PERIODIC TABLE OF THE ELEMENTS

Metals — Non-Metals

Legend / Key:

Group Number: New / Original

- Atomic Mass (g/mol)
- Open symbol elements are not stable in nature
- Atomic Radius* (A)
- Element Name

Atomic Number — 5
Symbol — **B** — 10.81
Common valence — B⁺³ 0.10
Boron

Transition Metals

Eu²⁺ is LIL
Eu³⁺ is HFS

Noble Gases

Alkali Metals
Alkaline Earth Metals

High Field Strength (HFS) Element
Large-Ion Lithophile (LIL) Element

Atomic Radius* is for typical valence and is only approximate, as it depends upon valence, coordination number, bond type, etc.

GROUP / PERIOD

Group 1 IA	2 IIA	3 IIIB	4 IVB	5 VB	6 VIB	7 VIIB	8	9 VIIIB	10	11 IB	12 IIB	13 IIIA	14 IVA	15 VA	16 VIA	17 VIIA	18 VIIIA
1 1.01 **H** H° 0.46 Hydrogen																	2 4.00 **He** Helium
3 6.94 **Li** Li⁺ 0.68 Lithium	4 9.01 **Be** Be²⁺ 0.25 Beryllium											5 10.81 **B** B³⁺ 0.10 Boron	6 12.01 **C** C° 0.77 Carbon	7 14.01 **N** N³⁻ 0.16 Nitrogen	8 16.00 **O** O²⁻ 1.30 Oxygen	9 19.00 **F** F⁻ 1.23 Fluorine	10 20.18 **Ne** Neon
11 22.99 **Na** Na⁺ 1.24 Sodium	12 24.305 **Mg** Mg²⁺ 0.80 Magnesium											13 26.98 **Al** Al³⁺ 0.61 Aluminum	14 28.09 **Si** Si⁴⁺ 0.34 Silicon	15 30.97 **P** P⁵⁺ 0.55 Phosphorus	16 32.065 **S** S⁶⁺ 0.37 Sulfur	17 35.45 **Cl** Cl⁻ 1.72 Chlorine	18 39.95 **Ar** Argon
19 39.10 **K** K⁺ 1.68 Potassium	20 40.08 **Ca** Ca²⁺ 1.20 Calcium	21 44.96 **Sc** Sc³⁺ 0.83 Scandium	22 47.87 **Ti** Ti⁴⁺ 0.69 Titanium	23 50.94 **V** V³⁺ 0.72 Vanadium	24 52.00 **Cr** Cr³⁺ 0.70 Chromium	25 54.94 **Mn** Mn²⁺ 0.80 Manganese	26 55.845 **Fe** Fe²⁺ 0.74 Fe³⁺ 0.64 Iron	27 58.93 **Co** Co²⁺ 0.72 Cobalt	28 58.69 **Ni** Ni²⁺ 0.77 Nickel	29 63.546 **Cu** Cu²⁺ 0.81 Copper	30 65.39 **Zn** Zn²⁺ 0.83 Zinc	31 69.72 **Ga** Ga³⁺ 0.70 Gallium	32 72.64 **Ge** Ge⁴⁺ 0.62 Germanium	33 74.92 **As** As⁵⁺ 0.58 Arsenic	34 78.96 **Se** Se⁶⁺ 0.37 Selenium	35 79.90 **Br** Br⁻ 1.88 Bromine	36 83.80 **Kr** Krypton
37 85.47 **Rb** Rb⁺ 1.81 Rubidium	38 87.62 **Sr** Sr²⁺ 1.33 Strontium	39 88.91 **Y** Y³⁺ 0.98 Yttrium	40 91.22 **Zr** Zr⁴⁺ 0.80 Zirconium	41 92.91 **Nb** Nb⁵⁺ 0.72 Niobium	42 95.94 **Mo** Mo⁶⁺ 0.80 Molybdenum	43 (98) **Tc** Technetium	44 101.07 **Ru** Ru⁴⁺ 0.70 Ruthenium	45 102.91 **Rh** Rh³⁺ 0.75 Rhodium	46 106.42 **Pd** Pd²⁺ 0.80 Palladium	47 107.87 **Ag** Ag¹⁺ 1.26 Silver	48 112.41 **Cd** Cd¹⁺ 0.97 Cadmium	49 114.82 **In** In³⁺ 1.00 Indium	50 118.71 **Sn** Sn⁴⁺ 0.77 Tin	51 121.76 **Sb** Sb⁵⁺ 0.69 Antimony	52 127.60 **Te** Te⁴⁺ 0.60 Tellurium	53 126.90 **I** I⁻ 1.97 Iodine	54 131.29 **Xe** Xenon
55 132.91 **Cs** Cs⁺ 1.96 Cesium	56 137.33 **Ba** Ba²⁺ 1.50 Barium	57-71 La-Lu Lanthanide	72 178.49 **Hf** Hf⁴⁺ 0.79 Hafnium	73 180.95 **Ta** Ta⁵⁺ 0.72 Tantalum	74 183.84 **W** W⁶⁺ 0.68 Tungsten	75 186.21 **Re** Re⁴⁺ 0.71 Rhenium	76 190.23 **Os** Os⁴⁺ 0.71 Osmium	77 192.22 **Ir** Ir⁴⁺ 0.71 Iridium	78 195.08 **Pt** Pt⁴⁺ 0.80 Platinum	79 196.97 **Au** Au³⁺ 0.78 Gold	80 200.59 **Hg** Hg²⁺ 1.10 Mercury	81 204.38 **Tl** Tl³⁺ 1.08 Thallium	82 207.2 **Pb** Pb²⁺ 1.57 Lead	83 208.98 **Bi** Bi³⁺ 1.19 Bismuth	84 (209) **Po** Po⁴⁺ 1.16 Polonium	85 (210) **At** Astatane	86 (222) **Rn** Radon
87 (223) **Fr** Francium	88 (226) **Ra** Ra²⁺ 1.56 Radium	89-103 Ac-Lr Actinide															

Lanthanide series (6)

| 57 138.91 **La** La³⁺ 1.25 Lanthanum | 58 140.12 **Ce** Ce³⁺ 1.22 Cerium | 59 140.91 **Pr** Pr³⁺ 1.22 Praseodymium | 60 144.24 **Nd** Nd³⁺ 1.20 Neodymium | 61 (145) **Pm** Promethium | 62 150.36 **Sm** Sm³⁺ 1.17 Samarium | 63 151.96 **Eu** Eu³⁺ 1.15 Europium | 64 157.25 **Gd** Gd³⁺ 1.14 Gadolinium | 65 158.93 **Tb** Tb³⁺ 1.12 Terbium | 66 162.50 **Dy** Dy³⁺ 1.11 Dysprosium | 67 (251) **Ho** Ho³⁺ 1.10 Holmium | 68 (257) **Er** Er³⁺ 1.08 Erbium | 69 (258) **Tm** Tm³⁺ 1.07 Thulium | 70 (259) **Yb** Yb³⁺ 1.06 Ytterbium | 71 174.97 **Lu** Lu³⁺ 1.05 Lutetium |

Actinide series (7)

| 89 (227) **Ac** Ac³⁺ 1.18 Actinium | 90 232.04 **Th** Th⁴⁺ 1.12 Thorium | 91 231.04 **Pa** Pa³⁺ 1.09 Proactinium | 92 238.03 **U** U⁴⁺ 1.08 Uranium | 93 (237) **Np** Neptunium | 94 (244) **Pu** Plutonium | 95 (243) **Am** Americium | 96 (247) **Cm** Curium | 97 (247) **Bk** Berkelium | 98 (251) **Cf** Californium | 99 (252) **Es** Ensteinium | 100 (257) **Fm** Fermium | 101 (258) **Md** Mendelevium | 102 (259) **No** Nobelium | 103 (262) **Lr** Lawrencium |

SECOND EDITION

An Introduction to
IGNEOUS
AND
METAMORPHIC
PETROLOGY

John D. Winter
Department of Geology
Whitman College

Prentice Hall
New York   Boston   San Francisco
London   Toronto   Sydney   Tokyo   Singapore   Madrid
Mexico City   Munich   Paris   Cape Town   Hong Kong   Montreal

Library of Congress Cataloging-in-Publication Data

Winter, John D. (John DuNann)
An introduction to igneous and metamorphic petrology / John D. Winter.—2nd ed.
p. cm.
Includes bibliographical references and index.
ISBN-13: 978-0-321-59257-6
ISBN-10: 0-321-59257-3
1. Rocks, Igneous. 2. Rocks, Metamorphic. I. Title.

QE461.W735 2010
552'.1—dc22   2008043585

*This effort is dedicated to the faculty and my fellow graduate students
at the University of Washington in the early 1970s.
They made geology interesting, and often downright fun.*

*It is also dedicated to my wife, Deborah,
and all the families deprived by those who attempt to write a book.*

Acquisitions Editor: *Drusilla Peters*
Editor in Chief, Science: *Nicole Folchetti*
Project Manager: *Crissy Dudonis*
Assistant Editor: *Sean Hale*
Editorial Assistant: *Kristen Sanchez*
Marketing Manager: *Amy Porubsky*
Managing Editor, Chemistry and Geosciences: *Gina M. Cheselka*
Project Manager, Production: *Ed Thomas*
Art Director: *Jayne Conte*

Senior Operations Supervisor: *Alan Fischer*
Art Editor: *Connie Long*
Composition: *GGS Higher Education Resources, A Division of Premedia Global, Inc.*
Production Editor: *Saraswathi Muralidhar, GGS Higher Education Resources, A Division of Premedia Global, Inc.*
Artist: *Jay McElroy*
Cover Design: *Suzanne Behnke*

Cover Images. Top: ASTER (Advanced Spaceborne Thermal Emission and Reflection Radiometer) satellite image of the Archean Pilbara craton, Western Australia. The area consists of ovoid granite-gneiss domes and folded greenstones of approximately 3.6 to 2.8 Ga. Deformation is not evenly distributed at this scale (50 km across), concentrating in the weaker (hence folded) supracrustal rocks rather than the more competent granites. Photo courtesy of NASA/GSFC/METI/ERSDAC/JAROS and U.S./Japan ASTER Science Team. **Bottom:** Photomicrograph under crossed polars of a pelitic schist from the Picuris Range, New Mexico, USA. Sedimentary bedding is probably responsible for the subhorizontal quartz-rich and mica-rich layers. The larger masses are porphyroblasts of staurolite, biotite, and garnet. Deformation imparts a crenulation cleavage (oriented NNW-SSE in the photo), which, at this scale (1.8 cm across), and in a fashion similar to the ASTER image, is not evenly distributed, but concentrated in zones of weakness. Photo from *A Practical Guide to Rock Microstructure*, provided courtesy of Ron Vernon and Cambridge University Press. As the cover suggests, petrologists address features and problems ranging from the global to the sub-microscopic scale.

© 2010, 2001 by Pearson Education, Inc.
Pearson Prentice Hall
Pearson Education, Inc.
Upper Saddle River, New Jersey 07458.

Printed in the United States of America
10 9 8 7 6 5 4 3 2 1

ISBN-13 978-0-32-159257-6
ISBN-10   0-32-159257-3

Prentice Hall
is an imprint of

www.pearsonhighered.com

BRIEF CONTENTS

CONTENTS

PREFACE

This text is designed for use in advanced undergraduate or early graduate courses in igneous and metamorphic petrology. The book is extensive enough to be used in separate igneous and metamorphic courses, but I use it for a one-semester combined course by selecting from the available chapters. The nature of geological investigations has largely shaped the approach that I follow.

Geology is plagued by the problem of *inaccessibility*. Geological observers really see only a tiny fraction of the rocks that compose the Earth. Uplift and erosion expose some deep-seated rocks, and others are delivered as xenoliths in magma, but their exact place of origin is vague at best. As a result, a large proportion of our information about Earth is indirect, coming from melts of subsurface material, geophysical studies, or experiments conducted at elevated temperatures and pressures.

The problem of inaccessibility has a temporal aspect as well. Most Earth processes are exceedingly slow. As a result, we seldom are blessed with the opportunity of observing even surface processes at rates that lend themselves to ready interpretation (volcanism is a rare exception for petrologists). In most other sciences, theories can be tested by experiment. In geology, as a rule, our experiment has run to its present state and is impossible to reproduce. Our common technique is to observe the results and infer what the experiment was. Most of our work is thus inferential and deductive. Rather than being repelled by this aspect of our work, I believe most geologists are attracted by it.

The nature of how geology is practiced has changed dramatically in recent years. Early geologists worked strictly in the observational and deductive fashion described above. The body of knowledge resulting from the painstaking accumulation of data observable with the naked eye or under a light microscope is impressive, and most of the theories concerning how the Earth works that were developed by the mid-20th century are still considered valid today, at least in broad terms. Modern postwar technology, however, has provided geologists with the means to study the Earth using techniques borrowed from our colleagues in the fields of physics and chemistry. We have mapped and sampled much of the ocean basins; we have probed the mantle using variations in gravity and seismic waves; we can perform chemical analyses of rocks and minerals quickly and with high precision; we can also study natural and synthetic specimens at elevated temperatures and pressures in the laboratory to approximate the conditions at which many rocks formed within the Earth. These and other techniques, combined with theoretical models and computing power, have opened new areas of research and have permitted us to learn more about the materials and processes of Earth's interior. These modern techniques have been instrumental in the development of plate tectonic theory, the encompassing paradigm that guides much present geologic thought. Given the limitations of inaccessibility mentioned above, it is impressive how much we have learned about our planet. Modern petrology, because it addresses processes that occur hidden from view, deep within the Earth, must rely heavily on data other than simple observation.

In the pages that follow I explain many of the techniques employed and the resulting insights they provide into the creation of the igneous and metamorphic rocks now found at the Earth's surface. The reader should be aware, however, that the results of our investigations, however impressive and consistent they may appear, are still based in large part on indirect evidence and inferential reasoning. I'm sure that the many researchers whose painstaking work we shall review would join me in urging a healthy skepticism, lest we become too dogmatic in our perspective. Ideas and theories are always in a state of flux. Many of today's ideas may be discarded tomorrow as new information becomes available and/or other ideas take their place. Petrology is not exempt from this process. If so, it would be far too dull to pursue.

The term *petrology* comes from the ancient Greek *petra* ("rock") and *logos* ("speak" or "explanation"), and means the study of rocks and the processes that produce them. Such study includes description and classification of rocks, as well as interpretation of their origin. Petrology is subdivided into the study of the three major rock types: sedimentary, igneous, and metamorphic. At the undergraduate level in most colleges and universities, sedimentary petrology is taught as a separate course, usually with stratigraphy. Igneous and metamorphic petrology are commonly combined, due to the similarity of approach and principles involved. I intend this book for either a combined igneous/metamorphic course or two separate ones. In the interest of brevity, I henceforth use the term *petrology* to mean the study of igneous and metamorphic rocks and processes. I hope not to offend sedimentary petrologists by this, but it would prove burdensome to continually redraw the distinction.

I shall concentrate on the processes and principles involved in the generation of igneous and metamorphic rocks rather than dwell upon lists of details to be memorized. Certainly facts are important (after all, they compose the data upon which the interpretations are based), but when students concentrate on the *processes* of geology, and the processes by which we investigate them, they get a deeper understanding, learn more lasting knowledge, and develop skills that will prove valuable beyond the classroom.

As mentioned above, modern petrology borrows heavily from the fields of chemistry and physics. Indeed, a student taking a petrology course should have completed a year of chemistry and at least high school physics. Calculus, too, would help but is not required. Some students who were attracted to geology for its field bias are initially put off by the more rigorous chemical and theoretical aspect of petrology. I intend this text to give students some exposure to the application of chemical and physical principles to geological problems, and I hope that some practice will give them confidence in using quantitative techniques. At the same time, I do not want to so burden them that they lose the perspective that this is a course in *geology*, not chemistry, physics, or computer science. We must bear in mind that the Earth itself is the true proving ground for all the ideas we deal with. Even the most elegant models, theories, and experimental results, if not manifested in the rocks in the field, are useless (and probably wrong as well).

All textbooks need to balance brevity, breadth, and depth. Whole books are dedicated to such subjects as thermodynamics, trace elements, isotopes, basalts, and such specialized subjects as kimberlites, lamproites, and mantle metasomatism. When distilling this sea of wisdom to an introductory or survey level, vast amounts of material must necessarily be abbreviated or left out. Of course it is up to the author to decide what is to be selected. We each have our own areas of interest, resulting in a somewhat biased coverage. To those who object to the light coverage I give to some subjects and my overindulgence in others, I apologize. The coverage here is not intended to replace more specialized classes and deeper levels of inquiry for those proceeding on to graduate studies in petrology. There is no attempt to develop theoretical techniques, such as thermodynamics or trace elements, from first principles. Rather, enough background is given for a degree of competence with using the techniques, but the direction is clearly toward application. We gain from our more general perspective a broad overview of the Earth as a dynamic system that produces a variety of igneous and metamorphic rocks in a wide range of settings. We will not only learn about these various settings and the processes that operate there, but we will develop the skills necessary to evaluate and understand them. Once again, I urge you to be critical as you progress through this text. Ask yourself whether the evidence presented to support an assertion is adequate. This text is different from texts only 10 years old. We might all wonder what interpretations will change in a text published 10 years from now.

Following the traditional approach, I have divided the book into an igneous section and a metamorphic section. Each begins with an introductory chapter, followed by a chapter on the description and classification of appropriate rock types and a chapter on the development and interpretation of textures. The chapters on classification and textures are intended as a laboratory supplement, and not for lecture-discussion. I have tried to explain most petrologic terms as they are presented, but you will invariably run across terms with which you are unfamiliar. I usually place a new term in **bold** typeface. If you forget a term, it can usually be found in the index, but a dictionary of geological terms is also a good companion. The inside covers contain some useful supplementary material.

Chapter 4 is a review of the field relationships of igneous rocks. It is relatively simple and intended to supply a background for the more detailed concepts to follow. Students may simply read it on their own. Chapters 5 through 9

are intensive chapters, in which I develop the theoretical and chemical concepts that will be needed to study igneous systems. By the time many students reach Chapter 9, they may fear that they are in the wrong course or, worse, the wrong major! Fortunately, things slow down after this and are oriented more toward application of the techniques to real rocks. Chapter 10 addresses the generation of basaltic magma in the mantle, and Chapter 11 deals with the evolution of such magmas once they are created. Chapters 12–20 explore the common igneous associations, using the techniques developed in Chapters 5–9 to develop models for their genesis. Few combined igneous–metamorphic petrology courses have the time to explore all of the igneous associations covered in Chapters 12–20, and instructors will commonly choose among their favorites (perhaps assigning selected sections rather than entire chapters if time is limited). Students can explore other chapters on their own or refer to them later, if the need arises. One can also get a decent review of these chapters by reading the final section in each, which discusses a petrogenetic model. The models, however, are based on the petrological and chemical data developed in the chapter, so many of the conclusions will have to be taken on faith using this approach. I teach a year-long mineralogy–petrology sequence and have found it advantageous to cover Chapters 5, 6, and 7 (which may be considered transitional between mineralogy and petrology) in the fall in mineralogy, leaving more time to focus on petrology during the second semester.

The metamorphic section is shorter than the igneous section, because there are not as many specific tectonic associations. The approach I follow is to consider metamorphic rocks as chemical systems at equilibrium, manifested as stable mineral assemblages. The mineral assemblages vary both spatially and temporally due to variations in pressure, temperature, composition, and the nature of associated fluids. Changes in mineral assemblage are achieved by chemical reactions and are controlled by the variables just mentioned. Qualitative approaches to assessing the equilibria are developed in Chapters 24–26. Chapter 27 addresses the quantitative approach using thermodynamics and geothermobarometry. Chapters 28 and 29 apply the techniques to specific common rock types: pelites, carbonates, and ultramafic rocks. Finally, Chapter 30 explores metasomatism. Less rigorous courses, or ones that run short of time, can drop Chapters 27 and 30 without rendering the other chapters incomprehensible.

I often make references in the text to other sections and figures where a concept, an approach, or a technique is introduced or developed more fully. These references are intended to assist the reader should a concept be slightly unfamiliar or if more information is desired. They do not imply that the reader must follow their lead in order to understand the discussion at hand.

To give students a better understanding of the processes and principles involved, I have integrated a number of problems into the text. The problems are an important part of the text, and working through them rather than simply scanning them will make an enormous difference in the student's understanding. The occasional problem integrated into the reading as a "worked example" should be done at the time it is encountered, as it is intended to illustrate the concept being presented. Problems available from my web site are intended as review and to bring together the material discussed in the particular chapter. Problems not only provide a deeper understanding of the principles involved, they give students some practice at data analysis and the tools that geologists commonly employ. Many problems can be done with just a hand calculator, but most would be done more efficiently with a computer, and some require one. Spreadsheets and other computer programs permit us quickly to get past the drudgery of handling moderate to large amounts of data or creating graphs, and into the more interesting aspects of interpreting the results. The ability to use a computer, and particularly a spreadsheet, is necessary for all science students. On my web page (www.prenhall.com/winter), you can download a number of files supplementary to this text. Included is a brief introduction to the use of Excel (Excel.doc, a Word document). If you are not familiar with the use of a spreadsheet, I suggest you read this and try the exercise as soon as you can. Also on the web site are some programs and data compilations for a number of the problems. Other problems will assume that particular petrologic programs are available on a campus computer. Among these are the following: IgPet and GCDkit, which are programs for manipulating and displaying chemical data for igneous rocks; TWQ, THERMOCALC, PERPLEX, and SUPCRT, which are thermodynamic databases that calculate mineral equilibria at elevated temperatures and pressures; and THERMOBAROMETRY, which calculates pressures and/or temperatures of equilibrium from mineral compositions. A list of those programs and pertinent information for their acquisition is on the web site. Of course, the basic law of computers must be kept in mind: To be useful, the data must be representative, relatively complete, and of good quality. Computers can output beautifully crafted diagrams that can make even wrong data look deceptively good.

Finally, a word on units. I have used SI (*Système International*) units throughout the text. Although most petrologists are more familiar with calories and bars for energy and pressure, respectively, the SI units are gradually becoming the norm. I think it best that you get your exposure to these units as early as possible because I have an awful time with them after so many years thinking in kilobars and calories. Calculations are also easier because the units are standardized. The only deviation that I make from the strict SI terminology is to commonly refer to temperature in degrees Celsius rather than in kelvin, but this is only a difference with regard to reference and not the magnitude of the units themselves. The inside front cover lists the units and prefixes used. Please look at this list as soon as possible for a general familiarity, and particularly for the prefixes of magnitude, so that you will be able to interpret such common terms as Ma (million years), or GPa (10^9 pascals pressure). Because minerals commonly have long names and are mentioned so often, at times I abbreviate the names in the text and on figures. The inside back cover lists the mineral abbreviations

and several of the other acronyms that petrologists commonly employ. I have also included a periodic table of the elements with some useful information on each.

If you have any comments, questions, corrections, or suggestions for future editions, please let me know. My e-mail address is winterj@whitman.edu.

SECOND EDITION NOTES

This second edition gave me the opportunity to correct errors and update several concepts. I also added to each chapter sections to aid students, including a brief introductory series of questions the chapter addresses, a summary, review questions, and some important "first principle" concepts of a fundamental nature. Several areas of petrology have advanced since the first edition (or perhaps just my understanding of them has). This book has been updated to address those areas. I have added a brief section on plate tectonics to Chapter 1 and expanded descriptions of mid-ocean ridge segmentation, dynamics, and melt production models in Chapter 13. I have added a section on mantle re-enrichment to Chapter 16 and a section on textural geochronology to Chapter 23 (with added detail in Chapter 27). One area in which we have *less* certainty than we did a decade ago is in regard to mantle layering and compositional stratification or diversification. In the 1990s, petrologists were relatively convinced that the mantle was layered, with a melt-depleted upper layer and a non-depleted or even enriched lower layer. Recent data indicate a more complex situation, and the debate over the nature and geometry of compositionally diverse mantle components is presently quite lively. Several chapters address this issue, and Chapter 14 deals with it most directly and has been significantly expanded as a result. I consider this issue important enough to address it in detail, but courses with time constraints may elect to address only the models and summary toward the end of Chapter 14.

As stated in our sustainability initiative, this book is carefully crafted to minimize environmental impact. To add to this effort, the decision was made to utilize technology to save on the amount of paper used in this publication. All review questions and problems for each chapter and the entire list of references are available for download from my website (www.prenhall.com/winter) instead of printing them in the book. Interested students and fellow researchers may print the reference list and keep it with the text. Electronic references are also searchable. Providing this material online also enables me to update review questions and problems. I encourage colleagues who have useful problems that they are willing to share to send them to me.

As previously, I have made all text figures available in color as PowerPoint™ presentations from my web site. My e-mail indicates that many colleagues have downloaded and used them. The presentations are fully editable and may be modified to suit individual tastes and uses. I don't expect other instructors to follow my approach exactly. Most of my versions, in fact, are laden with text because I now use them mostly for student review. I wouldn't presume to tell my colleagues how to teach, and many methods I now employ may merely reflect some inadequacy as a lecturer, but I have found it advantageous to vary what I do. The PowerPoints I actually use in class employ images with little text. One of my favorite approaches is to assign reading and let students look at the PowerPoints online (both before and after class), and we work on the review questions in class (usually with students working in pairs). I then walk around and listen, occasionally offering assistance. When this approach works well, my lectures become short reviews or on-demand mini-lectures addressing concepts with which the students seem to be having difficulty as they handle the question sheets. Perhaps making the review questions available on the web site as editable Word documents might even be an advantage in this way, as it allows instructors to adapt their own question sheets from mine.

As with the first edition, I hope students and colleagues alike find this work useful.

ACKNOWLEDGMENTS

No text is an individual endeavor. My deepest gratitude goes to those workers whose efforts in the field and laboratory I have summarized in what follows. Special thanks to colleagues and students who have reviewed early drafts and corrected errors or helped refine my thinking when it was muddled or in error. I owe more than I can say to the expertise and generosity of Paul Hoornbeek, Bernard Evans, Frank Spear, Spencer Cotkin, Rick Conrey, Peter Crowley, John Brady, and Jack Cheney, who reviewed substantial portions of the text with great care and patience. Many thanks also to colleagues who unselfishly reviewed and commented on individual chapters: Stu McCallum, Roger Mitchell, John Gittins, William Scott, Jack Rice, Barrie Clark, Lawford Anderson, Tracy Vallier, Bill Bonnischen, and Andrew Wulff. Thanks to the many many colleagues who made data or figures available and provided useful and encouraging assistance: Peter Kelemen, David Pattison, Roger Powell, Jamie Connolly, Julie Baldwin, Brennan Jordan, Tim Grove, and Susanne Kay. I am also very grateful to Amy Kushner, Nadine McQuarrie, and the staff at Prentice for their efforts in handling copyright permissions, a job for someone more organized than I. All of the editorial and production staff at Prentice Hall (and subsidiaries) were very helpful and cooperative in producing this text. Thanks again to everyone.

John D. Winter
Whitman College

About Our Sustainability Initiatives

This book is carefully crafted to minimize environmental impact. The materials used to manufacture this book originated from sources committed to responsible forestry practices. The paper is FSC certified. The binding, cover, and paper come from facilities that minimize waste, energy consumption, and the use of harmful chemicals.

Pearson closes the loop by recycling every out-of-date text returned to our warehouse. We pulp the books, and the pulp is used to produce items such as paper coffee cups and shopping bags. In addition, Pearson aims to become the first carbon neutral educational publishing company.

Pearson is also supporting student sustainability efforts through our Sustainable Solutions Awards, our Student Sustainability Summits, and our Student Activity Fund.

The future holds great promise for reducing our impact on Earth's environment, and Pearson is proud to be leading the way. We strive to publish the best books with the most up-to-date and accurate content, and to do so in ways that minimize our impact on Earth.

© **Mixed Sources**

Product group from well-managed forests, controlled sources and recycled wood or fiber
www.fsc.org Cert no. SCS-COC-00648
© 1996 Forest Stewardship Council

FSC

Prentice Hall
is an imprint of

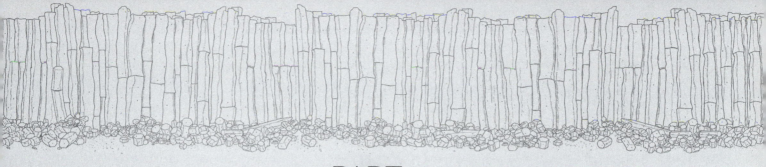

PART

I

Igneous Petrology

1

Some Fundamental Concepts

Questions to be Considered in this Chapter:

1. What is igneous petrology and what techniques are used to study igneous rocks?
2. What is the Earth made of, and what are the major subdivisions of the Earth's interior?
3. How did the Earth form?
4. How do the conditions of pressure and temperature vary within the Earth?
5. If we can constrain what composes the Earth and the conditions at depth, what can we initially conclude about melting and the generation of igneous rocks?

In this initial chapter, I will take the opportunity to generalize and set the stage for the more detailed chapters to come. It will be assumed that you, the student, are familiar with the most basic concepts. You may want to briefly review the chapter on igneous rocks and processes in your introductory geology text, as doing so will refresh your memory and provide an initial "big picture" as we proceed to refine the ideas. Reviewing a chapter on plate tectonics would also help in this regard.

1.1 INTRODUCTION TO IGNEOUS PETROLOGY

Igneous petrology is the study of melts (magma) and the rocks that crystallize from such melts, encompassing an understanding of the processes involved in melting and subsequent rise, evolution, crystallization, and eruption or emplacement of the eventual rocks. Origin by crystallization from a melt seems a simple enough criterion for considering a rock to be igneous. But we can only rarely observe the formation of igneous rocks directly, and then only for some surface lavas. The history of the study of igneous rocks is a tribute to the difficulties involved (see Young, 2003, for a comprehensive review). Many humans, including the Greeks, Romans/Italians, Japanese, Icelanders, and Indonesians, have lived with volcanoes. In the Western world, early biblical and Greek references to volcanic phenomena are generally attributed to angry gods. The Greek Hephaestus and later the Roman god Vulcan were gods of fire, and Romans considered volcanoes (particularly the island of Vulcano north of Sicily) as the chimneys of Vulcan's forge. The Greeks and Romans recognized that volcanoes emitted fiery-hot lava (or explosive ash) and that lava cooled to stone. Several ancient scholars even proposed theories of volcanism. Aristotle, apparently impressed by the explosive nature of some Greek eruptions, attributed volcanism to the movement and expulsion of subterranean winds, or "exhalations," that also gave rise to earthquakes. Others, impressed with the heat, suggested that volcanic subterranean fires required fuel as well, such as sulfur, alum, or asphalt (later extended to include coal and pyrite in 17th and 18th century Europe). Recognizing ancient volcanic deposits and extending the concept of volcanism beyond the very recent past was a much more difficult endeavor. This was probably first done in the Auvergne in south-central France in the mid-18th century, where although volcanism was no longer active, several conical mountains were still evident, and the summit craters, with flow-like and cindery deposits, were identical to those associated with active volcanoes. Similar extinct volcanic terranes were recognized in Germany and the United Kingdom.

Resistance to the idea of a volcanic origin of basalt was unusually strong in 18th and early 19th century Europe, led primarily by the influential German geologist Abraham G. Werner. He was impressed by numerous examples of flat-lying stratified layers of basalt with no associated volcanic mountain and the crystalline nature of those basalts, which resembled chemical precipitates such as limestone or salt. Thus was born the *Neptunist* school, that, while recognizing the existence of true volcanics, considered most basalt to be deposited by seawater. The word *magma*, from the Greek μαγμα ("paste"), was actually introduced in a Neptunist context, not igneous, by Dolomieu in 1794, in the belief that the rocks originating from it were reduced to paste by evaporation. Another school of thought, the Plutonists, following James Hutton in the late 18th century, agreed that basalt was igneous, but intrusive, not volcanic. It was not until the early 19th century that sufficient field and experimental work was accumulated to lead to a general consensus on the volcanic origin of basalt. A similar controversy raged in the 19th and early 20th centuries over the origin of granite. The Neptunists believed it, too, was an aqueous precipitate, and later arguments involved those who thought granites originate from a dry melt, those who preferred a water-saturated solution/melt, and those who thought them to be metamorphosed sediments.

These historical controversies are now essentially resolved. Let's begin, then, by considering some observational criteria for determining that a rock is indeed of igneous origin. Such criteria will be developed further later on, but, by way of introduction, they include:

1. *Field criteria* Intrusive igneous bodies commonly crosscut the "country rocks" into which they intrude, thereby truncating external structures, such as bedding or foliation. They may also exhibit some types of contact effects resulting from the sudden juxtaposition of hot magma and cooler country rocks. When developed, a narrow, fine-grained chilled margin (or "chill zone") within the igneous body margin or localized baking of the country rocks are good indicators of an igneous origin for plutonic (intrusive) bodies. In addition, we have come to associate certain specific forms of rock bodies with an igneous origin. For example, a strato-volcano, a pahoehoe flow, a sill or laccolith, etc. have become associated with igneous processes, either by direct observation of an igneous event or by the application of some of the criteria mentioned above. Field aspects of igneous rocks will be discussed further in Chapter 4.

2. *Textural criteria* Petrography is the branch of petrology that deals with the description and systematic classification of rocks. By observing thin sections of igneous rocks under the petrographic microscope, we have come to associate a specific *interlocking* texture with slow crystallization from a melt (Figure 1.1). When crystals are forming

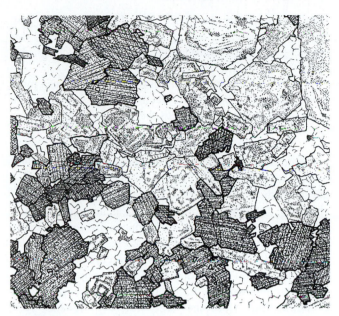

FIGURE 1.1 Interlocking texture in a granodiorite. From Bard (1986). Copyright © by permission Kluwer Academic Publishers.

in a cooling melt, they usually develop a nearly perfect crystal form, as the melt provides no obstruction to the preferred crystal shape. As the melt continues to cool, and more crystals form, they eventually begin to interfere with one another and intergrow. The resulting interlocking texture shows interpenetrating crystals, much like a jigsaw puzzle. As will be discussed in Chapter 3, the relative development of crystal form, inclusions, and interpenetration can be used to infer the sequence in which different mineral species crystallized. Henry Sorby introduced the polarizing microscope to the geological world in the 1850s, and it quickly quelled many of the controversies mentioned above.

Because liquids cannot sustain substantial directional stresses, **foliations** rarely develop in igneous rocks. A common textural criterion for distinguishing an igneous from a high-grade metamorphic crystalline rock in hand specimen is thus based on the *isotropic* texture (random orientation of elongated crystals) of the former. One must use caution, however, when applying this criterion, as some igneous processes, such as crystal settling and magmatic flow, can produce mineral alignments and foliations in igneous rocks.

Pyroclastic deposits (those resulting from explosive eruptions) can perhaps be the most difficult to recognize as igneous. Usually the magmatic contribution to these deposits has solidified and cooled considerably before being deposited along with variable proportions of pulverized preexisting rocks caught in the explosion. The actual deposition of pyroclastic material is in large part a sedimentary process, and hence the difficulty in recognition. There is still some debate among geologists as to whether pyroclastics should be considered igneous or sedimentary. They are igneous in the sense that nearly all of their matter crystallized from a melt, even though a proportion may have been earlier volcanic deposits. This is the *pyro* part. They are sedimentary as well, in the sense that they represent solid particles deposited by a fluid medium: air or sometimes water. This is the *clastic* part. Some geologists have wisely suggested that we avoid this hopeless debate by considering pyroclastics to be igneous going up and sedimentary coming down.

In Chapter 3, we will discuss igneous textures in more detail, including both those seen in hand specimen and those seen in thin section with the aid of the petrographic microscope.

As we initially consider the study of igneous rocks and processes, perhaps we should consider what exactly it is that we want to know. The types of very "broad-brush" questions that we would expect to have answered might include the following: How are melts generated? What is melted, and where? What is produced by this melting? How do the melts so produced crystallize to igneous rocks, and what processes accompany this crystallization? In what way(s) do the liquid and solid portions evolve (change composition) during the process of melting or crystallization? Does the large variety of igneous rock compositions now found at the Earth's surface result from different sources of melts, or can it be attributed to variations in the processes of melting and crystallization? Is there a relationship between igneous rock type and tectonic setting? If so, what controls this? Finally we might ask, What do we need to know to assess these? In other words, what background and approach does a good modern petrologist need? I would suggest the following as imperative background:

1. A petrologist needs experience looking at rocks and textures. One cannot begin to study rocks without knowing how to recognize, describe, organize, and analyze them. As H. H. Read (1957) quipped, the best geologist is the one "who has seen the most rocks."

2. Experimental data (based on synthetic and natural samples) are necessary. We can best understand the generation and crystallization of melts by re-creating these processes in the laboratory, simulating the conditions found at depth, and analyzing the results. This also allows us to place some constraints on the physical conditions under which igneous processes may have taken place.

3. Some theory is required, so we can organize and understand the experimental results better and apply those results beyond the exact compositions and conditions of the experiments. A bit of chemistry is necessary, encompassing major elements, trace elements, and isotopes, as is some thermodynamics. As we shall see, these techniques also help us characterize rocks and evaluate source regions and evolutionary processes. A knowledge of physics is also helpful because it permits us to place reasonable constraints on magmatic processes. It is useful to know something about the viscosity, density, heat capacity, thermal conductivity, and other properties of materials if we are to understand the processes of melting, cooling, crystallization, rise, and emplacement of magmatic systems.

4. We need a knowledge of what comprises the Earth's interior and the physical conditions that exist there. Melts are created deeper than we can directly observe. If we want to know what is melted and how, we must review what is known about the Earth's interior and how the constituents and conditions vary with tectonic setting.

5. Finally, we need some practical experience with igneous activity. A literature-based survey of common igneous rocks and processes in nature can provide a framework for all of the above and give a more complete picture.

I hope that you will acquire the requisite skills and experience in the chapters that follow. Although it would perhaps be preferable to develop each of the above skills sequentially, this is not practical, and may be impossible, as these skills are integrated in the scientific process. The student will get direct observational experience in the class laboratories, with the aid of Chapters 2 (classification) and 3 (textures). Chapter 4 is designed to provide a survey of igneous rocks in their field setting. This is largely a review of fundamental concepts. Next, we get some theory on melt systems and use that theory to look at some simple experimental systems. Then we will proceed to more complex natural systems and the (largely chemical) tools required to study them. Finally, we will embark upon an overview of the most common igneous provinces and review the current state of modern theories on their development. This latter portion will require considerable application and even amplification of the theoretical skills acquired.

On a grand scale, igneous, metamorphic, and sedimentary processes all contribute to a differentiated planet. Igneous processes are by far the most dominant in this regard, as they are largely responsible for the segregation of the crust from the mantle and for the origin of many natural resources. It will benefit us all to keep this large-scale differentiation in mind as we explore the more focused components in what follows.

1.2 THE EARTH'S INTERIOR

As mentioned above, virtually all igneous rocks originate by melting of material at some depth within the Earth. All terrestrial rocks that we now find at the Earth's surface were derived initially from the mantle, although some have since gone through one or more cycles of subsequent sedimentary, metamorphic, and/or igneous processes. If these rocks have an ultimate origin at depth, it follows that we need to know what makes up the Earth if we want to understand their origin more fully. From a *compositional* perspective, the Earth's interior is subdivided into three major units: the *crust*, the *mantle*, and the *core* (Figure 1.2). These units were recognized decades ago, during the early days of seismology, because they were separated by major discontinuities in the velocities of *P* (compressional) and *S* (shear) waves as they propagate through those layers in the Earth (Figure 1.3).

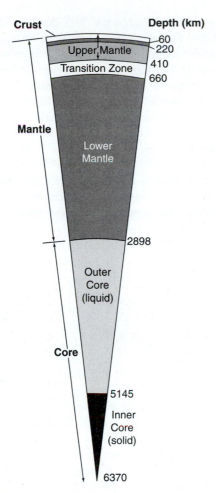

FIGURE 1.2 Major subdivisions of the Earth.

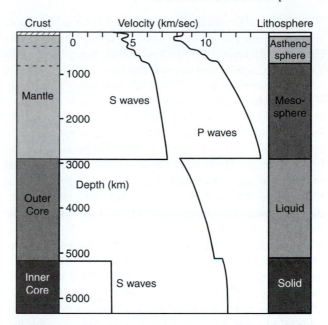

FIGURE 1.3 Variation in *P* and *S* wave velocities with depth. Compositional subdivisions of the Earth are on the left, and rheological subdivisions are on the right. After Kearey and Vine (1990). Reprinted by permission of Blackwell Science, Inc.

The **crust** comprises about 1% of the volume of the Earth. There are two basic types of crust—*oceanic* and *continental*—and both are too thin to represent accurately on Figure 1.2 (even the thickest continental crust would be thinner than the uppermost line). The thinner of the two, the **oceanic crust** (about 10 km thick), has an essentially basaltic composition. We will discuss the composition, structure, and origin of the oceanic crust in Chapter 13. Because plate tectonics is creating oceanic crust at mid-ocean ridges and consuming it at subduction zones, the oceanic crust is continually being renewed and recycled. The oldest oceanic crust is in the southwest Pacific and is about 160 Ma old. The **continental crust** is thicker: typically 30 to 45 km beneath stable areas but generally 50 to 60 km thick in orogenic areas and extending up to 90 km at a few localities. It is also more heterogeneous, including all sorts of sedimentary, igneous, and metamorphic rocks. A *very* crude average composition of the continental crust would be represented by a granodiorite. Continental crust covers about 40% of the Earth's surface. Unlike the oceanic crust, it is too buoyant to subduct far. The amount of continental crust has been increasing over the past 4 Ga. Some continental crust is thus very old, whereas some is quite new. The stable continents (**cratons**) consist of more ancient crystalline **shields** and stable **platforms**. Platforms typically have basement crystalline rocks (typically younger than shields, but not necessarily so) overlain by a

few kilometers of sedimentary rocks. Several marginal **orogenic belts** may also flank cratons, typically associated with subduction, which add to the continental crust over time. The lower crust is believed to be depleted in many of the more mobile elements and to have a more mafic character than the upper crust. A seismic discontinuity in the lower continental crust (the Conrad discontinuity) is recognized in some areas and may mark the transition between this deeper crust and the shallower sialic portion. It was once believed that the Conrad discontinuity was continuous and separated a more granitic shallow crust from a basaltic lower crust (similar to the oceanic crust) that formed the continental base, but we now realize that the upper and lower continental crust is much more heterogeneous. The base of the continents is not always sharply defined seismically and may locally be gradual and even have a layered transition into the subcontinental mantle. There will be lots more to say about the nature of the continental crust and the igneous and metamorphic processes associated with it throughout this book. For a good summary of continental structure, see Sleep (2005), and for the lower portion, see Fountain et al. (1992).

Immediately beneath the crust, and extending to nearly 3000 km, is the **mantle**, comprising about 83% of the Earth's volume. The boundary, or discontinuity, between the crust and mantle is called the Moho, or M discontinuity (shortened from Mohorovičić, the name of the Balkan seismologist who discovered it in 1909). At this discontinuity, most readily observed beneath oceanic crust, the velocity of *P* waves increases abruptly from about 7 to over 8 km/sec. This results in refraction, as well as reflection, of seismic waves as they encounter the discontinuity, making it relatively simple to determine the depth. The mantle is composed predominantly of Fe- and Mg-rich silicate minerals. We will discuss the petrology of the mantle in Chapter 10.

Within the mantle, several other seismic discontinuities separate layers that are distinguished more by physical than chemical differences. The shallowest such layer, between 80 and 220 km, is called the **low-velocity layer** because within it, seismic waves slow down slightly, as compared to the velocity both above and below the layer (Figure 1.3). The slowness of seismic waves is unusual because seismic velocities generally increase with depth because they propagate more readily through more compacted (hence more incompressible and rigid) materials. The reason seismic waves slow down in the low velocity layer is attributed to 1 to 5% partial melting of the mantle. The melt probably forms a thin discontinuous film along mineral grain boundaries, which retards the seismic waves. The melt also weakens the mantle in this layer, making it behave in a more ductile fashion. The low-velocity layer varies in thickness, depending on the local pressure, temperature, melting point, and availability of H_2O. We will discuss the origin of the low-velocity layer further in Chapter 10.

Below the low-velocity layer we encounter two more seismic discontinuities within the mantle. The 410-km discontinuity is believed to result from a phase transformation in which olivine (the major mineral constituent of the upper mantle) changes from the well-known ("α-phase") structure to *wadsleyite* ("β-phase") and then to *ringwoodite* ("γ-phase") with an isometric spinel-type structure. At 660 km, the coordination of Si in mantle silicates changes from the familiar IV-fold to VI-fold, and the dominant silicate becomes an $(Mg,Fe)SiO_3$ magnesium silicate with a perovskite-like structure, and the excess Mg and Fe form an $(Mg,Fe)O$ oxide called either *magnesiowüstite* or *ferropericlase*. This latter transition, of course, is not a simple a → b phase transformation, but an a → b + c reaction. Both the 410-km and 660-km transitions result in an abrupt increase in the density of the mantle, accompanied by a jump in seismic velocities.

Below the 660-km discontinuity, the velocities of seismic waves increase fairly uniformly with depth (Figure 1.3). At the very base of the mantle is a ~200 km thick heterogeneous layer of anomalously low seismic velocity called the **D″ layer**. A thin (~40km), apparently discontinuous layer with even lower velocities has also recently been resolved at the mantle–core boundary, most clearly beneath the central Pacific (Garnero and Helmberger, 1995, 1996). The nature of the D″ and 40-km sublayer are not entirely clear, but their properties are sufficiently anomalous to require more than a thermal boundary perturbation, and they probably represent a layer of different composition (and hence greater density) than the overlying mantle. A popular proposition is that they represent an accumulation of dense "dregs" of subducted oceanic crust that has settled to the base of the mantle (Christensen and Hofmann, 1994). We will explore this idea further in Chapters 14 and 16.

Beneath the mantle is the **core**. The mantle–core boundary is a profound chemical discontinuity at which the silicates of the mantle give way to a much denser Fe-rich metallic alloy with minor amounts of Ni, S, Si, O, etc. The outer core is in the liquid/molten state, whereas the inner core is solid. The composition of the inner core and outer core is probably similar. The transition to a solid is a response to increased pressure with depth, which favors the solid state. *S*-waves cannot propagate through a liquid because liquids cannot resist shear. Although *S*-waves are only slowed by the thin liquid films in the low-velocity layer, they disappear entirely as they reach the outer core (Figure 1.3). *P*-waves slow in the liquid core and refract downward, resulting in the seismic "shadow zone," a ring-like zone in which earthquake *P*-waves don't reach the surface of the Earth on the side away from which they originated.

The two types of crust, the mantle, and the core are distinguished on the basis of composition. An alternative way to consider the subdivisions of the Earth is based on *rheological* properties (right side of Figure 1.3). Using these criteria (how materials respond to deformation), we can consider the crust plus the more rigid portion of the uppermost mantle above the low-velocity layer to behave as a strong, coherent unit, collectively called the **lithosphere**. Oceanic lithosphere is thin (~50 km) near warm mid-ocean ridges and thickens to about 110 km when cool and mature. The lithosphere is 200 to 250 km thick under the stable continental shields (McKenzie and Priestly, 2008). The more ductile mantle immediately below the lithosphere is called the **asthenosphere** (from the Greek *asthenes*, "without strength"). The asthenosphere is important to plate tectonics because the ductility is thought to provide the zone of dislocation upon which the rigid lithospheric plates may differentially move. From a rheological standpoint, the mantle below the asthenosphere is called the **mesosphere**. The lithosphere–asthenosphere–mesosphere boundaries are all within the mantle and correspond to the transition from rigid to ductile and back to less ductile material with depth. The transitions are somewhat gradual and difficult to resolve seismically, particularly the bottom of the ductile layer. The asthenosphere is probably about 150 km thick. The rheological nature of the mesosphere is not well known, but seismic waves are not greatly attenuated, suggesting that this layer is relatively strong. The liquid outer core and solid inner core are of course distinguishable on a mechanical basis.

1.3 ORIGIN OF THE SOLAR SYSTEM AND THE EARTH

Now that you have some idea of what comprises the Earth, it is interesting to speculate on how it got that way. The following scenario summarizes the most generally accepted theories on the origin of the solar system. It will be presented as fact only to avoid the constant use of disclaimers and indefinite phrases. Remember, however, that this is only a collection of internally consistent ideas by which we explain what we now observe, although the extent of this consistency does lend credence to the models.

The most popular model for the origin of the universe has the Big Bang occurring between 12 to 15 Ga before present (b.p.). According to radiometric dating of meteorites, the solar system began to form about 4.56 Ga b.p. as a huge cloud of matter called the **solar nebula** (Figure 1.4).

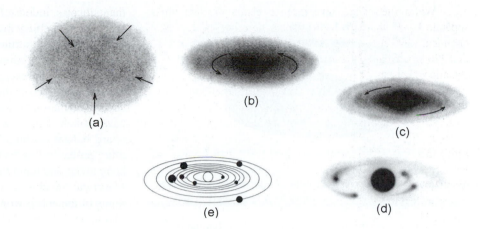

FIGURE 1.4 Nebular theory of the formation of the solar system. **(a)** The solar nebula condenses from the interstellar medium and contracts. **(b)** As the nebula shrinks, its rotation causes it to flatten to a disk **(c)**, with most of the matter concentrated toward the center as the primordial sun **(d)**. Outer solid particles condense and accrete to form the planets **(d)** and **(e)**. From Abell et al. (1988). Copyright © by permission Saunders.

The nebula consisted mostly of molecular H_2 plus some He and minor Be and Li (the only products of the Big Bang). A further 2% comprised heavier elements, including some other gases and fine solid particles ("dust"), presumably created by nuclear synthesis reactions in earlier nearby stars and supernovae. The nebular cloud began to collapse slowly because of the gravitational interactions of its constituents. Because it was rotating, it flattened to a disk-like shape as a result of centrifugal forces, with 1 to 10% of the mass constituting the central disk. The balance between gravitational collapse, centrifugal force, and conservation of angular momentum resulted in the majority of the mass losing angular momentum and falling to the center of the disk, eventually to form the sun. In the area that eventually became the inner terrestrial planets, small particles collided and aggregated, gradually forming larger bodies that swept up more material, growing to form meter- to several kilometer-sized bodies, called **planetesimals**. The gravitational collapse of the mass and its compression generated considerable heat, eventually reaching the stage where nuclear synthesis (fusion) of hydrogen to helium became possible and the sun became a star.

The first 100,000 years witnessed a very rapid evolution of the "proto-sun," accompanied by high luminosity caused by the heat generated by the initial contraction. When the compression was nearly over, the sun entered the **T-Tauri stage**, characterized by less vigorous activity, lasting up to 10 Ma. The **solar wind**, a stream of charged particles, changed character during the T-Tauri stage and began to emanate radially outward from the sun rather than spirally from the poles. The nebula may have lost about half of its initial mass during this stage.

Of the remaining material, 99.9% of the mass collapsed to form the sun, and the other 0.1%, with the majority of the angular momentum, remained in the disk. The disk material had sufficient mass to contract to the median plane, where it eventually separated into localized accumulations that formed the planets. The process of planetary accretion took place within a strong temperature and pressure gradient generated by the early sun. As a result, the more volatile elements within the solid particles of the nebula vaporized in the inner, hotter portion of the solar system. The vapor particles were then stripped off by the intense T-Tauri solar wind and condensed directly to solids further outward, where the

temperature was lower. Only the larger planetesimals survived this intense activity in the inner solar system. The actual condensation temperatures (and hence the distance from the sun at which condensation took place) depended upon the particular elements or compounds involved. Only the most refractory elements survived or condensed in the innermost zone, whereas the more volatile constituents were moved further outward. As a result, then, primarily due to the temperature gradient and solar wind, the nebula experienced a chemical differentiation based on condensation temperatures. Refractory oxides such as Al_2O_3, CaO, and TiO_2 either failed to volatilize at all or condensed quickly in the innermost portions of the solar system. Fe-Ni metal alloys, Fe-Mg-Ni silicates, alkali metals and silicates, sulfides, hydrous silicates, H_2O, and solids of ammonia, methane, etc., condensed and concentrated progressively outward. The distance beyond which the very volatile compounds such as water and methane condensed has been referred to as the **snow line**. Apparently, a gradient of decreasing pressure outward from the center of the nebula also had an effect, principally on the relative condensation temperatures of Fe metal versus silicates, and thus on the Fe/Si ratio (and oxygen content) of the planets.

The condensed solids continued to accrete as planetesimals. In the inner portion of the solar system, the more refractory planetesimals further accumulated and formed the **terrestrial** (Earth-like) planets (Mercury, Venus, Earth, and Mars) as well as the parent bodies that produced the present asteroids and meteorites. In the outer portions, beyond the snow line, the large **gaseous** planets formed. Pluto was considered a planet until August 2006, when it was demoted following the discovery of a similar and even larger object (subsequently named Eris) orbiting the sun. Pluto and Eris are two of many recently discovered objects with orbits beyond Neptune. They are now considered part of the *Kuiper belt* of icy objects (frozen methane, ammonia, and H_2O). As astronomers considered alternatives to increasing the number of planets to accommodate these bodies, they wanted to call the larger of them "plutons," but geologists objected loudly to the theft of our term. They are now called "minor solar system bodies," and poor Pluto, no longer a planet, is reduced to "number 134340."

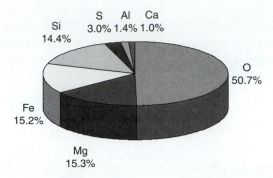

FIGURE 1.5 Relative atomic abundances (by mass) of the seven most common elements that comprise 97% of the Earth's mass.

From this very brief sketch, it seems clear that the composition of a planet is in large part the result of rather specific conditions that existed at a particular radial distance from the center of the solar nebula during the first 10 Ma of stellar evolution. The composition of the Earth is largely a result, then, of the nature of the ancient supernova that "seeded" the solar nebula with solid particles and the evaporation/condensation processes associated with the temperature at Earth's particular distance from the T-Tauri sun. Thus we would not expect the Earth's composition to be equal to that of other planets or to that of the solar nebula as a whole.

The differentiation process that produced the chemical variation across the solar system was not perfectly efficient. The composition of the Earth is complex, including some of every stable element, and not just those that could condense at our distance from the sun. Some of the varied constituents of the Earth, including the volatiles, were contained in the early planetesimals that were large enough to resist complete vaporization during the hot T-Tauri stage of solar evolution, whereas others may have been added later via impact of comet-like bodies from the outer solar system. Nonetheless, the process described above strongly favored the concentration of certain elements, and only seven elements now account for 97% of the mass of the Earth (Figure 1.5). These elements are consistent with the solar abundances and condensates that we expect to have formed at the pressures and temperatures at the Earth's position in the nebular gradients described above.

1.4 DIFFERENTIATION OF THE EARTH

The planetesimals that now form the Earth probably accumulated in a sequential fashion caused by the gravitational accretion of denser ones first, concentrating Fe-Ni alloys and denser oxides toward the Earth's center. Whether or not the Earth got this "head start" toward further differentiation is hard to say, but it differentiated more extensively soon thereafter (probably about 50 Ma after the beginning of the solar system). This extensive differentiation resulted from heating, caused by gravitational collapse, impacts, and concentrated radioactive heat. Eventually the planet heated sufficiently to initiate melting at some shallow depth, probably beneath a solid crust that stayed cooler by radiating heat to space. Because iron-nickel alloys melt at lower temperatures

than silicates, a dense Fe-Ni-rich liquid probably separated first. Once melting began, mobility within the Earth increased. Denser portions of the melts moved downward, whereas lighter portions rose. The gravitational energy released by this process, plus late impacts, probably generated enough heat to melt the entire Earth, with the possible exception of the outermost chilled layer. This layer may also have melted if there was sufficient gaseous atmosphere to retard radiant cooling. The Moon also formed at this early time, probably due to the impact at a glancing angle of a body about one-tenth the size of the Earth (about the size of Mars) and traveling at a velocity of approximately 2 km/sec. Some of the debris from this collision coalesced to form the Moon, and the rest fell back to the Earth (Taylor and Esat, 1996; Cameron, 1997).

The result of the early differentiation process was the Earth separating into layers controlled by density and the chemical affinities of the elements that comprise it. The concept of chemical affinity will be developed further in Chapters 9 and 27, but, in simplest terms, we can say for now that element behavior is controlled by the configuration of electrons in the outermost shells, which affects their bonding characteristics.

Goldschmidt (1937) proposed that the elements of the Earth tended to incorporate themselves into separate phases, analogous to the layers in ore smelting pots. Although his notion was simplistic by present standards, we have inherited his terms:

- **Lithophile** ("stone-loving") elements form a light silicate phase.
- **Chalcophile** ("copper-loving") elements form an intermediate sulfide phase.
- **Siderophile** ("iron-loving") elements form a dense metallic phase.

A separate phase of **atmophile** elements may also have formed in early Earth as a very minor ocean and atmosphere, but most of these light gaseous elements were not held by the Earth during the earliest stages and escaped into space. Most of the oceans and atmosphere probably accumulated slowly later.

It is simple enough to determine the affinity of every element empirically and use the results to predict the size (thickness) of each layer/reservoir in the early differentiated Earth, but this approach doesn't work very well. For instance, Fe, which should be siderophile, occurs in all three phases. To explain this, we must remember that the atoms are typically ionized, so the requirement of *electric neutrality* must be satisfied as well. We usually concentrate on cations, but anions are equally important. For example, sulfur is obviously required to create a sulfide, so the amount of sulfur dictates the size of the chalcophile layer in smelting pots. Because there was not enough sulfur to satisfy all the chalcophile cations in the Earth, excess chalcophile cations had to go elsewhere. Oxygen is the principal anion in silicate minerals. It combined with silicon for the lithophile layer, but other cations were required before neutrality was achieved. The most common minerals in the lithophile layer of the early Earth were probably olivine ($(Fe,Mg)_2SiO_4$), orthopyroxene ($(Fe,Mg)SiO_3$), and clinopyroxene ($Ca(Fe,Mg)Si_2O_6$). The relative abundance of oxygen thus determined the thickness of the lithophile layer. The inner siderophile layer was determined by the excess of siderophile cations (mostly Fe) left over after neutrality was achieved with O and S. All the other elements, accounting for the remaining 3% of the Earth's mass, went preferentially into one of these layers, in accordance with a particular atom's affinity.

As with the differentiation of the solar system, the Earth's differentiation was certainly not perfectly efficient: not all of the elements are restricted to the predicted layer. Otherwise, we would never find such elements as gold (siderophile), copper (chalcophile), etc., at the Earth's surface today. This may be caused, in part, by a lack of complete equilibrium during the differentiation process, but (as we shall see in Chapters 9 and 27) even if equilibrium is attained, elements typically partition themselves into different reservoirs in less than the most extreme proportions (not all into one reservoir).

After a few hundred million years, this molten, differentiated Earth cooled and mostly solidified to a condition similar to the planet we now inhabit, having a distinct temperature and pressure gradient with depth.

The lithophile, chalcophile, and siderophile layers are not to be confused with the present layers of the Earth: crust, mantle, and core. The core of the modern Earth is the siderophile layer, but the chalcophile component was probably dissolved in the siderophile core and never separated as a distinct phase. Although such a phase does form in smelters, it is much less likely to do so at the high pressures associated with the core. The Earth is not a smelting pot. If a separate chalcophile phase did form, it might be an outermost layer of the outer core.

The mantle certainly represents the early lithophile segregation, but it is unlikely that either the oceanic or the continental crust formed at this point by a large-scale differentiation event in the early Earth (although this probably did happen for the plagioclase-rich highlands of the Moon). If any crust formed from a primordial surface magma ocean, no samples have yet been found. The Earth's crust is believed to have formed later and more progressively. The processes by which the mantle differentiates to produce the crust are predominantly igneous in nature and are occurring to this day. As outlined above, the basaltic oceanic crust is created by partial melting of the mantle at divergent plate boundaries (Chapter 13), and most is eventually consumed by subduction (Chapters 16 and 17) and recycled. The oldest non-subducted oceanic crust is only about 160 Ma old. Most of the heterogeneous continental crust was probably created during the Archean via partial melting of mafic source rocks in subduction zone (Chapters 16 and 17) and rift (Chapters 13, 15, and 19) settings, typically with intermediate to silicic products. Archean island arcs and micro-continents are believed to have assembled to form larger stable continental masses by the early Proterozoic. Early continents have assembled to larger "supercontinents" and rifted apart several

times. The most recent breakup is that of "Pangea," beginning in the Jurassic. Continents grow by collision, arc accretion, and other orogenic processes at continental margins. Due to its thickness, buoyancy, and high viscosity, continental crust is not recycled at subduction zones, so the amount has been increasing over the past 4 Ga.

1.5 HOW DO WE KNOW ALL THIS?

If you are now asking yourself how we can possibly know what has just been presented, you're approaching petrology with the right attitude. Theories, such as those concerning the origin of the universe, the solar system, and the Earth, represent the best inferences we can make based on our interpretation of the data. The simplest explanation of all data, without violating physical "laws," is preferred. The more varied the nature of the phenomena a theory explains, the more confidence we place in it. The scenario described above is consistent with the physical "laws" of celestial mechanics, gravity, nuclear synthesis, and so on. It is also consistent with our observations of seismic waves and the nature and composition of the solar system. But rigorous evaluation of these criteria is well beyond the scope of this book. The scenario is intended only as background information, however. The information on the composition and layering of the Earth's interior in Figures 1.2 and 1.3 is the final result of the process and is presented as fact. This information is very important to the material that we will address in the pages ahead. After all, if igneous rocks are the products of melting at depth, it might be nice to know with some confidence what is being melted. For petrologic concerns, let's focus our skepticism here for a moment. We have not yet drilled a hole to the mantle (and will never do so to the core) in order to directly sample these materials. And our hypothetical mantle and core are far different from the materials we find at the Earth's surface. What evidence do we have to support the alleged composition and structure of our planet?

First, from careful measurements, we can accurately determine the gravitational constant, and use that, plus the measured moment of inertia of the Earth, to calculate its mass and, from that, the average density. This places several constraints on the materials that make up the Earth. For instance, the average density of the Earth is approximately 5.52 g/cm^3. It is relatively easy to observe and inventory the chemical composition of the rocks exposed at the surface of the Earth. But the density of surface rocks is rarely greater than 3.0 g/cm^3. The Earth must thus contain a large proportion of material that is much denser than can be accomplished by compression of surface-type rocks due to the increased pressure at depth.

One could come up with a variety of recipes for the dense material at depth, by mixing proportions of atoms of various atomic weights. However, such a random approach would better be guided by having some idea of which elements are more naturally abundant. The Earth must have formed from the solar nebula, so the composition of the nebula must provide us with significant clues to the makeup of our planet. The material that makes up the solar system can be analyzed from a distance by spectroscopic means. Atoms can be excited by heat or particle interactions and emit characteristic light spectra when they return to their lower-energy "ground state." The wavelength of light that reaches the Earth can be determined and related to the type of element or compound that emits it. By comparison with spectra of elements measured in the laboratory, the emitting atoms or molecules can be identified. The intensity of the spectral lines is proportional to their concentrations at the source. We thus get a good idea of what elements constitute the sun, other stars, and even other planetary surfaces, and by analogy with these, our own planet.

Figure 1.6 illustrates the estimated concentrations of the elements in the solar nebula (estimated from certain meteorites, as discussed below). Note the logarithmic scale for concentration, which makes it easier to show the full range of abundances. Hydrogen is by far the most abundant element, as it made up most of the original nebula. Other elements (except He) were synthesized from H in the sun and other stars. The decrease in abundance with increasing atomic number (Z) reflects the difficulty of synthesizing progressively larger atoms. Another interesting feature that is clear from Figure 1.6 is the "sawtooth" nature of the curve. This is in accordance with the Oddo-Harkins rule, which says that atoms with even atomic numbers are more stable, and hence more abundant, than their odd-numbered neighbors.

We must assume that the elements that compose the Earth are among the more common elements in Figure 1.6. For example, Fe, and to a lesser extent Mg and Ni, are much more abundant in the solar system than in the Earth's crust, so we might infer that these elements are concentrated elsewhere in the Earth. Fe is also dense enough to satisfy the Earth's high density requirement. In other words, using the data in Figure 1.6 as a starting point to model a planet with an average density of 5.52 g/cm^3 should lead us in the direction of the concentrations in Figure 1.5. Of course, the process is complicated by such inhomogeneities as the radial differentiation of the solar nebula and density variations and phase changes associated with increasing pressure in the Earth.

Seismic studies place further constraints on the materials that constitute the Earth. The velocities of *P* and *S* waves in various materials at elevated pressures and temperatures can be measured in the laboratory and compared to seismic velocities within the Earth, as determined from earthquakes or human-made explosions (Figure 1.3). In addition, reflection and refraction of seismic waves at discontinuities within the Earth provide direct evidence for the Earth's internal structure and the depths of the discontinuities that subdivide it into crust, mantle, outer core, and inner core, as well as other more detailed features.

Finally, although we haven't visited the mantle or core for samples, we have had samples delivered to us at the surface (or so we believe). (We will talk more about mantle samples in Chapters 10 and 19.) There are a number of rocks found at the surface that we believe to be of mantle origin. In many active and fossil subduction zones, slivers of oceanic crust and underlying mantle are incorporated into the accretionary prism. Thickening of the prism, followed by uplift and erosion, exposes the mantle-type rocks. Xenoliths of presumed mantle material are occasionally carried to the surface in some basalts. Deeper mantle material is believed to come to the surface as xenoliths in diamond-bearing kimberlite pipes (Section 19.3.3). The vast majority of samples found in all these situations are olivine- and pyroxene-rich ultramafic rocks. When many of these samples are partially melted in the laboratory, they produce melts similar to natural lavas that we believe to be mantle derived.

Because of the great density and depth of the Earth's core, not a single sample of the core has reached the surface. However, we do believe that pieces of the core of other planetesimals have reached Earth in the form of some meteorites. We will briefly discuss meteorites in the next section, as they are quite varied and provide important information about the composition of the Earth and solar system.

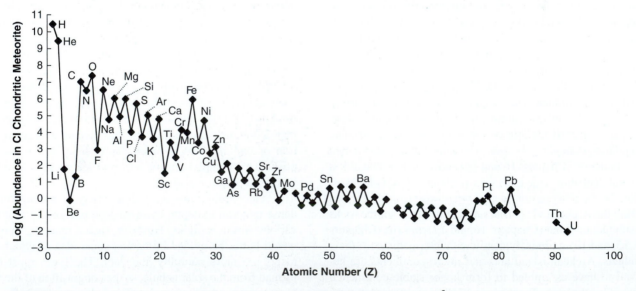

FIGURE 1.6 Estimated abundances of the elements in the solar nebula (atoms per 10^6 Si atoms). After Anders and Grevesse (1989). Copyright © with permission from Elsevier Science.

1.6 METEORITES

Meteorites are solid extraterrestrial objects that strike the surface of the Earth after surviving passage through the atmosphere. Most of them are believed to be fragments derived from collisions of larger bodies, principally from the asteroid belt between the orbits of Mars and Jupiter. They are very important because many are believed to represent arrested early to intermediate stages in the development of the solar nebula that have not undergone subsequent alteration or differentiation like the Earth. They thus provide valuable clues to the makeup and development of the solar system. Meteorites have been classified in a number of ways. Table 1.1 is a simplified classification, in which I have combined several subclasses to give a general indication of the more important types and the percentages of each from observed falls.

Irons (Figure 1.7b) are composed principally of a metallic Fe-Ni alloy, **stones** are composed of silicate minerals, and **stony-irons** (Figure 1.7a) contain subequal amounts of each. Because stones look much like terrestrial rocks,

TABLE 1.1   Simplified Classification of Meteorites

Class	Subclass	# of Falls	% of Falls
Irons	All	42	5
Stony-irons	All	9	1
Stones Achondrites	SNC's	4	8
	Others	65	
Chondrites	Carbonaceous	35	86
	Others	677	

After Sears and Dodd (1988).

they are seldom recognized as meteorites, so irons (quickly recognizable by their density) tend to dominate museum collections. When we consider only specimens collected after an observed fall, however, stones comprise 94% of meteoritic abundance. Iron meteorites are believed to be fragments of the core of some terrestrial planets that have undergone differentiation into concentrations of silicate,

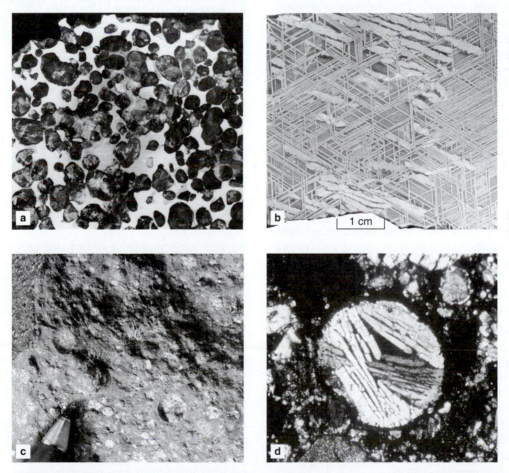

FIGURE 1.7   Meteorite textures. **(a)** Polished section of the stony-iron Springwater (Saskatchewan, Canada) meteorite (light is iron, dark is ~5mm olivines). © Courtesy Carleton Moore, Arizona State University. **(b)** Widmanstätten texture in the Edmonton (Kentucky) iron meteorite. © Courtesy John Wood and the Smithsonian Astrophysical Observatory. **(c)** Chondrules in the Allende chondrite meteorite (Mexico). Ball pen tip for scale. **(d)** Photomicrograph of a 0.5 mm diameter chondrule from the Dhajala (India) chondrite meteorite. The chondrule is composed of skeletal plates of olivine in a finer matrix, including crystal fragments and glass. Other chondrules may be dominated by orthopyroxene, or mixtures of several minerals, including sulfides. Chondrules may be very fine grained or even totally glassy in meteorites that have suffered little or no reheating. Chondrite photographs © courtesy of the Geological Museum, University of Copenhagen.

sulfide, and metallic liquids, in the manner discussed above for the Earth. These meteorites contain portions of siderophile (Fe-Ni alloy) and chalcophile (segregations of troilite: FeS) phases. The Fe-Ni alloy is composed of two phases, kamactite and taenite, which exsolved from a single, homogenous phase as it cooled. The two phases are commonly intergrown in a crosshatched pattern of exsolution lamellae called Widmanstätten texture (Figure 1.7b). Stony-irons are similar to irons but include a significant proportion of the silicate (lithophile) segregation mixed in. All irons and stony-irons are considered "differentiated" meteorites because they come from larger bodies that underwent some degree of chemical differentiation. Meteorites, however, display a large variation in the degree to which they represent differentiated portions of a planet. The parent bodies, most believed to be present in the asteroid belt, were of various sizes and thus capable of different degrees of differentiation. In addition, the collisions that disrupted the parent bodies into meteoritic fragments have remixed and even brecciated the material that we find in many meteorites.

Stones are further subdivided on the basis of whether they contain **chondrules** (Figure 1.7c and d), nearly spherical silicate inclusions between 0.1 and 3.0 mm in diameter. At least some chondrules appear to have formed as droplets of glass that have subsequently crystallized to silicate minerals. Stones with chondrules are called **chondrites**, whereas those without are called **achondrites**. As with irons and stony-irons, achondrites are differentiated meteorites. Chondrites, on the other hand, are considered "undifferentiated" meteorites because the heat required to initiate melting and differentiation of a planet would certainly have destroyed the glassy chondrules. The small size of the chondrules indicates rapid cooling (< 1 hr), requiring a cooler nebula at the time of their formation. The chondrules probably formed after condensation but before formation of the planetesimals. Chondrites are thus considered to be the most "primitive" type of meteorites, in the sense that they are thought to have compositions closest to the original solar nebula (hence their use in estimating solar abundances in Figure 1.6). It has been suggested that all of the inner terrestrial planets formed from a material of average chondritic composition. This has led to the Chondritic Earth Model (CEM), which provides a close fit to the composition of the Earth for most elements, but with a few important differences. For example, the Earth is much denser and must have a higher Fe/Si ratio than chondrites. Models such as the one presented above, based on condensation temperatures as a function of distance from the sun, are much better for explaining the chemical composition of the planets (particularly their variations) than is assuming that some meteorite represents them all.

Further subdivision of meteorites is based on their textures and/or mineral content. There is considerable variety in the overall ("bulk") composition, as well as in the mineralogy. More than 90 minerals have been found among the stony meteorites, some of which are not found elsewhere on Earth. Some meteorites appear to come from the moon and neighboring planets. The SNC meteorites, for example, appear to be from Mars. Given this variety, the study of meteorites can provide us with valuable information on the chemical composition of the solar system and its constituents.

1.7 PRESSURE AND TEMPERATURE VARIATIONS WITH DEPTH

We now have a good idea of what comprises the Earth and how it got there. If we are to proceed to an understanding of melts (and later of metamorphism), we should next attempt to understand the physical conditions (pressure and temperature) that occur at depth, so that we can appreciate how these materials respond and behave. As depth within the Earth increases, both pressure and temperature increase as well. Pressure increases as a result of the weight of the overlying material, whereas temperature increases as a result of the slow transfer of heat from the Earth's interior to the surface.

1.7.1 The Pressure Gradient

The pressure exerted in a ductile or fluid medium results from the weight of the overlying column of the material. For example, the pressure that a submarine experiences at depth is equal to the weight of the water above it, which is approximated by the equation:

$$P = \rho g h \qquad (1.1)$$

where: P = pressure

ρ = the density (in this case, that of water)

g = the acceleration caused by gravity at the depth considered

h = the height of the column of water above the submarine (the depth)

Because water is capable of flow, the pressure is equalized so that it is the same in all directions. The horizontal pressure is thus equal to the vertical pressure (the axis along which the imaginary column of water would exert itself). This equalized pressure is called **hydrostatic** pressure. Near the surface, rocks behave in a more brittle fashion, so they can support unequal pressures. If the horizontal pressures exceed the vertical ones (or vice versa), rocks may respond by faulting or folding. At depth, however, the rocks also become ductile and are capable of flow. Just as in water, the pressure then becomes equal in all directions, and is termed **lithostatic** pressure. Equation (1.1) will apply then, too, with ρ being the density of the overlying rock.

The relationship between pressure and depth is complicated because density increases with depth as the rock is compressed. Also, g decreases as the distance to the center of the Earth decreases. A more accurate approach would be to use a differential form of the P–depth relationship, complete with estimates of the variation in g and ρ, and integrate it over the depth range. However, the changes in g and the density of a given rock type are relatively minor in the crust and upper mantle, and they also tend to offset each other, so Equation (1.1) should suffice for our needs. Only when the rock type changes, as at the Moho, would a different value of ρ be required. One need only calculate the pressure to the base of the crust, using an appropriate average crustal density, and continue with depth using a density representative of the mantle.

For example, a reasonable estimate of the average density of the continental crust is 2.8 g/cm^3. To calculate the pressure at the base of 35 km of continental crust, we need only substitute these data into Equation (1.1), being careful to keep the units uniform (refer to the front inside cover for units and constants):

$$P = \frac{2800 \text{ kg}}{\text{m}^3} \cdot \frac{9.8 \text{ m}}{\text{s}^2} \cdot 35,000 \text{ m}$$
$$= 9.6 \times 10^8 \text{ kg/(m s}^2)$$
$$= 9.6 \times 10^8 \text{ Pa} \approx 1 \text{ GPa}$$

This result is a good average pressure gradient in the continental crust of 1 GPa/35 km, or about 0.03 GPa/km, or 30 MPa/km. Because of (upward) round-off, this gradient is also suitable for oceanic crust. A representative density for the upper mantle is 3.35 g/cm^3, resulting in a mantle pressure gradient of about 35 MPa/km. These are numbers worth remembering, as they provide a good way to interpret pressures in the phase diagrams to come. Figure 1.8 shows the variation in pressure with depth using the Preliminary Reference Earth Model (PREM) of Dziewonski and Anderson (1981).

1.7.2 Heat Transfer and the Temperature Gradient

Determining the **geothermal gradient**, the temperature variation with depth, is much more difficult than doing so for pressure, as there is no simple physical model analogous

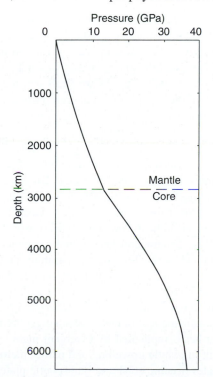

FIGURE 1.8 Pressure variation with depth. From Dziewonski and Anderson (1981).

to Equation (1.1). There are models, however, based on methods of heat transfer, that must be constrained to conform to measured heat flow at the surface. There are two primary sources of heat in the Earth.

1. **Secular cooling** Primordial heat developed early in the history of the Earth from the processes of accretion and gravitational differentiation described in Section 1.4 has been gradually escaping since that time. This set up an initial temperature gradient once the planet solidified and began to cool. Some continued gravitational partitioning of iron in the inner core may contribute some heat as well. Estimates of the primordial contribution to the total surface heat flux is 10 to 25%.

2. **Heat generated by the decay of radioactive isotopes** For reasons that will be discussed in Chapters 8 and 9, most of the more radioactive elements are concentrated in the continental crust. Radioactive decay produces 75 to 90% of the heat that reaches the surface.

Once generated, heat is transferred from hotter to colder regions by any of four processes, depending on the nature of the material involved in the transfer:

1. If a material is sufficiently transparent or translucent, heat can be transferred by **radiation**. Radiation is the movement of particles/waves, such as light or the infrared part of the spectrum, through another medium. This is the principal way a lamp loses heat, or how the Earth loses heat from its surface into space. It is also the way we receive heat energy from the sun. Heat transfer by radiation is not possible within the solid Earth except possibly at great depth, where silicate minerals may become hot enough to lose some of their opacity to infrared radiation.

2. If the material is opaque and rigid, heat must be transferred through **conduction**. This involves the transfer of kinetic energy (mostly vibrational) from hotter atoms to adjacent cooler ones. Heat conduction is fairly efficient for metals, in which electrons are free to migrate. This is why you can burn yourself if you handle an iron bar that has one end in a fire. Conduction is poor for silicate minerals but relatively efficient in the core.

3. If the material is more ductile, and can be moved, heat may be transferred much more efficiently by **convection**. In the broadest sense, convection is the movement of material due to density differences caused by thermal or compositional variations. For the present purposes, we shall consider the type of convection that involves the expansion of a material as it heats, followed by the rise of that material due to its gain in buoyancy. This convection explains why it is hotter directly above a candle flame than beside it. The air is heated, it expands, and it rises because it is now lighter than the air around it. The same thing can happen to ductile rocks or liquids. Convection may involve flow in a single direction, in which case the moved hot material will accumulate at the top of the

ductile portion of the system (or a cooled density current will accumulate at the bottom). Convection may also occur in a cyclic motion, typically in a closed cell above a localized heat source. In such a *convection cell*, the heated material rises and moves laterally as it cools and is pushed aside by later convective matter. Material pushed to the margin cools, contracts, and sinks toward the heat source, where it becomes heated and the cycle continues.

4. **Advection** is similar to convection but involves the transfer of heat with rocks that are otherwise in motion. For example, if a hot region at depth is uplifted by either tectonism, induced flow, or erosion and isostatic rebound, heat rises physically (although passively) with the rocks.

Convection works well in the liquid core and in the somewhat fluid sub-lithospheric mantle. Convection is also a primary method of heat transfer in hydrothermal systems above magma bodies or within the upper oceanic crust, where water is free to circulate above hot rock material. Beyond these areas, however, conduction and advection are the only available methods of heat transfer. The transfer of heat is a very important concept in petrology, as it controls the processes of metamorphism, melting, and crystallization, as well as the motion, mixing, and mechanical properties of Earth materials. Several petrologic processes, from explosive volcanism to metamorphism to lava flows and pluton emplacement, are critically dependent upon maintaining a heat budget. We will consider heat transfer again later, in association with the cooling of a body of magma. The rate at which a body cools depends upon a number of variables, including the size, shape, and orientation of the body; the existence of a fluid in the surroundings to aid in convective heat transfer; and the type, initial temperature, and permeability of the country rocks. An example of heat models applied to a cooling magma body will be discussed in Section 21.3.1.

Figure 1.9 schematically illustrates heat flow in the upper few hundred kilometers of the Earth's interior. The passage of seismic waves through the mantle demonstrates that it reacts as a solid to abrupt stress changes. The mantle is therefore solid, but the sub-lithospheric mantle is capable of flow when subject to slower stress changes, such as those associated with heating. Heat is thus transported more effectively by convection at depth (steep gradient in Figure 1.9) but only by conduction across the stagnant rigid lithosphere (shallow gradients in Figure 1.9). Both the conductive and convective geothermal gradients are nearly constant (linear) but of different magnitudes. Deep convection is sufficiently rapid that no heat is lost by conduction to the surroundings, so heat is retained within the rising material. Processes in which heat is neither lost nor gained are called **adiabatic** processes. Between the convecting and conducting layers is a "boundary layer" a few tens of kilometers thick across which occurs the transition in rheology and heat transfer mechanism. Thinner lithosphere (such as in the ocean basins and illustrated on the right side of Figure 1.9) allows convective heat rise to shallower depths than does thicker

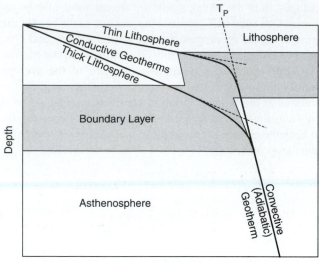

FIGURE 1.9   Diagrammatic cross-section through the upper 200 to 300 km of the Earth, showing geothermal gradients reflecting more efficient adiabatic (constant heat content) convection of heat in the mobile asthenosphere (steeper gradient) and less efficient conductive heat transfer through the more rigid lithosphere (shallower gradients). The boundary layer is a zone across which the transition in rheology and heat transfer mechanism occurs. The thickness of the boundary layer is exaggerated here for clarity: it is probably less than half the thickness of the lithosphere. Notice that thinner lithosphere (on right) allows convective heat transfer to shallower depths, resulting in a higher geothermal gradient (greater change in temperature for a given pressure increment) across the boundary layer and lithosphere. T_P is the *potential temperature*: the temperature that the deep solid mantle would attain at atmospheric pressure as extrapolated adiabatically to the surface (McKenzie and Bickle, 1988).

lithosphere (continental, on the left in Figure 1.9). This results in a higher geothermal gradient (greater temperature change for a given increment in pressure) across the oceanic boundary layer and lithosphere, across which it must drop to ambient surface conditions. As we shall see, most melting occurs within the upper few hundred kilometers of the Earth's interior, so the temperature distribution at these levels is of great interest to petrologists.

Geothermal gradients are typically calculated using models based on measured heat flow at the surface or in drill holes or mines. Estimates of the steady-state heat flow from the mantle range from 25 to 38 mW/m^2 beneath the oceans to 21 to 34 mW/m^2 beneath the continents. Heat flow is commonly expressed in heat flow units (HFU), where 1 HFU = 41.84 mW/m^2. Continental geothermal gradients are fairly well constrained by equilibration temperatures and pressures determined for mantle xenoliths from various depths (e.g., Rudnick and Nyblade, 1999). Because xenoliths are rarely found in the open ocean, oceanic geothermal gradients rely strictly on mathematical models. Oceanic geothermal gradients are further complicated by divergent plate boundaries, which result in mantle upwelling and partial melting, which in turn creates oceanic crust and lithospheric plates at ocean ridges (Chapter 13). The geothermal gradient in the ocean basins thus depends on the age of the lithosphere (and hence

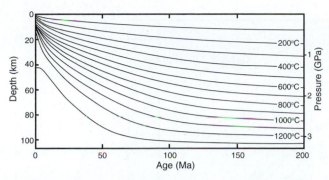

FIGURE 1.10 Temperature contours calculated for an oceanic plate generated at a mid-ocean ridge (age 0) and thickening as it cools. The 1300°C isotherm is a reasonable approximation for the base of the oceanic lithosphere. The plate thus thickens rapidly from zero to 50 Ma and is essentially constant beyond 100 Ma. From McKenzie et al. (2005).

distance from the ridge and spreading rate). Figure 1.10 shows the temperature contours calculated by McKenzie et al. (2005) for depth versus age of oceanic lithosphere. The transition from rigid (lithospheric) to ductile (asthenospheric) behavior is in the 1300 to 1400°C range (1375°C, according to McKenzie), so the 1300°C isotherm can be used as an approximation for the base of the oceanic lithospheric plate. The plate thus thickens from about 50 km at the ridge to ~110 km at "maturity" of ~100 Ma, beyond which it remains essentially constant in thickness and thermal structure. In the continental lithosphere heat flow is high in rifts and orogenic belts but settles down to a steady state after about 800 Ma in the platforms and shields (Sclater et al., 1980).

Several estimates of geothermal gradients for continental shields and mature oceanic intraplate settings are shown in Figure 1.11. Because the thermal conductivity of shallow crustal rocks is very low, heat transfer is slow, and the shallow geothermal gradient is correspondingly high. Simply extrapolating the shallow gradient on the basis of heat flow measurements leads to an impossibly high geothermal gradient (as shown). As expected, the geothermal gradient is higher in the ocean setting than in the continental cratons because the oceanic lithosphere is thinner. Shallow gradients in old oceanic lithosphere in Figure 1.11 range from about 17 to 20°C/km, gradually steepening to about 7°C/km with depth. Shallow shield gradients range from about 12 to 18°C/km, transitioning to about 4 to 6°C/km below about 30 km. The oceanic versus shield temperature differences are essentially restricted to the lithosphere and boundary layer, so the oceanic and continental curves converge by about 250 to 300 km where convection has a homogenizing influence (see Figure 1.9). Nearly adiabatic (constant heat content) convective heat flow deeper than about 300 km results in a linear geothermal gradient of approximately 0.3°C/km (~10°C/GPa), as shown in Figure 1.12. The gradient shallows across the D″ layer and then becomes steep in the metallic core, where the thermal conductivity and convection (in the liquid portion at least) is very high. The density contrast across the core–mantle boundary prohibits core material from rising into the mantle (convection), so heat can only be

transferred upward (across the boundary) by conduction, resulting in a thermal boundary layer across which the thermal gradient is estimated to be 300 to 1000°C.

1.7.3 Dynamic Cooling of the Earth: Geodynamics and Plate Tectonics

Figure 1.11 is intended to illustrate the thermal structure at shallow levels of the Earth's interior in stabilized areas, similar to a static, chemically and physically stratified Earth with continents and oceans. The Earth is far more dynamic, however, behaving like a viscous solid in a gravitational field, heated from within and below, cooled from above, and expanding when heated. Convection in such a fluid occurs when the thermal **Rayleigh number** (a dimensionless ratio representing the potential vigor of convection) exceeds 1000. With a Rayleigh number about 10,000 times this, the lower mantle not only convects but probably does so vigorously. Laboratory and model studies demonstrate that the temperature variations that drive convective flow are concentrated in thin boundary layers that are much smaller than the overall circulation pattern. Given the temperature-dependent viscosity of mantle material, the upper boundary of this system is a cool thermal boundary layer with low viscosity: the lithosphere. Because of its high density, the lithosphere is gravitationally unstable (negatively buoyant). In laboratory models of convecting fluids, cool dense upper thermal boundary layers descend as either cylindrical downwellings or as larger networks of partially connected tabular-shaped downwelling slabs, depending on the rheology of the material and the ratio of internal to basal heating. The mantle is perhaps 80 to 90% internally heated, largely by radioactive decay, with roughly 10% of its heat coming from the base/core. This ratio and the stiff rheology of the lithosphere clearly lead to slab-like downwelling, resulting in plate tectonics. Early debates as to whether mantle convection drives plate tectonics or plate instabilities lead to mantle convection now appear misguided. Mantle convection and plate tectonics are inseparable manifestations of the heat-driven dynamic cooling process of the Earth, given its present thermal state and rheological properties. For the Earth, plate tectonics *is* mantle convection.

Several forces are at work, however, and their relative contribution has been debated since plate tectonics was first recognized. The negative buoyancy of the descending plates has been called **slab pull**, with the regrettable implication that the dense subducting slab pulls the rest of the plate after it. The motion of the lithospheric plates down and away from the elevated mid-ocean ridges has also received an unfortunate name: **ridge push**. Both, however, are actually *body* forces, affecting the entire plate (not just a pulled or pushed end). Slab pull may be considered a horizontal pressure gradient associated with the sinking of a slab. As it descends in a subduction zone, the slab can't pull significantly on the rest of the plate, which would simply fault and break. Instead, as it sinks, the slab sets up circulation patterns in the mantle that exert a sort of suction force. As the slab descends from the surface, the pressure behind it is lowered, which is immediately compensated by feeding more plate

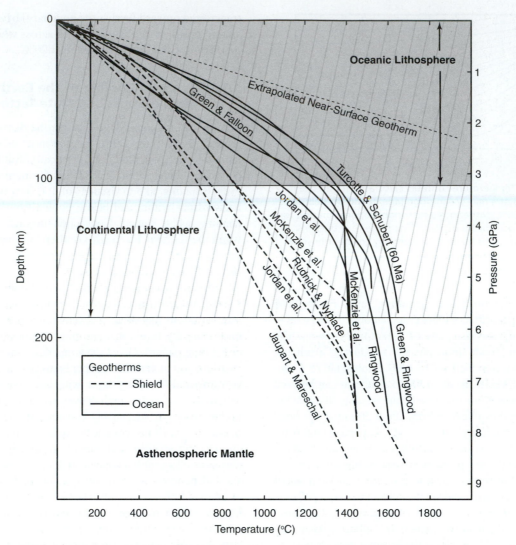

FIGURE 1.11   Estimates of oceanic (solid curves) and continental shield (dashed curves) geothermal gradients to a depth of 300 km. The thickness of mature (> 100Ma) oceanic lithosphere is shaded, and that of continental shield lithosphere is hatched. Data from Green and Falloon (1998), Green and Ringwood (1963), Jaupart and Mareschal (1999), McKenzie et al. (2005 and personal communication), Ringwood (1966), Rudnick and Nyblade (1999), and Turcotte and Schubert (2002).

into the subduction zone. Ridge push is the gravitational sliding of a plate off the elevated ridge, not an active push by ascending mantle flow, which would buckle the plate. Forsyth and Uyeda (1975) found that plate motion velocities are roughly proportional to the percentage of a plate's perimeter that corresponds to subduction zones and not to ridges, strongly implying that the negative buoyancy associated with slab pull is the principal force. Lithgow-Bertelloni and Richards (1998) estimate that ridge push constitutes only 5 to 10% of the driving force of slab subduction. The predominance of slab pull and the lack of any significant gravity anomaly at mid-ocean ridges suggest that mantle upwelling at divergent plate boundaries, which moves heat upward, is essentially a passive response to plate separation and descent. Upward heat transfer is thus essentially *advective* in this situation (carried upward in material that is rising for reasons other than thermal expansion). We shall investigate the petrologic consequences of this upwelling in Chapter 13.

From this brief description, we can conclude that the scale of the plates effectively controls the scale of mantle convection (i.e., the location of major upwelling and downwelling circulation), and not the reverse.

But what of the other boundary layer at the base of the mantle? The density more than doubles from the lower mantle to the core, thus preventing convection across the boundary, regardless of thermal differences. The liquid outer core must be internally convecting in order to create the Earth's magnetic field, and this convection delivers heat to the base of the mantle. The D" layer may be the resulting mantle-side thermal boundary layer, across which the temperature gradient is estimated to be 300 to 1000°C. Convection within the mantle results from thermal instabilities as the heated D" layer becomes positively buoyant. The viscosity is much lower than that of the plates at these elevated temperatures, and the instabilities take the alternative form of rising cylindrical plumes rather than slabs. These **plumes** are more of an

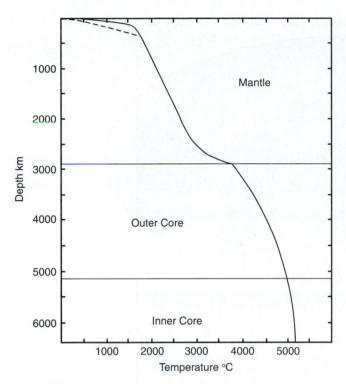

FIGURE 1.12 Estimate of the geothermal gradient to the center of the Earth (after Stacey, 1992). The shallow solid portion is very close to the Green and Ringwood (1963) oceanic geotherm in Figure 1.11, and the dashed geotherm is the Jaupart and Mareschal (1999) continental geotherm.

active upwelling, in comparison to the more passive shallow mantle rise in response to plate separation. The plumes appear to be largely independent of plate tectonics, but in many cases are sufficiently vigorous to penetrate the lithosphere and reach the Earth's surface, where they result in **hotspot** volcanism (Chapter 14) and large elevated lithospheric swells. Calculations of the excess heat required to create such swells have yielded estimates of the total heat transported by plumes, which agree with estimates of the heat flux from the core to the mantle (approximately 10% of the total heat flux to the surface). Plate tectonics thus appears to be the method by which a largely internally heated and viscous mantle cools by convection, whereas plumes essentially cool the core and have an increasing effect on the lower mantle.

In addition to plate tectonics and plumes, Earth rotation, melt migration, frictional drag, compositional heterogeneity, phase transformations, "torroidal" (strike-slip) plate motion, and the insulating effects of continents, among other factors, further complicate the heat flow and dynamics of the mantle. All of these factors can modify the simple static Earth model depicted in Figures 1.9 and 1.11. Figure 1.13 illustrates the variation in surface heat flux based on a smoothing of data from surface measurements from more than 20,000 measuring stations. The darker patterns clearly indicate the role of mid-ocean ridges as heat "leaks" with high geothermal gradients. Pollack et al. (1993) also note that fully half of the Earth's heat loss is associated with the cooling of relatively young Cenozoic oceanic lithosphere. Cratons are cooler, but not uniformly so. The mathematical

smoothing process used hides small-scale features such as hotspot plumes. A technique called *seismic tomography* uses powerful computers to model three-dimensional seismic wave velocity distributions in the Earth. Velocity anomalies are usually associated with temperature variations, so the technique has great promise for yielding a clearer picture of detailed mantle dynamics. For a general discussion of seismic tomography, see Lowrie (1997). Schubert et al. (2001) also provide a comprehensive review of thermal and seismic models of the Earth. Comparison of mantle temperatures from one locality to another is complicated by the natural variation in temperature with depth. The problem can be alleviated by using a pressure reference frame. McKenzie and Bickle (1988) suggested a 1-atmosphere reference they called the *potential temperature* (T_P): the temperature that the solid mantle would attain if it could reach the surface adiabatically without melting (see Figure 1.9).

Even a general model of mantle convection, plumes, etc. depends on the influence of the 660-km phase transition on mantle dynamics. Some investigators consider the density contrast across the boundary insufficient to impede convection, so that mantle convection cells span the full vertical extent of the mantle. Such models are called *whole-mantle* models. Others believe the 660-km transition can impede convection, so that warm, buoyant mantle material in rising portions of deep convection cells would be less buoyant than the material above the transition, causing the rising material to spread laterally instead. If so, then heat can transfer across the boundary only by conduction, perhaps inducing convection in the upper mantle. Flow in the upper layer can also be induced by plate motion, rising at divergent boundaries and sinking with subduction. Models with a 660-km barrier are called *two-layer* mantle models. Figure 1.14 combines aspects of both model types. In it, the 660-km boundary impedes most lower mantle convection from rising further, but subducted oceanic lithosphere (although impeded somewhat) is dense enough to sink through the boundary layer and accumulate as the D″ layer. According to this model, when heated at the core–mantle boundary, this material regains positive buoyancy, resulting in ascending plumes that rise vigorously enough to penetrate the boundary again and reach the surface (e.g., Hawaii). The 660-km boundary layer in such models allows some material transfer but impedes wholesale mantle convection from homogenizing the full vertical extent of the mantle and thus permits the composition of the shallow and deep mantle to evolve independently. Mantle flow dynamics and heat transfer play critical roles in magma genesis and evolution, so we will be exploring many of these processes (and the mantle layering controversy) more fully in subsequent chapters.

1.8 MAGMA GENERATION IN THE EARTH

We next address the problem of magma generation. **Petrogenesis** is a good general term in igneous petrology for the generation of magma and the various methods of diversification of such magmas to produce igneous rocks. Most magmas originate by melting in the Earth's mantle, but some show evidence of at least a partial crustal component.

Heat Flow

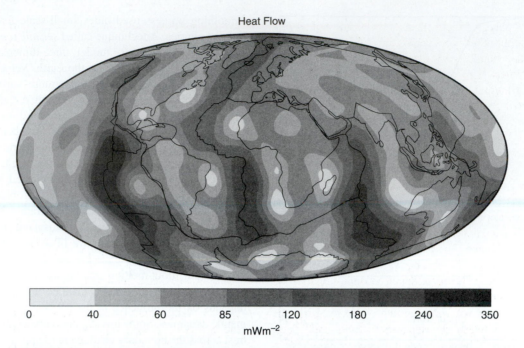

mWm⁻²

FIGURE 1.13   Pattern of global heat flux variations compiled from observations at more than 20,000 sites and modeled on a spherical harmonic expansion to degree 12. From Pollack et al. (1993). © AGU with permission.

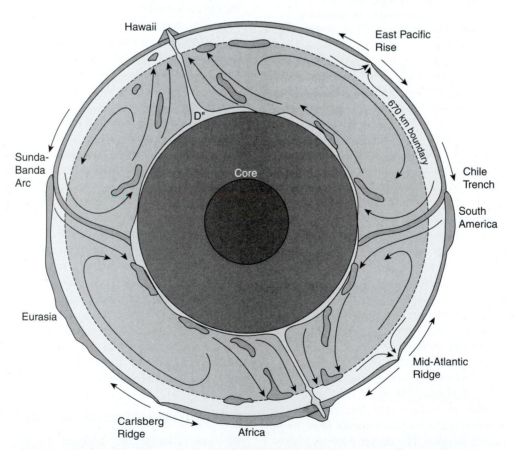

FIGURE 1.14   Schematic diagram of a two-layer dynamic mantle model in which the 660-km transition is a sufficient density barrier to separate lower mantle convection (arrows represent flow patterns) from upper mantle flow, largely a response to plate separation. The only significant things that can penetrate this barrier are vigorous rising hotspot plumes and subducted lithosphere (which sink to become incorporated in the D″ layer, where they may be heated by the core and return as plumes). After Silver et al. (1988).

Plate tectonics plays a major role in the generation of several magma types, but other types seem to result from processes at depths in the mantle greater than are influenced directly by plate tectonics. Figure 1.15 is a very generalized summary of the principal types of magmas and their geologic setting. We will study the processes and products in much more detail in Chapters 13 through 20.

The most voluminous igneous activity occurs at divergent plate boundaries. Of these, mid-ocean ridges (location 1 in Figure 1.15) are the most common (Chapter 13). As introduced above, the shallow mantle beneath the ridge undergoes partial melting, and the resulting basaltic magma rises and crystallizes to produce the oceanic crust. If a divergent boundary is initiated beneath a continent (location 2), a similar process takes place. The resulting magmatism, particularly at the early stages of continental rifting, is commonly alkaline and typically shows evidence of contamination by the thick continental crust (Chapter 19). If the rift continues to develop, oceanic crust will eventually be created in the gap that forms between the separating continental fragments. The result will be a new ocean basin and igneous activity similar to that in location 1.

An oceanic plate created at mid-ocean ridges moves laterally and eventually is subducted beneath a continental or another oceanic plate. Melting also takes place at these subduction zones. The number of possible sources of magma in subduction zones is far more numerous than at ridges and may include various components of mantle, subducted crust, or subducted sediments. The types of magma produced are correspondingly more variable than for divergent boundaries, but andesites are the most common. If oceanic crust is subducted beneath oceanic crust (location 3), a volcanic island arc forms (Chapter 16). If oceanic crust is subducted beneath a continental edge (location 4), a continental arc forms along the "active" continental margin (Chapter 17). A continental arc is generally more silica rich than is an oceanic arc. Plutons are also more common in continental arcs, either because the melts rise to the surface less efficiently through the lighter continental crust or because uplift and erosion is greater in the continents and exposes deeper material.

A different, and slower, type of plate divergence typically takes place behind the volcanic arc associated with subduction (location 5). Most geologists believe some sort of "back-arc" extension is a natural consequence of subduction, probably created by frictional drag associated with the subducting plate. Such drag pulls down part of the overlying mantle, requiring replenishment from behind and below. Back-arc magmatism is similar to mid-ocean ridge volcanism. Indeed, a ridge also forms here, and oceanic crust is created and spreads laterally from it. Back-arc spreading, however, is slower, volcanism is more irregular and less voluminous, and the crust created is commonly thinner than in the oceans. At times, rifting occurs behind a continental arc, and the volcanic portion separates from the continent as a marginal sea forms through back-arc spreading. Such a process is believed to have separated Japan from the Asian mainland. At other times, such a process seems to initiate, and then it mysteriously ceases. The result may just be a graben structure, or plateau-type basalts may form prior to cessation of activity.

Although magmatism is certainly concentrated at plate boundaries, some igneous activity also occurs within the plates, both oceanic (location 6) and continental (location 7). Ocean islands such as Hawaii, the Galapagos, and the Azores all form via volcanism within the oceanic plates (Chapter 14). The products are usually basaltic but are commonly more alkaline than ridge basalts. The reason for this type of igneous activity is much less obvious than it is for plate margins because our plate tectonic paradigm is of little use in these mid-plate regimes. The source of the melts is also less clear but appears to be deep, certainly well into the asthenosphere. Several of these occurrences exhibit a pattern of igneous activity that gets progressively younger in one direction. The direction correlates well with plate motion in a manner which suggests that the plate is moving over a stationary "hotspot," or mantle, "plume" (as described above) with the most recent activity occurring directly over the plume. Intraplate activity within continental plates is much more variable than that within the oceans (Chapter 19). It is compositionally variable but usually alkaline—and occasionally extremely so. This reflects the more complex and heterogeneous continental crust and subcontinental mantle as well. Some of the most unusual igneous rocks, such as kimberlites and carbonitites, occur within continental provinces. The term **igneous–tectonic association** refers to these broad types of igneous occurrence, such as mid-ocean ridge, island arc, and intra-continental alkalic systems. We will address these associations in the later chapters of the igneous section.

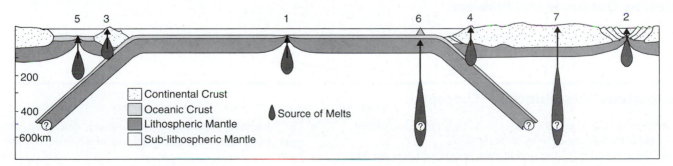

FIGURE 1.15   Generalized cross section illustrating magma generation associated with various plate tectonic settings.

Summary

Petrology is the study of the nature and origin of rocks. Igneous petrology addresses the processes that produce melts (magma), how those melts then rise, the chemical and mineralogical evolution as they cool and crystallize, and the eventual eruption or emplacement of the resulting rock bodies. Field, chemical, and textural criteria (the latter best observed using the polarized-light microscope) are the principal ones used to evaluate these processes. The Earth comprises the ~10 km thick Mg-Fe-rich (mafic) oceanic crust and the ~40–60 km thick Si-Al-rich (sialic) continental crust, both underlain by ultramafic mantle (~3000 km thick), as well as the central metallic Fe-rich core. From a rheological perspective, the Earth comprises the rigid lithospheric plates, underlain by the slightly molten ductile asthenosphere, the more rigid mesosphere, and the outer liquid and inner solid core. The Earth originated by accretion of matter in the solar nebula to form small planetesimals, which further collapsed under the influence of gravity to form the Earth itself. Concentration of heavy elements (principally Fe and lesser Ni) toward the center of the Earth released sufficient potential energy to melt at least most of the Earth and differentiate it into the metallic core and silicate mantle (and perhaps a thin ocean and atmosphere). The crust evolved later and more gradually by plate tectonic processes: oceanic crust at divergent plate boundaries and continental crust as small island-arc terranes at convergent (subduction) boundaries, which accreted by collisions to form larger continental masses.

If we are to evaluate melting and crystallization in the Earth, we must understand the pressure and temperature gradients. Because melts that reach the surface are generated in the crust and upper mantle, we are most interested in the gradients across the upper few hundred kilometers. Pressure increases with depth due to the weight of the progressively greater overlying material. A reasonable shallow pressure gradient is ~35 km/GPa. Because the Earth is still cooling following its early accretion and due to radioactive heat, it gets hotter with increasing depth. The increase in temperature with depth is called the geothermal gradient. Heat is transferred upward by conduction across the rigid lithosphere and more effectively by convection below that. Because the oceanic lithosphere is thinner, convection transfers heat to shallower levels, and the geothermal gradient is thus higher in the upper 200 to 300 km in oceanic areas than in continental shields.

Magma generation is largely controlled by plate tectonic processes. Magmas are generated in six principal settings:
- Mid-ocean ridges (oceanic divergent plate boundaries)
- Continental rifts (incipient continental plate boundaries)
- Island arcs (intra-oceanic subduction zones)
- Continental arcs (ocean-continent subduction zones)
- Back-arc basins (divergence behind the leading edge of the overriding plate at subduction zones)
- Hotspots (rising plumes that can penetrate either oceanic or continental lithospheric plates)

Key Terms

Magma 2	Lithosphere 6	Lithostatic pressure 12
Lava 2	Asthenosphere 6	Conduction of heat 13
Volcanic 2	Low-velocity layer 6	Convection of heat 13
Plutonic 3	Lithophile 8	Radiation of heat 13
Oceanic crust 5	Chalcophile 8	Advection of heat 14
Continental crust 5	Siderophile 8	Geothermal gradient 14
Shield 5	Iron meteorites 11	Slab-pull/ridge-push 15
Platform 5	Stone meteorites 11	Petrogenesis 19
Mantle 5	Stony-iron meteorites 11	Igneous-tectonic association 19
Core 6	Chondrite meteorites 12	

Review Questions and Problems

Review Questions and Problems are located on the author's web page at the following address: **http://www.prenhall.com/winter**

Important "First Principle" Concepts

- Igneous rocks crystallize from a melt. They can be recognized either by field setting or by textural criteria.
- The oceanic crust, continental crust, mantle, and core are the four principle *compositional* subdivisions of the Earth.

- The lithosphere, asthenosphere, mesosphere, outer core, and inner core are the five principal *mechanical* subdivisions of the Earth.

- The lithosphere contains both the crust and some upper mantle, so the moho (which separates the crust from the mantle) lies *within* the lithosphere. Lithospheric plates are about 110 km thick, and their base is called the low-velocity layer, where the mantle becomes more ductile (probably due to small proportions of partial melt).
- The Earth probably accreted from a gravitationally collapsing cloud of planetesimals and quickly separated into an Fe-rich metallic core and a silicate mantle. The oceanic and continental crust formed over long time periods (and continue to do so today).
- Basaltic oceanic crust is generated at mid-ocean ridges and is consumed at subduction zones. Therefore, the crust in the ocean basins is relatively young (< 160 Ma) and recycles.
- Heterogeneous continental crust is more silicic, thicker, and more buoyant than oceanic crust. It is thus not recycled at subduction zones but continues to increase, typically as a result of those very subduction zone processes.
- The composition of chondrite meteorites is considered a close approximation of the primordial Earth (before it differentiated).

- The most common elements in the Earth are, in decreasing order of abundance, Fe, O, Si, Mg, S, Ca, and Al.
- A good average pressure gradient in the crust is about 30 MPa/km (yielding 1 GPa at the base of typical 35-km thick continental shield crust). In the denser upper mantle the pressure gradient is about 35 MPa/km.
- Plate tectonics is the process that convectively cools the Earth's mantle. Convection involving descending slabs (as compared to cylindrical plumes) is the natural method, given the physical properties of the mantle at present (positive thermal expansion, temperature-dependent viscosity, etc.). Plume-style convection is more appropriate for the deeper and hotter boundary layer at the core–mantle boundary, however, and such rising plumes probably cool the core and rise through the mantle, largely independently of plate tectonics.
- Plate tectonics plays a pivotal role in most igneous processes, particularly those processes at divergent and consuming plate margins.

Suggested Further Readings

History of Igneous Petrology

Young, D. A. (2003). *Mind Over Magma. The Story of Igneous Petrology*. Princeton University Press. Princeton, NJ.

The Interior of the Earth and Global Dynamics

Anderson, D. L. (1992). The Earth's interior. In: *Understanding the Earth* (eds. G. C. Brown, C. J. Hawkesworth, and P. C. L. Wilson). Cambridge University Press. Cambridge, UK.

Brown, G. C., and A. E. Mussett. (1993). *The Inaccessible Earth*. Chapman & Hall. London.

Davies, G. F. (2001). *Dynamic Earth: Plates, Plumes, and Mantle Convection*. Cambridge University Press. Cambridge, UK.

Gurnis, M., M. E. Wysession, E. Knittle, and B. A. Buffett (eds.). (1998). *The Core–Mantle Boundary Region*. Geodynamics Series **28**. American Geophysical Union. Washington, DC.

Jackson, I. (1998). *The Earth's Mantle. Composition, Structure, and Evolution*. Cambridge University Press. Cambridge, UK.

Jacobs, J. A. (1975). *The Earth's Core*. Academic Press. London.

Jones, J. H., and M. J. Drake. (1986). Geochemical constraints on core formation in the Earth. *Nature*, **322**, 221–228.

Karato, S.-I., A. M. Forte, R. C. Liebermann, G. Masters, and L. Stixrude. (eds.). (2000). *Earth's Deep Interior: Mineral Physics and Tomography from the Atomic to the Global Scale*. Geophysical Monograph **117**. American Geophysical Union. Washington, DC.

Montagner, J.-P., and D. L. Anderson. (1989). Constrained reference mantle model. *Phys. Earth Planet. Sci. Lett.*, **44**, 205–207.

Palme, H., and H. St. C. O'Neill. (2003). Cosmochemical estimates of mantle composition. In: *The Mantle and Core* (ed. R. W. Carlson), *Vol. 2 Treatise on Geochemistry* (eds. H. D. Holland and K. K. Turekian). Elsevier-Pergamon. Oxford, UK. pp. 1–38.

Richards, M. A., R. G. Gordon, and R. D. van der Hilst. (eds.). (2000). *The History and Dynamics of Global Plate Motions*. Geophysical Monograph **121**. American Geophysical Union. Washington, DC.

Schubert, G., D. L. Turcotte, and P. Olson. (2001). *Mantle Convection in the Earth and Planets*. Cambridge University Press. Cambridge, UK.

Silver, P. G., R. W. Carlson, and P. Olson. (1988). Deep slabs, geochemical heterogeneity, and the large-scale structure of mantle convection: Investigation of an enduring paradox. *Ann. Rev. Earth Planet. Sci.*, **16**, 477–541.

Turcotte, D. L., and G. Schubert. (2002). *Geodynamics*. Cambridge University Press. Cambridge, UK.

Wyllie, P. J. (1971). *The Dynamic Earth: Textbook in Geosciences*. John Wiley and Sons. New York.

The Origin of the Earth and Solar System

Brown, G. C., and A. E. Mussett. (1993). *The Inaccessible Earth*. Chapman & Hall. London.

Cameron, A. G. W. (2001). From interstellar gas to the Earth–Moon system. *Meteorit. Planet. Sci.*, **36**, 9–22.

Ernst, W. G. (2007). Speculations on evolution of the terrestrial lithosphere–asthenosphere system—Plumes and plates. *Gondwana Research*, **11**, 38–49.

Lauretta, D., L. Leshin, and H. Y. McSween, Jr. (eds.). (2006). *Meteorites in the Early Solar System II*. University of Arizona Press. Tucson, AZ.

Newsom, H. W., and J. H. Jones. (1990). *Origin of the Earth*. Oxford University Press. New York.

Taylor, S. R. (1992). *Solar System Evolution: A New Perspective*. Cambridge University Press. Cambridge, UK.

Taylor, S. R. (1992). The origin of the Earth. In: *Understanding the Earth* (eds. G. C. Brown, C. J. Hawkesworth, and M. S. Matthews). Cambridge University Press. Cambridge, UK.

Walther, M. J., and R. G. Trønnes. (2004). Early Earth differentiation. *Earth Planet. Sci. Lett.*, **225**, 253–269.

Weatherill, G. W. (1989). The formation of the solar system: Consensus, alternatives, and missing factors. In: *The Formation and Evolution of Planetary Systems* (eds. H. A. Weaver and L. Danly). Cambridge University Press. Cambridge, UK.

Weaver, H. A., and L. Danly (eds.). (1989). *The Formation and Evolution of Planetary Systems*. Cambridge University Press. Cambridge, UK.

Wood, J. A. (1979). *The Solar System*. Prentice Hall. Englewood Cliffs, NJ.

Meteorites

Goldstein, J. I., and J. M. Short. (1967). The iron meteorites, their thermal history and parent bodies. *Geochim. Cosmochim. Acta*, **31**, 1733–1770.

Kerrige, J. F., and M. S. Matthews (eds.). (1988). *Meteorites and the Early Solar System*. University of Arizona Press. Tucson, AZ.

MacSween, H. Y. (1987). *Meteorites and Their Parent Planets*. Cambridge University Press. Cambridge, UK.

Wasson, J. T. (1985). *Meteorites*. W.H. Freeman. New York.

Pressure and Temperature Gradients in the Earth

McKenzie, D., J. Jackson, and K. Priestly. (2005). Thermal structure of oceanic and continental lithosphere. *Earth Planet. Sci. Lett.*, **233**, 337–349.

Sclater, J. G., C. Jaupart, and D. Galson. (1980). The heat flow through oceanic and continental crust and the heat loss of the Earth. *Rev. Geophys. Space Sci.*, **18**, 269–311.

Williamson, E. D., and L. H. Adams. (1923). Density distribution in the Earth. *J. Wash. Acad. Sci.*, **13**, 413–428.

2

Classification and Nomenclature of Igneous Rocks

Questions to be Considered in this Chapter:

1. On what basis are igneous rocks classified?
2. What general terms are used to describe the basic textural and compositional parameters by which igneous rocks are classified?
3. What is the generally accepted classification scheme?
4. How do we deal with the plethora of igneous rock names that we might run across in the literature?

2.1 INTRODUCTION

This chapter is designed to assist the student in laboratories with hand specimen and thin-section description and identification. The preferred method for classifying any rock type (igneous, sedimentary, or metamorphic) is based on *texture* and *composition* (the latter usually in terms of mineral proportions). Textural criteria are commonly considered first, as textures provide the best evidence for rock origin and permit classification into the broadest genetic categories. The first step in igneous rock description should be to determine whether the rock falls into one of the following three categories:

Phaneritic The majority of crystals that compose the rock are readily visible with the naked eye (> ~0.1 mm). If a rock exhibits phaneritic texture, it typically crystallized slowly beneath the surface of the Earth and may be called **plutonic**, or **intrusive**.

Aphanitic Most of the crystals are too small to be seen readily with the naked eye (< ~0.1 mm). If a rock is aphanitic, it crystallized rapidly at the Earth's surface and may be called **volcanic**, or **extrusive**.

Fragmental The rock is composed of pieces of disaggregated igneous material, deposited and later amalgamated. The fragments themselves may include pieces of preexisting (predominantly igneous) rock, crystal fragments, or glass. Fragmental rocks are typically the result of a volcanic explosion or collapse and are collectively called **pyroclastic**.

The grain size of *phaneritic* rocks may be further subdivided as follows:

Fine grained	< 1 mm diameter (< sugar granules)
Medium grained	1–5 mm diameter (sugar to pea sized)
Coarse grained	5–50 mm diameter
Very coarse grained	> 50 mm diameter (the lower size limit is not really well defined)

Pegmatitic is an alternative term for very coarse grain size but has compositional implications for many geologists because pegmatites have historically been limited to late-stage crystallization of granitic magmas (Section 11.2.2). Please notice the distinction between *aphanitic* (too fine to see individual grains) and *fine grained* (grains are visible without a hand lens but less than 1 mm in diameter).

Some rocks classified as phaneritic and aphanitic are relatively *equigranular* (of uniform grain size), whereas others exhibit a range of grain sizes because different minerals may experience somewhat different growth rates. The grain size usually varies over only a modest range, and it does so somewhat gradually. If, on the other hand, the texture displays two dominant grain sizes that vary by a significant amount, the texture is called **porphyritic**. The larger crystals are called **phenocrysts**, and the finer crystals are referred to as **groundmass**. Whether such rocks are considered plutonic or volcanic is based on the grain size of the *groundmass*. Because the grain size is generally determined by cooling rate, porphyritic rocks generally result when a magma experiences two distinct phases of cooling. This is most common in, although not limited to, volcanics, in which the phenocrysts form in the slow-cooling magma chamber, and the finer groundmass forms upon eruption.

2.2 COMPOSITIONAL TERMS

The composition of a rock may refer either to its chemical composition or the proportions of minerals in it. These different compositional aspects are related but may lead to some confusion at times. Nearly all igneous rocks are composed principally of silicate minerals, which are most commonly those included in Bowen's Series: quartz, plagioclase, alkali feldspar, muscovite, biotite, hornblende, pyroxene, and olivine. Of these, the first four (and any feldspathoids present) are **felsic** minerals (from *fel*dspar + *si*lica), and the latter four are **mafic** (from *ma*gnesium + *fer*ric iron). Generally, felsic refers to the light-colored silicates (feldspars, quartz, feldspathoids), whereas mafic refers to the darker ones, but composition has precedence (e.g., smoky quartz and dark feldspars are felsic). In addition to these principal minerals, there may also be a number of *accessory* minerals, present in small quantities, usually consisting of apatite, zircon, titanite, epidote, an oxide or a sulfide, or a silicate alteration product such as chlorite.

Most geologists agree that the best way to form the compositional basis for a classification of igneous rocks is to use the exact principal mineral content, as will be described shortly. A number of general descriptive terms, however, are not meant to name specific rocks but to emphasize some compositional aspect of a rock. Unfortunately, many of these terms address similar, but not equivalent, compositional parameters, resulting at times in confusion. For example, the terms in the previous paragraph are commonly applied not only to minerals but also to the rocks that they compose. **Felsic**, then, describes a rock composed predominantly of felsic minerals, whereas **mafic** describes a rock with far more mafic minerals. The term **ultramafic** refers to a rock that consists of over 90% mafic minerals. Similar, but not equivalent, terms are **leucocratic**, indicating a light-colored rock, and **melanocratic**, indicating a dark-colored rock. The first two terms are based on mineral content,

whereas the latter ones are based on rock color, but the relationship between these two parameters is obvious: rocks composed principally of light-colored minerals (felsic) should be light-colored themselves (leucocratic). Color, however, is not a very reliable measure of the composition of a rock. Thus terms such as mafic, which are defined rather loosely by color and composition, can be confusing. For example, when plagioclase becomes more calcic than about An_{50}, it is commonly dark gray or even black. Smoky quartz is also quite dark. Should these minerals be considered mafic? Most geologists would resist this, as the mnemonic roots of the felsic and mafic terms refer to their chemical composition, even though color may have replaced composition as the common distinguishing feature. A rock composed of 90% dark feldspar would thus be considered both felsic and melanocratic. The color of a rock has been quantified by a value known as the **color index (M′)**, which is taken here to simply mean the volume percentage of dark minerals. (It is more specifically defined for advanced work by Le Maitre et al., 2002.)

Purely chemical terms such as **silicic**, **magnesian**, **alkaline**, and **aluminous** that refer, respectively, to the SiO_2, MgO, ($Na_2O + K_2O$), and Al_2O_3 content of a rock may also be used, particularly when the content of some particular component is unusually high. Silica content is of prime importance, and the term **acidic** is synonymous with silicic. Although based on the outdated concept that silicic acid is the form of silica in solution, even in melts, the term is still in use. The opposite of acidic is **basic**, and the spectrum of silica content in igneous rocks has been subdivided as follows:

Acidic	> 66 wt. % SiO_2
Intermediate	52–66 wt. % SiO_2
Basic	45–52% wt. % SiO_2
Ultrabasic	< 45 wt. % SiO_2

Because the concept behind "acidic" and "basic" is not accurate, many petrologists consider these terms outdated as well, whereas others consider them useful. Naturally, basic rocks are also mafic, so we see some of the unnecessary complexity involved in simply describing the most general compositional properties of igneous rocks.

Yet a further problem arises when we attempt to refer to the composition of a melt. How, for example, do we refer to the magma that, when crystallized, becomes a basalt? Because there are few, if any, minerals in a melt, the mineral-based terms *felsic* and *mafic* are technically inappropriate. Color, as mentioned above, is an unreliable compositional measure, and it is notoriously poor for magmas and glasses, as the common occurrence of black obsidian, that is quite silicic, can testify. *Basic* would apply to melts, and it is here that the term may best be used. Many North American geologists consider the terms *mafic* and *felsic* appropriate compositional terms and prefer to call a dark magma *mafic* rather than *basic*, which they consider outmoded. British geologists prefer *basic* and *acidic* to refer to magmas because they are appropriately based on composition and not mineralogy. Others escape the problem altogether by naming the magma after the rock equivalent. The literature is thus full of descriptions of "basaltic magmas," "mafic magmas," "basic magmas," etc. As a true-blue North American, I tend to use the term *mafic* for both magmas and rocks. We will find some other descriptive chemical terms that are useful in the classification of melts and igneous rocks in Chapter 8.

2.3 THE IUGS CLASSIFICATION

Over time, a number of classification schemes have been applied to igneous rocks, resulting in a plethora of equivalent or overlapping rock names (see Table 2.1 at the end of the chapter for a partial list). Young (2003, Chapters 14, 21, and 27) provided an excellent review of the tumultuous history of igneous rock classification. In the 1960s, largely at the initiative of Albert Streckeisen, the International Union of Geological Sciences (IUGS) formed the Subcommission on the Systematics of Igneous Rocks to attempt to develop a standardized and workable system of igneous rock nomenclature. The latest version of this classification was recently published by Le Maitre et al. (2002).

2.3.1 Calculations and Plotting

The IUGS system requires that we determine the mineral components of a rock and plot the percentages of three of those components on appropriate triangular diagrams to determine the proper name. Figure 2.1 shows how triangular diagrams are used.

In Figure 2.1, the three components are labeled X, Y, and Z. The percentage of X (at the upper apex) is zero along the Y–Z base and increases progressively to 100% at the X apex. Any horizontal line represents a variation in the Y/Z ratio at a constant value of X. Such lines (at 10% X increments) have been shown on the left diagram. Likewise, lines of constant Y and constant Z have been added. These lines can be used like graph paper to plot a point, and a few of these lines have been labeled. In order to plot a point on a triangular

diagram using particular values of X, Y, and Z, they must total 100%. If they do not, then they must be **normalized** to 100%. This is accomplished by multiplying each by 100/(X + Y + Z). As an example, point A has the components X = 9.0, Y = 2.6, Z = 1.3. We can normalize these values to 100 by multiplying each by 100/(9.0 + 2.6 + 1.3) = 7.75. That gives the normalized values X = 70%, Y = 20%, and Z = 10%. If we count up 7 lines from the Y–Z base, we get a line representing a constant 70% X. Next, counting 1 line from the X–Y base toward Z, we get a line representing 10% Z. Their intersection (point A) is also intersected by the line representing 20% Y because the sum must be 100%.

If the proper grid lines are available, this technique is simple and direct. However, on the diagrams in Figures 2.2 and 2.3, the lines are not supplied, and for these diagrams, an alternative method is used to locate point A, as illustrated on the right diagram in Figure 2.1. Because Y = 20% and Z = 10%, the ratio 100Y/(Y + Z) = 2000/30 = 67. If we move 67% of the way along the Y–Z base from Z toward Y, we have a point with the proper Y/Z ratio of our point A. Any point along a line from this point to the X apex will also have this same Y/Z ratio. If we proceed along such a line to the 70%X position, we may plot point A at the same position as on the left diagram. Although this method is not as direct for locating a point exactly, it quickly determines the field in which a point falls in the figures that follow.

To classify and name most rocks using the IUGS system, one should use the following procedure. The full spectrum of igneous rocks is complex, however, and no single system will work for all. For example, it may be impossible to determine the mineralogy for many fine-grained or glassy volcanic rocks; pyroclastic rocks are classified on a separate basis; and some rocks, such as ultramafic rocks and some unusual and highly alkaline rocks, simply do not have the common mineralogy to fit in the usual context. (We will deal with the highly alkaline rocks in Chapter 19.) A wise approach would be to determine whether a rock fits one of these special categories before proceeding to the general IUGS classification. Le Maitre et al.

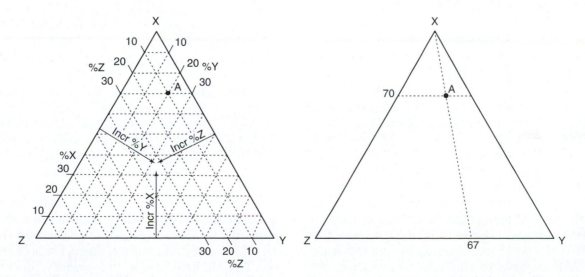

FIGURE 2.1 Two methods for plotting a point with the components 70% X, 20% Y, and 10% Z on triangular diagrams.

(2002) provide a detailed approach to do this, but it applies to a much broader range of rock types than we are likely to encounter at this introductory level, so I provide a somewhat simplified approach below, which should suffice for most general purposes. Advanced students with igneous rocks containing significant carbonate, melilite, or kalsilite, or with kimberlites, lamproites, or charnockites, are encouraged to use Le Maitre et al. (2002) directly.

1. Determine whether the rock is pyroclastic (comprised of rock fragments, ash, etc.). If so, go to Section 2.5.
2. If the rock is aphanitic (volcanic), and the volume percentages of the minerals cannot be determined, go to Section 2.4. If you can determine aphanitic minerals using a microscope or if the rock is phaneritic, proceed to Step 3.
3. Determine the **mode** (the percentage of each mineral present, based on volume). The mode is estimated on the basis of the cumulative area of each mineral type, as seen on the surface of a hand specimen or in a thin section under the microscope. A more accurate determination is performed by "point counting" a thin section. Point counting involves a mechanical apparatus that moves the section along a two-dimensional grid on the petrographic microscope stage. With each shift, the mineral at the crosshair of the microscope is identified and counted. When several hundred such points are counted, the count for each mineral is summed, and the totals are normalized to 100% to determine the mode. All these methods determine relative *areas* of the minerals, but these should correlate directly to volume in most cases.
4. From the mode, determine the volume percentage of each of the following:

   Q' = % quartz (or other SiO_2 polymorph)

   P' = % plagioclase (An_5–An_{100}) The compositional restriction is meant to avoid confusing the nature of nearly pure albite, which should be considered an alkali feldspar

   A' = % alkali feldspar (including albitic plagioclase: An_{0-5})

   F' = total % feldspathoids ("foids")

   M' = total % mafics and accessories

5. The majority of igneous rocks found at the Earth's surface have at least 10% $Q' + A' + P'$ or $F' + A' + P'$. Because quartz is not compatible with feldspathoids, they will never occur in equilibrium together in the same rock. *If a rock to be classified has at least 10% of these constituents, ignore M and normalize the remaining three parameters to 100% (once again, by multiplying each by $100/(Q' + P' + A')$ (or $100/(F' + P' + A')$)).* From this we get $Q = 100Q'/(Q' + P' + A')$, and similarly for P, A, and F (if appropriate), which sum to 100%. It may seem strange to ignore M, but this is the procedure (unless M > 90%). As a result, a rock with 85% mafic minerals can have the same name as a rock with 3% mafics, if the ratio of P:A:Q is the same. If the rock is phaneritic, and M' is > 90, see Section 2.3.4.

6. Determine whether the rock is phaneritic (plutonic) or aphanitic (volcanic). If it is phaneritic, proceed to Figure 2.2. If it is aphanitic, use Figure 2.3.
7. To find in which field the rock belongs, first determine the ratio $100P/(P + A)$. Select a point along the horizontal P–A line (across the center of the diamond) on Figure 2.2a (or Figure 2.3) that corresponds to this ratio. Next proceed a distance corresponding to Q or F directly toward the appropriate apex. Because quartz and feldspathoids can't coexist, there should be no ambiguity as to which triangular half of the diagram to select. The resulting point, representing the Q:A:P or F:A:P ratio, should fall within one of the labeled subfields, which provides a name for the rock. *If P > 65 and Q < 20, see page 27 (for phaneritic rocks) or 28 (for aphanitic rocks).*
8. If the rock does not fit any of the criteria above, you have either made a mistake or have an unusual rock that requires the more detailed approach of Le Maitre et al. (2002).

2.3.2 Phaneritic Rocks

Let's try an example. Upon examining a phaneritic rock, we determine that it has the following mode: 18% quartz, 32% plagioclase, 27% orthoclase, 12% biotite, 8% hornblende, and 3% opaques and other accessories. From this we get $Q' = 18$, $P' = 32$, and $A' = 27$. $Q' + P' + A' = 77$, so, we multiply each by 100/77 to get the normalized values $Q = 23$, $P = 42$, and $A = 35$ that now sum to 100. Because the felsic minerals total over 10%, Figure 2.2a is appropriate. To determine in which field the rock plots, we must calculate $100P/(P + A)$, which is $100(42/(42 + 35)) = 55$. By counting along the P–A axis from A toward P in Figure 2.2a, we find that it plots between the 35 and 65 lines. Then we move upward directly toward point Q. Because 23 falls between 20 and 60, the appropriate name for this rock is granite.

A rock with 9% nepheline, 70% orthoclase, 2% plagioclase, and the rest mafics and accessories would be a nepheline syenite. Try the calculation yourself. The term *foid* is a general term for any feldspathoid. *Don't use the term "foid" in a rock name.* Rather, substitute the name of the actual feldspathoid itself. The same applies for *alkali feldspar* in the fields for alkali feldspar granite and alkali feldspar syenite. Use the true feldspar name, if you can determine it, such as *orthoclase granite*.

For rocks that plot near P, a problem arises. Three relatively common rock types—gabbro, diorite, and anorthosite—all plot near this corner and cannot be distinguished on the basis of QAPF ratios alone. Anorthosite has greater than 90% plagioclase in the un-normalized mode and is thus easily distinguished. Diorite and gabbro have more than 10% mafics and are distinguished on the basis of the average composition of the plagioclase (which can be estimated on the basis of extinction angle in thin section). The plagioclase in gabbros is defined as more anorthite rich than An_{50}, whereas the An-content of plagioclase in diorite is less than An_{50}. In hand sample, this generally correlates with color. As mentioned earlier, plagioclase more calcic than An_{50} is usually dark gray to black, whereas it is whitish when

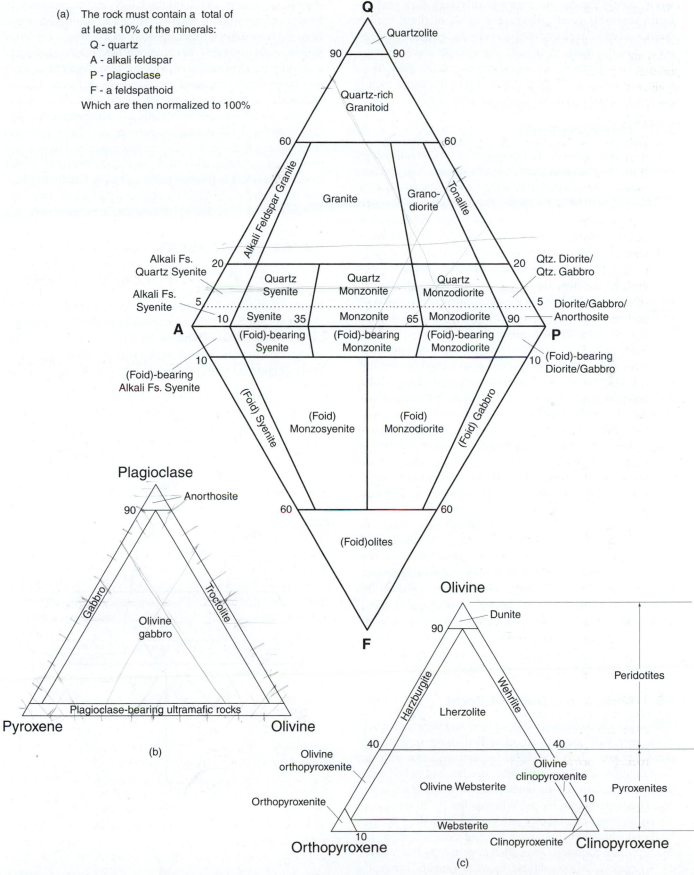

(a) The rock must contain a total of
at least 10% of the minerals:
   Q - quartz
   A - alkali feldspar
   P - plagioclase
   F - a feldspathoid
Which are then normalized to 100%

Q

Quartzolite

90   90

Quartz-rich
Granitoid

60   60

Granite   Grano-
diorite   Tonalite

Alkali Feldspar Granite

Alkali Fs.
Quartz Syenite   20   20   Qtz. Diorite/
Qtz. Gabbro

Quartz
Syenite   Quartz
Monzonite   Quartz
Monzodiorite

Alkali Fs.
Syenite   5   5   Diorite/Gabbro/
Anorthosite

10   Syenite   35   Monzonite   65   Monzodiorite   90

A   P

(Foid)-bearing
Syenite   (Foid)-bearing
Monzonite   (Foid)-bearing
Monzodiorite

10   10   (Foid)-bearing
Diorite/Gabbro

(Foid)-bearing
Alkali Fs. Syenite

(Foid) Syenite   (Foid) Gabbro

(Foid)
Monzosyenite   (Foid)
Monzodiorite

60   60

(Foid)olites

F

Plagioclase

Anorthosite

90

Gabbro   Troctolite

Olivine
gabbro

Plagioclase-bearing ultramafic rocks

Pyroxene   Olivine

(b)

Olivine

Dunite

90

Harzburgite   Wehrlite

Lherzolite

Peridotites

40   40

Olivine
orthopyroxenite   Olivine
clinopyroxenite

Olivine Websterite   10   Pyroxenites

Orthopyroxenite

Websterite

10   Clinopyroxenite

Orthopyroxene   Clinopyroxene

(c)

FIGURE 2.2 A classification of the phaneritic igneous rocks. **(a)** Phaneritic rocks with more than 10% (quartz + feldspar + feldspathoids). **(b)** Gabbroic rocks. **(c)** Ultramafic rocks. After Le Maitre et al. (2002).

more sodic. Gabbros are thus typically very dark (and the mafic mineral is usually a pyroxene and/or olivine), whereas diorites have a black-and-white ("salt-and-pepper") appearance (and the mafic is usually hornblende and/or biotite). Further complication may arise from compositionally zoned plagioclase (Section 3.1.3). If you determine that a rock is gabbro, proceed to Section 2.3.4 and address Figure 2.2b.

2.3.3 Modifying Terms

It is acceptable under the IUGS system to include mineralogical, chemical, or textural features in a rock name. The goal here is to impart some descriptive information that you consider important enough to put in the name. This is a matter of judgment and is flexible. If the rock is unusually light colored *for its category*, you may want to add the prefix *leuco-*, as in leucogranite. If it's unusually dark, add the prefix *mela-*, as in melagranite. You may also use textural terms (see Chapter 3), such as porphyritic granite, rapakivi granite, graphic granite, etc. when they are very obvious. When naming a rock, *always* try to find the correct name from the proper IUGS diagram. Names such as pegmatite, aplite, and tuff are incomplete. Rather, use these *textural* terms to modify the rock name, as in pegmatitic orthoclase granite, aplitic granite, and rhyolite tuff. If you want to convey some important mineralogical information, you can add that to the name as well. Naturally, quartz, plagioclase, and alkali feldspar are already implicit in the name so are redundant if mentioned specifically. However, you may want to describe a rock as a riebeckite granite or a muscovite biotite granite. If more than one mineral is included, they are listed in the order of *increasing* modal concentration. In the previous example, then, there should be more biotite than muscovite in the rock. At times, it may also be desirable to add a chemical modifier, such as alkaline, calc-alkaline, aluminous, etc. A common example is the use of the prefix *alkali-*. High contents of alkalis can stabilize an alkali amphibole or an alkali pyroxene. We usually don't think of pyroxene-bearing granites, but some "alkali granites" may indeed contain a sodium-rich pyroxene. As we shall see in Chapter 8, some chemical characteristics are manifested throughout a whole series of cogenetic magmas in some igneous provinces. The chemical terms are thus more commonly applied to "suites" of igneous rocks (i.e., groups of rocks that are genetically related).

2.3.3 Mafic and Ultramafic Rocks

Gabbroic rocks (plagioclase + mafics) and ultramafic rocks (with over 90% mafics) are classified using separate diagrams (Figures 2.2b and 2.2c, respectively). As with any classification, the IUGS subcommittee has had to find a delicate balance between the tendencies for splitting and lumping. The same is true for us. Whereas the IUGS must serve the professional community and guide terminology for professional communication, we must find a classification suitable for more common use in student petrology laboratories.

Figure 2.2b for gabbroic rocks is simplified from the IUGS recommendations. When one can distinguish pyroxenes in a gabbro, there is more specific terminology in Le Maitre et al. (2002) (e.g., an orthopyroxene gabbro is called a norite).

Figure 2.2c is more faithful to the IUGS recommendations. In hand specimen work, it may be difficult to distinguish ortho- from clinopyroxene in black igneous rocks. Hence the terms peridotite and pyroxenite are commonly used for ultramafics because they are independent of pyroxene type. When the distinction can be made, the more specific terms in Figure 2.2c are preferred. The presence of over 5% hornblende further complicates the nomenclature of both mafic and ultramafic rocks. I believe that the IUGS distinction between an olivine-pyroxene hornblendite, an olivine-hornblende pyroxenite, and a pyroxene-hornblende peridotite adds more detail than necessary at this point. The student is referred to the complete IUGS classification (Le Maitre et al., 2002) for proper names if it becomes important to make more detailed distinctions in nomenclature.

2.4 APHANITIC ROCKS

Volcanic rocks for which a mineral mode can be determined are treated in the same way as plutonics in the original IUGS classification. One determines the mode, normalizes to find P, A, and Q or F, and plots the result in Figure 2.3, in a manner identical to that described for Figure 2.2. Because the mode is commonly difficult to determine accurately for volcanics, Figure 2.3 is a simplification, modified from the more

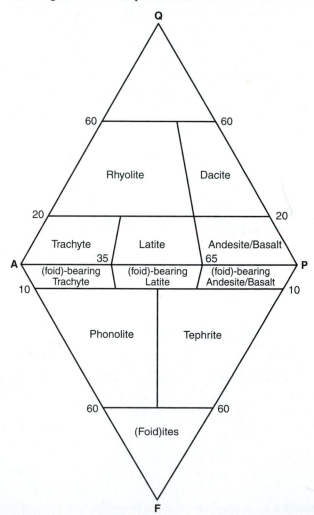

FIGURE 2.3   A classification and nomenclature of volcanic rocks. After Le Maitre et al. (2002).

detailed diagram published by the IUGS (Le Maitre et al., 2002). The matrix of many volcanics is composed of minerals of extremely fine grain size and may even consist of a considerable proportion of **vitreous** (glassy) or amorphous material. Thus it is commonly impossible, even in thin section, to determine a representative mineralogical mode. If it is impossible to recognize the mineralogy of the matrix, a mode must be based on phenocrysts. The IUGS recommends that rocks identified in such a manner be called **phenotypes** and have the prefix *pheno-* inserted before the name (e.g., pheno-latite). As we shall soon see, minerals crystallize from a melt in a sequence (as indicated by, but certainly not restricted to, Bowen's Series), so the first minerals to crystallize do not necessarily represent the mineralogy of the rock as a whole. If based on phenocrysts, the position of a rock on Figure 2.3 will be biased toward the early-forming phases and usually erroneous for the rock as a whole.

Again, **rocks that plot near P** in Figure 2.3 present a problem in the volcanic classification, just as in the plutonic one. One cannot distinguish andesite from basalt using Figure 2.3. The IUGS recommends a distinction based on color index or silica content (see below) and not on plagioclase composition. An andesite is defined as a plagioclase-rich rock either with a color index below 35% or with greater than 52% SiO_2. Basalt has a color index greater than 35% and has less than 52% SiO_2. Many andesites defined on color index or silica content have plagioclases of composition An_{65} or greater.

The most reliable way to avoid the matrix problem discussed above is to analyze the volcanic rock chemically and use a classification scheme based on the analytical results (as is implicit using % SiO_2 in the IUGS distinction between andesite and basalt discussed above). The IUGS has subsequently recommended a classification of volcanics based on a simple diagram comparing the total alkalis with silica, also called a "TAS" diagram (Le Bas et al., 1986). The diagram (Figure 2.4) requires a chemical analysis and is divided into 15 fields. To use it, we normalize a chemical analysis of a volcanic to a 100% nonvolatile basis, combine $Na_2O + K_2O$, and plot the total against SiO_2. Results are generally consistent with the QAPF diagram when a good mode is available. The shaded fields in the TAS diagram can be further subdivided by considering the concentrations of Na_2O and K_2O independently, if desired, according to the lower box. Further refinements are presented for Mg-rich volcanics, and the IUGS (Le Maitre et al., 2002) recommends the name *picrite* for volcanics containing less than 3% alkalis and 12 to 18% MgO and *komatiite* ($TiO_2 < 1\%$) or *meimechite* ($TiO_2 > 1\%$) for volcanics with similar alkali content but richer in MgO. The term *boninite* is recommended for an andesite or basaltic andesite with more than 8% MgO and less than 0.5% TiO_2.

The diagrams shown in Figures 2.2 to 2.4 provide you with the names of most common igneous rocks, but a number of important rock types classified by the IUGS are not included in the figures. For instance, the classification shown does not cover any **hypabyssal** (shallow intrusive) rocks such as **diabase** (or **dolerite** in Britain), nor does it cover the less common rock types such as **carbonatites** (igneous carbonates), **lamproites/lamprophyres**, (highly alkaline, volatile-rich mafic flow/dike rocks), **spilites** (sodic basalts), or **keratophyres** (sodic intermediate volcanics), etc.

Highly alkaline rocks, particularly those of continental origin, are varied, both mineralogically and chemically.

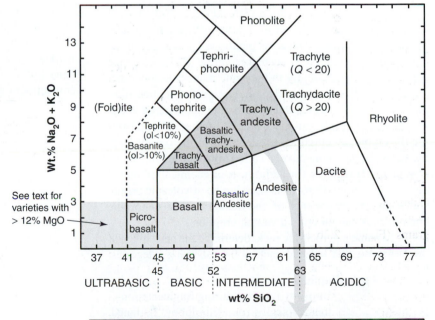

FIGURE 2.4 A chemical classification of volcanics based on total alkalis versus silica ("TAS"). After Le Bas et al. (1986) and Le Maitre et al. (2002). The line between the (Foid)ite field and the Basanite-Tephrite field is dashed, indicating that further criteria are needed to distinguish these types: if normative *ne* > 20%, the rock should be called a nephelinite, and if *ne* < 20% and normative *ab* is present but < 5%, the rock is a melanephelinite. Abbreviations: *ol* = normative olivine and Q = normative 100 * $q/(q + or + ab + an)$. See Section 8.4 and Appendix B for an explanation and calculation of norms. Reprinted by permission of Cambridge University Press.

Further subdivisions of shaded fields	Trachybasalt	Basaltic Trachyandesite	Trachyandesite
$Na_2O - 2.0 \geq K_2O$	Hawaiite	Mugearite	Benmoreite
$Na_2O - 2.0 \geq K_2O$	Potassic Trachybasalt	Shoshonite	Latite

The composition of highly alkaline rocks ranges to high concentrations of several elements present in only trace amounts in more common igneous rocks. The great variety results in a similarly complex nomenclature. Although these alkaline rocks comprise less than 1% of igneous rocks, fully half of the formal igneous rock names apply to them. Such an intricate nomenclature is far beyond the intended scope of this chapter. Chapter 19 will deal with several of the more common or interesting alkaline rock types.

I have tried to avoid the cumbersome detail that a comprehensive classification requires and have attempted to provide a useful compromise between completeness and practicality. Table 2.1 (at the end of the chapter) lists a much broader spectrum of igneous rock names that can be found in the literature. Those in **bold** are recommended by the IUGS and can be found either in Figures 2.2 to 2.4 or in Le Maitre et al. (2002). The other terms are not recommended by the IUGS because they are too colloquial, too restrictive, inappropriate, or obsolete. I have attempted to provide a *very* brief definition of each term, including the IUGS-approved term that most closely approximates it. It is impossible to do this with precision, however, as the chemical, mineralogical, and/or textural criteria seldom coincide perfectly. Table 2.1 is intended to provide you with a quick reference to rock terms that you may encounter in the literature and not a rigorous definition of each. For the latter, you are referred to the *AGI Glossary* or Le Maitre et al. (2002). For the sake of brevity, I have included only the root IUGS terms, such as *granite*, *andesite*, or *trachyte*, and not compound terms such as *alkali-feldspar granite*, *basaltic trachy-andesite*, etc.

2.5 PYROCLASTIC ROCKS

The initial IUGS classification (Streckeisen, 1973) did not cover pyroclastic rocks, but they were addressed in a later installment (Schmid, 1981). As mentioned previously, if the chemical composition is available, these rocks could be classified compositionally in the same manner as any other volcanics, but they commonly contain significant impurities, and only those for which the foreign material is minimal can a meaningful compositional name be applied. Pyroclastics are thus typically classified on the basis of the *type* of fragmental material (collectively called **pyroclasts**) or on the *size* of the fragments (in addition to a chemical or modal name, if possible). Pyroclasts need not be of volcanic origin: some may be fragments of sedimentary or metamorphic country rock caught up in a violent eruption.

To name a pyroclastic rock, determine the percentage of the fragments that fall into each of the following categories:

> 64 mm diameter
    Bombs (if molten during fragmentation—thus typically rounded/blobby, flattened, or stretched)
    Blocks (if not molten during fragmentation—thus typically angular or broken)
2–64 mm **Lapilli**
< 2 mm **Ash**

The IUGS maintains that a true pyroclastic rock must contain at least 75% pyroclasts. The fragments may be individual crystals, glass, or rock fragments. Individual crystals in pyroclastic rocks are referred to as "crystal fragments," not "phenocrysts," because their origin is uncertain. Glass may occur as pumice, ash-sized fragments of shattered thin pumice vesicle-walls, or as dense angular or rounded droplet-shaped pieces. The relative proportions of fragments in the size categories above are then plotted on Figure 2.5 to determine the rock name. Tuffs and ashes may be further qualified by the type of fragments they contain. **Coarse (ash) tuffs** contain particles predominantly in the 1/16 mm to 2 mm range, whereas the particles in **fine (ash) tuffs** or **dust tuffs** are generally < 1/16 mm. **Lithic tuff** would contain a predominance of rock fragments, **vitric tuff** a predominance of pumice and glass fragments, and **crystal tuff** a predominance of crystal fragments. Again, *it is good practice to include a compositional name whenever possible*, based on a chemical analysis, the color index, or crystal fragment mineralogy. A name such as rhyolitic lapilli tuff is thus a complete and descriptive name for a light pink, tan, or very light gray pyroclastic rock dominated by lapilli-sized fragments.

For rocks containing both pyroclasts and sedimentary clastic material (**epiclasts**), the IUGS subcommission (Le Maitre et al., 2002) suggests the general term **tuffite**, which may be subdivided further by adding the prefix *tuffaceous-* to the normal sedimentary name, such as shale, siltstone, sandstone, conglomerate, or breccia. **Epivolcaniclastics** are secondary deposits, meaning that they are not deposited directly by eruptive activity. These may occur due to volcanic flank collapse or as marine aprons around volcanic islands, volcanic mudflows (lahars), or reworked epiclasts. An **aquagene tuff** is a waterborne accumulation of ash. It may result from a subaqueous eruption, or it may be an airborne accumulation that has been reworked by water. **Hyalotuff** or **hyaloclastite** is an aquagene tuff that is created when magma is shattered upon contact with water.

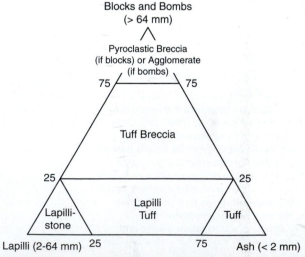

FIGURE 2.5 Classification of the pyroclastic rocks. After Fisher (1966). Copyright © with permission from Elsevier Science.

Summary

Most common igneous rocks may be classified and named on the basis of their texture and mineral content. The International Union of Geological Sciences has developed a standardized igneous rock nomenclature for classifying and naming igneous rocks based on the modal (volume) percentages of the constituent minerals. To name a typical rock, one determines the modal percentages of quartz minerals (Q'), Alkali feldspars (A'), Plagioclase (P'), Feldspathoids (F'), and Mafics (M'). If the rock has 90% or more Q' + A' + P' + F', the rock is named by plotting the relative proportions of these mineral constituents (normalized to 100%) on the appropriate phaneritic or aphanitic diagram. Volcanic rocks in which a mode is impossible to determine should be classified on the basis of chemical composition on a total alkali versus silica (TAS) diagram. Pyroclastic rocks are classified separately on the basis of the size and nature of the fragments (pyroclasts) that compose them, including a compositional name, whenever possible.

Key Terms

Phaneritic 23	Groundmass 24	Pyroclast 30
Aphanitic 23	Felsic 24	Ash 30
Fine, medium, coarse grained 23	Mafic 24	Lapilli 30
Fragmental 23	Silicic 24	Blocks 30
Pyroclastic 23	Acidic 24	Bomb 30
Porphyritic 24	Basic 24	Epi-volcaniclastic 30
Phenocryst 24	Mode 26	Epiclast 30
	Q, A, P, F, M 26	

Review Questions and Problems

Review Questions and Problems are located on the author's web page at the following address: **http://www.prenhall.com/winter**

Important "First Principle" Concepts

- Igneous rocks are dominated by silicate minerals and typically range in composition from ultramafic through mafic and intermediate to silicic varieties. More specific (and generally more unusual) varieties may be described chemically, using appropriate modifiers, such as alkalic, potassic, calcic, aluminous, etc.

- Texturally, igneous rocks are either phaneritic (intrusive rocks), aphanitic (extrusive or volcanic rocks), or fragmental (pyroclastic rocks).
- Igneous rocks may be classified and named on the basis of their mineral content (which is a reflection of their composition) and their texture.

Suggested Further Readings

The following are IUGS publications:

Le Bas, M. J., R. W. Le Maitre, A. L. Streckeisen, and B. Zanettin. (1986). A chemical classification of volcanic rocks based on the total alkali-silica diagram. *J. Petrol.*, **27,** 745–750.

Le Maitre, R. W., A. Streckeisen, B. Zanettin, M. J. Le Bas, B. Bonin, P. Bateman, G. Bellieni, A. Dudek, S. Efremova, J. Keller, J. Lameyre, P. A. Sabine, R. Schmid, H. Sørensen, and A. R. Wooley (eds.). (2002). *Igneous Rocks: A Classification and Glossary of Terms.* Cambridge University Press. Cambridge, UK.

Streckeisen, A. L. (1967) Classification and nomenclature of igneous rocks. *Neues Jahrbuch für Mineralogie Abhandlungen,* **107,** 144–240.

Streckeisen, A. L. (1973). Plutonic rocks, classification and nomenclature recommended by the IUGS subcommission on the systematics of igneous rocks. *Geotimes,* **18,** 26–30.

Streckeisen, A. L. (1974). Classification and nomenclature of Plutonic rocks. *Geol. Rundschau,* **63,** 773–786.

Streckeisen, A. L. (1976). To each plutonic rock its proper name. *Earth-Sci. Rev.,* **12,** 1–33.

Streckeisen, A. L. (1979). Classification and nomenclature of volcanic rocks, lamprophyres, carbonatites, and melilitic rocks: Recommendations and suggestions of the IUGS subcommission on the systematics of igneous rocks. *Geology,* **7,** 331–335.

Schmid, R. (1981). Descriptive nomenclature and classification of pyroclastic deposits and fragments: Recommendations and suggestions of the IUGS subcommission on the systematics of igneous rocks. *Geology,* **9,** 4–43.

TABLE 2.1   Common Rock Names, with IUGS Recommended Terms in Bold Print

Name	Approximate Meaning	Name	Approximate Meaning
Adakite	Mg-rich andesite	Glimmerite	Biotite-rich ultramafic
Adamellite	Quartz Monzonite	**Granite**	Figure 2.2
Alaskite	Leuco-(alkali feldspar) granite	**Granitoid**	Felsic Plutonic (granite-like) rock
Alnöite	Melilite lamprophyre (Table 19.6)	**Granodiorite**	Figure 2.2
Alvekite	Med.- to fine-grained calcite carbonatite	Granophyre	Porphyritic granite with granophyric texture
Andesite	Figures 2.3 and 2.4	Grazinite	Phonolitic nephelinite
Ankaramite	Olivine basalt	Grennaite	Nepheline syenite
Anorthosite	Figure 2.2	Harrisite	Troctolite
Aplite	Granitoid with fine sugary texture	**Harzburgite**	Figure 2.2
Basalt	Figures 2.3 and 2.4	Hawaiite	Sodic trachybasalt (Figure 2.4)
Basanite	Olivine tephrite (Table 19.1)	**Hornblendite**	Ultramafic rock > 90% hornblende
Beforsite	Dolomite carbonatite	Hyaloclastite	Pyroclastic rock = angular glass fragments
Benmoreite	Trachyte (Figure 2.4)		
Boninite	Mg-rich andesite or basaltic andesite (wt. % MgO > 8%, TiO_2 < 0.5%)	Icelandite	Al-poor, Fe-rich andesite
		Ignimbrite	Applied to welded tuffs
Camptonite	Hornblende lamprophyre (Table 19.6)	**Ijolite**	Clinopyroxene nephelinite (Table 19.1)
Cancalite	Enstatite sanidine phlogopite lamproite (Table 19.5)	**Italite**	Glass-bearing leucititolite
		Jacupirangite	Alkaline pyroxenite
Carbonatite	> 50% carbonate	**Jotunite**	Orthopyroxene monzonorite
Cedricite	Diopside leucite lamproite (Table 19.5)	Jumillite	Olivine madupitic lamproite (Table 19.5)
Charnockite	Orthopyroxene granite		
Comendite	Peralkaline rhyolite	Kalsilitite	Kalsilite-rich mafic volcanic (Table 19.1)
Cortlandite	Pyroxene-olivine hornblendite	Kamafugite	Collective term for the group: KAtungite-MA Furite-Ugandite
Dacite	Figures 2.3 and 2.4		
Diabase	Medium-grained basalt/gabbro	Katungite	K-rich olivine melilitite
Diorite	Figure 2.2	Keratophyre	Albitized felsic volcanic
Dolerite	Medium-grained basalt/gabbro	**Kersantite**	Biotite-plag. lamprophyre (Table 19.6)
Dunite	Figure 2.2	**Kimberlite**	Volatile-rich ultramafic (Table 19.1)
Enderbite	Orthopyroxene tonalite	**Komatiite**	Ultramafic volcanic (mostly Archean): wt. % MgO > 18%, TiO_2 < 1%
Essexite	Nepheline monzo-gabbro/diorite		
Felsite	Microcrystalline granitoid	**Kugdite**	Olivine melilitolite
Fenite	Alkali feldspar-rich metasomatic rock associated with carbonatites	Ladogalite	Mafic alkali feldspar syenite.
		Lamproite	A group of K-, Mg-, volatile-rich volcanics (Table 19.1)
Fergusite	Pseudolucite (foid)ite		
Fitzroyite	Leucite-phlog. lamproite (Table 19.5)	**Lamprophyre**	A diverse group of dark, porphyritic, mafic to ultramafic K-rich hypabyssal rocks (Table 19.1)
(Foid)ite	Figure 2.3; replace "foid" with the actual feldspathoid name		
(Foid)olite	Figure 2.2; replace "foid" with the actual feldspathoid name	**Larvikite**	Augite syenite-monzonite
		Latite	Figures 2.3 and 2.4
Fourchite	Mafic analcime lamprophyre	**Leucitite**	Volcanic rock that is nearly all leucite
Fortunite	Glassy olivine lamproite (Table 19.5)	**Lherzolite**	Figure 2.2
Foyaite	Nepheline syenite	**Limburgite**	Volcanic = pyroxene + olivine + opaques in a glassy groundmass
Gabbro	Figure 2.2		
Gabbronorite	Gabbro with subequal amounts of clinopyroxene and orthopyroxene.	**Liparite**	Rhyolite
		Luxulianite	Porphyritic granite with tourmaline

TABLE 2.1 *Continued*

Name	Approximate Meaning	Name	Approximate Meaning
Madupite	Lamproite with poikilitic phlogopite ground mass (Table 19.5)	**Phono-tephrite**	Figure 2.3
Mafurite	Alkaline ultramafic volcanic	**Picrite**	Olivine-rich basalt (Figure 14.2)
Malchite	Lamprophyre	**Picrobasalt**	Figure 2.4
Malignite	Aegirine-augite nepheline syenite	**Pitchstone**	Hydrous volcanic glass
Mamilite	Leucite-richterite lamproite (Table 19.5)	**Plagioclasite**	Anorthosite
Mangerite	Orthopyroxene monzonite	**Plagiogranite**	Leucotonalite
Marianite	Mg-rich andesite	**Polzenite**	Melilite lamprophyre (Table 19.6)
Marienbergite	Natrolite phonolite	**Pyroxenite**	Figure 2.2
Masafuerite	Picrite basalt	**Quartzolite**	Figure 2.2
Meimechite	Ultramafic volcanic (mostly Archean): wt. % MgO > 18%, TiO$_2$ > 1%	Rauhaugite	Coarse dolomitic carbonatite (Table 19.2)
Melilitite	Ultramafic melilite-clinopyroxene volcanic (Table 19.1)	Reticulite	Broken-up pumice
Melilitolite	Plutonic melilitite	**Rhyolite**	Figures 2.3 and 2.4
Melteigite	Mafic ijolite	**Sannaite**	Na-amph-augite lamprophyre (Table 19.6)
Miagite	Orbicular gabbro	Sanukite	Mg-rich andesite
Miaskite	Felsic biotite-nepheline monzosyenite	Scoria	Highly vesiculated basalt
Minette	Biotitic lamprophyre (Table 19.6)	**Shonkinite**	Alkaline plutonic with Kfs, foid, and augite
Missourite	Mafic type of leucite (foid)ite	**Shoshonite**	K-rich basalt (Table 19.1) and Figure 2.4
Monchiquite	Feldspar-free lamprophyre (Table 19.6)	**Sövite**	Coarse calcite carbonatite (Table 19.2)
Monzonite	Figure 2.2	**Spessartite**	Hbl-di-plag. lamprophyre (Table 19.6)
Monzonorite	Orthoclase-bearing orthopyroxene gabbro	**Spilite**	Altered albitized basalt
Mugearite	Mafic sodic trachyandesite (Figure 2.4)	**Syenite**	Figure 2.2
Natrocarbonatite	Rare sodic carbonatite volcanic	Tachylite	Basaltic glass
Nelsonite	Ilmenite-apatite dike rock	Tahitite	Haüyne tephri-phonolite
Nephelinite	Nepheline-bearing basalt (Table 19.1)	**Tephrite**	Figure 2.3 and Table 19.1
Nephilinolite	Plutonic nephelinite (a foidite)	**Tephri-phonolite**	Figure 2.3
Nordmarkite	Quartz-bearing alkali feldspar syenite	**Teschenite**	Analcime gabbro
Norite	Orthopyroxene gabbro	**Theralite**	Nepheline gabbro
Obsidian	Volcanic glass (typically silicic)	**Tholeiite**	Tholeiitic basalt or magma series (Section 8.7)
Oceanite	Picrite	**Tonalite**	Figure 2.2
Odinite	Lamprophyre	**Trachyte**	Figures 2.3 and 2.4
Opdalite	Orthopyroxene granodiorite	**Troctolite**	Olivine-rich gabbro (Figure 2.2)
Orbite	Lamprophyre	**Trohdhjemite**	Leuco-tonalite
Orendite	Di-sanidine-phlog. lamproite (Table 19.5)	**Uncompahgrite**	Pyroxene melilitolite
Orvietite	Volcanic near the tephri-phonolite– phono-tephrite border (Figure 2.4)	**Urtite**	Felsic nephilinolite (Table 19.1)
Oachitite	Ultramafic lamprophyre	**Verite**	Glassy ol-di-phlog. lamproite (Table 19.5)
Pantellerite	Peralkaline rhyolite (Table 19.1)	Vicolite	*See* orvietite
Pegmatite	Very coarse-grained igneous rock	**Vogesite**	Hbl-di-orthocl. lamprophyre (Table 19.6)
Peridotite	Figure 2.2	**Websterite**	Figure 2.2
Perlite	Volcanic glass that exhibits fine concentric cracking	**Wehrlite**	Figure 2.2
		Woldigite	Di-leucite-richterite madupidic lamproite (Table 19.5)
Phonolite	Figures 2.3 and 2.4	Wyomingite	Di-leucite-phlog. lamproite (Table 19.5)

3

Textures of Igneous Rocks

Questions to be Considered in this Chapter:

1. What textures may be produced as magma cools and crystallizes to form igneous rocks?

2. What physical variables control the development of igneous textures, and how do they do so?

3. What recrystallization textures may result as high-temperature igneous minerals, once formed, cool further toward near-surface conditions?

4. How can we work backward from the knowledge we have and use the textures we see to interpret the developmental history of the rock exhibiting them?

As mentioned in Chapter 1, **petrography** is the branch of petrology that deals with the description and classification of rocks. In Chapter 2, we discussed how to categorize and name an igneous rock, and, because the classification scheme is now largely developed, most modern petrography involves the detailed study of rocks in thin section, using the polarizing light ("petrographic") microscope. Thin sections are cut from rock samples, cemented to microscope slides, and ground down to 0.03 mm thickness so that they readily transmit light. From a purely descriptive standpoint, a good rock depiction should include the mineralogy, a proper name, and a good description of the rock's texture in hand sample and from thin sections. But textures are much more important than mere descriptive aids. The texture of a rock is a result of various processes that controlled the rock's genesis and, along with mineralogy and chemical composition, provides information that we may use to interpret the rock's origin and history. It is thus important for us to be able to recognize and describe the textures of a rock and to understand how they are developed. In Chapter 1, for example, I mentioned that **interlocking** texture (Figure 1.1) was produced by crystallization from a melt and can be used to infer the igneous origin of a rock. In this chapter, we will explore igneous textures in more detail, seeking to discover what controls those textures so that we can use textural criteria to aid us in understanding the crystallization history (and perhaps also some of the post-crystallization history) of a particular rock. I have supplied a glossary of textural terms at the end of the chapter, where you will find the definitions of terms you may encounter. In some of the references listed at the end of the chapter, you will also find excellent color photographs and line drawings (based mostly on thin sections) that further illustrate many of the textures.

The textures that you observe in an igneous rock result from a number of processes that can be grouped into two principal categories. **Primary textures** occur during igneous crystallization and result from interactions between minerals and melt. **Secondary textures** are alterations that take place after the rock is completely solid. The following is a very general discourse on how a number of the most common textures develop. I will concentrate on thin section study, but many of the textures described can be recognized in hand specimens also.

3.1 PRIMARY TEXTURES (CRYSTAL/MELT INTERACTIONS)

The formation and growth of crystals, either from a melt or in a solid medium (metamorphic mineral growth), involves three principal processes: (1) initial *nucleation* of the crystal, (2) subsequent crystal *growth*, and (3) *diffusion* of chemical species (and heat) through the surrounding medium to and from the surface of a growing crystal.

Nucleation is a critical initial step in the development of a crystal. Very tiny initial crystals have a high ratio of surface area to volume and, thus, a large proportion of ions at the surface. Surface ions have unbalanced charges because they lack the complete surrounding lattice that balances the charge of interior ions. The result is high surface energy for the initial crystal and, therefore, low stability. The clustering of a few compatible ions in a cooling melt will thus tend to spontaneously separate, even at the saturation temperature when conditions are otherwise suitable for crystallization of a particular mineral. Under such conditions, crystallization would be possible, but the prerequisite nucleation isn't. Before crystallization can take place, a critically sized "embryonic cluster" or "crystal nucleus" must form, with a sufficient internal volume of fully bonded ions to overcome the surface-related instability. This typically requires some degree of **undercooling** (cooling of a melt below the true crystallization temperature of a mineral) or supersaturation before a sufficient number of ions to be stable can spontaneously cluster together ("homogeneous nucleation"). Alternatively, a preexisting crystal surface may be present: either a "seed crystal" of the same mineral or a different mineral with a similar structure on which the new mineral can easily nucleate and grow ("heterogeneous nucleation"). For reviews of the kinetics of nucleation, see Dowty (1980a), Kirkpatrick (1981), and Cashman (1990).

Several experimental studies have indicated that crystals with simple structures tend to nucleate more easily than those with more complex structures. Oxides (such as magnetite or ilmenite) generally nucleate more easily (with less undercooling) than does olivine, followed by pyroxene, plagioclase, and alkali feldspar, with progressively more complex Si-O polymerization. This may explain why oxides are typically small and numerous, whereas alkali feldspars generally grow quite large, seemingly regardless of the degree of undercooling.

Crystal growth involves the addition of ions onto existing crystals or crystal nuclei. In a simple structure with high symmetry, faces with a high density of lattice points ($\{100\}$, $\{110\}$) tend to form more prominent faces (the "Law of Bravais"). Different faces also grow at different rates. As a rather simplistic generalization, fast-growing faces tend to be those with smaller interplanar lattice spacings (and higher surface energies). If the c-axis unit cell spacing is particularly small, for example, a crystal may be expected to become elongated in the c-axis direction. Fast-growing faces thus tend to grow themselves out of existence. In general, *faces with low surface energy become more prevalent*. When low-energy faces predominate over high-energy faces, the overall energy of the system is lower and, hence, more stable. In more complex silicates, this tendency may be superseded by preferred growth in directions with uninterrupted chains of strong bonds. Pyroxenes and amphiboles thus tend to be elongated in the direction of the Si-O-Si-O chains, and micas tend to elongate plate-like in the directions of the silicate sheets. Defects such as screw dislocations may also aid the addition of new ions to a growing face, and impurities may inhibit growth in some directions. The surface energy on different faces of a crystal may vary disproportionately with changing conditions, so the shape of a particular mineral may vary from one rock to another. Discussions of crystal growth based on crystal defects, the nature of lattice-building elements, the nature of the crystal–melt interface, and structural coherence between the melt and growing faces may be found in Kirkpatrick (1975, 1981), Dowty (1980a), Lofgren (1980), and Cashman (1990).

In most situations, the composition of a growing crystal differs considerably from that of the melt. Only in simple chemical systems, such as water–ice, is this not true. In the general case, then, the growth of a mineral will gradually deplete the adjacent melt in the constituents that the mineral preferentially incorporates. For growth to proceed, new material must **diffuse** through the melt, cross the depleted zone, and reach the crystal surface. In addition, the formation of a crystal from a melt produces heat (the latent heat of crystallization, which is merely the opposite of the latent heat of fusion; see Section 6.3). This heat must also be able to diffuse away from the crystal, or the temperature at the growing surface may become too high for crystallization to proceed.

3.1.1 Rates of Nucleation, Growth, and Diffusion

Because there are three main processes involved in mineral development, and not just one, their relative rates have considerable influence on the ultimate texture of the resulting rock. We shall see that, as with the weakest link in a chain, whichever rate is the *slowest* will be the overall rate-determining process and exert the most control on crystallization. There is a further rate that we must also address: the **cooling rate** of the magma. If the cooling rate is very slow, equilibrium is maintained or closely approximated. If the cooling rate is high, significant undercooling can result because there is seldom time for nucleation, growth, or diffusion to keep pace. The cooling rate is an important externally controlled variable that influences the rates of the other crystal-forming processes. Much of the textural information that we observe is thus used to interpret the cooling rate of a rock.

The rates of both nucleation and crystal growth are strongly dependent on the extent of undercooling of the magma. Initially, undercooling enhances both rates, but further cooling decreases kinetics and increases viscosity, thus inhibiting the rates. As illustrated in Figure 3.1, the maximum growth rate is generally at a higher temperature than is the maximum nucleation rate because it is easier to add an atom with high kinetic energy onto an existing crystal lattice than to have a chance encounter of several such atoms at once to form an embryonic cluster. Further undercooling inhibits growth because atoms have to diffuse farther to add onto a few existing crystals, and it is easier for the slowed atoms to nucleate in local clusters than to move far.

We can use Figure 3.1 to understand why the rate of cooling so profoundly affects the grain size of a rock. As it relates to Figure 3.1, "undercooling" is the degree to which temperature falls below the melting point (which, of course, is also the crystallization temperature when we consider cooling) before crystallization occurs. For example, if the cooling rate is low, only slight undercooling will be possible (such as at temperature T_a in Figure 3.1). At this temperature, the nucleation rate is very low, and the growth rate is high. Fewer crystals thus form, and they grow larger, resulting in the coarse-grained texture common among slow-cooled plutonic rocks. Quickly cooled rocks, on the other hand, may become significantly undercooled before crystallization begins. If rocks are undercooled to T_b in Figure 3.1, the nucleation rate exceeds the growth rate, and many small crystals are formed, resulting in the very fine-grained texture of volcanic rocks. Very high degrees of undercooling (T_c in Figure 3.1) may result in negligible rates of nucleation and growth, such that the liquid solidifies to a glass with very few or no crystals.

Two-stage cooling can create a bimodal distribution of grain sizes. Slow cooling followed by rapid cooling is the only plausible sequence and might occur when crystallization began in a magma chamber, followed by the opening of a conduit and migration of magma to the surface. Initially, the magma would be only slightly undercooled, and a few coarse crystals would form, followed by volcanism and finer crystals. When there is a distinctly bimodal distribution in

grain size, with one size considerably larger than the other, the texture is called **porphyritic**. The larger crystals are called **phenocrysts**, and the finer surrounding ones are called **matrix** or **groundmass**. A porphyritic rock is considered plutonic or volcanic on the basis of the *matrix* grain size. If the phenocrysts are set in a glassy groundmass, the texture is called **vitrophyric**. If the phenocrysts contain numerous inclusions of another mineral that they enveloped as they grew, the texture is called **poikilitic**. The host crystal may then be called an **oikocryst**.

The growth rate of a crystal depends upon the surface energy of the faces and the diffusion rate. For a constant cooling rate, the largest crystals will usually be those with the most plentiful or fastest-diffusing components. The diffusion rate of a chemical species is faster at higher temperature and in material with low viscosity. Diffusion rate is thus low in highly polymerized viscous melts. (Such melts are generally silica rich and also tend to be cooler than mafic melts.) Small ions with low charges diffuse best, whereas large polymerized complexes diffuse poorly. In general, diffusion in a fluid is better than in a glass, and it is better in glass than in crystalline solids. H_2O dramatically lowers the degree of polymerization of magma (Chapter 7), thereby enhancing diffusion. Alkalis have a similar effect, although less extreme. The very coarse grain size of many pegmatites may be attributed more to the high mobility of species in the H_2O-rich melt from which they crystallize than to extremely slow cooling.

The rates of nucleation and growth vary with the surface energy of the minerals and the faces involved, the degree of undercooling, and the crystal structure. These values can be different for different minerals, even in the same magma. Different minerals can be undercooled to differing extents because the melting point in Figure 3.1 is specific to each mineral. We shall see in Chapters 6 and 7 that minerals develop sequentially in a cooling magma as the melting point of each is progressively reached. The temperature may thus be lower than the melting point of one mineral (undercooled) and higher than that of another. Many stable nuclei of one mineral may thus form, while only a few of another may form, resulting in many small crystals of the former and fewer, larger crystals of the latter. The popular notion that the large crystals in a porphyritic rock must have formed first or in a slower-cooling environment is thus not universally valid. The sudden loss of an H_2O-rich fluid phase from a melt will quickly raise the crystallization temperature (Section 7.5) and can also produce porphyritic texture in some plutonic rocks, as we shall see in Chapter 11.

When the diffusion rate is not the limiting (slowest) rate, crystals growing free and unencumbered in a melt will tend to be euhedral and nicely faceted. Different crystal faces have different atomic environments and surface energies. As discussed above, faces with low surface energy will generally be most stable and manifest themselves when growing freely in a liquid; other factors may influence growth rates in some directions, with considerable effect on crystal shapes.

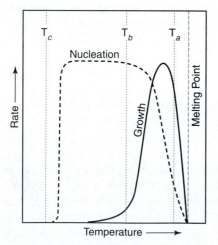

FIGURE 3.1 Idealized rates of crystal nucleation and growth as a function of temperature below the melting point. Slow cooling results in only minor undercooling (T_a), so that rapid growth and slow nucleation produce fewer coarse-grained crystals. Rapid cooling permits more undercooling (T_b), so that slower growth and rapid nucleation produce many fine-grained crystals. Very rapid cooling involves little if any nucleation or growth (T_c), producing a glass.

When the rate of diffusion is slower than the rate of growth (as in quickly cooled, or "quenched," lavas with substantial undercooling), the crystals take on an increasingly radiating form, or a tree-like, branching form termed **dendritic** (Figure 3.2). When diffusion is slower than growth, a zone of depleted liquid builds up at the crystal–liquid interface, as described above. Some propose that the crystals reach out in thin tendrils beyond the zone to tap a supply of appropriate elements or cooler melt. Others suggest that the perturbations in the surface shape toward dendritic forms help to eliminate the local heat buildup that accompanies crystallization. Perhaps both processes contribute to dendritic or spherulitic growth.

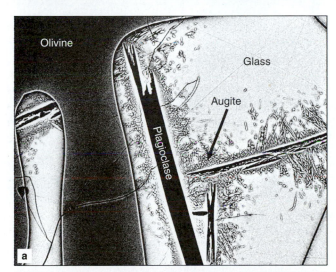

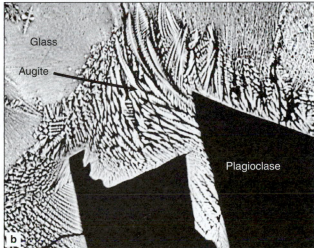

FIGURE 3.2 Backscattered electron image of quenched "blue glassy pahoehoe," 1996 Kalapana flow, Hawaii. Brightness is proportional to back-scattering ability and, hence, to mean atomic number. Black minerals are felsic plagioclase, and gray ones are mafics. **(a)** Large, embayed olivine phenocryst with smaller plagioclase laths and clusters of feathery augite nucleating on plagioclase. Magnification ca. 400×. **(b)** ca. 2000× magnification of feathery quenched augite crystals nucleating on plagioclase (black) and growing in a dendritic form outward. Augite nucleates on plagioclase rather than preexisting augite phenocrysts, perhaps due to local enrichment in mafic components as plagioclase depletes the adjacent liquid in Ca, Al, and Si. Photographs taken with the assistance of Jack Rice, Kathy Cashman, and Michael Schaeffer, University of Oregon.

Ultramafic lavas, such as Precambrian komatiites, when quenched, may develop spectacular elongated olivine crystals, in some cases up to 1 m long, called **spinifex** texture. The unusual size may be caused by rapid growth of the simple olivine structure in a very low-viscosity magma, not by slow cooling. Spinifex pyroxenes over 5 cm long have also been described.

Crystal corners and edges have a larger volume of nearby liquid to tap for components (or to dissipate the heat of crystallization) than do crystal faces (Figure 3.3). In addition, corners and edges have a higher proportion of unsatisfied bonds. Thus we might expect the corners and edges to grow more rapidly than the faces in such quench situations. When this occurs, the resulting forms are called **skeletal** crystals. In some cases, the extended corners may meet to enclose melt pockets at the recessed faces (Figure 3.4a). The corners of quenched plagioclase tend to grow straighter, creating a characteristic **swallow-tailed** shape (Figures 3.2a and 3.4b). Of course any *motion* of the liquid or crystals rehomogenizes it and reduces the limiting effects of slow diffusion.

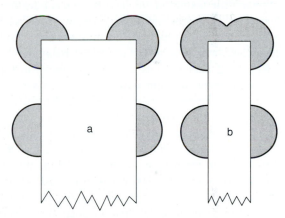

FIGURE 3.3 **(a)** Volume of liquid (shaded) available to an edge or corner of a crystal is greater than for a side. **(b)** Volume of liquid available to the narrow end of a slender crystal is even greater. After Shelley (1993). Copyright © by permission Kluwer Academic Publishers.

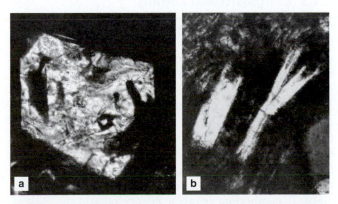

FIGURE 3.4 **(a)** Skeletal olivine phenocryst with rapid growth at edges enveloping melt at ends, Taupo, New Zealand. **(b)** "Swallow-tail" plagioclase in trachyte, Remarkable Dike, New Zealand. Length of both fields ca. 0.2 mm. From Shelley (1993). Copyright © by permission Kluwer Academic Publishers.

3.1.2 Nucleation at Preferred Sites

Epitaxis is the general term used to describe the preferred nucleation of one mineral on another preexisting mineral, thereby avoiding problems associated with slow nucleation. Similarity of the crystal structures of the mineral substrate and the new phase is a prerequisite for epitaxial growth. The atomic constituents of the new mineral thereby find favorable spots to accumulate, and a stable nucleus forms. The growth of sillimanite on biotite or muscovite in metamorphic rocks, rather than as a direct replacement of available crystals of its polymorph kyanite, is a common example. The Si-Al-O structures in both sillimanite and mica are similar in geometry and bond lengths, so that sillimanite (which nucleates poorly) tends to form in areas of mica concentration. **Rapakivi texture** (Rämö and Haapala, 1995), involving plagioclase overgrowths on orthoclase, occurs in some granites where the plagioclase preferentially forms on the structurally similar alkali feldspar rather than nucleating on its own. A crystal nucleus may also form epitaxially in a twin orientation on a preexisting grain of the same mineral, leading to the formation of growth twins.

Spherulitic texture in silicic volcanics is a texture in which needles of quartz and alkali feldspar grow radially from a common center. This and **variolitic** texture of radiating plagioclase laths in some basalts are probably the result of nucleation of later crystals on the first nuclei to form. Both are considered to form during the devitrification of glass, which will be discussed in Section 3.2. Nucleation of minerals on dike (or even vesicle) walls is also common. Growth of elongated crystals (generally quartz), with c-axes normal to vein walls, results in a structure called **comb structure** because the parallel columns resemble the teeth of a comb. **Crescumulate** texture is similar and describes the parallel growth of elongated, non-equilibrium arrangements of olivine, pyroxene, feldspar, or quartz crystals that appear to nucleate on a wall or layer and may grow up to several centimeters long. Crescumulate texture typically occurs in layered mafic plutons (where it may appear in multiple layers) and in the margins of granites.

3.1.3 Compositional Zoning

Compositional **zoning** is common and occurs when a mineral changes composition as it grows during cooling. The composition of most solid-solution minerals in equilibrium with other minerals or liquid is temperature dependent. The reasons for this will become clear in Chapters 5, 6, 7, and 27. Compositional zoning can be observed petrographically only when the color (Figure 3.5a), birefringence, or extinction position varies with composition. In the case of plagioclase, the extinction angle is highly composition dependent, and the compositional variations show up as concentric bands of varying brightness in cross-polarized light (Figure 3.5b). If equilibrium between the crystal and the melt is maintained, the composition of the mineral will adjust to lowering temperature, producing a compositionally homogeneous crystal. Chemical zoning, on the other hand, occurs when equilibrium is *not* maintained and a rim of the new

FIGURE 3.5 **(a)** Compositionally zoned hornblende phenocryst with pronounced color variation visible in plane-polarized light. Field width 1 mm. **(b)** Zoned plagioclase twinned on the carlsbad law. Andesite, Crater Lake, Oregon. Field width 0.3 mm.

composition is added around the old. Compositional re-equilibration in plagioclase requires Si-Al exchange, and this is difficult due to the strength of the Si-O and Al-O bonds. Diffusion of Al is also slow. Zoning in plagioclase is therefore very common.

As Figure 6.8 shows, the composition of plagioclase in equilibrium with a melt becomes more Na rich as temperature drops. The expected zonation in cooling igneous plagioclase would thus be from a more anorthite-rich core toward a more albite-rich rim. This type of zoning is called **normal zoning**. It is common in igneous rocks, although it is typically interrupted by reversals. **Reverse zoning** is the opposite of normal zoning, with more sodic inner and calcic outer zones. It is common in some metamorphic plagioclase, where growth is accompanied by rising temperature. Reverse zoning is rarely a long-term trend in igneous plagioclase; rather, it is typically a short-term event where it contributes to localized reversals as a component of oscillatory zoning. **Oscillatory zoning** is the most common type of zoning in plagioclase because a regular decrease in An content rarely dominates the full crystallization period. The sample illustrated in Figure 3.5b is oscillatory, and Figure 3.6 shows some typical oscillatory zoning profiles, extending from plagioclase cores to rims. Abrupt changes in zoning, such as the reversal in Figure 3.6a, require abrupt changes in the conditions of the magma chamber. Most

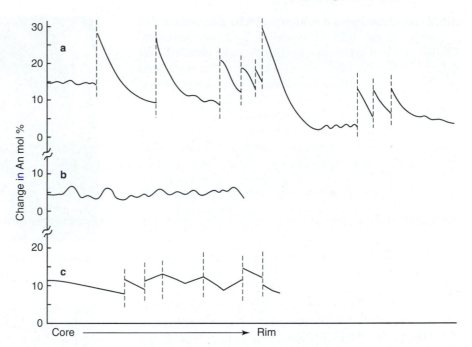

FIGURE 3.6 Schematic examples of plagioclase zoning profiles determined by microprobe point traverses. **(a)** Repeated sharp reversals attributed to magma mixing, followed by normal cooling increments. **(b)** Smaller and irregular oscillations caused by local disequilibrium crystallization. **(c)** Complex oscillations due to combinations of magma mixing and local disequilibrium. From Shelley (1993). Copyright © by permission Kluwer Academic Publishers.

petrologists believe that the injection of hotter, more juvenile magma into a cooling and crystallizing chamber effects this change. The common occurrence of corroded or remelted embayments of the crystal rim accompanying many reversals supports this conclusion. The more gradual oscillations illustrated in Figure 3.6b are more likely to result from diffusion-dependent depletion and re-enrichment of the liquid zone adjacent to the growing crystal in an undisturbed magma chamber. The depleted components could be the anorthite molecule or a constituent such as H_2O that can lower the melting point and thus shift the equilibrium composition of the plagioclase (Loomis, 1982).

Most other minerals are not as conspicuously zoned as plagioclase. This may be because the zoning is simply less obvious in thin section because it may not affect the color or extinction. Most minerals apparently maintain equilibrium with melt because ion exchange does not involve disruption of the strong Si-Al-O bonds. Fe-Mg exchange is also easier because these elements diffuse more readily than Al-Si. Microprobe analysis, however, reveals chemical zonation in several igneous and metamorphic minerals. A color element map of garnet (courtesy of Jack Rice, University of Oregon) is shown on the back cover of the text.

3.1.4 Crystallization Sequence

As a rule, early-forming minerals in melts that are not significantly undercooled are surrounded completely by liquid and develop as **euhedral** crystals, bounded on all sides by crystal faces. As more crystals begin to form and fill the magma chamber, crystals will inevitably come into contact with one another. The resulting mutual interference impedes the development of crystal faces, and **subhedral** or **anhedral** crystals form. In some cases, one can infer the sequence of mineral crystallization from these interferences. Early minerals tend to have better forms, and the latest ones are **interstitial**, filling the spaces between the earlier ones

(Figure 3.7). Phenocrysts in an aphanitic groundmass are typically euhedral and thus clearly formed early in the sequence. Some compositionally zoned minerals may show euhedral cores that formed when the crystals were suspended in melt and anhedral rims that formed later when the crystals were crowded together (see Figure 3.14c).

Unfortunately, the simple principle that a crystal that molds itself to conform to the shape of another must have crystallized later is not as reliable as we might wish. Whether or not a crystal grows with well-developed faces depends largely upon the surface energy of the faces. Minerals with very low surface energy may form euhedral crystals even in metamorphic rocks, where all growing crystals are necessarily in contact with neighboring grains. Garnet and staurolite, for example, are nearly always euhedral in pelitic schists. Igneous accessory minerals such as zircon, apatite, and titanite likewise tend to be euhedral, even though they commonly form during the later stages of crystallization. Metamorphic

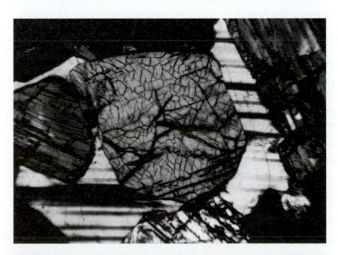

FIGURE 3.7 Euhedral early pyroxene with late interstitial plagioclase (horizontal twins). Stillwater Complex, Montana. Field width 5 mm.

petrologists have noted a tendency toward euhedralism that diminishes in the order of increasing Si-O polymerization. Olivine and pyroxenes thus tend to be more euhedral than quartz or feldspar. Furthermore, Flood and Vernon (1988) noted that, although there is certainly a sequence of mineral formation, there is considerable overlap, and most crystallization takes place via simultaneous crystallization of several mineral types. We shall see why this is true in Chapters 6 and 7. Molding relationships thus develop after most, if not all, of the minerals have begun to crystallize. When growth is simultaneous, the relative surface energy of mutually impinging minerals may have considerable influence on which mineral develops good crystal faces and which molds to the other. Hunter (1987) demonstrated that, although crystals suspended in melt tend to form euhedral grains, once they touch each other, they are likely to dissolve at areas of high surface curvature (interfacial edges) and crystallize at areas of low curvature, thus becoming more rounded. Except for minerals with very low surface energy, then, euhedral crystals should be rare in cases of simultaneous crystallization.

Geologists have often appealed to grain size as another indicator of crystallization sequence. In porphyritic volcanics, the large phenocrysts are generally considered to have formed before the groundmass phases. Although this is commonly true, grain size depends upon nucleation and growth rates, as discussed above, and some of the minerals in the groundmass may have formed early, with a faster nucleation and slower growth rate than the phenocrysts. The large, euhedral, K-feldspar megacrysts found in many granitic rocks, for example, are generally believed to form *late* in the crystallization sequence of those rocks, not early.

Another sequence indicator is based on inclusion relationships. Igneous inclusions should have formed at an earlier stage than the host that enveloped them. One must be aware, however, that a thin section is a two-dimensional slice through a three-dimensional rock, and a mineral that may *appear to be* surrounded by another could be jutting into it from above or below the plane of the section. One should thus note whether a mineral is *consistently* included in another throughout the section before concluding that it is truly an inclusion. In the case of the K-feldspar megacrysts mentioned above, they are commonly poikilitic, and the numerous inclusions of other minerals in them are taken as important indicators of their late formation, overruling arguments for early formation based on grain size. But even when a mineral is consistently included in another, this is not always unequivocal evidence that the included phase ceased to crystallize before the host crystallization began. For example, **ophitic** texture (Figure 3.8) refers to the envelopment of plagioclase laths by larger clinopyroxenes and is commonly interpreted to indicate that the clinopyroxene formed later. McBirney and Noyes (1979), however, noted a case in the Skaergård intrusion of Greenland in which the size of the plagioclase inclusions increases steadily from the clinopyroxene core to the rim. This suggests that plagioclase and clinopyroxene crystallized simultaneously. The clinopyroxene nucleated less readily, so fewer crystals formed, and they grew more rapidly and enveloped the more numerous

FIGURE 3.8   Ophitic texture. A single pyroxene envelops several well-developed plagioclase laths. Width 1 mm.

and smaller plagioclases. The later plagioclase grains that were included toward the host rims had longer to grow and were therefore larger.

As Flood and Vernon (1988) concluded, none of the classical criteria for determining the sequence of crystallization are entirely satisfactory. Inclusions are perhaps the most reliable. Cases in which one mineral forms a rim around another, or in which a mineral is included only in the core areas of another mineral, provide the strongest evidence for one mineral ceasing to crystallize before the other forms (or at least before the other ceases to crystallize). When one mineral occurs commonly as inclusions in another, and not vice versa, it strongly implies, but does not prove, that the included mineral crystallized earlier. Vernon (2004; see Section 3.6) provided a good summary of textural criteria and the sequence of crystallization in igneous rocks.

Although it may be difficult to unequivocally establish that one mineral formed entirely before another, some textures provide a clear testimony of the opposite: simultaneous mineral growth. The ophitic example of simultaneous plagioclase and clinopyroxene described by McBirney and Noyes (1979), cited above, is one example. In some shallow, H_2O-rich, granitic systems, a single alkali feldspar might form (as discussed in association with Figures 6.16 and 6.17). If the H_2O is suddenly lost, the melting point will rise quickly (Section 7.5.1), resulting in undercooling (even at a constant temperature) and rapid simultaneous crystallization of the alkali feldspar and quartz. Under these conditions, the two minerals do not have time to form independent crystals but rather form an intergrowth of intricate skeletal shapes referred to as **granophyric** texture (Figure 3.9a). A rock dominated by such texture may be called a **granophyre**. The intergrowth may nucleate epitaxially on preexisting phenocrysts or dikelet walls. Granophyric texture looks like branching quartz rods set in a single crystal of feldspar. The quartz rods go extinct at the same time, indicating that they are all parts of the same larger crystal. A coarser variation of granophyric texture is called **graphic**, where the cuneiform nature of the quartz rods in the feldspar host is readily seen in hand specimen (Figure 3.9b).

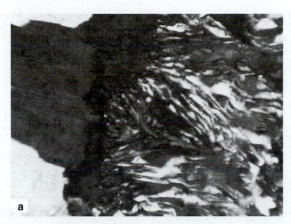

FIGURE 3.9 **(a)** Granophyric quartz–alkali feldspar intergrowth at the margin of a 1-cm dike. Golden Horn, Washington, granite. Width 1 mm. **(b)** Graphic texture: a single crystal of cuneiform quartz (darker) intergrown with alkali feldspar (lighter). Laramie Range, Wyoming.

3.1.5 Magmatic Reaction and Resorption

In some systems, early crystals react with the melt as crystallization proceeds. The reaction of olivine with melt to form pyroxene in the SiO_2-Mg_2SiO_4 system (seen in Figures 6.12 and 6.14) is a prime example. Figure 3.10 illustrates an olivine phenocryst mantled by orthopyroxene, produced at the olivine–melt interface by such a reaction.

Other reactions may result from dropping pressure as a magma rapidly approaches the surface or from magma mixing or other compositional changes. Another common type of reaction occurs when a hydrous magma reaches the surface, where a sudden loss of pressure may release volatiles and cause hydrous phenocrysts, such as hornblende or biotite, to dehydrate and oxidize, developing rims of fine iron oxides and pyroxenes (Figure 3.11c).

Resorption is the term applied to re-fusion or dissolution of a mineral back into a melt or solution from which it formed. Resorbed crystals commonly have rounded corners or are embayed. Some have attributed **sieve texture** (Figure 3.11a), or deep and irregular embayments (Figure 3.11b), to advanced resorption, but others argue that it is more likely to result from rapid growth enveloping melt due to undercooling (Figure 3.4a).

3.1.6 Differential Movement of Crystals and Melt

Flow within a melt can result in alignment of elongated or tabular minerals, producing **foliated** (planar) or **lineated** mineral textures. If lath-shaped microlites (typically plagioclase) in a volcanic rock are strongly aligned (commonly flowing around phenocrysts), the texture is called **trachytic** (Figure 3.12a). Random or non-aligned microlites are called **pilotaxitic** or **felty** (Figure 3.12b). Mingling of two magmatic liquids (either in a chamber or as flows) can create **flow banding** (alternating layers of different composition, (Figure 3.13). Banding and mineral alignment can also result from flow near magma chamber walls, as discussed in Chapter 4.

Suspended phenocrysts may cluster together and adhere by surface tension, a process that many call **synneusis** (Vance, 1969). Synneusis may be a prime mechanism for the production of growth twins because the twin orientation may be an energetically favorable orientation for two crystals of the same mineral to adhere to each other. Multiple-grain clusters of adhering phenocrysts is called **cumulophyric** texture. If the clusters are essentially of a single mineral, some petrologists distinguish the texture as **glomeroporphyritic**.

FIGURE 3.10 Olivine mantled by orthopyroxene in plane-polarized light **(a)** and crossed nicols **(b)**, in which olivine is extinct and the pyroxenes stand out clearly. Basaltic andesite, Mt. McLaughlin, Oregon. Width ~5 mm.

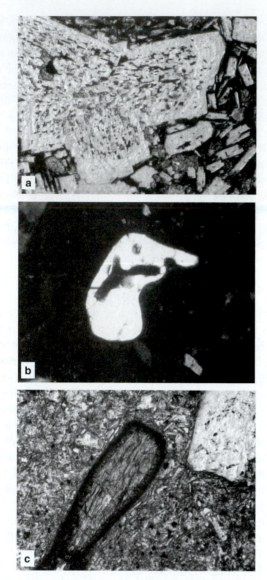

FIGURE 3.11 **(a)** Sieve texture in a cumulophyric cluster of plagioclase phenocrysts. Note the later non-sieve rim on the cluster. Andesite, Mt. McLoughlin, Oregon. Width 1 mm. **(b)** Resorbed and embayed olivine phenocryst. Width 0.3 mm. **(c)** Hornblende phenocryst dehydrating to Fe-oxides plus pyroxene due to pressure release upon eruption of andesite, Crater Lake, Oregon. Width 1 mm.

3.1.7 Cumulate Textures

The development of igneous cumulates will be studied in conjunction with layered mafic intrusions in Chapter 12. Cumulate texture is a hallmark of these fascinating bodies (although it is not restricted to them). For the moment, we will avoid the complex question of how the crystals accumulate. Suffice it to say that, historically, the crystals were considered to accumulate by sinking or floating due to density contrasts with the liquid.

In the ideal case, early-forming crystals of a single mineral accumulate (somehow) to the extent that they are in mutual contact, with the remaining liquid occupying the **interstitial** spaces between the crystals (Figures 3.7 and 3.14a). Mutual contact, however, is not a rigorous requirement for cumulate texture, and close approximation will

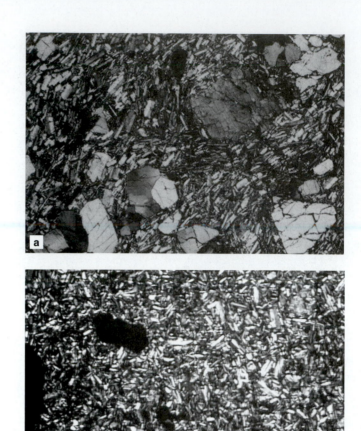

FIGURE 3.12 **(a)** Trachytic texture in which microphenocrysts of plagioclase are aligned due to flow. Note flow around phenocryst (P). Trachyte, Germany. Width 1 mm. From MacKenzie et al. (1982). **(b)** Felty or pilotaxitic texture in which the microphenocrysts are randomly oriented. Basaltic andesite, Mt. McLaughlin, Oregon. Width 7 mm.

suffice. The principal types of cumulates are distinguished on the basis of the extent to which the early-formed crystals, once accumulated, grow prior to ultimate solidification of the interstitial liquid. It would be unusual if the interstitial liquid had the same composition as the accumulated crystals because most magmas are chemically more complex than any single mineral. So if the liquid crystallizes essentially in-place, without exchange with the larger magma reservoir in the interior of the chamber, it should produce some of the initial mineral (assumed to be plagioclase in Figure 3.14, but it could be olivine, pyroxene, chromite, etc.) plus any other minerals that together constitute the interstitial magma. There may thus be some modest additional growth of the early minerals, together with formation of other, later-forming minerals in the interstitial spaces. The result is called an **orthocumulate** texture (Figure 3.14b).

If the interstitial liquid can escape and exchange material (via diffusion and/or convection) with the liquid of the main chamber, the early-forming cumulate minerals may continue to grow as rejected components in the interstitial liquid escape. The result is **adcumulate** texture (Figure 3.14c): a

FIGURE 3.13 Flow banding in andesite. Mt. Rainier, Washington.

nearly monomineralic cumulate with perhaps a few other minerals caught in the last interstitial points. Hunter (1987) concluded that compaction and expulsion of some of the intercumulus liquid *must* accompany the formation of adcumulates because adcumulate texture can be observed in areas too far from the open melt for material to diffuse through the limited porosity associated with late-stage growth. He also concluded that textural equilibrium, resulting in **polygonal** texture (see the polygonal mosaic in Figure 3.17) may be approached or maintained as the grains change shape during late growth and compaction. If true, this readjustment of mineral shapes in contact with each other and melt has important bearing on the earlier discussion of crystallization sequence and molding of one crystal to accommodate another. Hunter (1987) further noted that dihedral angles (see Figure 11.1) between minerals in contact and the late stages of melt trapped at mineral edge boundaries are nearly constant in adcumulates. As crystals grow and impinge, he maintained, surface energy considerations (and hence dihedral angles) control the shape of the last

liquid pockets, and few euhedral terminations are preserved. Thus early euhedral minerals will not dominate the final textures of plutonic rocks.

If the later minerals have a slow nucleation rate, they may envelop the cumulus grains, as described in Section 3.1.4. The result is **poikilitic** texture, but the host oikocryst may be so large and interstitial in some instances that it may be difficult to recognize it as such in the small area of a thin section. A large oikocryst also requires exchange between the interstitial liquid and the main magma reservoir in order to provide enough of its components and to dispose of excess components that would lead to the formation of other minerals. It is thus considered to be a type of adcumulate phenomenon and is termed **heteradcumulate** (Figure 3.14d). Finally, **mesocumulate** is a term applied to cumulate textures that are intermediate between ortho- and adcumulates.

3.1.8 Primary Twinning

A **twin** is an intergrowth of two or more orientations of the same mineral with some special crystallographic relationship between them. Primary (or growth) twins are twins that form because of mistakes during crystallization from a melt. An example is the simple (two-part) carlsbad twins in feldspars shown in Figures 3.5b and 3.18a. Nucleation error of twin-oriented domains is probably the predominant process involved in primary twinning, but synneusis adsorption of two constituents in twin orientation may also occur. Nucleation error is most likely to occur during rapid growth, as might immediately follow nucleation, which should quickly reduce supersaturation or undercooling. For a detailed discussion of synneusis, see Vance (1969) and Dowty (1980a). Repetitive albite twinning (Figure 3.18b) is also believed to result from nucleation errors during growth.

3.1.9 Volcanic Textures

Volcanic rocks cool quickly and tend to form numerous small crystals, as discussed above. Phenocrysts are an exception, resulting from slower cooling beneath the surface prior to eruption. Upon eruption, the remaining liquid crystallizes to

FIGURE 3.14 Development of cumulate textures (using plagioclase as an example). **(a)** Crystals accumulate by crystal settling or simply form in place near the margins of the magma chamber. In this case, plagioclase crystals (white) accumulate in mutual contact, and an intercumulus liquid (gray) fills the interstices. **(b)** Orthocumulate: intercumulus liquid crystallizes to form additional plagioclase rims plus other phases in the interstitial volume (dark). There is little or no exchange between the intercumulus liquid and the main chamber. **(c)** Adcumulates: open-system exchange between the intercumulus liquid and the main chamber (plus compaction of the cumulate pile) allows components that would otherwise create additional intercumulus minerals to escape, and plagioclase fills most of the available space. **(d)** Heteradcumulate: intercumulus liquid crystallizes to additional plagioclase rims, plus other large minerals (hatched and shaded) that nucleate poorly and poikilitically envelop the plagioclases. After Wager and Brown (1967).

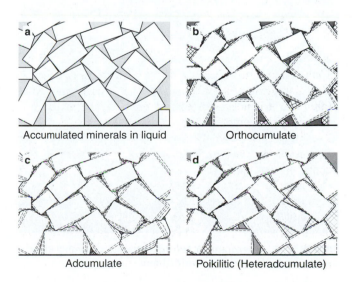

Accumulated minerals in liquid

Orthocumulate

Adcumulate

Poikilitic (Heteradcumulate)

fine tabular or equaint crystals comprising the **groundmass**. The groundmass crystals are called **microlites** (if they are large enough to be birefringent) or **crystallites** (if they are not). Microlites that are significantly larger than the groundmass, yet still microscopic, are called **microphenocrysts**. They are formed upon eruption and represent minerals with a higher ratio of growth rate to nucleation rate than the finer groundmass phases.

Basalts crystallize readily because they are very hot and are dominated by minerals with simple structures. The common result is a texture with a dense network of elongate plagioclase microphenocrysts and granular pyroxenes, with smaller magnetite crystals. Glass may solidify as a late interstitial material. The amount of glass in basaltic rocks is generally less than in more silicic volcanics, but it can vary considerably, from virtually none to highly glassy when basaltic lava comes into contact with water. The lexicon of basaltic textures reflects the variation in glass content, which is crudely correlated with decreasing pyroxene size in a fretwork of intergrown plagioclase laths. **Ophitic** texture (Figure 3.8) refers to a dense network of lath-shaped plagioclase microphenocyrsts included in larger pyroxenes, with little or no associated glass. This grades into **subophitic** (smaller pyroxenes that still partially envelop the plagioclase) and then into **intergranular** texture (Figure 3.15), in which the plagioclase and pyroxene crystals are subequal in size, and glass (or its alteration products) is still relatively minor. Intergranular texture grades into **intersertal** texture when interstitial glass or glass alteration is a significant component. When glass becomes sufficiently plentiful that it surrounds the microlites or microphenocrysts, the texture is called **hyalo-ophitic**. Hyalo-ophitic grades into **hyalopilitic** as the glass fraction becomes dominant, and crystals occur as tiny microlites. The textural terms described above are usually applied to randomly oriented crystals, but they may grade into **trachytic** texture as flow causes alignment of the microlites. (Don't basalt petrographers have fun with nomenclature?)

FIGURE 3.15 Intergranular texture in basalt. Columbia River Basalt Group, Washington. Width 1 mm.

Holohyaline (glassy) texture is most common in silicic rhyolite and dacite flows. If a rock is > 80% glass, it is called **obsidian**. Most investigators prefer to restrict the term to relatively silica-rich glasses and refer to basaltic varieties as **tachylite** or simply **basaltic glass**. Obsidian is very dark colored, despite its commonly silicic nature, because glass is readily tinted by very minor amounts of impurity. Glass in silicic lavas is not necessarily caused by very rapid cooling because some obsidian flows are too thick for the interiors to cool so quickly. Motion and/or the characteristically slow diffusion and nucleation of highly polymerized and viscous silicic flows may impede crystallization and produce these highly glassy rocks.

Trapped bubbles of escaping gas create subspherical voids in volcanics, called **vesicles**. Bubbles tend to rise in less viscous basaltic magmas and thus concentrate near the surface of basaltic flows. There is a complete gradation from basalt to **vesicular basalt** to **scoria**, with increasing vesicle content. Vesicles filled with later mineral growth, typically secondary zeolite, carbonate, or opal, are called **amygdules**. The silicic counterpart of scoria is **pumice**. Pumice is typically light and frothy, and fresh samples float in water. Frothy pumice is typically light gray, even though its corresponding vesicle-free obsidian may be black. The reason for this contrast is that the bubbles expand the glass to a thin film between the bubbles, which refracts and diffuses light, just as breaking waves form whitecaps on otherwise dark seawater.

3.1.10 Pyroclastic Textures

Pyroclastic rocks are fragmental, generally produced by explosive volcanic activity. The classification of pyroclastic rocks is based on the nature of the fragments (**pyroclasts**, or **tephra**), as discussed in Section 2.5. Pyroclastic eruptive and emplacement modes are discussed in Chapter 4. The ash component of pyroclasts is typically a mixture of pulverized rock and primary glass (including shattered pumice and aerosol liquid). The vesicles in pumice expand rapidly upon explosive eruption and are usually destroyed. The interstitial glass then forms cuspate- or spicule-shaped three-pointed shards in thin section (Figure 3.16a). Because these shards are commonly warm in a pyroclastic flow, they deform in a ductile fashion and fold over to the shape indicated on the right of Figure 3.16a and in Figure 3.16b. This type of bending, and other structures caused by compression and deformation resulting from settling in hot ash accumulations, are collectively referred to as **eutaxitic** textures. Larger pieces of pumice may accumulate intact and have the gas squeezed from them, eliminating the bubbles. If all the gas is expelled, the pumice returns to the black color of obsidian, and the squashed fragments are called **fiamme**. In fluid lavas, such as basalts, bursting bubbles hurl fine spray aloft, and it falls as glassy pellets called "Pele's tears" (after the Hawaiian volcano god, Pele), or the magma may be stretched to form delicate glass threads ("Pele's hair"). Ash falling through very moist air may accumulate successive layers on a single ash nucleus, forming spheroidal balls called **accretionary lapilli**. Consolidated deposits of such lapilli are called **pisolitic tuffs**.

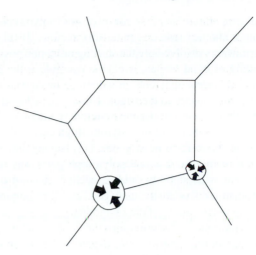

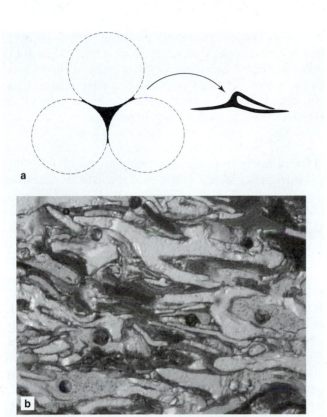

FIGURE 3.16 **(a)** The interstitial liquid (black) between bubbles in pumice (left) become three-pointed-star-shaped glass shards in ash containing pulverized pumice. If they are sufficiently warm (when pulverized or after accumulation of the ash), the shards may deform and fold to contorted shapes, as seen on the right and in the photomicrograph **(b)** of the Rattlesnake ignimbrite, southeastern Oregon. Width 1 mm.

3.2 SECONDARY TEXTURES: POSTMAGMATIC CHANGES

Secondary textures are those that develop after the igneous rock is entirely solid. These processes do not involve melt and are thus really *metamorphic* in nature. The process of crystallization does not necessarily cease when the magma becomes solid. As long as the temperature is high enough, recrystallization and both chemical and textural re-equilibration take place. (Otherwise, metamorphic petrologists would have very little to do.) Large cooling plutons may remain at temperatures equivalent to high-grade metamorphism for thousands of years, so there is ample opportunity for such processes to occur. Solid-state processes that occur as a result of igneous heat (even though waning) are called **autometamorphic** and are covered in this section. Because solid-state crystallization processes are truly metamorphic, they will be discussed more fully in Chapter 23.

Ostwald ripening is a process of annealing (or textural maturing) of crystals in a static environment. As Hunter (1987) noted, differences in grain–boundary curvature drive grain growth by Ostwald ripening until straight boundaries result (Figure 3.17). In such recrystallization, grain boundaries migrate toward their centers of curvature. Small grains with convex outward curvature are thus eliminated as the surfaces of neighboring larger grains with convex

FIGURE 3.17 Ostwald ripening in a monomineralic material. Grain boundaries with significant negative curvature (concave inward) migrate toward their center of curvature, thus eliminating smaller grains and establishing a uniformly coarse-grained equilibrium texture with 120° grain intersections (polygonal mosaic).

inward curvature encroach upon them. If the process attains textural equilibrium in a solid, there will be similarly sized grains having straight, approximately 120° triple-grain intersections (Figure 3.17). This equilibrium texture is most common in monomineralic metamorphic rocks (quartzite and marble), particularly if metamorphosed in a nearly static stress regime. We will discuss annealing further in Chapter 23. Most igneous rocks are not monomineralic, however, and rarely attain a good equilibrium texture. Relative differences in surface energy of contrasting mineral types and the coarse grain size of plutonics serve to establish and maintain interlocking textures in most cases. Ostwald ripening, however, may eliminate smaller grains in favor of larger neighbors at an early stage of growth, producing a more uniform distribution of grain sizes. Volcanic rocks with small initial grain size are far less stable than plutonics, and the groundmass recrystallizes readily. Glass is particularly unstable and readily devitrifies to fine mineral replacements, as described in Section 3.2.4. But volcanics cool quickly to low temperatures, and kinetic restrictions to recrystallization develop early. The retention of magmatic textures is thus surprisingly good in igneous rocks, but some types of solid-state recrystallization are well known, and I shall review them here.

3.2.1 Polymorphic Transformation

As you undoubtedly learned in mineralogy, many natural substances have more than one crystal structure. Alternative structural forms of the same chemical substances are called **polymorphs**. Familiar polymorphs are graphite–diamond, calcite–aragonite, kyanite–andalusite–sillimanite, and the several polymorphs of SiO_2. As we shall see in Chapter 6, a given structure is most stable over a particular range of pressure and temperature conditions, so that one polymorph will transform to another when the conditions change from its stability range to that of an alternative structure. As they cool and rise, partly or wholly crystalline igneous rocks may leave the pressure–temperature stability range of one polymorph and enter that of another, resulting in transformations. **Displacive transformations** involve only the shifting of atomic positions and bending of bond angles. A classic example is the high-quartz to low-quartz transition, in which the hexagonal high-quartz structure inverts to the trigonal structure of low quartz upon cooling (Figure 6.6). **Reconstructive transformations**, such as graphite–diamond or tridymite–high-quartz, involve the breaking and re-forming of bonds. Displacive transformations occur readily, so that one polymorph gives way to another as soon as its stability field is reached. Reconstructive polymorphism is less easily managed, and one polymorph may remain in the stability field of another.

Polymorphic transformations are common in many minerals, including quartz and feldspars, but may be difficult to recognize texturally because evidence for the initial phase might be completely lost, and only the replacement polymorph remains. If the crystal form of the earlier phase is distinctive, however, the replacement polymorph may form a **pseudomorph** of the original rather than assume its own characteristic form. For example, high quartz may crystallize as early phenocrysts in some rhyolites. The crystal form of high quartz is characteristically a hexagonal dipyramid, without the predominant prism faces of low quartz. Because high-to-low quartz is a displacive transformation, the high quartz in such crystals must invert to low quartz at 573°C (if at atmospheric pressure). The original phenocryst shape is usually preserved, however, thus providing evidence for the initial form. Another possible result of polymorphic transformations is the development of secondary twins, which can also provide a clue to the transformation process. Recognition of such features can provide some very useful information concerning the thermal history of a rock.

3.2.2 Secondary Twinning

In addition to the primary twinning discussed above, twinning may occur by secondary processes in preexisting minerals. Secondary twins can occur as a result of polymorphic transformation or deformation. **Transformation twins** are caused when a high-temperature crystal structure inverts to a low-temperature polymorph. As high-temperature structures have more vibrational energy, they generally exhibit a higher degree of symmetry than the low-temperature alternative. Because symmetry is lowered with cooling, the high-temperature form typically has a choice of two or more alternative low-symmetry orientations. If a whole crystal assumes one of the alternatives, no twinning will result. If, on the other hand, different portions of the same crystal are displaced into each of the alternative choices, the portions will generally be in twin relationship with the other portion. This is the origin of the crosshatched, or "tartan," twins of microcline (Figures 3.18c and 3.18d), which are produced when the high-temperature monoclinic form inverts to the low-temperature triclinic microcline structure. The familiar multiple "albite" twins of some plagioclases (Figure 3.18b) are attributed to a similar monoclinic to triclinic transformation, but this does not occur in plagioclase of intermediate

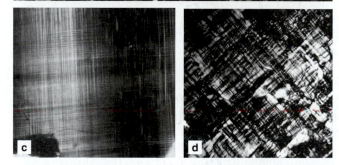

FIGURE 3.18   Feldspar twins. **(a)** Carlsbad twin in orthoclase. Wispy perthitic exsolution is also evident. Granite, St. Cloud, Minnesota. **(b)** Very straight multiple albite twins in plagioclase, set in felsitic groundmass. Rhyolite, Chaffee, Colorado. **(c–d)** Tartan twins in microcline. All field widths ~1 mm.

composition, and these twins are typically primary twins resulting from nucleation error during growth. Cyclic twins in quartz and olivine are other examples of transformation twins.

Twins can also be generated by deformation of solidified igneous rocks. Twinning is an important aspect of a rock's response to deformation because the shifting of a portion of a grain into the twin orientation is an easier response to effect than is grain rupture. **Deformation twins** in plagioclase can occur on the albite twin law, but they usually lack the extremely straight lamellar form of their primary counterparts. Deformation twins in plagioclase are most easily recognized when they are wedge shaped and

FIGURE 3.19 Polysynthetic deformation twins in plagioclase. Note how they concentrate in areas of deformation, such as at the maximum curvature of the bent cleavages, and taper away toward undeformed areas. Gabbro, Wollaston, Ontario. Width 1 mm.

bent (Figure 3.19). Calcite also readily develops deformation twins in response to shear. Deformation twins enhance the ductile response of rocks such as marble and are uncommon in other minerals.

3.2.3 Exsolution

Exsolution, discussed further in Section 6.5.4, involves chemical mixing that becomes increasingly limited in some solid-solution minerals as they cool. Perhaps the most common example occurs in alkali–feldspars, where unmixing results in separation of Na-rich and K-rich segregations. Because the unmixing in this case involves partitioning only of K and Na ions, and not strongly bonded Si and Al, it takes place relatively easily, and the segregations appear as a coherent intergrowth of long, wispy lamellae. "Coherent," in this sense, means that the lattices of the lamellae have a specific crystallographic relationship to the host and are not randomly oriented. When the alkali feldspar is potassic, the result is exsolved albite lamellae in a K-feldspar host, called **perthite** (Figure 3.18a). When the alkali feldspar is sodic, the lamellae are K-feldspar in an albite host, referred to as **antiperthite**. Exsolution also occurs in plagioclase at times, but albite–anorthite unmixing involves Si-Al, as well as Na-Ca exchange, and the process, when it occurs at all, produces much finer lamellae.

Exsolution also occurs in pyroxenes. For example, a low-Ca orthopyroxene may separate from high-Ca clinopyroxene. Thin lamellae of one of these in a host of the other are common. Pigeonite, an intermediate mixture, is found principally in volcanic rocks that cooled too quickly to allow such unmixing to occur.

Unmixing does not have to be coherent, and irregular patches of exsolved phases are commonly found, as are instances of complete expulsion of the exsolved phase to form separate grains immediately outside the host. Exsolution also occurs in amphiboles and a few other common minerals.

Some mafic silicates may even exsolve an Fe-Ti-oxide. In some anorthosites, high-temperature and high-pressure pyroxenes can dissolve a considerable amount of Al, and they exsolve lamellae of plagioclase as they cool at lower pressures. The red or pink color common to some feldspars is caused by the exsolution of fine hematite.

Ocelli are spherical or ovoid bodies a few millimeters to a few centimeters across that occur in some igneous rocks. Some appear to result from *liquid* immiscibility, a primary exsolution phenomenon discussed in Chapters 6, 11, and 19. Others are probably amygdule fillings, and yet others may be isolated blobs of mingled magmas.

3.2.4 Secondary Reactions and Replacement

Solid–solid and solid–vapor reactions are processes that dominate during metamorphism, and they will be left largely to Part II of this text. As stated above, however, igneous rocks cool through a temperature range appropriate to this realm, and plutonic rocks remain in that realm for a considerable length of time. Secondary mineral reactions that occur in igneous rocks as they cool, and that are not products of a distinct later metamorphic event, are commonly called **autometamorphic** processes rather than metamorphic ones because they are a natural part of igneous cooling. Autometamorphic processes are more common in plutonic rocks than in volcanics because they remain at elevated temperatures for a longer time. Diagenetic and weathering processes are not considered to be autometamorphic (a rather arbitrary distinction). Most, but not all, autometamorphic reactions involve minerals at moderate temperatures in an environment in which H_2O is either liberated from residual melt or externally introduced. Such alterations are a subset of autometamorphism, which involves hydration, and they are called **deuteric** alterations. Some of the principal alteration processes follow.

Pyroxene is a common primary mafic mineral in a variety of igneous rocks. If H_2O penetrates at modest temperatures, a deuteric alteration of pyroxene to amphibole results, called **uralitization** (Figure 3.20a). Any gradation from amphibole rims on pyroxene cores to multiple patches of pyroxene in an amphibole to complete replacement is possible. The amphibole may be a single crystal of hornblende or a fibrous actinolite or hornblende aggregate. Either, when demonstrated to result from pyroxene alteration, may be called **uralite**, but the term is more commonly applied to the aggregates.

Biotitization is a similar process of hydration/deuteric alteration that produces biotite, either directly from pyroxene, or, more commonly, from hornblende. Because biotite contains little Ca, epidote may be produced as Ca is released during the alteration of hornblende to biotite.

Chloritization is the alteration of any mafic mineral to chlorite. Chlorite is a very hydrous phyllosilicate and typically replaces the less hydrous mafics at low temperature when water is available. Pyroxenes, hornblendes, and biotites are commonly observed in thin section in various stages of alteration to chlorite. As in other deuteric alterations, hydration attacks the outer margin of a mineral, so that chlorite

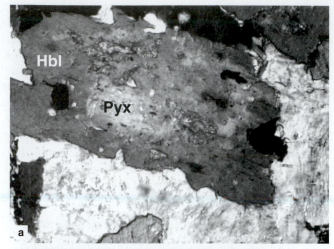

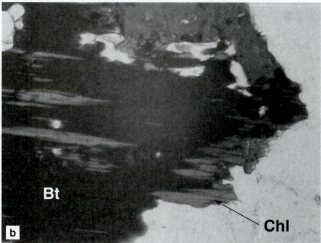

FIGURE 3.20 (a) Pyroxene largely replaced by hornblende. Some pyroxene remains as light areas (Pyx) in the hornblende core. Width 1 mm. (b) Chlorite (light) replaces biotite (dark) at the rim and along cleavages. Tonalite, San Diego, California. Width 0.3 mm.

generally replaces the initial mafic from the rim inward. In the case of biotite, H_2O may work its way along the prominent cleavages, and chlorite can be seen to replace biotite margins as well as along cleavage planes (Figure 3.20b).

Sericite is a term applied to any very fine-grained white mica. **Seritization** is the process by which felsic minerals (usually feldspars of feldspathoids in igneous rocks) are hydrated to produce sericite. Incipient stages can be recognized by a fine, dusty appearance of feldspars in plane-polarized light. In more advanced stages of alteration, the feldspars appear speckled, with fine micas having yellowish birefringence and then larger clots with coarser crystals and higher birefringence. K^+ ions are required for plagioclase to be altered to the common forms of sericite. Potassium may be released by chloritization of nearby biotite. K-feldspar requires no additional K^+ and may be more sericitized than associated plagioclase.

Saussuritization is the alteration of plagioclase to produce an epidote mineral. Higher-temperature plagioclase tends to be more calcium-rich (Figure 6.8), which is less stable than its sodic counterpart at low temperatures. The calcium-rich types thus break down to more nearly pure albite,

releasing Ca and Al to form an epidote mineral (± calcite and/or sericite). In zoned plagioclase, one can occasionally observe the products of saussuritization or seritization concentrated in the Ca-rich core or some of the more Ca-rich oscillatory bands in the original grain.

Olivine is easily altered in cooling mafic rocks, even in volcanics. It is commonly rimmed or replaced by serpentine or dark brown iddingsite.

Symplectite is a term applied to fine-grained intergrowths resulting from the combined growth of two or more minerals as they replace another mineral. As in any of the replacements described above, replacement may be partial or complete. Complete replacement and pseudomorphs are common. The fibrous actinolite–hornblende "uralite" aggregates replacing pyroxene are an example. Biotite and epidote replacing hornblende is another. **Myrmekite** is an intergrowth of dendritic quartz in a single crystal of plagioclase (Figure 3.21). The quartz appears rod-like in thin section, and numerous adjacent rods go extinct in unison, indicating that they are all parts of a single quartz crystal. Myrmekites are very common in granitic rocks and occur preferentially where plagioclase is in contact with K-feldspar. Myrmekites appear to have grown from the plagioclase–K-feldspar boundary into the K-feldspar. As the plagioclase replaces the K-feldspar, SiO_2 is released (the anorthite component of plagioclase contains less SiO_2 than the K-feldspar), thereby producing the quartz.

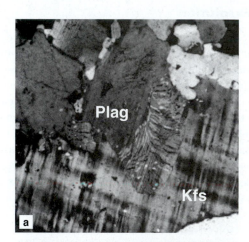

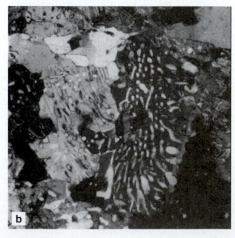

FIGURE 3.21 Myrmekite formed in plagioclase at the boundary with K-feldspar. Photographs courtesy L. Collins.

Myrmekite commonly forms during cooling of granitic rocks but may also occur in metamorphics. For details of the process, see Shelley (1993, pp. 144–147) or Vernon (2004; see Section 4.10).

Devitrification is the secondary crystallization of glass to fine-grained mineral aggregates. Glass is an inherently unstable material and is readily replaced by more stable minerals when kinetics permits. Water-quenched basaltic glass may be replaced by brown, optically isotropic oxidation--hydration products collectively known as **palagonite**. More silicic glassy rocks commonly devitrify to produce a microgranular mass of small, equidimensional grains of interlocking feldspar and silica minerals called **felsitic** texture (Figure 3.18b). Crystal form is entirely suppressed in felsitic rocks, and the texture looks very much like that of chert. Devitrification of glass may also produce radial aggregates of crystals (commonly cristobalite or tridymite plus feldspar) called **spherulites**. **Lithophysae** are large cavities bordered by spherulitic growth. Common in rhyolites, they probably represent late volatile releases that created a gas bubble in the glass. Spherulites may be found imbedded in a felsite matrix. The minerals produced by devitrification are generally too fine to identify under the polarizing microscope, and x-ray analysis may be required.

3.2.5 Deformation

Deformation of solid rock can result in a variety of textures, most of which we will cover in Chapter 23. For now, we are concerned only with common types of post-crystallization deformation of igneous rocks. The result is bent, broken, and crushed crystals or fragments. Foliations are created in many igneous rocks that remain at depths and temperatures where they are sufficiently ductile to deform readily. Compaction of pyroclastic deposits creates a flattening effect and typically imparts an enhanced foliation. Deformation can also produce **undulose extinction**, a waviness in the optical extinction pattern due to minor bending of the crystal lattice. It is not to be confused with compositional zoning, which results in a distinctly *concentric* pattern of varying extinction. Quartz is notoriously susceptible to the development of undulose extinction and may exhibit it when other minerals do not. Undulose extinction is commonly used to distinguish quartz from non-undulatory feldspars when observing a thin section under low power.

3.3 A GLOSSARY OF IGNEOUS TEXTURAL TERMS

Table 3.1 is a brief glossary of igneous textural terms, designed to assist you in describing hand specimens and thin sections. The terms may seem strange to you now, but they serve to describe most of the common characteristics of igneous rocks. Terms are grouped into categories to help you find a name for a texture you observe. Unfortunately, this makes it more difficult if you simply want to find the definition of a term. Petrography is an iterative process. It is not enough to look at a hand specimen or thin section once and then work through a checklist of terms. Rather, at each step, it is advantageous to reassess all previous information

---

TABLE 3.1   Common Igneous Textural Terms

Crystallinity

Holocrystalline	Consisting entirely of crystals (default term, not commonly used).
Hypocrystalline	Containing both crystals and glass.
Holohyaline, vitric	Consisting entirely of glass.

Grain Size

Aphanitic	Having minerals too fine grained to see with the naked eye.
Phaneritic	Having minerals coarse enough to see with the naked eye.
Cryptocrystalline	Having minerals too fine grained to distinguish microscopically.
Fine grained	Having an average crystal diameter less than 1 mm.
Medium grained	Having an average crystal diameter 1–5 mm.
Coarse grained	Having an average crystal diameter greater than 5 mm.
Very Coarse Grained	Having an average crystal diameter greater than 50 mm.
Pegmatitic	Being very coarse grained (historically associated with very coarse granitoid rocks: pegmatites).
Saccharoidal	Being fine- to medium-grained xenomorphic and equigranular (looking like sugar).
Aplitic	A synonym for saccharoidal, but typically restricted to leucocratic granitoid rocks.
Equigranular	Having grains that are all approximately the same size.
Inequigranular	Having grains that vary considerably in size.

Porphyritic Textures

Porphyritic	Having approximately bimodal size distribution (usually requires a great difference).
Megaporphyritic	Having a porphyritic texture that can be seen in hand specimen (rarely used).
Microporphyritic	Having a porphyritic texture that is visible only under the microscope.
Phyric (-phyric)	An adjective (or suffix) referring to porphyritic texture.

(continued)

TABLE 3.1 Common Igneous Textural Terms *(Continued)*

Porphyritic Textures (*cont.*)

Phenocryst	A large crystal set in a fine matrix.
Microphenocryst	A microscopic crystal that is larger than the remainder of the groundmass.
Megacryst	An unusually large crystal, either a phenocryst or a xenocryst.
Poikilitic	The state of a host phenocryst (oikocryst) containing many inclusions of other minerals.
Oikocryst	A host phenocryst in poikilitic texture.
Groundmass (matrix, mesostasis)	The glassy or finer-grained element in the porphyritic texture.
Cumulophyric	Having phenocrysts of the same or different minerals occurring in clusters (Figure 3.11a).
Glomeroporphyritic	Synonymous with cumulophyric (used by some to specify that only one mineral is involved).
Hiatial porphyritic	Having a pronounced difference in size between the phyric and groundmass phases (the default term: rarely used).
Seriate	Having a continuous gradation in size.
Aphyric	Not porphyritic (having no phenocrysts).

Form of Individual Grains

Euhedral (idiomorphic)	Completely bounded by crystal faces.
Subhedral (subidiomorphic)	Having crystal faces that are only partially developed.
Anhedral (allotriomorphic)	Having crystal faces that are entirely absent.
Crystal habits: equiant, prismatic, columnar, accicular, fibrous, tabular, platy, micaceous, lath-shaped, etc. (see any mineralogy text).	
Corroded (embayed)	Subhedral or anhedral and produced by partial melting (resorption) of phenocrysts by the melt.

Forms of Grains in the Rock as a Whole

Panidiomorphic	Having a majority of euhedral grains. Rare.
Hypidiomorphic	Consisting predominantly of subhedral grains. Common in many granitic rocks.
Allotriomorphic	Having a majority of anhedral grains (common).
Sutured	Characterized by articulation along highly irregular interpenetrating boundaries. Common in recrystallized deformed rocks.
Mosaic	A texture of polygonal equigranular crystals.

Intergrowths

Host (Oikocryst)	A large mineral that includes others in poikilitic texture.
Guest	The included mineral in poikilitic texture.
Poikilitic	Containing several small discrete crystals of another mineral. Refers to growth phenomena, not exsolution or replacement.
Graphic	Having an intergrowth in which the guest shows angular wedge-like forms. Usually occurs with quartz in microcline (Figure 3.9b).
Micrographic	Having graphic texture that is visible only under the microscope.
Granophyric	Having a texture in which the quartz and feldspars penetrate each other as feathery irregular intergrowths. Resembles micrographic texture but is more irregular.
Exsolution lamellae	Lamellar bands of a phase exsolved from a host phase (Figure 3.9a).
Perthitic	Having irregular veins, patches, lenses, etc., of sodic plagioclase in an alkali feldspar host. Usually results from exsolution (Figure 3.18a).
Antiperthitic	Having exsolution lamellae of alkali feldspar in a plagioclase host. Usually much thinner than perthite.
Symplectite	A replacement texture in which a mineral is replaced by an intergrowth of one or more minerals.
Myrmekite	A secondary texture consisting of irregular "wormy" blebs or rods of quartz in a plagioclase host adjacent to alkali feldspar grains (Figure 3.21).
Spherulitic	Having a radial intergrowth of fibrous minerals. Commonly alkali feldspar and quartz in devitrified silicic volcanics.
Axiolitic	Similar to spherulitic, but with fibers occurring in a layer and oriented normal to its walls.
Variolitic	Spherulitic and consisting of divergent plagioclase fibers. Applies to certain basalts.
Coalescent	Having anhedral texture developed by simultaneous growth of two mineral grains in contact.

(continued)

TABLE 3.1   *(Continued)*

Textures of Mafic Igneous Rocks

Ophitic	Having large pyroxene grains enclosing small, random plagioclase laths (Figure 3.8).
Subophitic	Having plagioclase laths that are larger and only partially enclosed by the pyroxene.
Nesophitic	Having a plagioclase that is larger, with interstitial pyroxenes.
Intergranular	Having small, discrete grains of pyroxene, olivine, etc., filling the interstices in a random network of larger plagioclase laths (Figure 3.15).
Intersertal	Having glass, cryptocrystalline material, or alteration products occupying the spaces between plagioclase laths.
Hyalo-ophitic	Having an intersertal texture in which a larger amount of glass is present than pyroxene.
Hyalopilitic	Having a large amount of glass, with plagioclase occurring only as tiny, random microlites.
Diktytaxitic	The texture of certain volcanics in which bounding crystals protrude into abundant angular interstitial gas cavities.
Cumulate	Displaying interstitial growth of a mineral between earlier ones that are all in contact and give the distinct impression that they accumulated at the bottom of a magma chamber (Figure 3.14).
Orthocumulate	Having cumulate texture, with other minerals occupying the interstitial areas (Figure 3.14b).
Adcumulate	Having cumulate texture in which the early cumulate minerals grow to fill the pore space (Figure 3.14c).
Mesocumulate	Having a texture that is intermediate between ortho- and adcumulate.

Replacement Textures

Pseudomorph	A replacement texture in which one or more minerals replace another, retaining the form of the original mineral.
Symplectite	A replacement texture in which a mineral is replaced by an intergrowth of one or more minerals.

Specific Mineral Replacements

Uralitization	Replacement of pyroxene by amphibole (Figure 3.21a).
Saussuritization	Replacement of plagioclase by epidote.
Biotitization	Replacement of pyroxene or amphibole by biotite.
Chloritization	Replacement of any mafic mineral by chlorite (Figure 3.20b).
Seritization	Replacement of feldspar or feldspathoids by fine white micas.

Miscellaneous Terms

Interstitial	Having one mineral filling the interstices between earlier crystallized grains (Figure 3.7).
Crystallites	Minute, inchoate crystals in the earliest stages of formation. They are isotropic and cannot be identified under the microscope.
Microlites	Tiny needle- or lath-like crystals of which at least some properties are microscopically determinable.
Felty	Consisting of random microlites (Figure 3.13b).
Pilotaxitic	A synonym for felty.
Trachytic	Consisting of (feldspar) microlites aligned due to flow (Figure 3.12a).
Embayed	Having embayments due to reaction with the melt (resorption) (Figure 3.02).
Skeletal	Having crystals that grew as, or have been corroded to, a skeletal framework with a high proportion of internal voids (Figure 3.4).
Sieve	Crystals filled with channelways (appearing as holes) due to resorption (Figure 3.11a).
Epitactic	Oriented nucleation of one mineral on another of a different kind.
Rapakivi	Overgrowths of plagioclase on alkali feldspar.
Vesicular	Containing gas bubbles.
Scoriaceous	Highly vesicular.
Pumiceous	Having a frothy vesicular structure characteristic of pumice.
Miarolitic	Having gas cavities into which euhedral minerals protrude. Applies to certain plutonic rocks.
Pipe vesicles	Tubelike elongate vesicles that result from rising gases.
Vesicular pipes	Cylindrical bodies that are highly charged with vesicles.

(continued)

TABLE 3.1   Common Igneous Textural Terms *(Continued)*

Miscellaneous Terms *(cont.)*

Amygdaloidal	Having vesicles that are completely or partially filled with secondary minerals.
Lithophysae	Large ovoid structures representing gas bubbles in devitrified rhyolitic glass.
Foliation	Planar parallelism.
Banding	Alternating planar layers.
Lineation	Linear parallelism.
Xenolith	An inclusion of country rock.
Xenocryst	A single-crystal foreign inclusion.
Perlitic	Having a concentric fracture pattern resulting from contraction of some volcanic glasses upon cooling.
Pyroclastic	Composed of fragments.
Ocelli	Ovoid blobs created by liquid immiscibility, mingled magmas, or filled vesicles.
Orbicules	Ovoid masses of radiating crystals, commonly concentrically banded, found in some granites.
Spinifex	A centimeter-scale texture subparallel to dendritic growth of olivine crystals in some quenched ultramafics.

Pyroclastic Terms

See Section 2.5.

Pyroclastic Glass textures

Pele's tears	Glassy lapilli.
Pele's hair	Hair-like strands of glass.
Fiamme	Compressed pumice fragments in a tuff.

accumulated on a specimen in light of the latest observations. During this process, be sure to check for possible relations between pieces of information, as well as for consistency of data.

A good approach to characterizing a rock is to describe the hand specimen in terms of the most general attributes, such as color, crystallinity, felsic or mafic character, and the most general textures (foliation, porphyritic, etc.). Then go on to determine the mode and describe individual minerals, their grain size (and variance in grain size), as well as shape, intergrowths, and specific textures. Be *descriptive* here but make deductions concerning origin of textures when you feel it is appropriate. Avoid textural terms that make unsupportable assumptions regarding the genesis of a specimen. Some textures (e.g., foliations) may occur in a number of ways. If you choose a genetic term (e.g., flow banding), be sure you can support your choice. Just be clear on the distinction between *observation* and *interpretation* when you do so. Good observations shouldn't change, but our interpretations may. One day, you may return to your descriptive notes and reinterpret a texture in a different (hopefully better) way.

Finally, make any general deductions you like and name the rock. Be sure you use a compositional term from Chapter 2 in the name (e.g., basalt or granite) and any textural features that you consider important or want to emphasize (e.g., vesicular *basalt*, *rhyolitic* tuff, porphyritic *monzonite*).

Summary

The study of igneous textures in thin section using the polarized-light petrographic microscope allows petrologists to readily identify minerals and observe important textures that greatly aid the interpretation of the rock's cooling and crystallization history. The relative rates of nucleation, growth, and diffusion have a profound effect on the grain size and texture of the resulting rock. Mineral inclusions, and the degree to which one mineral develops excellent crystal faces or interferes with the ability of an adjacent mineral to do so can aid in interpreting the sequence in which the minerals crystallized from the molten state. Minerals that permit significant solid solution typically vary systematically in composition as they crystallize over a temperature interval, and if equilibrium is not maintained, they may develop chemical zoning as successive growth layers are added. Differential motion of crystals and melt may lead to foliations, banding, or cumulate textures. Volcanic rocks cool quickly, typically resulting in fine microlites or crystallites and/or glass. Pyroclastic rocks typically result from explosive eruptions and thus exhibit a fragmental texture.

Igneous rocks typically solidify at temperatures in the 700 to 1200°C range (depending on composition), and as the resulting rock continues to cool, a number of postmagmatic processes may further modify the mineralogy and texture. These include annealing, devitrification of glass, polymorphic transformations of certain high-temperature minerals to lower-temperature structural forms, transformation twinning, exsolution, and replacement of igneous minerals with other (metamorphic) minerals that are more stable under the cooler conditions.

Key Terms

Primary textures *34*
Secondary textures *34*
Crystal nucleation *35*
Crystal growth *35*
Diffusion *35*
Undercooling *35*

Porphyritic *36*
Phenocryst *36*
Groundmass *36*
Epitaxis *38*
Compositional zoning *38*
Twinning *43*

Euhedral, anhedral, subhedral *39*
Interstitial *42*
Cumulate *42*
Autometamorphism *47*
Polymorphism *46*
Exsolution *47*

Review Questions and Problems

Review Questions and Problems are located on the author's web page at the following address: **http://www.prenhall.com/winter**

Important "First Principle" Concepts

- Crystals form by first nucleating (as a critical cluster of constituent atoms gather together) and then growing on that crystal nucleus.
- Crystal growth typically requires diffusion of the constituents through the matrix (in igneous systems, this would be a melt) to the surface of the growing crystal.
- The rate of cooling of a magmatic system plays a critical role in governing the extent of undercooling, which in turn affects the nucleation and growth rates of crystals and, ultimately, the texture of the final rock.
- Different minerals form sequentially in natural magmatic systems, so that a magma will completely crystallize over a range of temperatures.

- Early-forming crystals tend to be more euhedral than later-forming crystals because development of their crystal faces is not impeded by adjacent crystals. Later, when many crystals of several mineral types are occupying sufficient space as to be mutually interfering with each other's growth, the shapes become more subhedral to anhedral. The last mineral(s) to form crystallize from the residual melt in the interstices between previously formed minerals and take on the interstitial shape.

Suggested Further Readings

Bard, J. P. (1986). *Microtextures of Igneous and Metamorphic Rocks*. Reidel. Dordrecht, The Netherlands.

Hibbard, M. J. (1995). *Petrography to Petrogenesis*. Prentice Hall. Englewood Cliffs, NJ.

MacKenzie, W. S., C. H. Donaldson, and C. Guilford. (1982). *Atlas of Igneous Rocks and Their Textures*. John Wiley & Sons. New York.

McBirney, A. R. (1993). *Igneous Petrology*. Jones and Bartlett. Boston. Chapter 4: Igneous minerals and their textures.

Shelley, D. (1993). *Igneous and Metamorphic Rocks Under the Microscope*. Chapman & Hall. London.

Vernon, R. H. (2004). *A Practical Guide to Rock Microstructure*. Cambridge University Press. Cambridge, UK.

Williams, H., F. J. Turner, and C. M. Gilbert (1982). *Petrography: An Introduction to the Study of Rocks in Thin Section*. W.H. Freeman. San Francisco.

4

Igneous Structures and Field Relationships

Questions to be Considered in this Chapter:

1. In what types of field settings are volcanic and plutonic rocks found?

2. How do the properties of magmas, such as composition, volatile content, and viscosity, affect the style of volcanic eruption?

3. What are the types of volcanic landforms, and how do they form?

4. What are the typical shapes of plutonic rock masses, and how are they related to intrusion style?

5. What are the outcrop-scale features of volcanic and plutonic rocks, and how do they form?

6. What features develop at the contact between hot intruded magma and the adjacent wall rock, and how do these features vary with depth and size of intrusion?

7. How are plutons emplaced, and how do they make sufficient room for themselves?

Igneous rocks can be studied on scales from the microscopic to the global. Chapters 2 and 3 dealt with the hand specimen and microscopic scales. The present chapter explores the middle scale: that of the outcrop, landform, and deposit. We shall begin with the volcanic products, perhaps because they are the most dramatic and familiar. Then we shall proceed to the intrusive bodies and their structure, shape, and relationship to the country rocks. There is a tremendous accumulation of literature on these subjects, particularly on the volcanic side, and the present discussion can only be a condensed summary. For more detailed descriptions, turn to some of the excellent works listed at the end of the chapter.

4.1 EXTRUSIVE, OR VOLCANIC, PROCESSES, PRODUCTS, AND LANDFORMS

As I sit here in eastern Washington State, writing this textbook, I sit atop flows of basaltic lava that cover over 165,000 km^2, with a cumulative average thickness of more than 1 km. The great bulk of these outpourings was extruded over the land surface 14 to 16 Ma ago, and it must have been absolutely catastrophic. Mt. St. Helens erupted in 1980, darkening the skies and covering a quarter of the state in ash. Many here remember the distant rumble and the ominous cloud that imposed an eerie darkness by midmorning. In the American West, volcanism is much more than an academic subject. It created much of our landscape and provides a great deal of recreation and scenic beauty. It also remains a constant threat.

4.1.1 Properties of Magma and Eruptive Styles

The style of volcanic eruption, and the resulting deposits, are determined by the physical properties of the magma, particularly the viscosity and volatile (gas) content. Viscosity (i.e., the resistance to flow) is determined by the composition and temperature of the magma. The strong Si-O and Al-O bonds in silicate melts can link together (i.e., **polymerize**) to form extensive molecular networks. Recent work (see Section 7.5.1) has indicated that melts, at least near their melting point, have structures vaguely similar to the minerals that would crystallize from them (although less ordered and less polymerized). Remember that each Si atom is bonded to four oxygens in silicate minerals. Each oxygen can be bonded to another Si (called "bridging" oxygens, which create -Si-O-Si-O polymers) or, more weakly, to some other cation ("non-bridging" oxygens). Because melts approximate the structures of the corresponding minerals, basaltic melts tend to have more isolated Si-O complexes, with no bridging oxygens (as in olivine) and partially bridged -Si-O-Si-O- chains (as in pyroxene), than fully bridged three-dimensional -Si-O-Al-O- networks (as in plagioclase). Because they yield much more feldspar and quartz, rhyolitic melts tend to be dominated by three-dimensional -Si-O-Al-O- networks. The more polymerized -Si-O-Al-O- networks create strong interconnected bonds throughout the rhyolitic melt, and thus higher viscosity. Greater viscosity thus generally correlates with higher silica content. Also, for a given SiO_2 content, polymerization is weaker at high temperatures and increases toward the melting point and with increasing crystallinity. Viscosities range from about 1 Pa s (Pascal second) for anhydrous olivine basalt at 1400°C to about 10^4 Pa s for anhydrous rhyolite melt at the same temperature (Figure 4.1a). Rhyolite viscosity increases to about 10^7 Pa s at 1000°C. See Appendix A for methods of calculating estimated magma viscosities on the basis of chemical analyses.

When crystals form, the viscosity of resulting melt–crystal mixtures increases abruptly as they cool, owing perhaps to the rigid crystals or surface adsorption effects (Figure 4.1b). When stationary, such crystal–melt mixtures also develop a certain resistance to induced flow. This resistance, called **yield strength**, must be overcome before the material can deform and behave as a viscous fluid again.

H_2O and alkalis have the ability to break down some of the polymerized networks into shorter segments (Section 7.5.1), thereby reducing viscosity. For example, adding 2 weight % (wt. %) H_2O to rhyolite melt at 1000°C lowers the viscosity from 10^7 to 10^5 Pa s (Figure 4.1c). It requires about 8 wt. % H_2O to lower the viscosity to 10^3 Pa s, so the effect of adding H_2O gradually tapers off. Volatiles are an important constituent of magmas for yet another reason. As the magma rises, the pressure is reduced, and the volatile constituents escape from solution in the magma and expand. Gas pressures can become higher than the confining pressure of the surrounding rocks at shallow levels. This can contribute to very explosive eruptions. H_2O and CO_2 are the dominant volatile species, but there may also be SO_2, H_2, HCl, Cl_2, F_2, and a number of other constituents. Volatile content may range from less than 0.5 wt. % in some basaltic magmas to over 5 wt. % in some rhyolitic ones, and considerably more in carbonatites and some other exotic melts. As explained more completely in Section 7.5.1, 5% may not seem like much, but most volatiles have low molecular weights, so their *molecular* proportions may be several times the percentages reported by *weight*. Volatile pressure can thus be substantial.

The combination of viscosity and volatile content determines whether a volcanic eruption will be violent or quiescent. Volatiles diffuse fairly easily through silicate magmas, especially if polymerization is low. Because volatiles are also of low density, they concentrate at or near the top of shallow magma chambers. As a result, the initial stages of most eruptions are more violent than the later stages because the volatile-rich upper portions are the first to be expelled. Pressure keeps the volatiles dissolved in the magma while at greater depths, but they are released when the magma approaches the surface and pressure is relieved, in a manner similar to the uncapping of carbonated soda. If the viscosity is low enough, the volatiles can escape easily. Basaltic eruptions in Hawaii

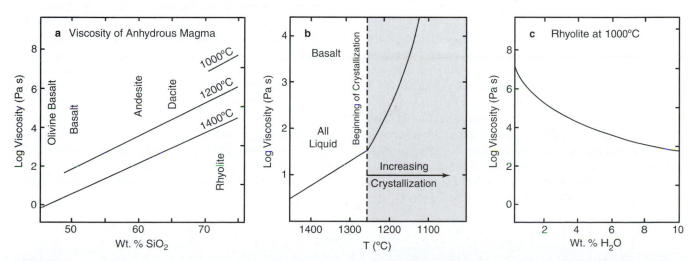

FIGURE 4.1 **(a)** Calculated viscosities of anhydrous silicate liquids at one atmosphere pressure, calculated by the method of Bottinga and Weill (1972) by Hess (1989). **(b)** Variation in the viscosity of basalt as it crystallizes (Murase and McBirney, 1973). Copyright © The Geological Society of America, Inc. **(c)** Variation in the viscosity of rhyolite at 1000°C with increasing H_2O content (Shaw, 1965). Reprinted by permission of the *American Journal of Science*.

typically exhibit early volatile-driven lava fountaining, followed by a calmer phase in which a lava lake forms and latent gases escape gradually, or locally and suddenly as large bubbles or bursts, throwing incandescent blobs of **spatter** into the air. At the other extreme, the high viscosity of many rhyolitic and dacitic magmas resists volatile escape until the pressure of the expanding gases overcomes the high resistance, resulting in tremendous explosive eruptions. The 1883 explosion of Krakatau in Indonesia was heard near Mauritius, over 4800 km away in the Indian Ocean, and spread a cloud of ash around the globe, which reflected sunlight and lowered atmospheric temperatures everywhere for several years.

Although most of the dissolved volatiles are expelled during pressure release associated with the initial eruption, some remain in the magma. The soda pop analogy continues to be a useful illustration. If you shake the can and open it, the pressure release will make a rather messy initial eruptive episode. If you pour the remaining drink into a glass, you will be able to watch more carbonated bubbles form and rise to the surface for several minutes. In a similar fashion, any dissolved gases remaining in the lava after the first eruption will leave solution and form bubbles, which rise and concentrate toward the surface of lava flows, where they may remain as voids, called **vesicles**. Highly vesicular basalt (called **scoria**) usually results from rapid vesiculation during violent eruptions. More viscous magmas, such as rhyolite (both because they generally contain more volatiles and because they more effectively trap the bubbles), can become so vesicular that blocks of the cooled rock actually float on water. This light-colored frothy glass is called **pumice**, and it generally forms as bits of magma included in explosive ejecta but also at the top of some rhyolitic flows. Long after solidification of lava flows, vesicles may fill with later minerals (e.g., opal, prehnite, calcite, zeolites) deposited by hydrothermal solutions. Such filled vesicles are called **amygdules** and are common in mafic flows.

4.1.2 Central Vent Landforms

Magma can issue from either a central vent or a linear fissure. In **vent** eruptions, lava issues from a generally cylindrical, pipe-like conduit through a subcircular surface hole (the vent). There may be a bowl- or funnel-shaped depression at the vent, called a **crater**. The rising magma follows weakened zones, fractures, or fracture intersections in the brittle, shallow crust. Magma solidifies in the vent and plugs it, strengthening the old weakness and forcing later eruptive phases to seek other conduits nearby. There may thus be more than one vent associated with a single volcano, particularly if we consider successive eruptions over a sufficient time span.

Figure 4.2 illustrates some examples of the major types of volcanic landforms associated with a single vent and their approximate relative sizes. **Shield volcanoes** range in size from a few kilometers across to the largest vent landforms (Figure 4.2a). The lavas that comprise a shield volcano are predominantly basalts. Because of the relatively low viscosity, flows predominate, and cover a large area, producing a landform with a convex upward profile and low slope (generally less than $10°$ and commonly closer to 2 to $3°$). Over some hotspots (Chapters 14 and 15), basalts issue forth in vast quantities, producing the largest single landforms on the Earth. The Hawaiian shield volcanoes, Mauna Loa and Mauna Kea, rise nearly 9 km from the floor of the Pacific Ocean, Mauna Loa to an elevation of 4169 m above sea level. The total volume of basalt in these shields exceeds 40,000 km^3. Individual flows have traveled up to 50 km from the source vent and are typically only 5 to 10 m thick. Although many flows in shield volcanoes come from the central vent, there may also be marginal **flank eruptions**. **Fissure** or **rift** eruptions, in which low-viscosity lavas issue from cracks in the swelling shield, are also common. Several small satellite vents may occur along a fissure as local effusive concentrations. Fissures are usually the predominant eruptive centers of plateau or flood basalts (see Section 4.1.3 and Chapter 15).

Another common volcanic landform is the **composite volcano**, or **stratovolcano** (Figures 4.2b and 4.3). These steep-sided cones are usually slightly concave upward and have slopes up to $36°$. They are the volcanic landform that we typically associate with volcanoes represented by such famous examples as Mt. Fuji, Mt. Rainier, Mt. Vesuvius, and Mt. St. Helens. They average about 2 km in height and are only $1/100$ the volume of a large shield.

As we shall see in Chapters 16 and 17, stratovolcanoes are typically composed of a fairly large range of magma compositions, even at a single locality. They are generally more silicic than shields, with andesite as the most common magma type. Their eruptive histories are generally complex, having repeated eruptions of both flows and pyroclastics, the layering of which gives the volcanic form its name. The ratio of flows to pyroclastic deposits varies considerably among volcanoes and may also vary over time at a given volcano. Flows are generally associated with the more mafic and hotter magmas, whereas the more silicic and cooler types produce explosive eruptions. The 1980 eruption of Mt. St. Helens, for example, was dacitic and locally devastating. Whereas some volcanic centers may be rather monotonous, either in terms of composition, eruptive style, or both, others may show considerable variety. The underlying reasons for this difference are not clear.

Each eruptive phase will leave some solidified magma filling the central vent and conduit. Successive magmas

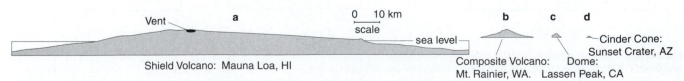

FIGURE 4.2 Volcanic landforms associated with a central vent (all at the same scale).

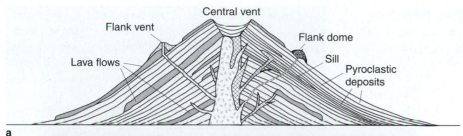

FIGURE 4.3 **(a)** Illustrative cross section of a stratovolcano. After Macdonald (1972). **(b)** Deeply glaciated north wall of Mt. Rainier, Washington, a stratovolcano, showing layers of pyroclastics and lava flows.

must move through, or around, earlier plugs. Shallow magma chambers may send offshoots to feed flank eruptions, forming parasitic cones, as illustrated in Figure 4.3.

Although the volcano in Figure 4.3a looks symmetrical and is strictly aggradational (continuously built up, or added to), real composite volcanoes are far more complex. Mass wasting, glacial erosion, and explosive destruction of an earlier edifice are important elements in the history of these volcanoes (Figure 4.3b). Many present-day composite volcanoes are merely the most recent edifices in a series of cones that were constructed and subsequently destroyed at the same general site. At some localities, the "site" may be somewhat distributed, and volcanism is spread out over a larger area, with sporadic activity lasting thousands to hundreds of thousands of years. Such occurrences are commonly called **volcano complexes**. Figure 4.4 is a schematic cross section of the Lassen Peak area in northeastern California. Note the removed mass of the original andesitic Brokeoff Mountain and the later occurrence of several dacitic and rhyolitic domes.

Smaller volcanic landforms are associated with more limited eruptive events and are not really comparable to the large shields or composite cones that build up over time as a result of successive eruptive phases. Examples of some of these landforms are illustrated in Figure 4.2 and in Figure 4.5 (where they are not at the same scale). **Pyroclastic cones**, such as **scoria cones** or **cinder cones** (Figures 4.2c, 4.5a, and 4.6c), result from the collection of airborne ash, lapilli, and blocks as they fall around a central vent in association with weak explosive activity. They are typically less than 200 to 300 m high and 2 km in diameter, and they generally last only a few years to tens of years. Parícutin, for example, a scoria cone in central Mexico, grew in a cornfield to 410 m in a few years beginning in 1943. Activity ceased in 1952. These small cones are usually basaltic and straight-sided, with slopes at approximately 33°, the angle of repose of the loose scoria. These small cones are typically asymmetrical, either elongated along a fissure or larger on the side downwind at the time of the eruption. They have a central, bowl-shaped crater,

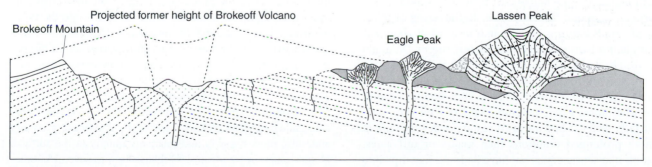

FIGURE 4.4   Schematic cross section of the Lassen Peak area. After Williams (1932).

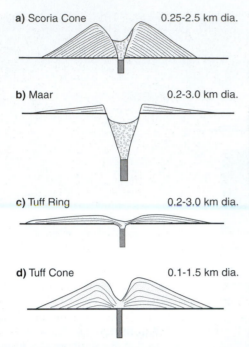

a) Scoria Cone   0.25-2.5 km dia.

b) Maar   0.2-3.0 km dia.

c) Tuff Ring   0.2-3.0 km dia.

d) Tuff Cone   0.1-1.5 km dia.

FIGURE 4.5 Cross-sectional structure and morphology of small explosive volcanic landforms with approximate scales. After Wohletz and Sheridan (1983). Reprinted by permission of the *American Journal of Science*.

which gradually fills and flattens as a result of mass wasting. A **maar** (Figures 4.5b and 4.6a) is typically lower than a scoria cone and has a much larger central crater relative to the deposited ring of debris. Maars result from the explosive interaction of hot magma with groundwater, which flashes to steam. Such explosions are called **hydromagmatic**, or **phreatic** eruptions. Note that the explosive power here is supplied by groundwater, not water contained in the melt. Geologists commonly use the terms *meteoric* and *juvenile* to make the distinction clear. **Meteoric** water refers to surface water or groundwater, and **juvenile** refers to either water or other constituents that are products of the magma itself. A maar is primarily a negative feature, in that the phreatic eruption excavates a crater into the original substrate. Tuff rings and tuff cones also form as a result of magma–H_2O interactions, but neither have deeply excavated craters. **Tuff rings** (Figure 4.5c) form when rising magma (usually basaltic) comes closer to the surface than with a maar before interacting explosively with shallow groundwater or surface water. They also involve a higher ratio of magma to H_2O than maars, forming a subdued ring of scoria and ash that has a low rim, and layering of pyroclastic material that dips inward and outward at about the same angle. Perhaps the most famous tuff ring is Diamond Head, which stands above the southeast end of Waikiki Beach (Figure 4.6b). **Tuff cones** (Figure 4.5d) are smaller than tuff rings, with steeper sides and smaller central craters. They form where magma interacts with very shallow surface water. They seem to result from less violent and more prolonged eruptions than maars or tuff rings. They look like scoria cones but have bedding that dips into the craters as well as outward.

FIGURE 4.6 **(a)** Maar: Hole-in-the-Ground, Oregon. Courtesy of USGS. **(b)** Tuff ring: Diamond Head, Oahu, Hawaii. Courtesy of Michael Garcia. **(c)** Scoria cone: Surtsey, Iceland, 1996. Courtesy of Bob and Barbara Decker.

Most classifications, including the one above, give the impression that the "pigeonholes" are distinct, reflecting discontinuities in the processes that generate the forms. Although there may be some clustering of forms into the types described above, there is a continuous spectrum of these small pyroclastic landforms, reflecting variations in viscosity, composition, the ratio of magma to meteoric water, and the depth of the explosive interaction. Even shields and composite cones are not completely distinct.

Domes (Figure 4.7) form when largely degassed, viscous, silicic magma, such as dacite or rhyolite (less commonly andesite), moves slowly and relatively quietly to the surface. Domes range in size from less than 100 m to several kilometers in diameter. Domes may form early or late in an eruptive

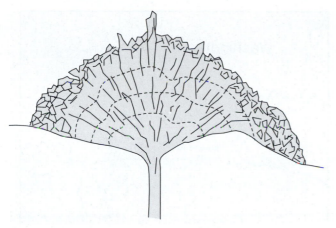

FIGURE 4.7 Schematic cross section through a lava dome.

FIGURE 4.8 Pressure ridges on the surface of Big Obsidian Flow, Newberry Volcano, Oregon. Flow direction is toward the left.

cycle but typically late. After an early phase of explosive activity, the final (gas-depleted) magma may inflate to a dome in the central crater. The process is usually **endogenous**, in that the dome inflates by the injection of magma from within. **Exogenous** dome eruptions are events in which the later additions break through the crust and flow outward. The crust that forms on the dome brecciates as the dome inflates, giving the surface a very rough texture. Brecciated blocks fall off and accumulate at the base of the dome as an apron of talus. The inflation may fracture the outer zones in a radial fashion and even push some sections outward or upward, forming a **spine**, as illustrated in Figure 4.7.

There is a gradation from steep-sided domes, to domes that flatten and flow downhill (called **coulées**, a name also applied to a type of canyon), to thick flows. Viscosity, slope, and rate of magma production are important factors in the final form. Coulées and thick rhyolite or obsidian flows are viscous, and don't travel far from the source. They typically have a surface that folds into "pressure ridges" that are usually convex toward the advancing front (Figure 4.8). How some domes, coulées, and thick flows can be glassy throughout is an interesting problem. Certainly they didn't cool that quickly. Perhaps the viscous and highly polymerized melt inhibits crystal nucleation, or the motion retards nucleation (as in taffy candy making). Some domes are inflated beneath the ground surface and are called **cryptodomes**. The inflation that caused the spectacular "bulge" on the flank of Mt. St. Helens in 1980 was a cryptodome. It was a bit unusual in that it was certainly not late in the eruptive cycle,

nor was it depleted in volatiles, as became clear when the bulge oversteepened, the lower portion slid away, and the unroofed and depressurized section of the dome released the gas in a very explosive event. For more on domes and coulées, see Fink and Anderson (2000).

Calderas are large-scale collapse features that typically form at a central vent fairly late in an eruptive episode. They form when the denser solid strata above a shallow magma chamber founder into the draining chamber. The magma may drain as a flank eruption, or it may move up the fractures separating the blocks of the collapsing roof. On basalt shields, the caldera may fill with magma from below to create a **lava lake**. In such an occurrence, the cool, dense solid roof founders into the less dense magma chamber beneath, displacing the magma upward.

In more silicic cases, fractures reaching down to the magma chamber reduce the pressure at the top of the chamber, inducing rapid vesiculation, and a substantial proportion of the magma may then escape along these fractures in the roof in the form of pyroclastic activity. Figure 4.9 illustrates the explosive eruption of Mt. Mazama to form Crater Lake 6850 years ago. According to Bacon (1983) and Bacon and Lanphere (1990), Mt. Mazama was a large composite volcano, consisting of several overlapping basaltic andesite shields and andesitic to rhyolitic composite cones that stood ~3600 m high. The first material from the large magma chamber that led to the climactic eruption was a rhyodicite pumice fall and lava flow dated at ~7015 b.p. The climactic eruption began with a single-vent phase that produced a

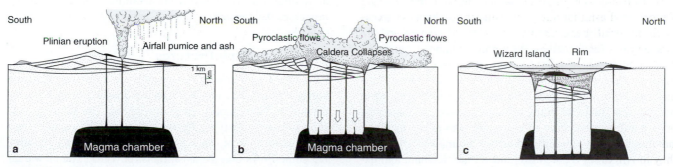

FIGURE 4.9 Development of the Crater Lake caldera. From Bacon (1988).

Plinian column (Figure 4.9a), producing a widespread air-fall tephra, then collapsed to a valley-hugging series of ign-imbrites (now welded). Immediately thereafter, a ring-dike phase began, characterized by caldera collapse and fairly high pyroclastic columns issuing from the marginal faults (Figure 4.9b). The columns produced lithic volcanic brec-cias near the vent and on nearby hilltops and poorly to non-welded pumiceous ignimbrites in the valleys. A later ring-dike phase was characterized by pyroclastic basaltic an-desite and andesite scoria. The total volume of ejected mate-rial is estimated at ~50 km^3. A few hundred years following the climactic eruption, small andesitic vents erupted sporad-ically within the caldera, producing small cones, such as Wizard Island (Figure 4.9c). Whether they represent new magma or remnants from the older chamber is unknown.

The largest calderas, some over 100 km across, are as-sociated with tremendous pyroclastic eruptions of rhyolite. The resulting calderas are too large to visualize at ground level, and some have been recognized only recently from satellite imagery. Examples of large calderas in the western United States are those at Yellowstone, Valles Caldera in New Mexico, and Long Valley in California. Yellowstone contains several partially overlapping calderas. Such occur-rences are called **caldera complexes**. The first eruption, 2 Ma ago, produced approximately 2500 km^3 of rhyolitic material in a single huge ash flow. The third eruption, 600,000 years ago, produced over 1000 km^3 of material. Following the collapse of the third caldera, the center has risen again (perhaps as new magma refilled a chamber below) to become what is called a **resurgent caldera**.

Smith (1979) has shown a strong correlation between caldera area and volume of the expelled ash flow, which holds for caldera sizes spanning five orders of magnitude. He used caldera *area* because volume is less easy to determine in such collapsed, ash-covered, and eroded features. If this cor-relation reflects a similar correlation between the ash flow volume and the caldera volume, it suggests that calderas drop approximately the same distance (about 0.5 km), regardless of diameter, suggesting that some process limits the depth of collapse. Perhaps at greater depths, the lithostatic pressure is sufficient to impede volatile separation, which might disrupt the magma, lower the viscosity, and permit rapid escape.

4.1.3 Fissure Eruptions

In contrast to central vent volcanism, **fissure eruptions** occur as magma erupts to the surface along either a single fracture or a set of fractures. The planar conduits, when ex-posed by erosion, are filled with solidified magma and are referred to as **feeder dikes**. As mentioned above, some fis-sure eruptions occur on the flanks of central vents, where fractures typically accompany inflation of the edifice as the magma chamber fills. Such fractures may form singly or in multiple sets (in a concentric or radial pattern about the vent or as parallel sets). Fissure eruptions also occur in larger areas undergoing regional extension. Examples include the African rift valleys, the Basin and Range, Iceland, and ex-tensional basins behind volcanic arcs. Figure 4.10 illustrates

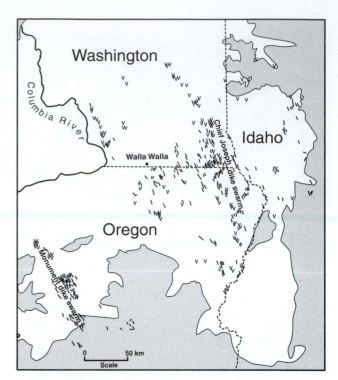

FIGURE 4.10 Location of the exposed feeder dikes (heavy lines) and vents (V's) of the southeastern portion of the Columbia River Basalts. Unshaded area covered by Columbia River Basalts. After Tolan et al. (1989). Copyright © The Geological Society of America, Inc.

the location of some of the exhumed feeder dikes and vents of the Columbia River Basalts, a major plateau-type, or flood basalt province, in eastern Washington. Although we think of these great flood basalts as being fed exclusively by dikes, a number of localized vents also feed some flows or even form local cones. The amount of basalt in plateau or flood basalts is tremendous, and the displacement of such a large volume to the surface commonly results in sagging of the crust below, producing a structural basin. These great outpourings will be discussed in detail in Chapter 15.

The most common type of fissure eruption is never seen by most humans. These fissures typically occur be-neath the sea, at mid-ocean ridges, and produce the most common and voluminous rock type on Earth: mid-ocean ridge basalts (MORBs), which constitute the oceanic crust. Here, two plates diverge, resulting in normal faults and ex-tensional tectonics, an ideal situation for fissure eruptions. Only in Iceland is this type of fissure eruption exposed above the sea. We will further examine MORBs and the processes that generate them in Chapter 13.

4.1.4 Lava Flow Features

Lava flows are more quiescent than the dramatic explosive volcanic eruptions, but they are the dominant form of vol-canism on Earth, and perhaps throughout the solar system. Flows occur most typically in lavas with low viscosity and low volatile content. They are thus most common in basalts, but some flows may be as silicic as rhyolite. The Hawaiian flows you may have seen generally travel fairly slowly and

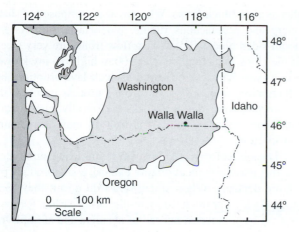

FIGURE 4.11 Aerial extent of the N2 Grande Ronde flow unit (approximately 21 flows). After Tolan et al. (1989). Copyright © The Geological Society of America, Inc.

over broad, but limited, areas in any single event. Flows rarely kill people, but they have engulfed a lot of property, including the occasional town. Fortunately, they are usually slow enough that people can evacuate their homes before they are lost to the advancing flow.

Some flows, however, particularly those associated with flood basalts, are of enormous size. The lack of phenocryst

alignment in many suggests that these flows typically travel in a turbulent fashion and come to rest before they have cooled enough to be viscous and preserve any internal flow features. Some individual flow units of the Columbia River Basalt Group cover nearly 120,000 km^2 (Figure 4.11) and approach 3000 km^3 in volume (Tolan et al., 1989). Self et al. (1997) concluded that large individual flows required 5 to 50 months for emplacement and averaged ~4000 m^3/s of lava effusion. These are formidable flows, and if one were to occur in modern times, it would have considerable impact on property, lives, and even the global climate.

In the early stages of a basaltic eruption, such as in Hawaii, magma emerges as incandescent lava at about 1200°C. This lava has a very low viscosity and runs down slope in rivers with initial velocities as high as 60 km/hr. This runny lava cools and forms a smooth black surface, which may develop a corrugated, or ropy, appearance (Figure 4.12a and the left side of Figure 4.12b). The corrugations are usually less than 2 cm high, with axes perpendicular to, or convex to, the flow direction. Such lavas are called **pahoehoe**. As the lava cools further and the viscosity increases, the flows begin to move more slowly and develop a thicker, scoriaceous crust. As the fluid interior continues to move, the crust breaks up into blocks of clinkery scoria, which ride passively on the top. Pieces also tumble down the advancing front. The motion is

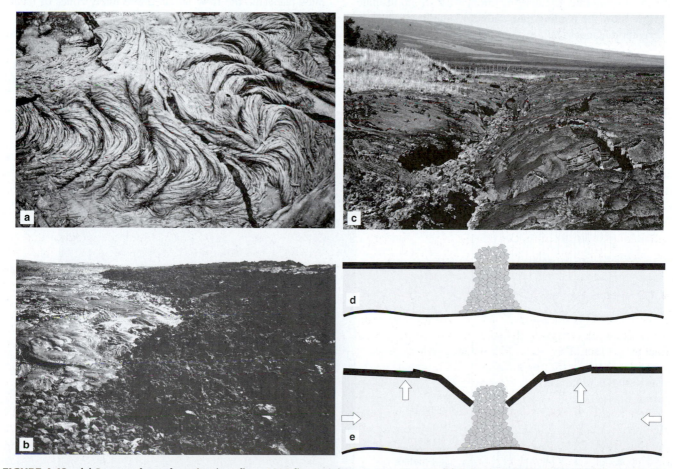

FIGURE 4.12 **(a)** Ropy surface of a pahoehoe flow, 1996 flows, Kalapana area, Hawaii. **(b)** Pahoehoe (left) and aa (right) meet in the 1974 flows from Mauna Ulu, Hawaii. **(c)** Inflated lava flow north of Kilauea (in background). **(d–e)** Illustration of the development of an inflated flow. In **(d)**, a thin flow spreads around a rock wall. In **(e)**, the flow is inflated by the addition of more lava beneath the earlier crust. An old stone wall anchors the crust, keeping it from lifting. The wall can be seen in the low area in part **(c)**.

like a conveyor belt, in which the surface slides beneath the front of the advancing flow. Thus the blocks are found both at the top and the base of the flow. The rubble-like lava flows that result are called **aa** (right side of Figure 4.12b). Occasionally the fluid core material escapes and flows as tongues through the carapace. **Lava tubes** also form as conduits within many basalt flows as an efficient means of conserving heat and delivering lava to the advancing front. Lava typically drains from the conduit, leaving a tube-like tunnel. Aa and pahoehoe are end members of a continuous series of flow-top characteristics. Pahoehoe is restricted to *basalts* of low viscosity, but aa can occur in flows spanning a range of compositions. Numerous basalt flows begin as pahoehoe and turn to aa further from the vent as they cool and slow down.

Holcomb (1987) described the **inflated** flow, applying the term to some Hawaiian basaltic flows (Figures 4.12c–e). These flows begin as a thin pahoehoe flow, perhaps as thin as 20 to 30 cm. After the crust sets, lava is added beneath it and inflates the flow internally to thicknesses as great as 18 m. As a result, the crust rises, cracks, and tilts in complex patterns. Inflated flows have been described in several places, including Hawaii, Oregon, the Juan de Fuca Ridge, and the Columbia River Basalts. Inflation of the Columbia River Basalts would avoid the incredibly rapid flow emplacement rates of about a week, estimated by some on the basis of cooling rate, because the initial crust would insulate the flow, slow the cooling rate, and allow the basalt to flow for greater distances (Self et al., 1997).

Andesite flows are more viscous than basalts, and, although they may occur as flows, they more typically form either as aa or **block lavas**. Whereas aa is composed of sharp cindery and scoreaceous rubble, block lavas have larger, smooth-sided blocks. Block lava flow fronts are very steep, and the block piles may be over 100 m high. To my knowledge, no one has ever observed a block lava eruption, but such eruptions appear to result from a process in which the lava is nearly solid, and the blocks are literally pushed out a vent. In larger flows, a more massive flow may form behind a blocky front.

Dacitic and rhyolitic lava flows are less common, as these magmas are typically more explosive and produce pyroclastic deposits. When silicic flows do occur, they form domes, coulèes, or thick flows, as discussed earlier. Rhyolite flows are commonly composed of obsidian and are **aphyric** (lacking in phenocrysts), indicating that they were hot (and hence less viscous).

Thick intermediate-to-silicic lavas typically exhibit **flow foliation**, which may consist of aligned phenocrysts, bands of different color (Figure 3.13), or pumice bands. These layers could have been different batches of mingled magmas or portions of the same magma with a different temperature, composition, or content of crystals, H_2O, or oxygen. The layers were then stretched, sheared, and/or folded during flow. In the case of pumice layers in obsidian, these layers may have been frothy surface layers incorporated back into the flow but are more likely zones of localized shear that induced vesiculation.

Subaerial lava flows (those that flow on land) and some shallow sheet-like intrusions may develop a characteristic jointing pattern called **columnar joints**. Figure 4.13

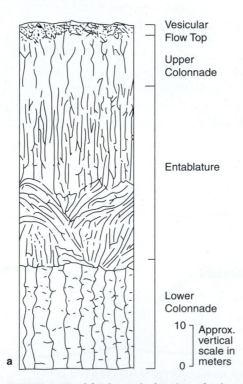

FIGURE 4.13 **(a)** Schematic drawing of columnar joints in a basalt flow, showing the four common subdivisions of a typical flow. The column widths in **(a)** are exaggerated about 4×. After Long and Wood (1986). **(b)** Colonnade–entablature–colonnade in a basalt flow, Crooked River Gorge, Oregon.

illustrates the structures within a typical basalt flow (in this instance, one of the Columbia River Basalt flows). In the ideal case, there are four subdivisions: a thin vesiculated and brecciated flow top, an upper and lower **colonnade** with fairly regular straight columns, and a central **entablature** that has more irregular columns that are typically curved and skewed. The three main subdivisions are not uniformly developed, in that each may vary greatly in thickness, be absent entirely, or occur repeatedly within a single flow. Although models have been proposed for the development of columnar joints based on convection currents (Sosman, 1916; Lafeber, 1956) or diffusion (Kantha, 1981; Hsui, 1982), the most widely accepted mechanism is based on contraction of the flow as it cools (Tomkeieff, 1940; Spry, 1962; Long and Wood, 1986; Budkewitsch and Robin, 1994). Because the top and bottom of the flow cool before the central layer, the outer areas contract, but the center doesn't. This results in tensional stresses that create regular joint sets as blocks pull away from one another to create polygons separated by joints. The joints propagate down from the top and up from the bottom as cooling progresses toward the center, forming the column-like structures in the colonnades. Columns form perpendicular to the surfaces of constant temperature, which are usually parallel to the surfaces of the flow (i.e., horizontal). Columns may be four, five, six, or seven sided, but five- and six-sided ones predominate.

The irregular entablature is more difficult to explain, with ideas generally relying on disturbed cooling surfaces or post-joint deformation of the still-ductile flow center. Petrographic textural differences between the colonnade and the entablature (Swanson, 1967; Long, 1978; Long and Wood, 1986) suggest that the difference results from primary crystallization effects. The greater proportion of glassy mesostasis and the feathery Fe-Ti oxides in the entablature suggested to Long and Wood (1986) that the entablature cooled more rapidly than the colonnades, contrary to what one would expect for the more insulated flow interior. They proposed a model in which water infiltrated along the joints (perhaps associated with lakes or floods) to cool the entablature portion by convective water circulation through the upper joint system. Variation in the development of the subdivisions might then be explained by the erratic nature of flooding, and the fanning of columns in the entablature may be related to the preferential percolation of water down large, early-formed joints.

When basaltic lava flows enter standing water, they form either tongues or more equidimensional blob-like structures, both of which are called **pillows**. The resulting **pillow lavas** (Figure 4.14) contain numerous packed pillows, usually with glassy rinds, concentric bands, and radial fractures within. Later pillows commonly flatten and conform to the shape of the pillows upon which they settle, providing geologists with a method of determining the original top of the deposit in folded terranes. The pillows may be embedded in fine glassy debris called **hyaloclastite**, produced by fragmentation of the hot lava as it contacts the water. The hyaloclastite commonly devitrifies quickly to an orangish-brown material called **palagonite**.

FIGURE 4.14 Subaqueous pillow basalts (hammer handle for scale in center foreground). Olympic Peninsula, Washington.

4.1.5 Pyroclastic Deposits

The term **volcaniclastic** refers to any fragmental aggregate of volcanic material, regardless of how it formed. **Autoclastic** refers to volcanics that suffer a quiescent self-imposed breakup, such as aa and block flows, dome talus aprons, or gravitational collapse features. **Pyroclastic** deposits are a subset of volcaniclastics that consist of fragmented material formed from explosive volcanic activity or aerial expulsion from a volcanic vent. Other volcaniclastic deposits include volcanic mudflows, or **lahars**, which form when volcanic debris mixes with sufficient water (rain runoff, melted snow/ice, groundwater, or lake water) to mobilize (Vallance, 2000). Lahars may be eruption associated and hot, or they may occur later and be cold debris flows off volcanoes. Hot dome collapse features are generally considered pyroclastic as well, though they do not technically qualify as such. Pyroclastic deposits may be deposited in water also. They may then mix with the water to become lahars or water-laid deposits (and are thus no longer pyroclastic by our definition). Collectively, the pyroclastic *particles* that comprise the deposits are called **pyroclasts**, and a collective term for the deposited *material* is **tephra**. Further classification of pyroclasts is based on particle size, as discussed in Chapter 2.

Pyroclastic deposits are classified on the basis of the mode of transport and deposition. They are divided into **falls** and **flows**. **Surges** are distinguished by some geologists as a third category of pyroclastic deposit, but they are really a type of flow.

4.1.5.1 PYROCLASTIC FALL DEPOSITS **Pyroclastic fall deposits** are produced by fallout from an eruptive jet or the plume of an explosive eruption (Houghton et al., 2000). Pyroclasts may be propelled forcefully upward during an explosive eruption, or they may be carried aloft by convection and the buoyancy of the hot gases above the vent. **Plinian** eruptions (Figures 4.9a, 4.15a, and 4.18a) are a combination of the two. Particles are forced from the vent by powerful jetting of gases and are then carried aloft by hot convection. The simplest falls accompany the small basaltic eruptions that produce scoria

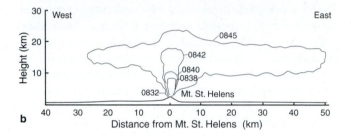

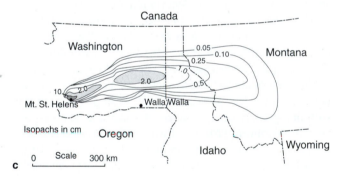

FIGURE 4.15 Ash cloud and deposits of the 1980 eruption of Mt. St. Helens. **(a)** Photo of the Mt. St. Helens vertical ash column, May 18, 1980. Courtesy of USGS. **(b)** Vertical section of the ash cloud, showing temporal development during the first 13 minutes. **(c)** Map view of the ash deposit. Thickness is in centimeters. After Sarna-Wojcicki et al. (1981).

cones and tuff rings/cones. The more dramatic and larger eruptions produce a high ash plume that may reach heights of 50 km. As the plume expands, the driving force dissipates, and the pyroclasts fall back to Earth under the influence of gravity. Larger and denser particles fall more quickly, and the finer ash remains aloft longer. As a result, fall deposits are well sorted, with grain size decreasing both vertically toward the top of a deposit and laterally away from the source vent. The size, thickness, and shape of the deposit, and the distribution of particle size, depend on the rate of expulsion, the volume erupted, the force of the explosion, and the direction and velocity of the prevailing winds at the time of eruption.

Fall deposits *mantle* the land surface. Thickness decreases gradually away from the source but is independent of topography (much like snow). Fall deposits have a better chance to cool while suspended, so they are rarely welded by their own heat after deposition, except, perhaps, near the vent. Figure 4.15b illustrates the time-sequence development of the 1980 Mt. St. Helens ash plume. In only 13 minutes (8:32 to 8:45 A.M.), the ash cloud rose to nearly 25 km and spread 50 km laterally, as monitored from ground stations and by satellite. The thickness of the resulting ash fall deposit is shown in Figure 4.15c, with lines of constant thickness, or **isopachs**, expressed in centimeters. Over 1 cm of ash was deposited as far as 500 km away. The ash blanket from the 6950-year-old eruption of Mt. Mazama (now Crater Lake) covered portions of seven states and part of Canada (Figure 4.16). Ash deposits 50 cm thick are found as far as 100 km from Crater Lake.

The most astonishing ash fall deposits are associated with huge rhyolitic eruptions, such as those associated with the Yellowstone (Wyoming), Long Valley (California), and Valles (New Mexico) calderas. For example, the area covered with ash from the Long Valley eruption, some 700,000 years ago, covered much of the western United States (Figure 4.17). Ash was 1 m thick over a 75-km radius and 1 cm thick in Kansas. It is estimated that devastation was total over a 120-km radius (the circle in Figure 4.17). Ash from three Yellowstone eruptions over the past 2.2 Ma produced 3800 km^3 of ash and pumice, some of which reached Minnesota and Louisiana (2200 km away).

As awful as these ash deposits are, the thinner extensions, which cover the largest area, are commonly transient. For example, the ash deposits of the May 18, 1980, Mt. St. Helens eruption created quite a mess in eastern Washington, but it is very difficult to find remnants of the ash blanket there

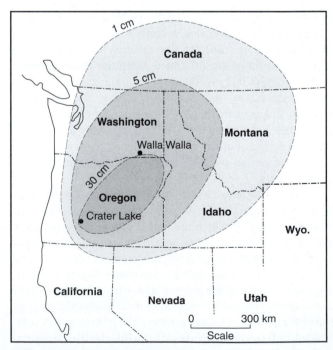

FIGURE 4.16 Approximate aerial extent and thickness of Mt. Mazama (Crater Lake) ash fall, which erupted 6950 years ago. After Young (1990).

FIGURE 4.17 Maximum aerial extent of the Bishop ash fall deposit from the eruption at Long Valley 700,000 years ago. After Miller et al. (1982).

today. The thick deposits near the mountain are still there, but the blanket that covered much of eastern Washington is practically gone. It would appear that the deposits in the subaerial environment are quickly eroded and accumulate in sheltered lowlands, particularly those with standing water. Thus, although they originally mantle the ground surface, the deposits may eventually concentrate in depressions.

4.1.5.2 PYROCLASTIC FLOW DEPOSITS   Pyroclastic flow

deposits are left by dense ground-hugging clouds of gas-suspended pyroclastic debris (mostly pumice and ash, with variable amounts of lithic and crystal fragments). They are generated in several ways (Figure 4.18). The largest flows result from the collapse of a vertical explosive or Plinian column that falls back to Earth and continues to travel along the ground surface (Figure 4.18a). An alternative explosive mechanism is a lateral blast, such as occurred at Mt. St. Helens in 1980 (Figure 4.18b). The flows are **fluidized** because the magmatic gases and air trapped within and beneath the advancing flow cannot easily escape from the high concentration of suspended particles (see Reynolds, 1954; or Wilson, 1984, for discussion). Other pyroclastic flows result from steady fountaining (ranging from relatively forceful to low-force "boiling over") of a highly gas-charged magma from a vent (Figure 4.18c), or as hot avalanches due to the gravitational collapse and disintegration of an expanding dome (Figure 4.18d) or of an unstably perched recent flow deposit (Figure 4.18e). In all cases, the hot particle-laden cloud is (or becomes) denser than the surrounding atmosphere, so it flows downward and deposits the pyroclasts as it loses momentum.

Pyroclastic flows are controlled by topography, and the deposits concentrate in valleys and depressions, unlike the uniform mantling of ash falls. Pyroclastic flows are hot (400 to 800°C). Velocities vary from 50 to over 200 km/hr (the Mt.

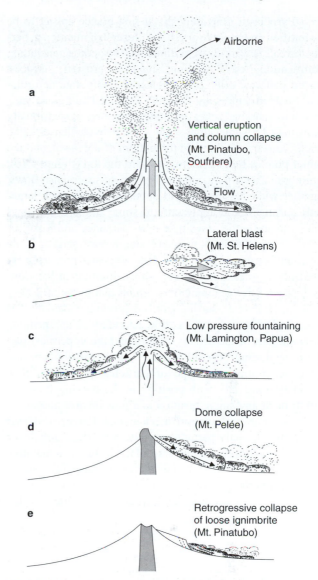

FIGURE 4.18 Types of pyroclastic flow deposits. After MacDonald (1972), Fisher and Schminke (1984), and Branney and Kokelaar (2002).

St. Helens blast was estimated at a maximum of 540 km/hr). They can also travel long distances (typically a few kilometers, but some deposits are found over 100 km from the source). The temperature and velocity make them very dangerous. Dense, hot, ground-hugging flows produce a buoyant ash/gas component, known as a **phoenix cloud**, or co-ignimbrite gas cloud, that rises above the descending flow. Deposits vary in thickness from centimeters to hundreds of meters, and reported volumes range from a few thousand cubic meters to several thousand cubic kilometers and may cover up to 45,000 km². Pyroclastic flow terminology can be a bit confusing. The moving flow itself is called either a *pyroclastic flow* or an *ash flow*. The deposits are called **ignimbrites** (Latin: *ignis* = "fire" + *imber* = "rain"). Some authors reserve the term *ignimbrite* for the pumice-rich varieties, using the term **block-and-ash deposits** for the unvesiculated-clast-bearing types (generally associated with lava dome collapse). The rock name for a sample taken from the deposit is called a **tuff**. (See Section 2.5 for more specific rock names.)

Pyroclastic flow deposits at first glance appear to be unstratified and poorly sorted, but closer examination generally reveals a variety of sedimentary structures, including stratification, cross-bedding, internal erosion surfaces, graded patterns, oriented particles, and soft-state deformation. Such structures are typically localized and grade variably into massive zones or other structures. Occasionally there may be some large-scale sorting, with denser lithic blocks concentrated toward the bottom of a deposit, and the lighter pumice blocks floating toward the top (Figure 4.19). Ignimbrite sheets may contain deposits of several ash flows, together with interlayered fall deposits, so any such large-scale sorting would apply only to some single-event flows that pond and settle as a single mass. Branney and Kokelaar (2002) expressed doubt that such single-mass settling is the norm and proposed that most flow deposits progressively build up (aggrade) during the (short) lifetime of the flow. Such progressive accumulation implies that pyroclastic flow sedimentation (e.g., sorting and bed-form characteristics) is controlled by processes in a "flow boundary zone" spanning the basal part of the density current and the uppermost part of the aggrading deposit. Velocity, shear distribution, and the concentration and type of particles within this flow boundary zone vary with position in the moving flow and with time at any single position and thus involve a range of processes and clast support mechanisms. Larger or denser blocks may be only partially supported by fluid turbulence and probably roll or saltate into place in the accumulating pile rather than settle through the entire suspended mass. High shear at the base of a flow may even result in local erosion or inverse grading as larger particles are lifted by dispersive forces.

Ignimbrites are emplaced at high temperatures, and the lower portions typically become **welded** by the internal heat in the settled pile to a very hard rock called a **welded tuff**. Hotter portions of a deposit are ductile for a period and are typically compressed by the weight of the overlying mass, becoming dense and foliated. Pumice shards commonly have the vesicles squeezed out and take on the black color of obsidian. The light gray color of pumice results from the vesicles, much like whitecaps on the sea. The flattened black bits of compressed pumice are called **fiamme**.

Although pyroclastic flow deposits generally cover a smaller area than fall deposits, due to their velocity and heat they can be absolutely devastating. Several such eruptions have occurred in historic times, but, for obvious reasons, eyewitness accounts are rare. On August 25, 79 A.D., ash falls and several hot pyroclastic flows descended from Mt. Vesuvius and buried the cities of Pompeii and Herculaneum, near Naples, Italy. The inhabitants were either burned or suffocated, and their bodies were buried in the soft ash, preserved to this day as molds. On May 18, 1902, an incandescent pyroclastic flow from Mt. Pelée descended suddenly on the city of St. Pierre, on the island of Martinique, and instantly killed approximately 28,000 people. Only 2 survived, 1 a prisoner. The pyroclastic flows associated with the Long Valley eruption (Figure 4.17) traveled at speeds over 200 km/hr and are estimated to have covered over 1500 km^2 to depths of tens to hundreds of meters (Miller et al., 1982). The origin of ignimbrites was somewhat mysterious until a glowing cloud was observed descending the flanks of Mt. Pelée in 1902. Still, no eruption of the truly huge ignimbrites, such as the Bishop Tuff (Long Valley) or Yellowstone Tuff has been observed, nor do I expect any volunteers. These deposits are the only volcanic deposits that approach the magnitude of flood basalts, and their origin is still speculative.

4.1.5.3 PYROCLASTIC SURGE DEPOSITS
Pyroclastic surge deposits are left by surges, which are a type of pyroclastic flow. The common occurrence of dunes and antidunes indicates that surges result from more turbulent flow, with a lot of gas but a lower concentration of particulates than the other flows. Surges hug the ground surface, but because of their low density and high velocity, they are not as topographically constrained, as are other flows. The resulting deposits thus both mantle topography and concentrate in low areas. Because of their low density, surges lose momentum rather quickly, so they tend to accumulate near the vent. The deposits are usually stratified and may show a number of current-bedding features. Valentine and Fisher (2000) provided a good summary of surges, and Burgisser and Bergantz (2002) discussed models governing flow versus surge behavior.

4.1.5.4 COMPARING PYROCLASTIC DEPOSITS
As with so many other classification schemes, the three categories of pyroclastic deposits are not totally distinct. There are gradations between them in terms of eruptive style and resulting

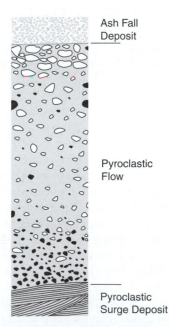

Ash Fall Deposit

Pyroclastic Flow

Pyroclastic Surge Deposit

FIGURE 4.19 Idealized section through an ignimbrite, showing the basal surge deposit, ponded middle flow that settled as a single mass, and upper ash fall cover. White blocks represent pumice, and black blocks represent denser lithic fragments. After Sparks et al. (1973). Copyright © The Geological Society of America, Inc.

deposit. Most pyroclastic deposits have features associated with a combination of two mechanisms or even of all three. As illustrated in Figure 4.19, an ignimbrite may have a stratified basal surge layer, followed by a heterogeneous flow deposit, which is then covered by an ash-fall mantle. The sequence of events that such a deposit represents might involve a lateral surge laying down the initial deposit, followed by a flow caused by the collapse of the ash plume. The flow may then be covered by a blanket of the slow-settling ash fall. Lateral variations will also be apparent because the surge will be restricted to the area near the vent, and the ash fall deposit may be the only layer found a great distance away. Branney and Kokelaar (2002) proposed an alternative to the flow–surge classification of pyroclastic density currents. They envisioned a continuous spectrum of currents between two end-member types. At one extreme are *fully dilute* pyroclastic density currents, in which particle transport and support is maintained by turbulence of a (dusty gas) fluid phase. Deposits of fully dilute currents are typically stratified and occur within ignimbrite sheets and surge-type deposits. At the other extreme are *granular fluid-based* pyroclastic density currents, with high clast concentrations in the critical flow boundary zone, within which collisional momentum transfer between grains and escaping fluid contribute significantly to particle support. Deposits from this type of current may be massive or bedded with various grading patterns.

Many descriptions of pyroclastic deposits, and generalized sections such as Figure 4.19, give a simplified account of pyroclastic flows, which is quickly dispelled when one encounters the deposits in the field. Most pyroclastic deposits reflect an eruptive event that is episodic and heterogeneous and that changes character as it proceeds. The composition, viscosity, and volatile content of the source magma for vertical eruption and fountaining flows are likely to vary as progressively deeper levels of the emptying magma chamber are tapped. Such deposits are a complex accumulation of varying materials, textures, and structures that cools as a single unit.

4.2 INTRUSIVE, OR PLUTONIC, PROCESSES AND BODIES

The generic term for an intrusive igneous body is a **pluton**, and the rocks outside the pluton are called the **country rocks**. The size and shape of plutons is generally somewhat speculative because erosion exposes only a small portion of most bodies. Nonetheless, we have managed to accumulate considerable data from more deeply eroded plutons, geophysical studies, and mining works. From these we have classified plutonic bodies into a few common forms. These forms are grouped into **tabular**, or sheet-like, bodies and **non-tabular** ones. Further classification is based on specific shapes and whether a body cuts across the fabric (usually bedding) of the country rocks or whether it follows the external structures. Crosscutting bodies are called **discordant**, and those that are intruded parallel to the country rock structure are called **concordant**.

4.2.1 Tabular Intrusive Bodies

Tabular intrusive bodies are simply magma that has filled a fracture. A concordant tabular body is called a **sill**, and a discordant one is called a **dike**. A sill occurs when magma exploits the planar weaknesses between sedimentary beds or other foliations, and is injected along these zones (Figure 4.20). A dike is a magma-filled fracture that cuts across bedding or other country rock structures. A fracture is an ideal conduit for magma because fractures can penetrate deeply and form easily, particularly in areas affected by extension or the force of a rising magma diapir. Clearly, magma could not have been generated between two bedding layers, so a sill must be fed by a dike somewhere along its length (unless the bedding dips steeply, and the sill is nearly vertical). Dikes and sills are typically shallow and thin, occurring where the rocks are sufficiently brittle to fracture.

Although most dikes and sills are emplaced in a single event, some may have a history of multiple injections. More than one stage of injection may occur because the dike or sill contracts as it cools, leaving a weakened zone for later magmas. Alternatively, the ductility contrast between a dike or sill and the country rock might make the contacts susceptible to localized deformation and later magmatic injection. A body is described as **multiple** if all phases of injection are of the same composition and **composite** if more than one rock type is represented.

Dikes and sills can occur as solitary bodies, but dikes, at least, more typically come in sets, reflecting the tendency for fractures to form in sets as a brittle response to imposed stresses over an area. Genetically related sets of numerous dikes or sills are called **swarms**. Note the Chief Joseph and Monument basaltic **dike swarms** in Figure 4.10. These swarms consist of subparallel dikes filling a set of fractures oriented NW–SE in response to SW–NE extension in the area during the Miocene. Dike swarms can consist of very large numbers of individual dikes. Figure 4.21 is a map showing two early Proterozoic dike swarms in the Søndre Strømfjord region of southeastern Greenland. Both swarms are mafic but are now highly metamorphosed. The principal

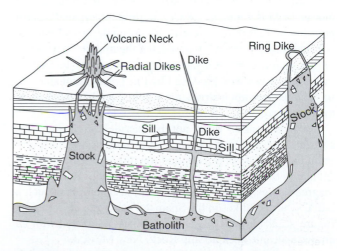

FIGURE 4.20 Schematic block diagram of some intrusive bodies.

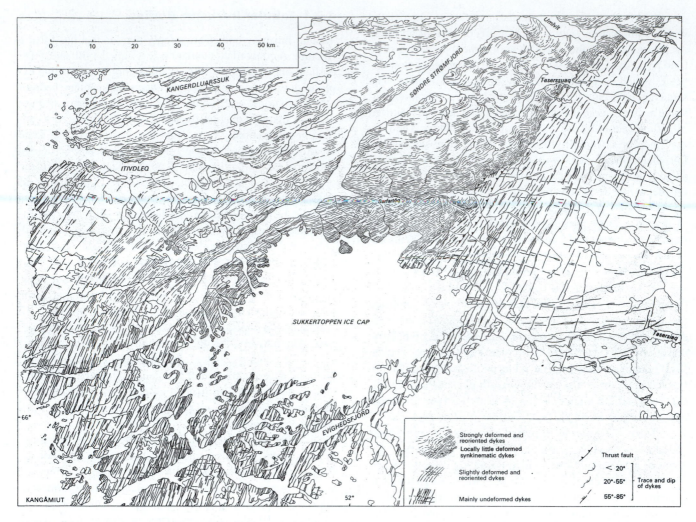

FIGURE 4.21   Kangâmiut dike swarm in the Søndre Strømfjord region of southeastern Greenland. From Escher et al. (1976). Copyright Geological Survey of Denmark and Greenland.

NNE-striking swarm is the 1.95 Ga Kangâmiut swarm, and the E–W swarm is somewhat older. These dikes represent up to 15% of the crust over large areas, indicating 15% crustal extension. The Kangâmiut dikes are dramatic in their own right but have also proved useful to delineate the margin of the area affected by the late phases of the Nagssugtoqidian mobile belt (as young as 1.8 Ga). In the southern and eastern portions of Figure 4.21, the dikes are undeformed but become highly deformed upon entering the mobile belt. Over 1 billion years of erosion has removed the great mountains that would have been produced by this deformational event, so we must rely upon structures within the rocks themselves to characterize it.

Dike swarms need not be parallel. A *radial* dike swarm about a volcanic neck is illustrated in the block diagram in Figure 4.20. Such radial fractures are a common response to the stresses imposed upon the rocks above rising magma bodies. Figure 4.22a is a famous example of such a radial swarm of dikes in the Spanish Peaks area of Colorado. Figure 4.22b shows two radiating dikes associated with an eroded volcanic neck at Ship Rock, New Mexico.

In addition to radial dikes over plutons, *concentric* ones may also form. There are two principal types of concentric dikes: ring dikes and cone sheets (Figure 4.23). **Ring dikes** occur when the pressure exerted by the magma is less than the weight of the overlying rocks. In this case, cylindrical fractures will form, as in Figure 4.23a. If the overlying rock is slightly more dense than the magma, cylinders of the roof will drop, and the magma will intrude along the opening fracture, as in Figure 4.23b. This is more likely to occur with less dense, silicic magmas. Erosion to the level X–Y will result in a ring dike exposed at the surface (similar to Figure 4.23c). More than one concentric fracture may form, resulting in a series of dikes. A fracture may also penetrate to the surface, in which case a ring dike would feed a volcanic event, as in caldera collapse (Figure 4.9). The block that drops into the chamber may consist of country rock, an earlier phase of the pluton itself, or associated volcanics. Ring dikes are either vertical or dip steeply away from the central axis. Figure 4.24 illustrates the classic Tertiary ring dikes and cone sheets on the Island of Mull in western Scotland. Note that there are two centers and several intrusive phases with rings of different compositions. Most of the blocks within a ring consist of earlier intrusives.

Cone sheets form when the pressure of the magma is greater than the confining pressure of the overlying rocks. In

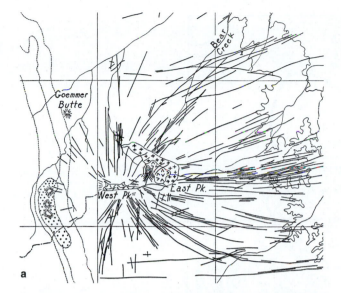

FIGURE 4.22 **(a)** Radial dike swarm around Spanish Peaks, Colorado. After Knopf (1936). Copyright © The Geological Society of America, Inc. **(b)** Eroded remnant of a volcanic neck with radial dikes. Ship Rock, New Mexico. From Shelton (1966).

fracture orientations and stress trajectories associated with ring dikes and cone sheets above magma chambers, see Anderson (1936) and Roberts (1970).

Illustrations of dikes and sills, such as Figure 4.20, generally leave the impression that dikes are close to vertical and sills are nearly horizontal. Although commonly true, this is not necessarily the case. Dikes, by definition, are discordant with the country rock structures, and sills are concordant. If the bedding is vertical, sills will be vertical, and dikes may be horizontal.

The term **vein** refers to a small tabular body, whether or not it is discordant or concordant. This term is typically used in association with ore bodies, where numerous small offshoots from a pluton penetrate the adjacent country rocks. These offshoots (veins) are typically quartz rich and may contain ore minerals. The term is *not* recommended for other igneous bodies. Rather, the terms *dike* and *sill* are preferable, no matter how small. Dikes and sills can range in thickness from less than 1 mm to over 1 km, although most are in the 1 to 20 m range.

Tabular bodies are generally emplaced by **injection**, associated with dilation of the dike walls, as shown in Figure 4.25a. In this case, the walls of the NE–SW dike open, usually in a direction normal to the margins as the dike is injected. All features across the dike match up if we mentally remove the dike and close the walls in a direction normal to the margins. An alternative geometry is illustrated in Figure 4.25b. In this case, the features cross the dike as though the dike were a passive feature, requiring no movement of the dike walls. In fact, any movement of the walls in an effort to close the dike will fail to match both the bedding and the smaller dike. The distinction between these two types is strongest when there are at least two nonparallel structures in the country rocks, such as bedding and the small dike. This is a rare occurrence. The type of geometry in Figure 4.25b implies that the dike *replaced* the rock volume that it now occupies. This is most readily accomplished by chemical permeation and replacement of the country rock. Stoping of the country rock and injection of a dike is an alternative interpretation, but this requires a fortuitous lack of any motion whatsoever across the stoped block. If the dike is a product of replacement, it is no longer technically a dike.

this case, inwardly dipping fractures form, as in Figure 4.23d, usually as a nest of concentric cones. Intrusion along these fractures produces a set of rings at the surface that, in contrast to ring dikes, dip *inward*. Cone sheets and ring dikes can occur together, where they result from different phases of a single intrusion. For a more detailed discussion of the

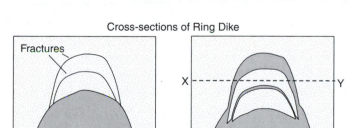
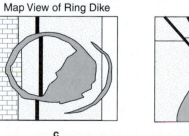
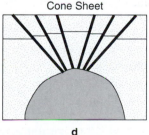

FIGURE 4.23 The formation of ring dikes and cone sheets. **(a)** Cross section of a rising pluton causing fracture and stoping of roof blocks. **(b)** Cylindrical blocks drop into less dense magma below, resulting in ring dikes. **(c)** Hypothetical map view of a ring dike with N–S striking country rock strata as might result from erosion to a level approximating X–Y in **(b)**. **(d)** Upward pressure of a pluton lifts the roof as conical blocks in this cross section. Magma follows the fractures, producing cone sheets. The original horizontal bedding plane shows offsets in the conical blocks. **(a)**, **(b)**, and **(d)** after Billings (1972), **(c)** after Compton (1985). Copyright © and reprinted by permission of John Wiley & Sons, Inc.

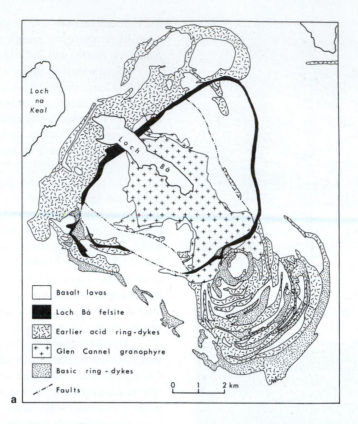

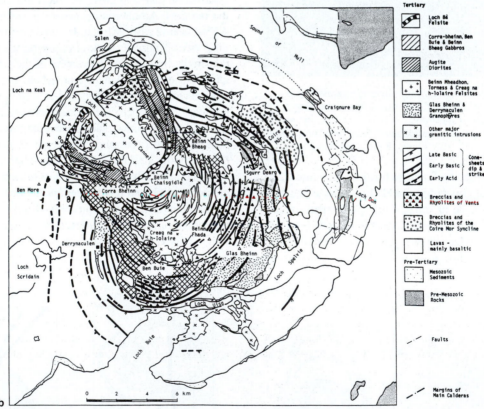

FIGURE 4.24 **(a)** Map of ring dikes, Island of Mull, Scotland. After Bailey et al. (1924). **(b)** Cone sheets in the same area of Mull, after Ritchey (1961). Note that the black felsite ring dike in part **(a)** is shown as the striped ring in the northwest of part **(b)**. Copyright © by permission British Geological Survey.

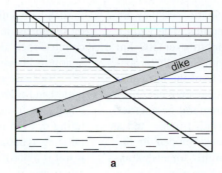

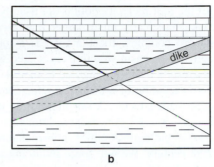

FIGURE 4.25 Types of tabular igneous bodies in bedded strata, based on method of emplacement. **(a)** Simple dilation (arrows) associated with injection. **(b)** No dilation associated with replacement or stoping.

Such replacements are necessarily small, and perhaps the term *vein* would be preferable for such features. There may be a symmetrical mineral zonation parallel to the vein axis and some traces of the original structures of the replaced country rocks within the vein. There may also be a central joint, which acted as a conduit for the fluids that carried the replacing elements. We will talk more about replacement when we discuss metasomatism in Chapter 30. Field evidence indicates that the vast majority of dikes and sills form by injection and dilation. Dikes and sills cool progressively from the walls inward and, in some cases, develop columnar joints in the process. Subvertical dikes develop columns oriented nearly horizontally, which helps to identify them in places where the dikes intrude volcanic rocks of a similar composition, such as the feeder dikes to flood basalts.

4.2.2 Non-Tabular intrusive bodies

The two most common types of non-tabular plutons are illustrated in Figure 4.20; they are called stocks and batholiths. The shape of these bodies is irregular and depends upon the depth of emplacement, the density and ductility of the magma and country rocks, and any structures existing in the country rocks at the time of emplacement. A **stock** is a pluton with an exposed area less than 100 km^2, and a **batholith** is a pluton with an exposed area larger than 100 km^2. Distinguishing the two on the basis of exposed area is very simple and useful in the field, but it is also unfortunate in that it is not based on the size of the body itself but, in large part, on the extent of erosion. It is possible that a small pluton that is deeply eroded would have an exposed area greater than 100 km^2 and be a batholith, whereas a much larger pluton, only barely exposed, would be a stock. For that matter, the "batholith" illustrated in Figure 4.20 is not exposed, except as a volcanic neck and a ring dike, so it would not technically be a batholith at all by the common definition. Also, several smaller stocks in an area

may be so similar in age and composition that they are believed to be connected at depth, forming parts of a larger body. These separate extensions are called **cupolas** (see Figure 4.30). When geophysical evidence or mapping suggests a larger pluton beneath the surface, it would be better to base the type of pluton on the estimated volume of the body and not on the exposed area. With the exception of drill data, however, subsurface interpretations are still somewhat speculative, and the classification based on exposed area does have the advantage of being definite. One way out of this dilemma was proposed by Pitcher (1993), who used the general term *pluton* for any large non-tabular body and reserved the term *batholith* for the large composite arrays of multiple plutons developed in orogenic belts. Further complication stems from the central European (and, until recently, French) practice of calling smaller bodies batholiths and larger ones plutons.

Some types of stocks are genuinely smaller bodies and not simply limited exposures of larger batholiths. Some stocks represent the cylindrical conduit and magma chamber beneath volcanoes. This type of stock is called a **plug**. The exposed portion of a plug, commonly remaining after the more easily eroded volcanics of the cone have been removed, is called a **volcanic neck** (Figures 4.20 and 4.22). Alkaline igneous rocks, which form in mid-continent regions (typically associated with rifts), tend to rise along vertical conduits of limited cross-sectional area. Thus they typically form small plugs.

In addition to the two "generic" types of plutons, stocks and batholiths, there are a number of special types, based on shape rather than size. A **laccolith** is a concordant stock with an arched roof and ideally a flat floor (Figure 4.26a). In many cases, however, the floor may sag somewhat due to the weight of the injected magma. A **lopolith** is another concordant type of pluton intruded into a structural basin (Figure 4.26b). Both are essentially sills. A laccolith is sufficiently viscous (and silicic) to limit magma flow along the horizontal plane, and it is shallow enough to physically

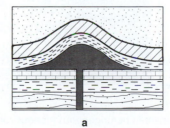

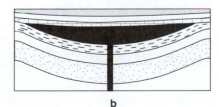

FIGURE 4.26 Shapes of two concordant plutons. **(a)** Laccolith with flat floor and arched roof. **(b)** Lopolith intruded into a structural basin. The scale is not the same for these two plutons; a lopolith is generally much larger than a laccolith.

lift the roof rocks. Lopoliths are usually mafic, and they are characteristically much larger than laccoliths. The Duluth gabbro complex, for example, is a lopolith over 300 km across. The basin structure associated with lopoliths probably results from sagging due to the weight of the intrusion, but evacuation of a deeper magma chamber may also be a contributing factor. A fair number of structural geometries are possible for concordant plutons in folded rocks, and a correspondingly complex nomenclature developed in the 20th century. We will forgo indulging in this somewhat arcane classification. Perhaps the following excerpt, defining *cactolith*, as described by Hunt et al. (1953), from the AGI *Glossary of Geology*, would serve to illustrate the degree of specialization attained:

> *Cactolith:* a quasi-horizontal chonolith composed of anastomosing ductoliths whose distal ends curl like a harpolith, thin like a sphenolith, or bulge discordantly like an akmolith or ethmolith.

4.2.3 Contact Relationships of Plutons

The emplacement of a pluton involves the juxtaposition of a hot, viscous, commonly fluid-saturated liquid in motion against a cooler, stationary solid, generally of a much different composition. Such contrasts in properties and relative motion are certain to result in chemical and mechanical interactions that will impart to the contact zone some diagnostic structures and textures but may also obscure the simple relationship of igneous material abutting country rock. The structures and textures, both within the pluton and in the adjacent country rocks, can provide important clues to the processes involved during emplacement. The contact itself may be sharp, with igneous rock against relatively unaltered country rock, or it may have a gradational border zone.

The border zone may be strictly mechanical (injected), and, when broad, exhibit a gradation from undisturbed country rock through a zone of increasing dikes, veins, or tongues (collectively called **apophyses**) extending from the pluton. As shown in Figure 4.27a, the ratio of igneous to wall rock may gradually increase across such a zone, beginning with country rock containing a spaced network of small injected dikes through an **agmatite** zone (a rock with a high concentration of country rock fragments, or **xenoliths**, in an igneous matrix), to spaced xenoliths of country rock floating in igneous material, until even the xenoliths become rare. Any degree of gradation is possible.

Because many of the more silicic plutons are fluid saturated and chemically distinct from the country rocks, fluids emanating from a pluton may permeate the surroundings, altering or even partially melting the country rocks and crystallizing igneous-type minerals in the exterior matrix. The result is a border zone across which the igneous rock gradually passes into the country rock without a distinct boundary (Figure 4.27b). We will return to this type of alteration in Chapter 30.

A third possible gradational border zone would combine the two processes of injection and permeation, as in Figure 4.27c. The various wall rocks and injected and/or locally melted components can become so intimately associated as to result in a hybrid rock of mixed character.

Along sharper contacts at shallow depths, an intrusion may have thermal and chemical effects on the country rock. A gradational zone may result from strictly thermal processes, particularly if the intrusion is both hot and dry. In such cases, unmetamorphosed country rock is heated and recrystallized by the pluton across a narrow interval at the contact. The grain size and degree of recrystallization would decrease rapidly away from the contact, and the contact itself should still be distinct. The thermal effects are more commonly combined with a chemical gradient, established by the silica-saturated fluids released from a fluid-saturated pluton. The result is a **contact metamorphic aureole**, which is usually gradational but affects the country rocks more than the pluton itself, leaving the original contact distinct. Contact metamorphism will be discussed in more detail in the metamorphic sections of this text.

The dynamics at the contact with the wall rocks may impart some lasting features *within* the pluton as well. In addition to xenoliths, there may be a **chill zone**. A chill zone is a zone of finer grain size resulting from the rapid solidification of the pluton where it comes in contact with the cooler country rock. Because the igneous material being intruded is in motion against a stationary country rock, there may also be a zone of distributed shear in the viscous magma. The more viscous the magma, the more pronounced the shear and thicker the sheared zone. The result of the shear will be

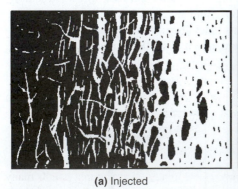

(a) Injected

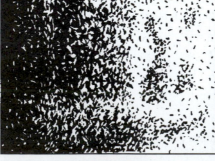

(b) Permeated

(c) Combination

FIGURE 4.27 Gradational border zones between homogeneous igneous rock (light) and country rock (dark). After Compton (1962).

FIGURE 4.28 Marginal foliations developed within a pluton as a result of differential motion across the contact. From Lahee (1961). Copyright ©, with permission of the McGraw-Hill Companies.

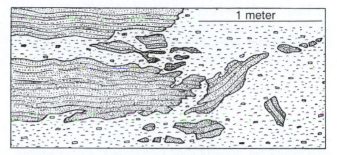

FIGURE 4.29 Continuity of foliation across an igneous contact for a pre- or syn-tectonic pluton. From Compton (1962).

the development of an imparted lineation and/or foliation parallel to the contact in the marginal portion of the pluton (Figure 4.28). This will be most evident in any elongate or platy minerals, such as hornblendes, pyroxenes, and micas. There may also be some elongated or flattened mineral aggregates or ductile heated xenoliths that produce disc-shaped masses called **schlieren**. The intensity of magmatic foliations typically increases toward the margins where shear is greatest (Paterson et al., 1991). Some magmatic foliations also form at an angle to the contact (Berger and Pitcher, 1970; Courrioux, 1987).

If the magma is sufficiently viscous and the country rock is hot enough, the contact shear may also affect the country rock, rotating the foliation *outside* the pluton, into parallelism with the contact, as in Figure 4.28 (Castro, 1986). Such transposition of the exterior foliation causes an otherwise discordant contact to become concordant.

The overall result of the shearing process is the loss of a clear, sharp, discordant igneous contact and the loss of the characteristically isotropic igneous texture. The igneous rock looks more like a metamorphic gneiss, and painstaking field and petrographic work may be required to interpret the contact correctly.

Because of the great difference in mechanical properties that typically exists between many crystallized plutons and the country rock, later deformation is commonly concentrated along the pluton margins. As a result, many igneous contacts suffer post-emplacement shearing, superimposing a tectonic imprint over the original igneous contact. This shear may have only minor effects, or it may obliterate any original igneous contact textures. If the shear is great enough, the pluton may be faulted into juxtaposition with a completely different rock, with the contact becoming strictly tectonic. It may be impossible to distinguish an extensively sheared igneous contact from a strictly tectonic one, or even from a contact that is sheared during emplacement of the pluton, as described above. As pointed out by Paterson et al. (1991), the structural patterns within and around plutons reflect the superposition of deformation related to emplacement and regional stresses. A variety of structures are possible, ranging from ductile to brittle.

4.2.4 Timing of Intrusion

Most batholiths and stocks are emplaced in mountain belts as part of the overall subduction/orogenic process, and they typically play an important and active role in the evolution of those belts (see Chapters 18 and 19, as well as Paterson

et al., 1991). Most orogenic belts are complex, involving multiple deformational, metamorphic, and igneous events (Section 23.4). Any attempt to relate deformation to pluton emplacement must demonstrate that deformation (and metamorphism) is closely related to intrusion not only spatially but also *temporally*. Textures within plutons and in country rocks reflect the timing of emplacement with respect to tectonic activity.

Post-tectonic plutons are emplaced after an orogenic/metamorphic episode, and the igneous rocks thus lack any deformational features, such as foliations (other than those related to intrusion). The regional deformational fabrics and structures of the country rocks will either be cut discordantly by the pluton or curve into parallelism with the contact. They will also be overprinted by any emplacement-related structures.

Syn-tectonic plutons are emplaced during the orogenic episode. Any regional foliation will be continuous with emplacement-related foliations in the pluton. The pluton will also be affected by any later continuing orogeny.

Pre-tectonic plutons are emplaced prior to the orogenic episode. Both pre- and syn-tectonic plutons suffer the imprint of the deformational and metamorphic processes associated with the orogeny. They therefore have an internal foliation that parallels, and is continuous with, that of the country rocks (Figure 4.29). The regional foliation may curve around a non-foliated pluton because of ductility contrasts. It is commonly difficult or impossible to distinguish whether a pluton is pre- or syn-tectonic because both share this attribute. Syn-tectonic intrusions are usually more ductile at the time of deformation, so they may be more elongated in the foliation direction, with more concordant contacts. Pre-tectonic plutons are cooler and likely to be more resistant to deformation, which would be concentrated at the pluton margins. Because plutonism and orogeny are related in most orogenic belts, purely pre-tectonic plutons are rare. Most plutons that can be characterized as pre-tectonic are commonly syn- or post-tectonically associated with an earlier orogenic event in a multiply deformed belt.

4.2.5 Depth of Intrusion

In addition to timing, the depth of emplacement affects many structural and textural features of plutons. Many of these characteristics were summarized by Buddington (1959) on the basis of the level of emplacement with respect to the three **depth zones** first distinguished by Grubenmann (1904). These zones—the epizone, mesozone, and catazone—are based on the characteristics of the country rocks, and the actual depth limits are only approximate (and variable as well), due to differences in the geothermal gradient between deformed belts.

The **epizone** is characterized by relatively cool (less than 300°C) country rocks of low ductility, at depths less than about 8 km. Some plutons, such as the Boulder Batholith, Montana (Hamilton and Myers, 1967), may be so shallow as to intrude their own volcanic carapace and thus be emplaced above the previous ground surface (Figure 4.36). Plutons of the epizone are usually post-tectonic and have sharp, discordant contacts. The wall rocks are typically brecciated, and there may be numerous dikes and offshoots from the main igneous body. The top of a pluton commonly penetrates the roof rocks in an irregular manner, with several lobes extending upward and outward.

Figure 4.30 illustrates several of the relationships between a barely exposed epizonal pluton and the country rocks composing the roof of the body. Contacts are generally sharp and discordant, crosscutting the structures in the country rocks. An **offshoot** is a general term for any lobe of the main plutonic body that protrudes into the country rock. A **cupola** is a non-tabular offshoot that is isolated from the main body, as observed in map view. A **septum** is a peninsula-like projection of the country rock into the pluton so that it separates two igneous lobes. A **roof pendant** is a projection of the roof rocks into the pluton that has become isolated *by erosion*. The structures of the roof pendant are parallel to those of the country rocks outside the pluton, signifying that the pendant was connected to the country rocks and prevented from rotating. If the structures of the isolated block were at a different angle than those of the country rocks, the block would be free floating and rotated, and thus it would be a xenolith. Large xenoliths are commonly called **rafts**.

Although some large, shallow batholiths are found in the epizone, most epizonal plutons are fairly small. This is to be expected because erosion is rather limited if the country rocks are epizonal. Many of the small stocks may themselves be cupolas of a larger body at depth, such that progressive erosion would expose larger bodies and deeper country rocks.

Emplacement of an intrusive body may fracture the country rocks as it forces its way toward the surface. Hydrothermal alteration and ore mineralization are common and concentrated along such fractures. A convective hydrothermal system may be established above the pluton, driven by the heat of the intrusion, making the alteration intense and pervasive (see Section 4.3). Contact metamorphism may be dramatic where the country rocks are previously unmetamorphosed or poorly metamorphosed. The principal limitations on the size and development of a contact aureole are the rapid cooling of small plutonic bodies and the loss of fluids along fracture channelways.

Fabrics in an epizonal pluton are typically isotropic, and this commonly includes the areas near the contacts that have experienced some shear against the stationary wall rocks. **Miarolitic cavities**, representing "bubbles" of fluid (presumably liberated at low pressure) with euhedral mineral projections inward, are common. The epizone is the typical environment for laccoliths, lopoliths, plugs, ring dikes, and cone sheets.

The **mesozone** is the depth interval of about 5 to 15 km. The country rocks are low-grade regional metamorphic rocks at temperatures of 300 to 500°C. Plutons in this zone have characteristics that are transitional between those of the epizone and the catazone (Figure 4.31a). They may be either syn- or post-tectonic. The contacts may be sharp or gradational and discordant or concordant—because the country rocks are more ductile than in the epizone. A contact metamorphic aureole is typically well developed because the plutons are larger and cool more slowly; and the country rocks are metamorphosed to only a low grade prior to emplacement and thus are susceptible to modification by the thermal and compositional contrasts imposed by the pluton. The rocks of the contact aureole commonly have a foliated fabric because they experience regional as well as contact metamorphism. "Spotted" slates and phyllites are thus common. The spots are larger contact metamorphic minerals that typically overprint the foliated regional metamorphic minerals (Section 23.2). A chill zone is minor or absent. Fabrics in the pluton may be directionless (isotropic) but are commonly foliated or lineated near the contact.

The deepest zone, the **catazone**, is deeper than about 10 km. Here the country rocks are undergoing medium to high-grade regional metamorphism in the 450 to 600°C range. Plutons are usually syn-tectonic, with gradational

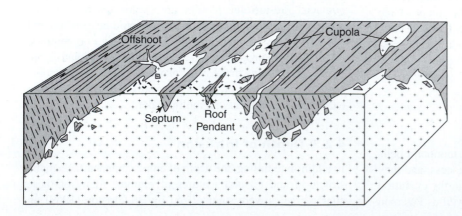

FIGURE 4.30 Block diagram several kilometers across, illustrating some relationships with the country rock near the top of a barely exposed pluton in the epizone. The original upper contact above the surface is approximated by the dashed line on the front plane. After Lahee (1961). Copyright ©, with permission of the McGraw-Hill Companies.

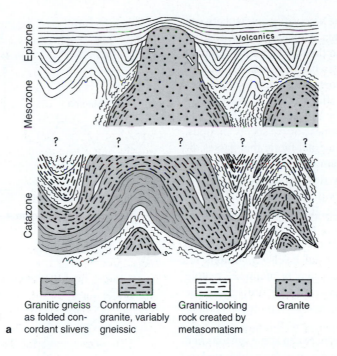

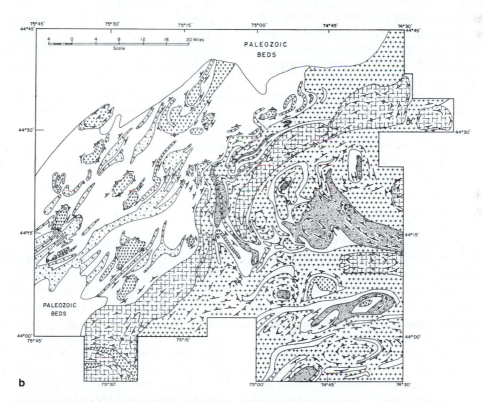

FIGURE 4.31 (a) General characteristics of plutons in the epizone, mesozone, and catazone. (b) Shape of concordant igneous bodies typical of the catazone, northwest Adirondacks, New York. The blank areas are metamorphic country rocks, and the patterned areas are various plutons. From Buddington (1959). Copyright © The Geological Society of America, Inc.

contacts and no chill effects. The viscosity contrast between the country rocks and the magma is relatively low, so contacts are generally concordant in the sense that the foliation of the fairly ductile country rocks has been sheared and rotated to become parallel to the contact. There are seldom noticeable contact metamorphic effects because the country rocks are already at fairly high metamorphic grades. The plutons commonly occur as domes, or as sheets or pods, with a foliated and/or lineated internal fabric that passes directly into the fabric of the metamorphic country rocks (see Figure 4.31). In addition to any flow foliation, the igneous body may have been imprinted with a metamorphic foliation during or after final solidification. It is difficult to recognize such foliated plutonic rocks as igneous because they have the characteristics of high-grade metamorphic gneisses. In the deep catazone and lower crustal levels, conditions for partial melting of the crustal rocks may be attained, creating localized melts from the country rocks that further blur the distinction between the igneous and metamorphic components.

Because of the similarities between some igneous and high-grade metamorphic components and the gradational relationships between them in the deep crust, it is common in exposed high-grade metamorphic terranes to be able to traverse from rock that is metamorphic, through a mixed zone, into igneous-looking rocks in a completely gradational manner without ever encountering a distinct contact.

The boundaries between the depth zones are neither sharp nor static. Rocks in any of these zones can enter the next deeper zone by burial or thrust stacking or the next shallower zone by uplift and erosion. Thus neighboring plutons in a given exposed terrane can be emplaced at different times in different zones.

4.2.6 Multiple-Injection and Zoned Plutons

As with volcanic centers, many plutons show complex histories of multiple intrusion of magmas of varying composition. Large batholith belts (Chapter 17) are composed of numerous compound intrusions. For example, although early descriptions of the Sierra Nevada Batholith in California refer to it as a single batholith, it has long been recognized as a broad zone of repeated intrusion of numerous plutons over a considerable time span. The number of separate intrusions is now known to be in the hundreds. Even smaller intrusive bodies, if mapped carefully, commonly show multiple intrusive events.

One well-documented example of multiple emplacement is the Tuolumne Intrusive Series in Yosemite National Park, a pluton that is part of the Sierra Nevada Batholith. Bateman and Chappell (1979) showed this small batholith to be a series of related intrusions, ranging in composition from an outer diorite through a granodiorite and finally a porphyritic granite at the core (Figure 4.32). A general concentric zonation, with later magmas toward the center, suggests that fresh magma surges took place within a solidifying outer shell. This concentric pattern is not perfect, however, because some portions of the later intrusives either abraded, assimilated, or

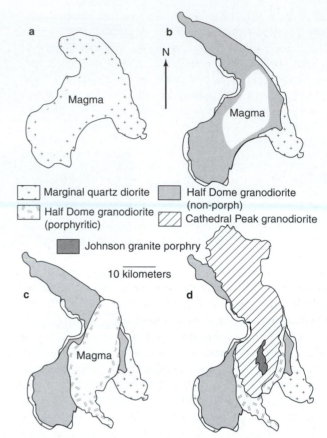

FIGURE 4.32 Developmental sequence of intrusions composing the Tuolumne Intrusive Series (after Bateman and Chappell, 1979). **(a)** Original intrusion and solidification of marginal quartz diorite. **(b)** Surge of magma followed by solidification of Half Dome Granodiorite. **(c)** Second surge of magma followed by solidification of porphyritic facies of Half Dome Granodiorite. **(d)** Third surge of magma followed by solidification of Cathedral Peak Granodiorite and final emplacement of Johnson Granite Porphyry. Copyright © The Geological Society of America, Inc.

fractured their carapace to crosscut the older crystallized margins. A number of chemical and mineralogical trends associated with this sequence of magmas are generally consistent with the evolution of a deeper magma chamber undergoing some combination of crystal fractionation, assimilation of wall rocks, and/or mixing of several magma injections. In an even more detailed study, Coleman et al. (2004) cited isotopic age dates and field evidence which reveal that some of the relatively homogeneous phases of the Tuolumne Series of Bateman and Chappell (1979) must have represented incremental assemblies of smaller components, typically dike-like sheets, over several million years. Because successive increments within a larger phase are similar in composition and commonly injected into a still-warm body, contacts between them may be very subtle and easily overlooked.

The Tuolumne Series is probably an example of a common phenomenon. Many plutons of various sizes display multiple injections and chemical variations across zones. Glazner et al. (2004) suggested that many large plutons, even apparently continuous bodies, have grown by amalgamation from many small, probably dike-fed, increments. Even where distinct magma types are not apparent, there is commonly a systematic variation in the mineralogy or composition of some minerals. For example, the amount of quartz or potassium feldspar may increase, or the plagioclase composition may become progressively more sodic toward the center of a pluton. These trends are compatible with a progressively cooling and evolving magma, as will be addressed in Chapter 8.

4.2.7 The Process of Magma Rise and Emplacement, and the "Room Problem"

Intrusive igneous rocks are simply magmas that have not reached the surface. Plutonism lacks the drama of volcanism because no one has ever died from pluton emplacement, and there has never been a pluton alert. The volume of igneous rock in plutonic bodies, however, is considerable, and there is a pluton of some sort beneath every volcano. In a sense, we can think of these bodies as liquid that crystallized, or "froze," in the plumbing system on its way toward the surface.

When magma forms through some melting process at depth, it segregates from the residual unmelted solids to form a discrete liquid mass. This mass is typically less dense than the surrounding solid (see Chapter 5), so it becomes *buoyant*. The buoyant magma body tends to rise, and, if the surrounding material is sufficiently ductile, it is generally considered to do so as a diapir. A **diapir** is a mobile mass that rises and pierces the layers above it as it does so. Whenever a sufficiently ductile rock mass is covered by denser rocks that are also ductile, diapirs are capable of forming.

A well-documented example of diapirism is the formation of salt diapirs in areas such as northern Germany, Iran, and the American Gulf Coast. In these cases, the salt beds (evaporite deposits) are covered by subsequent sediments that compact and become denser than the salt beneath. Halite (salt) not only has a low density but will begin to behave in a ductile fashion and flow under conditions of low confining pressure. After 5 to 10 km of overburden develops, the salt beds begin to flow, and the upper surface develops irregular swells (Figure 4.33). Salt flows upward into these swells, forming diapirs that rise toward the surface. If a diapir rises sufficiently above its source and is not supplied by enough salt flow from the evaporite bed, the column feeding the diapir stretches and constricts until the body separates from the source bed. The inverted raindrop shape, with bulbous leading edge and tapered tail, is characteristic of rising diapirs. Because buoyancy is the driving force that causes a diapir to rise, once it reaches a level where the density of the surrounding rock is the same as that of the salt, it will stop rising and spread out laterally.

Magma diapirs have long been the classic concept of rising magma and are believed to behave in a fashion similar to the description above: those forming in the mantle rise through the ductile mantle, just as the salt does through the overlying sediments. As long as the viscosity of the magma and country rock are fairly similar, they rise by radially distending the overlying rock as they force their way upward. In the less ductile areas of the upper mantle and crust, rising

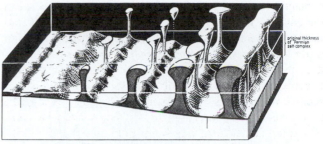

Progressive development of diapirs ⟶

FIGURE 4.33 Block diagram of subsurface salt diapirs in northern Germany. After Trusheim (1960). Copyright © American Association of Petroleum Geologists, reprinted by permission of the AAPG.

magma can no longer rise in diapiric fashion. Rather, it must exploit fractures or other weaknesses in the rocks through which it rises.

The method by which a large intrusive body moves upward through the crust and creates sufficient room for itself is far from clear. This **room problem** has been the subject of considerable debate for decades and is still an enigmatic problem. For excellent reviews of the room problem, see Newell and Rast (1970) and Paterson et al. (1991).

Open fractures and voids are limited to the very shallow near-surface environment (a few tens of meters). Below this, the rise of magma by simply filling such open voids is not an option. Rising magma can follow preexisting (closed) fractures, however, by forcibly displacing the rocks that form the fracture walls and following these planar conduits. At depth, the ability of magma to force a fracture open is limited because the injecting pressure of the magma is seldom great enough to significantly displace rigid rock walls that are forced together by lithostatic pressure below a depth of a few kilometers. Of course, if an area is undergoing regional extension, the walls of any fractures will not be under compression, and magma can force them apart. The number and width of dikes or sills in such filled conduits depends upon the rate of extension and should be commensurate with typical plate tectonic rates of less than ~3 cm/yr.

The room problem becomes more troublesome for larger intrusive bodies, which occupy a significant volume, which means much more rock must be displaced for them to move upward. Figure 4.34 summarizes the proposed mechanisms by which a pluton could make room and rise. Plutons, such as laccoliths, can **lift the roof** (number 1 in Figure 4.34) by either folding or block elevation along faults. Controversy exists over whether the lifting force of plutons is restricted to magma buoyancy, which limits the ability of a pluton to lift the roof when it reaches the level at which its density equals that of the country rocks. Lifting may be facilitated in such cases by **magmatic overpressure**, which might supply additional depth-derived pressure. Roof doming to form laccoliths is limited to depths less than about 2 to 3 km, where the magmatic pressures can exceed the weight and strength of the overburden (Corry, 1988). Similar restrictions apply to block uplift along faults.

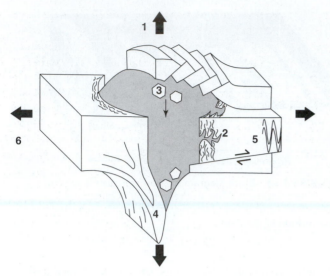

FIGURE 4.34   Diagrammatic illustration of proposed pluton emplacement mechanisms. **(1)** doming of roof; **(2)** wall rock assimilation, partial melting, zone melting; **(3)** stoping; **(4)** ductile wall rock deformation and wall rock return flow; **(5)** lateral wall rock displacement by faulting or folding; **(6)** (and **1**) emplacement into extensional environment. After Paterson et al. (1991). © The Mineralogical Society of America.

Alternatively, magma may melt its way upward (a process called **assimilation**; number 2 in Figure 4.34). The ability to melt the walls using magmatic heat is limited by the available heat of the magma—and the amount of magma that can penetrate in the first place. As we shall discuss in Chapter 10, the intruded magmas are themselves most typically generated by *partial* melting, so they are seldom appreciably "superheated" (i.e., heated above the temperature at which the melt coexists with solids). Thus the heat available for melting the country rocks does not exist in excess and must be supplied by the latent heat of crystallization of some portion of the magma, making it at least partially solid and, therefore, less mobile. This type of heat is necessarily limited.

If the country rocks are sufficiently brittle, blocks of the roof over a rising pluton could become dislodged, fall, and sink through the magma (number 3 in Figure 4.34). This process is called **stoping**, after the mining practice of quarrying the roof of an underground working. Considerable evidence for stoping is to be found in the upper portions of many plutons, where blocks of country rock are suspended in the crystallized igneous rock as rafts or xenoliths. The process is evident where agmatites form by injection along a fracture network (see Figure 4.27a). Stoping requires that the country rock be denser than the magma. Also, for stoping to be effective, the stoped blocks must be large enough to sink rapidly enough in a viscous magma, which can require blocks as large as tens of meters in granitic melts. Stoping may be effective only in the shallow crust where the rocks can be fractured. Cauldron subsidence (Figure 17.16) and caldera formation (Figure 4.9) are examples of large-scale stoping at shallow depths. Of course, the pluton must get to this level of the crust before this method can be possible, so stoping may be an emplacement mechanism at shallow depths but cannot make the original room for the pluton

where there was none before. The lack of field evidence for significant quantities of stoped xenoliths calls into question the effectiveness of stoping as a major process in magma emplacement.

A combination of assimilation and stoping, called **solution stoping** or **zone melting**, may operate at depths where the country rock is near the melting point. In this process, the minerals of the roof rock melt, and an equivalent amount of magma at the floor crystallizes, as proposed by Ahren et al. (1981). This process mitigates the heat loss from the intrusive magma, a major impediment to assimilation. Such a process may be effective in the mantle and deep crust, where the magma rises as a diapir anyway, but if the country rocks are well below their melting point, more minerals would have to crystallize than melt, and the pluton would quickly solidify. This process will be discussed further in Chapter 11.

Ductile deformation and **return downward flow** (number 4 in Figure 4.34) are the mechanisms associated with the rise of diapirs at greater depths, and are probably efficient where the country rocks are initially ductile or have been softened by the heat of the pluton. Recently Glazner et al. (2003) noted that many batholiths exhibit structural features indicative of sinking and suggested that shallow plutons (and their volcanic cover) may be accommodated by isostatic sinking of the thickening crustal pile into its substrate, which may then be displaced toward the back-arc region (Chapter 16) by deeper crustal flow. At any depth, **ballooning**, or radial expansion of the magma chamber by addition of magma from below, may also take place. The mechanism of roof yielding to this stretching would be ductile at depth, and more brittle near the surface. Ballooning may also physically compress the wall rocks at the sides of the diapir, forcing the walls apart with accompanying aureole deformation (number 5 in Figure 4.34). Theoretical and experimental studies suggest that for diapirism and ballooning to be effective, the viscosities of the pluton and wall rock must be similar (Marsh, 1982; Arzi, 1978; Van der Molen and Paterson, 1979). The lack of textural evidence for wall rock softening around most mid- and upper-level plutons argues against diapirism as an emplacement mechanism at these crustal levels (Paterson et al., 1991). As in the case for dikes mentioned above, extensional environments (number 6 in Figure 4.34) would facilitate this process, but, as mentioned above, the rate of extension places limits on the degree to which it can contribute.

Although magma is believed to rise by some or all of the above processes, they all have limits to the extent and the circumstances under which they may operate. The room problem is still a problem to this day, and it is particularly vexing when we consider the giant batholiths, which occupy vast volumes in the crust. Our concept of these great igneous bodies, however, is based on exposure at the surface, and the shape of a batholith at depth is not certain. Because these bodies are intruded from below, there is a natural tendency to think of them as extending outwards with depth, and occupying a considerable volume beneath the surface that we can observe (as in Figure 4.20). This is the way batholiths

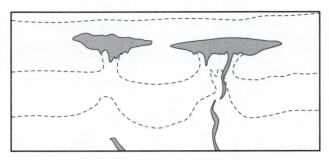

FIGURE 4.35 Sketches of diapirs in soft putty models created in a centrifuge by Ramberg (1970).

have been drawn for decades in most geologic cross sections involving them. They simply get larger downward, which certainly leaves the impression of much more igneous material below.

The shape of experimental diapirs, on the other hand, is different, at least as diapirs near the surface. Figure 4.35 is a drawing of some results of the experiments by Ramberg (1970), in which he modeled diapirs using soft materials of low density covered by layers of higher-density strata and placed them in a centrifuge to drive the diapiric motion. He found that the low-density material rises as diapirs due to its buoyancy and the ductility of all the material in the centrifuge. Upon reaching a level where the surroundings are less dense, however, the buoyancy of the diapir is reduced or lost entirely, at which point the diapir tends to spread laterally and become more broad and thin. If this behavior is typical, even in the more brittle epizone, batholiths may be much thinner than we have imagined, and the amount of room required for their emplacement is greatly reduced. Based on gravity models and fieldwork in eroded batholiths, several are now suspected of being shallow-floored intrusions. Figure 4.36 is a proposed cross section of the Boulder Batholith in Montana, which suggests that it is less than 10 km thick. Other plutons are now suspected of being this thin or even thinner. The Lilesville Granite, North Carolina (Waskom and Butler, 1971), the Katahdin Pluton, New Hampshire (Hodge et al., 1982), and the Flamanville Granite, NW France (Brun et al., 1990) have gravity-modeled floors at depths less than 3 km.

To summarize: several recent observations on the shape and internal structure of plutons are helping to solve the room problem. In his recent summary, Cruden (2005) concluded that most plutons are either tabular or wedge shaped (with a relatively flat roof and angled floor). Tabular sheets certainly alleviate the need for great amounts of room. If recent concepts of batholiths as multiple injections over long time periods (Coleman et al., 2004; Glazner et al.,

2004) prove to be generally true, one could imagine a large batholith growing incrementally by multiple dike–sill complexes (Figure 4.37). Magma ascent via dikes is thermally and mechanically very efficient, particularly in orogens where the wall-rocks are being dilated or sheared. At shallow levels, the dike-fed magma may then spread laterally as tabular sills when the magma becomes neutrally buoyant, when the magma encounters a permeable or rheological barrier, or when magma/country-rock instabilities are amplified. If sill propagation and inflation occur at shallow levels, the thin roof rocks may be lifted for each new increment (Figure 4.37a). At deeper levels, floor depression (Figure 4.37b) in response to the displacement of magma from beneath to above may prevail, forming a wedge-like lopolith shape. Some composite plutons may develop less systematically (Figure 4.37c). Barker (2006) noted several examples of incremental pluton growth, which he subdivided into *sheeted* and *nested* types. Time intervals separating successive small-volume magma batches in *sheeted* types are sufficient for nearly complete solidification, resulting in coalesced dikes and/or sills (as in Figure 4.37). *Nested* plutons (such as the Tuolumne) grow by successive magma injections into the interiors of largely liquid chambers. From today's perspective, the combination of tabular shape, incremental dike–sill additions over millions of years, and roof elevation or floor depression may eventually provide the best overall solution to the room problem for large batholiths.

4.3 HYDROTHERMAL SYSTEMS

The fractured and permeable sedimentary and volcanic rocks overlying many hot, shallow intrusions are ideal sites for the development of extensive hydrothermal systems. A typical hydrothermal system associated with a silicic volcanic terrane is shown in Figure 4.38. Stable isotope studies have shown that meteoric water predominates over juvenile in most hydrothermal systems (Mazor, 1975), but the ratio is variable from one locality to another. The heat of the shallow magma chamber typically associated with recent volcanism heats the groundwater (plus some added juvenile component) so that it expands and rises through the permeable material above, commonly resulting in fumaroles and hot springs at the surface. The water then cools, moves laterally as more hot water rises from below, and descends again as it becomes denser in a typical convective system. The result is the recirculation of the groundwater system above the magma body and perhaps in the upper solidified portion of the body itself, which may also be extensively fractured (Henley and Ellis, 1983; Hildreth, 1981).

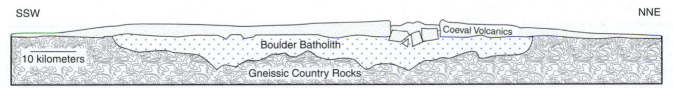

FIGURE 4.36 Diagrammatic cross section of the Boulder Batholith, Montana, prior to exposure. After Hamilton and Myers (1967).

Downward-building   Upward-building   Irregular and
    laccolith   lopolith   unsystematic

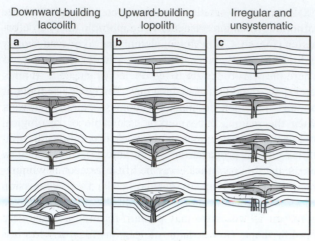

FIGURE 4.37 Possible methods by which a large batholith may grow by successive small increments over millions of years. Magma rises initially as a series of dikes in an extensional terrane. Each dike spreads laterally as a thick sill upon reaching a level at which it is no longer significantly buoyant. Room may be created by **(a)** lifting the roof rocks if the overburden is small, **(b)** depressing the chamber floor as magma is displaced upward and withdrawn from below (Cruden and McCaffrey, 2001; Cruden, 2005), or **(c)** some more irregular and sporadic process. Image courtesy of John Bartley.

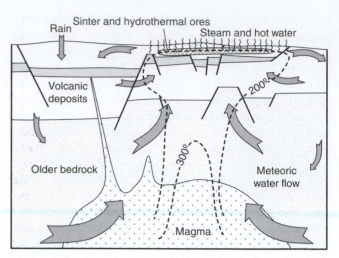

FIGURE 4.38 Schematic section through a hydrothermal system developed above a magma chamber in a silicic volcanic terrane. After Henley and Ellis (1983). Copyright © with permission from Elsevier Science. Oxygen isotopic studies have shown that most of the water flow (dark arrows) involves recirculated meteoric water. Juvenile magmatic water is typically of minor importance.

Hydrothermal systems over shallow batholiths can affect a considerable volume of rock. Aqueous fluid flow is controlled by the permeability of the overlying rocks. Extensive fracture systems are common above shallow intrusions, and these act as particularly effective conduits for hydrothermal fluids. Caldera structures are therefore common centers for hydrothermal activity. Drill holes in some systems extend to depths of up to 3 km, where they have encountered nearly neutral pH saline waters (brines) at temperatures up to 350°C. Below this, the character of the solutions is imprecisely known, but higher temperatures are certainly to be expected. If boiling occurs in the shallow portion of the system, CO_2 and H_2S are typically concentrated in the steam, which may reach the surface as fumarolic activity or condense and oxidize to form a distinctive acidic sulfate/bicarbonate solution common in many geothermal fields.

The hydrothermal fluids evolve through chemical (and isotopic) exchange with the silicate melt and/or the solidified portion of the pluton, whether the majority of the water is juvenile or not. It will thus contain a number of dissolved igneous constituents. As this fluid interacts with the surrounding rocks, it can cause a number of chemical, mineralogical, and textural changes to take place, depending on temperature, permeability, chemical composition and nature of the fluid and the rocks, the fluid/rock ratio, and the longevity

of the hydrothermal system. Fluid/rock ratios range from 0.001 to about 4, based on oxygen isotope exchange (Taylor, 1974). The great variation in the physical and chemical nature of the hydrothermal system results in a similar variety of alteration products, including quartz, feldspars, clay minerals, chlorite, calcite, epidote, a number of sulfides and ores, zeolites, and biotite, actinolite, diopside, and/or garnet at higher temperature. Volcanic glass and mafic minerals in the country rocks are particularly susceptible to alteration. The alteration and ore mineralogy is typically zoned and reflects a temperature gradient as well as chemical gradients and kinetic effects (Henley and Ellis, 1983). Such zonation may occur on a small scale, perhaps about a fracture, and/or on a large scale, as concentric zones that span the entire cap of the pluton. Considerable proportions of economic ore deposits are associated with present and ancient hydrothermal systems. These deposits are an important source of gold, silver, copper, lead, zinc, molybdenum, etc.

Analogous sub-ocean floor systems have been recognized at ocean ridge spreading centers in which ocean and juvenile water are convectively circulated in the thin oceanic crust. Submersible research vessels have even permitted scientists to see and document sea floor hot springs ("smokers") actively depositing metal sulfides and supporting a new and unique biological community (Francheteau et al., 1979; Hekinian et al., 1980).

Summary

The style of a volcanic eruption depends upon the eruption temperature, the volatile content, and the composition of the magma. SiO_2 content is of particular importance because it

dramatically affects the viscosity. H_2O and alkalis reduce viscosity, although to a lesser extent. Volcanic eruptions may issue from a linear fissure or a more restricted central vent. Low-viscosity (typically mafic) eruptions typically produce lava flows and broad-shield volcanoes with gentle slopes or volcanic plateaus. Blocky aa and smoother ropy pahoehoe flow features are common. High-viscosity (typically silicic) eruptions are generally more explosive, resulting in composite volcanoes with steeper slopes and layers of lava flows and pyroclastic deposits. Hydromagmatic eruptions occur when rising hot magma encounters groundwater, creating explosive landforms such as maars, tuff cones, etc. Post-eruption magma chamber collapse under a volcano may result in a broad caldera. Pyroclastic *fall* deposits settle from a vertical explosive eruption and typically produce extensive ash deposits that blanket the topography. Pyroclastic *flow* deposits are left by dense ground-hugging clouds and are generally hotter and denser, concentrating in valley areas. Pyroclastic *surges* are a type of pyroclastic flow that travel rapidly with more gas and fewer particles. Pyroclastic flows may result from collapse of the denser portion of a vertical eruption column, a laterally directed blast, low-pressure fountaining, or dome collapse.

Tabular intrusive bodies are either sills (which are concordant to the surrounding sedimentary strata) or dikes (which discordantly cut across the grain of the surrounding country rocks). Both are formed when magma is injected along a planar fracture. Ring dykes are cylindrical, and cone sheets are conical curved sheet-like intrusive bodies that typically form over shallow magma chambers. The most common forms of non-tabular plutonic bodies are stocks (with exposed surface area < 100 km^2) and batholiths (> 100 km^2 exposed). The contact zone between hot magma and cooler wall/country rocks may exhibit a chill zone and wall rock xenoliths in the igneous portion and contact metamorphism and injection of small dikes on the wall rock side.

Larger plutons are usually emplaced at subduction zones, where they are associated with tectonism/orogeny (mountain building) and metamorphism, typically representing several events. Such plutons probably accumulate incrementally as smaller episodes over millions of years. Pre-tectonic plutons/episodes are emplaced prior to a deformational event, syn-tectonic plutons are emplaced during such deformation, and post-tectonic plutons are emplaced after deformation. Plutons may also be emplaced at various depths, where uplift and erosion eventually expose them at the Earth's surface. Plutons crystallizing at shallow levels of the *epizone* (< 8 km) typically exhibit sharp crosscutting contacts and obvious contact metamorphic and chill zone effects. Those crystallizing at the greater depths of the *catazone* (>10 km) typically occur in high-grade metamorphic rocks and are more concordant to country rock foliations and exhibit only minor contact effects. Plutons emplaced into the mesozone (5 to 15 km) show intermediate features. Many plutons, particularly the larger orogenic batholiths, comprise several phases of injection in which successive phases may represent progressively more chemically evolved liquids from a deeper chamber.

The *room problem* addresses the question of what happened to the often considerable volume of material displaced by intrusion of a plutonic body. This, in turn, poses the problem of pluton rise and emplacement mechanism. A mass of magma may rise as a diapir if the surrounding rock is sufficiently ductile. In more competent surroundings at shallow levels, the molten mass rises by stoping, lifting, or assimilation of the roof rock, mobilization and downward flow of wall rock, or ballooning as the melt expands radially. Our classical concept of batholiths as huge subspherical bodies has recently been questioned. Gravity and seismic surveys and theoretical considerations suggest that batholiths are actually relatively broad and thin, occupying much less volume than originally thought and greatly reducing the room problem. Hot batholiths emplaced at shallow levels are capable of driving convective overturn of hydrothermal groundwater systems through permeable and fractured roof rocks, resulting in geothermal systems and hydrothermal alteration and ore deposits.

Key Terms

Viscosity *55*
Vent *56*
Fissure *56*
Shield volcano *56*
Composite (strato-) volcano *56*
Scoria cone *57*
Maar *58*
Meteoric versus juvenile water *58*
Tuff cone *58*
Tuff ring *58*
Dome *58*
Caldera *59*
Aa *62*
Pahoehoe *61*

Columnar joints (colonnade, entablature) *62*
Pillow lava *63*
Lahar *63*
Pyroclastic fall *63*
Pyroclastic flow *65*
Pyroclastic surge *66*
Ignimbrite *65*
Dike *67*
Sill *67*
Batholith *71*
Stock *71*
Laccolith *71*
Lopolith *71*

Xenolith *72*
Chill zone *72*
Pre-, syn-, and post-tectonic plutons *73*
Epizone, mesozone, catazone *74*
Cupola *74*
Septum *74*
Roof pendant *74*
Diapir *77*
Room problem *77*
Assimilation *78*
Stoping *78*
Zone melting *78*
Ballooning *78*
Hydrothermal *79*

Review Questions and Problems

Review Questions and Problems are located on the author's web page at the following address: **http://www.prenhall.com/winter**

Important "First Principle" Concepts

- Silicate liquids near the crystallization temperature have a structure similar to that of the minerals that would form with further cooling. Therefore, high-silica liquids, from which quartz and feldspars would crystallize, are more polymerized (and hence more viscous) than more mafic magmas.
- All lavas contain dissolved volatiles upon eruption. The viscosity, more so than the volatile content, controls the violence of an eruption. Mafic lavas thus typically form lava flows, whereas rhyolitic lavas form ignimbrites.
- Batholiths are typically complex, involving multiple intrusive phases. Their buoyancy wanes at shallow depths, where the country rock is not so dense, so they are now believed to spread laterally, developing a shape more like a pancake than a balloon.

Suggested Further Readings

General Field Relationship of Igneous Rocks

Compton, R. R. (1985). *Geology in the Field*. Wiley. New York.

Lahee, F. H. (1961). *Field Geology*. McGraw-Hill. New York.

Thorpe, R. S., and G. C. Brown. (1985). *The Field Description of Igneous Rocks*. Halsted. New York.

Volcanism and Volcanic Landforms

Bullard, F. M. (1976). *Volcanoes of the Earth*. University of Queensland Press. St. Lucia, Australia.

Decker, R. W., and B. Decker. (1998). *Volcanoes*. W.H. Freeman. New York.

Decker, R.W., T. L. Wright, and P. H. Stauffer. (1987). *Volcanism in Hawaii*. USGS Prof. Paper **1350**.

Francis, P. (1993). *Volcanoes, a Planetary Perspective*. Oxford Clarendon Press. Oxford, UK.

Sigurdsson, H., B. Houghton, S. R. McNutty, H. Rymer, and S. Stix. (eds.). (2000). *Encyclopedia of Volcanoes*. Academic Press. San Diego.

Simkin, T., and R. S. Fiske. (1983). *Krakatau 1883: The Volcanic Eruption and Its Effects*. Smithsonian Institution. Washington, DC.

Williams, H., and A. R. McBirney. (1979). *Volcanology*. Freeman, Cooper. San Francisco.

Pyroclastics

Branney, M. J., and P. Kokelaar. (2002). *Pyroclastic Density Currents and the Sedimentation of Ignimbrites*. Memoir **27**. The Geological Society. London.

Cas, R. A. F., and J. V. Wright. (1988). *Volcanic Successions, Modern and Ancient*. Unwin Hyman. London.

Fisher, R. V., and H. -U. Schminke. (1984). *Pyroclastic Rocks*. Springer-Verlag. Berlin.

Freundt, A., C. J. N. Wilson, and S. N. Carey. (2002). Ignimbrites and block-and-ash flow deposits. In: *Encyclopedia of Volcanoes* (eds. H. Sigurdsson, B. Houghton, S. R. McNutty, H. Rymer, and S. Stix). Academic Press. San Diego, pp. 581–600.

Leyrit, H., and C. Montenat. (2000). *Volcaniclastic Rocks from Magmas to Sediments*. Gordon and Breach. Amsterdam.

Intrusions

Bergantz, G. W. (1991). Chemical and physical characterization of plutons. In: *Contact Metamorphism* (ed. D. M. Kerrick). *Reviews in Mineralogy*, **26**. Mineralogical Society of America.

Buddington, A. F. (1959). Granite emplacement with special reference to North America. *Geol. Soc. Am. Bull.*, **70**, 671–747.

Clarke, D. B. (1992). *Granitoid Rocks*. Chapman & Hall, London. Chapter 2: Field relations.

Crittenden, M. D., and P. J. Coney (1980). *Cordilleran Metamorphic Core Complexes*. Geol. Soc. Amer., Memoir, **153**.

Didier, J. (1973). *Granites and Their Enclaves*. Elsevier. New York.

Marsh, B. D. (1982). On the mechanisms of igneous diapirism, stoping, and zone melting. *Am. J. Sci.*, **282**, 808–855.

Newell, G., and N. Rast. (eds.). (1970). *Mechanisms of Igneous Intrusion*. Gallery Press. Liverpool, UK.

Paterson, S. R., R. H. Vernon, and T. K. Fowler. (1991). Aureole tectonics. In: *Contact Metamorphism* (ed. D. M. Kerrick). *Reviews in Mineralogy*, **26**. Mineralogical Society of America.

Paterson, S. R., and T. K. Fowler. (1993). Reexamining pluton emplacement mechanisms. *J. Struct. Geol.*, **15**, 191–206.

Hydrothermal Systems

Burnham, C. W. (1967). Hydrothermal fluids at the magmatic stage. In: *Geochemistry of Hydrothermal Ore Deposits* (ed. H. L. Barnes). Holt, Rinehart and Winston. New York. pp. 38–76.

Goff, F. and C. J. Janik (2002). Geothermal systems. In: *Encyclopedia of Volcanoes* (eds. H. Sigurdsson, B. Houghton, S. R. McNutty, H. Rymer, and S. Stix). Academic Press. San Diego, pp. 817–834.

Henley, R. W., and A. J. Ellis (1983). Geothermal systems ancient and modern: A geochemical review. *Earth Sci. Rev.*, **19**, 1–50.

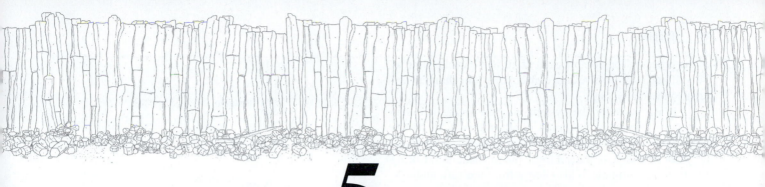

5

An Introduction to Thermodynamics

Questions to be Considered in this Chapter:

1. How can we determine the stability range of a mineral or of the mineral assemblages that constitute rocks so that we can judge whether a given rock (or some chemically equivalent alternative) is stable under a particular set of physical conditions?

2. How can we evaluate or predict the effects on a stable geologic system of changing some parameter, such as pressure or temperature, or of adding some chemical constituent?

3. How can we qualitatively evaluate phase diagrams and use them to understand the nature of the systems involved?

Petrologists use a knowledge of thermodynamics in two principal ways. First, thermodynamic principles can be applied *qualitatively*, to assess a geologic system or predict the effects that a change in pressure (P), temperature (T), or composition (X) may have on a stable rock/fluid/melt assemblage. Qualitatively, we might answer important questions such as, "What would be the general effect of increased pressure on half-molten rock?" or "What would happen if H_2O were added to a rock near its melting temperature?" Second, if some fundamental basic data can be experimentally determined, we can calculate *quantitatively* whether a certain assemblage of minerals (i.e., a rock), with or without a fluid phase or melt, is stable at some particular combination of P, T, and X. Both of these approaches are based on an understanding of equilibrium and energy.

Through years of experiments, we have compiled sufficient data for minerals and gases that we can quantitatively calculate the stability ranges of many mineral and mineral–fluid assemblages. Because the compositions of melts are so complex and variable compared to those of minerals and gases, we are only in the early stages of being able to quantitatively treat igneous systems. I will thus postpone developing the quantitative approach until we deal with metamorphism and have occasion to apply the results. For the present, I will develop the general thermodynamic basics and then concentrate on the qualitative approach, which will help us assess a variety of igneous phenomena.

5.1 ENERGY

A **system** is some portion of the universe that one might isolate (either physically or mentally) in order to study it. The **surroundings** are the adjacent portions of the universe outside the system in question. All natural systems are governed by energy. Any macroscopic change in a system is accompanied by the conversion of energy from one form to another. For instance, the dropping of an object, such as a rock, involves the conversion of **potential energy** (associated with its height) to **kinetic energy** (motion). Lifting the rock involves the transfer of **chemical energy** (stored in your body) to kinetic energy (the motion of your muscles, and eventually the rock). Likewise, the mixing of a strong acid and base is the conversion of stored chemical energy to **thermal energy** as they react to and neutralize each other, heating up the beaker in the process. The compression of air in a bicycle pump as you inflate a tire involves the conversion of **mechanical energy** to thermal energy.

If we consider the system plus the immediate surroundings, energy is conserved in all processes. If, on the other hand, we consider only the system, such as a rock, energy can be lost to, or gained from, the surroundings. Lifting the rock, for example, adds energy from the surroundings, thereby increasing the potential energy of the rock system. Of course, if we remove support from the rock, it will fall to Earth, spontaneously losing potential energy as it is converted to kinetic energy, which in turn is converted to heat (friction) and mechanical energy as it hits the Earth and deforms it a bit. The original potential energy is a useful commodity. It is capable of doing work, if we attach it to a pulley and let a string turn a generator. One could describe a mass such as a rock at a high elevation as having a higher energy content than a similar mass at a lower elevation. It should be obvious that an unsupported mass will spontaneously fall to the lowest point that it can. An unsupported rock falls until it hits the Earth, not stopping halfway down. This leads us to an important and fundamental property of natural systems: *Systems naturally tend toward configurations of minimum energy.* Such minimum energy configurations, such as the rock on the ground, are referred to as **stable**. A rock hurled aloft is in an **unstable** configuration (or *state*) because it will quickly fall to Earth under the influence of gravity.

Fortunately, not all natural systems change spontaneously to the minimum energy state. Some systems may exist in a state that is low in energy but not the lowest possible. There may be some energy barrier that must be overcome before the true minimum energy state can be attained. Returning to our falling rock analogy, suppose that the rock hits a sloping area of the ground and stops in a depression (Figure 5.1). Clearly, this is not the most stable configuration possible, as there are lower elevations nearby to which it would roll if it could. However, this elevation is lower than all immediately adjacent possibilities. It would have to roll upward briefly, thereby *increasing* its potential energy momentarily, before it could roll to the lowest elevation available. There thus exists an energy barrier that prevents the rock from easily reaching the lowest energy state. Such states, neither stable (lowest energy possible) nor unstable (capable of spontaneous change), are called **metastable** (Figure 5.1). We can also think of systems as being either at **equilibrium** or in a state of **disequilibrium**. Any system that is not undergoing some form of transition is said to be at equilibrium. It can be either stable equilibrium (such as the rock at the bottom of the slope) or metastable equilibrium (such as the perched rock).

The energy barriers that maintain metastable states may be potential energy, as in the rock in Figure 5.1, or kinetic in nature. Kinetic barriers keep many familiar materials from reverting to the most stable state. When your car is scratched, the iron doesn't immediately become oxidized to rust. Diamonds do not invert to graphite, and rocks at the Earth's surface do not instantly weather to clays. This is because the energy barriers involved in breaking bonds for the

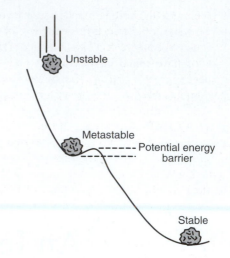

FIGURE 5.1   Stability states.

reactions to take place are too high for the low kinetic (vibrational) energies of the bonds at the low temperatures near the Earth's surface. It is indeed fortunate that metastable states exist, or most of the materials and energy sources that we use today would not be available.

5.2 GIBBS FREE ENERGY

Addressing systems in terms of energy is a useful approach. The crucial point is to find the proper energy expressions and the variables that control them (**variables of state**). **Mechanics** is the study of motion, such as projectiles or our falling rock. By dealing with potential, kinetic, and gravitational energy, as well as with mass, velocity, and momentum, mechanics has permitted us to understand the motion of projectiles and planets sufficiently well to place a person on the moon and predict the collision of a comet with Jupiter. The field of **thermodynamics**, as the name implies, deals with the energy of heat and work. Thermodynamics can help us understand a number of things, from steam engines to refrigeration. The work of J. Willard Gibbs elegantly related thermodynamic concepts to the understanding of chemical systems. Gibbs formulated an energy parameter, **Gibbs free energy**, that acts as a measure of the energy content of chemical systems. The Gibbs free energy at a specified pressure and temperature can be defined mathematically as:

$$G = H - TS \qquad (5.1)$$

where: G = Gibbs free energy

H = **enthalpy** (or heat content)

T = temperature in kelvins

S = **entropy** (most easily perceived as randomness)

As a simple example of enthalpy, consider heating water on a stove. The enthalpy of the water increases as it is warmed because you are adding heat to it. When it boils, the steam has a higher enthalpy than the water, even when both are at the same temperature (the boiling point), because you had to add heat to the water to convert it to steam. As for entropy, imagine the ordered structure of a crystal lattice. This is low entropy. Liquids have much less ordered arrangement of atoms and, hence, greater entropy. Gases have even more entropy than liquids because the atoms or molecules are much more widely distributed.

Using the Gibbs free energy parameter, we can assess the stability of chemical systems, just as we used potential energy for the rock in Figure 5.1. In a manner analogous to the rock situation, under a particular set of conditions, a chemical system should proceed to a state that minimizes the appropriate energy parameter. In other words:

> *Stable forms of a chemical system are those with the minimum possible Gibbs free energy for the given conditions.*

The Gibbs energy parameter is ideal for petrologists and geochemists because it varies as a function of pressure, temperature, and composition, the most important determining variables (variables of state) in nature. We will use Gibbs free energy to analyze the behavior of a variety of igneous and metamorphic systems. We begin with some relatively basic concepts, applying them to very simple chemical systems. We can then add new thermodynamic principles and techniques later (principally in Chapter 27) as we require them to treat more complex systems.

A variety of petrologic processes, from polymorphic transformations, to metamorphism, to crystallization and melting, can be expressed as chemical *reactions* of the type $A + B + \ldots = Q + R + \ldots$, where each letter represents a chemical species, such as a mineral, liquid, gas, or ionic species in solution. The species on the left side of the reaction are called the **reactants**, and those on the right are called the **products**. Of course, the reaction must balance stoichiometrically, in that the number of atoms of each element must be the same on both sides of the reaction. Because the stable form of a system at any value of pressure (P), temperature (T), and composition (X) is the form with the lowest Gibbs free energy, our object is to determine the Gibbs free energy for the alternative forms (reactants vs. products) of the system at the P, T, and X of interest. For a reaction, this comparison can be accomplished for *any* variable (in this case G) by defining:

$$\Delta G = \Sigma (n_{products}G_{products} - n_{reactants}G_{reactants}) \quad (5.2)$$

where: Σ = sum

n = stoichiometric coefficient for each phase in the reaction

For example, consider the hypothetical reaction:

$$3A + 2B = 2C + D$$

ΔG can be expressed as:

$$\Delta G = G_D + 2G_C - 3G_A - 2G_B$$

If ΔG is negative, then the products have a lower total free energy than the reactants (meaning that they are more stable), and the reaction should run from left to right, as written. Thus, if we knew G for each **phase** (mineral, gas, liquid) and ionic species (if included) in a reaction at some pressure and temperature, we could compare the values, summed for the reactants and for the products, to evaluate which side of the reaction is stable at the conditions of interest. Similarly, we could use the Gibbs free energies of the constituents to determine the *P-T-X* conditions under which the reactants and products are equally stable: the **equilibrium** condition for the reaction. In order to do all of this, we must first be able to determine a value for the Gibbs free energy of any phase (or ionic species) at any temperature, pressure, and composition of interest. It is then simply a matter of combining them using Equation (5.2) to do the rest.

5.3 THE GIBBS FREE ENERGY FOR A PHASE

Let's deal for now with a simple system composed of a substance of *fixed composition* (putting off the rather complex relationships between free energy and composition until later). For a *single phase*, such as a mineral or a liquid, we would like to be able to determine the Gibbs free energy at any specific pressure and temperature. Although we can measure volume, temperature, pressure, and various other properties for a phase, it is impossible to measure an absolute value of chemical free energy of any phase, compound, ion, etc. We can, however, determine *changes* in the free energy of phases as the variables of state change. If we can measure the changes, we need only choose some arbitrary state (a "reference state") of a phase and assign any value of G that we choose. We can then use the changes in G as conditions vary to assign values of G for any other state.

The most common reference state is to consider pure elements in their natural (stable) form at 25°C (298.15 K) and atmospheric pressure (0.1 MPa), the conditions in a typically overheated laboratory, and assign a Gibbs free energy of zero joules (0 J) to that state. The reference Gibbs free energy for oxygen, for example, is 0 J for pure O_2 gas, and for silicon it is 0 J for pure Si metal, both at 298.15 K and 0.1 MPa. Note that G is an **extensive variable of state**, in that it is dependent upon the quantity of material in the system (the *extent* of the system). We can avoid this problem by expressing G in terms of **molar** free energy, or the number of joules per mole of the substance. The molar Gibbs free energy of Si, then, is an intrinsic constant property of the element. For a compound such as quartz, we can measure the heat (enthalpy) *change* (ΔH) associated with the reaction of 1 mole Si + 1 mole O_2 to 1 mole SiO_2 (by a technique called **calorimetry**). We can also calculate the entropy of quartz based on the assumption that the entropy of any substance is zero at 0 K (based on the third law of thermodynamics), and calculate the change in entropy between 0 K and 298.15 K (as discussed in the following section). From these values of H and S, we can compute the Gibbs free energy for low quartz using Equation (5.1). The result is known as the **molar Gibbs free energy of formation** (from the elements) and given the symbol $\Delta \overline{G_f^\circ}$, where the subscript stands for *formation* (from the elements), the superscript refers to the **reference state** of 298.15 K and 0.1 MPa, and the bar above the G indicates that it is a molar quantity.

For quartz, $\Delta \overline{G_f^\circ}$ is −856.3 kJ/mol (Robie and Hemingway, 1995). This represents ΔG for the reaction Si (metal) + O_2 (gas) = SiO_2 (quartz). Note that the large negative free energy value tells us that the product (quartz) is much more stable than the reactants (Si and O_2), which is why we find quartz, and not silicon metal, in our oxygen-rich environment. The free energies for other compounds, including minerals, gases, ions, etc., are determined in a similar fashion. $\Delta \overline{G_f^\circ}$ is the value used for the Gibbs free energy (G) of various phases at 298.15 K and 0.1 MPa, and it is the basis for most thermodynamic calculations. Thermodynamic data are tabulated and available from a number of sources, including published compilations, such as Robie and Hemingway (1995), or computer databases (commonly as part of a program that performs the calculations as well). Minor variations in the values reported reflect experimental inaccuracies in the calorimetry or even the technique used. Lately we have devised ways to extract thermodynamic data from high-temperature and high-pressure experiments at equilibrium.

5.3.1 Variations in the Gibbs Free Energy for a Phase with Pressure and Temperature

Once we have the reference state data for geological phases of interest, we can determine the value of the Gibbs free energy (G) of a phase at elevated temperatures and pressures. We can do this by using the following differential equation:

$$dG = VdP - SdT \qquad (5.3)$$

where: G = Gibbs free energy of the phase
V = volume
S = entropy

The equation also holds true for molar properties ($\overline{G}$, $\overline{V}$, and $\overline{S}$). I will henceforth treat all these parameters as molar and dispense with the bar symbols. Equation (5.3) thus formalizes the variation in G with P and T. G also changes with composition, and parameters for this variation can be added to Equation (5.3), but I have decided to hold composition constant for the time being to ease our initiation process. As changes in temperature (dT) and/or pressure (dP) occur, G will also change (dG) in a determinable way. Thus G for a phase, such as forsterite, will be different at different temperatures and pressures. We solve for G at different pressures and temperatures by integrating Equation (5.4):

$$G_{P_2, T_2} - G_{P_1, T_1} = \int_{P_1}^{P_2} VdP - \int_{T_1}^{T_2} SdT \qquad (5.4)$$

where: P_2 = pressure of interest
T_2 = temperature of interest
P_1 = initial pressure (e.g., 0.1 MPa)
T_1 = initial temperature (e.g., 298 K)

To perform the integration accurately, we need to know how V varies with P and how S varies with T. The variation in V with respect to P (called the **isothermal compressibility**) is sufficiently small for solids that V can be treated as a constant for a fairly large range of pressures, but the volume of liquids and particularly gases will certainly change with pressure, so, for them, calculations that assume a constant V will be in error. S varies appreciably with T for most phases, whether solid, liquid, or gas. The relationship can be expressed as $dS = (C_p/T)dT$, where C_p is the **heat capacity** (the amount of heat required to raise 1 mole of the substance 1°C). Substances with a high heat capacity can absorb considerable heat with only a small temperature change. Consider the amount of heat you must add to a pot of water (high heat capacity) to raise the temperature 50° versus the much smaller quantity of heat required to do so for the same volume of air (low heat capacity). The process is complicated in that the heat capacity is itself a function of T and varies in a nonlinear fashion. Polynomial equations, usually of the form $C_p = a + bT - c/T^2$, are empirically determined for each phase and are reported in most sources (including Robie and Hemingway, 1995), allowing us to calculate S accurately at any temperature.

If the variation in pressure and temperature (dP and dT) are small, we can assume that V and S are constant as a first approximation, and Equation (5.4) reduces to a simple algebraic form. We will make this assumption (for solids at least) in the problems and exercises in this text and avoid calculating the integrals that include compressibility and polynomial heat capacity functions. Fortunately, some computer programs do the integration for us, so we can still derive more accurate results if we need to. Some programs have V, S, H, and C_p data for many common minerals, liquids, and gases so that the calculations, including corrections for compressibilities and changes in S and H with T are performed instantly. We can understand enough of the theory and application by using the algebraic form and let the computer perform the more complex mathematics. In the following sample problem, we will get some experience with handling the algebraic form of Equation (5.3) and compare our results for quartz to the integrated form in order to get some idea of the magnitude of the errors that our simplification introduces.

SAMPLE PROBLEM: Calculating the Gibbs Free Energy of Quartz at Elevated *P* and *T*

As an example, we shall use Equation (5.3) to calculate G for quartz at 500°C and 500 MPa. Because Equation (5.3) deals with *changes* in G with P and T, we must first know the value of G_{quartz} at some initial temperature and pressure. We will use the common reference state of 298.15 K and 1 atmosphere pressure, at which G_{quartz} is −856.3 kJ (± ~1 kJ: the analytical error) per mol (from Robie and Hemingway, 1995). T in Celsius (C) equals T in kelvins (K) minus 273.15, but we commonly round to the nearest degree, which is precise enough for geological work. Thus 298 K is

25°C, or "room temperature." To calculate G_{quartz} at higher pressures, say 0.5 GPa, we also need to know the molar volume of quartz, which is 22.69 cm³/mol. Because G is expressed in joules (m²kg/sec²mol) and P in pascals (kg/sec²m), V must be expressed in m³/mol if the units are to remain consistent. However, $cm^3 = 10^{-6}$ m³ and $MPa = 10^6$ Pa, so we can avoid the very small and very large numbers by using joules for G, cm³/mol for V, and MPa for P, and the exponents will cancel when we multiply $V \cdot P$. The choice is up to you, but always remember to be careful with the units in your calculations.

We begin with the pressure correction, calculating G_{quartz} at 0.5 GPa and 298.15 K. Because T is constant, $dT = 0$, and Equation (5.3) reduces to $dG = VdP$. We can integrate this to get:

$$G_{P_2} - G_{P_1} = \int_{P_1}^{P_2} Vdp \qquad (\text{constant } T) \qquad (5.5)$$

If V is constant as pressure changes, V can be removed from the integral, and Equation (5.5) becomes algebraic:

$$G_{P_2} - G_{P_1} = V\int_{P_1}^{P_2} dP = V(P_2 - P_1) \qquad (5.6)$$

where $P_2 = 500$ MPa and $P_1 = 0.1$ MPa at constant T:

$G_{0.1}$ is $-856,300$ J, so $G_{500,298} = G_{0.1,298} + V(500 - 0.1)$
$= -856,300 + 22.69(499.9) = -844,957$ J
(or -845.0 kJ).

If we wanted to correct for temperature first, we could use a similar process to calculate G at 500°C (773 K) and 0.1 MPa. Because P is constant, Equation (5.3) becomes:

$$G_{T_2} - G_{T_1} = \int_{T_1}^{T_2} - SdT \qquad (\text{constant } P) \qquad (5.7)$$

and, if we assume that S is also constant over this temperature range, this reduces to:

$$G_{T_2} - G_{T_1} = -S(T_2 - T_1) \qquad (5.8)$$

$S_{0.1,298} = 41.46$ J/K mol (Robie and Hemingway, 1995), yielding $G_{0.1,773} = G_{0.1,298} - 41.46(773 - 298) = -876.0$ kJ.

Finally we can perform both operations in sequence (either one first) to get G at any temperature and pressure. If we use Equation (5.6) to get $G_{500,298}$ and then use Equation (5.8) at a constant pressure of 500 MPa, we get $G_{500,773} = -844,957 - 41.46(773 - 298) = -864.6$ kJ.

Now we can compare our results with the computer method that integrates the equations to see how our assumptions of constant V and S worked. Table 5.1 lists the results for low quartz generated by the computer program SUPCRT (Helgeson et al., 1978).

The first thing to notice is that SUPCRT uses its own database, so even the reference state value of G_{quartz} differs

TABLE 5.1 Thermodynamic Data for Low-Quartz. Calculated Using Equation (5.1) and SUPCRT

Low Quartz		Eq (5.6) (5.8)	SUPCRT		
P(MPa)	T(C)	G(J)	G(J)	V(cm³)	S(J/K)
0.1	25	−856,288	−856,648	22.69	41.36
500	25	−844,946	−845,362	22.44	40.73
0.1	500	−875,982	−890,601	23.26	96.99
500	500	−864,640	−879,014	23.07	96.36

from the Robie and Hemingway (1995) value (although by less than 400 J, which is less than 0.04% of G, or half the reported analytical error). Next, we can see that our constant V assumption is not perfect (column 5), but apparently it is good enough because our calculated G at 500 MPa and 25°C agrees with the integrated SUPCRT value (again within about 400 J). In fact, if we had used the same standard-state value for G in Equation (5.3) that SUPCRT used, the agreement would have been excellent, differing by about 50 J. So, for solid phases at least, the assumption that volume remains constant is fine for pressure changes less than 1 GPa (the thickness of the crust). This will be true for most solids, as long as no pressure-induced phase transitions take place. As we will see in Chapter 6, low quartz inverts to coesite at about 2 GPa, with a significant volume change.

Our assumption that S remains constant as temperature changes is not as good. The difference between the free energy extrapolated linearly (our hand calculation at constant S) and as a curve (integrated by SUPCRT) over 475°C temperature change is 14,619 J. In column 6 of Table 5.1, we see that the entropy has more than doubled over this range. Nonetheless, the relative error in G is only about 2%. If we calculate G at 800°C, the difference increases to 35,247 J, or 4%.

In summary, we can use Equation (5.3) to calculate the change in Gibbs free energy for a mineral, liquid, or gas phase with changes in temperature and pressure. If we can determine G for a phase at some initial P–T state (tabulated in several sources), we can then determine G at other pressures and/or temperatures. Pressure calculations are relatively easy for solid phases because we can assume that V is a constant and use a simple algebraic form [Equation (5.6)] without sacrificing much accuracy. Temperature corrections using the algebraic form [Equation (5.8)] are less precise, and the integration may be necessary if dT is large.

The result of these calculations is just a number. What does the number mean? Remember, G is a measure of the stability of a phase or a system. If there are two (or more) alternative forms that a system may have (e.g., SiO_2 may occur as low quartz, high quartz, tridymite, cristobalite, a melt, etc.), the form with the lowest G, at a given pressure and temperature, will be the stable form. We could determine this by calculating G for each form at any P and T in question and then compare them. Alternatively, we might treat the difference between the free energies of two competing forms directly, obviating the need to check every conceivable alternative form. The transition from one form to another is a *reaction*, so we would then be dealing with the changes in G involved during such a reaction.

5.4 GIBBS FREE ENERGY FOR A REACTION

Now that we can calculate the Gibbs free energy of a phase at any temperature and pressure, we proceed to considering a reaction. Because we are embarking on igneous petrology, I have chosen a simple melting reaction. As an example, the schematic **phase diagram** in Figure 5.2 shows the limits of stability of a solid mineral phase (S) and its corresponding liquid melt (L) in terms of the variables P and T. The reaction in question can be written $S = L$. I use the equality sign instead of $S \rightarrow L$ to imply that the two sides of the equation are chemically equivalent, and the reaction could run either way, depending on the conditions. Only when the direction in which a reaction progresses is important will I use an arrow.

We must now compare and consider the free energies of both phases involved in the reaction. If we follow the logic of the boxed statement above (that all natural systems tend toward the lowest energy state), it should be clear that the solid must be more stable than the liquid of equivalent composition anywhere in the field labeled "Solid" in Figure 5.2, whereas the liquid is the more stable phase anywhere in the field labeled "Liquid." In other words, the value of G for the solid must be lower than that of the liquid at any point (such as point A) in the solid field. Likewise, the free energy of the liquid must be lower than that of the solid at any point (such as B) in the liquid field. We can summarize the stability concept for a reaction with the statement:

The side of a reaction equation with lowest G under a given set of conditions is the most stable.

5.4.1 Variation in the Gibbs Free Energy for a Reaction with Pressure and Temperature

In the case for the reaction $S = L$, we have two phases, each with the same composition, and each of which has a Gibbs free energy as defined by Equations (5.1) to (5.4). Will our system be in the liquid or the solid state, and how will the answer vary with temperature and pressure? This may be best illustrated if we isolate a single variable in Equation (5.3), so let's address only temperature. For any phase at a constant pressure, we can take the partial differential of Equation (5.3) with respect to temperature for a phase and get:

$$\left(\frac{\partial G}{\partial T}\right)_P = -S \tag{5.9}$$

where the subscript P denotes constant pressure. Equation (5.9) expresses the manner in which G for a given phase varies with changes in T (at constant P). In other words, $-S$ is the *slope* of a line representing the value of G on a graph of G versus T. Entropy can be considered a measure of randomness, and there is no such thing as negative randomness. According to the third law of thermodynamics, entropy drops to zero at 0 K (complete order, not even vibration), and it cannot go lower. Because S must be positive so far above 0 K, the slope of G versus T for any phase must be negative according to Equation (5.9), as shown in Figure 5.3. The slope is steeper for the liquid than for the solid because $S_L > S_S$ (because liquids have a more random atomic structure than crystalline solids). Points A and B in Figure 5.3 represent the endpoints of the **isobaric** (constant pressure) temperature increase shown in Figure 5.2. At point A, the solid phase has a lower G than the liquid and is thus more stable. As T is increased, G of both phases decreases (as in Table 5.1), but G of the liquid decreases more rapidly than G of the solid. By the time we reach point B, the liquid has a lower G and is thus stable. Because S varies (increasing with T), the slopes of the curves in Figure 5.3 are not linear: they increase somewhat with increasing T.

An identical approach can be taken with an **isothermal** (constant temperature) traverse from low P to high P. If we take the partial differential of Equation (5.3) for a phase with respect to pressure at constant temperature, we get:

$$\left(\frac{\partial G}{\partial P}\right)_T = V \tag{5.10}$$

Because V is positive, so is the G versus P slope, and V_{liquid} is greater than V_{solid}, and the slope of G versus P for

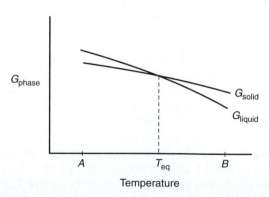

FIGURE 5.3 Relationship between Gibbs free energy and temperature for the solid and liquid forms of a substance at constant pressure. T_{eq} is the equilibrium temperature.

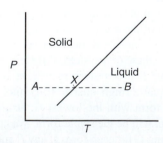

FIGURE 5.2 Schematic *P-T* phase diagram of a melting reaction.

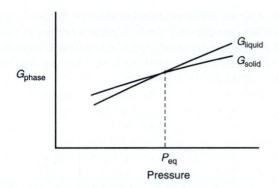

FIGURE 5.4 Relationship between Gibbs free energy and pressure for the solid and liquid forms of a substance at constant temperature. P_{eq} is the equilibrium pressure.

the liquid is greater than that for the solid (Figure 5.4). The liquid is more stable than the solid at low pressure, but it is less stable at high pressure, in agreement with Figure 5.2.

Next we apply Equation (5.3) directly to our reaction $S = L$. *It is important to clearly write the reaction before we can deal with it quantitatively.* It is customary to place the high entropy side of a reaction on the right. Applied to a reaction, Equation (5.3) becomes:

$$d\Delta G = \Delta V dP - \Delta S dT \qquad (5.11)$$

Δ and d both denote changes, but they denote different types of changes. d represents a finite change in the variables of state, such as pressure and temperature, whereas Δ represents the change in some variable as a result of the reaction [as defined by Equation (5.2)]. For our simple reaction, $S = L$, therefore, $\Delta V = V_L - V_S$. In other words, ΔV is the change in volume that occurs when the reaction progresses—in this case, when the solid melts. The solid form of most materials is more dense than the corresponding liquid at the same temperature because the ordered crystal lattice is more compact than the disordered liquid. Imagine the difference between a neatly stacked array of bricks and a random pile of the same bricks. The random pile, including the air spaces between the bricks, would occupy a larger total volume than the stacked pile. A notable exception to this general rule is the case for water and ice. The polar nature of the H_2O molecule permits closer packing of the molecules in the liquid than in the hexagonal ice lattice. Assuming that we are not melting water in our reaction, but are melting some mineral, ΔV must be positive for the reaction $S = L$ as written. In a similar fashion, we can deduce that because $S_S < S_L$, ΔS is also positive.

ΔG, the difference in G between the reactants and the products, is not as easy to predict as are ΔV and ΔS. That is why we have Equation (5.2), so that we can calculate the change in G from the more easily understood variables. Remember that dG represents the change in G for a phase as T and/or P change (dT and/or dP). $d\Delta G$ is therefore the change in ΔG for the reaction as T and/or P change. Be-

cause G for each phase varies differently with T and P, the value of ΔG for a reaction must vary with T and P as well. Compare Figures 5.2 and 5.3, and you can determine that ΔG, as defined for the reaction ($G_{liquid} - G_{solid}$), must be positive at point A in the solid field and negative at point B in the liquid field. ΔG, in this case, thus has to decrease with increasing T.

5.4.2 The Equilibrium State

At point x in Figure 5.2, or at any point on the curve that separates the solid and liquid fields, both phases are equally stable, so that they *coexist at equilibrium*. The curve is thus called the (stable) **equilibrium curve**. Because both phases are equally stable at equilibrium, they must have the same value of G. Hence $\Delta G = 0$, *which must be true anywhere along the equilibrium curve*. This is another fundamental axiom:

ΔG for a reaction at equilibrium = 0

The concept of equilibrium, as it applies to our system of equally stable reactants and products, is extremely important. Such systems might appear to be static, but they certainly are not. Equilibrium represents a *dynamic* state of flux, but the fluxes in this case cancel one another. For our coexisting liquid and solid, the reactions $S \rightarrow L$ and $L \rightarrow S$ both happen continuously, but the rates at which they proceed are equal. The amount of each phase will thus remain the same over time if the conditions are unchanged. The dynamic nature of this equilibrium state can be observed macroscopically because the shapes of the crystals will slowly change as they exchange atoms with the liquid.

Le Châtelier's Principle tells us how such a state of dynamic equilibrium will react to changes imposed upon it. *A system will react in a fashion that acts so as to mitigate the changes.* For instance, if we heat the system, the $S \rightarrow L$ process will take place at a faster rate than the $L \rightarrow S$ one, so that there will gradually be more liquid and less solid (some of the crystals appear to melt). This process absorbs the added heat and maintains the system at the same temperature, thereby mitigating the change imposed by adding heat. If our system is in a compressible container, and we press on a piston to increase the pressure, the $L \rightarrow S$ process will outpace the $S \rightarrow L$ one, thereby reducing the volume in an attempt to offset the change. Le Châtelier's Principle can be stated as follows:

Le Châtelier's Principle

If a change is imposed on a system at equilibrium, the position of the equilibrium will shift in a direction that tends to reduce that change.

This is an eclectic principle and has applications beyond simple chemical systems (e.g., marriages, freeway traffic, even first dates).

5.4.3 Thermodynamic Evaluation of Phase Diagrams

Although thermodynamic data are sparse for melts, we can still use Equation (5.3) to *qualitatively* evaluate practically any phase diagram such as Figure 5.2. Doing so will help us understand the results of any calculations that we may perform, and greatly improve our ability to interpret phase diagrams we may create and those we encounter in the rest of the text.

There are five interrelated variables in Equation (5.3): G, S, V, P, and T. We can use these variables to understand various aspects of phase diagrams, such as the slope of an equilibrium curve or why the solid is on the high pressure–low temperature side. The approach is simple and qualitative but provides us with a powerful tool to assess reactions and equilibria.

For example, as pressure increases, volume naturally decreases, as you could imagine with a piston-and-cylinder apparatus, such as Figure 5.5. Applying pressure on the piston will force it downward, and the pressure in the cylinder below the piston will increase. Similarly, if there are two (or more) possible configurations or states for a given chemical system, the one with the lowest volume will be favored at higher pressure because the lower volume is a preferred response to the pressure increase. This is an application of Le Châtelier's Principle. Consider the transition of quartz to the polymorph coesite and then to stishovite as pressure increases. These transitions are crystallographic changes by which SiO_2 responds to increased pressure. The structure of a polymorph will compress only so far before inverting to different polymorphs with successively more compact structures. Lower-volume phases are thus favored at higher P. Following a vertical line in Figure 5.2, representing increasing P at constant T (isothermal), the system passes from a liquid to a solid, confirming that $V_S < V_L$ (as we previously established).

In a similar fashion, increasing T at a constant P (isobaric heating) should create more kinetic and vibrational motion in atoms and thus greater entropy. Note that a horizontal line in Figure 5.2, representing such an isobaric T increase, moves us from the solid to the liquid field above,

confirming our previous assertion that $S_S < S_L$. Because both decreasing pressure and increasing temperature can cause a solid to melt, the curve separating the solid and liquid fields must have a *positive* slope. This is true for any material of geological interest (except H_2O).

Equation (5.11) can be applied to Figure 5.2 in the same way. In a manner similar to the G versus T treatment for Equation 5.3, we can take the partial differential of ΔG with respect to T in Equation (5.11) and get:

$$\left(\frac{\partial \Delta G}{\partial T}\right)_P = -\Delta S \qquad (5.12)$$

which is negative for the reaction $S = L$ because we have already determined that ΔS ($S_L - S_S$) is positive ($S_S < S_L$). ΔG therefore decreases with increased T, so increasing T from equilibrium (where $\Delta G = 0$) results in negative ΔG, meaning that the reaction product (liquid) has a lower G than the reactant (solid), and the reaction ($S = L$) runs toward the right (liquid). This is true, as you can see from Figure 5.2. A path of increasing T moves from the solid into the liquid field. Likewise:

$$\left(\frac{\partial \Delta G}{\partial P}\right)_T = \Delta V \qquad (5.13)$$

which is positive (liquid has a higher V than solid), and increasing P from equilibrium makes ΔG positive (the reactant has a lower G) and drives the reaction toward the reactant (solid).

Finally, let's choose any two points on the equilibrium curve in Figure 5.2. ΔG at both points must be zero. Thus:

$$d\Delta G = 0 = \Delta V dP - \Delta S dT \qquad (5.14)$$

and thus $\Delta V dP = \Delta S dT$, so:

$$\frac{dP}{dT} = \frac{\Delta S}{\Delta V} \qquad (5.15)$$

Equation (5.15) is called the **Clapeyron equation**, which gives the slope of the equilibrium curve that separates the L and S fields in a pressure versus temperature phase diagram, such as Figure 5.2. Although in our example we do not know the exact values of ΔV and ΔS, we do know their signs. In this case, because both ΔS and ΔV have the same sign, the slope is positive, agreeing with our previous conclusion.

Thermodynamic data for melts are only beginning to become available (see the references at the end of this chapter) and are less reliable than for minerals. We will thus settle for a qualitative to semi-quantitative approach to igneous thermodynamics, use this technique to estimate the slopes of melting reactions in phase diagrams, and assess the relative stabilities of reactants and products. In Chapter 27, we will proceed from this point to develop quantitative methods to

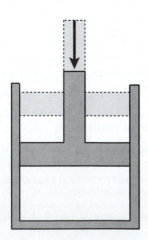

FIGURE 5.5 Piston-and-cylinder apparatus to compress a gas.

calculate the *P-T* and *P-T-X* conditions for metamorphic reactions based upon thermodynamic data for the mineral and gas phases involved in common metamorphic reactions. This approach is routinely performed on metamorphic minerals and fluids because relatively accurate data are available for these phases.

Summary

Thermodynamics deals with the stability of various chemical and mechanical systems and how they respond to their surroundings. All systems tend toward states with minimum energy. Gibbs free energy (*G*) is a measure of energy content of chemical systems and, hence, a measure of the relative stability of minerals, fluids, melts, etc. $G = H - TS$, where H = enthalpy (heat content), T = temperature (in *K*) and S = entropy (most easily thought of as randomness). Because we can only measure *changes* in thermodynamic properties such as G, elements at some reference (standard) state (typically 298 K and 0.1 MPa) are arbitrarily assigned zero values of G. We can then determine the molar Gibbs free energy of formation ($\Delta \overline{G_f^\circ}$) for any phase (i.e., mineral, fluid, melt) from this reference by addressing the reaction by which the phase is created from the stable elemental constituents. Quartz, for example, may be created by the reaction Si (metal) + O_2 (gas) = SiO_2 (quartz). The change in any variable resulting from a reaction (in this case, the change in G) is $\Delta G = \Sigma (n_{products} G_{products} - n_{reactants} G_{reactants})$ so that $\Delta \overline{G_f^\circ} = G_{Quartz} - G_{O_2(gas)} - G_{Si(metal)}$ (the latter two values = 0), which in this case = -856.3 kJ/mol, determined from measured enthalpies as the reaction proceeds in the lab and from theoretical (third law of thermodynamics) entropies. The negative value of ΔG indicates that the product of the reaction (quartz) is more stable (lower G) than the reactants, which is why quartz is so common and metallic silicon does not occur in nature where O_2 is present. The value of G for any phase (actually $\Delta \overline{G_f^\circ}$, but we shall accept the shorthand), once determined, can be extended to any pressure and temperature by using Equation (5.4):

$$G_{P_2,T_2} - G_{P_1,T_1} = \int_{P_1}^{P_2} V dP - \int_{T_1}^{T_2} S dT$$

Determining the change in Gibbs free energy of *reactions* between phase assemblages is a very powerful extension of the free energy concept, allowing us to determine the relative stability of a reactant phase or phase assemblage and the compositionally equivalent products under a particular set of physical conditions. This permits us to know the pressure–temperature stability range of a particular mineral assemblage and the *P-T* conditions of the equilibrium state under which the reactant and product assemblages coexist (the *equilibrium curve* separating the stability field of the reactants from that of the products).

Le Châtelier's Principle is a powerful concept that allows us to qualitatively assess the effect of changing some variable of state on a system at equilibrium. Because equilibrium is a dynamic, rather than static, state, the rates of the forward and reverse reaction will vary in such a way as to offset or absorb any change imposed. Simple thermodynamic considerations can also be used to qualitatively assess phase diagrams, allowing us to deduce some information about the phases involved with a particular equilibrium curve or even to detect errors in diagrams. For example, the phase assemblage (reactants or products) on the high-pressure side of an equilibrium curve must occupy less volume than the assemblage on the low-pressure side, and the phase assemblage on the high-temperature side of an equilibrium curve must have higher entropy and enthalpy than the assemblage on the low-temperature side. The slope of the equilibrium curve can also be determined using the Clapeyron equation [Equation (5.15)].

Key Terms

System *83*
Surroundings *83*
Phase *85*
Stable/unstable/metastable *84*
Equilibrium *84*
Thermodynamics *84*
Gibbs free energy *84*

Enthalpy *84*
Entropy *84*
Reactants/products *85*
Reference state *86*
Extensive variable of state *85*
Heat capacity *86*
Gibbs free energy of reaction *88*

Phase diagram *88*
Equilibrium curve for a reaction *89*
Le Châtelier's Principle *89*
Clapeyron equation *90*

Review Questions and Problems

Review Questions and Problems are located on the author's web page at the following address: **http://www.prenhall.com/winter**

Important "First Principle" Concepts

- All natural systems tend toward states of minimum energy. Stable systems, then, are at equilibrium at the lowest possible energy state under a particular set of physical conditions.
- Practically everything in this chapter addresses systems at equilibrium. Unstable and metastable systems cannot easily be dealt with using the concepts of thermodynamic equilibrium.
- For chemical systems, the governing energy parameter to be minimized is the Gibbs free energy (G).
- G can be determined for virtually any mineral or fluid phase at 298 K and 0.1 MPa from compiled standard-state thermodynamic data and can then be extended to any pressure and temperature, using appropriate compressibility and heat capacity values.
- Reactions are conventionally written with the *reactants* (low enthalpy) on the left and the *products* (high enthalpy) on the right.
- For the change in any parameter (for example, G) due to a reaction, $\Delta G = \Sigma (n_{products}G_{products} - n_{reactants}G_{reactants})$, where n represents the stoichiometric coefficients of the phases in the reaction.
- The side of a reaction that is most stable under a particular set of conditions is the side with the lower G.
- It follows from above that if $\Delta G < 0$ the products are more stable than the reactants, if $\Delta G > 0$ the reactants are more stable than the products, and if $\Delta G = 0$ the reactants and products are equally stable, so the system is at equilibrium with all of them coexisting.

- Equilibrium is generally a dynamic thing, with the forward reaction occurring at the same rate as the reverse reaction so that neither the reactants nor the products are totally consumed.
- Le Châtelier's Principle is an elegant concept that addresses the effect of an imposed change on the dynamic equilibrium state. It states that *a system will react in a fashion that acts so as to mitigate the changes*.
- We can extract useful information from phase diagrams. They typically indicate the stability limits of the reactants, the products, and the equilibrium curve (conditions under which the reactants and products coexist) for one or more reactions. When we address an equilibrium curve, the phase or assemblage on the high-pressure side has the lower volume, and the phase or assemblage on the high-temperature side has higher entropy and enthalpy. As a general rule, solids have lower molar volumes, entropies, and enthalpies than liquids, and both have lower molar volumes, entropies, and enthalpies than gases.
- The slope of the equilibrium curve at any pressure and temperature is given by the Clapeyron equation:

$$\frac{dP}{dT} = \frac{\Delta S}{\Delta V}$$

Suggested Further Readings

General Thermodynamics

Fletcher, P. (1993). *Chemical Thermodynamics for Earth Scientists*. Longman Scientific. Essex, UK.

Powell, R. (1978). *Equilibrium Thermodynamics in Petrology*. Harper Row, London.

Saxena, S. (1992). *Thermodynamic Data: Systematics and Estimation*. Springer-Verlag, New York.

Spear, F. S. (1993). *Metamorphic Phase Equilibria and Pressure–Temperature–Time Paths. Mineral. Soc. Am. Monograph* **1**. Mineralogical Society of America.

Wood, B. J., and D. G. Fraser. (1976). *Elementary Thermodynamics for Geologists*. Oxford University Press, Oxford, UK.

Thermodynamic Models for Melts

Berman, R. G., and T. H. Brown. (1987). Development of models for multicomponent melts: Analysis of synthetic systems. In: *Thermodynamic Modeling of Geological Materials: Minerals, Fluids, and Melts* (eds. I. S. E. Carmichael and H. P. Eugster). *Reviews in Mineralogy*, **17**. Mineralogical Society of America.

Ghiorso, M. S. (1987). Modeling magmatic systems: Thermodynamic relations. In: I. S. E. Carmichael and H. P. Eugster (eds.), *Thermodynamic Modeling of Geological Materials: Minerals, Fluids, and Melts* (eds. I. S. E. Carmichael and H. P. Eugster). *Reviews in Mineralogy*, **17**. Mineralogical Society of America.

Ghiorso, M. S. (1997). Thermodynamic models of igneous processes. *Ann. Rev. Earth Planet. Sci.*, **25**, 221–241.

6

The Phase Rule and One- and Two-Component Systems

Questions to be Considered in this Chapter:

1. How do crystallization and melting of chemically complex natural systems differ from simple systems such as water–ice?

2. How might we simplify natural systems in laboratory studies sufficiently to understand the complexities?

3. How can we formally analyze the behavior of systems in phase diagrams in order to make the dynamics most clear and to understand the effects of changing intensive variables?

4. Why do minerals crystallize (or melt) in repeatable sequences, and what controls the sequence?

5. How do liquids and associated mineral solids vary in composition during crystallization or melting?

6.1 INTRODUCTION: CRYSTALLIZATION BEHAVIOR OF NATURAL MAGMAS

In this chapter, we address the behavior of simple chemical systems as analogues of more complex natural ones. To see why it is advantageous to do so, let's begin by observing what happens when a natural melt crystallizes. In a work that combined natural samples of a cooling magma with laboratory analysis, Wright and Okamura (1977) studied the crystallization behavior of the Makaopuhi lava lake in Hawaii. They drilled through the thin crust of the lava lake and sampled the magma beneath, using stainless steel and ceramic probes. Because the upper portion of the ponded basaltic magma cools from the surface downward, by inserting the probe deeper into the liquid, just beneath the crust, one can sample progressively hotter portions of the magma. Thermocouples were also inserted into the drilled holes to determine the temperature gradient in the magma, in order to estimate the temperature at which each sample was collected. The result is a series of samples of uniform basaltic composition collected at a range of known cooling temperatures.

Once extracted, the small samples cool quickly and solidify. Fortunately, this process is so rapid that the solidification of the liquid portion of the sample has no time to form crystals. Rather, it rapidly solidifies ("quenches") to a glass (a solid phase with no ordered arrangement of the atoms). If any crystals were present in the original liquid at depth, they remain embedded in the newly formed glass because they, too, have no time to grow or react with the melt during quenching. Wright and Okamura took these types of quenched samples to the laboratory for chemical and microscopic analysis.

The results of the Makaopuhi study are summarized in Figures 6.1 to 6.3. Figure 6.1 shows that the amount of glass (representing liquid magma at the time of sample acquisition) decreases continuously from 100% at ~1200°C to 0% at ~950°C. Liquid is progressively replaced by crystals over this temperature range. Contrast this to some familiar simple substance, such as H_2O. At atmospheric pressure, water solidifies to ice at a constant 0°C. The Makaopuhi lava, on the other hand, began to crystallize at 1205°C and only became completely solid when it got 250° cooler.

Figure 6.2 shows that a specific sequence of solids formed as the magma cooled. Olivine began to crystallize first, followed by pyroxene, then plagioclase, and finally opaque iron–titanium oxide minerals (ilmenite and titanomagnetite).

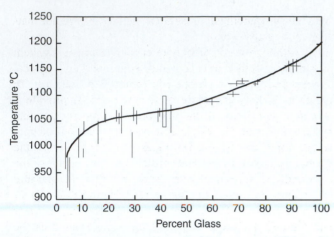

FIGURE 6.1   Percent melt (glass) as a function of temperature in samples extracted from the cooling of basalt of the Makaopuhi lava lake, Hawaii. Lines represent ranges observed. After Wright and Okamura (1977).

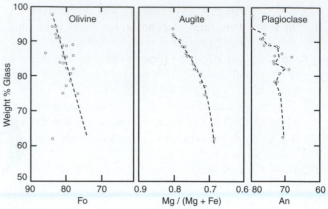

FIGURE 6.3   Model composition of minerals in Makaopuhi lava lake samples. From Wright and Okamura (1977).

If you remember Bowen's Reaction Series from previous classes (or see Figure 7.14), it might help a bit as you consider basaltic melt crystallization. The formation of olivine, followed by pyroxene, is exactly what Bowen's Series predicts. Plagioclase crystallization at Makaopuhi, however, begins to form after pyroxene and not along with olivine, as Bowen's Series indicates. Another unusual feature in Figure 6.2 is that the amount of olivine increases as crystallization of the magma proceeds from 1205 to 1180°C, and then *decreases* as the melt cools and crystallizes further. Microscopically, the early-forming olivines first grew and then began to appear embayed and corroded below 1180°C, indicating that the olivine began to be resorbed (consumed by reacting with the melt) as cooling progressed.

Figure 6.3 shows that the composition of the minerals also varies with temperature. The mafic phases get more Fe

rich, whereas the plagioclase, although somewhat irregular, gets less calcic and more sodic (in agreement with Bowen's Series). Although not shown in Figure 6.3, the composition of the glass also changed progressively during crystallization, with the remaining glass becoming preferentially depleted in Mg, Fe, and Ca.

Instances in which we can observe the crystallization behavior of natural melts are rare. We can study crystallization indirectly, however, by using sequential textures (Section 3.1.4) or by creating melts in the laboratory. From such textural and experimental criteria, we have confirmed that melts do indeed crystallize over a temperature range, a sequence of minerals forms over that range, and the composition of most minerals varies across the range as well. But there are many variations on this theme. Clearly, the minerals that form in a granite are not the same as those that form in a basalt, nor, we have discovered, is the temperature range over which that crystallization takes place the same. More silicic melts crystallize at lower temperatures than basalts, and the mineral sequence in silicic magmas may begin with biotite or amphibole and end with alkali feldspar or quartz. The actual sequence of minerals that crystallizes varies with composition and pressure. Parts of the sequence may even be reversed from one rock type to another.

From the accumulated textural and experimental data, we can make the following general observations about the complex crystallization behavior of natural melts (following Best, 1982):

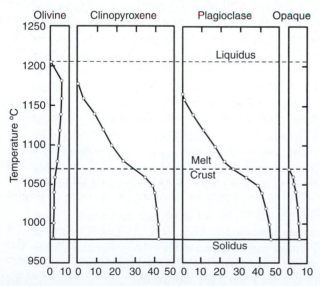

FIGURE 6.2   Weight percent minerals in Makaopuhi lava lake samples as a function of temperature. From Wright and Okamura (1977).

1. Cooling melts crystallize from a liquid to a solid over a range of temperature.
2. Several mineral phases crystallize over this temperature range, and the number of minerals tends to increase as temperature decreases.
3. Minerals usually crystallize sequentially, generally with considerable overlap.
4. Minerals that involve solid solution change composition as cooling progresses.
5. The melt composition also changes during crystallization.

6. The minerals that crystallize, as well as the sequence in which they form, depend on the temperature and composition of the melt.
7. Pressure can affect the temperature range at which a melt crystallizes. It may also affect the minerals that crystallize.
8. The nature and pressure of any volatile components (such as H_2O or CO_2) can also affect the temperature range of crystallization and the mineral sequence.

At this point, it may seem that magmas are simply too complex to understand. The chemical complexity of natural melts makes it difficult to focus on the various factors that control the behaviors described above. A basaltic melt doesn't behave the same as a granitic one, but why? Which of the many chemical variables is responsible for what aspect of the difference? A successful approach is to simplify the systems we study. By doing so, we reduce the complexity and make it possible to assess the effects of individual chemical constituents and minerals during crystallization and melting.

Of course, there is a price for this benefit, and that price is that the simplified systems are not really the natural ones that interest us. We may understand the simple "model" systems better, but the results may not apply directly to the more complex natural phenomena. Nonetheless, we shall see that this approach has been of great benefit to petrologists. With a bit of theory and some experimental results from simplified systems, we can understand the basis for the behaviors listed above. Upon finishing the next two chapters, you should be able to return to the list and understand how all these eight properties of melts are controlled. Application of model systems to real rocks is indeed possible, and we shall do so effectively.

6.2 PHASE EQUILIBRIUM AND THE PHASE RULE

If we are to understand the simplified systems that follow, we need a bit of theoretical preparation. We want to be able to analyze systems in a way that allows us to grasp the dynamics of each and to account for the contribution of each chemical constituent to the variations in those dynamics. If we understand how the introduction of additional constituents affects a system, we can not only understand each new system better, but we become prepared to apply the systems we study to the more complex systems in nature. The **phase rule** is a simple, yet rigorous and elegant, theoretical treatment for this approach. To develop the phase rule, we must first define a few terms.

As mentioned in Section 5.1, a **system** is some portion of the universe that you want to study. In the lab, we get to choose the system, but in the field, the system may be forced upon us. The **surroundings** can be considered the bit of the universe just outside the system. A system may be **open** (if it can transfer energy and matter to and from the surroundings), **closed** (only energy, such as heat, may be exchanged

with the surroundings), or **isolated** (neither energy nor matter may be transferred).

Although we commonly refer to the *state* of a system as simply whether it is a liquid, solid, or gas, physical chemists have a far more specific definition of the term. To them, to specify the **state** of a system is to provide a *complete* description of the macroscopic properties of that system. For example, consider a system composed of pure water. It may be contained in a glass, but we can define the system as only the water and consider the glass a part of the surroundings. We can measure the temperature (T), pressure (P), volume (V), mass (m), density (ρ), composition (X), or any of a number of other possible parameters of the water and thereby determine each. All these variables must be known if the state of the system is to be *completely* described. However, once a critical number of these variables is known, the others become fixed as a result because many of the properties are interdependent. For example, if we know the mass and the volume of our water, we know its density. But just how many of these variables must we specify before the others are determined? The phase rule was formulated to address this question.

A **phase** is defined as a type of physically distinct material in a system that is mechanically separable from the rest. A phase may be a mineral, a liquid, a gas, or an amorphous solid such as glass. A piece of ice is a single phase, whereas ice water consists of two phases (the ice and the water are separable). Two pieces of ice are mechanically separable, but because they are equivalent, they are considered different pieces of the same phase, not two phases. A phase can be complex chemically (such as a tequila sunrise), but as long as you cannot separate it further by mechanical means, it is a single phase.

A **component** is a chemical constituent, such as Si, H_2O, O_2, SiO_2, or $NaAlSi_3O_8$. We can define individual components as we please, but, for purposes of the phase rule treatment, we shall define the *number of components* as the *minimum* number of chemical species required to completely define the system and all of its phases. For example, ice water, although two phases, has but one component (H_2O). We could define it as H and O, but because H_2O describes both ice and water as a single component, not two, it is preferred for phase rule purposes. A pure mineral, such as albite, has a single component ($NaAlSi_3O_8$). Minerals that exhibit solid solution, however, are commonly treated as multicomponent systems. Plagioclase is commonly a single phase that comprises two components: $NaAlSi_3O_8$ and $CaAl_2Si_2O_8$. Why? Because we need to vary the proportions of these two components to determine the state of plagioclases of varying composition.

The proper choice of the number of components for the application of the phase rule is not always easy. The choice commonly depends on the behavior of the system and the range of conditions over which it is studied. For example, calcite may be considered a single-component system ($CaCO_3$). Although this is true at relatively low temperatures if we heat it to the point that it decomposes to solid CaO and gaseous CO_2, it would be a two-component system because we would have to use both CaO and CO_2 to describe the composition of the solid and gaseous phases. Compare this to our ice water: a single component suffices to describe the chemical composition of each phase. But for calcite, lime (CaO), and CO_2, we require two components, separately or in combination, to describe them all. This concept will become clearer with practice.

The **variables** that must be determined to completely define the state of a system can be either extensive or intensive in nature. **Extensive** variables depend on the quantity of material (the extent) in the system. Mass, volume, number of moles, etc. are all extensive variables. Such variables are not intrinsic properties of the substances in the system. In other words, it is possible to have 10 g of water, or 100 g. Either way, it's still the same water. Although it is nice, perhaps, to have more of some things (money, influence . . .) and less of others (debts, nose hairs . . .), such extensive variables are of little concern to us now, as they do not affect the macroscopic properties (the state) of matter in a system.

Intensive variables, on the other hand, don't depend upon the size of the system and are properties of the substances that compose a system. Intensive variables include pressure, temperature, density, etc. If we divide any *extensive* variable by another one, the extent cancels, and the ratio is an *intensive* variable. For example, the volume of a phase divided by the number of moles is the *molar* volume, an intensive variable. Density (mass divided by volume) is another example. These latter two intensive variables are certainly properties that can change for substances in a closed system, and therefore must be specified if we are to determine the state of that system. The molar volume and density of water is different at 10°C than at 50°C, and it is also different than the corresponding values for CO_2 at the same temperature. Because materials expand with increasing heat and contract with increasing pressure, the pressure and temperature must also be specified when describing the state of a system. Another important intensive variable is the composition of the phases present. Although the number of moles of Fe and Mg in olivine are extensive variables, the ratio of Fe/Mg is intensive and affects such properties as molar volume, density, etc.

There are a large number of possible intensive variables, and we've seen that many are interdependent. We now return to the question stated above: How many must we specify before the others are fully constrained and the state of the system is known? The phase rule of Gibbs (1928) is designed to do this. If we define F, the **number of degrees of freedom** (or the **variance**) of a system, as the minimum number of intensive variables that need to be specified to completely define the state of the system *at equilibrium*, the phase rule can be expressed as:

$$F = C - \phi + 2 \qquad \text{(the Gibbs phase rule)} \qquad (6.1)$$

Where ϕ is the number of phases in the system, and C is the number of components. A rigorous derivation of the phase rule (Gibbs, 1928) is based upon the number of variables (one for each component, plus P and T) minus the number of equations relating those variables [one for each phase, each similar to Equation (5.3)]. Hence, $F = C + 2 - \phi$. The mathematics of Equation (6.1) is simple enough and tells us that for each component we add to a system, we must specify one additional intensive variable to completely constrain the state of the system. For each additional phase, there is one fewer variable that needs to be specified. Once we have specified this critical number of independent intensive variables, all other intensive variables are fully constrained (invariable).

The phase rule applies only to systems in chemical equilibrium. One cannot count disequilibrium assemblages, such as incomplete replacement of biotite by chlorite in a cooled granitic rock (Figure 3.20b), as separate phases for most phase rule applications. The reaction of biotite to chlorite is typically arrested because the rock cooled too quickly for the reaction to run to completion. As a more extreme example, consider the great number of mineral phases that can coexist in a clastic sediment such as a graywacke. These minerals are collected together and deposited by clastic processes, but they are not in chemical equilibrium at low, near-surface temperatures, and to apply the phase rule to such a system would be useless.

6.3 APPLICATION OF THE PHASE RULE TO THE H_2O SYSTEM

Let's see how the phase rule works by applying it to a very simple system: the heating of ice on a hot plate. The system is defined by a single component, H_2O, so $C = 1$. If we begin with ice at equilibrium at some temperature below 0°C (we left the window open in January), then our system is completely solid, and $\phi = 1$ as well. The phase rule [Equation (6.1)] at this point would tell us:

$$F = 1 - 1 + 2 = 2$$

meaning that we must specify only two intensive variables to define the system completely. In the natural world, pressure and temperature are the most common independent variables, so if we were to specify P and T, the state of the system would be completely defined. If we were to specify −5°C and 0.1 MPa (atmospheric) pressure, all the other intensive parameters of the ice would necessarily also be fixed (density, molar volume, heat capacity . . . everything). By fixed, I mean that they are measurable parameters that are constant properties of ice under the conditions specified.

We can think of F as the number of variables that we must specify, or we can think of it as the number of variables that we are free to change *independently*. The fact that F variables can vary independently explains why we need to specify each. In the present case, we are free to change two intensive variables *as long as the value of the other parameters in the phase rule (C and ϕ) remain the same.* With our ice, we have $F = 2$ and have chosen P and T as the ones that we shall specify. Alternatively, we could say that we can change P and T independently in our pan of ice (either or both), and still have only ice.

Let's heat the system at constant pressure (turn on the hot plate beneath our pan of ice). We can heat it initially with no change in the parameters of the phase rule (i.e., F remains equal to 2 as long as ϕ and C are both equal to 1). At each new temperature we can specify T, and the other intensive parameters also have new values (e.g., the ice expands, so the density changes). The phase rule still holds, however, telling us we need to specify two intensive variables if we want to fix the others.

Eventually we heat the ice until a new phase appears: the ice begins to melt, and ice and water coexist stably at equilibrium in the pan. Now $\phi = 2$ and $F = 1 - 2 + 2 = 1$. We need to specify only *one* intensive variable to completely define the state of the system. Which variable do we choose? Pressure or temperature? *The phase rule cannot make this choice for us.* It tells us about the variance, but it cannot choose the variables for us. In other words, the phase rule is a tool for the analysis of chemical systems; it does not attempt to tell a system how it should behave. The responsibility is ours to apply the phase rule appropriately and interpret the results.

If we look at a pressure–temperature **phase diagram** for the H_2O system in Figure 6.4, we can interpret the phase rule more clearly. We began at point A in the field labeled "Ice" and moved along the dashed path to the melting point (point B on the ice–water boundary). Initially in the ice field we could vary pressure and temperature independently and still have only one phase. Thus we had to specify both variables to define the state of the phase. When we heated the ice to point B, we encountered the line separating the ice and water fields, meaning that both phases can coexist at equilibrium ("ice water") under P-T conditions anywhere along this line. Because $\phi = 2$, then $F = 1$, meaning that we must specify only *one* variable now (pressure *or* temperature), and all the other intensive variables for *both* phases are then determined. This may seem counterintuitive at first because we now have twice as many phases, each with its own density, molar volume, etc. to be determined, but the phase rule tells us that for each new phase, the number of *independent* variables is actually decreased. For example, if we specify that ice and water are at equilibrium at a pressure of 0.1 MPa (atmospheric pressure, represented by our dashed line in Figure 6.4), then the temperature *must be* 0°C (point B). We thus know both T and P, and the density, molar volume, etc. of each phase are therefore fixed.

Alternatively, $F = 1$ means that we cannot vary pressure *and* temperature independently anymore without changing the parameters of the phase rule (i.e., losing a phase). Consider point B in Figure 6.4 again. If we were to vary pressure independently vertically upward, we would leave the water–ice equilibrium boundary curve and enter the water field, losing ice and changing the phase rule parameter ϕ to one. If we were to raise the pressure and still maintain $\phi = 2$, we would have to change temperature in a sympathetic fashion so as to remain on the ice–water equilibrium line. So if $\phi = 2$, P and T are not independent. More generally, we can

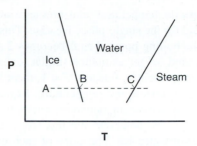

FIGURE 6.4 Schematic pressure-temperature phase diagram of a portion of the H_2O system.

say that, at equilibrium in a one-component, two-phase system (such as ice water), *there must be a relationship between pressure and temperature*. This relationship is expressed as the *slope* of the equilibrium curve that separates the ice and water fields on the pressure–temperature diagram (Figure 6.4). If we change either variable, we would also have to change the other along the curve to maintain two coexisting phases.

As mentioned above, the phase rule tells us the variance but can't choose which variable(s) are determinative. If we run our experiment in a lab on a hot plate at atmospheric pressure, the phase rule is not aware that pressure is not a possible variable. We can simply remember that one of our variables is fixed, or we can modify the phase rule to account for such restricted operating conditions. If all of our experimentation is conducted at constant pressure, we have forced the removal of a degree of freedom from our analysis. Under these conditions, the phase rule (Equation 6.1) would be reduced to:

$$F = C - \phi + 1 \qquad \text{(isobaric)} \qquad (6.2)$$

We could use a similarly reduced phase rule for any system with one fixed variable (e.g., constant temperature, constant volume).

In our constant 0.1 MPa pressure system, having ice water at equilibrium on our hot plate in the lab, Equation (6.2) tells us that $F = 1 - 2 + 1 = 0$. Such a system is completely fixed. The temperature must be 0°C, and every intensive variable of both phases is determined. At a constant pressure, we cannot change the temperature of our ice water as long as we have two phases. *We can add heat, but this won't change the temperature!* The hot plate is hot, pumping heat into the system, but the phase rule tells us that the temperature must remain constant as long as ice and water are both in the pan. Here we see an example of the difference between *heat* and *temperature*. In thermodynamics, heat has the symbol q. Heat is supplied to the ice water and has the effect of melting the ice *at a constant temperature* until the ice is consumed. This heat, which drives the transition from the solid to the liquid form, is called the **latent heat of fusion**.

Once we have melted all the ice, we again have one phase (water). Returning to our general system with variable pressure, Equation (6.1) tells us that $F = 1 - 1 + 2 = 2$, and we must specify temperature and pressure independently, just as we did for the single-phase ice case. This is true until we get to the boiling point; then ϕ becomes 2 again, and F returns to 1, and we get a situation similar to that of water + ice discussed previously. Because $F = 1$, there is a relationship between pressure and temperature, expressed by the slope of the water–steam boundary curve in Figure 6.4. There is no reason, however, that this slope has to be the same as that of water–ice. The slope of each is determined not by the phase rule but by the molar volume and entropy of the coexisting phases, as expressed by the Clapeyron equation [Equation (5.15)].

If the pressure is fixed, as in our lab, the modified phase rule [Equation (6.2)] for $\phi = 2$ reduces F to 0, meaning we can only have boiling water (coexisting water and steam at equilibrium) at a single specific temperature (100°C at 0.1 MPa). The heat supplied at this constant temperature converts the water to steam and is now called the **latent heat of vaporization**.

Some cooks think that briskly boiling water is hotter than slowly boiling water. This is clearly impossible as long as liquid water and water vapor coexist (a prerequisite for boiling). The added (wasted) energy used for brisk boiling simply supplies the latent heat to make more steam. If your cooking goal is to reduce the quantity of liquid, go ahead and boil away briskly. If you're only trying to cook some pasta, however, boil it as slowly as you can and save some money.

Smart cooks know how to get boiling water hotter than 100°C. They use a pressure cooker, which frees them from the 1 atmosphere restriction. Because water and steam coexist in the pressure cooker, the temperature and pressure must change along the water–steam boundary curve in Figure 6.4. Because heat is added, the pressure must increase because the boundary has a positive slope. This situation could get dangerous because the pressure could explosively overcome the strength of the container. That's why a valve with a weight is placed on the top of a pressure cooker. Once the pressure reaches a specific value, the vapor lifts the weight, and steam is released. Thus pressure cookers operate at a constant, though elevated, pressure (and therefore constant elevated temperature). Incidentally, the steam released by the valve instantly drops to atmospheric pressure while cooling only slightly in the one-phase steam field of Figure 6.4. Released steam is thus *very* hot and can cause severe burns. Most cooks know of another way to elevate the temperature of boiling water a little: They add salt. But this violates our premise of a single-component system (constant composition) and will be dealt with later.

Let's now apply the phase rule to some simple model systems of geological significance. The phase diagrams we shall use have been derived empirically (by experiments on simple mineral systems). As you are, I hope, beginning to recognize, the phase rule is a theoretical treatment that helps us understand the dynamics of the systems represented by these diagrams. Remember that the phase rule is $F = C - \phi + 2$ for the general case (not artificially fixing pressure or tem-

perature) and that the final term is decreased by 1 for each variable that we fix externally. We shall look at a number of synthetic and natural rock systems.

Experiments on silicate systems require furnaces capable of melting rocks (or chemically simplified rock analogs) at high pressure. Figure 6.5 is a schematic cross section through a typical high-pressure furnace. Note that the size of the cylindrical sample in Figure 6.5 is less than 1 cm in diameter. In furnaces that are combined with high-pressure hydraulic rams, small samples can be heated to temperatures sufficient for complete melting at pressures equivalent to those attained in the upper mantle. The sample is prepared and inserted into the furnace, and the furnace is then closed and gradually heated as the pressure is increased by the ram. The pressure is loaded vertically in Figure 6.5, but the furnace is confined radially, so that the horizontal pressure quickly approaches that of the vertical load. A thermocouple inserted just above the sample records the temperature attained and permits external temperature control. The result is that we can expose a sample to a variety of temperatures and pressures. In the ensuing discussion, we shall begin with the simplest one-component experimental systems and gradually explore the effects of added chemical complexity.

6.4 ONE-COMPONENT SYSTEMS

Figure 6.6 is the pressure–temperature phase diagram for the SiO_2 system. The upper limits of 10 GPa and 1900°C reflect generous limits of pressure and temperature to which a pure SiO_2 phase would typically be subjected in nature (remember, 1 GPa represents the approximate pressure at the base of 35 km of continental crust, and SiO_2 minerals are not common in the mantle or core). There are a number of solid silica polymorphs, and there is a liquid phase, each with a stability field shown on the diagram.

For conditions that fall within any one field in Figure 6.6, only one phase is stable, hence $\phi = 1$ and $F = 1 - 1 + 2 = 2$. These areas are called **divariant** fields because the variance in them is two. Both pressure and temperature are variable for these single-phase situations, and both must be specified to determine the state of any one-phase system. Curves separating the fields represent conditions under which two phases coexist at equilibrium. Because $\phi = 2$ and $F = 1 - 2 + 2 = 1$, the curves are called **univariant** curves. Along these curves, because two phases coexist, one need only specify pressure *or* temperature. Specifying one of these permits us to determine the other from the location on the curve, and thus the state of the whole system (both phases) is defined. This is true for any two phases in stable equilibrium, such as water and ice in Figure 6.4, cristobalite and liquid in Figure 6.6, or any two coexisting silica polymorphs. For example, if I were to ask at what temperature α- and β-quartz coexist at 1 GPa, we could look along the curve separating the fields of low quartz and high quartz at the pressure specified and determine the temperature: approximately 810°C. Note the positive slope for any solid/liquid equilibrium curve, as predicted in Chapter 5.

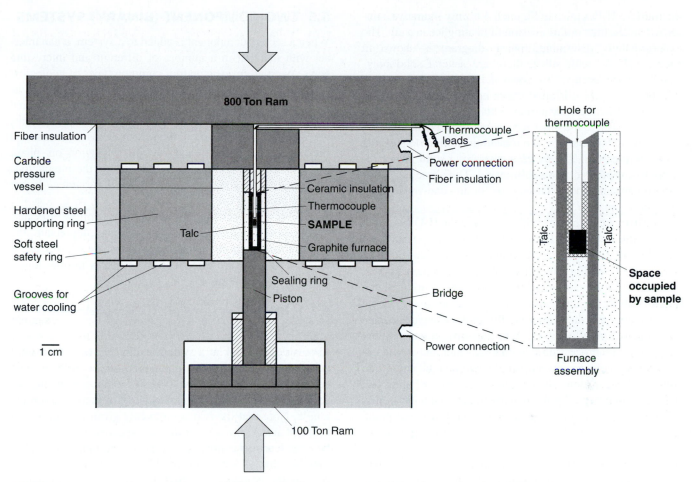

FIGURE 6.5   Cross section through a typical furnace for the experimental study of natural and synthetic rocks and minerals at pressures equivalent to depths as great as 150 km. Diagonally hatched areas are steel, and stippled areas are typically ceramic. After Boyd and England (1960). Copyright © AGU with permission.

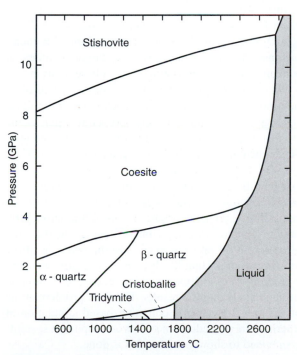

FIGURE 6.6   Pressure–temperature phase diagram for SiO_2. After Swamy et al. (1994). Copyright © AGU with permission.

Note that there are also points where univariant lines intersect. At these points, *three* phases coexist. When $\phi = 3$, $F = 1 - 3 + 2 = 0$. Whenever three phases coexist at equilibrium in a one-component system, the system is completely determined, and these points are called **invariant** points. Such points are obviously possible only at specific temperatures and pressures. Try this yourself. Under what conditions are low quartz, high quartz, and coesite stable together? What about high quartz, cristobalite, and liquid?

The phase rule and phase diagrams are very closely related. You have probably noticed that the variance of a system corresponds directly to the dimensions of the appropriate assemblage in the diagram. Divariant assemblages ($F = 2$) exist as two-dimensional fields (areas) on the phase diagram. Univariant assemblages ($F = 1$) are represented by the one-dimensional curves that act as boundaries between the divariant fields, and invariant assemblages are represented by the zero-dimensional points where three fields meet.

Figure 6.6 is a fairly representative one-component P-T phase diagram for minerals, showing a high-T liquid field and a solid field that may be subdivided into fields for various polymorphs. For substances with lower melting points and higher vapor pressures (natural liquids, gases, and some minerals), there may also be a vapor phase field. The phase diagram for H_2O shown in Figure 6.4 is only figurative, simplified to illustrate our discussion of heating ice in a lab. The experimentally determined phase diagram is shown in Figure 6.7. As with silica, there are several solid polymorphs of ice, each with its own stability range. The negative slope of the ice I/liquid curve is very rare. Ice I has a greater volume than the liquid. This irregularity, however, does not extend to the other ice polymorphs. Note the vapor field and the boundary for the transition from ice to steam at very low pressure. The process whereby a solid passes directly to a vapor is called **sublimation**. The phase rule treatment for sublimation is the same as for any one-component, two-phase situation.

There are several invariant points in the H_2O system—for example, where ice I, water, and steam all coexist. Note also that the liquid/vapor curve ends in a **critical point** at 374°C and 21.8 MPa. You can see from the diagram that it is possible to begin with liquid water at 0.1 MPa, increase the pressure above the critical pressure, heat it above the critical temperature, and then decrease the pressure again to produce steam. The effect is to create steam from water, but at no point in the process do two phases, water and steam, coexist. When increasing pressure is applied to coexisting water and vapor along the equilibrium curve in Figure 6.7, the vapor compresses more than the liquid, and their properties (density, etc.) gradually converge. At the critical point, they become identical. At pressures and temperatures above the critical point (called the **supercritical** region), there is no distinction between the liquid and vapor phases. The conditions required to create a supercritical fluid in aqueous systems are readily attainable in igneous and metamorphic processes. The terms *liquid* and *vapor* lose their meaning under these conditions, and we call such phases **supercritical fluids**, or simply **fluids**.

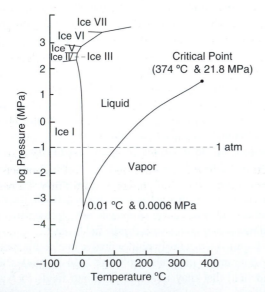

FIGURE 6.7 Pressure–temperature phase diagram for H_2O (after Bridgman, 1911, 1935–1937). Ice IV was created using D_2O and is not stable for H_2O.

6.5 TWO-COMPONENT (BINARY) SYSTEMS

When a second component is added to a system, it can interact with the first in a number of different and interesting ways. We shall investigate four common types of geological binary systems. Because $C = 2$, the variance can be as high as 3 in one-phase systems, requiring three-dimensional diagrams to illustrate properly. Rather than attempt this, we simplify most two-component igneous systems and illustrate their cooling and melting behavior on phase diagrams by fixing pressure and discussing the interactions of temperature and the compositional variables. If we restrict pressure, the phase rule becomes $F = C - \phi + 1$ [Equation (6.2)] in the discussions that follow. Because temperature–composition (T-X) diagrams depreciate the importance of pressure in natural systems, we will occasionally discuss pressure effects on the systems in question in this chapter and more fully in Chapter 7.

6.5.1 Binary Systems with Complete Solid Solution

First, we shall look at a system exhibiting **complete solid solution**, in which both components mix completely with each other. The plagioclase system, composed of the two components $NaAlSi_3O_8$ and $CaAl_2Si_2O_8$, is a common example. Note that the two components given are equivalent to the phases albite and anorthite, respectively. There might be a tendency to confuse components and phases here. Remember that C (the number of components in the phase rule) is the *minimum* number of chemical constituents required to constitute the system and all of its phases. It is most convenient to treat this system as the *two* components $CaAl_2Si_2O_8$-$NaAlSi_3O_8$, corresponding to the composition of the two phases. This is not a coincidence because this choice of C is the easiest one we can make to represent the phases in the system. Using the simple oxides, CaO-Na_2O-Al_2O_3-SiO_2 may seem like a more logical choice for chemical components, but it results in a larger number of components than is necessary (thus violating the definition of C). For the remainder of Chapters 6 and 7, I shall often use the mineral abbreviation (inside back cover) to indicate components and the mineral name proper to refer to the phases in a system. Thus the abbreviation Ab will be used to indicate the component $NaAlSi_3O_8$, and An will indicate $CaAl_2Si_2O_8$. From your mineralogy course, you may remember that the solid solution for this system involves the *coupled substitution* of $(Na^{1+} + Si^{4+})$ for $(Ca^{2+} + Al^{3+})$ in a constant $AlSi_2O_8$ reference frame. Figure 6.8 is an isobaric temperature–composition (T-X) phase diagram at 0.1 MPa pressure. Before we proceed to analyze the behavior of the system, I must stress that all the phase diagrams we shall cover are *empirically* determined by melting real mineral samples and analyzing the results. The final problem at the end of Chapter 5 involves a computer-based simulation of the experimental procedure on two two-component systems that is intended to show you how this is done.

At each end of the horizontal axis in Figure 6.8, we have a one-component system, representing each pure

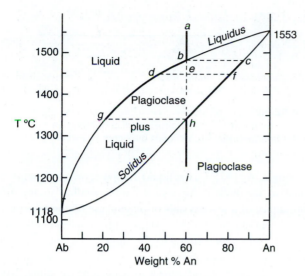

FIGURE 6.8 Isobaric *T-X* phase diagram for the albite–anorthite system at 0.1 MPa (atmospheric) pressure. After Bowen (1913). Reprinted by permission of the *American Journal of Science*.

"end-member" of the solid-solution series, pure albite on the left and pure anorthite on the right. Each of these pure systems behaves like a typical isobaric one-component system, in that the solids melt at a single fixed temperature, at which solid and liquid coexist in equilibrium ($\phi = 2$), just as in our ice–water example. Applying the isobaric phase rule [Equation (6.2)] with $C = 1$ and $\phi = 2$ yields $F = 1 - 2 + 1 = 0$. Albite melts at 1118°C, and anorthite melts at 1553°C.

Now we proceed to the effects of an added component on either pure system. First, the addition of the Ab component to pure anorthite lowers the melting point (just as adding salt lowers the melting point of ice on a frozen sidewalk). Adding An to pure albite raises its melting point. But this is not the only effect. Crystallization of multicomponent melts becomes much more interesting than pure one-component melts.

To understand this, let's use the phase rule to analyze the behavior of a melt of intermediate composition. Consider cooling a melt of composition *a* in Figure 6.8. We refer to the composition of the *system* as the **bulk composition** (X_{bulk}). The composition in question is 60% anorthite and 40% albite, by weight. This composition may be referred to as An_{60}. Note: An_{60} is usually a reference to *mole* % An. The notation doesn't preclude its use for weight % (wt. %), and because Figure 6.8 has been created for weight %, we will use it this way. At point *a* in Figure 6.8, at about 1600°C, we have a single liquid of composition An_{60}. In this case, the liquid composition is equal to the bulk composition because the system is entirely liquid. Because $C = 2$ and $\phi = 1$, Equation (6.2) yields $F = 2 - 1 + 1 = 2$. There are thus 2 degrees of freedom for a single two-component liquid at constant pressure. What are they? Once again, any two intensive variables will do, but they should be geologically realistic. Because the diagram is a temperature–composition diagram, it might seem appropriate to choose these two. What are the possible compositional variables? They must be intensive,

so the choices are the weight (or mole) *fraction* of any component in any phase. We can define the weight fraction of the An component in the liquid phase:

$$X_{An}^{liq} = n_{An}/(n_{An} + n_{Ab}) \text{ in the liquid} \qquad (6.3)$$

where: n = number of grams of any component

If the system weighs 100 g, and $n_{An} = 60$ g, then $X_{An}^{liq} = 60/(60 + 40)$, or 0.60. The phase rule thus tells us that, if we have a liquid in the Ab-An system *at fixed pressure*, we must specify T and a single compositional variable to completely determine the system. If we choose to specify T as 1600°C, and $X_{An}^{liq} = 0.60$, *all* the other variables, such as density and the other compositional variables, are fixed. Under the present circumstances, the only remaining intensive compositional variable is X_{Ab}^{liq}. Because the system is binary, it follows that $X_{Ab}^{liq} = 1 - X_{An}^{liq} = 0.40$.

If we cool the system to point *b* in Figure 6.8, at about 1475°C, plagioclase begins to crystallize from the melt. *However, the plagioclase that first forms has a composition at c (An_{87}), a different composition than that of the melt.* How does the phase rule help us understand what's going on at this point? Because $C = 2$ and $\phi = 2$, $F = 2 - 2 + 1 = 1$. Now we must specify only *one* intensive variable to completely determine the system. If we specify any one of T, X_{An}^{liq}, X_{Ab}^{liq}, X_{An}^{plag}, or X_{Ab}^{plag}, the others must be fixed. From Figure 6.8 we can see that this is true. Whereas the one-component systems has a single curve separating the liquid and solid fields, there are now *two* curves that specify *a relationship between the composition of both the liquid and the solid with respect to the temperature.* The upper curve is called the **liquidus**. It specifies the composition of any liquid that coexists with a solid at a particular temperature. The lower curve is the **solidus**, which specifies the composition of any solid that coexists with a liquid phase at some particular temperature. Remember that these diagrams are determined *empirically*. There is no way to theoretically predict the actual compositions, and the phase rule merely tells us about the variables, not what values they should have. We can thus specify one variable, such as $T = 1475°C$. If plagioclase and liquid coexist, $\phi = 2$, then we specify a horizontal line at 1475°C that intersects the liquidus at point *b* and the solidus at point *c*. Points *b* and *c* represent the composition of the liquid and solid, respectively, which we could determine from the abscissa. The system is therefore fully determined. The dashed line connecting *b* and *c* is called a **tie-line**, and it connects the composition of coexisting phases, by definition at a particular temperature. Try choosing another temperature and determine the composition of the phases that coexist at that T. Next, choose any other variable from the list above. If your choice of temperature meets the requirement that $\phi = 2$ on Figure 6.8, you can determine all the remaining variables. If $\phi = 2$ for this system at fixed pressure, we need only specify one intensive variable to determine the full state of the system. In a practical sense, temperature behaves as a determining variable most commonly in nature.

What the phase rule then says for this situation is: *For a two-component–two-phase system at a fixed pressure, the composition of both phases (in this case the liquid and the solid) depend only upon temperature.* This follows directly from the liquidus and solidus curves on Figure 6.8. The phase diagram, which is determined by experiment, is a manifestation of the relationships predicted by the phase rule. Picture this situation in a magma chamber at some particular depth (pressure) in the crust. The composition of the plagioclase that crystallizes from a melt is a function of the temperature of that melt and thus changes as temperature changes.

As we continue to cool our original bulk mixture of 60% An below 1475°C (point *b* in Figure 6.8), the compositions of both coexisting phases (liquid and solid) vary. The liquid composition changes along the liquidus from *b* toward *g*, whereas the plagioclase changes from *c* toward *h*. This process is one in which the solid *reacts with* the liquid, enabling the exchange of components between them, resulting in a compositional change in the phases. Such reactions that have at least 1 degree of freedom, and thus occur by exchange over a range of temperatures (and/or pressures), are called **continuous reactions**. In this case, the generalized reaction may be represented by:

$$Liquid_1 + Plagioclase_1 = Liquid_2 + Plagioclase_2 \quad (6.4)$$

By cooling, new $Liquid_2$ becomes incrementally more Na rich than old $Liquid_1$, and new $Plagioclase_2$ becomes more Na rich than old $Plagioclase_1$.

We can use the length of the tie-lines at any specified temperature to calculate the *relative* amounts of the phases. At 1445°C and a bulk composition of An_{60}, for example, we have the tie-line *d-f* connecting the liquid and solid phases in Figure 6.8. The bulk composition = *e*, whereas the composition of the liquid = *d* (An_{49}), and plagioclase has composition *f* (An_{82}). The relative amount of liquid versus solid is calculated geometrically by reference to Figure 6.9, in which the quantities of each phase must balance on the bulk composition fulcrum point.

Using the lengths of the tie-line segments:

$$\frac{amt_{liq}}{amt_{plag}} = \frac{\overline{ef}}{\overline{de}} \quad (6.5)$$

where: $\overline{ef}$ = length of the line segment *e-f*
$\overline{de}$ = length of the line segment *d-e* in Figures 6.8 and 6.9.

This approach is called the **lever principle**, and it works like a fulcrum, with the amounts of the phases balanced at the fulcrum point represented by the bulk composition in Figure 6.9. According to Equation (6.4), the amount of a given phase is proportional to the length of the segment on the *opposite* side of the fulcrum. The closer a phase is to the fulcrum point (bulk composition), the more predominant it is. At 1445°C $\overline{ef} = An_{82} - An_{60} = 22$ and $\overline{de} = An_{60} - An_{48} = 12$. Thus the proportion of liquid/solid is 22/12, or 1.83, or 65% liquid

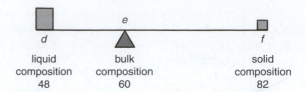

FIGURE 6.9 Use of the lever principle to determine the relative quantities of two phases coexisting along an isothermal tie-line with a known bulk composition.

by weight. At 1475°C, the liquid composition was essentially equal to the bulk composition, corresponding to the first appearance of a few crystals of plagioclase of composition *c*. As we continue to cool the system, with a constant bulk composition, $\overline{ef}$ gets progressively larger, whereas $\overline{de}$ gets smaller, corresponding with a decreasing ratio of liquid to solid, just as we would expect to occur upon cooling. Note that as cooling of the two-phase system in Figure 6.8 continues, *the composition of both the solid and liquid becomes richer in the Ab (low melting point) component.* As the temperature approaches 1340°C, the composition of the plagioclase reaches *h*, which is equal to the bulk composition (An_{60}). Obviously, there can be only a tiny amount of liquid present at this point. This last liquid has composition *g* (An_{22}) in Figure 6.8. Continued cooling consumes this immediately. We then lose a phase and gain a degree of freedom. We have only a single solid phase below 1340°C (plagioclase of composition An_{60}) that cools along the line *h-i*. With a single phase, $F = 2 - 1 + 1 = 2$, so we must specify both *T* and a compositional variable of the plagioclase to specify the system completely.

Crystallization of any liquid of a composition intermediate between pure Ab and pure An will behave in an analogous fashion.

Equilibrium **melting** is simply the opposite process. The divariant one-phase solid system of composition *i* in Figure 6.8 heats up until melting begins. The partially melted system is univariant, and the first liquid formed has composition *g*. The first liquid to form is not the same as the solid that melts. As heating continues, the compositions of the solid and liquid are constrained to follow the solidus and liquidus, respectively (via a continuous reaction). The liquid moves to composition *b* as the plagioclase shifts to composition *c* (the composition of the last plagioclase to melt). Whether the process is crystallization or melting, the solid is always richer in An components (Ca and Al) than the coexisting liquid. Ca is thus more **refractory** than Na, meaning that it concentrates in the residual solids during melting.

Notice how the addition of a second component affects the crystallization relationships of simple one-component systems:

1. There is now a *range* of temperatures over which a liquid crystallizes (or a solid melts) at a given pressure.
2. Over this temperature range, the compositions of both the liquid melt and the solid mineral phases change.

Compare these to the list of eight observations on the crystallization behavior of natural melts in Section 6.1. Even though we have studied only a simple model system, the processes responsible for observations 1, 4, and 5 should be getting clear.

The above discussion considers only **equilibrium crystallization** and **equilibrium melting**, in which the plagioclase that crystallizes or melts remains in chemical equilibrium with the melt. It is also possible to have *fractional* crystallization or melting. Purely **fractional crystallization** involves the physical separation of the solid from the melt as soon as it forms. If we remove the plagioclase crystals as they form (perhaps by having them sink or float), the melt can no longer react with the crystals. The melt composition continues to vary along the liquidus as new plagioclase crystallizes along the solidus. Because the crystals are removed from the system, however, *the melt composition continuously becomes the new bulk composition*, thus shifting inexorably toward albite. As a result, the composition of both the final liquid and the solids that form from it will be more albitic than for equilibrium crystallization and will approach pure albite in efficiently fractionating systems. Fractional crystallization implies that a range of magma types could be created from a single parental type by removing varying amounts of crystals that have formed in a magma chamber.

Fractional melting is another important geologic process. Purely fractional melting refers to the nearly continuous extraction of melt increments as they are formed. If we begin to melt a plagioclase of An_{60} composition in Figure 6.8, the first melt has composition g (An_{20}). If we remove the melt, the residual solids become progressively enriched in the high-melting-temperature component and continuously become the new bulk composition of the remaining solid system. The final solid, and the liquid that may be derived from it, shift toward anorthite.

Most natural magmas, once created, are extracted from the melted source rock at some point before melting is completed. This is called **partial melting**, which may be fractional melting or may involve equilibrium melting until sufficient liquid accumulates to become mobile. For example, suppose we begin with An_{60} in Figure 6.8 and melt it at equilibrium to 1445°C, at which point there is 65% melt (according to the lever principle) with a composition d (An_{49}). If melt d rises to a shallow magma chamber and cools, the bulk composition in the chamber is now An_{49}, as is the final plagioclase to crystallize from this melt (assuming equilibrium crystallization). *Partial melting, then, increases the concentration of the low-melting-point component in the resulting melt system* (An_{49} rather than An_{60}). Likewise, it increases the concentration of the high-melting-point component in the residual solids (point f, An_{82}, at the time of melt extraction).

Partial melting processes have some important implications for the source of melts. Suppose we partially melt the mantle to produce a basaltic liquid. If only small quantities of melt are produced, the remaining solid mantle must be more refractory (enriched in high-temperature components) than the melt produced. The mantle source will also be progressively depleted in the low-melting-point components and gradually become more refractory as partial melting continues over time, requiring successively higher temperatures in order to melt. Unless the melted source rocks are replenished by mixing with unmelted mantle, they may become sufficiently refractory that further melting is inhibited.

As soon as a system permits the solid and liquid portions to have different compositions, separating these phases can have a profound effect on the composition of the derivative systems. The ability to change the composition of magmas and the resulting rocks by fractional processes of melting and crystallization are thus prime methods for the production of the range of compositions of igneous rocks found at the surface of the Earth. Indeed these fractional processes are probably more common in nature than their equilibrium counterparts. We shall encounter numerous examples of these phenomena, both experimental and natural, as we proceed in our study of igneous processes.

Compositional zoning (Section 3.1.3) is another disequilibrium process that can occur in solid-solution systems. Rather than react with the melt and re-equilibrate, a mineral may simply add a rim with a composition equal to the solidus composition. Plagioclase, for example, may add a rim of new growth rather than react to maintain a single composition throughout. This results in a more calcic core and a progressively more sodic rim. Plagioclase (Figure 3.5b) is noted for this characteristic because the re-equilibration exchange is not simply Na for Ca but requires Al for Si as well, and this involves breaking the strong Si-O and Al-O bonds, which inhibits re-equilibration. Figure 6.8 implies that uniform cooling would produce successive rims of progressively more albitic composition ("normal" zoning).

Solid solutions are common in natural minerals. The most common substitution in the mafic minerals is that between Fe and Mg. This occurs in all mafic minerals and has an effect on the melting relationships similar to that in plagioclase. The olivine system, Mg_2SiO_4 (Fo, forsterite) − Fe_2SiO_4 (Fa, fayalite), is illustrated in Figure 6.10. Mg and Fe have the same valence and similar size. Mg is slightly smaller and thus forms a stronger bond in the mineral phase. As a result, the Mg-rich end-member typically has a higher melting point in olivine and other mafic minerals, and Mg is thus enriched in the solid as compared to the liquid at intermediate compositions. A melt of composition a (Fo_{56}), for example, will first produce a solid at c (Fo_{84}) at about 1700°C, and it will completely crystallize at 1480°C, when the final liquid (point d, Fo_{23}) is consumed. The behavior of the olivine system is entirely analogous to that of plagioclase.

6.5.2 Binary Eutectic Systems

Adding a second component certainly has a profound effect on a one-component system, but the effects are not limited to solid-solution behavior. In a great number of

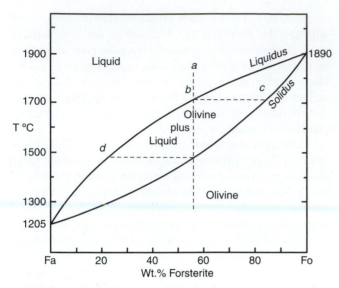

FIGURE 6.10 Isobaric *T-X* phase diagram of the olivine system at atmospheric pressure. After Bowen and Schairer (1932). Reprinted by permission of the *American Journal of Science*.

binary systems, the additional component does not enter into a solid solution but changes the melting relationships nonetheless. As an example of a binary system with no solid solution, let's turn to a system with considerable natural applicability. The system $CaMgSi_2O_6$ (Di, diopside)–$CaAl_2Si_2O_8$ (An, anorthite) is interesting in that it provides a simplified analog of basalt: clinopyroxene and plagioclase. The system is illustrated in Figure 6.11 as another isobaric (atmospheric pressure) *T-X* phase diagram. In this type of system, there is a low point on the liquidus, point *d*, called the **eutectic point**. Such systems are thus called **binary eutectic systems**. Because there is no solid solution, there is no solidus (although some petrologists refer to the line *g-h* as a type of solidus).

Let's discuss equilibrium cooling and crystallization of a liquid with a bulk composition of 70 wt. % An from point *a* in Figure 6.11. This *T-X* phase diagram is also isobaric, so Equation (6.2) with a single liquid yields $F = 2 - 1 + 1 = 2$. We can thus specify *T* and X_{An}^{liq} or X_{Di}^{liq} to completely determine the system. Cooling to 1450°C (point *b*) results in the initial crystallization of a solid that is *pure* An (point *c*). $F = 2 - 2 + 1 = 1$, just as with the plagioclase system. If we fix only one variable, such as *T*, all the other properties of the system are fixed (the solid composition is pure anorthite, and the liquid composition can be determined from the position of the liquidus at the temperature specified).

As we continue to cool the system, the liquid composition changes along the liquidus from *b* toward *d* as the composition of the solid produced remains pure anorthite. Naturally, *if anorthite crystallizes from the melt, the composition of the remaining melt must move directly away from An* (on the left in Figure 6.11) as it loses matter of that composition. The crystallization of anorthite from a cooling liquid is another **continuous reaction**, taking place

over a range of temperature. The reaction may be represented by:

$$Liquid_1 = Solid + Liquid_2 \qquad (6.6)$$

We can still apply the lever principle [Equation (6.5)] at any temperature to determine the relative amounts of solid and liquid, with the fulcrum at 70% An. If we do so at a number of temperatures, we would see that the ratio of solid to liquid increases with cooling, as we would expect.

At 1274°C, we have a new situation: diopside begins to crystallize along with anorthite. Now we have *three* coexisting phases, two solids and a liquid, at equilibrium. Our horizontal (isothermal) tie-line connects pure diopside at *g*, with pure anorthite at *h*, and a liquid at *d*, the eutectic point minimum on the liquidus. $\phi = 3$, so $F = 2 - 3 + 1 = 0$. This is a new type of invariant situation, not represented by any specific invariant point on the phase diagram. Because it is invariant, *T* and the compositional variables for all three phases are fixed (points *g*, *d*, and *h*). The system is completely determined and remains at this temperature as heat is lost and crystallization proceeds (just as with our ice water and boiling water, as discussed above). The amount of liquid decreases, and both diopside and anorthite are produced. Because the *amounts* (extensive variables) of all three phases change at a constant temperature, it is impossible to determine the relative amounts of them geometrically using the lever principle. The lever principle can be applied, however, to determine the ratio of diopside to anorthite that is being crystallized at any instant from the eutectic liquid. If the liquid composition is the fulcrum (about 42% An), and the solids are pure (0% An and 100% An), the ratio of diopside to anorthite crystallizing at any moment must be 58/42. Removing this ratio keeps the liquid composition from changing from the eutectic as crystallization proceeds.

The fact that the compositions of diopside, anorthite, and liquid are collinear is an example of an important relationship that we encounter often in petrology. It is one type of geometric relationship that implies a possible reaction. When three points are collinear, the central one can be created by combining the two outer compositions (in the proportion determined by the lever principle). In the present case the reaction must be:

$$Liquid = Diopside + Anorthite \qquad (6.7)$$

because liquid is in the middle. This type of reaction is a **discontinuous reaction** because it takes place at a *fixed* temperature until one phase is consumed. When crystallization is complete, the loss of a phase (liquid, in this case) results in an increase in *F* from 0 to 1, and thus temperature can once again be lowered, with the two phases diopside and anorthite coexisting at lower temperatures. Because the composition of the two solids is fixed, we have a unique opportunity to determine exactly which of our intensive variables is free to vary; temperature is the only variable left.

A discontinuous reaction involves one more phase than a corresponding continuous reaction in the same system, and

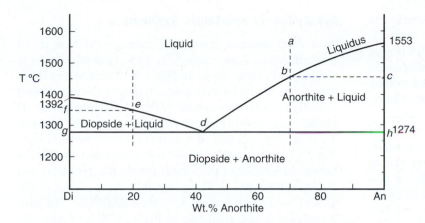

FIGURE 6.11 Isobaric (0.1 MPa) *T-X* phase diagram of the diopside–anorthite system. After Bowen (1915b). Reprinted by permission of the *American Journal of Science*.

because this decreases the variance, the compositions of the reacting phases do not vary as the reaction progresses. Only the proportion of the phases changes (usually until one phase is consumed). Such reactions are discontinuous in the sense that the phase assemblage changes at a single temperature due to the reaction. In this case, diopside + liquid gives way to diopside + anorthite as the system is cooled through the reaction temperature (1274°C).

Let's see what happens on the left side of the eutectic point. Cooling a liquid with a composition of 20 wt. % An results in the crystallization of pure diopside first, at 1350°C as the liquidus is encountered at point *e* in Figure 6.11. Diopside continues to crystallize as the liquid composition proceeds directly away from diopside toward point *d*. At point *d* (1274°C again), anorthite joins diopside and the eutectic liquid in the same invariant situation as above. The system remains at 1274°C as the discontinuous reaction, liquid = Di + An, runs to completion, and the liquid is consumed.

In these eutectic systems note that, for any binary bulk composition (not a pure end-member), *the final liquid to crystallize must* always *be at the eutectic composition and temperature.* The final cooled product of a binary liquid with no solid solution must contain both anorthite and diopside. To get there, we must have them both coexisting with a melt at some point, and that melt has to be at the eutectic point. Remember that solid-solution systems do not behave this way. In them, crystallization is complete when the composition of the solid becomes equal to the bulk composition, so the final liquid, and the temperature, depends on the bulk composition.

Equilibrium melting is the opposite of equilibrium crystallization. *Any* mixture of diopside and anorthite begins to melt at 1274°C, and the composition of the first melt is *always* equal to the eutectic composition *d*. Once melting begins, the system is invariant and will remain at 1274°C until one of the two melting solids is consumed. Which solid is consumed first depends on the bulk composition. If X_{bulk} is between Di and *d*, anorthite is consumed first, and the liquid composition will follow the liquidus with increasing temperature toward Di until the liquid composition reaches X_{bulk}, at which point the last of the remaining diopside crystals will melt. If X_{bulk} is between An and *d*, diopside is consumed first, and the liquid will progress up the liquidus toward An.

Note the discontinuities between either one-component end-member and the binary mixture. For example, pure anorthite melts at a single temperature of 1553°C. If we add just a tiny amount of Di to this, the first melt occurs at 1274°C and has a composition equal to *d*. Of course, there won't be much of this melt. (Use the lever principle if you don't see why.) As temperature increases in this An_{99} mixture, the amount of melt increases gradually and becomes rapidly more anorthitic. Melting is extensive and complete just below 1553°C.

Fractional crystallization has no effect on the path followed by the liquid in eutectic systems without solid solution. Unlike in the plagioclase or olivine systems, removing a solid of constant composition does not affect the composition of either of the two final minerals or of the last liquid. The compositions of the minerals are fixed, and the liquid must reach the eutectic composition whether or not the solids are removed. Only the composition of the final *rock* is affected. Following equilibrium crystallization, the final rock composition is the same as the bulk composition. If fractional crystallization is efficient, the final rock composition is equal to the eutectic because earlier crystals are lost and the last liquid is always the eutectic liquid.

Partial melting, however, does affect the path that the liquid follows. Perfect fractional melting (removal of any melt increment as soon as it forms) should not occur in nature. As we shall see later, a critical amount of melt (perhaps 1 to 10%) must form before it can be physically removed from the solid. A smaller amount will merely wet the mineral grain boundaries and remain adsorbed to the crystal surfaces. Nonetheless, if a few percent partial melt were almost continuously removed from an initially solid sample of diopside + anorthite, and the first melts were (necessarily) of the eutectic composition *d*, the melt increments being removed would continue to be of composition *d*, until one of the solid phases was finally consumed by melting. Then the remaining solid would be a one-component system. Therefore, no melting would occur between 1274°C and the melting point of the remaining pure phase, so the composition and the temperature of the melt being extracted would *jump discontinuously* from *d* at 1274°C to either pure diopside at 1392°C or pure anorthite at 1553°C, depending on the initial bulk composition and which phase is consumed first. Thus a partially melting mantle source, if it becomes depleted in one mineral phase, may require a significantly higher temperature in order to create further melts. If only sufficient heat were available to initiate melting at the eutectic temperature, consumption of any single mineral could raise the melting point of the residual solid by several hundred degrees, thereby shutting off the supply of magma feeding a volcanic area.

Furthermore, suppose we have **equilibrium partial melting**, and an intermediate melt separates from the solids as a single event at any point during an equilibrium melting process. Then the melt has a bulk composition different than that of the original system. Extracting such a melt and crystallizing it in some shallower magma chamber will produce a rock with a different anorthite/diopside ratio than would have resulted from crystallizing the original bulk composition. In other words, partial melts should not have the same composition as their source, and they should be enriched in the low-melting-temperature components (higher Fe/Mg, Na/Ca, etc.).

Using these simplified systems can be useful because they permit us to avoid chemical complexities (Ca-Na in plagioclase, Fe-Mg in clinopyroxene) and thus let us focus on some particular property without considering other complicating variables. Once we can isolate some property and understand it, we can then add other components to approach more realistic magmatic systems. What does this simplified Di-An basalt system tell us? It explains once again how liquids of more than one component crystallize over a *range* of temperatures, even without solid solution (observation 1 concerning the behavior of natural melts in Section 6.1). Second, it explains how we can get a *sequence* of minerals crystallizing as a basalt cools and that the sequence varies with composition (observations 2, 3, and 7). If the composition is to the right of the eutectic, anorthite will form, followed by diopside (+ anorthite). Figure 3.8 illustrates ophitic texture in a basalt, in which earlier euhedral to subhedral plagioclase crystals are surrounded by later interstitial augite. Note that the single augite crystal in Figure 3.8 conforms to the shape of the earlier plagioclases. If the bulk composition is to the left of the eutectic, augite crystals form first. Figure 3.7 illustrates a gabbroic cumulate in which earlier euhedral augite crystals are embedded in poikilitic late plagioclase. Also note that the Di-An system suggests that the composition of initial partial melts, even of more complex natural plagioclase–clinopyroxene rocks, ought to be concentrated around a specific composition (the eutectic).

Figure 6.11 illustrates the common tendency for an added component to lower the melting point of the complementary one-component system. Adding a diopside component to pure anorthite or an anorthite component to pure diopside results in a lower melting point of the mixture than in the nearby pure system. In contrast to the plagioclase or olivine systems, the second component here results in a second mineral, not a single mineral with a variable composition. We can thus distinguish two different types of additional components: those that mix into the phase with the original component(s) and those that don't and thus require a new phase.

6.5.3 Binary Peritectic Systems

As a third example of binary systems, we'll look at the forsterite–silica system (Mg_2SiO_4-SiO_2) in another isobaric *T-X* phase diagram, shown in Figure 6.12. In addition to the eutectic minimum on the liquidus (point *c*), there is another inflection point in this system (point *i*), called the **peritectic point**. Such systems are thus called **peritectic systems**. There are still only two components in binary peritectic systems, but an intermediate phase, in this case enstatite (En), is located between the end-member phases (forsterite, Fo, and a silica polymorph, S). The actual SiO_2 phase present in this system varies with temperature (see Figure 6.6). The two-liquid field on the right side of Figure 6.12 is another unusual feature, but this is not an essential feature of peritectic systems. We will discuss it further below.

Following an earlier discussion, because the composition of enstatite plots between forsterite and cristobalite, it is possible to combine forsterite and cristobalite in some proportions to produce enstatite. In other words, a **reaction** is possible. In this case the reaction is:

$$Mg_2SiO_4 + SiO_2 = 2MgSiO_3 \tag{6.8}$$
$$\text{forsterite} \quad \text{crist.} \quad \text{enstatite}$$

If we were to use thermodynamics to calculate the conditions under which this reaction takes place, we would find that under all realistic conditions, the reaction runs to the right, as written. In other words, *Mg-rich olivine and quartz can never coexist in equilibrium in igneous rocks!* If together, they would react to form orthopyroxene until one or the other were consumed. As a result, there is no field labeled "forsterite + quartz" (or any silica polymorph) anywhere in

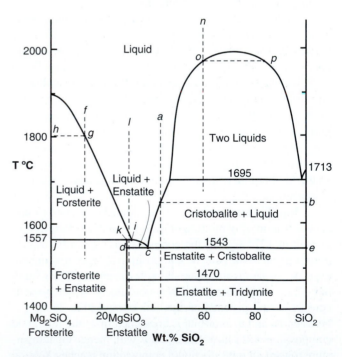

FIGURE 6.12 Isobaric *T-X* phase diagram of the system Fo-silica at 0.1 MPa. After Bowen and Anderson (1914) and Grieg (1927). Reprinted by permission of the *American Journal of Science*.

Figure 6.12. Only at the extreme Fe-rich end of the olivine series can fayalite coexist with quartz in some alkaline granites/rhyolites and uncommon iron-rich rocks.

On the right side of the *eutectic* point in Figure 6.12 (point *c*), the behavior of the Fo-SiO$_2$ system is similar to that of a eutectic system such as Di-An (if we avoid, for the moment, the two-liquid field). For example, suppose we cool a melt of composition *a*. Above the liquidus, Equation (6.2) tells us that $F = 2 - 1 + 1 = 2$, so we can vary temperature and the liquid composition independently, and we must specify *T* and either X_{An}^{liq} or X_{Fo}^{liq} to determine the state of the system. If we draw a line straight down from point *a*, emulating decreasing temperature, it would intersect the liquidus at about 1660°C, at which point cristobalite begins to crystallize (point *b*). Because $\phi = 2$, $F = 2 - 2 + 1 = 1$, and the liquid composition is temperature dependent and is constrained to follow the liquidus with continued cooling and crystallization of more cristobalite via a continuous reaction until it reaches point *c*, the eutectic, at 1543°C. At this temperature, enstatite (point *d*) joins cristobalite (point *e*) and liquid in a manner analogous to the diopside–anorthite eutectic situation. Now, because $C = 2$ and $\phi = 3$, $F = 2 - 3 + 1 = 0$, the temperature must remain constant as the liquid is consumed by a *discontinuous* reaction to form enstatite and cristobalite. Once again, we can use our geometric relationship to see that, because the composition of the liquid lies between those of enstatite and quartz, the reaction must be:

$$\text{Liquid} \rightarrow \text{Enstatite} + \text{Cristobalite} \qquad (6.9)$$

Once the liquid is consumed, the system has only two solid phases (enstatite and cristobalite) and $F = 1$, so we can continue to lower the temperature. At 1470°C, there is a phase transition in the SiO$_2$ system, and cristobalite inverts to tridymite (see Figure 6.6). Similar transitions to high and low quartz occur at temperatures below that of Figure 6.12.

Let's next explore the left side of the eutectic in Figure 6.12 with a liquid of composition *f*. At high temperatures, we begin with just liquid, and $F = 2$. At 1800°C, forsterite begins to crystallize (point *h*). With two coexisting phases, $F = 2 - 2 + 1 = 1$, so the composition of the liquid depends upon temperature. With further cooling, the liquid composition varies along the liquidus as the amount of forsterite increases and liquid decreases. At 1557°C, enstatite (point *k*) joins forsterite (point *j*) and liquid (point *i*). Because $\phi = 3$, Equation (6.2) for this isobaric system yields $F = 2 - 3 + 1 = 0$. This peritectic is a new type of invariant situation for us. Because $F = 0$, all the intensive variables are fixed, including *T* and the composition of all phases. Just as in the eutectic situation (Di + An + liquid), the cooling system must remain at constant *T* as heat is lost. Eventually a phase will be lost, the system will regain a degree of freedom, and the temperature can again change.

While invariant, we note that there are again three phases in a two-component system, and the three phases are collinear, implying that a reaction must take place, and we can determine the reaction geometrically. This time, however,

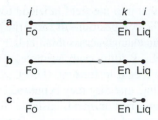

FIGURE 6.13 Schematic enlargements of the peritectic tie-line from Figure 6.12, showing the coexisting phases Fo, En, and Liq. The gray dot indicates possible bulk compositions.

notice that the composition of enstatite (point *k*) lies between those of liquid (point *i*) and forsterite (point *j*). This is different from the eutectic situation, where the liquid falls in the center. This geometry tells us that the reaction must be:

$$\text{Forsterite} + \text{Liquid} \rightarrow \text{Enstatite} \qquad (6.10)$$

This is somewhat new: we have a liquid reacting with a solid to produce another solid as it cools. Like Reactions (6.7) and (6.9), this reaction is *discontinuous* ($F = 0$), so it runs at constant temperature. Reaction (6.10) involves *two* reactants as written. It will thus continue (with cooling) until *one* of the reactants is consumed. This is an important point but is often overlooked. We tend to look at reactions such as Reaction (6.10) and think that both reactants are completely consumed as the product is produced. *This will really happen only in the rare circumstances that the reactants exist in exactly the correct proportions so that they are exhausted simultaneously.* One will typically be exhausted first. The other, with no remaining co-conspirator with which to react, remains perfectly stable in the system with the product(s). In the present situation, which reactant is consumed first by the reaction? We can answer this question geometrically on Figure 6.12 by applying the lever principle in a slightly different way. Figure 6.13a illustrates the geometry of the peritectic tie-line. I have shifted the positions of the phases from Figure 6.12 a little to aid visualization and used mineral abbreviations to save space. Reaction (6.10) involves the two line-end phases (Fo + Liq) combining to form the phase between them (En). The lever principle tells us that Fo and Liq must be combined (react) in the proportions Fo:Liq $= \overline{kj} : \overline{jk}$ in order to produce En. This is reasonable because En plots closer to Liq than to Fo, so it must contain less Fo than it does liquid components. If the bulk composition of our melt were exactly equal to En, then Fo and Liq would both be exhausted at the same instant. If the bulk composition plots between Fo and En, however (the gray dot in Figure 6.13b), there is an excess of Fo, and the liquid would be consumed first, leaving forsterite and enstatite. Alternatively, if the bulk composition plots between En and Liq (the gray dot in Figure 6.13c), there would be an excess of liquid, and forsterite would be consumed first.

In the present situation, the bulk composition (*f*) lies between enstatite and forsterite, so the liquid must be consumed first. Once the liquid is used up, the remaining forsterite will coexist with enstatite, $\phi \rightarrow 2$, $F \rightarrow 1$, and the temperature drops into the field labeled "forsterite + enstatite." Because the compositions of forsterite and enstatite are fixed, only *T* can change across this field.

Looking back at this path, we note an unusual feature. Olivine crystals begin to form at 1800°C, they continue to grow with cooling, and then they begin to be consumed back into the melt as a new mineral, enstatite, forms. This phenomenon is observed in some basalts, in which the olivine phenocrysts are ragged and embayed, suggesting that they were partially consumed (**resorbed**) by the melt after they had initially formed. This is exactly what we observed in the Makaopuhi lava lake data in Figure 6.2. The enstatite produced may (but does not have to) occur as a **reaction rim**, or mantle, on the olivine, where the two reactants were in contact (Figure 3.10). You might also recognize this olivine → pyroxene transition as the first step in the "discontinuous" left side of Bowen's Reaction Series (Figure 7.14). The Ab-An system is, of course, the "continuous" right side.

An even more interesting scenario happens for a composition such as *l* (as shown enlarged in Figure 6.14). Here we get a similar first appearance of olivine when the liquid first encounters the liquidus, and then enstatite (*k*) + liquid (*i*) + olivine (off to the left) coexist as an invariant assemblage at the peritectic temperature. Because our bulk composition (*l*) is now between liquid and enstatite, as in Figure 6.13c, forsterite is consumed before the liquid by Reaction (6.10). Here the first-forming olivine crystals are *completely* resorbed back into the melt, and the remaining liquid composition then moves further along the liquidus, coexisting with enstatite as *F* = 1, until it eventually reaches the eutectic composition (*c*), and cristobalite (off to the right) begins to crystallize along with enstatite (*d*) and liquid (*c*). We now have the invariant *eutectic* situation, and the system remains at 1543°C until the liquid is consumed by Reaction (6.9). The final rock contains enstatite and a silica polymorph. Compositions that fall between the peritectic and enstatite thus behave in an unusual fashion. Olivine is the first phenocryst phase, and it is joined by pyroxene at a lower temperature. Olivine then disappears, and quartz appears later as a substitute. Such phenomena are observed in some silica-saturated basaltic compositions.

Fractional crystallization, to the left of the peritectic in Figure 6.12, involves the isolation of the crystallizing olivines, and the bulk composition shifts (being equal to the liquid composition). If olivine fractionation is effective, the final liquid will *always* reach the eutectic, even if the original composition is to the left of the enstatite composition. This is an unusual situation in which fractional crystallization does affect the liquid evolution path in a system without solid solution.

Let's return to the two-liquid field in the silica-rich portion of Figure 6.12. If we cool a melt of composition *n*, it intersects a loop in the liquidus. This loop is called a **solvus** and represents the **exsolution** process (the separation of once-mixed phases). In this case, we have *liquid* exsolution, but we shall see some *solid* exsolution examples shortly. The initial liquid (φ = 1 and *F* = 1) cools to about 1980°C and separates into two **immiscible** liquids as a

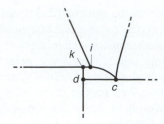

FIGURE 6.14 Enlargement of the peritectic area of Figure 6.12.

second liquid of composition *p* forms. The separation resembles oil and vinegar in salad dressing. Now φ = 2, so *F* = 2 − 2 + 1 = 1, so the compositions of both liquids are temperature dependent and follow the two limbs of the solvus curve with continued cooling. One liquid thus gets more silica rich, whereas the other becomes enriched in Mg. At 1695°C, the silica-rich liquid reaches a eutectic, and cristobalite crystallizes. The system now contains two liquids plus a solid and is invariant (*F* = 2 − 3 + 1 = 0). The temperature remains constant as a discontinuous reaction involves these three collinear phases. Because the silica-richer liquid plots between the Mg-richer liquid and cristobalite, the reaction must be:

$$\text{Silica-rich liquid} \rightarrow \text{Mg-rich liquid} + \text{Cristobalite}$$
$$(6.11)$$

which proceeds with cooling until the silica-rich liquid is consumed. Further cooling will be identical to the case for liquid *a*, discussed earlier. The exact location of the solvus in this system is poorly constrained.

Let's consider **melting relations** for a moment. If we begin with pure enstatite and melt it, we produce a liquid of composition *i* (the peritectic composition) and forsterite via Reaction (6.10) in reverse (with increasing *T*). The process by which a solid melts to a liquid and another solid, both of different compositions than the original, is called **incongruent melting**. Enstatite is one of several minerals that melt incongruently. All peritectic systems behave this way, and a peritectic phase diagram may also be called an *incongruent melting diagram*. The system leucite–silica is similar to forsterite–silica and exhibits the incongruent melting of the intermediate compound, sanidine.

Another interesting process begins with a mixture of solid olivine and enstatite. If we melt this, we get an initial melt of composition *i*, which is more silica rich than either of the two solids. If we remove the melt (partial melting again) and crystallize this melt elsewhere, the result is a mixture of enstatite + *quartz*. So we began with pyroxene and olivine and ended with pyroxene and quartz. Once again, by observing these simple systems, we get important keys to more complex melt behavior. By extracting a partial melt from a solid mineral assemblage, the new melt system will have a different composition, which, when crystallized, may produce a very different final rock. In the case of the peritectic diagram, the passing of the liquidus above and across the enstatite composition permits the liquid composition to cross over from olivine-rich to olivine-depleted compositions. Removal of such liquids can produce systems of very different composition than the original.

We will discuss **pressure effects** in more detail in Chapter 7, after we discuss three-component systems. For now, however, note the following points. Changing pressure can cause changes in these phase diagrams. As you may remember from Chapter 5 and the one-component diagrams, increasing pressure causes the melting point of most phases to increase. From this, we may predict that the effect of rising pressure on these two-component diagrams would be to increase the liquidus temperatures. But not all phases respond to the same extent as pressure increases. In a eutectic system, for example, the melting point of one mineral may rise more than another. As a result the eutectic point generally shifts laterally away from the phase for which the melting point rises the most.

For the Fo–En–silica system, the peritectic shifts toward Fo with increasing pressure. It eventually becomes coincident with the enstatite composition (at a specific pressure) and then moves further toward forsterite. The diagram then resembles the upper one in Figure 6.15, which is essentially two eutectic diagrams connected at enstatite. At these high pressures, enstatite melts *congruently* to a liquid of the same composition. The temperature maximum at enstatite now forms a **thermal barrier** for liquids in the system. Once enstatite forms, any liquid to the left of enstatite will cool toward forsterite, whereas liquids to the right will cool toward quartz. There are no fancy shifts across the enstatite composition, as in the peritectic system. At these high pressures, it becomes impossible to extract a silica-rich melt from which cristobalite can crystallize from an original mixture of olivine and enstatite. Also, at high pressures, there should be no peritectic reaction upon cooling between olivine and liquid to produce enstatite, so embayed olivine crystals, or orthopyroxene rims on olivine, should be absent.

6.5.4 The Alkali Feldspar System

As a final example of binary systems, we address the $NaAlSi_3O_8$-$KAlSi_3O_8$ system (Ab-Kfs, or the alkali feldspar system). The somewhat simplified *T-X* phase diagram (at a

H_2O pressure of 0.2 GPa) is illustrated in Figure 6.16. The system at this low pressure is like a cross between the plagioclase solid solution system and the Di-An eutectic system. Complete solid solution is possible in Figure 6.16, so that there is a looped liquidus–solidus pair (as in the plagioclase or olivine systems). The loop, however, shows a minimum temperature, thus forming two loops on either side of a eutectic minimum point, *f*. The cooling behavior is similar to that for the plagioclase system. Cooling a melt of composition *a* in Figure 6.16 to the liquidus results in the crystallization of a potassium-rich alkali feldspar (orthoclase) of composition *b* at about 1100°C. The feldspar coexists with a more sodium-rich melt (point *c*). Because $F = 2 - 2 + 1 = 1$, the compositions of both melt and feldspar are temperature dependent, following the liquidus and solidus curves, respectively, with continued cooling. Cooling is accompanied by a continuous reaction, similar to Reaction (6.4), and the relative amount of liquid decreases and solid increases, as can be determined at any temperature by the lever principle. Unlike in the Di-An eutectic system, the liquid will *not* reach the eutectic via equilibrium crystallization. When the composition of the feldspar reaches *d*, the solid composition becomes equal to the bulk composition, so there is only a tiny quantity of liquid (of composition *e*) remaining. The final drop of liquid is thus used up at this point, and there is a single feldspar at lower temperatures as $\phi \rightarrow 1$ and $F \rightarrow 2$ in the divariant field labeled "single feldspar" in Figure 6.16. With a single feldspar, we must define T and X_{Ab}^{feld} or X_{Or}^{feld} to determine the system.

Cooling a melt of composition *i* in Figure 6.16 would result in crystallization of a Na-rich feldspar of composition *j* at about 1000°C, coexisting with a more K-rich melt. Cooling would cause both the liquid and feldspar to become *less* sodic as the compositions follow the liquidus and solidus, respectively. The final liquid would have the composition *k*, and a single feldspar of composition *i* would exist below this.

Because there is solid solution, **fractional crystallization** would affect this system, shifting the last liquid (and solid) compositions closer to the eutectic point, regardless of which side of the eutectic the bulk composition lies.

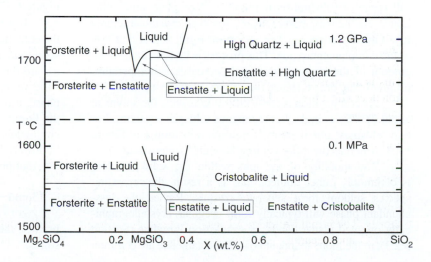

FIGURE 6.15   The system Fo-SiO₂ at atmospheric pressure and 1.2 GPa. After Bowen and Schairer (1935), Chen and Presnall (1975). Copyright © Mineralogical Society of America.

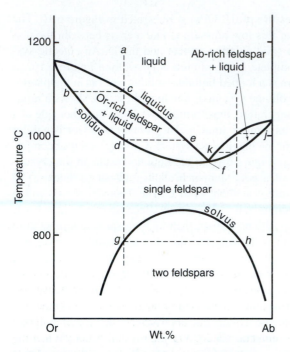

FIGURE 6.16 *T-X* phase diagram of the system albite–orthoclase at 0.2 GPa H_2O pressure. After Bowen and Tuttle (1950). Copyright © by the University of Chicago.

There is also a **solvus** in this system, one that involves the separation of two *solid* phases from a homogeneous solid solution. The solvus is caused by differences in the size of K^+ ions (ionic radius = 1.59 Å) and Na^+ ions (ionic radius = 1.24 Å). In the pure end-members, this size difference is accommodated by a slightly larger unit cell in orthoclase than in albite. Both structures are stable, but when some of the larger K^+ ions are introduced into the smaller albite unit cell, and vice versa, some distortion results. If the distortion is great enough, each end-member would be able to accept only a limited quantity of the foreign ion before the distortion created by the mixed sizes caused the structure to reject any further additions. This would place a limit on the amount of Na^+ that orthoclase could accept and of K^+ that albite could accept.

At high temperatures, the vibrational energy in the crystal structure permits minerals to accept more of the foreign ion. In the case of the alkali feldspars in Figure 6.16, the full range of substitution is possible ("complete solid solution"). As the temperature drops, however, crystal structures lose vibrational energy, become more rigid, and accept less of the complementary ion. This process is incremental, so the amount of the impurity tolerated decreases as temperature progressively drops. Hence the solvus is convex upward.

As composition *a* cools until it intersects the solvus at about 780°C, a single, homogeneous feldspar separates into two feldspars; one is more K rich (composition *g* in Figure 6.16), whereas the other is more Na rich (composition *h*).

The mobility of the ions within the solid structure is commonly rather limited, and, as a result, the separating species seldom form separate crystals. Rather, the less-abundant phase will typically form irregular, crystallographically oriented planar bands, or **exsolution lamellae** (Section 3.2.3), in the more abundant host. These lamellae are common

in alkali feldspars. When the bulk composition is K rich (as in composition *a* in Figure 6.16), lamellae of Na-rich feldspar form in a K-rich host, and the texture is called **perthite** (Figure 3.18a). You have probably seen this texture in orthoclase hand specimens. These are the thin, wispy layers seen on a number of cleavages. When the bulk composition is Na rich (as in composition *i* in Figure 6.16), lamellae of K-rich feldspar form in a Na-rich host, and the texture is called **antiperthite**. Exsolution is less common in plagioclases (probably due to the Ca-Al component and the requirement of coupled Ca-Al for Na-Si exchange). There are three known solvi that occur at low temperatures in the plagioclase system but rarely develop as lamellae and have been left out of Figure 6.8.

When two feldspars coexist in Figure 6.16, $\phi = 2$, so $F = 1$. This means that the composition of both feldspars is a function of temperature. As the temperature decreases, the compositions of both feldspars follow the solvus limbs, with the sodic feldspar becoming more sodic and the potassic feldspar becoming more potassic. This is an example of the technique of "geothermometry," by which we can calculate the temperature of equilibration from the composition of analyzed coexisting minerals. We will discuss this technique in Chapter 27.

The alkali feldspar system provides another important example of the effects of pressure on mineral systems. The effect of increased H_2O pressure will be greater on the solid–liquid equilibrium than for the solvus because the liquid is both more compressible than either solid and is the only phase that can accept some H_2O. The addition of H_2O would enter the liquid, thereby stabilizing it at the expense of the solid (Le Châtelier's Principle). Increasing the H_2O pressure, therefore, *lowers* the melting point but will have little effect on the solvus.

Figure 6.17 shows that, as H_2O pressure increases, the liquidus and solidus move to lower temperatures, whereas the solvus is largely unaffected. The liquidus and solidus eventually intersect the solvus as H_2O pressure approaches 500 MPa (Figure 6.17c). The area of a single homogeneous phase, either liquid or solid, has been shaded in all three diagrams. Notice that the area of a single feldspar solid solution gets progressively smaller with increasing water pressure, and above 500 MPa the range of solid solution is no longer complete. Systems such as Figure 6.17c are called **limited solid solution**.

If we cool a liquid with a composition to the right of the eutectic point in Figure 6.17c, the first solid to form is a sodic feldspar. The liquid composition follows the liquidus toward the eutectic point *a*, whereas the composition of the solid moves toward *b*. At around 600°C, we get a third phase, a potassium-rich feldspar of composition *c*, joining the liquid and the sodic feldspar. *F* is now 0, so we have a completely determined system, and the temperature cannot be lowered further until a phase is consumed. Upon cooling, that phase would be the melt, and the reaction must be:

$$Liquid \rightarrow Na\text{-rich feldspar} + K\text{-rich feldspar} \quad (6.12)$$

because the melt composition lies between those of the two solids.

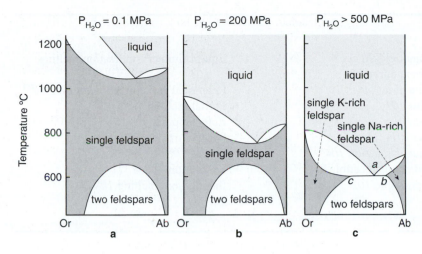

FIGURE 6.17 The albite–K–feldspar system at various H_2O pressures. Shaded areas represent possible single-phase compositions. **(a)** and **(b)** after Bowen and Tuttle (1950), copyright © by the University of Chicago. **(c)** after Morse (1970). Reprinted by permission of Oxford University Press.

In natural Ca-Na-K systems, the solvus separates a K-rich alkali feldspar (usually orthoclase) and a Na-Ca-rich plagioclase feldspar. The behavior of this system is familiar to us, in that it results in two coexisting feldspars, a plagioclase and an orthoclase, such as we find in many granitic rocks. Figure 6.17c is called a **subsolvus** feldspar system because the solvus is truncated, and no single midrange feldspar can form above it. The addition of calcium expands the solvus, resulting in subsolvus behavior at even lower H_2O pressures than in Figure 6.17.

Figures 6.16, 6.17a, and 6.17b are called **hypersolvus** feldspar systems because a melt crystallizes completely to a *single* alkali feldspar above the solvus, followed by solid-state exsolution. Granitic rocks formed in shallow chambers (low H_2O pressure) may exhibit single intermediate feldspars rather than separate crystals of orthoclase and plagioclase.

Once the melt is completely crystallized, further cooling causes the two solids to change composition via a continuous reaction along the solvus limbs and further expel some of the foreign component. Subsolidus reactions, such as along solvi, are very sluggish because they must occur in solid crystal structures at low temperature. If the system cools fairly quickly, equilibrium will not be attained, and the solids will not exsolve. Volcanic feldspars, and those in some shallow or small intrusives, may cool quickly enough to suppress perthitic textures.

Summary

Natural melts crystallize over a range of temperatures. Over that temperature interval, several mineral phases form in sequences that vary, depending upon the composition of the melt. The composition of the liquid also changes, as do the compositions of minerals that exhibit solid solution. We can understand many of these complex features by performing experiments on simplified analog systems (typically having three or fewer components). The Gibbs phase rule is a simple yet elegant tool for analyzing and understanding the variables that govern the behavior of these systems. The phase rule is $F = C - \phi + 2$, where F is the variance (the number of intensive variables that must be specified to fully determine the state of a system), ϕ is the number of phases in the system, and C is the number of components. Intensive variables are those that are not dependent on the amount of the constituents, and those most important to geologic systems are T (temperature), P (pressure), and X (composition of the system or phases in it).

Phase diagrams are graphic depictions of the stable phases or phase assemblages, as determined by experiments and typically shown on graphical $P–T$ or $T–X$ diagrams. We can use the phase rule to study in detail the behavior of experimental systems displayed on phase diagrams. One-component systems (e.g., SiO_2, H_2O) are readily depicted on $P–T$ phase diagrams, whereas binary (two-component) diagrams require isobaric $T–X$ (or isothermal $P–X$) diagrams if we are to visualize them adequately. On these diagrams, divariant ($F = 2$) mineral assemblages occupy two-dimensional fields, univariant ($F = 1$) mineral assemblages adhere to one-dimensional curves, and invariant ($F = 0$) mineral assemblages occur at points.

By selecting various hypothetical bulk compositions in binary systems and drawing descending paths on $T–X$ phase diagrams (representing cooling), we can determine the sequence of minerals that form, variation in the composition of the liquid and solid phases, and the dependence of these factors on the intensive variables under any particular set of conditions. We can also construct tie-lines connecting coexisting phases at any temperature and determine the relative proportions of the phases or infer the reactions between them as the system continues to cool. Fractional crystallization can also be understood if we consider removing solids as they form, thereby causing the bulk composition to shift with the composition of the liquid. Fractional crystallization both extends the range of temperature over which a melt crystallizes and shifts the composition of solution phases toward the low-temperature end-member. We can similarly understand melting processes (both equilibrium and partial melting) with increasing temperature. The phase rule is entirely general, and can be applied to subsolidus (metamorphic) systems as well. One example is the exsolution behavior associated with the solvus in the Or-Ab binary system.

Key Terms

Phase rule *95*
System (open, closed, isolated) *95*
Surroundings *95*
State *95*
Phase *95*
Component *95*
Variable (intensive, extensive) *96*
Variance (degrees of freedom) *96*
Latent heat of fusion/vaporization *98*

Divariant field *98*
Univariant curve *98*
Invariant point *99*
Liquidus *101*
Solidus *101*
Tie-line *101*
Continuous reaction *104*
Discontinuous reaction *104*
Lever principle *102*

Equilibrium crystallization/melting
103
Fractional crystallization/melting *103*
Partial melting *103*
Eutectic point *104*
Peritectic point *106*
Solvus *108*
Exsolution *108*
Incongruent melting *108*

Review Questions and Problems

Review Questions and Problems are located on the author's web page at the following address: **http://www.prenhall.com/winter**

Important "First Principle" Concepts

- The phase rule, $F = C - \phi + 2$, is fundamentally important.
- Intensive variables, those that are independent of the size/extent of a system, are really the variables that govern the behavior of systems. The ones that are most important to systems of geological interest are T (temperature), P (pressure), and X (the composition of a system or phase).
- The phase rule applies only to systems at equilibrium. Any number of phases can coexist at low temperatures where chemical equilibrium is not readily attained (e.g., in a graywacke).
- The initial melt (and the final liquid to crystallize) in eutectic systems is always the eutectic composition.
- Chemical reactions generally occur when the variance is less than 2.
- *Continuous* reactions occur when $F > 0$ and take place over a range of temperatures (in isobaric systems). The compositions of solution-type phases vary continuously across the temperature interval of the reaction. Examples include $Liq_1 + Solid_1 = Liq_2 + Solid_2$ in systems exhibiting solid solution (the subscripts indicate compositions) or $Liq_1 = Liq_2 + Solid$ in systems with fixed-composition solids.
- *Discontinuous* reactions occur when $F = 0$ and thus cannot involve compositional or temperature changes. The entire reaction must occur at a single temperature, so that one phase assemblage exists above that temperature (however small an increment) and a different phase assemblage exists below it.

- *Discontinuous* reactions can be inferred simply by observing the relative positions of the phases on the tie-line connecting them at a particular temperature. For isobaric binary systems (such as those in this chapter), this involves three phases, and the reaction is thus: *the two phases at the ends of the tie-line combine to form the phase between them.* Examples include $Di + An = Liq_d$ at 1274°C in Figure 6.11 and $Fo + Liq_i = En$ at 1557°C or $En + SiO_2 = Liq_c$ at 1543°C in Figure 6.12.
- A reaction involving two or more reactants will proceed only until *one* of the reactants is consumed. The remaining reactants, because they no longer have all the requisite phases to proceed, will remain stable under conditions beyond the conditions of equilibrium. We can also use the position of the bulk composition on a tie-line to determine which phase is consumed. For example, if the bulk composition plots between Di and Liq_d in Figure 6.11, An will be consumed first with progressive heating, and Di will continue to be stable with liquid at higher temperatures. If the bulk composition plots between An and Liq_d, Di will be consumed first, and An will be stable with liquid at higher temperatures.
- We must not overlook pressure, which may have significant modifying effects on crystallization and melting behavior. We shall further explore the effects of pressure and fluids in Chapter 7.

Suggested Further Readings

Bowen, N. L. (1928). *The Evolution of the Igneous Rocks*. Princeton University Press. Princeton, NJ. (1956 reprint by Dover. New York.)

Ehlers, E. G. (1972). *The Interpretation of Geological Phase Diagrams*. W.H. Freeman. San Francisco.

Morse, S. A. (1994). *Basalts and Phase Diagrams. An Introduction to the Quantitative Use of Phase Diagrams in Igneous Petrology*. Krieger. Malabar, FL.

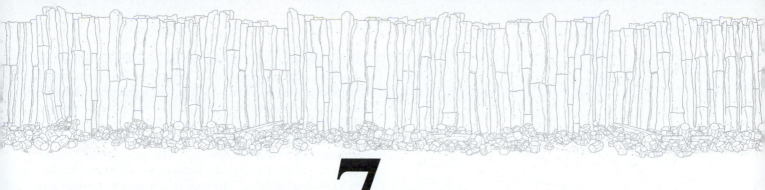

7

Systems with More Than Two Components

Questions to be Considered in this Chapter:

1. In what ways do systems with three or more components differ from the two-component systems described in Chapter 6?

2. What can we learn from melting experiments with natural igneous rocks?

3. Given the crystallization behavior that we have investigated, particularly the variation in crystallization sequence with changes in bulk composition of magmas, how are we to interpret Bowen's reaction series?

4. What are the principal effects of varying pressure on crystallization and melting behaviors?

5. What are the principal effects of volatile constituents, particularly H_2O and CO_2 on crystallization and melting behaviors?

7.1 THREE-COMPONENT (TERNARY) SYSTEMS

The addition of each successive component makes it increasingly difficult to visualize the variables in a system. Ideally, we should have a dimension for each variable in F, but we are limited to two-dimensional graphical portrayal. Pressure–temperature diagrams are easy to depict for one-component systems. For two-component systems, we have to rely on isobaric T-X diagrams, which are really isobaric sections through three-dimensional T-P-X diagrams. As we move to three-component systems, we can use isobaric three-dimensional T-X-X diagrams or can explore other ways to further simplify to two dimensions. We shall see that effective solutions ensue. Beyond three components, the difficulty involved in visualization commonly outweighs the gains of using models as simplified analogs of igneous systems. In this chapter, we shall analyze three relatively simple three-component systems that provide enough variety to illustrate the physical processes and the analytical techniques.

7.1.1 Ternary Eutectic Systems

The simplest three-component systems are eutectic systems with no solid solution. We shall thus return to the Di-An eutectic system (Figure 6.11) and add a third component: Mg_2SiO_4 (Fo, forsterite). Because olivine occurs in many basalts, the Di-An-Fo system is a more comprehensive basaltic model than Di-An or Fo-SiO_2 (Figure 6.12) separately. Of course, there is solid solution in the natural basalt system, particularly Fe-Mg exchange, but we shall ignore this for now because we only desire to investigate three-component systems. The principal effect of adding Fe is to induce solid solution in diopside and forsterite and to lower the crystallization temperatures of the mafic minerals (see Figure 6.10).

The compositional variation in the Di-An-Fo system can be represented by a triangle, with each of the components at a corner of the triangle. (See Figure 2.1 for a review of plotting on triangular diagrams.) If we add temperature as a variable, the system is then composed of three binary eutectics, making up the sides of a three-dimensional temperature-composition triangular prism (Figure 7.1). The familiar Di-An system (Figure 6.11) is the left-rear face in Figure 7.1. The

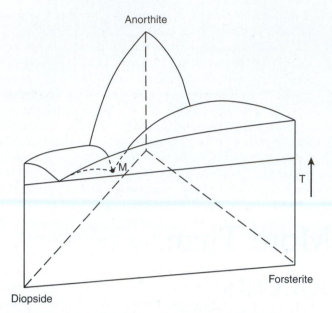

FIGURE 7.1 Three-dimensional representation of the diopside–forsterite–anorthite system versus temperature at atmospheric pressure. *M* is the ternary eutectic point. After Osborn and Tait (1952). Reprinted by permission of the *American Journal of Science*.

Di-Fo binary eutectic (front face in Figure 7.1) was explored by Bowen (1914), and the eutectic point was located at $Di_{88}Fo_{12}$ and 1387°C. The An-Fo system was studied by Kushiro and Schairer (1963) and is complicated somewhat by the participation of spinel. I have chosen to ignore the small spinel field in the Di-An-Fo system and treat An-Fo as another simple binary eutectic (right-rear face in Figure 7.1). Students interested in the true Di-Fo diagram are referred to Morse (1994).

When the number of components increases by one, the phase rule tells us that the variance increases by one. Thus,

the one-dimensional liquidus curve in an isobaric *T-X* binary diagram, such as Figure 6.11, becomes a two-dimensional liquidus *surface* in Figure 7.1. Thus the top surface of the diagram is represented by a complex curved liquidus surface, similar to three hills with summits at Di, An, and Fo, respectively. The hills slope down to valley bottoms, which are one-dimensional ternary extensions of the invariant binary eutectic points. These valley bottoms are similar to streams in a topographic sense. The streams converge to a ternary eutectic point *M* (analogous to a sinkhole) at the lowest point on the liquidus surface of the diagram.

The topographic analogy is apt for studying ternary systems, and we can simplify our perception of the system in the same way we simplify topography: by plotting the compositional variation in two dimensions, similarly to map areas, and contouring the liquidus surfaces with lines of constant temperature, just as with topographic elevation contours. The resulting diagram is shown for the Di-An-Fo system in Figure 7.2. This figure shows the compositional plan of the system as though the observer is looking down the temperature axis. The liquidus surface is shown, and it is contoured with lines of constant temperature (isotherms). The extensions of the binary eutectics, called **cotectic** curves, are shown, with arrows to indicate the downslope (down-temperature) direction to the ternary eutectic minimum, point *M*.

The cotectic curves separate the liquidus surface into three areas, just as a binary eutectic point separates the liquidus in Figure 6.11 into two sections, each sloping down-temperature toward it. In the binary system, each section of the liquidus corresponds to liquids coexisting with a different solid. The same is true for the ternary system, and each area is labeled with the name of the solid mineral phase that coexists with a liquid in that particular portion of the liquidus. The three areas are thus labeled Forsterite + Liq, Diopside + Liq, and Anorthite + Liq. Now let's use the

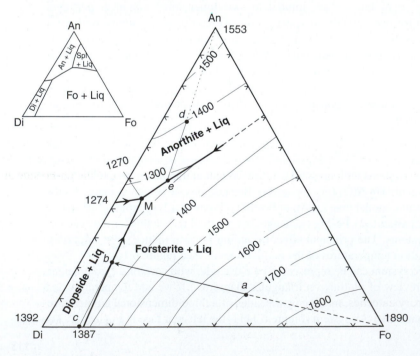

FIGURE 7.2 Isobaric diagram illustrating the liquidus temperatures in the Di-An-Fo system at atmospheric pressure (0.1 MPa). The dashed part of the Fo-An cotectic is in a small spinel + liquid (insert) field that is ignored here. After Bowen (1915b), reprinted by permission of the *American Journal of Science*, and Morse (1994) copyright © with permission from Krieger Publishers.

phase rule to analyze the crystallization behavior of some melts in this ternary system.

We begin with a liquid of composition *a* in Figure 7.2 (approximately $Di_{36}An_{10}Fo_{54}$, by wt. %). When a melt of this composition is above 1700°C, there is only a single melt phase present. Because $C = 3$ and $\phi = 1$, the *isobaric phase rule* [Equation (6.2)] is $F = C - \phi + 1$, so that F equals $3 - 1 + 1 = 3$. We must thus specify three of the possible intensive variables, which, in a natural system, are probably T, X_{Di}^{liq}, X_{Fo}^{liq}, and X_{An}^{liq} (P is constant). Only two of the three compositional variables are independent because the three sum to 100%. The variance of three corresponds to the three-dimensional volume occupied by the liquid field above the liquidus (see Figure 7.1).

By cooling to 1700°C, the system encounters the liquidus in the forsterite + liquid field in Figure 7.2, and forsterite begins to crystallize. Now $\phi = 2$ and $F = 3 - 2 + 1 = 2$. Because the composition of pure forsterite is fixed, we still have the same variables to choose from as we did for the liquid-only situation. But now we must specify only two intensive variables to determine the state of the system. As the temperature is lowered, forsterite continues to crystallize from the liquid, depleting it in the Fo component. Thus the liquid composition changes in a direction *directly away from the Fo corner* of the triangle from point *a* toward point *b* in Figure 7.2. This involves a *continuous reaction* of the type:

$$liquid_1 = forsterite + liquid_2 \qquad (7.1)$$

By cooling, new $liquid_2$ becomes incrementally more Ca-Al-rich than old $liquid_1$ by losing Mg_2SiO_4.

At any temperature, the relative amounts of liquid and solid forsterite can be calculated by applying the lever principle, with the three collinear points representing the liquid composition, forsterite, and using point *a* as the bulk composition and fulcrum point (see the discussion accompanying Figure 6.9). Note that the divariant nature of the system with two phases agrees with the two-dimensional nature of the liquidus surface in the three-dimensional diagram, Figure 7.1.

As the liquid cools to point *b* in Figure 7.2 (at approximately 1350°C), diopside joins forsterite and liquid in the system. Now there are three phases, and $F = 3 - 3 + 1 = 1$. Specifying *one* intensive variable now completely determines the system. Theoretically this could be any variable, but because the solids in this system are of fixed composition, it suggests that the composition of the liquid (coexisting with two solids) is now determined by temperature and is constrained to follow the one-dimensional cotectic curve between the binary Di-Fo eutectic and the ternary eutectic (*M*) with continued cooling. This will be accompanied, of course, by continued crystallization of forsterite and diopside via a continuous reaction:

$$liquid_1 \rightarrow liquid_2 + diopside + forsterite \qquad (7.2)$$

Determining the relative amounts of Di, Fo, and liquid at any temperature is also possible, but it is more complicated than for two phases. Imagine that the system has cooled, and the liquid composition has moved partway down the cotectic trough from point *b* toward *M* in Figure 7.2, while still coexisting with Di and Fo. At any temperature, we can construct an isothermal planar triangle with the corners at the composition of the three coexisting phases (Figure 7.3). If we were to place a mass on each corner that is proportional to the weight of the phase at that corner (because the diagram is in wt. %), the triangular plane must balance on point *a* (the bulk composition) as the fulcrum. The simplest way to determine the relative amounts of the three phases is to construct a working line from any one of the phases, through the bulk composition, to the opposite side of the triangle. This has been done for the line *Liq-a-m* in Figure 7.3. In this case, point *m* represents the *total solid* content at the temperature in question (diopside + forsterite). The ratio of the total solids to liquid is identical to the linear lever rule explained in Figure 6.9, so the weight ratio *m/Liq* is equal to the ratio of the length of the lines $\overline{Liq - a}/\overline{a - m}$. Then the ratio of diopside/forsterite within *m* equals $\overline{Fo - m}/\overline{Di - m}$.

We can readily see that the triangle Fo-Di-*b* in Figure 7.2 is the appropriate triangle for the first appearance of diopside with forsterite and liquid *b*. Because point *a* lies between Fo and *b* along an edge of the triangle, a qualitative application of our weight-on-triangle balancing analogy tells us that there must be an infinitesimally small amount of diopside on the far corner, which is consistent with the first appearance of that phase.

At any point during cooling of the liquid along the cotectic, we can draw a curve tangent to the cotectic (at the point of the liquid composition) back to the Di-Fo base of the Di-An-Fo triangle in Figure 7.2. The lever principle can then be used to determine the ratio of diopside to forsterite that is crystallizing from the liquid *at that instant* (referred to as the **bulk mineral extract**). For example, suppose diopside has just joined forsterite as the liquid reached point *b* in Figure 7.2. The liquid composition then begins to follow the cotectic toward point *M*. At the instant the liquid moves from point *b*, a tangent to the cotectic at *b* is drawn and intersects the Di-Fo base at point *c* (in Figure 7.2). For the liquid composition to move along the cotectic from *b*, a mass of solid corresponding to composition *c* must be removed from it (just as the liquid moved from *a* to *b* by losing forsterite). Point *c* must be composed of forsterite and diopside. The diopside/forsterite ratio being crystallized *at this instant* is then equal to the ratio of the distances $\overline{Fo - c}/\overline{Di - c}$ in Figure 7.2. As the cooling liquid follows the cotectic curve toward *M*, the tangent extrapolates back to the Di-Fo side at

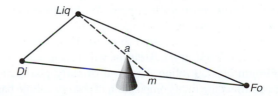

FIGURE 7.3 Diagram to illustrate the derivation of the relative amounts of three phases coexisting with a given bulk composition (*a*).

a point progressively closer to Di, meaning that the ratio of diopside to forsterite crystallizing at any instant slowly increases with cooling. This *instantaneous* ratio that is crystallizing at any one time is not to be confused with the *cumulative* amount of any phase that has crystallized since the inception of crystallization. The latter involves the amounts of solid phases that balance on the triangle, as shown in Figure 7.3. Because forsterite has been crystallizing since 1700°C, the total amount of forsterite crystallized is much greater than the instantaneous ratio of diopside to forsterite being added to the accumulated solids at any given time.

When the liquid reaches point *M*, at 1270°C, anorthite joins forsterite, diopside, and liquid. Now $F = 3 - 4 + 1 = 0$, and we have an invariant, completely determined situation at fixed temperature (and pressure) and composition of all phases, now including liquid. Continued cooling results in the *discontinuous reaction* involving the crystallization of all three solids at a constant temperature:

$$\text{liquid} \rightarrow \text{diopside} + \text{anorthite} + \text{forsterite} \quad (7.3)$$

until a phase (in this case liquid) is consumed. Only then do we gain a degree of freedom, and the temperature can again drop. Below 1270°C, three solid phases coexist with a single degree of freedom. Because the composition of all three phases is fixed, we need to specify only the temperature to determine the state of the system.

The path $a \rightarrow b \rightarrow M$ is called the **liquid evolution curve**, or the **liquid line of descent**, and it describes the way the liquid composition changes as crystallization proceeds. Paths of the type just described apply to compositions in any other liquidus field in Figure 7.2. For example, if we cool composition *d* to 1400°C, anorthite crystallizes first. The liquid composition moves directly away from the An apex of the triangle with continued cooling [via a continuous reaction similar to Reaction (7.1), but producing anorthite] until it reaches point *e* on the cotectic, at which point forsterite joins anorthite and liquid. The liquid then follows the cotectic, crystallizing both forsterite and anorthite (in a ratio that can be deduced at any point by erecting a tangent to the cotectic and extrapolating it back to the Fo-An side of the triangle). At the ternary eutectic, point *M*, diopside joins the system, and crystallization continues via the *discontinuous* Reaction (7.3) at a constant 1270°C until the liquid is consumed. As an exercise, try to find different bulk composition points on Figure 7.2 that result in every possible sequence of crystallization of the three phases.

As in the binary eutectic systems, the last liquid to crystallize for *any* bulk composition that plots within the triangle Fo-An-Di *must* occur at point *M*, the ternary eutectic composition (approximately $Di_{50}An_{43}Fo_7$).

Because there is no solid solution, the liquid evolution *path* for **fractional crystallization** will be the same as for equilibrium crystallization. Removal of early phases, however, can affect the final composition of the crystalline basalt. If earlier phases are removed by crystal fractionation (settling or floating), and the liquid is then separated from the crystallized solids at any point along the liquid evolution curve, the rock that forms by equilibrium crystallization from that derivative liquid will have a composition equal to that of the evolved liquid at the point of separation. In other words, the liquid, at separation, becomes the new bulk composition. The final rock could then have a bulk composition ranging from the original melt bulk composition to point *M* (43% plagioclase, 50% clinopyroxene, and 7% olivine, by weight). Fractional crystallization may thus be an important mechanism by which igneous rocks diversify because a range of rock types can be derived from a single parent magma.

Equilibrium melting is simply the reverse of equilibrium crystallization. Melting of *any* mixture of diopside, anorthite, and forsterite, regardless of proportions, produces a first melt at composition *M*. The liquid remains at *M* as the discontinuous Reaction (7.3) proceeds (in reverse), until *one* phase is consumed. The phase that is consumed first depends on the bulk composition and will determine which cotectic curve the liquid follows from *M*. If forsterite is consumed first, for example, the liquid follows the diopside–anorthite cotectic curve, driven along by a continuous reaction similar to Reaction (7.2), only in reverse and involving diopside and anorthite. At what point will the liquid leave the cotectic? That depends, again, on the bulk composition. The liquid evolution path leaves the curve when one of the two remaining phases is consumed by the continuous melting reaction. This happens when the liquid composition reaches a point that is a linear extrapolation of a line drawn from the remaining phase, through the bulk composition, to the cotectic. Returning to our original example (bulk composition *a*), the liquid leaves the cotectic at point *b*, which is the point on the cotectic where the line forsterite $\rightarrow a$ extrapolates to the cotectic curve. At that point, diopside must be eliminated because *b-a*-Fo are collinear, implying that the bulk composition (*a*) consists of liquid (*b*) and Fo only. Then the liquid composition moves directly from the cotectic curve toward the bulk composition, driven by Reaction (7.1), in reverse. When the liquid composition reaches point *a*, all of the forsterite must be consumed, and only one phase remains: the liquid.

Partial melting in the ternary eutectic system also behaves like binary eutectics. The first melt of a mixture of Di + An + Fo, in any proportions, is produced at the eutectic point *M*. Suppose melts of this eutectic composition are produced via invariant Reaction (7.3), in reverse, and extracted in small increments (fractional melting) until one phase is consumed. If we begin with bulk composition *a* in Figure 7.2, for example, anorthite must be consumed first. The remaining solid system is then binary (Fo + Di), and no further melting can occur until the temperature rises from 1270°C to 1387°C (the temperature of the binary Fo-Di eutectic). Binary invariant fractional melting then produces melts at the Fo-Di eutectic composition at a constant temperature of 1387°C until diopside is consumed. The remaining solid is forsterite, one component, and no further melting is possible until 1890°C, when pure forsterite melts. Partial melting thus occurs in three discrete episodes.

Partial melting may also occur as a single event, involving equilibrium melting until the melt is removed from the residual crystals at some point along the continuous liquid evolution path. Extraction results in a melt with a new bulk composition equal to that of the liquid composition at the time. As with fractional crystallization, we are thus able to vary the composition of derivative magmas (and hence rocks) over a fairly wide spectrum of liquid compositions by varying the percentage of partial melting of a single source rock.

7.1.2 Ternary Peritectic Systems

As an example of a ternary system with a peritectic, consider the Fo-An-silica system shown in Figure 7.4, which is a combination of the familiar Di-An binary eutectic system (Figure 6.11) and the Fo-silica binary peritectic system (Figure 6.12), which form two of the *T-X* sides of the triangle in Figure 7.4. The third binary, An-silica, is an eutectic system with the liquidus minimum at 52 wt. % An and 1368°C. (The isothermal contours have been omitted in Figure 7.4 to avoid clutter.)

Behavior in the silica + liquid and anorthite + liquid fields of Figure 7.4 is similar to that in eutectic systems, such as Figure 7.2. Our time is best spent if we focus on the characterization of ternary peritectic behavior, which is most evident in the forsterite and enstatite fields. For example, a liquid of bulk composition *a* in Figure 7.4, upon cooling to the liquidus surface, produces forsterite first. In an isobaric three-component system with two phases, $F = 3 - 2 + 1 = 2$, and the liquid composition moves directly away from the Fo corner of the triangle, via the continuous Reaction (7.1). When the liquid composition reaches the forsterite–enstatite boundary curve at point *b*, enstatite forms, and because $F = 3 - 3 + 1 = 1$, the liquid is constrained to follow the boundary curve toward point *c*. Note that a tangent projected backward from any liquid point along this peritectic curve to the Fo-SiO$_2$ edge of the triangle, for the purpose of determining the ratio of

instantaneously crystallizing solids, falls *outside* the forsterite–enstatite segment. In other words, the bulk solid composition being removed from the melt cannot be a combination of forsterite + enstatite, which are the only solids available. This is reflected in the crystallization reaction, which is *not* Liquid$_1$ → Liquid$_2$ + Forsterite + Enstatite, as it would be if the projected line did fall between Fo and En, but rather:

$$\text{liquid}_1 + \text{forsterite} \rightarrow \text{liquid}_2 + \text{enstatite} \qquad (7.4)$$

which is a peritectic-type reaction. Thus the boundary curve between the forsterite + liquid and enstatite + liquid fields in Figure 7.4 is a *peritectic* curve. For our original bulk composition *a*, there is sufficient forsterite that it is not consumed entirely by Reaction (7.5) by the time the liquid reaches point *c*, and anorthite joins forsterite, enstatite, and liquid in an invariant situation ($F = 3 - 4 + 1 = 0$). At this invariant, a discontinuous reaction takes place. Because the liquid composition in this case doesn't lie within the Fo-En-An triangle, the reaction cannot be the same as Reaction (7.3) but is:

$$\text{liquid} + \text{forsterite} = \text{enstatite} + \text{anorthite} \qquad (7.5)$$

We can always tell the final solid mineral assemblage to which a melt eventually solidifies via equilibrium crystallization by noting the sub-triangle in which the bulk composition plots (see the insert in Figure 7.4). Point *a* is in the forsterite–enstatite–anorthite sub-triangle of the system, so these three minerals must compose the final rock. This is how we know that forsterite was not consumed before the liquid reached point *c*. We can also deduce that, of the four phases, forsterite + enstatite +anorthite + liquid, that coexist at point *c*, the *liquid* must be consumed first at this invariant ternary

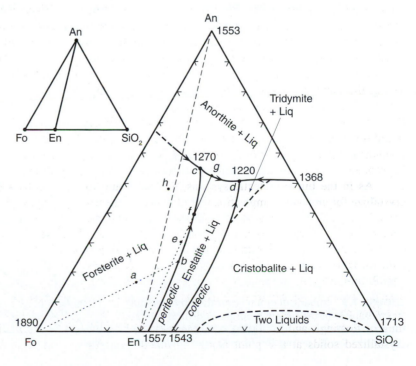

FIGURE 7.4 Isobaric diagram illustrating the cotectic and peritectic curves in the system forsterite–anorthite–silica at 0.1 MPa. After Anderson (1915), reprinted by permission of the *American Journal of Science*, and Irvine (1975). The inset is a chemographic diagram showing the position of all solid phases, with tie-lines connecting phases that stably coexist together. It can be used to determine the equilibrium mineral assemblage of entirely solid rocks. For example, most bulk compositions in the Fo-An-SiO$_2$ system will fall within the En-An-Q triangle or the Fo-En-An triangle. These two assemblages compose any rock plotting within the corresponding triangle. Only rarely will a bulk composition plot on a two-phase tie-line or a one-phase point (correspondingly reducing the components).

peritectic point. This is analogous to bulk compositions be-tween Fo and En at the peritectic temperature in the binary Fo-silica system (Figures 6.12 and 6.13). The system thus remains at 1270°C until the liquid is consumed by Reaction (7.5). The final mineral assemblage is then forsterite + enstatite + anorthite in relative proportions that can be determined by the bulk composition and the lever principle in the Fo-An-En tri-angle (as in Figure 7.3). Had the bulk composition been just to the right of the En-An join in Figure 7.4, a silica mineral would be required in the final assemblage, and the liquid line of de-scent would have left point c (because forsterite would be the first phase consumed and $F \rightarrow 1$) and reached point d (the ternary eutectic where $F = 0$ again), where tridymite would have formed. The last liquid would then have been consumed at this point via another invariant discontinuous reaction:

$$\text{liquid} \rightarrow \text{enstatite} + \text{anorthite} + \text{tridymite} \quad (7.6)$$

An interesting variation happens for compositions in a narrow band for which bulk composition e is representative. Forsterite forms first, and the liquid composition moves di-rectly away from Fo to the peritectic curve, where enstatite forms. As with composition a, the liquid follows the univariant peritectic curve as Reaction (7.4) progresses. At point f, how-ever, the bulk composition plots at an intermediate point on a line between enstatite and the liquid (dashed in Figure 7.4), meaning that the system can comprise only these two phases, and thus the forsterite must be consumed by the peritectic Reaction (7.4) when X_{liq} reaches point f, not at point c, as in the previous example. At f, the liquid composition must *leave the peritectic curve*, moving directly away from enstatite now, as enstatite forms by a continuous reaction similar to Reaction (7.1), X_{liq} proceeds along the f-g line to point g, where anorthite crystallizes, and the liquid then continues along the cotectic to point d, where Reaction (7.6) takes place until the liquid is consumed. It is unusual for a liquid to leave a curve in this manner, but the curve in question is a

peritectic, not a cotectic ("valley bottom"), and it is thus pos-sible to move off the curve with descending temperature.

Fractional crystallization associated with $X_{\text{bulk}} = a$ involves the removal of forsterite crystals, causing the bulk composition to migrate with the liquid composition away from Fo. There will be no forsterite to participate in the peri-tectic Reaction (7.4) when the liquid reaches the peritectic. Thus all liquids produce enstatite directly and leave the peri-tectic curve as soon as it is reached, moving directly away from enstatite toward either the enstatite–anorthite or the enstatite–silica cotectic curves, depending upon the geometry (the An content of the liquid).

7.1.3 Ternary Systems with Solid Solution

As an example of solid-solution behavior in three-component systems, once again with applicability to basalts, we can use the system Di-An-Ab. In this system, there is complete mis-cibility between two components (Ab and An), whereas the third (Di) is insoluble in either of the others. This system at atmospheric pressure is illustrated in Figure 7.5, once again projected down the temperature axis with the liquidus surface contoured in isothermal increments. We are already familiar with two of the T-X sides of the diagram: the diopside–anorthite eutectic (Figure 6.11) and the albite–anorthite solid solution (Figure 6.8). The diopside–albite system, like diopside–anorthite, is also not truly *binary*, in that diopside incorporates some Al from albite, and albite also takes a bit of Ca from diopside. We shall ignore this small effect and treat the system as a simple binary eutectic, with the minimum at $Ab_{91}Di_9$ and 1133°C (Schairer and Yoder, 1960). Because there is no eu-tectic in the albite–anorthite system, no cotectic descends from that side of the triangle in Figure 7.5. Rather, the cotectic runs from the binary eutectic on the diopside–anorthite side to the one on the diopside–albite side. In this case, it slopes continuously to a minimum at the diopside–albite edge, reflecting the slope of the albite–anorthite liquidus. This

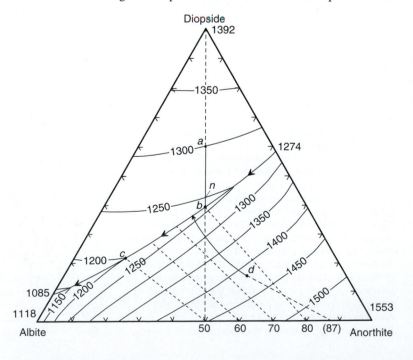

FIGURE 7.5   Isobaric diagram illustrating the liquidus temperatures in the system diopside–anorthite–albite at atmospheric pressure (0.1 MPa). After Morse (1994). Copyright © with permission from Krieger Publishers.

continuous slope is not a requirement for solid solutions, of course, and other systems may have a thermal minimum within the three-component triangle.

Unfortunately, we have only the liquidus contours on the diagram, and when solid solution is involved, we need to know about the solidus in order to know the plagioclase composition that is in equilibrium with a particular liquid. Only when we know what solid is interacting with the liquid can we quantitatively analyze the liquid evolution path during crystallization or melting. One might think that we can simply use the binary solidus in Figure 6.8 because the plagioclase that coexists with ternary liquids in the Di-An-Ab system is a binary mixture and plots along the albite–anorthite edge of the triangle. Regrettably, this is not the case, and the composition of the plagioclase that coexists with any ternary liquid at a given temperature differs from the binary values and must be empirically determined in the full three-component system. We can approximate this information from the ternary experimental data, but it is not sufficiently comprehensive to provide accurate constraints for any liquid. A few tie-lines showing the composition of the plagioclase that coexists with *cotectic liquids only* have been added to Figure 7.5 (dashed lines) as a partial aid.

Let's begin on the side of the cotectic that is easier to analyze: the "diop-side." If we cool a melt of composition *a* to 1300°C, diopside begins to crystallize as the first solid phase. When diopside and liquid coexist, $F = 3 - 2 + 1 = 2$, and pure diopside crystallizes from the liquid via the continuous reaction:

$$\text{liquid}_1 = \text{diopside} + \text{liquid}_2 \qquad (7.7)$$

and the liquid composition moves directly away from the diopside apex of the triangle toward the cotectic curve. X_{liq} reaches the cotectic at point *b*, at 1230°C, at which point plagioclase, of a composition An_{80}, begins to crystallize with diopside. Note that we *can* use the tie-lines on our diagram to determine the plagioclase composition because we are only considering plagioclases that coexist with diopside and cotectic liquids. Now, with continued crystallization of diopside and plagioclase, the liquid composition moves down the cotectic curve, whereas the plagioclase composition moves down the solidus as both become progressively more albitic. Because $\phi = 3$ and $C = 3$, $F = 3 - 3 + 1 = 1$, the composition of both plagioclase and liquid is dependent upon temperature via the continuous reaction:

$$\text{liquid}_1 + \text{plagioclase}_1 \rightarrow \text{liquid}_2 \\ + \text{diopside} + \text{plagioclase}_2 \qquad (7.8)$$

Because there is solid solution, the liquid does *not* reach the cotectic minimum, as it would in eutectic systems. Instead, the bulk composition determines when crystallization is complete, as it did in the binary plagioclase solid-solution system. When the liquid reaches point *c*, at about 1200°C, the plagioclase composition reaches An_{50}, and the bulk composition lies at an intermediate point on a line between the coexisting plagioclase and diopside.

Because these three points are collinear, plagioclase of composition An_{50} plus diopside can be combined to produce X_{bulk}, so the amount of liquid must approach zero. The last liquid is thus consumed at about 1200°C, and crystallization is complete.

Again, the sequence of plagioclase compositions associated with cotectic liquids and the tie-lines in Figure 7.5 is *not* the same as the binary solidus in the albite–anorthite (plagioclase) system. Figure 7.6 illustrates the difference. The plagioclase (An_{80}) that coexists with liquid *b* in the ternary system at 1230°C can be located on the lower loop in Figure 7.6 (point *x*). In the binary system, An_{80} coexists with a liquid at 1440°C (point *y*). The effect of adding diopside to the plagioclase system is to flatten the loop and lower the solidus and liquidus temperatures. For any noncotectic liquid composition on the ternary liquidus surface between the binary Ab-An system and the cotectic, some intermediate loop applies. Thus Figure 7.6 supplies us with limiting values but not with every plagioclase composition required if we want to rigorously determine solid and liquid crystallization or melting paths on Figure 7.5. We need appropriate experimental data for more accurate analysis. As with the binary systems and the ternary liquidi shown to date, the experimental data are only sufficient to constrain the approximate positions of the liquidus and solidus, which are drawn by interpolating between the known points.

Let's try an example of the crystallization behavior of a liquid on the plagioclase side of the cotectic with a liquid of composition *d*, which is $Di_{15}An_{55}Ab_{30}$ [the final plagioclase composition = 55/(55 + 30) = An_{65}]. In this case, the liquidus is encountered at approximately 1420°C, and a plagioclase of composition $\sim An_{87}$ crystallizes. Note that

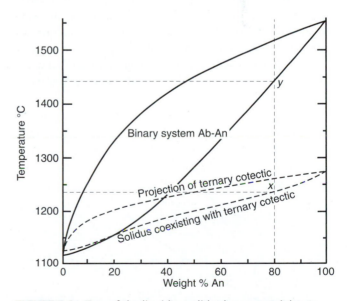

FIGURE 7.6 Two of the liquidus–solidus loops pertaining to the system diopside–anorthite–albite. The solid loop is the An-Ab binary liquidus–solidus loop from Figure 6.8, and the dashed lower loop represents the solidus that coexists with ternary cotectic liquidus (the curve from 1274 to 1133°C in Figure 7.5). Both are projected onto the albite–anorthite side. From Morse (1994). Copyright © with permission from Krieger Publishers.

this composition is not predictable from Figure 7.6 because it is neither a binary system nor a cotectic liquid, so it is not one of the dashed tie-lines in Figure 7.5. We can only estimate the plagioclase composition. Because $\phi = 2$, $F = 3 - 2 + 1 = 2$, and crystallization proceeds via a continuous reaction along the divariant liquidus surface:

$$\text{liquid}_1 + \text{plagioclase}_1 \rightarrow \text{liquid}_2 + \text{plagioclase}_2 \quad (7.9)$$

which is the ternary equivalent of Reaction (6.4). The liquid composition must evolve directly away from this solid composition being removed from it, but the plagioclase composition also shifts toward albite with cooling and progress of Reaction (7.9). Thus the liquid moves away from a shifting point, creating a curved path in Figure 7.5. This path is illustrated more clearly in Figure 7.7, where several tie-lines are shown connecting liquid compositions to the coexisting plagioclase, all passing through point d. (The bulk composition must always lie between the phases it comprises.) At each such condition, the liquid composition must be moving directly away from the coexisting solid. The curvature of the liquid evolution path has been slightly exaggerated in the figures. The liquid reaches the cotectic at point e, at about 1220°C, which coexists with a plagioclase of composition An_{75} (which *is* predictable from Figures 7.5 and 7.6 because this is now a cotectic liquid). At this point, diopside joins plagioclase as a crystallizing phase, and the liquid path turns abruptly to follow the liquidus via Reaction (7.8). Because the situation is isobarically univariant, the plagioclase and liquid compositions are dependent only upon temperature. In other words, in an isobaric three-component system, only one plagioclase composition can coexist with liquid and diopside at a given temperature. Crystallization continues until the plagioclase composition migrates to a point where the bulk composition is on a line connecting diopside and plagioclase, which for $X_{\text{bulk}} = d$, is An_{66}. The last liquid, in this case, is point f in Figure 7.7.

At any given temperature, the composition of all coexisting phases can be shown on an **isothermal section**. How

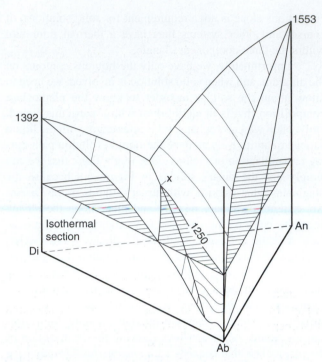

FIGURE 7.8   Oblique view illustrating an isothermal section through the diopside–albite–anorthite system. From Morse (1994). Copyright © with permission from Krieger Publishers.

such a section is derived is illustrated at 1250°C in Figure 7.8, and the section itself is shown in Figure 7.9. An isothermal section shows how the coexisting phase assemblages present at equilibrium vary with bulk composition at the temperature in question. Any bulk composition that plots in the shaded region in Figure 7.9 is above the liquidus, as can be seen in Figure 7.8. Therefore, the composition of this single liquid phase is equal to the bulk composition chosen. Point x is the intersection of the isothermal plane and the cotectic curve, and it therefore represents the composition of the only liquid

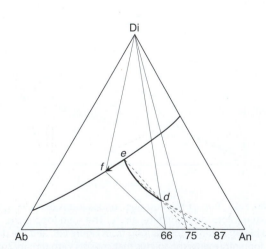

FIGURE 7.7   Liquid evolution path of bulk composition d ($\text{Di}_{30}\text{An}_{55}\text{Ab}_{15}$) in the diopside–anorthite–albite system at 0.1 MPa. The curvature of path d-e is exaggerated.

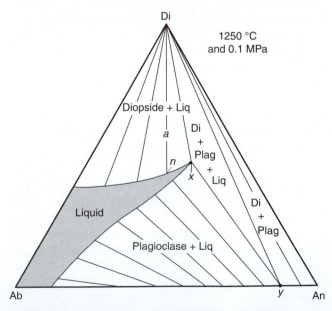

FIGURE 7.9   Isothermal section at 1250°C (and 0.1 MPa) in the system Di-An-Ab. From Morse (1994). Copyright © with permission from Krieger Publishers.

that can coexist with *both* plagioclase and diopside at 1250°C. The phase rule tells us that $F = C - \phi + 0 = 3 - 3 + 0 = 0$ at fixed pressure and temperature, requiring that the composition of all three coexisting phases also be fixed. The composition of the coexisting plagioclase is point y, or An_{89} in Figure 7.9, and diopside is pure. Any bulk composition falling within the triangle Di-x-y comprises these three phases at 1250°C in a fashion similar to Figure 7.3. The triangle is accordingly labeled with these three phases in Figure 7.9.

Any bulk composition in the field labeled diopside + liq will contain only these two phases at 1250°C, in a manner similar to point a in Figure 7.5. At 1250°C in Figure 7.5, bulk composition a consists of pure diopside and a liquid equal to the composition at the point where the line Di-a-b intersects the 1250°C isotherm on the liquidus. This point is labeled n in both Figures 7.5 and 7.9. Note that the bounding curve of the shaded portion in Figure 7.9 represents the liquidus, which, by definition, is the locus of all liquid compositions that coexist with a solid. Thus, for any bulk composition that plots in the Diopside + Liq field in Figure 7.9, the composition of the liquid that coexists with diopside at 1250°C can be determined as the point where a line drawn from diopside through the particular bulk composition intersects the liquidus curve. Several such tie-lines are included in Figure 7.9.

The same arguments hold for the plagioclase + liquid field in Figure 7.9. If we begin with composition d in Figure 7.5, the liquid composition that coexists only with plagioclase can be located by the point where the curved liquidus path crosses the 1250°C isotherm. Several other tie-lines for coexisting plagioclase + liquid at 1250°C are included in Figure 7.9 as well. Note that the tie-lines are appropriate only for the temperature corresponding to the isothermal section, in this case 1250°C. Only the x-y tie-line is of the same type as those in Figure 7.5 because plagioclase y coexists with diopside and a cotectic liquid. The other plagioclase–liquid tie-lines relate plagioclases to non-cotectic liquids (with no diopside) at 1250°C. The orientation of plagioclase–liquid tie-lines shifts with temperature, as should be clear from the dashed tie-lines that pass through point d at several temperatures in Figure 7.6. Also note in Figure 7.5 that the An_{87}-d tie-line (at 1420°C) crosses the An_{80}-b tie-line (at 1230°C).

Finally, any bulk composition in the triangle labeled Di + Plag in Figure 7.9 plots to the right of the limit of any solids that coexist with a liquid at 1250°C (the line Di-y). These compositions are thus associated with a completely solid mineral assemblage, diopside + plagioclase, and the composition of the appropriate plagioclase can be determined by extrapolating the tie-line connecting the bulk composition to diopside.

These isothermal sections are useful for determining quickly and easily how the phases present in a sample vary with bulk composition at a particular temperature and pressure. Diagrams of this type are commonly used by metamorphic petrologists, as we shall see in the metamorphic section of this text.

Once again, **equilibrium melting** is just the opposite of equilibrium crystallization. **Fractional crystallization,** because it involves solid solution, affects the composition

of the final liquid to crystallize. In the case of composition d in Figure 7.7, for example, perfectly efficient removal of crystallizing plagioclase means that the bulk composition of the residual melt system will always equal the liquid composition along the evolution path. Crystallization will therefore not be complete at point f but will continue to lower temperatures and produce more albitic plagioclase. **Partial melting**, as always, creates new melt compositions that differ from the bulk composition, if extracted at any point along a liquid evolution curve.

For **systems with more than one solid-solution series**, it is far more difficult to depict the evolution paths because the liquidus and solidus surfaces are more complex and cannot both be contoured as in Figures 7.2 and 7.5. Figure 7.10 is a somewhat schematic oblique view of the ternary feldspar system. Because of the obvious complexity, it is suitable only for a qualitative description of the system rather than a quantitative analysis of the crystallization or melting behavior. The left-front plane is the Ab-Or binary system at low H_2O pressure (Figure 6.16), and the rear plane is the now-familiar

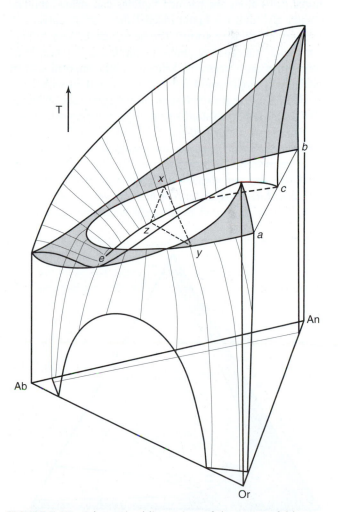

FIGURE 7.10 Schematic oblique view of the ternary feldspar system versus temperature. The ternary liquidus and solvus surfaces have the hatch pattern, and the solidus is shaded. Line e-c is the cotectic minimum, and the curve from a to b is the trace of solids that coexist with the cotectic liquids. After Carmichael et al. (1974). Copyright © with permission of the McGraw-Hill Companies.

Ab-An system (Figure 6.8). The remaining An-Or binary side of the system has a large solvus that limits solid solution, as in Figure 6.17c. This solvus continues through the ternary space to the Ab-Or side, but not to the Ab-An side. The cotectic curve runs from the An-Or eutectic (point *c* in Figure 7.10) toward the Ab-Or side but stops at point *e* before reaching that side because of the complete Ab-Or solid solution. At higher H_2O pressures, however, the Ab-Or system does form a eutectic (Figure 6.17c), and the ternary cotectic continues across to this side.

Many facets of feldspar crystallization and melting can be deduced from Figure 7.10, but it is admittedly a complex figure, and I bring it into the discussion here only to focus on the crystallization of coexisting plagioclase and alkali feldspar, so common in felsic rocks. Cooling of most melt compositions results in the crystallization of one feldspar as the liquidus is encountered. If the melt composition is to the Or side of the *c-e* cotectic, alkali feldspar will form first. If the melt composition is on the other side, plagioclase will form first. Either way, the liquid and coexisting solid both follow curved paths along the hatched liquidus and shaded solidus, respectively, until the liquid reaches the cotectic curve, *c-e*, and a second feldspar forms. The heavy curved line running from point *a* through *y* and *x* in the ternary system back around to point *b* traces the locus of solid compositions that coexist with another solid and a cotectic liquid (as with the *a-c-b* tie-line in the An-Or system on the right face of Figure 7.10). In the ternary system, one feldspar is a plagioclase, and the other is an alkali feldspar (see also Figure 7.11, which shows several isothermal sections through the system). Because the cotectic curve in Figure 7.10 slopes toward albite with decreasing temperature, the liquid is more albitic than either coexisting feldspar (except in some very Ca-poor magmas). This is illustrated by a single isothermal three-phase

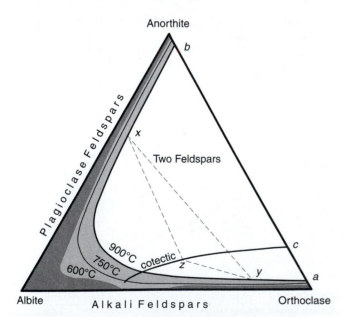

FIGURE 7.11 Traces of the ternary feldspar solvus at 600 and 750°C. The solvi separate areas in which two feldspars form (blank) from one in which a single feldspar forms (shaded). Included are the cotectic curve and the trace of the polythermal curve *a-y-x-b*, from Figure 7.10.

triangle *x-y-z* in Figures 7.10 and 7.11, where *x* represents the plagioclase, *y* the alkali feldspar, and *z* the cotectic liquid. Most compositions in the ternary system develop two separate feldspars such as this. Only outside this loop, in the shaded portions of Figure 7.11, would a rock develop only a single feldspar. The size and shape of the single-feldspar region varies with temperature (as shown) and with pressure.

In the ternary systems that we have seen, the two-component behavior is generally maintained when a third component is added. Binary eutectic systems retain the eutectic behavior as ternary cotectic curves; solid solutions, such as Ab-An, continue to be so when diopside or orthoclase is added; and peritectic behavior also continues into three-component systems (e.g., Fo-An-SiO₂). One can safely predict that these fundamental relationships will be maintained with continued compositional diversity, although with some modification because of the added variance. Thus the lessons we learn from these simplified systems should be broadly applicable to natural situations, as we observe patterns of partial melting, fractional crystallization, etc.

7.2 SYSTEMS WITH MORE THAN THREE COMPONENTS

Each additional component adds a potential degree of freedom to a system, as well as either an additional phase or a solid-solution component in existing phases. The addition of a third component, for example, changes the three-phase (pyroxene + plagioclase + liquid) eutectic *invariant point* in the two-component Di-An system (point *d* in Figure 6.11) into a three-phase cotectic *univariant curve* in the three-component Di-An-Fo system (Figure 7.2). The ternary situation returns to an invariant one when a fourth phase (forsterite) joins the other three at point *M* in Figure 7.2. Each additional component can successively turn a three-component eutectic point into a four-component univariant curve and a five-component divariant surface, a six-component volume, etc. There is thus no fixed temperature limitation for a four-phase assemblage such as clinopyroxene, olivine, plagioclase, and liquid in natural systems, for example, as there is in the three-component analog system shown in Figure 7.2.

Visualizing the compositional space of a four-component system requires a tetrahedron, which can be illustrated two-dimensionally only in perspective view. One-phase four-component systems have five degrees of freedom ($F = 4 - 1 + 2$), which is quadravariant in isobaric sections. The variance of such systems is impossible to depict on paper. Isothermal–isobaric systems with two or more phases are two dimensional or less and are easier to visualize, but they are less informative because of these restrictions. The simplest and most common approach to this problem is to take isobaric systems and restrict the analysis to assemblages with at least three phases present. An example of this approach is illustrated in Figure 7.12 for the system diopside–anorthite–albite–forsterite.

The diopside–anorthite–albite and diopside–anorthite–forsterite systems are familiar (Figures 7.5 and 7.2, respectively). The three-phase cotectic curves from these

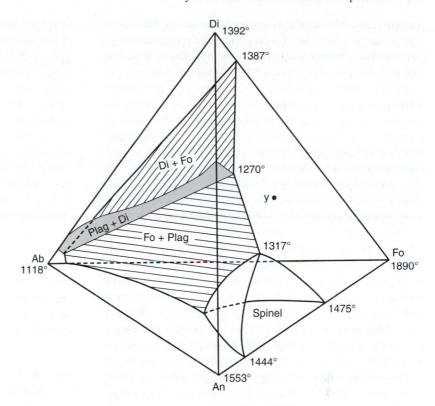

FIGURE 7.12 The four-component system diopside–anorthite–albite–forsterite. After Yoder and Tilley (1962). Reprinted by permission of Oxford University Press.

systems gain a degree of freedom with the additional component and become surfaces within the quaternary system. Naturally, such curves and surfaces cannot be isothermal, so the figure represents a polythermal projection in which it is impossible to depict temperature, except at a few invariant points. An additional problem is that we cannot easily determine depth in the diagram. Does point *y* lie on the Di-Fo-An face, on the Di-Ab-Fo face, or somewhere between? As a consequence of these limitations on what can be depicted, it is very difficult, if not impossible, to quantitatively trace liquid evolution curves and, thus, the sequence of minerals to form. Despite the more comprehensive basalt models that these four-component (or greater) systems afford us (they can be used effectively with some practice), we have now passed the point at which the difficulty in visualization and analysis has surpassed the benefits to be derived from model systems, at least for our present purposes.

For systems with a greater number of components, it is perhaps just as easy, and certainly more applicable, to perform melting experiments directly on natural substances. For these more complex systems, rather than struggle with multidimensional composition space, the melting relationships are projected back to the pressure–temperature diagrams, which we used for one-component systems. For example, melting experiments on a tholeiitic basalt from the Snake River Plain by Thompson (1972) are shown in Figure 7.13. We regain some simplicity in these diagrams by having all the information on a *P-T* phase diagram. For instance, we get a clear idea of the temperature interval of melting (or crystallization) at any pressure, represented by the shaded solidus–liquidus interval.

Within this interval, the curves representing the final disappearance of a particular phase with increased temperature can be added so that we can determine the sequence of minerals

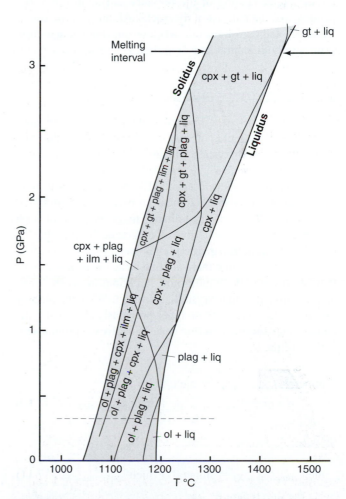

FIGURE 7.13 Pressure–temperature phase diagram for the melting of a Snake River (Idaho, USA) tholeiitic basalt under anhydrous conditions. After Thompson (1972).

that melt. In our simpler three-component model systems, these curves involve the loss of a phase and the movement of the liquid from a univariant curve to a divariant surface, or the transition from a divariant surface to a trivariant field, etc. We need only compare adjacent fields in Figure 7.13 to determine the phase lost (with heating) or gained (with cooling) that each curve represents. For example, at pressures below 0.5 GPa (dashed traverse), melting begins at the solidus, and we progress across the field via an unspecified continuous reaction that consumes the solids and produces liquid. Ilmenite is the first solid to be consumed by the reaction, and it is lost for this particular basalt at the first curve above the liquidus. In the next field, olivine + plagioclase + clinopyroxene + liquid are the remaining phases, and a continuous reaction across this field consumes some of the remaining solids and produces more liquid. Clinopyroxene is lost next, then plagioclase, and finally olivine at the liquidus, where melting is complete. We can also evaluate the effect of pressure on the melting–crystallization relationships directly from this diagram. Although these pressure–temperature diagrams are informative and simple to understand, they cannot provide us with such information as the relative quantity of each phase, or their compositions, as melting or crystallization progresses. Thus we trade the ability to understand the compositional evolution of the liquid (or solids) for the understanding of the pressure and temperature dependence of the melting of real rocks. All this information is valuable, and we draw our wisdom and knowledge from studies of both natural and simplified systems.

7.3 REACTION SERIES

In the analysis of experimental systems, we have encountered several examples of reactions between the liquid and previously formed crystals. The importance of such reaction relationships was clearly recognized by N. L. Bowen, a pivotal figure in the development of experimental petrology and its application to geological problems. He considered the reaction aspect of crystallizing magmas to be of such significance in the evolution of magmatic liquids that he called it the **reaction principle** (Bowen, 1922). He recognized, as we have, that there are two basic types of reactions that can occur under equilibrium conditions between a melt and the minerals that crystallize from it. We have already discussed the basis of these types previously. The first type, called a **continuous reaction series**, involves continuous reactions of the type:

$$\text{mineral}_{(\text{composition A})} + \text{melt}_{(\text{composition X})}$$
$$= \text{mineral}_{(\text{composition B})} + \text{melt}_{(\text{composition Y})} \quad (7.10)$$

or:

$$\text{melt}_{(\text{composition X})} = \text{mineral} + \text{melt}_{(\text{composition Y})} \quad (7.11)$$

Any of these may involve more than one mineral when $C > 2$. Examples of this type of reaction, in which the

composition of the melt, the mineral, or both varies across a range of temperature, abound in solid solution series. Examples include Reactions (6.4, 6.6, 6.11, 7.1, 7.2, 7.7, 7.8, and 7.9). By such reactions, plagioclase feldspars and coexisting melts both become more sodic; and mafic minerals, such as olivines or pyroxenes, and their coexisting melts become more iron rich (see Figures 6.8 and 6.9). Such reactions are continuous in the sense that the compositions of some phases, if equilibrium is maintained, adjust in a smooth, continuous fashion over the full crystallization temperature interval of the mineral.

The second type of reaction is the **discontinuous reaction**, such as:

$$\text{melt} = \text{mineral}_1 + \text{mineral}_2 \quad (7.12)$$

or the peritectic type of reactions:

$$\text{mineral}_1 + \text{melt} = \text{mineral}_2 \quad (7.13)$$

Examples of discontinuous reactions of the first type include Reactions (6.7, 6.9, 6.12, 7.3, 7.5, and 7.6). The classic peritectic reaction is forsterite + liquid = enstatite [Reaction (6.10), Figures 6.13 and 6.14]. Although Reaction (6.10) is discontinuous in the binary Fo-SiO$_2$ system, it becomes continuous in the three-component system Fo-An-SiO$_2$ [Figure 7.4, Reaction (7.4)]. A similar peritectic reaction exists in the Fo-Di-SiO$_2$ system in which a Ca-poor pyroxene reacts with the liquid to form a Ca-rich clinopyroxene (Osborn, 1979).

Such reaction relationships are common in crystallizing magmas, and more than one reaction or continuous series may take place simultaneously (or sequentially) in multicomponent melts (Osborn, 1979). One such series, the now-famous reaction series proposed by Bowen (1928), has the form illustrated in Figure 7.14.

Perhaps no other concept in petrology is as well known in general but misunderstood in detail by petrology

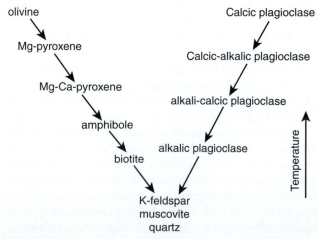

FIGURE 7.14 Bowen's Reaction Series (Bowen, 1928). Copyright © reprinted by permission of Princeton University Press.

students (at all levels of experience) as Bowen's reaction series. Certainly the right (continuous) side is clear to us [Reaction (6.4)], as is the discontinuous reaction of olivine to Mg-pyroxene on the left [Reaction (6.10)], but we should consider these classic examples more closely before proceeding. Does the right side mean that a pure anorthite gradually reacts to form a pure albite? What compositions are involved? Figure 6.8 tells us that the composition of the first and last plagioclase to crystallize depends upon the bulk composition of the system. It also depends upon the degree to which disequilibrium and zoning occur. Bowen's series gives us the proper trend but does not attempt to supply details or limits.

On the left side of Figure 7.14, the questions and processes are even more complex. Do all melts begin by crystallizing olivine? Will they all then proceed to pyroxenes and amphiboles, etc.? How far down the series will a given system proceed? Must they all eventually produce quartz? Will earlier phases be consumed by discontinuous reactions, or will they remain and coexist with the later phases in the final solidified rock? The answers to these questions also depend upon the bulk composition and the degree to which equilibrium is maintained, and they can be found by referring to appropriate phase diagrams of the types previously illustrated. Incidentally, simultaneous Mg-Fe continuous reactions take place in the mafic phases as well (Figure 6.10).

We have discussed scenarios involving equilibrium crystallization associated with Figures 6.12 and 7.4, in which olivine is either completely consumed or remains in coexistence with enstatite after the liquid is consumed during peritectic reactions. Figure 6.13 explains how bulk composition controls which scenario occurs. It is not a requirement, therefore, that an earlier phase *must* disappear in order to produce a phase below it on the discontinuous series. Likewise, for bulk compositions between points *i* and *c* in Figure 6.12, enstatite forms directly from the liquid, without involving olivine. Not all phases in the discontinuous series need thus be formed by discontinuous reactions. Clinopyroxenes may crystallize more commonly by cotectic-type crystallization than reaction between the melt and orthopyroxene. As anhydrous phases continue to form, the H_2O content of the dwindling melt volume increases to the point that hydrous phases such as amphiboles and biotite become stable. These phases may form as reaction rims around less hydrous phases (hornblende on pyroxene or biotite on hornblende) as a result of mineral–melt reactions, but they more commonly form as separate crystals directly from the melt.

How far down Bowen's series will a melt proceed? The common mineral assemblage olivine + pyroxene + plagioclase in basalts and gabbros tells us that, for most basaltic bulk compositions, crystallization is complete with the formation of plagioclase and Ca-Mg pyroxene so that they can proceed no further. Fractional crystallization, however, if conditions permit, can cause a system to proceed further down the series than under equilibrium conditions. Granitic liquids, on the other hand, seldom contain sufficient mafic components to form olivine, so they do not begin to crystallize until further down

the series, and they continue until quartz is produced. Clearly, *bulk composition* is the principal determining factor. Mafic liquids become solid before they get very far down the series, and granitic liquids may not begin to crystallize until a more sodic plagioclase and hornblende are stable.

By now it should be clear to you that Bowen's reaction series is not of universal application to silicate melts. Many budding young petrologists are dismayed to find this is so. Others with a bit more experience accept this but tend to see Bowen as being rather naive for having proposed it in the first place. If we turn to Bowen's original work, however, we discover that Bowen never intended his reaction series to be some Holy Grail of melt behavior. Bowen (1928) stated:

> An attempt is made to . . . arrange the minerals *of the ordinary sub-alkaline rocks* [author's emphasis] as a reaction series. The matter is really too complex to be presented in such simple form. Nevertheless the simplicity, while somewhat misleading, may prove of service in presenting the subject in concrete form. (p. 60)

Bowen further stated:

> The impression seems to have been gained by some petrologists that the postulated reaction relation between, say pyroxene and hornblende, carries with it the implication that all hornblende is secondary after pyroxene. Nothing is farther from fact. Just as there are many liquids in the system anorthite–forsterite–silica that precipitate pyroxene as a primary phase without any previous separation of olivine, so many magmas may precipitate hornblende directly without any previous precipitation of pyroxene. (p. 61)

It is in this vein that we should interpret and apply Bowen's reaction series, particularly the discontinuous section. It is entirely possible for each of these minerals to form by a discontinuous reaction involving melt and the mineral above it on the series. But it may also, and perhaps even more commonly, form as a result of direct crystallization from the melt. As with many other principles in science, Bowen's reaction series should be used as a guide and not as an overlay that restricts our observations and thinking. Some of the reactions expressed in it are well documented, both in nature and in the laboratory. The amphibole-, biotite-, and K-feldspar-forming ones are less common or reliable, as strictly interpreted.

Some generalities can be deduced from Bowen's series that provide useful guides when we consider igneous processes. Note that the minerals of the discontinuous arm of the series get progressively more Mg-Fe rich, and the continuous arm gets more calcic toward the high-temperature end. Thus, initial crystallization involves phases more mafic and calcic than the bulk composition of the system, and the later liquids will be more silicic and alkalic. We can thus expect fractional crystallization to produce evolved liquids that are enriched in SiO_2, Al_2O_3, and alkalis, as well as early solid fractionates enriched in CaO, FeO, and particularly MgO. Conversely, partial melts involve the lower minerals on the series, and we can expect them to be more silicic and alkalic than the bulk chemical composition of the system being melted. These aspects of melt systems are common.

The sequence of crystallization for any silicate liquid will commonly conform, in very generalized terms, to *a portion of* the series. We occasionally turn to Bowen's reaction series to guide our interpretations of melt behavior, but we can see from the experimental results above that compositional variation results in a wide variety of melting and crystallization paths. We should not be surprised when a natural system appears to violate the sequence in Bowen's series, and we can usually explain the behavior by reference to an appropriate experimental system, where we can observe and analyze the effects of composition on the path or sequence of crystallization. Even commonly overlooked components can have an important effect on the minerals that form. For example, free O_2 gas, typically present in exceedingly minute quantities (the partial pressure of oxygen ranges from 10^{-10} to 10^{-40}), can have a profound effect on the silicate phases that crystallize. If the partial pressure of oxygen is low, Fe remains as Fe^{2+} and mixes with Mg in mafic silicates. If the partial pressure of oxygen is high, some Fe is oxidized to Fe^{3+} and forms Fe-(Ti-)oxide phases, lowering the Fe^{2+} content and thus the total Mg + Fe available for mafic silicates. This and the increased effective Mg/Fe ratio may inhibit the formation of mafic silicates and change the sequence of minerals that forms in a reaction series.

7.4 THE EFFECTS OF PRESSURE ON MELTING BEHAVIOR

In order to simplify the graphical analysis of systems that have more than one component, we have usually ignored pressure as a variable. We justify this by noting that pressure generally has a much smaller effect on the stability of minerals than does temperature and that crystallization can be an isobaric process. The effects of pressure are not negligible, however, and we should take a moment to explore these effects.

As discussed in Chapter 5, because the entropy change and the volume change associated with melting of practically any solid both have a positive sign, the slope of the melting curve is positive on a *P-T* phase diagram, meaning that the melting point increases with increased pressure. Figure 7.15 schematically illustrates this point, showing the increase in melting temperature, from T_1 to T_2, that corresponds to a pressure increase from P_1 to P_2. The amount by which the melting point is elevated for a given pressure increase depends upon the slope of the equilibrium curve on a pressure–temperature diagram, which in turn depends on the relative values of ΔS and ΔV for the reaction, as expressed by the Clapeyron equation [Equation (5.15)]. Increasing lithostatic pressure raises the melting point of virtually all solid phases (except ice) and, thus, the liquidus in general.

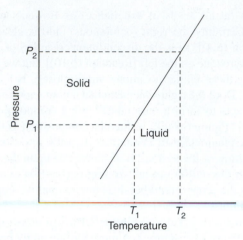

FIGURE 7.15 Schematic diagram illustrating the elevation of the melting temperature caused by an increase in pressure.

For example, raising the pressure by 1 GPa, corresponding to a depth change of about 35 km (from the surface nearly to the base of the continental crust), raises the melting point of basalt about 100°C (see Figure 7.13).

As the Clapeyron equation indicates, the magnitude of the pressure effect is not the same for all minerals, so the melting point elevation is different for each. If the liquidus surface is raised differentially in temperature–composition diagrams as pressure increases, the effect will generally shift the position of the eutectic. Note in Figure 7.16 that increasing the pressure from 1 atm to 1 GPa raises the diopside liquidus much more than that of anorthite. This causes the eutectic point to shift toward anorthite. Note also the change in nature from peritectic to eutectic behavior in the forsterite–silica system, as shown in Figure 6.15. For certain bulk compositions near the eutectic or peritectic point, the shift in such points with pressure from one side of the bulk composition to the other could alter the sequence of minerals that form. Similar eutectic shifts occur in ternary and higher-order systems.

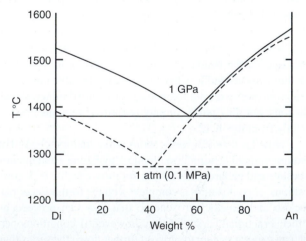

FIGURE 7.16 Effect of lithostatic pressure on the liquidus and eutectic composition in the diopside–anorthite system. 1 GPa data from Presnall et al. (1978). Copyright © with permission from Springer-Verlag.

Increased pressure may also cause some mineral phases to become unstable and be replaced by others. For example, feldspars become unstable at high pressure, where they break down to Na-Al pyroxenes and/or Ca-Al garnets. Pressure can also influence the composition of some minerals (e.g., pyroxenes are generally more aluminous at high pressure). As a result of these changes, pressure can have an effect on melting and crystallization besides simply raising the liquidus temperature. It can affect the very nature of the minerals that crystallize and hence the liquid evolution curves, as we shall see in Chapter 10. An example of the effect of pressure on the sequence of minerals that crystallize in a basalt can readily be seen in Figure 7.13. At low pressure, the sequence of minerals that form in the basalt as it cools is olivine → plagioclase → clinopyroxene → ilmenite. Above 0.5 GPa, it changes to plagioclase → clinopyroxene → ilmenite. At about 1 GPa, it becomes clinopyroxene → plagioclase → garnet → ilmenite. At the highest pressures, it is garnet → clinopyroxene.

7.5 THE EFFECTS OF FLUIDS ON MELTING BEHAVIOR

The large volume increase associated with the release of a dissolved volatile in solution to a free vapor phase makes the solubility of a volatile species in a melt susceptible to pressure variations. In other words, increasing pressure can effectively force a volatile to dissolve in a silicate melt. The effect is greatest at low pressures, where the free gas volume would be largest. We discussed some dramatic examples of the pressure dependence of volatile solubility in Chapter 4 when volatiles were released with pressure loss in explosive volcanic events.

A **fluid-saturated** melt is one that contains the maximum amount of a dissolved volatile species possible under the existing P-T-X conditions. Any volatiles present in excess of that amount must be present as a separate coexisting fluid phase. The term **fluid pressure** (P_f) is commonly used to describe the combined effect of pressure and fluid content in fluid-bearing systems. A melt system at a specific pressure can vary from fluid saturated ($P_f = P_{total}$) to fluid free ($P_f = 0$, also called "dry"), depending on the amount of fluid species available. P_f rarely exceeds P_{total}, at least not for long, because the result would be expulsion of the excess volatiles, often explosively.

By the time we collect and analyze a rock sample, the fluids contained in it are mostly gone, so we commonly overlook their importance. Occasionally we can observe and analyze **fluid inclusions** (fluids released from solution and trapped before they can escape; Figure 7.17), but they are tiny, and the analytical process is complex. Many fluid inclusions form during postmagmatic stages and thus do not represent magmatic fluids. Alternatively, we can collect and analyze volcanic gases as they escape. From these studies, we learn that the volatile component of magma comprises gases predominantly in the C-O-H-S system. H_2O and CO_2 dominate, with lesser amounts of CO, O_2, H_2, S, SO_2, and H_2S. There may also be some minor amounts of

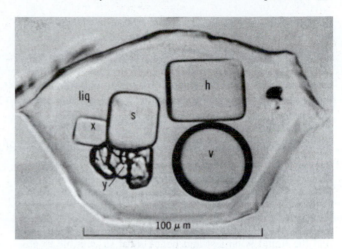

FIGURE 7.17 Fluid inclusion in emerald. The inclusion contains aqueous liquid (liq), plus a vapor bubble (v), and crystals of halite (h), sylvite (s) and two other crystalline phases (x and y). From Roedder (1972).

other components, such as N, B, Cl, and F. We shall limit our discussion of volatiles to the two major species, H_2O and CO_2. Although the volatile constituents are present in relatively small amounts, we have discovered from experimental melting studies with gas components that they can have a profound effect on melting temperatures, the sequence of minerals that crystallize, and liquid evolution paths.

7.5.1 The Effects of H_2O

In Chapter 5 and in Figure 7.15, we discussed the slope of the generic idealized anhydrous solid → liquid equilibrium melting curve on a P-T phase diagram. Let's take a moment to address from a simple qualitative standpoint the effects that we might expect if H_2O is added to this anhydrous system. If we add H_2O, the reaction itself changes. Most minerals don't accept much H_2O. The only common igneous minerals that do are micas and amphiboles, which are usually subordinate. Melts, as we shall shortly see, accept more H_2O. For anhydrous minerals (those with no H_2O content), the one-component melting reaction:

$$\text{solid} = \text{liquid} \qquad (7.14)$$

becomes:

$$\text{solid} + H_2O = \text{liq}_{(aq)} \qquad (7.15)$$

The subscript (aq) means *aqueous*, or an H_2O-bearing phase. H_2O must be on both sides of the reaction for Reaction (7.15) to balance. It occurs as a separate fluid phase on the left and as H_2O that is dissolved in the liquid on the right. Because the phase(s) on the high-temperature side of the equilibrium boundary accommodate H_2O better than those on the low-temperature side, Le Châtelier's Principle tells us that adding H_2O to this new equilibrium causes the high-T side to expand at the expense of the low-T side. In other words, if we begin at equilibrium in an initially anhydrous system (on the equilibrium curve in Figure 7.15), adding H_2O causes aqueous melt to become more stable with respect to the solid. *The result of the addition of H_2O to an anhydrous system, then, is to lower the melting point at a given pressure.* Because more H_2O can be forced into solution at higher pressure, the melting point depression becomes progressively greater as pressure increases.

Now we turn to this effect in more detail. The solubility of H_2O in a few silicate melts at 1100°C is shown in Figure 7.18, as a function of H_2O pressure. As predicted above, the amount of H_2O that dissolves in a melt increases with pressure, from 0 at atmospheric pressure to 10 to 15 wt. % at 0.8 GPa. Also, the rate of solubility increase is greatest at low pressures, where ΔV of the reaction between a free gas and the liquid is largest and hence most susceptible to pressure changes. The amount of H_2O in the melts is somewhat understated by using wt. % in the figure. In the case of the albite curve, we can easily convert to mol %. The molecular weight of H_2O is 18 g/mol, and that of albite is 262 g/mol. If our melt has 10 wt. % H_2O, it has 10 g H_2O and 90 g albite. This is 0.56 mol H_2O and 0.34 mol albite—or 62 mol % H_2O! Similar ratios apply to the other melts.

A principal mechanism by which H_2O dissolves in silicate melts is a process in which the H_2O molecule is involved in a hydrolysis reaction with the "bridging" oxygen that links (polymerizes) adjacent SiO_4 tetrahedra. By this hydrolysis reaction, H_2O dissociates to H^+ and OH^- and interacts with the -Si-O-Si- link to form -Si-OH and HO-Si- (Wasserburg, 1957; Burnham, 1979; Mysen et al., 1980). The H^+ ion satisfies the negative charge on the once-bridging tetrahedral oxygen, and the depolymerization reduces the viscosity of highly polymerized melts. The extent to which this process occurs depends upon the structure of the melt (which determines the initial degree of polymerization). A model for melt structures proposed by Weyl and Marboe (1959; see also Burnham, 1979; Mysen, 1988; Mysen et al., 1982) proposed that the structure of a silicate melt is similar

to its mineralogical equivalents, having essentially the short-range order of silicate bonds but lacking the long-range order necessary to diffract x-rays. This implies that the more polymerized the aluminosilicate structure of the minerals that would crystallize from it, the more polymerized the corresponding melt. Melted olivine is not polymerized, whereas melted quartz or feldspar is extensively polymerized. Basalts are then less polymerized, and rhyolites are more so. This, of course, agrees with the correlation between viscosity and silica content as it is related to the explosiveness of volcanic eruptions in Chapter 4. If this notion is true, we could correlate the relative extent of H_2O solubility by this mechanism in silicate melts to the degree of polymerization of the minerals corresponding to the melt. Such an approach has proven fruitful in explaining the greater solubility of H_2O in melts of framework silicates than in olivine or pyroxene melts (see Figure 7.18). The actual mechanism of H_2O solution and hydrolysis of silicate melts, however, must be more complex than this single reaction involving bridging oxygens. In the works cited above by Burnham and by Mysen (see also Mysen 1990, 1991), hydroxyl groups also appear to form complexes with alkalis, alkaline earths, and some transition elements.

The effect of H_2O in lowering the melting point in silicate systems has been substantiated in numerous experiments under hydrous conditions. We shall look first at the albite-H_2O system because it is relatively simple and well documented. Figure 7.19 shows the effect of H_2O on the melting of albite. The high-temperature curve is the "dry" melting of albite under H_2O-absent ($P_{H_2O} = 0$) conditions. This is the type of melting that we have addressed in the model systems up to this point. The other curve represents H_2O-saturated conditions ($P_{H_2O} = P_{total}$). The depression of the melting point is dramatically lowered by the addition of H_2O. The rate of lowering is rapid at lower pressures, and

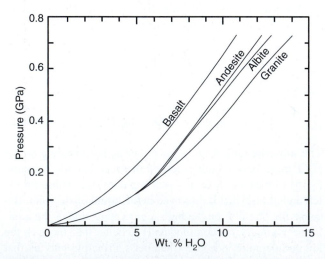

FIGURE 7.18 The solubility of H_2O at 1100°C in three natural rock samples and albite. After Burnham (1979). Copyright © reprinted by permission of Princeton University Press.

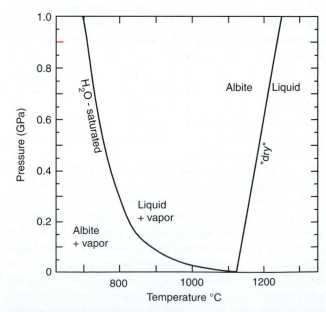

FIGURE 7.19 The effect of H_2O saturation on the melting of albite, from the experiments by Burnham and Davis (1974). Reprinted by permission of the *American Journal of Science*. The "dry" melting curve is from Boyd and England (1963).

decreases to a nearly linear relationship above ~0.4 GPa. This reflects the large negative ΔV of the reaction:

$$H_2O_{(vapor)} + albite = liquid_{(aq)} \qquad (7.16)$$

because of the large volume of the vapor phase at low pressure. Thus a slight increase in pressure drives the reaction to the right and stabilizes the liquid at the expense of the solid + vapor, thereby lowering the melting point. We can also use the Clapeyron equation to come to the same conclusion. Remember Equation (5.15):

$$\frac{dP}{dT} = \frac{\Delta S}{\Delta V} \qquad (5.15)$$

As described in Chapter 5, ΔV and ΔS are both positive for the "dry" melting of a solid such as albite, so the slope (dP/dT), according to Equation (5.15), is positive. This is true for the "dry" melting of albite in Figure 7.19. For the H_2O-saturated system, ΔS of melting is positive (because of the very low entropy of the solid phase compared to the liquid or vapor), but ΔV is *negative* (because of the very large volume of the vapor compared to the liquid or solid). Thus the slope is negative and of low magnitude because the denominator in the Clapeyron equation, ΔV, is much larger than the numerator, ΔS. At higher pressure, however, ΔV decreases dramatically (because the vapor phase is so compressible), whereas ΔS changes much less. The slope thus gets steeper. At very high pressure (beyond that of Figure 7.19), the vapor phase occupies a small volume, and ΔV becomes positive again (i.e., the volume increase associated with melting becomes larger than the volume lost as the vapor phase dissolves). The slope thus returns to a positive value, and there is a temperature minimum on the melting curve. At pressures approaching zero, it is impossible to maintain a dissolved vapor phase, as it immediately escapes into the surroundings. The curves for vapor-saturation and "dry" conditions therefore meet at a single point, close to 1118°C, the atmospheric pressure melting point.

The two bounding condition curves in Figure 7.19 are typical for silicate melting under "dry" and H_2O-saturated conditions. The amount of H_2O that can dissolve in a melt increases with pressure, and the melting point depression increases as well (although the *rate* of reduction decreases). Any H_2O present in excess of that required for saturation exists as a separate vapor phase and has no further effect on melting. The saturation curve is thus the limit of the melting point reduction.

The effect of H_2O on melting of a natural silicate systems is illustrated in Figure 7.20, which shows the melting (or crystallization) interval for both H_2O-free ("dry") and H_2O-saturated conditions for a basaltic composition. The dry system is similar to that of Figure 7.13. (The differences probably reflect compositional variables resulting from the choice of sample material.) The individual mineral melting curves have been omitted to simplify the diagram. The saturated system behaves like the albite system, with both the solidus and liquidus depressed. Up to about 1.5 GPa, the

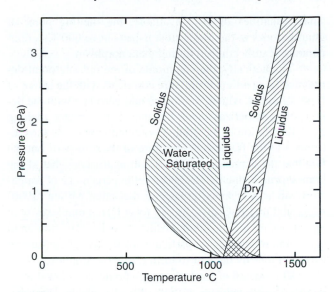

FIGURE 7.20 Experimentally determined melting intervals of gabbro under H_2O-free ("dry") and H_2O-saturated conditions. After Lambert and Wyllie (1972). Copyright © by the University of Chicago.

solidus is depressed further than the liquidus, resulting in a greater melting temperature interval. Above 1.5 GPa, we see the reversal in slope of the solidus. The liquidus slope reverses at about 3.5 to 4.0 GPa.

The solidus curves in Figure 7.20 represent the beginning of melting with rising temperature. Figure 7.21 illustrates these same solidus curves for three common igneous rock types covering a range of composition from ultramafic to silicic. In all cases, the effect of adding H_2O is to greatly lower the melting point of silicate rocks. Thus H_2O could play a very important role in the generation of melts, particularly in the crust, where H_2O is more abundant. With even

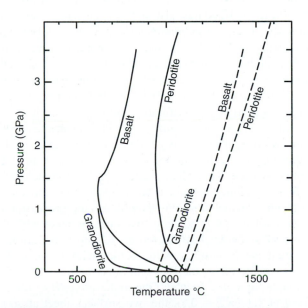

FIGURE 7.21 H_2O-saturated (solid) and H_2O-free (dashed) solidi (beginning of melting) for granodiorite (Robertson and Wyllie, 1971), basalt (Lambert and Wyllie, 1968, 1970, 1972), and peridotite (H_2O-saturated: Kushiro et al., 1968a; dry: Hirschman, 2000).

small quantities of H_2O available, the melting point of granitic rocks in the lower crust is just above 600°C, a temperature readily attained during metamorphism.

Theoretically, the magnitude of the melting point depression at a particular pressure correlates with the extent of H_2O solubility, which, in turn, should correlate with the degree of polymerization of the melt. The greater melting point depression in the granodiorite and gabbro are thus attributed to the framework structure of the feldspars, but the fact that the depression of the basalt approaches that of the granodiorite is anomalous because the percentage of framework silicates is much greater in the latter. Mysen (1990) suggested that H_2O dissolves to form OH^- complexes with *all* cations in calcium–aluminosilicate melts. Ca^{2+} is high in basalt and might then be particularly effective at consuming hydroxyl ions by the pair to form $Ca(OH)_2$ complexes.

The amount of melt that can be produced below the dry solidus depends upon the quantity of H_2O available. This is because the two situations—H_2O saturated and dry—represent the boundary conditions for rock–melt H_2O contents. Natural systems typically contain some intermediate amount of H_2O. The amount may be fixed, either a fixed percentage of the rock or melt or a fixed proportion of the coexisting fluid phase (controlled by some external reservoir). The quantity of H_2O in a rock in the deep crust or mantle may be small, probably less than 2 to 3 wt. %, and the first melt to form may absorb all of the H_2O available. The remaining solid must then melt under H_2O-undersaturated conditions, requiring significantly higher temperatures.

To illustrate this last point, let's return to our simpler and better-constrained albite-H_2O system. Figure 7.22 is a projection of the system in pressure-temperature-$X_{H_2O}^{melt}$ space to the pressure–temperature face. Think of it as a three-dimensional block with x = temperature, y = pressure, and z (toward you) = composition with pure albite at the near end of the block and pure H_2O at the far end. This figure was calculated using the solution model for Ab-H_2O of Burnham and Davis (1974). The dry and H_2O-saturated curves agree well with the experimental data from Figure 7.18. The sub-horizontal curves represent contours (similar to the temperature contours in Figure 7.2) of the H_2O content (X_w^m in *mole* fraction) of H_2O-saturated molten albite (which can contain only a certain quantity of H_2O at a particular pressure and temperature). Any H_2O present in excess of this value exists as a free fluid phase coexisting with an H_2O-saturated melt. The other, steeper lines are contours for X_w^m on the albite melting surface, representing the conditions under which albite melts with a fixed proportion of H_2O available. Where these two surfaces meet with the same value of X_w^m (the conditions for H_2O saturation of the melt and the conditions for melting with the same H_2O content) determines the H_2O-saturated melting curve. For example, in a case in which the molar albite/H_2O ratio is fixed at 1:1 ($X_w^m = 0.5$), the two surfaces meet at about 0.2 GPa and 830°C (the intersection of the two $X_w^m = 0.5$ curves). Naturally, this must be on the H_2O-saturated melting curve as well so that all three curves intersect at this point.

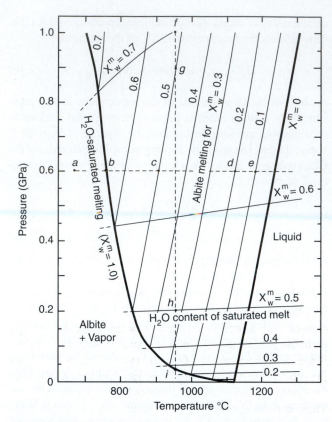

FIGURE 7.22 Pressure–temperature projection of the melting relationships in the system albite-H_2O. X_w^m is the mole fraction of H_2O dissolved in the melt. From Burnham and Davis (1974). Reprinted by permission of the *American Journal of Science*.

If the limitation on H_2O content is appropriate for a particular case, *the H_2O-saturated melting curve is valid only at pressures less than this point*. At pressures greater than this point, more H_2O is required to attain saturation of the melt and induce saturated melting. If that H_2O is not available, melting cannot proceed until higher temperatures are attained.

I realize that these curves and the concept stated above can be confusing. The processes involved, and their important ramifications for real melt behavior, are best illustrated by using some examples. First, let's try a simple scenario involving the isobaric heating of albite with a fixed quantity of H_2O. We shall assume that the system contains 10 *mol* % H_2O (perhaps a generous estimate for rocks in the lower crust) and heat it from point *a*, at 0.6 GPa and 670°C. At point *a*, the system is composed of solid albite coexisting with a fluid phase. When heated to about 770°C (point *b*), the system reaches the H_2O-saturated melting curve and begins to melt. At this point, an *H_2O-saturated* melt could contain about 64 *mol* % H_2O. There is only sufficient H_2O available, however, for a melt with 10% H_2O (if all of the albite were to melt). Thus only 10/64, or 16%, of the albite reverts to an H_2O-saturated (64% H_2O) melt at this point. As the temperature is raised, the amount of H_2O in the melt that coexists with albite and a fluid phase decreases, so a bit more albite will melt. At point *c* (~905°C), the system would be completely molten if it were to contain 50% H_2O, but it contains only 10%, so essentially 10/50, or 20%, will

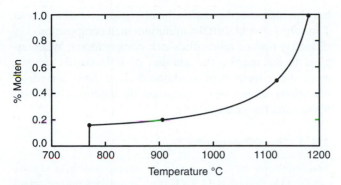

FIGURE 7.23 Percentage of melting for albite with 10 mol % H_2O at 0.6 GPa as a function of temperature along traverse *a-e* in Figure 7.22.

be melted. At point *d* (~1120°C), it would be completely molten if it contained 20% H_2O, but it is only 10/20, or 50%, melted. Only at point *e* (~1180°C) is albite with 10% H_2O completely molten. The ratio of melt to solid thus slowly increases with temperature. At any point along the melting path, the ratio of melt to solid is approximately *x/y*, where *x* is the mole fraction of H_2O in the system, and $y = X_w^m$ for the albite liquidus contours. This ratio increases slowly at first, until the value of *y* approaches that of *x* (Figure 7.23). Thus, at point *d*, over 400°C above the point of initial melting, as the $X_w^m = 0.2$ contour is crossed, the albite is only half molten. The final 50% of the solid is melted between $X_w^m = 0.2$ and 0.1 (point *e*), an interval of less than 15°C that is close to the dry liquidus. We may conclude that, in spite of the ability of H_2O to dramatically lower the melting point of silicate phases, *the amount of melt generated at the lowered temperatures is limited by the availability of that H_2O.*

To further illustrate the implications of this diagram, consider yet another simple scenario. We begin with a system that contains $X_w^m = 0.5$ at 1 GPa (point *f*, at approximately 925°C). The initial conditions are above the H_2O-saturated melting curve, so the H_2O available has caused some of the albite to melt, but the system is not H_2O saturated. It requires $X_w^m = 0.52$ (52 mol %) to be saturated at this *P* and *T*, which would result in complete melting. This system is thus only partially molten but almost completely so (50/52, or 96% melt). Suppose a diapir of this crystal–melt mixture were to rise rapidly enough to remain essentially isothermal. At point *g*, the melting curve for $X_w^m = 0.5$ is reached, and the last of the albite melts. Below this pressure, the system is at a temperature above the liquidus for $X_w^m = 0.5$. As the liquid diapir continues to rise, it thus becomes progressively more **superheated** (further above the liquidus), with no other significant changes until point *h*. At this point, the melt reaches the conditions for which a melt with $X_w^m = 0.5$ is an H_2O-saturated melt. *Now a separate fluid/vapor phase begins to form.* This phase is H_2O fluid that is in turn saturated in dissolved albite. With further rise, the amount of H_2O that the melt can contain decreases, as the curves for $X_w^m = 0.4$ and 0.3 are crossed, resulting in the generation and release of more of the fluid phase. Finally, it reaches point *i*, where the melt, containing

about 25% H_2O (halfway between the $X_w^m = 0.2$ and the $X_w^m = 0.3$ curves) at 0.3 GPa, crystallizes to albite, which is anhydrous, and the remaining H_2O is set free. *Thus we can expect rising melts containing H_2O to reach H_2O saturation and expel a fluid phase as it rises and expel even more as it crystallizes.* We shall see in Chapter 11 that these fluids can create separate systems such as pegmatites or hydrothermal ores, and they may play a major role in contact metamorphism of the country rocks. A further implication of this scenario is that rising hydrous melts will intersect the H_2O-saturated solidus and crystallize at some relatively shallow depth prior to reaching the surface. Granitic melts, which are generally hydrous, thus tend to form plutons, whereas dry mafic melts, which are not likely to intersect their positively sloping anhydrous solidus, are prone to reach the surface as basaltic lava flows.

More realistic natural processes are neither isothermal nor isobaric but heat or cool along curves, usually with a positive slope, reflecting the diffusion of heat, such as a geothermal gradient or temperature drop in a rising melt diapir. Such processes, however, differ more in detail than in principle from the isothermal or isobaric ones just described. More complex natural chemical systems also have different saturation and melting curves than the ones shown for albite, but they should have similar shapes, and the principles developed for albite at intermediate H_2O contents are applicable.

In the situations described above, H_2O was considered to be present in a fixed quantity. At high pressure, H_2O was the only constituent dissolved in the fluid portion of the melt, but the fluid was usually present in quantities less than required for saturation ($P_{H_2O} = P_{fluid} < P_{total}$). An alternative way that the P_{H_2O} can be less than P_{total} is for the system to be fluid saturated, but H_2O comprises only a portion of the fluid ($P_{H_2O} = P_{fluid} < P_{total}$). This occurs when the fluid contains a mixture of H_2O and other species. Instead of a constant mol % H_2O in the system, we have a constant *activity* of H_2O in the melt ($a_{H_2O}^{melt}$; see Section 27.3). In such cases, the melting curve will not follow the saturated curve and then shift toward the dry curve when the H_2O is consumed by melting, as it does in Figure 7.22. Rather, it will have a shape intermediate between the saturated and dry curves. The calculated melting curves for the albite-H_2O system are shown in Figure 7.24. Lines of constant $a_{H_2O}^{melt}$ are calculated using an ideal solution model based on the solution mechanism of Burnham (1979). These curves represent the initiation of melting for various H_2O activities. In this case, the magnitude of the temperature depression effect for first melting is related to the activity.

Because the solubility of H_2O in the melt varies for different minerals being melted, the addition of H_2O to rock systems will lower the liquidus differentially and thus affect the composition of the liquids that coexist with minerals at a particular pressure and temperature. Figure 7.25 illustrates this effect of P_{H_2O} on the eutectic point for the diopside–anorthite system (see also Figure 7.27). Note that, compared to the effect of lithostatic pressure, H_2O pressure depresses the liquidus temperature, and this depression is greater for anorthite than for diopside, as we would predict from the

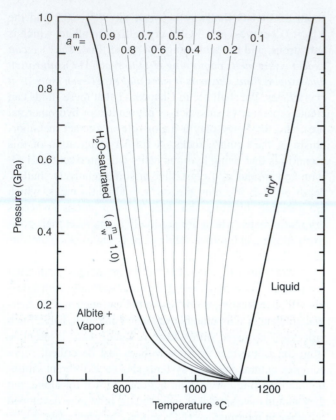

FIGURE 7.24 Pressure–temperature projection of the melting relationships in the system albite-H_2O, with curves representing constant activity of H_2O. From Burnham and Davis (1974). Reprinted by permission of the *American Journal of Science*.

more abundant bridging oxygens in the former. This causes the eutectic to shift toward more anorthitic compositions. Mysen et al. (1980) pointed out that the addition of H_2O to melting systems containing pyroxenes plus feldspars or silica minerals not only favors melting of the framework minerals

but also increases the chain-silicate component of the melts. This effect should shift the minimum melt compositions significantly toward more silica-rich compositions. When applied to the mantle, the addition of H_2O should produce more silicic melts than would be derived under anhydrous conditions, as has been confirmed by Kushiro (1972) and Mysen and Boettcher (1975).

7.5.2 The Effects of CO_2

The solubility of CO_2 in silicate melts contrasts significantly with that of H_2O. CO_2 does not dissociate or attack bridging oxygens because the C^{4+} ion behaves much differently than the H^+ ion. C^{4+} is small and highly charged, so will not stably bond to an oxygen adjacent to the Si^{4+} cation. Consequently, CO_2 does not dissolve appreciably in melts, particularly highly polymerized silicic melts. Several experimental investigations have found that the addition of CO_2 to systems has resulted in little solution and little change in the melting point (Holloway and Burnham, 1972; Eggler, 1972). Many investigators have accordingly treated CO_2 as an inert component of the fluid phase and have used it as a dilutant of H_2O to reduce a_{H_2O} in the fluid (as in Figure 7.24). Figure 7.26 shows the results of some mixed-volatile experiments on the melting of albite. Whereas the H_2O reduces the melting point of albite, the addition of CO_2 appears to mitigate the effect of H_2O, as though it were simply diluting H_2O and reducing its capacity to dissolve in the melt. (Compare the middle curve in Figure 7.26 to the curve for $a_{H_2O} = 0.5$ in Figure 7.24.) By lowering a_{H_2O} in the fluid phase, Le Châtelier's Principle tells us that the reaction solid + H_2O = melt$_{(aq)}$ should shift the equilibrium toward the reactants, stabilizing them more than the aqueous melt, and shifting the H_2O-saturated melting curve toward higher temperatures.

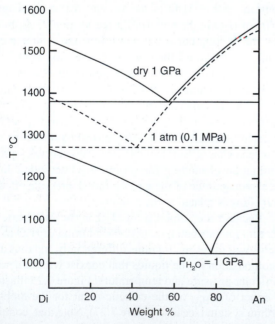

FIGURE 7.25 The effect of H_2O on the diopside–anorthite liquidus. Dry and 1 atm from Figure 7.16; $P_{H_2O} = P_{total}$ curve for 1 GPa from Yoder (1965).

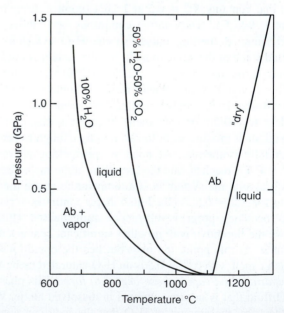

FIGURE 7.26 Experimentally determined melting of albite: dry, H_2O-saturated, and in the presence of a fluid comprising 50% H_2O and 50% CO_2. From Millhollen et al. (1974). Copyright © by the University of Chicago.

CO_2 does dissolve to some extent in silicate melts, however, particularly at pressures greater than 1 GPa. Eggler (1973) found that CO_2 dissolved to the extent of about 4 to 5% in diopside, enstatite, and albite melts at high pressure. He stated that the solubility of CO_2 strongly depends on the composition of the melt, dissolving more extensively in mafic, less polymerized melts (in marked contrast to the behavior of H_2O). He found that a little H_2O dramatically increased the solubility of CO_2 to as much as 35% in diopside at 2 GPa and to 18% in enstatite, but only to 5 to 6% in albite. Mysen and Virgo (1980) suggested that CO_2 dissolves to form carbonate (CO_3^{2-}) complexes in silicate melts, particularly with calcium to form $CaCO_3$ complexes. In order to form CO_3^{2-}, CO_2 reacts with two *non-bridging* oxygens to take one and make a bridging oxygen of the other: CO_2 + -Si-O O-Si- = CO_3^{2-} + -Si-O-Si- (Eggler and Rosenhauer, 1978). For more recent experiments and models of H_2O-CO_2 mixtures in silicate melts, see Blank et al. (1993), Papale (1999), Newman and Lowenstern (2002), and Liu et al. (2005).

Because CO_2 dissolves in melts to some extent, it will lower the melting point of silicate systems, although considerably less so than does H_2O. Beyond that, CO_2 has effects that are different from those of H_2O. CO_2 tends to make a *more* polymerized melt, whereas H_2O breaks the bridging bonds. CO_2 should thus dissolve to a greater extent in more mafic, less polymerized melts, and it should *raise* the viscosity of those melts as it dissolves. This also explains why CO_2 dissolves to a greater extent if H_2O is present because H_2O creates less polymerized melts, which, in turn, attract CO_2. The addition of CO_2 should thus shift the diopside–anorthite eutectic (Figure 7.25) to lower temperature and toward diopside. In mafic melt systems, then, CO_2 should favor *less* siliceous melt compositions, in direct contrast to the effect of H_2O. This prediction is clearly illustrated in Figure 7.27, based on experiments in the basalt system at 2 GPa. E_{dry} is the position of the

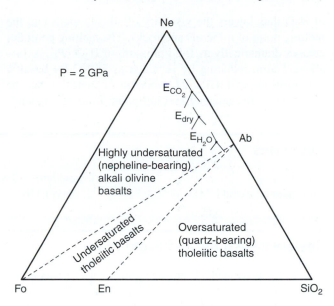

FIGURE 7.27 Effect of volatiles on the ternary eutectic (minimum melt composition) in the system Fo-Ne-SiO$_2$ (base of the "basalt tetrahedron") at 2 GPa. Volatile-free curve from Kushiro (1968), H$_2$O-saturated curve from Kushiro (1972), and CO$_2$-saturated curve from Eggler (1974).

ternary eutectic under volatile-free conditions, whereas E_{CO_2} and E_{H_2O} are the CO_2 and H_2O saturated eutectic positions.

Because CO_2 solution is greater at high pressure and in more mafic compositions, the effect of CO_2 should be greatest in the mantle. Indeed, CO_2 can be influential in lowering the solidi of mantle rocks and causing incipient melting of alkaline basalt. Of course, the amount of alkaline melt created by this process depends upon the quantity of CO_2 available in the mantle. We will return to the creation of mantle melts and the variables that control it in Chapter 10. CO_2 is influential in peridotite melting and the formation of carbonatites, as discussed in Section 19.2.

Summary

The addition of another component doesn't drastically modify the behaviors observed in binary systems, but it permits us to explore the dynamics of systems a step closer to natural ones. Ternary magmatic systems are more difficult to portray graphically and are typically represented by isobaric *T-X* triangular diagrams. The liquidus surface can be represented by using isothermal contours. As in Chapter 6, we can apply the phase rule to multicomponent systems and phase diagrams to understand the dynamics of equilibrium or fractional melting–crystallization processes. We can determine the sequence of minerals that crystallize from a cooling melt, the dependence of the sequence on the bulk composition of the initial melt, and the variation in composition of liquids and solids across the crystallization interval. Equilibrium melting is simply the reverse process of equilibrium crystallization.

Systems with more than three components are very difficult to depict graphically, although the phase rule may still be applied effectively to understand the variance and dynamics of such systems. Melting relationships for natural samples are generally shown on *P-T* phase diagrams with relevant univariant reaction curves within the melting interval. Laboratory experiments and phase diagrams illustrate the complexities of melting and crystallization of natural materials, particularly the variation in the sequence of minerals that are consumed or formed in systems of differing composition. These variations show us that any broad generalization, such as Bowen's Reaction Series, although useful as a guide for magmatic crystallization sequences, should not be treated as law.

Increased pressure generally raises the melting temperature of minerals (and therefore of rocks) but affects various minerals to different extents. As a result, the eutectic point shifts, affecting the composition of the first melt (or final liquid to crystallize).

H_2O dissolves readily in most silicate melts, where the molecule breaks up polymerized Si-O bonds. Addition of H_2O thus lowers the viscosity of silicate melts and the melting point of minerals and rocks. The melting point decreases dramatically as fluid pressure of H_2O (P_{H_2O}) rises above 1 atm, reducing by as much as 500°C for basaltic compositions at 1 GPa. The *amount* of melt that can be generated at the reduced melting point, however, depends on the amount of H_2O available. CO_2 has a less dramatic effect. Addition of H_2O and CO_2 also shift the composition of the eutectic point in natural systems. H_2O shifts the first melt of basaltic systems toward more SiO_2-rich compositions, and CO_2 shifts it toward more alkaline SiO_2-undersaturated compositions.

Key Terms

Cotectic *114*

Bulk mineral extract *115*

Liquid line of descent (evolution curve) *116*

Bowen's reaction series *124*

Fluid pressure *127*

Review Questions and Problems

Review Questions and Problems are located on the author's web page at the following address: **http://www.prenhall.com/winter**

Important "First Principle" Concepts

- We can use a chemographic diagram of mineral assemblages (see inset in the upper left of Figure 7.4) to determine the final equilibrium mineral assemblages that correspond to any solidified bulk composition. This can be helpful in determining which phase is consumed by invariant reactions, particularly peritectic ones.

- The first mineral to form from a cooling magma is the one on the same side of the eutectic point (in binary systems) or the cotectic curves (in ternary systems) as the bulk composition (X_{bulk}).

- The composition of a cooling liquid shifts directly away from the mineral from that liquid (or from the bulk mineral extract if more than a single mineral is removed).

- Evolving liquids in cooling ternary eutectic systems move from divariant fields to univariant (cotectic) curves to invariant (eutectic) points as each additional new mineral crystallizes (and thus lowers the variance).

- As in binary systems, the initial melt (and the final liquid to crystallize) in eutectic systems is always the eutectic composition.

- The melt evolution during equilibrium melting in ternary eutectic systems follows a continuous path that is the opposite of equilibrium crystallization. Fractional melting, however, involving the removal of melt increments as they are generated, occurs in discontinuous steps. The first melt composition is at the ternary eutectic point. Melting ceases when a phase is consumed and begins again at the composition and temperature of the remaining binary system eutectic (if heated sufficiently).

- Continued heating will consume another phase, halting melting again until the melting point of the final phase is reached. This suggests that melting in natural systems will cease when a phase is consumed and involve a discontinuous jump in temperature if it is to proceed.

- We can use an isothermal section through a ternary *T-X* phase diagram (Figure 7.9) to determine the equilibrium phase assemblage corresponding to any particular X_{bulk}. This is an important tool used most by metamorphic petrologists.

- Bowen's reaction series is a very useful generalization of magma crystallization (and melting) behavior, but, like any other generalization, it cannot explain the many variations in crystallization sequence. Appropriate phase diagrams are much more useful in this regard.

- Pressure raises the melting point of silicate systems and also generally changes the composition of the eutectic point.

- Addition of H_2O can dramatically reduce the melting point of silicate systems at elevated pressures, but the amount of melt so generated depends on the amount of H_2O available.

- A rising H_2O-bearing (but undersaturated) magma will gradually approach H_2O saturation. If it contains enough H_2O, it will reach saturation at a fairly shallow depth and expel a hydrous vapor phase (contributing to hydrothermal systems above the magma body and perhaps also pegmatites and ores).

- A rising H_2O-bearing magma will probably encounter its H_2O-saturated solidus and crystallize at a shallow depth before reaching the surface of the Earth.

Suggested Further Readings

Bowen, N. L. (1928). *The Evolution of the Igneous Rocks*. Princeton University Press. Princeton, NJ. (1956 reprint by Dover. New York.)

Burnham, C. W. (1979). The importance of volatile constituents. In: *The Evolution of the Igneous Rocks. Fiftieth Anniversary Perspectives* (ed. H. S. Yoder). Princeton University Press. Princeton, NJ.

Ehlers, E. G. (1972). *The Interpretation of Geological Phase Diagrams*. W.H. Freeman. San Francisco.

Morse, S. A. (1994). *Basalts and Phase Diagrams. An Introduction to the Quantitative Use of Phase Diagrams in Igneous Petrology*. Krieger. Malaber, FL.

8

Chemical Petrology I: Major and Minor Elements

Questions to be Considered in this Chapter:

1. How do we analyze rocks to determine their chemical composition?

2. How are rock analyses reported, and what do they tell us about the composition of igneous rocks?

3. How can we use the compositions of rocks and of groups of associated rocks to investigate the processes involved in their genesis?

4. How can we display the compositions of many associated rocks so that we can investigate trends in the data?

5. What quantitative methods are used to model observed compositional trends?

6. How can we use chemical criteria to characterize or classify igneous rocks and entire sequences of igneous rocks that appear to be related in the field?

Geology borrows heavily from other disciplines, applying the principles and techniques of physicists, chemists, and materials scientists, among others, to geological problems. Petrologists borrow most heavily from the field of chemistry, where the application of **geochemistry** to petrologic problems has proved so fruitful that modern petrology simply cannot be accomplished properly without it. In this chapter and Chapter 9, I shall lay the groundwork for an understanding of the principles of chemistry that we apply to igneous and metamorphic rocks. This chapter begins with the chemical applications that address elements usually present in concentrations greater than 0.1 weight % (wt. %) in igneous rocks. Trace elements and isotopes are treated in Chapter 9. The material in these two chapters is critical to our ability to interpret and understand the processes of magma generation and evolution in the various igneous–tectonic settings that we shall investigate in Chapters 13 to 20. At times you may want to refer to the periodic table inside the front cover of this text.

For convenience in what follows, elements are considered to be either major, minor, or trace elements, based on their concentration in rocks. Limits on the groups are arbitrary, but the following is a guide to the common behavior:

Major elements >1.0 wt. % (expressed as an oxide)
   Typical examples: SiO_2, Al_2O_3, FeO, MgO, CaO, Na_2O, and K_2O
Minor elements 0.1 to 1.0 wt. % (expressed as an oxide)
   Typical examples: TiO_2, MnO, P_2O_5, and perhaps the volatiles H_2O and CO_2
Trace elements <0.1 wt. % (usually expressed as an element)

Major elements, because they are present in high concentrations, control to a large extent the mineralogy and crystallization–melting behavior in igneous systems. They also control such properties as viscosity, density, diffusivity, etc. of magmas and rocks. **Minor elements** typically substitute for a major element in a mineral. (Mn, for example, substitutes for Fe or Mg in most mafic minerals.) If they reach a sufficient concentration, however, minor elements may form a separate mineral phase, present in minor amounts (called an **accessory mineral**). For example, if sufficient Zr is present, it causes the mineral zircon to form, sufficient P will generate apatite, and Ti may form titanite, rutile, or a Fe-Ti oxide, such as ilmenite. **Trace elements** are too dilute to form a separate phase, so they act strictly as substitutes for major or minor elements. Although major elements, minor elements, and trace elements are classified in accordance with the guidelines above, most petrologists consider TiO_2, MnO, P_2O_5, and H_2O minor elements, even in uncommon instances when they are present in concentrations greater than 1.0 wt. %. Likewise, K_2O, which may be present in concentrations below 0.1 wt. % in some mafic rocks, is more abundant in most rocks and is thus considered a major element by many investigators, regardless of concentration.

As a result of the different roles played by major and trace elements, each provides distinct insights into various igneous processes. Major elements may be used to classify igneous rocks and study the chemical control on the physical properties of crystal–melt systems. They are also used to study the chemical evolution of melts (and minerals) during the crystallization or melting process. The concentration and distribution of trace elements may also be used to study the evolution of magmas. They have proved to be particularly effective as tracers that help to constrain magma sources or to discriminate between some magmatic processes.

8.1 ANALYTICAL METHODS

Chemical techniques were first applied to rocks in 16th-century Europe in order to understand and exploit ore veins. It was only in the 18th and 19th centuries, however, that the majority of the elements were recognized and organized on the basis of similar properties (giving rise to the periodic table of the elements in 1869). The techniques for the accurate chemical analysis of rocks, minerals, volcanic gases, and aqueous solutions were developed in the past 150 years. The only available method prior to the 1920s and 1930s was classical volumetric and gravimetric analysis (commonly called **wet chemical analysis**). Preparation of liquids and gases for wet chemical analysis was relatively easy compared to that for rocks and minerals, which had to be dissolved in strong acids and diluted before the resulting liquid could be analyzed. Mineral analyses were particularly formidable because a reasonable quantity of the pure mineral (free of inclusions and other mineral fragments) had to be physically separated before the mineral could be dissolved and analyzed accurately. Wet chemical analyses involved analysis by the tedious and time-consuming processes of

titrating, or adding certain chemicals that would combine with a particular ion in solution to form an insoluble precipitate. The concentration of the ion in question could then be calculated either from the accurately determined volume titrated or the weight of the precipitate. When this was done for each of the principal elements in the sample, the analysis was complete. A single analysis could require over a day, and years of experience were needed to master the techniques to produce high-quality results. Needless to say, chemical analyses of rocks and minerals were not common when these techniques were the only ones available. Even then, analysis could be performed only for major and minor elements.

As technology advanced, wet chemical analysis was gradually augmented by, and eventually replaced by, **instrumental** techniques of analysis. Techniques useful to the analysis of rocks and minerals are based on the ability of atoms to either emit or absorb radiation with frequencies characteristic of the element responsible. Because these techniques involve emission or absorption of some particular portion of the electromagnetic spectrum, they are called **spectroscopic** techniques. They permit much more rapid analysis of geological materials, and they can determine the concentrations of various trace elements as well. Figure 8.1 illustrates, in a simplistic fashion, the general principles of emission and absorption upon which these instruments are based. All of them require an energy source that bombards the sample to be analyzed (also called the "unknown") with energy and a detector capable of distinguishing the energy of the emitted or transmitted photons and determining the intensity of the radiation corresponding to a particular energy. The detector output provides a signal that is a measure of the emitted or absorbed radiation from the sample.

Emission requires that atoms in the sample being analyzed be "excited," or raised to an unstable state by the absorption of some form of energy. The excited state usually involves the transfer of an electron from a lower energy orbital about the nucleus to one of higher energy. The excited atom then returns spontaneously to the unexcited ("ground") state. Because energy must be conserved, the return is accompanied by emission of a photon of light, the energy (or wavelength) of which is related directly to the energy of the electron transfer (the difference between the energy levels of the two orbitals). Because the energy of electron orbitals are quantized, the radiated energy is characteristic of the element. Not every atom in the sample will continuously emit characteristic photons, but the number of photons with a particular energy emitted in a given period of time is proportional to the number of atoms of that element present in the sample (i.e., the concentration of the element). The raw data, then, are the rate of emission from the sample (usually in "counts per second," [cps]) at a particular frequency or energy. Equating cps to concentration requires comparison of the emission rate of the "unknown" (sample) to the emission rate of a set of "standards" of known concentration. These standards may be natural samples of known composition (previously analyzed using wet chemical techniques) or

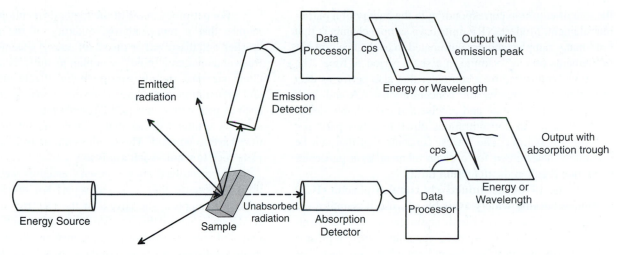

FIGURE 8.1 Diagram illustrating the geometry of typical spectroscopic instruments. "Data processor" represents possible electronic signal amplifiers, filters, and data analysis electronics. Output is illustrated as a plot of intensity (typically in counts/second [cps]) of the emitted or unabsorbed or absorbed radiation (*y*-axis) versus energy, frequency, or wavelength (*x*-axis).

prepared by dissolving and diluting pure analytical reagents to specific concentrations.

The simplest (and therefore the first) instrument for spectral detection and analysis was the **optical spectrometer**. It uses a simple prism or a diffraction grating to disperse the visible-light spectrum emitted by an excited substance and photographic film to record the light of a given color (frequency). The intensity of the spot or line of a particular color on the developed film is proportional to the concentration of a certain element. Later, electronic detectors with photomultipliers were developed to determine the intensity more accurately and quantitatively and to permit detection of radiation emitted beyond the visible portion of the spectrum. One could excite a sample in the optical spectrometer by using intense heat or simply point the detector at a self-excited source, such as the sun and stars.

Absorption techniques work in a fashion similar to emission but determine the amount of energy *absorbed* by a sample. The energy is absorbed by atoms in the sample in order to reach an excited state (generally associated with an electron jump between orbitals), so that a particular element may absorb energy at characteristic frequencies as well (equal to the energy difference between the ground and excited states). In absorption instruments, electromagnetic radiation (usually light) is passed through a sample into a detector that measures the final intensity (Figure 8.1). The extent of absorption is compared to standards in order to compute the concentration of an element (or compound) in an unknown. Of course, the light source has to be of variable frequency in order to provide the energy that correlates with the electron shifts for a variety of elements.

A number of instrumental techniques, both emission and absorption, have been developed over the years, some of which are more suitable than others for certain elements, concentrations, and materials. The first fully instrumental techniques to be developed were flame photometry, emission spectroscopy, and colorimetry, none of which is used much today. These deal with the ultraviolet–visible–near-infrared portions of the spectrum. They can also be applied to elements in far lower concentrations than can be analyzed by using wet chemical analysis. The **flame photometer** involves the aspiration of a solution into a flame (the energy source). The flame can provide only enough energy to excite the outer (valence) electrons of easily ionized elements, primarily the alkali metals. For example, the yellow-colored light observed when you strike a match is the 589 nm emission of sodium in the match head when excited by the heat of the flame. The photometer is capable of dispersing the emitted spectrum, focusing on the (sodium) emission line, and quantifying the emission. **Emission spectrometry** is similar to flame photometry, but the energy is supplied by a graphite electrode. The bright-white light emitted by the emission spectrometer testifies to a far more energetic source that can excite a much greater number of elements than can flame photometry. **Colorimetry** is based on the absorption of visible light associated with the color imparted to an aqueous solution by the addition of a chemical that forms a colored complex with the element to be analyzed.

Later, **atomic absorption (AA)** and **x-ray fluorescence (XRF)** spectroscopy were developed. In AA spectroscopy, a solution is aspirated into a flame or a graphite furnace, and a beam of light (of a predetermined wavelength) is also passed through the flame or furnace. The absorption (the reduction in intensity of the resultant light) is measured and compared to the reduction in standards. XRF irradiates a *solid* sample with x-rays that are sufficiently energetic to excite electron transitions in the *inner* electron shells (not just among the valence electrons) for a variety of elements in a sample. The ensuing return to the ground state results in emission of **fluorescent x-rays**, characteristic spectra in the higher-energy x-ray range. The emitted x-ray spectrum can no longer be dispersed by a prism or a light diffraction grating to isolate a particular frequency but requires an oriented crystal lattice to diffract it (in accordance with Bragg's law). Once the spectrum is dispersed, an x-ray detector can be tuned to a particular frequency (based on the angle with reference to the diffracting crystal) and measure the rate of emission corresponding to the energy of a particular element. Modern XRF units have sample chambers that hold many samples and are automated to evaluate a number of elements on each sample in a short period of time. The process is controlled by a computer that also performs data reduction and computes the concentrations. AA and XRF are reasonably accurate and precise and can reliably detect many elements in concentrations down to a few parts per million (see below). The lowest concentration that can be accurately determined for a given element by a particular technique is called the **detection limit**.

In the 1970s the **inductively coupled plasma (ICP)** technique became widely available. To use it, samples are dissolved and then mixed with argon gas as they are aspirated into a tiny radio frequency generator, where a plasma (similar to the gas in a fluorescent light tube) is created. This is a particularly effective means of exciting atoms. The emission line relating to each element is isolated by a diffraction grating and detected by a series of photomultipliers, each on a separate channel tuned for a specific element. All channels are simultaneously counted and fed to an on-board computer, where the data are reduced and the concentration of each element is calculated. A good modern ICP can produce a chemical analysis of more than 60 elements in a matter of minutes (once the sample has been prepared).

In addition to these spectroscopic methods that rely upon *electron* interactions, there are methods that rely upon *nuclear* interactions. As we shall see in Chapter 9, excited or unstable particle configurations within the nucleus of an atom may release either a photon or a particle when they *decay*, or revert to the ground state. We commonly refer to these releases as "radioactivity." We can detect and measure the natural radioactivity of rocks and minerals by using a Geiger or scintillation counter. Of course this works only for isotopes of elements that are naturally unstable and emit radioactivity as they decay. Nuclear techniques have the advantage of determining the concentration of *isotopes* and not just elements. In a technique called **instrumental neutron activation analysis (INAA)**, the nuclei of a number of otherwise stable nuclides are excited by bombarding samples with an intense neutron flux in a nuclear reactor. The sample that is removed is "hot" in the radioactive sense, and the particles emitted as the atoms within it return to stable configurations can be analyzed spectroscopically in a fashion similar to XRF. This method is very accurate, with low detection limits for many elements.

In **mass spectrometry**, a sample is heated and ionized, and the ions are introduced into an evacuated chamber with a strong magnetic field. The stream of charged ions follows a curved path in the magnetic field. Due to their greater momentum, heavier ions follow a path with a larger radius of curvature. Ions of each mass are counted as they pass through slits in a charged detector. This instrument is not like the others in that it does not rely on emission or absorption, but it physically dissociates the sample and physically counts the particles of a particular mass. This is an expensive method, but it is the only reliable way to determine isotopic ratios for petrology and geochronology.

For petrology, most of the methods mentioned thus far require that a representative quantity of the sample be crushed and dissolved or fused. Of course a sample must be fresh and unaltered (unless you plan to study the alteration). Grain size and sample heterogeneity affects the sampling and preparation procedure. For very coarse-grained rocks, such as pegmatites, several kilograms may be required, and a sample splitter, a device that produces an unbiased portion, may have to be used. The result is an analysis of the complete rock (a **whole-rock analysis**).

As mentioned above, *mineral* analyses were far more difficult because they required separation of clean monomineralic samples. The advent of the **electron microprobe (EMP)** has changed this; the EMP provides an efficient, accurate, and rapid method of mineral analysis. With this instrument, a polished thin section of the sample is bombarded with a beam of electrons only ~2 μm in diameter. The electrons excite the atoms in a tiny spot on the sample (perhaps 10 μm in diameter), which emit fluorescent x-rays, much as with XRF. In both XRF and EMP, the x-rays can be dispersed either on the basis of wavelength, using oriented crystals and Bragg's law (**wavelength dispersive spectrometers [WDS]**), or on the basis of energy, using silicon semiconductor wafers (**energy dispersive spectrometers [EDS]**). WDS has better resolution of the elements but requires a separate crystal and detector channel for each element to be analyzed. Most WDS "probes" have four or more channels, each of which, after finishing the counting for one element, are re-tuned by computer to another wavelength and begin collecting again for a different element. After two to four such passes per channel, the excited sample spot has been analyzed for the 8 to 14 principal elements found in most minerals. EDS systems are less accurate than WDS systems, but they are compact and require only one small detector port. An EDS system can be mounted on a scanning electron microscope (SEM), and it adds to existing imaging devices the ability to do chemical analyses. The main advantage of EMP is that it can provide very rapid and reasonably accurate mineral analyses. Further, we can use it to analyze a traverse across a mineral, from core to rim, and determine the nature of any chemical zoning, if present. With advanced computer-driven electron beams, it is possible to create element distribution "maps" that portray the relative concentration of various elements of interest in a small area of the thin section (two color maps for a garnet are shown on the back cover of this text). The EMP has the further advantage that it is nondestructive, in that the thin section is undamaged (other than a few <20 μm pits if the electron beam intensity is high). The drawbacks are that it is not appropriate for whole-rock analysis, and the detection limits are high enough that it is not good for many trace elements.

A variation on the EMP idea is the **ion microprobe (IMP)**. This device bombards a sample surface with a stream of oxygen ions, literally blasting a crater and ionizing the target area. The ions are released into a mass spectrometer and analyzed by mass. This produces both an elemental analysis and an isotopic analysis. Ion probes are very expensive, and there are only a handful of them in operation, mainly in large government-supported laboratories.

Today, whole-rock analysis is done quickly and routinely, usually using a combination of XRF and ICP techniques. Some elements are more reliably analyzed by XRF and others by ICP. AA and colorimetry are also used in some laboratories. For a list of elements and the appropriate technique(s) for each, see Rollinson (1993). Mineral analyses are done on the EMP, and isotopic analysis requires a mass spectrometer. Oxidation states cannot be determined by using spectroscopic techniques, so Fe^{3+}/Fe^{2+}, when desired, must still be determined by titration.

8.2 ANALYTICAL RESULTS

Raw data from instrumental techniques are in counts per second of an emitted spectral line (or the decrease in emission associated with absorption), and there is always some statistical variation in the output. Any two consecutive counting intervals produce slightly different total counts, both for the sample and the standards. When combined with electronic fluctuations in the instruments, a certain level of statistical uncertainty is invariably associated with the determined concentration for each element. Because the elements are analyzed separately, the resulting analysis would not be expected to total exactly 100.000%. In addition, not all elements are routinely determined. For example, electron microprobes require that the samples be coated with an electrically conductive material (typically carbon) to disperse the electron buildup resulting from the beam, so it is not possible to analyze for carbon (unless another coating is used). Also, the windows on the detectors, required to maintain a vacuum in them, absorb the weak x-rays produced by light elements (usually below atomic number 9), so these elements are not determinable using XRF or the microprobe. The concentration of oxygen is not determined directly either because it has an atomic number of 8, and it is present in the atmosphere and in the water in which samples are dissolved for AA and ICP. Rather, oxygen is calculated on a charge-balance basis and added to the analysis in the proportion required to balance the cations. If an analysis is less than 100%, it may be because of analytical uncertainty, unanalyzed constituents, or both.

Even though *elemental* concentrations are determined using instrumental techniques, the major and minor elements for silicate rocks and minerals are routinely reported as *wt. % oxides* (grams of the oxide per 100 g of the sample). This practice is a holdover from the days of wet chemical analysis, when the precipitates were typically oxides and were weighed to determine the concentration. Certainly, most cations in our oxygen-rich atmosphere are bonded to oxygen, and this is also true for cations in silicate minerals, so reporting the analyses as oxides makes a certain degree of sense. Oxides were so commonly addressed that most common ones have specific names (SiO_2 = "silica," Al_2O_3 = "alumina," MgO = "magnesia," CaO = "lime," Na_2O = "soda," K_2O = "potash," P_2O_5 = "phosphate;" oxides of iron and manganese have no names . . . unless you consider "rust"). Our interest in rock and mineral compositions, however, is focused on the flow, exchange, and distribution of atoms (or ions), and these atomic (or molecular) proportions may be obscured by atomic mass differences when reported in wt. %. The first step in most chemical calculations is thus to convert wt. % to atomic proportion. Some works report analyses or chemical ratios in molecular proportions directly, and care must be taken to notice the difference. Trace element concentrations are typically reported in *parts per million (ppm)* of the *element*, not the oxide. This is still on a weight basis, so ppm equals grams of the element per million grams of sample. A convenient conversion to remember is *1 wt. % is equivalent to 10,000 ppm.*

The first numeric column of Table 8.1 represents a typical analysis taken directly from the literature, in this case of a basalt. The major and minor element *oxides* are usually combined and listed in order of decreasing valence. Error estimates are seldom reported, but most analyses are good to three significant figures, with an error in the range of 1 to 5 relative percent. **Relative percent** means the percentage relative to the total *reported for that element*. For example, a relative error of 3% in the value for MgO in Table 8.1 equals 3% of 6.44, or 0.19 **absolute percent**. Thus the analysis might be expressed as MgO = 6.44 ± 0.19 wt. %. Many authors simply report analyses to two decimal places, so the values for silica (SiO_2), alumina (Al_2O_3), and lime (CaO) are given to four significant figures, which is more precise than can be justified by analytical uncertainty.

H_2O is the most common volatile constituent in most rocks and minerals and may be expressed as H_2O^+ and H_2O^-. H_2O^+ represents **structural water**, present as OH^- bonded in

TABLE 8.1 Chemical Analysis of a Basalt (Mid-Atlantic Ridge)

Oxide	Wt. %	Mol Wt.	Atom Prop.	Atom %
SiO_2	49.2	60.09	0.82	17.21
TiO_2	2.03	79.88	0.03	0.53
Al_2O_3	16.1	101.96	0.32	6.64
Fe_2O_3	2.72	159.70	0.03	0.72
FeO	7.77	71.85	0.11	2.27
MnO	0.18	70.94	0.00	0.05
MgO	6.44	40.31	0.16	3.36
CaO	10.5	56.08	0.19	3.93
Na_2O	3.01	61.98	0.10	2.04
K_2O	0.14	94.20	0.00	0.06
P_2O_5	0.23	141.94	0.00	0.07
H_2O^+	0.70	18.02	0.08	1.63
H_2O^-	0.95	18.02	0.11	2.22
(O)			2.82	59.27
Total	99.92		4.76	100.00

	ppm			ppm
Ba	5	137.33	0.04	0.8
Co	32	58.93	0.54	11.4
Cr	220	52	4.23	88.9
Ni	87	58.7	1.48	31.1
Pb	1.29	207.2	0.01	0.1
Rb	1.14	85.47	0.01	0.3
Sr	190	87.62	2.17	45.6
Th	0.15	232.04	0.00	0.0
U	0.16	238.03	0.00	0.0
V	280	50.94	5.5	115.5
Zr	160	91.22	1.75	36.9

Data from Carmichael et al. (1974), p. 376, col. 1.

hydrous minerals such as amphiboles and micas. H_2O^- is adsorbed, or trapped along mineral grain boundaries. Both are too light to be detected by spectrographic means. H_2O^- can be driven off by heating the powdered sample to about 100°C and determined by the weight loss accompanying this process. **Loss on ignition (LOI)** is the weight loss that occurs when the powdered sample is heated to about 800°C, at which point all of the remaining volatiles, including structural volatiles (H_2O, CO_2, etc.), are released. If LOI is determined without first weighing the post-100°C sample, LOI will include H_2O^- as well as the bonded volatile elements. Therefore, heating is usually done in two steps.

The total of 99.92 in Table 8.1 suggests that the analysis is good. If all major and minor elements are determined, an analysis is generally considered to be acceptable if the total falls between 99.8 and 100.2%. A total of 100.00%, on the other hand, does not mean that an analysis is error free but that the statistical errors for each element happened to offset in such a way as to coincidentally produce this total. On the other hand, a total of 94% may not be significantly in error, either, if a major constituent is not analyzed. For example, because CO_2 is rarely analyzed in rocks or minerals, a good analysis for a limestone may total only 56% (the wt. % of CaO in $CaCO_3$). Because microprobed amphiboles are hydrous, they commonly have totals near 98%, and micas may not have totals much higher than 96%.

Trace elements, when determined, are listed (typically in ppm) after the major/minor elements. Even when considered cumulatively, they are unlikely to compose a significant proportion of the sample, so they are generally not included in the total. It would be very time-consuming and expensive to determine *all* of the trace elements present in a rock or mineral, so one must choose the ones to be analyzed based on the purpose for which the analysis is intended. Trace elements are determined only when necessary, and many published whole-rock analyses do not include them. ICP and XRF analyses have become common, and the number of analyses that include trace elements is increasing because it requires little extra time to produce them. Isotopic analysis is still slow and expensive, so it is done only for specific applications. Because the electron microprobe does not analyze sparse trace elements or isotopes, the number of mineral analyses that include either is miniscule.

I have added the remaining two numeric columns in Table 8.1 to illustrate the conversion from wt. % oxides to atomic % Column 3 contains the molecular weights of the oxides (or elements for the trace elements). You can determine these by adding the atomic weights in the periodic table on the inside front cover. Column 4, atomic proportions, is calculated by dividing column 2 by column 3 (weight to moles) and multiplying each by the number of cations in the oxide formula. For Si, this number is 49.2/60.09, for Al it is 2(16.09/101.96), etc. Oxygen is totaled in a similar fashion for each oxide (2 for each Si atom + 3/2 for each Al, etc.). These atom proportions are then added for the major and minor elements only (assuming that the trace elements have a negligible effect). Each value in column 5 is the value in column 4 divided by the sum of column 4 multiplied by 100, which converts the atom proportions

to atomic %. Converting to a molecular basis but retaining the expression as mol % oxides is easier. We need only divide each wt. % by the molecular weight and then normalize each to 100.

8.3 MAJOR AND MINOR ELEMENTS IN THE CRUST

Based on compilations of published rock analyses and estimates of the relative proportions of the rocks represented, a number of investigators have attempted to produce a representative analysis of the continental crust. Table 8.2 is one such example (for major elements only). Column 2 is an average of the estimates (expressed as oxides) from Poldervaart (1955) and Ronov and Yaroshevsky (1976). I calculated the approximate atomic % from these values in the manner discussed above. Note that these eight elements constitute nearly 99% of the total crust, so it should not be surprising to find that these elements are the major elements of most rocks and minerals that we shall encounter. Certainly, O and Si dominate, as reflected in the high percentage of silicate minerals in the crust. When expressed as oxides, seven major oxides would be expected to compose most crustal materials (column 2 in Table 8.2).

Table 8.3 shows some examples of the major and minor element compositions of some common igneous rocks, ranging from ultramafic to granitic and alkalic. These examples are taken from a study by LeMaitre (1976b), in which he determined average compositions based on more than 26,000 published analyses. Note that the seven major element oxides in Table 8.2 make up essentially all of the rocks listed. If the peridotite is representative of the mantle, these oxides are the dominant mantle constituents as well. TiO_2, H_2O, MnO, and P_2O_5 are the other oxides typically present in minor concentrations, and they compose the minor elements of most igneous rocks. Cr_2O_3 may be a minor element in ultramafics, but it is typically a trace element in most rocks. Fe is the only major or minor element that occurs widely in two different valence states. The ratio of Fe^{3+}/Fe^{2+} increases with the **oxygen fugacity** [Equation (27.7)] in the rock or melt at equilibrium. Because Fe^{3+} is concentrated in Fe-Ti oxides, rocks with relatively high oxygen fugacities have less Fe available for silicates, which will reduce the mafic silicate content of a rock. When the titration required to determine oxidation state has been performed, Fe is reported as Fe_2O_3

TABLE 8.2 Estimated Relative Abundances of the Major Elements in the Continental Crust

Element	Wt. % Oxide	Atomic %
O		60.8
Si	59.3	21.2
Al	15.3	6.4
Fe	7.5	2.2
Ca	6.9	2.6
Mg	4.5	2.4
Na	2.8	1.9
K	2.2	1.0
Total	98.5	98.5

After Poldervaart (1955) and Ronov and Yaroshevsky (1976).

TABLE 8.3   **Chemical Analyses of Some Representative Igneous Rocks**

Oxide	Peridotite	Basalt	Andesite	Rhyolite	Phonolite
SiO_2	44.8	49.2	57.9	72.8	56.2
TiO_2	0.19	1.84	0.87	0.28	0.62
Al_2O_3	4.16	15.7	17.0	13.3	19.0
Fe_2O_3	1.36	3.79	3.27	1.48	2.79
FeO	6.85	7.13	4.04	1.11	2.03
MnO	0.11	0.20	0.14	0.06	0.17
MgO	39.2	6.73	3.33	0.39	1.07
CaO	2.42	9.47	6.79	1.14	2.72
Na_2O	0.22	2.91	3.48	3.55	7.79
K_2O	0.05	1.10	1.62	4.30	5.24
H_2O^+	0.0	0.95	0.83	1.10	1.57
Total	99.36	99.02	99.27	99.51	99.20

Peridotite: average for Lizard (Green, 1964); other averages from LeMaitre (1976a).

(Fe^{3+}, or *ferric*, iron) *and* FeO (Fe^{2+}, or *ferrous*, iron). When it hasn't, most authors calculate all Fe as FeO* (or FeO_T: the asterisk and T subscript indicate that all Fe has been converted mathematically to FeO). Fe_2O_3* may also be used; it means that all Fe is converted and reported as ferric. If we want to compare the Fe content of analyses using different conventions, Fe_2O_3 can be converted to FeO by multiplying by 0.8998. FeO can be converted to Fe_2O_3 by multiplying by 1.1113. For example, we can convert Fe_2O_3 to FeO for the peridotite in Table 8.3, and add it to the remaining FeO:

$$FeO^* = FeO + 0.8998 \cdot Fe_2O_3 = 6.85 + 0.8998 \cdot 1.36 = 8.07$$

Heating to determine LOI oxidizes some of the Fe, so Fe^{2+}/Fe^{3+} ratios should either be determined before LOI, or Fe should be reported as total Fe.

Note in Table 8.3 that MgO and FeO* decrease as silica and alkalis increase, going from a peridotite to a rhyolite. This is a common mafic-to-silicic trend in igneous rocks. Of course, this variation is reflected in rock mineralogy as well. The more silicic rocks generally contain more alkali feldspars plus quartz and fewer mafic minerals. The relationship between the composition of a rock and the compositions of its constituent minerals should be obvious: If we were to analyze all of the minerals that are in a particular rock, we could calculate the composition of the rock by combining the mineral analyses, factored for their percentage in the rock. Of course, it is much easier (and more accurate) to simply obtain a whole-rock analysis.

For volcanics, which may have a considerable glassy or amorphous component, a chemical analysis may be required if we are to discern the rock type. This problem was discussed in Section 2.4. Many of the volcanics used for data in the averages presented in Table 8.3 were not entirely crystalline. Most of the rhyolites were either tuffs or glassy rocks, with very few minerals. They can best be identified as

rhyolites based on their chemical composition. For example, if you take the silica content of the rhyolite in Table 8.3 (72.82%) and total alkalis ($Na_2O + K_2O = 7.85\%$), and plot these values on Figure 2.4, they plot in the rhyolite field.

The chemical composition of rocks also permits the comparison of igneous rocks with their altered and metamorphosed equivalents. Although the mineral composition may vary with changes in pressure and temperature, the bulk rock chemical composition should remain largely unaltered (at least with respect to some critical, or immobile, elements). This may allow us to identify the original igneous precursor to a number of metamorphic rocks and perhaps correlate modern and ancient igneous provinces (Section 9.6). In many cases, we may also be able to document the chemical changes that accompany alteration and metamorphic processes (Chapter 30).

8.4 NORMATIVE MINERALS

Because many volcanic rocks are too fine grained for us to recognize their mineral constituents, even microscopically, and may have a significant glassy component, a method exists to calculate an idealized mineralogy for such rocks so they can be compared with coarse-grained rocks. The mineralogy of coarse-grained rocks may also vary with pressure and temperature, making direct comparison difficult. The **norm** is an attempt to reconcile these differences to a consistent and limited set of minerals that reflect compositional variables only, facilitating the direct comparison of a broad spectrum of rocks. Norms can also be used to calculate an approximate mineralogy from published whole-rock analyses when the mineralogy is not reported. Because the norm is calculated on an anhydrous basis, it can be used to compare the mineralogy of rocks with different H_2O contents. Because they reflect the chemical composition of a rock, norms have also been used in a number of classification schemes.

The norm was first developed by C. W. Cross, J. P. Iddings, L. V. Pirsson, and H. S. Washington (Cross et al., 1902) at the beginning of the 20th century, as part of an elaborate igneous rock classification scheme. Their norm is called the **CIPW norm**, from the first letters of their last names. Although the norm-based classification scheme never met with popular approval, the norm itself lives on. They based their normative minerals on the typical minerals that might be expected to crystallize from an anhydrous melt at low pressure. Since that time, a number of variations and alternative norms have been proposed, some for special circumstances (such as high pressure). The original CIPW norm is still the one in standard use in the United States.

The *norm* is not to be confused with the *mode*. The **mode** is the actual mineral composition of a rock, based on the observed volume % (see Section 2.3.1 for a description). The norm is the idealized mineralogy calculated from the chemical composition of a rock. Because the chemical composition is reported in wt. % of the oxides, the CIPW norm is expressed in wt. % of the normative minerals. The mode and the norm may differ for a number of reasons, but the volume versus weight proportion is a consistent difference that distorts the relationship between a mode and a norm. The norm exaggerates the denser minerals compared to the mode. We could convert from wt. % to volume % by using mineral densities, but this is seldom done. An alternative norm, the **cation norm** or **Barth-Niggli norm**, expresses the normative minerals on an atomic basis. This superior method is popular in Europe.

The CIPW norm is calculated by following a rigidly prescribed set of rules that allocates the various oxides to a set of end-member and solid-solution minerals. To accomplish this, it combines the thrilling processes of mineral stoichiometry and business accounting. The step-by-step technique is described in detail in Appendix B, so you can calculate a norm by hand (or spreadsheet). Such a rigid formal methodology is an ideal application for a computer program, and a number of norm programs are available. Oxygen fugacity and the oxidation state of Fe are generally not known, and estimates must usually be supplied for reasonable normative mafic mineral contents. Irvine and Baragar (1971) discussed a method to estimate the Fe^{3+}/Fe^{2+} ratio for a norm. Once computerized, the conversion between CIPW and cation norms should be an easy process, as is the conversion to approximate volume % (if mineral densities are available) for comparison to modes.

The norm simplifies and organizes the chemical composition of a rock in a way that emphasizes certain subtle chemical characteristics, particularly (but not limited to) **silica saturation**. A "silica-oversaturated" rock is one that contains quartz (or another silica polymorph) in the mode as a stable phase. "Silica-undersaturated" rocks contain a mineral that is incompatible with quartz, such as olivine or a feldspathoid. A rock that is just silica saturated may contain a trace of quartz but no undersaturated phase. If you look at the analyses in Table 8.3, which are silica saturated? Certainly, the degree of silica saturation must be correlated with the silica content in the analysis, but there is no simple silica

concentration that will determine this. Whether quartz (or a polymorph) forms depends upon the silica content *but also on what other elements are competing for silica to form a variety of silicate minerals.* For example, a rock with 100% silica will be pure quartz. If we then add 20% MgO, this combines with the silica to form enstatite. We could then determine whether free quartz would remain by calculating how much silica, if any, is present after a molecule of SiO_2 was apportioned to each molecule of MgO to create $MgSiO_3$. If rocks contained only MgO and SiO_2, then there would be a critical silica concentration that we could relate to silica saturation: Any rock with over 60 wt. % silica (the fraction of silica in enstatite) has excess SiO_2, and would thus have quartz in it. But if we add Na_2O and Al_2O_3, they also combine with silica to form albite. Each atom of Na consumes three atoms of Si (and one of Al) to create albite ($NaAlSi_3O_8$). Now the relation between silica concentration and saturation gets complicated and depends on the Si:Mg:Na:Al ratios. During the norm calculation, silica is sequentially apportioned to the various silicate minerals in a fashion similar to that just discussed for enstatite. The last mineral to be determined is quartz, which represents any excess silica left after the other oxides have had SiO_2 allocated to them. Thus a norm can be used for a variety of rock compositions to estimate silica saturation. If quartz appears in the norm, the rock is considered oversaturated. Other normative minerals can be used to indicate silica undersaturation, high alkalinity, excess alumina content, etc. Many other aspects of rock chemical composition can be indicated by the various normative minerals.

The norm is held in varying levels of esteem by petrologists. Some value it highly and use it both to compare and classify igneous rocks. Others think it is outdated and adds little to the discussion of chemical characteristics that is not already apparent in a chemical analysis. Whether or not we like the norm, its common use in the literature requires that we be familiar with it. You will also see that it has some uses in this chapter. All of us "old timers" have calculated norms by hand, and many of us also require our students to do so. Whether this is because of the "you can't understand it well until you do it" philosophy or "misery loves company" is difficult to tell. You can calculate a norm from an analysis such as those in Table 8.3 by following the procedure in Appendix B. By doing so, you can become familiar with the process and thereby get a better understanding of how to interpret the results. Problem 3 of this chapter is an alternative way to understand how the norm works without having to perform any arduous calculations.

8.5 VARIATION DIAGRAMS

Suppose you are engaged in a research project mapping and evaluating a sequence of volcanic rocks in a volcanic terrane, consisting of several small vents, cones, and small flows. You suspect the rocks to be genetically related to some local volcanic process or event of limited duration. You carefully map individual flows and collect a **suite** of rocks (a collection that is either genetically related or representative of an area) with samples from each flow and cone layer. Now that it is possible to generate chemical (or mineralogical, geophysical, structural, etc.) data rather rapidly, and in quantity, it is possible to become swamped by your own data. Let's say you analyze all of your rocks and end up with more than 100 analyzed samples. Using a table of chemical analyses for a suite of rocks, similar to Table 8.3, is an excellent way to organize the data, but it is often difficult to see important trends, particularly if you have many analyses. When data for a series of cogenetic volcanic or plutonic rocks are analyzed, they usually show significant chemical variation. It thus becomes critically important to be able to display the data in a fashion that allows you to recognize trends in the variation so that you can describe and interpret them. There is no single best way to display data, and indeed it is something of an art. The objective is to find the parameters that show systematic variation so that you can investigate the underlying causes. Diagrams that do this are called **variation diagrams**.

There are three common formats for variation diagrams of chemical data in petrology. In the first, the **bivariate (Cartesian, or x-y) plot**, two parameters are plotted, one vertically (called the ordinate, or y-axis) and one horizontally (called the abscissa, or the x-axis). In the second, the **triangular diagram**, we can represent three parameters, one at each corner, but they can show only *relative* proportions, not absolute quantities, because the three parameters must be normalized so that they sum to 100% in order to plot as a unique point (as explained in Figure 2.1). The third common type of diagram, the *normalized multi-element diagram*, will be introduced in Sections 9.3 and 9.4.

Any correlations or trends in bivariate or triangular diagrams show up in the pattern of plotted points (if we choose the parameters wisely). More dimensions can be represented by contouring (as was done for temperature in the three-component phase diagrams, such as Figure 7.2), or projecting (as we shall see in Section 24.3.2), but the diagrams can quickly become so complex that the benefits gained by adding meaningful data are quickly overcome by the loss in simplicity and the ability to visualize the correlations.

Additional chemical data can also be represented by combining chemical constituents that behave in a similar fashion (such as FeO + MgO + MnO as a single component). Of course, a compromise is involved when we do this as well. We can get some additional data, but we lose the ability to focus on effects attributable to individual constituents. There is no unique way to display data, and researchers are continually inventing creative ways to do it. Computers are useful in this process, allowing us to rapidly select and display data in a variety of formats, searching for correlations or patterns. Variation diagrams not only help us to recognize trends in geochemical data, but they can also help us to interpret recognized trends and evaluate the process or processes responsible.

8.5.1 Bivariate Plots

Any chemical constituents, such as major elements, trace elements, or even combinations of elements and element ratios, can be compared on bivariate diagrams. Perhaps the first such

diagram applied to petrologic chemical data was the **Harker diagram**, first used by Iddings (1892b) on a molecular basis but named after British petrologist Alfred Harker, who used weight percents and advocated their broad usage (Harker, 1900). This simple x-y diagram, which plots wt. % silica as the abscissa against the other major oxides, is still one of the most common variation diagrams in use today. Figure 8.2 is an example of a Harker diagram for the volcanic rocks of Crater Lake/Mt. Mazama. The first things that one notices in Figure 8.2 are that the rocks from Crater Lake span a significant compositional range (from basalts to rhyolites) and that there is a smooth trend in the variation of each of the major oxides. It would certainly require considerable study of a table of analytical results before one could recognize these trends. The diagram makes this obvious. Of course, there is some scatter, a combination of analytical error and the type of variation generally found in nature. Rarely in geology do data fall exactly on a nice line or curve. Nonetheless, the trends in Figure 8.2 are remarkably clear. Curves have been drawn for the trends of each oxide as a best fit to the data.

The smooth trends strongly suggest that the lavas at Crater Lake are genetically related in some fashion and that some process is at work, perhaps in a shallow magma chamber, that causes the continuous variation. **Primary magmas** are those derived directly by partial melting of some source, and they have no characteristics that reflect the effects of subsequent differentiation. Unless specified otherwise, the source is presumed to be the mantle. Magmas that have experienced some form of chemical differentiation along the trends in Figure 8.2 are referred to as **evolved magmas** or

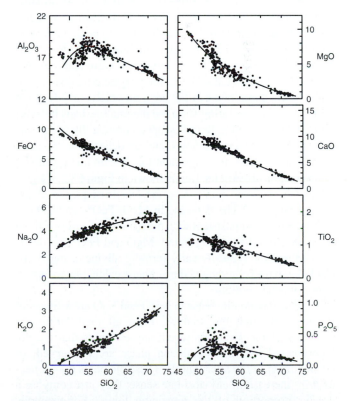

FIGURE 8.2 Harker variation diagram for 310 analyzed volcanic rocks from Crater Lake (Mt. Mazama), Oregon Cascades. Data compiled by Rick Conrey (personal communication).

derivative magmas. Magmas that are not very evolved are called **primitive**. The **parental magma** corresponds to the most primitive rock found in an area and, thus, the magma from which we suppose the others are derived. We shall return to the definition of parental again in Section 11.2.1.

Harker (1909) proposed that SiO_2 increased steadily with magmatic evolution, and he thus used it as the abscissa to indicate the extent of evolution. The magma with the lowest silica content in Figure 8.2 is thus accepted as the *parental* magma. It may be impossible, however, to demonstrate conclusively that it is a true *primary* magma because it may also have evolved during ascent (see Section 10.4).

Although we cannot directly observe the dynamics of a magma chamber, we can gain considerable insight into the processes by studying the chemical characteristics of the various products of natural igneous systems, such as at Crater Lake or the Tuolumne Intrusive Series discussed in Chapter 4. One test is to determine whether the chemical and/or mineralogical trends are consistent with some process, such as crystal settling, mixing of two magmas, assimilation of the wall rock, etc.

In Chapters 6 and 7, we had the opportunity to observe the manner in which a liquid varies in composition as minerals form during cooling of some simple experimental systems. If the crystals are removed from the melt, the process is called **fractional crystallization** (or **crystal fractionation**), and the composition of the remaining liquid system follows the **liquid line of descent**, usually along cotectic curves toward the eutectic minimum-temperature liquid composition, as explained and traced in several of the experimental phase diagrams in Chapter 7. For most of the 20th century, in large part because of the great influence of N. L. Bowen, petrologists considered fractional crystallization to be the predominant method by which magmas **differentiate**, or change composition, in nature. If such a process were applicable to the rocks at Crater Lake, we could imagine a magma chamber at some depth beneath the area, in which early-forming minerals were separating from the magma (perhaps by sinking), and the evolving liquid escaped periodically to the surface, forming the cones, flows, and pyroclastics in the area.

Let's assume for a moment that fractional crystallization is responsible for the trends shown in Figure 8.2. Can we relate the observed trends to a reasonable sequence of extracted minerals? The decrease in MgO, FeO*, and CaO as SiO_2 increases is consistent with the removal of early-forming minerals from the cooling liquid. MgO and FeO* are incorporated into the typically early-forming olivine or pyroxene. CaO may have been removed by either a calcic plagioclase, a clinopyroxene, or both.

The apparent increase in Na_2O and K_2O are artifacts of the necessity that the analyses must all total 100%. As the other elements are removed, any elements that are not incorporated into a crystallizing phase are **conserved**, or concentrated in the later liquids. These elements do not necessarily increase in any absolute sense; they just compose a greater proportion of the remaining liquid. For example, imagine that you have a bag containing an equal number of red and green candies. If you prefer the red ones, and you eat

10 of them for every 1 green candy you eat, the ratio of green to red candies remaining in the bag increases with time. Without adding any more candy, you may have increased the *proportion* of green ones from 50 to 95%. On a *percentage* basis, it would appear that the green candies have increased, whereas they have actually *decreased* on an *absolute* basis (because you ate 1 for every 10 red ones). The fact that analyses must total 100%, and the effect this has on apparent trends in variation diagrams, is referred to as the **closure problem**, and it has been discussed in detail by Chayes (1964). The increasing Na_2O and K_2O trends in Figure 8.2 suggest that the albite component in any plagioclase that may have formed and settled was low and that potassium feldspar either did not form or was very minor. Although the closure problem tells us that alkalis and other components that increase with progressive magmatic evolution need not be added to increase in percentage, this does not preclude their being added by assimilation or mixing processes either.

The Al_2O_3 curve shows an interesting trend. First it increases, and then it decreases. Because CaO decreases continuously, we can reconcile these trends by speculating that clinopyroxene was removed early on, removing Ca, but not Al, and plagioclase began to crystallize later, removing both Ca and Al.

Many geologists believe that Harker diagrams amply demonstrate the crystal fractionation process. When we make analyses of igneous processes based on variation diagrams, however, we should be careful to distinguish *observations* from *interpretations*. Only then can we be clear on what is interpretive, and perhaps then we can explore the assumptions upon which the interpretations are based. The interpretation that relates the trends to a crystal fractionation process assumes that these analyses represent *consanguineous* lavas (i.e., lavas with a common ancestor) erupted from a magma chamber beneath the volcano at various stages of progressive evolution. This assumption is supported, but not proved, by the close spatial and temporal association of the rocks. Our interpretation further assumes that the silica content is related to the evolutionary process in such a way that the wt. % SiO_2 increases as the magma evolves and that crystal fractionation is the sole process involved. By recognizing these assumptions, we can return to the chemical data, or even the petrography, or the field, and evaluate them in light of our assumptions and interpretations.

For example, if a crystal fractionation process is indeed responsible for the trends, care should be taken if porphyritic or coarse-grained rocks are included because these rocks may still include minerals that should have been removed by fractionation. In the case of porphyritic rocks, they may include some early-forming minerals and not others, if, perhaps, mineral density controlled the phenocryst separation (sinking) process. Some porphyritic rocks may actually have *accumulated* phenocrysts as they settled from higher levels or floated. Such rocks would plot off the true liquid line of descent, and the fit on a Harker diagram might improve if these rocks were excluded.

Phenocrysts of the minerals that are inferred to be fractionating should either be found in some of the lavas or be shown by experiments to be stable phases in the magmatic systems in question. The trends in variation diagrams should also be consistent with experimental cotectic liquid lines of descent. Further, the more evolved rocks should be younger than the less evolved ones. Likewise, the trends should be amenable to *quantitative* simulation based on extraction of specific proportions of minerals that are naturally, experimentally, or theoretically compatible with the types of magmas present, as we shall discuss shortly. Can this be substantiated in the field? If not, how must the model be amended?.

In many systems, including layered mafic intrusions (where we can document the fractionation process; see Figure 12.12) and several volcanic series, the silica content does not increase during most of the differentiation process. In such cases, a different index should be used, one that is more sensitive to the particular process at hand. Table 8.4 lists several chemical parameters that have been applied by various investigators as measures of progressive differentiation for some igneous province. Most of the indices are based on major elements, but a number of trace elements are also not incorporated into early crystallizing phases and thus concentrate in late melts. The high precision of modern analytical techniques has led several investigators to prefer some of these trace elements, such as Zr, Th, or Ce, as differentiation indices.

Any of the indices in Table 8.4 can be used as the abscissa in a bivariate diagram if the objective is to document magmatic evolution in an igneous series. Because different systems evolve in different ways, we might not expect a single parameter, such as wt. % SiO_2, to work equally well in them all. Some may work better than others throughout a particular area, or some may be more sensitive at a particular stage of evolution. For example, the indices based on Mg/Fe ratios are most effective in the early evolution of mafic systems (where SiO_2 varies little), whereas the more alkaline parameters generally work best in the late stages of igneous evolution.

TABLE 8.4   Some Indices of Differentiation Proposed in the Literature

Name	Formula
Felsic Index	$100 (Na_2O + K_2O)/(Na_2O + K_2O + CaO)$
Larson Index	$\frac{1}{3} SiO_2 + K_2O - (FeO + CaO + MgO)$
Nockolds Index	$\frac{1}{3} Si + K - (Mg + Ca)$
Mafic Index (MI)	$(Fe_2O_3 + FeO)/(Fe_2O_3 + FeO + MgO)$
MgO	Wt. % MgO
M (or Mg#, Mg', Mg*)	$100 Mg/(Mg + Fe^{2+})$ (can be fraction or %)
Solidification Index	$100 MgO/(MgO + Fe_2O_3 + FeO + Na_2O + K_2O)$
% Normative Plag.	Normative *pl*
Differentiation Index	Normative $q + or + ab + ne + ks + lc$
Normative Felsic Index	Normative $100(ab + or)/(ab + or + an)$
Conserved Trace Element	Zr, Th, Ce

Variation diagrams are extremely useful, both descriptively and as a basis for interpretation, but they should be used as the *first step* toward rigorous interpretation. As we shall see in Chapter 11, crystal fractionation is not the only method by which magmas differentiate. The wealth of recent geochemical data, in fact, suggests that fractional crystallization may not be the principal process responsible for subduction-related magma evolution, such as at Crater Lake. Conrey (unpub. manuscript), in a review of extensive chemical data, concluded that, although fractional crystallization may be important at the primitive end of the spectrum, mixing of mafic magma (mantle melts) and silicic magma (crustal melts) is the dominant process in the evolution of subduction-related magma series. We shall discuss the evolution of subduction-related magmas more fully in Chapters 16 and 17.

In addition, the differentiation process may not be restricted to a single shallow magma chamber, which is merely the last place of residence of a rising magma. New influx of mafic parental magma from a deep source may alter, or even reverse, some of the trends. Assessing the various processes that may be responsible for the evolutionary trends apparent in a particular igneous province is an important and difficult job, and the chemical trends are the principal method by which we attempt to distinguish and document them. *Variation diagrams work best when they are carefully designed to test a specific hypothesis.* Because several processes may be contributing to the evolution of a magmatic series, several types of variation diagrams may be employed to analyze the same data.

Careful observations of variation diagrams can provide further benefits as we focus on details in the data. For example, the peaked Al_2O_3 curve in Figure 8.2 may simply be a result of the mathematical fit of the curve to data that are widely scattered at the low-silica end. We may want to analyze more rocks with low silica and carefully observe the Al-bearing minerals to provide some better constraint on the curve.

The gap between 62 and 66% SiO_2 is also provocative. Is this simply an accident of sampling, or is it real? If the latter, why? Was there simply a quiescent period during this stage of magmatic evolution, resulting in no eruption of lavas, or is there some developmental or mineralogical reason why no representative rocks exist from this interval? Based on more than 350 analyses from Crater Lake, Bacon and Druitt (1988) found a similar gap in whole-rock analyses but noticed that the *glass* compositions fill this gap and display a more continuous spectrum. Thus the whole-rock gap may be a discontinuity in *crystal* content (the mafics are richer in crystals than the felsics), which, Bacon and Druitt (1988) proposed, reflects a recharge period in the magma chamber, during which eruptions were rare. On the other hand, volcanics with SiO_2 contents between 48 and 58 wt. % are less common than basalts (<48%) and trachytes–rhyolites (>58%) in many subduction-related volcanic areas, not just Crater Lake. This compositional gap in volcanic series is commonly called the **Daly gap** (after Reginald Daly). The gap may be more apparent than real, reflecting the dynamics of fractional crystallization and the phases involved.

The abrupt appearance on an oxide mineral, for example, may cause the liquid line of descent to cross the Daly gap in terms of SiO_2 content, with only a minor amount of fractionation. When a fractionation index other than SiO_2 is used, the gap is usually lessened. If magma mixing is responsible for the trends in the more evolved end of the spectrum of subduction-related magma series, the Daly gap may easily be explained by the simple notion that magmas closer to the two mixed end-members are likely to be more common than intermediate mixtures.

In order to have interpretive significance, variation diagrams are best restricted to a single igneous locality or a somewhat broader set of igneous centers that are still apparently related to the same process. The term **petrogenetic province** (or petrographic province) is commonly used to refer to a geographic region in which the igneous rocks are related in space and time and are presumed to have a common genesis. The scope of this term is purposefully left rather vague: a single volcano is generally considered to be a bit restrictive, but the term can range from a specific phenomenon such as Crater Lake to the High Cascade volcanic arc or the Oregon High Lava Plateau, to the Jurassic volcanic arc of the western United States.

8.5.2 Triangular Plots: The AFM Diagram

The most common triangular variation diagram used by igneous petrologists is the AFM diagram (Figure 8.3), in which A (alkalis: $Na_2O + K_2O$), F (FeO + Fe_2O_3), and M (MgO) plot as the corners of the triangle. The AFM diagram used by igneous petrologists should not be confused with the different AFM diagram used by metamorphic petrologists (Chapter 24). Igneous AFM diagrams are generally cast on a wt. % basis, but they can also be created on a cation basis. Note that the AFM diagram actually accounts for *four* cation elements and that the relative proportions of Na and K are obscured due to the conviction (or hope) that the total alkali content is more informative than either one separately. Figure 8.3 includes data for both Crater Lake and the

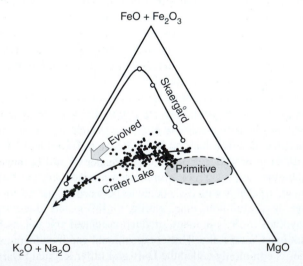

FIGURE 8.3 AFM diagram for Crater Lake volcanics (data from Figure 8.2) and the Skaergård intrusion, eastern Greenland.

Skaergård mafic intrusion of East Greenland (Section 12.2.3). Here also, trends are noticeable, and the Crater Lake and Skaergård trends are clearly distinguished. Remember from the phase diagram of the olivine system (Figure 6.10) that, for most mafic minerals, Mg/Fe is higher in the solid phase than the coexisting melt. Removal of the solid by fractional crystallization thus depletes MgO preferentially from the melt and enriches it in FeO, a trend evident in the Skaergård data. Alkalis are typically enriched in evolved liquids and enter into solid phases only during the late stages of crystallization. We can thus recognize the evolution curves on the AFM diagram. The parental magmas, if present, are closer to the MgO corner, and the most evolved ones are closer to the alkali corner. Note that, although the Skaergård trend shows a pronounced Fe enrichment in the early stages of magmatic evolution, the Crater Lake trend does not. We will discuss possible reasons for this in Chapter 16.

8.6 USING VARIATION DIAGRAMS TO MODEL MAGMATIC EVOLUTION

So far, our interpretation of variation diagrams has been inferential and qualitative. We have sought minerals capable of extracting certain components that could produce the increases or decreases in the evolutionary trends on variation diagrams. A more satisfactory evaluation would be to quantify the evolutionary process and test the proposed minerals to see if they can really produce the observed patterns. Two methods are commonly employed to assess the effects of mineral fractionation, using variation diagrams for magmatic suites. One, called Pearce element ratios, uses the slopes of variation trends based on fractionating mineral stoichiometry. The other is far more rigorous and varied in approach: direct mass-balance modeling that can be done graphically or by computer. This latter approach goes far beyond simply using variation diagrams, and sophisticated models have been developed.

8.6.1 Pearce Element Ratios (PERs)

Pearce (1968) proposed a method for using chemical data to indicate phases extracted from evolving liquids. This method is entirely empirical and uses element *ratios* to test hypotheses of mineral fractionation in a set of cogenetic analyses. The technique involves plotting on bivariate diagrams the ratios of certain elements, designed to test for the fractional crystallization of a particular mineral or minerals. The denominator of the ratio is always the same for both axes and is usually a single element (but may be more) and *not* contained in the fractionating minerals, but conserved in the remaining melt. This minimizes the closure problem discussed above. The numerators are linear combinations of elements that reflect the stoichiometry of the proposed fractionating mineral or minerals.

Because most minerals have a fairly simple stoichiometry, separation of a specific mineral will remove certain elements from the remaining melt in the proportion in which they are contained in that mineral, thereby leaving an imprint of that stoichiometry on the chemical variation in the series of remaining melts that eventually comprise the suite of rocks. Trends on a properly devised PER diagram have slopes that give an immediate and quantitative indication of the mineral(s) that may have fractionated and thereby controlled the chemical variations in the set.

For example, olivine, $(Fe, Mg)_2SiO_4$, contains $(Fe + Mg)/Si$ in the atomic ratio 2/1, so olivine fractionation removes $(Fe + Mg)/Si$ from the remaining melts in that proportion, defining a trend with a slope of +2 on a plot of $(Fe + Mg)/K$ as the ordinate versus Si/K as the abscissa (on an *atomic* basis) of the derivative liquids. Either elements or oxides can be used in PER diagrams, as long as molecular proportions are used and not weight percents. Figure 8.4 is a plot of $0.5(Fe + Mg)/K$ versus Si/K for two sets of Hawaiian basaltic magmas. Because the mafic component is halved in Figure 8.4, a slope of 1.0, not 2.0, would be in accordance with olivine fractionation. Note that the two sets have distinct original $(Fe + Mg)/Si$ ratios, so they plot as two separate lines, but each fits well to a slope of 1.0, supporting the notion of olivine fractionation within each set.

Like any variation diagram, PERs do not *prove* that a particular mineral is fractionating, or even that fractional crystallization is at work. They merely indicate whether the chemical composition of a suite of lavas is consistent with such a process. PERs are strictly empirical and may be better at excluding mineral extracts than at proving them. When the pattern of points does *not* conform to the slope one would expect from the fractionation of a particular mineral, it provides compelling negative evidence that the magmatic suite is *not* the result of fractionation of that mineral. If the data plot on the predicted trend, then the data are compatible with hypothesis of fractionation of the mineral, but, again, do not prove it.

As a more complex example of the application of PERs, we turn to another set of Hawaiian basalts erupted from Kilauea from 1967 to 1968 (Nicholls, 1990). In this set, Ti, K, and P appear to be conserved, such that Ti/K and

P/K remain essentially constant for the set of analyses. This suggests that the set of rocks is genetically related because they have these consistent ratios. Figure 8.5 is a set of 4 PER diagrams for the Kilauean volcanic rocks. Figure 8.5a is a plot similar to Figure 8.4 (except that the mafics are not halved). A slope of 2 is thus compatible with olivine fractionation. Indeed, the picrites (light circles) fall on such a line, but the basalts appear to fall on a line with a shallower slope. This suggests that another mineral is fractionating in the basalts, either along with or instead of olivine.

Figure 8.5b is a PER diagram designed to test for the fractionation of plagioclase. In anorthite, the ratio of Ca to Si is 1:2. In albite the ratio of Na to Si is 1:3. If we plotted Ca/K versus Si/K, a slope of 0.5 would be compatible with anorthite fractionation. By plotting 2Ca/K versus Si/K, a slope of 1 would then be compatible. Because the plagioclase composition also varies as fractionation proceeds, so would the Ca/Si ratio of the plagioclase, the linearity in a 2Ca/K versus Si/K diagram would be lost, and a curve would result. If we plot 3Na/K versus Si/K, a slope of 1 would also result if albite were fractionating. So if we now plot (2Ca + 3Na)/K versus Si/K, a slope of 1 would result if *any* plagioclase composition between pure anorthite and pure albite were to be fractionating. This has been done in Figure 8.5b. The dashed line with a slope of 1 closely fits the basalt data, meaning that they are compatible with plagioclase fractionation. The data actually show a slightly shallower slope than the line, suggesting that some other phase is also fractionating. The picrites have a slope of zero, suggesting no plagioclase fractionation in them.

Figure 8.5c combines Figures 8.5a and 8.5b. A vertical line is compatible with plagioclase-only fractionation (no change in Fe or Mg), and a horizontal line is compatible with mafic-only fractionation (no change in Ca or Na). The non-vertical slope of the basalts suggests that a *mafic* phase is fractionating with plagioclase in the basalts because Fe and Mg are affected. Figure 8.5d is a more comprehensive plot. This repeats the Ca-Na process described above for a variable plagioclase stoichiometry but includes the $(Fe + Mg)/Si$ stoichiometry of olivine to create a combined parameter $(0.5(Fe + Mg) + 2Ca + 3Na)/K$ that should vary in a 1:1 ratio with Si/K if any combination of anorthite, albite, and olivine is removed. Figure 8.5d is compatible with such a process for both the basalt and picrite data. This suggests that olivine fractionates with the plagioclase in the basalts because pyroxene would remove $0.5(Fe + Mg)$ versus Si in a 2:1 ratio rather than a 1:1 ratio.

PERs, and the resulting diagrams, can be used to critically evaluate the minerals that might be fractionating to relate a set of cogenetic lavas. The ratios that can be plotted are limited only by the stoichiometry of the minerals and the ingenuity of the investigator. The reader should be aware, however, that any technique involving ratio correlation can be misleading. As pointed out by Chayes (1971), Butler (1982, 1986), and Rollinson (1993), even when a set of variables shows no correlation, *ratios* of those variables with components in common tend to be correlated, sometimes even highly correlated. In other words, because the denominators are the same in both the ordinate and abscissa of PER diagrams, they may indicate a correlation where none really

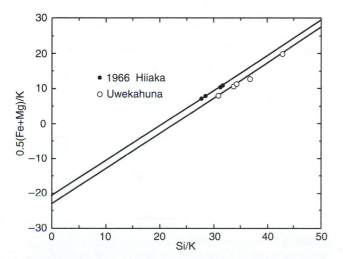

FIGURE 8.4   Pearce element rato (PER) diagram of 0.5(Fe + Mg)/K versus Si/K for two Hawaiian picritic magma suites. From Nicholls and Russell (1990).

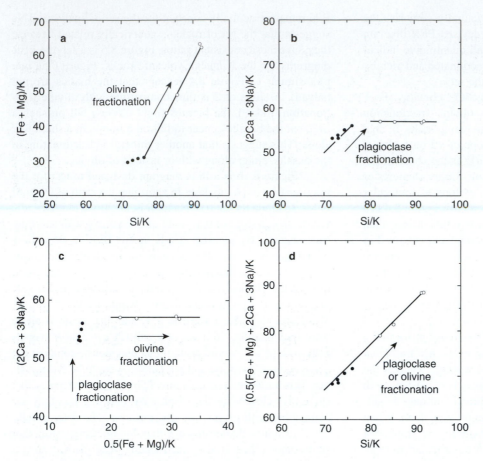

FIGURE 8.5 PER diagrams for basalts (dark circles) and picrites (light circles) erupted from Kilauea, Hawaii, between November 1967 and August 1968. After Nicholls (1990).

exists. The PER approach is a good way to test hypotheses and can serve to eliminate a bad hypothesis, but it can only support, not *prove*, a good one. When scattered data on Harker-type diagrams suddenly become linear on PER diagrams, one should not jump to the conclusion that this correlation necessarily demonstrates that a particular mineral is fractionating. Rather, it should be treated as a suggestion, to be evaluated using other textural and chemical criteria.

8.6.2 Graphical and Mathematical Models of Magmatic Evolution

If some process of magmatic evolution is responsible for the trends on variation diagrams, we should be able to quantitatively model the process by subtracting some components to cause the resulting melt to follow the given path. Variation diagrams, particularly bivariate diagrams, provide an excellent basis for these models. In this section, I shall present the methodology, with emphasis on the graphical analysis of crystal fractionation.

The basic method by which a particular model of magma evolution, such as fractional crystallization, is tested is an extension of the simple lever principle introduced in Chapters 6 and 7 to determine the relative proportions of coexisting phases in experimental phase diagrams. The principles involved are illustrated in Figure 8.6, which shows five different plots on a Harker-type variation diagram, using the hypothetical components *X* and *Y* (as either weight or mol %). In all of the diagrams, *P* represents the parental

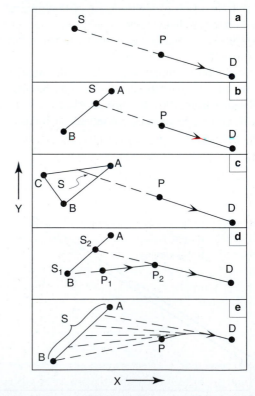

FIGURE 8.6 Stacked variation diagrams of hypothetical components *X* and *Y* (either wt. % or mol %). *P* = parent, *D* = daughter, *S* = solid extract, *A, B, C* = possible extracted solid phases. For explanation, see text. From Ragland (1989). Reprinted by permission of Oxford University Press.

sample, and D is the derived one. S represents the solid bulk composition removed from the parent to produce the derived liquid (the "bulk mineral extract" of Section 7.1.1). A, B, and C represent the composition of individual minerals that may be in the extract.

In Figure 8.6a only a single mineral (S) forms and is removed from the parent magma (P). As discussed in Chapters 6 and 7, the composition of the melt, upon loss of S, must move *directly* away from S, so that S-P-D must define a straight line. A derived melt of composition D can be formed when sufficient S has been extracted from P. The D/S ratio is readily calculated by using the lever principle:

$$\frac{D}{S} = \frac{\overline{SP}}{\overline{PD}} \qquad (8.1)$$

where $\overline{SP}$ is the length of the line S-P, etc. Likewise, the percentage of D and S are derived by:

$$\%D = 100\overline{SP}/\overline{SD} \qquad (8.2)$$

and:

$$\%S = \overline{PD}/\overline{SD} = 100 - \%D$$

The line S-P-D is commonly called the **control line** for mineral S fractionation.

In Figure 8.6b, *two* minerals (A and B) are removed from P to create D. The bulk mineral extract, S, must fall somewhere on the line connecting the two minerals that compose it. S can be determined by extrapolating the line D-P back to where it crosses the line A-B. We can determine the S/D ratio by using Equations (8.1) and (8.2). We can also determine the A/B ratio in S by using:

$$\frac{A}{B} = \frac{\overline{BS}}{\overline{AS}} \qquad (8.3)$$

An equation similar to Equation (8.2) can be used to recast the ratios as percentages. Note that Equations (8.1) and (8.3) (or their % counterparts) can be used in conjunction to determine the relative amounts of all three phases, D, A, and B.

In Figure 8.6c, three minerals (A, B, and C) are extracted. The bulk mineral extract, S, in this case cannot be uniquely determined because the extrapolation of the line D-P intersects the triangle ABC in a *line* (the solid line in Figure 8.6c). S must lie on this line, but we cannot know where because any point on the line could, when extracted, produce D from P. As a result, we cannot determine the A:B:C ratios, nor the S/D ratio, without some further information.

Figure 8.6d represents a **sequence** of two minerals being extracted from P. This is analogous to a binary eutectic-type situation. First, bulk mineral extract S_1 (mineral B) crystallizes and is removed, driving the parental melt from point P_1 directly away from B toward P_2. At P_2, mineral A joins B in such a way that the A/B ratio in the extract equals S_2. Now the melt at that time (P_2) moves directly away from S_2, toward the final derived melt D. The resulting liquid line

of descent on the variation diagram shows a definite kink at P_2 and is not a straight line, as in the previous cases that involve only a single bulk mineral extract. At any point along either path, the relative proportions of the coexisting phases can be determined by using Equations (8.1) to (8.3).

Figure 8.6e illustrates the effect of extracting a solid solution, or two minerals in which the ratio varies continuously (as would occur with a curved cotectic). In this case, the bulk mineral extract moves along the line from B toward A. As it does so, the melt composition must move directly away from a shifting bulk extract point, resulting in a curved liquid line of descent, similar to those for Al_2O_3, MgO, and Na_2O in Figure 8.2

This simple graphical example is the basis for more rigorous numeric solutions. Complex combinations of mineral sequences and solid solutions can be specifically modeled mathematically using the computer, by extending this treatment. We can analyze the simple processes in Figure 8.6 graphically or mathematically. If the number of phases gets beyond three, the mathematical method is far superior.

The technique described can be used equally well to model **crystal accumulation**. In this case, D in Figure 8.6 would be the initial parent magma, and P would be the accumulative rock that formed by the addition of cumulate crystals $S = A$, B, etc. The only way to tell the difference between fractional crystallization and crystal accumulation would be based on textures. **Fractional melting** could be modeled with P as the parent, D the extracted melt, and S, A, B, etc. the solid assemblage left behind. Assimilation of wall rocks and magma mixing can also be modeled, or any combination of processes. Sophisticated computer models have been developed to quantitatively model observed trends in terms of single or combined processes.

As an example of the graphical treatment of fractional crystallization, I offer one developed by Ragland (1989), which, along with Cox et al. (1979), should be consulted for further amplification of the methods. The example begins with a suite of cogenetic rocks, ranging from a basalt to a rhyolite, from a typical subduction-related volcano. The analyses (selected from samples with <5% phenocrysts) are given in Table 8.5. These data are then plotted on a set of Harker diagrams in Figure 8.7. Smooth curves are fitted to the trends for each oxide. Note that three curves are linear, three are curved, and two show a maximum. We can now proceed to analyze these rocks based on the following three assumptions:

1. The rocks in Figure 8.7 are related to each other by a process of crystal fractionation.
2. The trends in Figure 8.7 represent liquid lines of descent.
3. The basalt is the parental magma from which the others are derived.

There are two methods to evaluate the hypothesis that fractional crystallization is responsible for the trends in Figure 8.7. Both involve stepwise analysis from each rock type to the next, more evolved one. This avoids the complexities involved with curved liquid descent lines and continuously shifting extract compositions required to produce them. Rather, they are

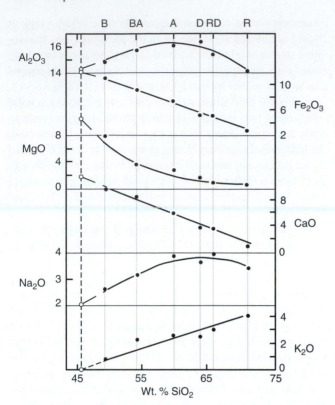

FIGURE 8.7 Stacked Harker diagrams for the calc-alkaline volcanic series of Table 8.5 (dark circles). From Ragland (1989). Reprinted by permission of Oxford University Press.

treated as a series of straight-line segments connecting each analysis. We shall do this only for the basalt (*B* in Table 8.5 and Figure 8.7) to basaltic andesite (*BA*) increment, which should suffice for the purpose of illustration.

The first method is the most general. It assumes that the point representing *B* (the parent) must lie on a straight line between *BA* (the derived melt) and *S* (the bulk mineral extract), as was demonstrated in Figure 8.6. If we want to find *S*, then, it must lie on the low-SiO$_2$ extrapolation of the line drawn from *BA* through *B*. Of all the analyses in Figure 8.7, three have positive correlations with SiO$_2$ in the *B-BA* range: Al$_2$O$_3$, Na$_2$O, and K$_2$O. If we extrapolate lines from

BA back through *B* to more primitive compositions, all three will eventually extend to zero. K$_2$O proves to be the limiting value in this example because the *BA-B* line extrapolates to zero potassium in Figure 8.7 first, at 46.5% SiO$_2$ (dashed line), the highest silica value of the three. At lower silica values, the concentration of K$_2$O in the extract would be negative, which is clearly impossible for any analysis. If we next assume that there was no potassium in the bulk mineral extract (a reasonable assumption for a basalt), this indicates that 46.5% is the silica concentration for our extract. Using a vertical line on Figure 8.7 at 46.5 % SiO$_2$, we can graphically determine the concentrations of the other oxides in the bulk mineral extract by the intersection of their variation curves with this line (unfilled circles in Figure 8.7).

A more accurate approach to this method is a mathematical one using Table 8.5 and the linear equation applied to chemical variables *X* and *Y* and rocks 0, 1, and 2:

$$(X_2 - X_1)/(Y_2 - Y_1) = (X_1 - X_0)/(Y_1 - Y_0) \quad (8.4)$$

If we perform the calculation for $1 = B$, $2 = BA$, and $0 =$ bulk mineral extract (in which the wt. % K$_2$O = 0) and substitute SiO$_2$ for *X* and K$_2$O for *Y*, we can calculate X_0, the value of SiO$_2$ when K$_2$O $= Y_0 = 0$. From Table 8.5 we get:

$$(54.3 - 50.2)/(2.1 - 1.0) = (50.2 - X_0)/(1.0 - 0)$$

for which $X_0 = 46.5$ wt. % SiO$_2$.

If we take this value for X_0, we can use Equation (8.4) to get any other oxide as Y_0. For example, MgO:

$$(54.3 - 50.2)/(3.7 - 7.4) = (50.2 - 46.5)/(7.4 - Y_0)$$

for which $Y_0 = 10.8$% MgO.

Using either the graphical or mathematical approach for all major oxides, we get the chemical composition for a bulk mineral extract reported in Table 8.6. Note that all oxides are positive, except for K$_2$O = 0, as was our aim. At this point, a norm calculation comes in handy (columns 3 and 4 of Table 8.6), so we can express the chemical composition as an approximate mineralogy. (We may prefer another norm scheme, such as a high-pressure norm, if we suspect a deep fractionation

---

TABLE 8.5 Chemical Analyses (wt. %) of a Hypothetical Set of Related Calc-Alkaline Volcanics

Oxide	B	BA	A	D	RD	R
SiO$_2$	50.2	54.3	60.1	64.9	66.2	71.5
TiO$_2$	1.1	0.8	0.7	0.6	0.5	0.3
Al$_2$O$_3$	14.9	15.7	16.1	16.4	15.3	14.1
Fe$_2$O$_3$*	10.4	9.2	6.9	5.1	5.1	2.8
MgO	7.4	3.7	2.8	1.7	0.9	0.5
CaO	10.0	8.2	5.9	3.6	3.5	1.1
Na$_2$O	2.6	3.2	3.8	3.6	3.9	3.4
K$_2$O	1.0	2.1	2.5	2.5	3.1	4.1
LOI	1.9	2.0	1.8	1.6	1.2	1.4
Total	99.5	99.2	100.6	100.0	99.7	99.2

B = basalt, BA = basaltic andesite, A = andesite, D = dacite, RD = rhyo-dacite, R = rhyolite. Data from Ragland (1989).

TABLE 8.6 Bulk Mineral Extract Required for the Evolution from B to BA in Table 8.5

Oxide	Wt. %	Cation Norm	
SiO$_2$	46.5	*ab*	18.3
TiO$_2$	1.4	*an*	30.1
Al$_2$O$_3$	14.2	*di*	23.2
Fe$_2$O$_3$*	11.5	*hy*	4.7
MgO	10.8	*ol*	19.3
CaO	11.6	*mt*	1.7
Na$_2$O	2.1	*il*	2.7
K$_2$O	0.0		
Total	98.1		100

Data from Ragland (1989).

process.) Note that olivine, diopside, and plagioclase dominate the bulk mineral extract, which is reasonable for a basalt phenocryst assemblage. Because the norm is a cation norm, the composition of the plagioclase = 100 $an/(an + ab)$ = An_{62}, perhaps a bit low for a basalt (for which plagioclase is usually in the An_{70} to An_{85} range), but it's close. The *hy* component may not require orthopyroxene but may be included as a component in the clinopyroxene (especially if it's a pigeonite).

A more precise method can be used if there is a phenocryst assemblage associated with the basalt. If the compositions of the phenocrysts have been analyzed using a microprobe, they can be plotted on the same variation diagram as *B* and *BA* and used to solve the liquid line of descent, either graphically (as in Figure 8.6) or mathematically. As stated previously, the trends in Figure 8.7 are considered reliable examples of the evolved liquids only if the rocks contain few phenocrysts. Thus only samples with <5% phenocrysts were used. However, these phenocrysts can now be useful. Let's say we have phenocrysts of olivine, augite, and plagioclase in the basalt *B*, and we have only a few augite and plagioclase phenocrysts in the basaltic andesite *BA*. Figure 8.8 is a variation diagram created by Ragland for this example. In it, *B* and *BA* are plotted, along with the three phenocryst compositions. For variety, the compositions have been recast as molecular %, but this is not necessary. Using wt. % oxides would work equally well. Ragland combined Fe + Mg to minimize errors associated with Fe oxidation state. K + Na were also combined because they are both susceptible to subsolidus alteration, and combining them may serve to minimize these effects. It is possible to do this only when both elements behave in a similar fashion. In this case, both Na and K increase. If we were to model the *RD* to *R* path, instead of *B* to *BA*, Na would decrease, and K would increase (Figure 8.7), and combining them should then be avoided.

If you compare Figures 8.6c and 8.8, you will see the similarities. Figure 8.8 produces four triangles (for Al, Fe +

Mg, Na + K, and Ca), with four lines on which possible values for *S*, the bulk mineral extract, may lie. Because the triangles are different sizes and shapes, they cannot be directly compared. If the four triangles are converted to equilateral triangles of the same size, however, they can be superimposed. The four solid extract lines can be added to the equilateral composite (Figure 8.9) by noting the relative intercepts of each line on the two sides of the triangle that it pierces in Figure 8.8. For example, in the Al_2O_3 section of Figure 8.8, the *ratios* of the line lengths $\overline{Py}/\overline{PC}$ and $\overline{Px}/\overline{PO}$ define the line *x-y* in *any* shape of triangle POC. Ideally, when added to the triangle, the four lines intersect at a point equal to *S*, the bulk extract. If they don't intersect at a point, the area bounded by the intersecting lines defines the zone containing *S*, which should be relatively small (shaded in Figure 8.9). The ratios of olivine,

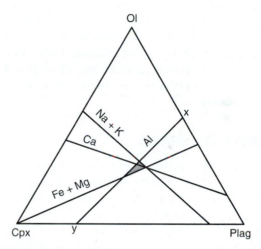

FIGURE 8.9 Equilateral triangle showing the solution to the bulk mineral extract (shaded area) best fitting the criteria for the variation diagrams in Figure 8.8. From Ragland (1989). Reprinted by permission of Oxford University Press.

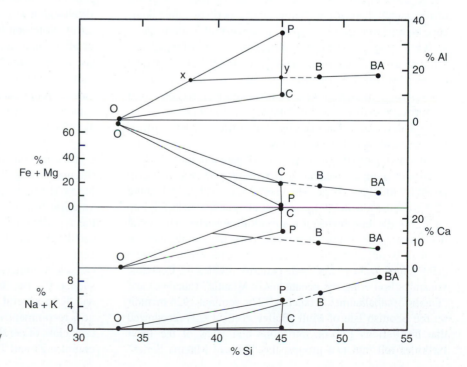

FIGURE 8.8 Variation diagram on a cation basis for the fractional crystallization of olivine, augite, and plagioclase to form BA from B (Table 8.6). From Ragland (1989). Reprinted by permission of Oxford University Press.

plagioclase, and augite can then be determined from S, using the method described in association with Figures 7.3 or 2.1.

This graphic example illustrates the process by which fractionation processes can be modeled. The graphic method is more instructive, but the mathematical process is more efficient, for it can be extended to several components and phases, and it can address combinations of processes, such as fractional crystallization and magma mixing, or wall–rock assimilation (Bryan et al., 1969; Wright and Doherty, 1970; DePaolo, 1981c; Nielsen, 1990). The methods involve the best least-squares fit of analyzed minerals to a regression line of successive analyzed lavas. Care must be taken when several simultaneous processes are considered, for more than one unique solution may be possible. Like PERs, the graphic and numeric models provide a test for a hypothesis, eliminating some hypotheses effectively, but they cannot prove that a particular process was operating. Later quantitative approaches have attempted to model mineral–melt equilibria based on thermodynamic data extracted from experimental melting data (see Ghiorso, 1985; Nielsen, 1990; Ghoirso and Sack, 1995). A "forward" approach may be employed, using a model to predict a liquid line of descent, which is then compared to a natural suite.

8.7 MAGMA SERIES

We have already seen in Chapter 2 how chemical composition can be used to classify and name individual igneous rocks, a method that is particularly useful for volcanic rocks that may have few identifiable minerals. The different trends in Figure 8.3 for two sets of samples, Crater Lake and Skaergård, each show a fairly continuous, yet distinct, chemical variation that strongly implies a genetic relationship or an evolutionary process. This is an invitation to attempt a different type of chemical classification, one that distinguishes whole *families* of magmas (and the resulting rocks). A group of rocks that share some chemical (and perhaps mineralogical) characteristics and shows a consistent pattern on a variation diagram, suggesting a genetic relationship, can be referred to as a **magma series**. Synonymous terms include *association, lineage, magma type,* and *clan.* The American Commission on Stratigraphic Nomenclature (1961) preferred the term *group* to *series,* but this usage is no longer common.

The concept that many igneous rocks fall into distinct kindred series and that each series follows some characteristic evolutionary path from a unique type of parent magma through a series of more evolved silicic derivative types was first proposed by Iddings (1892a). He recognized the chemical nature of this distinction and proposed that all igneous rocks fell into either an **alkaline series** or a **subalkaline series**. The initial distinction of most series was made in the field on the basis of mode of occurrence and mineralogical characteristics. Harker (1909), intrigued by large-scale patterns, divided the Cenozoic volcanics bordering the oceans into "Atlantic" (alkaline) and "Pacific" (subalkaline) "branches." The seminal 1924 memoir on the Scottish Isle of Mull (Bailey et al., 1924) recognized that the various Eocene intrusions and lavas of the Inner Hebrides fell into two groups, the "Plateau Magma Series"

(alkaline) and the "Main Magma Series" (subalkaline). Kennedy (1933) considered these two types to be worldwide in scope and proposed the "tholeiitic" (Plateau, or subalkaline) and "olivine-basalt" (Main, or alkaline) "magma types." Basalts, with their high liquidus temperatures, are considered to be the parental magmas from which the more evolved types within a series are derived. Bowen (1928) provided sound experimental support for this concept, and, under his great influence, the process of crystal fractionation was accepted as the dominant process by which the various series evolved. As mentioned above, there is reason to doubt that fractional crystallization is the sole process of magmatic evolution, and it may not even be the dominant one in many provinces, giving way to magma mixing or assimilation in some types of magmatism.

Although many series were distinguished on field and mineralogical characteristics, the chemical characteristics and influence were equally obvious, particularly the alkalinity and silica content, because these features were distinctive throughout a series. The influence of silica and alkali content are reflected in the names of the two original major series. Subsequent work has only reinforced the early series concept, but we now consider the large-scale groupings of Harker and Kennedy as overly simplified (i.e., that subalkaline rocks occur in the Atlantic, and alkaline rocks in the Pacific). Alkaline rocks are richer in alkalis and are commonly silica undersaturated, whereas subalkaline rocks are silica saturated to oversaturated. If the series are indeed unique, they should be distinguished by their evolutionary patterns on variation diagrams, which might also permit further analysis of the evolutionary processes.

Peacock (1931) used a plot of CaO and $(Na_2O + K_2O)$ versus SiO_2 (Figure 8.10a) to somewhat arbitrarily distinguish four chemical classes, based on a single parameter, the "alkali–lime index" (the wt. % SiO_2 at which the increasing alkali curve met the decreasing CaO curve on a Harker diagram). He called the classes **alkalic** (alkali–lime index <51), **alkali-calcic** (51 to 56), **calc-alkalic** (56 to 61), and **calcic** (>61). Note that the Crater Lake data in Figure 8.10a yield an alkali–lime index of ~60.7, indicating that the series is just to the calc-alkalic side of the boundary between Peacock's calcic and calc-alkalic classes. Shand (1927) grouped igneous rocks based on the total *molar* alkali versus alumina content as either **peralkaline** $[Al_2O_3 < (Na_2O + K_2O)]$, **peraluminous** $[Al_2O_3 > (CaO + Na_2O + K_2O)]$, or **metaluminous** $[Al_2O_3 < (CaO + Na_2O + K_2O)$ but $Al_2O_3 > (Na_2O + K_2O)]$, a classification that is useful mostly for very felsic rocks (Figure 8.10b). The molar alkali–alumina ratio, $(Na_2O + K_2O)/Al_2O_3$, is called the **peralkalinity index**. The term **agpaitic** has been used as a synonym for peralkaline (and the agpaitic index, or coefficient, as initially proposed, indicated the same thing as the peralkalinity index). More recently, however, the agpaitic character of alkaline rocks considers mineralogical and other chemical parameters, such as Ca, Ba, Sr, Rb, and Cs, which may substitute for Na and K in several alkaline silicate minerals, and the IUGS now recommends that agpaitic rocks be restricted to a specific class of peralkaline nepheline syenites characterized by complex Zr and Ti minerals (see Sørensen, 1974, 1997).

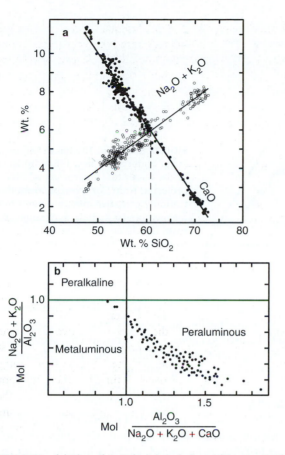

FIGURE 8.10   (a) Plot of CaO (solid circles) and (Na$_2$O + K$_2$O) (open circles) versus SiO$_2$ for the Crater Lake data used in Figures 8.2 and 8.3. Peacock (1931) used the value of SiO$_2$ at which the two curves crossed as his "alkali–lime index" (dashed line). **(b)** Alumina saturation indices (Shand, 1927), with analyses of the peraluminous granitic rocks from the Achala Batholith, Argentina (Lira and Kirschbaum, 1990).

Since these pioneering works, a number of investigators have attempted to identify and classify magma series based on chemical characteristics. Other series have been proposed, based, among other things, upon the dominant type of alkali. Some of the terms were considered to be true magma series, whereas others served more to describe the chemical characteristics of an igneous rock or province. Such different applications of the same terms led to obvious confusion. Most investigators continue to recognize Iddings's (1892a) original two series, alkaline and subalkaline, and accept that the subalkaline can be further divided into the tholeiitic series and calc-alkaline series. Beyond that, there is less agreement as to what constitutes a series and how series are distinguished.

In 1971, Irvine and Baragar attempted to systematize the growing diversity of nomenclature of magmatic rocks and series, and they recommended a classification to the National Resource Council of Canada. Using major element composition and Barth-Niggli (cation) norms, they attempted to provide a reasonable classification of magma series and methods by which they could be distinguished. They accepted the original alkaline and subalkaline series and the subdivision of the subalkaline series into the tholeiitic and calc-alkaline series. They also recognized a **peralkaline series**, based on the

alumina content, as defined above, though it is less common. They also provided a chemical definition for conventional rock names (as parts of the series), but, because it overlaps with the chemical classification by the IUGS presented in Chapter 2, we shall disregard that aspect of the work.

The alkaline and subalkaline series are distinguished in a total alkali versus silica diagram in Figure 8.11. In this diagram, the alkaline rocks plot distinctly above and the subalkaline rocks below the dividing lines chosen by MacDonald (1968) and Irvine and Baragar (1971).

Figure 8.12a is the Ne-Di-Fo-Qtz tetrahedron, called the **basalt tetrahedron** because it is so useful in characterizing basalts. The Di-Ab-En plane is called the **plane of silica saturation** because to the right of it, a silica polymorph is stable (indicating SiO$_2$ oversaturation), whereas to the left, the silica-undersaturated phase olivine is stable without a silica polymorph (see Figures 6.12 and 7.4). The Di-Ab-Fo plane is called the **critical plane of silica undersaturation**, and to the left of this plane, the alkaline and silica-undersaturated feldspathoid mineral nepheline is stable. Figure 8.12b is the Ne-Fo-Qtz base of the tetrahedron (compare to Figure 7.4, which has anorthite instead of albite). The alkaline and subalkaline rocks, when plotted on this diagram using the (cation) normative minerals *ne*, *ol*, and *q*, are distinguished by the dividing line shown (proposed by Irvine and Baragar, 1971). This line is close to the critical plane of silica undersaturation. Alkaline rocks plot to the left of this plane and are thus silica undersaturated. These two series should be distinct, at least at low pressures, because of the **thermal divide** along the line Ab-Ol (see Figure 8.13) that prevents liquids from crossing it as they cool. Rather, liquids on the flanks of this divide descend away from it as they cool, evolving toward either the silica-saturated or the alkaline and silica-undersaturated eutectic. Subalkaline rocks can be olivine bearing or quartz bearing, depending upon which side of the plane of silica saturation they occupy. The common evolutionary sequence in the alkaline series begins with an alkali olivine basalt and proceeds through trachybasalts and trachyandesites, to trachytes or

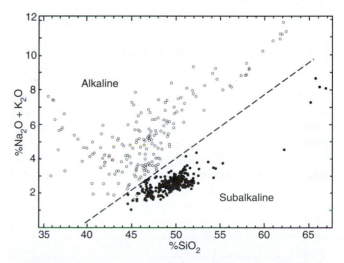

FIGURE 8.11   Total alkalis versus silica diagram for the alkaline (open circles) and subalkaline rocks of Hawaii. After MacDonald (1968). Copyright © The Geological Society of America, Inc.

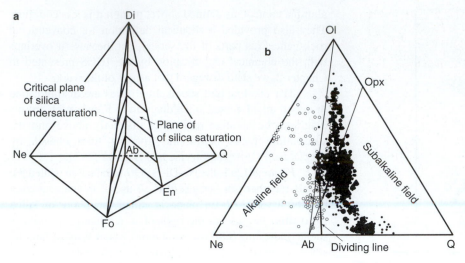

a

Di

Critical plane of silica undersaturation

Plane of of silica saturation

Ne   Ab   Q

En

Fo

b

Ol

Opx

Alkaline field

Subalkaline field

Ne   Ab   Q

Dividing line

FIGURE 8.12 **(a)** The basalt tetrahedron (after Yoder and Tilley, 1962). **(b)** The base of the basalt tetrahedron on which are projected from Cpx the compositions of (cation) normative minerals, determined from whole-rock analyses of subalkaline rocks (black) and alkaline rocks (gray) from Figure 8.11. After Irvine and Baragar (1971).

phonolites (see Figure 2.4). The common sequence for the subalkaline series is the more familiar basalt → andesite → dacite → rhyolite family.

The subalkaline series was further subdivided into a **tholeiitic** and a **calc-alkaline series** by Tilley (1950). Although these two subdivisions cannot be distinguished in either the alkali-silica or the *ne-ol-q* diagrams (despite Peacock's 1931 attempt), they do plot as distinct fields in the AFM diagram (Figure 8.14) and on a plot of Al_2O_3 versus the composition of the (cation) normative plagioclase (Figure 8.15).

If we compare on AFM diagrams in Figures 8.3 and 8.14, we see that the Skaergård trend is clearly tholeiitic, and the Crater Lake trend is calc-alkaline. Both series progress along the basalt–andesite–dacite–rhyolite trend, but there are distinctive mineralogical and chemical differences between the two series that are most evident for intermediate compositions. The parental end-members converge toward the M corner, making distinction among them on an AFM diagram problematic. Likewise, Irvine and Baragar (1971) found it impossible to consistently distinguish the more siliceous members of the two series because they converge after the different iron enrichment paths that characterize the intermediate stages (Figure 8.14). As pointed out by Sheth et al. (2002), Figures 8.14 and 8.15 distinguish tholeiitic from calc-alkaline series on the basis of criteria other than CaO versus alkalis. Although

they may effectively do so, and those distinctions may lead us toward understanding the responsible processes, neither should really be used to *define* a rock series as calc-alkaline without addressing both CaO and alkalis (as the original Peacock criteria do). We will investigate the trends and distinctions of alkaline versus subalkaline series more fully in Chapter 14 and calc-alkaline in Chapter 16.

Irvine and Baragar (1971) further divided the alkali series into a **sodic series** and **potassic series**, and Middlemost (1975) recommended adding a **high-K series** as well (Figure 8.16). The remaining terms, such as *peraluminous*, *metaluminous*, etc., although they may yet be shown to characterize some distinctive magma series, are more useful as descriptive terms when some chemical characteristic is emphasized for a particular igneous rock or province (as in Section 18.1).

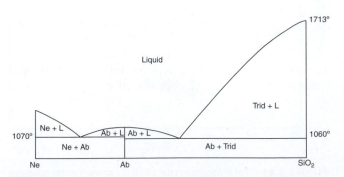

FIGURE 8.13 The thermal divide at the albite composition on the Ne-Q system.

Liquid

1713°

Trid + L

Ne + L

Ab + L Ab + L

1070°   1060°

Ne + Ab   Ab + Trid

Ne   Ab SiO_2

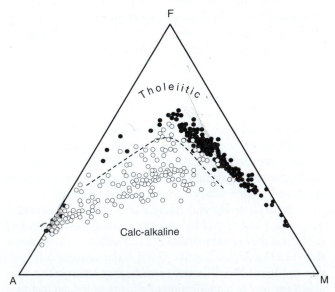

F

Tholeiitic

Calc-alkaline

A   M

FIGURE 8.14 AFM diagram showing the distinction between selected tholeiitic rocks from Iceland, the Mid-Atlantic Ridge, the Columbia River Basalts, and Hawaii (solid circles) plus the calc-alkaline rocks of the Cascade volcanics (open circles). From Irving and Baragar (1971).

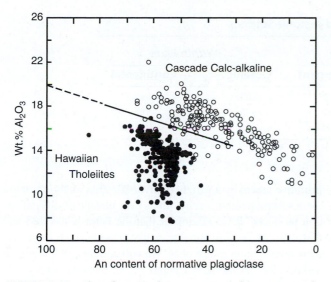

FIGURE 8.15   Plot of wt. % Al_2O_3 versus anorthite content of the normative plagioclase, showing the distinction between the tholeiitic and calc-alkaline series. From Irvine and Baragar (1971).

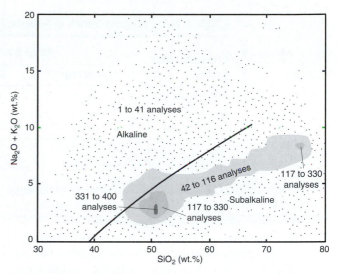

FIGURE 8.17   Plot of more than 41,000 igneous rock analyses on an alkali versus silica diagram. The alkaline–subalkaline dividing line of Irvine and Baragar is included. After LeMaitre (1976b). Copyright © with permission from Springer-Verlag.

Although the alkaline, tholeiitic, and calc-alkaline magma series predominate the igneous history of the Earth, transitional types can be found. Diagrams such as Figures 8.11 to 8.16 suggest a clear separation of the series types, but when additional data are added, the distinction becomes less clear. Figure 8.17 is a plot of more than 41,000 analyses of igneous rocks compiled from the literature on an alkali versus silica diagram, such as that used in Figure 8.11 to distinguish the principal alkaline and subalkaline magma series. Besides noting that the subalkaline rocks are more common than the alkaline, we see that any supposed gap separating these two dominant series is missing. In fact, a modest concentration of analyses lie on a portion of the boundary line of Irvine and Baragar (1971).

In addition, the commonly accepted series are too narrowly defined to accommodate all magmas, or even all

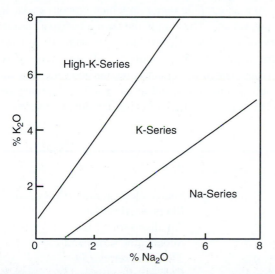

FIGURE 8.16   Wt. % K_2O versus Na_2O diagram subdividing the alkaline magma series into High-K-, K-, and Na-sub-series. After Middlemost (1975). Copyright © with permission from Elsevier Science.

magma series, particularly among the highly variable alkaline rocks of the continental interiors (Chapter 19). Thus, although the classification into magma series may be useful, it is far from perfect. Classifications, with their convenient "pigeonholes," are commonly difficult to apply unambiguously to nature. Classifications also focus our attention on *similarities*, when individual *differences* may be equally important, if not more so. Although the basalts of Mauna Kea, the Columbia River Plateau, and the Mid-Atlantic Ridge are all tholeiitic, they do not have identical origins, chemistries, or evolutionary paths. However, it is still convenient to use the three main magma series (at least). Despite the somewhat gradational distinctions at times, a significant number of magmas fall into one of these series. Each series is characterized by a particular parent basalt and shows a sequence of derivative magmas that follow a liquid line of descent from the parent. This leads us to the conclusion that there must be some important physical and chemical controls on the development of a number of primary and derivative magmas. The search for patterns and correlations, however general, is a fundamental scientific approach, and the patterns represented by magma series are real, even if they are not universal. These patterns must reflect some basic Earth processes that we cannot afford to overlook. If we look at the tectonic setting of the various series in Table 8.7, we find some other correlations.

First, calc-alkaline magmas are essentially restricted to subduction-related plate tectonic processes. This must carry some genetic significance and will certainly be a factor in any model for subduction zone magma genesis (Chapters 16 and 17). Recognition of calc-alkaline characteristics in the composition of ancient volcanic rocks may help determine their tectonic framework and aid the historic interpretation of an area. Second, tholeiitic magmas are practically the exclusive magma type associated with divergent boundaries. Although some alkaline rocks are found, they play a

TABLE 8.7   Magma Series in Specific Plate Tectonic Settings

Characteristic Series	Plate Margin		Within Plate	
	Convergent	Divergent	Oceanic	Continental
Alkaline	yes	no	yes	yes
Tholeiitic	yes	yes	yes	yes
Calc-alkaline	yes	no	no	no

After Wilson (1989), p. 11.

subordinate role, usually in the initial stages of continental rifting. This too, must provide some useful clues to the genesis of magmas at mid-ocean ridges (Chapter 13).

Magma series are important to our understanding of igneous petrogenesis. Earlier, as we first regarded the spectrum of igneous rocks that occur on Earth, I asked a basic question: Does the observed diversity derive from a single primitive parent by a variety of diversification methods, or are there several parents, each with its own lineage of diversification products? It would appear that the latter is the case and that there are at least three principal magma series, and perhaps several other minor ones. We will continue to investigate and develop this theme in Chapters 12 to 20 as we survey several igneous provinces.

Summary

Petrologists use the chemical composition of rocks and minerals to characterize and categorize individual types, to identify series of related rocks, and to model processes that relate the rocks in a consanguineous series. Analyses were initially performed by using tedious gravimetric methods but are now quickly and routinely accomplished by using instrumental spectroscopic techniques based on emission or absorption of energy when directed at prepared samples. Analyses are typically reported in wt. % oxide for the major elements (SiO_2, Al_2O_3, FeO, MgO, FeO, CaO, Na_2O, and K_2O) and minor elements (TiO_2, MnO, P_2O_5, H_2O, and CO_2), and they are reported as parts per million (ppm) of the element for trace elements. A norm is a method of recasting a chemical analysis into a set of hypothetical minerals. Norms are useful in categorizing volcanic/glassy rocks that have not fully crystallized, and they also have other uses in igneous rock classification and modeling.

Graphical display of accumulated chemical data in suitable variation diagrams may prove helpful in discerning trends that may discriminate between possible evolution mechanisms relating members of a suite of rock samples collected in an area. Bivariate (x-y) diagrams are commonly used, particularly ones in which various components are plotted versus some chemical parameter that is considered a measure of chemical fractionation (e.g., SiO_2, Zr, Mg#). Ternary diagrams, such as the AFM diagram, are also common. By using bivariate diagrams, such as Harker or Pearce element ratio diagrams, we can evaluate specific differentiation mechanisms (e.g., fractional crystallization of certain minerals) deemed responsible for the chemical evolution within a suite of supposedly consanguineous rocks. Quantitative graphical and computational methods extend the power of this approach and can determine such things as the exact amounts of minerals of specific compositions that could, if fractionated, relate the rocks within an evolving suite.

Petrologists use the term *petrogenetic province* to refer to a geographic region in which igneous rocks are apparently related. They apply the term *series* to a broader category of characteristic evolutionary path from a type of parent magma through a sequence of more evolved derivative types. The most popularly recognized series are alkaline and subalkaline series, and the subalkaline type may be subdivided further into tholeiitic and calc-alkaline series.

Key Terms

Major, minor, and trace elements *136*
Spectroscopic analysis *136*
Detection limit *138*
Loss on ignition (LOI) *140*
Norm (CIPW) *142*
Mode *142*
Silica saturation *142*
Suite *142*
Variation diagram *143*

Primary magma *143*
Derivative magma *144*
Primitive *144*
Parental *144*
Conserved elements *144*
Closure problem *144*
Consanguineous *144*
Daly gap *145*
Petrogenetic province *146*

Pearce element ratio (PER) *146*
Magma series *152*
Alkaline series *152*
Subalkaline series *152*
Tholeiitic series *154*
Calc-alkaline series *154*
Thermal divide *153*
Peralkaline/agpaitic *152*
Peraluminous *152*

Review Questions and Problems

Review Questions and Problems are located on the author's web page at the following address: **http://www.prenhall.com/winter**

Important "First Principle" Concepts

- Major element behavior is different from that of trace elements. Because of their high concentrations, major elements (and, to a lesser extent, minor elements) determine the physical properties of magmas and the minerals that crystallize from them. Differentiation processes resulting in the evolution of magmas within a petrogenetic province are largely controlled by major element concentrations and may be detected by noting variations in major elements across a suite of cogenetic igneous rocks.

- We can qualitatively and quantitatively model fractionation processes to evaluate the potential of specific processes to produce the compositional evolutionary trends observed in such an igneous suite.

- Trends revealed in variation diagrams are best used to critically evaluate a specific hypothesis. They may provide grounds to reject or support the hypothesis, but cannot prove that it has occurred.

- A rock containing sufficient SiO_2 to combine with the other elements to create silicates (such as feldspars and pyroxene) and yet have excess SiO_2 left over will manifest that excess as quartz (either in the mode or norm). Such a rock is considered *silica oversaturated*. Rocks with insufficient SiO_2 will develop minerals unstable with quartz, such as olivine or a feldspathoid. Just-saturated rocks will develop feldspars and mafics such as pyroxene but neither quartz nor olivine/feldspathoid.

- *Primary* magmas are magmas derived directly from a source (typically the mantle) without evolving after separation from that source. Most suites of cogenetic (consanguineous) rocks display a range of compositions from *primitive* to *evolved* (or *derivative*) compositions. *Parental* magmas are the most primitive ones found in an area and are considered the local type from which the others evolved.

- Three common magma series are popularly recognized: alkaline, calc-alkaline, and tholeiitic. Numerous other series have been proposed by investigators desiring more detail.

Suggested Further Readings

Analytical Methods and Major Element Analysis

Cox, K. G., J. D. Bell, and R. J. Pankhurst. (1979). *The Interpretation of the Igneous Rocks*. Allen & Unwin. London.

Johnson, W. M., and A. Maxwell. (1981). *Rock and Mineral Analysis*. John Wiley & Sons. New York.

Ragland, P. C. (1989). *Basic Analytical Petrology*. Oxford University Press. New York.

Rollinson, H. R. (1993). *Using Geochemical Data: Evaluation, Presentation, Interpretation*. Longman/Wiley. Harlow/New York.

Variation Diagrams, Magma Series, and Classification

Carmichael, I. S. E., F. J. Turner, and J. Verhoogen. (1974). *Igneous Petrology*. McGraw-Hill. New York. pp. 32–50.

Irvine, T. N., and W. R. A. Baragar. (1971). A guide to the chemical classification of the common volcanic rocks. *Can. Journ. Earth Sci.*, **8**, 523–548.

LeMaitre, R. W. (1976). Some problems of the projection of chemical data in mineralogical classifications. *Contrib. Mineral. Petrol.*, **56**, 181–189.

Russell, J. K., and C. R. Stanley (eds.). (1990). *Theory and Application of Pearce Element Ratios to Geochemical Data Analysis*. Volume 8. Geological Association of Canada. St. John's.

Wilcox, R. E. (1974). The liquid line of descent and variation diagrams. In: *The Evolution of the Igneous Rocks. Fiftieth Anniversary Perspectives* (ed. H. S. Yoder). Princeton University Press. Princeton, NJ.

Young, D. A. (2003). *Mind over Magma. The Story of Igneous Petrology*. Princeton University Press. Princeton, NJ. (Several chapters cover the historical development of many concepts and methodologies.)

See also Ragland (1989) and Cox et al. (1979), above.

9

Chemical Petrology II: Trace Elements and Isotopes

Questions to be Considered in this Chapter:

1. In what ways do the properties of trace elements and isotopes differ from those of major elements?

2. How can we capitalize on those differences and use trace element and isotopic data to further our understanding of the processes involved in the genesis of rocks and rock suites?

Trace elements and isotopes have a number of uses in the geological sciences. In Chapter 8, we discussed ways that major elements can be used to classify rocks and test hypotheses concerning the origin and evolution of magmatic systems. Because different phases selectively incorporate or exclude trace elements with much greater selectivity than they do major elements, trace elements are far more sensitive to igneous fractionation processes. As a result, the origin of melt systems and their evolutionary processes can usually be constrained better by using trace elements.

It would be prohibitively laborious and expensive to analyze for *all* possible trace elements in a rock or mineral sample. Experience has shown that some are particularly useful in petrology, although new applications are continuously being found. Trace elements are classified on the basis of their geochemical behavior. Among the most useful trace elements are the transition metals (particularly Sc, Ti, V, Cr, Mn, Co, Ni, Cu, and Zn), the lanthanides (more commonly called the rare earth elements [REE]), and Rb, Sr, Y, Zr, Nb, Cs, Ba, Hf, Ta, Pb, Th, and U. These groupings are shown in the periodic table on the inside front cover.

Unlike trace elements, which fractionate on the basis of *chemical* affinity for various phases, isotopes of any particular element can fractionate only on the basis of *mass* differences. Isotopic distributions may also result from radioactive decay from elements that chemically fractionated at some earlier time. Isotopes can place further constraints, including temporal ones, on the history of rock systems.

9.1 ELEMENT DISTRIBUTION

As discussed in Chapter 1, different elements have different affinities for specific crystallographic sites or other physical/chemical environments in which to reside. For example, as we have seen in the last three chapters, potassium tends to be concentrated in late melts, whereas magnesium is usually concentrated in early-forming minerals during the crystallization of a melt. Although most students are familiar with the major elements, trace elements are typically strangers. How can we understand and predict their behavior? In addition to his classification of elements into siderophile (preferring a native metallic state), chalcophile (preferring a sulfide phase), and lithophile (preferring a silicate phase) types, presented in Chapter 1, Goldschmidt (1937) also advanced some simple rules for the qualitative prediction of trace element affinities, based solely on the ionic radius and valence:

1. Two ions with the same radius and valence should enter into solid solution in amounts proportional to their concentration. In other words, they should behave about the same. Using this rule, we can predict the general affinity for

some trace elements by analogy with a major element with similar charge and radius. This type of substitution is often called **camouflage**. For example, Rb might be expected to behave as does K and concentrate in K-feldspars, micas, and evolved melts. Ni, on the other hand, should behave like Mg and concentrate in olivine and other early-forming mafic minerals.

2. If two ions have a similar radius and the same valence, the *smaller* ion is preferentially incorporated into the *solid* over the liquid. Because Mg is smaller than Fe, it should be preferred in solids, as compared to liquids. This is clearly demonstrated by noting the Mg/Fe ratio in olivine versus liquid in the Fo-Fa system (Figure 6.10).

3. If two ions have a similar radius but different valence, the ion with the *higher charge* is more readily incorporated into the *solid* over the liquid. Thus Cr^{+3} and Ti^{+4} are almost always preferred in solids as compared to liquids.

Goldschmidt's approach is rather simplistic and has many exceptions. The substitution of a trace element for a major element requires not only similar radius and valence but electronegativity as well, a factor that affects the bonding characteristics of an ion in minerals and was overlooked by Goldschmidt (see Ahrens, 1953; Ringwood, 1955; and Whittaker, 1967). The real affinity of an ion has a lot to do with *crystal field* effects and the valence electron shell configurations as they relate to the electric and energetic geometry of prospective mineral sites. Such an approach has proved fruitful, but it is beyond the scope of this text (for a discussion, see Henderson, 1982). The approach of Goldschmidt, however flawed, has the allure of simplicity, and it suffices for our present purposes.

Practically all elements distribute themselves unevenly between any two phases at equilibrium. This effect is known as **chemical fractionation**. For example, note that Ca/Na is always greater in plagioclase than in the coexisting melt in Figure 6.8, and Mg/Fe is always greater in olivine than in the coexisting melt in Figure 6.10. In Chapter 27, when we apply thermodynamics to mineral equilibria in a quantitative manner, we shall see that the distribution of an element between any two phases at equilibrium at a particular temperature, pressure, and compositional range is fixed and can be expressed using an "equilibrium constant," *K*.

If the reaction between two phases, such as a solid and a liquid, is an exchange reaction of some component *i*:

$$i_{(liquid)} \rightleftarrows i_{(solid)} \qquad (9.1)$$

We can define a simple empirical **distribution constant**, K_D, as:

$$K_D = \frac{X_i^{solid}}{X_i^{liquid}} \qquad (9.2)$$

where:

X_i = mol fraction of component *i* [the mol equivalent of Equation (6.3)] in the solid or liquid phase

As long as the concentrations of the components are relatively dilute:

$$K_D = \frac{C_S}{C_L} \qquad (9.3)$$

where:

C_S = concentration of a trace element in the solid (in ppm or wt. %)
C_L = concentration of a trace element in the liquid (in the same units as C_S)

K_D, although qualitatively predictable, is determined empirically, and Equations (9.2) and (9.3) simply state that a component has a tendency to be distributed in coexisting phases at equilibrium in a consistent and reproducible fashion.

When referring to trace elements, K_D is often replaced by *D* and is called the **distribution coefficient**, or **partition coefficient**. Distribution coefficients can be compiled by measuring the concentration of trace elements in the glass and coexisting mineral constituents from experimental runs (or from equilibrium matrix/phenocryst pairs in natural volcanic rocks), as has been done in Table 9.1 for a number of useful trace elements, as they partition themselves between some common minerals and a basaltic to andesitic melt.

The distribution coefficients in Table 9.1 should be considered only approximations because they vary with temperature, pressure (only slightly), and composition of the melt (often considerably). Some distribution coefficients vary by one or two orders of magnitude between basaltic and rhyolitic compositions. For an extensive compilation of partition coefficients, go to the database on the Geochemical Earth Reference Model (GERM) web page, at http://earthref. org. *Major elements* do not fractionate extremely, so major element K_D values between common phases are usually within an order of magnitude of 1.0. K_D values for trace elements, on the other hand, can range over several orders of magnitude (note, for example, the range of *D* values in the garnet column of Table 9.1). As a result, trace elements can be very sensitive to distribution and fractionation processes.

By convention, **incompatible trace elements** are concentrated in the melt more than the solid, and K_D (or *D*), as defined by Equation (9.3), is considerably less than 1. **Compatible trace elements** concentrate in the solid and K_D (or *D*) $\gg$ 1. This, of course, depends on the minerals involved, as we can see from Table 9.1, but we commonly standardize to mantle minerals (olivine, pyroxenes, and perhaps garnet). Of the major elements, then, Mg and Fe are generally considered compatible, whereas K and Na are incompatible. Incompatible elements are commonly subdivided into two subgroups based on the ratio of valence to ionic radius. The smaller, more highly charged **high field strength (HFS)** elements include the REE, Th, U, Ce, Pb^{4+}, Zr, Hf, Ti, Nb, and Ta. The low-field-strength **large ion lithophile (LIL)** elements (including K, Rb, Cs, Ba, Pb^{2+}, Sr, and Eu^{2+}) are generally considered to be more mobile, particularly if a fluid phase is involved. Small, low-valence elements are usually compatible, including the trace elements Ni, Cr, Cu, W, Ru,

TABLE 9.1   Partition Coefficients (C_S/C_L) for Some Commonly Used Trace Elements in Basaltic and Andesitic Rocks

		Olivine	Opx	Cpx	Garnet	Plag	Amph	Magnetite
Rb		0.010	0.022	0.031	0.042	0.071	0.29	
Sr		0.014	0.040	0.060	0.012	1.830	0.46	
Ba		0.010	0.013	0.026	0.023	0.23	0.42	
Ni		*14*	5	7	0.955	*0.01*	6.8	29
Cr		0.70	10	34	1.345	0.01	2.00	7.4
La		0.007	0.03	0.056	0.001	0.148	0.544	2
Ce		0.006	0.02	0.092	0.007	0.082	0.843	2
Nd	*Rare Earth Elements*	0.006	0.03	0.230	0.026	0.055	1.340	2
Sm		0.007	0.05	0.445	0.102	0.039	1.804	1
Eu		0.007	0.05	0.474	0.243	0.1/1.5*	1.557	1
Dy		0.013	0.15	0.582	3.17	0.023	2.024	1
Er		0.026	0.23	0.583	6.56	0.020	1.740	1.5
Yb		0.049	0.34	0.542	11.5	0.023	1.642	1.4
Lu		0.045	0.42	0.506	11.9	0.019	1.563	

Data from Rollinson (1993) and http://earthref.org
*Eu^{3+}/Eu^{2+}

Italics indicate estimates.

Rh, Pd, Os, Ir, Pt, and Au. HFS and LIL elements are indicated in the periodic table on the inside front cover (as are typical charges and ionic radii).

For a *rock*, we can determine the distribution coefficient for any element, i, by calculating the contribution for each mineral that comprises the rock. The result is called the **bulk distribution coefficient**, $\overline{D}_i$, and is defined by the equation:

$$\overline{D}_i = \sum W_A D_i^A \qquad (9.4)$$

where:

W_A = weight fraction of mineral A in the rock
D_i^A = distribution coefficient for the element i in mineral A

For example, if we take a hypothetical garnet lherzolite, containing 60% olivine, 25% orthopyroxene, 10% clinopyroxene, and 5% garnet (all by *weight*, not volume, as in the mode), the bulk distribution coefficient for erbium, using the data in Table 9.1, is:

$$\overline{D}_{Er} = (0.6 \cdot 0.026) + (0.25 \cdot 0.23) + (0.10 \cdot 0.583)$$
$$+ (0.05 \cdot 6.56)$$
$$= 0.459$$

Notice how the distribution coefficient for a mineral that is much different from the others (especially if it is very high) can significantly affect the bulk distribution coefficient for an element. In this case, garnet, comprising only 5% of the peridotite, caused the bulk distribution coefficient

to be greater than 0.4, when 60% of the rock contained olivine, with a distribution coefficient of only 0.013.

Next consider the partial melting of this peridotite. Just by browsing Table 9.1, we can readily see that Rb (calculated $\overline{D}_{Rb} = 0.016$), Sr ($\overline{D}_{Sr} = 0.025$), Ba ($\overline{D}_{Ba} = 0.008$), and the REE are incompatible elements for typical mantle minerals, and are concentrated in the melt, whereas Ni ($\overline{D}_{Ni} = 10.4$) and Cr ($\overline{D}_{Cr} = 6.39$) are compatibles and remain principally in the solid peridotite residua. Likewise, crystal fractionation of a basaltic magma increases the enrichment of Rb and the other incompatibles in the late liquids, whereas Ni and the compatibles will be selectively removed into the early crystals of olivine or pyroxene. Once again, as a result of the more extreme values of D for many trace elements as compared to major elements, the fractionation and concentration effects will be much greater for trace elements, and thus they provide a better measure of partial melting and crystallization processes.

Moreover, trace elements that are strongly partitioned into a single mineral may provide a sensitive measure of the amount of the mineral that has fractionally crystallized, provided that analyses are available for a cogenetic suite of rocks (representing a range of liquid compositions developed via fractional crystallization). For example, Figure 9.1a is a Harker-type diagram of Ni versus SiO_2 for the Crater Lake volcanic suite. The abrupt drop in Ni below 55% SiO_2 indicates that crystal fractionation of olivine occurred over this interval (note the high D value for Ni in olivine in Table 9.1). The loss in MgO above 55% SiO_2 in Figure 8.2 must be caused by the removal of some other mineral or by some process other than fractional crystallization. Note also that the variation in Ni ranges from about 250 ppm to essentially 0, or two orders of magnitude.

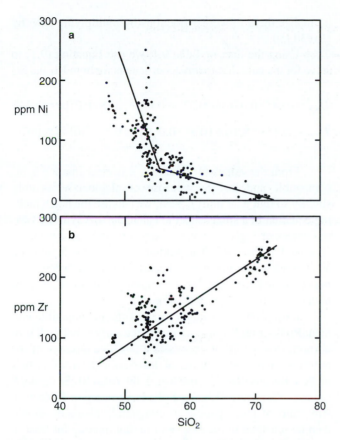

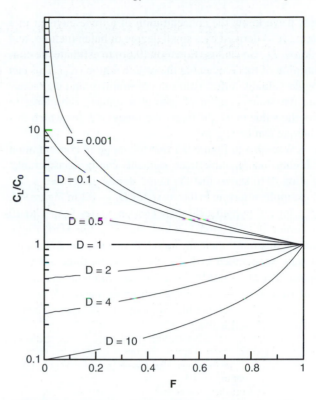

FIGURE 9.1 Harker diagrams of **(a)** Ni versus SiO_2 and **(b)** Zr versus SiO_2 for the Crater Lake suite of volcanic rocks in Figure 8.2.

Conversely, a trace element with a very small $\overline{D}_i$ is preferentially concentrated in the liquid and will reflect the proportion of liquid at a given state of crystallization or melting because its concentration varies inversely with the amount of liquid diluting it. The plot of Zr versus SiO_2 in Figure 9.1b is an illustration. Some relatively simple but useful quantitative models have been developed that can treat multiple fractionating phases and various crystallization or melting schemes using a variety of trace elements. We shall discuss a few of these next.

9.2 MODELS FOR SOLID–MELT PROCESSES

9.2.1 Batch Melting

The simplest model for an equilibrium process involving a solid and a liquid is the **batch melting model**. In this model, the melt remains in equilibrium with the solid, until at some point, perhaps when it reaches some critical amount, it is released and moves upward as an independent system. Shaw (1970) derived the following equation to model batch melting:

$$\frac{C_L}{C_o} = \frac{1}{\overline{D}_i(1 - F) + F} \quad \text{Batch Melting} \quad (9.5)$$

where:

C_o = concentration of the trace element in the original assemblage before melting began

F = weight fraction of melt produced [= melt/(melt + rock)]

FIGURE 9.2 Variation in the relative concentration of a trace element in a liquid versus source rock as a function of $\overline{D}$ and the fraction melted, using Equation (9.5) for equilibrium batch melting.

Figure 9.2 shows the variation in C_L/C_o with F for various values of $\overline{D}_i$, using Equation (9.5). Many petrologists consider values of F greater than 0.4 to be unlikely for batch melting in the mantle because such large amounts of melt should separate and rise before higher values are reached. The left side of Figure 9.2 is thus of most interest.

When $\overline{D}_i = 1$, there is (by definition) no fractionation, and the concentration of the trace element in question is the same in both the liquid and the source (hence the horizontal line at $\overline{D}_i = 1$ in Figure 9.2). The concentration of a trace element in the liquid varies more as $\overline{D}_i$ deviates progressively from 1. This is particularly true for small values of F (low degrees of partial melting) and for highly incompatible elements ($\overline{D}_i \ll 1$). Such incompatible elements become greatly concentrated in the initial small fraction of melt produced by partial melting, and they subsequently get diluted as F increases. Naturally, as F approaches 1, the concentration of every trace element in the liquid must be identical to that in the source rock because it is essentially all melted. This can be shown in Equation (9.5), for, as F approaches 1, Equation (9.5) becomes:

$$\frac{C_L}{C_o} = 1 \quad \text{as } F \rightarrow 1 \quad (9.6a)$$

On the other hand, as F approaches 0, Equation (9.5) reduces to:

$$\frac{C_L}{C_o} = \frac{1}{\overline{D}_i} \quad \text{as } F \rightarrow 0 \quad (9.6b)$$

Thus, if we know the concentration of a trace element in a magma (C_L) derived by a small degree of batch melting, and we know $\overline{D}_i$, we can use Equation (9.6b) to estimate the concentration of that element in the source region (C_o). This can provide valuable information in constraining and characterizing the source region of natural magmas. The closer to unity the value of $\overline{D}_i$ the larger range of F for which this technique can be applied.

Note also in Figure 9.2 that the range in concentration for highly incompatible trace elements can be considerable. Equation (9.6) shows that the range should not exceed $1/\overline{D}_i$. For example, referring to the curve for $\overline{D}_i = 0.1$ in Figure 9.2, C_L/C_o for our hypothetical trace element varies from 1 to 10 (one order of magnitude), which is $1/\overline{D}_i$. On Harker-type variation diagrams, then, the variation in some trace elements can vary by as much as $1/\overline{D}_i$, which can be three orders of magnitude if $\overline{D}_i = 0.001$. As can be seen in Figure 8.2, major elements tend to vary by a single order of magnitude or less.

For very incompatible elements, as D_i approaches 0, Equation (9.5) reduces to:

$$\frac{C_L}{C_o} = \frac{1}{F} \quad \text{as } \overline{D}_i \to 0 \tag{9.7}$$

This implies that if we know the concentration of a very incompatible element in both a magma and the source rock, we can determine the fraction of partial melt produced. This is another useful way trace elements can be used to evaluate melting processes at depth.

WORKED EXAMPLE 1: Batch Melting

Suppose a gabbroic source rock with a *mode* of 51% plagioclase, 33% clinopyroxene, and 18% olivine undergoes batch melting. We can use the batch melting Equation (9.6) to calculate C_L/C_o for Rb and Sr at values of $F = 0.05, 0.1, 0.15, 0.2, 0.3, 0.4, 0.5, 0.6, 0.7, 0.8,$ and 0.9. Next, we can plot C_L/C_o versus F for each (on the same graph), connecting the points for each with a line.

The first step is to calculate $\overline{D}_{Rb}$ and $\overline{D}_{Sr}$, but first we must convert the mode (volume %) to weight fraction. To do this, we must multiply each volume percent by the density, to get a weight basis, and then we normalize to 1.0. Based on some crudely estimated densities (g/cm^3), this can be done as shown in Table 9.2. The mode is in cm^3, so multiplying each by the density gives the weight proportion. If we get the sum of the weight proportions (= 303.9), we can

normalize them to weight fractions by multiplying each by (1.0/303.9).

Using the data in Table 9.1, we use Equation (9.5) to solve for the bulk distribution coefficients for both Rb and Sr:

$$\overline{D}_{Rb} = (0.45 \cdot 0.071) + (0.37 \cdot 0.031) + (0.18 \cdot 0.010) = 0.045$$

$$\overline{D}_{Sr} = (0.45 \cdot 1.830) + (0.37 \cdot 0.060) + (0.18 \cdot 0.014) = 0.838$$

From the values for $\overline{D}$ above, it is clear that Rb is incompatible and that Sr, because of plagioclase, is only slightly so (but near unity). Next, we can use the batch melting equation in a spreadsheet to calculate C_L/C_O at the various values of F given above. A spreadsheet that does this is presented in Table 9.3. The bold numbers are the input data, and the other columns were calculated using the batch melting equation [Equation (9.5)]. Graphing F versus C_L/C_O for both elements is easily done (Figure 9.3).

From Figure 9.3, it is clear that the incompatible element Rb is strongly concentrated in the early small melt proportions (low F). It thus provides a sensitive measure of the progress of partial melting, at least until the rock is half melted. Because $\overline{D}_{Sr}$ is close to 1.0, the **ratio** Rb/Sr versus F is nearly the same as Rb alone (the last column in Table 9.3). Any ratio of incompatible to compatible elements should then be sensitive to the degree of partial melting (at least in the initial stages). When $\overline{D}_i$ for the compatible element is very small, the ratio varies even more.

Once again, note that highly incompatible elements are strongly partitioned from the source rock and are concentrated in the earliest melts. As melting progresses, the fractionation is less extreme because the other elements are gradually added to the melt and dilute the incompatibles already there. That melts with highly varied Rb/Sr ratios can be derived from a source with a fixed Rb/Sr ratio is of great interest in isotopic studies and will be discussed further in Section 9.7.2.2.

TABLE 9.2   Conversion from Mode to Weight Percent

Mineral	Mode	Density	Wt. Prop.	Wt. %
Ol	15	3.6	54.0	18
Cpx	33	3.4	112.2	37
Plag	51	2.7	137.7	45
Sum	—	—	303.9	100

TABLE 9.3   Batch Fractionation Model

	$C_L/C_O = 1/(D(1 - F) + F)$		
	D_{Rb}	D_{Sr}	
F	**0.045**	**0.838**	**Rb/Sr**
0.05	10.78	1.18	9.12
0.1	7.12	1.17	6.08
0.15	5.31	1.16	4.58
0.2	4.24	1.15	3.69
0.3	3.02	1.13	2.67
0.4	2.34	1.11	2.11
0.5	1.91	1.09	1.76
0.6	1.62	1.07	1.51
0.7	1.40	1.05	1.33
0.8	1.24	1.03	1.20
0.9	1.11	1.02	1.09

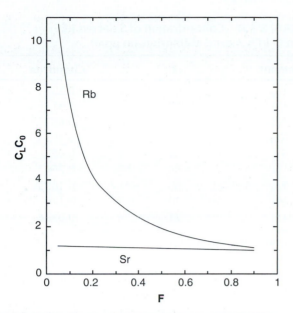

FIGURE 9.3 Change in the concentration of Rb and Sr in the melt derived by progressive batch melting of a basaltic rock consisting of plagioclase, augite, and olivine.

the chamber, and are essentially removed and isolated from further reaction with the remaining liquid. Using such a model, the concentration of some element in the *residual* liquid, C_L, is:

$$\frac{C_L}{C_o} = F^{(\overline{D}_i - 1)} \quad \text{Rayleigh crystal fractionation} \quad (9.8)$$

where:

> C_o = concentration of the element in the original magma
>
> F = fraction of melt *remaining* after removal of crystals as they form

The Rayleigh equation can also be applied to the melting process. A model for perfect fractional melting, or **Rayleigh fractional melting**, is:

$$\frac{C_L}{C_o} = \frac{1}{\overline{D}_i}(1 - F)^{(1/\overline{D}_i - 1)} \quad \text{Rayleigh fractional melting}$$

$$(9.9)$$

where:

> F = fraction of melt *produced*

Removal of every tiny melt increment, however, is not considered a likely process because it is very difficult to extract small amounts of melt from the source. As mentioned in Chapter 4, the initial melt occupies the intergranular space in a rock and is adsorbed to the grain surfaces. Some critical melt quantity is probably required before melt can be extracted.

Other models are used to analyze mixing of magmas, wall rock assimilation, zone refining, and combinations of some of these processes (see Cox et al., 1979, for a review). These models are capable of predicting and analyzing trace element distribution for a variety of igneous processes. Using these models and published partition coefficients, we can not only assess crystallization and melting processes, but we can occasionally identify specific minerals involved and constrain the source area of some melts.

Our experience with ternary experimental systems in Chapter 7 tells us that it is unrealistic to expect the ratio of minerals in the solid residue to remain constant throughout the melting process. Rather, we can expect the ratio to change as melting progresses, and a sequence of minerals will be consumed as well until melting is complete. We can apply Equation (9.5) repeatedly to deal with melting as increments of F, each increment with a different mineralogy or mineral ratio and, hence, a different value of $\overline{D}_i$. Such a model is called **incremental batch melting**. If the increments are few, it can be done readily by hand or spreadsheet, but if the increments are more continuous, a computer program is much better. Figures 9.2 and 9.3 show that the model is most sensitive to $\overline{D}_i$ at low values of F, so it is most important to work with small increments in this area. Above $F =$ 0.4, the increments need not be finely adjusted, as batch melts in this range vary less and are unlikely anyway.

9.2.2 Rayleigh Fractionation

As a second model, we turn to an idealized model for crystal fractionation. If all of the crystals that form remain in equilibrium with the melt, the batch melting equation [Equation (9.5)] would apply because equilibrium processes are reversible. The only difference would be that F would be the proportion of the liquid remaining after extraction, not the amount formed by melting. Such equilibrium crystallization may not be likely, except perhaps in highly viscous silicic melts where crystal settling is impeded. At the other extreme would be the separation of each crystal as it formed. This model for perfectly continuous fractional crystallization in a closed reservoir (such as a magma chamber) is called **Rayleigh fractionation** after the Rayleigh equation that is used to model it. In this situation, crystals form and accumulate, presumably on the floor of

9.3 THE RARE EARTH ELEMENTS: A SPECIAL GROUP OF TRACE ELEMENTS

The rare earth elements (REE), the series from lanthanum to lutetium (atomic numbers 57–71), are members of Group IIIA of the periodic table (see the inside front cover). They all have similar chemical and physical properties, causing them to behave as a coherent series (the "lanthanide series"). They all have a 3+ oxidation state, as a rule, and their ionic radius decreases steadily with increasing atomic number (called the **lanthanide contraction**, also apparent in the ionic radii listed in the periodic table on the inside front cover). The decrease in ionic radius causes the heavier REEs to be progressively less compatible (Goldschmidt's rule #2), so some fractionation does occur within the series. Because of crystal field effects, the fractionation for some minerals is more pronounced than for others. This may be clear from the $\overline{D}_i$ values in Table 9.1. Note that plagioclase, for example, is insensitive to the effect of ionic radius, whereas garnet strongly favors the heavy rare earth elements (HREE). This slight fractionation, commonly mineral specific in an otherwise coherent series, makes the REE a valuable tool for petrologists.

There are two exceptions to the 3+ valence norm for the REE. At low values of oxygen fugacity (f_{O_2}), Eu can have a 2+ valence, and Eu^{2+} may be more abundant than Eu^{3+} over a common range of f_{O_2} in igneous systems. Eu^{2+} substitutes for Ca in plagioclase (but is too large to do so in clinopyroxene or most other Ca-bearing phases). Thus D_{Eu}^{2+} for plagioclase is inordinately high for the REE series (Table 9.1). Under oxidizing conditions, Ce can also have a valence of 4+.

The REE are usually treated as a group, and REE diagrams are plots of concentration as the ordinate (y-axis) against increasing atomic number, so that the degree of compatibility increases from left to right across the diagram. All 15 REE are seldom, if ever, determined, but the trends are clear when using about 9 or 10 of them. A direct plot of concentration versus atomic number suffers from the jagged, or "sawtooth," Oddo-Harkins effect, whereby atoms with an even atomic number are more abundant than their immediate neighbors with odd atomic numbers (Figure 1.6). The effect is eliminated by normalizing, or dividing the concentration of each REE by the concentration of the same REE in a standard. Regrettably, there is no standard standard (if you will), but the most commonly used standards are estimates of primordial mantle or chondrite meteorite concentrations. As discussed in Chapter 1, some consider chondrites to be the least altered samples inherited from the primordial solar nebula, so they probably approximate the chemical composition of the early Earth. By normalizing to a chondrite standard, not only is the Oddo-Harkins effect eliminated, but the resulting REE diagram can also be compared directly to (approximate) primordial Earth values. Table 9.4 lists the average concentrations in CI chondrite meteorites of those REE in Table 9.1. For an excellent discussion of the various normalization schemes, see Rollinson (1993). Some confusion can result when REE concentrations or ratios are reported because some values may be normalized whereas others are not. It is a good practice to use the subscript N for normalized values and to state the particular normalization used.

REE diagrams are commonly used to analyze igneous petrogenesis, and we shall see several of these in the ensuing chapters. Before we do so, it would be beneficial to know how to interpret them. To help us do this, let's use the data in Tables 9.1 and 9.4 to create REE diagrams for the magmas in some of the models we have just considered.

TABLE 9.4   Concentration of Selected REE in C-1 Chondrite Meteorites (in ppm)

Element	Z	Chondrite
La	57	0.237
Ce	58	0.612
Nd	60	0.467
Sm	62	0.153
Eu	63	0.058
Dy	66	0.254
Er	68	0.1655
Yb	70	0.170
Lu	71	0.0254

Data from Sun and McDonough (1989).

WORKED EXAMPLE 2: REE Diagrams

Consider the garnet lherzolite, for which we calculated the distribution coefficient for erbium using Equation (9.4) in Worked Example 1. Let's now apply the batch melting model [Equation (9.5)] to get C_L/C_O for each of the REE in Table 9.1 at $F = 0.05, 0.1, 0.2, 0.4$, and 0.6 (5 to 60% melting). We shall assume that C_O for the present mantle is the same as the primordial mantle, which, in turn, can be approximated by CI chondrites. From this we can calculate C_L and create an REE diagram for the melts by plotting $C_L/C_{chondrite}$ as the ordinate versus the atomic number for each REE as the abscissa.

Given the chondrite model, Equation (9.5) becomes $C_L = C_{chondrite}/[\overline{D}_{REE}(1 - F) + F]$ for each REE. A spreadsheet saves us an inordinate amount of work to create this REE diagram. The spreadsheet shown in Table 9.5 has the data from Table 9.1 in the heavy-bordered cell range from A16 to G25 and the data from Table 9.4 in the heavy-bordered cell range from A3 to C11. Row 26 contains the wt. % of each mineral in the lherzolite, which completes the input data for the problem. Using the mineral % plus the individual D values, we can calculate $\overline{D}_i$ for each element by using Equation (9.4). The results are in column H. Next, we pick a value of F (cell I15) and use Equation (9.5) to calculate C_L (using C3:C11 as C_O) in column I. For ease of graphing, I copied column I to D3:D11 and finally normalized to chondrite by dividing D3/C3, putting the result in E3, and so on down the column. Finally, I plotted B3:B11 versus E3:E11 to make the REE diagram. The file REE.XLS at www.prenhall.com/winter is the Excel spreadsheet I used.

The resulting REE diagram for the selected values of F is reproduced as Figure 9.4. If $F = 1.0$, all of the source rock would be melted, and the liquid must have the same REE concentration as chondrite (assuming that the chondrite model is correct). Because the concentration of each REE is then divided by the chondrite values as a normalization, the result is a horizontal line at 1.0 on the REE diagram. This gives a

TABLE 9.5 Spreadsheet to Calculate an REE Diagram for the Batch Melting Model

	A		B	C	D	E	F	G	H	I
1	Rare Earth Element Diagram									
2	Element		Atom #	chondr	sample	samp/chon				
3	La		57	0.237	1.75	7.40				
4	Ce		58	0.612	4.86	7.94				
5	Nd		60	0.467	3.37	7.22				
6	Sm		62	0.153	0.98	6.43				
7	Eu		63	0.058	0.15	2.62				
8	Dy		66	0.254	1.00	3.94				
9	Er		68	0.1655	0.44	2.65				
10	Yb		70	0.17	0.37	2.17				
11	Lu		71	0.0254	0.05	2.00				
12										
13				in ppm	in ppm					
14									Batch Melting	
15				Distribution Coefficients					*F* =	0.1
16			**Olivine**	**Opx**	**Cpx**	**Garnet**	**Plag**	**Amph**	Bulk D	C_L
17	La	Rare Earth Elements	0.007	0.03	0.056	0.001	0.148	0.544	0.039	1.75
18	Ce		0.006	0.02	0.092	0.007	0.082	0.843	0.029	4.87
19	Nd		0.006	0.03	0.230	0.026	0.055	1.340	0.038	3.48
20	Sm		0.007	0.05	0.445	0.102	0.039	1.804	0.060	0.99
21	Eu		0.007	0.05	0.474	0.243	1.5	1.557	0.316	0.15
22	Dy		0.013	0.15	0.582	3.17	0.023	2.024	0.214	0.87
23	Er		0.026	0.23	0.583	6.56	0.020	1.740	0.371	0.38
24	Yb		0.049	0.34	0.542	11.5	0.023	1.642	0.600	0.27
25	Lu		0.045	0.42	0.506	11.9	0.019	1.563	0.626	0.04
26	Weight fract.		0.50	0.21	0.08	0.04	0.17	0.00		
27	Min. prop.		60	25	10	5	20		120	

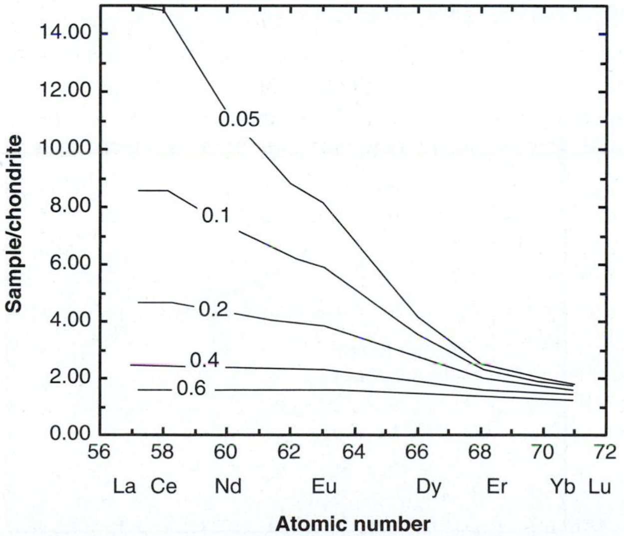

FIGURE 9.4 Rare earth element concentrations (normalized to chondrite) for melts produced at various values of F via melting of a hypothetical garnet lherzolite, using the batch melting model [Equation (9.5)].

convenient mental reference: complete melting of the unaltered mantle should produce a horizontal REE plot at 1.0. For progressively lower values of *F*, representing successively smaller melt fractions, the REE will be fractionated, but not equally so. The increasing $\overline{D}_i$ values as atomic number (Z) increases in Table 9.1 (column H) reflect the lanthanide contraction and the progressively greater proportion of the heavy REE (HREE) in the solid and the light REE (LREE) in the liquid, resulting in a negative slope for the curves in an REE diagram. The effect is enhanced for small values of *F*, as we would expect, and the increasing negative slope with decreased *F* is clearly illustrated in Figure 9.4. You are encouraged to experiment with REE.XLS on a computer and change any of the parameters—*F* and/or the weight proportion of any mineral. I have a normalization in rows 26 and 27 to be sure that the weight fraction totals 1.00 for $\overline{D}_i$ to be realistic. You can thus change the values in row 27 without having to worry about whether they total 100%. The resulting chart on the spreadsheet will instantly reflect your choices. You can get a variety of slopes on your REE diagrams, but *a positive slope is impossible*, whatever you may try, because the LREE will always favor the liquid compared to the HREE. Note that garnet, with its very high distribution coefficients for the HREE, will, when added to the minerals of the hypothetical source rock, result in the highest negative slopes for a given value of *F*.

The slope on an REE diagram can easily be approximated mathematically by the ratio of the normalized concentration of an element on the left side, such as La or Ce, divided by an element on the right, such as Yb or La. This value obviously increases with the slope. A $(La/Lu)_N$ ratio of 1.0 is a horizontal line, and a ratio below 1.0 indicates a positive slope. Similarly $(La/Sm)_N$ or $(La/Eu)_N$ can be used to measure enrichment within the LREE, whereas $(Tb/Yb)_N$ will do the same for the HREE.

The process described above can be performed, and REE diagrams created, for any of the melting or crystallization models discussed. The resulting REE diagrams are different for each model, but, given the uncertainty and variance in the D values, it is debatable (and certainly debated) whether they are sufficiently different to be used to distinguish between the modeled processes when applied to natural rocks.

One interesting change you can make on the spreadsheet and graph used in our example is to include plagioclase in the source rock. The distribution coefficient in the spreadsheet table for europium in plagioclase is for a high-Eu^{2+} component. If you assign 20% plagioclase to the source rock, there will be a pronounced dip in the REE pattern at Eu (Figure 9.5). This is referred to as a **europium anomaly**, reflecting the Eu^{2+} substitution for Ca in plagioclase. The anomaly may be either negative, as in Figure 9.5, or positive, depending on whether plagioclase was removed or accumulated, respectively. The magnitude of the Eu anomaly is commonly expressed as **Eu/Eu***, where Eu* is the hypothetical value of Eu if no Eu^{+2} was captured by plagioclase (the value on a straight line between the nearest neighboring elements; Figure 9.5). A negative europium anomaly is a good indicator that the liquid was at one time in equilibrium with now-absent plagioclase, but it cannot easily determine whether the reason for this is a plagioclase-bearing melt source or the removal of plagioclase phenocrysts from the melt at a later time. As I have suggested, this is a common problem for most of our models. Using them, we might be able to identify

the participation of a specific mineral, but we cannot necessarily distinguish the exact process.

In Chapters 10 to 20, we shall see the important role REE diagrams play in the interpretation of the petrogenesis of several types of igneous rocks. The overall shape of the REE patterns and individual element anomalies can be used to constrain the source of a melt or the participation of specific minerals in the evolution of a magma. REEs become considerably more complicated in very silicic rocks, such as granites, where a number of minor and accessory minerals, such as apatite, zircon, monazite, and allanite, have very *high* REE distribution coefficients, concentrating them and having a disproportionate influence on REE patterns.

9.4 NORMALIZED MULTIELEMENT (SPIDER) DIAGRAMS

The use of a reference, such as chondrite normalization, in REE patterns has led to an expansion of the technique to applications with a broader range of trace element data, collectively called **normalized multielement diagrams**. Thompson (1983) was the first to use in print the more colloquial term **spider diagram** (often shortened to **spidergram**), considering plots with several samples reminiscent of a cluster long-legged spiders (R. N. Thompson, personal communication, although he credits Mike Norry with prior verbal usage). Figure 9.6 is an example of a spider diagram for an alkaline basalt from Gough Island in the southern Atlantic. The diagram is similar to a typical REE diagram but has a larger range of trace elements plotted. In a spider diagram, the abundances of a number of mostly incompatible trace elements are normalized to estimates of their abundances in some uniform primitive reservoir, such as the primordial Earth. Whereas the absolute abundances of these elements in the bulk Earth may approximate chondrite values, those in the primordial mantle may be greater because of the concentration effects of early core formation, which rejected these elements.

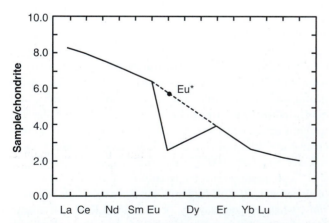

FIGURE 9.5 REE diagram for 10% batch melting of a hypothetical lherzolite with 20% plagioclase. Because Eu^{2+} is retained in the plagioclase of the source, the extracted melt is depleted in Eu, resulting in a pronounced negative europium anomaly.

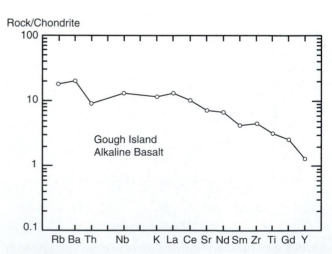

FIGURE 9.6 Chondrite-normalized multielement (spider) diagram showing the trace element patterns of an alkaline island basalt from Gough Island (Sun and McDonough, 1989).

Several variants of the spider diagram have been used in the literature with different elements and different normalization schemes (hypothetical primordial mantle, chondrite meteorite, etc.). In addition, the order of the elements along the abscissa may vary slightly with different authors (Wood et al., 1979; Sun, 1980; Thompson, 1982c, 1984). The order is usually based on the author's estimate of increasing incompatibility of the elements from right to left in "typical" mantle undergoing partial melting. The elements selected are almost always incompatible during mafic to intermediate partial melting and fractional crystallization processes. The main exceptions are Sr, which may be compatible if plagioclase is involved, Y and Yb with garnet, and Ti with an Fe-Ti oxide. Troughs at these elements are taken to indicate involvement of these minerals.

In general, the more incompatible elements on the left-hand side of the spider diagram should be more enriched in the melt during partial melting (particularly for small degrees of it), resulting in a negative slope. Any fractional crystallization subsequent to magma segregation from the source should increase the slope even further. The dynamics are the same as for REE diagrams, but a spider diagram permits us to extend the technique to a broader range of trace elements. Once again, no known process can produce a liquid with a positive slope directly from a non-sloping source for either an REE or a spider diagram. Slopes on a spider diagram can also be estimated by the ratio of two elements of contrasting compatibility, such as $(Rb/Y)_N$.

Spider diagrams are flexible, and a variety of elements and normalizations have been used. For example, Figure 9.7 illustrates a type of spider diagram used by Pearce (1983), normalized to an average mid-ocean ridge basalt. This diagram compares the incompatible trace elements of a sample to the most abundant igneous rock on the planet. The LIL elements are on the left side of the diagram, and the HFS elements are on the right. Each set is arranged in order of increasing incompatibility away from the margins, so that the most incompatible elements are just left of the center of the diagram. An average of many ocean island basalt analyses is plotted in Figure 9.7 and has Y and Yb contents nearly equal to MORB (Y_N and $Yb_N \sim 1.0$), but ocean island basalts are enriched in all other incompatible trace elements on the diagram in proportion to the degree of incompatibility, resulting in the broadly arched pattern that peaks at Ba-Th.

Of course, spider diagrams portray a much more heterogeneous array of trace elements than the more systematic lanthanide series in REE diagrams, and they are likely to show more peaks and troughs, reflecting the different behavior of the elements involved. The LIL elements (particularly the more mobile ones) may be mobilized by a hydrous fluid phase, whereas variations in the HFS elements are more likely to be controlled by the source region and mineral/melt fractionation processes during magma evolution. High Ba and Rb (the more mobile LIL elements) may thus suggest metasomatism, or contamination by a crustal component, because the LIL are easily extracted from the mantle and eventually become concentrated in the continental crust. Some individual elements may be strongly influenced by particular minerals, such as Zr by zircon, P by apatite, Sr by plagioclase, and Ti, Nb, and Ta by ilmenite, rutile, or titanite. If rocks from a particular petrogenetic province display similar patterns of peaks and troughs, this strongly suggests that they share a common parent, process, or contaminant.

9.5 APPLICATION OF TRACE ELEMENTS TO IGNEOUS SYSTEMS

The simplest application of trace elements is to use them in variation diagrams (Figure 9.1) in the same fashion as major elements were used in Chapter 8. As mentioned earlier in this chapter, the high distribution coefficients for many trace elements result in a larger variation with partial melting or fractional crystallization. They can thus be used to assess the extent to which these processes may have operated because their concentrations tend to vary considerably as these processes continue.

A further use of trace element partitioning is the identification of the source rock or a particular mineral involved in either partial melting or fractional crystallization processes. For example, the REE can commonly be used to distinguish between high pressure and low pressure sources of mantle-derived melts. In the deep continental crust, and at depths over about 70 km in the mantle, garnet and clinopyroxene are important phases (see Figure 10.3) and remain as residual solids during the generation of up to 15 to 20% partial melts. As a result, particularly due to the presence of garnet, the bulk distribution coefficient will be high for the HREE (Table 9.1), and up to 10% partial melts will be highly HREE depleted (with highly negative slope on an REE diagram, as you can test using REE.XLS). Because the slope on an REE diagram is also a function of F, the fraction of melt generated, we must be aware of the differences

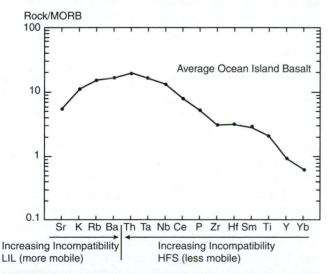

FIGURE 9.7 Ocean island basalt plotted on a mid-ocean ridge basalt (MORB) normalized multielement (spider) diagram of the type used by Pearce (1983). Data from Sun and McDonough (1989).

between HREE depletion by garnet and LREE enrichment due to low degrees of partial melting. Garnet extraction, however, typically imparts a negative slope *within* the HREE, whereas LREE enrichment at low F results in only minor HREE variation. At depths shallower than 40 km, plagioclase is an important phase (Figure 10.3) and can be detected via an Eu anomaly in the melt. Thus the shape of the REE pattern of some mantle-derived basalts may yield important information on their depth of origin. We shall use this technique in later chapters.

The concentration of a *major* element in a phase (mineral or melt) is usually buffered by the system, so that it varies little in a phase as the system composition changes. For example, consider the olivine system, Fo-Fa, in the phase diagram in Figure 6.10. Let's specify that olivine and liquid coexist at 1445°C. If the Mg/Fe ratio of the system increases from 20 to 50%, this will have no effect whatsoever on the composition of either phase because the composition of each is fixed by temperature, as the phase rule predicts. Only the ratio of solid to liquid changes. In marked contrast to this behavior, *trace* element concentrations are in the Henry's law region of concentration (Figure 27.3), and their activity varies in direct relationship to their concentration in the system. Thus the concentration of Ni in all phases will double if the Ni concentration in the system doubles. This does not mean that the concentration of Ni is the *same* in all phases because trace elements do fractionate. Rather the concentration within each phase varies in proportion to the system concentration. If, for example, the Ni concentration in olivine is 200 ppm, and 70 ppm in orthopyroxene, doubling it in the system will result in 400 and 140 ppm, respectively.

Because of this proportionality, the *ratios* of trace elements are commonly superior to the concentration of a single element in identifying the role of a specific mineral. For example, in the case of garnet above, the ratio of the HREE Yb divided by the LREE La would be a good indicator of the slope of the REE diagram. The *absolute* values of either La or Yb would vary with the garnet effect, but they also vary with the overall concentration of REEs in the source, and we would be unable to distinguish between these effects in a single rock sample on the basis of the concentration of La or Yb alone. A low value of Yb in a volcanic rock could result from either a garnet-bearing source (which held Yb in the source) or simply a source with a low REE content. Because La and Yb should behave similarly, except with respect to garnet, a low La/Yb ratio is more likely to reflect the influence of garnet. Likewise, the ratio of Eu to the adjacent Sm would indicate an Eu anomaly and, thus, plagioclase participation. Although these REE examples serve to illustrate the point, the full REE diagram is usually more informative.

As a more practical example, the K/Rb ratio has been used to indicate the importance of amphibole in an ultramafic source rock, such as a hornblende peridotite. In mafic assemblages, K and Rb behave similarly, so their ratio should be nearly constant for all mafic rocks. Olivine and pyroxene contain little of either of these elements, so their

contribution to the bulk distribution coefficient is negligible. Almost all of the K and Rb must therefore reside in the amphibole, which has a D of about 1.0 for K and 0.3 for Rb. Because amphibole has D_{Rb} less than D_K, melting of a hornblende-bearing assemblage (as long as some hornblende is left behind) results in a decrease in the K/Rb ratio in the melt over what it was in the original rock. Other factors being equal, magma produced by partial melting of an amphibole-bearing source rock would have lower K/Rb than would magma derived from a source without amphibole. Naturally, high absolute K or Rb contents would also indicate an amphibole-bearing source but may result from other causes, such as phlogopite, or an alkali-enriched fluid. The ratio is more indicative of amphibole because of the different D values particular to this mineral. Fractional crystallization of an amphibole would also result in a low K/Rb ratio in the evolved liquid.

Another example involves the incompatible pair Sr and Ba. These incompatible elements tend to be enriched in the first products of partial melting or the residual liquids following fractional crystallization. The effect is selective, of course, according to the mineral phases involved in the process. Sr is excluded from most common minerals except plagioclase (Table 9.1), and Ba is similarly excluded from all but alkali feldspar. The Ba/Sr ratio thus tends to increase with crystallization of plagioclase, but it levels off and may even decrease when orthoclase begins to crystallize.

As an example of the use of *compatible* element ratios, Ni is strongly fractionated into olivine but less so into pyroxene. Cr and Sc, on the other hand, enter olivine only slightly but are strongly fractionated into pyroxenes. The ratio of Ni to Cr or Sc would then provide a way of distinguishing the effects of olivine and augite in a partial melt or a suite of rocks produced by fractional crystallization.

In all of the above cases using ratios, the idea is to find a mineral with a unique pair of elements for which it alone has a relatively high value of D for one element and a relatively low value of D for the other. The ratio of these elements is then sensitive only to liquid/crystal fractionation associated with that particular mineral.

There are myriad applications of trace elements to petrology, including some that are not mineral specific. For example, the ratio of two highly incompatible trace elements should be the same throughout a magma series developed at a volcanic center by fractional crystallization because the crystallizing minerals remove little of either. If volcanics were derived from distinct parents or sources, however, the ratio would be expected to be more variable.

These are but a few examples of specific trace element uses. Table 9.6 provides a summary of some important trace elements used as petrogenetic tracers in attempts to identify minerals involved in differentiation or partial melting (see also Taylor, 1969). These elements can tell us something about a single whole-rock analysis, but they are much more reliable if used in variation diagrams for a suite of related rocks with a significant compositional range in a single area. A decrease in these elements for a rock series implies the

TABLE 9.6   A Brief Summary of Some Particularly Useful Trace Elements in Igneous Petrology

Element	Use as a Petrogenetic Indicator
Ni, Co, Cr	Highly compatible elements. Ni and Co are concentrated in olivine, and Cr is concentrated in spinel and clinopyroxene. High concentrations indicate a mantle source, limited fractionation, or crystal accumulation.
Zr, Hf	Very incompatible elements that do not substitute into major silicate phases (although they may replace Ti in titanite or rutile). High concentrations imply an enriched source or extensive liquid evolution.
Nb, Ta	High-field-strength elements that partition into Ti-rich phases (titanite, Ti-amphibole, Fe-Ti oxides). Typically low concentrations in subduction-related melts.
Ru, Rh, Pd, Re, Os, Ir, Pd	Platinum group elements (PGEs) that are siderophile and used mostly to study melting and crystallization in mafic–ultramafic systems in which PGEs are typically hosted by sulfides. The Re/Os isotopic system is controlled by initial PGE differentiation and is applied to mantle evolution and mafic melt processes.
Sc	Concentrates in pyroxenes and may be used as an indicator of pyroxene fractionation.
Sr	Substitutes for Ca in plagioclase (but not in pyroxene), and, to a lesser extent, for K in K-feldspar. Behaves as a compatible element at low pressure, where plagioclase forms early, but as an incompatible element at higher pressure, where plagioclase is no longer stable.
REE	Myriad uses in modeling source characteristics and liquid evolution. Garnet accommodates the HREE more than the LREE, and orthopyroxene and hornblende do so to a lesser degree. Titanite and plagioclase accommodate more LREE. Eu^{2+} is strongly partitioned into plagioclase.
Y	Commonly incompatible. Strongly partitioned into garnet and amphibole. Titanite and apatite also concentrate Y, so the presence of these as accessories could have a significant effect.

fractionation of a phase in which they concentrate. High concentrations of a trace element in the parental magmas may reflect the high concentration of that element in the source rock, helping to constrain the source area mineralogy.

9.6 GEOCHEMICAL CRITERIA FOR DISCRIMINATING BETWEEN TECTONIC ENVIRONMENTS: DISCRIMINATION DIAGRAMS

Some of the trace element patterns that we now recognize for igneous rocks show distinct trends or ratios that correlate empirically with a particular tectonic setting, such as mid-ocean ridges, ocean islands, or subduction zones. Of course, modern examples are readily characterized based on field criteria and location, but the chemical characteristics, when recognized based on rocks of known affinity, can be applied to much older igneous rocks, which may be considerably deformed, faulted, displaced, and isolated from their original setting (Pearce and Cann, 1971, 1973).

Figure 9.8 shows some attempts to use the ratios of several minor and trace elements to indicate the original source of mafic volcanic rocks that now occur as greenschists and amphibolites in deformed and metamorphosed terranes where their source is no longer recognizable. If we analyze older rocks in deformed terranes, we can plot them on one of numerous such **discrimination diagrams** found in the literature (see Rollinson, 1993, Chapter 5, for a comprehensive review) and infer the original tectonic/igneous

setting. Vermeesch (2005) recently tested several discrimination diagrams in a statistically more rigorous way than the customary method of drawing boundaries by eye. He concluded that the Ti-Si-Sr system provided the best linear fit to his database of 756 oceanic basalts of known tectonic affinity and that the Na-Nb-Sr system provided the best quadratic fit (Figure 9.8d and e).

Please note that these techniques are strictly empirical and are used to the extent that the composition of igneous rocks whose histories we can infer on other grounds continue to support these conclusions. The usage and results are at times a bit ambiguous because there are so many variables involved: the source rocks; the extent of partial melting, fractional crystallization, magma mixing, and wall–rock assimilation; and the effects of subsequent metamorphism. The effects of metamorphism, however, can be minimized by choosing trace elements generally considered to be immobile during metamorphism (usually including Ti, Cr, Zr, Hf, and Y). The effects of fractional crystallization, assimilation, and mixing can be minimized by applying the technique to mafic volcanic rocks only.

Rocks may plot in contradictory fields in different diagrams, leading several investigators to question the validity of this approach (e.g., Wang and Glover, 1992). Different tectonic environments, however, do have some distinctive geochemical signatures, which suggests that careful application of these techniques may yield useful information not otherwise available. Again, plotting suites of related rocks is far superior to plotting a single sample. Of course, wise investigators don't rely on any single technique or plot.

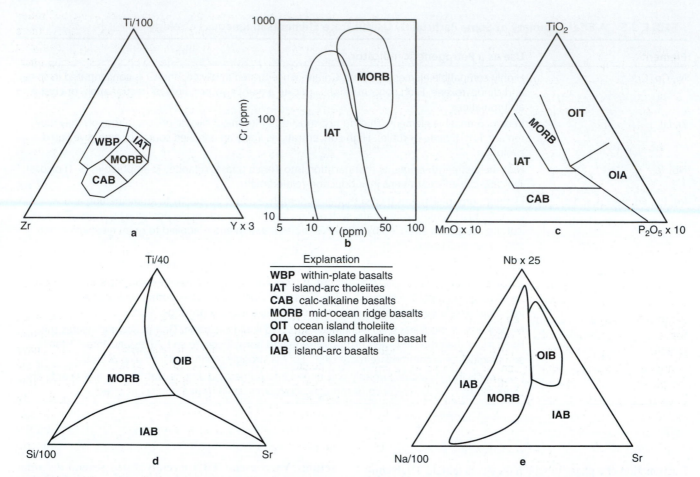

FIGURE 9.8 Examples of discrimination diagrams used to infer tectonic setting of ancient (meta-) volcanics. **(a)** After Pearce and Cann (1973), **(b)** after Pearce (1982), Coish et al. (1986). Reprinted by permission of the *American Journal of Science*, **(c)** after Mullen (1983) Copyright © with permission from Elsevier Science, **(d)** and **(e)** after Vermeesch (2005) © American Geophysical Union with permission.

Trace elements have become important tools for petrologists, and they have myriad applications to igneous rocks. We shall apply and expand upon the principles introduced above in Chapters 10 to 20, when we evaluate the generation and development of various magmatic systems. New trace element uses are continually being developed.

9.7 ISOTOPES

Elements are characterized by the number of protons in the nucleus. All atoms with 6 protons in their nucleus are carbon atoms, those with 7 are nitrogen, etc. Because neutrons have no charge, the nucleus of an element can contain variable amounts of them without affecting the atom's chemical properties. Thus carbon may have 6, 7, or 8 neutrons, resulting in carbon atoms with different masses. These variants, all of the same element, are called **isotopes**. Neutrons do affect the nucleus, however, and only a limited range of neutron capacities are stable for any particular element.

The general notation for the proton and neutron content of **nuclides** (atoms) is conventionally written $^{12}_{6}C$, where C is the element symbol (carbon, in this case), the subscript 6 is the atomic number (Z, the number of protons

in the nucleus), and the superscript 12 is the mass number (A, the number of protons plus neutrons in the nucleus). The three natural carbon isotopes are then $^{12}_{6}C$, $^{13}_{6}C$, and $^{14}_{6}C$. Because Z is characteristic of the element, the subscript is redundant with the element symbol, and it is commonly omitted. Isotopes have become important petrogenetic tracers. The isotopes most commonly used by petrologists are H, C, O, S, K, Ar, Rb, Sr, U, Pb, Th, Sm, and Nd, but, as with trace elements, new useful isotopic systems are continually being explored.

Isotopes can be classified as either stable or radioactive. **Stable isotopes** are those that remain indefinitely. **Radioactive isotopes** are unstable and undergo a process of radioactive decay to produce another nuclide, plus a particle or gamma ray and considerable energy. The original unstable isotope is referred to as the **parent**, the resulting isotope is the **daughter**, and the particle may be either an alpha particle (essentially a $^{4}_{2}He$ atom) or a beta particle (an electron). Some radioactive isotopes decay by **nuclear fission**, in which they split into two daughter isotopes, some of which may also be unstable and decay further. Daughter isotopes, because they are created by radioactive decay, are called **radiogenic** isotopes. Radioactive isotopes do not decay instantaneously but do so at rates that depend on their relative stability, so decay rates are specific to each nuclide. $^{24}_{11}\text{Na}$, for example, is created in nuclear reactors and is very unstable, lasting only a few days to weeks. $^{238}_{92}\text{U}$, on the other hand, is more stable, and only about half of the Earth's original ^{238}U has decayed so far. As we shall see, because the rate of decay for a particular radioactive isotope is constant, these isotopes can be used to determine the age of rocks as well as having uses as petrogenetic tracers.

9.7.1 Stable Isotopes

Most elements have more than one isotope. That's why the atomic weight of an element is not a whole number: it represents an average mass of the isotopes for that element in a typical natural sample. Because stable isotopes of a particular element are *chemically* identical (all isotopes of carbon are still carbon), they cannot fractionate *chemically* between two phases, as would two trace elements such as Rb and Sr, as shown in Figure 9.3. Rather, **mass fractionation** is the only process that can separate the existing isotopes of a single element. In other words, during some reaction, such as melting, crystallization, or evaporation, isotopes of the same element can only fractionate among the phases as a function of their difference in mass. If any mass fractionation does take place, the *light isotope always fractionates, preferably into the phase with weaker bonding, and is generally favored in the vapor over the liquid and in the liquid over the solid*. These mass differences are usually small compared to chemical differences, so the mass fractionation is usually small and can be documented only with the use of very sensitive analytical equipment. *The efficiency of mass fractionation is a function of the mass difference divided by the total mass.* Thus ^{204}Pb and ^{205}Pb do not mass fractionate appreciably because the mass difference is only about 0.5% of the total. ^1H and ^3H, on the other hand, fractionate well because ^3H has essentially three times the mass of ^1H.

As an example of the use of stable isotopes, consider the three stable isotopes of oxygen:

^{16}O 99.756% of natural oxygen

^{17}O 0.039% of natural oxygen

^{18}O 0.205% of natural oxygen

Isotope fractionations are usually expressed as ratios and are referenced to some standard in order to make them more easily handled and understood, as well as to calibrate the results of various laboratories. The most common international standard for oxygen isotopes is *standard mean ocean water* (SMOW). ^{18}O and ^{16}O are the commonly used isotopes, and their ratio is expressed as δ:

$$\delta(^{18}\text{O}/^{16}\text{O}) = \frac{(^{18}\text{O}/^{16}\text{O})_{\text{sample}} - (^{18}\text{O}/^{16}\text{O})_{\text{SMOW}}}{(^{18}\text{O}/^{16}\text{O})_{\text{SMOW}}} \times 1000$$

(9.10)

The factor 1000 yields results expressed in thousandths, or *per mil* (‰), not hundredths, or per*cent* (%). $\delta(^{18}\text{O}/^{16}\text{O})$ for SMOW as a sample would be zero, according to Equation (9.10). Phases enriched in ^{18}O with respect to SMOW have positive δ values, whereas ^{18}O-depleted phases have negative values.

What would $\delta^{18}\text{O}$ be for meteoric water? Remember, the light isotope is favored in vapor over liquid, so evaporation from seawater preferentially selects ^{16}O, so clouds have a negative δ. Condensation to rain should take most of the vapor as rain, so there is little reverse mass fractionation when virtually all of the vapor is converted to liquid. We would thus expect to find $(^{18}\text{O}/^{16}\text{O})_{\text{rain}} < (^{18}\text{O}/^{16}\text{O})_{\text{SMOW}}$ and thus negative δ values in meteoric waters. Figure 9.9 shows this is indeed true, but δ is a function of climatic temperature as well. Why?

As can be seen for Rb in Figure 9.3, fractionation is most effective when F, the amount of material converted from one state to another, is small. In most hot climates, much of the moisture in the air remains as humidity, and a lower percentage of the water vapor condenses back to rain. Although oxygen mass fractionation during vaporization is still high in warm climates, reverse mass fractionation during condensation also takes place because only a small proportion of the vapor condenses. As a result, the liquid condensate tends to concentrate the heavy isotope, and it returns closer to SMOW values. Where the temperature is lower, and a higher fraction of the cloud vapor condenses to liquid, the mass fractionation during condensation is less efficient. (If all of the water condensed, no fractionation at all

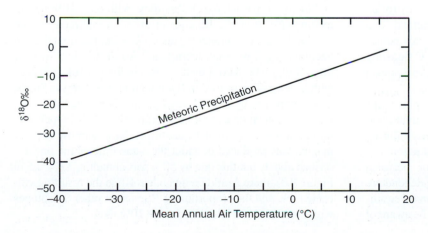

FIGURE 9.9 Relationship between $\delta(^{18}\text{O}/^{16}\text{O})$ and mean annual temperature for meteoric precipitation. After Dansgaard (1964).

would occur.) δ thus remains low in cooler climates, closer to its value in the vapor. Once again, mass fractionation of oxygen is fairly effective because the mass difference is approximately one-eighth of the total mass. We have thus found that the fractionation in rainwater of oxygen, a light isotope, is dependent upon climate temperature. This is useful in a number of ways, such as paleoclimate estimates from snow accumulations and cores from ice caps or even from the $^{18}O/^{16}O$ ratios in carbonates (which, incidentally, have a different standard than SMOW).

Stable isotopes are useful in assessing the relative contribution of various reservoirs, each with a distinctive isotopic signature. For example, the oxygen and hydrogen isotopic ratios of meteoric and juvenile (magmatic) water are different, as are those of brines. O and H isotopes have been used to evaluate the juvenile/meteoric/brine characteristics of water (including hydrothermal systems above intrusions) and the type of water responsible for rock alteration. As discussed in Section 4.3, most works indicate that hydrothermal systems above plutons are dominated by recirculating meteoric water.

Oxygen is also an important constituent in most minerals. The oxygen isotopic compositions of igneous, sedimentary, and metamorphic rocks display systematic variations in $\delta^{18}O$ that contain important information about their formation and history. For example, $\delta^{18}O$ of most igneous rocks is between +5 and +15‰, which differs from that of surface-reworked sediments, which exchange oxygen and equilibrate with meteoric water. $\delta^{18}O$ is thus different for mantle rocks/melts and sediments, and can be used to evaluate the extent to which mantle-derived magmas are contaminated by crustal sediments. Other igneous rocks show lowered ^{18}O and deuterium (^2H, or D) values, interpreted to result from exchange between the rocks and infiltrating meteoric water. O'Neil et al. (1977) used the $\delta^{18}O$ and H/D values of some granites in Australia to subdivide those plutons derived from igneous sources and those derived by melting of clay-rich sedimentary rocks. Oxygen and hydrogen isotopes have also been used to study the genesis of hydrothermal ores, and the related wall-rock alteration, concentrating largely on the source of associated water.

Carbon isotopes of carbonates and fluids in igneous rocks (or of diamonds and graphite in some) yield important information concerning the source and alteration of the rocks. $^{13}C/^{12}C$ ratios for most carbonates in typical igneous rocks, for example, indicate that the carbonate is nonmagmatic and is generated by circulating hydrothermal fluids. Carbonatites (Section 19.2) are a rare exception. The low-$^{13}C/^{12}C$ signature of many hydrothermal ores, on the other hand, indicates a deep-seated source for most of the carbon in them.

The stable isotopic composition of metamorphic rocks is also a good indicator of the pre-metamorphic parent rock. It can also indicate the nature of the fluids present during metamorphism, and the extent of fluid–rock interaction.

Isotopic exchange between minerals, or between minerals and intergranular fluids, is temperature dependent. In principle, one can use stable isotopes to estimate equilibration temperatures of minerals, but we should be aware of re-equilibration and alteration, which can readily upset the original equilibrium values.

Because the efficiency of mass fractionation is a function of the mass difference divided by the total mass, stable isotope studies are usually limited to sulfur and lighter elements. For a further discussion of the many petrological applications of stable isotopes, see the suggested readings at the end of the chapter.

9.7.2 Radioactive and Radiogenic Isotopes

The isotopic ratios of elements with radioactive (parent) or radiogenic (daughter) isotopes among their isotopic components can be affected by mass fractionation processes, if they are light enough, but also have isotopic ratios that vary with time. For heavy elements the mass fractionation effects are insignificant, so we shall concentrate on these, and focus on the way that the isotopic ratios reflect the time-dependent decay processes.

Isotopic ratios of elements that include *radioactive* isotopes reflect losses of the unstable isotopes due to decay. Variations in the isotopic ratio of elements with *radiogenic* isotopes are more complex. Their isotopic ratios reflect additions of radiogenic isotopes due to the gradual radioactive decay of a parent nuclide, but the parent is a different element. In the case of radiogenic isotopes, then, the isotopic ratio of an element depends upon the *ratio* of the parent to the daughter element, which is usually the result of *chemical* fractionation during some earlier process, such as partial melting, fractional crystallization, etc. For example, assume that some parent element ("Pn") has an isotope that decays to a daughter element ("Dr") and that Dr has 4 isotopes, one of which (Dr*) is radiogenic. Suppose that after a specific time, half of the unstable isotopes of Pn decay to radiogenic isotopes of Dr*. Clearly, the more initial Pn in the sample, the more radiogenic Dr* produced, resulting in a greater percentage of Dr* among the isotopes of Dr. Also, if there is very little Dr in the original sample, the greater will be the effect of the addition of a fixed quantity of Dr* isotope on the overall Dr isotopic ratios. In other words, the addition of 100 atoms of Dr* will have a minor effect on the isotopic ratio of 1 million preexisting Dr atoms, but it will have a profound effect if there were initially only 10 Dr atoms.

Figure 9.3 provides an example of this effect. Suppose we begin with a solid rock with about equal amounts or Rb and Sr. If we partially melt this rock, with only 10% partial melting ($F = 0.1$), the Rb content of the partial melt will be several times the Sr content. This is *chemical* fractionation. In Section 9.7.2.2 we shall learn that ^{87}Rb decays to ^{87}Sr over time. If this 10% partial melt cools rapidly to a volcanic rock with little chance for further fractionation of Rb versus Sr, the ^{87}Sr isotope produced by decay from the large quantity of Rb will later constitute a significant proportion of Sr (which was initially present in smaller quantities). Now suppose another magma was produced at about the same time as the one described above, but this one by 50% partial melting ($F = 0.5$ in Figure 9.3); there would be much less Rb in the resulting volcanic rock, and the proportion of ^{87}Sr versus other Sr isotopes with time would be less than for the 10% melt.

According to the theory of radioactive decay (Rutherford and Soddy, 1903), the rate of decay of an unstable parent atom in a sample at any time (t) is proportional to the number of parent atoms existing at that time. Mathematically, this means:

$$-\frac{dN}{dt} \propto N \quad \text{or} \quad -\frac{dN}{dt} = \lambda N \qquad (9.11)$$

where:

> N = number of parent atoms
>
> t = time
>
> λ = proportionality constant, called the **decay constant**, an empirical constant specific to a particular isotopic system (α means "is proportional to")
>
> dN/dt = change in N with respect to time, or the rate of decay at any particular time

We can rearrange and integrate this equation (see Faure, 1986) to derive:

$$N/N_o = e^{-\lambda t} \qquad (9.12)$$

where:

> N_o = original number of atoms of the radioactive nuclide
>
> N = number after some time t (in years) has passed

If the rate of decay is proportional to the number of parent atoms remaining in the sample at any time, it follows that a constant proportion of the remaining parent atoms decay in a fixed time period. Half of the unstable atoms in a sample thus decay during a specific time interval that is some function of λ, and half of the remaining atoms decay in the next time interval of the same length, etc. We can define the **half-life** ($T_{1/2}$) as the time required for half of the unstable atoms to decay. If we begin with N_o unstable atoms, after one half-life, $\frac{1}{2}N_o$ will remain. $\frac{1}{4}N°$ atoms will remain after two half-lives, then $\frac{1}{8}$, $\frac{1}{16}$, etc. We can substitute $T_{1/2}$ into Equation (9.12) to solve for the relationship between $T_{1/2}$ and λ:

$$\frac{1}{2}N_o = N_o e^{-\lambda T_{1/2}}$$
$$\frac{1}{2} = e^{-\lambda T_{1/2}}$$
$$\ln\left(\tfrac{1}{2}\right) = -\lambda T_{1/2} \qquad (9.13)$$
$$\ln 2 = \lambda T_{1/2}$$
$$T_{1/2} = \ln 2/\lambda = 0.693/\lambda$$

If we know N, the number of atoms of the radioactive parent isotope in a rock at the present time, N_o, the original number of such atoms, and λ, we can use Equation (9.12) to solve for t, the age of the rock (the number of years that has passed since the decay process began). It is impossible, however, to measure both N and N_o in a modern rock. We can only measure N, the amount in the sample today. We can, however, determine N_o by adding D^*, the radiogenic daughter isotope produced, to N, the remaining parent isotope. Because:

$$D^* = N_o - N \qquad (9.14)$$

then we can substitute ($D^* + N$) for N_o in Equation (9.12) to get:

$$D^* = Ne^{\lambda t} - N = N(e^{\lambda t} - 1) \qquad (9.15)$$

We thus need to know the amount of the parent remaining, the amount of the daughter produced, and λ in order to determine the age of the rock. The difficult part is distinguishing the radiogenic daughter isotopes from any of the same daughter-type isotopes that are not a product of radioactive decay since the time of the geologic event we are attempting to date.

Most radiometric dating techniques are practical only over a certain range of ages. The limitation results from a combination of analytical accuracy and statistical validity. The lower age limit is imposed by the difficulty in accurately determining the tiny amounts of the daughter produced and the validity of the decay equation for so few product atoms. As an example of the statistical problem, consider the following. If we begin with 10^8 parent atoms, and the half-life is 100 years, it is statistically reasonable to assume that, after 100 years, there will be $5 \cdot 10^7$ atoms left (at least within our measuring accuracy). If, on the other hand, there are only two initial atoms, exactly when will each decay, and how sure can we be that only one will be around in 100 years? We need a statistically valid sample and, thus, a certain critical number of daughter isotopes. The practical upper age limit is imposed for the same reason on the few remaining parent atoms. A good example of this is the ^{14}C dating technique. Because the half-life of ^{14}C is only 5730 years, very little ^{14}C remains after 100,000 years, and decay becomes sporadic. The technique is thus limited to dating Holocene events and archaeology.

As a result of the variations in decay rates, element concentrations, and daughter stabilities, different isotopic systems are used, depending on the rocks and ages involved. Examples of the most common petrological isotopic systems follow.

9.7.2.1 THE K-Ar SYSTEM We begin with the potassium-argon system because it is (conceptually, at least) the most straightforward. The method is based on the branching decay of ^{40}K to either ^{40}Ca or ^{40}Ar. Because ^{40}Ca is so common, we run into the problem just mentioned of distinguishing radiogenic ^{40}Ca from nonradiogenic ^{40}Ca. ^{40}Ar, on the other hand, is an inert gas. Although Ar can be trapped in many solid phases, virtually all of the Ar escapes when a rock is sufficiently hot. The escape of Ar effectively "resets the radiometric clock" because all of the daughter is removed, and the remaining ^{40}K produces ^{40}Ar again from a fresh start. Thus when magmas form and crystallize, the clock is reset, and any ^{40}Ar in the rock after this time must be the daughter from the decay of ^{40}K.

The equation for the K $\rightarrow$ Ar decay is slightly complicated because it is a two-step process, involving first electron capture and then positron decay. The equation is:

$$^{40}\text{Ar} = {}^{40}\text{Ar}_o + \left(\frac{\lambda_e}{\lambda}\right) {}^{40}\text{K}(e^{\lambda t} - 1) \quad (9.16)$$

λ_e is the decay constant for the electron capture process ($\lambda_e = 0.581 \cdot 10^{-10} \, a^{-1}$), and λ is the decay constant for the entire decay process ($\lambda = 5.543 \cdot 10^{-10} \, a^{-1}$). Normally, $^{40}\text{Ar}_o$ will be zero because all of the Ar escapes during an igneous event, but subsequent ^{40}Ar produced by K decay is trapped in the solidified igneous rock. The attractiveness of this technique is that a single rock sample provides a value for ^{40}Ar and ^{40}K and, thus, an age. For this reason K-Ar has been a popular technique, producing many "dates" in a relatively simple and straightforward process (call it a computer dating service).

Recently we have realized that some complications affect this simple system. The release of Ar is a complex process, and slowly cooled rocks may not cease to release Ar until well after their initial crystallization. The temperature below which a particular mineral will no longer release Ar is called the **blocking temperature** for that mineral. If all of the ^{40}Ar were quickly removed soon after an igneous rock formed, and none was leaked later or added subsequently, and similarly no ^{40}K was introduced or removed, the calculated K-Ar age from Equation (9.16) should be an accurate measure of the time since crystallization of the rock. However, K and Ar are both mobile elements. Subsequent metamorphism may leach or add K or release Ar.

Blocking temperatures vary for different minerals. For amphiboles, it is in the vicinity of 600°C, for micas closer to 300°C, and for apatite as low as 100°C. The simple one-rock $\rightarrow$ one-age relation thus becomes less than a simple and reliable process. These considerations do not render the K-Ar technique useless. On the contrary, the K-Ar dates from Equation (9.16) may be reliable for some rock types, such as unmetamorphosed volcanics. In addition, if the "true" age of a K-Ar-upset rock can be determined based on another isotopic system, the K-Ar age may provide additional information, such as the age of a later metamorphic event that released Ar but did not reset the other systems.

An extension of the K-Ar technique that builds on the notion of delayed Ar release is the **^{40}Ar-^{39}Ar method**. To analyze a K-bearing mineral, such as mica or amphibole, it is irradiated in a nuclear reactor, in which neutron bombardment converts some of the nonradioactive ^{39}K to ^{39}Ar. Then the sample is heated incrementally in a vacuum, and the ^{40}Ar/^{39}Ar ratio determined at each step as various minerals release Ar gas. Because the ^{39}Ar content is proportional to ^{39}K, and ^{39}K is proportional to the *original* ^{40}K (because the isotopes did not mass fractionate when the mineral formed, and the ^{39}K has been in the mineral since that time), the ^{40}Ar/^{39}Ar ratio can be used to calculate the ^{40}Ar/^{40}K ratio and, hence, the age.

In the geological environment, Ar escapes more readily from some crystallographic sites than from others, and it must diffuse from the interior of a crystal to the surface before it can escape along intergranular boundaries. Figure 9.10a

illustrates the theoretical evolution of radiogenic ^{40}Ar* in a spherical K-bearing mineral grain in a rock that experienced a single episode of Ar loss at some intermediate period during its history. When the rock first crystallized from a melt, there was no ^{40}Ar*, but after time passes, the ^{40}Ar*/^{40}K ratio gradually increases uniformly throughout the grain. The dashed curve x in the figure represents the uniform ^{40}Ar*/^{40}K at some time following crystallization. During a heating event, Ar diffuses from the mineral grain to the grain surface. Ar is readily lost from the boundary area, but diffusion within the crystal is slow, and a typical diffusion profile develops (curve y), varying from zero ^{40}Ar* at the grain rim to the initial value at some point in the interior. Curve y gets progressively lower with time, but at some point, the thermal event ceases, and curve y is frozen in place when the temperature drops below the blocking temperature for diffusion. In Figure 9.10a, we assume that curve y did not lower to the point at which Ar was lost from the mineral core area. Curve z represents a profile in ^{40}Ar*/^{40}K that would result as more ^{40}Ar* is added to curve y due to continued ^{40}K decay with time following the Ar release event. If Ar is reduced to zero at the grain rim and no Ar is lost

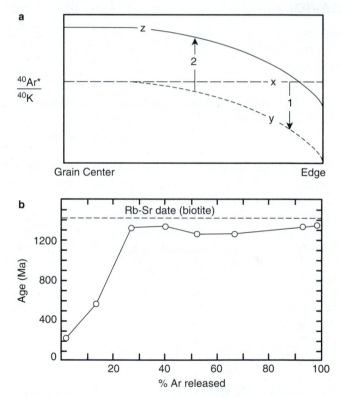

FIGURE 9.10 (a) Illustration of the idealized evolution of radiogenic ^{40}Ar* in a spherical grain experiencing a single event of diffusive Ar loss. Curve x is the original ^{40}Ar/^{40}K at some time after initial mineral formation. Curve y represents partial loss of Ar from the grain due to diffusion during a thermal event. Curve z represents the evolution of curve y by addition of ^{40}Ar from ^{40}K decay after the passing of more time without further Ar loss. After Faure (1986) and Turner (1968). Copyright © reprinted by permission of John Wiley & Sons, Inc., and Elsevier Science, respectively. (b) The spectrum of dates calculated from ^{40}Ar/^{39}Ar ratios of gas released by the incremental heating of biotite from the Precambrian Marble Mountains of southeastern California. After Lanphere and Dalrymple (1971). Copyright © with permission from Elsevier Science.

subsequently, the $^{40}Ar*/^{40}K$ ratio at the rim yields the age of the thermal event, whereas the ratio in the deep interior of the grain yields the overall time that has passed since the rock was formed.

Heating of biotite separated from the sample releases Ar from the margin inward. Ar in the marginal areas is released at low temperature, whereas Ar from progressively deeper into the grain interior is released at incrementally higher temperatures. If the sample is first irradiated to convert some ^{39}K to ^{39}Ar, the $^{39}Ar/^{40}Ar$ ratio of the released Ar gas yields an age. Incremental heating yields constant ^{39}Ar (because ^{39}K was uniformly distributed in the grain), but variable ^{40}Ar that corresponds to the profile of curve z in Figure 9.10a. Figure 9.10b shows the result of incremental heating of biotite separated from Precambrian gneisses of the Marble Mountains in southeastern California. The Rb-Sr age (see below) was determined as 1410 to 1450 Ma, but the conventional K-Ar age was only 1152 ± 30 Ma. Stepwise heating of biotite yielded $^{39}Ar/^{40}Ar$ ages that indicated that Ar was partially lost during a thermal event ~200 Ma ago, and the plateau indicates a grain-interior age of ~1300 Ma. The conventional K-Ar age is a mixture of these events. The age at the plateau is less than the Rb-Sr age, which may reflect loss of Ar from the grain core.

$^{39}Ar/^{40}Ar$ ages are controlled by the mineralogy and thermal history of a rock and can be related to such processes as uplift rates in eroded orogenic belts. For more complete descriptions of the ^{40}Ar-^{39}Ar technique, see Faure (1986), York (1984), and McDougall and Harrison (1999).

9.7.2.2 THE Rb-Sr SYSTEM

One of the most common isotopic systems used by petrologists for both determining age and constraining the source of magmatic rocks is the rubidium–strontium system. The system has the following characteristics:

- ^{87}Sr can be created through the breakdown of ^{87}Rb → ^{87}Sr + a beta particle ($\lambda = 1.42 \cdot 10^{-11} a^{-1}$).
- ^{86}Sr is a stable isotope and not created by breakdown of any other element.
- Rb behaves like K so it concentrates in micas, amphiboles, and, to a lesser extent, K-feldspar.
- Sr behaves like Ca, so it concentrates in plagioclase and apatite (but not clinopyroxene).
- ^{88}Sr:^{87}Sr:^{86}Sr:^{84}Sr in the average natural sample = 10 : 0.7 : 1 : 0.07.

Because ^{87}Sr is stable and is present in all Sr-bearing rocks, the amount of it in any particular rock will reflect the original ^{87}Sr, plus any radiogenic ^{87}Sr added from the decay of ^{87}Rb over time. Thus a single sample cannot provide an unambiguous age because it is impossible to distinguish the radiogenic and nonradiogenic components of the daughter isotope. This problem is common for isotopic systems, and K-Ar is a fairly unique exception because of the inert-gas nature of the daughter, allowing it to escape during thermal or magmatic events. The problem is overcome in an ingenious fashion by using the **isochron** technique, which uses two or more samples and normalizes the isotopes that vary with

time to ^{86}Sr, which is neither radiogenic nor radioactive and is thus constant.

For example, the value of $^{87}Sr/^{86}Sr$ in a sample at the present time is equal to the original $^{87}Sr/^{86}Sr$ ratio of the sample at the time it first crystallized, $(^{87}Sr/^{86}Sr)_o$, plus the radiogenic ^{87}Sr formed since that time. Once again, this latter amount is determined by the Rb concentration in the original sample and time. It is important to realize that these heavy Sr isotopes do not *mass fractionate* during melting or crystallization, but Sr and Rb may *chemically fractionate* during these processes.

When applied to the Rb/Sr system, Equation (9.15) can be recast by dividing through by the constant ^{86}Sr to get:

$$^{87}Sr/^{86}Sr = (^{87}Sr/^{86}Sr)_o + (^{87}Rb/^{86}Sr)(e^{\lambda t} - 1) \tag{9.17}$$

λ for the breakdown of Rb to Sr is $1.42 \cdot 10^{-11} a^{-1}$. For values of λt less than 0.1, $(e^{\lambda t} - 1) \cong \lambda t$. Thus Equation (9.15), for ages less than 70 Ga (which certainly ought to cover terrestrial rocks) reduces to:

$$^{87}Sr/^{86}Sr = (^{87}Sr/^{86}Sr)_o + (^{87}Rb/^{86}Sr)\lambda t \tag{9.18}$$

This is the equation for a straight line ($y = b + mx$) in a $^{87}Rb/^{86}Sr$ versus $^{87}Sr/^{86}Sr$ plot, as illustrated in Figure 9.11. In this figure, three straight lines (**isochrons**) represent three different times: t_0, t_1, and t_2. Consider first the horizontal line, t_0. The three dots a, b, and c on this line represent the $^{87}Sr/^{86}Sr$ and $^{87}Rb/^{86}Sr$ analyses of either three minerals in a single rock or three cogenetic rocks with a spread of Rb and Sr concentrations. Let's assume that the samples are rocks resulting from partial melting of some uniform source (perhaps in the mantle). Remember from Figure 9.3 that it is possible, by varying F (the fraction of the source being melted), to produce melts with different Rb/Sr ratios from the same source. A similar spectrum of Rb/Sr ratios could also be generated in a series of late liquids due to fractional crystallization. We can use Equation (9.3) and the decay constant ($\lambda = 1.42 \cdot 10^{-11} a^{-1}$) to derive the half-life ($4.95 \cdot 10^{10} a$) for Rb decay. Therefore, any melting, rise, and crystallization processes occurring within a few hundred thousand years can be considered "instantaneous" and need not affect the subsequent development of the isotopic system.

Because Sr isotopes don't mass fractionate, the $^{87}Sr/^{86}Sr$ values will be the same for all three samples at the time of separation from the mantle and crystallization, regardless of the process(es) involved. The line joining the three dots is thus horizontal at t_0 (the time of formation). In other words, t_0 is the time at which the rocks or minerals crystallized and began to act as independent isotopic systems.

Following t_0, ^{87}Rb in each sample continuously breaks down to form ^{87}Sr. ^{87}Rb thus decreases as ^{87}Sr increases. Naturally, the more Rb in the rock, the more that is lost in a particular time period, and the more ^{87}Sr created. Thus the three points at t_0 move in the direction of the arrows to the

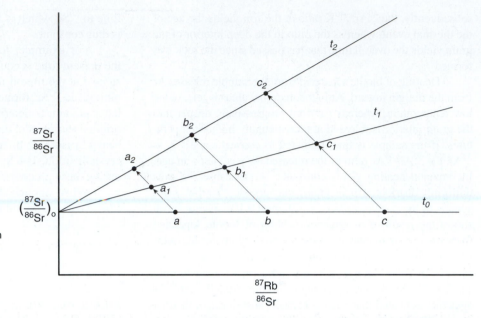

FIGURE 9.11 Schematic Rb-Sr isochron diagram showing the isotopic evolution over time of three rocks or minerals (*a*, *b*, and *c*) with different Rb/Sr ratios after their derivation from a homogeneous source at time t_o.

next set of points, a_1, b_1, and c_1, at time t_1. Because the breakdown rate of Rb is constant, the three points continue to be collinear but define a new line, or isochron, with a positive slope. Such an isochron can give us *two* pieces of information. First, Equation (9.18) tells us that the slope equals λt, and by knowing λ, we can calculate the age of the rock ($t_1 - t_0$) at time t_1. Second, the line connecting the three dots at t_1 can be extrapolated to zero ^{87}Rb. Naturally, if ^{87}Rb = 0, no new ^{87}Sr will be created over time, so the ^{87}Rb = 0 intercept will be the original ^{87}Sr/^{86}Sr ratio, or (^{87}Sr/^{86}Sr)$_o$ of all three rocks, *as well as that of the solid source of the melts at the time of their separation from it.* After another time interval (at t_2), the three samples plot along a new line, t_2, at points a_2, b_2, and c_2. The new slope yields the age ($t_2 - t_0$) at that time. The line t_2 is steeper, reflecting the higher value of t in λt, the slope. Extrapolating the line back to ^{87}Rb = 0 yields the same value for (^{87}Sr/^{86}Sr)$_o$.

Only two samples are required to produce an isochron because two points define a line. It is common practice, however, to use at least three widely spaced points, and more if possible, because the degree of fit that several points make to a line is considered an indication of the accuracy of the results.

Figure 9.12 is an Rb-Sr isochron for analyzed whole-rock and hornblende separates from the Eagle Creek Pluton, part of the Mesozoic Sierra Nevada Batholith of California. The data fall on a good linear isochron that yields a slope of 0.00127. Using Equation (9.18) with the slope = λt, the age = slope/λ = 0.00127/1.4 $\cdot$ 10^{-11} = 91 Ma.

The isochron intercept also gives us the initial ^{87}Sr/^{86}Sr ratio, (^{87}Sr/^{86}Sr)$_o$, which can be an excellent petrogenetic tracer. The reason for this may become evident if we turn to Figure 9.13, which illustrates a simplified model for the long-term Sr isotopic evolution of the upper mantle, beginning with the primordial Earth (modeled on chondritic meteorites) back some 4.6 Ga before present (b.p.). The ^{87}Sr/^{86}Sr ratio of the mantle at that early time was about 0.699 (based on extrapolation back in time from present

chondritic meteorite values). Since that time, the ^{87}Sr/^{86}Sr ratio of the upper mantle has slowly increased, following the **growth curve** of Figure 9.13, as the small quantity of ^{87}Rb in the mantle decays (assuming an initial Rb/Sr = 0.027, based on chondrites).

If at any time the mantle should be partially melted, and the melts so derived become continental crust, Rb will be preferentially incorporated into the crust (because it is incompatible and acts similarly to potassium). Figure 9.13 shows a rather massive hypothetical melting event about 3.0 Ga b.p. to illustrate the process in a simple way. After that event, the mantle growth curve has a lower slope because Rb has been selectively removed by the melts so that less ^{87}Sr is subsequently generated. The *crustal* growth curve, on the other hand, has a dramatically greater slope because Rb will be greatly concentrated from a large mantle reservoir into a smaller volume of sialic crust (F is low as in Figure 9.3). The crustal rock evolution curve is based on an assumed Rb/Sr

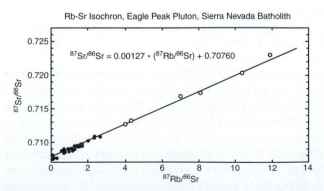

FIGURE 9.12 Rb-Sr isochron for the Eagle Peak Pluton, central Sierra Nevada Batholith, California. Filled circles are whole-rock analyses, and open circles are hornblende separates. From Hill et al. (1988). The regression equation for the data is also given. Reprinted by permission of the *American Journal of Science.*

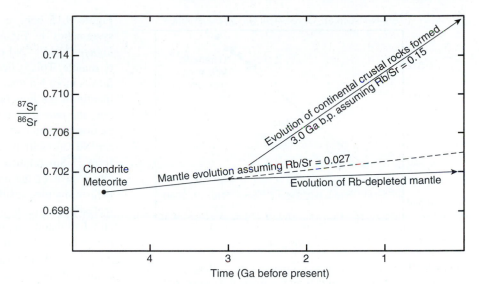

FIGURE 9.13 Estimated Rb and Sr isotopic evolution of the Earth's upper mantle, assuming a large-scale melting event producing granitic-type continental rocks at 3.0 Ga b.p. After Wilson (1989). Copyright © by permission Kluwer Academic Publishers.

of 0.15. The extrapolated original mantle growth curve evolves to a modern mantle with $^{87}Sr/^{86}Sr = 0.704$.

Although Figure 9.13 shows a single crust-forming event, a more likely scenario involves a more continuous process by which melts are derived from the mantle, perhaps punctuated by some larger episodic events. Such a process would result in a series of nearly parallel silicic melt paths and a kinked and curved mantle evolution path, but the final $^{87}Sr/^{86}Sr$ ratio for the mantle, because of the low Rb content, will be close to the values in Figure 9.13. The present-day $^{87}Sr/^{86}Sr$ value of the residual upper mantle is estimated to be about 0.703. Crustal values are higher, but the exact value naturally depends on the Rb/Sr ratio and the age of the rocks in question. The higher the Rb content and age, the higher will be $^{87}Sr/^{86}Sr$.

Perhaps now it is apparent how useful $(^{87}Sr/^{86}Sr)_o$, commonly called the **initial ratio**, can be. With it, we may be able to judge whether an igneous rock or magma is derived by melting of the mantle or of the continental crust. As a general rule, an initial ratio (i.e., corrected for age by extrapolating to $^{87}Rb = 0$) of less than 0.706 indicates that the rock was derived by partial melting of the mantle. The number 0.706 allows for some inhomogeneity and provides a slight margin for error. If the initial ratio is greater than 0.706, the rock may have had a source with higher Rb/Sr (crust), or perhaps it was a mantle-derived magma that was contaminated during transit to the surface by assimilating some very Rb/Sr-rich rocks, such as old granitic continental crust. Note that the intercept for the Eagle Creek Pluton in Figure 9.12 is 0.7076, indicating that there are some crustal components involved. This may not be surprising for a granitic pluton emplaced in thick Proterozoic continental crust, even if the ultimate melt were produced within the mantle. Once again, the intercept approach works because these heavy isotopes do not mass fractionate during melting or crystallization processes, so $^{87}Sr/^{86}Sr$ is the same in the melt and solid portions at the time of separation. To repeat: very low initial ratios probably result from Rb depletion in areas of the mantle that have been extensively melted to extract crustal rocks. Progressively higher ratios indicate greater input from Rb-richer sources, probably old, K-Rb-rich crust.

9.7.2.3 THE Sm-Nd SYSTEM As another example of radiogenic isotopes, we shall look at the samarium–neodymium system. Because both Sm and Nd are LREE, they are incompatible elements and tend to preferentially fractionate into melts. Because Nd has a lower atomic number, it is a little larger than Sm and is thus concentrated slightly more in liquids relative to Sm. As a result, the Sm/Nd ratio *decreases* in partial melts (compared to the source) or in late liquids resulting from progressive fractional crystallization. Among the numerous isotopes of both Sm and Nd, two are related by radioactive decay: $^{147}Sm \rightarrow\ ^{143}Nd$ by alpha decay for which $\lambda = 6.54 \cdot 10^{-12}\ a^{-1}$. An isochron-type decay equation, equivalent to Equation (9.18), can be derived by reference to the nonradiogenic ^{144}Nd isotope:

$$^{143}Nd/^{144}Nd = (^{143}Nd/^{144}Nd)_o + (^{147}Sm/^{144}Nd)\lambda t$$

$$(9.19)$$

Once again, the λt approximation for $(e^{-\lambda t} - 1)$ is reasonable for ages less than $1.5 \cdot 10^{12}$ years. Because Sm and Nd are separated on the periodic table by only promethium (which does not occur in nature), the fractionation between them is minor, and Sm-Nd isotopic work requires extremely precise and accurate analysis to be useful and reliable.

Figure 9.14 is an example of an Sm-Nd isochron for whole-rock analyses of volcanics from the Archean Onverwacht Group, South Africa. The data define a good linear trend with a slope of 0.02135, which equals λt, yielding an age of 3.54 Ga. The intercept gives an initial $(^{143}Nd/^{144}Nd)_o$ ratio of 0.50809.

Figure 9.15 is analogous to Figure 9.13 and shows the $^{143}Nd/^{144}Nd$ evolution of the upper mantle over time. CHUR (chondrite uniform reservoir) is an estimate of the average chondrite composition by DePaolo and Wasserburg (1976).

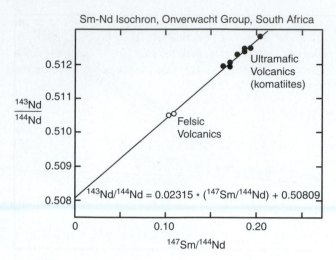

Sm-Nd Isochron, Onverwacht Group, South Africa

$^{143}Nd/^{144}Nd = 0.02315 * (^{147}Sm/^{144}Nd) + 0.50809$

FIGURE 9.14 Sm-Nd isochron for whole-rock analyses of ultramafic and felsic volcanics from the Archean Onverwacht Group of South Africa. Age calculated as 3.540 Ga ± 30 Ma. After Hamilton et al. (1979).

From the chondrite model, the CHUR bulk Earth line shows the $^{143}Nd/^{144}Nd$ evolution of the mantle if it were a closed system. Because ^{144}Nd is not a radiogenic product, it is constant over time. As ^{147}Sm breaks down to ^{143}Nd, the $^{143}Nd/^{144}Nd$ ratio gradually increases, just as $^{87}Sr/^{86}Sr$ did in Figure 9.13. The CHUR bulk Earth line is derived by applying Equation (9.19) to CHUR using an estimate of the present-day value of $^{143}Nd/^{144}Nd = 0.512638$ (I told you this stuff was precise) and $^{147}Sm/^{144}Nd = 0.1967$ in chondrite. The CHUR bulk Earth line is thus equal to $0.0512628 - 0.1967 \cdot (e^{\lambda t} - 1)$.

Next, we postulate a partial melting event at 3 Ga b.p. (point *a* in Figure 9.15), as we did in the Rb-Sr system in Figure 9.13. An alternative to partial melting would be a mantle-enrichment event. As we shall see in Chapters 15 and 19, there is evidence that such an enriched mantle is present in some locations, particularly beneath the continents. Qualitatively at least, partial melts (eventually becoming incorporated in the crust) and enriched mantle behave similarly. The enriched and depleted portions in

Figure 9.15 behave in an opposite manner to the Rb-Sr systematics in Figure 9.13. The depleted mantle shows *higher* $^{143}Nd/^{144}Nd$ ratios with time than the enriched melt or mantle. This is because partial melting of the mantle removes more Nd than Sm (as noted above). This depletes the mantle in the *daughter* isotope, whereas in the Rb-Sr system, the *parent* is depleted. As a result of the daughter depletion in partially melted areas of the mantle, the higher Sm/Nd ratio will, over time, generate more radiogenic ^{143}Nd from ^{147}Sm relative to the original ratio of $^{143}Nd/^{144}Nd$. Other areas of the mantle may become enriched (either by melts remaining resident and locally concentrating Nd or by metasomatic fluids concentrating and depositing Nd). Such enriched mantle areas (if they exist), or melts derived from the mantle, follow the enriched trend in Figure 9.15, reflecting a lower Sm/Nd ratio. The low Sm/Nd results in the generation of smaller quantities of ^{143}Nd and has a diminished effect on the larger amount of Nd initially in the system. Naturally, basalts derived from the mantle have the same $^{143}Nd/^{144}Nd$ as the mantle source at the time of partial melting because Nd does not mass fractionate during melting or crystallization processes. Equation 9.19 thus becomes:

$$\left(\frac{^{143}Nd}{^{144}Nd}\right)_{CHUR,t} = \left(\frac{^{143}Nd}{^{144}Nd}\right)_{CHUR,today} \quad (9.20)$$
$$- \left(\frac{^{147}Sm}{^{144}Nd}\right)_{CHUR,today} (e^{\lambda t} - 1)$$

Because the differences in Nd isotope ratios are small, DePaolo and Wasserburg (1976) introduced the term ε (epsilon) to express the degree of Nd enrichment. ε is defined as:

$$\varepsilon_{Nd} = \left[\frac{(^{143}Nd/^{144}Nd)_{initial}}{I^t_{CHUR}} - 1\right] \times 10,000 \quad (9.21)$$

where:

I^t_{CHUR} = the $^{143}Nd/^{144}Nd$ ratio for CHUR *at the time* (t) *of formation of the rock*

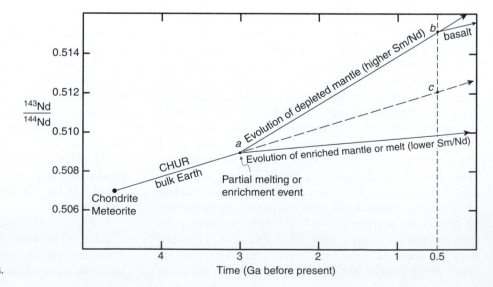

FIGURE 9.15 Estimated Nd isotopic evolution of the Earth's upper mantle, assuming a large-scale melting or enrichment event at 3.0 Ga b.p. After Wilson (1989). Copyright © by permission Kluwer Academic Publishers.

A positive ε_{Nd} value is depleted (higher ^{143}Nd), and a negative ε_{Nd} is enriched (lower ^{143}Nd), both with respect to the CHUR standard (corrected for time along the CHUR bulk Earth line in Figure 9.15). Thus a positive ε_{Nd} for a rock implies that it was derived from a depleted mantle source, and a negative ε_{Nd} indicates that the rock was derived from either an enriched mantle or a crustal source enriched over time. A similar calculation can be used to derive ε_{Sr} in the Rb-Sr system.

For example, consider a basalt derived from the depleted mantle at point b in Figure 9.15, equivalent to 500 Ma b.p. This enriched basalt evolves along the line labeled "basalt" in the figure. If we had several samples with a range of Sm/Nd ratios, we could derive an isochron and use it to derive the age and $(^{143}\text{Nd}/^{144}\text{Nd})_{initial}$ for this basalt (point b). We could then compare this to I^t_{CHUR} at the time of formation of the basalt (point c on the CHUR bulk Earth evolution curve). From this we can use Equation (9.20) to determine $\varepsilon_{Nd} = (0.515/0.512 - 1) \cdot 10^4 = 5.86$, a positive value, supporting our model that the rock was derived from a depleted source. As a more quantitative example, consider the volcanics of the Onverwacht Group in Figure 9.14. $(^{143}\text{Nd}/^{144}\text{Nd})_o$ is the intercept of the regression line, = 0.50809. I^t_{CHUR} at 3.54 Ga can be calculated from Equation (9.20) by substituting $t = 3.54 \cdot 10^9$, yielding 0.508031. Substituting this value into Equation (9.21) yields $\varepsilon_{Nd} = [(0.50809/0.508031) - 1] \cdot 10,000 = 1.16$, suggesting a slightly depleted mantle source.

9.7.2.4 U-Th-Pb SYSTEM
Our final example of isotopic systems is the uranium–thorium–lead system. This system is complex, involving three radioactive isotopes of U (^{234}U, ^{235}U, and ^{238}U) and three radiogenic isotopes of Pb (^{206}Pb, ^{207}Pb, and ^{208}Pb). Naturally-occurring uranium today is 99.2745% ^{238}U, 0.720% ^{235}U, and 0.0055% ^{234}U. Only ^{204}Pb is strictly nonradiogenic. U, Th, and Pb are all incompatible

elements, and they concentrate in early melts to become incorporated in the crust (particularly the continental crust). In addition to any original lead, the isotopic composition of Pb in rocks is a function of three decay reactions involving the breakdown of U and Th to Pb:

$$^{238}\text{U} \rightarrow {}^{234}\text{U} \rightarrow {}^{206}\text{Pb} \; (\lambda = 1.5512 \times 10^{-10} a^{-1}) \quad (9.22)$$

$$^{235}\text{U} \rightarrow {}^{207}\text{Pb} \; (\lambda = 9.8485 \times 10^{-10} a^{-1}) \quad (9.23)$$

$$^{232}\text{Th} \rightarrow {}^{208}\text{Pb} \; (\lambda = 4.9475 \times 10^{-11} a^{-1}) \quad (9.24)$$

With three concomitant decay schemes, the U-Pb-Th system can be rather complex. Each system can be treated independently, using the standard isochron technique. A common alternative is to treat Equations (9.22) and (9.23) simultaneously. Figure 9.16 illustrates the ^{206}Pb and ^{207}Pb isotopic development of a hypothetical Precambrian rock. Figure 9.16a shows the development of the Pb system for the first 2.5 Ga of the rock's history. If radiogenic ^{206}Pb (^{206}Pb*) and radiogenic ^{207}Pb* both evolve in unison via Equations (9.22) and (9.23), the isotopes (when standardized by dividing by their parent concentrations) follow the curve shown, called the **concordia**. All natural samples with coherent U-Pb systems must develop along this concordia curve. Because of its smaller decay constant, ^{235}U decays faster, so ^{207}Pb*/^{235}U is always larger than ^{206}Pb*/^{238}U at any time, and the difference gets greater with time, resulting in the characteristic concave downward shape of the concordia.

Suppose that after 2.5 Ga of evolution, the package of rock is disturbed by some event that causes some Pb, a mobile element, to be lost. This may be a melting event, a thermal event such as metamorphism, or infiltration by fluids that preferentially scavenge LIL elements, including Pb. Because Pb isotopes do not mass fractionate during depletion, all of

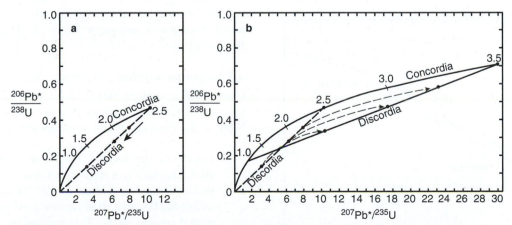

FIGURE 9.16 Concordia diagram illustrating the Pb isotopic development of a 3.5-Ga-old rock with a single episode of Pb loss. **(a)** Radiogenic ^{206}Pb* and ^{207}Pb* evolve simultaneously along the *concordia* curve for the first 2.5 Ga, at which time a thermal or fluid infiltration event causes lead to be lost. Both isotopes of Pb are lost in the proportions in which they exist in the rock at the time, so that the isotopic compositions of the depleted rocks trend along the *discordia* directly toward the origin (arrow). The filled circles represent hypothetical rocks with variable degrees of depletion due to the event. **(b)** Continued evolution of the Pb system for a further 1 Ga causes undepleted rocks to follow the concordia for a total of 3.5 Ga of evolution. Depleted rocks follow separate concordia-type curves (dashed) to the new positions shown. The final discordia intersects the undepleted concordia at two points, one yielding the total age of the rocks, and the other yielding the age of the depletion event. After Faure (1986).

the Pb isotopes are depleted in proportion to their concentrations in the rock. As a result, depletion causes the Pb isotopic system to move directly toward the origin from the 2.5 Ga point on the concordia, along a line called the **discordia**.

Let's assume that we have four different rocks, or four different grains of U-Pb-bearing zircons in a single rock, and that each of the four becomes depleted in Pb to differing degrees. One rock (or zircon) is not depleted at all, and it remains on the concordia, whereas the others move to three different points along the discordia, represented by the dots.

Following the depletion event, the system continues to evolve for another 1 Ga (Figure 9.16b). The undepleted rock or zircon continues to follow the concordia for a total of 3.5 Ga. The other three samples follow their own concordia-type curves (dotted) from their point of origin on the 2.5 Ga discordia. After 1 Ga, they are still collinear and define a new discordia. The final discordia intersects the concordia at two points. The one on the right intersects the concordia at a point that yields the total age of the system (3.5 Ga). The intersection on the left is at 1.0 Ga, the age of the depletion event.

Figure 9.17 is a concordia diagram for three zircons from some Archean gneisses in Minnesota. The three points define a good linear discordia. When the (universal) concordia curve is added to the diagram, the discordia intersects the concordia at 3.55 Ga (the U-Pb age of the granite) and 1.85 Ga (the age of the Pb-loss episode).

Analyzing single zircon grains isotopically is a difficult procedure. Zircons with different histories may be found in the same rock (see Section 18.1) and can generally be recognized by differences in color, morphology, or crystallinity. Aggregates of similar zircons can be analyzed by conventional mass-spectrographic means if enough of the material can be accumulated. A newer method utilizes the **ion microprobe**, similar to the electron microprobe, but it bombards the sample with a focused beam of oxygen ions instead of electrons. The ions blast a tiny crater in the sample, sending ions of the sample into a mass spectrometer, which analyzes the liberated material. An ion microprobe can analyze tiny areas within single zircon crystals, so that individual crystals, or even zoned zircon growth, can be studied. We will return to the dating of single crystals and events in Section 23.7.

Due to the mobility of both U and Pb, systematics in the U-Pb system are complex. We shall not explore the many intricate details of this system, but will return to it in when needed in several subsequent chapters. For more information, see Faure (1986).

You might be interested to know the record-holders for oldest crustal rocks at the time of this printing. The oldest dated rocks are Acasta gneisses, from the Acasta River area of the Slave Craton, Northwest Territories, Canada. They were initially dated via ion microprobe U-Pb dating of zircons within tonalitic gneisses at 3.96 Ga (Bowring et al., 1989a). Three Acasta samples yielded Sm-Nd ages of 3.85, 3.92, and 4.1 Ga (Bowring et al., 1989b). Detrital zircons from the Jack Hills conglomerate of the Archean Yilgarn Craton, Western Australia (see front cover), yielded an ion microprobe U-Pb age of 4.40 Ga (Wilde et al., 2001), only ~150 Ma younger than the estimated age of the Earth! Because the zircons are detrital, they predate the conglomerate, but they imply that there was some very old, presumably continental crustal rock being eroded nearby.

Care must be taken in all of the above systems that they are not upset by later metamorphic, metasomatic, or alteration events. The usual criterion for the reliability of a date is the approximation that the data points make to a linear fit to the isochron. The errors reported with an age determination are the statistical errors of the analysis and do not reflect any systematic errors imposed by metamorphism or alteration. It is indeed possible to upset an isotopic system so that the results are still linear, producing an "errorchron" and a similarly erroneous initial ratio. For a discussion of the possible upsets in isotopic systems, see Faure (1986) and Bridgwater et al. (1989).

9.7.2.5 OTHER ISOTOPIC SYSTEMS

There are a number of other useful isotopic systems. All are useful for geochronology, and some have uses as petrogenetic tracers as well. Some of the most commonly used systems in petrology are:

- **Lu-Hf** Lu (parent) is a heavy rare earth element, and Hf (daughter) is a dispersed Group IV element (replacing Zr and, to a lesser extent, Ti). Lu-Hf is similar to Sm-Nd because Hf is concentrated more than Lu in partial mantle melts. This method has been used to date rocks and study mantle differentiation.
- **Re-Os** Re and Os are both chalcophile and have been used to date Mo-Cu-sulfide minerals in hydrothermal ore deposits. In silicate systems, Re (parent) is highly incompatible during mantle melting, whereas Os is compatible. The small quantity of Os in the melt (and eventually in the crust) is thus very sensitive to radiogenic ^{187}Os addition over time. $^{187}Os/^{186}Os$ has been used in oceanic

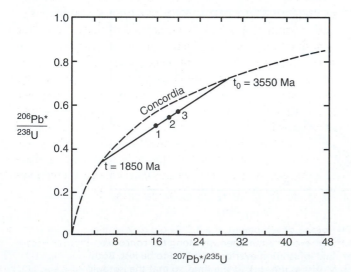

FIGURE 9.17 Concordia diagram for three discordant zircons separated from an Archean gneiss at Morton and Granite Falls, Minnesota. The discordia intersects the concordia at 3.55 Ga, yielding the U-Pb age of the gneiss, and at 1.85 Ga, yielding the U-Pb age of the depletion event. From Faure (1986). Copyright © reprinted by permission of John Wiley & Sons, Inc.

basalts to infer the involvement of subducted crustal material or contamination by the core (Chapter 14). Os ratios in continental ultramafic xenoliths may provide a clearer view of the timing and nature of melt extraction from the continental lithosphere and its relation to continent formation (e.g., Walker et al., 1989).

- **Hf-W** Hf (parent) is lithophile, whereas W is siderophile, so the Hf remained in the mantle while W was preferentially incorporated into the core during early Earth differentiation. ^{182}Hf produced ^{182}W with a very short (9 Ma) half-life. ^{182}W/^{183}W ratios in the mantle and in meteorites have been used to date core formation in the Earth and planetary bodies. Recent data (Yin et al.,

2002; Kleine et al., 2002) suggest that core separation in the terrestrial planets took place within 30 Ma of the birth of the solar system (before all the ^{182}Hf became extinct).

- **U/Th-He** Decay of ^{235}U, ^{238}U, and ^{232}Th produce ^{4}He, which diffuses readily in association with shallow mantle melting. High ^{3}He/^{4}He ratios are usually correlated with deeper, less outgassed mantle (more primordial ^{3}He).

As we survey the tectonic–igneous associations in later chapters, we will continue to explore the uses of trace elements and isotopes as we turn to them for help in constraining the source of melts as well as any processes that may have modified them prior to final solidification.

Summary

Due to their low concentrations, *trace elements* only rarely create separate mineral phases. They typically distribute themselves as impurities in all phases present, where they generally substitute for major elements with similar ionic radius and valence (most likely in the same group on the periodic table). The distribution of any particular trace element between coexisting phases at equilibrium is generally unequal, often highly so, and occurs in fixed, reproducible proportions. The *distribution* (or *partition*) *coefficient* (D) is simply equal to C_S/C_L, the ratio of the measured concentrations in a solid and the liquid with which it equilibrated. Because trace elements are so ubiquitous, yet selective, their concentrations can be much more sensitive than major elements to fractionation processes such as partial melting or crystallization. Trace elements may thus be very good indicators of the source of a melt, the extent of partial melting or fractional crystallization, or even the participation of a particular mineral in these processes. We can mathematically model the variation in concentration of any element in a liquid (C_L) versus the original concentration in that liquid (C_o) as a function of the fraction of melt that has formed (F), using various assumptions (typically the amount of liquid generated before removal). Crystallization may be modeled as the reverse process.

The rare earth elements (lanthanides) are a particularly useful group of trace elements. Because they have the same valence (in most cases) and their compatibility increases gradually with atomic number, they behave similarly, but not identically, to one another. REE are typically plotted on normalized variation diagrams, in which the concentration of each REE determined is divided by a reference standard

(e.g., chondritic concentrations) and plotted along the x-axis from left to right, in order of increasing atomic number (and hence compatibility). REE plots of partial melts of a primordial source would thus have a negative slope. Slopes and other patterns on normalized REE diagrams are used regularly for interpreting the history of magmatic systems. The normalization technique has spawned a plethora of "normalized multielement" ("spider") diagrams, extending the range of useful trace elements. Trace element ratios (typically as plotted on bivariate or ternary diagrams) have also been used empirically to infer the genetic environment of igneous rocks no longer found in their original setting.

Different *isotopes* of the same element cannot fractionate chemically, but those of the lighter elements can mass fractionate during some natural processes, such as evaporation, crystallization, weathering, etc. Geologists use such *stable isotope* fractionation (most commonly ^{18}O versus ^{16}O and ^{3}H versus ^{1}H) to assess the involvement of various reservoirs with distinctively fractionated isotopic ratios (e.g., hydrothermal fluids, seawater, sediments, rocks) in the genesis of other rocks, minerals, or fluids. Radiogenic isotopes are particularly useful to petrologists. Not only can several isotopic systems be employed to determine the age of various events in the genesis of igneous and metamorphic rocks, but the *initial ratios* (the ratio of a radiogenic isotope to a nonradiogenic isotope of the same element extrapolated to zero radiogenic on an isochron diagram)—because the isotopes do not fractionate during melting and crystallization processes—are in many cases the most convincing indicators of the ultimate source material of melts.

Key Terms

Review Questions and Problems

Review Questions and Problems are located on the author's web page at the following address: **http://www.prenhall.com/winter**

Important "First Principle" Concepts

- Trace elements generally act as impurities, distributed in consistent proportions in all phases.
- The concentration of any element in a melt depends upon the nature of the melted source, the percentage of the source melted, the proportions of various minerals that may have fractionally crystallized from the melt at any particular depth, and any assimilation or exchange with wall rocks.
- Trace element distributions between two coexisting phases can be extreme, so the concentrations of certain trace elements can be sensitive indicators of the progress of partial melting or fractional crystallization.
- Highly incompatible elements are strongly concentrated in low-percentage initial partial melts and are diluted with continued melting. The reverse is also true: highly incompatible elements concentrate in late residual liquids upon cooling.
- Trace elements that partition strongly into only a single phase can be sensitive indicators of the fractionation of that phase.
- Trace element concentration *ratios*, if the elements behave differently and are selected wisely, are generally more sensitive to specific processes than is the concentration of any single trace element.
- Trace element patterns can retain a record of the deep source of melts when major elements have lost most of their record due to crystallization or wall–rock exchange upon ascent.
- *Isotopes* can fractionate only on the basis of *mass* (not chemical differences), and only isotopes of S and lighter mass fractionate significantly.
- Variations in isotopic ratios of heavier elements with a radiogenic component result from initial *chemical* fractionation affecting the relative concentration of the parent and daughter elements and daughter production over time.
- Fractionates with high parent/daughter elemental ratios will have much higher radiogenic/nonradiogenic daughter isotope ratios as time passes.
- High parent/daughter elemental ratios can be achieved by parent enrichment, daughter depletion, or both.

Suggested Further Readings

Trace Elements

Cox, K. G., J. D. Bell, and R. J. Pankhurst. (1979). *The Interpretation of Igneous Rocks*. George Allen & Unwin. London.

Ragland, P. C. (1989). *Basic Analytical Petrology*. Oxford University Press. New York.

Rollinson, H. R. (1993). *Using Geochemical Data: Evaluation, Presentation, Interpretation*. Longman/Wiley. New York.

Rare Earth Elements

Henderson, P. (ed.). (1984). *Rare Earth Element Geochemistry*. Elsevier. Amsterdam.

Lipman, B. R., and G. A. McKay (eds.). (1989). *Geochemistry and Mineralogy of the Rare Earth Elements. Reviews in Mineralogy* 21. Mineralogical Society of America. Washington, DC.

See also Rollinson (1993), above.

Isotopes

Basu, A., and S. R. Hart (eds.). (1996). *Earth Processes: Reading the Isotopic Code*. AGU Monograph 95. American Geophysical Union. Washington, DC.

Faure, G. (1977). *Principles of Isotope Geology*. John Wiley & Sons. New York.

Faure, G. (2001). *Origin of Igneous Rocks: The Isotopic Evidence*. Springer-Verlag. Berlin.

Hoefs, J. (1980). *Stable Isotope Geochemistry*. Springer-Verlag, Berlin.

Taylor, H. P., Jr., J. R. O'Neil, and I. R. Kaplan. (1991). *Stable Isotope Geochemistry: A Tribute to Samuel Epstein*. Special Publ. No. 3. The Geochemical Society. San Antonio, TX.

Valley, J. V., H. P. Taylor, Jr., and J. R. O'Neil (eds.). (1986). *Stable Isotopes in High Temperature Geological Processes. Reviews in Mineralogy* 16. Mineralogical Society of America. Washington, DC.

See also Rollinson (1993), above.

10

Mantle Melting and the Generation of Basaltic Magma

Questions to be Considered in this Chapter:

1. Is mantle melting a "normal" process in the sense that it is a natural consequence of the typical shield and oceanic plate geotherms? If not, what is required to initiate partial melting?

2. When the mantle melts, what governs the nature of the diverse partial melts produced?

3. What are the implications of observed magma diversity on the nature of the mantle?

Let's begin by considering the Earth as a simple magma-generating machine, reducing the problem of melt generation to its simplest parameters. As the ensuing chapters will reveal, the Earth's crust is generated over time by melting of the mantle (although, once created, crustal rocks may experience several episodes of sedimentary, igneous, and metamorphic reworking). We shall thus focus on the earliest process in this sequence, the one responsible for generating the crust and allowing the other processes to occur: partial melting of the mantle. We shall discover that the common product of this melting is basalt, by far the most common type of volcanic rock generated today. As we shall see in Chapter 11, much of the spectrum of igneous rock types *can be* derived from primitive basaltic material by some evolutionary process, such as fractional crystallization, assimilation, etc. The generation of *basaltic* magma from the mantle is thus a critical first step in developing a comprehensive understanding of magma genesis. Because basalts occur most commonly in the ocean basins, and the Earth's structure is simplest in the oceanic areas, lacking the physical and chemical complexities imposed by the continents, we shall begin our investigation in the ocean basins.

Generating basalt, however, is not sufficient for our model. Table 8.7 lists the three most common magma series, each with its own parental basalt type. Because the calc-alkaline series is restricted mostly to convergent plate boundaries in complex settings, we shall set it aside for later consideration in Chapters 16 and 17. For the moment, we shall focus on tholeiitic and alkaline basalts as the principal types of basaltic magmas generated beneath the oceans. The main petrographic characteristics of each are summarized in Table 10.1. Although these characteristics are common, you will rarely find all of them in a single specimen. Tholeiites are considerably more voluminous, being generated at the mid-ocean ridges (mid-ocean ridge basalts [MORBs]) as well as in scattered intraplate volcanic centers creating oceanic islands. Alkaline basalts are more restricted to the intraplate occurrences. Both of these magma types are distinct, and each evolves to more silicic types along separate paths, as discussed in Chapter 8 and developed further in Chapter 14. Our model for magma generation from the mantle, then, must not only be able to generate basalt but must be able to generate at least these two parental types of basalt.

With this perspective, we might begin by asking if the generation of magma in our magma machine is a "normal" process. In other words, will the natural geothermal gradient result in the melting of the material comprising the Earth at depth? If so, at what depth will melting occur, what is melted, and what is produced? If the natural geothermal gradient will not result in the melting of the material comprising the Earth at depth, we must ask what can be done to melt things before we address any other questions.

TABLE 10.1   Summary of the Characteristics of Tholeiitic and Alkaline Basalts

	Tholeiitic	Alkaline
Groundmass	Usually fine-grained, intergranular.	Usually fairly coarse, intergranular to ophitic.
	No olivine.	Olivine common.
	Clinopyroxene = augite (plus possibly pigeonite).	Titaniferous augite (reddish).
	Hypersthene common.	Hypersthene rare.
	No alkali feldspar.	Interstitial alkali feldspar or feldspathoid may occur.
	Interstitial glass and/or quartz common.	Interstitial glass rare, and quartz absent.
Phenocrysts	Olivine rare, unzoned, and may be partially resorbed or show reaction rims of orthopyroxene.	Olivine common and zoned.
	Orthopyroxene relatively common.	Orthopyroxene absent.
	Early plagioclase common.	Plagioclase less common and later in sequence.
	Clinopyroxene is pale-brown augite.	Clinopyroxene is titaniferous augite, reddish rims.

After Hughes (1982) and McBirney (1993).

Because volcanoes and igneous rocks are relatively common phenomena, you might be tempted to answer our initial question in the affirmative and say that melts are a product of the normal geothermal gradient. Every student who has had an introductory course in geology, however, has the background to answer this question more fully. The study of seismic waves has allowed us to evaluate the Earth's interior (see Chapter 1) and can tell us something about the state of the material at depth. For example, S-waves are shear waves, so they cannot propagate through a liquid, which doesn't resist shear. Because these waves pass through the mantle, we can deduce that the mantle is essentially solid, and does not melt under normal circumstances. Only the outer core is liquid and fails to transmit S-waves, but material from this layer is too dense and too deep to reach the surface. Melts that do reach the surface must then be derived from the mantle or as remelts of the crust. Crustal melts will be addressed in Chapters 17 and 18 and, because the crust is ultimately mantle-derived anyway, we shall now address the *mantle* as the ultimate source of magmas.

Eruption temperatures of most basalts are in the 1100 to 1200°C range. Because basaltic magma must cool as it rises (even if adiabatic), these temperatures, when compared to the geotherm in Figure 1.11, provide a *minimum* depth of origin that indicates a mantle source at least 100 km deep. This conclusion is supported by the deep seismic activity preceding many volcanic events. If the upper mantle is the source of basalts, we must know in some detail what comprises it before we can address how to melt it and what is produced when it melts.

10.1 PETROLOGY OF THE MANTLE

In Chapter 1, we learned that the mantle is composed mostly of mafic silicate minerals (olivine and pyroxenes). Although seismic and gravity/angular momentum data can place physical constraints on the composition and mineralogy of the mantle, and our theories of the origin of the Earth can add some geochemical constraints, only by observing direct samples can we get a more accurate idea of what is there.

Surface samples generally accepted to be of mantle origin come from the following sources:

1. Ophiolites These are large sheet-like mafic to ultramafic masses, presumed to be ancient oceanic crust and upper mantle thrust onto the edge of continents and/or incorporated into mountain belts (see Coleman, 1971, 1977; Peters et al., 1991). Erosion then exposes a characteristic section of sedimentary, mafic, and ultramafic rocks. Refer to Figure 13.4 for a typical section through an ophiolite. Ophiolites show a considerable range in size, thickness, and degree of structural integrity. We will discuss ophiolites more fully in Chapter 13, but for now we are primarily interested in the ultramafic rocks in the lower portions because these are believed to represent significant portions of the upper mantle now exposed at the surface of the Earth.

Smaller slivers of presumed ophiolitic ultramafics, now dismembered and incorporated into deformed mountain belts, are commonly referred to as **alpine peridotites**. The ultramafic portions of ophiolites and alpine peridotites contain a variety of peridotites, predominantly harzburgite and dunite, with subordinate wehrlite, lherzolite, and pyroxenite. The original mineralogy is dominated by olivine, orthopyroxene, and clinopyroxene, with lesser amounts of plagioclase and oxide minerals, including magnetite, ilmenite, and chrome-rich spinel. Hornblende and serpentine appear to be later hydrous replacement minerals. The larger, more intact ophiolites allow us to see the geometric relationships between various rock types, but our observations are limited to the very shallowest mantle (less than 7 km).

2. Dredge samples from oceanic fracture zones Differences in ridge elevation can result in significant scarps at many ridge-offsetting fracture zones (transform faults). We will also discover in Chapter 13 that slow-spreading ridges have extensive detachment zones exposing deeper mantle (e.g. Figure 13.24). It was a favorite mantle sampling practice in the 1960s and 1970s to drag bottom-sampling dredges along the scarps. As with ophiolites, dredged samples represent only the uppermost mantle beneath this oceanic crust. It is also impossible to know the original location of a dredged sample or the relationships between any two samples. Dredged samples are varied (Melson and Thompson, 1970; Bonatti et al., 1970; Thompson and Melson, 1972), but the types are nearly identical to those exposed in ophiolites, providing strong evidence that ophiolites are indeed samples of oceanic crust and upper mantle.

3. Nodules in basalts Ultramafic xenoliths, called *nodules*, are occasionally carried to the surface by basalts, usually basanites or alkali basalts (White, 1966; Irving, 1978; Boyd and Meyer, 1979; Menzies, 1983; Frey and Rodin, 1987). They are usually fist sized or smaller, and the most common rock types are gabbro, dunite, harzburgite, spinel lherzolite, plagioclase lherzolite, wehrlite, garnet lherzolite, and eclogite (a high-pressure metamorphic garnet-pyroxene rock, chemically equivalent to basalt). Some lower crustal xenoliths are also found in many basaltic lavas. Many of the nodules are **autoliths**, or **cognate xenoliths**, meaning they are genetically related to the magma and not picked up from the wall rocks far from the magma source. Some of these may be cumulates (particularly the gabbros and pyroxenites), and others are **restites**, a refractory (high-melting-point) residuum left behind after partial melts have been extracted. Because the basalts have a mantle source, restites are olivine rich. In order to carry such dense olivine-rich nodules in suspension to the surface, transport was apparently rapid. These nodules are restricted to alkaline basalts and basanites and are not found in the more common tholeiites, implying that the former travel more rapidly and have had less time to crystallize or interact with the wall rocks. This has led several petrologists to conclude that alkali basalts and basanites are more primitive than tholeiites. The high-pressure garnet-bearing lherzolites occur only in the most alkaline and silica-deficient basalts, suggesting that perhaps these basalts have a deeper origin than the less alkalic and more silica-saturated basalts. We will explore these ideas more fully below.

4. Xenoliths in kimberlite bodies Kimberlites are unusual igneous phenomena that will be described in Section 19.3.3. Several lines of evidence suggest that kimberlites tap an upper mantle source as deep as 250 to 350 km and travel rapidly to the surface, bringing a variety of mantle and crustal samples to the surface as xenoliths. Kimberlites thus give us the only glimpse we have of mantle samples below the uppermost layers. All known kimberlites occur in continental areas, so the xenoliths represent continental crust and subcrustal mantle. Nonetheless, the mantle at depth, at least, is believed to be similar in both continental and oceanic areas. Ultramafic xenoliths are diverse, suggesting a heterogeneous upper mantle, but garnet lherzolite and spinel lherzolite are dominant among the unaltered deeper kimberlite samples (Boyd and Meyer, 1979; Mitchell, 1986; Dawson, 1980).

Data from the above sources (plus stony meteorites) lead us to believe that the mantle is composed of ultramafic rocks, as discussed in Chapter 1. Spinel and garnet lherzolites stand out from the array of mantle samples as prime suspects for pristine mantle material because they have the composition (particularly the calcium, aluminum, titanium, and sodium content) that,

when partially melted, can yield a basaltic liquid. Such material also has the appropriate density and seismic properties to match those determined for the mantle (Boyd and McCallister, 1976; Green and Lieberman, 1976; Jordan, 1979). Copious experimental evidence indicates that basalts can be generated by partial melting of such lherzolitic material.

For the reasons noted above, most students of the mantle believe that "typical" mantle is composed of peridotite. More specifically, it is a four-phase lherzolite, composed of olivine, orthopyroxene, clinopyroxene, and a subordinate aluminous phase, such as garnet, spinel, or plagioclase. Much of the shallowest oceanic mantle, now represented by dunite and harzburgite in ophiolites and many nodules in basalts, appears to be related to the lherzolites as refractory residuum after basalt has been extracted, as illustrated in Figure 10.1. Note that the compositions of dunite, harzburgite, lherzolite, and tholeiitic basalt are collinear, and that lherzolite composition lies at an intermediate position between the basalt and the other two rock types. If the tholeiite is created by partial melting of the lherzolite, extraction of the liquid will shift the composition of the remaining material directly away from the tholeiite toward harzburgite and dunite.

We can apply the lever rule to Figure 10.1 and determine that 20 to 25% partial melting of the lherzolite will produce tholeiitic basalt, leaving 75 to 80% as an essentially Al_2O_3-free dunite residuum. 15 to 20% partial melting will leave a low-alumina harzburgite. Some dunite/harzburgite samples have cumulate textures, however, and these may have formed by the accumulation of the olivine and orthopyroxene during fractional crystallization of the basaltic magma in magma chambers at or near the top of the mantle. The aluminous lherzolite represents undepleted mantle, also called **fertile mantle**, with a composition presumed to be close to that of the original mantle. It is a prime source for generating basaltic partial melts. Table 10.2 summarizes the mineralogy

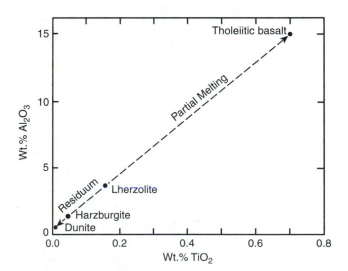

FIGURE 10.1 Relationship between TiO_2 and Al_2O_3 for garnet lherzolite, harzburgite, and dunite, as well as tholeiitic basalt, showing how the extraction of a basaltic partial melt from a garnet lherzolite can result in the creation of a solid refractory harzburgite or dunite residue. From Brown and Mussett (1993). Copyright © by permission Kluwer Academic Publishers.

and chemical composition of the most common lherzolite samples from kimberlites: garnet lherzolites and spinel lherzolites. Note the great similarity in composition between the two rock types. They are essentially identical. For the first time, we address a common metamorphic question: How can two *chemically* identical rocks have different *mineralogy*?

We can not only answer this question, but we can gain considerable insight into mantle petrology by choosing some mantle material and subjecting it to high pressure and temperature experimentation. This has been done by several investigators for both natural and synthetic mantle rocks (Ito and Kennedy, 1967; Green and Ringwood, 1967; Wyllie, 1970). The best-known example of a synthetic mantle analog is **pyrolite**, a *py*roxene–*ol*ivine material proposed and synthesized by A. E. Ringwood (1966). The composition of pyrolite, and several other estimates of primitive early mantle compositions, are included in Table 10.2.

Figure 10.2 is a pressure–temperature phase diagram created by experimental studies on mantle-type rocks of aluminous lherzolite composition. I have inverted the pressure axis from its conventional orientation to reflect its increase with depth. The diagram shows the liquidus and solidus, with the melting interval shaded (compare to Figure 7.13 for basalt), as well as some subsolidus metamorphic reactions. The shallower metamorphic reactions determine which aluminous phase is stable. The shallow slopes of the reactions suggest that ΔV plays a more important role than ΔS [from the Clapeyron equation, Equation (5.15)], and thus the reactions are more sensitive to pressure than temperature. Included in the figure is an estimated geotherm

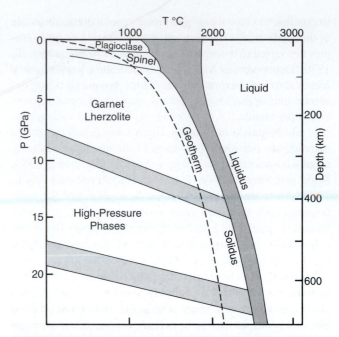

FIGURE 10.2   Phase diagram of aluminous lherzolite with melting interval (gray), subsolidus reactions, and oceanic geothermal gradient. Note that pressure increases downward, to reflect the trend in the Earth. After Wyllie (1981).

beneath the ocean basins. Where this geotherm intersects a particular reaction, the pressure–temperature conditions become appropriate for the reaction to take place. Thus at low pressure (below 30 km), plagioclase is stable, then spinel from 30 to

---

TABLE 10.2   Mineralogy and Chemical Composition of Spinel and Garnet Lherzolites, "Pyrolite," and Estimates of Primitive Mantle

| Mode/ Oxide | Spinel Lherzolite | | Garnet Lherzolite | | | Primitive Mantle | | | |
	Range	Average	Range	Average	Pyrolite	H&Z '86	M&S '95	P&O '03	L&K '07
Olivine	55–90	67	55–80	63	56				
Opx	5–35	24	20–40	25	26				
Cpx	3–14	8	0–10	2	16				
Spinel	0.2–3	2							
Garnet			3–15	10					
SiO$_2$	42.3–45.3	44.2	43.8–46.6	45.9	42.7	46.0	44.9	45.5	45.0
TiO$_2$	0.05–0.18	0.13	0.7–0.18	0.09	0.47	0.18	0.20	0.21	0.16
Al$_2$O$_3$	0.43–3.23	2.05	0.82–3.09	1.57	3.3	4.06	4.43	4.51	3.52
Cr$_2$O$_3$	0.23–0.45	0.44	0.22–0.44	0.32	0.45	0.47	0.38	0.37	0.39
FeO*	6.52–8.90	8.29	6.44–8.66	6.91	7.92	7.54	8.04	8.12	7.97
MnO	0.09–0.14	0.13	0.11–0.14	0.11	0.13	0.13	0.13	0.13	0.13
NiO	0.18–0.42	0.28	0.23 -0.38	0.29	0.42	0.28	0.25	0.24	0.25
MgO	39.5–48.3	42.2	39.4–44.5	43.5	41.4	37.8	37.8	36.9	39.5
CaO	0.44–2.70	1.92	0.82–3.06	1.16	2.11	3.21	3.53	3.66	2.79
Na$_2$O	0.08–0.35	0.27	0.10–0.24	0.16	0.49	0.33	0.36	0.35	0.30
K$_2$O	0.01–0.06	0.06	0.03–0.14	0.12	0.18	0.03	0.03	0.03	0.02

After Maaløe and Aoki (1977), pyrolite after Green and Ringwood (1967), H&Z: Hart and Zindler (1986), M&S: McDonough and Sun (1995), P&O: Palme and O'Neill (2003), L&K: Lyubetskaya and Korenaga (2007a, b).

80 km, and finally garnet from about 80 to 400 km. At greater depths, high-pressure phases occur. The ~600-km transition appears to represent the upper limit of Si in IV-fold coordination, and silicate structures similar to the mineral perovskite, with Si in VI-fold coordination, probably exist beyond this depth (as discussed in Section 1.2).

This sequence of reactions explains how we can have compositionally equivalent spinel and garnet lherzolites and also tells us that plagioclase lherzolites are a low-pressure alternative, also with the same chemical composition. It further explains why plagioclase, spinel, and garnet are rarely found together in the same sample and why plagioclase lherzolites are found only in shallow mantle samples (ophiolites and some oceanic basalts), whereas garnet lherzolites occur more commonly in kimberlites that tap a deeper mantle source. Because plagioclase peridotites are limited to depths less than about 30 km, which is less than the thickness of much of the continental crust, we would expect plagioclase peridotite to be absent in most of the subcontinental mantle, the top of which is commonly deeper than 30 km. This explains why it is so rare in kimberlites. The transitions from plagioclase to spinel peridotite and from spinel to garnet peridotite are accomplished by the following idealized metamorphic reactions:

$$CaAl_2Si_2O_8 + 2\,Mg_2SiO_4 = 2\,MgSiO_3 + CaMgSi_2O_6$$

Plagioclase   olivine   opx   cpx

$$+ MgAl_2O_4 \quad (10.1)$$

spinel

$$MgAl_2O_4 + 4\,MgSiO_3 = Mg_2SiO_4 + Mg_3Al_2Si_3O_{12}$$

Spinel   opx   olivine   garnet

$$(10.2)$$

10.2 MELTING OF THE MANTLE

Now that we have an idea of the chemical and mineralogical nature of the mantle, let's return to our original question regarding the feasibility of mantle melting. The geotherm shown in Figure 10.2 does not intersect the solidus for fertile mantle lherzolites, which supports our earlier contention, based on seismic data, that melting of the mantle does not occur under normal circumstances. There are a number of estimates for the average oceanic geotherm (Figure 1.11), but none of them approach the solidus. So our first problem, because we know that basalts are indeed generated, is to figure out how the mantle can be melted. There are three basic ways to accomplish this goal, following the three principal natural variables. We can either raise the temperature, lower the pressure, or change the composition. Of course, these must be done in a geologically feasible manner. Let's analyze each of these possible mechanisms in turn.

10.2.1 Raising Temperature

Figure 10.3 shows how melting could be accomplished by simply heating the mantle above the normal geotherm. Perhaps the simplest way we might do this is to accumulate

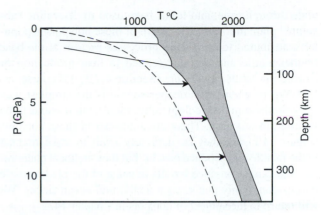

FIGURE 10.3   Melting by raising the temperature.

enough heat by the decay of radioactive elements because this is the only known source of heat other than that escaping from the primordial differentiation process (Section 1.8.2). The prime radioactive elements (K, U, and Th) occur in such low concentrations in the mantle that they are capable of producing less than 10^{-8} J $g^{-1}a^{-1}$. A typical rock has a specific heat (the heat required to raise 1 g of rock 1°C) on the order of 1 J g^{-1} deg^{-1}. It would thus require over 10^7 years for radioactive decay to raise the temperature of peridotite 1°C. The thermal conductivity of rocks is pretty low, but it is certainly high enough to allow this heat to dissipate long before any rocks would even approach melting. In fact, this radioactive generation *and conduction* of heat is exactly the process responsible for up to half of the heat flow reaching the surface that creates the geotherm in the first place. Local concentration of radioactive elements to increase heat production would require highly unrealistic concentration factors, with no driving mechanism to compensate for the entropy loss. If, in the unlikely event that we did somehow accomplish this concentration and managed to heat the mantle to its solidus, the heat required for melting must supply the latent heat of fusion of the minerals, which is about 300 times greater than the specific heat required to bring the minerals up to the melting point. The job of producing sufficient melts to be extractable thus becomes even more difficult. It is widely believed that the sudden jump in heat required to further raise the temperature of rocks already at their solidus plays a significant role in moderating unusual thermal fluxes in the mantle. Finally, if a percent or two of melt were produced, the K, U, and Th, which are highly incompatible, would concentrate in the melt and escape, leaving the peridotite too depleted to produce further melts equal to the surface volcanism we observe.

The heat flux from the lower mantle or core is not well understood. In general, it is constrained by the geothermal gradient itself in Figure 10.2, but this gradient is an average, and local perturbations may be possible. The most obvious manifestations of locally high heat flow are the **hotspots**, such as Hawaii, which are above narrow pipe-like conduits of basaltic magma that appear to have a stationary source in the mantle (Crough, 1983; Brown and Mussett, 1993). The motion of plates over these stationary hotspots results in the apparent migration of the volcanic activity across the plate through time, and it has been used to determine "absolute" plate motions. The origin of these hotspots is popularly attributed to very deep processes present at the base of the mantle and perhaps related to heat production or convection in the liquid core (Section 1.7.3). Of course, attributing problems to a deeper source is the simplest way to avoid them in the geological literature, but it seems to be justified in this case. There are a number of these hotspots (Figure 14.1), and they are definitely a way to add extra heat to the mantle and produce basalts, but they are local phenomena and cannot produce basalts in some of the places where basalts commonly occur, such as the mid-ocean ridges. We will return to the subject of hotspots in Chapter 14.

10.2.2 Lowering Pressure

Because the dry peridotite solidus has a positive P/T (Clapeyron) slope (it appears negative in Figures 10.2 and 10.3 because I have inverted the pressure axis), we could achieve melting by reducing the pressure at constant temperature. The simplest idea, local zones of lower pressure (the pressure equivalent of hotspots?) is untenable in ductile material such as the mantle because high-pressure material would quickly flow to the low-pressure areas until lithostatic equilibrium was restored. A more plausible way to lower pressure is to raise mantle rocks to shallower levels while maintaining their stored heat content. When material moves upward, the pressure is reduced and the volume increases slightly, resulting in a slight temperature reduction (10 to 20°C/GPa or 0.3 to 0.6°C/km for mantle rocks; Ahren and Turcotte, 1979). The upwelling mass would also move into cooler areas and lose heat by conduction to the surroundings, so it would then simply follow the geotherm, never approaching its melting point. If, on the other hand, the rise were sufficiently rapid to minimize heat loss to the surroundings, the only temperature difference would be due to expansion. If conductive heat loss were zero, the process is referred to as **adiabatic**, and any rising rock material would follow a path with the ~10°C/GPa (0.3°C/km) slope (Figure 10.4), called the **adiabat**. As discussed in Section 1.7.2, convective rise in the sub-lithospheric mantle may be

adiabatic until it reaches the less ductile boundary layer of the lower lithosphere, which leads to the curved shallow geotherms in Figures 1.11 and 10.4. Continued upwelling of mantle material beneath extensional areas may allow flow and adiabatic rise to shallower levels, departing from the shallow geotherm and following a P-T path with a much steeper slope than the solidus (~130°C/GPa), eventually intersecting the solidus and initiating melting. Once melting begins, the latent heat of fusion will absorb heat from the rising mass, causing the adiabatic path to follow a shallower temperature/pressure slope closer to the solidus curve, thus traversing the melting interval more obliquely. As a result, upwelling mantle material will diverge from the solidus slowly, producing limited quantities of melt. The process is called **decompression partial melting**.

Upwelling of mantle material occurs at divergent plate boundaries (Chapter 13), where two plates are pulling apart, and mantle material must flow upward to fill in. Langmuir et al. (1992) calculated that about 10 to 20% melting will occur per GPa (about 0.3%/km) of continued pressure release above the solidus. Thus, if the upwelling material began at its solidus temperature, it would have to rise about 35 to 100 km to attain the 20 to 30% melting estimated to produce mid-ocean ridge basalt. Of course, this would have to be added to the rise necessary to bring it to its solidus temperature in the first place, which would be on the order of 150 km.

10.2.3 Adding Volatiles

Because we have specified a mantle composition, we are not permitted a great deal of creative adjustment when we consider changing the composition. Although the content of the immobile components may be constrained, volatile species are mobile, and we might speculate as to their effect on the melting of mantle lherzolite. In some mantle xenoliths we find phlogopite or an amphibole. They are minor phases, but they do attest to the presence of some H_2O in the mantle. Wyllie (1975) estimated that the amount of H_2O in normal mantle material is unlikely to exceed 0.1 wt. % and suggested that it is not uniformly distributed. (The mantle is probably much more hydrated in subduction zones, however.) Microscopic examination reveals fluid inclusions (Figure 7.17) up to 5 μm in diameter, most of which contain H_2O but many of which are filled with dense liquid CO_2. This, and the occurrence of carbonate inclusions in some mantle minerals (McGetchin and Besancon, 1973) and in the matrix of kimberlites, suggests that CO_2 is also present in the mantle.

In Chapter 7, we explored the effects of H_2O and CO_2 on melting relationships in silicate systems. Figure 10.5 shows the dry peridotite solidus of Ito and Kennedy (1967) and several determinations of H_2O-saturated peridotite solidi. As predicted theoretically in Chapter 5 and demonstrated experimentally in Chapter 7, the effect of adding H_2O is to dramatically lower the solidus temperature, especially at higher pressure, where more of the volatile species can be accommodated in the melt. The geotherm intersects all of the H_2O-saturated solidi, so melting of the mantle is certainly possible in a hydrous environment, without needing to

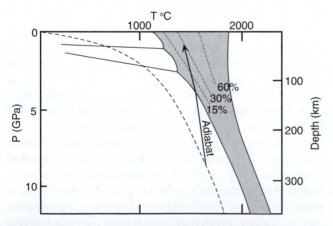

FIGURE 10.4 Decompression melting by (adiabatic) pressure reduction. Melting begins when the adiabat crosses the solidus and traverses the shaded melting interval. Dashed lines represent approximate percentage of melting.

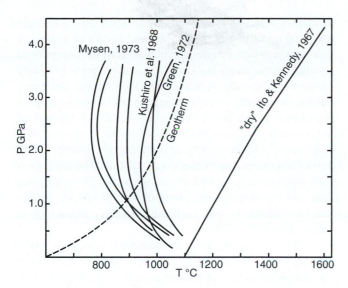

FIGURE 10.5 Dry peridotite solidus compared to several experiments on H_2O–saturated peridotites.

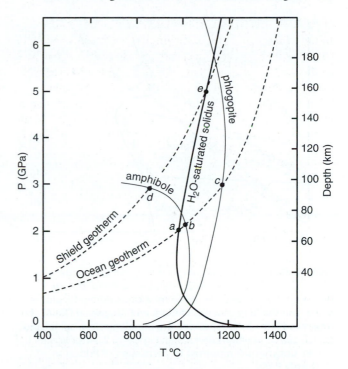

FIGURE 10.6 Phase diagram (partly schematic) for a hydrous mantle system, including the H_2O–saturated lherzolite solidus of Kushiro et al. (1968a), the dehydration breakdown curves for amphibole (Millhollen et al., 1974) and phlogopite (Modreski and Boettcher, 1973), plus the ocean and shield geotherms of Clark and Ringwood (1964). After Wyllie (1979). Copyright © reprinted by permission of Princeton University Press.

perturb the geotherm in any way. This method alone would solve the problem of melting the mantle, except the mantle is clearly not saturated in H_2O (or CO_2). As mentioned previously, the H_2O content of "normal" mantle is probably on the order of 0.1 to 0.2 wt. %, and it is probably bonded in hydrous mineral phases. The three common hydrous phases stable in ultramafic systems are phlogopite, an amphibole, and serpentine. The latter, however, is not stable above about 600°C and is thus strictly a low-temperature alteration of mantle rocks brought near the surface.

Figure 10.6 shows the combined phase diagram for a hydrous mantle system. It includes the Kushiro et al. (1968a) H_2O-saturated solidus for lherzolite, the dehydration reactions for amphibole and phlogopite, and estimated ocean and shield geotherms. All three equilibrium curves have the shapes predicted by the Clapeyron equation, given the variation in ΔV as pressure increases (Chapter 5). As pressure increases, ΔV decreases because of the higher compressibilities of the fluid phases, so the slope steepens. At very high pressure, the slope becomes vertical and reverses as the fluid-bearing side of the reaction begins to occupy less volume, and ΔV changes sign. The requirements for melting in this system are to satisfy both of the following:

1. Free H_2O, unbound in minerals, and
2. Temperature/pressure conditions sufficient to melt the lherzolite under H_2O-saturated conditions

If we assume "normal" sub-oceanic geothermal conditions, we must follow the ocean geotherm in Figure 10.6. At a depth of nearly 70 km, the geotherm intersects the H_2O-saturated solidus at point a, but there is no free H_2O, and the system is below the dry solidus. As a result, no melting occurs at this point because we have satisfied criterion (2) above, but not criterion (1). At point b, if there is any amphibole present, the amphibole will break down, releasing H_2O, which will immediately be free and available to produce some melt because the conditions are above the H_2O-saturated solidus, and

both criteria are thus met. If phlogopite is present, rather than amphibole, no melt will be produced until point c, at 90 km depth, when the phlogopite breaks down and releases H_2O. If we consider a different P/T path, such as the shield geotherm, the system will cross the amphibole breakdown first at point d, where amphibole dehydrates to an assemblage of anhydrous minerals and releases H_2O. Because this happens at a pressure and temperature below the H_2O-saturated peridotite solidus, we have satisfied criterion (1) but not criterion (2), so no melting will occur at this point. Only when the solidus is crossed at point e can the H_2O be used to melt the peridotite (if it doesn't migrate from the system between points d and e).

Because the amount of H_2O in either case is very small, the amount of melt that can be produced in this fashion is also very small, less than 1% (see Figure 7.23 for an example of the effect of limited H_2O on melting). Thus, in spite of an attractive method to produce melts without having to invoke unusual P/T conditions, the amount of melts that can realistically be produced in a hydrous mantle is probably less than 1%, which will be adsorbed to crystal surfaces, probably as a discontinuous film, and cannot be extracted from the peridotite source (see Section 11.2). Although we may not have found a way to generate the melts that reach the surface, we may have found a good way to explain the occurrence of the seismic **low-velocity layer** (Chapter 1). The small quantity of melt retained in the peridotite still permits the transmission of

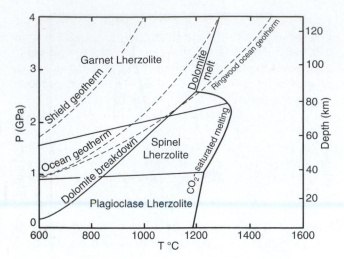

FIGURE 10.7 Phase diagram for a carbonated mantle system, including the CO_2-saturated lherzolite solidus, the decarbonation breakdown curve for a calcic dolomite, plus the ocean and shield geotherms of Clark and Ringwood (1964) and Ringwood (1966). Also included are the plagioclase–spinel and spinel–garnet reaction curves from Figure 10.2. After Wyllie (1979). Copyright © reprinted by permission of Princeton University Press.

S-waves but can be expected to slow both S- and P-waves. The depth of the layer (60 to 220 km) agrees well with those expected in Figure 10.6. For a discussion of this interpretation for the low-velocity layer, see Green and Lieberman (1976) or Solomon (1976); and see Priestly and McKenzie (2006) for an alternative view. We shall see in Chapters 16 and 17 that the H_2O content of the mantle in subduction zones is probably much greater and is an important factor in subduction-related magmatism.

Experiments on CO_2-bearing peridotite systems can also produce small amounts of melt at pressures in the vicinity of 2 to 4 GPa (Huang and Wyllie, 1974; Wyllie and Huang, 1975; Eggler, 1976). In Figure 10.7, the Ringwood-estimated ocean geotherm intersects the dolomite breakdown curve, which liberates CO_2 and then the CO_2-saturated solidus, initiating melting. The Clark and Ringwood geotherm intersects the direct dolomite-melting curve. Because the mantle samples suggest that the carbonate content is small, this should result in the production of a very small quantity of melt in the 75- to 120-km depth interval, depending on the geotherm chosen. Therefore, the addition of CO_2 can also cause small quantities of melt to be produced between 75 and 120 km with normal heat flow. We shall return to the effects of CO_2 on mantle melting in Chapter 19.

10.2.4 A Brief Summary of Mantle Melting

So far, we have an idea of what comprises the typical mantle, and we have found several reasonable ways to initiate melting. All of the methods explored above for mantle melting are probably at work in one area or another. Abnormally high geothermal gradients may occur in hotspots, ascending areas of convection cells, or places where magmas are rising

as diapirs. Pressure reduction may be associated with near-adiabatic upwelling of mantle material at rifts or in association with any rising material, such as those just mentioned. Volatiles, particularly H_2O and CO_2, can reduce solidus temperatures sufficiently to create melts with normal geotherms, but the low mantle volatile content severely limits the quantity of melt so produced. In Chapters 16 and 17, we will address subduction zone magmatism, where the role of H_2O (subducted in hydrous crust or sediments) is thought to play a more important role in magma generation. The methods of melt generation described above, supported by geophysical data and petrological indicators of igneous rocks, confine mantle melting to the shallowest mantle, extending no deeper than 150 to 200 km.

We also know from experimental work that partial melting of mantle lherzolite will generally produce basalt. There is some debate at present as to the significance of the **cusps**, the small low-T dimples where the plagioclase–spinel and spinel–garnet reactions intersect the liquidus in Figure 10.2. These cusps are required on theoretical grounds (as described in Section 26.10), but their magnitude is not known. Some authors speculate that they may lower the solidus enough to facilitate melting, thereby favoring the generation of melts at specific depths (about 40 and 90 km). More complete melting of the mantle to produce ultramafic magma is rare, probably because the geothermal gradient is not great enough to produce such high melt proportions and because the partial melts, once they reach 10 to 20%, tend to separate and rise, leaving a refractory residue that is unlikely to melt further. **Komatiites**, ultramafic volcanic rocks, are almost completely restricted to the Archean, when the geothermal gradient was much higher, and a larger fraction of partial melting (F) was apparently easier to generate (see Nesbitt et al., 1979).

It remains to be seen if we can generate a variety of basalt types, particularly tholeiites and alkaline basalts, from our typical mantle material. As a first approximation, it is useful to consider the mantle as a reservoir of uniform composition from which we wish to extract more than one type of melt. The "law of parsimony," generally referred to as "Ockham's razor," asserts that the simplest solution to a problem is most likely the correct one. If the spectrum of basalts found at the surface can be generated from a chemically uniform mantle, we have solved our problem in the simplest possible way. If not, we may have to move on to some more complex model. We shall now review some of the experimental data related to melting of typical mantle material and see if the types of basalts we find at the surface in the ocean basins can be generated.

10.3 GENERATION OF BASALTS FROM A CHEMICALLY UNIFORM MANTLE

If we are to generate a variety of products from a chemically uniform starting material, we will have to vary certain parameters. If we are presently prohibited from varying the composition, we must begin by testing the effects of varying pressure and temperature. Pressure variation at a particular temperature implies different geothermal gradients. High *T/P* gradients reach the solidus and initiate melting at high pressure, whereas low *T/P* gradients initiate melting at lower pressure. Temperature variation at a particular pressure translates to the extent to which the solidus is overstepped, which affects the percentage of partial melt (*F*) produced. This is an important variable at any depth, as we found from the experimental systems in Chapters 6 and 7, as well as from the partial melting models discussed in Chapters 8 and 9. What effects will varying the depth and extent of melting have on the partial melts produced from aluminous lherzolite?

We begin by looking again at the effects of pressure on melting. Figure 10.2 and Reactions (10.1) and (10.2) show us one important pressure effect that causes pressure-sensitive changes in the mineralogy of mantle lherzolite. Thus although the *chemical* composition of the mantle may be constant, the *mineralogical* composition is variable with depth. Because *minerals* are being melted, the first melt of a garnet lherzolite would not be the same as the first melt of a plagioclase lherzolite of the same composition. Naturally, *complete* melting of each would produce an identical melt because all of the components are then converted.

In Section 7.4, we learned that pressure changes have a differential effect on the melting of minerals (as a result of their different compressibilities). This causes a shift in the position of the eutectic minimum, resulting in different eutectic melt compositions. Figure 10.8 shows how the ternary eutectic minimum varies with pressure in the system Ne-Fo-Q. Notice that the eutectic minimum (the composition of the first melt) moves with increasing pressure from silica-saturated (tholeiitic) to highly undersaturated and alkaline melts. This shift reflects both the differential effects of compression on the minerals and pressure-controlled mineralogical changes. At 1 atm pressure, albite is involved in melting, but at 3 GPa

(as we shall calculate in Chapter 27), albite is no longer stable, and a jadeitic pyroxene is being melted instead, with an associated effect on the liquidus and the eutectic. The implication from this simplified basalt system is that tholeiites are favored by shallow melting, and silica-undersaturated alkaline basalts are favored by deeper melting.

We now turn to partial melting experiments on mantle-type rocks. Figure 10.9 illustrates the results of the experiments by Green and Ringwood (1967) on Ringwood's synthetic pyrolite mantle material. The size of each patterned block corresponds to the amount of a given mineral present in the experimental charge, thus giving a clear appraisal of the changes in mineralogy with depth and melting. Note that olivine and two pyroxenes dominate the sub-solidus mineralogy (hence a lherzolite), and plagioclase is the aluminous phase below 1 GPa. At higher pressure, the mineralogy of the pyrolite mantle differs from that of the lherzolite in Figure 10.2. In the experiments of Green and Ringwood (1967), spinel did not form a separate phase but dissolved as a component in the aluminous pyroxenes before the solidus temperature was reached. This results in a garnet-producing reaction that is different than Reaction (10.2), does not involve spinel, and occurs at about 3 GPa.

Figure 10.9 shows both pressure and % partial melting effects (the sizes of the blank blocks are proportional to the amount of liquid produced). Here we see that the first melt is more undersaturated (alkaline) at 60 km than at 25 km, which agrees in general with Figure 10.8. Also note that several different runs were made at 1.8 GPa (60 km), each at a different temperature, producing a different fraction of melt. Lower fractions of partial melting at this pressure result in more alkaline basalts. This is because alkalis are highly incompatible and enter the early melts, whereas successive melt increments slowly dilute the early alkali concentrations (see Figures 9.2 and 9.3), resulting in a more silica-saturated tholeiitic character. The solid residuum beyond about 20% partial melting is a harzburgite. It may require over 60% partial melting to create a dunite residuum. See also Figure 19.14 for a *P* versus *F* (degree of melting) grid of Green (1970), showing that more alkaline partial melts at high *P* and low *F* give way to more tholeiitic compositions as *P* decreases or *F* increases.

Included in Figure 10.9 is a brief summary of the possible effects of fractional crystallization in shallow crustal magma reservoirs. In such reservoirs, the usual trend is toward more silicic evolved liquids (andesites, rhyolites, etc.). Note the low-pressure thermal divide separating tholeiitic and alkaline magma series (Figure 8.13), which makes these series distinct as they evolve in shallow magma chambers to more silicic members, as discussed in Chapter 8. The nature of fractional crystallization also varies with depth. Just as the phases in the mantle vary with depth in Figure 10.2, likewise the phases that crystallize from the melt as it rises will vary.

Figure 10.10 shows Wyllie's (1971) summary of the melting and fractional crystallization products of the Green and Ringwood (1967) and Green (1969) schemes. Fractional crystallization, if accomplished at the same depth as

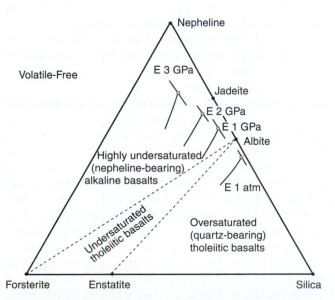

FIGURE 10.8 Change in the eutectic (first melt) composition with increasing pressure from 1 to 3 GPa, projected onto the base of the basalt tetrahedron. All but the low-*T* ends of cotectic curves have been omitted to avoid clutter. After Kushiro (1968).

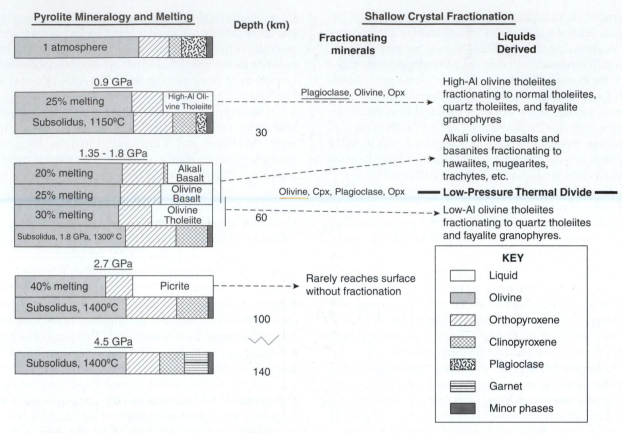

FIGURE 10.9   Variation in the nature of the liquids and refractory residua associated with partial melting of pyrolite at various pressures. Included is the near-solidus mineralogy of pyrolite. After Green and Ringwood (1967). Copyright © with permission from Springer-Verlag.

melting, is simply the reverse of the partial melting process, as you can see at the 1.8 GPa (60 km) level in Figure 10.10. Removal of an aluminous enstatite (Al-En) from the olivine tholeiite produced by 30% partial melting results in an alkaline basalt that could have been created simply by 20% partial melting in Figure 10.9. By creating partial melts at one depth and fractionating them at a shallower depth, however, a larger variety of magma types can be generated. Also, the thermal divide separating alkaline and tholeiitic series (D in Figure 10.10) does not exist at high pressure, so it is possible to go from a tholeiitic to an alkaline magma via fractional crystallization at high pressure, as discussed above and shown at 1.8 GPa in Figure 10.10.

Notice the changes with depth in the type of minerals that fractionate from liquids. Of particular importance is the solubility of aluminum in pyroxenes, which is more extensive at high pressure. Al-poor pyroxenes are removed with shallow fractionation, leaving more Al behind, resulting in high-Al liquids. Higher-Al pyroxenes fractionate at depth, producing highly SiO_2- and Al_2O_3-undersaturated liquids (nephelinites).

O'Hara (1965, 1968) also investigated mantle melting and crystal fractionation from basaltic liquids at various depths. He also found that silica-saturated tholeiites are generated at shallow pressures (below 0.5 GPa), giving way to silica-undersaturated alkaline basalts with increasing depth. Tholeiites are also favored by increasing degrees of melting,

extending their generation range to greater depths. O'Hara believed that melts fractionate continuously as they rise, with at least olivine crystallizing, and that the magmas reaching the surface are residual liquids of such fractionation and not truly primary (in the sense that they reflect the unadulterated partial melts produced at depth).

In his comprehensive summary of over 50 years of experiments directed toward mantle-melting products, Kushiro (2001) also showed that a variety of common basalt types can be generated by varying the depth of melting and/or the amount of partial melt produced (Figure 10.11). His results agreed that generation of alkaline basalt is favored by low melt fractions and higher pressures, whereas tholeiitic basalt is favored by higher melt fractions and lower pressures. Low-pressure partial melting produces rather SiO_2-rich initial melts. For example, Kushiro's (1996) experiments at 0.5 GPa yielded melts with 55.9, 52.8, 52.3, and 53.7 wt. % SiO_2 at 6.5, 18.5, 21.2, and 25.5% melting, respectively. All are quartz-normative basaltic andesites (becoming boninitic basaltic andesites with increasing melt fraction due to increasing MgO content). These results agree with those of Hirose and Kushiro (1991) on dry natural spinel lherzolite and the models of Kinzler and Grove (1992a, 1992b, 1993), based on their experiments on natural MORBs: low melt fractions lead to alkaline basalts, whereas higher melt fractions lead to more tholeiitic compositions. This is also shown in Figure 10.12 (a modification of the basalt tetrahedron base

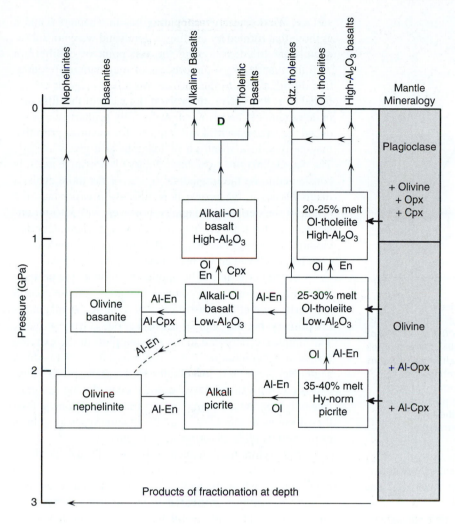

FIGURE 10.10 Schematic representation of the fractional crystallization scheme of Green and Ringwood (1967) and Green (1969) relating several basaltic magma types at moderate to high pressures. Minerals fractionating are listed near the arrows. After Wyllie (1971). Copyright © reprinted by permission of John Wiley & Sons, Inc.

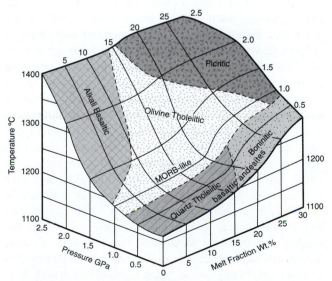

FIGURE 10.11 Extent of partial melting of fertile garnet lherzolite with increasing temperature at various pressures with the composition of partial melts produced (from Kushiro, 2001, based on the results of Kushiro, 1996). Liquids in the fields labeled "Olivine Tholeiitic" and "Boninitic" are basaltic andesites, and the transition into the latter field is due to increasing MgO with greater melt fraction.

shown in Figure 8.12) by the trends away from normative *ne** and toward *opx*. Melting at higher pressures leads to picrites that are nepheline normative (alkaline and silica undersaturated) at low degrees of partial melting and silica saturated at higher melt fractions.

Although there is some disagreement in the details between the schemes of Green and Ringwood (1967), O'Hara (1968b), Kushiro (2001), and Kinzler and Grove (1993), it is clear that a substantial variety of melts can be derived from a chemically homogeneous mantle and that this variety includes all of the basic magma types we see at the surface. These generalizations are substantiated by other extensive works (Yoder and Tilley, 1962; Yoder, 1976; Hess and Poldervaart, 1968).

Finally, we should remember that fluids in the mantle can also affect the type of melts generated. Figure 7.27 shows the effects of dry melting, H_2O-saturated melting, and CO_2-saturated melting on analog basaltic systems at 2 GPa. Eutectic point E_{dry} in Figure 7.27 is the same point as E_{2GPa} in Figure 10.8. H_2O shifts the eutectic toward silica saturation (tholeiites), whereas CO_2 shifts it toward more alkaline compositions. Although the quantity of fluids in the mantle as a whole is generally considered to be low, as

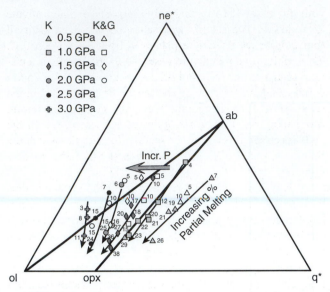

FIGURE 10.12   The normative *ol*-ne*-q** projection diagram of Irvine and Baragar (1971), showing the trends in partial melts of spinel lherzolites. *ne* = ne + 0.6 ab, q* = q + 0.4 ab + 0.25 opx* from the experiments described by Kushiro (2001). Open symbols refer to experimentally based isobaric batch melting calculations by Kinzler and Grove (1992a, 1992b, 1993). Numbers refer to percentage of lherzolite melted. The Kinzler and Grove model results in slightly greater ol* content at a given pressure and more gentle slopes (ol* enrichment) as partial melting progresses at higher pressures. Compare to Figure 10.8.

discussed in conjunction with the low-velocity layer above, we know from kimberlites and carbonatites that H_2O and CO_2 can locally be much higher.

To summarize, most models of basalt petrogenesis indicate the following (modified from Wyllie, 1971):

1. The composition of *primary* basalts is controlled by the depth of partial melting and segregation from mantle peridotite, the degree of partial melting, and the amount and type of the volatile phase, if present.
2. The composition of the basalt reaching the Earth's surface is also controlled by any subsequent crystal fractionation during post-segregation ascent.
3. Tholeiites may be formed by shallow melting or by olivine fractionation during rise of deep-seated picritic liquids. Tholeiites are also favored by H_2O-rich volatiles. Silica-poor alkaline basalts are derived by low degrees of partial melting, deeper sources, and CO_2-rich volatiles.
4. Tholeiites may also be formed when olivine fractionates during rise of deep-seated picrites, and alkaline basalts may be formed by deep-seated fractionation of Al-rich silicate phases.

10.4 PRIMARY MAGMAS

Figure 10.10 suggests that fractional crystallization may be common during ascent of basaltic magma. O'Hara (1965, 1968a) considered it inevitable. This brings into question exactly how common primary magmas are at the Earth's

surface. We defined **primary magmas** in Chapters 6 and 8 as those that formed by melting at depth and were not subsequently modified, following the last point of equilibrium with the mantle, by some process of magmatic differentiation during ascent to the surface. Modified magmas are referred to as **derivative**, or **evolved**. Because of the difficulty in determining whether a magma is truly primary, we introduced the term **parental** in Chapter 8 for the most primitive magma type in a spectrum of magmas at a given locality. The parental magma, whether primary or derivative itself, is considered to be the immediate source of the more evolved magma types in the series. It is difficult to understand the origin of the derivative magmas, however, without first understanding the truly primary ones that foster them. Our discussion above suggests that a variety of primary magmas are possible, but how can we recognize a primary magma?

Several criteria can be applied to evaluate the primary nature of a magma. As is common in geology, these criteria are fairly good at demonstrating that a particular magma is *not* primary, but they cannot prove that one *is* truly primary. The simplest criteria are that a magma plots at the extreme end of a differentiation index (such as low % SiO_2, high Mg# [Mg/(Mg + Fe)], low alkalis, etc.) and that it has a high extrusion temperature. These criteria are useful but can indicate only a parental magma, not a primary one. Nonetheless, they do show us that basalts are the magma types to address as the best candidates for primary magmas.

Let's return for a moment to Figure 6.11: the anorthite–diopside eutectic. As we cool composition *a* to 1455°C, we reach point *b*, where the melt, a complex solution of ions and molecules, becomes **saturated** in the anorthite component. At this point, anorthite begins to crystallize, much like the precipitation of salt from a salt solution during cooling or evaporation. At point *d*, the eutectic, the melt is saturated in both anorthite and diopside, as both phases crystallize. In the ternary system, when the melt reaches the ternary eutectic (point *M* in Figures 7.1 and 7.2), the melt becomes saturated in three phases, and so on as more components are considered. We refer to a melt that is saturated in several phases at once as a **multiply saturated** melt. Most basalts, when erupted, are close to their liquidus temperature (they either contain phenocrysts or phenocrysts form with little additional cooling). Furthermore, all of the major mineral types begin to crystallize within a narrow temperature interval, which suggests that they are close to being multiply saturated. Take the example from the Makaopuhi lava lake in Chapter 6. The eruption temperature was estimated at 1190°C (Wright and Okamura, 1977), at which point olivine had already begun to crystallize (Figure 6.2). Thus the magma was already at the liquidus temperature (not superheated) and saturated in olivine. The other major phases, plagioclase and clinopyroxene, began to crystallize within 20°C of cooling (Figure 6.2). Thus the Makaopuhi basalt, although certainly a high-temperature magma, was erupted at the liquidus temperature and was close to being multiply saturated.

At first we might consider a multiply saturated melt to be a good candidate for a primary magma. After all, if a primary magma is produced by *partial* melting of mantle peridotites, at least at low melt fractions, it must be a eutectic melt and is therefore multiply saturated. This is true, however, only at the pressure of *formation* at depth, not at the pressure of *eruption*. Figure 10.8 shows that there may be considerable difference between the 1 atm eutectic and the higher-pressure eutectic points. If a magma is multiply saturated at low pressure, it is nearly impossible to be so at high pressure as well. Low-pressure multiple saturation thus suggests that the magma was in equilibrium with the solid phases at very shallow depths, too shallow for magma formation. This requires that minerals were crystallizing, keeping the magma at the shifting eutectic as it rose. Such a magma must have been modified by fractional crystallization and is *not* a primary magma.

Consider Figure 10.13, which shows the results of melting experiments at a range of pressures on a mid-ocean ridge basalt glass suspected of being a primary melt. This method of melting a surface rock at various pressures to look back at the phases with which it may have been in equilibrium is commonly referred to as the **inverse method** because it begins with the product of partial melting as an indicator of the source character. Beginning with possible mantle samples and melting them to assess the results, as was done in Figures 10.9 and 10.10, is called the **forward method** because it attempts to mimic the original process. See Myers and Johnston (1996) for more on the forward versus inverse methods.

Back to our sample: unlike most mid-ocean ridge basalts, this sample was not multiply saturated at low pressure. As can be seen in Figure 10.13, at 1 atm, it was saturated

with olivine at 1215°C, and only plagioclase joined olivine 75°C lower. The sample *was* saturated, however, with all four plagioclase lherzolite phases at 0.8 GPa and 1250°C. This implies that the melt represented by the sample is a primary magma, once in equilibrium with the mantle at 25 km depth. This may or may not be the depth of "origin," in the sense of the partial melting event that created the basalt. More likely, it was the last point during rise at which the melt was well mixed with the mantle, the point at which the liquid separated from the mantle solids and behaved as a separate system.

A melt that is multiply saturated at the surface is thus *not* a likely candidate for a primary magma. A multiply saturated melt would have the saturation curves, such as those illustrated in Figure 10.13, intersect at near-surface pressures, and the curves would diverge at greater depths, indicating a lack of equilibrium with the mantle. Unfortunately, the criterion of non-multiple saturation at surface pressure is only a negative one because a magma that is not multiply saturated isn't proved to be primary. Only by subjecting a non-multiply saturated natural basalt to experimental melting and recrystallization at elevated pressures, as in Figure 10.13, can we see if it may be multiply saturated at high pressure.

If partial melting is extensive, however, as each successive mantle phase is consumed by progressive melting, the melt is no longer saturated in that phase, until finally it is only saturated with olivine, the last residual mantle phase. Primary melts resulting from extensive partial melting, then, may not be multiply saturated at any pressure. At the very least, however, the melt must be saturated with olivine at high pressure because olivine is the dominant phase of the mantle and the last mineral to melt. Figure 7.13, which illustrates the melting experiments on a Snake River tholeiitic basalt, shows that olivine is only a liquidus phase at pressures below 0.5 GPa. Either this basalt was formed at very low pressure (not even in the mantle because the Moho is deeper than this beneath the Snake River Plain) or it lost so much olivine by fractional crystallization that it can no longer remain olivine saturated at high pressure. The latter is more likely, so the basalt is not a suitable candidate for a primary basalt.

If experiments show that a basalt corresponds to a high-pressure eutectic composition and not a low-pressure one, they indicate high-pressure equilibrium with the mantle. This can be interpreted to mean that it is a primary melt at that pressure or that it formed deeper, fractionated last at the indicated pressure, and then rose essentially unaltered thereafter. In such a case, the support for a primary magma will have improved, but proof is elusive. Complications may also arise in applying experimental techniques when a reaction relationship exists for one of the minerals. If the source contains an orthopyroxene, it may not appear as a liquidus phase at all in the experiments because it will have a reaction relationship: opx = olivine + liquid (Figure 6.12).

As pointed out in Chapter 8, aphyric lavas, because they are untainted by previously formed crystals, are commonly chosen for variation diagrams as examples of good derivative melts. Some petrologists also consider the more primitive aphyric samples to be good possibilities for primary magmas

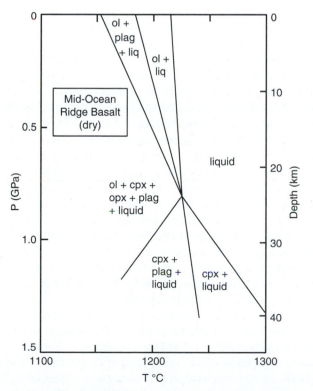

FIGURE 10.13 Anhydrous *P-T* phase relationships for a mid-ocean ridge basalt suspected of being a primary magma. From Fujii and Kushiro (1977).

because they show no evidence for having formed and lost phenocrysts. Others question this wisdom for multiply saturated magmas (see McBirney, 1993). They point out that multiply saturated magmas have probably lost their early phenocrysts. They suggest that some phenocryst-rich lavas, particularly the olivine-rich picrites, may still contain their early-forming phases and that the total assemblage may reflect a primary melt. Unfortunately, it is impossible to determine whether some of the phenocrysts have been lost. In addition, crystal-rich picrites may represent phenocryst accumulations at the bottom of a magma chamber and may thus in a sense be *more* primitive than a primary melt. Many investigators have looked to glass inclusions in early-forming phenocrysts as possible trapped primary magmas.

Two other criteria have also been advanced to support the idea that a magma may be primary. First, magmas containing dense dunite and peridotite nodules are presumed to have risen rapidly in order to keep the nodules suspended. Such magmas would have less time to undergo fractionation and may be primary. Second, the olivine in residual mantle peridotite nodules is usually in the narrow composition range Fo_{86} to Fo_{91}. Basaltic liquids in equilibrium with such olivines should have a ratio of $MgO/(MgO + FeO)$ in the range 0.66 to 0.75 (Roeder and Emslie, 1970; Green, 1971). Thus we should be able to quickly judge whether a basalt is a good candidate for a primary magma from its major element chemical composition or its position on an AFM diagram. Further chemical characteristics of a primary magma include high contents of Cr (>1000 ppm) and Ni (>400 to 500 ppm).

Remember, the criteria above are more negative criteria than positive, in that they are better at demonstrating that a particular magma is *not* primary than that it is. Because this is the case, a good candidate for a primary magma should meet as many of the above criteria as possible. Herzberg et al. (2007) and Herzburg and Azimov (2008) provided a tutorial (and spreadsheet) for a technique that can be used to estimate a primary liquid composition from an analysis (if available) of a primitive basalt from which only olivine and clinopyroxene are believed to have fractionated (see also the discussion associated with Figure 11.2).

10.5 A CHEMICALLY HETEROGENEOUS MANTLE MODEL

We have seen that, by varying the conditions of pressure, percentage of partial melting, fluid composition, and fractional crystallization in feasible ways, it is possible to generate a broad spectrum of basalts from a single mantle composition (including both tholeiitic and alkali basalts, as well as high-alumina basalts and picrites). Our quest for the parental magmas for the derivative series found at the surface would thus appear to be at an end, as was believed by most petrologists in the 1970s. However, modern techniques of trace element and isotopic analysis have changed our perspective on both basalt genesis and the nature of the mantle from which basalts are derived. This is a subject that we will explore more fully in Chapters 13 and 14, but a brief introduction is in order now. Although a chemically homogeneous mantle

has provided a simple solution to the variety of melts that occur in the oceanic areas, and Ockham's razor tells us that the simplest solution is most probably the correct one, there are a few ugly little facts that we cannot explain this way.

First, although the mantle samples fall into the broad categories of lherzolite, harzburgite, and dunite, the variety is too important to overlook. Although the major elements appear to be rather uniform, their variation permits the samples to be separated into two main groups (Nixon et al., 1981). **Fertile** or **enriched** xenoliths have somewhat higher contents of Al, Ca, Ti, Na, and K and lower Mg/(Mg + Fe) and Cr/(Cr + Al) than **depleted** xenoliths. Fertile xenoliths thus have more incompatible elements, which can be correlated with their potential to yield magma before becoming more refractory like the depleted samples. Garnet and spinel lherzolites are the most fertile samples, and dunites are the most depleted. There is a subtle, yet important, difference between the terms *fertile* and *enriched*, in that the latter implies that something has actually been added to them, whereas the former merely implies that little has been removed. We have not yet discussed ways to assess the contention that a sample is truly enriched, a subject developed more fully in Chapter 14 and Section 19.4. I will only mention in passing at this point that some of the mantle xenoliths show evidence of having been altered at depth. Although phlogopite and amphiboles are usually primary, some show textures suggesting an origin associated with volatiles and some other fluid-born components, added via an external fluid phase. Other alteration minerals include apatite, titanite, carbonate, etc.

Second, some trace element patterns for mantle-derived basaltic magmas present a complication. In one of the Problems in Chapter 9, you were asked to create a REE and spider diagram for two rock types. The rocks were average compositions of "typical" ocean island basalt (OIB) and mid-ocean ridge basalt (MORB), and the diagrams are reproduced as Figure 10.14. The negatively sloping OIB pattern in each diagram is a typical enriched pattern and can readily be explained via a model of either partial melting of peridotite or fractional crystallization of a peridotite-derived melt, in which the more incompatible elements are concentrated in the liquid (OIB) fraction. The positively sloping MORB trends, on the other hand, cannot be reconciled with any process of partial melting or fractional crystallization of chondrite-like mantle that incorporates the HREE and other relatively compatible elements into the liquid in preference to the less compatible elements. The only way a partial melt can have a pattern with a positive slope on the diagrams in Figure 10.14 is to melt a significant proportion of a solid that is already LREE and incompatible element depleted and thus has a positive slope to begin with.

Because these diagrams are chondrite normalized, and undepleted mantle should be similar to chondrite, it should plot at rock/chondrite = 1.0 and have a slope of 0. In order to become depleted in LREE and incompatible elements, these elements must be extracted from the mantle and incorporated into melts prior to the formation of the MORB. In other words, *the most common magma on the planet must be derived from a mantle that has been previously depleted* (probably by the

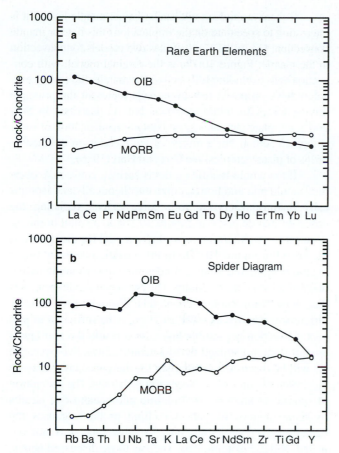

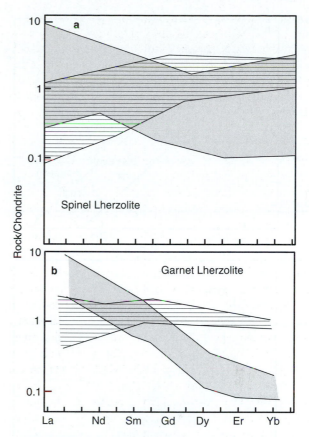

FIGURE 10.14 Chondrite-normalized **(a)** REE and **(b)** multielement (spider) diagrams of a typical alkaline ocean island basalt (OIB) and tholeiitic mid-ocean ridge basalt (MORB). Data from Sun and McDonough (1989).

FIGURE 10.15 Chondrite-normalized REE diagrams for **(a)** spinel and **(b)** garnet lherzolites. After Basaltic Volcanism Study Project (1981).

earlier extraction of melts to form the oceanic crust and continental crust). The other common magma of the ocean basins shows no such pattern and appears to be derived from a nondepleted (fertile) mantle source. We are faced with the conclusion that, despite our earlier success in generating different magma types from a singular mantle source, the mantle is not homogeneous and contains at least two principal reservoirs—one depleted and the other fertile.

Trace element data for the lherzolite nodules supports this conclusion. Figure 10.15 shows both positive and negative slopes on REE diagrams for both garnet and spinel lherzolites, showing that the mantle source rocks are certainly not a uniform reservoir. Because these samples represent potential source rocks, and not the derived partial melts, the LREE-enriched rocks with a negative slope and rock/chondrite ratios up to 10 must be truly *enriched* and not simply nondepleted. In other words, if our assumption that chondrite meteorites represent primordial mantle characteristics is correct, some process, perhaps the addition of melts or fluids from below, must have added incompatible LREE to some of these mantle samples.

Finally, we turn to the isotopic data. Figure 10.16 shows the variation in the ^{143}Nd/^{144}Nd and ^{87}Sr/^{86}Sr ratios for a variety of both oceanic basalts and mantle xenoliths. Remember from Chapter 9 that depleted mantle will, over time, evolve to *lower* ratios of ^{87}Sr/^{86}Sr (Figure 9.11) and to

higher ^{143}Nd/^{144}Nd (Figure 9.13). Because these are isotopes and heavy, they will not fractionate during melting or crystallization processes, so the initial isotopic ratio (corrected for age) of a melt will be identical to that of the melted source. The broadly linear belt with a negative slope on Figure 10.16a shows a correlation between the Nd and Sr systems consistent with a spectrum of enriched to depleted mantle material. This was called the **mantle array** by DePaolo and Wasserburg (1977) and Zindler et al. (1982). The upper-left part of the array (where MORB plots) has the high-^{143}Nd/^{144}Nd and low-^{87}Sr/^{86}Sr characteristics of a depleted source, and the lower-right part of the array is progressively less depleted. The large star represents the isotopic ratios of chondritic meteorites, the values we ascribe to the primitive Earth (neither depleted nor enriched). Note, then, that most of the "mantle array" reflects variable depletion, with MORB being derived from the most depleted source.

Rather than propose a mantle with continuous variability along the linear array, a much simpler interpretation of the array of isotopic data is that it represents a *mixing line*. Intermediate values along the line would then represent mixing of different proportions of only two components, represented by the endpoints of the line. One end represents a depleted, MORB-like reservoir, and the other is either nondepleted (primordial, or chondrite like) or slightly enriched. In order for the end-members to have developed their isotopic signatures independently, they must have been separate (unmixed) for at least 2 Ga.

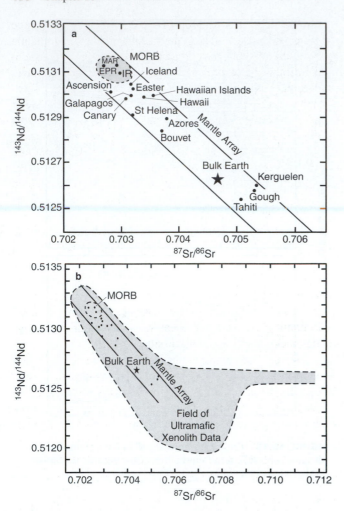

FIGURE 10.16 Initial $^{143}Nd/^{144}Nd$ versus $^{87}Sr/^{86}Sr$ for **(a)** oceanic basalts and **(b)** ultramafic xenoliths from the subcontinental mantle. From Wilson (1989). Data from Zindler et al. (1982) and Menzies (1983). MAR = Mid-Atlantic Ridge, EPR = East Pacific Rise, IR = Indian Ocean Ridge. Copyright © by permission of Kluwer Academic Publishers.

Oceanic basalts only rarely show truly enriched characteristics (more enriched than chondrite), but in Figure 10.16b we see that the mantle xenoliths show a much larger variation than the basalts, particularly with respect to enrichment (especially for the Sr data). Because this enrichment occurs for xenoliths originating beneath the continents (kimberlite xenoliths), it appears to be related to the subcontinental mantle. The fact that some oceanic islands also show enrichment beyond chondrite, however rare, is still significant and suggests that some portion of the sub-oceanic mantle has also been the recipient of incompatible components. We will return to the subject of mantle enrichment in Chapters 14, 17, 18, and 19.

The trace element and isotopic data led many petrologists and geochemists to believe that the mantle is stratified into two major levels: an upper depleted level and a lower one that is less depleted, if not enriched. The boundary is thought to be at about 660 km, which on Figure 1.2 is a deep seismic discontinuity that we believe represents a phase change to perovskite-like structures in which Si is in VI-fold coordination.

If there are indeed two such distinct mantle layers, it is interesting to speculate on the implications this has for mantle convection. Figure 10.17 illustrates two models for convection in the mantle. Figure 10.17a is the original model, with convection cells extending fully across the mantle. Several investigators have proposed that these cells represent the principal driving forces for plate tectonics, but, as described in Section 1.7.3, plate tectonics is probably a manifestation of more complex cooling. For a review of the theories on the mechanisms of plate tectonics, see Cox and Hart (1986).

If the whole-mantle model is correct, convective overturn should mix and homogenize the chemically and isotopically distinct reservoirs to a large extent, reducing the observed mantle array in Figure 10.16a to a small homogeneous area. An alternative two-layer model for the mantle is illustrated in Figure 10.17b. In this model, a depleted upper mantle is separated from a nondepleted lower mantle by the 660 km seismic discontinuity. The two-layer model proposes that convection cannot cross this discontinuity. Density differences perhaps as small as 2% may be sufficient to prevent convection across this layer. As a result, the two layers are not re-homogenized by full-mantle convective overturn. As will be discussed in Chapter 13, the upper layer may be the depleted source of mid-ocean ridge basalts. The generation of significant amounts of tholeiitic mid-ocean ridge basalts by large degrees of partial melting probably causes the depletion over geologic time. The lower layer is not depleted, and the isotopic data, such as those discussed above,

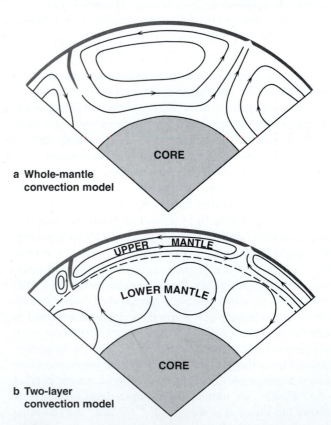

FIGURE 10.17 Simplified versions of two possible models for mantle convection. After Basaltic Volcanism Study Project (1981).

suggest that at least some of it is actually slightly enriched. This layer is tapped by deep hotspot "plumes" (Chapter 14) that feed some oceanic islands.

Debate over whole-mantle versus two-layer mantle convection continues. Whole-mantle convective models tend to homogenize the mantle, degrading long-term distinct mantle reservoirs required to explain the mantle isotopic array and other OIB signatures (Chapter 14). On the other hand, recent seismic tomography models provide convincing evidence that subducted slabs penetrate the 660-km discontinuity (e.g., Jordan, 1977; Grand et al., 1997; van der Hilst et al., 1997; Bijwaard et al., 1998) and accumulate in the lower mantle, where they may nourish the D" layer (Section 1.2 and Figure 1.14). Hotspot plumes (Chapter 14) originate in the deep mantle (probably at the D" layer) and are sufficiently buoyant to penetrate the 660-km discontinuity and return much of the subducted material upward. It thus appears that neither rigorous two-layer nor whole-mantle convection is tenable, leading investigators to appeal to more exotic models in an attempt to reconcile the differences.

The model of Silver et al. (1988) illustrated in Figure 1.14 is a modified two-layer mantle which retains the notion that the 660-km discontinuity is a sufficient viscosity or density barrier to impede wholesale convective mixing of the upper mantle and lower mantle. The discontinuity accommodates penetration of subducted slabs and plumes but imparts density stratification and separate depleted and fertile source regions, thereby satisfying the geochemists. Upper mantle flow in this model is largely a passive upwelling response to plate separation (driven by lithospheric density), and lower mantle upwelling is associated largely with plumes. Objections to a 660-km density barrier persist, and several other models have been proposed, some involving whole-mantle convection and others proposing deeper-layer boundaries.

Note the importance and impact of trace element and isotopic studies. Those nasty little positive slopes on REE and multielement (spider) diagrams could not be explained by any known process involving partial melting of a chondritic and chemically uniform mantle. Along with the isotopic trends, they forced us to reassess our models of the mantle and the nature of mantle convection. There is still uncertainty as to the nature of the mantle and its convection, but an exciting debate is now taking place, and interesting new avenues of inquiry are being opened. We shall return to this debate in Chapters 13 and 14 and use the more detailed geochemical data for mid-ocean ridge and ocean island igneous rocks to refine our discussion.

Finally, let's return to our earlier question of basalt generation in light of the possibility of a heterogeneous mantle. Figure 10.18 shows the results of melting experiments on both depleted and fertile lherzolite samples. The steep dashed lines indicate the extent of partial melting, contoured in increments

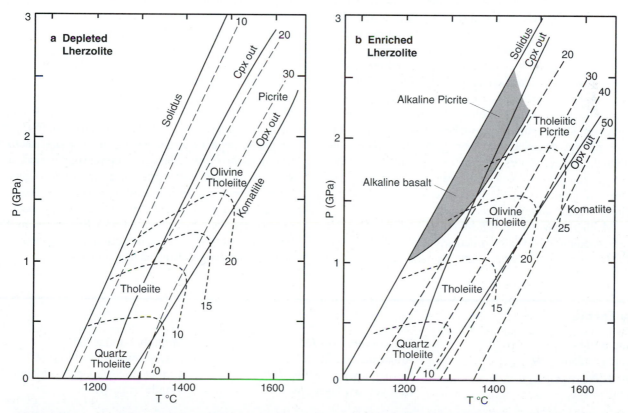

FIGURE 10.18 Results of partial melting experiments on **(a)** depleted and **(b)** fertile lherzolites. Dashed lines are contours representing percentage of partial melt produced. Strongly curved lines are contours of the normative olivine content of the melt. "Opx out" and "Cpx out" represent the degree of melting at which these phases are completely consumed into the melt. The shaded area in **(b)** represents the conditions required for the generation of alkaline basaltic magmas. After Jacques and Green (1980). Copyright © with permission from Springer-Verlag.

of 10%. The strongly curved contours represent the normative olivine content in the melt. Silica-saturated tholeiites can be generated by 10 to 40% partial melting of *either* a depleted or enriched source at depths of 30 to 40 km. Alkali basalts are generated at greater depths by 5 to 20% partial melting of fertile lherzolite. These findings are in general agreement with the earlier data, in that increased alkalinity correlates with greater depths and a lower percentage of partial melting. However, it now appears that generating alkali basalts from depleted mantle would be difficult under any conditions.

So it now looks as though we have a heterogeneous mantle, with at least a shallow depleted component and a less depleted mantle, probably concentrating at depth. The shallow source has been depleted by melt extraction and, perhaps, devolatilization. If partial melting processes have been operating throughout geologic time, then the upper mantle may be heterogeneous, comprising both fertile and depleted lherzolite as well as more refractory harzburgite and dunite. In addition, there must be numerous bodies of frozen-in partial melts that failed to reach the surface. The process of convection should re-homogenize these differences to a large degree in the sub-lithospheric (convecting) upper mantle. In contrast, the rigid lithospheric mantle should preserve many of these irregularities imposed upon it since it was formed. As we shall see in Chapter 19, this is of particular importance in the older continental lithosphere.

Recent high-pressure melting experiments (Wyllie, 1992) indicate that the compressibility of the melt may cause ΔV of dry lherzolite melting to approach 0 and to reverse sign at pressures greater than 7 GPa (just over 200 km depth). This would cause the liquidus (and solidus) to reverse slope (but remain steep) in Figure 10.2. The solidus and liquidus may remain above the geotherm at these depths (because S-waves still propagate), but the implications for magma genesis may be profound. If ΔV of melting is very low or negative, the buoyancy of the melt will likewise be low or negative, and melts created at these depths may never rise to the surface. This might help to explain why some deep reservoirs have not become depleted over time. The generation of alkaline basalts from the enriched layer below 660 km may not involve melting at these depths. Rather, *solid* diapirs of enriched material may rise from deep levels and undergo decompression melting at shallower levels to produce enriched alkaline basalt, as will be discussed in Chapter 14.

Summary

Mantle-derived samples may be delivered to the Earth's surface as ophiolites (slivers of oceanic crust and upper mantle caught up in orogenesis), nodules in basalts, or xenoliths (particularly noteworthy in kimberlites). Partial melting of the mantle is probably the original igneous rock-producing process on Earth, and it is still the predominant one occurring today. We infer from samples believed to be delivered to the surface that primitive undepleted mantle has a lherzolitic composition with sufficient aluminum to produce a subordinate Al-rich phase (plagioclase at very shallow levels, giving way to spinel and then garnet with increasing depth in the upper mantle). Melting of the mantle is not considered a natural consequence of increasing depth, considering the *P-T* trajectory of the lherzolite solidus and normal shield or oceanic intraplate geotherms.

Melting may be realistically achieved by hotspot plumes, addition of volatiles (particularly H_2O), or plate divergence followed by adiabatic mantle rise and decompression partial melting. *Primary* magmas may be recognized on the basis of high Mg# (>0.66), Cr (>1000 ppm), and Ni (>400 ppm) concentrations, as well as through multiple saturation with several mantle phases at some high pressure and temperature in melting experiments.

It is possible to generate the two most common types of primitive basalts, tholeiitic and alkaline, from a chemically homogeneous mantle. Alkaline basalts are favored over tholeiites by a lower percentage of partial melting of the lherzolite, medium-high pressure fractional crystallization from a tholeiite, greater depth of partial melting, and less H_2O and/or more CO_2 in an associated fluid phase. In spite of this, trace element and isotopic variations suggest that the mantle is chemically heterogeneous, requiring at least two mantle reservoirs developing distinct isotopic signatures without having been mixed and homogenized for at least 2 Ga. A few whole-mantle and two-layer mantle convection models may satisfy both seismic/geodynamic and geochemical requirements.

Key Terms

Ophiolite *184*	Hotspot (hotspot) *187*	Multiply saturated *194*
Alpine peridotite *184*	Adiabatic/adiabat *188*	Forward/inverse method *195*
Restite *185*	Decompression melting *188*	Fertile/enriched/depleted mantle *196*
Pyrolite *186*	Low velocity layer *189*	Mantle array *197*

Review Questions and Problems

Review Questions and Problems are located on the author's web page at the following address: **http://www.prenhall.com/winter**

Important "First Principle" Concepts

- Even if the composition of the mantle were relatively constant from top to base, the mineralogy, density, and viscosity would vary. Density and viscosity may vary gradually over some depth ranges, but abrupt phase (mineral) transformations produce discontinuities.

- Significant mantle melting is not the normal consequence of increased temperature and pressure with depth. A vertical column from Moho to core should be essentially solid beneath shields and old oceanic crust.

- The mantle adiabat is steeper than the solidus, so even near-adiabatic rise, if sufficiently great, should result in partial melting at depths less than 150 to 200 km.

- Extensional tectonics invites mantle rise, which, if sufficiently fast and great, will result in partial melting.

- There is little evidence for mantle melting at depths greater than 200 km.

- Typical mantle is greatly undersaturated in H_2O. Any H_2O present (perhaps 0.1 to 0.2 wt. %) probably occurs in hydrous silicates (amphibole or phlogopite). This quantity of H_2O would only be enough to produce around 1% partial melting by lowering the dry solidus; this is insufficient to separate and rise but may be enough to create the low-velocity layer.

- The mantle is heterogeneous, having components with distinctly depleted and fertile trace element and isotopic signatures. Isotopic contrasts require that the separate components have been isolated for at least 2 Ga.

Suggested Further Readings

Mantle Petrology

Anderson, D. L. (1977). Composition of the mantle and core. *Ann. Rev. Earth Planet. Sci.,* **5**, 179–202.

Boyd, F. R., and H. O. A. Meyer (eds.). (1979). *The Mantle Sample: Inclusions in Kimberlites and Other Volcanics.* American Geophysical Union. Washington, DC.

Brown, G. C., C. J. Hawkesworth, and R. C. L. Wilson. (1992). *Understanding the Earth.* Cambridge University Press. Cambridge, UK.

Brown, G. C., and A. E. Mussett. (1993). *The Inaccessible Earth: An Integrated View to Its Structure and Composition.* Chapman & Hall. London.

Hart, P. J. (ed.). (1969). *The Earth's Crust and Upper Mantle.* Geophysical Monograph 13. American Geophysical Union. Washington, DC.

Pearson, D. G., C. Canil, and S. Shirey. (2003). Mantle samples included in volcanic rocks: xenoliths and diamonds. In *Treatise on Geochemistry. Volume 2: The Mantle and Core* (ed. R. W. Carlson). Sec. 2.05. Elsevier, Amsterdam.

Ringwood, A. E. (1966). Mineralogy of the mantle. In: *Advances in Earth Sciences* (ed. P. M. Hurley). MIT Press. Cambridge, MA.

Basalt Genesis

Asimow, P. D. (2000). Melting of the mantle. In: *Encyclopedia of Volcanoes* (eds. H. Sigurdsson, B. Houghton, S. R. McNutty, H. Rymer, and S. Stix). Academic Press, San Diego. pp. 55–68.

Dick, H. J. B. (ed.). (1977). *Magma Genesis.* Oregon Department of Geology and Mineral Industries Bulletin **96**.

Hess, H. H., and A. Poldervaart (eds.). (1968). *Basalts: The Poldervaart Treatise on Rocks of Basaltic Composition.* Vol. 2. John Wiley & Sons. New York.

Kushiro, I. (2001). Partial melting experiments on peridotite and origin of mid-ocean ridge basalts. *Ann. Rev. Earth Planet. Sci.,* **29**, 71–107.

Morse, S. A. (1994). *Basalts and Phase Diagrams.* Krieger. Malabar, Florida.

Ragland, P. C., and J. J. W. Rogers (eds.). (1984). *Basalts.* Van Nostrand Reinhold. New York.

Wyllie, P. J. (1971). *The Dynamic Earth: Textbook in Geosciences.* John Wiley & Sons. New York.

Yoder, H. S. (1976). *Generation of Basaltic Magma.* National Academy of Science. Washington, DC.

Mantle Convection

Davies, G. F. (1999). *Dynamic Earth: Plates, Plumes and Mantle Convection.* Cambridge University Press. Cambridge, UK.

Davies, G. H., and M. A. Richards. (1992). Mantle convection. *Jour. Geology,* **100**, 151–206.

Peltier, W. R. (ed.). (1989). *Mantle Convection Plate Tectonics and Global Dynamics.* Gordon & Breach. New York.

Silver, P. G., R. W. Carlson, and P. Olson. (1988). Deep Slabs, Geochemical Heterogeneity, and the Large-Scale Structure of Mantle Convection: Investigation of an Enduring Paradox. *Ann. Rev. Earth Planet. Sci.,* **16**, 477–541.

Trace Element and Isotopic Systematics of the Mantle

Hart, S. R. (1988). Heterogeneous mantle domains: Signatures, genesis, and mixing chronologies. *Earth Planet. Sci. Lett.,* **90**, 273–296.

Rollinson, H. R. (1993). *Using Geochemical Data: Evaluation, Presentation, Interpretation.* Longman. Essex, UK.

Wilson, M. (1989). *Igneous Petrogenesis: A Global Tectonic Approach.* Unwin Hyman. London.

Zindler, A., and S. Hart. (1986). Chemical geodynamics. *Ann. Rev. Earth Planet. Sci.,* **14**, 493–571.

11

Magma Diversity

Questions to be Considered in this Chapter:

1. How can compositional diversity be developed in magmatic systems, and what are the products?

2. Once a parental magma is created, how can its composition change so as to yield a sequence of evolved compositions in a magma series?

3. What methods do we have to evaluate diversification processes?

In Chapter 10, we concluded that a range of basaltic magmas could be created by partially melting the mantle. Now we shall explore the dynamics of magmatic diversification. We do this for two reasons. First, differentiation is an important natural process, capable of modifying the composition of magma and leading to a spectrum of igneous rocks. As students of petrology, we should understand these processes so we can interpret the rocks and their associations. Second, we now understand the generation of primary basalts. If we are correct in our interpretation that the continental crust evolved over time, then such basalts (and komatiites) must once have represented the only types of primary magmas possible on Earth. This is certainly a restricted range of magma types. There is obviously a much wider range of magma and igneous rock compositions, extending from basalt to myriad types showing considerable variation in silica, alkalis, alumina, etc. (Chapter 2 lists many.) Is it possible to produce all of the other igneous rock types from primary basaltic magmas (as various magma series) by realistic diversification processes, or must we turn to other sources and primary magmas? In order to evaluate this fundamental question, we must first know what processes are possible. After that, we shall see to what extent those processes are capable of producing the other igneous rocks.

Diversification involves separating different phases of contrasting composition. This separation can happen during melting or during crystallization, when phases in different states coexist. The main subject of this chapter involves crystallization processes, but first we shall review and develop the concept of partial melting because it is also a method by which a variety of magmas can be produced, and it is familiar to us from earlier discussions.

11.1 PARTIAL MELTING

Separation of a liquid from the partially melted solid residue is a form of diversification because it involves partitioning and separation of chemical constituents, and it can produce a variety of melt compositions from a single source. This process has already been discussed in several chapters, so we shall only briefly review it here. In Chapters 6 and 7, we studied the phase relationships of several simplified experimental systems and discussed, among other options, the effects of removing the liquid fraction at various stages of melting (fractional melting). One important point to remember from that discussion is that, in systems involving eutectic behavior, the first melt is *always* produced at the eutectic composition, regardless of the relative proportions of the phases in the melting rock (as long as all relevant phases are present). The major element composition of this melt remains constant as more is produced, until one of the source mineral phases is consumed by the melting process. Once a phase is consumed, the next melt increment will occur at a different composition and temperature because it now corresponds to the minimum melt in a reduced system with one fewer component and phase (see Section 7.1). Obviously, different melts can be extracted from an initially uniform source rock.

In Chapter 9, we looked at several models for trace element behavior during crystallization and melting, including models for Rayleigh fractional melting and batch melting. The melting conventions discovered for simplified systems were applied to more complex natural systems in Chapter 10. We found that the nature of the melt produced by partial melting of a particular source is a function of the source composition, pressure, the nature of the fluid phase, and the extent of melting. This last factor is limited by the available heat and the ease with which the melt can be extracted from the crystalline residue.

As mentioned in Chapters 4 and 9, when a rock begins to melt, a tiny fraction of initial melt forms discrete liquid drops at the junctions of mineral grains, usually at the points where three or four grains meet (Figure 11.1). Only when a critical quantity of melt is produced will there be a sufficient liquid volume that:

1. The liquid forms an interconnected network.
2. The interior body of the liquid can be free from the restraining effects of crystal surface adsorption.

Only when conditions 1 and 2 are met can some of the melt be separable from the solids. The critical melt fraction required to form an interconnected network depends upon the **dihedral angle**, θ, formed between two solid grains and the melt (defined graphically in Figure 11.1a). When the interfacial energy (or surface tension) of the melt is similar to that of the minerals, the dihedral angle is low, and the melt forms an interconnected network at a low melt fraction. If θ < 60° the melt may form a network with as little as ~1% melt. As the dihedral angle increases above 60°, the amount of melt required to establish connectivity also increases (Beere, 1975). This theory was developed for structurally isotropic solids, and the predicted melt distributions in rocks containing several types of minerals may deviate from such simple geometric models, especially if the minerals have low symmetry. In mafic systems, θ has been found to be less than 50°, allowing very small melt fractions to be extractable. Some experimental evidence suggests that rhyolitic melts have higher angles, but still in the range of 50 to 60° (Laporte et al., 1997). Laporte et al. (1997) also found that a permeability threshold, the fraction of melt required for an interconnected network, was low, generally less than 1%.

Separation of a melt may require higher melt fractions, however, than indicated by θ and the permeability threshold. **Viscosity** is an important factor in melt segregation, once a continuous network has been formed. High-viscosity silicic melts, such as granitic–rhyolitic liquids, are less easily extracted. The **critical melt fraction**, or **rheological critical melt percentage (RCMP)** (Wickham, 1987), is the percentage of melt at which a crystal-dominated, more rigid granular framework gives way to a melt-dominated, fluid suspension,

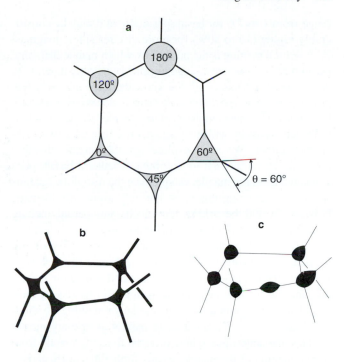

FIGURE 11.1 (a) Illustration of the dihedral angle (θ) of melt droplets that typically form at multiple grain junctions. Low angles of θ occur when the surface energy of the melt is similar to that of the minerals. When a melt has a low dihedral angle, it tends to "wet" the grain surfaces and form a continuous interconnected network (b). Higher surface energy contrasts result in higher θ and isolated melt droplets (c). After Hunter, 1987. Copyright © by permission Kluwer Academic Publishers.

commonly called a **crystal mush**. For a theoretical system of spheres, the RCMP is 26% melt, but for irregular shapes and variable sizes, this amount may vary between 30 and 50% for *static* situations involving viscous granitic compositions.

Liquid separation is commonly motivated by gravitational effects as the buoyant liquid seeks to rise and escape the crystal residue. Because melt source regions are generally deep and under pressure, the separation may be aided by a process known as **filter pressing**, or **compaction**, in which the crystal–liquid system is squeezed like a sponge, and the liquid migrates from the compacted solids (McKenzie, 1987). For systems being deformed or compacted, the RCMP drops considerably from the static theoretical value of 26%.

Rushmer (1996) stated that classic models based on dihedral angles of melt pockets at multiple-grain intersections are of limited use in rocks because the interfacial energy is anisotropic, resulting in planar solid–melt interfaces as well. Rushmer's experiments also suggest that the critical melt fraction is not reliable and that melt segregation in natural systems is controlled by several variables, including the depth of melting, the type of reaction involved, the volume change upon melting, and the tectonic setting (which controls the rate of melting and the style of contemporaneous deformation). The critical fraction required for separation of natural melts thus varies with several factors, also including temperature, viscosity, composition, and volatile content. It is not known with any precision, therefore, and estimates range from 1 to 7% for basalt/peridotite and may be considerably higher (15 to 30%) for more viscous silicic magmas.

Partial melting is the process by which mantle lherzolite fractionates to produce a range of primary basaltic magma, as we explored in Chapter 10. We know there are relatively few primary magma types, although there is debate as to exactly how many. The main emphasis of this chapter addresses potential mechanisms by which a magma, once formed, may differentiate further. We shall explore several magma chamber processes as we attempt to determine whether a basalt, once generated from the mantle, can produce the rest of the igneous spectrum. Expecting this to be the only avenue to diversity, however, would be asking too much, and partial melting should not be overlooked as a component of diversification. Most Icelandic rhyolites, for example, are considered to be formed by a two-stage partial melting process. The first stage produces the ubiquitous Icelandic basalts from mantle lherzolite, and the second stage produces rhyolite (the second most abundant rock type in Iceland) by partially remelting the solidified basaltic crust. In addition to basalts as primary mantle melts, some andesites are now believed to result directly from partial melting of the mantle, as are kimberlites, carbonatites, and some other alkaline rocks (Chapter 19). We also consider many silicic magmas to have formed by partial melting of sialic crustal rocks (some of which are differentiated by sedimentary processes) or of mafic rocks solidified from basaltic magmas trapped at the base of the crust (Chapters 17 and 18). Some subduction zone andesites may also result from a two-stage mantle melting process: stage one produces basaltic oceanic crust at mid-ocean ridges (Chapter 13), and this crust is partially melted as it is subducted to produce andesite (Chapter 16). Nonetheless, differentiation plays a great role in creating a wide range of chemical and mineralogical diversity in magmas and rocks.

11.2 MAGMATIC DIFFERENTIATION

Magmatic differentiation is defined as any process by which magma is able to diversify and produce a magma or rock of different composition. Differentiation (and partial melting) involves two essential processes:

1. Creation of a compositional difference between one or more phases as elements partition themselves in response to a change in an intensive variable, such as pressure, temperature, or composition. This determines the *trend* of the differentiation process.
2. Preservation of the chemical difference created in part 1 by segregating the chemically distinct portions, which then evolve as separate systems. **Fractionation** is the physical process by which different portions (usually distinct phases) are mechanically separated. The effectiveness of the fractionation process determines the *extent* to which differentiation proceeds along a particular trend.

By far the most common forms of magmatic differentiation involve the physical separation of phases in multiphase systems. The effectiveness of this separation depends upon contrasts between the phases in physical properties such as density, viscosity, and size/shape. The energy providing the force for the separation is usually thermal or gravitational. The fractionated phases in magmatic systems can be either liquid–solid, liquid–liquid, or liquid–vapor.

11.2.1 Fractional Crystallization

Fractional crystallization has traditionally been considered the dominant mechanism by which most magmas, once formed, differentiate. We first encountered fractional crystallization in Chapters 6 and 7, when we noted the effects of removing crystals as they formed, compared to equilibrium crystallization, on the *liquid line of descent*. Fractional crystallization was a popular mechanism (even discussed by Darwin, 1844) by the time Bowen was able to reproduce the process experimentally. Bowen (1915a) created olivine crystals on the forsterite-rich side of the diopside–forsterite eutectic (Figure 7.1), most of which sank 1 to 2 cm and accumulated at the bottom of the platinum crucible after 15 minutes at 1405°C. This was observed as small crystals of forsterite concentrating near the bottom of a eutectic glass once the charge was removed from the furnace and quenched. In other experiments, Bowen also demonstrated the sinking of pyroxenes and the *floating* of tridymite.

The discussion in Section 10.4 on the scarcity of primary magmas provides a qualitative assessment of magmatic differentiation between the time of melt formation and eruption. The predominance of magmas found to be multiply saturated at low pressure implies that most mantle-derived magmas have equilibrated to low-pressure conditions by crystal fractionation en route to the surface.

A volcanic series can be evaluated for the effects of fractional crystallization by using variation diagrams. In Section 8.7 we developed the theoretical method to evaluate the evolution of a suite of lavas based on fractional crystallization. Figure 11.2 illustrates a particularly clear and simple relationship among a series of Hawaiian lavas that can be related by the crystallization of a single phase. Because these lavas are rather primitive, the wt. % MgO is a more useful differentiation index than silica, which typically varies little in the early stages of magmatic evolution. MgO *decreases* toward the right in Figure 11.2, so it conforms to most bivariate differentiation diagrams, in which evolution increases from left to right. All the lavas in Figure 11.2 plot along linear paths connecting the proposed parental magma and extrapolating to the proposed olivine phenocrysts. The parent was considered to be represented by the most mafic *glass* found because only glasses can be assured to represent liquid compositions. The vertical "extracted olivine" line in Figure 11.2 corresponds to an olivine analysis that is compatible with observed phenocrysts. TiO_2 and Na_2O project to slightly negative olivine extract values, the result either of analytical error and the "heroic" extrapolation across over 25 wt. % MgO or crystallization of an unaccounted-for minor phase that formed in addition to olivine. The excellent linear fit of all data in Figure 11.2 provides strong support for the rocks being related by variation

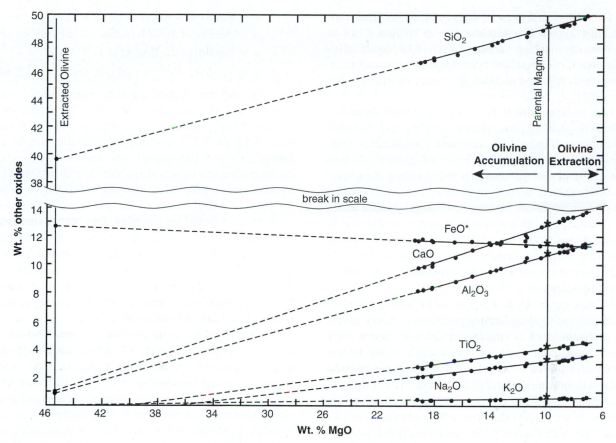

FIGURE 11.2 Variation diagram using MgO as the abscissa (sometimes called a "Fenner" diagram) for lavas associated with the 1959 Kilauea eruption in Hawaii. The parent melt (asterisks) was estimated from the most primitive glass found. All the variation can be accounted for by the extraction and accumulation of olivine phenocrysts (after Murata and Richter, 1966b, as modified by Best, 1982). Reprinted by permission of the *American Journal of Science*.

in olivine content alone. Note that fractionation, the physical separation of olivine and liquid, was probably occurring, but it was not particularly effective in creating highly evolved liquids in this example. Less than half of the analyses plot to the low-Mg (right) side of the parent, consistent with their being derivative liquids created from the parental magma by the removal of olivine. These also account for only ~3 wt. % variation in MgO. The remainder plot to the left and represent liquids in which the olivine had accumulated but was still suspended as phenocrysts in the magma.

Sometimes the most mafic phenocrysts found are too Mg rich to be in equilibrium with any lavas present in an area, suggesting that there is a more primitive, yet unexposed, "parent" liquid that gave rise to the suite exposed by shallow fractionation of olivine. A creative way to estimate such a parental melt composition is to select a primitive glass from which only olivine has probably fractionated (as in the example above) and then to "unfractionate" by mathematically adding olivine. For example, Eggins (1992) added 1% increments of *equilibrium* olivine to a primitive glass composition from Kilauea. (These additions cause the bulk composition to follow an olivine-controlled fractionation trend similar to that shown Figure 11.2.) Equilibrium olivine compositions were computed using an olivine–liquid Fe-Mg K_D value of 0.30 (Roeder and Emslie, 1970). After each addition, the bulk composition was recalculated, and another 1% equilibrium olivine increment

was added until the bulk composition eventually became compatible with the most Mg-rich phenocryst found (which for Eggins was Mg# = 90.5, a value too high to be in equilibrium with any exposed glass composition). The result would be a hypothetical parental magma capable of crystallizing the Mg-rich olivine and evolving to the volcanic compositions observed. This rather stretches our definition of a "parental" magma from Chapter 8 (the most primitive rock found in cogenetic suite), and if this technique is adopted, we might use the definition proposed by Herzberg et al. (2007): "the most magnesian liquid that can be inferred from a given rock suite in the crust." In other words, the most primitive magma to arrive at the shallow magma chamber and leave a trace in terms of phenocrysts carried to the surface by derivative liquids.

How might fractional crystallization occur? **Gravity settling**, such as the sinking of Bowen's olivines, has long been considered the dominant mechanism by which fractional crystallization is accomplished. It involves the differential motion of crystals and liquid under the influence of gravity due to their differences in density. Considerable evidence for the process has also been gathered from field studies of mafic plutonic rocks, such as thick sills and the dramatic layered mafic intrusions (Chapter 12). One particularly clear example is the Duke Island ultramafic intrusion in southeastern Alaska, where one can see numerous sedimentary-type structures, including bedding of different lithologic layers and slump structures (Irvine, 1974, 1979). The common texture known as **cumulate** texture (Figure 3.14), in which mutually touching phenocrysts are embedded in an interstitial matrix, is a result of crystal fractionation and accumulation (although not necessarily always by sinking; see Irvine, 1982).

Due in large part to N. L. Bowen's great influence, including his famous text (Bowen, 1928) and Bowen's Reaction Series (Figure 7.14), fractional crystallization (particularly resulting from the process of gravity settling) achieved its dominant role among petrologic theories on magmatic differentiation. Some considered it the only method by which silicic magmas are created. A wealth of detailed recent work, however, has shown that fractional crystallization is only one of several mechanisms responsible for evolved igneous rocks, and many trends once attributed to fractional crystallization may result from other processes. These alternatives will be addressed in ensuing sections.

In concept, at least, the process by which fractional crystallization and gravity settling produce a sequence of layers in a plutonic rock is straightforward. Let's return for a moment to the discussion of crystallization in the system forsterite–anorthite–silica (Figure 7.4) for a somewhat simplified and idealized example. If we cool bulk composition h, olivine forms first. If the olivine sinks and accumulates on the bottom of the magma chamber, a layer of dunite forms at the base. Any melt trapped between olivine crystals as an intercumulus phase may cool and crystallize under equilibrium conditions if no further crystal fractionation occurs within the very small liquid pockets. The composition of the melt remaining in the main chamber progresses directly away from the Fo corner as olivine forms and sinks until it reaches the olivine–anorthite cotectic, at which point anorthite and olivine crystallize together. If both are heavier than the liquid, they will sink to the bottom as a layer of troctolite above the dunite. The liquid progresses next to the isobaric invariant point c as plagioclase and olivine continue to accumulate. At c, orthopyroxene joins the plagioclase and olivine, so an olivine gabbro (norite) layer forms above the troctolite. As the liquid progresses toward point d, we get an olivine-free norite layer and, finally, a quartz-bearing norite layer at the top when point d is reached.

Gravity settling of minerals in a magma can be quantitatively modeled if we make a few simplifying assumptions. If we simplify the geometrical shape of the mineral to a spherical particle and then assume that the magma is a Newtonian fluid (a fluid with no yield stress, deforming as soon as a differential stress is applied), the settling under the influence of gravity is governed by **Stokes' Law**:

$$V = \frac{2gr^2(\rho_s - \rho_l)}{9\eta} \quad \text{(Stokes' Law)} \quad (11.1)$$

where:

V = settling velocity (cm/sec)

g = acceleration due to gravity (980 cm/sec^2 for the Earth)

r = *radius* of a spherical particle (cm)

ρ_s = density of the solid spherical particle (g/cm^3)

ρ_l = density of the liquid (g/cm^3)

η = viscosity of the liquid (1 g/cm sec = 1 poise)

We can use Stokes' Law to determine the settling velocity of a spherical olivine in a theoretically Newtonian basaltic liquid. Consider fairly typical values for olivine ($\rho_s = 3.3$ g/cm^3, $r = 0.1$ cm) and basaltic liquid ($\rho_l = 2.65$ g/cm^3, $\eta = 1000$ poise). See Appendix A for empirical methods using chemical analyses to estimate the viscosity of magmas. From this, $V = 2 \cdot 980 \cdot 0.1^2 (3.3 - 2.65)/(9 \cdot 1000) = 0.0013$ cm/sec. To non-geologists this may seem like a number approaching zero, but it translates to 4.7 cm/hr, or over 1 m per day. In the five years that the cooling of the Makaopuhi lava lake was studied (and it was largely liquid at the end of that period), these olivines could have settled over 2 km! Plutons solidify over time periods of 10^4 to 10^6 years, permitting considerable gravity settling, if Stokes' Law is any proper measure.

Next consider a rhyolitic melt, with $\eta = 10^7$ poise and $\rho_l = 2.3$ g/cm^2. A 0.1-cm radius hornblende crystal ($\rho_s = 3.2$ g/cm^3) would now settle at a rate of $2 \cdot 10^{-7}$ cm/sec, or 6 cm/year, and feldspars ($\rho_s = 2.7$ g/cm^3) would settle at 2 cm/year. For the feldspar, this equates to about 200 m in the 10^4 years that a stock might cool. If the feldspar were 0.5 cm in radius (1 cm diameter), it would settle at a rate of 0.65 m/yr, or 6.5 km in the 10^4 year cooling interval for a stock. From this analysis, it appears that gravity settling of crystals should be possible for both basaltic and granitic liquids, but it is much more effective in the former because of its low viscosity, explaining why mafic plutons display more obvious textural features of the process. Notice also that plagioclase crystals ($\rho = 2.7$ g/cm^3) would not sink in a slightly Fe-rich basaltic melt ($\rho = 2.7$ g/cm^3) and should even *float* in more Fe-rich liquids.

This analysis using Stokes' Law is overly simplified for a number of reasons. First, the assumption of spherical-shaped crystals is unrealistic. Tabular, accicular, and platy minerals are common and settle with slower velocities, but it is difficult to determine exactly how much slower they are. A far more serious problem involves the assumption of Newtonian fluid behavior. McBirney and Noyes (1979) pointed out that only basaltic magmas near or above their liquidus temperatures behave as Newtonian fluids. Once these begin to crystallize, they develop a significant **yield strength** that must be overcome before any motion is possible. McBirney and Noyes (1979) found that a Columbia River Basalt heated to 1195°C had a yield strength of 60 Pa. In order to overcome this resistance, an olivine crystal must be several centimeters in diameter, well above realistic sizes. The yield strength must be considerably higher for cooler and more silicic liquids. The implication of this study is that gravity settling may be a viable process only within a short temperature interval near the liquidus of a mafic magma. Cooling rate also affects fractional crystallization because it controls the duration of time that a system remains within the optimal non-Newtonian temperature interval.

The previous analysis indicates that only mafic intrusions are susceptible to the effects of differentiation by gravity settling of crystals. There is evidence, however, that many silicic bodies have evolved along a liquid line of descent. Several investigators have used the approach to the ternary eutectic in the systems albite–orthoclase–silica (Figure 11.3) to indicate the evolution of late granitic liquids toward the thermal minimum, or eutectic, composition. Others have used the same clustering of analyses around the minimum to indicate eutectic partial melts of sialic material in the continental crust. Certainly, a eutectic magma could result from either process. Included in Figure 11.3 are the cation norm components of the successive intrusive phases of the Tuolumne Intrusive Series (Figure 4.30). Note the progressive approach of these magmas to the eutectic composition along a trend that follows the decompression eutectic path. Whatever the origin of the parental magma in this case, this series appears to have evolved toward the low-pressure thermal minimum. Harker-type bivariate variation diagrams of intrusive sequences also indicate evolutionary trends. Bateman and Chappell (1979) interpreted the trends for the Tuolumne Intrusive Series to be the result of fractional crystallization. Similar fractional crystallization-based interpretations for Sierran zoned granitoids have been proposed by Bateman and Nokelberg (1978) and Noyes et al. (1983). Although the compositional trends may be compatible with fractional crystallization, crystal settling has been considered very problematic in such viscous silicic magmas (Brandeis and Jaupart, 1986; Sparks et al., 1984). Harper et al. (2005), however, cited viscosity and field criteria supporting crystal settling in hydrous granites. Later in this chapter, we shall explore methods by which fractional crystallization can occur without crystal settling. As we shall see in Section 11.4, mixing of silicic (crustal melts) and mafic (mantle melts) is a popular alternative interpretation to the evolutionary trends in some of these systems.

In addition to gravity settling, three other mechanisms may facilitate the separation of crystals and liquid. **Filter pressing (compaction)**, mentioned earlier in reference to partial melting, is also possible in crystal mushes that form as cumulates or crystal suspensions. The amount of trapped intercumulus liquid between cumulate minerals may be as high as 60 vol. % (Irvine, 1980b). With the added weight of further accumulation, the crystal mush may be compacted (McKenzie, 1984), squeezing much of the liquid out into the main magma body. Another method of filter pressing involves the movement of a phenocryst-laden crystal mush. Any constriction in the conduit causes the crystals to interfere and slow with respect to the liquid.

Another similar mechanism by which crystals may be segregated from the liquid occurs when crystal-rich magmas flow in a laminar fashion near the walls of the magma body. The process is known as **flow segregation** (or flow[age] separation, or flow[age] differentiation). The motion of the magma past the stationary walls of country rock (Figure 11.4) creates shear in the viscous liquid as a result of the velocity gradient near the walls. The resulting differential motion forces the magma to flow around phenocrysts,

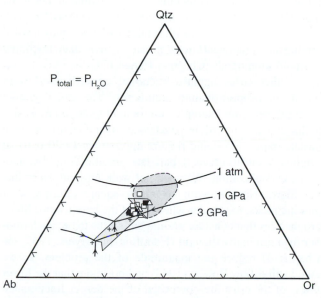

FIGURE 11.3   Position of the H_2O-saturated ternary eutectic (minimum melt composition) in the albite–orthoclase–silica system at various pressures. The shaded portion represents the composition of most granites. Included are the compositions of the Tuolumne Intrusive Series (Figure 4.32), with the arrow showing the direction of the trend from early to late magma batches. Experimental data from Wyllie et al. (1976).

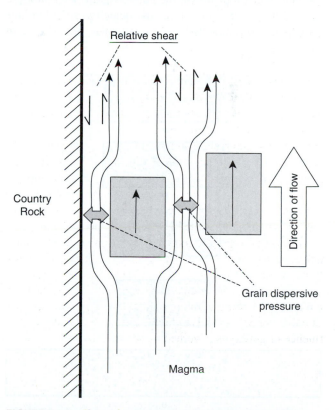

FIGURE 11.4   Flow of magma adjacent to a wall of country rock results in differential motion and shear in the magma. Where such shear is constricted, as between adjacent phenocrysts or between phenocrysts and the contact, a force (called grain-dispersive pressure) is generated and pushes the phenocrysts apart and away from the contact.

thereby exerting pressure on them at constrictions where phenocrysts are near one another or near the contact itself. The pressure, called **grain dispersive pressure** (Komar, 1972), forces the grains apart and away from the contact. This effect is greatest near the walls, and it drops off quickly toward the magma interior, where the flow becomes uniform. Phenocrysts thus concentrate away from the walls to mitigate the pressure buildup. This concentration is most apparent in dikes and sills, where the volume affected by the contact comprises a substantial proportion of the body, resulting in a distinct concentration of coarse phenocrysts toward the center (Figure 11.5). Flow segregation is an interesting, though localized, phenomenon and cannot be responsible for the evolution of more than a small proportion of igneous rocks.

A third mechanism involves the separation and rise of buoyant liquids from boundary layers in which crystals form without themselves moving. This relatively new model has become popular recently and will be introduced in Section 11.5.

The majority of fractional crystallization models assume that fractionation has taken place in a stationary magma chamber at constant pressure. The rise of basaltic magmas, as pointed out by O'Hara (1968b), may involve fairly continuous fractional crystallization as it rises, which must obviously be a *polybaric* fractionation process. One result is that the fractionating minerals vary as their stability fields are crossed (e.g., garnet to spinel to plagioclase). Another is that the shift in the eutectic point with pressure (as in Figure 7.16) also causes the quantity of the liquidus phases that crystallize to vary. In particular, the increase in

the size of the field for olivine with decreasing pressure requires that a lot of olivine must form as the melt composition follows the liquidus away from the olivine side of the diagram in a rising basaltic melt (see Problem 1). Thus, the relative amount of olivine that crystallizes with a rising basaltic magma will be far greater than the amount that forms during isobaric crystallization.

The broad acceptance of Bowen's (1928) contention that fractional crystallization is the predominant mechanism of magmatic differentiation is now being questioned. This one process cannot account for all of the diversity in the broad spectrum of natural igneous rocks, even if we allow for variations due to the influence of changes in pressure or associated fluids. Some observed chemical trends simply cannot be accomplished by fractional crystallization. Other classical examples of fractional crystallization have not withstood more critical analysis and the test of time. For example, the 300-m thick Triassic diabase Palisades Sill, on the eastern banks of the Hudson River, is commonly cited as an example of a vertically differentiated sill with layers formed by gravity settling. The overall composition of the sill is tholeiitic basalt, as demonstrated by the upper and lower chill zones. The 10- to 20-m thick olivine-rich layer at the base is commonly attributed to differentiation by settling and accumulation of early-forming dense olivine crystals. Although vertical chemical trends in the sill are compatible with fractional crystallization of pyroxene and pyroxene accumulation zones occur near the bottom of the sill, the striking olivine layer is not compatible with the trends, and olivine is far too rare elsewhere in the sill to be consistent with the concentration in the layer. The layer has recently been reinterpreted as one of several late intrusions of magma into the crystallizing tholeiitic liquid of the sill. This injected pulse was olivine rich and dense, so it accumulated near the base (Husch, 1990). Some magma series, such as the calc-alkaline series associated with subduction zones, may involve mixing of components to a greater extent than fractional crystallization trends (Chapters 16 and 17).

Other cases against fractional crystallization were based on proportionality arguments. The great granite batholith belts, for example, are thought to be too extensive to have been created by fractional crystallization from a basaltic parent. It would require approximately 20 parts of original basalt to create 1 part late granitic liquid by fractional crystallization. We need not walk long in places like the Sierra Nevada, with so many square kilometers of granitic rocks, to wonder where all of the basalt went! Modern theories that consider granite batholiths to be much thinner than originally thought (Hamilton and Myers, 1967, see Chapter 4) reduce the magnitude of the problem, so we could still follow Bowen (1948) and postulate that the lower levels of the crust are composed of the denser fractionated gabbros. Seismic and gravity surveys, however, argue against this possibility, making the origin of granite batholiths via fractionation from a basaltic parent untenable (Presnall, 1979). These arguments serve only to demonstrate that fractional crystallization cannot lead to *all* of the magmatic rocks now exposed at the surface of the Earth. It is still a

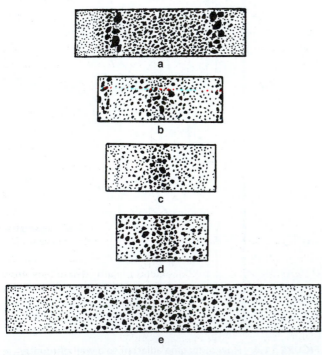

FIGURE 11.5 Increase in size and concentration of olivine phenocrysts toward the center of small dikes by flow differentiation. Isle of Skye, Scotland. After Drever and Johnston (1958). Reproduced by permission of the Royal Society of Edinburgh.

common and important process, particularly in the early crystallization of mafic liquids, but there are other important differentiation processes and other primary magmas.

11.2.2 Volatile Transport

Chemical differentiation can also be accomplished when a separate vapor phase coexists with a magma and liquid–vapor fractionation takes place. A vapor phase may be introduced in any of three principal ways. First, a fluid may be released by heating of hydrated or carbonated wall rocks. We shall discuss some ramifications of this process in later sections of this chapter.

Second, as a volatile-bearing but undersaturated magma rises and pressure is reduced, the magma may eventually become saturated in the vapor, and a free vapor phase is released (see the discussion of the H_2O-albite system in association with Figure 7.22). Because the vapor phase has a lower density than the melt, it rises, diffusing through the magma, and concentrates near the top of the magma chamber. Such concentrated fluid may even permeate into the roof rocks. This process usually involves an H_2O-rich fluid, and it produces a variety of hydrothermal alteration effects. For example, the alkali metasomatism known as **fenitization** above nephelinite–carbonatite bodies has been attributed to alkali-rich fluids derived from the highly alkaline intrusives (Section 19.2.3).

A third mechanism for generating a separate fluid phase is a result of late-stage fractional crystallization. Most early-formed igneous minerals are anhydrous (even hydrous minerals are less so than associated melts), so their segregation from a hydrous melt enriches the melt in H_2O and other volatile phases. Eventually the magma reaches the saturation point, and a hydrous vapor phase is produced. This somewhat paradoxical "boiling off" of water as a magma *cools* has been called **retrograde** (or **resurgent**) **boiling.**

Of course, the three processes by which a vapor can be produced need not be entirely separate, and all three may contribute to saturation and volatile release from a magma, depending upon the composition of the original magma, the rates of cooling and rise, the initial volatile content, the extent of fractional crystallization, the temperature, the nature of the wall rocks, etc. As a separate vapor is produced, the chemical constituents in the system partition themselves between the liquid and vapor phases in appropriate equilibrium proportions, some remaining preferentially in the melt and others becoming enriched in the vapor phase. The result is a silicate-saturated vapor phase in association with a vapor-saturated silicate liquid phase.

The cation sites in minerals are much more constrained and selective than in melts, so the chemical constituents in minerals are generally much simpler. As a result, the process of fractional crystallization tends to remove only a few elements from the liquid in significant quantities, and a number of incompatible, LIL, and non-lithophile elements become concentrated in the latest liquid fraction. Many of these will further concentrate in the vapor, once formed. This is particularly true in the case of resurgent boiling because the melt already is evolved by the time the vapor

phase is released. The vapor phase may contain unusually high concentrations of volatile constituents such as H_2O, CO_2, S, Cl, F, B, and P, as well as a wide range of incompatible and chalcophile elements.

The volatile release and concentration associated with pluton rise or resurgent boiling may momentarily increase the pressure at the top of the intrusion and fracture the roof rocks in some shallow intrusions (it may also initiate volcanic eruptions). Both the vapor phase and some of the late silicate melt are likely to escape along a network of these fractures as dikes of various sizes. The silicate melt commonly crystallizes to a mixture of quartz and feldspar. It is typically found in small dikes with a sugar-like texture, which is informally called **aplite**. The vapor phase is typically concentrated as dikes or pods in, or adjacent to, the parental granitic pluton, where it crystallizes to form a characteristically **magmagenic** form of **pegmatite**.

Although pegmatite is used as a textural classification term for very coarse grain size (Chapter 2), and there are other methods of creating large crystals, the type of pegmatite described above is the most common. The large grain size in magmagenic pegmatites is not due to a slow cooling rate but is a result of poor nucleation and very high diffusivity in the H_2O-rich phase, which permits chemical species to migrate readily and add to rapidly growing minerals. The size of crystals in pegmatites can occasionally be impressive, such as spodumene, microcline, or mica crystals 6 to 10 m across. Most pegmatites are "simple," essentially very coarse granites. Others are more complex, with a tremendous concentration of incompatible elements and a highly varied mineralogy, commonly displaying a concentric zonation (Jahns and Burnham, 1969; Černý, 1991; Simmons et al., 2003), as shown in Figure 11.6. Because the late fluid segregation concentrates several unusual elements, pegmatites are important economic resources and are mined for Li, Be, the rare earths, W, Zr, and a host of others elements that are rarely concentrated in other environments. They are also a major source of gems.

Vapors that completely escape the magma and move to higher levels may cool further and precipitate low-temperature minerals, such as sulfides in a hydrothermal system (commonly mixed in part with meteoric water).

Miarolitic pods, or **cavities**, are smaller fluid segregations trapped in the plutonic host. When finally exposed at the surface, they are coarse mineral clusters (usually a few centimeters across), the centers of which are typically hollow voids from which the fluid subsequently escaped. The hollow cavities have euhedral crystals (of the same minerals comprising the pluton) that extend inward, where they grew into the fluid, unimpeded by other minerals. Like complex pegmatites, some miarolitic cavities or pods have a concentric structure consisting of layers of different mineralogy (Jahns and Burnham, 1969; McMillan, 1986.

Because the addition of H_2O lowers the melting point of magmas, the release of hydrous fluid into the country rocks causes the liquidus temperature in the main magma body to rise suddenly, resulting in rapid crystallization of much of the liquid remaining with the previously formed

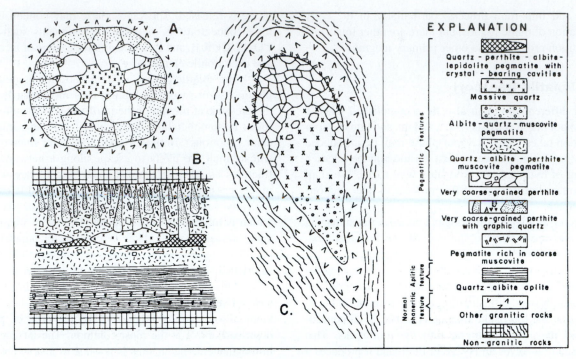

FIGURE 11.6   Schematic sections of three zoned fluid-phase deposits (not at the same scale). **(a)** Miarolitic pod in granite (several centimeters across). **(b)** Asymmetric zoned pegmatite dike with aplitic base (several tens of centimeters across). **(c)** Asymmetric zoned pegmatite with granitoid outer portion (several meters across). From Jahns and Burnham (1969).

minerals. This is an alternative way of generating porphyritic texture and is common in many silicic plutons.

11.2.3 Liquid Immiscibility

Two liquids that don't mix seems an unlikely occurrence, and it is. Yet most of us are familiar with salad oil and oil slicks, so we have some concept of the phenomenon. Many oils do not mix with water, and, because they are less dense, the oil floats to the top of the water, forming a distinct layer. Most immiscible phases, whether liquids or solids, homogenize at elevated temperatures due to the increased entropy and molecular vibrational energy, although for oil–water at atmospheric pressure, the homogenizing temperature is above the boiling point of water. The solvus, representing liquid or solid immiscibility on a phase diagram, is therefore convex upward on a temperature–composition diagram (Figure 6.16). We have already encountered immiscible liquids in the forsterite–silica system (Figure 6.12), where, on the high-silica side of the diagram, a highly silica-rich liquid separates from a less silica-rich one.

Throughout the 20th century, geologists appealed to liquid immiscibility as a mechanism for magmatic differentiation, thinking that it might be responsible for the separation of a granitic liquid from an evolving system (presumably from an initial basaltic parent). Such a separation into contrasting liquid systems was also used to explain enigmatic cases of bimodal volcanism, such as the basalt–rhyolite occurrences of the Snake River–Yellowstone area, or the Basin-and-Range of the southwestern United States.

There are two problems with applying the forsterite–silica liquid immiscibility gap to natural magmas. First, the temperature of liquid immiscibility is far too high (over 1700°C) to represent a reasonable crustal process. Of course, the Mg-Si-O system is rather restricted, leading one to ask whether the addition of other components, required to create more natural magmas, would lower the temperature of the solvus. The effect, however, of adding alkalis, alumina, etc. is to eliminate the solvus completely (see Figure 7.4). When this was experimentally demonstrated, liquid immiscibility was relegated to the compost pile of magmatic processes.

Interest was renewed when Roedder (1951) discovered a low-temperature immiscibility gap in the central portion of the fayalite–leucite–silica system (Figure 11.7) at temperatures and compositions that are conceivable for some Fe-rich natural magmas. Roedder (1979) provided a review of liquid immiscibility in silicate magmas, citing dozens of references in which natural occurrences of immiscible liquids were described, including a significant proportion of the lunar samples returned by the Apollo program.

Three natural magmatic systems are widely recognized as having immiscible liquids in some portion of their compositional range. The first is the system mentioned above, which most commonly translates to natural **Fe-rich tholeiitic basalts**, which experience an initial trend toward iron enrichment (Figure 8.14). In the later stages of fractionation, a "granitic" melt ($>75\%$ SiO_2) separates from a basaltic melt ($\sim40\%$ SiO_2). Once separated, the silicic liquid must have a much lower density than the Fe-rich mafic liquid, and we would expect it to rise and collect near the top of the magma

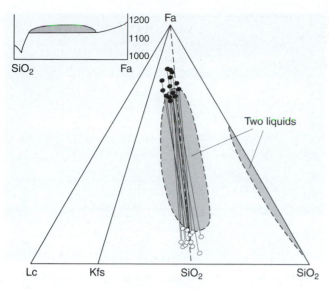

FIGURE 11.7 Two immiscibility gaps in the system fayalite–leucite–silica. The central one is of a composition, and at a low enough temperature (see the section in the upper left) to be attainable in some Fe-rich natural magmas (after Roedder, 1979, copyright © the Mineralogical Society of America). Projected into the simplified system are the compositions of natural immiscible silicate pair droplets from interstitial Fe-rich tholeiitic glasses (Philpotts, 1982).

chamber. Crystallization of the magma must be advanced by the time liquid separation occurs, however, and both liquids are likely to become trapped in the already-formed crystal network. Philpotts (1982) described the textures of some Fe-rich Hawaiian basalts in which small droplets of the two immiscible liquids are mingled in the interstitial glass trapped between plagioclase and augite crystals. The separate droplet compositions may be determined by microprobe and are projected into the Fa-Lc-silica system in Figure 11.7, along with the liquid immiscibility gap of Roedder (1951). The actual liquid compositions plot slightly outside the experimental gap, probably because of the effects of Fe_2O_3, TiO_2, and P_2O_5, which expand the immiscible field. The low oxygen fugacity of the lunar basalts is the probable reason that immiscible liquids are so common in them.

Observing immiscible droplets is clear evidence of the process, but evidence is far less obvious that immiscible granitic liquids have separated and formed substantial segregations from Fe-rich tholeiites that are over 70% crystallized. Perhaps filter pressing may aid the process, and the granophyric layers and lenses at the top of many mafic intrusions, including the Palisades Sill and the Skaergård intrusion (McBirney, 1975; see also Chapter 12) may be the products of immiscible liquids. In such cases, liquid immiscibility is a late-stage addition to a more extensive process of fractional crystallization in these mafic intrusions. Granitic bodies and other large-scale evolved liquids, however, are unlikely products of immiscible liquids.

A second system displaying immiscible liquid behavior is the separation of a sulfide-rich liquid from a sulfide-saturated silicate magma. Less than one-tenth of a percent of sulfur is sufficient to saturate a silicate magma and release

an iron–sulfide melt that is also rich in Cu, Ni, and other chalcophile elements. Small, round, immiscible sulfide droplets in a silicate glass matrix, similar to Philpotts' (1982) granitic–tholeiitic examples above, have been observed in a number of quenched ocean basalt glasses. Economically important **massive sulfide** segregations in large, layered mafic complexes have formed by separation and accumulation of immiscible sulfide melts.

A third liquid immiscibility gap occurs in highly alkaline magmas that are rich in CO_2. These liquids separate into two fractions, one enriched in silica and alkalis and the other in carbonate. These give rise to the nephelinite–carbonatite association, discussed further in Section 19.2.5.2.

Although these are the three generally recognized occurrences of immiscible liquids, other magmas might separate into two liquid phases under certain circumstances. These possibilities include lamprophyres (Philpotts, 1976; Eby, 1980), komatiites, lunar mare, and various other volcanics (see Roedder, 1979, for a summary).

The close spatial and temporal association of contrasting liquids may result from a number of processes in addition to liquid immiscibility. We can apply three tests to juxtaposed rocks to evaluate them as products of immiscible liquids. First, the magmas must be immiscible when heated experimentally, or they must plot on the boundaries of a known immiscibility gap, as in Figure 11.7. Second, immiscible liquids are in equilibrium with each other, and thus they must also be in equilibrium with the same minerals. If the two associated liquids crystallized different minerals or the same mineral with different compositions, they cannot be an immiscible pair. Finally, we may be able to use the pattern of trace element fractionation between the two liquids to evaluate them as immiscible. Partitioning of minor and trace elements between Fe-rich mafic liquids and granitic liquids, for example, can be distinctive when compared to the more common mafic magmas with less Fe. Some incompatible elements (P, for example) are preferentially incorporated into an Fe-rich mafic liquid over the complimentary silicic one. A granitic rock relatively depleted in these incompatible trace elements may be a product of liquid immiscibility. Of course, a low concentration in a particular trace element can also result if the liquid was derived from a similarly depleted source. It is far more reliable, then, if rocks representing *both* of the immiscible liquids can be evaluated. This has been accomplished for some mixed dike rocks (Vogel and Wilband, 1978), but no one has yet succeeded in identifying a medium-sized or larger granite as derived from an immiscible liquid.

Although liquid immiscibility is now widely accepted as a phenomenon in natural magmas, the extent of the process is still in question, and its importance in generating large bodies or a significant proportion of evolved magmatic rocks is doubtful.

11.3 MAGMA MIXING

Magma mixing is a bit like liquid immiscibility in reverse, and so was some of the reasoning behind its historical origins. The reigning paradigm of fractional crystallization implies gradually decreasing proportions of evolved liquids stemming from basaltic parent magmas. The common occurrence of rhyolites/granites, resulting in a bimodal basalt–rhyolite–dominated distribution of composition in many areas, invited a wide variety of explanations. One school of thought at the beginning of the 20th century considered basalts and rhyolites/granites to be the *two* main primitive magmas—one produced by partial melting of the mantle and the other derived similarly from the crust—with the intermediate magmas created by mixing these two in various proportions. Interest in magma mixing waned after Bowen (1915, 1928) demonstrated fractional crystallization experimentally and so forcefully proposed it as a natural mechanism. Although the variety of natural magmas is much broader than can be explained by the mixing of only two "end-members," the literature is becoming increasingly rich with examples of magma mixing.

For a suite of rocks from a particular province, we should be able to test for the effects of **end-member mixing** that result from the simple combination of two magma types. If we plot the components of the suite on Harker-type variation diagrams, the variation in each element or oxide must lie on a straight line between the values representing the two most extreme compositions (presumably representing the two mixed parental magmas). As we have seen in Chapter 8, some (indeed nearly all) volcanic suites have a number of curved lines that cannot be explained by strict binary end-member mixing. As we saw in Figures 8.6d and 8.6e, curved trends can result from the sequential extraction of two or more minerals, or the extraction of minerals of varying composition, and fractional models have usually been based on these premises. Mixing of hotter basaltic liquids and cooler silicic ones, however, invariably leads to cooling and crystallization of the former, so magma mixing plus crystallization can result in more complex curved patterns on variation diagrams. Mixing of more than one parental type can also result in nonlinear variations. Conversely, linear patterns on variation diagrams are no guarantee of magma mixing. The cause could be either magma mixing or fractional crystallization of a single phase (as in Figure 11.2). Usually textural or field evidence can differentiate between the two possibilities. In the case of Figure 11.2, the guilty perpetrators, the olivine phenocrysts, are easily observed.

The dynamics of magma mixing depend on the contrasting magma properties, such as the temperature, composition, density, volatile content, and viscosity, as well as the location and the turbulence with which one magma injects into the chamber containing the other. Recent studies by Huppert and Sparks (1980), Campbell and Turner (1986), Sparks and Marshall (1986), and Tait and Jaupart (1990) have attempted to assess and quantify the effects of these factors on the type and degree of mixing that takes place in magma chambers. Gas-charged or pressure-induced fountaining may mix magmas considerably, whereas quiescent injection of dense magmas may pond at the bottom of a silicic magma chamber and result in stratification with little mixing at all. If basalt composes most of the mixture, mixing occurs readily under a variety of conditions.

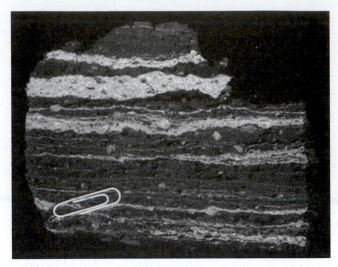

FIGURE 11.8   Commingled basaltic and rhyolitic magmas, Mt. McLoughlin, Oregon. Sample courtesy Stan Mertzman.

There is ample evidence for the mixing process taking place. It is most evident in cases in which the magmas are very different, such as basaltic and intermediate or silicic liquids. Due to the large differences in the physical properties of the contrasting magmas, the degree of mixing of these magmas may be limited. Two magmas can commonly be seen as **commingled** swirls of contrasting colors (Figures 3.13 and 11.8) on the hand sample or outcrop scale, or even as intimate mixtures of contrasting glass in thin section. As mentioned above, because basaltic is initially at a higher temperature than silicic magma, their commingling would tend to chill the basalt and superheat the silicic magma. Basaltic magmas entering granitic chambers commonly form pillow-like structures with curved boundaries and glassy quenched marginal textures that accumulate at the bottom of the chamber (Figure 11.9).

FIGURE 11.9   Pillow-like blobs created as basaltic magma is injected into a granitic magma chamber. The hotter, denser basalt is partially quenched, and the blobs accumulate at the bottom of the chamber, trapping some of the interstitial silicic liquid. Vinalhaven Island, Maine.

Some magma mixing occurrences may appear like immiscible liquids, but the mixed magmas are not in equilibrium, so disequilibrium assemblages (such as plagioclase or pyroxenes of radically different compositions, or juxtaposed olivine and quartz, and corroded and partially resorbed phenocrysts) are common. For similar magmas, the mixing may be more extensive, resulting in homogeneous mixtures, but the evidence for such mixtures is also obscured.

Replenishment of a differentiating magma chamber by reinjection of more primitive parental magma from below is probably a fairly common process. Such cases are far from the original concept proposed for dissimilar magmas mixing to produce an intermediate result and are really a type of genetic variation (although backward) within a single magmatic province. Replenishment can be documented structurally (cross-cutting dikes and layers), mineralogically (reversing back up a liquid line of descent and perhaps the *loss* of a crystallizing liquidus phase), texturally (some cases of phenocryst resorption; and zoning reversals in plagioclase or other minerals, as illustrated in Figure 3.6), or geochemically (see O'Hara and Matthews, 1981) and isotopically (DePaolo, 1985; Palacz, 1985). It is not yet known how common is the process of replenishment of a magma chamber in general, but it is probably common. It is considered to be a particularly important part of the evolution of long-lived magma chambers at mid-ocean ridges, as we shall see in Chapter 13.

Marsh (2002) suggests that when fresh magma invades a chamber in which the center has reached an advanced stage of crystallization, the new liquid pushes out the later-stage fractionated interstitial melt in a process called *flow fractionation*. The expelled liquid is free to erupt, and the new liquid, now out of equilibrium with the older crystals, will dissolve and perhaps disrupt them.

Magma mixing has experienced a modern renaissance, beginning perhaps with Eichelberger and Gooley (1975). A number of studies have been published documenting large-scale mixing processes in volcanic fields and plutonic complexes, including Iceland on the Mid-Atlantic Ridge (Sigurdsson and Sparks, 1981; McGarvie, 1984), Hawaii (Wright, 1973; Garcia et al., 1989, 1992), the Columbia River plateau basalts (Hooper, 1985), layered mafic intrusions (Chapter 12), continental arc volcanics (Eichelberger, 1975; Grove et al., 1982, Smith and Leeman, 1993), and granitic plutons, to name just a few. We shall see in Chapters 16 and 17 that mixing of mantle and crustal melts may be an important process in the evolution of many subduction-related calc-alkaline magmas. Reid et al. (1983), Frost and Mahood (1987), and Sisson et al. (1996) interpreted a significant proportion of the evolutionary trends of Sierran intrusives to be a product of mixing of silicic (crustal) melts and mafic (mantle) melts, as did Kistler et al. (1986) for the now-familiar Tuolumne Intrusive Series.

11.4 ASSIMILATION

Assimilation is the incorporation of chemical constituents from the walls or roof of a magma chamber into the magma itself. Assimilation may be capable of significantly altering the composition of a magma. Evidence for partial assimilation can be found in variously altered and resorbed contacts or xenoliths suspended in igneous rocks. Historically, there have been several zealous proponents of assimilation who have argued that many of the compositional variations in igneous rocks result from extensive assimilation of country rocks by more primitive magmas.

The degree to which a magma can assimilate the country rock by melting is limited by the heat available in the magma. As pointed out by Bowen (1928), the country rock must first be heated to the melting point and then at least partially melted in order to be assimilated, and this heat must be supplied by the magma itself. Let's evaluate the process, using the same estimates we used in Chapter 10 for the specific heat (the energy required to raise the temperature of 1 g of rock 1°C) of 1 J/g °C and heat of fusion (the energy required to melt 1 g of rock at the solidus) of 400 J/g. In order to bring 1 g of granitic country rock from an ambient temperature of 200°C to a hypothetical melting point at 800°C would require 600°C · 1 J/g °C = 600 J (per gram). To melt a gram would require a further 400 J. The magma would have thus to expend 1000 J of energy to heat, melt, and assimilate one gram of country rock. If the magma is at its liquidus temperature (as most magmas are), the only energy it can supply is its own heat of crystallization, so 2.5 g of magma would have to crystallize in order to melt and assimilate 1 g of wall rock. It is thus theoretically possible for a magma to assimilate 40% of its weight in country rock, and more if the country rock is initially closer to its liquidus temperature. In a more rigorous approach, using Mark Ghiorso's thermodynamic modeling program MELTS, Reiners et al. (1995) calculated that 3 to 7% fractional crystallization of olivine from basalt could, in the early stages, efficiently assimilate 5 to 18% felsic crust via isenthalpic (heat balanced) processes. Once plagioclase and/or pyroxene began to crystallize, the process became one-quarter to one-half as efficient, but significant assimilation with even small degrees of crystallization appear to be possible on the basis of heat budget.

The portion of the magma that crystallizes in order to supply the heat for assimilation occurs at the cool walls where this heat energy is consumed. Here it likely forms a barrier to inhibit further exchange with the wall rocks (unless turbulent flow or gravitational effects continuously sweep the contact clean, as proposed for some situations by Huppert and Sparks, 1985). Assimilated components must then diffuse through the marginal barrier. As diffusion of heat is much faster than diffusion of mass, formation of such a boundary should inhibit chemical exchange in most cases, and the magma would solidify before appreciable assimilation has occurred.

Stoping of wall and roof rocks (Chapter 4) to create xenoliths can enhance the assimilation process by increasing the surface area of country rocks exposed directly to hot interior magma. **Zone melting** (Section 4.2.7), a process that has been proposed as a method of pluton emplacement, involves considerable assimilation of country rock (Ahren et al., 1981; Huppert and Sparks, 1988a). In this process, a magma melts its way upward by crystallizing an amount of igneous material at the base of the pluton equivalent to the amount melted at the top, transferring heat between the two zones by convection (Harris, 1957; Cox et al., 1979). The process is similar to **zone refining**, an industrial process in which a material (usually a metal bar of some type) is passed through a small furnace, where a very thin cross section melts and solidifies again as the bar passes the hot spot. As a result, a melted zone passes through the bar, from one end to the other, picking up impurities that partition themselves into the melt rather than the solid. The melt thus sweeps the impurity out of the material. In a similar fashion, a rising magma can concentrate an unusual amount of incompatible trace elements as it zone melts its way upward (major elements do not concentrate in this manner; see Section 9.5). Geologists do not yet agree on the importance of zone melting as a method of magmatic rise or diversification. If it works extensively at all, it is most likely to occur with large mafic magmas in the deep crust, where the country rocks are already at or near the liquidus. Marsh (1982) has argued against it as a major factor in arc magmas because magmas that pass through continental crust while assimilating a large component of wall rocks should tend toward rhyolites, and those restricted to oceanic island arcs should tend toward tholeiites, which is not generally the case. Wilson (1989) pointed out that, in the case of basaltic magmas rising through the continental crust, the assimilated roof rocks are likely to be less dense than the rising magma and pond at the top of the chamber, protecting it from further erosion.

Although partial melting of the wall rock is considered the dominant method by which a magma can assimilate constituents, reaction and diffusive exchange, perhaps aided by absorption of volatiles, can also incorporate components from solid country rock into the magma (Patchett, 1980; Watson, 1982). Exchange between rising mantle-derived magmas and the upper mantle or oceanic crust is difficult to detect or evaluate because of the chemical similarity of the melt and the wall rocks. The continental crust is another matter, however, and there are still many adherents who think it unlikely that mantle-derived magmas pass through the sialic continental crust without assimilating at least a modest contribution, composed predominantly of the components with low melting temperature (silica, alkalis, etc.). The effects of small degrees of assimilation are still unlikely to be conspicuous because the major elements are the same ones in the magma and the country rocks. There may be a shift in bulk composition, but in most cases the melt still progresses toward the eutectic. Only the relative proportions of the solids change. In the case of solid solutions (Figures 6.9 and 6.10), assimilation of more sialic wall rocks causes the bulk composition to shift toward the low-temperature component, and the final solid will be somewhat enriched in that component (an effect identical to that of fractional crystallization). Only in the case of a thermal divide (as in Figure 8.13) can the composition be shifted across it by assimilation so that the liquid would evolve to a different eutectic. Similarly, in peritectic systems, assimilation can shift the bulk composition across the peritectic, resulting in a different final mineral assemblage. For example, addition of silica to a liquid, such as

f in Figure 6.12, can shift the liquid from the left of enstatite to the right of it, and the final assemblage will contain quartz, not olivine.

Although major elements are relatively insensitive to the effects of assimilation of limited amounts of country rock, this is not true for trace elements and isotopes. As mentioned in Section 9.5, trace elements, because they are present in such small quantities, are not "buffered" by the solid–liquid equilibrium in the same manner as major elements. Figure 6.10 again illustrates this point. Suppose we had a system equivalent to Fo_{30} at 1480°C, with $liquid_d$ coexisting with $olivine_a$, and added enough pure Fo to bring the bulk composition to Fo_{45}. This would be equivalent to assimilating solid forsterite amounting to half the original liquid. The effect would be to add considerable Mg to the overall system, but the addition of pure Fo would have to cause crystallization of equilibrium $olivine_a$ in this ideal isothermal example. This follows directly from the lever principle: bulk composition Fo_{30} requires a higher $liquid_d/solid_a$ ratio at 1480°C than does bulk composition Fo_{45}. Adding 50% assimilant in this case has no effect on the major element composition of either phase at equilibrium. The system effectively buffered the addition of Fo by crystallizing olivine of composition a. If the assimilated forsterite contained 200 ppm Ni, however, and the original system had only 100 ppm Ni, assimilation of 50% forsterite would change the trace element concentration of the combined system to 133 ppm and raise the Ni concentration of both the $liquid_d$ and $olivine_a$ by a proportionate 50%. Thus the addition of a particular trace element simply adds to the overall trace element content of the melt and minerals in the magmatic system. Some trace elements are much more abundant in the continental crust than in mantle-derived magmas, and the assimilation of a modest amount of crustal material rich in that element may have a considerable effect on a magma initially containing little of it (see McBirney, 1979, and Wilson, 1989 for reviews).

The effects of assimilation of continental crust might best be illustrated using a simple mixing approach. Figure 11.10 is a spider diagram from Wilson (1989) for a typical mid-ocean ridge tholeiite (chosen to represent mantle-derived basaltic magma free of contaminants) and the patterns resulting from assimilation of 15% typical deep-crustal gneiss or 15% typical upper to mid-crustal gneiss. Assimilation in this model involves mixing of 85% basalt and 15% gneiss, as well as trace element partitioning between the phases using appropriate partition coefficients. The resulting patterns are different and appear to indicate not only that assimilation has occurred but at what crustal levels. This approach may allow us to select certain trace elements for different purposes in the analysis of crustal contamination of basalts. For example, the more compatible elements Ti, Tb, Y, Tm, and Yb do not partition from the crust into the melt and thus do not modify the original basalt pattern. They are useful if we wish to see through the contamination effects to the trace element characteristics of the unmodified basalt. Ba and the group Nb to Hf are enriched in the contaminated magmas and might be useful as general indicators of crustal contamination. The Nb-Ta trough appears to be a distinctive feature of contaminated magmas. The elements Rb, Th, and K are also enriched in

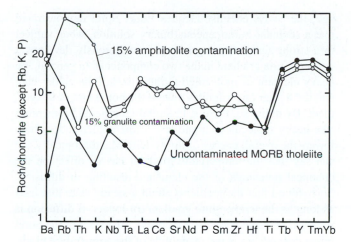

FIGURE 11.10 Chondrite-normalized multielement (spider) diagram of a typical mid-ocean ridge basalt and the effects of mixing 15% of typical deep crustal granulite facies gneiss and 15% typical mid-crustal amphibolite facies gneiss. From Wilson (1989). Copyright © by permission Kluwer Academic Publish.

11.5 BOUNDARY LAYERS, IN SITU CRYSTALLIZATION, AND COMPOSITIONAL CONVECTION

Magmatic differentiation involves partitioning elements between two phases (one a liquid and the other a solid, liquid, or vapor) and the subsequent differentiation that results when those phases are physically separated. Recent studies of magma chambers have shown that many are diversified in ways not adequately explained by the classical mechanisms of crystal settling. This has led several researchers to reevaluate historical ideas of magmatic differentiation and to propose alternative methods in which diversification takes place by **in situ** (in place) crystallization and compositionally induced convective processes within an initially stationary liquid or liquid–solid boundary layer.

Since the 1960s, the geological literature has contained several works addressing the processes involved with boundary-layer effects and convective fluid motion in multicomponent systems with gradients in temperature, density, and composition (Shaw, 1965; Bartlett, 1969; Turner, 1980; Sparks et al., 1984; Huppert and Sparks, 1984a; Brandeis and Jaupart, 1986; Langmuir, 1989; Tait and Jaupart, 1990, 1992, 1996; Jaupart and Tait, 1995). Layered mafic intrusions provide excellent natural laboratories for the study of magmatic differentiation and have fostered many of our ideas on the nature of fractional crystallization and gravity settling. A more thorough discussion of these processes as they apply to magma chambers requires a better look at the structures, textures, and chemical trends of these fascinating bodies. I shall thus postpone the study of fractional crystallization in convecting systems until Chapter 12, when we address layered mafic intrusions and have a better basis for evaluation. For now, by way of introduction to in situ processes, I present a work from a well-documented field area: the 0.76 Ma Bishop Tuff, erupted at Long Valley, California.

Hildreth (1979) proposed a novel model for the well-documented vertical compositional variation in the stratified Bishop Tuff. He proposed that the magma chamber beneath what is now the Long Valley caldera was progressively emptied from the top downward during the eruption, so the bottom-to-top variation in the tuff should correlate with top-to-bottom variations in a compositionally stratified upper chamber. Using mineral geothermometry (Chapter 27), Hildreth was able to demonstrate that the earliest eruptive materials (now at the base of the flow but representing the top of the chamber) erupted at 720°C, whereas the latest material (lower in the chamber) erupted at 780°C. He pointed out a number of trends based on mineral densities and the relationship between mineralogical and chemical enrichment that could not have resulted from classical fractional crystallization. Due to the high viscosity of such silicic magmas, crystal settling and thermal convective motion is expected to be minimal anyway (Brandeis and Jaupart, 1986; Sparks et al., 1984).

Instead of fractional crystallization, Hildreth (1979) proposed a model based on the in situ convection–diffusion model of Shaw et al. (1976). According to Hildreth's theory, the magma near the vertical contacts became enriched in

contaminated magmas, but they may also be able to distinguish upper and lower crustal effects. Relatively high proportions of these elements suggest an upper crustal contaminant, whereas low relative proportions in a basalt that appears to be contaminated based on the Ba-Hf criteria are more probably contaminated by deep crustal materials.

Probably the best way to detect the effects of assimilation, particularly the contamination of mantle-derived magmas by continental crust, is isotopically. As we saw in Chapter 9, the continental crust becomes progressively enriched over time in $^{87}Sr/^{86}Sr$ (Figure 9.13) and depleted in $^{143}Nd/^{144}Nd$ (Figure 9.15). Primitive magmas with unusually high values of $^{87}Sr/^{86}Sr$ and low values of $^{143}Nd/^{144}Nd$ are thus probably contaminated by ancient continental material. We learned in Section 9.7.2.2 that $^{87}Sr/^{86}Sr$ values below 0.706 would be appropriate for relatively unmodified mantle melts, whereas ratios above that value are probably contaminated by old continental components. The continental crust is greatly enriched in U, Pb, and Th relative to the mantle and the oceanic crust. Given the decay schemes in Section 9.7.2.4, we can see that, over time, the continental crust becomes enriched in ^{207}Pb and ^{206}Pb by the breakdown of U in the crust. Because ^{204}Pb is nonradiogenic, we can thus expect the $^{207}Pb/^{204}Pb$ and $^{206}Pb/^{204}Pb$ ratios to be considerably higher in the older continental crust than in the mantle or in mantle-derived melts (because no isotopic fractionation occurs during melting). Initial $^{87}Sr/^{86}Sr$ and $^{143}Nd/^{144}Nd$ ratios or $^{207}Pb/^{204}Pb$ or $^{206}Pb/^{204}Pb$ can be plotted against some differentiation index on Harker-type variation diagrams for suites of cogenetic magmas to determine the extent of crustal contamination. Linear enrichment trends in isotopic ratios with increasing differentiation suggest either continuous assimilation or mixing of mantle and crustal magmas, although I suspect that the latter is more common. Increased crustal isotopic signatures in only the most evolved magmas is more likely with assimilation of felsic crust in the marginal portions of the chamber. Field criteria, such as commingled magmas or partially melted crustal xenoliths, may also help us choose between the alternatives.

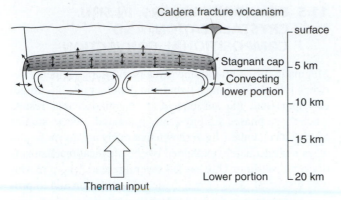

Caldera fracture volcanism

Stagnant cap
Convecting
lower portion

Lower portion

Thermal input

surface
5 km
10 km
15 km
20 km

FIGURE 11.11 Schematic section through a rhyolitic magma chamber undergoing convection-aided in situ differentiation. The upper high-silica cap is stagnated by an H_2O gradient. Because of this gradient, as well as thermal and density gradients, the cap zone may experience internal diffusion and exchange with the wall rocks and the convecting system below. Single-headed arrows indicate convection, and double-headed arrows indicate diffusion. After Hildreth (1979). Copyright © The Geological Society of America, Inc.

H_2O from the wall rocks. This H_2O-enriched boundary layer, although cooler, was less dense than the interior magma, and it rose under the influence of gravity to concentrate at the top of the chamber. This resulted, he proposed, in a growing density-stabilized boundary-layer cap that inhibited convection in the top portion of the magma chamber (although convection is likely to continue in the main portion of the chamber below the cap; see Figure 11.11). Although the cap rock was relatively stationary, there were initial gradients in temperature and H_2O content, with the most H_2O-rich, low-density liquids increasing upward. The H_2O gradient, Hildreth reasoned, should affect the structure of the melt and the degree of polymerization (Section 7.5.1). Higher H_2O content decreases the polymerization, and, because polymerized molecules should end in O atoms, not Si or Al, this increases the $O/(Si + Al)$ ratio in the liquid.

Hildreth (1979) further postulated that the resulting compositional gradients, combined with the temperature gradient, induced further diffusional mass transfer within the cap, resulting in vertical compositional gradients in the other components. There may also have been an exchange of matter with the walls and roof, as well as with the convecting lower chamber (which provided components upward to the stagnant cap that would otherwise be depleted by the slow rates of diffusion and resupply). The result, according to Hildreth (1979), was a compositionally stratified uppermost magma chamber that developed the stratification on the order of 10^3 to 10^6 years (depending upon the size of the chamber), which is much faster than the rates attainable by diffusion alone and might explain how some large rhyolitic calderas can regenerate chemical gradients within the 10^5- to 10^6-year intervals between eruptions.

Hildreth (1979) attributed at least part of the diffusional effects responsible for the compositional gradients at the top of the magma chamber beneath Long Valley to the **Soret effect** (also called **thermal diffusion**). Over a century

ago, the Swiss chemist Jacques Louis Soret demonstrated that a stagnant homogeneous binary solution, when subject to a strong temperature gradient, spontaneously develops a concentration gradient in the two components. In general, the heavy element or molecule migrates toward the colder end, and the lighter one migrates toward the hotter end of the gradient. By doing so, there is less total vibrational momentum (the heavy molecules vibrate less at the colder end), which lowers the overall energy of the system. Macroscopically, we can say that the temperature gradient has an effect on the chemical potentials of the elements, resulting in diffusion. Soret found that the gradients attain a steady state that lasts as long as the temperature gradient (or longer, if diffusion is impeded by rapid cooling). Of course, the extent of the Soret effect depends upon the magnitude of the temperature gradient and the nature of the system. The Soret effect was the most frequently invoked mechanism of magmatic differentiation in the 19th century. It was demonstrated experimentally for more complex natural systems on a basaltic magma by Walker et al. (1981) and Walker and DeLong (1982). They subjected two basalts to thermal gradients of nearly 50°C/mm (!) and found that the samples reached a steady state in a few days. They found the heavier elements (Ca, Fe, Ti, and Mg) concentrated at the cooler end and the lighter ones (Si, Na, and K) at the hot end of the sample, as predicted by Soret's theory. The chemical concentration in the results of Walker et al. (1981) and Walker and DeLong (1982) is similar to that expected from fractional crystallization, only with the temperature reversed. Here, the hot end was more andesitic, whereas the cool end was a low-silica basalt.

In the wake of the results of Walker et al. (1981) and Walker and DeLong (1982), it was thought that the density and compositional gradients in stagnant layers, such as at the top of the Long Valley magma chamber, enhanced Soret-type thermal diffusion. These combined gravity and thermal processes were given the somewhat imposing term **thermogravitational diffusion**. The thermal gradients required for thermal diffusion are far from realistic, however, and most of Hildreth's vertical thermal and chemical trends, derived from the stratigraphy of the Bishop Tuff and translated to the stagnant cap of the magma chamber, do not correlate well with the experimental trends of Walker et al. (1981) and Walker and DeLong (1982) for purely thermal diffusion. Thermal diffusion is unlikely in nature anyway because the unusually steep thermal gradients should diffuse and equilibrate in natural systems more rapidly than slower chemical diffusion can respond. Thermal gradients, however, can create density instabilities, which may then produce convective motion (as compared to diffusion). Thermogravitational diffusion, in which thermal gradients directly induce Soret-type chemical diffusion, was the most frequently invoked mechanism of differentiation until the mid-1890s, but it is no longer considered viable and is of little more than historical note today.

Most of Hildreth's (1979) work was correct, however. His observations concerning the chemical stratification of thick tuff sequences are not unique. Similar compositional variations have been observed in several localities, suggesting that silicic magma chambers are commonly stratified. Hildreth (1979) was also correct in proposing that boundary layers formed along the walls and cap of the magma chamber at Long Valley, and that these boundary layers resulted in compositional gradients in the cap. Observation of compositionally zoned margins of silicic plutons (Michael, 1984; Sawka et al., 1990; Mahood et al., 1996; Srogi and Lutz, 1996) provides direct support for such zonation along both the walls and cap of many magma chambers in which the viscosity and yield strength of the liquid should preclude crystal settling. We are realizing that chemically zoned magma chambers are likely to be the rule rather than the exception. Only Hildreth's appeal to thermogravitational diffusion was in error. As Michael (1983) pointed out, the gradients that Hildreth proposed for the compositional stratification in the stagnant cap correlate much better with liquid–crystal fractionation and the trends expected for partitioning between the phenocrysts observed in the tuff and the rhyolitic magma. Dunbar and Hervig (1992) came to a similar conclusion, finding that the trace element trends were compatible with ~40% fractional crystallization of quartz and alkali feldspar, although H, B, Li, F, and Cl decoupled from the others, and their distribution was related to variations in H_2O content in the stratified cap.

Figure 11.12 illustrates a possible mechanism for the development of compositional stratification along the walls and top of a magma chamber that may work even in relatively viscous silicic chambers. Because the magma cools from the margins inward, thermal gradients occur in the marginal areas where the magma is in contact with cooler wall and cap rocks (although the cap rock may be heated more by convection). If equilibrium is maintained, the magma will crystallize to a progressively increasing extent toward the cooler margins: a crystal/liquid "mush zone" or "solidification front" (Marsh, 2000, 2002). The gradient in the extent of crystallization produces a corresponding gradient in the liquid composition. If the interior of the chamber is entirely molten, the temperature ranges from the solidus temperature at the outer margin of the mush zone to the liquidus temperature at the inner margin, so the entire compositional spectrum of the liquid line of descent is present simultaneously, locally in equilibrium with the solids, but obviously not across the whole zone. Interstitial glasses from the drilling profiles through the caps of the Hawaiian lava lakes we visited in Section 6.1 range from parental olivine tholeiitic to rhyolitic (Helz, 1987). This **boundary layer** crystallization process, in which the crystals remain in situ, can produce gravitational instabilities where highly evolved interstitial liquids near the walls are less dense than the crystals and the liquid in the interior of the chamber. As a result, the marginal liquids rise liquids rise toward the top and are replaced by liquid moving toward the boundary layer from the interior. This process is called **compositional convection**, convective rise of low-density material, not because it is hot and expanded but because it loses the heavy elements, such as Fe, Ca, and Mg, to the crystals as it solidifies. The convecting liquid rises to the cap area, where it spreads laterally and stagnates, as Hildreth (1979) proposed. The stagnant stratification at the cap is thus not a result of diffusion

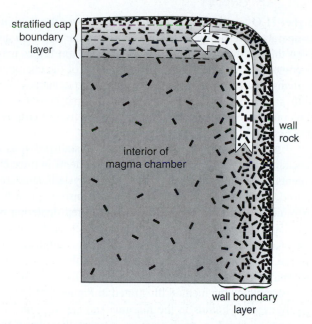

FIGURE 11.12   Formation of boundary layers along the walls and top of a magma chamber. Gradients in temperature result in variations in the extent of crystallization, which, in turn, creates a compositional gradient in the interstitial liquid. The liquid composition varies from that of the original magma in the interior to highly evolved at the margin, where crystallization is advanced. The evolved liquid is less dense and may rise to spread out at the top, forming a stagnant cap layer.

generated by an initial H_2O gradient only, as Hildreth (1979) proposed, but is inherited from fractional crystallization along the walls because the most evolved liquids are the least dense and occupy the highest portions of the cap boundary layer. A process similar to that at the walls may also occur at the cap, enhancing the inherited gradients. Assimilation of felsic wall or cap rocks may also contribute. Sawka et al. (1990) found the vertical and horizontal variations in 50 elements (and in mineralogy) in two zoned "granitic" plutons in the Sierra Nevada to be directly relatable to side-wall crystallization and upward migration of the buoyant melt to the chamber roof. Bachmann and Bergantz (2004) proposed a somewhat similar model for fractionation of rhyolites from intermediate magmas. Crystals were kept in suspension by convective stirring in their model until the chamber became about half crystallized. Convection in such a high yield–strength mush came to a halt and, as the mush compacted, the interstitial rhyolitic melt rose to a layer beneath an upper solidification front, perhaps aided by gas-driven filter pressing in volatile-rich systems. The layer may then have crystallized as a granitic phase of the pluton, or some then have escaped as a rhyolitic ignimbrite.

Intake of H_2O from the wall rocks, as Hildreth (1979) proposed, may also aid convective rise. Because H_2O diffuses more rapidly than nonvolatile species, it may also rise from the main magma body and become stratified in the cap area (Shaw, 1974). The density gradients in the cap area because H_2O and the more evolved liquids rise to the top, are gravitationally stable, and the stratification should remain (until eruption at least, in the case of Long Valley). The higher H_2O levels toward the top should lower the liquidus temperature, resulting in fewer crystals in the upper strata, even though the layers may be cooler. The top layers may become H_2O saturated long before the rest of the chamber, perhaps resulting in the development of pegmatites and aplites (Mahood et al., 1996) or hydrothermal processes in the country rocks above, or this may even trigger eruptive processes (Wolff et al., 1990).

A major problem with fractional crystallization as a mechanism for producing the observed gradients in trace elements is the high degree of fractional crystallization required. For example, Hildreth (1979) demonstrated that the extreme depletion in Sr, Ba, and Eu required depletion of over 50% feldspar, and Michael (1983) demonstrated the need for over 65 to 70% fractionation. This is a rather high degree of solidification in light of the volume of liquid magma subsequently erupted as the Bishop Tuff. Notice that it is possible by the method illustrated in Figure 11.12 to get highly evolved liquids in the margins and cap of a magma chamber with only a small fraction of crystallization of the total magma body. Such liquids have been recognized in several natural systems. Because fractional crystallization may produce nonlinear fractionation trends (curves on variation diagrams) as new phases join those crystallizing, linear mixing of convecting liquids with nonlinear compositional variations can produce liquid compositions that differ from liquid lines of descent for the magma body as a whole (Langmuir, 1989; Nielsen and DeLong, 1992). For example, Srogi and Lutz (1996) developed a model based on variable mixing of melts representing different stages of differentiation that explained the compositional variations producing granites in the zoned Arden pluton (of southeastern Pennsylvania and northern Delaware) that could not be modeled by regular fractional crystallization or magma mixing.

Marsh (2000) noted that the cooler and partially crystallized solidification fronts at the tops of plutons, if sufficiently thick, are denser *en masse* than the underlying uncrystallized melt. He suggested that tears develop as the internal strength is locally overcome by the weight of the leading portion of the front, which sinks, inviting local interstitial melt to be drawn into the openings to form the silicic pods that are commonly observed in many plutonic bodies. Some thick plutons may develop sufficiently thick tears to free eruptible volumes of silicic segregations. Solidification front instability, he notes, should produce bimodal rock suites: high volumes of mafic parent and silicic segregations but few intermediate compositions. Such a process, he proposes, may be responsible for the bimodal basalt–rhyolite volcanism in Iceland, particularly when new magma pulses remelt and disrupt older pod-bearing crust. Tait and Jaupart (1996) also discussed liquid segregation from boundary layers and both upward convection from the floor and downward convection from the roof of magma chambers.

As mentioned earlier, due to their high temperature and low viscosity, exposed layered mafic intrusions make ideal natural laboratories for studying the effects of magmatic differentiation processes. We will look further into these processes in Chapter 12.

11.6 MIXED PROCESSES

Two or more of the processes discussed above may work simultaneously or in sequence during the generation, migration, and solidification of magmatic systems. A number of magmas (and the resulting rocks) may thus be complex hybrids reflecting the combined effects of crystal fractionation, magma mixing, assimilation, volatile transport, and/or liquid immiscibility.

In some cases, a combined process may be more than coincidence, and two processes may operate cooperatively. For example, mixing of two magmas of different compositions and temperatures commonly produces thermal instabilities, resulting in quenching of the hotter magma and heating of the cooler one. Combinations of magma mixing, fractional crystallization, and convection are thus possible. Other novel effects will be explored in Chapter 12. Another example is the combination of assimilation and the accompanying fractional crystallization required to supply the necessary heat, as proposed above. Because many of the processes have similar effects, it may be difficult to distinguish the relative contributions of each with any degree of confidence. One approach to the problem is to devise mathematical models (similar to the Rayleigh fractionation and the other models presented in Chapter 9) for the behavior of certain trace elements and isotopes (or ratios) based on a combination of processes. This has been done by DePaolo (1981c) for the assimilation + fractional crystallization process (which he called AFC), fractional crystallization + recharge of more primitive magma (O'Hara and Matthews, 1981), and all three combined (Aitcheson and Forrest, 1994), using iterative techniques to model the ratio of contaminant to original magma. Nielsen and DeLong (1992) have also developed a numeric model to simulate in situ crystallization and fractionation in boundary layers. As mentioned in Chapter 9, because of the large number of variables, the uncertainties in the initial concentrations and partition coefficients, and now the addition of a number of concurrent mechanisms, we can be assured that at least one model, and probably several, can be derived to replicate the chemical variations of practically any suite of samples representing a particular petrogenetic province. The fact that a particular model can explain the chemical trends of the rocks in a province does not then guarantee that the process so modeled actually took place. Nonetheless, the models may prove useful or helpful in comparing the contrasting effects of different models or eliminating impossible processes. They may even help us refine our thinking on the complex dynamics of magmatic diversification. At times, a model may even be correct!

All of the processes described in this chapter probably occur in nature. An important field of modern igneous petrology evaluates the contributions of the various processes in the chemical characteristics and temporal evolutionary trends in many petrogenetic provinces. From these studies we can then evaluate the relative importance of the processes on a global scale. Which of them are principally responsible for the diversity of igneous rocks on Earth? In my opinion, fractional crystallization and magma mixing are probably the two most important magmatic differentiation processes responsible for magmatic diversity within a suite of related igneous rocks. Fractional crystallization modifies primary magmas to varying extents and is probably most effective for hot mantle-derived mafic magmas, few of which reach the surface without having evolved somewhat along their liquid line of descent. Reinjection of parental magma and mixing of independently generated magmas are common magma chamber processes, and the latter may be responsible for many igneous rocks of intermediate composition. The other processes may be important in particular instances and fundamental for some magma types, but they are not as common in general.

11.7 TECTONIC–IGNEOUS ASSOCIATIONS

With this chapter, we end our preparatory phase of igneous petrology. We have acquired the necessary background, and it is now time to embark on a survey of the major igneous occurrences and use our skills to evaluate them and the processes that shape them. Conversely, such a survey will help shape and refine our petrologic concepts. In Chapter 12, we shall take a look at layered mafic intrusions because they provide an excellent natural laboratory to shed some light upon the speculations on magmatic diversity just explored. After that, I have subdivided the various major igneous phenomena into groups commonly called **tectonic–igneous associations**. These associations are on a larger scale than the petrogenetic provinces discussed earlier. Where a petrogenetic province is a geographically limited and presumably cogenetic phenomenon, tectonic–igneous associations are an attempt to address global patterns of igneous activity by grouping provinces based upon similarities in occurrence and genesis. Of course, there is bound to be disagreement on the distinctions and

number of such groups, particularly on the minor, more esoteric ones. I have tended to be a "lumper" again, rather than a "splitter," and have selected large, comprehensive groupings. A list of these associations is:

1. Mid-ocean ridge volcanism
2. Ocean intraplate or ocean island volcanism
3. Continental plateau basalts
4. Subduction-related intra-oceanic island arcs
5. Subduction-related volcanism and plutonism of continental arcs
6. Granitoid rocks
7. Mostly alkaline igneous associations of stable craton interiors
8. Anorthosites

This grouping may not please all petrologists. Some would like to subdivide one or more of the associations into distinctive subgroups, and others might prefer to combine some of them or even omit some as being less global, and yet others would add some associations to the list. The associations I have chosen are certainly a compromise, and although the list may not include all igneous phenomena on the planet, it covers perhaps 99%. The remaining 1% would fill volumes, and we must perform some sort of "academic triage" if we ever hope to find some closure on the subject of igneous petrology. Each of the ensuing chapters addresses one of these associations in the order listed above. In them we shall review the mode of occurrence and the igneous products and then proceed to use petrographic, chemical, and perhaps geophysical and other data to interpret the association and come up with a petrogenetic model. This provides us with some application of our newly acquired techniques and gives us some exposure to global igneous phenomena, the final requirement for good petrologists listed in Chapter 1.

Summary

Simply put, diversification in geologic systems requires:

1. Two (or more) phases in which components are unequally distributed
2. A physical process in which the phases are separated

When combined, these two processes allow geologic materials, such as rocks, melts, fluids, etc., to change composition.

Partial melting is one such situation, during which the chemical constituents are distributed unequally between the melt being formed and the solid residuum, so that the melt, when finally separated, and the residuum both have different compositions than the original rock melted. The compositions of the melt and residuum depend upon the initial host composition, the conditions of melting, and the fraction melted (F). The fraction melted must reach some critical value before it can form a continuous intergranular network and escape the host as a result of buoyancy gained by expansion upon melting.

Fractional crystallization has been invoked as the principal process by which a magma, once formed, may differentiate

toward more "evolved" compositions. The most important mechanism for fractional crystallization is gravity settling: the sinking (or floating) of crystals in a liquid due to contrasting density. We can calculate ideal rates of settling using Stokes' Law, but natural systems are complicated by non-spherical grain shapes and the non-Newtonian behavior (particularly yield strength) of silicate liquids. Gravity settling may be enhanced by compaction (filter pressing) in which the interstitial liquid is squeezed out from a crystal-laden mush. Flow segregation may also play a minor role in differentiating crystal–liquid suspensions flowing through a narrow conduit such as a dike.

Crystallization and rising to lower pressure (with perhaps some wall-rock dehydration) may result in fluid saturation of an originally H_2O-undersaturated melt. The late stages of crystallization in a rising (typically hydrous granitic) magma body may thus lead to retrograde boiling: the release of a silicate-saturated fluid phase that is capable of fracturing the roof rocks and escaping to form pegmatites and/or hydrothermal veins and ores.

Liquid immiscibility, once believed to be a major mechanism of diversification and a pathway to granites, is now relegated to a minor role in basaltic evolution, producing a small quantity of late silicic liquids in some Fe-rich tholeiites, typically trapped interstitially with the late basaltic melt. Liquid immiscibility may be important in the segregation of a late sulfide liquid to form massive sulfide deposits and in the development of carbonatites.

Magma mixing can be observed as commingling of silicic and mafic liquids in some volcanic hand specimens and as basaltic pillows at the base of some crystallized granitic magma chambers. Chemical and textural evidence for replenishment of differentiating magma chambers with reinjected primary magma has also been documented at many localities, particularly at mid-ocean ridges. Mixing of disparate magmas in shallow magma chambers is increasingly recognized as an important process in the diversification of magmas in a variety of igneous provinces.

Assimilation of wall rocks is another source of contamination of magmas and can best be detected by trace element or isotopic patterns. Assimilation is limited by the amount of heat available in the magma, which is generally restricted to the latent heat of crystallization.

Recent studies of magma chambers and volcanic sequences reveal features that the traditional mechanisms above fail to explain. In situ crystallization within an initially stationary liquid or within a stationary crystal–liquid suspension boundary layer may produce an evolved liquid with reduced density. Thermal gradients near the chamber walls and roof may produce gradients in the degree of crystallization of the magma and, hence, in the composition of the liquid. Compositional convection, the rise of less dense liquids from the solid suspension, may result in highly evolved liquids segregating toward the top of the chamber, perhaps forming a stagnant, density-stratified cap boundary layer.

None of the above processes need work in isolation, and any combination may be possible in natural systems.

Key Terms

Dihedral angle *203*	Stokes' Law *206*	Zone refining *214*
Critical melt fraction or rheological critical melt percentage (RCMP) *203*	Newtonian/non-Newtonian *206*	In situ process *215*
	Yield strength *206*	Soret effect *218*
Compaction/filter pressing *203*	Flow segregation *207*	Thermogravitational diffusion *216*
Magmatic differentiation *204*	Retrograde or resurgent boiling *209*	Boundary layer *217*
Fractionation *204*	Miarolitic pods/cavities *209*	Compositional convection *217*
Gravity settling *205*	Commingling *212*	Tectonic–igneous association *219*
	Assimilation *213*	

Review Questions and Problems

Review Questions and Problems are located on the author's web page at the following address: **http://www.prenhall.com/winter**

Important "First Principle" Concepts

- A certain critical amount of melt must be created to form an interconnected grain–boundary film or network before it can be removed from the melting source. This amount may be as little as 1 to 7% for low-viscosity, low-dihedral-angle mafic systems to 15 to 30% for more viscous, high-dihedral-angle silicic systems.

- Fractional crystallization, magma mixing, and in situ processes are now believed to be the major processes of magmatic differentiation. The other processes probably play only a restricted role.

Suggested Further Readings

Bowen, N. L. (1928). *The Evolution of the Igneous Rocks*. Princeton University Press. Princeton, NJ.

Daines, M. J. (2000). Migration of melt. In: *Encyclopedia of Volcanoes* (eds. H. Sigurdsson, B. Houghton, S. R. McNutty, H. Rymer, and S. Stix). Academic Press. San Diego. pp. 69–88.

Hildreth, W. (1979). The Bishop Tuff: Evidence for the origin of compositional zonation in silicic magma chambers. In: *Ash-flow Tuffs* (eds. C. E. Chapin and W. E. Elston. Geological Society of America Special Paper **180**. pp. 43–75.

Jaupart, C., and S. Tait. (1995). Dynamics of differentiation in magma reservoirs. *J. Geophys. Res.,* **100**, 17,615–17,636.

Marsh, B. D. (2000). Magma chambers. In: *Encyclopedia of Volcanoes* (eds. H. Sigurdsson, B. Houghton, S. R. McNutty, H. Rymer, and S. Stix). Academic Press, San Diego. pp. 191–206.

McBirney, A. R. (1979). Effects of assimilation. In: *The Evolution of the Igneous Rocks. Fiftieth Anniversary Perspectives* (ed. H. S. Yoder). Princeton University Press. Princeton, NJ.

Presnall, D. C. (1979). Fractional crystallization and partial fusion. In: *The Evolution of the Igneous Rocks. Fiftieth Anniversary Perspectives* (ed. H. S. Yoder). Princeton University Press. Princeton, NJ.

Roedder, E. (1979). Silicate liquid immiscibility in magmas. In: *The Evolution of the Igneous Rocks. Fiftieth Anniversary Perspectives* (ed. H. S. Yoder). Princeton University Press. Princeton, NJ.

Tait, S., and C. Jaupart. (1990). Physical processes in the evolution of magmas. In: *Modern Methods of Igneous Petrology: Understanding Magmatic Processes* (eds. J. Nichols and J. K. Russell). *Reviews in Mineralogy,* **24**. Mineralogical Society of America. Washington, DC.

Turner, J. S., and I. H. Campbell. (1986). Convection and mixing in magmas. *Earth Sci. Rev.,* **23**, 255–352.

12

Layered Mafic Intrusions

Questions to be Considered in this Chapter:

1. What do large mafic intrusions, with their high temperatures and relatively low viscosity, reveal about the ways that magma chambers evolve as they cool and crystallize?

2. What types of layering develop in large mafic magma chambers?

3. To what extent can the classical explanation of crystal settling govern the layers that develop?

4. What other processes may be involved?

Our understanding of magmatic differentiation is based largely on experimental studies of natural and simplified systems in the laboratory (Chapters 6 and 7). Geochemical trends on variation diagrams, developed principally in volcanic suites, are also useful (although indirect) indicators of differentiation processes. All our concepts must be tested against observations in the field. The most informative natural examples of crystallization and differentiation are the large layered mafic intrusions (LMIs). Uplift and erosion of these great bodies have provided us with excellent natural laboratories in which to observe the products of the dynamic processes that accompany cooling and crystallization in magma chambers, where the viscosity is low enough to permit significant motion (hence differentiation) of crystals and/or liquids. Classical studies (Hess, 1960; Jackson, 1961; Wager and Brown, 1968) have recently been challenged or augmented by new interpretations based mainly on principles and experiments in fluid dynamics, making this an interesting time to study these occurrences. Young (2003) provided an excellent history of the study of layered mafic intrusions and discussed many of the theories proposed for their development.

Mafic intrusions come in all sizes, from thin dikes and sills to the huge (66,000 km^2 and up to 9 km thick) Bushveld intrusion of South Africa. They can occur in any tectonic environment where basaltic magma is generated (Chapters 13 to 15), and the mid-ocean ridges may be the most common site for their development (Chapter 13). The larger intrusions that are the real subject of this chapter, however, require a substantial volume of magma in a short period of time, and they must occur within the continental crust in order to become exposed at the surface by erosion. The most common environment for copious mafic magma production in a continental setting is associated with continental flood basalts (Chapter 15), where rifting may be associated with mantle hotspots. The majority of LMIs are associated with coeval flood basalts, although a few are not. Areas of unusually copious (mostly mafic) igneous activity (other than normal seafloor spreading) have recently been called **large igneous provinces (LIPs)** (Mahoney and Coffin, 1997). For more on LIPs, see Sections 14.1 and 15.1. Most of the larger LMIs are Precambrian, occurring perhaps in aborted rifts with a high percentage of mantle melting due to the high Precambrian geothermal gradient. There are probably a large number of mafic plutonic bodies deep in the continental crust that never made it to the surface because of their high density.

TABLE 12.1	Some Principal Layered Mafic Intrusions		
Name	Age	Location	Area (km²)
Bushveld	Precambrian	South Africa	66,000
Dufek	Jurassic	Antarctica	50,000
Duluth	Precambrian	Minnesota	4,700
Stillwater	Precambrian	Montana	4,400
Muskox	Precambrian	Northwest Territories, Canada	3,500
Great Dike	Precambrian	Zimbabwe	3,300
Kiglapait	Precambrian	Labrador	560
Skaergård	Eocene	Eastern Greenland	100

Although some thicker dikes and sills show stratifica-
[ ] that occurred during cooling, the compositional varia-
[ ] them is usually subdued. A certain critical thickness,
[ ]der of 400 to 500 m, seems to be required before
[ ] and crystallization processes generate the spec-
[ ]ring for which LMIs get their name. Cooling
[ ]zation times vary with the size and shape of the
[ ] well as the capacity for circulating water in the
[ ]ect away heat. Estimates vary from tens of
[ ] years to a million or more. Table 12.1 lists
[ ] largest and/or best-studied occurrences of
[ ] intrusions. The form of LMIs is typically a
[ ]uton, and many are either a lopolith (Figure
[ ]nnel (Figure 12.1), the difference being that the
[ ]opolith is considered to be conformable to the
[ ]ock strata, whereas a funnel is cross-cutting to
[ ] external strata and the internal layering. A chill
[ ] is common, but due to significant assimilation of
[ ] rocks, it can rarely be used to accurately determine the
composition of the parent magma. Nonetheless, LMIs are
overwhelmingly gabbroic and usually tholeiitic, exhibiting
the classic Fe-enrichment characteristic of tholeiitic magma
series (see the "Skaergård trend" in Figure 8.3 and compare
it to Figure 8.14).

12.1 IGNEOUS LAYERING

Crystal concentration and layering distinguish LMIs, and
they provide fascinating material for our contemplation. If
your grasp of cumulate textures is a bit rusty, it may help to
review the subject in Section 3.1.7. An excellent summary
of cumulates and layering is provided by Irvine (1982). Several types of layers can be developed in LMIs. Multiple
types can be developed in close proximity or even occur simultaneously. Layers occur as a consequence of changes in
mineralogy, texture, or mineral composition. In the context
of LMIs, a **layer** constitutes any conformable sheet-like unit
that can be distinguished by its compositional and/or textural features. Layers may either be **uniform** (mineralogically and texturally homogeneous) or **non-uniform** (varying
either along or across the layering). The most common type
of non-uniform layer is a **graded** layer, which shows gradual variation in either mineralogy (Figure 12.2a) or in grain
size (Figure 12.2b). The latter is rare in gabbroic LMIs, occurring more commonly in ultramafic ones.

Layering (or stratification), on the other hand,
deals with the structure and fabric of sequences of multiple layers. **Modal** layering is characterized by variation in
the relative proportions of constituent minerals. Modal
layering may comprise uniform layers (Figure 12.3b),
graded layers (Figure 12.2), or a combination of both. The
scale of modal layering is in the range of a centimeter to a
few meters.

Phase layering refers to layered intervals defined on
the basis of the appearance or disappearance of particular minerals in the crystallization sequence developed in modal layers.
Phase layering can be said to transgress modal layering, in the

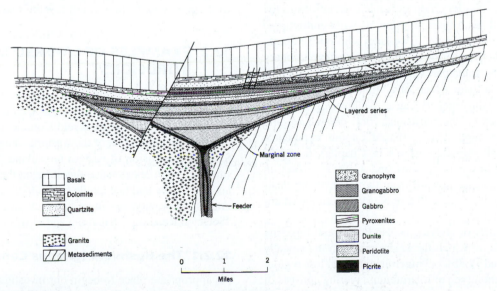

Basalt
Dolomite
Quartzite

Granite
Metasediments

Granophyre
Granogabbro
Gabbro
Pyroxenites
Dunite
Peridotite
Picrite

Layered series

Marginal zone

Feeder

0   1   2
Miles

FIGURE 12.1 Simplified cross section of the Muskox intrusion of Canada. From Irvine and
Smith (1967). Copyright © reprinted by permission of John Wiley & Sons, Inc.

FIGURE 12.2 Two types of graded layers. **(a)** Modal layering resulting from the concentration of mafic minerals toward the base (Skaergård intrusion, eastern Greenland). **(b)** Size layering resulting from the concentration of larger crystals at the base (Duke Islan[d] Alaska). From McBirney an[d] Noyes (1979). Reprinted [by] permission of Oxford [University] Press.

sense that it characterizes larger subdivisions of a layered sequence. A single phase layer, therefore, commonly consists of numerous modal layers. For example, thin repeated modal layering of olivine and plagioclase may be joined by augite at some stratigraphic level in a layered intrusion, defining a new phase layer. Three-phase modal layers may continue, until perhaps olivine is lost at some higher level, and another phase layer is begun. The introduction of augite and the loss of olivine can each be used to define boundaries of phase layering. By analogy with sedimentary stratigraphy: layers are like sedimentary beds, layering is like bedding sequences, and phase layering is like formations. As with formations, phase layering is commonly used to delineate major subdivisions in the layered sequences of differentiated LMIs.

Cryptic layering, as the name implies, is not obvious to the naked eye. It is based on systematic variations in the chemical composition of certain minerals with stratigraphic height in a layered sequence. If we keep with our sedimentary analogy, cryptic layering is rather like the evolutionary changes shown by some particular fossil genus as time progresses. Cryptic layering occurs on a broader scale than both modal and phase layering. Because the layering in LMIs must be produced as the intrusions cool, crystallize, and differentiate, we might expect (based on our experience with experimental mineral systems in Chapters 6 and 7) $Mg/(Mg + Fe)$ in mafic phases and $Ca/(Ca + Na)$ in plagioclase to decrease upward from the floor, if our analogy with progressive sedimentation of crystals on the chamber floor is correct.

The *regularity* of layering is also of interest. Layering can be **rhythmic**, in which the layers repeat systematically (Figures 12.3a and 12.3b), or **intermittent**, with less regular patterns. A common type of intermittent layering consists of rhythmic graded layers punctuated by occasional uniform layers (Figure 12.4). Rhythmic layering that is several meters

thick is commonly called **macrorhythmic**, whe[re] only a few centimeters thick is called **micror**[hythmic]. [Al]though the terms *rhythmic* and *intermittent* are [typi]cally applied to modally graded sequences, ther[e] they cannot be applied to any kind of prominent [modal] layers.

Because of the many obvious similarities [between] layering in LMIs and sedimentary layers, classica[l interpre]tations of the origin of the igneous layering have b[een based] on sedimentary processes: crystal settling as a resul[t of den]sity differences between growing crystals and coe[xisting] liquid. As mentioned above, this classical concept has been challenged recently, but before we can evaluate the relative merits of competing hypotheses, we should take a brief look at the evidence. We shall therefore have a look at three well-documented examples of layered mafic intrusions.

12.2 EXAMPLES OF LAYERED MAFIC INTRUSIONS

I have selected three gabbroic LMIs to illustrate what is common to these bodies, as well as the variability they can display. These are the Bushveld Complex of South Africa, the Stillwater Complex of the western United States, and the Skaergård intrusion of eastern Greenland. The descriptions summarize some extensive literature on these complex bodies, and I have had to exclude many interesting details in the interest of brevity. For more complete descriptions, see the "Selected Readings" list at the end of this chapter.

12.2.1 The Bushveld Igneous Complex

An appropriate place to begin our investigations of LMIs is with the largest: the Precambrian ($\sim$2.06 Ga) Bushveld Igneous Complex of South Africa. The main mafic/ultramafic

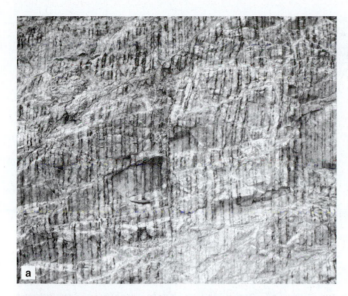

FIGURE 12.3 Rhythmic layering in LMIs. **(a)** Vertically tilted centimeter-scale rhythmic layering of plagioclase and pyroxene in the Stillwater Complex, Montana. **(b)** Uniform chromite layers alternate with plagioclase-rich layers, Bushveld Complex, South Africa. From McBirney and Noyes (1979). Reprinted by permission of Oxford University Press.

FIGURE 12.4 Intermittent layering showing graded layers separated by nongraded gabbroic layers. Skaergård intrusion, eastern Greenland. From McBirney (1993). Jones and Bartlett, Sudbury, MA. Reprinted with permission.

body (now called the Rustenburg Layered Suite) is a flat, funnel-shaped intrusion of initially gabbroic magma in a broad basin, complicated somewhat by folding and faulting. It is about 300 to 400 km wide and 9 km thick (Figure 12.5), and it somewhat discordantly cuts the local strata. The roof is composed of sediments and felsic volcanics that have been lifted, metamorphosed, and in places partially melted by the main intrusion. Some of the roof volcanics of the Rooiberg Group are also found beneath the pluton, and one felsic flow yielded a zircon age virtually identical to that of the layered mafics. The roof rocks are preserved mainly in the central depression of the basin. The central portion of the body was also intruded by the Rashoop and Lebowa granitic rocks very soon after the mafics were emplaced (2.054 Ga). As a result, the Rustenburg Suite crops out as a discontinuous ring, with main limbs on the east and west that are pinched out on the north and south between the granites and the older Transvaal Supergroup sediments below. Gravity and drilling data suggest that there is also a large buried

southeastern ("Bethal") limb (Irvine et al., 1983; Cawthorn et al., 2006). Local irregularities between limbs have suggested to some investigators that there may be four or more separately fed centers, with limited mixing between them, but others are impressed more with the similarities between exposed limbs than with the differences. It would not be surprising if thin layers in a single intrusion of such magnitude were laterally discontinuous, but the overall stratigraphy of the limbs is strikingly similar. Some well-mapped layers are remarkably regular. For example, the famous sulfide-bearing pyroxenite of the Merensky Reef (the world's chief source of platinum) is 1 to 5 m thick and can be followed for nearly 200 km in the western limb and 150 km in the eastern limb, some 200 km apart! It is thus more likely, although far from certain, that the Bushveld Complex formed not as several separate chambers but as a single large, shallow intrusion, and perhaps irregularities in the floor and roof within the basin and more than one pulse of magma led to the development of semi-isolated pockets that evolved somewhat differently. The similar timing of the mafics and felsics indicate that the Bushveld is a bimodal association, suggesting that the felsics are the products of crustal fusion due to localized heat input from below at this major mafic province.

The greater part of the Rustenburg Suite is well layered, with layers usually dipping 10 to 15° toward the center of the intrusion. There is also a broad compositional variation from ultramafic at the base through mafic and intermediate, to more felsic rocks upward. Figure 12.6 is a simplified stratigraphic column for the Bushveld. In addition to the major subdivisions, it illustrates both the phase and cryptic layering patterns. The cryptic layering is indicated only on a very coarse scale. In detail, there are irregularities and local reversals of the overall trends, especially in the central sections.

The lowest unit, the **Marginal Zone**, is an unlayered heterogeneous zone of norites about 150 m thick (but locally extending up to 800 m), occurring only at the bottom of the

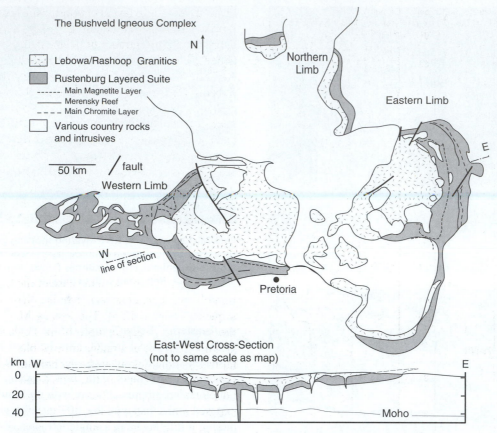

FIGURE 12.5 Simplified geologic map and cross section of the Bushveld Igneous Complex. After Willemse (1964), Wager and Brown (1968), and Irvine et al. (1983).

complex. It is complicated by floor irregularities, variable assimilation of the country rocks, and the presence of numerous composite mafic sills that are difficult to distinguish from the chill margin proper. Analyses of some fine-grained norites from the margin correspond to a high-aluminum tholeiitic basalt. See Eales (2002) and Cawthorn (2007) for summaries of proposed parental magmas of the Bushveld. The layered rocks above the Marginal Zone are subdivided into four principal units. LMI subdivisions are usually based on the appearance or disappearance of some cumulate mineral (phase layering), but, for practical reasons, some subdivisions in the Bushveld are separated by distinct, mappable marker horizons. Fortunately, the difference is relatively minor in this case because the markers chosen are stratigraphically close to important mineralogical changes.

The ultramafic **Lower Zone** comprises thin dunite cumulates alternating (often macrorhythmically) with thicker orthopyroxenite and harzburgite layers, nearly all of which are uniform. The intercumulus phases are augite and plagioclase, occurring commonly as large poikilitic crystals. In the far western limb, the Lower Zone contains nine cyclic units of dunite → harzburgite → pyroxene, totaling ~1050 m thick. The top of the Lower Zone has been variably defined. Teigler and Eales (1992) proposed the top of the uppermost thick olivine-rich interval as the boundary.

The next zone upward is the **Critical Zone**, composed of norite, orthopyroxenite, and anorthosite layers, with subordinate dunite and harzburgite plus significant chromite layers. The Lower Critical Zone contains an 800-m thick succession of pyroxenite cumulates with an olivine-bearing layer and up to seven chromite layers. The base of the Upper Critical Zone is defined by the first appearance of cumulus plagioclase. The transition from intercumulate plagioclase to cumulate plagioclase near this transition indicates that plagioclase becomes an early liquidus phase, and not one that forms later on in the trapped intercumulus liquid. Fine-scale layering is impressively developed, as can be seen in Figure 12.3b, which shows repeated strongly contrasting and sharply bounded chromite and plagioclase cumulate layers. The layers are parallel over remarkable distances but eventually pinch out laterally in most cases. The rhythmic nature of thin, homogeneous, or graded layers is certainly the most striking feature of LMIs. The layering of cumulus minerals can be fairly straightforward when there is a consistent trend in density or falling liquidus temperatures, but the *repetition* of lithological layers is a feature that has challenged the interpretive capabilities of petrologists for half a century.

The **Merensky Reef** and **Bastard Cyclic Units** are thin rhythmic units at the top of the Critical Zone that contain layers of cumulus anorthosite, pyroxenite, norite, and chromitite. The famous Pt-Pd and sulfide-bearing **Merensky Reef** sub-unit is an orthopyroxenite–olivine–chromite cumulate layer. The unit is unique in the Bushveld. Several investigators, beginning with Hess (1960), have proposed that it marks the horizon at which one or more major fresh surges of

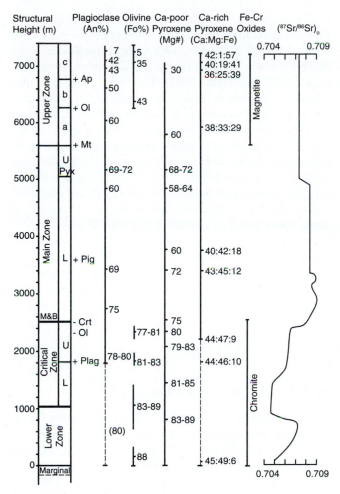

Structural Height (m) | Plagioclase (An%) | Olivine (Fo%) | Ca-poor Pyroxene (Mg#) | Ca-rich Pyroxene Oxides (Ca:Mg:Fe) | Fe-Cr Oxides | $(^{87}Sr/^{86}Sr)_o$

FIGURE 12.6 Generalized stratigraphic sequence of layering in the Rustenburg Layered Suite, Bushveld Igneous Complex. *Maximum* thicknesses of units are indicated. Compositions of cumulus phases are given on the right (dashed lines indicate intercumulus status). M&B = Merensky and Bastard Cyclic Units, Pyx = Pyroxenite Marker. After Wager and Brown (1968) and Cawthorn et al. (2006).

magma were introduced into the chamber. The Pt and Pd are then derived from the residual liquid of the earlier stage and the sulfides from an immiscible sulfide liquid associated with the latter stage.

The **Main Zone** is the thickest zone, above the uppermost cyclic unit, and contains thick, monotonous sequences of norite and gabbro–norite, with minor anorthosite and pyroxenite layers. Cumulus olivine and chromite are absent, and magnetite occurs only near the top of the zone. The layering is poorly developed compared to the lower units, but it is still present. About 2200 m above the base of the Main Zone in the west, and 3100 m in the east, pigeonite is supplanted by orthopyroxene in the Pyroxene Marker interval. This marks the boundary between the Lower and Upper Main Zones. The Lower, Critical, and Main Zones abut and terminate against the Marginal Zone in an onlap fashion toward the south, but the full sequence is developed farther north.

The base of the **Upper Zone** is defined by the appearance of cumulus magnetite. As the presence of magnetite indicates, the Upper Zone is enriched in Fe, following the typical tholeiitic trend in Figure 8.14. The zone is well

layered, including layers of anorthosite, gabbro, and ferrodiorite. Subzone A comprises ∼700 m of gabbro and anorthosite. The entrance of olivine defines the base of Subzone B, whereby troctolite joins the assemblages of Subzone A. The introduction of cumulus apatite defines the base of Subzone C. Determination of the exact nature and quantity of late Bushveld differentiates is complicated by the presence of metamorphosed, melted, and assimilated roof rocks and the almost immediately following Rashoop and Lebowa granites, microgranites, and granophyres. The felsic intrusives not only resemble possible differentiates, they commonly obliterate the original layers. Most, if not all, are interpreted as crustal melts. The Rustenburg Layered Suite proper comprises rock types ranging from dunite and pyroxenite through norite, gabbro, and anorthosite, to diorite, a comprehensive differentiation sequence for an initial mafic magma.

About 300 m above the base of the Upper Zone, iron-rich olivine *reappears* as a cumulus phase in the Bushveld sequence (Figure 12.6). The loss of Mg-rich olivine, and the subsequent reappearance of an Fe-rich one with progressive cooling, is a rather unusual phenomenon for crystallization behavior as we have come to understand it from the systems examined in Chapters 6 and 7. The experimental forsterite–fayalite–silica system investigated by Bowen and Schairer (1935) shows that this behavior is a natural consequence of Fe enrichment in olivine-bearing basaltic systems. With increasing Fe, the incongruent melting behavior of Mg-rich orthopyroxene no longer takes place, and the liquidus field of orthopyroxene eventually pinches out (Figure 12.7). Suppose we begin with a liquid of composition *a* on Figure 12.7. Mg-rich olivine is the first mineral to crystallize, and the liquid follows the schematic path from point *a* toward point *b* as an increasingly less Mg-rich olivine forms. At the peritectic, orthopyroxene joins olivine and liquid, and the liquid line of descent follows the peritectic as the reaction olivine + liquid = orthopyroxene takes place if equilibrium crystallization is maintained. If fractional crystallization predominates, however, there will be no olivine available for this reaction, and, as discussed by Osborn (1979), the liquid leaves the olivine–pyroxene peritectic boundary curve and crosses the pyroxene liquidus field (directly away from the Mg-rich orthopyroxene that forms) to point *c*, where tridymite forms. The liquid then follows the pyroxene–silica cotectic as fractional crystallization continues and the Fe content of the liquid increases. When the pyroxene field pinches out at point *d*, an Fe-rich olivine begins to crystallize in equilibrium with orthopyroxene and tridymite. Thus we have not only the odd situation in which olivine ceases to crystallize and then reappears, but we also have stable Fe-rich olivine + quartz in the final solid assemblage. In our discussion of the forsterite–silica binary system (Figure 6.12), we concluded that Mg-rich olivine and a silica mineral could never stably coexist. This generalization for olivine and quartz holds true for most rocks, except for very iron-enriched rocks (usually felsic differentiates) such as those we encounter near the top of the Bushveld.

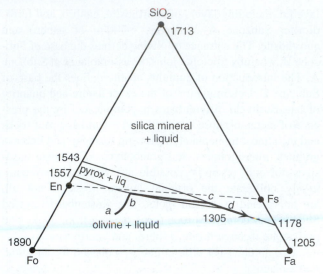

FIGURE 12.7 The Fo-Fa-SiO$_2$ portion of the FeO-MgO-SiO$_2$ system, after Bowen and Schairer (1935). Reprinted by permission of the *American Journal of Science*. The pyroxene + liquid field is bounded by the 1557 to 1305°C ternary peritectic curve on one side and the 1543 to 1178°C pyroxene–silica cotectic curve on the other. The field pinches out toward higher Fe content. Perfect fractional crystallization of a cooling liquid of bulk composition *a* follows the schematic path *a-b-c-d*. Fe-poor olivine forms at point *a*, and the liquid evolves toward point *b*. Because olivine is separated as it forms, the peritectic reaction (Ol + Liq = Opx) is not possible. Rather, Fe-poor orthopyroxene forms, and the liquid progresses directly away from the pyroxene composition to point *c*, where tridymite forms. Fe-rich olivine *reappears* with pyroxene and tridymite at point *d*. Temperatures of invariant points are in degrees Celsius. The immiscible liquid solvus in the silica-rich field is ignored.

Apatite enters as a significant cumulus phase (up to 5%) 400 m higher than Fe-olivine. Quartz and alkali feldspar occur within the uppermost 200 m. Cryptic zoning is continuous and commonly pronounced. Nearly pure fayalite occurs in some of the uppermost differentiates.

The Bushveld Complex serves as a stunning example of the problems we face in attempting to interpret the cooling and magmatic differentiation processes of what is probably the optimal natural situation: a large, slow-cooling chamber of hot, relatively low-viscosity magma. Of particular interest is the conspicuous development of rhythmic layering of (commonly) sharply defined uniform or graded layers. Our study of experimental (and natural) systems has shown that a cooling magma will first crystallize one liquidus phase, typically then joined by progressively more phases as the liquid line of descent reaches the cotectic and higher-order eutectics. The repetition of mineral sequences in LMIs, however, requires either some impressively periodic reinjection of fresh magma or cyclic variation in one or more physical properties if it is to be produced by gravitational crystal settling alone. Furthermore, the overall pattern of cryptic layering indicates a progressive differentiation spanning the full vertical height of the intrusion (Figure 12.6). A brief and gradual reversal in many mineral compositions

(and a change in $^{87}Sr/^{86}Sr$) occurs over ~200 m at the Pyroxene Marker in the Main Zone, interpreted as representing the addition of a large volume of new magma into the Bushveld mafic chamber (Eales and Cawthorn, 1996; Nex et al., 2002; Cawthorn, 2007). The initial $^{87}Sr/^{86}Sr$ ratios (Figure 12.6) relate to plagioclase cryptic variations because that is the only significant Sr-bearing phase. The irregular $(^{87}Sr/^{86}Sr)_0$ trends in the lower half of the complex and the sudden drop at the Pyroxenite Marker suggest multiple injections of magma with variable isotopic compositions but only modest major element contrasts, resulting in little offset in the main cryptic mineral trends. Significant magmatic inputs have been suggested for the Lower, Critical, and Main Zones and again at the Pyroxenite Marker. Despite these inputs, the otherwise smooth mineral composition trends and the far more intricate layering preclude any model for repetitive layering based solely on replenishment. Let's look at the remaining two examples of LMIs (both of which have been studied in considerable detail) with these problems in mind before we attempt to develop a model for the layering.

12.2.2 The Stillwater Complex

The Precambrian (2.7 Ga) Stillwater Complex in southwestern Montana outcrops in a belt about 45 km long and up to 6 km wide (Figure 12.8). It was uplifted, tilted, and eroded prior to the Middle Cambrian, and sediments of this age lie unconformably on the eroded top surface, so an unknown thickness of the upper sequences is lost. Later tilting and erosion now exposes an excellent cross section of the steeply northward-dipping remaining portions of the intrusion. A large positive-gravity anomaly to the north indicates that a considerably larger portion of the body (up to nine times the exposed section, and perhaps including the upper layers) extends beneath the Paleozoic sedimentary cover.

The exposed stratigraphy of the Stillwater is similar to that of the Bushveld and has been subdivided into a Lower Zone, an Ultramafic Series, and a Banded Series. Detailed sections have been compiled by Raedeke and McCallum (1984) and McCallum et al. (1980), a somewhat simplified version of which is shown in Figure 12.9. The **Lower Zone** is a thin (50 to 150 m) layer of norite and gabbro that is discontinuous due to faulting. Some of the rocks are sufficiently fine grained to indicate that they may represent chilled parental magma, but they are either cumulates or altered by assimilation or fluid interactions.

The base of the **Ultramafic Series** is defined as the first appearance of copious olivine cumulates. It is subdivided into a lower Peridotite Zone and an upper Orthopyroxenite Zone. The **Peridotite Zone** consists of 20 cycles, from 20 to 150 m thick, of macrorhythmic layering with a distinctive sequence of lithologies (Raedeke and McCallum, 1984). Each series begins with dunite (plus chromite), followed by harzburgite and then orthopyroxenite. Considering both cumulus and later intercumulus phases (intercumulus orthopyroxene giving way

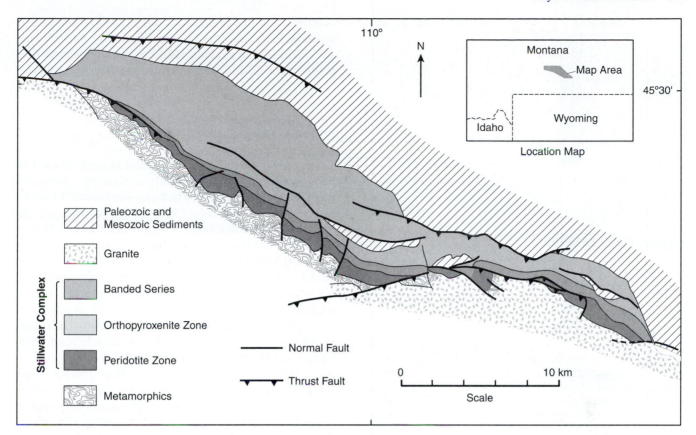

FIGURE 12.8 Simplified geologic map of the Stillwater Complex. After Wager and Brown (1968).

upward to plagioclase and then augite), the crystallization sequence in each rhythmic unit, with rare exception, is:

olivine + chromite →
olivine + orthopyroxene →
orthopyroxene →
orthopyroxene + plagioclase →
orthopyroxene + plagioclase + augite

This is a common basaltic crystallization sequence, which, when combined with the sharpness of the base of the dunite layer, suggests that each sequence is initiated by some major change in crystallization conditions, followed by a period of cooling and crystal accumulation. Once again, we are faced with the problem of explaining the *repetition* of these cycles.

The interesting upward *increase* in Mg# noticed in the cryptic layering for cumulus orthopyroxene and olivine in the lower 400 m of the section (Figure 12.9) has been attributed to subsolidus re-equilibration between the cumulus and intercumulus phases. According to this theory, the more rapid cooling near the floor resulted in a progressively larger proportion of trapped intercumulus liquid (which was more differentiated, thus having a lower Mg#). Later (post-igneous) re-equilibration between cumulus phases and the low-Mg intercumulus material reduced the Mg# of the lower cumulates.

I should stress at this point that the bulk composition of cumulate rocks is never equivalent to any liquid composition. Regardless of the specific process, cumulates form by the accumulation of early-forming minerals and the displacement/expulsion of associated liquids. As a result of the motion of the solids and/or liquid, neither the cumulate minerals, the intercumulus liquid, nor the sum of the two at any stratigraphic level correlate to a composition along the true liquid line of descent. This makes the identification of true liquid compositions difficult.

The **Orthopyroxenite Zone** is a single, thick (up to 1070 m), rather monotonous layer of cumulate orthopyroxenite with Mg# of 84 to 86. Some layers are graded from coarser at the base to finer upward.

The **Banded Series** begins with the sudden occurrence of cumulus plagioclase, marking a significant change from ultramafic rock types (Figure 12.9). This may readily be explained by the evolving magma reaching the point where plagioclase begins to crystallize. The most common lithologies are anorthosite, norite, gabbro, and troctolite (olivine-rich and pyroxene-poor gabbro). The Banded Series has been subdivided into three major zones that are generally consistent across the intrusion on a large scale, although subdivisions may show significant lateral variations (McCallum et al., 1980). The **Lower Banded Zone** consists of well-layered norite and gabbro with 1- to 20-m thick anorthosite layers. Augite is a common cumulate phase that begins at the 2500-m level. Olivine is rare but does occur

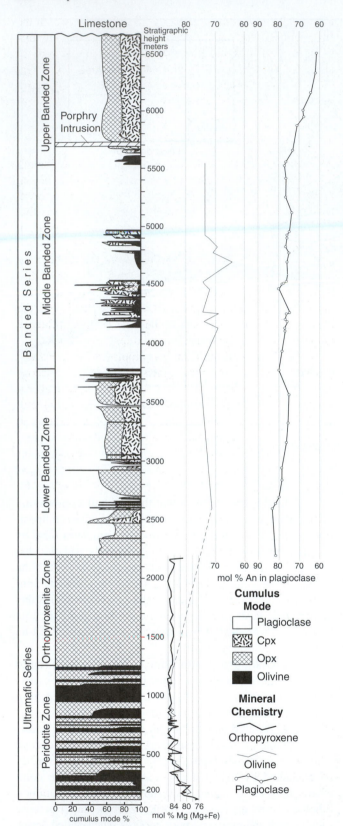

FIGURE 12.9 Composite stratigraphic column of the Stillwater Complex. After McCallum et al. (1980) and Raedeke and McCallum (1984).

perturbation of the chamber, as olivine-saturated magma is introduced. This initiates five repetitions over the next hundred meters of the rhythmic sequence troctolite–anorthosite–norite–gabbro, corresponding to the crystallization sequence:

$$\text{olivine} + \text{plagioclase} \rightarrow$$
$$\text{plagioclase} + \text{orthopyroxene} \rightarrow$$
$$\text{plagioclase} + \text{orthopyroxene} + \text{augite}$$

A Pt-Pd-sulfide–bearing horizon (known as the J-M Reef) has been discovered recently, similar both chemically and stratigraphically to the Merensky Reef of the Bushveld. Above this occurs nearly 250 m of norite, including an anorthositic norite that displays the fine-scale orthopyroxene–plagioclase layering shown in Figure 12.3a.

The **Middle Banded Zone** is plagioclase rich (averaging 82% of the zone), including numerous anorthosite layers, one reaching 570 m thick. Some rhythmites in olivine-bearing subzones are well layered, commonly on a centimeter scale, comprising complex sequences of troctolite, anorthosite, and gabbro, but orthopyroxene is a relatively minor and late phase in most of them. The crystallization sequence in the Middle Banded Zone differs from that of the Lower Banded Zone (clinopyroxene comes in before orthopyroxene). Such a change in crystallization sequence suggests the introduction of a different magma type into the Stillwater magma chamber at the time of this transition.

The **Upper Banded Zone** contains a lower olivine-bearing subzone and an upper one of uniform gabbro. Once again, the olivine-bearing subzone contains several rhythms of cyclic (troctolite–)anorthosite–norite(–gabbro).

The Stillwater Complex has many of the characteristics of the Bushveld. It has a similar overall composition, and it displays well-defined rhythmic layering as well as phase and cryptic layering (Figure 12.9); it grades from ultramafic to gabbroic and anorthositic compositions (before the unconformity at the exposed top, which has removed any later differentiates). Because of the detailed stratigraphic columns, we can see better evidence for the sequence of crystal fractionation that produces many of the lithologic successions and for intermittent influxes of new magma.

The simplest explanation for the Ultramafic and Banded Series, and the profound increase in plagioclase at the base of the latter, is that there are two principal magma types. The first is mafic and olivine saturated. The other is gabbroic or noritic. The Peridotite Zone crystallization sequence of olivine, orthopyroxene, plagioclase, and augite described above and the common Banded Zone gabbro/ norite sequence or orthopyroxene, plagioclase, and augite are in accordance with the sequences predicted for such magmas in the quaternary forsterite–silica–anorthite–diopside system.

On the other hand, some lithologic sequences, particularly those in the Middle Banded Zone, do not follow a progression consistent with phase equilibria. For example, the common sequence of olivine-bearing rhythmites in the Banded Series is troctolite → anorthosite → norite → gabbro. Injection of a contrasting magma into a chamber has been proposed for the development of rhythmic layering. The

periodically at several horizons. Modal grading, scour-and-fill, and slump structures in the lower few hundred meters of the zone attest to the activity of currents and/or slumping at the floor of the magma chamber. Reappearance of olivine (not Fe rich) at the 2600-m level suggests another major initial troctolite can be achieved by injection of the ultramafic type of magma into the norite/gabbro magma chamber, which also explains why these rhythmites are always associated with olivine-bearing subzones. The sawtooth pattern of cryptic layering is also suggestive of reinjection of a contrasting magma, although the overall trend is still one of advancing differentiation with time. The relatively pure anorthosite layers are more difficult to explain. Single-phase crystallization (anorthosite) should be the *first* step in any crystallization process, before other minerals join it with further cooling. Why, then, should plagioclase crystallization follow the crystallization of olivine + plagioclase? As I said in Chapter 11, however, linear mixing of magmas that evolve along curved trajectories can produce liquids that vary from the normal liquid line of descent. Note in Figure 7.4 that, because of the curvature of the cotectic, mixing of magmas of compositions *c* and *d* results in a hybrid directly between these two points that is within the anorthite + liquid field. The result is crystallization of anorthite, which returns the liquid to the cotectic. Because liquids *c* and *d* are invariant, they are likely to be rather common. The extensive amount of anorthosite in the Stillwater Complex, however, is difficult to generate by such a process. Deviations in sequence and non-cotectic mineral proportions are generally associated with plagioclase, which, unlike the mafic phases, has a density close to that of the coexisting magma (McCallum et al., 1980). Gravity would then have little effect on its motion relative to the liquid.

The mechanisms of crystallization and layer formation in LMIs are complex and extend beyond even combined replenishment and crystal settling. Reinjection of magma into the chamber almost certainly played a role in both the Bushveld and Stillwater complexes, and it is commonly difficult to distinguish the effects of magma additions from internal layering processes. We turn now to an intrusion where the role of reinjection is believed to be minor or absent, to see what layering is still possible.

12.2.3 The Skaergård Intrusion

A discussion of LMIs would be incomplete without including the Skaergård intrusion of eastern Greenland. (The old Danish spelling, used commonly in the geological literature, is "Skaergaard.") One of the proclivities of science is that good pioneering work spawns further work and comparison. Thus a small (only 10 km across) layered intrusion in an isolated and desolate stretch of Greenland, following the comprehensive early work of Wager and Deer (1939), has become the "type locality" for LMIs. It is now one of the most intensely studied igneous bodies on Earth, and the ideas generated by those studies have had a great influence on our concepts of igneous differentiation.

The Skaergård intrusion was emplaced in the Eocene (56 Ma) through Precambrian gneisses and thin Cretaceous sediments into the base of Tertiary flood basalts. Glacial erosion and (partial) recent recession have left over a kilometer of relief with excellent exposures. Ice and fjords cover some of the body, and the thick Basistoppen sill was intruded into the upper portion (Figure 12.10), but a good section, about 3200-m thick, is exposed across the southerly dipping layers. Only the lowermost section is still buried beneath the

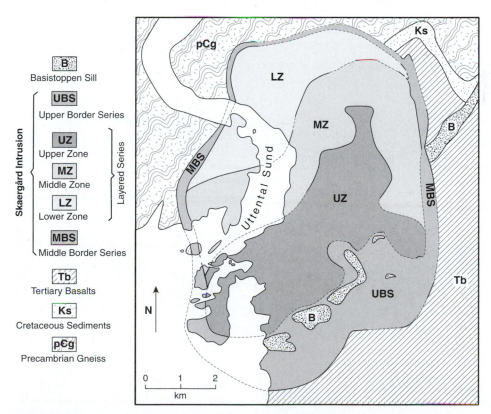

FIGURE 12.10 Simplified geologic map of the Skaergård intrusion, eastern Greenland. Contacts are dashed where under water or ice. After Stewart and DePaolo (1990). Copyright © with permission from Springer-Verlag.

gneisses to the north. The magma was apparently intruded in a single surge, so the Skaergård serves as a premier natural example of the crystallization of a mafic pluton in a single-stage process. There is a fine-grained chill margin, about 1 m thick, but, like most LMI margins, it has been contaminated and is no longer representative of the original magma.

The Skaergård has been subdivided into three major units: the Layered Series, the Upper Border Series, and the Marginal Border Series (Figures 12.10 and 12.11). The base is not exposed, and an unknown buried thickness is classified as the Hidden Zone. It is generally agreed that the Layered Series crystallized from the floor upward, the Upper Border Series from the roof downward, and the Marginal Border Series from the walls inward. The major cooling units, the Upper Border Series and the Layered Series, meet at what is called the "Sandwich Horizon," the level at which the most differentiated liquids developed.

The lowest exposed zones of the **Layered Series** begin with cumulates of olivine and plagioclase set in poikilitic (hence interstitial) augite crystals. Augite joins olivine and plagioclase as a cumulus phase 200 m higher (Figure 12.12), followed by pigeonite (a hypersolvus low-Ca pyroxene that inverts on cooling to orthopyroxene) and magnetite. The disappearance of olivine marks the top of the Lower Zone (800 m above the exposed base), and the reappearance of a more iron-rich olivine defines the base of the Upper Zone (nearly 1600 m above the base). This behavior of olivine is the same as observed in the Bushveld Complex, discussed above. Alkali feldspar and quartz form as late interstitial phases in the Upper Zone. The crystallization sequence manifested in the overall phase layering is then:

olivine + plagioclase → + augite → + pigeonite →

+ magnetite → − olivine → + Fe-rich olivine →

− pigeonite → + apatite → + alkali feldspar + quartz

We thus see a liquid evolve from two liquidus phases at the lowest exposed layers (and presumably only one below in the Hidden Zone) to six (excluding minor phases) in the more differentiated upper levels.

The cryptic layering, measured in the interior of cumulus grains to minimize post-cumulus reaction effects, is also shown in Figure 12.12. There are orderly trends of decreasing An content of plagioclase and Mg# of the mafics. The phase and cryptic layering, on the hundred-meter scale, are consistent with crystal fractionation and settling in a progressively cooling chamber of tholeiitic basalt (Figure 12.12).

Even the whole-rock chemical composition shows fairly regular trends compatible with progressive fractionation of a basaltic liquid. Figure 12.13a shows the trends for SiO_2, MgO, and FeO for both whole-rock analyses and estimated liquid compositions for the Layered Series, whereas Figure 12.13b shows the trends in selected compatible (Cr and Ni) and incompatible (Rb and Zr) trace elements for the complete section. The trends shown are compatible with differentiation of a single surge of magma, and there is no evidence for any cyclical variations suggestive of repeated injections of fresh magma.

Methods for determining true liquid evolution trends are necessarily indirect, however, because cumulate rocks are dominantly solid fractionates and cannot be equivalent to any bulk liquid. Thus one must estimate the liquid line of descent by some indirect means. One method, tried by Wager (1960), uses mass balance calculations based on original (chill) liquids less the sum of the cumulate crystals up to any stratigraphic level. Another uses experimental studies of supposedly solidified trapped interstitial liquids (McBirney, 1975). Hunter and Sparks (1987) challenged both methods as being inconsistent with tholeiitic volcanic trends and common phase equilibrium trends, and they proposed another liquid line of descent that they consider more appropriate. All three models for liquid major element trends are

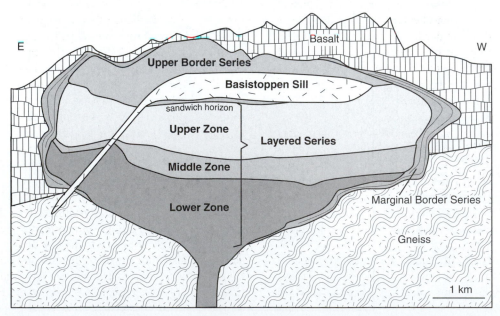

FIGURE 12.11   Schematic cross section of the Skaergård intrusion made by projecting the units at the surface to a plane normal to the regional dip. After Hoover (1978).

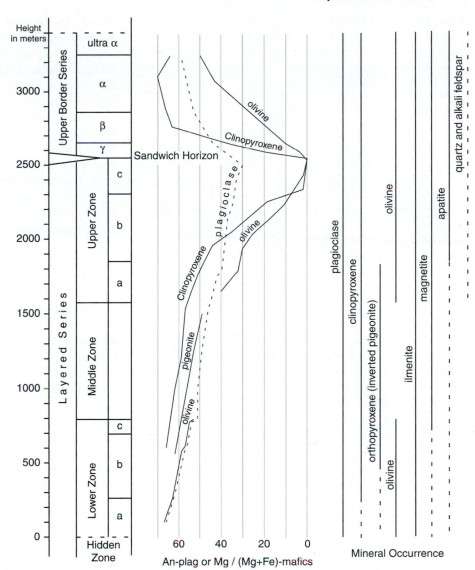

FIGURE 12.12 Stratigraphic column for the Skaergård intrusion. Solid vertical lines indicate cumulus minerals; dashed lines indicate intercumulus status. After Wager and Brown (1967) and Naslund (1983), reprinted by permission of Oxford University Press.

included in Figure 12.13a. Note among the major element trends that MgO decreases and FeO increases (until at least the last stages of differentiation). The behavior of SiO_2 in evolving tholeiitic liquids is controversial. If the models of Wager (1960) and McBirney (1975) are correct, SiO_2 varies only slightly for most of the evolutionary process. *This tholeiitic trend places into question the use of silica as a universal differentiation index (classic Harker diagrams) in all magma series.* If, on the other hand, Hunter and Sparks (1987) are correct, SiO_2 may yet work for later differentiation, although it is still relatively constant for at least the first half of the fractionation process. Recent experimental work by Toplis and Carroll (1995) suggested that SiO_2 remains nearly constant until the later stages of crystallization and then increases, in a fashion similar to the curve of Hunter and Sparks (1987) in Figure 12.13a.

The deceptively smooth and simple pattern for a single-stage mafic intrusion is complicated by the presence of layering on a finer scale. Thin (usually < 50 cm), rhythmic layering is well developed, as is intermittent layering that includes both uniform and graded layers (Figure 12.4). Graded layers in the Layered Series have mafics concen-

trated toward the well-defined base and more felsic minerals toward the top, as though they were gravity stratified. Gradations in grain size are rare. Repetition of such layers gives the rock a distinct dark–light rhythmically banded appearance. The smooth trends in mineral and whole-rock composition preclude the periodic reinjection of more primitive magma, leaving us to wonder how the rhythmites are formed. Further, if crystal settling under the influence of gravity created the layering, why are some layers graded and others, usually containing the same minerals, uniform?

The 960-m **Upper Border Series** is thinner but mirrors the 2500-m Layered Series in many respects. The upper border of Skaergård cooled from the top down, so the top of the Upper Border Series crystallized first. Plagioclase is more dominant and olivine less so than in the Layered Series, which may correspond to the low density of the former and high density of the latter. The most Mg-rich olivines and Ca-rich plagioclases occur at the top and grade to more Fe-rich and Na-rich compositions downward (Figure 12.12). Major element trends also reverse in the Upper Border Series as compared to the Layered Series (Figure 12.13). Olivine shows the hiatus between upper Mg-rich and lower

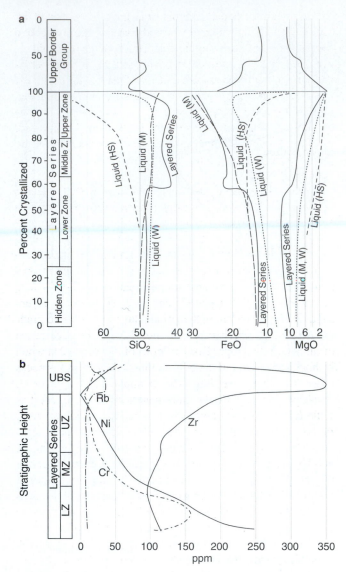

FIGURE 12.13 Variation in selected major **(a)** and trace **(b)** elements in the Skaergård intrusion. Major elements are determined for the Layered Series and Upper Border Series, as well as calculated for hypothetical liquids based on the models of McBirney, 1975 (M), Wager, 1960 (W), and Hunter and Sparks, 1987 (HS). After McBirney (1993). Jones and Bartlett, Sudbury, MA. Reprinted with permission.

Fe-rich olivine, orthopyroxene, apatite, and magnetite occur as cumulus phases in the predominant ferrogabbros in a granophyric mesostasis. Granophyric segregations of quartz and feldspar appear rooted in the ferrogabbro, and this pair is believed to represent immiscible liquids that evolve in the late stages of Fe-rich tholeiite differentiation (Section 11.2.3).

The **Marginal Border Series** is a 70- to 600-m thick shell around the intrusion. It contains a fine, relatively homogeneous chill zone for the outer 50 to 100 m, but in the rest of the Marginal Border Series, the rocks are also layered. The layers are thinner (a few centimeters to several meters) and roughly parallel to the wall contacts. Layering thus forms a concentric pattern for the Skaergård. The subzones based on phase layers of the Layered Series are also developed in a more compressed version, and most thicken upward (Wager and Brown, 1968; Hoover, 1978). Rhythmic layering is also developed due to grain size differences or variations in the proportions of light and dark minerals. Modally graded layers are commonly developed, with the dark minerals concentrated on the outer margin of a layer and light-colored minerals inward. This poses a further problem for any interpretation that attributes graded layers (or any vertical rhythmic layers) solely to crystal settling!

The Skaergård intrusion, despite its supposed single-stage development, is still a complex layered intrusion. The three major subdivisions signify that the Skaergård has crystallized inward from the margins, developing a concentric layering as a result. The Marginal Border Series is the thinnest, suggesting that either cooling was slowest along the sides or that much of the cooled liquid migrated away by convection. The bottom Layered Series is three times the thickness of the Upper Border Series, in spite of the obvious fact that the intrusion lost more heat through the roof (where convection of meteoric water in fractures above the intrusion sets up an efficient heat transfer process) than through the floor. The reason for the thicker Layered Series may be associated with convective transfer of heat upward from below and of the cooler liquid downward. The chamber is several kilometers thick, so the melting point of most minerals will be significantly (10 to 20°C) higher at the floor than at the roof. For the same reasons that decompression melting can occur in a rising diapir (Chapter 10), so might crystallization accelerate at the floor of an intrusion if crystal-laden magma is carried downward in a plume of cooled magma descending from the roof or walls, thus enhancing floor accumulation. Similarly, heat rising upward (as the lower magma cools plus a significant contribution from latent heat of fusion as crystals grow) will retard cooling at the roof and inhibit crystallization. It may even be sufficient to cause melting of the roof rocks in some LMIs.

Fe-rich types as well. The classic gravity-driven crystal settling model faces further problems when we attempt to explain the downward increase in differentiation if the first-formed crystals are to have settled in that direction. In this case, the higher-temperature crystals are near the roof. The occurrence of heavy olivines and pyroxenes in rhythmic layers along the chamber roof does not support the gravity settling hypothesis. The Upper Border Series is enriched in K_2O, SiO_2, P_2O_5, and H_2O in comparison to the Layered Series, suggesting that it received differentiated components from the crystallizing magma below (Naslund, 1983).

The Upper Border Series and Layered Series meet at what is called the **Sandwich Horizon**, where the latest, most differentiated liquids crystallized. The Sandwich Horizon is interpreted by Irvine et al. (1998) as the cumulate residue of a now-eroded last granophyre. Sodic plagioclase (An_{30}) plus

12.3 THE PROCESSES OF CRYSTALLIZATION, DIFFERENTIATION, AND LAYERING IN LMIs

Although LMIs may represent our most "simple and ideal" natural case of magmatic evolution upon cooling, the dynamics of cooling magma chambers are more complex than we had initially anticipated and are still only partly understood after a half century of study. It is impossible to come up with a single coherent model that explains all cases of differentiation and layering in these bodies, nor can we hope to cover all of the ideas in the rapidly expanding literature on the subject. Instead, I offer a brief and selective review of the hypotheses and discuss some alternative explanations for the development of the various types of layering that characterize most LMIs.

Any successful model for crystallization in LMIs must be able to explain the coeval phase, cryptic, and modal (including rhythmic) layering. A basaltic magma chamber is not a system in which minerals simply crystallize in the sequence of Bowen's Reaction Series and rain down upon the chamber floor to create a pile of cumulate rocks beneath progressively differentiated liquids in the shrinking magma volume remaining beneath the roof. Indeed this does occur to some extent. The large-scale patterns of phase and cryptic layering correspond well to the sequence of minerals that form, and the change in mineral compositions expected, as tholeiitic liquids cool. Phase and cryptic layering commonly span a coherent cooling unit, such as the upward-cooling Layered Series or the downward cooling Upper Border Series of the Skaergård.

Other features, however, are unusual, particularly the stratification on scales ranging from a few centimeters to several hundred meters, including cyclic repetition of layers, both homogeneous and graded. Our main job at this point, then, is not to explain phase and cryptic layering but to show how repetitive modal layering can occur and still be compatible with the phase and cryptic layering patterns. Here are some of the models proposed.

12.3.1 Gravity Settling

As discussed in Chapter 11, the oldest and most widely accepted mechanism for magmatic differentiation is **gravity settling** (or **crystal settling** under the influence of gravity). Wager (1963) described evidence for this mechanism of crystal growth and sinking to the bottom in the Skaergård intrusion. The theory and practicability of crystal settling for Newtonian fluids were developed in Section 11.3. There is ample evidence that crystal settling occurs in some magmas. The best textural evidence generally comes from the higher temperature and lower viscosity (more Newtonian) ultramafic varieties. Many picrites and komatiites show evidence for phenocryst accumulation. The ultramafic Duke Island intrusion in southern Alaska is another example where size sorting of graded layers has been convincingly attributed to crystal settling processes (Irvine, 1974, 1979). Slump structures, autoliths, xenoliths, and large crystal clusters that have dropped to the floor have also been described at Duke Island, Skaergård, and other LMIs.

The nature of cumulate rocks and graded layers has been compared to the textures of clastic sediments and invoked as further evidence for crystal settling. Indeed, the original definition of cumulate texture had the genetic implication of **accumulation** by crystal settling (Wager et al., 1960).

Rhythmic modal layering was initially explained by crystal settling interrupted by periodic large-scale **convective overturn** of the entire cooling unit. Thus a single rhythmic sequence would be deposited, involving settling of denser crystals beneath lighter ones and expulsion of the late-differentiated liquid by compaction under the weight of the accumulated crystals. Overturn would remove the late liquid and rehomogenize the system, and the process would then be repeated. Each cycle would be more evolved due to the removal of the phases in the rhythmic unit, resulting in the phase and cryptic patterns.

Despite its simple allure, a number of objections have been raised to the crystal settling theory. Many of the minerals found at a particular horizon in LMIs are not hydraulically equivalent in the sense that they would not be expected to settle at the same rate via Stokes' Law and thus concentrate at a given level by crystal settling. Size is more important than density in Stokes' Law [Equation (11.1)] because the radius term is squared, and thus size-graded layers should be more common than modal layering. Yet size grading is rare in the graded beds of most LMIs. The occurrence of olivine in the Upper Border Series of the Skaergård when it should have sunk to the bottom must result from a process other than crystal settling. The example that Wager et al. (1960) chose to support their genetic model for crystal-settled cumulates was plagioclase in the Skaergård intrusion. Interestingly, this may turn out to be one of the best arguments *against* the origin of cumulates by crystal settling. A number of investigators have determined that much of the plagioclase supposed to have settled out under the influence of gravity in the Skaergård should have been *less* dense than the liquid from which it crystallized. If gravity were the sole cause of the layering, why didn't plagioclase *float*?

In addition, the cryptic variations in the Upper Border Series, as mentioned above, are inverted, with the most An-rich plagioclase and Mg-rich mafics at the top, suggesting that the early-formed minerals, including heavy mafics, settled upward if they settled at all. The Marginal Border Series shows *vertical* layering, with mafic minerals concentrated toward the margin in graded layers. These relationships do not follow from simple gravitational settling.

As mentioned in Chapter 11, McBirney and Noyes (1979) also suggested that basaltic magmas become highly non-Newtonian, developing a high yield strength, slightly below liquidus temperatures. If this is true for LMIs, crystal settling would no longer be expected to occur soon after crystallization began, particularly if there is some convective motion to keep the crystals suspended, as is widely considered to be the case (Sparks et al., 1984).

12.3.2 Recharge and Magma Mixing

Other processes might work, alone or in conjunction with crystal settling, to produce some of the layer variations noticed in LMIs. As discussed above, the cryptic variations expected from a continuously evolving chamber are not observed in the Ultramafic Series in the Stillwater. Mg# of most mafics, for example, is irregular (Figure 12.9). Periodic **recharge** of the chamber with more primitive magma has been invoked to explain such situations. This would reset the differentiated cryptic trends back toward more primitive compositions, only to be followed by a forward event of settling and overturn. The result would be a sawtooth cryptic pattern like those in Figure 12.9. As discussed in conjunction with the Stillwater Complex, mixing of magmas along curved liquid lines of descent can result in non-cotectic liquids and the crystallization of monomineralic layers. Most volcanoes erupt intermittently due to recharge of fresh magma into the shallow chamber beneath the vent. Many LMIs are certainly subject to the same process, and major compositional shifts in the Stillwater and Bushveld are almost certainly due to recharge. Many cryptic trends, however, are like the Skaergård and show no cryptic spikes of regression indicative of recharge. Thus a model based on reinjection of fresh magma cannot produce all rhythmites, and we must find a way capable of producing them in a closed system.

12.3.3 Oscillations Across the Cotectic

Harker (1909), Wager (1959), and Maaløe (1978) suggested that well-layered rocks (such as macrolayering) can be attributed to transitory excursions across the cotectic. Boudreau and McBirney (1997) suggested that liquid compositions can be displaced from the cotectic liquid line of descent by a variety of events, such as convective overturn, magma mixing, assimilation of country rocks, gain or loss of volatiles, or changes in temperature or oxygen fugacity. If the composition of the liquid is displaced from the cotectic, crystallization of the phase corresponding to the liquidus adjacent to the cotectic will return the liquid to the cotectic again, producing a monomineralic layer. Subsequent displacement to the other side of the cotectic will result in a layer of another mineral. The idea is that oscillating changes in the parameters above cause the liquid path to zigzag across the liquidus during cooling, producing alternating layers.

12.3.4 Compaction

Compaction, expulsion, and convective rise of less dense intercumulus liquid has been proposed as an important process in the evolution of both cumulus assemblages and later liquids. The importance of compaction and exchange of intercumulus liquid with the larger magma reservoir to the development of adcumulus textures was discussed in Section 3.1.7.

Compression of a cumulate mass on the floor of a chamber by the weight of additional sedimentation of crystals on top can result in the expulsion of highly enriched residual evolved liquids into the reservoir (McBirney, 1995; Mathez et al., 1997). Some of these expelled intercumulus liquids can react with other liquid–mineral mixtures above, resulting in secondary replacement as they infiltrate upward (Boudreau and McBirney, 1997). Irvine (1980a, 1980b; Irvine et al., 1999) attributed the development of cross-cutting replacement anorthosites in some LMIs to the preferential resorption of mafic minerals by focused flow of expelled

H_2O-enriched liquids. McBirney (1995) attributed the enrichment of the Upper Border Series to the rise of compaction-released liquids from the floor.

Coats (1936) and Boudreau and McBirney (1997) proposed a mechanism by which layering can be initiated by mechanical segregation of mineral types during compaction. The segregations, when simulated experimentally by Boudreau and McBirney (1997), however, were very poorly defined.

12.3.5 In Situ Crystallization and Convection

For all of the above reasons, many investigators have turned to an alternative mechanism, initially proposed by Jackson (1961), emphasizing in situ crystal nucleation and growth, which operates in a similar fashion to evaporites or other chemical precipitates. In situ crystallization, as discussed in Section 11.5, involves nucleation and growth of minerals in a thin, stagnant **boundary layer** along the margins of the chamber. In situ is a bit of a misnomer, because differential motion of crystals and liquid is still required for fractionation by any in situ process. The dominant motion, however, is considered to be the migration of the depleted liquid from the growing crystals by a combination of diffusive and convective processes. For convection, this may seem a semantic distinction, but the crystals in this case only settle (or float) a short distance *within the boundary layer* as the melt migrates from the crystals in accordance with its density. The boundary layer interface inhibits material motion. Thus plagioclase will not easily rise out of a layer at the bottom of a chamber, and olivine will not easily sink from one at the top.

How does in situ crystallization and fractionation work, and how can it produce layering? This is a complex subject, governed by a number of variables, including the shape and size of the chamber, the composition and viscosity of the magma, and the processes at the boundary layer. I shall attempt to describe the in situ process in the most general way, and show how it may lead to layering. Our understanding of the processes is based mainly on the fluid dynamics in experiments on heat and mass transfer in multicomponent solutions, usually aqueous with dissolved salts (Huppert and Turner, 1981a, 1981b; Sparks et al., 1984; Huppert and Sparks, 1984b; Turner and Campbell, 1986; Brandeis and Jaupart, 1986).

There are numerous possible situations involving cooling, crystallization, magmatic evolution, recharge, and mixing in complex solutions. Of particular interest for layering are systems in which there are gradients in two or more properties (chemical or thermal) with different rates of diffusion. If a situation occurs in which these gradients have opposing effects on the density of the fluid in a vertical direction, a wide range of novel and complex convective phenomena may occur.

As an example, consider the following experiments. First, a container of pure water is heated from below. The result is a convecting system extending the full vertical length of the container, passing heat upward. Okay, so that is a dull experiment. Next, we try a container with an aqueous NaCl solution with an initial stable salinity gradient in which the concentration (and density) of the solution increases toward the base. The container is then heated from the bottom. The heating establishes a thermal gradient in the bottom layer of the container. The thermal gradient creates a thermal density (ρ_{temp}) gradient in the layer that increases upward (Figure 12.14a). This is now an example of the situation described in the previous paragraph. We have a vertical compositional density gradient (ρ_{salt}) that increases downward and a thermal density gradient that increases upward in the same layer. In addition, one gradient (in this case ρ_{temp}) is destabilizing because the density increases upward (although the total density gradient is stable) and the diffusivity of the destabilizing component (heat) is faster than the compositional diffusivity of the salt.

This creates a classic **double-diffusive convection** situation. A characteristic feature of double-diffusive convection is that opposed gradients with different diffusivities tend to produce a series of convecting layers rather than a single large overturning convection cell (Figure 12.14b). In our saline example, heat causes convection only in a thin bottom layer, which homogenizes both the salinity and thermal density gradients. Heat diffuses faster than the salt, so it is transferred across the *diffusive* interfaces between the convecting layers and causes a thermal density gradient to develop in the next layer upward. This creates convective overturn in that layer, which homogenizes the salinity and thermal gradients within it and overcomes the thermal density instability while transferring heat upward, where it will diffuse through the interface to the next layer and so on. The result is a series of layers, each with a different composition and temperature, and the gradual upward transfer of heat from the bottom of the container to the top by convection within layers and diffusion across the interfaces between them. The layers operate in the same fashion as the two-layer mantle model illustrated in Figure 10.17, with convection mixing the layers, but only diffusion (of heat much more than matter) across the boundary. Cooling from the top will have an effect similar to that of heating at the bottom because both situations impart a decreasing upward temperature gradient. The process described above is well known in the oceans, where it creates thermal- and salinity-stratified layers.

The model above is based on a single compositional variable and a temperature gradient, but similar effects can be created in more complex magmas by two (or more) compositional gradients as well, perhaps induced by crystallization, if their effect on density is opposing and their diffusibilities are different. Realistically, crystallization is initiated by decreasing temperature, and a thermal gradient would probably control initial nucleation and crystal growth, but, once begun, crystallization may create gradients in several components, resulting in double-diffusive convection and layering. Once formed, each of the layers can evolve as a chemically independent system and could crystallize to create an overall rhythmic sequence, including grading. Graded layers could be "normally" graded at the bottom of a magma chamber (high-temperature phases at the bottom), reverse graded at the top, and vertical layers with horizontal grading at the sides (all of which occur in the Skaergård) because diffusion in response to crystallization in a thermal gradient, not gravity, may be controlling the fractionation process.

Alexander McBirney, a strong proponent of double-diffusive convection as the method for the formation of graded rhythmic layering in LMIs, later raised doubts about its effectiveness in tholeiitic magmas at the floors of magma chambers (McBirney, 1985). Although it may still be effective along the walls, he concluded, his calculations indicated that the compositional density gradients along the floor were too large for the expected temperature gradients to destabilize.

12.3.6 Preferential Nucleation and Crystallization

As an example in which crystallization alone (without a thermal gradient) results in the development of rhythmic bands, consider the rhythmic oscillations found in Liesegang bands (Liesegang, 1896). These can be created by placing a crystal of silver nitrate in a gel impregnated with potassium chromate. As silver ions diffuse radially outward from the crystal, silver chromate precipitates after an area in the gel becomes supersaturated in both silver and chromate. Supersaturation seems to be required for initial nucleation and precipitation of

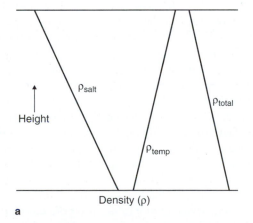

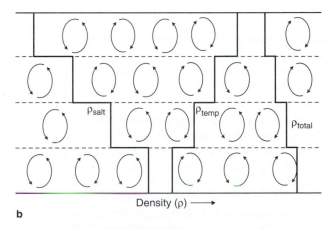

FIGURE 12.14   (a) Gradients in density created by salinity (ρ_{salt}) and temperature (ρ_{temp}) in a layer of fluid as it is initially heated from below. **(b)** The same system after it has broken up into a series of double-diffusive convecting layers. After Turner and Campbell (1986). Copyright © with permission from Elsevier Science.

silver chromate. The subsequent growth of silver chromate causes the depletion of chromate in the neighboring zone. Silver continues to diffuse outward (faster than chrome can recover from the depletion) until more chromate is encountered, and another band of silver chromate forms. The result is a series of bands of silver chromate separated by precipitate-free bands. Similar processes of **oscillatory nucleation** (McBirney and Noyes, 1979) and depletion caused by crystallization followed by diffusion or convection of the remaining magma could create rhythmic layers in magma chambers, including the fine rhythmic layers shown in Figure 12.3.

Boudreau and McBirney (1997) developed a model for layering involving competitive growth of crystals. Their model begins with the poorly developed layering that they attribute to compaction processes (see above). They proposed that expelled liquids from further compaction react preferentially with smaller crystals in the layers above (similar to Ostwald ripening, Section 3.2). They also proposed that the surface energy of a mineral is lower when in contact with the same mineral, so that reactions will preferentially scavenge subordinate minerals from indistinct layering and make it more regular. Boudreau (1995) developed a similar model for nucleation and preferential growth of early larger crystals that deplete neighboring layers of smaller crystals to produce layering.

12.3.7 Density Currents

Density currents of cooler, heavy-element-enriched, and/or crystal-laden liquid may descend (perhaps along the cool walls) and move across the floor of a magma chamber and deposit layers. Irvine (1980a) has made an excellent case for these gravity-driven currents in several LMIs, supported by such current features as scour-and-fill channels and cross-bedding (Figure 12.15). He discussed a number of mechanisms by which crystal-suspension density currents descend along the sloped walls of LMIs and surge across the floor to pond.

Irvine et al. (1999) presented a model for the Skaergård that draws from Neil Irvine's 30 + years studying LMIs. The model attempts to reconcile the many aspects of layering, particularly the modally graded layers (Figure 12.2a) and the intermittent layers (Figure 12.4), while reaffirming the importance of crystal sedimentation, density currents, and features associated with the thousands of displaced blocks evident in the Skaergård. The blocks (autoliths of crystallized material from the Upper Border Series and xenoliths of country rock stoped from the roof) have fallen through the magma of the Skaergård, impacting the floor and deforming the layering. Some blocks have shattered on impact. Other blocks have broken up and become entrained in density currents, eventually spreading horizontally in clusters across the floor. The evidence of these blocks led Irvine et al. (1999) to conclude that there was always a clear, sharp contact between the top of the cumulate pile and the bottom of the main magma body and that the pile must have been relatively coherent (perhaps 50% crystalline) in order for the layers to have been deformed by block impact. These conclusions argue against double-diffusive liquid layering at the floor of the chamber.

FIGURE 12.15   Cross-bedding in cumulate layers. **(a)** Duke Island, Alaska. Note also the layering caused by different size and proportion of olivine and pyroxene. From McBirney (1993). Jones and Bartlett, Sudbury, MA. Reprinted with permission. **(b)** Skaergård Intrusion, eastern Greenland. Layering caused by different proportions of mafics and plagioclase. From McBirney and Noyes (1979). Reprinted by permission of Oxford University Press.

Irvine et al. (1999) attributed the development of the modally graded layers to density currents that descend from the boundary layers along the walls of the chamber. Dense crystals are held in suspension by the agitation of the flow, and light crystals like plagioclase can also be trapped and carried downward in the currents. Using the flow patterns observed in flume experiments with flows of dense glycerin into water or silicone liquids, Irvine (1980a) proposed a number of mechanisms that might produce graded or thin, uniform layers within a flowing regime as well as after the density-driven flow has come to rest. Some investigators have objected to density currents as a mechanism for producing layering, pointing out that density currents descending along chamber walls are unlikely to produce thin layers of constant thickness extending for over 100 km across the floor of some large intrusions. Irvine's (1980a) flow experiments suggested that the flows can extend for a considerable distance, but Irvine et al. (1999) also pointed out that individual rhythmic modal layers in the Layered Series of the Skaergård never extend laterally for more than 300 m (although the larger *layering* sections extend much further).

Figure 12.16 illustrates the type of vortex that forms in density surge currents along boundary layers in many of Irvine's (1980a) flume experiments. The streamlines and arrows portray the persistent instantaneous flow directions relative to (stationary reference) points V, R, and S in the vortex. In the illustrated example, the vortex actually moves downslope (arrow at point V), but, because they are the figure reference, the flow lines in the cumulate floor, although truly stationary, appear to be flowing uphill. The material in the density current moves faster than the vortex (thereby causing it), so that the boundary layer flow separates from the floor at point S and reattaches at point R. Irvine et al. (1999) thus called the vortexes "S-R cells." This phenomenon is well known in fluid dynamics (Schlicting, 1968; Tritton, 1977). Laminar flow is illustrated, but turbulent flow may also occur in and around the vortex. Variations in velocity, viscosity, etc. can cause the vortex to remain stationary or migrate uphill as well. When it migrates uphill, it results in headward erosion, and when it advances, as in Figure 12.16, it deposits a layer. Irvine (1980a) scaled his experimental results to basaltic liquids and calculated that the currents should move at rates of 2 to 6 km/hr and could spread material far across the floor of the Skaergård in less than an hour.

Irvine et al. (1999) further proposed that the currents are associated with three forces: gravity, floor drag, and displacing the host liquid upward during advance. Floor drag, they proposed, is particularly effective at stripping the dense mafic minerals downward from the bottom front of the vortex, whereas the less dense plagioclase crystals are stripped from the top rear by host-liquid drag. The result is a graded bed in the wake of the vortex. The plagioclase is then trapped beneath the next uniform layer of cumulates deposited by the main chamber liquid soon after the current has passed. This model explains both the graded beds and the alternating uniform beds in the common intermittent layering of the Layered Series illustrated in Figure 12.4.

Similar flows, but more rapid, occur because of surges of debris slumping in from the walls and top. Many of these carry blocks that break up and spread in clusters across the floor, as described above. Irvine et al. (1999) argue that this process of density current vortexes in the boundary layer apply equally well to the Upper Border Series layers of the chamber roof and the Marginal Border Series along the walls. Figure 12.17 illustrates their model for the Skaergård layering, showing how the Upper, Marginal, and lower Layered Series, although of different thicknesses, are components of a system of concentric shells. Stoping of the roof releases the autoliths and xenoliths, which fall to the bottom and constitute the numerous blocks. Boundary-layer density currents form along the steep walls at the roof and deposit layers in the Upper Border Series. Graded layers are absent near the top, however, suggesting that the reinforcing combination of gravity and drag at the floor are required. Heavy minerals are captured at the top due to rapid heat loss through the roof, which freezes crystals in place.

Crystallization in a boundary layer next to the walls, due to the large vertical column, will be susceptible to convective magma migration. Turner and Campbell (1986) considered convective motion to be an "inevitable" consequence of fractional crystallization along vertical walls. If denser, the derivative magma may feed density currents like those described by Irvine (1980a). If lighter, it may rise to form a stagnant cap layer at the top (Figures 11.11 and 11.12). Experiments by Nilson et al. (1985) have shown that some solutions cooling adjacent to a vertical wall develop an outer downflowing boundary layer and an inner upflowing one.

Density currents may travel in other ways than along the walls. Suppose a layer of more dense material overlies less dense material, due either to cooling at the roof or to compositional differences. Rather than getting multiple convecting layers, as in Figure 12.14, a series of dense columnar **plumes** forms (Figure 12.18) that extend the height of the chamber where the cool fluid spreads laterally to form a layer across the floor. Discrete plume-type events could create the uniform layers that occur within sequences of graded beds comprising the intermittent type of layering shown in Figure 12.4. It is also a way to produce layers of uniform thickness at the bottom of large chambers that may be a problem for wall-generated density currents, which may not be able to travel as far across the floor.

Density variations caused by changes in magma composition resulting from crystallization commonly dominate over thermal density changes (Turner and Campbell, 1986). Tholeiitic magma, the type most common in LMIs, can thus add a further complexity to compositional convection. As illustrated in Figure 12.19, initial fractionation of mafic minerals lowers the density of the residual evolved liquid, as is most common. But when plagioclase joins the mafics, the Fe-enrichment trend in early tholeiite evolution may cause the magma density to *increase* with further fractionation.

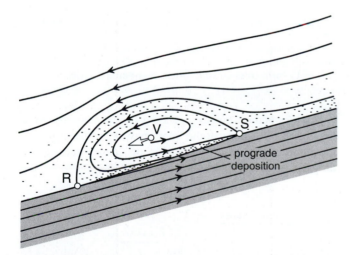

FIGURE 12.16 Cross section of a vortex cell in a density surge current along the boundary layer between floor cumulates (shaded) and magma. Streamlines and arrows portray the instantaneous flow relations relative to points V, R, and S. The floor is actually stationary, of course, and points V, R, and S move as shown with the arrow at V. Flow within the surge moves faster than the vortex (as in a tractor tread), so that the flow material separates from the floor at S and reattaches at R. Forward motion of the vortex results in deposition between it and the chamber floor. After Irvine et al. (1999). Copyright © The Geological Society of America, Inc.

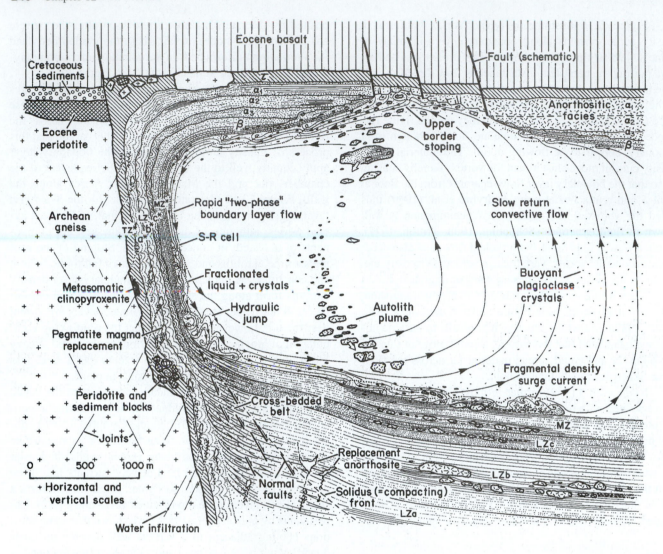

FIGURE 12.17 Schematic synthesis of the processes believed to be occurring in the Skaergård magma chamber early in the Middle Zone stage of development. Density-driven *S-R* vortex cells descend along the sloping roof and walls to spread across the floor, producing rhythmic layers that may be graded on the floor. Occasional larger density surge deposits result from slumps originating along the walls. Blocks of autoliths and xenoliths are also broken from the walls and roof, and impact the floor. Compaction of cumulates below the floor releases liquids that react with the overlying layers to form cross-cutting replacement anorthosite. From Irvine et al. (1998). Copyright © The Geological Society of America, Inc.

Once Fe-Ti oxides begin to fractionate the density again decreases. The density of calc-alkaline magmas, on the other hand, tends to decrease continuously as fractionation proceeds. Tholeiitic liquids, then, may rise or sink in a crystallizing boundary layer, depending on the stage of evolution. Crystallization at the roof may, at some stages of tholeiitic evolution, form plumes of denser magma that drop to the floor as in Figure 12.18.

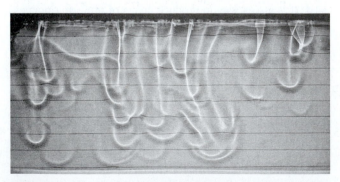

FIGURE 12.18 Cold plumes descending from a cooled upper boundary layer in a tank of silicone oil. Photo courtesy Claude Jaupart.

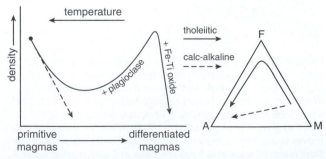

FIGURE 12.19 Schematic illustration of the density variation in tholeiitic and calc-alkaline magma series (after Sparks et al., 1984).

12.3.8 Combined Processes

The recharge and density current processes described earlier in conjunction with crystal settling need not be tied to fractionation by crystal settling in order to produce rhythmites. In his comprehensive work on density currents, Irvine (1980a) included models in which graded rhythmites can be created by in situ processes in a stagnant layer that results when a density current has come to rest.

Double-diffusive convective processes can readily be created in open systems where recharge is involved as well. Huppert and Turner (1981b) described an experiment in which a hot, dense layer of KNO_3 was slowly injected so that it flowed beneath a colder, lighter layer of $NaNO_3$ solution. The two layers exchanged heat (which diffused faster than the salts), causing crystallization of KNO_3 as the lower layer cooled, accompanied by vigorous convection driven by both heat and the release of light fluid as the crystals grew on the floor of the container. Eventually, the density of the residual KNO_3 solution became as low as that of the upper $NaNO_3$ solution, and a rapid overturn of the two layers took place.

Raedeke and McCallum (1984) applied such a model of combined reinjection and in situ fractionation (see also Huppert and Sparks, 1980) to the rhythmic dunite–harzburgite–orthopyroxenite layering of the Stillwater Ultramafic Series. In the model (illustrated in Figure 12.20), primitive olivine-bearing magma (magma A) is injected into a more differentiated magma (magma B) in which orthopyroxene is the sole liquidus phase. Rather than mixing to form a single homogeneous mixture, the experiments with aqueous solutions of various salts suggest that liquid–crystal suspensions of contrasting density will stratify according to their densities. The introduced magma is denser than the magma in the chamber, despite being hotter, because it has more mafic components. It would thus pond at the base of the liquid layer (Figure 12.20a). Olivine then crystallizes (solid dots in Figure 12.20) from the hotter primitive magma as it cools. Both layers may convect independently, maintaining a sharp interface as long as the material in the lower layer (combined crystals + liquid) is denser than in the upper one. Heat can diffuse (conduct) upward from the lower layer across the interface toward the cooler unit. The resulting cooling of the top of the lower layer and heating of the base of the upper layer drives the convection in each layer. It may even be possible for some of the orthopyroxene (open

circles) in the upper layer to melt because of the heat added from below. Convection may be sufficient to keep the crystals from settling to the base of their respective layers.

As the system cools further, convection slows, and the olivine and orthopyroxene crystals begin to settle to the base of their respective layers (Figure 12.20b). The density of the crystal-depleted and evolved top portion of the lower layer decreases, and that of the crystal-laden base of the upper layer increases. When the density of the base of the upper layer surpasses that of the liquid below it, the interface will break down and the layers will quickly mix (Figure 12.20c). Suspended olivine crystals above the accumulation at the base will be carried upward and mixed with the orthopyroxene as a hybrid magma (magma C) is created (Figure 12.20d). These crystals settle toward the bottom of the layer to create the harzburgite above the previously accumulated dunite. Continued crystallization of magma C will soon leave the peritectic and produce only orthopyroxene, creating the orthopyroxenite layer (Figure 12.20e). The process will be repeated with each injection of olivine-bearing magma, forming the rhythmites observed and the sawtooth cryptic patterns of Figure 12.9, where the repeated primitive injection resets the cryptic trends and retards any overall trend toward more differentiated liquids.

Fractionation processes based on compositional convection are also compatible with crystal settling processes, and both may occur, either concurrently or sequentially. Two of the principal champions of in situ fractionation published a model in which crystal settling plays an important role. Based on experiments with suspended grit particles in water or glycerin solutions cooled from the top, performed by Koyaguchi et al. (1993), Sparks et al. (1993) proposed a model for the origin of modal and rhythmic igneous layering by crystal settling in a convecting magma chamber. The experiments showed that when the solutions had low concentrations of particles ("phenocrysts"), convection kept the bulk of the particles suspended but simultaneous sedimentation occurred at the base, due to declining convection near the floor, so that the concentration of particles slowly decreased. Above some critical concentration of particles, however, a sharply descending interface developed between a particle-free convecting upper layer and a stagnant lower layer in which crystal settling took place in accordance with Stokes' Law.

Sparks et al. (1993) extended the concept to crystallizing systems, where they developed a model for layering.

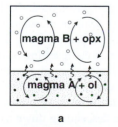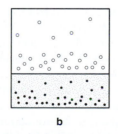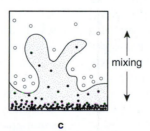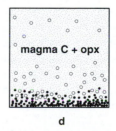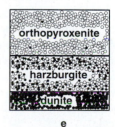

FIGURE 12.20 Schematic illustration of a model for the development of a cyclic unit in the Ultramafic Zone of the Stillwater Complex by influx of hot primitive magma into cooler, more evolved magma. From Raedeke and McCallum (1984). Reprinted by permission of Oxford University Press.

They envisioned a convecting chamber that is cooled from the top. Convection is both thermal and compositional, caused by crystals accumulating at the floor. As temperature decreases, crystals are formed, but most are kept suspended by convection as long as the concentration of crystals is low. As in the experiments, some crystals settle and accumulate on the floor. As crystallization progresses, a critical particle concentration is reached, producing a discrete sedimentation event. A descending interface separates a sedimenting lower layer from a particle-free convecting upper layer. The upper layer will eventually fill the chamber, when all the particles have settled. It will then cool and repeat the process. The discrete bed of crystals may be graded if there are crystal size or density differences because Stokes' Law controls sedimentation in the stagnant lower layer. In addition, each mineral phase will have a different critical concentration, resulting in substantial fluctuations in sedimentation rate and mineral proportions. Rhythmic phase graded layering is an expected result.

12.4 CONCLUSION

The examples discussed above are but a few of the myriad fractionation possibilities in cooling or recharged multicomponent systems in which crystallization and/or thermal gradients can create density variations in liquids or liquid–solid suspensions, convective motion, and even multiple horizontal layers. A variety of potential mechanisms are possible, depending on the size and geometry of the layers and the magma chamber, the nature of thermal and compositional gradients, the liquid viscosity, and the relative diffusibility of heat and chemical constituents. Such a complex set of variables means that an equally complex variety of processes may operate in cooling magma chambers, hence the bewildering array of structures in LMIs. Layering in a single intrusion may have different origins. McBirney and Nicolas (1997) noted that both "dynamic" (magmatic flow due to density currents) and "non-dynamic" (in situ) layers occur in the Skaergård.

Experimenting directly with crystallizing silicate magmas would provide some needed direct answers to our questions on magma chamber processes, but such experiments in the laboratory would require large systems into which we could see and are fraught with technical problems. We shall have to settle for indirect application of the results of fluid-dynamic experiments with solutions of various compositions, temperatures, and densities to the problem of magma chamber processes. The results appear to be fruitful, however, and it appears that a combination of density currents, crystal settling, reinjection, and in situ crystallization (accompanied by the separation of liquids as their density is altered, a process greatly facilitated by compaction) could lead to the types of layering observed in LMIs. The examples discussed above are but a small part of an extensive and growing literature on the subject but serve to show what types of processes may be possible. S. Sparks (personal communication) noted that in situ crystallization, crystal settling, and double-diffusive convection are commonly observed together in small-scale experiments.

In closing, I should note that the application of fluid dynamics to petrological problems is relatively new and in a state of flux. The fact that one of these proposed methods based on analogous solutions of salts, sugar, glycerin, etc. forms layered cells and *can* explain some type of layering in LMIs doesn't necessarily mean that it *does*. Nor does the fact that we have a set of possible solutions to the layering problem mean that we have the correct one (or ones). But the close analogies certainly indicate that we are on the right track, and the layering in LMIs in all probability results from processes similar to the ones described.

Summary

Due to their high temperature and low viscosity, large mafic intrusions should provide ideal natural laboratories for observing crystal–liquid fractionation processes during cooling and crystallization. Indeed they may do so, but the features they exhibit are far more complex than simple sequential crystal formation and gravity settling to form progressive accumulations from the bottom of an intrusion to the top. Rather, a bewildering array of layering occurs, including layers defined by varying proportions of minerals (modal layering), the appearance or disappearance of a phase across broader ranges of modal layering (phase layering), and variations in the composition of minerals (cryptic layering). Some of the layers are uniform in texture, whereas others may be graded. Layers are commonly repeated, often multiple times in a rhythmic fashion on a fine scale.

The cyclic repetition of layers, spanned by broader phase and cryptic trends, cannot result from the simple expedient of separate pulses of new magma injected into a larger chamber, causing the crystallization sequences we observed in Chapters 6 and 7 to repeat. Crystal settling and periodic convective overturn of the chamber system is another relatively simple alternative, but this cannot explain why some minerals in layers are not sorted by their density or Stokes' Law settling velocities, nor the inverted layering and cryptic trends in the upper portions of several LMIs. More complex processes are required to explain the layers and features of LMIs. Proposed explanations include oscillations across a eutectic path, crystallization in stagnant boundary layers and subsequent melt migration from the crystal/melt mush, the development of layers resulting from double-diffusive convection in systems with opposing gradients (typically in density) with different rates of diffusion, oscillatory nucleation in response to diffusion down composition gradients, and density currents descending down the sloping roof and/or walls of cooling chambers. Our understanding of such in situ processes and experiments using analog liquids is in an early stage.

Key Terms

Large igneous province (LIP) *222*
Layer *223*
Uniform layer *223*
Non-uniform layer *223*
Graded layer *223*
Layering *223*
Modal layering *223*

Phase layering *223*
Cryptic layering *224*
Rhythmic layering *224*
Intermittent layering *224*
Convective overturn *236*
Recharge *236*
Boundary layer *237*

Double-diffusive convection *238*
Oscillatory nucleation *238*
Density currents *238*
Descending plumes *239*
Vortex (*S-R*) cell *239*

Review Questions

Review Questions are located on the author's web page at the following address: **http://www.prenhall.com/winter**

Important "First Principle" Concepts

- Crystallization in cooling hot, viscous magmas differs substantially from an easily visualized concept of sand grains settling in water. Temperature gradients near the margins of a magma chamber and variable degrees of crystallization across these gradients result in compositional and density gradients.
- Liquids with density variations commonly result in convective motion: lighter liquids rise, and denser liquids sink.

- Crystals may also settle upward or downward in response to density contrasts with the liquid, or they may become entrained in viscous density currents.
- Compositional gradients (actually, *activity* gradients) may also cause components to migrate from areas of high activity toward areas of lower activity, thereby lessening the gradient.

Suggested Further Readings

Layered Intrusions

Cawthorn, R. G. (ed.). (1996). *Layered Intrusions*. Elsevier. Amsterdam. (A 25th anniversary volume commemorating the original Wager and Brown classic.)

Czamanski, G. K., and M. L. Zientek (eds.). (1985). *The Stillwater Complex, Montana: Geology and Guide*. Montana Bureau of Mines and Geology Special Publication **92**.

Journal of Petrology volume **20**, no. 3 (1979), and volume **30**, no. 2 (1989), are devoted to the Skaergård and Kiglapait intrusions.

Parsons, I. (ed.). (1987). *Origins of Igneous Layering*. NATO ASI series C. Reidel. Dordrecht.

Visser, D. J. L., and G. Von Gruenewaldt (eds.). (1970). *Symposium on the Bushveld Igneous Complex and Other Layered Intrusions*. Geol. Soc. South Africa Special Publication **1**.

Wager, L. R., and G. M. Brown (1968). *Layered Igneous Rocks*. W.H. Freeman. San Francisco.

Young, D. A. (2003). *Mind Over Magma*. Princeton University Press. Princeton, NJ. Especially see Chapters 17, 28, and 29.

Magma Chamber Processes

Brandeis, G., and C. Jaupart. (1986). On the interaction between convection and crystallization in cooling magma chambers. *Earth Planet. Sci. Lett.*, **77**, 345–361.

Huppert, H. E., and R. S. J. Sparks. (1984). Double-diffusive convection due to crystallization in magmas. *Ann. Rev. Earth Planet. Sci.*, **12**, 11–37.

Marsh, B. D. (1989). Magma chambers. *Ann. Rev. Earth Planet. Sci.*, **17**, 439–474.

McBirney, A. R., and R. M. Noyes (1979). Crystallization and layering in the Skaergaard intrusion. *J. Petrol.*, **20**, 487–554.

Sparks, R. S. J., H. E. Huppert, and J. S. Turner. (1984). The fluid dynamics of evolving magma chambers. *Phil. Trans. R. Soc. Lond.*, **A310**, 511–534.

Turner, J. S., and I. H. Campbell. (1986). Convection and mixing in magma chambers. *Earth-Sci. Rev.*, **23**, 255–352.

13

Mid-Ocean Ridge Volcanism

Questions to be Considered in this Chapter:

1. What igneous processes occur at divergent plate boundaries to create new lithosphere?
2. How does mantle melting occur beneath mid-ocean ridges, and what are the igneous products?
3. What variation is there in the igneous rocks at mid-ocean ridges, and what can we deduce from that variation?
4. What is the nature of the oceanic lithosphere, and how do igneous and tectonic processes interact to produce it?
5. Are exposed ophiolites representative of oceanic lithosphere?

The ocean floor covers over 70% of the Earth's solid surface and comprises two principal petrogenetic provinces, both of which are predominantly basaltic. The first, volcanism at divergent or constructive plate margins, is the most voluminous form of volcanism on the planet, and it is responsible for generating the oceanic crust. That is the subject of this chapter. The other, ocean island volcanism, is a sporadic and scattered intraplate form of volcanism at point sources throughout the ocean basins. It will be covered in Chapter 14. These two forms of volcanic activity are occasionally interrelated and together provide petrologists with an initial picture of the make-up and dynamics of the Earth's mantle beneath the oceans. A third oceanic province, intra-oceanic subduction-related island arcs, is of an entirely different nature and will be discussed in Chapter 16.

13.1 VOLCANISM AT CONSTRUCTIVE PLATE BOUNDARIES

Evidence from decades of dredging and from the Deep Sea Drilling Project (DSDP, then JOIDES, then JOI) indicates that the vast bulk of the oceanic crust is composed of a tholeiitic basalt of unique and relatively restricted chemical composition. This basalt type has acquired a number of names, including, most commonly, **ocean-floor basalt**, **abyssal basalt** (or abyssal tholeiite), and **mid-ocean ridge basalt (MORB)**. MORBs are generated at the mid-ocean ridges, where adjacent plates diverge. As two plates separate, the mantle flows upward to fill the potential gap. Rising mantle lherzolite undergoes adiabatic decompression (Section 10.2.2) and eventually reaches the solidus temperature, resulting in basaltic partial melts. The melts separate and rise to the crustal rift, where they collect and solidify and add to the trailing edge of the separating oceanic plates. As the magnetic poles periodically reversed, this plate accretion led to the familiar symmetric magnetic anomaly striping that was a fundamental component in the development of plate tectonic theory. We begin with a survey of the geophysical, petrographic, and geochemical data for ridge volcanism so that we can constrain and refine this simple model.

13.2 THE MID-OCEAN RIDGES

Modern technology has made the previously hidden ocean floor accessible to study. We now have extensive acoustic imagery, plus seismic, gravity, magnetic, and heat-flow data on much of the ocean bed, as well as direct sampling and observation using dredges, drilling, and submersibles. One of the most startling discoveries of postwar sonar seabed mapping was the existence of an extensive submarine mountain system. This system of **ridges** (or **rises**) stands 1 to 3 km above the abyssal plain. It is about 2000 km wide and forms a continuous globe-encircling submarine mountain range about 65,000 km long that covers approximately one-third of the sea floor. Figure 13.1 shows the location of the central spreading axes of the system. The term *mid-ocean ridge* is appropriate for the Mid-Atlantic Ridge (MAR) and the Indian Ocean segments of the system, but the East Pacific Rise (EPR) is clearly at the eastern margin of the Pacific Ocean. Accuracy, in this case, yields to convenience, and *mid-ocean* is generally applied to the ridge system as a whole.

Divergent plate boundaries are usually oceanic, a result of the density and thinness of the basaltic crust created by the resulting mantle melting. Where divergent boundaries are initiated within continents, such as the East African Rift (section 19.1), either they fail after the incipient formation of a rift valley, becoming aborted rifts (**aulacogens**), or the continent is split and the opposing portions separate with the creation of an intervening ocean basin (as in the case of the Atlantic Ocean separating the Americas from Europe and Africa since the Triassic).

Volcanism and topography are broadly symmetrical across mid-ocean ridge axes, but asymmetrical segments are not uncommon, particularly along slow-spreading ridges. A near-axis **neovolcanic zone** is a region along the immediate plate boundary within which volcanism is concentrated. The ridge axis is the locus of considerable earthquake activity. First-motion studies indicate that the earthquakes are associated with normal faulting, characteristic of extensional tectonic environments. The neovolcanic zone is also associated with very high heat flow, indicating an elevated geothermal gradient (probably resulting from hot upwelling mantle and partial melts). The high heat flow in a submarine environment creates an extensive hydrothermal system, composed largely of recirculated seawater percolating downward through fractured and porous upper crust, where it is heated and convects upward again. Dramatic visual evidence for the hydrothermal systems has been supplied by submersible dives along axial sections of the Galapagos Ridge, the EPR, and the Juan de Fuca and Gorda ridges. Hydrothermal fluids in excess of 350°C cool dramatically upon escaping from vents and precipitate a number of minerals, including barite, silica, and several valuable metal sulfides, forming **spires**, or **chimneys**. Sulfide precipitation results in **black smokers** where the hot fluid immediately precipitates fine black sulfide minerals upon contact with cold seawater (Figure 13.2). A unique and rich biological community has developed in these vent areas, sustained by the hydrothermal fluids. Among the odd creatures is a mouthless tube worm, sustained via symbiosis with internal bacteria that metabolize the sulfides.

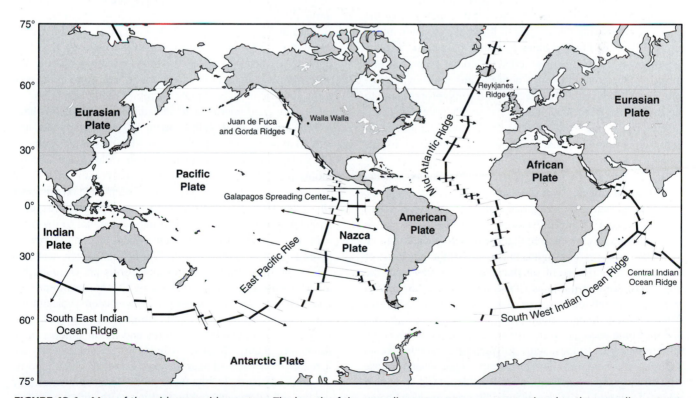

FIGURE 13.1 Map of the mid-ocean ridge system. The length of the spreading rate vectors are proportional to the spreading rate at that point (see Table 13.1). After Minster et al. (1974).

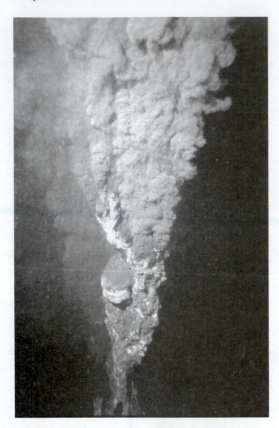

FIGURE 13.2 "Black smoker:" sulfide minerals precipitating from hot, mineral-laden water emanating from a chimney on the East Pacific Rise. Photograph courtesy of Dudley Foster, Woods Hole Oceanographic Institute.

TABLE 13.1   Spreading Rates of Some Mid-Ocean Ridge Segments

Category	Ridge	Latitude	Rate (cm/a)*
Fast	East Pacific Rise	21–23°N	3
		13°N	5.3
		11°N	5.6
		8–9°N	6
		2°N	6.3
		20–21°S	8
		33°S	5.5
		54°S	4
		56S	4.6
Slow	Indian Ocean	SW	1
		SE	3–3.7
		Central	0.9
	Mid-Atlantic Ridge	85°N	0.6
		45°N	0.5
		36°N	2.2
		23°N	1.3
		48°S	1.8

*Half spreading
From Wilson (1989). Data from Hekinian (1982), Sclater et al. (1976), and Jackson and Reid (1983).

Gravity studies indicate that ridges are essentially in isostatic equilibrium, from which we can conclude that the elevation of the ridge is a result of thermal expansion and is thus compensated by low-density mantle below. As the plate moves away from the ridge crest, it cools, contracts, and subsides. The amount of subsidence at any point on the distant flanks of the ridge has been found to be proportional to the square root of the age of the rocks at that point. This subsidence rate is in accord with the conductive heat loss equation, which predicts that cooling is proportional to the square root of time (Parker and Oldenberg, 1979).

The rate at which plate divergence (spreading) occurs at mid-ocean ridges is not the same for all ridge segments. The approximate spreading rates of various segments are given in Table 13.1 and illustrated in Figure 13.1 as relative vectors. From this we can see that the MAR spreads at a rate of about 1 to 2 cm/a, a figure that refers to the *half rate*, or the rate at which a point on one plate separates from the ridge axis. An alternative is to use the *full rate*, the rate at which two points on opposite plates separate. We shall restrict our usage to half-spreading rates, hoping to avoid confusion. Some sections of the EPR spread at half rates of up to 8 to 9 cm/a. Ridges with a spreading rate less than 3 cm/a are considered slow-spreading ridges, whereas those with a rate greater than 5 cm/a are considered fast-spreading ridges. Temporal variations in spreading rate are also known to have occurred (recognized on the basis of the widths and ages of the magnetic anomaly stripes). For more recent and detailed rate estimates, see Müller et al. (2008).

Side-scanning sonar surveys indicate that the ridge morphology for slow- and fast-spreading ridges is different (Basaltic Volcanism Study Project, 1981; Searle, 1992). Direct observation and sampling of individual ridge segments was made possible by deep-diving submersibles, and the first detailed study of a ridge segment was the FAMOUS (French American Mid-Ocean Undersea Study) project along the MAR near the Azores at 37°N (Moore et al., 1974; Ballard and Van Andel, 1977; Ballard et al., 1975). I shall present a brief introductory comparison between the morphology of fast-spreading and slow-spreading ridges. We will make a more detailed comparison later to help develop more specific models of plate accretion processes.

Slow- and intermediate-spreading ridges typically have a pronounced axial valley about 30 to 50 km wide and 1 to 5 km deep, with step-like inward-facing scarps, similar to rift valleys on land. Within this larger valley there is commonly a 3 to 9 km wide **inner rift valley** with a flat floor. Volcanism and crustal extension are concentrated on this inner rift valley floor, where fissures open, and pillow lavas, constrained by the scarp walls, flow mostly parallel to the ridge axis. Volcanic activity is not evenly distributed, and typically several volcanic mounds up to 300 m high occur, scattered across the floor. Elongated volcanic forms are common but are generally only 10 to 20 km long. Many may represent coalescing volcanic cones along a fissure system. Volcanic accumulations are not always split symmetrically but are irregularly cut by sporadic later fractures and fissures, imparting a rough topography. Because fissures and volcanic activity are generally concentrated near the axis, however, older deposits are gradually carried to the flanks of the inner rift, where they are typically dismembered by faulting. Although the process does not correspond exactly to our simple concept of adding successive dike-like layers to the plate on each side of the central rift, each being magnetized in accordance with the polarity at the time of injection, the coarse overall effect is the same. The result is the magnetic anomaly "striping" that is symmetric about the ridge on the broad scale, but the patterns are much more irregular in detail.

Fast-spreading ridges, such as the EPR, have also been studied by submersibles (CYAMEX, 1981; Francheteau and Ballard, 1983, Perfit et al., 1994; Macdonald et al., 1996). Fast-spreading ridges are smoother and less disrupted by large fault displacements. There is typically a narrow (2 to 5 km) **axial rise**, with a small (40 to 250 m wide by 5 to 14 m deep) **axial summit trough** (or caldera). The neovolcanic zone on the EPR is generally regarded as a zone of nearly continuous volcanism 0.5 to 2 km wide. In fast-spreading ridges, small pillow lava hills are flanked by smooth lava plains, attributed to sheet lavas formed by faster lava extrusion associated with the more rapid spreading and higher heat flow. Beyond the neovolcanic zone, the basalts become progressively (although not uniformly) older and more fractured. Recent detailed studies of small areas along the EPR indicate that young volcanics were erupted up to 4 km away from the axial summit caldera (Perfit et al., 1994). As a general rule, fissure-fed eruptions characterize fast-spreading ridges, and point-source volcanism (although probably fissure controlled) is more common at slow-spreading ridges.

Ocean floor samples collected by dredging and drilling in the southwestern Pacific are as old as Jurassic but are virtually identical to samples created recently at the ridges. This assures us that the processes we observe at the mid-ocean ridges have taken place, uninterrupted, for at least 140 Ma. Magma production at the FAMOUS area of the MAR is estimated to be 8,600 m^3/km $\cdot$ a (Moore et al., 1974). This is a modest rate for volcanism. Iceland, a portion of the MAR that has emerged above the surface of the Atlantic, has a productivity 10 times as great. Estimates for the whole mid-ocean ridge system range from 5 to 20 km^3/a (Basaltic Volcanism Study Project, 1981). This rate is not particularly impressive until it is multiplied by the time span over which it has taken place. If MORBs have been generated for only the 140 Ma we can observe, $1.4 \cdot 10^9$ km^3 have been created—an impressive amount. If this process has taken place for 10 times that long (probably a conservative estimate), cumulative MORB generation would represent a volume equivalent to 5% of the upper mantle (to the 660-km transition), which is capable of depleting the upper mantle to a considerable extent. Of course, subduction can be expected to reincorporate much of this material back into the mantle again. The fate of this material is poorly known, but we shall encounter some evidence for it later in this and subsequent chapters.

Mid-ocean ridges are **segmented** on a range of scales (Schouten et al., 1985; Macdonald et al., 1988; Macdonald, 1998; Dunn et al.; 2005). The **tectonic** (physical) **segments** have been classified on a four-tiered hierarchy (Figure 13.3), imparting perhaps a false clustering on what is probably more of a continuum in scale. *First-order* tectonic segments are the longest (300 to 900 km on fast-spreading ridges and 200 to 600 on slow-spreading ridges), are typically offset more

($>$30 km), and most enduring ($>$0.5 Ma). The seismically active portions of the offset between the first-order ridge segments are subparallel strike-slip faults (see Figures 13.1 and 13.3) called **transform faults** (Wilson, 1965). Where active spreading and volcanism at a ridge encounter a transform, the activity terminates abruptly against older, cooler, more stable lithosphere on the flank of a separate ridge segment across the offset. The nearly aseismic **fracture zone** extensions beyond the ridges are very linear features that appear to offset the magnetic anomaly patterns and extend in many cases across the ocean floor (Fox and Gallo, 1986, 1989). *Second-order* segments are typically 50 to 230 km long on fast-spreading ridges and 20 to 80 km long on slow-spreading ridges, with offsets of 2 to 30 km. Second-order discontinuities on fast-spreading ridges are usually associated with large **overlapping spreading centers (OSCs)** (Macdonald and Fox, 1983). An OSC occurs at a location on a ridge where two offset segments extend along the axis past each other, so that their tips overlap without a major fault (D2 in Figure 13.3a). Second-order offsets on the more tectonically disrupted slow-spreading ridges are generally associated with shear zones oriented obliquely to the axis or with kinks in the axial rift without obvious faulting (D2 in Figure 13.3b). *Third-order* segments are $\sim$20 to 80 km long on fast-spreading ridges, where they are offset 0.5 to 2 km by smaller OSCs, and they are perhaps 5 to 25 km long and offset by gaps between linear volcanic centers in slow-spreading ridges (D3s in Figure 13.3). *Fourth-order* segments (D4 in Figure 13.3) are $\sim$6 to 22 km long in fast-spreading ridges and offset less than 1 km by small axial strike changes, collectively termed **devals** (for **dev**iations from **a**xial **l**inearity) by Langmuir et al. (1986). Fourth-order segments in slow-spreading ridges are individual axial highs along a linear volcanic fissure, so the discontinuities are apparently gaps between volcanic vents.

The center of a ridge segment tends to be both higher and wider than the distal ends near discontinuities. Ridge segments also vary in length and position over time. Most notably, higher-order segments may extend in length along the ridge axis. When such a **propagating rift** is a first-order (fast-spreading) ridge segment (to which the term is typically applied), it may propagate into the lithosphere across a fracture zone, where it then runs parallel to the ridge already on that segment, resulting in complex accommodations of ridge-transform geometries over time (Hey, 1977; Hey and Wilson, 1982; Sinton et al., 1983; Hey et al. 1989). As OSC segments extend, small cracks and dikes typically form in the offset between them, and the local crust is then rotated and sheared. According to Macdonald et al. (1998), crack propagation is affected by the reoriented stresses, thereby deviating from the expected direction perpendicular to regional extension (parallel to the ridge). When overlap reaches about three times the offset distance, propagation stalls, and a new crack develops behind the tip of the previous crack, causing a propagating OSC segment to cut either toward or away from the opposing segment, repeatedly slicing off its own ridge tip. A more active segment may eventually cut off the tip of the opposing segment, joining the segments, perhaps in a deval-type ridge kink without overlap.

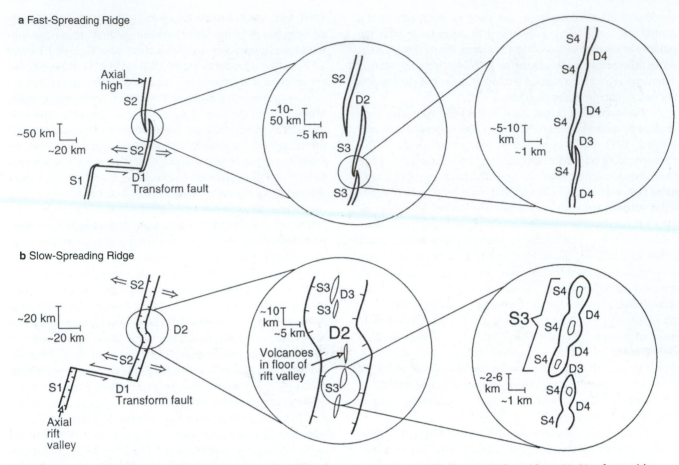

FIGURE 13.3 A hierarchy for ridge segmentation for **(a)** fast-spreading ridges and **(b)** slow-spreading ridges. S1–S4 refer to ridge segments of first- to fourth-order, and D1–D4 refer to discontinuities between corresponding segments. Solid lines associated with segments outline topographic highs. Lines with tick marks are faults, with ticks on the downthrown side (thus bounding grabens). After Macdonald (1998).

The greater size at segment centers and the tendency to propagate suggest that segments are related to upwelling warm mantle and melts, spaced along conduits at segment centers (Francheteau and Ballard, 1983; Macdonald and Fox, 1983; Whitehead et al., 1984; Langmuir et al., 1986; Macdonald et al., 1988; Solomon and Toomey, 1992; Batiza, 1996; Dunn et al., 2005), and that melt then migrates from the upwelling centers along the axis toward segment ends, driving propagation. Within each segment, the characteristics of magmatism, faulting, and hydrothermal circulation thus vary systematically with distance from the segment center, and each segment therefore has its own cycle of magmatic input, growth/propagation, and waning activity. The contrast in character between adjacent segments increases as segment order decreases.

13.3 STRUCTURE OF THE OCEANIC CRUST AND UPPER MANTLE

The four layers of the oceanic lithosphere were initially distinguished on the basis of discontinuities in seismic velocities. Most *direct* sampling of the oceanic lithosphere has recovered only the sedimentary veneer and uppermost volcanics. Even the Deep Sea Drilling Program has rarely penetrated the volcanics, and then only to a maximum depth of

about 1500 m. Dredging the base of fracture zone scarps has supplied samples from exposed deeper sources, but reliable stratigraphic control of such samples is lacking.

Our understanding of the petrologic nature of the oceanic lithosphere has been greatly enhanced by field studies in **ophiolite** terranes on land. As mentioned in Chapter 10, ophiolites are considered to be masses of oceanic crust and upper mantle thrust onto the edge of a continent or incorporated in mountain belts, where they are tectonically disrupted and now exposed by erosion. For a concise history of the ophiolite concept, see Dilek (2003). Figure 13.4 is a section through a "typical" ophiolite, in this case based primarily on the Semail ophiolite in Oman on the Arabian Peninsula, a particularly well-exposed and relatively intact ophiolite. Like most other ophiolites, however, it has undergone some disruption during tectonic emplacement, obscuring some of the internal details. There is a consistent layering, however, to which most other ophiolites broadly conform. How does this section compare with the oceanic lithosphere?

Although there is still some uncertainty as to how the seismic data are to be interpreted, most geologists accept the section of oceanic lithosphere illustrated in Figure 13.5, which includes characteristic seismic velocities and a comparison of the estimated thicknesses of the layers in both ophiolites and oceanic lithosphere. Plate accretion at mid-ocean

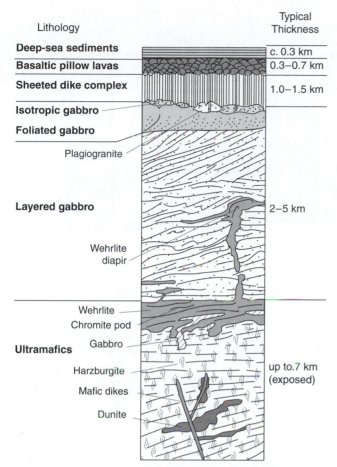

Lithology

Typical
Thickness

Deep-sea sediments — c. 0.3 km

Basaltic pillow lavas — 0.3–0.7 km

Sheeted dike complex — 1.0–1.5 km

Isotropic gabbro

Foliated gabbro

Plagiogranite

Layered gabbro — 2–5 km

Wehrlite
diapir

Wehrlite

Chromite pod

Gabbro

Ultramafics

Harzburgite

Mafic dikes — up to 7 km (exposed)

Dunite

FIGURE 13.4 Lithology and thickness of a typical ophiolite sequence, based on the Semail ophiolite in Oman. After Boudier and Nicolas (1985). Copyright © with permission from Elsevier Science.

their low-temperature metamorphic equivalents. Some investigators distinguish the porous and nonporous pillow flows, calling the porous zone Layer 2A and the nonporous zone Layer 2B. They then call the sheeted dikes **Layer 2C** (as adopted in Figure 13.5).

Layer 3 is more complex and a bit more controversial but is generally believed to comprise mostly gabbros, presumably crystallized from a shallow **axial magma chamber** that fed the dikes and basalts. **Layer 3A**, again by analogy with ophiolites, represents uppermost isotropic and lower, somewhat foliated ("transitional") gabbros, whereas **Layer 3B** is more layered, typically exhibiting cumulate textures. The layering may be horizontal but more commonly dips (toward the presumed ridge axis) at angles locally up to 90°. Both layers 3A and 3B are well foliated and lineated in the Semail (Oman) ophiolite.

At the top of the gabbros in the Oman are small discontinuous diorite and tonalite ("plagiogranite") bodies, presumed to be late differentiated liquids that are filter pressed and mobilized to rise and collect along the gabbro-sheeted dike contact, occasionally extending up into the pillow layer. The relationship between the dikes and gabbros is complex, reflecting a moderate time span of successive dike emplacement and gabbro crystallization in a steep thermal gradient. Only a few late dikes postdate and cut the upper gabbros and plagiogranites. Layer 3 appears to thicken from about 3 km near the ridge to nearly 5 km by the time the crust is 30 Ma old, correlating with a ~0.5 km thinning of Layer 2. McClain (2003) attributed this to rise of the boundary separating Layers 2 and 3 due to alteration at the base of Layer 2, which raises the seismic velocity, but the magnitude of Layer 3 thickening is greater than the change in Layer 2, so the difference may (also?) be a result of off-axis igneous activity and serpentinization of the top of Layer 4.

Layer 4 has seismic velocities that correlate well with ultramafic rocks. In ophiolites, the base of the gabbro grades into layered cumulate wehrlite and gabbro. Diapir-like bodies of wehrlite also appear to have moved upward into the layered gabbros. Cumulate dunite with harzburgite xenoliths and chromite lenses is usually found below the wehrlite layer. Below this is a tectonite harzburgite and dunite interpreted to be the unmelted refractory residuum of the source mantle left behind after basaltic magma was extracted. A few gabbroic dikes may also occur in this layer.

The boundary between Layers 3 and 4 is, broadly speaking, the Moho. The upper portion of Layer 4 is thought to be layered and of cumulate origin, as olivine and pyroxenes accumulate at the bottom of the axial magma chamber. Below this portion is original, unlayered, residual mantle material. This brings up an interesting question as to the nature of the crust–mantle boundary. Should we consider the top of the mantle to be the top of the *original* mantle, or is it the mafic–ultramafic transition, and hence the top of the ultramafic cumulates deposited in a shallow magma chamber? In other words, is the mantle defined by its petrogenesis or by its composition? A seismologist would argue that the Moho was originally defined by the seismic discontinuity where the *P*-wave velocity, V_p, jumps from about 7.3 km/s

ridges is a complex and variable process, so Figure 13.5 should be considered only a general representation. A brief description of the layers, beginning at the top, follows.

Layer 1 is a thin layer of pelagic sediment that is absent on the newly generated crust at ridge axes and thickens away from the axes as sediment accumulates on progressively older crust.

Layer 2 is basaltic. It can be subdivided into two sublayers. From submersible observations, drilling, and by analogy with ophiolites, **Layer 2A** is believed to comprise pillow basalts and sheet-lavas, and **Layer 2B** is believed to comprise vertical sheeted dikes emplaced in the shallow, brittle extensional environment at the ridge axis. Many dikes have only a single chill margin, implying that later dikes split and intruded earlier ones. The upper portion of Layer 2A has seismic velocities lower than predicted for basalts based on laboratory measurements of compressional wave velocities. Kirkpatrick (1979) interpreted this as resulting from fractures and cavities (13 to 41% porosity) in the upper part of the basalts. This porous layer thins and eventually disappears away from the ridge crest as the voids are filled by diagenetic mineralization (Houtz and Ewing, 1976). Below the porous zone, Layer 2 has seismic velocities commensurate with laboratory measurements for basalts and

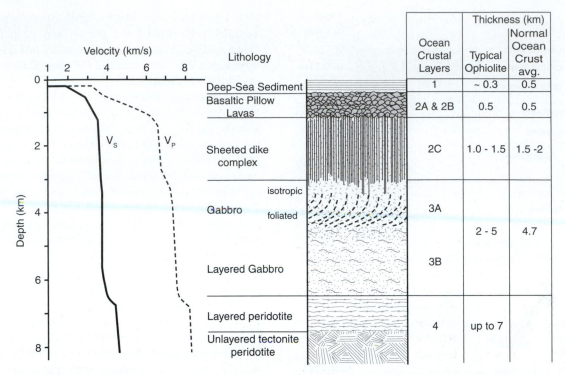

FIGURE 13.5   Schematic section of the oceanic lithosphere based on ophiolite data. Typical ophiolite thicknesses are compared to the average thickness of oceanic layers. After Moores (1982), Brown and Mussett (1993), and McClain (2003). Data from Gass (1982), Lewis (1983), Spray (1991), Basaltic Volcanism Study Project (1981), and Kennett (1982).

(characteristic of gabbro) to 8.1 km/s (characteristic of peridotite), and thus the top of the ultramafics, regardless of origin, is the top of the mantle. A petrologist might argue that the top of the mantle should be the top of the original ultramafics and should not include some later minerals sprinkled on top in a magma chamber. A number of authors distinguish a *seismic Moho* from a *petrological Moho* on this basis. Personally, I have to vote with the seismologists.

Although ophiolites are convincing analogs of oceanic lithosphere, a direct correlation is undermined by a few consistent differences. Seismic velocities in ophiolites are lower than those measured in oceanic lithosphere, and the magnetic anomaly striping that characterizes oceanic crust has not been found in ophiolites. These two differences may result from the extensive hydrothermal alteration and weathering affecting most ophiolites. Seismic velocities in ophiolites are further retarded by faults and joints, which must have accompanied their emplacement on land. Also, most layers in ophiolites are generally much thinner than their oceanic equivalents, and the chemical composition is usually more enriched and sialic. These latter differences are more fundamental and suggest that ophiolites may be associated with oceanic **back-arc** (marginal) **basins**, which form behind the volcanic arc at subduction zones (Chapters 16 and 17). The Sea of Japan is a good modern example of an oceanic back-arc basin. In these basins, the spreading is more erratic, and the crust that forms is thinner and somewhat more sialic. Many back-arc basins eventually close, and the arc plus segments of the intervening basin lithosphere are thrust (or **obducted**) onto the continental edge, producing ophiolites. There is no reason to preclude main-stage oceanic lithosphere from being obducted as well when ocean

basins are closed during continental collision, but mature oceanic lithosphere is probably a bit too old, thick, and dense to easily be obducted. It is also more likely to be highly deformed by the ensuing continental collision. We believe the crust of oceanic back-arc basins is nonetheless formed in much the same fashion as in the ocean basins, so the analogy, if correct, is considered reasonable.

We shall address the question of how these layers are formed shortly, when we develop a model for processes at mid-ocean ridges. But first we must understand a little more about the volcanic products at the ridges, which can supply us with some important data to focus our inquiry into layer development.

13.4 MORB PETROGRAPHY AND MAJOR ELEMENT GEOCHEMISTRY

A "typical" MORB is an olivine tholeiite, with low K_2O ($<0.2\%$) and low TiO_2 ($<2.0\%$) compared to most other basalts. This makes MORBs chemically distinctive from basalts of other petrogenetic associations. Textures range from glassy to phyric and, rarely, to gabbroic among seabed samples. Glass samples are very important chemically because they represent *liquid* compositions, whereas the chemical composition of phyric rocks can be modified by crystal accumulation processes. Common MORB phenocrysts are plagioclase (An_{40}–An_{88}), olivine (Fo_{65}–Fo_{91}), and a Mg-Cr spinel (Grove and Bryan, 1983). Ca-rich clinopyroxene phenocrysts are much less common and usually occur in rocks with abundant olivine and plagioclase, suggesting that clinopyroxene is typically a late crystallizing phase. The groundmass mineralogy of MORBs is dominated by plagioclase and clinopyroxene microlites and a Fe-Ti oxide. Textures and experiments on natural samples at low pressure reveal a common crystallization sequence: olivine ($\pm$ Mg-Cr spinel)$\rightarrow$ olivine + plagioclase ($\pm$ Mg-Cr spinel) $\rightarrow$ olivine + plagioclase + clinopyroxene. It is simple enough to visualize this sequence for compositions near, but to the forsterite side of, the center of the diopside–forsterite–anorthite system in Figure 7.2. Other sequences are possible, however, and depend upon bulk chemical composition, pressure, and magma chamber processes such as fractionation style or recharge of more primitive magma. Fe-Ti oxides are restricted to the groundmass and thus form only late in the MORB sequence (hence the early Fe-enrichment characteristic of the tholeiite trend on an AFM diagram, Figure 8.14).

Samples very rich in plagioclase phenocrysts and others rich in olivine phenocrysts (*picrites*) are fairly common. These samples were originally believed to represent two distinct parental magma types, but density studies show that plagioclase should float and that olivine and clinopyroxene should sink in magmas of MORB composition. Therefore most petrologists now consider such phenocryst-rich rocks to be accumulative rocks in which the phenocrysts concentrated. Once again, this underscores the importance of relying on *glasses* to represent true liquid compositions in work that attempts to deal with liquid evolution.

Larger megacrysts (>2 to 3 mm) may also occur, and are commonly rounded and embayed, with compositions that are not in equilibrium with the groundmass. For example, we can define a simple empirical **distribution constant**, K_D [see Equation (9.2)], for the Fe-Mg exchange between olivine and coexisting MORB glass (melt):

$$K_D = \frac{(FeO/MgO)_{ol}}{(FeO/MgO)_{liq}} \qquad (13.1)$$

where FeO and MgO are expressed as wt. % oxides. Such empirical distribution constants are entirely general and can be used for a number of element exchanges between any phases at equilibrium. They should hold for a modest range of magma compositions and temperatures. Measurements of olivine phenocryst/basalt glass pairs in both natural samples and experimental results have shown that K_D, as defined by Equation (13.1), is about 0.28 (Basaltic Volcanism Study Project, 1981). Olivine megacrysts commonly have compositions more Mg-rich than one would expect from the glass composition. Megacrysts with Mg# ($100 \, Mg/(Mg + Fe)$ on a *molecular* basis) equal to 90 have been observed in basalts with glass Mg# = 59 (Stakes et al., 1984). Such phenocrysts should be in equilibrium with a melt of Mg# = 71. This, and the corroded nature of the megacrysts, implies they formed in a more primitive magma that was injected into and mixed with a more evolved one that now dominates the groundmass. Similar observations have been made for plagioclase megacrysts based on An content.

The major element composition of MORBs was originally considered to be extremely uniform, which, in

TABLE 13.2 Average Analyses and CIPW Norms of MORBs

Oxide (wt. %)	All	MAR	EPR	IOR
SiO$_2$	50.5	50.7	50.2	50.9
TiO$_2$	1.56	1.49	1.77	1.19
Al$_2$O$_3$	15.3	15.6	14.9	15.2
FeO*	10.5	9.85	11.3	10.3
MgO	7.47	7.69	7.10	7.69
CaO	11.5	11.4	11.4	11.8
Na$_2$O	2.62	2.66	2.66	2.32
K$_2$O	0.16	0.17	0.16	0.14
P$_2$O$_5$	0.13	0.12	0.14	0.10
Total	99.74	99.68	99.63	99.64
Norm				
q	0.94	0.76	0.93	1.60
or	0.95	1.00	0.95	0.83
ab	22.17	22.51	22.51	19.64
an	29.44	30.13	28.14	30.53
di	21.62	20.84	22.50	22.38
hy	17.19	17.32	16.53	18.62
ol	0.0	0.0	0.0	0.0
mt	4.44	4.34	4.74	3.90
il	2.96	2.83	3.36	2.26
ap	0.30	0.28	0.32	0.23

All: Average of glasses from Atlantic, Pacific, and Indian Ocean ridges.
MAR: Average of MAR glasses.
EPR: Average of EPR glasses.
IOR: Average of Indian Ocean ridge glasses.

Basaltic Volcanism Study Project. (1981). *Basaltic Volcanism on the Terrestrial Planets.* Pergamon. New York. Table 1.2.5.2

conjunction with their great volume and restricted mode of occurrence, was interpreted to imply a simple petrogenesis. More extensive sampling and chemical work, however, has revealed that MORBs display a range of compositions. The range, however, is still considerably more restricted than for most petrogenetic associations. Table 13.2 lists some averaged chemical analyses of mid-ocean ridge basalts from the Atlantic, Pacific, and Indian Ocean ridges, as well as CIPW norms. All analyses are of glasses, so that only liquid compositions are represented and any possible effects of phenocryst accumulation is avoided. Note the very low content of K$_2$O (compared to the average basalt in Table 8.3) and that all analyses are quartz–hypersthene normative (although olivine is common in the mode).

Figure 13.6 is a Fenner-type variation diagram showing major element variations as a function of the concentration of MgO for MORB glasses from the Amar Valley of the MAR. MgO was chosen as the abscissa because the early stages of evolution of mafic liquids generally involve a more substantial decrease in Mg/Fe^{2+} than a change in SiO$_2$ (see Figure 12.13a). This should also be evident from Figure 13.6, where the SiO$_2$ content varies little and is practically

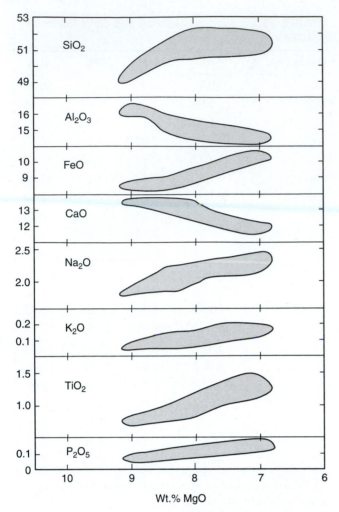

FIGURE 13.6   Fenner-type variation diagrams for basaltic glasses from the Afar region of the Mid-Atlantic Ridge. *Note different ordinate scales.* From Stakes et al. (1984). © American Geophysical Union with permission.

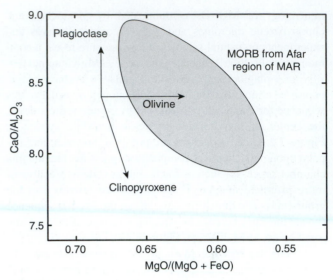

FIGURE 13.7   Variation in CaO/Al_2O_3 with Mg# for basaltic glasses from the Afar region of the Mid-Atlantic Ridge. The shaded region represents the range of glass analyses. Vectors show the expected path for liquid evolution resulting from fractional crystallization of the labeled phase. From Stakes et al. (1984). © American Geophysical Union with permission.

constant over most of the range, making it a poor index of liquid evolution. MgO *decreases* from left to right in Figure 13.6, to conform with typical Harker-type diagrams in which the more evolved liquids are on the right. Again, this decrease in MgO and relative increase in FeO is the characteristic early differentiation trend of tholeiites (see Figure 8.14 and the tholeiitic Skaergård trend in Figure 8.3). Only in the exposed ridge volcanic plateaus, such as Iceland and the Galapagos, are found more evolved rocks such as andesites and even rhyolites, which exhibit the subsequent alkali enrichment. Whether the occurrence of evolved samples in these plateaus is a result of better exposure and sampling density or the greater volume of melts produced (and hence larger magma chambers) is not presently known (although most Icelandic rhyolites are attributed to crustal remelting).

The major element patterns in Figure 13.6 are compatible with crystal fractionation of the phenocryst phases discussed above. Removal of Mg-rich olivine would raise the FeO/MgO ratio, and the separation of a calcic plagioclase would cause Al_2O_3 and CaO to decrease as well.

The CaO/Al_2O_3 *ratio,* however, also decreases with differentiation, from 0.90 to about 0.78 (Figure 13.7). Because

the CaO/Al_2O_3 ratio in plagioclase phenocrysts is about 0.55 (lower than in the melt), removal of this phase should *increase* CaO/Al_2O_3 in the remaining liquid rather than reduce it, as shown by the plagioclase vector in the diagram. Olivine contains neither Ca nor Al and should thus have no effect on the ratio. Some other Ca-rich phase containing little Al must also be removed from the MORB liquids in this portion of the MAR. Clinopyroxene is the likely candidate, as shown in Figure 13.7.

The behavior of clinopyroxene, however, is hardly straightforward. Data from a larger data set (*not* restricted to glasses), spanning a longer section along the MAR from 27 to 73°N (Schilling et al., 1983), found neither decrease in CaO/Al_2O_3 nor any compositional trends that require clinopyroxene fractionation (Figure 13.8a). Figure 13.8b is a Pearce element ratio diagram (see Section 8.6) for the same data set. A slope of $+1$ on a plot of $(0.5(Fe+Mg)+2Ca+3Na)/K$ versus Si/K suggests fractionation of only plagioclase and olivine for the MAR samples. Sinton et al. (1991) noted a slight decrease in CaO/Al_2O_3 for glasses from the EPR, and, although most of their samples were multiply saturated with olivine and plagioclase, some were also saturated with clinopyroxene. Clinopyroxene appears to form late in MORBs. If it is a phenocryst phase, it is commonly the last one to crystallize, and it is much more common as a matrix phase. How then can it be responsible for the Ca/Al reduction in some MAR and EPR magmas *across the whole chemical spectrum* and not just in the more evolved magmas? Rhodes et al. (1979) found that clinopyroxene removal is required for mass balance models of fractionation of many MORB suites in spite of the fact that clinopyroxene is not a liquidus phase in them. They called this strange conclusion the *pyroxene paradox.* Grove and Kinzler (1992) suggested that clinopyroxene is a eutectic phase at higher pressures, but the eutectic shifts toward clinopyroxene as

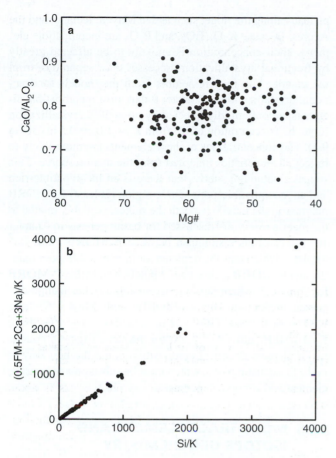

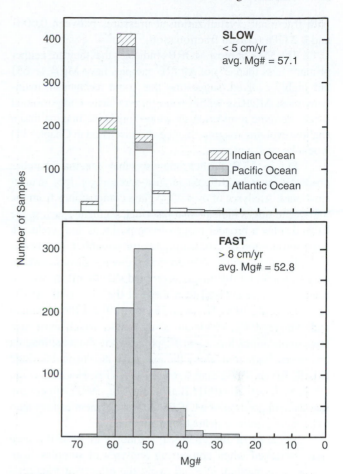

FIGURE 13.8 Variation diagrams for MORBs from the Mid-Atlantic Ridge. **(a)** Plot of CaO/Al_2O_3 versus Mg# /100 showing no discernible clinopyroxene trend. **(b)** Pearce element ratio plot with 1:1 slope for $(0.5 (FeO + MgO) + 2CaO + 3Na_2O)/K$ versus Si/K, suggesting olivine + plagioclase fractionation. Data from Schilling et al. (1983).

FIGURE 13.9 Histograms of more than 1600 glass compositions from slow and fast mid-ocean ridges. Note that the MAR is slow, the EPR is fast, and fast ridges show slightly more evolved melts on average. After Sinton and Detrick (1992). © American Geophysical Union with permission.

pressure is reduced. The melt composition, which was at the eutectic at depth, thus shifts into the olivine-plagioclase field as the melt rises, and is no longer clinopyroxene saturated.

Returning to the remaining fractionation trends in Figure 13.6: Na_2O, K_2O, TiO_2, and P_2O_5 all increase as MgO decreases. Na_2O is included in plagioclase, but, because the plagioclase is anorthitic, Na_2O is still enriched in the evolved liquids. The other three elements are totally excluded from the phenocryst phases, and the concentration of each increases by about 300% over the range of compositions shown.

From these data we can draw a few interesting conclusions about MORBs and the processes beneath mid-ocean ridges. First, we can confirm our earlier contention that MORBs are not the completely uniform magmas they were once considered to be. They show some chemical trends consistent with fractional crystallization of olivine, plagioclase, and perhaps clinopyroxene. This also tells us that the great bulk of MORBs *cannot be primary magmas*, but are derivative magmas resulting from fractional crystallization. Because the composition of most MORBs approximate a low-pressure cotectic for olivine + plagioclase + clinopyroxene + liquid, the fractional crystallization must have taken place near the surface, probably in shallow magma chambers. Primary magmas must be in equilibrium

with mantle phases, and therefore must have Mg# around 70 (Section 10.4), which corresponds to a MgO concentration of 10 to 11 wt. % for MORBs. Thus few, if any, of the MORB liquids that reach the surface are primary (as can be seen from Table 13.2 and Figures 13.6, 13.8, and 13.9).

To what extent does fractional crystallization take place in MORBs? The small variation in SiO_2, if we took it to be a measure of fractionation, is misleading. So too is MgO, which is contained in the mafic minerals and the melt. Rather, the *incompatible* elements should provide the variation that is directly related to the progress of fractional crystallization. If K_2O, TiO_2, and P_2O_5 all increase by 200 to 300% in the MORB suites, and they are concentrated by exclusion from crystallizing phases in an isolated magma chamber, then the liquid must be reduced by 50 to 67% to accomplish this. This is a surprising conclusion for such apparently uniform chemical composition.

Grove and Kinzler (1992) derived a model for MORB fractionation and applied it to several ridge segments. They concluded that fractional crystallization beneath some slow-spreading ridges occurs at pressures of 0.3 to 0.6 GPa (within the mantle), suggesting cooler mantle beneath them, allowing MORB melts to cool and crystallize as they rise. Data on a fast-spreading segment of the EPR were consistent with fractional crystallization at pressures between 0.0001 and 0.2 GPa (crustal fractionation).

The Mg# of most MORBs indicates that they are neither primary (less than 2% of MORB samples have Mg# > 65) nor highly evolved, suggesting that some mechanism maintains most MORBs within some intermediate compositional range. Periodic reinjection of a more primitive magma into a shallow evolving magma chamber (as discussed in Chapter 11) would be such a process.

Figure 13.9 further suggests that magma chamber processes may be different at fast-spreading ridges than at slow ones. Analyses of over 2200 glass compositions from the mid-ocean ridges indicate that fast ridge segments, such as the EPR, display a broader range of compositions and produce a larger proportion of evolved liquids (or possibly a greater degree of partial melting) than do slow segments. Several investigators also found that magmas erupted slightly off the axis of ridges are more evolved than those at the axis itself (MAR: Hekinian et al., 1976; Bryan and Moore, 1977; EPR: Hekinian and Walker, 1987; Hekinian et al., 1989), which must also have implications for magmatic processes in the axial magma chambers. Variations *along* the axis have also been described (MAR: Bryan, 1979; Stakes et al., 1984; EPR: Hekinian et al., 1989; Sinton et al., 1991; Batiza and Niu, 1992), suggesting that the composition of MORBs at ridge segment centers tend to be more primitive than at segment ends.

Because of the chemical variation in MORBs, care must be taken when comparing analyses of samples from one area to another. We can avoid the effects of fractional crystallization by comparing only analyses of samples with similar Mg# (as will be done more extensively below). We could minimize variations in shallow differentiation paths by comparing only the most primitive samples. Figure 13.10 shows the variation in K_2O with Mg# for the MAR data set of Schilling et al. (1983). Note the tremendous variation in K_2O, even for constant Mg# values between 65 and 70, indicating considerable variation in MORB parental liquids.

Similar variations independent of Mg# are found for TiO_2 and P_2O_5. The samples with high concentrations of incompatible elements in this data set come from the volcanic plateaus along the ridge, such as Jan Mayen, Iceland, and the Azores. Because K_2O, TiO_2, and P_2O_5 are incompatible elements, their concentration is unlikely to be affected greatly by fractional crystallization processes after separation from the mantle. You may remember from the models for fractional crystallization in Chapter 9 that most incompatible elements are only mildly affected by up to 50% crystallization (note the concentration of Rb from $F = 1.0 \rightarrow 0.5$ in Figure 9.3). The concentrations of these elements are more likely to reflect characteristics inherited from the mantle source. This suggests, although subtly, that there is an incompatible-rich and an incompatible-poor source region (at least) for MORB magmas in the mantle beneath the ridges, probably related to the mantle reservoirs discussed for basalt genesis in Chapter 10. We can thus distinguish between **N-MORB** ("normal" MORB), which taps the depleted, or incompatible-poor mantle, and **E-MORB** ("enriched" MORB, also called **P-MORB** for "plume"), which taps an incompatible-richer mantle discussed in Section 10.5. N-MORBs with Mg# > 65 have $K_2O < 0.10$, and $TiO_2 < 1.0$, whereas E-MORBs have $K_2O > 0.10$ and $TiO_2 > 1.0$ for the same Mg#. Of course, major element chemical composition is not the best way to make these distinctions, which must be substantiated by trace element and isotopic differences.

13.5 MORB TRACE ELEMENT AND ISOTOPE GEOCHEMISTRY

Variation diagrams for trace elements versus Mg# support the conclusions stated above for the major element trends. Highly compatible elements, such as Ni and Cr, decrease with decreasing Mg#, as we would expect with olivine fractionation. Highly incompatible elements, such as V, concentrate in the evolved liquids. If we want to address the distinction between P-MORB and E-MORB, however, we are best served by proceeding directly to the rare earth elements.

Figure 13.11 shows some selected REE patterns from the MAR data of Schilling et al. (1983). The patterns with open squares have negative slopes, resulting from LREE

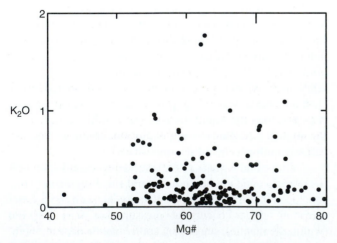

FIGURE 13.10 Variation in K_2O versus Mg# for MORBs from the Mid-Atlantic Ridge. Data from Schilling et al. (1983).

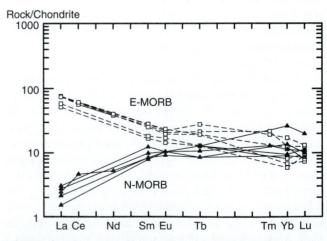

FIGURE 13.11 Chondrite-normalized REE patterns for selected samples from the Mid-Atlantic Ridge. Squares are E-type MORBs, and triangles are N-type MORBs. Data from Schilling et al. (1983).

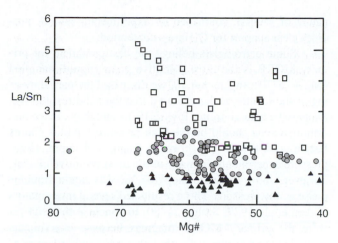

FIGURE 13.12 La/Sm versus Mg# for MORBs from the Mid-Atlantic Ridge. Open squares are E-type MORBs, and solid triangles are N-type MORBs. Gray circles are T-type MORBs. Data from Schilling et al. (1983).

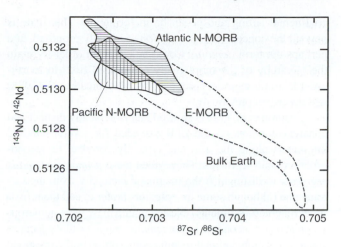

FIGURE 13.13 Sr and Nd isotopic ratios for Atlantic N-MORB (horizontal hatches), Pacific N-MORB (vertical hatches), and E-MORB of the southwestern Indian Ocean. Data from Ito et al. (1987), LeRoex et al. (1983), and Meyzen et al. (2005).

enrichment, similar to the enriched mantle xenoliths and basalts in Figure 10.15. The patterns with solid triangles have positive slopes (LREE depleted) considered characteristic of MORBs derived from a depleted mantle source in Figure 10.15. The LREE enriched samples are also enriched in other incompatibles, corresponding to the high K_2O and TiO_2 samples in Figure 13.10. As discussed in Chapter 10, these patterns of light rare earth enrichment and depletion cannot result from partial melting or fractional crystallization effects (because they are excluded from the solid phases) and must reflect different mantle reservoirs.

The HREE patterns are overlapping and similar, prompting some investigators to propose that the ratio of La/Sm (the slope within the LREE) should be used as a measure of LREE-enriched versus LREE-depleted magmas and not La/Lu or La/Yb (the slope across the entire REE diagram). Others consider only the most incompatible elements to be unaffected by differential fractionation due to crystal separation or partial melting effects, and suggest the La/Ce ratio is superior.

Figure 13.12 shows La/Sm versus Mg# for the data set of Schilling et al. (1983). Note that E-MORBs (open squares) always have a higher La/Sm ratio than N-MORBs (open triangles), regardless of Mg#. The lack of any distinct break between the enriched and depleted lavas led several investigators to distinguish *three* MORB types. For the data of Schilling et al. (1983), the selected subdivisions depended a bit on the particular ridge segment, but in general E-MORBs have La/Sm > ~1.7, N-MORBs have La/Sm < ~1.0, and **T-MORBs** (for "transitional MORBs") have intermediate values. The chemical data suggest T-MORBs form a continuous spectrum between N- and E-types, most simply explained by mixing of the two extreme magma types. Thus T-MORBs do not necessarily imply a third distinct source.

Because isotopes do not fractionate during partial melting or fractional crystallization processes, they should be the most useful indicators of source variations. Figure 13.13 shows the $^{143}Nd/^{144}Nd$ versus $^{87}Sr/^{86}Sr$ data for MORBs. N-MORBs plot as a relatively tight cluster with $^{87}Sr/^{86}Sr < 0.7035$ and $^{143}Nd/^{144}Nd > 0.5030$, both of which indicate a depleted

mantle source (see Section 10.5). Note that E-MORBs extend the MORB array to much more enriched values (higher $^{87}Sr/^{86}Sr$ and lower $^{143}Nd/^{144}Nd$), providing further strong support for distinct mantle reservoirs for N-type and E-type MORBs. T-MORBs also show intermediate (mixed) values.

The trace element and isotopic data thus provide confirming evidence that MORBs have more than one source region and that the mantle beneath the ocean basins is not homogeneous. Petrologists have preferred a model in which N-MORBs tap an upper depleted mantle and E-MORBs tap a deeper enriched source, as described in Sections 1.7.3 and 10.5, but this layered model is controversial. The time required for the isotopic systems to develop suggests that these reservoirs, whether layered or as distributed segregations, have been distinct for a very long time. T-MORBs are probably produced by mixing of the N-source and E-source magmas during ascent and/or in shallow chambers.

A further intriguing aspect emerges from the isotope studies of the ocean ridges, where there is evidence for a third, very enriched component. This component is especially noticeable in the data from the ridges in the Indian Ocean (Dupré and Allègre, 1983; Michard et al., 1986; Meyzen et al., 2005), as indicated in Figure 13.13. The evidence is stronger when we consider the U-Pb-Th system and the ocean island data, so we shall postpone discussion of this component and the implications for the sub-oceanic mantle until Chapter 14 on ocean island volcanism.

Increased sampling has revealed a correlation between some of the ridge *tectonic* segmentation described previously and chemical magmatic trends at several localities, leading to the concept of **magmatic segmentation**. Magmatic and tectonic segments must be related because the boundaries of zones exhibiting petrologic and geochemical similarities correspond to tectonic discontinuities. In his review of segments, Batiza (1996) proposed that the largest segment scale is based on regional isotopic patterns and corresponds to variations in mantle history and composition spanning a range of scales up to that of entire ocean basins or larger, which he related to similar-scale tectonic segmentation originating with initial continental rifting and ocean basin formation. This scale is beyond the concept of *segmentation* as initially proposed, and perhaps the term *domains* would be better, in order to retain the hierarchy of the more easily recognized first- to fourth-order tectonic segments proposed by Macdonald et al. (1988) and described previously.

Sinton et al. (1991) showed that isotopically distinct and coherent domains on the EPR extended for several hundred kilometers along the axis and typically correlate to second-order tectonic segments. They called these *primary magmatic segments* to distinguish the magmatic hierarchy from the tectonic one (although some investigators prefer to call them first-order magmatic segments). Nested within these were numerous *secondary* (or second-order) *magmatic segments* that appear to share a common parental magma composition and correspond to third-order tectonic segments. Langmuir et al. (1986) showed that small devals on the northern EPR commonly corresponded to changes in eruptive compositions best interpreted as resulting from fractionation or enrichment trends. Some detailed across-axis sampling studies have revealed systematic chemical fractional trends that correlate with distance from the ridge axis, suggesting temporal trends in recharge and evolution in axial magma chambers. Reynolds et al. (1992) showed that two deval-bounded segments near 12°N on the EPR had chemically coherent evolutionary trends, each of which independently varied with time. It appears that, for some ridge segments at least, the degree of chemical contrast correlates with the tectonic segment hierarchy, and that separate segments may recharge and evolve independently over time, further supporting the interpretation of segments as separate magmatic systems, fed from a conduit near the center of the segment and propagating as melt migrates parallel to the ridge toward the segment tips.

13.6 PETROGENESIS OF MID-OCEAN RIDGE BASALTS

We now have a fairly good idea of what constitutes MORB magmas and how they vary. It remains to develop a reasonable model for MORB petrogenesis. Such a model must be able to explain the chemical and mineralogical trends, the volcanic phenomena at the ridges, and the generation of the layers of oceanic crust, while remaining compatible with the geophysical data. Models of course change as more data become available, and any model presented at this date will most certainly be modified or changed in the future, just as previous models have evolved to our present concept.

13.6.1 Mantle Melting: The Generation of Mid-Ocean Ridge Basalts

Substantial controversy has arisen over the nature of primary N-MORB magmas. The great quantity and chemical uniformity of MORBs led early workers to conclude they were all primary melts delivered directly to the surface. As discussed above, the vast majority of, if not all, MORBs are now believed to be derivative liquids that have undergone at least some degree of fractional crystallization and magma mixing, as first proposed by O'Hara (1968a, 1977, 1982). As discussed earlier in this chapter, the low-P multiply-saturated

nature of MORB, confirmed by experimental studies, provides clear support for O'Hara's contention.

Some petrologists believe the best candidate for primary MORB is the most primitive *glass* (liquid) sampled (Mg# > 70). In Chapter 10, we discussed the partial melting of the mantle, and concluded that the solid residue was likely to be a harzburgite (Figure 10.1), and thus olivine and orthopyroxene should be the high-pressure liquidus phases last in equilibrium with primary magmas at depth. High-pressure experiments on many of the most primitive glasses, however, showed that orthopyroxene was not a liquidus phase at *any* pressure. These results led several investigators (Green et al., 1979; Stolper, 1980) to conclude that *picrites* were the primary MORB magmas because experiments show them to be saturated with both olivine and orthopyroxene at 1.0 to 1.2 GPa (30 to 35 km). Opponents of picrites as primary MORB magmas pointed to the lack of any glasses of picritic composition, which suggests picrites are accumulative with respect to olivine and do not represent true liquid compositions. Those who favored picrites replied that the density of picritic liquids inhibits their reaching the surface. Not all experiments on primitive MORB glasses, however, lacked orthopyroxene on the liquidus. Fujii and Kushiro (1977) and Bender et al. (1978) reported primitive MORBs saturated with orthopyroxene in the range of 0.8 to 1.2 GPa (see Figure 10.13). This controversy over the nature of MORB parent and the interpretation of experimental results extended for over a decade (Basaltic Volcanism Study Project, 1981; Stolper, 1980; Presnall and Hoover, 1987; Elthon, 1989; Fuji, 1989). Most recent experiments using new techniques to maintain small melt fractions in uncontaminated states (e.g., Hirose and Kushiro, 1991) suggest that primitive MORB with only 10 to 12% MgO (Mg# 63–70) can be generated at shallow pressures (0.1 GPa) in equilibrium with olivine and orthopyroxene, and that picrites are not necessary as MORB parent magmas.

Regardless of the parent, experimental data on the most likely parental MORBs (including picrites) indicate that they were multiply saturated with olivine, clinopyroxene, and orthopyroxene in the pressure range of 0.8 to 1.2 GPa, corresponding to about 25 to 35 km. This is in the spinel lherzolite field, which is compatible with the lack of both HREE depletion (expected if garnet were a residual phase) and a europium anomaly (expected if plagioclase were residual) in the REE data of Figure 13.11. We must be clear on what this means. Because MORB magmas are the product of partial melting of mantle lherzolite in a rising diapir, melting takes place over a range of pressures. The pressure of multiple saturation represents the point at which the melt was *last in equilibrium with the solid mantle phases*, as discussed in Chapter 10. Whereas the incompatible trace element and isotopic characteristics of the melt reflect the equilibrium distribution of those elements between the melt and the ultimate *source* reservoir, the major element (and hence mineralogical) character will be controlled by the equilibrium maintained between the melt and the residual mantle phases during its rise until the melt separates as an independent system with its own distinct character. Thus the depth of multiple saturation reflects the *separation*

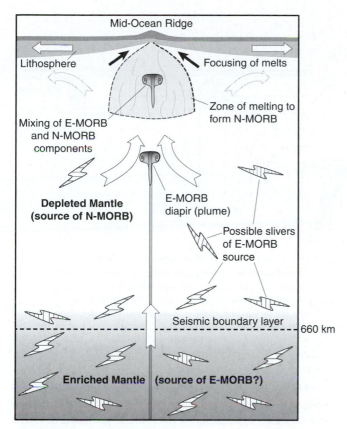

FIGURE 13.14 Schematic diagram illustrating a model for the rise and decompression melting of different source regions to produce N-type and E-type MORBs. After Zindler et al. (1984) and Wilson (1989). The gradational nature of the 660-km discontinuity and multiple slivers of enriched source material are intended to represent the uncertain nature of mantle heterogeneity and not the actual geometry, boundaries, or scale of mantle domains.

depth, which may be interpreted as the *minimum* depth of origin because the melt can separate from the solids at any point during the rise of the diapir from its *ultimate* source. Certainly the ultimate source is much deeper than 25 to 35 km as indicated by the experiments, perhaps as great as 80 km for N-MORB, and even deeper for plumes of E-MORB.

Our petrogenetic melting model, illustrated in Figure 13.14, thus begins with the separation of two lithospheric plates at a divergent plate boundary, resulting in the upward motion of mobilized mantle material into the extended zone, where it replaces the material shifted horizontally and undergoes decompression partial melting associated with near-adiabatic rise (Section 10.2.1) along a *P-T* path of ~0.3°C/km. The cause of plate motions is debated and described briefly in Section 1.7. In the reviews of Forsyth and Uyeda (1975) and Backus et al. (1981), the authors concluded that "slab pull" is the principal force, believing that the greater density of cool oceanic lithosphere results in slab descent at subduction zones. As a result, plate divergence at mid-ocean ridges is driven by gravitation and negative-buoyancy, so mantle upwelling at ridges is then largely a passive response to plate separation ("**passive rifting**") rather than an active cause of it ("**active rifting**"). Heat transport, therefore, is essentially *advective* (Section 1.7.2). Actively buoyant mantle flow may

occur on a local scale, however, where thermal gradients are strong and where melt formation and segregation occur (Forsyth, 1992). For N-MORB, melting is initiated in the 60 to 80 km depth range in uppermost depleted mantle, where it inherits its depleted trace element and isotopic character. Simple mathematical models suggest that the region of melting for N-MORBs is rather broad, approximately 100 km wide (Scott and Stevenson, 1989). The percentage of partial melting increases to approximately 15 to 40% (Plank and Langmuir, 1992; Forsyth, 1993) as the melting mantle ascends toward the surface (see Figure 10.4). Because melting removes Fe and Al preferentially, the residual mantle is *less* dense than the original mantle source, enhancing buoyancy and therefore further melting as it proceeds. Melting is terminated by heat loss to the surface near the top of the column, perhaps aided by the consumption of clinopyroxene, which, when gone, requires a discontinuous temperature jump before further fractional melting is possible (Section 7.1.1). Asimow et al. (1995) pointed out that solid-phase mantle transformations (Figure 10.4) affect the slopes of melting increments within the solidus–liquidus melting interval, which may also retard decompression partial melting. Their modeling using the MELTS algorithm (Ghiorso and Sack, 1995; Ghiorso et al., 2002) indicated that the garnet $\rightarrow$ spinel transition suppresses decompression melting and spinel $\rightarrow$ plagioclase may even reverse it.

An attractive model for melt production in rising mantle that traverses a simple triangular melting regime, turns a corner beneath the ridge, and then moves laterally, is illustrated in Figure 13.15. In such a "*cornerflow*" model (Ahren and Turcotte, 1979; Phipps Morgan, 1987; McKenzie and Bickle, 1988; Plank and Langmuir, 1992; Langmuir et al., 1992; Klein, 2003), the passively rising mantle intersects the solidus (see Figure 10.4, a simplified version of which is inserted in Figure 13.15) and progressively melts as it (adiabatically) traverses the lherzolite melting interval. Melting terminates when the flowing mantle moves horizontally from the ridge area, where it can then be cooled by hydrothermal circulation. Hotter mantle (Figure 13.15a) departs from the geotherm and intersects the lherzolite solidus at deeper levels (insert in Figure 13.15) and has the potential to traverse a greater fraction of the solidus $\rightarrow$ liquidus melting interval (represented by the shaded interval in the insert and the triangular melting region in Figure 13.15a). Cooler mantle (Figure 13.15b) departs from the geotherm and intersects the solidus at shallower levels with less melting potential (capable of traversing less of the solidus-liquidus interval, and hence a smaller triangular melting region). The extent of melting of any particular package of rising mantle material thus depends on the temperature of the upwelling mantle (presumably dependent on rate of plate separation) and the amount of near-adiabatic rise within the melting interval before turning the corner and exiting the melting region (related to proximity of a rising mantle package to the ridge axis). In her summary, Klein (2003) estimated global ranges in pressure of solidus intersection to be ~3.5 to 1.5 GPa ~(105 to 45 km) at temperatures of 1550°C to 1300°C and a mean extent of peridotite melting between 8 and 22%.

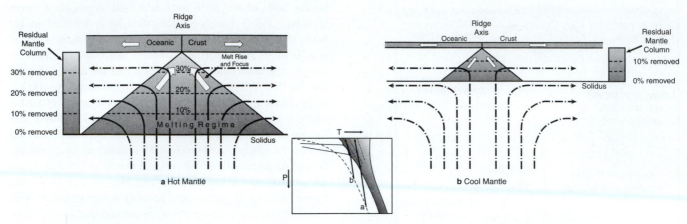

FIGURE 13.15 Idealized steady-state melting regimes produced by mantle flow in response to plate separation. Dash-dot curves represent mantle flow paths (solid within the triangular melting regime). Horizontal dashed lines represent the extent of melting within the regime (equivalent to the extent of melt removed in the residual mantle column upon exiting). **(a)** (right) Hot mantle from a deeper position on the geotherm [see **(a)** in lower center insert based on Figure 10.4] intersects the solidus deeper and has the potential to traverse a greater portion of the solidus → liquidus melting range. **(b)** (left) Cool mantle from a shallower position on the geotherm intersects the solidus shallower [**(b)** in insert] and has less chance to traverse the melting range. After Langmuir et al. (1992). © American Geophysical Union with permission.

Because the critical melt fraction for separation (Section 11.1) of basaltic liquid from rising host lherzolite is believed to be low ($<1\%$: McKenzie, 1984; Daines and Richter, 1988; Riley and Kohlstedt, 1991; Faul, 1997), small melt increments are considered to separate from the mantle host at a variety of pressures and temperatures. Simple batch melting models for parental MORBs have thus yielded to more sophisticated "incremental batch" models involving sequential polybaric melt production with separation of small (typically 1%) melt increments at progressively decreasing pressure steps (McKenzie and Bickel, 1988; Kinzler and Grove 1992a, 1992b; Kinzler, 1994). Rising mantle source peridotites in these models gradually change composition as melt increments are removed, and the next melt increment is then calculated for the new composition. Released melt increments are then focused into the 3- to 8-km wide neovolcanic zone at the mid-ocean ridges by processes that remain a subject of debate (Spiegelman and McKenzie, 1987; Phipps Morgan, 1987; Scott and Stevenson, 1989; Cordrey and Phipps Morgan, 1993; Sparks and Parmentier, 1994; Perfit et al., 1994; Kelemen et al., 1999).

Dasgupta and Hirschmann (2006) cited geophysical observations that indicate some melting occurs beneath mid-ocean ridges at depths as great as 300 km. Their experiments on peridotites with 2.5 wt. % CO_2 demonstrated that, if minor amounts of CO_2 were available, $< 0.3\%$ carbonate-rich liquids may be generated at depths as great as 330 km. These minor melts may affect the physical and chemical properties of the host. The effect of volatiles on melting may be more dramatic in other situations, and will be explored more fully in the next six chapters.

As indicated by the map in Figure 14.1, several mid-ocean ridge segments are associated with deep-seated mantle **plumes** (e.g., Iceland, Galapagos, Azores, Tristan da Cunha). It is debatable in such situations whether (a) plumes influence ridge formation, (b) mantle upwelling at ridges draws up or

"invites" plumes, or (c) the plume and ridge are of independent origins (but geographically coincidental). The deep source of plumes argues against notion (b), and the gradual drift of the Reykjanes segment of the Mid-Atlantic Ridge northwestward from the Iceland plume, only to jump and re-center over the plume every few million years (Hardarson et al., 1997), supports notion (a). But hotspot tracks indicate that some other plumes (perhaps smaller?) have jumped indifferently across ridges from one plate to another (e.g., Tristan da Cunha, Kerguelen, Reunion, Easter Island), supporting notion (c). Plumes originate deep in the mantle, probably at the core–mantle boundary (Figures 1.14 and 14.23) and reflect enriched trace element and isotopic character, perhaps of a deeper mantle reservoir. A rising plume also undergoes polybaric decompression partial melting at much shallower depths (Figure 14.20). This is the generally accepted mechanism for the formation of E-MORB magmas. Some E-MORBs occur at localities unrelated to noticeable plumes, however, perhaps involving a shallow (lithospheric) mantle source enriched by infiltration of low-degree melts, near either a mid-ocean ridge (Workman et al., 2004), or subduction zone (Donnelley et al., 2004) or the melts are ocean island alkali basalts (Hémond et al., 2006). The question of the layering of the mantle into a depleted upper mantle and a non-depleted (or perhaps enriched) lower mantle plume source, subdivided at the 660-km seismic discontinuity, remains in dispute, as discussed in Section 1.7.3, and the issue will be explored more fully in Chapter 14. Figure 13.14 attempts a somewhat neutral view, allowing for both dispersed enriched blobs (perhaps concentrated in the deeper mantle due to their greater density) and a 660 km layer boundary (made gradational to indicate its contested nature: seismically the phase transition is sharp, but geochemically perhaps not). As with N-MORB, plume melts should segregate at a variety of depths once the solidus has been crossed (generally shallower than 150 km), and each melt increment will become a semi-isolated system at relatively shallow depths where the major element and mineralogical character is determined.

MORB magmas are thus amalgamations of diverse polybaric melt increments from variable sources that focus to a shallow sub-ridge axial zone. Mixing of various E-MORB and N-MORB melts may take place over a range of mantle depths, and mantle-level fractional crystallization may also occur, either before or after mixing, particularly in some cooler slow-moving ridge segments (Grove et al., 1992). Erupted lavas at the ridge represent the aggregated product of these diverse melts, thereby explaining the difficulty in finding primitive MORB magma and the lack of MORB–mantle equilibrium. Two sources provide particularly strong evidence for MORB melt diversity. First, *near-axis seamounts* are believed to tap less aggregated melting products than do adjacent ridges (Klein, 2003, and references therein). Such seamounts exhibit a wider compositional range, including more depleted and more enriched compositions, presumably due to less mixing and homogenization. Second, *melt inclusions trapped within phenocrysts* represent liquids isolated at earlier stages of melt development and tend to exhibit a correspondingly more primitive character than the interstitial glasses described previously (e.g., Shimizu 1998; Sours-Page et al., 2002). Many trace element concentrations (particularly incompatible elements) and ratios in these glasses span a greater range than the major elements, attesting to a greater diversity of melt compositions that must have existed in the mantle prior to aggregation beneath the ridge. Such diverse melts (see also Figure 13.11) cannot all be in equilibrium with residual peridotites, leading to the general conclusion that most melts segregate from initial *porous flow* migration to *focused flow* in conduits, perhaps along melt-filled fractures or tubes that allow more rapid melt transport without maintaining equilibrium between the melts and their mantle surroundings. Kelemen et al. (1999) suggested that sharply discordant dunite segregations (commonly observed in the mantle section of ophiolites) mark the remains of these conduits.

13.6.2 Magma Chambers and the Creation of Oceanic Lithosphere

Mantle melts are eventually focused toward a shallow magma chamber beneath the ocean ridge crest. Originally the axial magma chamber was envisaged as a relatively large (approximately 5 km wide and 9 km deep) semi-permanent chamber that underwent fractional crystallization to produce the derivative MORB magmas (Figure 13.16). According to this theory, fractionation was moderated by periodic reinjection of fresh, primitive MORB from below. Dikes emanated upward through the extending and faulting roof to create the sheeted dike complex and feed the surface flows. Crystallization took place in this magma chamber near the top, along the floor, and along the sides, thereby adding successive layers of gabbro to create Layer 3. Cann (1974) called this model the "infinite onion" model because it resembled an infinite number of onion shells created continuously from within and added to the walls as they receded.

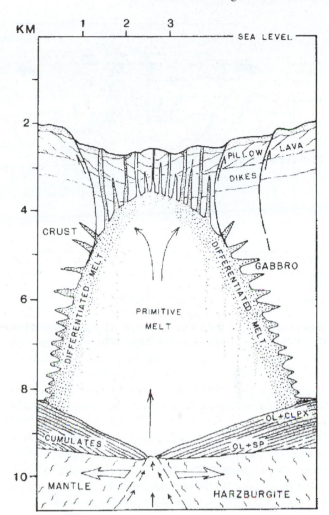

FIGURE 13.16 The early model of a semi-permanent axial magma chamber beneath a mid-ocean ridge. From Bryan and Moore (1977). Copyright © Geological Society of America, Inc.

Accumulation of dense olivine and pyroxene crystals on the chamber floor could have produced the ultramafic cumulates found in ophiolites and suspected to be present at the top of Layer 4. The layering in the lower gabbros (Layer 3B) may have resulted from density currents of suspended crystals flowing down the sloping walls and floor.

The large persistent chamber model gained wide acceptance for its simple and elegant explanations of mid-ocean ridge magmatism and the creation of the oceanic crust. The "open-system" periodic recharge of primitive magma and the continuous differentiation within the chamber explained a number of other features as well. First, it explained the narrow chemical range with a somewhat evolved character which reflected a near steady-state balance between differentiation and replenishment. That this moderating influence was not perfectly steady-state is reflected in the chemical variations shown in Figures 13.6 through 13.12 for the erupted volcanics. Of course, the evolved magma will no longer be in equilibrium with the mantle phases at depth, having undergone fractional crystallization in the shallow chamber. The more primitive nature of the volcanics toward the axis of the ridge and more evolved nature toward the flanks observed by some investigators could be explained by fresh

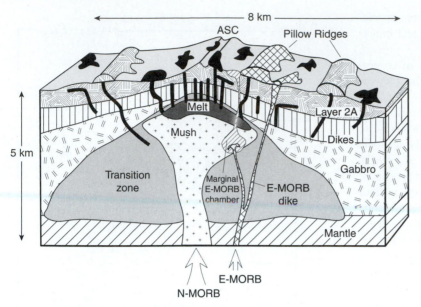

FIGURE 13.17 Schematic section through an axial magma chamber beneath a fast-spreading ridge such as the East Pacific Rise. The black zone is the liquid axial magma chamber. ASC = axial summit caldera. After Perfit et al. (1994). Copyright © Geological Society of America, Inc.

injections in the axial region and more advanced differentiation toward the cooler chamber walls. The "pyroxene paradox" in which chemical trends suggested clinopyroxene fractionation in magmas with no clinopyroxene may result from the mixing of primitive (clinopyroxene free) magma with a more evolved one saturated in clinopyroxene (Walker et al., 1979).

O'Hara (1977) proposed a model for the behavior of incompatible trace elements in open-system, periodically replenished magma chambers in which fractional crystallization plus discharge of evolved liquid to the surface would compensate more primitive recharge from below. Each recharge would partially reset the evolved magma in the chamber back toward more primitive compositions. Major elements approach a steady state of intermediate composition. Incompatible minor and trace elements, on the other hand, concentrate over time to values higher than they would under closed-system behavior because more is added with each recharge than is lost to the surface by volcanism, and the incompatibles are not removed by fractional crystallization. These elements eventually reach a steady state as well, but at elevated concentrations. This may explain the high concentrations of K_2O, TiO_2, and P_2O_5 that suggested 50 to 67% fractional crystallization in Section 13.4. Langmuir (1989) suggested an alternative model of **in situ crystallization** (Jackson 1961; McBirney and Noyes, 1979), involving a boundary layer of graduated solidification (as in Figure 11.12). Such a boundary layer would be a crystal-laden mush at a temperature near the solidus at the cooler solid walls and grade to nearly all liquid inward where the temperature is higher. The composition of the *liquid* mixed with the crystals thus varies from the low-temperature eutectic composition near the wall to the original bulk composition inward where fewer crystals are present. If this magma spectrum were mixed, melts of a range of derivative characteristics would be combined. Clinopyroxene would crystallize along the cool walls, and the melts in equilibrium with it might be expelled to mix with clinopyroxene-free melts in the interior, thus explaining the "pyroxene paradox," as well as the elevated concentration of incompatible elements.

The persistent chamber model of mid-ocean ridge magmatism has a beautiful simplicity and provides an elegant solution to the generation of the ophiolite sequence/oceanic crust layering. Despite the appeal of the model, however, an intensive search spanning over three decades of seismic imaging has failed to detect any chambers of this size at ridges, thus causing a fundamental shift away from this traditional view of axial magma chambers as large, steady-state, predominantly molten bodies of extended duration (Langmuir and Detrick, 1988). A more modern concept of the axial magma chamber beneath a *fast-spreading* ridge is illustrated in Figure 13.17, which combines the magma chamber geometry proposed by Sleep (1975) and Sinton and Detrick (1992) with the broad zone of volcanic activity noted by Perfit et al. (1994).

In the new model, the completely liquid body is a thin and narrow sill-like lens ($\sim$ 10 to 150 m thick and < 2 km wide). It is located 1 to 2 km beneath the seafloor and provides a high amplitude subhorizontal reflector noticed in detailed seismic profiles shot along and across sections of the EPR (Herron et al., 1978, 1980; Detrick et al., 1987; Kent et al., 1990; Toomey et al., 1990). The melt body is surrounded by a much wider zone of low seismic velocity that transmits shear waves, but may still have a minor amount of melt (the "mush" and "transition" zones in Figure 13.17). Petrologically and geochemically, the "magma chamber" may be considered to comprise both the well-mixed, convecting, liquid body and the partially solidified mush zone, as the liquid is continuous through them. The liquid gradually crystallizes to mush and the boundary moves progressively into the liquid lens as crystallization proceeds. The lens is maintained by reinjection of primitive magma, as in the "infinite onion" of Cann (1974). The mush zone is not seismically well constrained because the variation in seismic waves with melt proportion is not well known.

The crystal mush zone contains perhaps 30% melt (Detrick, 1991) and constitutes an excellent boundary layer for the in situ crystallization process proposed by Langmuir (1989) above, which is more consistent with recent laboratory and theoretical results concerning crystallization in

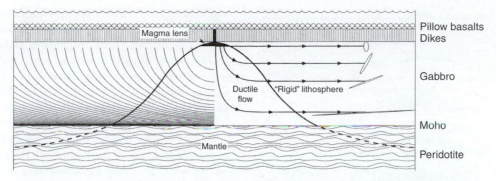

FIGURE 13.18 Diagram illustrating the theoretical crustal extension and ductile flow associated with a gabbro glacier model of magmatic accretion at a fast-spreading mid-ocean ridge. The heavy bell-shaped curve outlines the ductile mush region. Curves with arrows in the right half represent material flow lines away from the base of the magma lens, and ellipses represent accumulated strain along each flow line (deformed shapes of original circles). Strain is most intense in the lower levels. Curves on the left side indicate foliation in the gabbros, which conforms well to that believed to form in the Semail ophiolite. From Phipps Morgan et al. (1994).

magma chambers (Huppert and Sparks, 1980). Seismic velocities are still low beyond the mush, and are believed to result from a transition zone where the partially molten material grades to cooler solidified gabbro. The mush/transition boundary probably represents the **rigidus**, where crystallinity increases beyond 50 to 60%, and the magma becomes a crystal-bonded aggregate that behaves much like a solid and is no longer eruptable (Marsh, 1989). The relatively high seismic velocities throughout most of the transition zone limit the melt component to just a few percent.

The lack of any detectable large magma chamber in the most active igneous province in the world and the increasingly common detection of a small lens/mush zone at fast-spreading ridges have convinced most of the geological community that the small magma chamber model is correct. A shallow melt of any sort can readily erupt through cool crust in an extensional environment to produce sheeted dikes and volcanics, but the small ephemeral sill-like liquid chamber seems difficult to reconcile with traditional ideas of fractional crystallization and crystal settling to form the layered gabbros and cumulates, which correspond to over half the height of the gabbroic sequence and conform better to our concepts of a large liquid chamber, as in Figure 13.16. Debate has recently focused on the detailed nature of the lens/mush geometry and the processes by which such a limited chamber could create the observed layered and foliated gabbros and ultramafics.

Two principal types of "end-member" models attempt to reconcile the small melt lens with the thick sequence of, and structures within, the plutonic rocks. The first type, the so-called "gabbro glacier" (or "conveyor belt") models, are based on an idea first proposed by Sleep (1975). According to these models (Figures 13–18) crystals settle in a shallow melt-dominated lens, or sequence of ephemeral lenses in the same general place beneath the ridge axis. Ductile flow within the mush zone, as it compacts, solidifies, and recedes with plate separation, then imparts a (secondary, or nonmagmatic) foliation (Dewey and Kidd, 1977; Nicolas et al., 1988; Quick and Denlinger, 1993; Phipps Morgan and

Chen, 1993; Henstock et al., 1993; Phipps Morgan et al., 1994). Recent studies of oceanic gabbros, 1500 m of which were penetrated at Leg 735B of the DSDP in the southwest Indian Ocean (Meyer et al., 1989; Bloomer et al., 1991, Dick et al., 1991; Robinson et al. 2000) exhibit textural and chemical variations that are compatible with evolution in a crystal-rich zone at the margins of a small liquid chamber (Sinton and Detrick, 1992). Nicolas (1989) and Cannat et al. (1991) proposed that much of the layering of gabbros in ophiolites appears to be secondary, imposed during deformation of the spreading seafloor and not by crystal settling. In their summary of gabbroic foliation and layering in the Semail ophiolite, Quick and Denlinger (1993) recognized cumulate textures, but also secondary foliations and noted that both are typically discordant to the presumed Moho, exhibiting a generally concave-upward shape from low dips near the base to nearly vertical toward the top of the plutonic section. Many of the studies just mentioned have quantitatively modeled heat and strain considerations. The model of Phipps Morgan et al. (1994) shown in Figure 13.18 illustrates the flow directions and accumulated strain in the gabbros resulting from compaction and flow in the ductile region and the resulting gabbro foliation, which corresponds well to the concave-upward foliation trends observed in the Semial ophiolite. The portion of the lens that crystallizes laterally along the separating chamber walls have little strain and eventually constitute the upper isotropic gabbros of ophiolite sequences. Notice that the flow and deformation within the gabbros result largely from the weight of the accumulating crystals settling to the bottom of the magma lens, just like the ice in a glacier responds to accumulated snow (hence the model name).

The gabbro glacier model elegantly explains many features of ophiolites, particularly the geometry of the foliation developed. Detailed studies of the gabbros in ophiolites find little evidence for ductile deformation of individual crystals, however, but secondary gabbro flow and crystal alignment are still possible via lubricated grain boundary slip and pressure solution (Chapter 23) if a few percent or more

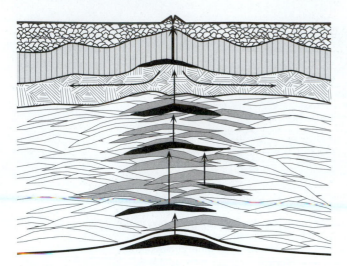

FIGURE 13.19 Sheeted sill model for the development of oceanic lithosphere at a fast-spreading ridge. The shallow melt lens feeds into only a minor fraction of the upper gabbros of Layer 3. Most of the lower gabbros crystallize in place as a series of sills. Crystallization and compaction in the sills result in release of residual melt that migrates upward to feed the shallow lens. After Kelemen et al. (1997).

melt is present (Nicolas and Idelfonse, 1996). The model is less successful at explaining the *layered* variations in mineral types, the correlated layering in mineral compositional variations, and the apparently primary near-vertical fabrics in the upper gabbros that appear to represent subvertical melt conduits (Korenaga and Kelemen, 1998; Boudier et al., 1996; Kelemen et al., 1997, Kelemen and Aharanov, 1998). It is unlikely that these features are secondary. Noting these features and the gabbroic sills in the mantle section of the Semail ophiolite, plus the similarity of those sills to many of the modally layered gabbros, Kelemen and coworkers concluded that most of the lower oceanic crust crystallized in place, and proposed a second model: the **sheeted sill** model for the development of the oceanic lithosphere (Figure 13.19). Kelemen and Aharanov (1998) suggested that these sills form as porous flow of rising basaltic liquids (or possibly small melt-filled fractures) encounter permeability barriers of earlier crystallized melts and pond to form the sills. These sills crystallize to form the modally layered cumulate gabbros in Layer 3 (and some deeper gabbroic sills in the mantle). Influx of new magma leads to increased pressure, which periodically exceeds the strength of the overlying cap and releases melt into vertical fracture-conduits (via hydrofracturing) that feed the shallow melt lens. The ophiolite gabbros are thus primary cumulates of early-forming minerals, the compositions of which are refractory and not equivalent to any basaltic liquids, but are complementary to the more evolved shallow gabbros, dikes and extrusives in the upper crust, suggesting that the solid cumulates and shallow liquids once coexisted in equilibrium at depth before the latter escaped. According to this model, the sills form the shallowly-dipping lower portions of Layer 3 in ophiolites and the dike-like conduits form the steeply-dipping components above. Kelemen et al. (1997) noted the abundance of dips <20° in the lower portion of the Semail ophiolite gabbros

and of dips >60° in the upper portion but claimed that there is essentially no continuity or transition between them.

A new geophysical imaging technique for constraining shallow crust-mantle structure uses *seafloor compliance* measurements, employing broad-band seismometers and pressure gauges placed on the seabed to measure tiny displacements of the seafloor in response to pressure fluctuations caused by long-period ocean waves (really!). Crawford et al. (1999), using seafloor compliance measurements to image regions of low shear velocity beneath the EPR at 9°48′N, detecting not only a fully molten shallow lens 1.4 km beneath the ridge axis of the seabed, but also a *second* on-axis melt lens (thus stacked lenses) with 3 to 18% melt at ~5.5 km depth (at or near the crust–mantle boundary). Dunn and Toomey (1997) and Dunn et al. (2000) also found a second deep melt accumulation elsewhere on the EPR using seismic tomography. According to Dunn et al. (2000) the deeper melt region near the crust/mantle transition beneath the EPR at 9°30′N has a lower proportion of melt (4 to 10%) than the shallow lens (and the deep lens imaged by Crawford et al., 1999), but, because of its greater size, it may contain up to 40% more total melt than the shallow lens. The accumulation of melt at this level should be an impediment to upward melt flow at the base of the crust, implying that a significant portion of the lower oceanic crust forms in-place. This melt may feed the sheeted-sill complex proposed by Kelemen and coworkers (notice the deep sub-axial sill near the crust/mantle boundary included in Figure 13.19). Supporters of the gabbro glacier model, acknowledging a second deeper melt body but still impressed with the gabbro foliation, proposed a hybridized model with two melt lenses (Figure 13.20a). The shallow melt lens feeds the upper gabbros, as in the original model, and the deeper lens periodically feeds the shallow lens and crystallizes more along the cooling top than along the hotter bottom so that crystallized material flows upward and outward to generate the lower foliated and layered gabbros (Schouten and Denham, 1995).

One objection to the sheeted sill model is the inefficient transfer of latent heat of crystallization from sills in the deep crust. The shallow lens cools largely by efficient hydrothermal circulation through the fractured overlying crust, but there is no such opportunity for the deeper sills. The parameterized modeling of Chen (2001), for example, indicates that crystallization of gabbro near the Moho in excess of 10% would generate sufficient heat to create a partially molten region much larger than detected seismically anywhere along the mid-ocean ridge system, so that Chen (2001) concluded that the lower lens imaged by Crawford et al. (1999) and Dunn et al. (2000) is short-lived. This prompted hybrid sill models in which most sills are emplaced off-axis in cooler crust (Figure 13.20b) or in which dense cumulates in the shallow lens periodically become unstable as they accumulate and cool, and the slurry descends into the lower mush region and spreads out to form the sills (Figure 13.20c). Cherkaoui et al. (2003) noted that the tomographic and compliance studies of Dunn et al. (2000) and Crawford et al. (1999) indicated steep-sided and narrow, low-velocity (partially molten) zones, which, they suggested, cool more

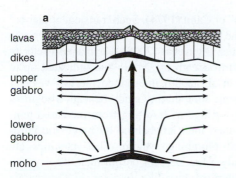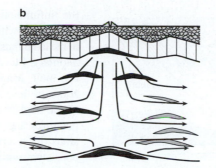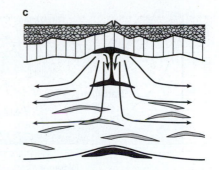

FIGURE 13.20 Hybrid models for development of oceanic lithosphere at a fast-spreading ridge (arrows represent material flow lines). **(a)** Ductile flow model incorporating a second melt lens at the base of the crust (e.g., Schouten and Denham, 1995). **(b)** Ductile flow with two melt lenses and off-axis sills (e.g., Boudier et al., 1996). **(c)** Sheeted-sill hybrid model in which lower sills are fed from above by descending dense cumulate slurries from the upper melt lens (Rayleigh-Taylor instabilities) into the lower mush region (Buck, 2000).

efficiently and do not conform to the parameterized models of heat transport (e.g., Chen, 2001), perhaps obviating the need for some of these hybrid modifications.

At the present, therefore, we are left with two end-member models and several hybrid models for the generation of oceanic crust and upper mantle at fast-spreading mid-ocean ridges. Perhaps all of the models have some validity, each indicating one or more processes that occur and contribute to lithospheric development in these complex dynamic systems.

Most models propose that diking and volcanism are limited to the immediate axial neovolcanic area. Perfit et al. (1994), however, reported young looking lava fields and pillow ridges that suggest recent volcanic activity up to 4 km from the axis of the EPR. This activity may be due to off-axis dikes from the central chamber, as well as smaller ephemeral magma chambers of both N-MORB and E-MORB, fed by rising blobs of melt (as illustrated in Figure 13.17). Heterogeneous chemical data (including compatible elements) for the off-axis volcanics, as compared to much more uniform magmas in the axial summit caldera, led Perfit et al. (1994) to suggest that several small spatially and temporally distinct magma bodies with lifespans on the order of 5 ka or less existed in the past 80 ka at the East Pacific Rise. Off-axis intrusions and extrusions add significantly to the oceanic crust, and are believed responsible, at least in part, for the thickening of

Layer 2A away from the ridge. The chemical heterogeneity of the axial magmas in their particular section of the EPR, although less than for off-axis, suggested fractional crystallization, plus recharge and mixing of at least two components within the N-MORB suite (in addition to E-MORB material), which Perfit et al. (1994) attributed to separate blobs of N-MORB that evolved during ascent to the axial area.

As mentioned previously, melt bodies can be traced as continuous reflectors for up to several kilometers *along* the ridge crest, but tend to occur as "magmatic segments" (e.g., Sinton et al., 1991) with notable gaps at fracture zones, and at smaller non-transform offsets (devals and OSCs), as illustrated in Figure 13.21. Chemical variations on the scale of hundreds of meters to a few kilometers indicate poor mixing along the axis, and/or intermittent liquid magma lenses, each fed by a source conduit. The character of the melt is more evolved at the distal ends of the chambers farther from the source of primitive magma.

Thermal calculations indicate that high heat flow at fast-spreading ridges can maintain relatively persistent magma chambers, whereas thermal constraints make such chambers beneath slow-spreading ridges highly unlikely (Sleep, 1975). Among the many investigations of *slow-spreading* ridge segments, only Sinha et al. (1999) have managed to detect a "significant" thin sill-like magma body in a larger low-velocity

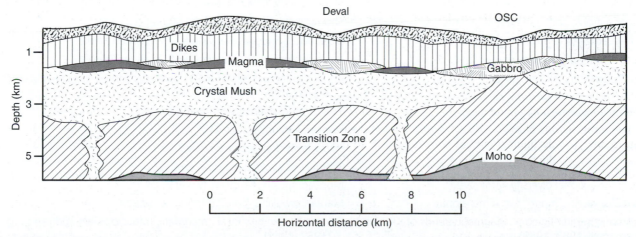

FIGURE 13.21 Schematic cross section along the axis of a fast-spreading ridge illustrating the lack of continuity of the axial magma chamber, crystal mush, and transition zones. Two first-order magmatic segments are separated by an OSC. After Sinton and Detrick (1992). © American Geophysical Union with permission.

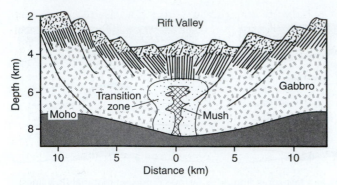

FIGURE 13.22 Schematic cross section of an axial magma chamber beneath a slow-spreading ridge such as the Mid-Atlantic Ridge. A persistent liquid axial magma chamber is typically absent. After Sinton and Detrick (1992). © American Geophysical Union with permission.

region 2 to 3 km beneath the axis of the Mid-Atlantic Ridge. Figure 13.22 illustrates a model for the nature of the magma chamber beneath a *slow*-spreading ridge, such as the Mid-Atlantic Ridge. With a reduced heat and magma supply, a steady-state eruptable magma lens is relinquished in favor of a dike-like mush zone and a smaller transition zone beneath the well-developed rift valley. With the bulk of the body well below the liquidus temperature, convection and mixing is far less likely than at fast ridges.

Nisbit and Fowler (1978) suggested that numerous, small, ephemeral magma bodies occur at slow ridges. They termed this model the "infinite leek" as a smaller variation of

the "infinite onion" of Cann (1974). Such transient bodies are compatible with the present model, and probably concentrate in the axial mush area. Dunn et al. (2005) also imaged a velocity anomaly 4 to 10 km beneath the MAR that indicated up to 5% melt. Magde et al. (2000) interpreted the low-velocity zones in their seismic tomography study beneath a segment of the Mid-Atlantic Ridge as resulting from increased temperatures and retained melt left behind by magmatic intrusions in a crustal plumbing system. The velocities are consistent with only a few percent partial melt. The system includes spaced vertical pipe-like features in the ductile asthenosphere, each of which probably defines a magmatic segment. As described earlier, rising magma in each segment, upon reaching the base of the thin lithosphere, interacts with extensional stresses and propagates laterally along-axis through the brittle layer and vertically as dikes. Magde et al. (2000) considered the low-velocity anomalies to represent the time-averaged signature of a number of short-lived injection episodes, rather than a single magma intrusion.

13.6.3 The Influence of Spreading Rate

Table 13.3 lists several principal contrasts between fast-spreading ridges and slow-spreading ridges. Most of these can be attributed to the differences in the **thermal budget** of individual ridge segments, which depends on the balance between magmatic heat input and hydrothermal cooling: percolating seawater that convectively transports heat

TABLE 13.3 General Differences Between Fast-Spreading Ridges ($> \sim 5$ cm/a) and Slow-Spreading Ridges

Fast-Spreading Ridge	Slow-Spreading Ridge
Ophiolite example: Semail (Oman)	Ophiolite example: Troodos (Cyprus)
Axial magma chambers are more steady state, volcanism more frequent	Axial magma chambers are more ephemeral and scattered, volcanism less frequent
Smoother flanks (less faulted)	Rougher flanks (highly faulted)
Symmetric and less tectonically disrupted	Commonly asymmetric, more listric faulting and low-angle detachments. Crustal layering is less uniform.
Ridge typically higher (shallower)	Ridge typically lower (deeper)
Longer tectonic and magmatic segments	Shorter tectonic and magmatic segments
Narrow axial rise with small axial trough	Deep discontinuous axial valleys with uplifted flanks
Wider low seismic velocity (partial melt) zone	Narrower low seismic velocity zone, melt lens rare
Narrow axial neovolcanic zone	Wider and more irregular axial neovolcanic zone with more distributed local sources → hills, seamounts
Thinner lithosphere (higher heat flow)	Thicker lithosphere (lower heat flow)
Thicker, more uniform crust	Thinner, less uniform crust
Extensive sheet lava flows	Pillow lavas dominate extrusives
Slightly more evolved magmas (avg. Mg# = 52.8).	Slightly less evolved magmas (avg. Mg# = 57.1).
Less compositional diversity within areas	More compositional diversity within areas
Mantle upwelling more "two dimensional"	Mantle upwelling more "three dimensional"
Commonly exhibit "global" magmatic trends of Klein and Langmuir (1987, 1989).	Commonly exhibit "local" magmatic trends of Klein and Langmuir (1987, 1989).

From Solomon and Toomey (1992), Mutter and Karson (1992), Small (1998), Perfit and Chadwick (1998), Karson (1998), Macdonald (1998), Thy and Dilek (2000), Karson (2002)

upward (a process far more efficient than conductive cooling). More robust magmatism along fast-spreading ridges is sufficient to sustain axial lenses/sills over longer time spans, and magmatic accretion onto the trailing edges of the separating plates overshadows tectonic processes. As a result, the crust is thicker (Figure 13.15), and the flanks are relatively smooth (Figure 13.23). Magma is also better focused to the shallow axial area at fast-spreading ridges, resulting in a narrower axial rise, where a small axial trough may be related to dike-induced grabens or collapse over lava drainback events (Perfit and Chadwick, 1998). Segments along fast-spreading ridges are longer, and the crust and mantle are warmer and less dense, resulting in fewer earthquakes, thinner rigid lithosphere, and a higher (shallower) ridge.

At slow-spreading ridges, the magmatism is more episodic, occurring perhaps as ephemeral magma chambers distributed farther from the axis, which completely solidify, allowing cooler extensional regimes between magmatic pulses. *Tectonic* extension thus becomes more important, leading to rough topography (Figure 13.23) with asymmetric listric faulting (allowing even more efficient hydrothermal cooling) and low-dip *detachment surfaces* between layers with different rheological properties (Mutter and Karson, 1992; Karson, 1998), similar to continental extension zones. Blockfault rotation and detachments disrupt the usual ophiolite sequence and expose serpentinized peridotites and gabbros on the seafloor, even in the axial area (Figure 13.24). DSDP drill core 735B *began* in Layer 3 in one such exposed area on an ultra-slow segment of the South West Indian Ocean Ridge (Robinson et al., 2000). Escarpments along normal faults and transforms, and exposed detachment surfaces (called "oceanic core complexes" by some investigators) provide "tectonic windows" into the deeper oceanic layers, from which several investigators (e.g., Karson, 1998, 2002) have noticed that the internal structure of oceanic lithosphere, particularly at slow-spreading ridges, is considerably more complex than the classical ophiolite layering of Figure 13.4. Major differences include missing ophiolite units (e.g., the sheeted dike complex is often missing in tectonic windows at slow-spreading ridges) and discontinuous non-horizontal lithologic contacts offset by complex structures. In several well-studied oceanic areas volcanic units appear to lie directly on variably deformed metagabbros or serpentinized peridotites. The remarkably uniform seismic layering of the oceanic crust seems all the more amazing in light of such variable structure and lithology. Either the tectonic windows and deeper drill cores are not representative of the oceanic lithosphere as a whole (windows are much more prevalent near major segment discontinuities), or the seismic layering is due largely to fracturing, hydrothermal mineralization, and other alteration/metamorphic processes than to original lithologic layering, a conclusion that agrees with the previously noted thinning of Layer 2 and thickening of Layer 3 away from ridge axes. Robinson et al. (2000) claimed that the gabbro–ultramafic boundary on the western edge of the Atlantis Bank near drill hole 735B is at a depth of ~2.5 km, yet the seismic Moho in the area is at a depth of ~6 km, leading them to infer that serpentinization of the uppermost mantle lowered seismic velocities, thereby depressing the Moho to a level *within* the ultramafics. Minshull et al. (1998) concluded that serpentinization of normal oceanic lithosphere does not extend deeper than 5 km, however (although H_2O might penetrate deeper along localized fractures under special circumstances), so the notion of the Moho representing a serpentinization front remains in question.

The increased complexity noted above and the greater variability observed with continued work in ophiolites has led to classification schemes of various ophiolite types. Nicolas and Boudier (2003) classified ophiolites into *harzburgite types* (most similar to Figure 13.4), *lherzolite types* (much thinner and lacking well-organized sheeted dikes), and *intermediate harzburgite lherzolite types*. Dilek (2003) went a

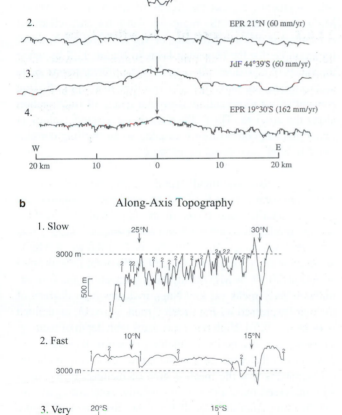

FIGURE 13.23 **(a)** Across-axis bathymetric profiles of selected ridge segments of the Mid-Atlantic Ridge, the East Pacific Rise, and the Juan de Fuca Ridge with different spreading rates. Slow ridges have rough topography, more normal faulting, and pronounced axial grabens. From Perfit and Chadwick (1998). **(b)** Along-axis profiles for slow, fast, and very fast spreading ridges. Numbers refer to segment discontinuity orders. Segments are highest in the center and deepest at discontinuities, probably reflecting mantle upwelling at segment centers. Faster ridges are higher and smoother, having longer segments with low-amplitude depressions at segment offsets. From Macdonald (1998). © American Geophysical Union with permission.

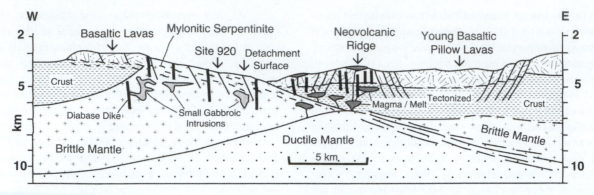

FIGURE 13.24 Interpretive cross section across the slow-spreading Mid-Atlantic Ridge near the Kane fracture zone. Tectonic extension results in a series of normal faults and exhumation along a shallow-dipping detachment surface, producing a disrupted and distinctly asymmetric architecture. From Thy and Dilek (2000).

step further, proposing a seven-fold ophiolite classification, each with its own "type locality" ophiolite and generic model.

Lavas at slow-spreading ridges are generally less differentiated than at fast ridges (Figure 13.9), but show complex chemical trends that can be ascribed to polybaric fractionation and/or plagioclase accumulation and reaction in separate rising magma blobs (Flower, 1980; Grove et al., 1992). If steady-state liquid lenses do not exist, magmas entering the axial area are more likely to erupt directly to the surface (and are thus more primitive), with some mixing of mush. Faster ridges with more persistent liquid chambers will, on average, undergo more advanced fractional crystallization in the longer-lived liquid portion.

Mantle upwelling appears to be more "two-dimensional" along fast-spreading ridges, meaning that there is more of a continuous vertical sheet of rising mantle under the spreading axis, leading to longer, more continuous volcanic centers and less along-axis variation. Thus any 2-D cross-section would be representative. Upwelling along slow-spreading ridges tends to separate more into individual diapirs ("three-dimensional" rise) that lead to spaced ridge volcanic segment centers. Mantle flow may be more buoyancy-driven along slow-spreading ridges (rather than a strictly passive rise in response to plate separation), which may tend to localize the upwellings.

The intermittent magma bodies at slow-spreading ridges may also remain separate more than those at fast-spreading ridges (as indicated by the multiple bodies in Figure 13.23), so that fractional crystallization will reflect closed-system behavior with less frequent recharge, as compared to the open-system behavior that dominates the more enduring reservoirs beneath fast ridges. Recharge and fractionation will thus be less advanced beneath slow ridges, which would explain the less differentiated character of slow-ridge magmas (Figure 13.9), as well as the more uniform compositions. Most of these magmas do not reach the surface, however, and the nature of the oceanic crust at slow ridges should be heterogeneous.

Klein and Langmuir (1987, 1989) and Langmuir et al. (1992) quantitatively analyzed extensive MORB chemical data sets to model partial melting and fractionation of mantle material beneath the mid-ocean ridge system. To avoid the effects of shallow magma chamber fractionation (presumably the latest process to affect magma composition) and focus on earlier genetic processes, they compared various chemical constituents adjusted to a constant degree of differentiation. They chose wt. % MgO as their differentiation index, and a value of 8.0 wt. % MgO as the standard. Their method was to use Fenner-type variation diagrams for series of magmas from ridge segments, regress the trends for each element versus MgO, and determine the regressed value for each chemical constituent at this standard value of MgO. This could be done, for example, for the Hawaiian lavas in Figure 11.2 by using the values of all oxides where their best-fit linear trends intersect a vertical line at 8.0% MgO, resulting in a series of oxide concentrations representing a singular extent of fractionation along the abscissa. Their "fractionation-corrected" values of major element compositions revealed some startling correlations. In brief, their conclusions were:

1. *Global* chemical trends represent averages for ~100-km long segments of ridges, designed to smooth out local irregularities and focus on the large-scale trends. One strong correlation found by Klein and Langmuir (1987) was between low-average $Na_{8.0}$ (wt. % Na_2O at 8.0 wt. % MgO) and high-average $Fe_{8.0}$ (also called $FeOT_8$: total Fe expressed as FeO at 8.0 wt. % MgO). Na_2O is incompatible and concentrates in early melts, so low $Na_{8.0}$ indicates high degrees of melting [expressed as $F = melt/(melt + rock)$, as defined in Equation (9.5)]. High $Fe_{8.0}$ correlates with depth of melting, as found by many peridotite melting experiments. They concluded from the correlation in these broadly averaged chemical signatures that the mean degree of melting, $\overline{F}$, increases with increased mean pressure of melting, both averaged over the "melting column." Recall from the discussion of Figure 13.15 that melts are generated in the shallow mantle as deeper material ascends adiabatically across the melting interval between the solidus and liquidus of Figure 10.4, equivalent to the triangular melting region in Figure 13.15. The melting column is thus the column of material along any flow line within this interval. The correlation that constitutes the global trend is a logical implication of the process of polybaric increments of fractional melting within this column. Hotter mantle (Figure 13.15a) results in earlier (deeper) initial melting. Because the melting column extends to greater depth, the mean pressure of melting (average for melt increments segregating across the column) is greater. This correlates with greater $\overline{F}$ because melting occurs across a larger range of pressures (the hotter mantle flow lines in Figure 13.15a traverse a larger triangular melting region than do the cooler mantle flow lines in the smaller triangle of part b). Langmuir et al. (1992) therefore concluded that *the global correlations are controlled by differences in thermal regime between major mantle segments* much more than by differences in mantle composition.

2. Because of correlations between the global chemical trend and physical ridge properties, Langmuir et al. (1992) concluded that the *thermal regime beneath a ridge segment exerts a major control on the quantity and composition of MORBs, and hence on crustal thickness and depth below sea level of the ridge axis* (all on very broad averages).

3. The fact that a correlatable mean pressure of partial melting signal can be extracted from the averaged MORB chemical composition requires that melts can be released at depth without re-equilibrating at low pressure as they rise. Low-*P* re-equilibration would destroy any high-*P* chemical characteristics. This supports *a process of fractional melting in which incremental melt blobs escape from the mantle matrix rapidly and efficiently as soon as they reach some small fraction sufficient to permit extraction,* as concluded previously. Equilibrium melting (in which melts remain resident and remain in equilibrium with the host mantle) does not produce the observed trends. *Different batches of melt, however, can easily mix at a variety of depths prior to eruption.*

4. *Local* trends diverge from the broadly averaged global trends. For example, when individual batch samples along a smaller ridge segment are compared, particularly for slow-spreading ridges, *F* varies *inversely* with the pressure of melting (in a manner opposite to the global trend). This indicates that individual melt batches occur as small degrees of melting at great depth (where melting is just beginning) and tend to increase in *F* as they rise. Langmuir et al. (1992) were unable to quantitatively model local trends on slow-spreading ridges with well-constrained fractional crystallization models and thus concluded that the *local trends must be controlled by mantle melting processes.* The correlation of global versus local trends with spreading rate is clear from Figure 13.25. Local trends associated with the fast-spreading East Pacific Rise differ from slow-spreading ridges, and are best modeled as due to small-scale mantle heterogeneities associated with an enriched component that is independent of depth, *indicating enriched veins, slivers, or pods distributed throughout the mantle.* Fast-spreading ridges also exhibit less temperature variation, which probably correlates with more efficient mantle processing. Slow-spreading ridges are less efficient, and thus more heterogeneous in terms of topography, morphology, and chemical variability as well. Niu and Batiza (1993) suggested that the local trends described above may be related to less efficient melt separation in rising mantle diapirs beneath slow-spreading ridges, so that melts continue to react with the matrix, obscuring higher-pressure signatures. If these melts periodically segregated, an ensemble of melts resembling the local trend would result. Deep melts at small F would be produced and escape early, whereas continued ascent and melting might release shallow melts with higher *F*. Kinzler and Grove (1992b) proposed that mixing of low-pressure fractionally crystallized magmas combined with

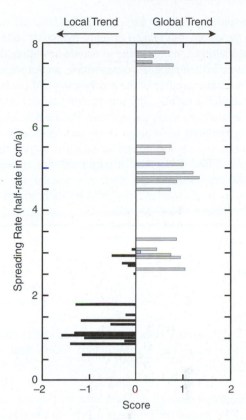

FIGURE 13.25 Geochemical systematics of Klein and Langmuir (1987, 1989), using the global trend versus local trend scoring system of Niu and Batiza (1993). Global trends predominate at spreading (half) rates greater than 5 cm/a, whereas local trends are more apparent at lesser rates. After Niu and Batiza (1993) and Phipps Morgan et al. (1994).

replenishment of more primary magmas (to reset MgO) may also produce the observed local trends.

5. Hotspots play an important role in the chemical composition of MORBs. This is particularly true for the slow-spreading Mid-Atlantic Ridge system, where more enriched E-MORB plume components correlate with ridge platforms, such as Iceland and the Azores. Enriched components are more dispersed and ubiquitous in the fast East Pacific Rise system.

Seismic measurements on the East Pacific Rise indicate that larger partially molten regions beneath the axis correlate with the swollen tectonic segment centers along the rise. Magmatic and tectonic segmentation are therefore linked and must in some way be related to mantle and melt temperatures, flow, and upwelling patterns. Macdonald et al. (1988, 1992) argued that first-order tectonic segments reflect separate large-scale asthenospheric upwelling cells, with second- and third-order segments related to periodic melt segregations from those cells that are spaced along the ridge ("three-dimensional" rise) and preferentially supply segment centers. Fourth-order segments they related to along-axis pinching and swelling of sub-axial magma accumulations as melt migrates from the centers (Figure 13.21). Lavas near lower-order segment centers tend to have higher MgO content, indicating that hotter and more primitive melt is focused there and differentiates somewhat as it propagates toward the segment ends. Batiza (2003) similarly proposed that magmatic segment patterns reflect a hierarchy in which mantle upwelling plays a more fundamental role than melting processes. He related primary magmatic segments to different melting conditions due to mantle upwelling domains with distinctive temperature, composition, and/or mineralogy.

Our understanding of the processes at mid-ocean ridges is now changing rapidly, and this is a particularly interesting time to be studying these phenomena. We are now aware that there are multiple upwellings from various mantle sources (including at least a normal and an enriched source) that have undergone differing degrees of partial melting, mixing, and fractional crystallization as they rise toward the surface. These melts are focused into the axial area beneath mid-ocean ridges. Here many of them mix further and undergo fractional crystallization in a liquid lens or crystal mush boundary layer before some of the liquid reaches the surface, whereas the rest crystallizes in the chamber(s). Ridge segments experience a complex interplay between magmatic and tectonic processes. Direct observation of exposed deeper portions of oceanic lithosphere at "tectonic windows" along detachment surfaces reveals far more complexity than the simple seismic layers and the layering of the ideal ophiolite sequence. More carefully collected samples and chemical data from ophiolites and ridges of various spreading rates will be necessary before a consistent model emerges that explains all of the variables.

Summary

The mid-ocean ridge system is a globe-encircling submarine mountain range centered on mature divergent plate boundaries. The popular consensus is that plate separation results principally from the negative buoyancy of dense oceanic lithosphere, and the mantle rises passively in response to plate subduction and divergence. Rising mantle retains heat, and the pressure reduction can result in (decompression) partial melting. The extent of mantle melting depends on the temperature of the initial rising mantle and how long the flowing mantle remains in the partial melting region beneath the ridge before flowing laterally into regions cooled effectively by convective circulation of seawater through the fractured crust. Upwelling normally originates in the isotopic and incompatible-element depleted upper mantle (probably above the 660 km seismic discontinuity), but *melting* generally occurs at depths less than 60 to 80 km. Non-depleted (and possibly even enriched) mantle appears to originate in deeper plumes, perhaps as deep as the core/mantle boundary. Depleted mantle melts at ridges produce normal or N-MORBs, and enriched mantle melts produce E-MORBs (also called P-MORBs). Partial melting beneath ridges is a process in which small melt increments (probably <1%) separate at various depths from the rising, melting source, and are focused toward a narrow zone beneath the ridge axis. These melts rarely maintain equilibrium with the mantle host and must therefore travel along some sort of semi-isolated conduits. The contrasting melts then accumulate and mix to varying extents in shallow magma chambers to form complex hybridized mixtures, ranging from N-MORB to E-MORB, including transitional T-MORB. Crystal fractionation in shallow chambers produces chemical trends on Fenner-Harker-type diagrams that indicate over 60% crystallization in some erupted lavas.

The shallow ridge-axis magma chambers release dikes upward into the cool roof, and the dikes in turn feed basalt sheet flows and pillow lavas onto the seabed. The chamber itself crystallizes along the floor and walls to produce gabbros and ultramafics. The sequence of abyssal sediments, basalts, sheeted dikes, gabbros, and ultramafics comprise the main layers of the oceanic lithosphere and of ophiolites, which are now considered to be slabs and slivers of oceanic lithosphere tectonically disrupted and emplaced into orogenic belts. The classical concept of sub-ridge-axis magma chambers as large, persistent bodies has recently been replaced by one in which thin melt lenses wax and wane near the top of a larger partially molten crystal mush.

Ridges are tectonically segmented on a number of scales and are offset by transforms, OSCs, or devals. Major segments seem to have a separate mantle upwelling and magmatic source near the segment center, and the magma then propagates parallel to the axis toward the segment ends. Lower-order segments have less contrasting origins.

Several competing models presently attempt to relate the small magma lens to the ophiolite layers (particularly the gabbros). The thermal structure of a ridge reflects a balance between magmatic heat input and hydrothermal cooling, and ridge segments differ in character, depending largely on the rate of spreading. Fast-spreading ridges are hotter, with more persistent magmatism. Magmatic accretion onto the receding plate edge thus dominates over tectonism, resulting in smoother, higher ridges. Slow-spreading ridges are cooler and tectonic disruption is more pronounced, resulting in pervasively normal-faulted rough topography, a distinct axial graben, and even detachment faulting. The complex interactions of magmatism and tectonism suggest that the petrological units of oceanic lithosphere, particularly if created at slow-spreading ridges, are really more complex than the deceptively simple horizontal seismic layers and the generalized ophiolite sequence.

Key Terms

Mid-ocean ridge basalt (MORB) *244*
Chimney/black smoker *245*
Inner rift valley *246*
Tectonic segment *247*
Magmatic segment *000*

Summit caldera *247*
Fracture zone *247*
Transform fault *247*
OSC *247*
Deval *247*

Ophiolite *248*
Layer 1, 2A, 2B, 2C, 3, 4 *249*
Obduction *250*
N-MORB *254*
E-MORB/P-MORB *254*

Review Questions and Problems

Review Questions and Problems are located on the author's web page at the following address: **http://www.prenhall.com/winter**

Important "First Principle" Concepts

- Plate separation is approximated by a "pulling" model in which dense oceanic lithosphere is susceptible to sinking onto the less dense mantle below at subduction zones.
- Mantle upwelling at divergent boundaries is more of a passive response to plate separation than an active force that drives it.
- The *major element* composition of a partial melt reflects the source composition and the last depth at which the melt was in equilibrium with the source (a rising source, in the case of MORBs).
- The *trace element* and (particularly) the *isotopic* characteristics of a partial melt are more likely than are the major elements to reflect the nature of the ultimate source at the depth of origin.
- Crystal fractionation can further modify major and trace element concentrations.

- Much of the crystal fractionation effects can be compensated by determining the ideal value of various component concentrations for a specified Mg# (or other differentiation index), using Harker-type variation diagrams.
- Variations in MORB composition indicate an isotopically distinct depleted mantle (probably shallow) and enriched mantle (probably deeper).
- Spreading rate controls the thermal state of a ridge segment, which, in turn, affects several chemical, magmatic, tectonic, and morphological characteristics.
- Despite exhaustive searching, no large, persistent magma chambers have been detected in the most prolific igneous–tectonic province on Earth. Perhaps large magma-filled chambers are more rare in general than we have thought.

Suggested Further Readings

Basaltic Volcanism Study Project. (1981). *Basaltic Volcanism on the Terrestrial Planets*. Pergamon. New York. Sections 1.2.5 and 6.2.1.

Buck, W. R., P. T. Delaney, J. A. Karson, and Y. Lagabrielle (eds.). (1998). *Faulting and Magmatism at Mid-Ocean Ridges*. Monograph **106**. American Geophysical Union. Washington, DC.

Cann, J. R., H. Elderfield, and A. Laughton. (eds.). *Mid-Ocean Ridges: Dynamics of Processes Associated with Creation of New Oceanic Crust*. Cambridge University Press. Cambridge, UK.

Dilek, Y., E. M. Moores, D. Elthon, and A. Nicolas (eds.). (2000). *Ophiolites and Oceanic Crust: New Insights from Field Studies and the Ocean Drilling Program*. Special Paper **346**. Geological Society of America. Boulder, CO.

Dilek, Y., and S. Newcomb (eds.). (2003). *Ophiolite Concept and the Evolution of Geologic Thought*. Special Paper **373**. Geological Society of America. Boulder, CO.

Klein, E. M. and C. H. Langmuir. (1987). Global correlations of ocean ridge basalt chemistry with axial depths and crustal thickness. *J. Geophys. Res.*, **92**, 8089–8115.

Klein, E. M. and C. H. Langmuir. (1989). Local versus global variations in ocean ridge basalt composition: A reply. *J. Geophys. Res.*, **94**, 4241–4252.

Langmuir, C. H., E. M. Klein, and T. Plank. (1992). Petrological systematics of mid-ocean ridge basalts: Constraints on melt generation beneath mid-ocean ridges. In: *Mantle Flow and Melt Generation at Mid-Ocean Ridges* (eds. J. P. Morgan, D. K. Blackman, and J. M. Stintin). Monograph **71**. American Geophysical Union. Washington, DC. pp. 183–280.

MacLeod, C. J., P. A. Tyler, and C. L. Walker (eds.). (1996). *Tectonic, Magmatic, Hydrothermal and Biological Segmentation of Mid-Ocean Ridges*. Special Publication **118**. Geological Society. London.

Mills, R. A., and K. Harrison (eds.). (1998). *Modern Ocean Floor Processes and the Geological Record*. Special Publication **148**. Geological Society. London.

Nicolas, A. (1989). *Structures of Ophiolites and Dynamics of Oceanic Lithosphere*. Kluwer Academic Publishers. Dordrecht, The Netherlands.

Nicolas, A. (1995). *The Mid-Oceanic Ridges*. Springer-Verlag. Berlin.

Parson, L. M., B. J. Murton, and P. Browning (eds.). (1992). *Ophiolites and Their Modern Analogs*. Special Publication **60**. Geological Society. London.

Phipps Morgan, J., D. K. Blackman, and J. M. Sinton (eds.). (1992). *Mantle Flow and Melt Generation at Mid-Ocean Ridges*. Monograph **71**. American Geophysical Union. Washington, DC.

Prichard, H. M., T. Alabaster, N. B. W. Harris, and C. R. Neary (eds.). (1993). *Magmatic Processes and Plate Tectonics*. Geological Society Special Publication **76**. Blackwell. Oxford, UK.

Ryan, M. P. (ed.). (1994). *Magmatic Systems*. Academic Press. San Diego.

Saunders, A. D., and M. J. Norry (eds.). (1989). *Magmatism in the Ocean Basins*. Geological Society Special Publication **42**. Blackwell. Oxford, UK.

Sinton, J. M. (ed.). (1989). *Evolution of Mid-Ocean Ridges*. Monograph **57** (IUGS vol. 8). American Geophysical Union. Washington, DC.

Thy, P., and Y. Dilek. (2003). Development of ophiolite perspectives on models of oceanic magma chambers beneath active spreading centers. In: *Ophiolite Concept and the Evolution of Geologic Thought* (eds. Y. Dilek and S. Newcomb). Special Paper **373**. Geological Society of America. Boulder, CO. pp. 187–226.

Wilson, M. (1989). *Igneous Petrogenesis: A Global Tectonic Approach*. Unwin Hyman. London. Chapter 5.

14

Oceanic Intraplate Volcanism

Questions to be Considered in this Chapter:

1. What is the nature of oceanic volcanism that occurs within plates and is thus not governed by plate tectonic (mid-ocean ridge or subduction zone) processes?
2. What variation does ocean island volcanism exhibit, and what is responsible for it?
3. What does the geochemical and isotopic variation in ocean island basalts tell us about the nature of the mantle?
4. What are hotspot plumes? Where do they originate, and what causes them?
5. What is the melting process in rising plumes?

14.1 INTRAPLATE VOLCANIC ACTIVITY

The processes responsible for intraplate activity are more enigmatic than those at plate margins because there are no obvious mechanisms we can tie to the plate tectonic paradigm. Mechanisms for generation of intraplate volcanics are poorly constrained by geophysical and field data, and the products are spatially scattered and chemically diverse. As with MORB, the dominant magma type for oceanic intraplate volcanism is basalt (commonly called **ocean island basalt (OIB)**), suggesting a mantle origin. As we might expect, intraplate volcanism is collectively much less voluminous than volcanism at plate margins. Schilling et al. (1978) estimated the volume ratio of MORB to OIB at 9:1. If we consider convergent margin igneous activity as well, the proportion of OIB falls to a few percent of all oceanic-related volcanism. Nonetheless, OIB volcanism is estimated at about 1.5 km^3/a, which is far from trivial.

The chemical character of oceanic intraplate volcanism is distinct from that of constructive or consumptive plate margins, permitting us to treat the products as a single petrogenetic province. The chemistry and petrography are highly variable, however, more so than for MORB. It is beyond the scope of this book to attempt a comprehensive survey of the varied products of oceanic intraplate volcanism. Space limitations dictate that we address the major themes by way of some well-studied and constrained examples, encompassing a reasonable compositional diversity of OIBs and their differentiation products. These examples serve as a basis for later speculation on petrogenetic processes and source regions of oceanic intraplate magmas.

The most familiar products of oceanic intraplate volcanism are the numerous islands that dot the world oceans. **Seamounts** (eroded or sunken islands or accumulations that never rose above sea level) also compose a significant (although poorly studied) proportion of the total intraplate igneous activity. Batiza (1982) estimated that there are between 22,000 and 55,000 seamounts dotting the ocean floor, of which only about 2000 are presently active or dormant. Seamounts appear to be concentrated along fracture zones, which supply convenient shallow conduits for magma rising toward the surface. There are also about 15 **oceanic plateaus** (Kerr, 2003): massive outpourings of basalt on the ocean floor, most similar to continental flood basalts on land (see Figure 15.1). We refer to all of the basalts from oceanic intraplate

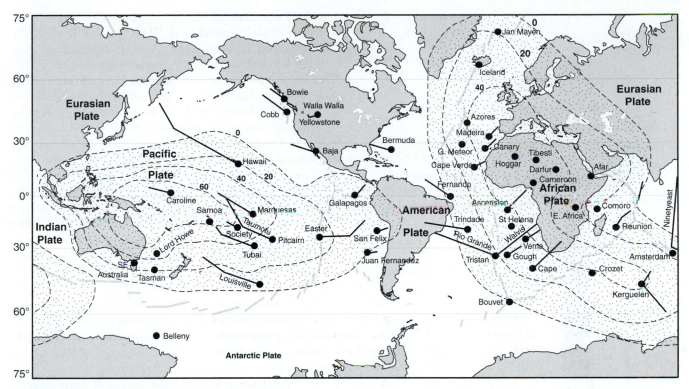

FIGURE 14.1 Map of relatively well-established hotspots and selected hotspot trails (island chains or aseismic ridges). Hotspots and trails from Crough (1983), with selected more recent hotspots from Anderson and Schramm (2005). Also shown are the geoid anomaly contours of Crough and Jurdy (1980, in meters). Note the preponderance of hotspots in the two major geoid highs (superswells). Walla Walla, incidentally, should not be confused with a hotspot of any kind.

settings as OIBs, regardless of whether the accumulations rise above sea-level, risking the occasional misnomer ("island") for the sake of simplicity.

Many (non-fracture-zone) seamounts and ocean islands define linear chains which show a progressive age relationship, commonly ending at an active volcanic island. Several of these chains are shown in Figure 14.1, perhaps the most famous of which is the Hawaiian-Emperor seamount chain. Island chains in the same plate follow subparallel paths and progress in age in the same direction, leading Wilson (1963) to conclude these chains were the result of volcanism generated by rising **plumes**. Most investigators believe that major plumes rise from a thermal boundary layer at the core–mantle boundary (see Nataf, 2000, and Thorne et al., 2004, for a summary of seismic evidence) and carry heat from the core (Section 1.7.3 and Figure 1.14). Figure 14.2 is a photograph of a newly formed laboratory thermal plume of dyed dense fluid initially ponded beneath an immiscible less-dense fluid. When heated from below, the denser fluid initially formed a stable thermal boundary layer across which heat could be transferred only by slow conduction. Upon further heating the lower fluid expanded and became sufficiently buoyant to rise through the overlying fluid as a plume. Because the plume meets resistance in the cooler and more viscous surroundings at its rising front, a flattened bulbous **plume head** forms, fed by continuing hot additions up the thinner conduit, or **tail**. The head spreads from the leading tip and entrains some of the surroundings in an eddy-like vortex

(Figure 14.2). Mantle plumes are thought to develop similarly. Hill et al. (1992) and Davies (2005) estimated that an incipient mantle plume head must reach a diameter of about 300 to 400 km before it has enough buoyancy to detach from the thermal boundary layer at the base of the mantle. As it rises, the head grows to a diameter of 800 to 1200 km. The arrival at the surface of the massive plume head leads to tremendous basaltic outpourings: continental flood basalts and oceanic plateaus. A general term for such huge outpourings is *large igneous province (LIP)*. Subsequent effusion from an established conduit tail (probably 50 to 100 km in diameter) marks the longer-lived **hotspots** (the surface expression of a plume). See Section 15.1 for more on LIPs and oceanic plateaus.

Estimates of the number of hotspots ranges from a conservative 16 (Morgan, 1971) to as many as 122 (Burke and Wilson, 1976). The more well-established hotspots identified by Crough (1983) and Anderson and Schramm (2005) are shown in Figure 14.1. Note that several hotspots occur at or near mid-ocean ridges (for example Iceland, Azores, Galapagos, Bouvet, and Tristan). These situations correspond to some of the ridge-associated E-MORB plumes discussed in Chapter 13. These plumes can be considered hybrids between true MORBs (N-MORBs) and OIBs. We shall see that their geochemical character reflects this dual (or intermediate) nature.

Hotspots are surrounded by broad elevated areas, called **swells**. The swell accompanying the Hawaiian chain, for example, is about 1 km high and 1000 km wide, far

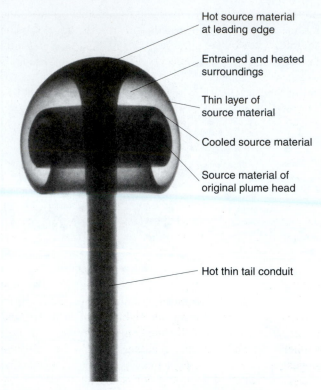

Hot source material
at leading edge

Entrained and heated
surroundings

Thin layer of
source material

Cooled source material

Source material of
original plume head

Hot thin tail conduit

FIGURE 14.2 Photograph of a laboratory thermal plume of heated dyed fluid rising buoyantly through a colorless fluid. Note the enlarged plume head, narrow plume tail, and vortex containing entrained colorless fluid of the surroundings. After Campbell (1998) and Griffiths and Campbell (1990).

larger than the narrow topography of the volcanoes themselves. Such swells may result from thickened oceanic crust, piling up of lava on top of the crust (and supported by the lithosphere), or from buoyant material thickening and thereby raising the lithosphere. Seismic surveys indicate normal crustal thicknesses, and the lithosphere is not strong enough to support a swell non-isostatically without sinking. The swell must therefore be supported by buoyant material at depth. This and the obvious volcanism are the strongest evidence supporting narrow columns of hot mantle (plumes) rising beneath hotspots. Geophysicists have used the swell sizes to estimate the heat flux of plumes, the sum of which is equivalent to estimates of heat loss from the core. Hence the contention in Section 1.7.3 that, whereas plate tectonics probably cools the mantle, major plumes originate at or near the D″ layer and cool the core. For detailed descriptions of plumes, see Wyllie (1988a), Sleep (1990, 1992), Griffiths and Campbell (1990, 1991), Hill et al. (1992), Gallagher and Hawkesworth (1994), White and McKenzie (1995), Davies (1999), and Condie (2001).

Plate motion over a plume gives the surface impression of hotspot migration and the production of linear volcanic chains. Active volcanic activity at the end of the chain marks the present position of the plume/hotspot. Hotspots were initially considered stationary, and the tracks were used to determine plate motions. More accurate surveys and correlations now suggest that hotspots move a little relative to one another (Basaltic Volcanism Study Project, 1981; Olson, 1987; Molnar and Stock, 1987; Müller et al., 1993;

Tarduno and Gee, 1995; Steinberger and O'Connell, 1998; Norton, 2000; Raymond et al., 2000; Tarduno et al, 2003). Most hotspots, however, move much less than plates do, consistent with their origin in the deep sub-asthenospheric mantle, where the viscosity is estimated to be two to three orders of magnitude higher than in the asthenosphere. Motion of hotspots is generally attributed to plumes being swept by convecting mantle flow (itself a response to plate motion, as described in Section 1.7.3). The models of Steinberger and O'Connell (1998) and Steinberger (2000) suggest that there is a tendency for plume conduits to be swept and inclined by a deeper mantle counterflow to plate motion (therefore toward mid-ocean ridges) or tilted from the core–mantle boundary toward large-scale upwellings. The two major upwellings of this type are beneath western Africa and the western Pacific, generally referred to as **superplumes**, or **superswells**, and are indicated by the broad elevated geoid features in Figure 14.1. Notice the concentration of hotspots within the superswells. These superswells may represent displaced deep mantle responding to subducting plates reaching great depth. Although hotspots aren't absolutely stationary, they are still useful as indicators of plate motion because they move much less than lithospheric plates.

Most familiar island chains consist of many separate islands and seamounts. They are separate because hotspot magma production is sporadic, and the shallow conduit is eventually abandoned as the plate carries it away from a deep hotspot source so that a new conduit must be established. Slower plate motion or more effusive magma production results in a more continuous volcanic ridge, called an **aseismic ridge**, which may or may not emerge locally above sea level to produce islands. Aseismic ridges are not to be confused with seismic mid-ocean ridges, as the former do not mark plate boundaries. The name similarity is unfortunate.

Some of the ridge-centered plumes leave V-shaped island/seamount chains with a limb on each side of the mid-ocean ridge as *both* plates are affected by the plume. For example, the Rio Grande and Walvis aseismic ridges stem from the Tristan hotspot in Figure 14.1. The Tristan plume, however, is now situated at the island of Tristan da Cunha, just east of the Mid-Atlantic Ridge, having been abandoned by the ridge as it migrated westward about 30 Ma ago. This ridge-associated creation and abandonment (described for Iceland in Chapter 13) may be true for a number of hotspots.

Note also the kink in a number of hotspot trails in Figure 14.1. The kink in the Pacific Hawaiian-Emperor and Tuamotu trails is duplicated by several smaller trails on the Pacific plate, and represents a change in plate direction that took place about 43 Ma ago.

Plumes can also rise beneath continents, causing a linear volcanic progression within continental areas. An example is the Snake River–Yellowstone hotspot (Figure 15.5), now located at Yellowstone National Park, Wyoming. Africa, probably because it is situated over a superswell, has several continental hotspots. Magmas from such hotspots commonly incorporate some of the low-melting-point continental crust, and have a much different volcanic character than oceanic occurrences (e.g., Chapter 15).

Alternatives to the popular theory that plumes originate from thermal perturbations at the core–mantle boundary include *compositional plumes* of less dense material (Anderson, 1975) and *volatile influx*, causing melting point reduction and rise of the buoyant melts ("wetspots"). The more traditional thermal plumes may originate at any thermal boundary layer, which, in addition to the obvious D″ layer, may include the 660-km transition or perhaps other mantle seismic boundaries (Section 14.6). Although the plume theory is very attractive, at present it is a challenge to accurately characterize plumes geophysically or in terms of temperature, size, rate of ascent, or initiating mechanism. In addition, not all oceanic intraplate volcanism is associated with an obvious hotspot. Several localized occurrences appear to be transient, and are not associated with island or seamount chains (e.g., Batiza, 1982). Many of these were probably erupted as off-axis volcanoes in near-ridge environments (Chapter 13), and others are concentrated near fracture zones and other ridge segment offsets, but some appear to be isolated (although ephemeral) events without adequate explanation.

14.2 TYPES OF OIB MAGMAS

Two principal magma series result from ocean intraplate volcanism. The more common type is a **tholeiitic** series, with a parental **ocean island tholeiitic (OIT)** basalt. This basalt is similar to MORB, but we shall see some distinct chemical and mineralogical differences. There is also a subordinate **alkaline** series (with parental **ocean island alkaline basalt (OIA)**). Hopefully you will recognize these two parental magma types from our discussion in Chapter 10 of how to create both by partially melting the mantle. The predominant characteristics of the two basalt types are compared in Table 10.1. There also appears to be two principal alkaline sub-series: one is silica undersaturated, and a less abundant series is slightly silica oversaturated. Modern volcanic activity at some islands is predominantly tholeiitic (e.g., Hawaii and Réunion), whereas other islands are more alkaline in character (e.g., Tahiti in the Pacific and a concentration of islands in the Atlantic, including the Canary Islands, the Azores, Ascension, Tristan da Cunha, and Gough).

The Hawaiian Islands are the best studied of all the ocean islands, and they present us with some interesting material for consideration. Hawaii reveals a sequential, or cyclic, pattern in eruptive history. A cycle is believed to begin in a **pre-shield stage** with submarine eruptions of alkaline basalt and highly alkaline basanites followed by tholeiitic basalt. Loihi Seamount, the newest volcanic center growing from the seabed about 35 km southeast of the big island of Hawaii is presently at this stage (Moore et al., 1982; Garcia et al., 1993, 1995). The initial alkaline basalts probably represent low degrees of partial melting as a new plume conduit starts up. This stage is immediately followed by tremendous outpourings of tholeiitic basalts in what is called the **shield-building stage**. The island of Hawaii, the largest island and the one located at or near the hotspot, comprises five overlapping shields. Kilauea and Mauna Loa (the two huge shields nearest the hotspot in the southern and southeastern part of the island) are presently in this stage of development. The quantity of basalt is impressive. Mauna Loa represents about 40,000 km^3 of basalt, and, because it rises from the seabed 5000 m below sea level to over 4000 m above sea level, it is the *tallest* mountain on Earth (although not the *highest*). The early shield-building stage produces 98 to 99% of the total lava in Hawaii.

The remaining three shields on Hawaii (Mauna Kea, Hualalai, and Kohala) have moved on to the next stage of development, a **post-shield stage** that typically follows caldera collapse. This stage is characterized by waning activity that is more alkaline, episodic, and violent. The lavas are also more diverse, with shallow fractionation producing rocks ranging from hawaiites to trachytes. This activity eventually fades, and, following a long period of dormancy (0.5 to 2.5 Ma), a late, **post-erosional** stage takes place. This stage is characterized by highly alkaline and silica-undersaturated magmas, including alkali basalts, basanites, nephelinites, and nepheline melilites. The two late alkaline stages represent ~1% of the total lava output. The alkaline stages are noted for the variety of **xenoliths** brought to the surface. Some are mafic and ultramafic cumulates from magma chambers of the early tholeiitic stage, whereas others are mantle materials representing various stages of depletion by melt extraction (including, perhaps, pristine mantle types). The samples include dunites, harzburgites, spinel lherzolites, and rare garnet lherzolites and garnet pyroxenites. These latter are high-pressure assemblages, which appear to have equilibrated at depths of 60 to 80 km.

The extent to which the Hawaiian sequence can be considered a general pattern of ocean intraplate volcanism is not known. Nor is it possible to evaluate the relative volumes of tholeiitic and alkaline rocks in most islands due to the limited exposure above sea level. A number of islands appear to evolve toward increased alkalinity and decreased output with time. If the Hawaiian stages are indeed universal, then the submerged portions of the many alkaline islands must be more tholeiitic, a proposition that has not been tested.

The Hawaiian pattern encompasses all three major OIB magma series in a single geographic occurrence. This pattern of early and voluminous tholeiites giving way to later, less extensive alkaline magmas has been traditionally related to (1) decreasing partial melting of the mantle (Figure 10.9) as the heat productivity wanes or (2) tholeiitic to alkaline evolution accomplished by fractional crystallization in a magma chamber (Figure 10.10). This latter proposition is difficult to reconcile with the overlap in Mg# and separate differentiation trends of the tholeiitic and alkaline lavas, as well as the low-pressure thermal divide separating the two series (Figure 8.13). It is still possible to evolve from OIT to OIA at high pressures, however, but we shall have to postpone such speculation until we have a better idea of the nature of the series involved, an idea greatly aided by trace element and isotope data.

14.3 OIB PETROGRAPHY AND MAJOR ELEMENT GEOCHEMISTRY

Let's review for a moment. We have learned that major element chemical composition is directly related to the mineralogy (or potential mineralogy) of magmas. It is thus of great use for characterizing magmas in a descriptive sense, in addition to its interpretive value. For *primary* magmas, the major element chemistry is related to the mineralogy of the source and the degree of partial melting (as well as the pressure at which this takes place). The rising melt may remain in equilibrium with the solid mantle, exchanging elements with it in accordance with the equilibrium constant (K), so that the residual mantle mineralogy at the depth of *segregation* (and not necessarily that of original melting) may leave a greater impression on the major elements. The pattern of major elements for derivative liquids is also affected by (or determines) the phases that crystallize and fractionate along the liquid line of descent. As a result of the continuous re-equilibration, major element chemical composition may be poor at constraining the ultimate source of initial melting, for which we turn to incompatible trace elements and isotopes.

OITs are similar to MORBs, but they do have some distinguishing characteristics. Table 14.1 lists some analyses of Hawaiian tholeiites. For the same Mg# OITs typically have higher K_2O, TiO_2, and P_2O_5, and lower Al_2O_3 than MORBs (compare to Table 13.2). There is more overlap in the other major elements. Magma types range from silica-saturated to slightly undersaturated olivine tholeiites to picrites. Olivine (Fo_{70-90}) is the dominant phenocryst phase in OITs, with Cr-spinel subordinate. Much of the evolution of tholeiitic series in OIBs can be modeled by fractional crystallization of olivine alone. The linear pattern for a Hawaiian tholeiite in Figure 11.2, for example, is different from the curved and kinked variation diagrams for MORBs (Figure 13.6) where plagioclase and clinopyroxene crystallize as well.

Plagioclase and clinopyroxene are usually only groundmass phases in OITs. Some clinopyroxene megacrysts are found, but these are more Mg-rich than the groundmass clinopyroxenes and are not in equilibrium with the liquid at low pressure. They probably represent cumulate minerals or phenocrysts from greater depths. Early olivine and Cr-spinel are eventually joined by later plagioclase and clinopyroxene in more evolved liquids. The order may be either plagioclase or clinopyroxene first because the composition of OIT is such that olivine fractionation drives the liquid composition toward the olivine–clinopyroxene–plagioclase minimum (similar to point M in the simplified Fo-Di-An system; see Figure 7.2). Small compositional differences in the initial liquid cause the liquid to reach the cotectic on either the clinopyroxene side or the plagioclase side of this point. The last phase to form is a Fe-Ti oxide. Remember, it is the late crystallization of a Fe-Ti oxide that permits the early Fe-enrichment so characteristic of tholeiite series (the Skaergård trend in Figure 8.3). This results from the removal of Mg-rich olivine. When the Fe-Ti oxide finally does crystallize, the liquid trend curves back from Fe-enriched derivative liquids and the alkali enrichment dominates. Some quartz-normative tholeiites crystallize a

TABLE 14.1   Analyses of Selected Hawaiian Tholeiites

Oxide	1	2	3
SiO_2	49.4	49.2	49.4
TiO_2	2.50	2.57	2.47
Al_2O_3	13.9	12.8	13.0
Fe_2O_3	3.03	1.50	2.32
FeO	8.53	10.1	9.16
MnO	0.16	0.17	0.18
MgO	8.44	10.0	9.79
CaO	10.3	10.8	10.2
Na_2O	2.13	2.12	2.24
K_2O	0.38	0.51	0.47
P_2O_5	0.26	0.25	0.24
Total	99.03	100.02	99.47
Mg#	57	61	61
Norm			
q	2.1	0.0	0.0
or	2.3	3.0	2.8
ab	18.2	18.0	19.1
an	27.6	23.8	24.1
di	18.0	22.6	20.2
hy	22.0	18.7	23.1
ol	0.0	6.2	2.1
mt	4.4	2.2	3.4
il	4.8	4.9	4.7
ap	0.6	0.6	0.6

1 = Average for all Hawaiian tholeiites (Macdonald, 1968).

2 = Most mafic glass, Kilauea summit (Murata and Richter, 1966a).

3 = Glassy tholeiite with 9% Ol, 1% Plag, 1% Aug (Moore, 1965).

Ca-poor pyroxene, either orthopyroxene (hypersthene) or pigeonite as a phenocryst phase (perhaps as a reaction rim on olivine) or in the groundmass. No amphibole or other hydrous phase forms.

The problem of identifying a primary OIT magma is similar to, only worse than, that for MORB. There are picrites with high Mg# (80) that would be in equilibrium with mantle phases at high pressure, but this high Mg content could be the result of olivine accumulation because no equivalent glasses have been found to substantiate a true picritic liquid. The most Mg-rich glasses contain about 10 wt. % MgO (Mg# about 62) and are even less primitive than the most primitive MORB glasses. They are thus not considered primary melts of primitive mantle. Clague et al. (1991) reported a glass from offshore of Kilauea with 15 wt. % MgO, but this is not considered primary either.

Because neither K, Ti, P, nor Al are included in olivine, the differences in these elements (and their ratios) between MORB and OIT cannot reflect shallow crystal fractionation. For example, the ratio of Al_2O_3/TiO_2 is close to 20 in MORB and only about 5 in OIT. This difference cannot be explained by fractional crystallization of olivine, the liquidus phase in both magmas at this early stage of development, and leads us to conclude there is a different source or generating process for the two types of oceanic tholeiites. Either OIT is a result of less extensive partial melting than MORB (Chapter 10), or the source is less depleted. We will return to this conjecture when we have had a chance to see the trace element and isotopic characteristics of OIBs.

As discussed in Section 8.7, alkaline basalts (OIAs) are characterized by higher alkali and lower silica content than tholeiites (see Figure 8.11). This should also be clear from a comparison of Tables 14.1 to 14.3 (accounting for the differences in Mg#). Although the alkaline series are highly variable compared to MORB and OIT, some fairly consistent mineralogical differences reflect the alkali/silica contrast. Because of the lower silica content, olivine is even more prevalent in OIAs, occurring in the groundmass as well as an ubiquitous phenocryst phase. Olivine also occurs over a broader range of the differentiated spectrum and has a greater compositional range (Fo_{35-90}) than it does in OIT. There is usually only one pyroxene in OIAs, a brownish Ti-rich augite. Amphibole is also an occasional phenocryst phase, indicating a higher volatile content. Due to the low-P curvature of dehydration curves in pressure–temperature space (e.g., Figure 10.6), hydrous phases such as amphiboles become unstable at low pressure and eruptive temperatures in excess of 1000°C, so groundmass amphiboles are rare, and phenocrysts may be resorbed, or develop reaction rims of fine anhydrous phases as the volatiles escape from the crystal perimeter (Figure 3.11c). The groundmass of alkali basalts usually contains all of the phenocryst phases plus an alkali feldspar as well as feldspathoids (such as nepheline, leucite, or sodalite).

As mentioned above, the composition and mineralogy of OIB is much more variable than that of MORB, and this is particularly true for the alkaline suites. A comprehensive survey is well beyond the scope (and intent) of this book, and we will content ourselves with a brief glimpse of two islands: Ascension and Tristan da Cunha. I chose these because each represents one of the two common types of alkali subseries, and available chemical data cover a broad range of compositions. Although both series are represented in Hawaii as well, it may benefit us to broaden our geographic perspective. For the location of these islands, see Figure 14.1.

TABLE 14.2 Representative Analyses and CIPW Norms of a Silica-Undersaturated Alkaline Series from Tristan da Cunha

Oxide	Basanite	Tephrite	Phon-Teph	Teph-Phon	Phonolite
SiO_2	42.8	46.0	49.4	54.6	60.2
TiO_2	4.14	3.41	3.19	1.75	0.80
Al_2O_3	14.3	17.1	18.5	19.5	20.2
Fe_2O_3	5.89	3.70	2.87	2.83	2.2
FeO	8.55	7.12	5.37	2.87	0.91
MnO	0.17	0.17	0.16	0.18	0.14
MgO	6.76	4.61	3.31	1.50	0.49
CaO	12.0	10.2	7.47	5.65	2.15
Na_2O	2.79	3.99	4.95	5.82	6.58
K_2O	2.06	3.02	3.68	4.85	6.23
P_2O_5	0.58	0.75	1.10	0.51	0.11
Total*	100.04	100.07	100.00	100.06	100.01
Mg#	46	44	42	32	22
Norm					
or	12.2	17.8	21.7	28.6	36.8
ab	6.7	11.4	24.3	31.1	42.0
an	20.4	19.8	17.4	12.8	7.2
ne	9.2	12.1	9.5	9.8	7.4
di	28.3	20.8	10.0	8.8	2.2
wo	0.0	0.0	0.0	0.3	0.0
ol	5.9	4.4	4.3	0.0	0.2
mt	8.2	5.4	4.2	4.1	1.1
il	7.9	6.5	6.1	3.3	1.5
hem	0.0	0.0	0.0	0.0	1.5
ap	1.3	1.7	2.5	1.2	0.3

Columns: 1 = Alkaline basalt; 2 = Average of 13 trachybasalts; 3 = Average of 3 Si-rich trachybasalts; 4 = Average of 9 trachyandesites; 5 = Average of 5 trachytes.

*Total recast to volatile-free.

Data from Baker et al. (1964).

TABLE 14.3 Representative Analyses and CIPW Norms
of a Silica-Oversaturated Alkaline Series from Ascension Island

Oxide	Basalt	Hawaiite	Benmoreite	Trachyte	Rhyolite
SiO_2	50.0	51.4	53.2	64.8	73.1
TiO_2	2.61	2.65	2.52	0.61	0.20
Al_2O_3	16.7	16.0	16.1	16.8	12.4
FeO	11.6	10.7	9.66	4.48	3.39
MnO	0.19	0.20	0.29	0.22	0.11
MgO	5.70	5.14	3.55	0.53	0.04
CaO	8.71	8.36	6.44	1.54	0.22
Na_2O	3.00	3.68	4.89	6.53	5.58
K_2O	1.16	1.35	1.91	3.63	4.70
P_2O_5	0.34	0.59	1.15	0.66	0.02
Total	100.01	100.01	99.67	99.82	99.75
Mg#	47	46	40	17	2
Norm					
q	1.3	1.6	1.5	9.6	24.4
or	6.8	7.9	11.3	21.4	27.8
ab	25.3	31.0	41.3	55.2	37.8
an	28.6	23.1	16.4	3.3	0.0
c	0.0	0.0	0.0	0.9	0.0
di	9.9	11.6	6.4	0.0	0.9
hy	16.4	12.4	9.9	3.7	2.9
ac	0.0	0.0	0.0	0.0	4.9
mt	5.9	6.0	5.8	3.1	0.0
il	4.9	5.0	4.8	1.2	0.4
ap	0.8	1.4	2.7	1.5	0.0
nms	0.0	0.0	0.0	0.0	0.9

Data from Harris (1983).

Tables 14.2 and 14.3 provide some representative analyses and CIPW norms for Tristan da Cunha and Ascension, respectively. Remember, the undersaturated series is far more common. Ascension (and the Azores) are well-known examples of the less common oversaturated type. The two alkaline series are plotted on an alkali–silica diagram in Figure 14.3, along with the Iceland OIT (or E-MORB) trend for comparison. Most of these alkaline rocks contain an alkali feldspar in addition to plagioclase. Note the significant difference in alkali/silica ratios between the two alkaline series in Figure 14.3. The highly alkaline and silica-undersaturated series from Tristan da Cunha evolves from alkaline basanites and alkaline basalts through tephrites and intermediates to phonolites. Most rocks are nepheline normative (Table 14.2) and many contain leucite. The silica-saturated Ascension series has a high $Na_2O:K_2O$ ratio and evolves from less alkaline basalts through hawaiites, mugearites, benmoreites to trachytes and eventually to quartz-bearing alkali rhyolites. The sudden drop in alkali content of the Ascension series in the rhyolite field is probably associated with fractional crystallization of alkali feldspar. The two alkaline trends correspond approximately to the two minima in the silica–nepheline–kalsilite system (see Figure 19.7). The path followed by a particular series is determined by initial compositional differences, such as silica content, alkalinity, or volatile content in the primary magma.

The great variety in alkalinity in ocean intraplate volcanics can be illustrated by the alkali/silica ratios presented in Table 14.4. These ratios are determined by least-squares linear regression of the alkali versus silica trends on variation diagrams for volcanic rock suites from each locality. For series containing rhyolites, analyses for the highly evolved samples showing alkali decreases were omitted in order to avoid the substantial effects on the average slope due to this relatively late process. The islands are listed in Table 14.4 in order of decreasing total alkalinity. I have included two tholeiitic series as well (Galapagos and Iceland) for comparison. These, of course, rank at the bottom of the list. The Na_2O/SiO_2 and K_2O/SiO_2 ratios are also listed separately for purposes of comparison. The slope of the line separating alkaline and subalkaline (including tholeiitic) series in Figure 8.11 is approximately 0.37. From Table 14.4, it should be abundantly clear that there is considerable

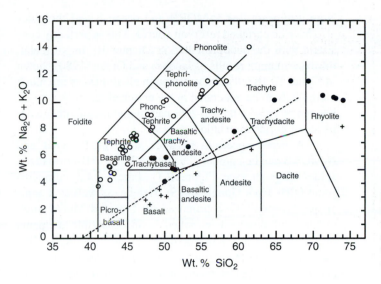

FIGURE 14.3 Alkali versus silica diagram of Cox et al. (1979), showing the differentiation trends for an ocean island tholeiitic series (Iceland), a silica-undersaturated ocean island alkaline series (Tristan da Cunha) and a slightly oversaturated OIA series (Ascension Island). After Wilson (1989). Copyright © by permission Kluwer Academic Publishers.

variability in the alkali content, both collectively and individually, for ocean intraplate volcanism.

Because alkalis are incompatible elements, only slightly affected by up to 50% shallow fractional crystallization, this again argues for distinct mantle sources or generating mechanisms. The variation in Na/K among the suites makes the former possibility much more likely and leads us to suspect that the mantle is more heterogeneous than we had previously thought.

From the major element chemical composition, we can conclude that OITs are distinct from MORBs and that the former are either a result of less extensive partial melting or melting of less depleted mantle, such as the depleted mantle reservoir deeper than 660 km. OIAs are also distinct, and heterogeneous, suggesting complex melting processes, a chemically heterogeneous mantle source, or both. We could continue to analyze the major element behavior of the various suites on variation diagrams, but I prefer that we move on to the most interesting aspect of the OIB story:

what they can tell us about the nature of the mantle. For this we must look at the trace element and isotope systems.

14.4 OIB TRACE ELEMENT GEOCHEMISTRY

The LIL trace elements (K, Rb, Cs, Ba, Pb^{2+}, and Sr) are incompatible (except for Sr and Ba in plagioclase), and are all enriched in OIB magmas with respect to MORBs. They can be used to evaluate the source composition, degree of partial melting (and residual phases), and subsequent fractional crystallization processes, although they may not always be able to discriminate well between them. As mentioned previously, the *ratios* of incompatible elements have been employed to distinguish between source reservoirs, and the K/Ba ratio is considered by many to be a particularly good source indicator. For N-MORBs K/Ba is high (usually >100), whereas for E-MORB it is in the mid-30s, OITs range from 25 to 40, and OIAs in the upper 20s. Thus all appear to have distinctive sources. HFS elements (Th, U, Ce, Zr, Hf, Nb, Ta, and Ti) are also incompatible and are enriched in OIBs over MORBs. Ratios of these elements have also been used to distinguish OIB mantle sources. The Zr/Nb ratio, for example, is generally high for N-MORB (>30) and low for OIB (<10). MORBs near ocean island plumes commonly show lower Zr/Nb and Y/Nb values that fit a mixing line between N-MORB and the adjacent plume OIB, implying that the two components are variably combined during ascent (LeRoex et al., 1983, 1985; Humphris et al., 1985). Compatible transition metals, such as Ni and Cr, are useful indicators of fractional crystallization of olivine and spinel, respectively. OIAs tend to be depleted in both relative to OITs and MORBs, which, along with the higher Mg#s, suggests they have experienced fractionation of these phases prior to eruption.

A number of REE curves for OIBs are illustrated in Figure 14.4. The bounding curves for N-MORB and E-MORB from Figure 13.11 are included for reference. Note that ocean island tholeiites (represented by the Kilauea and Mauna Loa samples) overlap with MORB and are not unlike E-MORB. The alkaline basalts have steeper slopes

TABLE 14.4 Alkali/Silica Ratios (Regression) for Selected Ocean Islands

Island	Alk/Silica	Na_2O/SiO_2	K_2O/SiO_2
Tahiti	0.86	0.54	0.32
Principe	0.86	0.52	0.34
Trinidade	0.83	0.47	0.35
Fernando de Noronha	0.74	0.42	0.33
Gough	0.74	0.30	0.44
St. Helena	0.56	0.34	0.22
Tristan da Cunha	0.46	0.24	0.22
Azores	0.45	0.24	0.21
Ascension	0.42	0.18	0.24
Canary Islands	0.41	0.22	0.19
Tenerife	0.41	0.20	0.21
Galapagos	0.25	0.12	0.13
Iceland	0.20	0.08	0.12

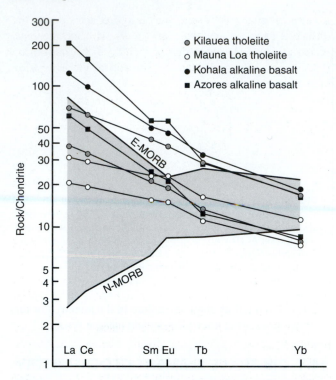

FIGURE 14.4 Chondrite-normalized REE diagram for Hawaiian tholeiites (light and gray circles), Hawaiian alkaline basalts (black circles), and Azores alkaline basalts (squares). Typical N-MORB and E-MORB are included for comparison, with the MORB interval shaded. After Wilson (1989). Copyright © by permission Kluwer Academic Publishers.

and greater LREE enrichment, although some fall within the upper MORB field.

The models for partial melting discussed in Chapter 9 showed that, if partial melting is more extensive than 10% ($F > 0.1$), there should be little inter-element fractionation of the REE (see Figure 9.4), and the pattern (the slope of La/Sm, La/Yb, or La/Ce ratios) should be similar to that of the source. Figure 14.4 thus suggests a heterogeneous source. La/Yb (the overall slope on the REE diagram) is crudely proportional to the degree of silica undersaturation in OIBs. Highly undersaturated magmas can have La/Yb in excess of 30, whereas OIA ratios are closer to 12, and OITs about 4. Note also that the *heavy* REEs are also fractionated in the OIB samples in Figure 14.4 (as compared to the flat HREE patterns in N- and E-MORB). This indicates that garnet was a residual phase because it is one of the few common minerals that differentially incorporates HREE. The implication is that these melts have segregated from the mantle at depths in excess of 60 km (Figure 10.2).

From the consistent negative slopes in Figure 14.4, we can deduce that E-MORBs and OIBs (OIAs and OITs) are distinct from N-MORBs (positive slope) and appear to originate in an enriched mantle reservoir, although very low degrees of partial melting may also produce LREE-enriched melts from a primordial or slightly depleted source. In Chapter 10, I argued that MORB tholeiites probably originated in the depleted upper mantle, and alkali basalts in an enriched mantle reservoir (perhaps in the lower mantle).

Now it appears that E-MORB and ocean island tholeiites also have an enriched reservoir source. This is perfectly consistent with the data presented in Chapter 10. In particular, the experimental results of Jacques and Green (1980) shown in Figure 10.18 tell us that, although alkali basalts are much more likely to be generated in the enriched reservoir, tholeiites can be created by 5 to 20% partial melting of either a depleted or enriched mantle source.

A broad spectrum of trace elements for OIB and N-MORB can be compared using the N-MORB-normalized spider diagram (see Figure 9.7). Figure 14.5 compares some OIBs from Gough and St. Helena, as well as a composite "average" OIB proposed by Sun and McDonough (1989). All three are enriched in incompatible elements over N-MORB (normalized values greater than one). They show the broad central hump in which both the LIL (Sr-Ba) and HFS (Yb-Th) element enrichments increase with increasing incompatibility (inward toward Ba and Th). This is the pattern we should expect in a sample enriched by some single-stage process (such as partial melting of a four-phase lherzolite) that preferentially concentrated incompatible elements. The hump pattern is regarded as typical of melts generated from relatively undepleted mantle in intraplate settings.

Because they shouldn't vary appreciably during modest ranges of partial melting and fractional crystallization, the ratios of a number of *similarly* incompatible trace elements have been used in place of La/Sm, La/Ce, etc. (REE slopes) to aid in the identification of sources. The K/Ba ratio was used above in this fashion. Hofmann (2003) suggested that variations in Th/U, Nb/U, Nb/La, Ba/Th, Sr/Nd, and Pb/Nd should reflect source differences more than melt extraction/evolution processes. Hofmann et al. (1986) noticed that the ratios of some highly incompatible trace elements are surprisingly uniform across a range of fresh ocean basaltic glasses, ranging from depleted N-MORBs to enriched OIBs. For example, Figure 14.6 illustrates the consistency of Nb/U over a substantial range of Nb concentrations.

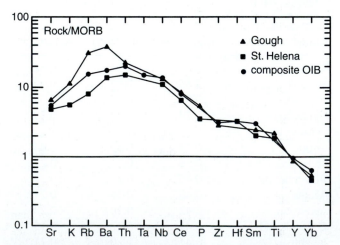

FIGURE 14.5 MORB-normalized multielement (spider) diagram for OIB magmas from Gough, St. Helena, and a "typical" OIB. Data from Sun and McDonough (1989).

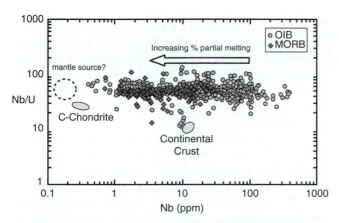

FIGURE 14.6 Nb/U ratios versus Nb concentration in fresh glasses of both MORBs and OIBs. The Nb/U ratio is impressively constant over a range of Nb concentrations spanning over three orders of magnitude (increasing enrichment should correlate with higher Nb). From Hofmann (2003). Chondrite and continental crust values from Hofmann et al. (1986).

It would appear that these two elements do not fractionate much with respect to each other during partial melting and subsequent crystal fractionation, and the ratio may thus be the same as in the mantle source, implying similar sources for MORBs and OIBs. Sims and DePaolo (1997) criticized this approach, arguing that it is somewhat circular and that constant ratios could result from a systematic relationship between partition coefficients and Nb/U source ratios, but such a relationship is also a bit fortuitous and the approach of Hofmann and colleagues is appealing. Perhaps even more striking in Figure 14.6 is the relationship between the Nb/U ratio of oceanic basalts and that of original primitive mantle (represented by C-type chondrites). The ratio for this obvious candidate for a melt source is lower than those of the partial melts, implying that the mantle must have evolved over time to become a source with higher Nb/U appropriate to the melts. Hofmann et al. (1986) argued that primitive mantle fractionated early in Earth history to produce continental crust, thereby yielding an enriched crust (higher Nb and lower Nb/U in Figure 14.6), and a complementary depleted mantle (lower Nb and higher Nb/U) from which modern MORBs and OIBs are derived (concentrating Nb and U, but not affecting the ratio). The complex process(es) by which continental crust is formed (Chapters 16–18) were presumably capable of affecting mantle Nb/U (and other ratios). The residual mantle was then rehomogenized over time, they speculate, yielding a more consistent mantle Nb/U ratio. Any subsequent mantle differentiation to depleted (MORB) and enriched (OIB) sources was considered sufficiently modest to leave Nb/U largely undisturbed. This reasoning makes the somewhat startling inference that *little, if any, primitive mantle presently exists* (no basalts have chondritic Nb/U).

Trace element data provide a clearer indication than the major elements of distinct source areas for the various oceanic magma types, but the data are still not conclusive. Differences in small degrees of partial melting, and perhaps deep-seated fractional crystallization of high-pressure phases,

such as garnet and pyroxene, may also produce some of the observed patterns. We thus turn to the isotopic systems, which do not fractionate during melting or crystallization processes, where a fascinating story behind the generation of oceanic magmas begins to emerge.

14.5 OIB ISOTOPE GEOCHEMISTRY

The isotope chemistry of OIBs is much more variable than shown for MORBs in Chapter 13. E-MORBs have recently been found to be more variable than originally thought as well, indicating that they tap a broader spectrum of mantle source material, a subject to be explored more fully in this chapter. The isotopic signature of melts believed to be derived directly from the mantle provides us with one of the best perspectives on the nature of the mantle itself. The isotopic variation shown in the ocean volcanics reveals that the mantle is far from a uniform reservoir, even in the simplest case of the sub-oceanic mantle.

Before we proceed further, let's discuss the effects of mixing. Suppose we have two distinct chemical reservoirs, A and B, that mix in varying amounts. The result would be a series of mixtures with compositions that vary along a linear array between the two mixed end-members (Figure 14.7a). The position of any single mixture depends on the relative proportions of the end-members in it. Figure 14.7 assumes some variability in each reservoir, so the mixtures define a band between them. This is the situation in the "mantle array" in Figure 10.16a for oceanic basalts. Melts derived from each reservoir apparently mix during transit to the surface, resulting in the mixed array. If mixing occurs between three reservoirs, mixtures will define a triangular field between them (as in Figure 14.7b), four-component mixtures define a tetrahedron, etc.

In order to characterize the complex mantle source areas of OIB melts, petrologists have turned to a broad range of isotopic systems. Because the behavior and partitioning of the parent/daughter trace elements leading to isotopic ratios of Pb, He, and Os differ from the familiar Rb/Sr and Sm/Nd systems, they provide additional information and constraints on the nature of the sub-oceanic mantle.

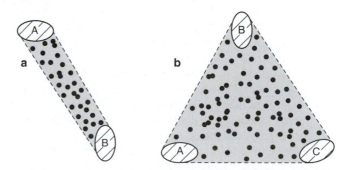

FIGURE 14.7 Idealized mixing among reservoirs. **(a)** Mixing of two components. Mixtures occur along a linear array between the two mixed end-members. **(b)** Mixing of three components. Mixtures occur in an area bounded by the end-members. In either case, the position of any mixture depends upon the relative quantities of the components being mixed.

14.5.1 Strontium and Neodymium Isotopes

Figure 14.8 illustrates the $^{143}Nd/^{144}Nd$ versus $^{87}Sr/^{86}Sr$ systematics for a spectrum of oceanic volcanism. I realize this is a rather busy diagram, but the point is not necessarily to discern Hawaii from the Azores but to notice the extensive variation and overall pattern in both $^{87}Sr/^{86}Sr$ and $^{143}Nd/^{144}Nd$ for the OIBs. The data for Figure 14.8 are largely from the comprehensive reviews by Zindler and Hart (1986) and Hart and Zindler (1989) and are expanded considerably over the material used in Figures 10.16 and 13.13. The "mantle array" from Figure 10.16 is included in Figure 14.8 for reference. MORBs are shaded and conform to the array, although E-MORBs (and particularly the Indian Ocean MORBs) extend the MORB data to more enriched values (low $^{143}Nd/^{144}Nd$ and high $^{87}Sr/^{86}Sr$). In Chapter 10, I proposed that the mantle array probably results from mixing of magmas derived from two principal mantle reservoirs, each with its own unique isotopic character. One reservoir was the upper, incompatible element depleted mantle, hypothesized to exist above the 660-km seismic transition. The other was a non-depleted or even enriched reservoir below the 660-km transition. The OIB data show us that this is a bit of an oversimplification, and Figure 14.8 requires mixing of material derived from at least three reservoirs. As we shall see, most mantle petrologists prefer at least *five* reservoirs.

The isotopic signatures of the most commonly cited mantle components are summarized in Table 14.5. The continental crust values in Table 14.5 should be viewed as least constrained because there are several possible sources and mantle mixing/transport processes (including sedimentary ones) that vary with crustal level and can differentially affect Rb/Sr, Nd/Sm, and U-Th/Pb.

The first major mantle reservoir is commonly called **DM** (for depleted mantle) or DMM (for depleted MORB mantle). This is the low-$^{87}Sr/^{86}Sr$ and high-$^{143}Nd/^{144}Nd$ reservoir end of the mantle array that is presumed to be the N-MORB source (Chapter 13). A second reservoir may be a non-depleted mantle, called **BSE** (bulk silicate Earth) by Zindler and Hart (1986) and primitive upper mantle (PUM) or primary uniform reservoir by others. BSE reflects the isotopic signature of the primitive mantle as it would evolve to the present without any subsequent fractionation of radioactive elements (which presumably were extracted to form continental and oceanic crust). Although a few oceanic basalts have this isotopic signature, it is rare, and the arguments of Hofmann et al. (1986) based on Nb/Ce and other incompatible element ratios (discussed above) suggest that little, if any, BSE still exists. The Sr-Nd isotopic data do not require that this reservoir now survives because it plots within the mixing area and has no distinctive end-member signature.

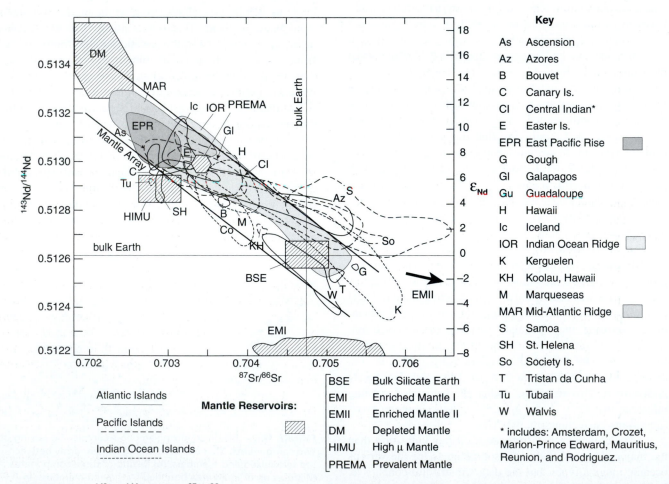

FIGURE 14.8 $^{143}Nd/^{144}Nd$ versus $^{87}Sr/^{86}Sr$ for ocean basalts, including MORB and OIB, after Zindler and Hart (1986), Staudigel et al. (1984), Hamelin et al. (1986), and Wilson (1989). Copyright © by permission Kluwer Academic Publishers.

TABLE 14.5 *Approximate Isotopic Ratios of Various Mantle Reservoirs*

Reservoir	$^{87}Sr/^{86}Sr$	$^{144}Nd/^{143}Nd$	$^{206}Pb/^{204}Pb$	$^{207}Pb/^{204}Pb$	$^{3}He/^{4}He$	$^{187}Os/^{186}Os$	$^{18}O/^{16}O$
End-Member Mantle							
DM	0.7015–0.7025	0.5133–0.5136	15.5–17.7	<15.45	7–9 R_A	0.123–0.126	low
HIMU	0.7025–0.7035	0.511–0.5121	21.2–21.7	15.8–15.9	2–6 R_A?	0.15	low
EMI	c 0.705	<0.5112	17.6–17.7	15.46–15.49	2–6 R_A?	0.152	low
EMII	>0.722	0.511–0.512	16.3–17.3	15.4–15.5	2–6 R_A?	0.156?	high
FOZO*	0.7030–0.7040	0.51280–0.51300	18.50–19.50	15.55–15.65	8–32 R_A	low	low
Other Mantle							
BSE	0.7052	0.51264	18.4	15.58	40–80 R_A?	0.129	low
PREMA	0.7033	<0.5128	18.2–18.5	15.4–15.5			
PHEM	0.7042–0.7052	0.51265–0.51280	18.4–19.0	15.5–15.6	>35 R_A	low	low
C	0.703–0.704	0.5128–0.5129	19.2–19.8	15.55–15.65	20–25+?	low	low
Continental Crust**	0.72–0.74	0.507–0.513	up to 28	up to 20	~0.1 R_A	high	high

*FOZO or some low-^{4}He reservoir, such as PHEM or C.

**Continental crust may be recycled back into the mantle via subduction.

DM, HIMU, EMI, EMII, and BSE from Rollinson (1993), pp. 233–236; PHEM from Farley et al. (1992); FOZO from Hauri et al. (1994); C from Hanan and Graham (1996); $^{187}Os/^{186}Os$ values from Shirey and Walker (1998) and van Keken et al. (2002a); and $^{18}O/^{16}O$ estimates based on Eiler (2001).

Because the Nd-Sr mantle array for OIBs extends beyond the primitive (BSE) values to truly enriched ratios, there must exist an enriched mantle reservoir capable of supplying such material. If such a reservoir exists, and it certainly appears that one does, BSE lies on the DM ↔ enriched reservoir mixing line, which is why it is not *required* as a mixing component. Zindler and Hart (1986) proposed *two* enriched reservoirs. Enriched mantle type I (**EMI**) has lower $^{87}Sr/^{86}Sr$ (near primordial), and enriched mantle type II (**EMII**) has much higher $^{87}Sr/^{86}Sr$ (>0.720), which is well above any reasonable indigenous mantle source. Both EM reservoirs have similar enriched (low) Nd ratios (<0.5124). Note that all of the Nd-Sr data in Figure 14.8 can be reconciled with mixing of *three* reservoirs—DM, EMI, and EMII—because the data are confined to a triangle with apices corresponding to these three components. The high Sr ratios in EMI and EMII require a high parental Rb content and a long time (>1 Ga) to produce the excess ^{87}Sr. This signature correlates well with continental crust (or sediments derived from it). Oceanic crust and sediment are other likely candidates for these reservoirs, but there is less Rb in the oceanic realm, and it would require more of that material to produce the same signature. The modest enrichment of EMI may be accomplished within the mantle, but EMII appears to require a crust/sediment contribution. The nature of the other reservoirs in Figure 14.8 will be discussed more fully below.

14.5.2 Lead Isotopes

In Chapter 9, we learned that Pb may be produced by the radioactive decay of uranium and thorium by the following three decay schemes:

$$^{238}U \rightarrow {}^{234}U \rightarrow {}^{206}Pb \qquad (9.22)$$

$$^{235}U \rightarrow {}^{207}Pb \qquad (9.23)$$

$$^{232}Th \rightarrow {}^{208}Pb \qquad (9.24)$$

All three of these elements are incompatible large-ion trace elements. As such, they tend to fractionate into the melt (or fluid) phase in the mantle, if available, and migrate upward, where they become incorporated in the oceanic or continental crust. Pb is scarce in the mantle, so Pb isotopic ratios for mantle-derived melts are particularly susceptible to contamination from U-Th-Pb-rich reservoirs, which can add a significant proportion to the mantle-derived Pb. As I described in Chapter 9, if you are given a dozen green jellybeans, they may not have much of an effect on the color proportion of your candy stash if you had a hundred or so variably colored jellybeans to begin with, but they certainly will if you initially had only three. U, Pb, and Th are concentrated in sialic reservoirs, such as the continental crust, which, over time, will develop high concentrations of the radiogenic daughter Pb isotopes. Addition of crustal Pb will thus have a large effect on the radiogenic lead proportions of low-Pb mantle systems (including young mantle melts). Remember, ^{204}Pb is nonradiogenic, so for materials with high concentrations of U and Th, $^{208}Pb/^{204}Pb$, $^{207}Pb/^{204}Pb$, and $^{206}Pb/^{204}Pb$ will increase as U and Th decay. Oceanic crust also has elevated U and Th content (compared to the mantle), as will sediments derived from oceanic and continental crust. Pb is thus a sensitive measure of crustal (including sediment) components in mantle isotopic systems. Because 99.3% of natural U is ^{238}U, Reaction (9.22) will dominate over Reaction (9.23), and thus $^{206}Pb/^{204}Pb$ will be most sensitive to a crustal-enriched component.

Figure 14.9 shows $^{207}Pb/^{204}Pb$ versus $^{206}Pb/^{204}Pb$ data for Atlantic and Pacific ocean basalts. The **geochron** is the line along which all modern *single-stage* (not disturbed or reset) Pb isotopic systems, such as BSE, should plot, regardless of their initial Pb contents. It works like the concordia in Figure 9.14, in that it represents the simultaneous evolution of Reactions (9.22) and (9.23). Note that practically *none* of the oceanic volcanics fall on the geochron. Nor do they fall

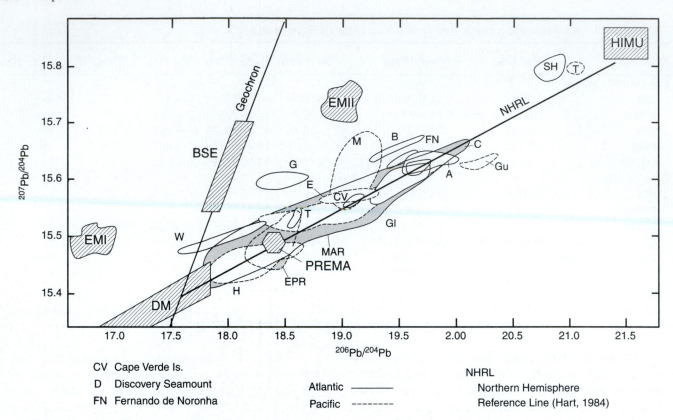

CV Cape Verde Is.
D Discovery Seamount
FN Fernando de Noronha

Atlantic ———————
Pacific --------

NHRL
Northern Hemisphere
Reference Line (Hart, 1984)

FIGURE 14.9 $^{207}Pb/^{204}Pb$ versus $^{206}Pb/^{204}Pb$ for MORBs and selected OIBs from the Atlantic and Pacific oceans. See Figure 14.8 for data sources and other abbreviations.

within the EMI–EMII–DM triangle, as they appear to do in the Nd-Sr systems. A different mantle reservoir: **HIMU** (high μ) has been proposed to account for this great radiogenic Pb-enrichment pattern. Incidentally, μ = $^{238}U/^{204}Pb$ and is used in U-Pb systems to evaluate parental uranium enrichment. Although intermediate in the Sr-Nd system (Figure 14.8), the HIMU reservoir is distinctive in the Pb system. It has a high $^{206}Pb/^{204}Pb$ ratio, suggestive of a source with high U, but it is not enriched in Rb (low $^{87}Sr/^{86}Sr$), and it is old enough (>1 Ga) to develop the observed isotopic ratios by radioactive decay over time. Several models have been proposed for this reservoir, including subducted and recycled oceanic crust (possibly contaminated by seawater), localized mantle lead loss to the core, and Pb-Rb removal by those commonly relied upon (but difficult to document) deep metasomatic fluids. The similarity of the HIMU reservoir to rocks from St. Helena Island in the southern Atlantic (where Napoleon was exiled and died) has led some workers to call this reservoir the "St. Helena component."

The $^{207}Pb/^{204}Pb$ data, especially from the northern hemisphere, show a close approximation to a linear mixing line between DM and HIMU, defining a line called the **northern hemisphere reference line (NHRL)** by Hart (1984). The data from the southern hemisphere, particularly from the Indian Ocean, depart from this line and appear to include a larger EM component (probably EMII). This should be apparent in Figure 14.10, which shows the $^{208}Pb/^{204}Pb$ data. Note that HIMU is also ^{208}Pb enriched, which tells us that this reservoir is enriched in Th as well as

U [Reaction (9.24)]. The ^{208}Pb content of the EM reservoirs is not as clearly defined as it is for the other isotopes, so the positions in Figure 14.10 are less well constrained. This highly enriched EM component has been called the **DUPAL** component, named for Dupré and Allègre (1983), who first described it. Note the "DUPAL Group" of high-$^{208}Pb/^{204}Pb$ islands in the Indian Ocean in Figure 14.10.

From geographic distribution of the isotopic data it is possible to construct maps of the various signatures. Figure 14.11 is an example of such a map created by Hart (1984) for the ^{208}Pb data. The contours are for Δ8/4, which, without going into the mathematics, is a quantitative way of estimating the distance that an isotopic data set for ^{208}Pb and ^{204}Pb plots above the NHRL (see Rollinson, 1993). Why this enriched anomaly extends as a band across the southern hemisphere at about 30° S is still a mystery. It has been proposed that this represents a dragging of Pb-enriched subcontinental lithospheric mantle from South America by mantle flow. Whatever the origin, the DUPAL map pattern suggests that the mantle is heterogeneous on a global scale (it includes very large domains).

14.5.3 Helium Isotopes

The noble gases are inert and volatile, making them useful alternative geochemical tracers, adding to the more conventional isotopic systems described above. See Farley and Neroda (1998) and Hilton and Porcelli (2003) for comprehensive reviews. For example, ^4He is produced in the solid

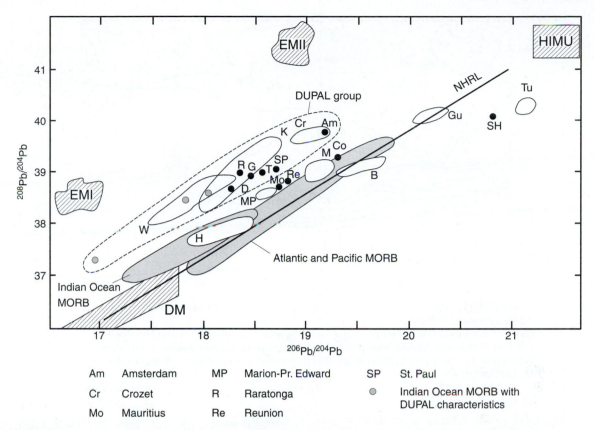

FIGURE 14.10 $^{208}Pb/^{204}Pb$ versus $^{206}Pb/^{204}Pb$ for MORBs and selected OIBs from the Atlantic, Pacific, and Indian Oceans. EM sources are approximate with respect to ^{208}Pb. Note Indian Ocean islands with high $^{208}Pb/^{204}Pb$ in the "DUPAL" group. Data from Hamelin and Allègre (1985), Hart (1984), and Vidal et al. (1984). See Figure 14.8 for other data sources and abbreviations.

Earth by the α-decay of U and Th (an α-particle is essentially a ^4He atom), whereas ^3He is largely primordial and nearly constant. The extent of radiogenic ^4He enrichment is typically expressed as $R = {}^3He/{}^4He$ (which, to my knowledge, is unique in petrologic isotope usage in that the radiogenic isotope is the *denominator*). *Low* $^3He/{}^4He$ values are thus more radiogenic. The common He reference is standard air, for which $({}^3He/{}^4He)_{Air} = "R_A" = 1.39 \cdot 10^{-6}$. Primordial "planetary" $^3He/{}^4He$ is estimated at 100 to 200 times R_A. He and the other noble gases are necessarily incompatible, concentrating in partial melts, from which they readily exsolve into a vapor phase such as CO_2 if the melts rise and decompress. Liberated noble gases are ultimately destined for the atmosphere, although most He escapes into space (after an atmospheric residence time of about 1 Ma). Atmospheric noble gases are not recycled back into the mantle, so mantle values are widely believed to reflect long-term degassing and production within the mantle of radiogenic noble gas isotopes such as ^4He from U and Th, or ^{40}Ar from K. The combined effects of degassing and ^4He production have served to *lower* the primordial $^3He/{}^4He$ ratio of the mantle over time.

N-MORB $^3He/{}^4He$ values are remarkably consistent with a sharp frequency peak at $8 \pm 1\ R_A$ (Farley and Neroda, 1998; Hilton and Porcelli, 2003). Outlying values range from 6 to 16 R_A with extreme ratios generally corresponding to E-MORB-like trace element and isotopic characteristics.

Such uniformity supports the notion of an extensive depleted DM-type N-MORB source reservoir, which appears to be rather uniformly degassed, lowering the present $^3He/{}^4He$ ratio from primordial ($>100\ R_A$) to about 8 R_A. OIBs exhibit much more variable $^3He/{}^4He$ ratios (Figure 14.12), ranging from about 5 R_A (two samples from the Galapagos reach 2 R_A) to as high as 43 R_A (some samples from Iceland and Hawaii yield the highest values reported thus far). Clearly, MORBs and OIBs sample different He reservoirs. Although some oceanic volcanoes exhibit considerable variation in $^3He/{}^4He$. Farley and Neroda (1998) noted that most localities tend to have ratios either higher or lower than N-MORB, suggesting mixing of melts derived from a MORB-type source material with a local reservoir having either higher or lower $^3He/{}^4He$.

A very low-$^3He/{}^4He$ mantle reservoir may result from ^4He addition by decay of U or Th in recycled crustal material. Ancient crust should have very low $^3He/{}^4He$ ($\sim0.1\ R_A$), and subduction recycling of only a little such material could significantly lower $^3He/{}^4He$ in partial melts of such contaminated mantle. The simplest and most plausible candidate for a *high* $^3He/{}^4He$ mantle reservoir is a deeper less-degassed mantle with values trending more toward primordial signatures. Deep-seated mantle plumes may tap such a reservoir in the lower mantle or even in the D″ layer at the core–mantle boundary. The D″ layer may also be He rich if the core has sequestered He (Porcelli and

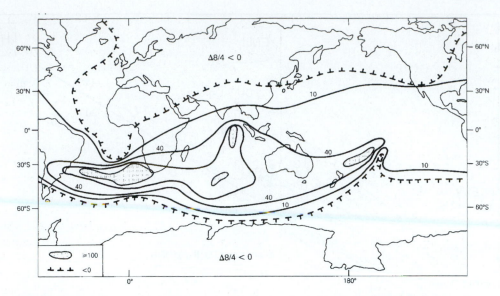

FIGURE 14.11 Map showing the global distribution of lead isotope anomaly contours for $\Delta 8/4$, a measure of the deviation in $^{208}Pb/^{204}Pb$ from the NHRL. The higher the value, the farther from the DM–HIMU mixing line and the greater the influence of the EMI–EMII ("DUPAL") source(s). Barbed lines point toward negative values and shaded areas have values over 100. From Hart (1984). Reprinted by permission from *Nature*, © Macmillan Magazines Ltd.

Ballentine, 2002) and slowly releases some as it progressively cools and solidifies (the solid inner core slowly grows at the expense of the liquid outer core). Such sources may also be responsible for the spread in MORB He ratios. Radiogenic isotopes of the other noble gases are also produced by U and Th decay, and their isotopic ratios, when corrected for atmospheric contamination, generally correlate with radiogenic He.

He isotopes do not correlate consistently with most of the nonvolatile isotopic reservoirs discussed earlier. Higher values of $^3He/^4He$ do tend to correspond (although poorly) to intermediate values of $^{87}Sr/^{86}Sr$ and $^{206}Pb/^{204}Pb$. This correlation may be due to the ease with which He diffuses in solids, or perhaps the proposed deep un-degassed reservoir is deeper than the other enriched isotopic reservoirs and is perhaps associated with a high $^3He/^4He$ core. Hilton and Porcelli (2003) also noted that HIMU-type mantle seems to correlate to low $^3He/^4He$ (5-8 R_A), which they attributed to ^4He produced by the elevated U concentration of the HIMU reservoir. Farley et al. (1992) proposed a high-$^3He/^4He$ mantle reservoir, which they called **PHEM** (**p**rimitive **he**lium **m**antle), to act as a mixing component and explain the spread of OIB data in the He-Sr-Nd-Pb systems (Figure 14.13). Some high-$^3He/^4He$ reservoir is required to explain the data spread. Other (similar) proposed high-$^3He/^4He$ mantle reservoirs are **FOZO** and C, which will be described below. PHEM, FOZO, and C appear to have moderately depleted Sr-Nd-Pb isotopic ratios between BSE and DM (but closer to BSE).

One nagging problem with the He data in Figure 14.12 is that OIBs have no obvious relationship between $^3He/^4He$ and total He concentration and typically have lower concentrations of He than do N-MORBs. If the high $^3He/^4He$ data reflect mantle that is less degassed than

MORB-source, why isn't high $^3He/^4He$ in a derivative partial melt correlated with high He content? Farley and Neroda (1998) suggested that noble gas concentrations are easily modified from their mantle values by partial melting, magmatic degassing, and solid-state diffusion. Hart and Zindler (1989) suggested that the higher degree of partial melting associated with MORBs results in low initial gas concentrations and correspondingly late gas saturation, resulting in relatively minor gas loss prior to eruption. Although OIBs may have higher initial gas contents, magmas may surface

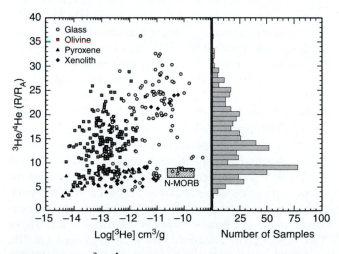

FIGURE 14.12 $^3He/^4He$ isotope ratios in ocean island basalts and their relation to He concentration. OIB $^3He/^4He$ values extend to both higher and lower values than N-MORBs (hatched box) but are typically higher. Fresh glasses (open circles) are generally required to avoid contamination. Some mineral inclusions, either fluid or melt/glass (other symbols), may also be good. Concentrations of ^3He are in cm³ at 1 atm and 298 K. After Sarda and Graham (1990) and Farley and Neroda (1998).

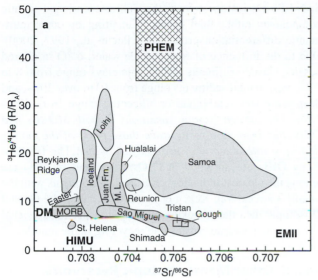

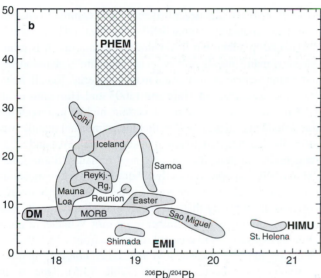

FIGURE 14.13 ^{3}He/^{4}He versus **(a)** ^{87}Sr/^{86}Sr and **(b)** ^{206}Pb/^{204}Pb for several OIB localities and MORB. The spreads in the diagrams are most simply explained by mixing between four mantle components: DM, EMII, HIMU, and PHEM. PHEM (**p**rimitive **he**lium **m**antle) is a hi-^{3}He/^{4}He reservoir with near-primitive Sr-Nd-Pb characteristics. After Farley et al. (1992).

^{3}He/^{4}He is some relatively dispersed pods or slivers in the mantle. Others (e.g., Albarede, 1998) attributed the high ^{3}He/^{4}He signal to old recycled oceanic lithosphere in the deep mantle that has been depleted in U and Th, so it develops less ^{4}He over time. Coltice and Ricard (2002) argued that only a small fraction of primitive material (~3%) is required to impart the observed primordial-like noble gas signal and that a partly degassed peridotite host may retain sufficient primordial ^{3}He, perhaps as pods of primitive mantle remaining in a depleted mantle host. Peridotite, recycled crust, etc. are better stirred in the lower-viscosity upper mantle, which, along with the larger melt fraction involved in MORB production, may explain the more homogenous noble gas signature of MORB. The less efficiently mixed lower mantle has more variable proportions of primitive versus depleted peridotite and recycled components in plume sources, they argued. Lower melt fractions homogenize smaller volumes of these sources, resulting in the heterogeneity observed in OIBs (including melts of sources with more of the primitive component).

14.5.4 Osmium and Oxygen Isotopes

^{187}Os is created by the breakdown of ^{187}Re with a half-life of about 42 Ga. Both Re and Os are highly siderophile, preferring metal and sulfide phases to silicates, so were concentrated in the core during early planetary differentiation. These elements are controlled far more by sulfide, oxide, and metal phases than by silicates. This contrasting behavior to lithophile Rb-Sr, Sm-Nd, and U-Th-Pb adds another significant new perspective to mantle isotopic studies. Shirey and Walker (1998) and Carlson (2005) provided excellent reviews. Mantle samples yield nearly chondritic Re/Os evolution values (^{187}Os/^{188}Os = 0.1296 ± 8), generally attributed to meteorite accretion to resupply the mantle following core formation. Because Os is compatible during partial melting of the mantle (concentrating in trace sulfides), and Re is moderately incompatible (distributed in silicates), the mantle is now enriched in Os (~3 ppb) relative to crustal rocks (~0.03 ppb). Because the Os content is low and the Re/Os ratio is relatively high in partial melts and crustal rocks, the ratio of radiogenic ^{187}Os to nonradiogenic ^{188}Os (^{187}Os/^{188}Os) is much higher in the crustal rocks. Of course, the quantities are minute, and this system has only been of use recently due to advances in analytical procedures.

The Re/Os ratio in peridotites cluster near 0.1, whereas in mafic rocks (including MORBs, OIBs, and eclogites) it is higher, with a peak near 10 and extending to about 1000. Unlike the other isotopic systems, there is practically no overlap between these populations. Over time these differences translate into large differences in ^{187}Os/^{188}Os, making elevated Os isotope ratios a direct tracer of the addition of mafic crust or melt to a mantle source (Hauri and Hart, 1993). Virtually all OIBs have elevated ^{187}Os/^{188}Os, as compared to peridotites, as shown in Figure 14.14. The OIBs in Figure 14.14 appear to follow three possible mixing arrays, each extending from a high-^{187}Os/^{188}Os mantle reservoir (EMI, EMII, and HIMU) and converging toward a common low-^{187}Os/^{188}Os reservoir

more slowly in the non-extensional OIB environment and pre-eruptive degassing of the more volatile-rich magma may thus be more extensive.

In summary, noble gas data suggest that the mantle is continually degassing, liberating primordial ^{3}He as well as primordial and radiogenic ^{4}He. The consistent 8 ± 1 clustering of N-MORB source data and the higher values for many deep-seated OIBs support the existence of significant mantle reservoirs, probably including a shallow He-depleted MORB source and a deep reservoir that has preserved high concentrations of volatiles since the formation of the mantle. Although most investigators find the layered mantle model most appealing for the noble gas data, others take a contrary view. Meibom et al. (2005), for example, suggested that the poor correlation between He and the nonvolatile isotopes and between ^{3}He/^{4}He and X$_{He}$ is fundamental and that the carrier of high

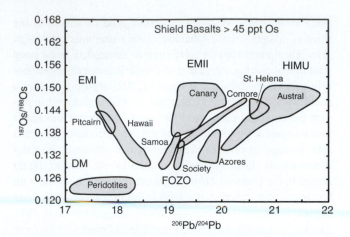

FIGURE 14.14 $^{187}Os/^{188}Os$ versus $^{206}Pb/^{204}Pb$ for mantle peridotites and several oceanic basalt provinces, all relatively rich in Os and for which quality Os data are available. All of the basalt provinces are enriched in ^{187}Os over the values in mantle peridotites and require more than one ^{187}Os-enriched reservoir to explain the distribution. The Os values for the various mantle isotopic reservoirs are estimates. After Hauri (2002) and van Keken et al. (2002a).

(FOZO). The $^{187}Os/^{188}Os$ ratios for the familiar enriched reservoirs in Table 14.5 are estimates, but clearly some Os-enriched reservoirs are required and, based on their Rb-Sr-Nd-Pb characteristics, they are good enriched-mantle candidates.

Brandon et al. (1998) found that both $^{187}Os/^{188}Os$ and $^{186}Os/^{188}Os$ are elevated in some Hawaiian lavas. ^{186}Os is produced by decay of ^{190}Pt, which is siderophile and should be concentrated in the core. The most viable mechanism for the correlated enrichments of both radiogenic Os isotopes, they conclude, is the addition of 0.5-1% of outer-core metal to a portion of the adjacent D'' layer, followed by upwelling of the mixture into the Hawaiian plume. Hauri and Hart (1997) used the Re/Yb ratio in OIBs to constrain such a core contribution to less than 0.2% in mantle plumes, and further proposed that elevated Os isotope ratios result from a ubiquitous recycled MORB component in the plumes.

As for oxygen, recall from Chapter 9 that ^{18}O and ^{16}O are both *stable* isotopes of oxygen that are sufficiently light and different in mass to mass-fractionate via exchange processes between coexisting phases. This fractionation is dominated by low-temperature processes near the Earth's surface, particularly water-rock interactions which strongly enrich altered/exchanged silicates in $\delta^{18}O$ (see the review by Eiler, 2001). Recent technical advances in heated laser-fluorination techniques have greatly improved the precision and accuracy of oxygen isotopic analysis of silicates, and I will only cite the more precise data from the last two decades produced using this technique. Oxygen fractionation during melting, crystallization, and gas exsolution of silicate systems is small, with partial melts enriched by only 1 to 2‰ over the source. $\delta^{18}O$ values in mantle peridotites cluster tightly near 5.5 ± 0.2‰. Variations in $\delta^{18}O$ in MORBs (5.3 to 6‰) and OIBs (up to 7‰ or more) are small, but higher values correlate with trace element and

Sr-Nd-Pb-Os isotope tracers that indicate enriched mantle components rather than extent of melting or other intra-mantle differentiation processes (Eiler et al., 1997, 2000). Due to the influence of near-surface water, $\delta^{18}O$ in altered basalts, clastic sediments and pelagic clays range from 8 to 25‰, continental sediments range from 10 to over 30‰, and deep-sea oozes (carbonate or siliceous) range from 30 to 40‰. *The oxygen isotope variations in both MORBs and OIBs thus seem to directly trace those parts of the mantle that have interacted with near-surface water.* The fact that many OIBs are enriched in $^{187}Os/^{188}Os$ and in $\delta^{18}O$ are among the clearest indicators of crustal components with the mantle sources. Van Keken et al. (2002a) called the Os and O isotopic data the "smoking guns" that testify to subducted sediment and oceanic crustal slab recycling in the mantle.

14.5.5 Other Mantle Isotopic Reservoirs

Most of the reservoirs described above are required to encompass the OIB data by mixing. BSE is not required because it is not an end-member, but lies within the data array. It is referenced by many investigators who are reluctant to abandon any remaining vestige of primordial mantle material. Recall, however, the arguments of Hofmann (2003 and Hofmann et al. (1986), based on the ratios of certain highly incompatible trace elements such as Nb/Ce, that the primordial mantle was probably depleted by extraction of continental crust and rehomogenized, so that little, if any, now remains. Some OIBs exhibit BSE isotopic values, but there is no great concentration of such values as one might expect for a major source component, implying that the values may simply result from fortuitous mixing of end-member reservoirs. Whether or not it remains today, it is still a useful reference from which mantle evolution may be evaluated. A few other non-end-member reservoirs have been proposed (Table 14.5). One is in Figure 14.8, from the early work of Zindler and Hart (1986), called **PREMA** (**pre**valent **ma**ntle). PREMA represents a restricted isotopic range that is common in ocean volcanic rocks. Although it lies on the mantle array, and could result from mixing of melts from DM and other primitive or enriched sources, the preponderance of melts with this restricted signature suggests it may be a distinct mantle source.

Hart et al. (1992) proposed **FOZO** (for **fo**cal **zo**ne) because the isotopic data from a number of OIB provinces, when plotted in Sr-Nd-Pb space, exhibit sublinear arrays that converge (or focus) toward it (Figure 14.15). FOZO is a high $^3He/^4He$ component lying approximately midway between DM and HIMU (Figure 14.15). Either it represents actual source material (a true mantle reservoir) or the partial melting process in rising OIB diapirs which sample a mixed DM-HIMU source in very consistent proportions. Such consistent proportions seem fortuitous, so FOZO may well be a distinct mantle source. Hanan and Graham (1996) found similar convergence of Pb-Nd-Sr isotopic data for MORBs and OIBs toward a restricted intermediate isotopic value, and called this the **C** (**c**ommon) mantle component (Table 14.5). FOZO and C are commonly called "convergence reservoirs," and C is a high-$^3He/^4He$ reservoirs as well. Several investigators have

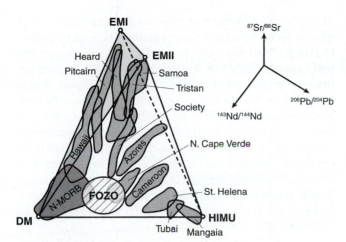

FIGURE 14.15 Projection of various oceanic isotopic data sets onto the base triangle of the tetrahedron defined by the four mantle end-members of Zindler and Hart (1986) in three-dimensional $^{87}Sr/^{86}Sr$ – $^{143}Nd/^{144}Nd$ – $^{206}Pb/^{204}Pb$ space. DM appears to be a mixing end-member only in the MORB array. Most OIB arrays converge toward a volume in isotopic space characterized by low $^{87}Sr/^{86}Sr$ and high $^{143}Nd/^{144}Nd$, $^{206}Pb/^{204}Pb$ (and $^{3}He/^{4}He$) ratios. Hart et al. (1992) proposed the name FOZO (**fo**cal **zo**ne) for this volume. After Hart et al. (1992).

suggested that FOZO, C, and PHEM (and perhaps PREMA) represent the same mantle source. These convergence trends and the notable lack of mixing arrays between EMI, EMII, and HIMU suggest that mantle mixing is not random.

The isotope data for ocean volcanic rocks have supplied some provocative information concerning the nature of the mantle. Geologists will continue to collect data and ponder the implications of the results for some time to come. At present we are compelled to accept the notion that the mantle is heterogeneous, and that there are several reservoirs within it, each with its own distinct isotopic (and trace element) signature. To the extent that there is a consensus among petrologists and geochemists, *five* mantle end-member components are at present the minimum number necessary to produce the OIB and MORB isotopic arrays by mixing: DM, EMI, EMII, HIMU, and a high-$^{3}He/^{4}He$ reservoir (FOZO is most popular). Clearly, there is a significant proportion of depleted mantle, due to the extraction of continental and oceanic crust over time. Other reservoirs are either less depleted (FOZO with $^{3}He/^{4}He$ values close to primordial) or enriched by some process. From the data amassed thus far, these reservoirs appear to be distinct. If they weren't, the mantle would become homogenized to a single reservoir and it would not be possible to derive magmas with such varied isotopic signatures from it. We are forced to conclude that the reservoirs have remained distinct for over 1 Ga, without rehomogenizing.

Perhaps the most interesting idea coming out of the study of MORBs and OIBs is that some of the isotopically enriched reservoirs (EMI, EMII, and HIMU) may be too enriched for any known mantle process, and they probably correspond to crustal rocks and/or sediments. EMII is highly enriched, especially in radiogenic Sr (indicating the Rb parent)

and Pb (U/Th parents), which correlates best with the upper continental crust, ocean island crust, or sediments. EMI, with its slightly enriched character, may be correlated with lower continental crust or oceanic crust. Hofmann (1997), however, found that HIMU basalts have trace element ratios similar to depleted mantle. Eiler (2001) also noted that $\delta^{18}O$ for EMII basalts is high, but EMI and HIMU basalt values are about normal for MORB, prompting him to agree with Sun and McDonough (1989) that HIMU and EMI may be products of magmatic or metasomatic differentiation within the mantle rather than addition of crustal components. Willbold and Stracke (2006) noted that the trace element ratios of HIMU basalts are remarkably uniform, with characteristics suggesting that HIMU may be recycled oceanic lithosphere (crust + mantle). EM basalts, they found, are much more variable and require incorporation of continental crustal materials to explain. Some OIA lavas have compositions unlike liquids derived from experimental melting of peridotites and have enriched isotopic signatures (e.g., Kogiso et al., 1997; Janney et al., 2002; Workman et al., 2004). Experiments involving melting of pyroxenites and eclogites (Hirschmann et al., 2003; Kogiso et al., 2003; Dasgupta et al., 2006) indicate that subducted oceanic crust may represent HIMU- and EMI-like sources capable of producing these lavas. Some lavas from these provinces also show elevated CO_2 (e.g., Dixon et al., 1997) and CaO (Kogiso et al., 2003), indicating perhaps a carbonated eclogite source (Dasgupta et al., 2006).

At present, then, EMII appears to have a clear crustal signature, and EMI and HIMU may be enriched either by crustal additions or by internal mantle processes (or both?). There are no specific (named) enriched Os or $\delta^{18}O$ reservoirs, but these isotopes appear to distinctly indicate crustal components in the mantle. Crustal or sedimentary material could only be introduced into the deeper mantle by subduction and recycling. To remain isotopically distinct, however, they could not have fully rehomogenized or re-equilibrated isotopically with the rest of the mantle. We have certainly been aware that oceanic crust, upper mantle, and associated sediments (both pelagic and in the fore-arc wedge, both of which have a continentally derived portion) have been subducting for eons, yet we have had little evidence of their fate other than the arc-related volcanism to be discussed in Chapter 16. Here is the first good indicator of their presence on a broad scale in the mantle.

14.6 NATURE OF THE MANTLE

The depths of plume-diapir origin, partial melting, and final equilibration of erupted basalts are poorly constrained, and estimates vary as to the extent of partial melting and mantle entrainment during diapir ascent. It thus becomes a complex issue to infer the size and three-dimensional distribution of mantle domains on the basis of basalt composition and surface distribution. What we can say with some assurance is the following:

- N-MORBs involve shallow melting of passively rising upper mantle and indicate a significant volume of depleted shallow upper mantle (that has lost lithophile elements and considerable He and other noble gases).

- OIBs typically originate from deeper levels than N-MORBs. The major- and trace-element data indicate that the deep source of OIB magmas, both tholeiitic and alkaline, is distinct from that of N-MORB. Trace element and isotopic data reinforce this notion and further indicate that the deeper mantle is relatively heterogeneous and complex, consisting of several domains of contrasting composition and origin. In addition to the depleted MORB mantle, there are at least four enriched components, including one or more containing recycled crustal and/or sedimentary material reintroduced into the mantle by subduction, and at least one (FOZO, PHEM, or C) that retains much of its primordial noble gases.

- MORBs are not as homogenous as originally thought, and exhibit most of the compositional variability of OIBs, although the variation is expressed in far more subordinate proportions. This implies that the shallow depleted mantle also contains some enriched components.

Since the initial trace element and isotopic indications of a shallow depleted mantle, petrologists and geochemists have favored a relatively simple two-layer mantle model (proposed in Chapter 10) with an upper mantle depleted by continental crust and MORB extraction, and an undepleted lower mantle, largely untapped and the recipient of subduction-recycled altered oceanic crust and sediments. Because Fe is heavier and a little less compatible than Mg, the density of depleted mantle is slightly lower than that of fertile mantle, which should contribute to the stability of this arrangement. The obvious choice for a boundary separating these two great layers is the 660-km phase transformation. As mentioned in Sections 1.7.3 and 10.5, debate over the effectiveness of this transition zone as a barrier to convective mixing and mantle homogenization remains unresolved.

As described in Chapter 1, the increased density of the colder oceanic lithosphere is believed to drive subduction and plate tectonics. But what about the crustal component of the slab? We generally correlate density with increasingly mafic character, which explains why both oceanic and continental crust "float" on top of the ultramafic mantle, and why the sialic continental crust stands higher than the mafic oceanic crust. Density, however, depends on more than composition, and mineral lattice packing is important. At moderate pressure, mafic rocks, such as the basalts and gabbros that compose the oceanic crust, contain more aluminum than peridotites and undergo a metamorphic reaction to become a garnet-pyroxene rock called **eclogite** (Section 25.3.4). As a result of its high garnet content, eclogite is denser than the olivine–pyroxene-dominated ultramafic mantle lherzolites at the same pressure, at least in the upper mantle. The oceanic crust in subduction zones may thus *facilitate* initial sinking of the subducted slab. But what happens at greater depth? Seismic evidence and mantle tomography indicate that, although some subducted slabs of oceanic lithosphere flatten

or fold at the 660-km transition, several penetrate and appear to break up in the lower mantle. The experiments of Ono et al. (2001) suggest that basaltic rocks are denser than peridotites until the 660-km transition but then become less dense until about 710 km, where they again become denser. Due to differences in density or rheology, some crustal component may delaminate from the lithospheric mantle, but components of each probably reach the lower mantle and feed the D″ layer at its very base (Christensen, 2001). Altered crust and sediments are less dense, of course, but these become dehydrated during subduction and compose only a minute fraction of the subducted lithosphere, and some is probably entrained with the rest of the lithosphere and carried to depth.

In addition to slab penetration of the 660-km transition, some vigorous plumes also appear to originate beneath the transition (perhaps also at D″) and rise through it. A strict 2-layer mantle model with isolated upper and lower mantle domains, however attractive from a petrological perspective, is therefore precluded, but the question remains as to the ability of such a phase transition to impede wholesale mantle remixing. Critical to its capacity to do so is the Clapeyron slope of the transition reaction.

As shown in Figure 14.16, a positive Clapeyron slope would enhance convective motion of descending cool material and rising warm material across the transition, and a negative slope would retard convection across the transition. Mathematical models and experiments on layered convecting fluids suggest that a Clapeyron slope of -4 MPa/°C is sufficient to retard penetration, at least for a time (see Davies, 1999, for a summary). Recall from Section 1.2 that the 660-km transition for peridotite involves the transformation of ringwoodite to a perovskite-like phase plus magnesiowüstite. The transformation is obviously pressure sensitive, and thus has a very low Clapeyron slope, but there are uncertainties as to the entropy and volume changes involved at such great pressures and temperatures. Most estimates of the slope are negative, with a magnitude of -2 to -3 MPa/°C, implying some resistance to penetration, but not enough to necessitate separately convecting mantle layers. Basaltic crust is a subordinate component of subducting lithosphere, and the persistence of majorite garnet in such crust deeper than 660 km imparts a slight buoyancy to the subducting slab beneath the transition, but the overall effect is considered minor. Liu (1979) and Anderson and Bass (1986) suggested that the 660-km transition may also involve a minor change in composition: increasing Fe and/or SiO_2 with depth, which could make the transition more of a barrier. Controversy over the 660-km transition therefore persists. Either it cannot inhibit mantle convection very well at all, resulting in whole-mantle convection, or it can partially inhibit it, allowing for some modified version of two-layer convection.

Convective **mixing**, whether affecting the entire mantle or partially isolated layers, involves two processes: **stirring and diffusion**. Stirring is the mechanical intermingling of matter and diffusion is the migration of chemical species through the host medium on the molecular scale. Suppose, for

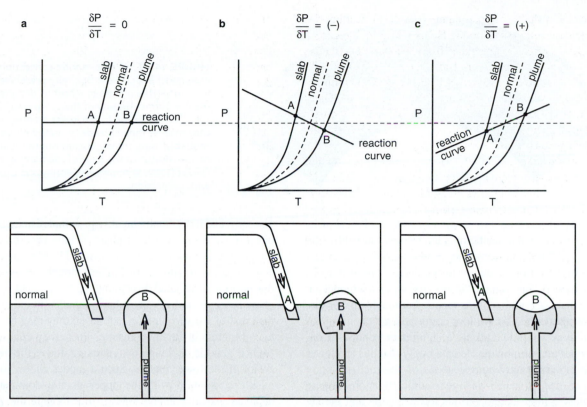

FIGURE 14.16   Effect of the Clapeyron slope of a phase transformation on the ability of material to pass through.
(a) Clapeyron slope = 0: material in a subducting slab with a cooler geotherm (A) and in a rising plume with a hotter geotherm (B) will undergo the transformation to a denser phase at the same depth as their "normal" surroundings. **(b)** Negative Clapeyron slope: cooler material in the slab undergoes the transformation at greater depth and warmer material in the plume does so at shallower depths. The slab at A is thus less dense than the immediate surroundings and the plume at B is more dense than the surroundings. The effect in both situations would retard penetration of the transition. **(c)** Positive Clapeyron slope: slab material undergoes the transformation at shallower depth and plume material does so at greater depths. The slab at A is thus more dense than the immediate surroundings and the plume at B is less dense than the surroundings. The effect in both situations would enhance penetration of the transition.

example, you add sugar to iced tea. You *stir* it to keep the sugar from collecting on the bottom of your glass, creating a suspension of small sugar crystals. The agitated mixture is homogenized on the large to intermediate scale but not on the very fine molecular scale. You may keep stirring as you wait for the crystals to dissolve and the molecules of sugar to *diffuse* throughout the tea, at which point the tea becomes a truly homogenized mixture. If you don't stir, or quit early, the sugar crystals accumulate at the bottom of the glass, and only the relatively slow and small-scale diffusion process can mix the solution further (with limited results). What you drink last is then much sweeter than what you drink earlier.

Mantle mixing is similar. Convection can *stir* the viscous material, stretching and folding any impurities (recycled crust or sediments, lithospheric mantle, stalled and solidified melts, metasomatized material, etc.). The high viscosity of the mantle limits flow to large scales and laminar (nonturbulent) behavior, which strongly affects the rate and effectiveness of stirring. Stirring iced tea is much more effective because it is turbulent, creating cascades of smaller and smaller eddies that intermingle the components much more intimately. Experiments using viscous fluids and numeric computer simulations (e.g., Gurnis, 1986; van Keken et al., 2002a, 2003) differ in detail but generally suggest that impurities can be distributed relatively well on timescales of 1 to 2 Ga, but only as stretched/folded pods on the scale of hundreds of meters or more. *Diffusion*, on the other hand, is limited to length scales of 0.1 to −10 m over the age of the Earth, so the stirred larger heterogeneities should last, each with its own geochemical and age characteristics. Localized streamlined mixing in high-shear areas (including hotspot plumes) may further stretch and thin heterogeneities. The mantle may thus resemble a "marble cake" (Allègre and Turcotte, 1986), with discontinuous commingled layers on the meter to multiple-kilometer scale. (I've also heard it referred to as "plum pudding" mantle, but perhaps fudge-ripple ice cream is a more familiar analogy to U.S. students.) Figure 11.8 illustrates the texture of commingled layers on a much finer scale resulting from magma mixing in a subvolcanic magma chamber. Several high-temperature peridotite massifs, believed to represent fault-emplaced mantle, exhibit centimeter- to meter-scale folded and deformed pyroxenite layers. The pyroxenites are mineralogically, geochemically, and isotopically diverse, and many are of a composition that may be expected of altered oceanic crust that has been somewhat depleted in a subduction zone (Allègre and Turcotte, 1986; Takazawa et al., 2000).

Whole-mantle convection can thus create a marble-cake mantle with considerable heterogeneity in terms of stretched

FIGURE 14.17 Whole-mantle convection model with geochemical heterogeneity preserved as blobs of fertile mantle in a host of depleted mantle. Higher density of the blobs results in their concentration in the lower mantle, where they may be tapped by deep-seated plumes, probably rising from a discontinuous D″ layer of dense "dregs" at the base of the mantle. From Davies (1984) © American Geophysical Union with permission.

and folded pods/layers of diverse ages and origins. Partial melting draws from such a mantle on a larger scale than diffusion could homogenize, and would thus produce melts representing mixtures of the various components within the volume of mantle being melted. The isotopic "reservoirs" in such a model are not large distinct portions of the mantle but much smaller dispersed constituents. The shallow origin and geochemistry of MORBs, however, still indicate a significant volume of depleted upper mantle, and the He data imply that this shallower reservoir is surprisingly homogeneous on the melt-sampling scale. The deeper-source and geochemically more diverse OIBs imply a more heterogeneous lower mantle and the He data suggest that this lower mantle retains more of its primitive noble gases. Also, the sum of continental and depleted mantle heat production (and plumes from D″) is insufficient to account for present-day surface heat flow (Albarede, 1998; Kellogg et al., 1999), suggesting a higher heat-producing lower mantle. These are among the more compelling features that cannot be explained by unrestrained whole-mantle convective stirring and require some form of layering.

A modified two-layer mantle convection model proposed by Silver et al. (1988) to address the contrasting requirements was presented in Figure 1.14. In this model, the 660-km transition is a sufficient density barrier to impede convective mantle stirring and homogenization, but subducted lithosphere and vigorous rising hotspot plumes represent sufficient density instability to pass through. The result is an upper and lower mantle that are compositionally distinct in several potential ways, but that share many components, including those recycled by subduction and carried by plumes. Stirring may be effective within each layer, so that each may develop a marble-cake heterogeneity, but the proportions of various constituents and the nature of the peridotite host matrix (e.g., depletion of incompatible elements and noble gases) may differ. Such a model could accommodate nearly any geochemical similarity or contrast within and between the layers, allowing for the greater diversity of OIBs without precluding similar contrasts in some MORBs.

Whole-mantle convection may also be modified to accommodate many objections. Figure 14.17 illustrates a whole-mantle model proposed by Davies (1984). As in the model of Silver et al. (1988), subducted lithosphere penetrates the 660-km transition, perhaps with some temporary deflection, and extends to the lowermost mantle. Descending lithosphere may buckle at greater depth due to the steadily increasing

viscosity, and it eventually breaks up, so that dismembered bits of crust, sediment, and lithospheric mantle become entrained into the marble-cake mantle by stirring. The 660-km transition is not a barrier to convection in these models, so the entrained bits become distributed throughout the mantle, but the density of mafic crust and of lithospheric mantle should be greater than that of the surrounding mantle, so they may be expected to concentrate in the lower mantle, where they escape shallow MORB genesis, and vigorous remixing due to the higher viscosity of the lower mantle. Such a model allows for whole-mantle convection with the upper mantle dominated by a depleted character and the lower mantle more fertile. An established plume (center of Figure 14.17) and new plume (right) rise from the core-mantle boundary area, where they are fed by a thermal boundary layer (beneath the dashed line) containing a discontinuous layer of dense settled crustal dregs (described in Section 1.2). OIB-MORB diversity may be accommodated by such a model, but the noble gas differences are more difficult. If the high $^3He/^4He$ reservoir is indeed a poorly degassed lower mantle, as popularly interpreted, whole-mantle convection should have dispersed it. Those favoring whole-mantle convection have proposed other explanations for the high-$^3He/^4He$ signal, as described in Section 14.5.3.

Whole-mantle and two-layer convection can both be shaped to conform to the geochemical and geophysical criteria. At the present state of uncertainty the 660-km transition may allow either, or practically any state between these two models, allowing for intermediate degrees of contrast between the upper and lower mantle. Other models have also been proposed to address various problems and objections. McNamara and van Keken (2000), for example, pointed out that density stratification at 660-km should produce significant topography along the boundary layer, as well as large (measurable) geoid and gravity anomalies. It also implies sufficient insulation to raise the top of the lower layer significantly above its melting temperature (unless radiative heat transfer significantly augments thermal conduction across the boundary). None of these features are observed. Kellogg et al. (1999) and van der Hilst and Kárason (1999) thus proposed deeper density-stratified boundaries (of 1600 km and 1700 to 2300 km) that allow for a deep heat-producing layer with more primordial helium and topography so as to avoid most of these problems (Figure 14.18). Mantle plumes may rise from the top of such a deep layer and perhaps also from the D″ layer below. Descending

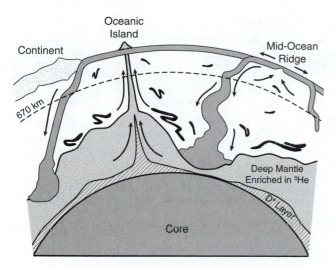

FIGURE 14.18 Two-layer mantle model with a dense layer in the lower mantle and less depletion in lithophile elements and noble gases. The top of the layer varies in depth from ~1600 km to near the core–mantle boundary. After Kellogg et al. (1999).

plates may also accumulate along and depress the top of this layer, and then become stirred into it, the D″ layer, and/or the upper layer (to create marble-cake layering).

Becker et al. (1999b) proposed a whole-mantle convection model in which large blobs of primitive mantle escape MORB depletion and reside mainly in the cores of convection cells, an obvious place (if they exist) to escape shallow MORB melting yet remain susceptible to plume penetration and incorporation. Calculations by Becker et al. (1999b) and by Manga (1996) show that such blobs, if sufficiently viscous, could persist in convective cores for very long times without being substantially mixed with the surrounding flow. The principal difference between the Davis (Figure 14.17) and Becker et al. (1999b) models is that the non-depleted trace element and isotopic source(s) result from subducted oceanic lithosphere (mostly partial mantle melts) in the former but from unmelted primitive mantle in the latter. Of course, neither of these models precludes the other, so both sources may be present. Some investigators have also proposed that a progressively cooling Earth may have experienced a transition from early two-layer convection to more recent whole-mantle convection (Honda, 1995). As a result, long-term mantle heterogeneity resulting from two separately convecting layers may have survived such a recent onset of mixing.

14.7 PETROGENESIS OF OIBS

Figure 14.19 illustrates a general model for oceanic basalt generation that allows for various possibilities of mantle reservoirs and sources. The shallow origin and depleted character of MORBs implies a depleted upper mantle and the deeper origin and broader distribution of geochemical heterogeneity in OIBs suggests a less depleted deeper mantle in which more enriched components are concentrated. The upper mantle in Figure 14.19 is therefore predominantly the depleted DM reservoir, the source of N-type MORB. The

lower mantle is the principal OIB and E-MORB source. Both layers, however, may contain isotopically and chemically distinct material, including enriched components, probably stretched and folded in a marble-cake fashion by convective stirring, but the deeper mantle probably contains significantly more. The OIB source may also contain non-depleted (BSE), partially depleted (PREMA), and non-degassed (FOZO, PHEM, and/or C) materials. The variably enriched EMI, EMII, and HIMU may have several origins. The high $^{87}Sr/^{86}Sr$ of EMII implies continental crust, most readily recycled into the deep mantle in the form of subducted sediments. EMI and HIMU are less enriched and may also represent a crustal component, either ancient oceanic crust or pelagic sediment, or they may possibly have been enriched by intra-mantle processes: stranded melts or metasomatizing fluids. As we shall see in Chapters 18 and 19, the subcontinental lithospheric mantle may be highly metasomatized, and some may detach from the continental crust and founder downward. The $^{187}Os/^{188}Os$ and $\delta^{18}O$ data in particular indicate that very shallow components have been incorporated in the deep mantle. To do so, they must be subducted and recycled back into the mantle but not completely re-homogenized (by stirring and diffusion) if their isotopic signature is to be recognizable, even in basalts generated far from contemporary subduction zones (Hofmann and White, 1982). Although Figure 14.19 rather tentatively retains a modified two-layer mantle model with the 660-km transition as the boundary separating the upper and lower regions, it could conform to other models discussed above with only minor modification. The lower density of residual mantle following melt extraction, for example, could result in more gradual density stratification that is independent of the 660-km transition, and the more dense recycled components could concentrate in the lower mantle in a whole-mantle convective model. Alternatively, the boundary might occur at a deeper level.

The plume model is most commonly invoked for the origin of most OIB magmas, but the true origin of plumes is still not clear. Certainly they are associated with a thermal-gravitational anomaly, which is generally attributed to heating of the D″ layer at the core–mantle boundary, but may also occur at the 660-km boundary layer or at 1700 km. An interesting correlation between times of intense volcanic activity in Hawaii and periods of normal magnetic polarity was pointed out by Moberly and Campbell (1984), which suggests that some plumes at least are related to core processes. Localized fluid concentrations may also decrease the density of mantle lherzolite and initiate diapiric rise of hotter material through cooler material above as a thermal plume. Similarly, the low density of subducted EM components with sialic continental character could also separate from their subducted plate host and initiate a "compositional" plume.

Whatever their origin, plumes at depth are characterized by upwelling *solid* diapiric masses (Figure 14.2). The source may also be a zone in which subducted material is concentrated. Rising plumes may also entrain some of the surrounding mantle, although experiments suggest that entrainment is most effective immediately above the source and becomes much less so as the plume rises (Hauri et al., 1994).

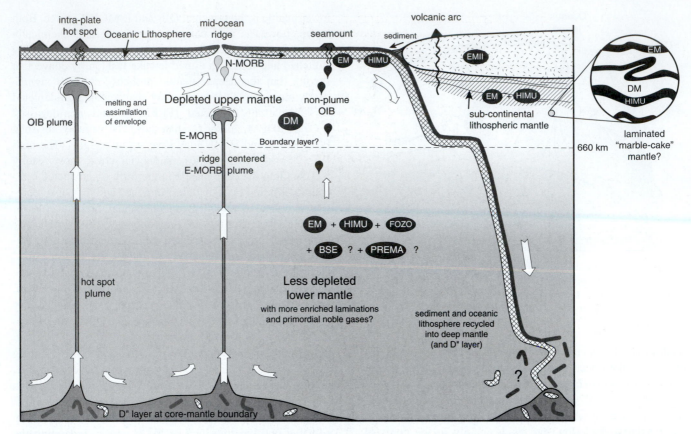

FIGURE 14.19   Schematic model for oceanic volcanism. The upper mantle is (variably) depleted (DM) and is the source of N-type MORB. The lower mantle is a major source for E-MORB and OIB magmas. Both regions are heterogeneous, containing stirred (stretched and folded) disrupted remnants of marble-cake crust, lithospheric mantle, and sediment recycled by subduction (as illustrated in the magnified insert for the upper mantle). The lower mantle contains more of these (denser) constituents: enriched mantle (EM), high-μ mantle (HIMU), high-^{3}He/^{4}He (FOZO), and perhaps primitive mantle (BSE) and "prevalent" mantle (PREMA) reservoirs as well. Nomenclature from Zindler and Hart (1986) and Hart and Zindler (1989). The 660-km seismic discontinuity is tentatively retained as the boundary layer between upper and lower mantle, but is left gradational to allow for other options, such as a deeper boundary near 1700 km or whole-mantle convection with progressively less depleted and more dense deeper mantle containing more recycled material.

Decompression partial melting occurs in the rising diapirs when the mass intersects the appropriate solidus at shallow depths. Plumes are hottest toward the inner axis and in the large head of new plumes. The hotter areas at least should begin to melt at greater depths than MORBs. Recall from Section 13.6.3 that Klein and Langmuir (1987, 1989) used the value of FeO at 8.0 wt. % MgO (called $Fe_{8.0}$ or $FeOT_8$) on Harker-type diagrams for MORB data sets to standardize for shallow fractionation and investigate earlier genetic processes. $Fe_{8.0}$ was correlated with depth of melting in that study. Lassiter and DePaolo (1997) found that values of $Fe_{8.0}$ for ocean island basalts overlap with the high-value (deep source) end of MORBs and extend toward deeper melting.

The onset of melting in a rising plume depends on material composition, temperature in the plume, and volatile content. Thermal models suggest that the axial zone is much hotter than the margin: *potential* temperatures (depth corrected temperatures based on some estimated value at depth and extrapolated adiabatically to the surface: see Figure 1.9) are about 1200°C at the plume margin and reach about 1550°C along the axis (Courtney and White, 1986; Watson and McKenzie, 1991). Mantle heterogeneity and the process of OIB generation is certainly more complex than we originally

envisioned in Chapter 10. A marble-cake mantle with pods and stringers of basaltic crust is perhaps a common source of many MORB–OIB melts, and melting affects (homogenizes) a volume larger than the scale of proposed marble-cake heterogeneity. Van Keken et al. (2002a) suggested that each hotspot exhibits an array of chemical and isotopic characteristics that extend from mantle peridotite components to enriched components related to recycled mafic crust and sediments. Hotspots, they concluded, do not simply produce random mixtures of five mantle components (the minimum number of mantle reservoirs required to encompass the OIB isotopic data); instead, each reflects a unique combination of components, including residual mantle and material reflecting different styles of near-surface water-rock alteration, subduction zone processing, in situ enrichment, and aging in the convecting mantle.

Figure 14.20 presents a general model for melting in an established deep-seated intra-plate plume tail such as Hawaii. It assumes the plume contains numerous stretched slivers and pods of recycled crustal components, mostly from the deep source area and concentrated toward the plume axis, where plume entrainment of the surrounding mantle during ascent is minimal. If the mantle is indeed stratified, the peridotite host is largely deep-seated (not depleted by MORB extraction

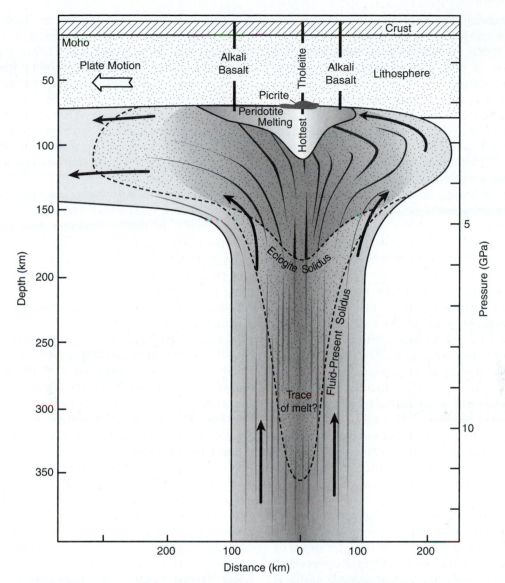

FIGURE 14.20 Diagrammatic plume-tail melting model. Rising plume material (heavy arrows are flow lines) is hotter toward the axis. Fluid-present melting of mantle lherzolite may begin at depths of about 350 km (stippled area), but the extent of such melts depends on the amount of fluid present and is probably minor. Melting of recycled crustal pods and stringers (dark streaks) may also begin near this depth, and such melting may be more extensive locally. Major lherzolite melting occurs at depths near 100 km. The melt fraction is greatest near the plume axis, producing picrites and tholeiites. The extent of plume asymmetry depends on plume flux and plate velocity. Plume-*head* melting is much more extensive (Chapter 15). Based on Wyllie (1988b).

and probably retaining most primordial noble gases), but the plume may contain some entrained depleted upper mantle toward the margins to account for the DM component evident in many OIBs. Kerr et al. (1995), Walker et al. (1999), and Kempton et al. (2000) suggested that the DM signature in deep plumes may also result from subducted oceanic lithospheric mantle that makes its way to greater depths. Whole-mantle convection models, of course, allow for the entire mantle host to be somewhat depleted (e.g., Lyubetskaya and Korenaga, 2007a, 2007b).

Models suggest that melting occurs at three principal levels in a rising plume. According to Wyllie (1988a, 1988b) vapor-present peridotite melting within a rising plume may begin at depths as great as 350 km (stippled area in

Figure 14.20), but volatiles (principally H_2O and CO_2) are generally quite limited and thus no more than a trace of volatile-charged, highly alkaline, low-SiO_2 melt is probably generated at this depth. Dasgupta and Hirschmann (2006) found that carbonate in peridotite is stabilized by about 5 ppm CO_2 and could lower the solidus temperature by 600°C at 6.6 GPa (compared to volatile-free peridotite). The melts produced in their experiments were minor in volume and carbonatitic (Section 19.2). Any melts produced at this deep stage probably fail to escape and they solidify or react with volatile-free peridotite to be incorporated into later melting events, although some (particularly in the cooler plume margins) may be swept away and back into the asthenosphere. Melting at this depth is unlikely to contribute a significant melt directly to OIB magmatism.

Recent "sandwich" type experimental techniques that melt thin layers of eclogitic (MORB-like ocean crust) material sandwiched within a peridotitic groundmass show that eclogite melts first. If present, eclogites would contribute to a second principal plume melting event. Various estimates for the onset of eclogite melting include ~190 km (Sobolev et al., 2005), ~175 km (Yasuda et al., 1997), and ~150 km (Kogiso et al., 2003). The onset of melting of entrained material in Figure 14.20 is chosen as an eclogite solidus at about 175 km depth (near the hot plume axis). Eclogites also have a higher melt productivity (% melt produced per degree of heating above the solidus) than peridotites and the resulting melts are enriched in FeO, TiO_2, and alkalis (similar to many OIBs). Such an olivine-free eclogite source could also account for the very high Ni content of many Hawaiian magmas because Ni would otherwise be concentrated in residual mantle olivine (Hauri, 1996).

At pressures greater than 2 GPa natural eclogites/garnet-pyroxenites apparently straddle a thermal divide separating quartz-normative melts of silica-saturated (typical MORB-like) eclogite and nepheline-normative melts of silica-poor eclogites (Hirschmann et al., 2003). SiO_2-deficient eclogites might be (1) oceanic or delaminated continental mafic cumulates, (2) created by reaction of common MORB-type eclogites with peridotite, or (3) be "re-fertilized" by deeper carbonatite metasomatism or other alkalic fluids. Yaxley and Green (1998) reported that the partial melts of re-fertilized pyroxenites are nepheline-normative picrites at the solidus and become tholeiitic picrites at higher melt fractions. Melting experiments on silica-deficient garnet pyroxenites by Kogiso et al. (2003, 2004) and Hirschmann et al. (2003) produce strongly nepheline-normative OIA basaltic compositions with more conventional (lower) Al_2O_3 than corresponding peridotite melts. Experiments on silica-deficient eclogite with 5 wt. % CO_2 by Dasgupta et al. (2006) produced strongly silica-undersaturated melts at 3 GPa (100 km) and 1225°C, indicating that CO_2 may play in important role in alkaline basalt genesis. Parental melts from eclogites on each side of the thermal divide, if they escape and rise, could thus be parental to both the predominant OIT (SiO_2-saturated) and OIA (SiO_2-undersaturated) magma series.

Yaxley and Green (1998) and Walter (2003) argued that melting of MORB-like eclogite produces siliceous partial melts that are not in equilibrium with the host mantle and therefore react with the peridotites (rather than escape), forming domains of fertile garnet pyroxenite. The darkest stringers in Figure 14.20 represent such pyroxenites (converted from the lighter-shaded deeper eclogites), along with refractory eclogite residue.

Many altered oceanic basalts contain several percent calcite in veins and vugs, most of which probably escapes removal in subduction zones and returns to the mantle (Section 16.8.4). Yaxley and Brey (2004) and Dasgupta et al. (2004, 2006) found that the presence of carbonate lowered the solidus temperature of eclogite and that carbonatite liquids may also form in carbonate-eclogite pods in the deeper first melting event described above. The viscosity of these carbonatite liquids is very low and they probably escape the pods, but then react with and are consumed by (and re-fertilize) the host peridotites. Hauri et al. (1993), for example, described samples of peridotite xenoliths from Samoa that have trace element characteristics indicative of equilibrium with carbonatite liquids and isotopic ratios consistent with EM and HIMU components. They ascribed their xenoliths to mantle plumes that may have contributed the EMII and HIMU signatures to many Samoan basalts. If the melts stall and re-fertilize their mantle surroundings, they may remelt later (as described below).

The third melting event, according to the model in Figure 14.20, is considered the most productive, and is probably limited to depths shallower than 100 km (Wyllie, 1988a, 1988b; Green and Falloon, 2005), where the volatile-free peridotite solidus is reached. OITs and OIAs may both be produced from this level as well. Greater depths of melting of peridotite produces more olivine-rich melts, and Wyllie (1988a, 1988b) proposed that picrites are generated in the hot plume axial region and alkali picrites at lower melt fractions toward the margin. In experiments on pyrolite with 1000 ppm H_2O and 400 ppm CO_2 Green and Falloon (2005) found that separation of <2% partial melts at depths near 90 km yields olivine nephelinites, melilitites, or leucitites; separation of ~4% partial melts at depths near 60 km yields olivine-rich basanites; and separation of 10% partial melts at depths near 40 km yields alkaline basalts that transition into tholeiites at shallower levels with higher melt fractions. Picrites were generated in these experiments with nearly 20% melting near 50 to 80 km. Picrites are favored as common primary plume magmas (Norman and Garcia, 1999; Green et al., 2001, Green and Falloon, 2005) but are rarely erupted. Wyllie (1988a, 1988b) proposed that the dense picrites and alkali picrites stall in lithospheric magma chambers and fractionate to produce the tholeiitic and alkaline basalts. The experiments of Takahashi and Kushiro (1983) indicated that melilitite OIA liquids should not be generated by partial melting of volatile-free peridotite, but they appear to be more compatible with carbonated peridotite (Eggler, 1978; Hirose, 1997).

Having eclogite or garnet pyroxenite in plumes facilitates melt production and may explain several OIB trace element and isotopic trends, as well as the major element concentrations of the highly alkaline end of the spectrum, but their presence is still debated. As mentioned at the end of the preceding section, eclogites (or peridotites altered by eclogite melts) may represent a potential source of the HIMU or EMI isotopic signature in many OIBs (Dasgupta et al., 2006). The experiments of Cordery et al. (1997) and Kogiso et al. (1998, 2003) suggest that mixtures of peridotite and eclogite domains in a marble-cake mantle can yield both tholeiitic and alkaline basalt melts similar to observed OIB magmas. Stracke et al. (1999), on the other hand, argued that Hf-Nd-Th isotopic ratios are sufficiently different for peridotites and garnet pyroxenites or eclogites, and their data suggest the latter two should be absent in the Hawaiian plume. I am attracted to the idea of stringers of eclogitized oceanic crust in the mantle, their potential to carry volatiles through subduction zones, and the effect of these lithologies and volatiles on partial melting. It will be interesting to see how this debate unfolds.

Whether primary or evolved from their respective picrites, erupted tholeiites should predominate toward the hotter central plume axis in the melting region where melt fractions are greater, and alkaline basalts should be more common toward the margins. Pyroxenite and peridotite partial melts *segregate* from the solid plume host at depths between 50 and 80 km, after which they begin to act as independent systems, which determines much of the *major* element character of the magma (trace elements and isotopes reflect more of the deep source characteristics). Fractional crystallization in shallow chambers produces more evolved differentiates for both the tholeiitic and alkaline series.

Models for magma genesis from mantle materials are best constrained using primary magma compositions. In many localities, including Hawaii, fractional crystallization has been sufficiently effective that no rocks with primary composition are exposed (see Section 10.4 for some criteria). As described in conjunction with Figure 11.2, we might circumvent this problem by selecting a parental glass from which one could infer that only olivine has fractionated, and then add small increments of equilibrium olivine until the bulk composition eventually becomes compatible with the most Mg-rich phenocryst found. For example, to estimate the primary magma composition for Loihi seamount in Hawaii, Garcia et al. (1995) added 1451 minute increments of equilibrium olivine (see Equation [13.1] for K_D between olivine and melt) and Cr-spinel (in 98.5:1.5 proportions, the observed average in Kilauea basalts) to primitive basalt compositions. After each increment, the new X_{bulk} and equilibrium mineral compositions are calculated and then added. This is repeated until MgO in the bulk composition reached 16 wt. %, which was deemed a reasonable value for Hawaiian primary magmas. Herzberg et al. (2007) provided a more robust method for estimating primary magmas based on a combination of the inverse method of adding increments of equilibrium olivine compositions and the forward method of calculating mantle melt compositions until the olivine compositions of both converge. They provided a nice tutorial and spreadsheet for doing so in the reference cited above (see also Herzberg and Asimov, 2008).

Magma mixing in shallow fractionating reservoirs can further complicate the volcanic products. Enriched and depleted components of a rising solid plume will inevitably mix to varying degrees during and after melting. Ridge-centered hotspots are associated with rising hot depleted mantle beneath the ridge, and therefore entrain more MORB component, resulting in E-MORBs.

We should also remember the *temporal* variations in plume activity and products. Plume heads generate significantly more magmatism when a new plume first reaches the surface (Chapter 15), and plume tails should eventually wane and die off. Recall also the early voluminous tholeiites giving way to less extensive late alkaline magmas in Hawaii. This trend of increasing alkalinity with time appears to be relatively common for volcanic islands on mature oceanic

crust, but it is not yet known how common. The transition from tholeiites to alkaline magmas by shallow fractional crystallization was rejected early on when the low-pressure thermal divide separating the two series was recognized (Figure 8.13). The customary explanation then relied upon early tholeiites created by a high degree of partial melting of a shallow source followed by less extensive partial melting of a deeper source as activity waned, an explanation that agrees with the experimental data reviewed in Chapter 10. The relatively recent discovery of alkaline melts at Loihi during the very early stages of hotspot activity adds a further complication. The model in Figure 14.20 may provide at least a partial answer. Note that plate motion interacts with rising plume material, dragging the plume into marked asymmetry and shifting lithospheric conduits across the plume melting regime. Loihi may tap low-melt-fraction alkaline partial melts near the advancing front of hotspot migration (right side of Figure 14.20), where temperatures are lower away from the plume axis. As the conduit migrates past the plume axis voluminous tholeiites bury the earlier alkaline basalts. Further migration taps the opposite ("downstream") plume margin and again produces alkaline liquids.

Major element, trace element, and isotopic differences between the Hawaiian tholeiitic and alkaline series preclude derivation from a single source, conforming more to a heterogeneous marble-cake plume. Figure 14.21 shows the Sr-Nd isotopic data for some Hawaiian tholeiitic and alkaline basalts. The data crudely fit a binary mixing line between enriched and depleted components, but notice that the alkaline magmas, which are major element and trace element *enriched*, are isotopically *depleted*, and the high-melt-fraction tholeiites are isotopically depleted. This odd decoupling between trace elements and isotopes at Hawaii

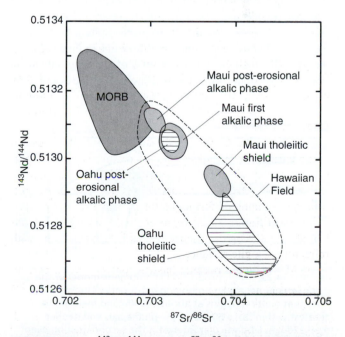

FIGURE 14.21 $^{143}Nd/^{144}Nd$ versus $^{87}Sr/^{86}Sr$ for Maui and Oahu Hawaiian early tholeiitic shield-building and, later, alkaline lavas. From Wilson (1989). Copyright © by permission Kluwer Academic Publishers.

has spawned some interesting and complex models for Hawaiian genesis involving variations in partial melting of a heterogeneous mantle and/or different degrees of lithospheric assimilation (Staudigel et al., 1984; Chen and Frey, 1983, Clague, 1987). The simplest explanations involve low degrees of partial melting (creating *chemically* enriched melts) of *isotopically* depleted mantle sources and higher degrees of partial melting (creating chemically more primitive melts) of isotopically enriched mantle sources. In Figure 14.20, this could be accommodated by more entrained (shallow?) depleted mantle in the plume margins (tapped at low melt fractions to produce alkaline basalts) and more enriched deep mantle and recycled materials toward the axis where tholeiites are produced.

The post-erosional highly alkaline magmatism at Hawaii poses another problem. Why would volcanism flare up again after a hiatus? A numeric melting model for the Hawaiian plume beneath moving lithosphere by Ribe and Christensen (1999), shown in Figure 14.22, suggests that plume melting (assuming homogeneous lherzolite) begins within 25 km of the axial area at about 150 km depth, producing basalts at a rate of ~0.1 g/cm^{3}s (assuming depleted mantle). The melting region extends upward and cools with gradually decreasing melt production but increasing radius (to ~50 km) until melting ceases at the base of the lithosphere at ~90 km depth. Plate motion causes the melt production region to become highly asymmetrical at depths less than 120 km, dragged nearly 200 km in the direction of plate motion. Less than 0.01 g/cm^{3}s of melt is produced in the final 100 km, however, and melting ceases because material descends slightly in the mantle flow-stream due to small-scale convective instability of the lowermost lithosphere as buoyant plume material spreads laterally. In all 45 of their numeric models, Ribe and Christensen (1999) noted a second minor melting zone about 300 to 550 km downstream from the vertical axis of the primary plume as material re-ascends slightly (notice the lower flow line in Figure 14.22) and partially melts again. Over 99% of the magmatism in their models occurs when the volcano is over the hot central

axis of the hotspot plume, followed by a hiatus of 4 to 5 Ma and a second weak episode, all in excellent agreement with the main shield stage of Hawaiian volcanism followed by a weak post-erosional phase. The main hot stage and higher degrees of partial melting should produce tholeiites and the later stage involves low melt fractions and alkaline basalts. Bianco et al. (2005) proposed an alternative model in which the post-erosional secondary magmatism is induced by flexural uplift of the lithosphere in a compensating ring marginal to the lithospheric depression caused by loading of the tholeiitic shields.

The mantle is clearly dynamic and heterogeneous. Over time, portions have undergone loss of incompatible elements and volatile components to melts and fluid phases. Trapped rising melts and metasomatic fluids may have locally enriched some pods or veins of the mantle at shallower levels (Phipps Morgan and Morgan, 1999; Frost, 2006). To this we can now add reintroduction of lost components by subduction (Chapter 16) and recycling, and sinking of portions of the altered subcontinental lithospheric mantle (Chapter 18). The resulting heterogeneous mantle is stirred by convective mixing, although diffusion is too limited to completely homogenize the chemically distinct reservoirs. The result is a locally laminated (marble-cake) mantle on a scale of centimeters to hundreds of meters that may also be stratified on the global scale. Local areas ascend (generally at mid-ocean ridges or as plumes), incorporating parts of the various reservoirs that they encounter in transit, and are partially melted. Many of the melting aspects presented in the previous chapter on MORBs, such as mixing of polybaric incremental melt fractions and channelized flow, may also apply to plumes. Conversely, many of the aspects of the plume model, such as marble-cake stringers of eclogite, carbonate, and so on, may also apply to MORB genesis.

Courtillot et al. (2003) proposed five criteria for deep plume origin:

- A long-lived track of islands or seamounts leading from the present hotspot
- A large igneous province at the initial head breakout
- Magma production in excess of 10^3 kg/s
- High ^{3}He/^{4}He (or primitive neon)
- Significantly low shear-wave velocities at the level of the 660-km transition

They reported that only 9 of 49 major hotspots they considered met three or more of the criteria and only seven met all five (although others may if more data were available). They noted that fluid mechanics suggests very large plume heads and enduring tails can only be produced at depths of origin far below the 660-km transition and proposed that the seven "primary" hotspots—Hawaii, Easter, Louisville, Iceland, Afar, Reunion, and Tristan (see Figure 14.1)—originate at or near the core–mantle boundary and rise to the surface.

Montelli et al. (2006) cataloged deep mantle plumes on the basis of new "finite-frequency tomography" and concluded that deep plumes exist beneath Ascension, the Azores, Canary, Cape Verde, Cook, Crozet, Ester, Kerguelen,

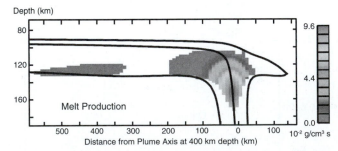

FIGURE 14.22 Melt production in a numeric model of the Hawaiian plume, assuming homogenous peridotitic material. Note that melting begins at about 160 km and melt flux is greatest within 30 to 50 km of the plume axis and deeper than 120 km. Of particular interest is the second melting event 300 km downstream of the primary melt zone, a result of the re-ascension of plume material that previously advected slightly downward beneath the lithosphere. Heavy black lines are mantle flow streamlines. After Ribe and Christensen (1999).

Hawaii, Samoa, and Tahiti. They concluded that Afar, the Atlantic Ridge, Bouvet, Cocos, Louisville, and Reunion had an origin at least as deep as 660 km. See also Boschi et al. (2007) for an assessment of deep plumes.

Courtillot et al. (2003) subdivided the lesser non-primary hotspots into two groups: *secondary* hotspots, which may originate at the 660-km transition zone, and *tertiary* hotspots, with a more superficial origin. As illustrated in Figure 14.23, the low-velocity superswells beneath the southwestern Pacific and Africa (Figure 14.1) also have deep origins but are sufficiently broad and diffuse that they rise very slowly. A thermal boundary layer at the upper interface between a superswell and the surrounding mantle, they argued, could generate secondary plumes. According to their model, superswells stall at the 660-km transition, at least for a time, so most secondary hotspot plumes originate there.

The tertiary hotspots of Courtillot et al. (2003) meet few, if any, of the deep-plume criteria and were assigned a shallow and superficial origin. The ocean basins are riddled with small, scattered seamounts and suggest a non-plume origin for many OIBs (Batiza, 1982; Anderson, 1998, 2005; Anderson and Schramm, 2005). Many may have originated near the spreading ridge as off-axis MORB-associated volcanoes. The mantle is near the peridotite solidus temperature and only minor variations may produce melts (particularly in recycled crustal pods). Proposed non-plume mechanisms for shallow melting include extensional stresses resulting in local near-adiabatic rise, low-melting-point material carried upward by mantle convection, channelized fluids, leaky transform faults and fracture zones, and convective rise deflected by local variations in lithosphere thickness.

As mentioned earlier, several investigators have recently proposed that not all low density plumes need be of strictly thermal origin, and that *chemical plumes* may also occur (e.g., Masters et al., 2000). The numeric modeling of Farnetani and Samuel (2005) abandoned the strictly thermal plume starting assumption and considered the effect of chemical heterogeneities in the deep mantle with no ad hoc temperature perturbation (which typically produces radial plume axis thermal symmetry). Their models indicated a range of plume types with two end-members. Plumes that

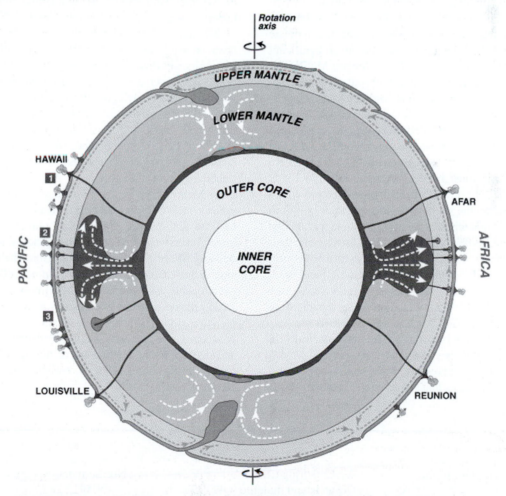

FIGURE 14.23 A schematic cross section through the Earth, showing the three types of plumes/hotspots proposed by Courtillot et al. (2003). "Primary" plumes, such as Hawaii, Afar, Reunion, and Louisville, are deep seated, rising from the D″ layer at the core–mantle boundary to the surface. Superplumes, or superswells, are broader and less concentrated, and they stall at the 660-km transition zone, where they spawn a series of "secondary" plumes. "Tertiary" hotspots have a superficial origin. From Courtillot et al. (2003).

sample a larger proportion of a deep mantle thermo–chemical boundary layer were hotter than the surrounding mantle through which they rise, and thus rise faster and develop the classic large bulbous head (Figure 14.2). At the other end of the plume spectrum are "spout"-like plumes that preferentially sample the upper part of the thermo–chemical boundary layer, develop less of a temperature excess, and rise more slowly with a smaller head. Both types were stalled momentarily at the 660-km transition from which head-type plumes released a secondary head-type plume to the base of the lithosphere and spout-type plumes released only a narrow headless tail. They also noticed that plumes were capable of stretching deep mantle heterogeneities into long-lasting filaments. Such filaments are randomly distributed in the plume and not simply concentrically zoned, and may be tapped by different volcanoes as an oceanic plate moves across the plume tail.

This discussion serves as a condensed summary of our knowledge of the mantle and its derivatives in the simplest situation: beneath the oceanic crust, where contamination by traversing the continental lithosphere does not complicate the products that reach the surface. In Chapter 15, we will look at what happens when basaltic magma, particularly a vigorous plume head, erupts within the continents.

Summary

Oceanic island volcanism occurs mainly within the lithospheric plates and is not governed by plate tectonic processes. The composition of ocean island basalts is more varied than that of MORBs. Three major oceanic magma series are recognized: a tholeiitic series (evolving to rhyolites), an alkaline series (evolving to trachytes and alkaline rhyolites), and a less common highly alkaline series (evolving to phonolites). Isotopes and some trace element data indicate that the mantle is heterogeneous, containing a significant proportion that has been depleted by prolonged MORB extraction and by extraction of melts to produce continental crust. It is unknown whether any primordial mantle remains, but there is ample evidence for enriched mantle components, which may include stalled partial melts, metasomatized mantle (including foundered lithosphere), and crust plus sediment recycled into the mantle by subduction. These mantle constituents must have evolved in isolation for at least 1 Ga for them to develop their distinct isotopic signatures. N-MORBs tap the shallow mantle, and their predominantly depleted characteristics suggest that the upper mantle is largely depleted. OIBs tap a variety of levels, and their greater heterogeneity suggests that the deeper mantle contains more enriched components. Petrologists and geochemists thus embraced a two-layer mantle convection model in which the 660-km transition zone separated an upper depleted mantle from a lower more enriched one.

Many ocean islands occur at the end of trails of seamounts of progressively younger age, indicating a persistent hotspot: the surface expression of a rising mantle plume. Seismic evidence indicates that some plumes originate in the deep mantle, perhaps the D″ layer at the core–mantle boundary. This and similar evidence that subducted slabs also penetrate the 660-km mantle transition zone suggest substantial material flux across the transition and undermine strict two-layer mantle convection models. Whole-mantle convection, on the other hand, should mix and largely re-homogenize the upper depleted and lower enriched materials. Recent mantle models attempt to reconcile these issues by modifying two-layer or whole-mantle convection to accommodate distinct reservoirs and flow across the 660-km transition. The transition is no longer considered an absolute barrier, but controversy persists as to whether it is capable of maintaining some degree of compositional stratification. Regardless of the large-scale variations, most models accept that the mantle is heterogeneous on the meter to kilometer scale, having enriched material as sheared and stretched pods in a less enriched matrix in a "marble-cake" fashion.

Mantle plumes may be of more than one type. Major plumes probably originate near the core–mantle boundary and rise to the surface. Secondary plumes may sprout from larger superswells that stall at a boundary layer, perhaps at 660 km. Minor plumes may result from shallow local extension, shear, fluid enrichment, etc. Plumes reflect their source, but probably also entrain material through which they rise. Rising plume material experiences adiabatic decompression partial melting, principally at shallow depths (<100 km), producing a variety of basalt types depending on the depth and extent of partial melting and on the type of material melted (depleted lherzolite, pods of enriched eclogite-pyroxenite, carbonate, etc.).

Key Terms

Ocean island basalt (OIB) *270*
Seamount *270*
Oceanic plateau *270*
Hotspot *271*
Plume *271*
Swell *271*
Superplume/superswell *272*

Aseismic ridge *272*
Ocean island tholeiite (OIT) *273*
Ocean island alkaline (OIA) basalt *273*
Depleted mantle (DM) *280*
Bulk silicate Earth (BSE) *280*
Enriched mantle (EMI and EMII) *281*
High-μ (HIMU) mantle *282*

Northern hemisphere reference line (NHRL) *282*
DUPAL component *282*
Primitive helium mantle (PHEM) *284*
Focal zone mantle (FOZO) *286*
Common mantle (C) *286*
Marble-cake mantle *289*

Review Questions and Problems

Review Questions and Problems are located on the author's web page at the following address: **http://www.prenhall.com/winter**

Important "First Principle" Concepts

- Mantle plumes are *solid* diapirs as they rise through most of the ductile mantle. They tap and entrain various mantle depths and partially melt only at very shallow depths, producing a variety of basalt types.
- Tholeiites are favored over alkaline basalts by shallower melting, higher partial melt fractions, and high H_2O/CO_2 (if present).
- OIBs are a good basis for inferring the nature of the mantle because they avoid potential contamination associated with traversing the thick, very-enriched continental crust.
- At least part of the mantle has been depleted by the prolonged extraction of continental and oceanic crust.
- Major elements are buffered to some extent by mineral–melt equilibrium and thus tend to reflect the conditions at which a melt was last in equilibrium with the mantle (plus any subsequent fractionation).

- Heavy isotopes do not fractionate during partial melting and crystallization processes, so the isotopic signatures of oceanic basalts reflect more the nature of their mantle source. Trace elements are transitional in nature between isotopes and major elements.
- Based largely on isotopic data, the mantle is revealed to be quite heterogeneous. Some is depleted, and some is enriched with respect to primordial mantle.
- Near-surface crustal material and sediments appear to have been recycled all the way to the core–mantle boundary.
- The mantle appears to be heterogeneous on a very large scale (e.g., DUPAL in the southern Indian Ocean) and a very small scale (meter- to several-kilometer-scale marble-cake layering).
- Mantle partial melting is odd in that BOTH products are less dense than the original. The melt is less dense because of the volume expansion, and the residue because if its higher Mg/Fe.

Suggested Further Readings

Ocean Island Basalts

Basaltic Volcanism Study Project. (1981). *Basaltic Volcanism on the Terrestial Planets*. Pergamon. New York. Sections 1.2.6 and 6.2.2.

Hofmann, A. W. (1997). Mantle geochemistry: The message from oceanic volcanism. *Nature,* **385**, 219–229.

Hofmann, A. W. (2003). Sampling mantle heterogeneity through oceanic basalts: Isotopes and trace elements. In: *Treatise on Geochemistry*. (ed. R.W. Carlson) Chapter 2.03. pp. 61–101. Elsevier, Amsterdam.

McKenzie, D., and R. K. O'Nions. (1995). The source regions of ocean island basalts. *J. Petrol.,* **36**, 133–159.

Prichard, H. M., T. Alabaster, N. B. W. Harris, and C. R. Neary (eds.). (1993). *Magmatic Processes and Plate Tectonics*. Special Publication **76**. Geological Society. London.

Saunders, A. D., and M. J. Norry (eds.). (1989). *Magmatism in the Ocean Basins*. Special Publication **42**. Blackwell. Oxford, UK.

Wilson, M. (1989). *Igneous Petrogenesis: A Global Tectonic Approach*. Unwin Hyman. London. Chapter 9.

Mantle Isotopes

Farley, K. A., and E. Neroda. (1998). Noble gases in the Earth's mantle. *Ann. Rev. Earth Planet. Sci.,* **26**, 189–218.

Hart, S. R., and A. Zindler. (1989). Constraints on the nature and development of chemical heterogeneities in the mantle. In: *Mantle Convection: Plate Tectonics and Global Dynamics* (ed. W. R. Peltier). Gordon and Breach. New York. pp. 261–387.

Hilton, D. R., and D. Porcelli (2003). Noble gases as mantle tracers. In: *Treatise on Geochemistry*. Elsevier, Amsterdam. pp. 277–318.

Porcelli, D., C. J. Ballentine, and R. Wieler (eds.). (2002). *Noble Gases in Geochemistry and Cosmochemistry*. Reviews in Mineralogy and Geochemistry **47**. Mineralogical Society of America/Geochemical Society. Washington, D.C.

Zindler, A. H., and S. R. Hart. (1986). Chemical geodynamics. *Ann. Rev. Earth Planet. Sci.,* **14**, 493–571.

Mantle Plumes

Condie, K. C. (2001). *Mantle Plumes and Their Record in Earth History*. Cambridge University Press. Cambridge, UK.

Courtillot, V., A. Davaille, J. Besse, and J. Stock. (2003). Three distinct types of hotspots in the Earth's mantle. *Earth Planet. Sci. Lett.,* **205**, 295–308.

Davies, G. F. (2005). The case for mantle plumes. *Chinese Sci. Bull.,* **50**, 1541–1554.

Foulger, G. R. (2005). Mantle plumes: Why the current skepticism? *Chinese Sci. Bull.,* **50**, 1555–1560.

Foulger, G. R., J. H. Natland, D. C. Presnall, and D. L. Anderson (eds.). (2005). *Plates, Plumes and Paradigms*. Special Paper **388**. Geological Society of America. Boulder, CO.

Nataf, H.-C. (2000). Seismic imaging of mantle plumes. *Ann. Rev. Earth Planet. Sci.,* **28**, 391–417.

Suetsugu, D., B. Steinberger, and T. Kogiso. (2004). Hotspots and mantle plumes. In: *The Encyclopedia of Geology* (eds. R. C. Selley, L. R. M. Cocks, and I. R. Plimer). Elsevier. Oxford, UK.

www.mantleplumes.org is a marvelous site for plume-related material.

Mantle Structure and Geodynamics

Albarède, F., and R. D. van der Hilst. (1999). New mantle convection model may reconcile conflicting evidence. *EOS,* **45**, 535–539.

Bennett, V. C. ((2003). Compositional evolution of the mantle. In: *Treatise on Geochemistry*. (ed. R.W. Carlson) Chapter 2.13. p. 493–519. Elsevier, Amsterdam.

Christensen, U. (1995). Effects of phase transitions on mantle convection. *Ann. Rev. Earth Planet. Sci.,* **23**, 65–87.

Davies, G. F. (1999). *Dynamic Earth: Plates, Plumes and Mantle Convection*. Cambridge University Press. Cambridge, UK.

Davies, G. F., and M. A. Richards. (1992). Mantle convection. *J. Geol.,* **100**, 151–206.

Davies, J. H., J. P. Brodholt, and B. J. Wood (eds.). (2002, November). *Chemical Reservoirs and Convection in the Earth's Mantle.* Philosophical Transactions of the Royal Society of London.

Garnero, E. J. H. (2000). Heterogeneity of the lowermost mantle. *Ann. Rev. Earth Planet. Sci.,* **28**, 509–537.

Garnero, E. J. H. (2004). A new paradigm for the Earth's core-mantle boundary. *Science,* **304**, 834–836.

Gurnis, M., M. E. Wysession, E. Knittle, and B. A. Buffett (eds.). (1998). *The Core–Mantle Boundary Region.* Geodynamics Series **28**. American Geophysical Union. Washington, DC.

Jackson, I. (ed.). (1998). *The Earth's Mantle. Composition, Structure, and Evolution.* Cambridge University Press. Cambridge, UK.

Kellogg, L. H. (1992). Mixing in the mantle. *Ann. Rev. Earth Planet. Sci.,* **20**, 365–388.

Peltier, W. R. (ed.). (1989). *Mantle Convection: Plate Tectonics and Global Dynamics.* Gordon and Breach. New York.

Schubert, G., D. L. Turcotte, and P. Olson. (2001). *Mantle Convection in the Earth and Planets.* Cambridge University Press. Cambridge, UK.

Tackley, P. J. (2000). Mantle convection and plate tectonics: Toward an integrated physical and chemical theory. *Science,* **288**, 2002–2007.

Turcotte, D. L., and G. Schubert. (2002). *Geodynamics* (2nd ed.). Cambridge University Press. Cambridge, UK.

van Keken, P. E., E. H. Hauri, and C. J. Ballentine. (2002b). Mantle mixing: The generation, preservation, and destruction of chemical heterogeneity. *Ann. Rev. Earth Planet. Sci.,* **30**, 493–525.

van Keken, P. E., C. J. Ballentine, and E. H. Hauri. (2003). Convective mixing in the Earth's mantle. In: *Treatise on Geochemistry.* (ed. R.W. Carlson) Chapter 2.12. pp. 277–318. Elsevier, Amsterdam.

Plume Melting

Green, D. H., and T. J. Falloon. (2005). Primary magmas at mid-ocean ridges, "hotspots," and other intraplate settings. In: *Plates, Plumes and Paradigms* (eds. G. R. Foulger,

J. H. Natland, D. C. Presnall, and D. L. Anderson). Special Paper **388**. Geological Society of America. Boulder, CO. pp. 217–247.

Kogiso, T., M. M. Hirschmann, and D. J. Frost. (2003). High-pressure partial melting of garnet pyroxenite: Possible mafic lithologies in the source of ocean island basalts. *Earth Planet. Sci. Lett.,* **216**, 603–617.

Kogiso, T., K. Horose, and E. Takahashi. (1998). Melting experiments on homogeneous mixtures of peridotite and basalt: application to the genesis of ocean island basalts. *Earth Planet. Sci. Lett.,* **162**, 45–61.

Leitch, A. M., and G. F. Davies. (2001). Mantle plumes and flood basalts: Enhanced melting from plume ascent and an eclogite component. *J. Geophys. Res.,* **106**, 2047–2059.

Phipps Morgan, J. (2001). Thermodynamics of pressure release melting of a veined plum pudding mantle. *Geoghem. Geophys. Geosystems,* **2**, doi2000GC000049.

Ribe, N. M., and U. R. Christensen. (1999). The dynamical origin of Hawaiian volcanism. *Earth Planet. Sci. Lett.,* **171**, 517–531.

Sobolev, A. V., A. W. Hofmann, S. V. Sobolev, and I. K. Nikogosian. (2005). An olivine-free mantle source of Hawaiian shield basalts. *Nature,* doi:10.1038/nature03411, 1–8.

Walter, M. J. (2003). Melt extraction and compositional variability in mantle lithosphere. In: *Treatise on Geochemistry.* (ed. R.W. Carlson) Chapter 2.08. pp. 61–101. Elsevier, Amsterdam.

Watson, S., and D. McKenzie. (1991). Melt generation by plumes: A study of Hawaiian Volcanism. *J. Petrol.,* **32**, 501–537.

Wyllie, P. J. (1988a). Solidus curves, mantle plumes, and magma generation beneath Hawaii. *J. Geophys. Res.,* **93**, 4171–4181.

Wyllie, P. J. (1988b). Magma genesis, plate tectonics, and chemical differentiation of the Earth. *Rev. Geophys.,* **26**, 370–404.

Yaxley, G. M., and D. H. Green. (1998). Reactions between eclogite and peridotite: Mantle refertilization by subducted oceanic crust. *Schewiz. Mineral. Petrogr. Mill.,* **78**, 243–255.

15

Continental Flood Basalts

Questions to be Considered in this Chapter:

1. What are large igneous provinces? Where are they, and what causes them?

2. How do hotspot plumes relate to the generation of flood basalts, and how do the processes and products differ on the continents as compared to ocean island basalts?

3. How do plumes affect extension and continental breakup?

Several times in the geologic past, tremendous outpourings of basaltic lavas have inundated areas, amounting to the largest igneous events on Earth. They have occurred within the ocean basins, on the continents, and along continental margins. These outpourings are generally associated with plume activity, and therefore overlap with processes discussed in the previous chapter. The great volume and short duration, however, suggest an association with the surfacing of the large head of a newly initiated plume (Figure 14.2). We begin with a general description of such provinces, and then concentrate on the well-exposed subaerial examples: continental flood basalts.

15.1 LARGE IGNEOUS PROVINCES

Large igneous provinces (LIPs) (Coffin and Eldholm, 1991, 1994; Mahoney and Coffin, 1997) are rather loosely defined. They were originally intended to include voluminous outpourings, predominantly of basalt, over very short durations, but the definition didn't specify minimum size, duration, petrogenesis, or setting. A recent attempt to refine and expand the classification by Sheth (2006), concentrating on the size issue, includes island arcs (Chapter 16), continental arcs and batholith belts (Chapter 17), and a number of other igneous occurrences. I prefer to adhere more closely to the initial definition and restrict LIPs to largely basaltic provinces, the origins of which lack an obvious plate tectonic control (unlike MORBs or subduction-related activity). LIPs characteristically cover large areas ($>10^5 \, \text{km}^2$), and the great bulk of the magmatism occurs in less than about 1 Ma, although activity may trail off for considerably longer. Figure 15.1 is a map showing the major basaltic LIPs on the Earth. Principal LIPs in the ocean basins include **oceanic volcanic plateaus (OPs)** and **volcanic passive continental margins** (seaward remnants of volcanism associated with continental rifting). **Oceanic flood basalts** are LIPs distinguished from oceanic plateaus by some investigators because they don't form morphologic plateaus, being neither flat-topped nor elevated more than 200 m above adjacent seafloor. Examples include the Caribbean, Nauru, East Mariana, and Pigafetta provinces. Other investigators also include aseismic ridges and large ocean island and seamount clusters in the category of LIPs.

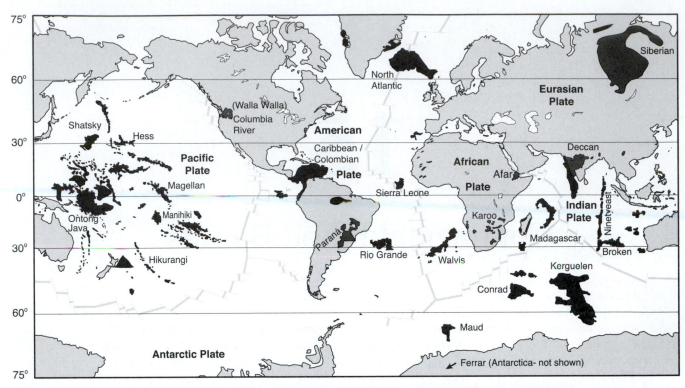

FIGURE 15.1   Map of the major large igneous provinces (LIPs) on Earth, including continental flood basalt provinces, volcanic passive margins, oceanic plateaus, aseismic submarine ridges, ocean basin flood basalts, and seamount groups. After Saunders et al. (1992) and Saunders (personal communication).

We shall concentrate here on the more easily observed continental manifestations, commonly called **continental flood basalts (CFBs)**, or **plateau basalts**. See Kerr (2003) for a detailed description of oceanic plateaus. A more ancient term, *traps* (meaning "steps," a reference to the step-like geomorphology of eroded flow layers, Figure 15.2) is also occasionally used for CFBs, and is part of the formal name applied to some occurrences, such as the Deccan Traps and Siberian Traps. The flows are unusually fluid and rapidly erupted, hence covering extensive areas with large volumes of magma (both cumulatively and as individual flows). CFBs are typically fissure fed, corresponding to extensional tectonics and continental rifting. Some examples of major CFB provinces are listed in Table 15.1, along with ages and approximate volumes. The volumes are difficult to estimate, and

FIGURE 15.2   Step-like erosion surface commonly formed on layered basaltic flows. Hell's Canyon, Idaho–Oregon border. Photo courtesy S. Reidel. From Geological Society of America Special Paper 239. Copyright © Geological Society of America, Inc.

TABLE 15.1   Major Flood Basalt Provinces

Name	Approximate Volume	Age	Location
CRB	$1.7 \cdot 10^5$ km³	Miocene	Northwestern United States
Keeweenawan	$4 \cdot 10^5$ km³	Precambrian	Lake Superior area
Deccan	$8.6 \cdot 10^6$ km³	Cretaceous–Eocene	India
Paraná	(area $> 10^6$ km²)	early Cretaceous	Brazil
Karroo	$> 2 \cdot 10^6$ km³	early Jurassic	S. Africa
Siberian	$3–5 \cdot 10^6$ km³	late Permian–early Triassic	Siberia

erosion of course removes significant proportions in many provinces, so these values are intended only as crude indicators of the magnitude of volcanic activity involved in these events.

The examples in Table 15.1 comprise most of the major recognized CFB provinces, but there are smaller and less well-documented ones in Antarctica, Australia, China, and elsewhere. The giant flood basalt accumulations grade downward in size to numerous smaller, rift-related volcanic provinces. Erosion in these volcanic piles has revealed cogenetic mafic intrusions, including sills, dikes, and the more massive layered mafic intrusions discussed in Chapter 11. Only one of the major provinces listed is distinctly pre-Mesozoic, and seven have occurred within the past 250 Ma. It is not yet clear how this is to be interpreted in terms of global dynamics.

15.2 THE TECTONIC SETTING OF CFBs

An important aspect of CFB (and most major LIP) volcanism is its correlation with the process of rifting. Although difficult to determine which leads to which, it is tempting to attribute both to some deep-seated thermal instability, such as a plume. The rifting can be relatively minor (as would be the associated volcanics as a rule), or it can lead to continental breakup and the formation of a new ocean basin. Rifts that do not lead to new oceans are called *failed rifts*, or *aulacogens*. The Keweenawan Province in the north-central United States is associated with what is termed the "mid-continent gravity high," a major geophysical anomaly that extends over 3000 km from Kentucky northwest to Lake Superior and then southwest into Kansas. This gravity high suggests thinning of the continental crust and the presence of more dense mafic crust or mantle at relatively shallow depths. The Siberian Traps is another example of a CFB associated with a failed rift (a major coeval basin on the west margin of the traps indicates extension). Is it coincidence that these are the two oldest large flood basalts on record, or may we attribute this to some earlier, more vigorous mantle dynamics?

CFBs are typically associated with the incipient stages of "successful" continental rifting. The Paraná, Karroo, North Atlantic, and Antarctic provinces occurred at sites undergoing initial continental fragmentation that led to the opening of the Atlantic and Indian oceans (Figure 15.3). These are now the classic volcanic passive margins, and CFBs are an early continental manifestation. Figure 15.3 illustrates the relationships between the Karroo, Paraná, Antarctic, and Etendeka (Namibia) provinces prior to the breakup of Gondwanaland during the Jurassic and Cretaceous. The earlier Karroo and Antarctic flood basalts erupted just before the Jurassic opening of the Indian Ocean, and the later Paraná–Etendeka were precursors to the initial rifting of the South Atlantic in the Late Jurassic and Cretaceous. Such correlations provide excellent criteria for plate tectonic reconstructions, as well as for models of continental fragmentation. Not all continental rifts are loci of CFB volcanism, however, suggesting that continental rifting is not, of itself, a sufficient mechanism to generate flood basalts.

Mantle plumes and hotspots are commonly associated with continental breakup, and plumes are now generally

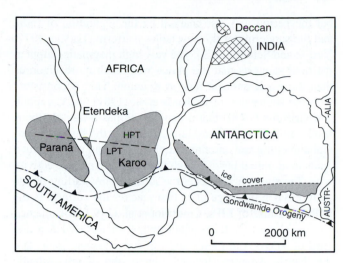

FIGURE 15.3 Flood basalt provinces of Gondwanaland prior to breakup and separation. After Cox (1978). The dashed line separates HPT and LPT (high and low P_2O_5–TiO_2) magmas, as described later in the chapter.

recognized as the principal cause of LIPs, including continental flood basalts (Morgan, 1972; Richards et al., 1989; Campbell and Griffiths, 1990; Gallagher and Hawkesworth, 1994; White and McKenzie, 1995; Turner et al., 1996; Mahoney and Coffin, 1997). Figure 15.4 shows that the Paraná and Etendeka provinces are of an age and in a position to have been located over the Tristan hotspot at the time of initial development and rifting in the South Atlantic. The combination of rifting and plume development results in exceptionally high thermal output, which probably resulted in the local development of CFBs.

In the case of the Deccan in India, the period of intense igneous activity postdated the rifting between India and Africa by about 50 Ma. Plate tectonic reconstructions for the period of flood volcanism put the Deccan area near or over the plume now beneath Reunion Island (Cox, 1980; Duncan et al., 1989). In this case, hotspot activity alone was apparently sufficient to generate this huge flood basalt province,

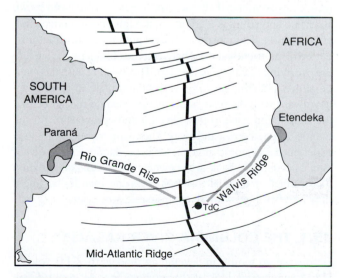

FIGURE 15.4 Relationship of the Etendeka and Paraná Plateau provinces to the Tristan hotspot. After Wilson (1989). Copyright © by permission Kluwer Academic Publishers.

but the period was also marked by the separation of India and the sundering of the Seychelles platform (Larson, 1977; Norton and Sclater, 1979). The very high magmatic output at the time suggests that the plume was much more vigorous than the weak remnant now at Reunion. This is consistent with the suggestion of Richards et al. (1989) and Campbell and Griffiths (1990) that a new plume has a large bulbous head followed by a narrow tail. The head is the site of considerable melting and entrainment of surrounding mantle which leads to an initial burst of high thermal and volcanic activity when the plume first reaches the surface. This starting-plume period, when the large head first reaches the surface, is the probable cause of LIPs. Courtillot et al. (1999) documented similar plume–rift–CFB relationships for the Red Sea and Gulf of Aden (Afar plume and Ethiopian/Yemen Traps) and the North Atlantic (Iceland plume and North Atlantic Province). Every major CFB is associated with both a plume and extension, and most principal oceanic plateaus appear to have developed near a spreading center at the time of their activity (Kerr, 2003). The peak volcanic activity for the Deccan around 65 Ma b.p. is tantalizingly close to the infamous Cretaceous–Tertiary boundary, marked by a mass extinction. This has led some authors to suggest that the intense Deccan volcanism may have put sufficient gas and aerosols into the atmosphere to have blocked the sun for years and to qualify as an alternative candidate to comet impact for the cause of the extinctions (Officer and Drake, 1985). Others have attempted to relate the Deccan Traps to melting caused by the impact itself (Hartnady, 1986). Neither seems likely now, as evidence for an impact in Central America of appropriate age to cause the extinctions grows ever more convincing.

Another possible setting for some CFBs is back-arc spreading. Many investigators consider the Columbia River Basalt Group to be a result of back-arc spreading behind the Cascade arc in a continental environment (Figure 15.5). This might be called a "failed" back-arc because it never fully separated to an offshore arc such as Japan with a marginal sea behind it. Another term for such a failed back-arc rift that retains most of its continental character is an *ensialic back-arc basin*. Back-arc spreading is a common, though poorly understood, phenomenon. The most widely accepted mechanism at present involves frictional drag between the subducting plate and the overlying mantle (Chapter 16). This would cause the mantle above and adjacent to the subducting plate to be dragged downward with the plate, requiring the mantle further behind the subduction zone to flow upward and compensate for the mantle dragged down and away. The rising mantle might then experience decompression melting, producing CFB volcanism, or eventually back-arc oceanic crust. CFBs are certainly the product of thermal instabilities in the mantle, but these may be associated with several possible processes.

15.3 THE COLUMBIA RIVER BASALTS

Before we go too far in developing models for CFB volcanism, I suggest we look at the products in a bit more detail. Once again, the variety is well beyond the scope of this text, and we shall content ourselves with a detailed look at one example of flood basalts, the one in my backyard. I will then compare it to others in the most general fashion to indicate the similarities and variations among the provinces. Just as OIBs provided us with a window of sorts on the nature of the suboceanic mantle, so too may CFBs, by comparison, contribute valuable data on the nature of the subcontinental mantle.

15.3.1 The Setting of the Columbia River Basalt Group

Figure 15.5 shows the aerial extent and present tectonic setting of the Columbia River Basalt Group (CRBG) in the northwestern United States. The Juan de Fuca Plate, presently being subducted beneath the Cascade Arc, is a remnant of the more extensive Farallon Plate that was subducting beneath the entire west coast of North America during the Mesozoic. As a result, the west coast experienced a substantial period of calc-alkaline igneous activity (Chapter 17) prior to the eruption of the CRBs. Note the curvature of the Cretaceous continental margin, represented rather vaguely by the dashed lines in Figure 15.5. This "Columbia Embayment" suggests that the margin curved abruptly eastward north of California (the Sierra Nevada Batholith and Klamath Mountains) to the site of the Idaho Batholith in western Idaho. This embayment may reflect the curvature of the original craton to the east, or it may represent removal of a wedge of continental crust by rifting and strike-slip motion, leaving some oceanic crust in the embayment. Lately it has been recognized that the curvature is greatly exaggerated by Basin and Range extension to the south, which moved the California coast westward relative to Idaho, Oregon, and Washington during the Cenozoic. In the embayment area several island-arc-like terranes of Triassic and Jurassic age were accreted to the continent in the Blue Mountains of northeastern Oregon and western Idaho prior to the Cretaceous.

Roughly 30 Ma ago, the East Pacific Rise collided with the North American Plate, converting much of the subduction regime to strike-slip motion in California (Atwater, 1970). At approximately the same time, the Basin and Range extension began in the southwestern United States, and calc-alkaline volcanism was replaced by a bimodal basalt–rhyolite type of activity (Lipman et al., 1972; Christiansen and McKee, 1978). The area of the Columbia Embayment experienced another strange event when subduction activity migrated abruptly from near the Idaho border westward to the present coast and the Cascade arc as the Blue Mountain terranes were accreted. This migration was probably a series of discontinuous westward jumps in Washington and Oregon to accommodate the docked terranes, but may have been more gradual to the south of the terranes. Such a gradual migration may have resulted from a rapid steepening of the dip of the subducting plate from a shallow (Laramide-type) dip to a more "normal" dip angle or from slab "rollback" (Figure 17.11). Migration of arc volcanism left the CRB area in a back-arc setting with subduction-altered mantle (and possibly an expanse of oceanic crust) beneath at least part of the jump interval.

The CRBG is composed of more than 300 individual flows, with an average volume of 500 to 600 km^3/flow

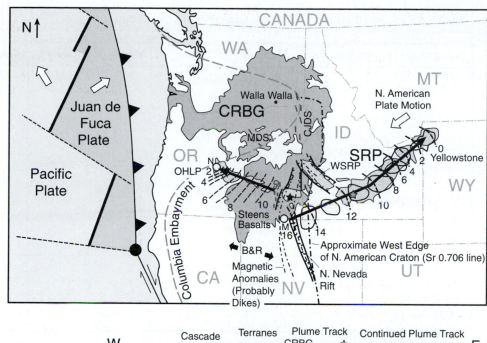

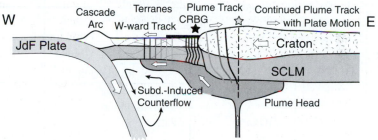

FIGURE 15.5 Setting of the Columbia River Basalt Group in the northwestern United States. Black star is the location proposed by Camp and Ross (2004) of the 16.6 Ma outbreak of the plume and plume-related basaltic volcanism. Gray star is the location of the deep plume conduit proposed by Jordan et al. (2004). Light shaded areas are Quaternary basalts and stippled areas are rhyolite centers. Heavy dashed curves represent the progressive younging of rhyolitic centers (with ages in Ma). Those on the east represent the proposed Yellowstone hotspot track (heavy arrow). Those on the west are the opposing westward track leading to Newberry Volcano (N), with ages reported by Jordan et al. (2004). Heavy solid lines are graben-bounding normal faults (ticks on downthrown side). Light dot-dash lines are gravity anomalies of Glen and Ponce (2002), believed to represent subsurface dikes. SRP = Snake River Plain basalt–rhyolite province. CJDS = Chief Joseph dike swarm, MDS = Monument dike swarm. SD = Steens dikes. Only a few dike orientations are indicated for each swarm. B&R = Basin and Range. OHLP = Oregon high lava plain. M = McDermitt Caldera. The lower cross section is diagrammatic, generally across southern Oregon and Idaho (south of the main CRBG), and illustrates the westward deflection of the plume head by the deep keel of the North American craton to beneath the thinner accreted terranes and the migration of the hotspot tracks both east and west. After Camp and Ross (2004) and Jordan et al. (2004) © American Geophysical Union with permission.

(Tolan et al., 1989). The CRBG is subdivided into several formations (Table 15.2). Although the Imnaha Basalts were considered the initial eruptive phase, recent mapping in eastern Oregon has revealed continuity with the basalts of the Steens Mountain shield volcano in the southeastern corner of the state (Camp et al., 2002; Camp and Ross, 2004; Hooper et al., 2008). Activity thus began about 16.6 Ma b.p. in the southeastern portion of the area with the eruption of the Lower Steens Basalts, which Hooper et al. (2008) tentatively proposed be included as the lowermost formation of the CRBG. Volcanic activity then moved rapidly northward and increased dramatically in flux. Nearly all (98%) of the CRBG basalts were erupted, mostly in eastern Washington, during the 2 Ma

period between 16.6 and 14.5 Ma, peaking at about 15.5 Ma with the Grande Ronde flows, which account for over 85% of the volume of the CRBG in less than 1 Ma (Figure 15.6). There was a brief hiatus in activity between the latest Grande Ronde and earliest Wanapum basalts at about 15.3 Ma. Activity declined rapidly following the Wanapum flows but extended to 6 Ma in the form of late reduced flows of the Saddle Mountains Basalt, which were topographically controlled, largely by river valleys and canyons at the time. The vast majority of flows issued from the nearly N–S Chief Joseph dike swarm fissure system (Figures 4.10 and 15.5). Based on geochemical similarities, the Imnaha, Grande Ronde, and Wanapum formations have been combined as the Clarkston

TABLE 15.2　Stratigraphy of the Columbia River Basalt Group

Sub-group	Formation	Member		Magnetic Polarity*	K/Ar Dates
Clarkston	Saddle Mountains Basalt	Lower Monumental		N	6 Ma
		Ice Harbor		N, R	8.5
		Buford		R	
		Elephant Mountain		R, T	10.5
		Pomona		R	12
		Esquatzel		N	
		Weissenfels Ridge		N	
		Asotin		N	13
		Wilbur Creek		N	
		Umatilla		N	
	Wanapum Basalt	Priest Rapids		R	14.5
		Roza		T, R	
		Frenchman Springs		N	
		Eckler Mountain		N	
	Grande Ronde Basalt	See Reidel et al. (1989) for Grande Ronde Units		N_2	15.0
		Picture Gorge		R_2	
				N_1	
				R_1	
	Imnaha Basalt	See Hooper et al. (1984) for Imnaha Units		R_1	
				T	
				N_0	
				R_0	16.5
	Lower Steens			R_0	16.6

*N = normal, R = reversed, T = transitional

Data from Reidel et al. (1989), Hooper and Hawkesworth (1993), Hooper (1997), and Hooper et al. (2008).

Basalt by Hooper and Hawkesworth (1993). The Picture Gorge Basalts are contemporaneous with part of the Grande Ronde, but erupted from their own fissure system, the NW–SE-oriented Monument dike swarm west-southwest of the Chief Joseph swarm (Figures 4.10 and 15.5). Camp and Ross (2002) proposed a second migration trend (in addition to the main northward trend) from the Steens Mountain area toward the northwest, to the Picture Gorge. The Picture Gorge is not only distinct geographically, it also has some geochemical characteristics that distinguish it from the main Columbia River Basalt Group, indicating a similarity to the younger basalts of the eastern Oregon High Lava Plateau. It is thus placed in a separate box in Table 15.2.

During peak activity, some massive individual flows may have exceeded $2000 \ km^3$ or even $3000 \ km^3$, which qualify them as the largest known terrestrial lava flows (Tolan et al., 1989). Typical flow thicknesses are in the 20 to 80 m range, but may exceed 150 m where ponded in low-lying areas. A major eruption occurred on average about every 100,000 years during the peak stage. Some Wanapum and Grande Ronde Basalts advanced over 600 km, following the Columbia River valley to the Pacific Ocean. Some of these flows were apparently invasive into the shelf sediments, and continued to flow offshore beneath the less dense sedimentary cover. Exposures are now found along the Washington and Oregon coasts as minor flows and dikes (Snavely et al., 1973).

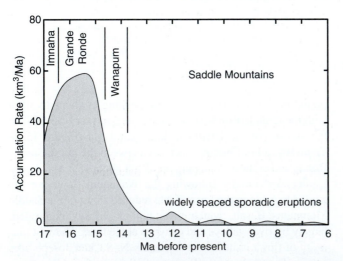

FIGURE 15.6 Time-averaged extrusion rate of CRBG basalts as a function of time, showing cumulative volume. After Hooper (1988a). Copyright © by permission Kluwer Academic Publishers.

TABLE 15.3 Selected Columbia River Basalt Analyses

	Imnaha		Grande Ronde			Wanapum			Saddle Mountains		
	Am. Bar	Rock Cr	High Mg	Low Mg	Prineville	Rob. Mtn	Fr. Spr.	Roza	Umatilla	Asotin	Goose Is.
SiO_2	51.1	49.5	53.8	55.9	51.6	50.0	52.3	51.2	54.7	50.7	47.5
TiO_2	2.24	2.41	1.78	2.27	2.71	1.00	3.17	3.13	2.80	1.45	3.79
Al_2O_3	15.1	16.3	14.5	14.0	13.9	17.1	13.2	14.1	14.1	16.2	12.5
FeO*	13.0	12.4	11.4	11.8	12.3	10.0	14.4	13.9	12.6	9.64	17.5
MnO	0.22	0.20	0.19	0.19	0.24	0.14	0.22	0.23	0.17	0.17	0.27
MgO	5.07	6.06	5.25	3.36	4.44	7.84	4.04	4.39	2.71	8.19	4.41
CaO	9.31	9.15	9.07	6.88	8.12	11.0	7.90	8.48	6.14	10.7	8.80
Na_2O	2.58	2.58	2.83	3.14	3.36	2.44	2.67	2.72	3.20	2.22	2.44
K_2O	0.91	0.93	1.05	1.99	2.02	0.27	1.41	1.22	2.68	0.51	1.23
P_2O_5	0.42	0.41	0.28	0.43	1.39	0.19	0.71	0.67	0.88	0.18	1.54
Total	99.25	99.94	100.15	99.96	100.08	99.99	100.02	100.04	99.98	99.96	99.98
Mg#	41	47	45	34	39	58	33	36	28	60	31
Norm											
q	5	2	6	10	4	0	9	7	9	2	4
or	5	5	6	12	12	2	8	7	16	3	7
ab	22	22	24	26	28	21	22	23	27	19	21
an	27	30	24	18	17	35	20	22	16	33	19
di	14	10	16	11	12	15	12	12	7	15	12
hy	17	20	15	12	13	20	14	14	11	21	18
ol	0	0	0	0	0	2	0	0	0	0	0
mt	5	6	5	5	6	4	7	7	6	4	8
il	4	5	3	4	5	2	6	6	5	3	7
ap	1	1	1	1	3	0	2	2	2	0	4
%an	55	58	50	41	37	63	47	50	37	64	49

Data from Basaltic Volcanism Study Project (1981), Table 1.2.3.2.

Many of the massive flows are remarkably uniform in composition throughout their extent of outcrop, which suggests amazingly little magmatic differentiation or chemical zonation in what must have been a substantial magma chamber (or chambers) beneath the Columbia Plateau. The occurrence of such gigantic sheet flows, rather than shield volcanoes, indicates an unusually large volume of magma erupted in a very short time from the chamber. Shaw and Swanson (1970) envisaged lava fronts 50 m high perhaps 100 km long, moving down the gentle slopes toward the Pacific at 3 to 5 km/hr. Ho and Cashman (1995) determined that the Gingko flow of the Frenchman Springs Member lost only 20°C as it traveled 550 km across the plateau. Self et al. (1997) proposed that the CRBG flows advanced as a succession of large lobes that were fed *internally*, each new magma pulse inflating the crust of older flows (see Figure 4.12c–e). The flows may thus have retained heat beneath the insulating crust and formed over a period of years rather than the days to weeks implied by the Shaw and Swanson (1970) model.

15.3.2 CRBG Petrography and Major Element Geochemistry

The Columbia River Basalts are tholeiitic and chemically similar to MORBs and OITs, although generally more evolved. Most CRBG lavas are aphyric, so the chemical composition might be considered representative of true liquid compositions. A few flows have sparse phenocrysts of plagioclase, and others may have additional subordinate olivine and/or clinopyroxene. The Imnaha Basalt is an exception, and is coarsely phyric, with plagioclase dominating over olivine and subordinate clinopyroxene. Selected analyses and CIPW norms are presented in Table 15.3. The basalts are hypersthene normative, and most are slightly silica oversaturated (quartz normative), though a few are undersaturated. SiO_2 ranges from 47.5 to 56 wt. %, which, along with the low Mg# values, indicates that the CRBG lavas are moderately evolved and cannot represent primary mantle melts. In fact, over 80% of the Columbia River basalts are actually basaltic andesites on the basis of silica content (Figure 2.4). The matrix usually consists of plagioclase, augite, pigeonite, an Fe-Ti oxide, and some glass.

The chemical range of the CRBG is somewhat restricted. Just as there are no primary melts or picrites found, more siliceous types, such as andesites or rhyolites, are also absent. The accumulated chemical data, however, reveal significant chemical diversity among the flows and individual provinces. Compositional jumps or shifts were the basis for distinguishing the individual formations, such as a significant increase in SiO_2 and K_2O between the Imnaha and Grande Ronde and a decrease in SiO_2 and increase in Fe between the Grande Ronde and Wanapum Basalts.

Figure 15.7 is a variation diagram showing the changes in selected major element oxides plotted against

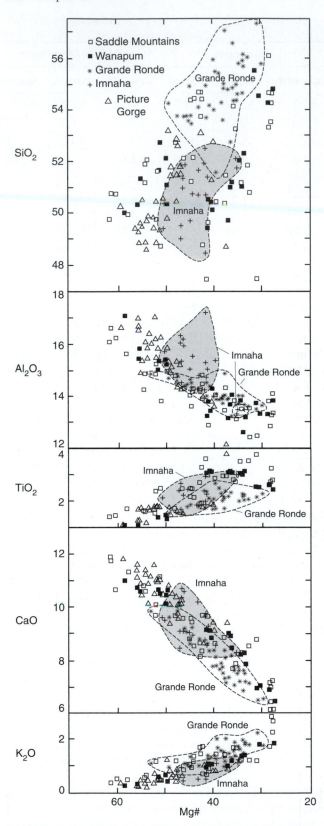

FIGURE 15.7 Variation in wt. % of selected major element oxides versus Mg# for units of the Columbia River Basalt Group. Data from Table 1.2.3.3 of Basaltic Volcanism Study Project (1981), Hooper (1988a), Hooper and Hawkesworth (1993).

Mg# for several representative analyses from the CRBG. The analyses conform to a broadly-defined evolutionary trend of increasing SiO_2, TiO_2, and K_2O along with decreasing CaO and Al_2O_3 as Mg# decreases (toward more evolved types). This trend is familiar by now, and is compatible with fractional crystallization of plagioclase plus clinopyroxene and/or olivine.

Despite this overall trend, however, there is no consistent chemical pattern related to stratigraphy or age within the Columbia River Basalts. The Imnaha and Grande Ronde formations vaguely follow the evolutionary trend in a temporal sense (Figure 15.7), but there is considerable overlap, and the trends do not hold within the formations. The Grande Ronde is the most evolved formation, on average, in the CRBG, and the earliest Wanapum flows that follow are among the most primitive to be found. The later Wanapum and Saddle Mountains formations are highly variable in their chemical composition.

Some limited sequences of flows have behaved in a sufficiently coherent fashion to be modeled on the basis of relatively simple evolutionary processes. For example, Hooper (1988b) successfully modeled the geochemical variation in the American Bar flows of the Imnaha Basalts based on a combined magma chamber recharge and fractional crystallization model. Takahashi et al. (1998) modeled the Grande Ronde basalts on the basis of addition and subtraction of high-pressure aluminous clinopyroxene. Such addition is consistent with melting at depths greater than 60 km of a *mafic* component, either deep mafic crust or eclogite at mantle depths. But the trends within most groups, and particularly between groups, have proven impossible to model on the basis of any one mechanism. In attempting to explain the poor fit of models based on fractional crystallization alone, most investigators have also required mixing of variable primary magma compositions (Wright et al., 1973; Reidel, 1983), crustal contamination (McDougall, 1976; Carlson et al., 1981), or a combination of both (Carlson, 1984; Hooper et al., 1984; Hooper, 1997). The evolved nature of even the most primitive CRBG flows indicates substantial fractionation of the primary melts or an unusually Fe-rich source.

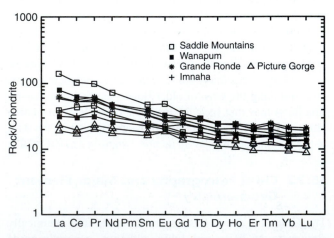

FIGURE 15.8 Chondrite-normalized rare earth element patterns of some typical CRBG samples. Data from Hooper and Hawkesworth (1993).

15.3.3 CRBG Trace Element Geochemistry

Concentrations of the compatible trace elements, such as Ni and Cr, are low in the CRBG and support the contention that they are not primary magmas. The focus of most trace element studies on the CRBG, however, is on the incompatible elements, which are generally enriched and much more revealing about the nature of CFB magmas. REE patterns have similar negative slopes, indicating similar LREE enrichment, regardless of concentration (Figure 15.8). The compatible element depletion and incompatible element enrichment in both major elements and trace elements holds true for CFBs in general, and clearly distinguishes tholeiitic CFBs from N-MORB tholeiites. This indicates that CFBs have experienced considerable fractional crystallization since separation from their peridotitic mantle source. CFB patterns are more like E-MORB and OIB.

Minor and trace element patterns, however, are considerably more complex than these simple statements indicate. Simple models based on fractional crystallization fail to explain the observed variations. Trace element enrichments typically require over 40 to 50% fractional crystallization plus crustal assimilation (assuming a peridotite parent). Individual trace elements generally require different enrichments, such that trace element ratios do not lend themselves to coherent solutions based on a single source and only one or two evolutionary processes.

A distinction between the CRBG and N-MORB is best shown by direct comparison of a wide variety of elements on the N-MORB-normalized multielement (spider) diagram (see Figures 9.7 and 14.5). Figure 15.9 is such a diagram for representative analyses of the CRBG. We can clearly see that all CRBG units are enriched in incompatible elements over MORB (normalized values greater than one). The Imnaha, Grande Ronde, and Wanapum (composing the Clarkston unit of Hooper and Hawkesworth, 1993) appear to be the most coherent, showing the broad central hump in which both the LIL (Sr-Ba) and HFS (Yb-Th) element enrichments increase with increasing incompatibility (inward toward Ba and Th).

Once again, this is the pattern to expect in a sample enriched by some single-stage mantle partial melting process that preferentially concentrated incompatible elements. This pattern is similar to OIBs generated from non-depleted mantle in intraplate settings. Some of the Saddle Mountains Basalt flows are similar, but others show very strong LIL (Sr-Ba) enrichments and a slight Ta-Nb trough. This negative Ta-Nb anomaly is also present in the Picture Gorge basalts. As we shall see in Chapter 16, this anomaly is a nearly ubiquitous feature of subduction-related magmas (see Figure 16.11), and is commonly attributed to mantle source material that has been preferentially enriched in LIL over HFS incompatible elements by fluids derived from the subducted slab below it. The high LIL/HFS pattern and negative Ta-Nb anomaly is also a feature of many intra-continental volcanics not associated with contemporary subduction, suggesting this pattern may also be a feature of the subcontinental lithospheric mantle (Fitton et al., 1988; Hawkesworth et al., 1990), perhaps caused by the effects of earlier subduction processes.

A comparison with OIB can also be made using the same multielement technique. Figure 15.10 is such a diagram normalized to a "typical" OIB analysis proposed by Sun and McDonough (1989). In this diagram, the incompatibility is considered to increase from right to left during partial melting processes, not toward the center of the diagram as in Figure 15.9. As can be seen, the basalts of the CRBG are closer to OIB, but there are still several discrepancies. The positive slopes on the right side of the diagram (Lu toward P) are curious and suggest a somewhat depleted mantle source with respect to that of OIB. The spikes at Pb, Ba, and K may reflect an enriched component, perhaps contamination by continental crust. The Nb trough is similar to the one in Figure 15.9 and can be attributed to a fluid-enrichment of the crust or mantle source.

The pronounced differences between the Clarkston, Picture Gorge, and Saddle Mountains units, as well as variations within units, certainly suggest there are multiple sources and/or more complex processes than simple partial melting of the mantle peridotite beneath the Columbia Plateau.

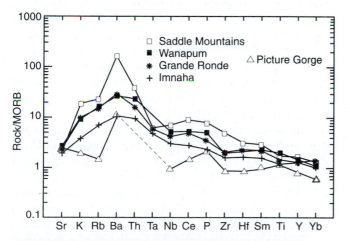

FIGURE 15.9   N-MORB-normalized multielement (spider) diagram for some representative analyses from the CRBG. Data from Hooper and Hawkesworth (1993), Picture Gorge from Bailey (1989).

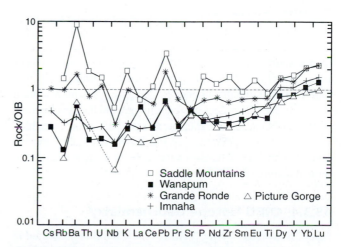

FIGURE 15.10   OIB-normalized multielement (spider) diagram for some representative CRBG analyses (data as in Figure 15.9).

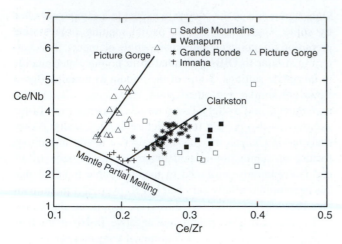

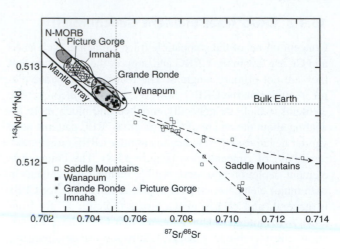

FIGURE 15.11 Ce/Zr versus Ce/Nb (un-normalized) for the basalts of the Columbia River Basalt Group. After Hooper and Hawkesworth (1993). Reprinted by permission of Oxford University Press.

FIGURE 15.12 $^{87}Sr/^{86}Sr$ versus $^{143}Nd/^{144}Nd$ for the CRBG. Data from Hooper (1988a), Carlson et al. (1981), Carlson (1984), McDougall (1976), Brandon et al. (1993), Hooper and Hawkesworth (1993).

Although REE and spider diagrams permit many elements to be plotted at the same time, it is difficult to plot a large number of rock analyses without considerable clutter. Hooper and Hawkesworth (1993) compared a larger array of CRBG samples on a diagram that still retains some critical features of Figure 15.9 by using a plot of Ce/Zr (which is a measure of the slope of the HFS elements on Figure 15.9), versus Ce/Nb (which is affected by the negative Nb anomaly). Such a diagram is shown in Figure 15.11. Note the two trends on the diagram. A negative trend separates the Picture Gorge, Clarkston (main CRBG), and the more variable late Saddle Mountains subgroups, whereas there is a positive trend within the subgroups. MORB, OIB, and primitive mantle lie along the negative trending line labeled "Mantle Partial Melting," which illustrates the approximate array of magmas associated with variations in partial melting of oceanic upper mantle. As the degree of partial melting increases, the proportion of a more incompatible element decreases with respect to a less incompatible element as it gets progressively diluted (Section 9.2 and Figure 9.3) so that Ce/Nb should decrease and the Cr/Zr ratio should increase. The extent of partial melting should thus increase from the upper left to the lower right along the Mantle Partial Melting curve in Figure 15.11. The distinct groupings of the Picture Gorge and Clarkston basalts, however, without intermediate values, suggest they are derived from distinct mantle sources, rather than by different degrees of partial melting of a single source. The positive trends within the Picture Gorge and the Imnaha–Grande Ronde groups are more continuous, and seem to reflect LIL enrichment and the development of the Ta-Nb troughs (Wanapum is slightly more variable). Figure 15.9 shows that the negative Ta-Nb anomalies are correlated with high LIL/HFS (and LREE) enrichment that is more a feature of subduction-related material.

15.3.4 CRBG Isotope Geochemistry

$^{87}Sr/^{86}Sr$ versus $^{143}Nd/^{144}Nd$ for the CRBG is presented in Figure 15.12. In this diagram, similar to Figures 10.16, 13.13, and 14.8, the Picture Gorge and Clarkston basalts show a consistent trend of enrichment (decreasing $^{143}Nd/^{144}Nd$ and increasing $^{87}Sr/^{86}Sr$) with time. These follow the Mantle Array, which has been related in Chapters 10 to 13 to mixing of depleted mantle (DM) and non-depleted mantle (BSE) or enriched components of the sort present in the sub-oceanic mantle. The Saddle Mountains samples appear to define two distinct trends, one toward very high $^{87}Sr/^{86}Sr$ values of the type associated with an EMII type reservoir. This was correlated in Chapter 14 with old crustal sediments. The somewhat steeper sloping trend is also directed toward EM but with a bit lower $^{87}Sr/^{86}Sr$ and $^{143}Nd/^{144}Nd$ ratios. Whether this is a mixture of EMI and EMII or a distinct type of enriched reservoir is not clear.

Figure 15.13 is a plot of $^{208}Pb/^{204}Pb$ versus $^{206}Pb/^{204}Pb$, similar to Figure 14.10. I have included the locations of EMI, EMII, the DUPAL group, the Atlantic and Pacific MORB array, and the NHRL (northern hemisphere reference line) connecting DM and HIMU mantle reservoirs from Figure 14.10 as references. Here the Picture Gorge and Clarkston Basalts plot near the NHRL, suggesting an HIMU component with radiogenic lead. The Clarkston data cluster more tightly than the oceanic array, suggesting perhaps a more limited enriched isotopic source than the mixing of DM and HIMU. The Saddle Mountains basalts are again more variable, and plot closer to the DUPAL material, suggesting a mixture of EMI and EMII.

Carlson et al. (1981), Carlson (1984), and Carlson and Hart (1988) proposed that three to five reservoirs could explain the isotopic and trace element data. Carlson's principal reservoirs, C1, C2, and C3, are shown on Figure 15.13. C1 is a DM-like source, perhaps altered by a small amount (~1%) of material (perhaps a fluid) derived from subducted sediment. This component is least diluted in the Imnaha (and Picture Gorge) basalts. C2 is similar mantle, but closer to HIMU, and it is considered a separate reservoir that is more contaminated by subduction-derived components. C3 has high radiogenic Sr and Pb and low radiogenic Nd, and it

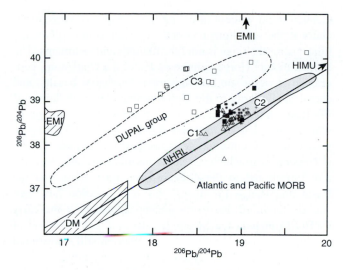

FIGURE 15.13 $^{208}Pb/^{204}Pb$ versus $^{206}Pb/^{204}Pb$ for the basalts of the CRBG. Included for reference are EMI, EMII, the DUPAL group, the MORB array, and the NHRL (northern hemisphere reference line) connecting DM and HIMU mantle reservoirs from Figure 14.8. Symbols as in Figure 15.12.

is either a crustal component or mantle containing significant (50 to 80%) crustal material. As can be seen in Figure 15.13, it is possible to produce the CRBG isotopic array using the mantle reservoirs proposed for oceanic basalts in Chapter 14. Carlson and colleagues, however, chose their reservoirs so as to provide a tighter bracket on the CRBG data, arguing that, if the oceanic mantle reservoirs were responsible, there should be a greater scatter of the data upon mixing. On the other hand, we must note that the CRBG array, particularly the Saddle Mountains data, is not constrained within the C1–C3 triangle, and seems to require at least two EM components.

The trace element and isotopic data for volcanics associated with continents are complicated by at least three components in addition to the "simple" mantle reservoirs in the ocean basins. The first of these is the continental crust itself, which is both chemically and isotopically heterogeneous. This crust has been derived from the mantle, probably by subduction, partial melting, and arc-related igneous activity over the course of at least 3.8 Ga (Chapters 16 and 17). The deep crust has probably experienced granulite facies metamorphism (Section 25.3.2), and would be somewhat isotopically depleted (due to partial melting and fluid loss), whereas the upper crust, as the recipient of this material, is further enriched. If we add sedimentary reworking, and a large variation in age, we come up with a very heterogeneous material indeed. The potential for contamination of any mantle-derived magma, as it traverses up to 70 km of this material, cannot be overlooked.

The second component is the wedge of mantle between the continental lithosphere and the subducting Juan de Fuca plate slab of oceanic crust and sediment, which has probably been altered by material rising from the slab: either melts, fluids, or both. This material should be enriched in incompatible elements, particularly the LIL elements, which are mobile in aqueous fluids. We shall see in Chapter 16 that

high ("decoupled") LIL/HFS ratios are characteristic of the subduction zone environment. This mantle material, if young enough, will exhibit enriched trace element characteristics without corresponding isotopic enrichment. As it gets older, of course, the isotopic character will change as well.

Yet a third component is the **subcontinental lithospheric mantle (SCLM)**. Presumably this mantle was initially depleted by the processes that produced the continental crust above it. Unlike the oceanic lithosphere, however, which is continuously subducted and recreated, this lithosphere has been part of the continental plate since the overlying continent was formed. This may be relatively recent in the case of accreted terranes, or as ancient as 4 Ga or more for the older portions of the cratons. During the period since it became part of the continental plate, the SCLM has probably been re-enriched, by mantle-derived fluids, melts, and/or the subduction-related processes described in the preceding paragraph (see also Section 19.4).

These three additional components complicate the major element, trace element, and isotopic signature of magmas in continental areas. The problem is particularly vexing for the isotopes, which have served us well up to this point in our study of igneous rocks and processes. Investigators have spent considerable time and effort trying to distinguish mantle reservoirs from crustal contamination based on trace element and isotopic characteristics of volcanics such as the CRBG. The problem is the difficulty, if not the impossibility, of distinguishing an enriched mantle reservoir from crustal contamination. Because the mantle enrichment is commonly affected by components derived from the crust anyway (such as the reworked crustal/sediment origin proposed for EMI, EMII, and HIMU in Chapter 14), the question is then not *whether* this component was introduced into the melt but *when* and *how*. These questions are much more difficult to answer, and the arguments over this issue are complex, detailed, and still inconclusive.

In addition to their three mantle reservoirs, Carlson and Hart (1988) argued that crustal contamination has also occurred in the CRBG, based largely on minor element ratios and oxygen isotope data. They admitted, however, that the oxygen isotope ratios are subject to alteration, and multiple sources make their conclusion tenuous. One clear way to substantiate contamination would be if the degree of isotopic enrichment by assimilation of continental crust correlated with some measure of fractional crystallization, as would result from a combined assimilation-fractional crystallization (AFC) process, in which the fractional crystallization provides the heat required for assimilation as described in Sections 4.2.7 and 11.10. There is, however, no obvious correlation between the isotopic characteristics and Mg# for the units of the CRBG. Brandon et al. (1993) modeled the Picture Gorge Basalts and found the chemical and isotopic data were compatible with a depleted mantle that has *recently* been enriched by subduction-related fluids and fractional crystallization plus assimilation (AFC) of 8 to 21% locally available accreted terrane material through which the magma traversed. The *recent* enrichment of the mantle source is because the isotopic signature is subtle, leaving the source of the Picture Gorge open to alternate interpretations. For example, Hooper and Hawkesworth (1993) found that, because of the subtle differences in isotopes, the enrichment of the Picture Gorge could equally well be accomplished by enrichment of the mantle source alone, without any crustal contamination.

In summary, because of the scatter in isotopic ratios, the Columbia River basalts must be derived from several sources. The likely candidates include a deep plume source (the basalts also yield high $^3He/^4He$ ratios), a depleted N-MORB-like source, subduction zone fluid enrichment, subcontinental lithospheric mantle, recycled sediments, and assimilation of continental crust.

15.3.5 Petrogenesis of the Columbia River Basalts

The CRBG comprise a great volume of tholeiitic basalt and basaltic andesite, which requires a mantle source (that may be at least as heterogeneous initially as the mantle beneath the oceans). The basalts were erupted into an ensialic back-arc basin above a subducting slab of oceanic lithosphere. The subducting slab was subjected to dehydration and perhaps partial melting, the products of which rise into the overlying mantle wedge. Also, the subcontinental lithospheric mantle can be expected to have been enriched during the >1 Ga period since the oldest continental crust in the area was created. In addition there is a heterogeneous assemblage of autochthonous and accreted sialic crust through which the magma must pass, a hotspot plume in the vicinity at the time of eruption, and coeval development of the extensional Basin and Range Province, with its own brand of basalt/rhyolite volcanism, developing along the southern margin. This makes for a very complex situation, both tectonically and geochemically, and allows for a plethora of models and petrologic conjecture.

Rather than review the various models proposed for the CRBG, all of which are variations on a central theme of multiple mantle and crustal sources, combined with fractional crystallization ± crustal contamination, I will rely principally on the recent models of Hooper and Hawkesworth (1993), Hooper (1997), and Camp and Ross (2004). Hooper and Hawkesworth (1993) suggested that the Picture Gorge, Clarkston, and Saddle Mountains represent three major units of the CRBG, each with its own source and isotopic/incompatible trace element characteristics (see Figure 15.11). The Picture Gorge Basalts, they noted, have a uniform depleted isotopic character (Figure 15.12), but a range of trace element enrichment (Figure 15.11), which includes the development of LIL enrichment and a Ta-Nb anomaly (Figure 15.10). They proposed that this is best explained by a depleted mantle source with a superimposed subduction-related enrichment. Other investigators have attributed the Picture Gorge characteristics to an enriched subcontinental lithospheric mantle (SCLM) source, which is also possible but would be expected to show a more variable isotopic signature. A lithospheric DM source with a more recent enrichment would be capable of producing the trace

element trend without affecting the isotopic systems, especially if the enrichment is derived from young subducted lithosphere (the ridge is not far offshore) and sediments. The limited Picture Gorge volume (<1% of the Clarkston), geographic location south of the main Clarkston basalts, and chemical similarity to the high-alumina olivine tholeiites of the Basin and Range Province, on the northern margin of which it lies, led Hooper and Hawkesworth (1993) to propose that the Picture Gorge is not part of the main CRBG (Clarkston) with which it is contemporaneous, but a separate unit transitional to the Oregon High Lava Plain and the Basin and Range Province. This interpretation has been slightly modified due to the recent extension of the early CRBG to include the Steens basalts, and consideration of the Picture Gorge basalts, which, although still transitional to the Basin and Range–type basalts, are probably a separate CRBG migration from the Steens area.

The main Clarkston basalts follow a more diverse but linear isotopic trend (Figure 15.12), which, along with the linear trace element patterns, indicates mixing of two source components. Hooper and Hawkesworth (1993) proposed that one (characterized by the Imnaha basalts) is similar to many OIBs and is probably in the deep (sub-660 km?) mantle. An obvious way to tap this deeper source is a hotspot plume. The other component has high LIL/HFS ratios, is LREE enriched, and also has enriched isotopic characteristics indicating this component has been enriched for over 1 Ga. An obvious choice for such a reservoir is the continental crust itself or the older SCLM beneath the edge of the craton. Hooper and Hawkesworth (1993) argued for the latter, based on a comparison of different isotopic ratios.

As can be seen in Figures 15.11 and 15.7, the Wanapum basalt is more varied, and does not fit as the enriched end of a simple model as might be implied in Figure 15.12. There is probably at least one other (enriched) component involved in this unit. The Saddle Mountains basalts have the widest range of chemical and isotopic values, which include the most enriched values of the CRBG. Hooper and Hawkesworth (1993) agreed with Carlson (1984) that the best source for this unit is to be found in the SCLM, and that the isotopic signature requires an enrichment age in the neighborhood of 2 Ga. The extended time span and limited erupted volumes allow for considerable variety with respect to sources, fractionation, and assimilation.

As mentioned previously, *within* each subgroup, the chemical effects of partial melting, fractional crystallization, assimilation, and magma mixing can be distinguished for a number of smaller flow sequences. This is simply not possible for the larger units.

The Grande Ronde basalts at the peak of volcanic activity produced 150,000 km^3 of basalt in about 1 Ma. This requires an average of 0.15 km^3/yr. The largest individual flows (>1200 km^3) imply flow rates approaching 0.1 to 1 km^3/day per linear km of vent (Shaw and Swanson, 1970). Such high flow rates are too rapid to permit appreciable assimilation of continental crust during transit. Shaw and Swanson (1970) used fluid dynamic models to suggest that the 100 km^3 average volumes of the Grande Ronde flows might represent only 1 to 10% of the volume of the available magma in the chamber at depth. This implies a huge magma chamber, with a volume approaching 1,000 to 10,000 km^3, feeding the Grande Ronde flows. The lack of any shallow collapse features suggests that the chamber was deep, supporting isostatic arguments that such dense basaltic magma would pond at the crust/mantle boundary.

Many models for the Columbia River flood basalts rely upon rifting in a back-arc spreading environment, based upon back-arc volcanism of the type occurring behind many modern arcs (see Section 16.1). The huge volume of the CRBG, and the general OIB/E-MORB affinity, as compared to N-MORB (Figures 15.8 to 15.10), has led a number of investigators to suggest a hotspot plume as an alternative explanation. The timing and proximity of the Snake River–Yellowstone hotspot plume makes such a connection very tempting. As discussed previously, plumes commonly accompany continental rifting, and some hotspot tracks lead directly away from CFB provinces (e.g., Figure 15.4). As discussed above, this has led several authors to propose that plumes, particularly the vigorous newly initiated ones, are an important mechanism for generating CFBs (Duncan and Richards, 1991; Richards et al., 1989; Campbell and Griffiths, 1990). The combination of back-arc spreading and initial hotspot plume impact may be responsible for CRBG genesis.

Historically, the principal problem with invoking the Snake River–Yellowstone hotspot for the generation of the CRBG is that the absolute motion of the North American Plate and the obvious track of the Snake River Plain, combined with the age determinations of rhyolitic volcanism by Armstrong et al. (1975) and others (Figure 15.5), place the location of the hotspot approximately 500 km south of the main Chief Joseph fissures at the time of their eruption. As mentioned above, however, recent mapping in NE Oregon has revealed links of the CRBG to the Steens Mountain basalts, suggesting that the original breakout of CRBG volcanism may have been much further south.

Pierce and Morgan (1993) proposed that a large plume head first approached the base of the continent ~17 Ma b.p. and spread out at the base of the crust when it encountered less dense material above. According to their model, the plume produced a large hot ponded magma chamber, which created an extending and thinning crustal welt, initiating Basin and Range extension. It also melted some of the sialic crust, producing rhyolitic liquids. Periodic escape of both rhyolite and basalt resulted in the bimodal volcanism that characterizes the Basin and Range. Pierce and Morgan (1993) further proposed a long north-south-striking rift system which funneled basaltic magma ~500 km to the north along the rift axis. Camp and Ross (2004) suggested that the thick cratonal lithospheric keel deflected the plume head to the west. Figure 15.5 shows the present distribution of the Columbia River Basalt Group, as extended southward by Camp and Ross (2004) and Hooper et al. (2008). The lower cross-section illustrates the deflection of the plume head and the black star shows the initial outbreak of deflected plume-related basalts near Steens Mountain proposed by Camp and Ross (2004). The gray star indicates the geographic location directly above the deeper plume tail axis at the time of outbreak, proposed by Jordan et al. (2004). Notice how the dike swarms and faults, as well as the magnetic anomalies of Glen and Ponce (2002), believed to represent subsurface dikes, all point toward the plume outbreak. Such radiating dike and fault patterns are thought to accompany plume impact (Ernst and Buchan, 2001). Plume-head deflection and rise toward the thin lithosphere beneath the Mesozoic accreted terranes was accompanied by extensive decompression partial melting (enhanced perhaps by back-arc upwelling and heat). Although the Lower Steens basalts are the first manifestation of this volcanism, magmatism migrated rapidly northward along fractures near the terrane/craton suture into the Chief Joseph dike swarm (this direction may also have led to even thinner lithosphere). Camp and Ross (2004) further proposed that the plume head spread along weaknesses or thin zones to the northwest to feed the Picture Gorge basalts along the Monument dike swarm. The sudden late waning of CRBG activity may reflect separation of the plume head from the tail as the North American craton progresses westward. Prevented from spreading eastward or further north by the craton, the plume head continued to thin and spread toward the west and south, encouraging Basin and Range extension and volcanism (including the late Tertiary to Quaternary Oregon High Lava Plains, with which the Picture Gorge shares many characteristics). The enigmatic *westward* progression of rhyolitic centers from Steens to the present site of Newberry volcano (N in Figure 15.5) may reflect the advancing front of the plume head, fed by convective rise of the remaining head mass and drawn westward perhaps by subduction-induced counterflow in the mantle wedge (Section 16.1) as proposed by Jordan et al. (2004). See also Xue and Allen (2006) for some alternative explanations for the westward migration.

Geist and Richards (1993) proposed a different model to explain the deflection of initial CRBG breakout, suggesting that that the plume was initially shielded from reaching the crust by the subducting Farallon Plate above the plume. About 17.5 Ma ago, according to their model, the plume finally ruptured the plate and passed through it, deflecting northeastward (in the direction of the subducting plate motion) to the location of the Chief Joseph vents. They argued that this would have released a large buoyant magmatic mass that had accumulated beneath the sinking plate, giving rise to the voluminous Clarkston phase of the CRBG. After this breakthrough, the lower portion of the Farallon Plate separated and sank, and the plume slowly returned to a vertical regime, migrating southward, and then following its natural eastward course to its present location beneath Yellowstone National Park in northwestern Wyoming. The recent discovery of magmatism linking southward to the Steens Mountain area obviates the need for such a mechanism, but the effect of an inclined subducting plate on a rising plume poses an interesting question.

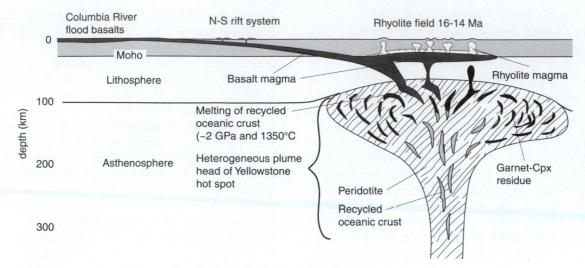

FIGURE 15.14   A model for the origin of the Columbia River Basalt Group, based on melting within a heterogeneous plume head (initial stages of the Yellowstone hotspot). The plume head contains recycled stringers of recycled oceanic crust that melts before the peridotite, yielding a silica-rich basaltic magma equivalent to the main Grande Ronde basalts and leaves a garnet–clinopyroxene residue. The large plume head stalls and spreads out at the base of the resistant lithosphere and the basaltic magma ponds (underplates) at the base of the crust, where it melts some crust to create rhyolite. Basalt escapes along a northward trending rift system to feed the CRBG. From Takahashi et al. (1998). Copyright © with permission from Elsevier Science.

Takahashi et al. (1998) have recently proposed a model to explain the silicic and enriched nature of the CRBG (Figure 15.14). Their experiments indicate the CRBG magmas can be generated by simple batch melting of MORB-like oceanic crust at ~2 GPa (~70 km). Following Cordery et al. (1997), they proposed that the plume head contains a mixture of peridotite and subduction-recycled basalt/eclogite. As described in Section 14.7, near-adiabatic rise and decompression partial melting preferentially melt the eclogite stringers first, producing a melt equivalent to the Grande Ronde basalts of the CRBG. Melting experiments show that such silica-rich basaltic andesite melts can be produced only when olivine is not in equilibrium with melts in the source rocks. This can occur only when the eclogites melt below the solidus temperature of the host peridotite and the eclogite partial melts escape, acting as independent systems. They proposed that this can occur when the plume head temperature is less than ~100°C above ambient. Once the peridotite begins to melt, they claimed, the eclogite melts must equilibrate with the host peridotite, producing basalts indistinguishable from peridotite partial melts.

Camp and Hanan (2008) noted that the Imnaha basalts appear to require a DM-dominated mantle plume source and that the Grande Ronde (with its higher SiO_2 and $^{87}Sr/^{86}Sr$, lower $^{143}Nd/^{144}Nd$, and a $^{187}Os/^{188}Os$ ratio that indicates a Mesozoic mafic component) is best interpreted as derived principally from melting of a mafic source, such as pyroxenite or eclogite. The model of Takahashi et al. (1998) may accommodate this, but the earlier effusion of plume-dominated Imnaha followed by a later and greater pulse of the latter does not quite fit the model, because the transition is quite abrupt and the mafic component should melt first (Section 14.7) and ought to produce less melt. Camp and Hanan (2008) thus proposed that the plume triggers a delamination of the sub-continental lithospheric mantle (see Figure 18.8c) and deep

dense crustal mafic underplate. Initial rise of the plume head into the delamination wedge may enhance partial melting of the plume, creating the Imnaha basalts, and partial melting of the delaminated material as it interacts with the plume produces the copious Grande Ronde outpouring.

15.4   OTHER CONTINENTAL FLOOD BASALT PROVINCES

Time and space considerations permit only a cursory review of other CFB provinces, intended only to highlight some of the variation among the occurrences rather than indulge in extensive descriptions. For more complete descriptions, the reader is referred to several sources listed at the end of the chapter.

The Proterozoic **Keweenawan Province**, as mentioned above, is associated with a gravity high, indicating thinned crust associated with an aborted rift, or aulacogen. The thinned (or absent) continental crust minimizes the effects of crustal contamination, and the basalts are similar to plume-related ocean tholeiites (OIT and E-MORB). Approximately 7% of the exposed basalts are alkaline basalts, as compared to a total absence in the CRBG. Some differentiation-produced rhyolites are extensive locally. $^{87}Sr/^{86}Sr$ ratios are generally high (0.7032 to 0.7141). These values are far from the mantle evolution curve for Proterozoic times, which is difficult to explain if no continental crust is involved. Several cogenetic mafic intrusions, including the Duluth Complex layered intrusion, are exposed by erosion.

The huge Permo-Triassic **Siberian Traps** represent the largest known subaerial eruptive sequence. They occur in a rifted continental basin with a thin continental crust and numerous N–S rifts and grabens. They cover an area of ~4 · 10^6 km^2 and are about 1 km thick on average. The bulk of the lavas erupted over about 1 million years ca. 248 Ma. b.p.

(the Permian–Triassic boundary). Approximately 3% of the basalts are alkaline. Numerous intrusive bodies, particularly sills, are associated with the volcanics. An unusual feature is the relative abundance of basaltic tuffs and breccias, greatly exceeding lavas in the lower sequences of the southern part of the province. One tuff layer is 15 to 25 m thick, covering about 30,000 km^2. Erosional breaks are very rare, and it appears that the area subsided evenly as the magma was displaced to the surface. The presence of aquatic fossils in the tuffs suggests the presence of widespread shallow lakes or lagoons, but subaerial conditions dominated later. The lack of substantial uplift features does not correlate well with a hot thermal plume, however. Both the Keweenawan and Siberian basalts are more magnesian on average and poorer in incompatible elements than most younger CFBs, perhaps as a result of the higher heat flow at that time. They are predominantly aphyric, but less so than the CRBG. The isotopic data of Sharma et al. (1991) suggested an only slightly depleted, nearly chondritic lower mantle source (initial $\varepsilon_{Nd} = +1.8$) with a component of crustal contamination consistent with deep granulites or granites. Contamination is greatest in the early lavas, perhaps reflecting stoping as the first magmas worked their way upward. They also detected an EMI-like source, which they interpreted as SCLM.

The Karoo and Paraná basalts are associated with the Juro–Cretaceous rifting of Gondwanaland (Figure 15.3), a successful rift attempt compared to the Keweenawan and Siberian failures. The **Karoo Province** basalts of South Africa lie in an intracratonal basin (as do practically all CFBs); Karoo is one of the larger provinces, initially covering perhaps 3 million km^2. It is most notable for its variety of rock types. Basalts certainly dominate, but picrites are locally abundant, as are nephelinites, acid lavas, and intrusives ranging from gabbros to granites and syenites. Rhyolites comprise up to 30% of the exposed volcanics. Interestingly, a plot of $^{87}Sr/^{86}Sr$ versus $^{143}Nd/^{144}Nd$ for the Karoo basalts shows essentially the same two diverging trends toward enriched compositions as the Saddle Mountains basalts in Figure 15.12 (Cox, 1988).

The **Paraná Province** and the **Etendeka Province** are associated with Gondwanaland rifting and the Tristan hotspot (Figures 15.3 and 15.4). Volcanism began at about 137 Ma, but most lavas were produced between 133 to 131 Ma. Thick "seaward-dipping reflectors" are seismically imaged offshore. First discovered off North America, they can reach thicknesses of 10 km. They are believed to accompany rifted continental margins, and are interpreted as the transition from large sub-aerial to submarine eruptions. Their presence offshore here suggests the lavas were associated with rifting and subsidence. As with the CRBG, both are mostly aphyric and strictly tholeiitic. Both are also strongly bimodal: basalt and rhyolite with few intermediate compositions. $^{87}Sr/^{86}Sr$ ratios are high (0.704 to 0.717) and in some sequences are related to the SiO_2 content, consistent with a common model for bimodal volcanism in which the basalts are of mantle origin whereas the rhyolites are produced by crustal melting induced by a large hot mafic magma chamber underplated beneath the crust. High $^{87}Sr/^{86}Sr$ ratios in some low-SiO_2 rocks also suggest an enriched subcontinental mantle source.

Cox (1988) recognized a high TiO_2/high P_2O_5 (HTP) magma type and a low TiO_2/low P_2O_5 (LTP) magma type in both the Karoo and Paraná provinces that are spatially related to the southern margin of Gondwanaland (Figure 15.2), along which subduction took place prior to rifting. He proposed that this chemical variation probably reflects the influence of subduction-altered mantle.

As mentioned above, the **Deccan Traps** appear to be related to the Reunion hotspot and rifting between India and the continental Seychelles block (where similar-age basalts are found). They now cover about $1.5 \cdot 10^6$ km^2 of India (similar in size to the Karoo and Paraná) and thicken westward to the Western Ghats near the coast at Mumbai, where they are about 2 km thick. Hotspot magmatism apparently created the N–S elongate Western Ghats ridge. Volcanism began in the north (Hooper, 1999) and migrated southward as the eruption waned, consistent with the rapid northward movement of the Indian plate over the plume. All of the lavas apparently erupted in <1 Ma at the Cretaceous–Tertiary boundary. Gravity and seismic data indicate that the basalts continue offshore into the seaward-dipping reflectors in Kerala basin, which apparently subsided as the plume head waned. The Chagos–Laccadive aseismic ridge and the Mascarene plateau extend to the south (Figure 15.1), linking the Deccan with the Reunion hotspot. The Carlsberg ridge passed over the plume, so the hotspot went from the Indian-Eurasian plate to the African plate (Figure 14.1). The Deccan flows are predominantly tholeiitic, and porphyritic lavas are the norm (some are highly phyric). Plagioclase crystals in excess of 5 cm are not uncommon. The major element chemical composition is uniform, although relatively evolved (Mg# generally less than 60, but a few picrites reach 80), but the trace element and isotopic variation is greater. Approximately 10% of the basalts are alkaline basalts. The $^{87}Sr/^{86}Sr$ versus $^{143}Nd/^{144}Nd$ also shows two diverging trends toward enriched values, but they are steeper than the CRBG and Karoo trends. Mahoney (1988) interpreted the isotopic data as indicating that some tholeiitic basalts were derived from the upper mantle (a mixture of Reunion plume and depleted Indian Ridge N-MORB source) without significant contamination, whereas others show contamination by upper crustal material and yet others by either deep crustal granulites or enriched SCLM. The mixing of the enriched components cannot be produced by simple assimilation because the lithospheric enrichments do not correlate with fractional crystallization trends. The sources must be melted by plume heat. $^{40}Ar/^{39}Ar$ ages on minor alkaline basalts in the NW indicate flows both older and younger than the main Deccan tholeiites and high $^3He/^4He$ values in the early alkali basalts indicate a deep mantle source (Basu et al., 1993). These basalts apparently were generated by smaller degrees of partial melting at the periphery of the plume and did not incorporate as much lithospheric component as the tholeiites.

The **North Atlantic Province** (also called the Brito-Arctic Province) extends from Canada to the British Isles and is associated with the Iceland hotspot and the early Tertiary rifting in the North Atlantic and the separation of Greenland from Europe. Saunders et al. (1997) noted that contamination by continental crust was prevalent during the early phase of continental breakup but less so in the second and final stage. The Iceland volcanics are also distinctly bimodal, with rhyolites the second most common rock type, significantly less than the basalts. The province includes the classic igneous localities of the Hebrides in which much of the pioneering work in igneous petrology was done on the many intrusives and volcanics. The famous Skaergård layered mafic intrusion of Greenland (Section 12.2.3) is also part of the province. Perhaps the most distinctive feature of the province is that approximately 45% of the exposed material is alkaline. Much of the early work in distinguishing the alkaline and sub-alkaline magma series, as discussed in Chapter 8, stemmed from the British Tertiary igneous province.

Korenaga and Kelemen (2000) calculated primary melt compositions that suggest the northern Atlantic source mantle is relatively fertile (Mg# = 86−87 as compared to normal mantle with values over 89). This implies a source with a significant basalt/eclogite component with a lower melting temperature than normal peridotite. They related this component to short-term recycling of the Iapetus Ocean crust (Iapetus was the pre-Atlantic ocean before the Appalachian–Caledonian orogen closed it). The Sr-Nd-Pb-Hf isotopic data of Kempton et al. (2000) require at least four mantle components for the North Atlantic Province. They recognized shallow N-MORB source mantle, both a depleted component and a small range of enriched components in the Iceland plume, and depleted material in the plume margin. They proposed that the depleted component within the plume is not shallow N-MORB source mantle but recycled ocean lithospheric mantle.

Hill (1991) noted that the Labrador Sea was already opening between Greenland and Canada as the Iceland plume head impacted beneath Greenland, but the plume was nearer Greenland's east coast. The plume, according to Hill's model, was deflected by the deep Greenland lithosphere both east and west. This added to the earlier Greenland west coast volcanics, but most plume material was deflected east, beneath the thin lithosphere of the Paleozoic Appalachian–Caledonian orogenic belts. The Labrador separation to the west thus stalled, and spreading between Greenland and Europe took over. Unlike the Reunion hotspot and the Carlsberg ridge, the northwestward drift of the Mid-Atlantic Ridge over the Iceland hotspot has not been effective. Every few million years, the ridge abandons its drifted position and re-centers as a new ridge segment over the plume.

15.5 PETROGENESIS OF CONTINENTAL FLOOD BASALTS AND LIPS

In summary, we can say a number of things about CFBs. They exist on every continent. Few remain from prior to the Mesozoic, and five have been generated in the last 250 Ma. They are characterized by extrusion of very high volumes of mafic magma (requiring a mantle source) into subsiding basins in extensional continental settings over a comparatively short period of time. Fissure-fed aphyric lava flows constitute the bulk of CFBs, and the combination of high eruption rate and low viscosity permits extensive lateral flows, leaving no near-source landforms such as a shield (although the Deccan may be a very broad shield). Intrusions, including the spectacular layered mafic intrusions, are subordinate but can be common locally. Tholeiitic basalts are the predominant magma type (exclusive in some occurrences), but alkaline types and more evolved differentiates are also represented.

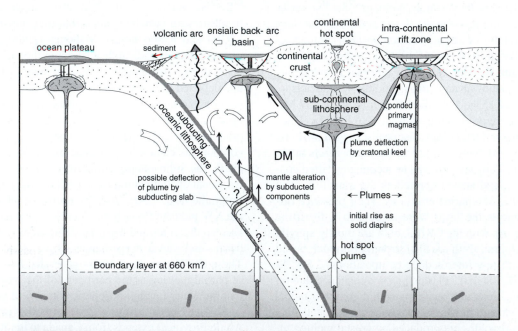

FIGURE 15.15 Diagrammatic cross section illustrating possible models for the development of continental flood basalts. DM is the depleted mantle (MORB source reservoir), and the gradient beneath 660-km depth represents a less depleted, or enriched, OIB source reservoir. See Section 14.6 for a discussion of the controversial layered nature of the mantle.

Although picrites occur in some CFBs (notably in the Karoo), the magmas are generally evolved. Non-picritic flows are characteristically high in Si, Fe, Ti, and K. Mg# is commonly <60. Compatible trace element contents, such as Ni and Cr, are low, suggesting that CFBs are not primary magmas, but have undergone substantial crystal fractionation prior to eruption. Derivation from an Fe-rich mantle material is an alternative possibility, but it is unlikely that an appropriate source is common.

Cox (1980) proposed a magma reservoir ponded at the base of the crust as a site for low pressure fractional crystallization of plagioclase, olivine, and pyroxene, which can explain the evolved chemical nature of CFBs. He suggested a sill complex, but a larger body, such as a plume head, is more likely. In areas of extension the crust is thinner and the moho higher, forming a natural collecting point for ascending magma. Alternatively, a plume head could *cause* extension. Which came first is difficult to tell, but if plumes are really deep mantle phenomena, they were probably first in most cases. Plumes may stall at the base of the more rigid lithospheric mantle, but the density of primary magmas, once produced, suggests that they should pond at the top of the mantle beneath the less dense continental crust. Shallow fractional crystallization would take place here, plus some assimilation of the lower crust (perhaps by an AFC process) either until the density changed enough so the magmas were once again buoyant enough to rise further, or extensional faults provided conduits for the escape and rise of dense mafic liquids.

CFBs are also characterized by high concentrations of incompatible trace elements, particularly the most incompatible ones, such as K_2O, LIL, and LREE. They have variable isotopic signatures, but are commonly enriched, and require multiple sources including substantial enriched components suggestive of continental crust and/or enriched mantle.

Figure 15.15 is a diagrammatic summary of models for the genesis of continental flood basalts, showing the principal hypotheses for CFB origin. The principal mechanism, illustrated on the right of the diagram, is the hotspot plume, initiating an intracontinental rift. This is called the **active** model of rifting because mantle upwelling is active and influences rifting. The alternative **passive** model of rifting proposes that intracontinental rifts are the cause of mantle upwelling and CFBs. The passive model has many of the characteristics of MORB genesis (Chapter 13). Extension causes faulting (probably listric) and thinning of the continental crust, providing ample conduits for magma to reach the surface. Thinning also initiates diapiric rise of sublithospheric mantle material, and this can be expected to induce decompression partial melting in the rising mantle diapir, hence the similarity between CFB and MORB magmas. CFB magmatism is commonly associated with the early stages in the development of such rifts. If the rifting is aborted, an aulacogen results with CFBs filling the basin, as in the Keweenawan or Siberian Traps. If the rifting continues, an ocean basin is created, which eventually separates two continental blocks, each perhaps with its own CFB, as in the case of the Karoo, Paraná, Etendeka, and North Atlantic provinces.

If rifting alone were sufficient to generate CFBs, however, we would expect to see CFBs associated with all rifts and ubiquitously occurring along the coasts of rift-separated continents. This is clearly not the case, and the high magma volumes in localized areas have led many investigators to conclude that a mantle hotspot plume is also a necessary component of CFB magmatism (particularly during the initial stages of development when the large plume head first reaches the surface). A deep plume tapping the deep enriched mantle could also explain the alkaline character of some CFB material. The active rift model in Figure 15.15 thus includes a rising hotspot plume, probably originating at the D'' layer at the core–mantle boundary. Because of the evolved nature of CFB magmas, the existence of a ponded magma chamber at the base of the crust probably accompanies any model of CFB genesis and is also included as a possibility in the models illustrated.

A model perhaps more appropriate to the Columbia River Basalt Group is one in which back-arc rifting creates an ensialic basin (toward the left in Figure 15.15). Although the exact cause of back-arc spreading is not clear, there are a number of models suggesting that back-arc extension is a necessary accompaniment of subduction (Chapter 16). One model discussed earlier is based on frictional drag between the subducting slab and the mantle wedge above the slab, requiring an upward flow of mantle to replace that dragged away. Either this mechanism or some other based on mantle convection (perhaps associated with slab-induced melting) is responsible for the ubiquitous grabens behind volcanic arcs, and even the development of marginal seas, such as the Sea of Japan, with its own (weak) spreading center. As mentioned above, the thermal output of such a back-arc system alone is not generally regarded as sufficient to generate the quantity of basalt found in CFB provinces such as the Columbia River Basalt Group, otherwise we would find more back-arc CFBs. Thus a plume is also suspected of taking part. The combination of back-arc rifting, accompanied by mantle rise and elevated heat flow, plus the Snake River–Yellowstone hotspot plume, is probably responsible for the creation of the CRBG.

The Siberian Traps and the Keweenawan basalts indicate that hotspots can produce mafic magmas in a continental interior with only limited rifting, as illustrated in the center of Figure 15.15. The Snake River–Yellowstone hotspot track in the northwestern United States also results in basaltic volcanism, but the volume of basalts in the narrow rifted graben of the Snake River Plain is considerably less than for CFBs proper. There is considerable doubt as to whether a *mature* plume (a plume tail) alone can produce a CFB. Either associated large-scale (passive) rifting and mantle flow or an active vigorous starting plume head appear to be required. The Deccan is another possible example of pure hotspot plume activity, although rifting between India and Madagascar may have accompanied the event. The considerably higher magmatic output of the Deccan is probably related to very high thermal activity associated with the initial surface activity of the highly productive Reunion plume. On the left of Figure 15.15 is an oceanic plateau created by the impact of a plume head beneath oceanic lithosphere.

Although the plume hypothesis has its critics, ample seismic, petrologic, geochemical, and modeling evidence supports it. Perhaps a brief review of the aspects of plumes that pertain to CFB-LIP genesis is in order. Plume tail conduits are believed to be 100 to 200 km in diameter. This is below the limits of resolution of typical seismology, but recent application of finite-frequency tomography has traced the Ascension, Azores, Canary, Easter, Samoa, and Tahiti plumes to the core–mantle boundary (Montelli et al., 2004). Numeric and fluid modeling indicates that plume head sizes depend largely on the depth of origin and should be ~1000 km in diameter if generated at the core–mantle boundary, ~750 km if at the 1700 km level, and ~250 km if at the 660-km transition zone (Campbell, 2001). We shall accept the deep origin and large head in what follows. As a newly-formed plume rises and encounters the more rigid lithosphere the head flattens further and expands to a disk about 2000 km in diameter. According to the active plume model for LIP generation (Campbell, 2001), this causes 500 to 2000 m of uplift due to thermal expansion and enormous flood basalts covering an area 2000 to 2500 km across in what he calls the *first phase* of CFB volcanism.

A deep-seated plume is believed to contain stretched pods of recycled oceanic lithosphere in a peridotite host: a mixture of mantle components including DM, EM, HIMU, and FOZO. Melting in the plume head is the result of shallow melting in accordance with the model described for plume tails in Section 14.7, except that the size of the head is much larger and the temperature may be somewhat hotter as well. Although volatile-induced melting of enriched stringers (mostly eclogite) may have occurred between 350 and 150 km in the tail model, little of this escapes the plume and the silicic eclogite melts probably react with the peridotite host to form solid garnet pyroxenites. Major melting occurs at about 100 km depth in the tail model, and the melt fraction increases as the plume head gets shallower.

Most of the enormous volcanism in major LIPs occurs very rapidly (<1 Ma), reflecting the great magmatic potential of the plume head, but may trail off for 10 to 100 Ma afterward. Early picrites and alkali basalts may erupt, but the majority of the volcanism is tholeiitic (shallow and high melt fraction). As the basaltic magma is displaced upward through the crust, the crust beneath subsides rapidly. Later, slower subsidence results from gradual cooling.

Most experiments and models indicate that peridotite melting is not extensive if it ascends and stalls beneath thick (>125 km) continental lithosphere. CFB generation therefore requires one or more of the following:

- The temperature excess in the plume head (the temperature over that of ambient mantle) is greater than the generally accepted 200°C.
- Considerable recycled eclogite and sediments are present in the plume, so that their melts, although not in equilibrium with the peridotite host, are more abundant and manage to escape (Figure 15.14).

- The lithosphere thins and allows the plume head to move to shallower levels where melting is more effective.

Any or all of the above may perhaps occur where a plume head approaches the Earth's surface. Farnetani and Richards (1994) proposed that the excess temperature in a plume is about 350°C greater than ambient, and the experiments of Yaxley (2000) indicate that a few tens of percent basaltic material in peridotite substantially enhances melt productivity. I believe that these may be important secondary effects in some plumes, but thinner lithosphere is a fundamental prerequisite to most LIP/CFB creation. Kent et al. (1992) stressed the importance of an incubation time between the arrival of a plume head beneath thick lithosphere and effusive magmatism. During this time, they proposed, the base of the lithosphere is removed (weakened) by conductive heating and melt injection. The delay time may be in the 10 to 20 Ma range and depends on the initial temperatures of the lithosphere and plume; the age, thickness, and composition of the lithosphere; thermal transfer and melt injection rates; and the presence of small-scale instabilities at the plume–lithosphere interface.

Hill (1991) noted that 1 to 2 km of uplift in the first phase of active plume arrival beneath continental lithosphere results in gravitationally induced horizontal stresses leading to extension. If the somewhat crude two-dimensional modeling of Houseman and England (1986) is any indicator, uplift-induced horizontal stresses could easily lead to "runaway" extension and the formation of a new ocean basin. According to their models the threshold from self-limiting extension to runaway extension depends on the elevation of the lithosphere and the temperature at the Moho, ranging from 800 m elevation at 750°C to 2500 m at 550°C. The continental geotherm estimates in Figure 1.11 and a crustal thickness of 40 km indicate a Moho temperature of 500 to 600°C, indicating that significant extension is to be expected when a plume head arrives beneath continental lithosphere. The formation of a new ocean basin, however, is a marginal case and CFBs do not require it. Campbell's (2001) *second phase* of CFB volcanism occurs when extension-related thinning of the lithosphere occurs and the plume head rises into the extending zone and further melts. If extension is limited, an aulacogen forms (e.g., Siberia, Keweenawan). If extension leads to the formation of a new ocean basin, magmatism during this phase may be as voluminous as in the first phase. Extensive basaltic flows and subsidence of the thinning–cooling continental margin results in the creation of volcanic passive continental margins and the creation of seaward-dipping reflectors that are evident in seismic surveys. Examples include the east coast of Greenland (Iceland plume), the west coast of India (Reunion plume), and the opposing coasts of South America and Africa at the Tristan da Cunha hotspot track. In all cases, the continuing trail of the less vigorous plume tail is evident as a wide trail of volcanics on the seafloor (typically as an aseismic ridge).

Although extension and rifting appear to be inevitable results of plume head arrival, most investigators believe that plumes alone (unless perhaps they are very energetic) are not enough to initiate runaway extension. Continental breakup and the formation of new ocean basins, they contend, requires extension of plate tectonic origin, and although plume head arrival may modify this somewhat, it cannot create it. Nearly all of the post-Paleozoic CFBs, however, seem to be associated with such breakup. Hill (1991) and Courtillot et al. (1999), among others, suggested that plumes exploit and accelerate situations in which extension is already in nascent stages, presumably related to the foundering effect of cool lithosphere ("slab-pull"). As the Columbia River Basalt Group indicates, thick cratonal keels may deflect rising plume heads toward thinner lithosphere: back-arc basins, incipient rifts, or old sutured orogenic belts that may be experiencing post-orogenic collapse (Section 18.4.2). This is included in the center of Figure 15.15 and may explain why some continental rifting centered on old orogens (e.g., the Atlantic opening along the Appalachian–Caledonian system). The Columbia River Basalt Group is an order of magnitude smaller than other CFBs, which may indicate a weak plume, perhaps explaining why extension has not led to runaway extension of the back-arc basin.

The classic Dewey and Burke (1974) model of continental breakup and rifting (Figure 15.16) suggests that a series of subcontinental plumes, each forming a three-rift triple junction (such as presently at Afar, Figure 19.2), link up along two arms of each, forming a continuous rift. The third arm of each hotspot eventually fails, becoming an aulacogen. But one might ask why does such a string of plumes develop beneath a continent? The reconstruction of Burke and Torsvik (2004) suggested that Pangea, at the time of breakup, sat above a large superswell, similar to the ones now in the central Pacific and under western Africa (Figure 14.1). If this superswell spawned several smaller plumes (Figure 14.23), some may have been close enough to connect as proposed by Dewey and Burke (1974). Alternatively, Bott (1992) proposed that subduction on both sides of a continent can induce tension in the interior, resulting in (passive) mantle upwelling and magmatism. Extensional rifts can certainly occur without hotspots and vice versa, but extension is much more vigorous when they coincide. On the basis of the number of hotspots today, White and McKenzie (1989) concluded that any point on the Earth's surface should drift near a plume every few hundred million years. If the stresses at some point are marginally extensional and a plume should pass beneath, the result may be a CFB and runaway extension. The opening of the Atlantic, for example, was sequential (the south opened long before north) and may thus reflect the need for this type of coincidence. The experiments and model of the effects of an insulating continental lid on mantle convection by Guillot and Jaupart (1995) indicate that a continent wider than the depth of mantle convection will create large thermal anomalies, resulting in large-scale organization of mantle convection, producing lines of (non-simultaneous) centrally located upwellings.

Oceanic lithosphere is thinner than continental. Rising asthenosphere in a hot plume can thus reach shallow levels suitable for decompression partial melting and LIP production in ocean basins without plume-induced or rift thinning of

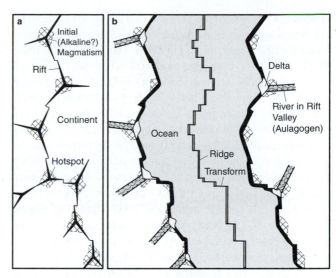

FIGURE 15.16 The Dewey and Burke model for the evolution of a continental rift by the concatenation of a series of three-rift triple junctions, each centered on a hotspot. Initial low melt-fraction alkaline magmatism is shown, but no subsequent effusive tholeiites. Two arms of each hotspot link up to adjacent hotspots, although generally not perfectly. The third arm fails and becomes a rift valley (aulacogen). The hotspots need not be coeval and different segments can form sequentially. From Dewey and Burke (1974).

the lithosphere. If the active plume model is correct, oceanic plateaus should form without significant rifting, and magmatism may be substantially greater than for CFBs, particularly in areas of thinnest oceanic lithosphere (near divergent plate boundaries). Magnetic anomaly reconstruction traces the plume head impact of the largest LIP on Earth, the Ontong Java Plateau, to very near the Pacific mid-ocean ridge at the time. The very thin lithosphere at such a location may explain the estimated $4.4 \cdot 10^7$ km^3 of material (an order of magnitude greater than any CFB and over 25% greater output than the present mid-ocean ridge system). Most oceanic plateaus appear to have been created near ridges (Kerr, 2003). The greater extent of source melting also explains the generally flatter REE patterns of oceanic plateaus. The crustal thickness of oceanic plateaus makes them resistant to subduction, and they may be peeled off from subducting lithosphere and accreted to island arcs or continents. Large plateaus may clog a subduction zone, resulting in back-stepping of subduction to behind the plateau. If an oceanic plateau is thrust beneath an island arc, subduction may even jump the opposite way: behind the arc and may reverse dip (Saunders et al., 1996; Kerr and Mahoney, 2006). The Caribbean plateau, for example, is believed to have collided with the east-dipping subduction zone along the west coast of Central America and cruised right through to form a west-dipping subduction zone in the Caribbean. Accretion of plateaus to continents and the direct emplacement of CFBs make LIPs a major contributor to the formation of continental crust.

The nature of possible source reservoirs for CFB and LIP volcanism is complex and controversial. Certainly a number of mantle and crustal components are required to explain the variable trace element and isotopic characteristics. Proposed sources include all the mantle reservoirs proposed for the generation of MORB and OIB (DM, BSE, HIMU, PREMA, and EM, as described in Section 14.5). The continentally derived sediments in the fore-arc wedge (Figure 15.15) are also a ready source for addition to the already enriched character of the subducted oceanic crust and pelagic sediments. As this material dehydrates and begins to melt, the incompatible element-enriched derivatives can be expected to alter the overlying mantle wedge, either as rising metasomatic fluids, partial melts, or both. A final mantle source, particularly for CFBs, is considered to be the subcontinental lithospheric mantle (SCLM) (shaded in Figure 15.15). The SCLM was probably initially depleted as the overlying crust was derived from it, but it was subsequently re-enriched from below, perhaps in excess of the initial depletion, during the 1 to nearly 4 Ga that this mantle material has remained beneath the crust. This lithosphere must be both geochemically complex and heterogeneous, having experienced most of the same tectonic, metamorphic, and magmatic events as the overlying crust. This is substantiated by the wide range of mantle xenoliths brought to the surface from beneath the continents by kimberlites and continental alkali basalts. Some of these are both old and enriched. White and McKenzie (1995) noted that one-quarter of the CFB suites they studied are too enriched to be modeled in terms of asthenospheric mantle alone, and they proposed that the subcontinental lithospheric mantle is the most likely source of the enrichment. We will take a closer look at the subcontinental lithospheric mantle in Chapter 18.

A large layer of continental crust overlies the CFB source, and the magmas must traverse this layer to reach the surface. It is thus difficult to ascertain whether the trace element and isotopic enrichment is a property of the mantle source, as it must be for OIB magmas, or whether it is due to assimilation of continental crust either at the base of the crust or during ascent toward the surface. The reason this is so difficult, even for isotopes, is that the enriched portions of the mantle are believed to have become enriched by addition of crustal material, probably in the form of oceanic crust and continentally derived sediments by subduction. If a mantle reservoir, such as EMI or EMII, was enriched by continental crust in the form of sediments, its isotopic signature would be a mixture of the original mantle and the continental materials. Magmas derived from a source with such a signature would be identical to ones derived from a more pristine mantle source that was subsequently mixed with an ancient continental reservoir by crustal assimilation because the only difference is when and how the two isotopic types were mixed.

I am attracted to the proposal by Cordery et al. (1997) that plumes may contain odd bits of ancient oceanic crust, recycled by convection (Figures 14.20 and 15.14). Their experiments, and the ones of Kogiso et al. (1998), Yaxley and Green (1998), and Yaxley (2000), suggest that melting of these eclogitic slivers or pods can generate basalts. The trace element and isotopic signatures of these basalts may also explain in part the enriched characteristics of CFBs. Geochemical variation in CFBs is a result of differences in the degree of partial melting, polybaric fractional crystallization, magma mixing, crustal contamination, and different source characteristics. No occurrence of CFBs can be simply modeled in a temporal sense by any one or any pair of the first four processes, and multiple sources are certainly required.

Some provinces show a more complete sequence of magmas than others, including andesites and rhyolites (and their intrusive equivalents) that can be attributed to shallow fractional crystallization. Although many of the rhyolites are products of differentiation of parental basalts, several areas (e.g., parts of the Deccan and Karoo) are characterized by a bimodal basalt–rhyolite association, with subordinate intermediate rock types. Other occurrences, such as the Snake River Plain and the Basin and Range in the United States, are almost exclusively basalt–rhyolite. In the bimodal occurrences, the trace element and isotopic signatures of the rhyolites are unrelated to the basalts and are attributed to melting of the more readily fusible continental crust by the copious high-temperature mafic magmas that pond at the base of the crust or form magma chambers at various crustal levels.

The generally accepted model of LIP formation is that of a vigorous plume head reaching shallow depths of less than 125 km beneath thin lithosphere. Hotspot tracks then typically record the location of the plume tail conduit from the location of initial surface impact as the overriding plate moves across the axis. Some hotspot tracks lack a LIP, however (e.g., Hawaii), and some LIPs lack a later hotspot track (e.g., Ontong Java). In the former, the LIP may have been subducted or accreted, but as pointed out in Chapter 14, not all plumes need be thermal in origin. *Chemical plumes* may also occur (e.g., Masters et al., 2000; Farnetani and Samuel, 2005) that may have significantly smaller heads. It is also possible that some plume heads have little or no tail, so lack of a post-LIP seamount chain may not rule out a plume-head origin for a LIP (Kerr and Mahoney, 2006). The modeling of Davaille et al. (2005) and Lin and van Keken (2006a, 2006b) suggests that a range of plume types can result from interaction between compositional heterogeneities and thermal convection near the core–mantle boundary. The type of upwelling depends on the balance between chemical and thermal buoyancy effects (Davaille et al., 2005). If the chemical effects are large, plumes from the core–mantle boundary can entrain dense material near the source and remain connected to the source by a plume tail. If the chemical effects are small, however, thermal "domes" or "megaplumes" develop, and the tail is generally severed. Smaller such plumes, if these models are appropriate to the Earth, may produce a LIP with no ensuing hotspot track.

As a bit of a postscript, LIP generation results in huge magmatic fluxes, and these may have considerable effect on the environment and extinction rates (Coffin and Eldholm, 1994; Kerr, 2003; White and Saunders, 2005; Saunders, 2005; Wignall, 2005). Subaerial CFBs can add both dust and gas to the atmosphere with dramatic climatic effects. As mentioned earlier, the Deccan CFB has been related by some to the Cretaceous–Tertiary boundary (although a bolide impact is a more popular alternative). Likewise, the Siberian Traps occurred at the Permian–Triassic boundary (and mass extinction), and the North Atlantic Province occurred at the Triassic–Jurassic extinction event. Several lesser extinction events have occurred in post-Paleozoic times that are tantalizingly close to LIP occurrences. Oceanic plateaus may release enough CO_2 to heat the atmosphere and oceans. This may trigger stored methane in permafrost and submarine methane hydrates, resulting in a runaway greenhouse effect. Sufficient planktonic activity may then lower O_2 levels in seawater to the point of massive ocean-life mortality. Extensive coeval black shales, indicating anaerobic conditions, support such a scenario. And who could resist a relationship between a LIP and the kiss of death?

Summary

The principal large igneous provinces (LIPs) are generally associated with massive decompression partial melting in the large head of a newly developing and rising mantle plume as it impacts the lithosphere. Continental flood basalts (CFBs) are LIPs that occur on continents. They typically represent over 1 million km^3 of basaltic magma generated in less than 1 Ma and constitute the largest subaerial rapid igneous events on the Earth. Some ocean plateaus are even larger. CFBs are typically associated with extension, either limited in scope ("failed rifts," or aulacogens) or "runaway" rifts that develop into new ocean basins.

The scatter in isotopic ratios suggests that CFBs, like OIBs, are derived from several mantle sources, including a deep plume source, a depleted N-MORB-like source, and enriched components that are probably recycled into the mantle by past subduction processes. Many CFBs also show evidence of the participation of enriched subcontinental lithospheric mantle, mantle enriched by released subduction zone fluids, and assimilation of continental crust. Major and trace element data indicate extensive fractional crystallization in magma chambers, presumably ponded at the base of the crust where rising basaltic magma loses its buoyancy. Several CFBs are bimodal basalt–rhyolite, in which the basalts represent mantle partial melts, whereas the rhyolites are typically the result of crustal melting, presumably induced by the large subcrustal magma underplates.

The common association of hotspots with rifts could result from either rift-induced mantle rise (passive upwelling/rifting) or plumes causing extension (active upwelling/rifting). Both probably occur at various places. Extension may be an inevitable result of plume head arrival, but plumes alone (unless perhaps very energetic) are not likely to initiate runaway extension. Continental breakup and the formation of new ocean basins may first need extension of plate tectonic origin, and a plume, should it then pass near the area, may then be deflected toward the thin lithosphere, where it could exploit and accelerate the process. Most post-Paleozoic CFBs involve initial uplift, followed by tremendous magmatism and extension, and then subsidence and formation of a new ocean basin. This suggests that the active model is more common, at least for the larger plume heads.

Key Terms

Large igneous province (LIP) *301*
Oceanic (volcanic) plateau (OP) *301*
Volcanic passive continental margin *301*

Continental flood basalt (CFB) *302*
Traps *314*
Failed rift/aulacogen *303*

Subcontinental lithospheric mantle (SCLM) *311*
Active/passive rift models *317*

Review Questions

Review Questions are located on the author's web page at the following address: **http://www.prenhall.com/winter**

Important "First Principle" Concepts

- Some plumes probably traverse all mantle depths, from the core–mantle boundary to the upper mantle.
- The large heads of newly rising plumes are capable of melting tremendous quantities of material brought from depth.
- Major melting of mantle peridotite in a rising plume occurs only at very shallow depths, probably less than 125 km, and the melt fraction increases considerably with continued rise above this depth.
- Melting is minimal in plumes rising beneath the lithosphere of thick continental cratons because they stall at depths greater than 125 km.

- Melting is most extensive beneath thin crust, such as at rifts, past orogenic belts, and at oceanic crust (particularly near spreading centers).
- Thick cratonic keels may deflect rising plumes to thinner areas.
- Melting may be enhanced greatly if basaltic material is included in the plume head.
- CFBs are generally associated with rifting, but plume heads are not believed to be capable of initiating "runaway" rifting and the creation of a new ocean basin. They may, however, exploit incipient rifting of plate tectonic origin and the combination may be able to do so.

Suggested Further Readings

Basaltic Volcanism Study Project. (1981). *Basaltic Volcanism on the Terrestrial Planets*. Pergamon. New York. Sections 1.2.2 and 1.2.3.

Campbell, I. H. (2001). Identification of ancient mantle plumes. In: *Mantle Plumes: Their Identification Through Time* (eds. R. E. Ernst and K. L. Buchan). Special Paper **352**, 5–21. Geological Society of America. Boulder, CO.

Campbell, I. H., and R. W. Griffiths. (1990). Implications of mantle plume structure for the evolution of flood basalts. *Earth Planet. Sci. Lett.,* **99**, 79–93.

Coffin, M. F., and O. Eldholm. (1991). Large igneous provinces: Crustal structure, dimensions, and external consequences. *Rev. Geophys.,* **32**, 1–36.

Coffin, M. F., and O. Eldholm. (2000). Large igneous provinces and plate tectonics. In: *The History and Dynamics of Global Plate Motions* (eds. M. A. Richards, R. G. Gordon, and R. D. van der Hilst). Monograph **121**, 309–326. American Geophysical Union. Washington, DC.

Condie, K. C. (2001). *Mantle Plumes and Their Record in Earth History*. Cambridge University Press. Cambridge, UK.

Courtillot, V., C. Jaupart, I. Manighetti, P. Tapponnier, and J. Besse. (1999). On causal links between flood basalts and continental breakup. *Earth Planet. Sci. Lett.,* **166**, 177–195.

Elements. (2005). Volume 1. An issue on large igneous provinces.

Ernst, R. E., and K. L. Buchan (eds.). (2001). *Mantle Plumes: Their Identification Through Time*. Special Paper **352**. Geological Society of America. Boulder, CO.

Kerr, A. C. (2003). Oceanic Plateaus. *Treatise on Geochemistry*, **3**, (ed. R.L. Rudnick) Chapter 3.16, pp. 537–565. Elsevier, Amsterdam.

Kerr, A. C., and J. J. Mahoney. (2007). Oceanic plateaus: Problematic plumes, potential paradigms. *Chemical Geology* **241**, 332-353.

Macdougall, J. D. (ed.). (1988). *Continental Flood Basalts*. Kluwer. Dordrecht, The Netherlands.

Mahoney, J. J., and M. F. Coffin (eds.). (1997). *Large Igneous Provinces: Continental, Oceanic, and Planetary Flood Volcanism*. Amer. Geophys. Union. Geophys. Monograph **100**. American Geophysical Union. Washington, DC.

Saunders, A. D., J. Tarney, A. C. Kerr, and R. W. Kent. (1996). The formation and fate of large igneous provinces. *Lithos,* **37**, 81–95.

Storey, B. C., T. Alabaster, and R. J. Pankhurst (eds.). (1992). *Magmatism and Causes of Continental Breakup*. Special Publication **68**. Geological Society. London.

White, R., and D. McKenzie. (1989). Magmatism at rift zones: The generation of volcanic continental margins and flood basalts. *J. Geophys. Res.,* **94**, 7685–7729.

White, R., and D. McKenzie. (1995). Mantle plumes and flood basalts. *J. Geophys. Res.,* **100**, 17,543–17,585.

Wilson, M. (1989). *Igneous Petrogenesis: A Global Tectonic Approach*. Unwin Hyman. London. Chapter 10.

Yaxley, G. M. (2000). Experimental study of the phase and melting relations of homogeneous basalt + peridotite mixtures and implications for the petrogenesis of flood basalts. *Contrib. Mineral. Petrol.,* **139**, 326–338.

Two useful web sites:

www.mantleplumes. org

www.largeigneousprovinces.org

16

Subduction-Related Igneous Activity, Part I: Island Arcs

Questions to be Considered in this Chapter:

1. Why does subduction lead to magmatism when the geothermal gradient is actually lowered by the sinking of a cool plate?

2. Why are island-arc magmas so much more diverse than other intra-oceanic igneous provinces? What is the nature of that diversity, and what is responsible for it?

3. What is the source of island-arc magmas? Is it the subducting lithosphere or the mantle?

4. How does island-arc crust differ from oceanic crust, and what are the ramifications of that difference for arc and continental crustal evolution?

5. What is responsible for the calk-alkaline magma series that is restricted to the subduction zone environment?

6. How does H_2O affect the melting temperature and composition of the melts in basalt and peridotite systems?

7. If island-arc magmas are generated at depth, how do they inherit an enriched isotopic component with a near-surface and continental signature?

8. How is subducted lithosphere reprocessed in subduction zones, and what is returned to the deeper mantle?

In this and the following chapter, we address the remaining great plate tectonic igneous regime: convergent margins. Igneous activity occurring along intra-oceanic arcuate volcanic island chains and some continental margins is distinctly different from the mainly basaltic provinces we have dealt with thus far. The composition of the volcanic suites is more diverse, with greater proportions of much more silicic types, so basalts compose a smaller proportion of the spectrum than in the other oceanic settings explored thus far. Arc volcanism is also typically more explosive, and stratovolcanoes (Figure 4.3) are a common volcanic landform.

Plate tectonics again provides us with an explanation for the generation of these distinctive magmas, as well as for their overall geological setting and geographic distribution. It is now well established that the island-arc type of igneous activity is related to convergent plate situations that result in the subduction of one plate beneath another (situations 3 and 4 in Figure 1.15). The initial petrologic model, and certainly the simplest, was that oceanic crust was partially melted as it was subducted and heated at depth. The melts then rose from the point of melting through the overriding plate to form volcanoes just behind the leading plate edge. As we have learned in several previous chapters, partial melts tend to be less mafic than their parent material (see, e.g., Figure 10.1). Just as partial melting of the *ultramafic* mantle produces mafic basalt, so might partial melting of basaltic oceanic crust produce the intermediate andesites, so common in the arc setting. The ongoing subduction process provides a continuous supply of basaltic crust to be processed into andesite in such a fashion. This two-step model by which mantle peridotites are processed first to basaltic oceanic crust, which then gives rise to arc andesites, is simple and elegant. It is commonly presented in introductory geology texts. We shall soon learn,

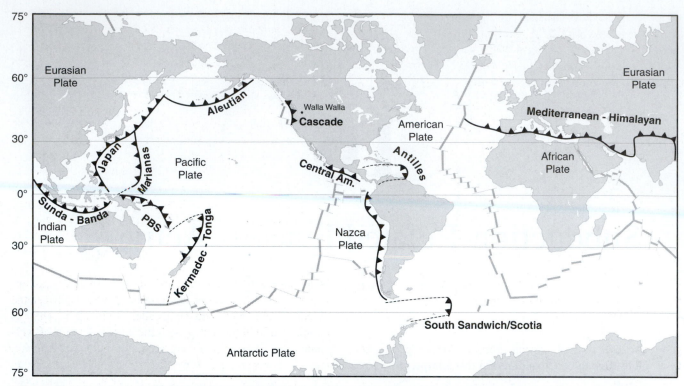

FIGURE 16.1 Principal subduction zones associated with orogenic volcanism and plutonism. Triangles are on the overriding plate. PBS = Papuan–Bismarck–Solomon–New Hebrides arc. Several zones are simplified on this scale. See Gill (1981) or Thorpe (1982) for detailed maps of individual arcs and Hamilton (1978) for the Indonesian-southwest Pacific region. After Wilson (1989). Copyright © by permission Kluwer Academic Publishers.

however, as we explore subduction zone magmatism, that this model cannot explain many of the characteristics of the melts produced.

Only *oceanic* crust can be subducted to any great extent because continental crust is too thick and buoyant. The *overriding* plate, however, is indifferent to density, and can thus be of either type. If the overriding plate is oceanic, the resulting magmatism forms an island arc (#3 in Figure 1.15). If the overriding plate is of continental character, the resulting igneous activity is referred to as a continental arc or an active continental margin (#4 in Figure 1.15). Figure 16.1 is a simplified map of the presently active subduction zones, including both island arcs and continental arcs. Although the overall process is similar for both types of subduction, the influence of continental crust does result in differences. We shall begin with the simpler intra-oceanic island arcs in this chapter and then proceed to continental arcs in Chapter 17.

Subduction is a complex process and produces not only some characteristic igneous associations, but also distinctive patterns of metamorphism (Figure 25.4). It is also responsible for the creation of mountain belts (orogeny). Subduction, magmatism, metamorphism, and orogeny are intimately yet complexly interrelated, and the terms *orogenic* and *subduction-related* are commonly used as synonyms when referring to the common volcanic association of basalts, basaltic andesites, andesites, dacites, and rhyolites produced at subduction zones. Many call it the *orogenic suite*.

Perhaps the greatest challenge I face in writing an introductory petrology text is to summarize and interpret petrologic processes and generalize enough to make the concepts comprehensible without grossly oversimplifying the complexities involved. Nowhere is this more difficult than in addressing subduction zones, certainly the most complex tectonic and magmatic environment on Earth. Subduction-related magmatism is a multistage, multicomponent, multilevel process with considerable variety within and between arcs. I shall attempt to provide as simple a model as possible, but for every generalization that I make, there are certain to be exceptions. I hope that a general model will provide a basic understanding of subduction-related magmatism from which more detailed interpretations can be developed, if necessary, on an individual basis.

16.1 ISLAND-ARC VOLCANISM

Intra-oceanic subduction results in an arcuate chain of volcanic islands. Island arcs are generally 200 to 300 km wide and can be several thousand kilometers long. Figure 16.2 is a schematic cross section through a typical subduction zone arc complex. The oceanic plate on the left, comprising oceanic crust and rigid lithospheric upper mantle, is shown subducting beneath the oceanic plate on the right. A trench, commonly deeper than 11 km, is the surface expression of the plate boundary. Gill (1981) summarized much of the

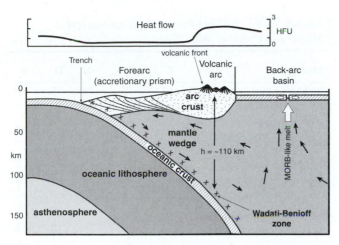

FIGURE 16.2 Schematic cross section through a typical island arc, after Gill (1981). Copyright © with permission from Springer-Verlag. Dashed arrows in the mantle wedge represent the flow directions of McCulloch and Gamble (1991). HFU = heat flow unit ($4.2 \cdot 10^{-6}$ joules/cm²/sec). X's mark earthquake foci of the Wadati-Benioff zone.

geophysical data for 28 volcanic arcs. Subduction rates vary from 0.9 to 10.8 cm/yr. Jarrard (2003) estimated that about 2.4 km² of oceanic crust is subducted annually. Subduction dip angles beneath the volcanic arc vary from 30° to nearly vertical and average about 45°. The dip angle is generally less with younger and hotter subducted lithosphere and with thicker subducted crust (due to aseismic ridges or oceanic plateaus, etc.) due to the greater buoyancy.

The geometry at depth can be inferred from seismic data. An inclined zone of earthquake foci was first described in the 1950s by Kiyoo Wadati of Japan and further elucidated by the American Hugo Benioff. Now commonly called the Benioff zone, or **Wadati-Benioff zone**, it represents the upper boundary of the cool, relatively brittle subducting slab where it slips against the overriding plate and mantle (the seismic zone is represented by X's in Figure 16.2). Seismicity extends to approximately 700 km depth, at which point the subducting plate is believed to soften to the point where brittle failure is no longer possible. The vertical depth from the **volcanic front** (the point of volcanic activity closest to the trench) to the Wadati-Benioff zone (symbolized h) places constraints on magma genesis. Several investigators have estimated h from compilations of arc volcanoes closest to the front, expressed either as ranges or as a single mean with standard deviation. Gill (1981) found h to range from 80 to 250 km and calculated a mean of 128 ± 38 km. The range of Syracuse and Abers (2006) was 72 to 173 km, with an average of 105 km. Tatsumi (2003) concluded that volcanic fronts are 108 ± 18 km (1σ) above the subducted slab surface. Schmidt and Poli (2005) based their compilation on slab surface tomography and the spatial extent of volcanic activity, finding a range in h of 50 to 250 km, with a mean around 105 km. Although there is obvious variation in h, for a given arc, this depth is relatively constant, regardless of dip angle, and averages about 110 km (Tatsumi, 1986). The horizontal distance from the trench to the volcanic front for any particular arc thus

largely depends on subduction zone dip. The width of most volcanic arcs also tends to vary with dip (wider arcs correspond to shallower dips). The close relationship between volcanism and depth to the subducting slab implies that magma generation may be largely controlled by pressure-dependent processes, such as dehydration of some hydrous mineral, which has a low dP/dT slope at high pressure (e.g., see the amphibole dehydration curve in Figure 10.6). The variation in h, however, suggests that arcs are more complex than any single generalization may adequately explain.

A volcanic arc is underlain by a thick mafic to sialic welt of accumulated lava, pyroclastic flows, and plutons (the arc crust). A typical heat flow profile is shown above the cross section in Figure 16.2. Naturally, heat flow is highest at the volcanic portion of the arc. Gill (1981) estimated that volcanism alone, however, accounts for less than 10% of the thermal anomaly, requiring mass transfer of heated material from below in the form of mantle convection and/or plutonism to account for the rest. The portion of the arc between the volcanic front and the trench is called the **fore arc**, which is composed of flows and pyroclastic material from the volcanic arc, immature sediments eroded from the growing arc, and oceanic sediments scraped from the subducting plate. The fore arc is typically highly deformed and imbricated by thrusting as a result of plate convergence. Slivers of oceanic crust and mantle (ophiolite) can be caught up in the thrusting and incorporated into the pile. The accumulation is often called an **accretionary prism**, or **accretionary wedge**.

Behind the arc, a **back-arc basin** is typically developed. This is a site of MORB-like volcanism that creates thin ocean-type crust in an extensional tectonic environment behind the volcanic arc. Having extension at a compressive plate boundary seems anomalous. Extension has been attributed to various processes, including differences in the rates of motion between the overriding and subducting plates (e.g., Scholz and Campos, 1995), "slab rollback" caused by sinking of the subducting plate and migration of the trench over it (see Figure 17.11), and "extrusion" of material perpendicular to the convergence resulting in localized separation of arc fragments (e.g., McKenzie, 1972; Mantovani et al., 2002). Although there may be several influences on extension (see Schellart and Lister, 2004, for a review), it is probably governed by rollback, and by frictional drag associated with the subducting slab, which pulls the adjacent mantle material in the wedge corner downward with it, as illustrated by the mantle flow arrows in Figure 16.2 (McCulloch and Gamble, 1991). Friction between the subducted lithosphere and mantle wedge requires that the dragged wedge material must be replaced, resulting in cyclic motion, convective upwelling, and spreading behind the arc. Notice also that this *induced corner flow* in the mantle wedge between the arc crust and subducted slab, if correctly interpreted, will stir the mantle, and, because it forms a nearly closed loop, it at least partially isolates much of the wedge from the rest of the mantle. As a result of the high heat flow and convective mixing, the lithospheric versus asthenospheric nature of this portion of the mantle above the subduction zone is the subject of considerable debate, and I have purposely left it ambiguous in Figure 16.2.

Volcanism is not uniformly developed in island arcs. Activity begins rather abruptly at the volcanic front, and in most arcs is fairly well focused at the surface along a zone parallel to the trench and less than 50 km wide. Volcanic activity declines more gradually toward the back arc, and arc widths of 100 km are not unusual and a few extend to over 300 km (Schmidt and Poli, 2005). Some arcs (including the Aleutians, Kamchatka, Kuriles, northeastern Japan, Sunda–Banda, and Scotia) have a **secondary arc**, which appears late and some 50 km behind the main arc (Marsh, 1979; Tatsumi and Eggins, 1995).

Volcanism is not continuous *along* the arc axis either. Like mid-ocean ridges (Chapter 13), many arcs are segmented (Stoiber and Carr, 1973; Marsh, 1979; Kay et al., 1982). Segments are nearly linear and 50 to 300 km long. Where they terminate, they typically offset to another segment, either closer to or further from the trench. These offsets usually occur at places corresponding to some structure in the subducting (or overriding) plate, such as a fracture zone. If *h* (the depth to the descending slab beneath the volcanic front) is nearly constant, as proposed above, at least some segments may correspond to different dipping sections of the subducting plate, separated by fracture zones.

Volcanic centers within a segment are often regularly spaced for a given arc (25 km for Central America to 70 km for the Cascade–Aleutian arc; Gill, 1981). Marsh (1979) discussed several hypotheses for this spacing, and concluded that gravitational instabilities produce spaced diapirs that rise from a ribbon of magma that runs parallel to most arcs.

The thickness of the arc-type (with apologies to C. G. Jung) crust ranges from 12 to 36 km and averages about 30 km. It typically thickens with age and consists almost exclusively of locally derived volcanic, plutonic, and volcaniclastic material. Some arcs, such as Japan and New Zealand, however, have a base of much older igneous and metamorphic rocks. Perhaps this material represents a detached continental fragment that separated from the edge of the continent behind it and migrated oceanward as a result of back-arc spreading or strike-slip faulting. If so, such occurrences apparently began as a continental arc, and now have characteristics intermediate between continental and island arcs.

Although arc eruption rates are highly variable and episodic, Reymer and Schubert (1984) estimated that about 30 km^3/Ma has been produced for each kilometer of arc, on average, during the Phanerozoic. This is perhaps 10% of the average rate of MORB production at the mid-ocean ridges.

16.2 ISLAND-ARC VOLCANIC ROCKS AND MAGMA SERIES

Arc complexity is often sadly reflected in the literature, where a plethora of terms arise, and even simple terms, such as *tholeiitic* and *calc-alkaline*, have different meanings to different investigators. It is difficult to find a proper balance between the need for descriptive terms for rock types or magma series and the tendency such classifications have to

TABLE 16.1 Relative Proportions of Analyzed Island-Arc Volcanic Rock Types

Locality	B	B-A	A	D	R
Mt. Misery, Antilles (lavas)[2]	17	22	49	12	0
Ave. Antilles[2]	17	(42	)	39	2
Lesser Antilles[1]	71	22	5	(3	)
Nicaragua, NW Costa Rica[1]	64	33	3	(0	)
W Panama, SE Costa Rica[1]	34	49	16	(1	)
Aleutians east of Adak[1]	55	36	9	0	0
Aleutians Adak and west[1]	20	31	38	(10	)
Ave. Japan (lava, ash falls)[2]	14	(85	)	2	0
Izu-Bonin/Mariana[1]	36	31	7	10	17
Kuriles[1]	34	38	25	(3	)
Talasea, Papua[2]	9	23	55	9	4
Scotia[1]	65	33	2	(0	)

[1] From Kelemen (2003a and personal. communication).
[2] After Gill (1981, Table 4.4) B = basalt, B-A = basaltic andesite, A = andesite, D = dacite, R = rhyolite

compartmentalize our thinking. I shall attempt here to provide a simple summary of the common arc-related magma types.

Table 16.1 illustrates the relative proportions of the main rock types in some quaternary volcanic island arcs, compiled from descriptions available in the literature. Although certainly not an exhaustive survey, these data are broadly representative of arcs. *Basalts still dominate most arcs* but can be subordinate in others. They may compose over 80% of the products from some volcanoes and be all but absent in others. More differentiated lavas compose 30 to over 80% of the rocks in various arcs in Table 16.1, which contrasts markedly with the occurrences previously described in Chapters 13 to 15. Dacites tend to be subordinate and rhyolites are rare. Andesite and basaltic andesite are very prominent in the island-arc rock spectrum, composing 27 to 82% of rocks in the arcs of Table 16.1. Andesite, of course, derives its name from the Andes, a subduction-related continental arc province. Gill (1981), in his comprehensive survey of andesites, pointed out that although andesites occur as minor evolved magmas in a variety of tectonic environments (e.g., Iceland on the Mid-Atlantic Ridge, or Hawaii in intraplate settings), the overwhelming majority occur at convergent plate boundaries.

The rock categories in Table 16.1 are based on wt. % SiO_2 (Figure 2.4), and there is considerable chemical variation within each. We can get a better sense of the arc magmas and their development if we adopt the concept of the magma series (Section 8.7). Since introducing the concept, we have dealt mainly with the tholeiitic and alkaline series. At subduction zones, we encounter the third major one: the calc-alkaline series. The calc-alkaline series became firmly associated with circum-Pacific arcs in the first half of the 20th century, well before plate tectonic theory helped explain the association. An "andesite line" around the Pacific was said to separate an oceanic province with tholeiitic and alkaline characteristics from the calc-alkaline orogenic zones that marked the margin of the sialic continental crust. Many petrologists consider calc-alkaline magmatism to be the hallmark of subduction zone magmatism, and some used the term as yet another synonym for the orogenic suite.

As more sampling and chemical data began to accumulate we realized that all three magma series are well represented at subduction zones (Table 8.7). The calc-alkaline series, however, is essentially restricted to convergent boundaries (it is also recognized in some continental rift areas), and is generally the dominant series in them. Certainly there must be some subduction-related process that is conducive to the development of calc-alkaline magmas.

The abundance of calc-alkaline rocks in volcanic arcs is indicated in Figure 16.3. The figure shows alkali-silica, AFM, and FeO*/MgO versus silica diagrams for 1946 analyses of the more mafic volcanics from about 100 volcanic centers in 30 arcs (both oceanic and continental). We are familiar with the alkali–silica and AFM diagrams from Chapter 8 (Figures 8.10 and 8.13, respectively) as attempts to discriminate alkaline and sub-alkaline (tholeiitic and calc-alkaline) magma series. The third diagram is another tholeiite/calc-alkaline discrimination diagram proposed for arc volcanics by Miyashiro (1974). From Figure 16.3a we can see that alkaline magmas are only a minor constituent of orogenic zones (they are more common above plumes in intraplate and ridge settings), and we shall not be discussing them much in this chapter. Figures 16.3b and 16.3c indicate that both tholeiites and calc-alkaline magmas are well represented in volcanic arcs (but more evolved rocks, toward the alkali corner of Figure 16.3b and toward the high-SiO$_2$ end of Figure 16.3c, are more calc-alkaline).

The terminology of arc magmas begins to get rather messy from this point. Problems arise when we note that the AFM and the FeO*/MgO versus SiO$_2$ diagrams in Figure 16.3 distinguish tholeiites from calc-alkaline series on similar, but not identical, grounds. The AFM diagram correctly implies that FeO*/MgO increases for tholeiites during the early stages of differentiation, whereas calc-alkaline magmas do so either to a lesser degree or not at all (recall Figure 8.3, with the tholeiitic Skaergård trend versus the calc-alkaline Crater Lake trend). The field boundary for the FeO*/MgO versus SiO$_2$ diagram has a positive slope, indicating that FeO*/MgO tends to increase with silica during differentiation within a series, but the distinction is made not on a differentiation trend but on the *magnitude* of the FeO*/MgO ratio for any single analysis (although factored for the silica content). Because the tholeiitic/calc-alkaline boundary is not the same for each diagram, Figure 16.3b indicates that tholeiites and calc-alkaline magmas occur in subequal amounts, whereas Figure 16.3c (for the same data) indicates that tholeiites are more common. Thus some points are tholeiitic by the criteria of Figure 16.3b and calc-alkaline by the criteria of Figure 16.3c. Of course, distinguishing tholeiitic and calc-alkaline rock types should be done for a *series* of cogenetic rocks, and not on the basis of a single analysis on Figure 16.3. Contamination and natural variations are always

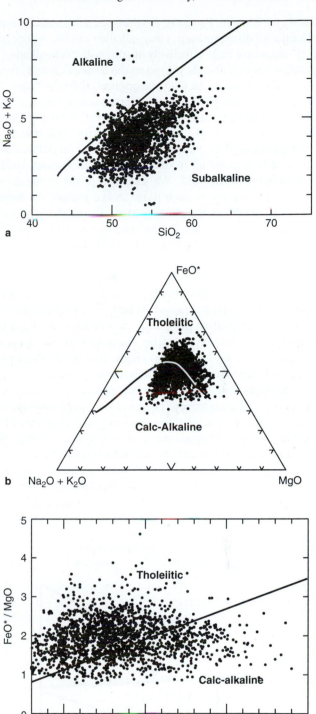

FIGURE 16.3 (a) Alkali versus silica, (b) *AFM*, and (c) FeO*/MgO versus silica diagrams for 1946 analyses from ∼ 30 arcs with emphasis on the more primitive island arc and active continental margin volcanics. Data compiled by Terry Plank (Plank and Langmuir, 1988).

possible for any selected rock, and by plotting a whole series we get not only a better statistical sampling, but we can observe important evolutionary trends that may be pivotal in distinguishing magma series. In addition, as pointed out by Sheth et al. (2002), the AFM diagram doesn't consider CaO, and distinguishes tholeiitic from calc-alkaline on the basis of

Fe/Mg enrichment. Figure 16.3c addresses neither alkalis nor calcium in distinguishing a calc-alkaline series. Although both diagrams may effectively distinguish "calc-alkaline" trends from tholeiitic, they do so on the basis of criteria other than CaO versus alkalis. Only the original criteria of Peacock (1931) in Section 8.7 use an appropriate calcic versus alkaline basis and should thus act as a suitable definition of *calc-alkaline* (he used the synonymous term *calc-alkalic*). Modern usage of the term, however, is based on differentiation trends conforming to diagrams such as Figures 16.3b and 16.3c or Figure 8.15. We shall accept the gentle misnomer and conform to such usage as we explore the basis for such trends in the next section. Paradoxically, most subduction-related "calc-alkaline" rock suites plot as "calcic" by Peacock's criteria (Middlemost, 1985). Alas, our mess is only just beginning.

Gill (1981) stressed the importance of K_2O in subduction-related rocks, calling it "the most significant variable in major element composition between andesites for tectonics." He adopted an alternative classification of orogenic magma series, first proposed by Taylor (1969), based on a combination of K_2O and SiO_2. Figure 16.4 shows the three principal K-based series: **low K**, **medium K**, and **high K**. Contours for more than 2500 analyses of "andesites" (as named in the source literature) are also plotted in the figure. From these data you can see that all three series are well represented in subduction-related magma suites. A fourth series, **very high K** (or **shoshonite**), also occurs, but it is relatively rare.

Any series based on a geochemical parameter such as K_2O cannot be considered separately from the tholeiitic and calc-alkaline series, so we must deal with both series-defining approaches simultaneously. From Figure 16.3c, we could

distinguish tholeiitic from calc-alkaline character on the basis of the value of FeO*/MgO at any chosen value of SiO_2. Gill (1981) chose 57.5% SiO_2 and determined that the value of FeO*/MgO for the boundary curve was 2.3 at that silica content. By this criterion, any series with FeO*/MgO over 2.3 at 57.5% SiO_2 would be tholeiitic, and if less, it would be calc-alkaline. Similar criteria for the value of K_2O at 57.5% SiO_2 could distinguish the three potassium series on Figure 16.4. Figure 16.5 is a diagram from Gill (1981) that plots these values and results in *six* series: the three potassium types (combining high-K and shoshonitic), each having both tholeiitic and calc-alkaline varieties. There should probably be a separate (rare) very-high-K seventh series as well.

At this point, the literature terminology becomes a morass because many investigators fail to distinguish clearly between the K-Si and tholeiitic–calc–alkaline parameters. Figure 16.5 suggests that, of the six common series, the low-K type is dominantly tholeiitic, the medium-K series is more calc-alkaline, and the high-K series is mixed. Thus most orogenic rock suites might most simply be described by *three* principal series: **low-K-tholeiitic**, **medium-K-calc-alkaline**, and **high-K (mixed)**. The first of these hybrid series corresponds to the "island-arc tholeiite" series of Jakes and Gill (1970), and the second and third correspond to the "calc-alkaline" and "high-K" series, respectively, of Peccerillo and Taylor (1976).

Lest we begin to believe in three, five, six, or seven truly *distinct* series, each evolving from a unique parent basalt toward evolved rhyolite along some fractionation path, we should examine Figures 16.3 to 16.5 more closely. First of all, rock analyses plot in all six fields in Figure 16.5, suggesting that there are both tholeiitic and calc-alkaline magma series belonging to each of the three potassium types

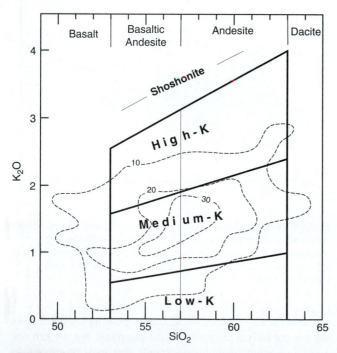

FIGURE 16.4 The three andesite series of Gill (1981). A fourth very high K shoshonite series is rare. Contours represent the concentration of 2500 analyses of andesites stored in the large data file RKOC76 (Carnegie Institute of Washington).

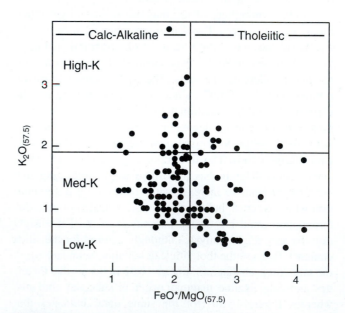

FIGURE 16.5 Combined $(K_2O)_{57.5}$ versus $(FeO*/MgO)_{57.5}$ diagram in which the low-K to high-K series are combined with the tholeiitic versus calc-alkaline types, resulting in six andesite series, after Gill (1981). The "57.5" subscript means the value of K_2O or FeO*/MgO of a series of cogenetic lavas at 57.5 wt. % SiO_2 The points represent the analyses in the appendix of Gill (1981). Copyright © with permission from Springer-Verlag.

(see Table 5.3 of Gill, 1981, for a list). Only the low-K calc-alkaline type appears to be rare. The shoshonitic series, also rare, occurs in several arc as well.

More importantly, one can clearly see from the data in Figures 16.3 to 16.5 that the compositions of arc rocks (and series) form a *continuum* rather than falling into any conveniently neat (yet artificial) groupings. In other words, convergent plate boundaries are complex places, with a variety of processes taking place and a similar variety of products. Perhaps we should not expect arc magmas to conform to a classification concept. Many rocks plot at or near the boundary curves, and it is not uncommon for some magma series to cross a boundary (usually from more tholeiitic primitive magmas to calc-alkaline derivative ones). Some confusion and disagreement is inevitable in a nomenclature for such rocks.

But a classification is still useful, at the very least as a descriptive aid to the investigation of arc magmatism. Subdividing and classifying rock types is useful in a descriptive sense, and is a natural course of scientific methodology (which is often one of finding patterns in nature). We can use the classifications to our advantage, but we should remember that there are differences in the chemical composition and genesis of magmas between various arcs, between igneous centers within

an arc, and even at a single center. Our simplified three-fold series classification provides us with a vocabulary for discussing the types of magmas, and, because they are based on demonstrably important chemical variations, they may also be indicators of trends in petrogenesis. As noted earlier, the tholeiitic series occurs in a variety of tectonic environments, where it usually results from shallow partial melting of rising mantle and even shallower differentiation (Chapters 13 to 15). Calc-alkaline magmas, on the other hand, are essentially restricted to subduction zones. It would be interesting to understand what controls the development of tholeiitic versus calc-alkaline magmas. But this is a matter of petrogenesis, and first we should address the chemical and field data that can give us more clarification on these series and constrain our models.

16.3 MAJOR ELEMENT GEOCHEMISTRY OF ISLAND ARCS

Table 16.2 lists some heroically averaged compositions and CIPW norms of basalts (B), andesites (A), dacites (D), and rhyolites (R) for each of the series in the three simplest groupings discussed above. The averages are calculated for a relatively wide variety of rocks from around the globe, and most are

TABLE 16.2   Average Compositions of Some Principal Rock Types in the Island-Arc Volcanic Suites

wt. % Oxide	Low-K (tholeiitic)				Meduim-K (mostly c-a)				High-K			
	B	A	D	R	B	A	D	R	B	A	D	R
SiO_2	50.7	58.8	67.1	74.5	50.1	59.2	67.2	75.2	49.8	59.4	67.5	75.6
TiO_2	0.8	0.7	0.6	0.4	1.0	0.7	0.5	0.2	1.6	0.9	0.6	0.2
Al_2O_3	17.7	17.0	15.0	12.9	17.1	17.1	16.2	13.5	16.5	16.8	16.0	13.3
Fe_2O_3	3.1	3.0	2.0	1.4	3.4	2.9	2.0	1.0	3.9	3.6	2.0	0.9
FeO	7.4	5.2	3.5	1.7	7.0	4.2	1.8	1.1	6.4	3.0	1.5	0.5
MgO	6.4	3.6	1.5	0.6	7.1	3.7	1.5	0.5	6.8	3.2	1.1	0.3
CaO	11.3	8.1	5.0	2.8	10.6	7.1	3.8	1.6	9.4	6.0	3.0	0.9
Na_2O	2.0	2.9	3.8	4.0	2.5	3.2	4.3	4.2	3.3	3.6	4.0	3.6
K_2O	0.3	0.6	0.9	1.1	0.8	1.3	2.1	2.7	1.6	2.8	3.9	4.5
P_2O_5	0.1	0.2	0.2	0.1	0.2	0.2	0.2	0.1	0.5	0.4	0.2	0.1
Total	99.8	100.1	99.6	99.5	99.8	99.6	99.6	100.1	99.8	99.7	99.8	99.9
CIPW Norm												
q	2.8	15.5	28.1	40.0	0.0	14.0	24.1	36.1	0.0	10.7	21.8	35.4
or	1.8	3.6	5.3	6.5	4.7	7.7	12.5	15.9	9.5	16.6	23.1	26.6
ab	17.0	24.5	32.3	34.0	21.2	27.2	36.5	35.5	28.0	30.6	33.9	30.5
an	38.5	31.6	21.3	13.3	33.2	28.6	17.6	7.3	25.6	21.5	13.6	3.8
di	13.9	6.0	1.9	0.0	14.9	4.5	0.0	0.0	14.5	4.7	0.0	0.0
hy	20.9	13.9	6.6	2.8	16.9	13.0	4.6	2.1	3.0	9.8	2.9	0.8
ol	0.0	0.0	0.0	0.0	3.1	0.0	0.0	0.0	10.8	0.0	0.0	0.0
mt	3.3	3.2	2.9	2.0	3.6	3.2	2.9	1.5	4.5	3.5	2.9	1.0
il	1.5	1.3	1.1	0.8	1.9	1.3	1.0	0.4	3.1	1.7	1.1	0.4
ap	0.2	0.5	0.5	0.2	0.5	0.5	0.5	0.2	1.2	0.9	0.5	0.2
Mg#	53	45	34	27	56	49	43	31	55	48	37	29

B = basalt, A = andesite, D = dacite, R = rhyolite, c-a = calc-alkaline

From Hess (1989, p. 152). Data from Ewart (1979, 1982).

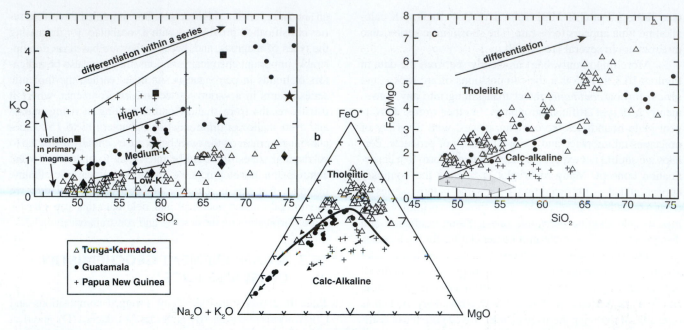

FIGURE 16.6 Variation in the Tonga–Kermadec, Guatemala, and Papua New Guinea volcanic suites. **(a)** K_2O-SiO_2 diagram (after Gill, 1981, copyright © with permission from Springer-Verlag) distinguishing high-K, medium-K, and low-K series. Large squares = high-K, stars = medium-K, diamonds = low-K series from Table 16.2. Smaller symbols are identified in the caption. Differentiation within a series (presumably dominated by fractional crystallization) is indicated by the arrow. Different primary magmas (to the left) are distinguished by vertical variations in K_2O at low SiO_2. **(b)** *AFM* diagram distinguishing tholeiitic and calc-alkaline series. Arrows represent differentiation trends within a series. **(c)** FeO^*/MgO versus SiO_2 diagram distinguishing tholeiitic and calc-alkaline series. The gray arrow near the bottom is the progressive fractional melting trend under hydrous conditions of Grove et al. (2003). Note that Tonga–Kermadec is both low-K and tholeiitic, PNG is high-K and calc-alkaline, whereas Central America is medium-K and calc-alkaline by the criteria of **(b)** but borderline by that of **(c)**.

phyric (a typical feature of orogenic volcanics), so that true liquid compositions are not necessarily represented (although they may be if the crystals remain with the parent liquid). These analyses should be treated only as *very* general indicators of the rock types and series in question and as being useful in a not-too-rigorous comparison to one another or to rocks from other petrogenetic associations, such as MORBs or OIBs. The basalts of all three island-arc series are similar to MORBs. Al_2O_3 and K_2O are higher than in MORBs, whereas MgO (and Mg#) is lower. Al_2O_3 is also higher than in ocean island basalts, but it is variable and can be as low as 13 wt. %, which overlaps with MORB. High Al_2O_3 content is particularly common in association with calc-alkaline magmas, and basalts with 17 to 21 wt. % Al_2O_3 are called **high-alumina basalts**. Several investigators have noted this association and have proposed that high-alumina basalts are parental to the calc-alkaline series. This association, however, may be due to the physical conditions prevalent in subduction zones producing both high-alumina basalts and calc-alkaline trends rather than a parent-derivative series (as will be discussed later).

The averages in Table 16.2 are much less useful for evaluating evolutionary processes because none represents a real rock, so they are not really members of a single natural series. As examples of true island-arc series, I have chosen three from the literature that range from low-K to high-K and tholeiitic to calc-alkaline:

- ***Tonga-Kermadec:*** low-K tholeiitic
- ***Guatemala:*** medium-K calc-alkaline
- ***Papua New Guinea Highlands:*** high-K (calc-alkaline)

The data for these series can be found in the Excel file IA.XLS available at www.prenhall.com/winter. When plotted on the K_2O-SiO_2 diagram (Figure 16.6a) the data for these three series (and the data in Table 16.2) fall within the appropriate areas as defined by Gill (1981). The data in these examples all come from more than one volcano and are thus rather scattered, but even when combined, they still serve to illustrate the chemical characteristics of each series. Variation within a series is controlled mostly by fractional crystallization resulting in progressively increasing SiO_2 and perhaps by magma mixing in the more evolved portions. The effects of crustal contamination are minimized in these intra-oceanic arcs because the crust is thin and not much different than the arc magmas themselves. Projecting back toward more primitive magmas the trends converge, so there is less variation among the parental basalts. True primary magmas are rare, and chemical differences are due to a number of factors, including different sources, depth and extent of partial melting of the source, fractionation during ascent, depth of magma segregation from the source, mixing of magmas from different series, etc.

Figure 16.6b is an AFM diagram for the three series. Note that the low-K Tonga–Kermadec series is indeed tholeiitic and shows the typical Fe enrichment in the early stages of differentiation. The medium-K Guatemalan and high-K PNG series fall within the calc-alkaline field and show less Fe enrichment. All three series show significant alkali enrichment as they evolve. Note that the Guatemalan data cross from more primitive tholeiitic affinities to more evolved calc-alkaline ones. This trend is not unusual for an orogenic series on AFM diagrams, implying that something in the subduction zone environment favors calk-alkaline evolution even if the parent is tholeiitic. In general, the higher the K_2O + content, the less Fe enrichment, so that the high-K series plots closest to the base of the AFM triangle. The most differentiated rhyolites in the high-K series also plot closest to the alkali corner, as we would expect.

Figure 16.6c is the FeO*/MgO versus SiO_2 diagram of Miyashiro (1974). Here again, the high-K series is calc-alkaline and the low-K series tholeiitic, but the medium-K series straddles the border between the two types, and appears slightly more tholeiitic. This confusion in tholeiitic versus calc-alkaline character in medium- to high-K series

is common, and the Fe-enrichment is probably related as much to H_2O and oxygen content, and their effect on Fe-Ti-oxide formation during shallow level fractional crystallization, as to intrinsic differences in parental magma type. We shall address these factors shortly.

Figure 16.7 shows a series of major element variation ("Harker") diagrams for the three series. Although these data represent more than one volcano in each arc and cannot be interpreted rigorously as cogenetic, the diagrams serve to illustrate the general differentiation trends for the three series. The decrease in Al_2O_3, MgO, FeO*, and CaO with increasing SiO_2 are all familiar by now and are compatible with fractional crystallization of plagioclase and mafic phases,

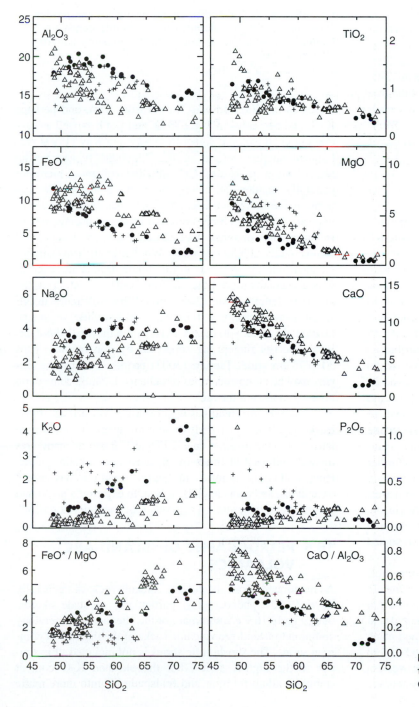

FIGURE 16.7 Harker variation diagrams for the three example volcanic series: Tonga–Kermadec, Guatemala, and PNG. Symbols as in Figure 16.6.

such as olivine or pyroxene. As discussed in Section 13.4, fractionation of plagioclase should cause CaO/Al_2O_3 to increase for these compositions, but this ratio decreases in the three series (Figure 16.7). Thus clinopyroxene probably also forms and removes calcium as well. The K_2O diagram is the same as in Figure 16.6. Because the three series were chosen on the basis of varying K_2O contents there is considerable spread in this diagram. K_2O and Na_2O are largely conserved, and thus have positive slopes as they concentrate in the more evolved melts. The decrease in TiO_2 is probably a result of fractionation of an Fe-Ti oxide. FeO*/MgO is also included in the figure, and increases in all three series, a fact obscured somewhat by the constant sum effect in the AFM diagram (A + F + M must be normalized to equal 1.0). The increase in FeO*/MgO is greatest for the low-K tholeiite series and least for the high-K series, as we saw in Figure 16.6b.

Although it does not show up on Harker diagrams very well because it is the abscissa, SiO_2 increases progressively throughout the calc-alkaline series. In tholeiites, SiO_2 remains relatively constant until the late stages (see the SiO_2 curve in Figure 12.13a). This silica enrichment and the iron enrichment and elevated H_2O content (resulting in a more explosive nature) are the most obvious and consistent differences between calc-alkaline and tholeiitic magma series.

16.4 SPATIAL AND TEMPORAL VARIATIONS IN ISLAND ARCS

When considered as a whole, most arcs show considerably more compositional variation than is shown in Figures 16.6 and 16.7, which concentrate on specific series within an arc. More than one series typically occurs in a single arc (or even in a single volcano). Figure 16.8, from the Sunda–Banda arc, is an illustration of the variety possible. This serves to underscore the complexity of the island-arc environment, and the resultant diversity of the magmas produced.

Some spatial and temporal patterns in the distribution of magma series are found in several island arcs. Kuno (1959) first described the relationship between the concentration of K_2O in a lava and the depth (h) from the volcano to the Wadati-Benioff zone (the so-called **K-h relationship**) in Japan. Japanese low-K tholeiites occur closer to the trench, and medium- and high-K, mostly calc-alkaline, magmas (and even shoshonitic types) occur progressively farther from the trench. This pattern of higher K_2O content (and lower SiO_2 saturation) has been described from several arcs, particularly continental ones (see Hatherton and Dickinson, 1969; Arculus and Johnson, 1978; Gill, 1981; Tatsumi and Eggins, 1995), but there are exceptions, and even reversed patterns, known.

Patterns are also evident *along* the axis of some arcs. In the Lesser Antilles, there is a variation, at approximately constant h, from low-K tholeiites at St. Kitts in the north to calc-alkaline magmas in the central area to alkaline types at Grenada in the south (Brown et al., 1977). Kelemen et al. (2003b) noted that primitive magmas in the central and eastern Aleutians tend to be basaltic whereas those in the western Aleutians are mainly andesitic. They also noticed systematic

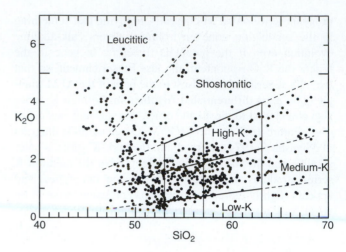

FIGURE 16.8 K_2O-SiO_2 diagram for nearly 700 analyses for Quaternary island-arc volcanics from the Sunda–Banda arc. From Wheller et al. (1987). Copyright © with permission from Elsevier Science.

increases in SiO_2 and Na_2O toward the west, and perhaps K_2O; and decreases in MgO, FeO, and CaO; as well as several enriched isotopic signatures. Kay and Kay (1994) recognized another pattern in the Aleutians related to the segmentation of the arc. Calc-alkaline volcanoes there are generally smaller and concentrated along the middle sections of the individual segments, whereas tholeiitic volcanoes are larger and more common at segment ends, nearer fracture zones or other large structures, such as seamount chains, on the subducting plate. Wheller et al. (1987) distinguished four arc segments in the Sunda–Banda arc, each of which becomes more potassic from west to east along the arc axis. Tamura et al. (2002) noted that volcanoes in NE Japan were concentrated in ten ~50-km-wide segments, elongated normally to the arc axis, which they called *hot fingers*. The hot fingers are separated by non-volcanic gaps 30 to 75 km wide. Tamura (2003) proposed that these fingers may be elongated zones of enhanced mantle convection up and out toward the arc.

A *temporal* trend has also been described (Gill, 1981; Baker, 1982), in which early tholeiitic volcanism gives way to later more calc-alkaline and K-rich volcanism as many arcs mature, followed at times by an even later alkaline phase. Ringwood (1977) pointed out that this trend also has several exceptions, and considered it, at best, to express "a broad and general trend rather than a rigorous sequential development".

16.5 PETROGRAPHY OF ISLAND-ARC VOLCANICS

As mentioned above, island-arc volcanic rocks are generally phyric (>20% phenocrysts), particularly the calc-alkaline ones. Figure 16.9 is a summary of the phenocryst mineralogy common to island-arc volcanic rocks of the three most common series. The shoshonitic series is not included. It is similar to the high-K series, but the elevated K_2O content stabilizes alkali feldspar and feldspathoids into more mafic compositions, so that either may occur in basalts of this series. Andesites and basaltic andesites are typically much more phyric than basalts. Surprisingly, dacites, and especially rhyolites, are less phyric than andesites, perhaps because of restricted diffusion in the highly polymerized liquid, which inhibits migration of constituents to, and hence the growth of, large crystals.

Plagioclase is a ubiquitous phenocryst phase in arc volcanics. Plagioclase phenocrysts typically exhibit complex zoning and resorption or reaction patterns, which may differ from one crystal to another in the same thin section. Zoning may exhibit combinations of normal, reverse, and oscillatory patterns, with either gradual or abrupt variations in An content. Plagioclase compositions are highly variable but typically more calcic than in MORBs or OIBs, often in the range An_{50} to An_{70}, but they may reach An_{90} or higher. The high An content is commonly attributed to the high H_2O content of arc magmas. Remember that H_2O tends to disrupt Si-O-Si-O polymers, so elevated H_2O makes for a less polymerized melt. Remember also that the melt structure is similar to the minerals that form from it. High H_2O content should thus suppress minerals with high Si-O-Si-O polymerization. Albite ($NaAlSi_3O_8$) has 50% more Si-O bonds than anorthite ($CaAl_2Si_2O_8$), so raising the H_2O content and suppressing Si-O-Si-O polymers should favor crystallization of anorthite over albite. Al is more affected by hydration than Si, as shown by the high number of minerals with Al-OH bonds and the paucity of ones with Si-OH bonds. The type of plagioclase that forms with more hydrous melts is thus more calcic. Most phenocrysts are more calcic than either the groundmass plagioclase or the normative plagioclase based on the whole-rock composition. Plagioclase phenocrysts often appear corroded or sieve like, and have frequently contained inclusions in certain zones. The various disequilibrium features suggest that magma mixing is common (Eichelberger, 1975; Eichelberger and Gooley, 1977) but may also reflect complex convection effects (plus recharge) in shallow magma chambers (Singer et al., 1995).

Plagioclase is accompanied by phenocrysts of olivine and augite in the mafic parts of all series, and in andesites of the low-K series. Mafic phases are usually rather Mg rich (olivine = $Fo_{70} - Fo_{85}$ and augite Mg# = $85 - 90$), even in dacites and rhyolites. Augite is second only to plagioclase as an arc-related phenocryst phase. Arc clinopyroxenes are typically more aluminous than their MORB counterparts. The high-Al effect can also be attributed to high H_2O content (increased tschermakite component results in more Al-O than Si-O bonds). Orthopyroxene also occurs as a phenocryst phase in some arc volcanics.

Black hornblende phenocrysts are more common in medium- to high-K (calc-alkaline) andesites. Hornblende is an important phenocryst phase, as it is stable only if the melt contains >3 wt. % H_2O at pressures in excess of 0.1 to 0.2 GPa. It thus provides information on both H_2O content and depth of magma segregation. Biotite occurs only in the more evolved rocks of the medium- to high-K series. Both hornblende and biotite phenocrysts are commonly replaced by, or develop rims of, fine anhydrous phases, such as pyroxene,

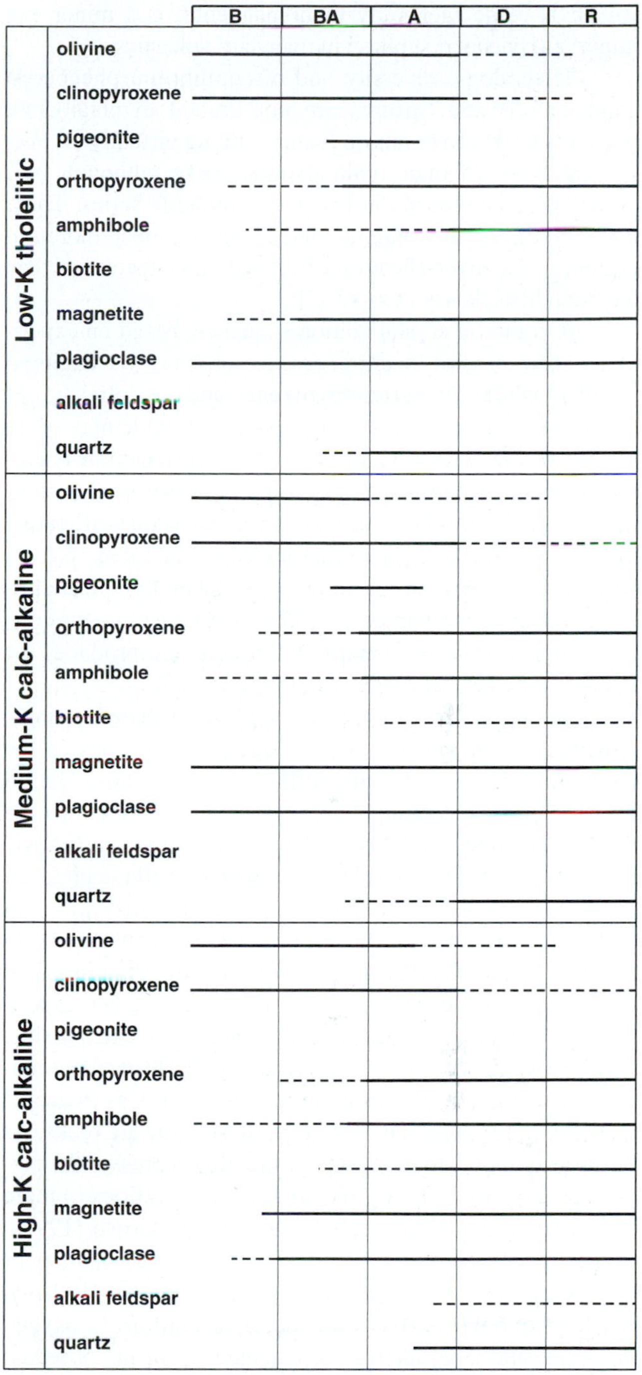

FIGURE 16.9 Major phenocryst mineralogy of the low-K tholeiitic, medium-K calc-alkaline, and high-K calc-alkaline magma series. B = basalt, BA = basaltic andesite, A = andesite, D = dacite, and R = rhyolite. Solid lines indicate a dominant phase, whereas dashes indicate only sporadic development. From Wilson (1989). Copyright © by permission Kluwer Academic Publishers.

plagioclase, magnetite, etc. The iron in them may also oxidize and produce dark brown colors in the remaining crystals. This dehydration and oxidation has been interpreted as partial to complete decomposition of the phase due to the sudden loss of H_2O pressure upon eruption. An alternative mechanism involves hydrous phenocrysts in rhyolitic or dacitic magma, which may be heated above their stability range when mixed with hotter, more primitive magma (Feeley and Sharp, 1996).

An Fe-Ti oxide (generally titanomagnetite) is a minor, but important, phenocryst phase in many arc volcanics.

Mineralogic diversity and disequilibrium phenocryst compositions and textures are not limited to plagioclase phenocrysts. It is a common feature among virtually all phenocryst types in many calc-alkaline rocks, although it is much less common in the low-K arc tholeiite series. These features suggest that magma mixing may be much more important in the diversification of the calc-alkaline series than has been historically recognized.

A common crystallization sequence, based on experiments and natural occurrences, is olivine $\rightarrow$ + augite $\rightarrow$ + plagioclase $\rightarrow$ + orthopyroxene and − olivine $\rightarrow$ + hornblende, biotite, and quartz. An Fe-Ti oxide may occur at almost any time depending on f_{O_2}. Most natural occurrences have olivine + augite + plagioclase in the most primitive volcanics because truly primary magmas are rarely found at the surface. (Note that the Mg#s in Table 16.2 are considerably less than the value of 70, taken to indicate primary magmas in Chapter 10.) This is a typical crystallization sequence that has been noted before and produces the trends on the Harker diagrams in Figure 16.7.

Of course, there are more complexities than this single sequence might indicate, which should be obvious in the tholeiitic versus calc-alkaline differentiation trends. The increasing SiO$_2$ content and lack of dramatic Fe enrichment in the calc-alkaline trend has been attributed to the early crystallization of an Fe-Ti oxide phase, which would deplete the remaining liquids in Fe. This simple explanation was first proposed by Kennedy (1955), who related it to the high H$_2$O content of calc-alkaline magmas in arcs, the dissolution of which produced high f_{O_2} (Osborn, 1959). Further experiments showed that magnetite formation and the calc-alkaline trend required oxygen fugacities too high for the arc environment. The amount of magnetite required is at least twice its modal proportion as phenocrysts in most mafic arc rocks. As mentioned above, high H$_2$O pressure also depresses the plagioclase liquidus, shifting plagioclase compositions to more calcic varieties. Experiments by Sisson and Grove (1993a) suggested that crystallization of olivine, An-rich, SiO$_2$-poor plagioclase, and Cr-spinel or magnetite from an H$_2$O-rich basalt (4 to 6 wt. % H$_2$O) can occur at reasonable oxygen fugacities and is an alternative mechanism for the observed calc-alkaline differentiation trend. Boettcher (1973, 1977) and Cawthorn and O'Hara (1976) proposed that fractionation of a low-silica high-Fe hornblende could also produce the calc-alkaline trend, but it is not sufficiently common as a phenocryst phase and the Fe-content is too low to make this a likely option.

16.6 ISLAND-ARC TRACE ELEMENT GEOCHEMISTRY

Recall from earlier chapters that major elements are typically buffered by magma–rock/mineral interactions and thus reflect host characteristics at the point of segregation and subsequent fractionation and mixing processes. Trace

elements, although also affected by crystal fractionation, can tell us more about ultimate source characteristics. Compatible trace element concentrations in more primitive island-arc basalts are in the ranges Ni: 75 to 150 ppm and Cr and V: 200 to 400 ppm. These are very low for considering arc basalts to be primary mantle partial melts. Because these elements are included in early-forming minerals, the concentrations drop during fractional crystallization. For andesites, the values are typically Ni: 10 to 60 ppm, Cr: 25 to 100 ppm, and V: 100 to 200 ppm. If andesites are partial melts of the mantle, as some have proposed in order to explain their great abundance at convergent plate boundaries, most are far from primary melts of normal mantle. V, which concentrates in magnetite, typically shows a strong decrease with increasing SiO$_2$ in calc-alkaline magmas, supporting the theory of magnetite fractionation in the calc-alkaline trend (Garcia and Jacobson, 1979).

Chondrite-normalized REE concentrations for basaltic andesites and andesites are shown for our three most common types of K series in Figure 16.10. Several points are worth mentioning. The overall shape and slope of the REE pattern varies from one series to another with the K$_2$O

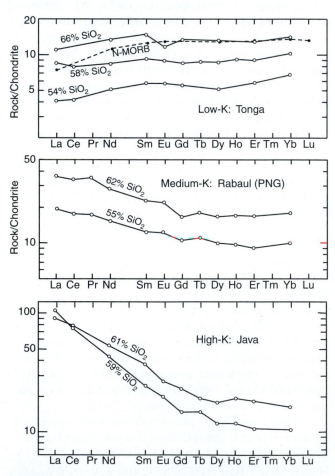

FIGURE 16.10 Chondrite-normalized REE diagrams for some representative low-K (tholeiitic), medium-K (calc-alkaline), and high-K basaltic andesites and andesites. N-MORB is included for reference (from Sun and McDonough, 1989; compare also to Figure 13.11). After Gill (1981). Copyright © with permission from Springer-Verlag.

content, but not with variations in SiO$_2$ within a series. Because the REE are incompatible elements, their absolute concentrations increase with SiO$_2$ content (presumably due to fractional crystallization), but they do so coherently, and the slope is relatively unchanged as the curves rise in Figure 16.10. Thus fractional crystallization must involve phases such as olivine, pyroxenes, and feldspars that do not selectively incorporate light versus heavy REEs. Only a small negative Eu anomaly develops in the high-SiO$_2$ curves, which can be related to the removal of plagioclase.

The low-K tholeiitic series have a low negative REE slope, similar to MORB, but not as steep. Because LREE depletion (negative REE slope) cannot be accomplished by any partial melting process of a chondritic (non-sloping) type of mantle, the source of low-K tholeiites must be *depleted mantle*, similar to that for MORB. The REE abundances of the low-silica samples are even less than that for MORB, however, suggesting that the source of arc magmas may be even *more depleted* than the MORB source. Medium-K and high-K series are progressively more LREE enriched. LREE enrichment is similar to that of other highly incompatible elements, such as K itself, and can be accomplished by low degrees of partial melting of primitive mantle. An alternative explanation, proposed by Thompson et al. (1984), is that the mantle source for island-arc magmas is a heterogeneous mixture of depleted MORB and enriched OIB mantle types. Either way, there appears to be more than one source with different incompatible element concentrations, reflecting variable depletion and perhaps enrichment.

It is unlikely that the three series can be derived from one another, or from a single parent, via shallow fractional crystallization because the silica contents overlap and fractional crystallization does not vary the REE slope within a series. Only at some greater depth, where different phases are stable (such as garnet) and may selectively incorporate the HREE over the LREE, would it be possible to vary the

slope significantly by fractional crystallization. Then, perhaps, the REE pattern for one series could be derived from that of another.

The HREE portion of all the curves shown in Figure 16.10 is relatively flat, implying that garnet, which strongly partitions among the HREE, was not in equilibrium with the melt at the time of segregation. This suggests that the three arc series are not related by deep fractionation processes. It also puts in doubt the idea that most island-arc magmas are derived from the oceanic crust of the subducted slab because the basalt should be converted to eclogite (a clinopyroxene-garnet rock) at depths of 110 km where the magmas are apparently derived. If any garnet were left in the residuum after partial melting, there should be a positive slope to the HREE portion of the curves.

Figure 16.11 shows two MORB-normalized spider diagrams for island arcs, which show a broader range of trace elements than the REE diagram. Only island-arc *basalts* are shown because they are more likely to reflect the trace element concentrations of the source and be less affected by fractional crystallization or contamination. Figure 16.11a is the MORB-normalized spider diagram of Pearce (1983), also used in Figures 9.7, 14.5, and 15.9. In every arc in the diagram, the LIL elements (Sr, K, Rb, and Ba) are enriched and behave differently than the HFS elements (Th-Yb), which show nearly MORB-like concentrations (rock/MORB ratio of 1.0).

This high-LIL/HFS pattern is now recognized as a distinctive feature of most subduction zone magmas. The large ionic radius and low valence of LIL elements make them very soluble in aqueous fluids, and are thus readily fractionated into a hydrous fluid phase, if one is available. HFS elements, with a higher valence, are much less H$_2$O soluble and for that reason are often called *immobile elements* (referring to mobilization in an aqueous pore fluid). Because the LIL and HFS elements are all incompatible and behave similarly in solid–melt exchange, the "decoupling" of these

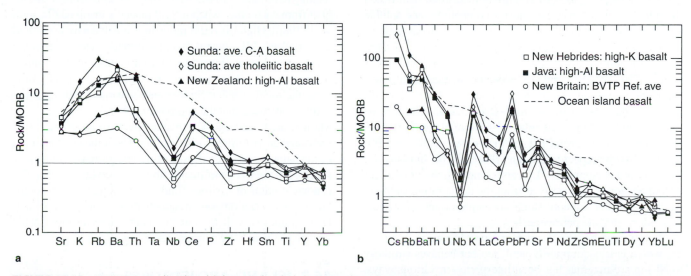

FIGURE 16.11 MORB-normalized multielement (spider) diagrams for selected island-arc basalts. **(a)** Using the normalization and ordering scheme of Pearce (1983), with LIL on the left and HFS on the right and compatibility increasing outward from Ba-Th. **(b)** Using the normalization and ordering scheme of Sun and McDonough (1989), with increasing compatibility to the right. The legend is split between the diagrams and applies to both. Data from Basaltic Volcanism Study Project (1981). OIB data from Sun and McDonough (1989).

two groups, and enrichment of the LILs, is most readily explained by the participation of H_2O-rich fluids in the genesis of subduction zone magmas. The obvious source for H_2O so deep in the Earth would be water contained in the sediments and hydrated oceanic crust of the subducted slab. These fluids, enriched in LIL elements scavenged from the sediment and crust, can both lower the melting temperature of the solid source rocks and concentrate LILs in the resulting hydrated magma. Such a model is also consistent with the common occurrence of hydrous minerals (such as hornblende), calcic plagioclase, and the explosive nature of arc volcanism.

Y and Yb are concentrated in garnet, and the lack of any negative Y-Yb anomaly in Figure 16.11 again suggests that the magma source was not deep and garnet-bearing. The relatively flat HFS element pattern near 1.0 means the concentration of these elements is similar to MORB, and again suggests that the source of island-arc basalt is similar to that of MORB: depleted mantle, and not subducted crust. The fact that the HREE and compatible HFS elements in both diagrams are lower than MORB (less than 1.0) may be because of the MORB standard concentration values chosen for normalization, but may also indicate, as mentioned previously, that the mantle source of many island-arc basalts is even *more* depleted than MORB source (Pearce and Peate, 1995).

Figure 16.11b is a variation on the MORB-normalized spider diagram used by Sun and McDonough (1989), which places the elements in order of ascending compatibility (most compatible on the right), similarly to the OIB-normalized diagram in Figure 15.10. This ordering considers only partial melting and does not take into account aqueous solubility, so the LIL elements (Cs, Rb, K, Ba, Pb, Sr, and Eu^{+2}) and HFS elements appear less coherent than in Figure 16.11a. The diagram also includes a broader range of trace elements than the Pearce (1983) type diagram in Figure 16.11a. Once again, all the arcs shown in Figure 16.11b have a broadly similar pattern that shows LIL enrichment, often manifested as spikes in Cs (the largest, most incompatible, and soluble element of the group), K, and Pb. An OIB analysis was included for reference in Figure 16.11 and has a humped (Figure 16.11a) or continuously sloping (Figure 16.11b), non-spiked pattern. This pattern is widely accepted as a coherent, non-subduction zone, or "intraplate" pattern. All of the spiked elements in the other patterns are H_2O soluble and likely to be concentrated in fluids. Leeman (1996) also noted that the concentrations of several LIL elements (As, B, Sb, and Pb) increased with respect to Ce (an HFS element) in many arc magmas and attributed the LIL enrichment to their mobilization in aqueous fluids derived from the dehydrating slab. We can conclude that aqueous fluids are an important component of subduction zone petrogenesis.

The large negative trough at Nb in Figure 16.11b (and Ta, which is less commonly analyzed but behaves similarly to Nb) in arc magmas has been interpreted in various ways. Some investigators have noted the similarity of the overall trace element pattern between island arcs and OIBs and propose that the source of island-arc magmas is a somewhat enriched one, similar to that of OIB. They then attribute the low Nb-Ta concentrations to the presence of a residual Nb-Ta-bearing mineral. Nb and Ta behave similarly to Ti, so rutile, ilmenite, titanite, and even hornblende are suspected of remaining and holding Nb and Ta in the source (Morris and Hart, 1983; Saunders et al., 1991). Others, such as McCulloch and Gamble (1991), noted that Nb and Ta have MORB-like concentrations (rock/MORB = 1.0 in Figure 16.11), comparable to many other more compatible elements (toward the right) that would not be incorporated in any Nb-Ta-concentrating phase. The deep Nb "troughs" in Figure 16.11 may thus result more from the additions of the neighboring elements on each side to a MORB-like source than from any actual depletion of Nb. In other words, the trough may be an artifact of the location of Nb (and Ta) on the abscissa rather than due to an abnormally low Nb concentration in the source. McCulloch and Gamble (1991) concluded that the immobile HFS element concentrations are similar to those of MORB and probably reflect overall mantle source characteristics, whereas the LIL element concentrations reflect the more water-soluble components from the slab. I am inclined to agree, but there could be considerable variation in the mantle character in a complex subduction zone. Jahn (1994) calculated that only 2% contamination of primitive mantle by an upper crustal component would produce a pronounced negative Nb-Ta anomaly (without significantly changing the major element composition).

In summary, trace element data suggest that the principal source of island-arc magmas is similar to MORB source, though perhaps even more depleted and is therefore dominated by mantle rather than subducted crust. Basalts are more common than was initially thought in these andesite-characterized arc provinces, which further indicates a mantle source. Thus the model presented at the beginning of this chapter, in which andesites are created as primary melts of the subducting oceanic crust, is beginning to look less attractive. LIL and HFS elements, which behave coherently during normal mantle melting to produce MORB–OIB, may become strongly decoupled from each other when a fluid phase is involved (Tatsumi et al., 1986). High LIL/HFS ratios indicate that a hydrous fluid and its excess LIL content, both presumably derived from the subducting slab, are introduced into a mantle magma source, where they may play a pivotal role in magma genesis by lowering the melting point. Stolper and Newman (1994), for example, concluded that the compositional variation in primitive Mariana arc basalts can be explained by mixtures of N-MORB melts and an H_2O-rich component in which the concentration of the latter controls the degree of mantle partial melting, presumably in the wedge of mantle above the subducted slab. The canonical status of hydrous fluids dominating the slab contribution to subduction zone magmatism has some recent challengers, and we will look at this controversy in more detail when we consider arc petrogenesis.

16.7 ISLAND-ARC ISOTOPES

The isotope systematics of island-arc volcanics is complex and reflects the heterogeneity of melt source(s) and a multiplicity of processes. I shall try to provide a simplified summary of this exciting and rapidly developing field that gives a suitable assessment of the general trends. Figure 16.12 shows the variation in $^{87}Sr/^{86}Sr$ versus $^{143}Nd/^{144}Nd$ for several island arcs. New Britain, Marianas, Aleutians, and South Sandwich volcanics plot within a surprisingly limited range of depleted values similar to MORB. This suggests that the principal source of island-arc magmas is the same mantle source that produces MORBs, although the trace element data for these four areas still require additional enriched components. How trace elements can be enriched without affecting isotopic ratios is a vexing problem (see Hawkesworth et al., 1991). The data for other arcs extend along the enriched "mantle array" (Figure 10.16) and yet further toward Nd- and Sr- enriched regions following the now familiar trends of the OIB data discussed in Chapter 14. Two enrichment trends, one for the Banda arc and the other for the Lesser Antilles, extend beyond the OIB field (Figure 14.8). These trends are similar to those in the Columbia River Basalts (Figure 15.12) and probably represent high concentrations of two different EM-type crustal components. In the preceding two chapters we concluded that these reservoirs most closely matched the characteristics of continental crust (or sediments derived from continental crust). One of the enriched reservoirs (such as EMII) has a higher radiogenic Sr content than the other. This may represent upper continental crust that was further enriched as the deeper crust (possibly EMI) was depleted during high grade metamorphism. I proposed in Chapter 14 that these reservoirs were introduced into the mantle by subduction of sediments derived from erosion of the continents and deposited in the ocean basins or in the fore-arc sedimentary wedge. Of course this source arrived in oceanic intraplate OIB volcanics by a rather circuitous route. In the present case, directly above the subducting slab, the route is much more direct.

In Figure 16.12, we can see that the Antilles trend, where the Atlantic Ocean crust is subducted, follows a mixing curve between depleted mantle (MORB source) and Atlantic sediments. The same is true for the Pacific data (Banda and

New Zealand), where the detrital sediment has $^{87}Sr/^{86}Sr$ around 0.5123 and $^{143}Nd/^{144}Nd$ around 0.715 (Goldstein and O'Nions, 1981). These arc magmas can therefore be explained by partial melting of a depleted mantle source with the addition of a component derived from the type of sediment that exists on the appropriate subducting plate. The increasing north-to-south chemical enrichment along the Antilles arc described above is also true for the isotopes, and is probably related to the increasing proximity of the southern end of the arc to the South American sediment source of the Amazon. The observations of Kelemen et al. (2003b), finding enriched isotopic signatures toward the west in the Aleutians, may also be correlated with a decreased continent-derived sediment input away from the North American margin.

Figure 16.13 shows $^{207}Pb/^{204}Pb$ versus $^{206}Pb/^{204}Pb$ for a number of island arcs. As argued in Chapter 14, lead in mantle-derived systems is very low, and thus readily shows the effect of contamination by enriched crustal sources. Included in Figure 16.13 are the principal isotopic reservoirs shown in Figure 14.9 as well as the field of MORB Pb ratios. The lead in some arcs overlaps with the MORB data, once again suggesting that a depleted mantle component would serve as a major source reservoir (end member) for subduction zone magmas. The majority of the arc data are enriched in radiogenic lead (^{207}Pb and ^{206}Pb), trending toward the appropriate oceanic marine sedimentary reservoir. Several arcs could represent mixing of DM, PREMA, and sedimentary sources. The Sunda data extend to EMII. The Aleutians data follow a nice mixing line between DM or PREMA and Atlantic sediments, perhaps extending beyond toward HIMU (Figure 14.9). The sources are somewhat varied, but the Pb data clearly indicate a sedimentary component in arc magmas.

The question is whether the enrichment is ancient, reflecting a heterogeneous and isotopically diverse mantle

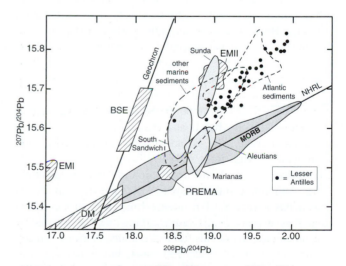

FIGURE 16.13 Variation in $^{207}Pb/^{204}Pb$ versus $^{206}Pb/^{204}Pb$ for oceanic island-arc volcanics. Included are the isotopic reservoirs and the northern hemisphere reference line (NHRL) proposed in Chapter 14 (see Figure 14.9). The geochron represents the mutual evolution of $^{207}Pb/^{204}Pb$ and $^{206}Pb/^{204}Pb$ in a single-stage homogeneous reservoir. Also shown are the Pb isotopic ranges of Atlantic and global marine sediments. Data sources listed in Wilson (1989, p. 185).

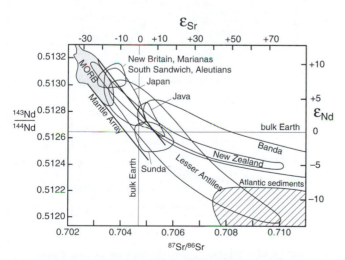

FIGURE 16.12 Nd-Sr isotopic variation in some island-arc volcanics. MORB and mantle array from Figures 13.12 and 10.16. After Wilson (1989), Arculus and Powell (1986), Gill (1981), and McCulloch et al. (1994). Atlantic sediment data from White et al. (1985).

source, or more recent, such as alteration of a homogeneous depleted mantle by the addition of sedimentary components in the subduction zone system. As we concluded in Chapter 15, there is no way to distinguish these two possibilities on the basis of Nd, Sr, and Pb isotopes, as the question addresses the timing of the mixing rather than the characteristics of the mixed components. The Nd, Sr, and Pb isotopic systems, which evolve over very long time periods and do not fractionate during melting, crystallization, or mixing, are indifferent to the issue of when the enriched component was introduced into the source region, as long as the two components were chemically different and remained separate long enough to develop unique isotopic ratios prior to the mixing that accompanied melting. If we now assume that the principal source of arc magmas is in the mantle wedge above the subducting slab, as the trace elements indicate, we can have either a heterogeneous mantle wedge (with depleted and enriched components that remain isotopically distinct over time frames on the order of 1 Ga or more) or a somewhat more homogeneous (though depleted) mantle with LIL and Sr-Nd enriched additions from the subducted slab and sediments. The proximity of a subducted sedimentary source (with the appropriate isotopic character) in the general area of magma generation makes the latter possibility attractive yet still unproved. The two possibilities are not mutually exclusive, by the way, and both may be true.

A technique based on [10]Be has shed some light on the problem of timing of the sedimentary contribution to island-arc magmas. [10]Be is created by the interaction of oxygen and nitrogen with cosmic rays in the upper atmosphere. It is carried to Earth by precipitation, where it is readily incorporated into clay-rich oceanic sediments. The unique contribution of [10]Be is its convenient half life of only 1.5 Ma. This is long enough to be subducted and enter directly into the arc magmatic system, but short enough to excuse itself from long engagements in the mantle systems. After about 10 Ma, [10]Be is no longer detectable. [9]Be is a stable natural isotope and is used as a normalization factor for Be isotopes so that [10]Be enrichment in samples with different beryllium concentrations can be compared directly. [10]Be averages $5 \cdot 10^9$ atoms/g in the uppermost oceanic sediments, and [10]Be/[9]Be averages about $5000 \cdot 10^{-11}$ (Morris et al., 1990). In mantle-derived MORB and OIB magmas, as well as in continental crust, [10]Be is below detection limits ($<1 \cdot 10^6$ atom/g) and [10]Be/[9]Be is $<5 \cdot 10^{-14}$.

Morris et al. (1990) used boron in a similar fashion to [10]Be. Although a stable element, B is analogous to [10]Be in that it has a very brief residence time deep in subduction zone source areas, in this case escaping quickly into shallow crustal and hydrospheric systems. B concentrations in recent sediments are high (50 to 150 ppm), but it has a greater affinity for altered oceanic crust (10 to 300 ppm). In MORB and OIB, it rarely exceeds 2 to 3 ppm.

Morris (1989) and Morris et al. (1990) found that lavas from many recent arc volcanoes have elevated [10]Be and B concentrations that could only result from a contribution from *recent* sediments becoming involved in the source. Here, finally, was an answer to the riddle regarding the timing of the

enriched component. At least some of the enriched component in arc magmas is derived from sediments about 2.5 Ma old in the subduction zone itself. Figure 16.14 shows the data on a [10]Be/Be$_{total}$ versus B/Be$_{total}$ diagram (Be$_{total} \approx$ [9]Be because [10]Be is so rare). The Be-B data for each arc studied formed linear arrays, each arc having a unique slope. Also shown are the Be-B characteristics of the other known reservoirs, including typical mantle (virtually no [10]Be or B), hydrated and altered oceanic crust (high B, low [10]Be), and young pelagic oceanic sediments (low B, but [10]Be/Be$_{total}$ extends well off the diagram up to 2000). The simplest explanation for the linear arrays is that each arc represents a mixing line between a mantle reservoir (near the origin) and a fluid (or melt) reservoir that is specific for each arc, and is itself a mixture of slab crust and sediment. For example, the Kurile data in Figure 16.14 follow a line between the mantle and a fluid (labeled "Ku") located somewhere in the broad field between the sediment and crust reservoirs. Hypothetical fields for each arc are illustrated, but the exact location along the extrapolated line of data cannot be determined. The ratio of sediment to slab crust is different for each arc.

For the data to be linear, the fluid in each arc must be of a restricted Be-B range and hence it must be homogenized throughout the arc! This is a surprising result of the study and awaits substantiation with additional data. For now, we shall accept that the participation of young sediment and altered oceanic crust in the subduction zone magmatic system is confirmed. Hawkesworth et al. (1991), on the basis of trace element and Pb-Nd-Sr isotope data, concluded that the slab-derived component that reaches arc magmas is but a small proportion of the overall slab, and there is a substantial quantity of enriched material that survives the subduction zone system and becomes recycled back into the mantle, where it reaches the oceanic magma systems discussed in Chapters 13 and 14. Hawkesworth

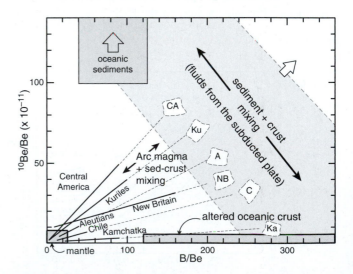

FIGURE 16.14 [10]Be/Be$_{(total)}$ versus B/Be for six arcs. Data for each arc follow linear mixing lines between a mantle-like reservoir with virtually no [10]Be or B, and various points representing slab-derived fluids. The fluids correspond to another mixing line between young oceanic sediments and young altered hydrous oceanic crust. After Morris (1989).

et al. (1991, 1993) also pointed out that the trace element enrichments of island-arc magmas suggest a much larger subducted slab/sediment component than do the isotope data and suggested that the hydrous fluids may selectively incorporate (or **scavenge**) water-soluble components from the mantle wedge, thus requiring less total material from the slab itself.

Several short-lived intermediate isotopes in the complex $U \rightarrow Pb$ chain have also recently been used to constrain the timing of addition of slab components to the mantle wedge. Each of these "chronometers" is sensitive to a different process and time scale (see Turner et al., 2000; Turner, 2002; Elliott, 2003; and Zellmer et al., 2005, for reviews). Because the initial U breakdown is long-lived, the concentrations of short-lived intermediate nuclides, if undisturbed, gradually reach a steady state, termed *secular equilibrium*. Scavanging by fluids or partial melts may selectively remove some of the intermediate elements in the chain, resulting in nuclide ratios that differ from the constant (equilibrium) secular ones. Disequilibrium, either of the depleted source or an enriched destination, remains discernible for about five times the half-life of the disturbed nuclide before secular equilibrium is again reestablished by further decay. Some nuclides have half-lives that make them very useful as subduction-related chronometers, notably ^{230}Th (half-life = 75,280 a), ^{226}Ra ($T_{1/2}$ = 1,600 a), and ^{231}Pa ($T_{1/2}$ = 32,760 a). Elevated values of ($^{230}Th/^{232}Th$) in many arc lavas have been attributed to ^{238}U addition to the wedge via fluids from the subducting crust, the occurrence of which ranges from 10,000 to 200,000 years before eruption (Turner et al., 2000). Radium is an alkaline earth element, behaving similarly to Ba, and thus mobilized in aqueous fluids. Very high ($^{226}Ra/^{230}Th$) enrichments in some arc lavas typically correlate with high Ba/Th. As discussed above, this decoupling of LIL/HFS elements implies aqueous fluid transfer of LIL elements from subducted sediments to the mantle wedge. The high ($^{226}Ra/^{230}Th$) and the very short half-life of ^{226}Ra indicate that transfer, melting, and magma rise to the surface all occurred in as little as 1000 years! (I think I've traveled on elevators slower than this.)

16.8 PETROGENESIS OF ISLAND-ARC MAGMAS

We began this chapter by outlining the initial model in which arc magma was generated by partial melting of the oceanic crust in the subducting slab, which produced the characteristic andesites directly. Since then, we have found that basalts are very common in arcs and that the trace element and isotopic evidence indicates that the dominant source component resembles MORB source (depleted mantle), along with some other constituents, such as oceanic sediment and altered ocean crust. The manner in which these components are added to the mix is complex and controversial, and it probably varies from one arc to another. We now return to the topic of magma generation in subduction zones and attempt to develop a reasonably comprehensive model.

16.8.1 Thermal Constraints

Subduction zone magmatism is a paradox in the sense that great quantities of magma are generated in regions where cool lithosphere is being subducted into the mantle and isotherms are depressed, not elevated. No adequate petrogenetic model can be derived without considering the thermal regime in subduction zones. Numerous thermal models have been published over the past two decades (see Davies and Stevenson, 1992; van Keken et al., 2002a; Peacock, 2003; van Keken, 2003; and van Keken and King, 2004, for reviews). Most models are in substantial agreement on the variables involved and differ principally in their estimates of the exact thermal regime and the extent to which the various mechanisms affect it. Of the many variables capable of affecting the distribution of isotherms in subduction zone systems, the five main ones are:

1. The rate of subduction
2. The age of the subduction zone
3. The age of the subducting slab

In addition to these "first-order" plate tectonic variables, but also believed to be important (although less easy to assess), are:

4. The extent to which the subducting slab induces flow in the mantle wedge and the vigor and geometry of that flow
5. The effects of frictional or shear heating along the Wadati-Benioff zone

Other factors, such as the dip of the slab, endothermic metamorphic reactions, and metamorphic fluid flow, are now thought to play only a minor role (Peacock, 1991; Furukawa, 1993).

Figure 16.15 illustrates a typical thermal model for a subduction zone, in this case by Furukawa (1993). Isotherms will be higher (i.e., the system will be hotter) if:

1. The convergence rate is slower
2. The subducted slab is young and near the ridge (hence warmer)
3. The arc is young (<50 to 100 Ma, according to Peacock, 1991)

In spite of these variables, we can consider Figure 16.15 to be sufficiently representative to serve as a basis for the overall thermal regime, and we can consider variations as we proceed.

All thermal models agree that the isotherms are greatly depressed in the bulk of the slab itself, as compared to the "normal" intraplate geothermal gradient away from any subduction zone. This simply means that the rocks of the slab, being such poor thermal conductors, heat up much more slowly than they sink. This certainly affects the melting behavior of the subducted material. Basaltic ocean crust, if sufficient H_2O is available, should begin to melt at about 700°C (Figure 7.20) or about 40-km depth with a normal ocean geotherm. In the subducted slab, however, it would not melt until it reaches nearly 200 km. It would not reach its *dry* solidus (>1100°C) until depths of several hundred kilometers. Remember, h, the depth beneath the volcanic

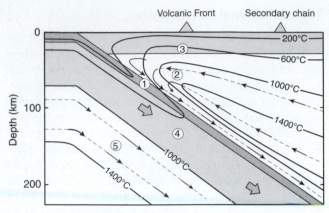

FIGURE 16.15 Cross section of a subduction zone showing isotherms (after Furukawa, 1993) and mantle flow lines (dashed and arrows, after Tatsumi and Eggins, 1995). Potential magma source regions are numbered and discussed in the text.

front to the Wadati-Benioff zone, which should approximate the area of magma genesis, is about 110 km.

Several potential source components for island-arc magmas are labeled in Figure 16.15. The principal ones, with numbers corresponding to circled numbers in the figure, are:

1. The crustal portion of the subducted slab, which includes three components:
   a. The altered oceanic crust itself, which is hydrated by circulating seawater, and partly metamorphosed to greenschist facies (including chlorite, actinolite, and albite)
   b. Subducted oceanic and fore-arc sediments
   c. Seawater trapped in pore spaces
2. The mantle wedge between the subducting slab and the arc crust
3. The arc crust
4. The lithospheric mantle of the subducting plate
5. The asthenosphere beneath the slab

The last three sources on the list are unlikely to play much of a role, at least in early arc development. The lithospheric mantle of the subducting plate (4) is already refractory, due to the extraction of MORB at the ridge, and it heats very little in the upper 200 km of the subduction zone. The asthenosphere beneath the slab (5) flows with it, but does not heat much (the isotherms are essentially parallel to the flow lines at these depths). Because it didn't melt in the ocean basins, it is unlikely to melt upon subduction. As for the arc crust, this crust is the product of subduction zone magmatism, so it cannot be a fundamental and necessary component for the generation of arc magmas or the process could never begin. It may be incorporated into rising melts by crustal contamination, but the composition is not much different than the melts themselves, and so the extent is hard to evaluate. The overriding crustal component can only be a minor factor in early arc development, but may be much more important in mature island arcs or continental arcs with a thicker crust. The isotherms in Figure 16.15 at the base of the (predominantly andesitic) arc crust indicate that the temperature is too low

for crustal melting, even under hydrous conditions. These isotherms are very broad averages, however, and rising melts, once created, can add considerable heat locally, perhaps sufficient to melt the base of the crust. We will consider this effect later, especially in Chapter 17.

We are left with the subducted crust and mantle wedge as the two principal sources of arc magmas. The trace element and isotopic data reveal a combination of depleted and enriched signatures, suggesting that both contribute to arc magmatism, but the question is how and to what extent. We know from Figure 10.5 that the dry peridotite solidus is too high for melting of *anhydrous* mantle to occur anywhere in the thermal regime in Figure 16.15. The high LIL/HFS ratios of arc magmas, however, suggest that H_2O plays a significant role in arc magmatism. Because we know the general composition of the constituents in Figure 16.15, we can model magma generation by combining this information with the pressure-temperature conditions to which the constituents are subjected as they move through the subduction zone (also shown in the figure) and considering the consequences. The sequence of pressures and temperatures to which a rock is subjected during a sequence of burial, subduction, metamorphism, uplift, etc. is called a **pressure–temperature–time path** (**P-T-t path**). For example, from Figure 16.15 we can see that the oceanic crust, as it subducts, begins to heat at about 50 to 70 km depth and will continue to heat with rising pressure, although slowly due to the depressed isotherms. The mantle wedge material will follow the path of drag-induced flow, also illustrated in Figure 16.15. This is less well constrained but should follow a path (like the arrows beginning near point 2 in the figure) of initial cooling from about 1100°C to about 800°C at nearly constant pressure and then heating toward 1000°C as pressure increases.

16.8.2 Dehydration and Melting in Subducted Slabs

P-T-t paths for materials moving through subduction zones depend on many variables. Direct analytical solutions are therefore difficult, and most paths have been produced by numeric modeling. Figure 16.16 shows some *P-T-t* paths for the *subducted crust* in a variety of arc scenarios modeled numerically by Peacock (1990, 1991). All curves are based on a subduction rate of 3 cm/yr, so the length of each curve represents about 15 Ma. The dotted *P-T-t* paths represent various arc ages. A newly formed arc will not have subducted sufficient cool slab material to fully depress the ocean geotherm that existed prior to the initiation of subduction. It will thus follow a path toward higher temperatures at low pressure than in a mature arc. After about 30 to 70 Ma a steady-state equilibrium is attained between subduction and heating of the slab, and *P-T-t* paths remain essentially constant. The age of the subducted slab also affects the path (dashed curves). Young oceanic lithosphere will be near the ridge and still warm. The lithosphere cools fairly quickly away from the ridge, so *P-T-t* paths first recede rapidly to lower *T/P* as the age of the slab increases, and then more slowly toward a near steady state at around 200 Ma.

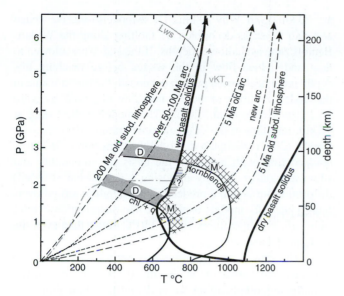

FIGURE 16.16 Subducted crust pressure–temperature–time (*P-T-t*) paths for various situations of arc age (dotted curves) and age of subducted lithosphere (dashed curves, for a mature approximately 50-Ma-old arc), assuming a subduction rate of 3 cm/yr (Peacock, 1991). Included are some pertinent reaction curves, including the wet and dry basalt solidi (Figure 7.20), the dehydration of hornblende (Lambert and Wyllie, 1968, 1970, 1972), chlorite + quartz (Delaney and Helgeson, 1978). Areas in which the dehydration curves are crossed by the *P-T-t* paths below the wet solidus for basalt are stippled and labeled D for dehydration. Areas in which the dehydration curves are crossed above the wet solidus are hatched and labeled M for melting (see text). See Figure 16.17 for curve vKT₀.

The temperature at the top surface of the slab at a given pressure, say 2 GPa, can thus vary (along a horizontal traverse in Figure 16.16) from 300°C (old subducted crust, mature arc) to nearly 1100°C (< 5 Ma old subducted crust or subduction system). The high-temperature paths are unusual situations, but they serve to indicate some of the variability in subduction zones, which may in turn be reflected in variations in magma generation and chemical composition. More on this in a moment. So we might now ask whether any of these *P-T-t* paths result in melting of the slab.

Included in Figure 16.16 are the solidus curves for dry and H_2O-saturated melting of basalt (from Figure 7.20). Note first that *none* of the *P-T-t* paths intersect the dry solidus, precluding substantial partial melting of dry basaltic crust (or peridotite). Melting of *hydrated* basalt depends upon the availability of "free" (non-bonded) H_2O released by the dehydration of minerals, as discussed in conjunction with Figure 10.6. Devolatilization of oceanic crust and the overlying sediments involves an assortment of hydrous and carbonate phases and a variety of potential prograde reactions, both continuous and discontinuous. (I introduced these reaction types in Chapters 6 and 7 and will develop the concepts further in Chapters 25, 26, and 28.) Because chlorite and hornblende are typically the predominant hydrous minerals in metamorphosed hydrated basalt (at least at lower pressures), the breakdown curves for chlorite + quartz and hornblende are included in Figure 16.16. Biotite, actinolite,

and other hydrous minerals dehydrate mostly between these curves, so the curves represent reasonable boundary conditions and serve to illustrate the process, although in a simplified fashion. Where a *P-T-t* path intersects a dehydration curve *below* the wet solidus, dehydration and liberation of H_2O takes place. This is the situation for mature arcs in Figure 16.16, in which the lithosphere is over about 25 Ma old. The areas above the dehydration curves in such cases are stippled in Figure 16.16, and labeled with a *D* (for dehydration). Only in young arcs, or arcs subducting young lithosphere, does dehydration of chlorite or amphibole (or some intermediate hydrous phase) release H_2O at temperatures *above* the wet solidus, directly initiating melting of the crustal portion of the slab to form Mg-rich andesites. Where *P-T-t* paths cross dehydration curves above the wet solidus, the area is cross-hatched and labeled with an *M* (for melting).

What happens to intermediate cases, such as the curve representing the 5-Ma-old arc? If dehydration occurs before the subducted material reaches the wet melting point, will it eventually melt when the *P-T-t* path crosses the wet solidus? This depends upon the fate of the H_2O released by dehydration. If it remains in place, melting will occur when the wet solidus is reached (labeled in Figure 16.16 with a question mark). Many investigators believe that H_2O rises and leaves the system as dehydration occurs so that the remaining crust no longer contains free water, and melting is inhibited at the wet solidus. If fluids readily escape, only when conditions reach that of the first dehydration reaction above the wet solidus will more fluid be generated and melting initiated. In the simplified system illustrated in Figure 16.16, this occurs when hornblende breaks down. In more realistic situations, other hydrous phases, such as actinolite, chloritoid, or an epidote mineral, may be present and break down before hornblende (but these reactions typically release less H_2O). Continuous reactions release diminished amounts of H_2O over a range of *P-T* conditions, and smaller quantities of melt may then form as soon as the wet solidus is crossed in such situations.

For mature arcs, however, hornblende breaks down before the wet solidus and, although other hydrous phases such as lawsonite and/or glaucophane may also be present (Chapter 25), an anhydrous eclogite eventually forms. Will the *P-T-t* path cross the wet solidus, and if so, will crust then melt, or will H_2O be released and escape first? Notice in Figure 16.16 that the *P-T-t* paths for mature arcs approach the solidus, running nearly parallel to it. The models suggest that they don't cross it, however, and melting is precluded. But real arcs need be only slightly hotter than these models indicate to cross the solidus (we will return to this point shortly). Slab melting is the subject of great controversy, and opinions diverge at this point into those that perceive the melting of the oceanic crust (eclogite plus H_2O) as the dominant source of arc magma and those that favor the overlying mantle wedge. Both types of models have strong arguments and adherents (Green, 1982; Wyllie, 1982; Sekine and Wyllie, 1982a, 1982b; Johnston and Wyllie, 1989; Wyllie and Wolf, 1993; Tatsumi and Eggins, 1995, Kelemen et al., 2003a, 2003c; Peacock, 2003; Tatsumi, 2003, 2005).

The high LIL/HFS trace element data underscore the importance of slab-derived H_2O and a MORB-like mantle wedge source, and the typically flat HREE pattern argues against a garnet-bearing (eclogite) source. This and the abundance of basalts have caused the majority opinion to swing toward the non-melted slab for most cases. In light of the trace element data, wholesale melting of the slab crust to produce primary andesites is generally considered to occur only in rare instances of high-Mg andesites. McCulloch (1993) suggested that young, warm subducted crust is more likely to melt before it dehydrates. Thus slab melting may be more common where a subduction zone is close to a ridge, or during the Archean, when heat production was greater.

Eruption temperatures for arc volcanics, including andesites, are typically in excess of 1100°C, which places further constraints on the melting process. According to the thermal models referred to above, nowhere *in the slab* beneath the volcanic front is the temperature high enough to produce such an eruption temperature (Figure 16.15). Only in the mantle wedge above the slab is the temperature high enough. In view of these arguments, most petrologists favored a model in which melting of the subducted crust is only minor, and that the mantle wedge is the principal magma source. Like most arc-related generalizations, this is neither conclusive nor all-encompassing, and the nature of the slab contribution to the wedge and to arc magmas is probably complex and variable. We continuously reevaluate our models as new information arises.

For example, numeric models have recently attempted to address the effects of variations in the viscosity of the mantle on slab-induced flow in the mantle wedge (variable 4 in the list at the beginning of Section 16.8.1). Most early models, including those determining the *P-T-t* paths in Figure 16.16, assumed an isoviscous (constant viscosity) mantle. Viscosity within the mantle wedge may be lowered by increases in temperature, stress, and fluid or melt content (van Keken et al., 2002a; van Keken, 2003; Peacock, 2003). Models with stress- and/or temperature-dependent viscosity exhibit enhanced mantle flow. The flow primarily affects the temperature of the wedge tip because wedge material is more readily dragged down by friction with the slab and replaced by hotter fluid mantle toward the wedge tip (where it is again dragged down). This raises the temperature of the wedge and of the upper slab surface by several hundred degrees in comparison with isoviscous models (Kincaid and Sacks, 1997; van Keken et al., 2002; Peacock, 2003; Kelemen et al., 2003b). Heat can descend further into the slab only by conduction, so the thermal effect is concentrated toward the top of the slab, with steep thermal gradients beneath. As an illustration, Figure 16.17 compares slab temperatures at the slab surface and at 7 km stratigraphic depth, based on several models using different assumptions of mantle rheology. Limited conduction moderates the temperature difference between the isoviscous and reduced-viscosity models at 7 km beneath the slab surface. The effect of varied rheology is much greater near the slab surface, resulting in a dramatic rise in temperature for the vKT_0 model with lowered wedge viscosity as compared to the isoviscous PW_0 model, beginning

at ~70 km depth. Why 70 km? These models force the wedge and slab to act as decoupled by faulting along the Wadati-Benioff zone at shallower depths. If the slab were allowed to drag and induce flow in the wedge before reaching this depth, more heat would penetrate beneath the fore-arc area in the models, contrary to the low heat flow observed in Figure 16.2 (van Keken et al., 2002). The models permit drag at 70 km, resulting in the dramatic temperature rise indicated. Conder (2005) attempted to model the transition from slip to coupled flow with depth on the basis of temperature and strain rate rather than by imposing a (somewhat artificial) dislocation surface to a predetermined depth or a rigid overlying plate. His models using that parameterization exhibit slab surface temperatures 100 to 150°C hotter than even the vKT_0 model above.

The effects of hydration and partial melting on the rheology and buoyancy of the mantle wedge add further challenging complications to modeling of subduction zone dynamics and temperatures. Not every model with enhanced ductility predicts higher slab surface temperatures, however. The models of Arcay et al. (2006), for example, suggest that enhanced flow in the softened hydrous wedge erodes the overlying fore-arc lithosphere. Cool removed material is then dragged to create a *cool* blanket at the slab surface. If the wedge is weakened further by H_2O in the models, it convects, and cool blobs then detach from the lithosphere and

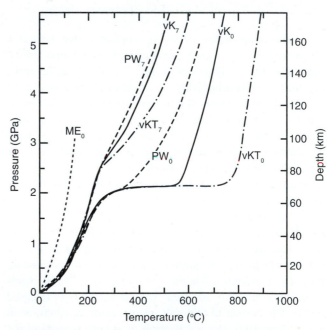

FIGURE 16.17 *P-T-t* paths at a depth of 7 km into the slab (subscript = 7) and at the slab/mantle-wedge interface (subscript = 0), predicted by several published dynamic models of fairly rapid subduction (9 to 10 cm/yr). ME = Molnar and England's (1992) analytical solution with no wedge convection. PW = Peacock and Wang (1999) isoviscous numeric model. vK = van Keken et al. (2002a) isoviscous remodel of PW with improved resolution. vKT = van Keken et al. (2002a) model with non-Newtonian temperature- and stress-dependent wedge viscosity. After van Keken et al. (2002a) © American Geophysical Union with permission.

sink to form a much thicker blanket above the slab, resulting in significantly cooler slab surface temperatures. All we can say at this point is that numeric modeling, once used to reject the notion of slab melting, can no longer do so with the same authority. It will be interesting to see how future models develop with more computing power, sophisticated algorithms, and more precise geophysical constraints.

The LIL/HFS decoupling in arc volcanics, which indicates the slab component is transported to the melting source via hydrous fluids, may not be as universal as once thought either. Johnson and Plank (1999) noticed that Th/Rb (HFS/LIL) for basalts in several arcs correlated with that of the sediments sampled on the yet-to-be-subducted seafloor bordering each arc. Their experiments on melting of pelagic red clay indicated that this coupling and Be concentrations in many arc lavas is best maintained if the sediment is melted rather than dehydrated. The geochemical modeling of George et al. (2003, 2005) also indicated element transfer to the wedge source by sediment partial melts in the Aleutians and Tonga–Kermadec. Elliott et al. (1997) proposed that the high Th/Nd at Arigan Island required sediment melting, but element ratios elsewhere in the Mariana arc indicated fluid transport. Kelemen et al. (2003b) also argued that the geochemical trends from several arcs required melting of sediment. In his global compilation of trace element and isotopic analyses from mafic island-arc lavas, Elliott (2003) found trace element and isotopic evidence for contributions from both sediment melts and fluids released from the mafic crust of subducted slabs.

The indicators of sediment melting agree with the suggestion of some of the newer dynamic models that the top of the slab may be considerably hotter than previously considered. I have included the vKT_o curve from Figure 16.17 in Figure 16.16 to show that these revised models no longer preclude partial melting of the uppermost slab surface in mature arcs. Recall that temperatures decrease rapidly below the slab surface in these models, leading to the recently popular aphorism that *sediments melt and the basaltic crust dehydrates*. Indeed, George et al. (2005), despite noting indicators for sediment melts, could find "no geochemical evidence for a contribution from melts of the oceanic crust" (p. 209). But others disagree. Kelemen et al. (2003b, 2003c) pointed out that the newer thermal models with a hotter upper slab allow mafic crustal melting in mature arcs. Kelemen et al. (2003b) noted that nonradiogenic MORB-like Pb isotope ratios (indicating a mafic crustal source rather than sediments in the slab) occur in Aleutian volcanics with low Pb/Ce (LIL/HFS) and attributed this to partial melting of the subducted crust. Kelemen et al. (2003a) analyzed geochemical data from a number of arcs and argued that LIL/HFS ratios in the volcanics are variable, and the relative enrichments in Ba, Th, Pb, Sr and LREE can be explained better if mafic slab melts are considered as a component.

Part of the LIL/HFS controversy may be our imperfect knowledge of the distribution coefficients (Section 9.1) of these elements between minerals, fluids, and melts, particularly at high pressures. Experimentally determined D values for a number of elements depend upon temperature,

pressure, the total solute content of the fluid, and the composition and H_2O-content of the melt (see, e.g., Brenan et al., 1995; Keppler, 1996; Stalder et al., 1998; Rapp et al., 1999, and Kessel et al., 2005a). Ryan et al. (1996) noted decreases in concentrations of typically H_2O-soluble elements with increasing slab depth in a number of arc transects. As pressure (and, to a lesser extent, temperature) increases, so does the solubility of H_2O in siliceous melts and the solubility of silicates in hydrous fluids. Aqueous fluids and hydrous melts get progressively more similar until the solvus disappears entirely and only a single aqueous–silicate phase exists beyond some critical point. This behavior mimics the liquid–vapor phase relations of H_2O at the critical point (Figure 6.7). Because this fluid–melt convergence occurs at a temperature and pressure greater than the liquid–vapor convergence of H_2O, it has been called the **second critical endpoint** by several investigators (for a detailed explanation, see Manning, 2004). Bureau and Keppler (1999) noted complete supercritical H_2O-melt miscibility in a wide range of experimental melt compositions, including nepheline, jadeite, and granitic systems. Kessel et al. (2005b) found supercritical behavior in H_2O-saturated basaltic melts above 6 GPa. Mibe et al. (2007) determined the second critical endpoint for peridotite-H_2O at 3.8 GPa and 1000°C. It is difficult at this time to say what effect this transition in behavior has on trace element patterns such as LIL/HFS. Arguments over the potential and pervasiveness of slab melting will continue. I think the evidence for slab-derived hydrous fluids is very strong. If melts are substantiated they should be considered in addition to fluids rather than as a substitute. For evaluations of melts versus fluids as the enriched component transport medium, see Miller et al. (1992), Kay and Kay (1994), Tatsumi and Eggins (1995), Pearce and Peate (1995), Class et al. (2000), Ulmer (2001), Grégorie et al. (2001), Elliott (2003), and Kelemen et al. (2003a).

Island arcs are magmatic, of course, and where slab melts do not occur, the source must be the mantle wedge above the slab. Before we explore the potential for wedge melting, let's consolidate what we have learned so far and consider wedge modification and melts in the context of the entire subduction process.

16.8.3 A Possible Model

Figure 16.18 attempts to distill the complex and varied dynamics of subduction into a possible model for subduction zone magmatism. It includes isotherms from Figure 16.15 (based on the newer temperature-dependent mantle viscosity models) for reference, as well as the mantle flow direction in the wedge. H_2O may be structurally bound in minerals and occupy the pore spaces in the oceanic crust and sediments. Fisher (1998) suggested that the upper kilometer of the oceanic crust is ~10% porous. H_2O stored in hydrous minerals amounts to 8 to 9% in the crust (zeolites, prehnite, pumpellyite, chlorite, etc.) and is also plentiful in the thinner sediments (13 to 14% in the pelagic clays of Johnson and Plank, 1999). Carbonates are also present. As described in Chapter 13, the mantle portion of the subducting plate is a partially hydrated, variably depleted peridotite. The degree of serpentinization is highly variable, depending in part on the intensity of fracturing and the ridge dynamics where the plate was generated. Slow spreading ridges, for example, are more broken up and the mantle may be exposed along detachment surfaces, becoming more hydrated to serpentine.

Fluid expulsion from sediments and oceanic crust begins almost immediately upon subduction, as compaction and mineral growth in cracks and voids reduce porosity. Let's concentrate on the further evolution of subducted lithosphere following the steep *P-T-t* paths for mature arcs and old subducted lithosphere in Figure 16.16 because these should represent the most common situations. For more detailed descriptions of the dehydration reactions for subducted crust, see Section 25.3 or Iwamori (1998), Hacker et al. (2003), Kawamoto (2006), and the reviews by Poli and Schmidt (2002) and Schmidt and Poli (2005). Minerals in the altered oceanic crust following the steep *P/T* paths begin to dehydrate at depths around 20 km as zeolites break down. As the slab goes deeper, chlorite, phengite, and other hydrous phyllosilicates then begin to decompose through various continuous and discontinuous reactions. The effect of continuous reactions, variations in X_{bulk}, the degree of hydration, and the effect of carbonates on the fluid composition (Section 29.1.1.2) will be to smear out dehydration over wider ranges than indicated in Figure 16.16. Initially abundant chlorite (containing up to 12 wt. % H_2O) should be entirely decomposed by 70 km depth. Lawsonite may contain over 11% H_2O and becomes abundant beyond this, as will sodic amphiboles

such as glaucophane. The newer temperature-dependent viscosity models (vKT_o in Figure 16.16) suggest significantly hotter slab surface temperatures, and sediment melting may be expected at depths greater than 50 km. Melting may also extend into the mafic crust, but it is still a matter of controversy as to how far, if at all, in mature arcs. Beneath any thin zones of melting, further dehydration takes place to greater depths as other hydrous phases break down (again via a series of continuous and discontinuous reactions), including amphibole at about 100 km (Figure 16.16). The slab crust is successively metamorphosed to blueschist, amphibolite if warm enough, and finally a nearly anhydrous eclogite as it reaches about 80 to 100 km depth. Most slab H_2O is lost by 100 km, and dehydration reactions (as well as the wet solidus) are nearly parallel to *P-T-t* paths in Figure 16.16, inhibiting the rate of further dehydration or of melting. Lawsonite may be the last phase to hold H_2O, and it breaks down at nearly 300 km (Schmidt and Poli, 2005).

Most liberated H_2O is believed to rise into the overlying mantle wedge, where it reacts with the lherzolite to form serpentine and chlorite at shallow levels beneath the fore arc (Iwamori, 1998), ± talc, a pargasitic amphibole, and perhaps phlogopite (hatched area in Figure 16.18). Any crustal (or sediment) melts that may have formed will be relatively silicic and probably react with the ultramafic mantle, as described in Section 14.7. Most melts therefore stall, solidify, and, like the fluids released previously, locally enrich the mantle wedge rather than reach the surface. The now hydrous and slightly enriched mantle immediately above the slab, carrying enriched fluid ± slab melt trace element and isotopic signatures in addition to its original depleted mantle one, is carried downward by slab drag, where it heats up. At this point we return to the question of wedge *P-T-t* paths.

P-T-t paths followed by the mantle wedge are quite different from those of the subducting slab. The hydrated peridotite of the wedge follows the flow lines illustrated in Figure 16.15 and discussed above. *P-T-t* paths for the wedge peridotite have also been modeled by Peacock (1990, 1991) and some typical paths are shown in Figure 16.19. Wedge *P-T-t* paths, beginning beneath the back arc, involve first cooling and then heating. Fluids rising from the dehydrating slab (arrows in Figure 16.19) rise into and hydrate the mantle wedge (hatched area in Figure 16.18). Following the *P-T-t* paths in Figure 16.19, the mantle is in the stability range of pargasite. As mantle material is dragged down, the path crosses the pargasite stability curve and the amphibole dehydrates at temperatures above the wet solidus of the peridotite (at *M* near the center of Figure 16.19). The presence of free water causes partial melting (point *A* in Figure 16.18) and initiates rise and further decompression melting of the peridotite. Note that the dehydration curve for amphibole has a very low *P/T* slope in Figures 16.16 and 16.19, which means that dehydration, and in the present situation the initiation of partial melting of the wedge peridotite, is more dependent upon pressure than temperature. The dehydration curve for pargasite in Figure 16.18 suggests that melting should occur at approximately 110-km depth over a range of temperatures, which agrees with the commonly observed height of the

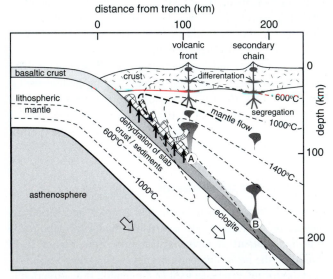

FIGURE 16.18   A proposed model for subduction zone magmatism with particular reference to island arcs. Dehydration of slab crust causes hydration of the mantle (hatched), which undergoes partial melting as amphibole (A) and phlogopite (B) dehydrate. Melt quantity increases as melts rise into hotter portions of the wedge (indicated by thickness of balloon and gradient). Alternatively, dehydration of hydrated peridotite at or before (A) may expel fluids upward into the wedge where they are above wet solidus ("fluid" arrow). After Tatsumi (1989) and Tatsumi and Eggins (1995). Reprinted by permission of Blackwell Science, Inc.

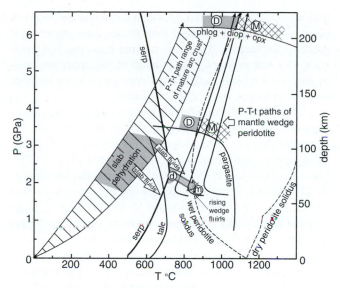

FIGURE 16.19 Calculated *P-T-t* paths for the peridotite in the mantle wedge as it follows paths similar to the flow lines in Figure 16.15. Included are dehydration curves for serpentine, talc, pargasite, and phlogopite + diopside + orthopyroxene. Also shown is the *P-T-t* path range for the subducted crust in a mature arc, and the wet and dry solidi for peridotite from Figures 10.5 and 10.6. The subducted crust dehydrates, and water is transferred to the wedge (labeled arrows). Areas in which the dehydration curves are crossed by the *P-T-t* paths below the wet solidus for peridotite are stippled and labeled D for dehydration. Areas in which the dehydration curves are crossed above the wet solidus are hatched and labeled M for melting. Note that although the slab crust dehydrates, the wedge peridotite melts as pargasite dehydrates (Millhollen et al., 1974) above the wet solidus. An alternative model involves dehydration of serpentine ± chlorite nearer the wedge tip (lowercase d), with H_2O rising into hotter portions of the wedge (gray arrow) until H_2O-exess solidus is crossed (lowercase m). A second melting may also occur as phlogopite dehydrates in the presence of two pyroxenes (Sudo, 1988). After Peacock (1991) and Tatsumi and Eggins (1995).

volcanic front above the Wadati-Benioff zone (*h* in Figure 16.2) with only minor variation due to thermal regime.

An alternative and equally attractive mechanism involves the mantle wedge nearer the wedge tip. The mantle is cooler there, and early fluids from the slab should initially hydrate the mantle to serpentine + chlorite. As this portion of the wedge is dragged downward and heated these minerals dehydrate and the fluids rise into the hotter wedge (gray arrow in Figure 16.19) where the temperature eventually reaches the H_2O-saturated peridotite solidus, resulting in fluid-fluxed partial melting (e.g., Manning, 2004; Iwamori et al., 2007). Pargasite is stable under these conditions, but fully serpentinized peridotite contains about 7 wt. % H_2O, whereas pargasite can hold about 1 wt. % and comprises a smaller proportion of the mode, so there will still be excess H_2O. Melting in the first model occurs at or near the slab–wedge interface and at 60 to 80 km in the second model, but the depth to the slab itself beneath the zone of melting is still ~110 km in both models. Melts in the first model (and perhaps the second) should also rise in temperature, maintain equilibrium with the mantle, and increase in volume, so it

would be quite challenging to distinguish between the models. Neither process excludes the other, so both melting and fluid rise may occur within the mantle wedge in subduction zones.

Melt generation in subduction zones must be a multiple-stage process. Dehydration of the slab (and sediments) provides the LIL, [10]Be, B, [226]Ra, and other highly incompatible element enrichments and often the enriched Nd, Sr, and Pb isotopic signatures. These components, plus other dissolved silicate materials, are transferred to the wedge in a fluid phase (perhaps accompanied by hydrous melt). The mantle wedge provides the HFS and other depleted and compatible element characteristics. Mass-balance calculations suggest that the contribution of the subducted sediments is only a few percent in most arc systems but can produce the high LIL/HFS trace element patterns (Figure 16.11) that characterize subduction zone magmas. Sediment melting and perhaps melting of the uppermost basaltic crust may also yield H_2O-saturated, or "second critical," melts that add some of the more moderate LIL/HFS signature found in some arc lavas. The nearly closed-cell induced flow in the wedge proposed by McCulloch and Gamble (1991; see Figure 16.2 and discussion) permits progressive depletion of the wedge as arc magmas are extracted. This provides an explanation for the HREE and compatible trace element data, which can be even more depleted than MORB (Figures 16.10 and 16.11). Melting of the hydrated wedge peridotite occurs when a hydrous phase (serpentine, chlorite, talc, and/or pargasite) releases H_2O, either at supra-wet-solidus conditions or that rises into portions of the wedge where such conditions exist.

Phlogopite is stable in ultramafic rocks beyond the conditions at which amphibole breaks down. The *P-T-t* paths for the wedge reach the phlogopite-2-pyroxene dehydration reaction at about 200 km depth (Figure 16.19). If this occurs above the wet peridotite solidus, a second phase of melting will occur at a position appropriate for the secondary volcanic chain that exists behind the primary chain in several island arcs. The *P-T-t* paths are nearly parallel to the solidus, and may be above it in some arcs and below it in others. Thus dehydration may or may not be accompanied by melting, so the development of a second arc will depend critically upon the thermal and flow regime of a particular arc. Melting initiated by the breakdown of potassium-rich mica will probably be more potassic, as is true in most secondary arc occurrences. The K-*h* relationship apparent in some arcs is probably more complex than this, reflecting the decreasing quantity of H_2O with depth and thus the degree of partial melting, as well as the depth of melting (which becomes more alkaline with depth, Chapter 10).

As noted above, young arcs, slow convergence, or arcs subducting young crust are hotter, so hydrous melting of subducted crust may be more important and many melts so produced may reach the surface as Mg-rich andesites, adakites, sanukites, etc. (see below). The extreme case of young subducted crust accompanies **ridge subduction**, in which a spreading system segment is subducted. Because plate separation is apparently passive at ridges (Section 1.7.3 and Chapter 13), subducted ridge-juxtaposed plates still separate and hot mantle still rises at the gap, but there is no surface cooling to form new plate. As a result, a **slab window** forms in the subducting plate and gradually expands (Thorkelson, 1996). The hot thinned edges of these windows can be expected to be important sites of slab melting, adakite generation, and mantle wedge enrichment (Thorkelson and Breitsprecher, 2005).

The product of partial melting of the mantle wedge is the subject of considerable experimentation, analysis, and speculation (Kelemen, 1995; Gaetani and Grove, 1998, 2003; Ulmer, 2001; Grove et al., 2002, 2003; Kelemen et al., 2003a; Tatsumi and Kogiso, 2003; Myers and Johnston, 2003; Davidson, 2003). We have learned in several previous chapters that the primary melt of depleted mantle, at least at shallow levels, is typically an olivine tholeiitic basalt. We have also learned the important role of H_2O in subduction zone melting, which explains the high LIL/HFS characteristic of many arc lavas, and because it dramatically reduces the melting temperature in silicate systems, solves the apparent paradox of melt generation at plate boundaries characterized by lower than average geotherms.

16.8.4 A Panoply of Arc Parental Magmas

Arc magmatism is complex, and arc lavas highly diversified. No single model of magma generation is capable of explaining all of this diversity. Although andesites are a hallmark of arc magmatism, basalts still predominate (Table 16.1). Several parental arc magmas have been recognized or proposed by various investigators. Due to the many opportunities for exchange between the subducted lithosphere and the overlying wedge, the subduction zone environment is chemically complex and mixing is probably common among melt sources and products. The brief survey of parental magmas that follows is necessarily incomplete and the distinctions somewhat artificial due largely to difficulties involved in attributing chemical characteristics of evolved liquids to different parental magmas versus different evolutionary paths from similar parents.

Among the earliest erupted lavas in many arcs are tholeiitic basalts (also called **island-arc tholeiites**). They tend to be primitive and are more common in island arcs than continental arcs (Chapter 17). Island-arc tholeiites have depleted REE, similarly to MORBs, but they are usually richer in FeO and K_2O and lower in Cr and Ni. They correspond to the low-K tholeiitic series in Section 16.3 and probably correlate to more typical depleted mantle partial melts formed by decompression partial melting of upwelling mantle, but with a somewhat more fertile and hydrated source. Arc tholeiites are typically more dry than other arc magmas, and they are more common in younger and otherwise hotter subduction zones. Mantle upwelling toward the wedge tip (illustrated by the flow lines in Figures 16.15 and 16.16) may enhance production of tholeiitic partial melts. This may also explain their tendency to concentrate closer to the trench than mainstream calc-alkaline rocks as well as their general correlation with arcs experiencing back-arc extension. In hotter subduction zones and in continental arcs the upwelling portion of the wedge rising from the back-arc area may be a favorable locus for arc tholeiites. Primary

tholeiitic melt is dense, and, although it can rise buoyantly in the mantle wedge, it is denser than the arc crust impeding it reaching the surface. Tholeiitic magma may lose H_2O as it rises. Dry low pressure fractionation in the upper arc crust produces the typical tholeiitic differentiation trend to more evolved andesites, dacites, and rhyolites.

The **high-alumina basalt** type is largely restricted to the subduction zone environment, and its origin is still controversial (Wyllie, 1982; Mysen, 1982; Grove and Baker, 1984; Grove and Kinzler, 1986; Crawford et al., 1987; Sisson and Grove, 1993b). High-alumina basalts have higher Al_2O_3 (>17 wt. %) than tholeiites and are also enriched in K, Rb, Sr, and Ba. Some high-Mg (>8 wt. % MgO), high-alumina basalts may be primary melts of peridotite. Myers et al. (1986) and Brophy and Marsh (1986), however, argued that their composition was most readily explained by partial fusion of quartz eclogite of the subducted oceanic crust and sediment. Others argue that most rocks representing high-alumina surface lavas have compositions too evolved to be primary. One idea is that the more common low-Mg (<6 wt. % MgO), high-Al (>17 wt. % Al_2O_3) rocks are the result of somewhat deeper fractionation of a primary tholeiitic magma that ponds at a density equilibrium position at the base of the arc crust in more mature arcs (Sisson and Grove, 1993b; Kay and Kay, 1994). Here fractional crystallization of olivine and augite (without plagioclase) in the presence of H_2O can produce the low-Mg, high-alumina basalts and basaltic andesites observed. According to this model, most high-alumina basalts would thus be derivative and not primary. Further rise of these hydrous basalts results in volatile loss and partial crystallization (see the discussion associated with Figure 7.22). Magmas that reach the surface are therefore highly phyric andesites and basaltic andesites.

Calc-alkaline rocks (particularly basaltic andesites and andesites) are a hallmark of subduction zones. Many investigators believe that there are parental **calc-alkaline basalts**, but this term has problems in my opinion. First, the calc-alkaline differentiation trend (Figure 16.6) is most easily recognized in evolved rocks, and converges with the tholeiitic trend toward the parental basalts, making it difficult to define any unique calc-alkaline character of primitive liquids. Second, there are probably several ways to generate calc-alkaline liquids from a variety of parental basaltic magma types, including tholeiitic and high-alumina ones (but H_2O seems to be important in virtually all of the ways). For example, the common calk-alkaline basaltic andesites and andesites can probably be derived by high-H_2O crystal fractionation, as described in Section 16.5. Crustal assimilation and magma mixing of crustal and deeper subduction zone melts may also be a pathway. The most commonly erupted (recognizable) calc-alkaline parent is a **basaltic andesite**, higher in Al_2O_3 than tholeiites, and enriched in Ba, Sr, REE, and LIL elements but low in Nb and Ta. They show the clear signature of fluid-mobilized element enrichment (derived from the slab via hydrous fluids and/or melts) superimposed on a depleted mantle wedge. Because H_2O can lower the melting point, the extent of partial melting of the mantle wedge is dependent in part upon the amount of H_2O

in the source (e.g., Stolper and Newman, 1994), a process commonly referred to as *(fluid-) fluxed melting*. Ulmer (2001) noted that adding 0.1 to 0.5 wt. % H_2O to a mantle source led to a 100 to 150°C lowering of the melting point and 5 to 20% partial melting below the dry solidus. Grove et al. (2002) and Gaetani and Grove (1998) added that increasing H_2O content led to increased $SiO_2/(FeO + MgO)$ in mantle wedge melts. They presented an interesting model in which the mantle wedge experiences adiabatic decompression partial melting to produce high-alumina tholeiites, leaving a depleted mantle residue. Slab-derived H_2O later invades this depleted wedge, and the fluxed melting produces a very hydrous initial melt. Addition of H_2O makes this melt very buoyant and of low viscosity. It therefore rises, they proposed, by porous flow as an interconnected intergranular network, initially into progressively hotter mantle, where it equilibrates with the peridotite and the melt fraction increases. Under these conditions orthopyroxene melts incongruently to olivine + a silica-rich liquid and the melt composition migrates from the tholeiitic to the calk-alkaline field in Figure 16.6c (following the broad gray arrow at the low-differentiation end). The result is a primitive basaltic andesite to Mg-rich andesite that Grove et al. (2002) consider to be parental to the calk-alkaline suite. As the source gets more hydrous, the initial melt gets less tholeiitic.

High-Mg# andesite is another magma type that has been argued as primary. Although high-Mg# andesites overlap with many of the calc-alkaline basaltic andesites described above, high-Mg# andesites cannot be formed by low-pressure fractional crystallization or partial melting of basaltic rocks because these processes lower the Mg# (Kelemen, 1995). Adakites, sanukites, and boninites are variations on the high-Mg# andesite theme (although the terms, adopted from the type-locality islands, are poorly constrained and commonly misused). **Adakites** (Kay, 1978; Stern and Kilian, 1996; Martin, 1999; Garrison and Davidson, 2003; Rollinson and Martin, 2005, Martin et al., 2005) are high-Mg# andesites with high alkalis, Ni, Sr, and Cr and low Nb, Ti, and HREE (hence high Sr/Yb). **Sanukites** (Stern et al., 1989; Tatsumi, 2006) are enstatite-bearing high-Mg andesites with $TiO_2 > 0.5$ wt. %. Both have depleted HREE indicative of a garnet-bearing source and may indeed represent warm-slab melts in some hot subduction zones. As Myers and Johnston (1996) pointed out, however, most high-Mg# andesites are not saturated with an eclogite assemblage, and thus cannot have been in equilibrium with the slab at the point of origin. Kelemen et al. (2003a) also noted that there were few primitive andesite lavas in their compilation with a clear eclogite melting signature (notably fractionated HREE). The low-HREE "garnet signature" simply implies that garnet was a fractionating phase. Partial melting of an eclogitic slab is probably the simplest mechanism for accomplishing this, but partial melting of deep arc or continental crust, or high-pressure crystal fractionation of garnet may also work. There is some indication that hornblende can also fractionate HREE. Thick crust can result from mafic magma accumulating and solidifying at the base of the crust, crustal thickening by compression, or "subduction erosion" of the lower crust that carries crustal material down with the slab (the

latter two processes are illustrated in Figure 17.11b). Kelemen (1995) and Tatsumi and Kogiso (2003) also cited several experiments which suggest that high-Mg# andesites can be generated by partial melting of the wedge peridotite under hydrous conditions. **Boninites** (Cameron et al., 1979; Hickey and Frey, 1982; Falloon and Danyushevsky 2000; Deschamps and Lallemand, 2003), another type of high Mg#, high silica, Ni, Cr, low TiO_2 and high LIL/HFS andesite, are also probably derived from the wedge. Ulmer (2001) and Gaetani and Grove (2003), however, argued that quartz-normative high-Mg# andesites are not in equilibrium with peridotite. Kelemen (1995) concluded that the high H_2O, Na_2O, and K_2O contents as well as the high LREE enrichment made the origin of high-Mg# andesites from depleted peridotite unlikely, claiming that their derivation from a uniquely enriched peridotite source would not explain their abundance. Kelemen (1995) and Kelemen et al. (2003a) preferred a multisource model in which H_2O-rich melt derived from the subducted crust (which imparts most of the trace element characteristics) reacts and re-equilibrates with the overlying mantle (which buffers the major elements, but may also obliterate any slab garnet signature). They believed the moderate LREE enrichment and higher Th concentrations are more in line with a hydrous slab melt than a fluid. They called this process **melt fluxing** of the peridotite. We encountered a similar process with melting eclogite pods in rising plumes in Chapter 14. Whether the slab component is a hydrated melt or a fluid/melt hybrid beyond the "second critical endpoint" is still an open question.

Finally, there are the high-K **alkaline** and **shoshonitic series**, typically developed from the deeper wedge and which rise closer to the back-arc. The alkaline character may be due to the greater depth of melting, lower melt fraction due to less H_2O, or the greater concentration of alkalis in the slab-derived fluid: perhaps a result of mantle phlogopite or slab phengite breakdown.

The various parental magmas described rarely reach the surface, which is dominated by variably evolved rock types. Fractional crystallization in island arcs takes place at a number of levels, as illustrated in Figure 16.18. In the shallower chambers, the well-known tholeiitic fractionation trend occurs where H_2O is limited. The calc-alkaline fractionation trend takes place in a hydrous magma with the fractionation of magnetite/Fe-Cr-Al spinel, a highly anorthitic plagioclase, and/or hornblende, as discussed in Section 16.3. The restriction of calc-alkaline magmas to subduction zones is probably a result of the uniquely high H_2O content, aided perhaps by the thickened arc crust that causes primary liquids, including some tholeiites, to pond and fractionate along the calc-alkaline trend.

The temporal trend from early tholeiite-dominated to later calc-alkaline-dominated magmas in several arcs may be attributed to thin crust in young arcs, providing ready conduits for more primitive liquids to the surface. Young arcs are also hotter at depth and drier, favoring tholeiitic melts. In the Aleutians, tholeiites also concentrate where major structures in the upper or subducted plate may provide similar conduits in mature arcs (Kay and Kay, 1994).

Mixing of primitive magma injected into evolving chambers (perhaps of different parentage) is also likely to be common, and is an important part of several models that attempt to explain phenocryst disequilibrium in calc-alkaline rocks as well as major- and trace-element trends. The process of **magma mixing** has only recently been given the attention that it deserves in subduction zone petrogenesis. Conrey (unpublished manuscript) argued that magma mixing is a common process in arc magmatism and is all but essential to the development of the calc-alkaline trend (see also Tatsumi and Kogiso, 2003). Conrey noted the ubiquity of disequilibrium and reaction textures among calc-alkaline phenocrysts, and the common presence of quenched mafic blobs in intermediate and silicic calc-alkaline rocks. The composition gaps and bimodal nature of many calc-alkaline volcanic suites (in which mafic and silicic rocks are more abundant than intermediate rocks) also suggest that mixing may be important. He also described some common element variation trends as incompatible with developing evolved calc-alkaline rocks from early high alumina basalts via crystal fractionation. For example, because apatite is rare in arc cumulates and is too soluble in mafic magmas to crystallize at such high temperatures, P_2O_5 must be conserved and increase in later liquids as fractional crystallization of other phases proceeds. The P_2O_5 content of many intermediate calc-alkaline rocks, however, overlaps with that of mafic rocks in the same area, and is too low to be a product of fractional crystallization from the mafic magmas. Conrey (unpublished manuscript) noted several such chemical anomalies and proposed that many calc-alkaline suites represent mixing of mantle-derived mafic magmas with dacitic or rhyolitic melts produced by partial melting of the arc crust due to thermal input from the rising mafic material. The mafic melts may evolve via fractional crystallization at various depths prior to mixing, but this process, Conrey asserted, is effective only in the early stages of evolution, and mixing with silicic melts is the dominant process beyond this early stage. It will be interesting to see how the relative importance of fractional crystallization versus magma mixing plays out in the light of future work and debate.

Given the range of possible source materials and *P-T* conditions beneath arcs, we should not be surprised by the great variety and gradational nature of the igneous products and series. The K_2O-based series may be related to the depth and degree of partial melting of a deeper source, the tholeiitic versus calc-alkaline nature to either melting processes, mixing of various melts, or fractionation at intermediate levels, and the final fractionation toward dacites and rhyolites occurs at shallow levels. There are many variables at each level.

The increasing alkalinity away from the trench (the K-*h* relationship) of some arcs may result from lower degrees of partial melting (concentrating alkalis and other incompatible elements in the smaller melt fraction) due perhaps to less H_2O available behind the main dehydration front or from deeper melt segregation.

16.8.5 Mantle Re-enrichment

We learned from ocean island basalts in Chapter 14 that the mantle apparently contains several types of enriched components (or "reservoirs") with isotopic signatures indicative of crustal and sedimentary affinities. From the discussions above, we can now see that these constituents were probably parts of the crustal and sedimentary portion of subducted lithosphere. They were partially depleted by solution of more mobile elements into escaping fluids and, in some situations, by melts, and then became part of the dehydrated dense slab dominated by eclogite and cool lithospheric mantle that subsequently sinks to the deep mantle. Subducted slabs may disaggregate, with parts then stirred into the mantle to become thinned stringers, or they may sink further and accumulate at the core–mantle boundary (Sections 1.7.3 and 14.6). Tatsumi (2005), assuming steady-state subduction of 7-km-thick crust for 3 Ga, calculated that enough basalt could be re-introduced to occupy ~10% of the volume of the lower mantle.

Xenoliths reveal that the mantle also contains volatile-bearing phases (Section 19.4), and is releasing fluids, particularly H_2O and CO_2, as part of many magmatic processes (see reviews by Luth, 2007, and Pearson et al., 2007, for more on mantle xenoliths and volatiles). To what extent might some of these volatiles be reintroduced into the mantle at subduction zones by retention in subducted slabs? We have noted that many hydrous phases break down during subduction, particularly in warmer zones. But some hydrous and carbonate phases persist in the cooler regions of subduction zones. Peacock (2003) calculated that about 10^{12} kg of bound H_2O is subducted with basaltic crust each year, of which at least 90 to 95% is lost to the mantle wedge. Kawamoto (2006) summarized the potential hydrous phases in subducting slabs, including many phases found only in high pressure experiments and not yet in nature (because they would probably decompose en route to the surface). In sediment compositions, these include (with wt. % H_2O content) phase egg (7.5), Topaz-OH (10), phase pi (9), and δ-AlOOH (15). Kawamoto (2006) considered lawsonite to be the most important hydrous phase in basaltic systems for transferring H_2O to great depth because it is common and stable to depths greater than 300 km (9 GPa) at temperatures below ~850°C along more "normal" *P-T-t* paths in Figure 16.16 (beyond the stability ranges of amphibole, zoisite, and other common hydrous mafic phases). Kerrick and Connolly (2001a) compared phase equilibria to published *P-T-t* paths and concluded that all H_2O was lost in the slab by 80 km for a hot path, but most (over 2 wt. %) may be retained past 180 km along cold paths. Schmidt and Poli (2005) concluded that a maximum of 1.5 wt. % H_2O could remain stable in subducted crust deeper than about 70 km (~2.4 GPa, when amphibole breaks down), contained in lawsonite, zoisite, and/or phengite.

The ultramafic portion of the subducted slab has an equal or even greater potential for high-pressure H_2O storage (Angel et al., 2001; Kerrick and Connolly, 2001a; Frost, 2006; Hirschmann, 2006). High-pressure ultramafic phases known only from experiments include (with wt. % H_2O) phase A (11.8), phase B (1.6–2.4), phase E (11.4), phase D (10.1), the 10-Å phase (7.6–13), phase X (1.7–3.5), Mg-sursassite (7.2), hydrous wadsleyite (3.3), and hydrous ringwoodite (3.3). (Who names these things?) Initially, serpentine is stable to ~6 GPa or 200 km (Ulmer and Trommsdorff, 1995, or see

Figure 16.19) and, if potassium is available, K-richterite is stable to 15 GPa, and K-bearing Phase X is stable to 22 GPa (nearly the 660 km transition zone). Only K-bearing phase X is stable at very high pressures with normal mantle geotherms; others require cooler regimes.

H_2O can apparently be accommodated in several low-P hydrous phases in ultramafic rocks and also in several candidates for hi-P hydrous phases, but the intermediate-P hydrous phases are limited to lower temperatures and can accommodate less H_2O than peridotites at lower and higher pressures. For normally low-K peridotite, serpentine is the principal low-P-T hydrous phase. For H_2O to be passed from serpentine to the next most likely low-K hydrous phase, phase A, at 200 km, the temperature would have to be below 700°C. According to models (see Figure 16.15), the temperature is likely to be higher than that throughout the subducted lithosphere under most circumstances. Kerrick and Connolly (2001a) concluded that virtually all H_2O in subducted mantle would be lost along hot and intermediate P-T-t paths by 65 and 105 km, respectively. If cool enough ($<$580°C at 6 GPa according to Schmidt and Poli, 2005), or if metastable, fully serpentinized peridotite could transport about 4 wt. % H_2O to phase A, but the slab should be too buoyant to make it down there if $>$10% serpentinite. Angel et al. (2001) therefore estimated that perhaps 0.4 wt. % H_2O is a reasonable maximum to be delivered to Phase A and hence to depth in altered peridotites. If the temperature were between 580 and 720°C at 6 GPa, Schmidt and Poli (2005) determined that serpentine could deliver $\sim$0.6 wt. % H_2O to the 10-Å phase. When conditions are such that the low-pressure phases ($<$6 GPa) break down before higher-pressure hydrous phases become stable, H_2O will probably be lost (upwards), unless it can be trapped as in intergranular film. Kawamoto et al. (1996) referred to such situations as **choke points** because they supposedly choke off transfer of H_2O to the deeper mantle.

The P-T-t path followed by any particular portion of a subducting slab or of the mantle wedge has a major role in controlling whether metamorphosed sediments, mafic crust, or peridotite escape a choke point and succeed in recycling some mantle H_2O. The high-pressure hydrous phases break down eventually, but peridotites at depth can still contain significant water in phases that are considered anhydrous at the surface (olivine, pyroxene, and garnet). Estimates range from 0.03 to 0.065 wt. % H_2O in those phases at high pressure, suggesting transport of $(1 \text{ to } 8) \cdot 10^{11}$ kg of H_2O per year to the deep mantle and that the mantle has a potential to store 4.6 to 12.5 times the present ocean mass of H_2O (Iwamori, 2007).

Carbonates are stable from low to very high pressures, so CO_2 appears to be easier to recycle. Kerrick and Connolly (2001b) computed that siliceous limestones should experience little devolatilization between 80 and 180 km along low-temperature P-T-t paths, unless infiltration of H_2O promotes sub-arc decarbonation. They estimated that about $5.4 \cdot 10^{12}$ moles of CO_2 are subducted annually, whereas only about $(2 \text{ to } 3) \cdot 10^{12}$ moles are expelled in arc magmatism. Thus about $(2.5 \text{ to } 3.5) \cdot 10^{12}$ moles may be carried deeper or dissolved in H_2O each year. Gorman et al. (2006) considered open-system fluid behavior and concluded that H_2O infiltration still permits considerable CO_2 return to the mantle. They calculated that nearly all of the original CO_2 should be lost under warm subduction zone arcs, but slabs under cool arcs may retain up to 80% of the original CO_2. Dasgupta et al. (2004), however, concluded that experimental results on partial melting of carbonated eclogite show that elimination of carbonate from the residue of eclogite requires temperatures $\sim$100°C hotter than any plausible subduction geotherm.

Despite some uncertainty over the details, most investigators agree that considerable CO_2 may be returned to the mantle via subduction. As discussed in Chapter 14, carbonated eclogite from subducted crust may be a possible source of some of the enriched isotopic mantle reservoirs, such as HIMU (Kogiso et al., 1997; Workman et al., 2003; Dasgupta et al., 2006).

Summary

Subduction zones are the most complex geologic features on Earth. Into intra-oceanic subduction zones come lithospheric slabs carrying heterogeneous mixtures of sediments and variably altered oceanic crust and upper mantle. These materials are variably dehydrated and perhaps melted in the subduction zone and the products rise and mix with the overlying mantle wedge peridotites, which, when dragged deeper, are partially melted in response. The products of this "subduction factory" include shallow- and deep-focus earthquakes along the Wadati-Benioff zone, a variety of magmas, island arcs with their explosive volcanoes, the arc crust (and, ultimately, the continental crust), and the processed remnants of the subducted slab and wedge which are recycled back into the mantle. Subduction zones are where old dense plates sink and are hence where plate tectonics is believed to manifest its main driving force ("slab pull").

Island-arc magmas can (rather artificially) be subdivided into a low-K tholeiitic, medium-K calc-alkaline, and high-K mixed series. Tholeiitic rocks occur in virtually any situation involving mantle melting, but calc-alkaline rocks are essentially restricted to subduction-related environments and are hence a hallmark of subduction-related magmatism (particularly the basaltic andesites and andesites). H_2O plays a crucial role in subduction zone processes. It explains the anomalously high LIL/HFS ratios of arc lavas and dramatically lowers the melting temperatures of silicate systems, thereby solving the fundamental paradox of magmatism in intrinsically cooler-than-average tectonic regimes. Warm plates actually corelate with reduced volcanism in subduction zones. This is probably because the subducting slab dehydrates earlier in warm zones and considerable H_2O escapes into the fore-arc region before it can be dragged into the hot wedge where melting occurs. H_2O also explains the occurrence of the calc-alkaline series by either shifting partial melts of the mantle wedge toward high-SiO_2 and low-Mg# or by fractional crystallization of calcic plagioclase, an oxide, or hornblende.

No single evolutionary process can produce the variety of arc magmas, and the present model provides only a basis for the principal features of arc magmatism. A general model involves the following:

1. Heating and dehydration of subducted oceanic crust and associated sediments at temperatures below the wet basalt solidus. Most petrologists believe the major contribution from the subducted slab is in the form of aqueous fluids and their transported solutes. Partial melting of subducted sediments is also possible, and perhaps of the uppermost mafic crust as well. Crustal melting may be particularly important in young arcs or arcs with young subducted crust (especially if a ridge is subducted). Dehydration of subducted lithospheric mantle may also transfer fluids upward.

2. Rise of released aqueous fluid (plus dissolved constituents, particularly LIL elements) into the overlying mantle wedge. If melts form they also rise and react with the wedge peridotite and probably solidify at depth under most arcs. Slab melts are believed by some investigators to impart some of the more moderate LIL/HFS trends found in some arc volcanics.

3. Dragging of the now hydrous and hybridized peridotite to greater depths where it then either dehydrates above the wet peridotite solidus or releases fluids upward into parts of the wedge above the wet solidus, either of which initiates partial melting to form an olivine tholeiitic basalt with 1 to 2 wt. % H_2O. Calc-alkaline basaltic andesites and high-Mg# andesites, adakites, sanukites, and boninites may also be generated by slab crustal melting or fluid/melt fluxing of the wedge peridotites. Enhanced slab melting is generally attributed to unusually hot subduction zones or Archean times of higher thermal gradients. The variation in h (depth to the slab beneath the volcanic front),

even along a single arc segment, underscores the complexity of arcs. The interdependence of thermal and mechanical processes, and local perturbations in each, results in variable reaction paths and complex patterns of fluid release and magma genesis.

4. Ponding (underplating) of tholeiitic magma at the base of the arc crust to form high-alumina basaltic magma by fractional crystallization. Other high-alumina basalts may be primary, as may some calc-alkaline liquids. Large hot mafic magma chambers may also partially melt the overlying arc crust, producing silicic melts that rise toward the surface and mix with mafic melt derivatives.

5. Differentiation of both tholeiitic (dry) and calc-alkaline (with H_2O) series at higher crustal levels (including fractional crystallization, assimilation, and magma mixing) to produce the spectrum of volcanic products now found at the surface. Magmas that do not reach the surface form plutons at depth.

6. More alkaline magmas may also form, perhaps well behind the arc where less subduction zone-related fluid is available, and probably contain more dissolved alkalis.

7. Induced mantle flow behind the arc can cause convective mantle upwelling and back-arc volcanism. This process may create marginal ocean basins and remnant arcs (Karig, 1974).

Subduction zones act as a *filter* in the sense that material from the mantle and oceanic realm is differentiated, selectively incorporating some into the arc and continental crust, with the remainder returned into the deep mantle, most notably as the enriched reservoirs (EMI, EMII, and HIMU). Subduction zones thus have a crucial role in the chemical evolution of the Earth. Material added to crust from the mantle at subduction zones is predominantly basalt, with subordinate basaltic andesite and andesite (Kushiro, 1990; Grove and Kinzler, 1986; Gaetani and Grove, 1998). This has some important implications for the development of continental crust, and will be explored further in the next two chapters.

Key Terms

Orogenic or subduction-related magma suite *324*
Wadati-Benioff zone *325*
Volcanic front *325*
h *325*
Fore arc *325*
Accretionary prism/accretionary wedge *325*
Back-arc basin *325*

Secondary arc *326*
Low-K/medium-K/high-K/shoshonitic series *328*
Calc-alkaline *330*
K-h relationship *332*
Secular equilibrium of U-series radioactive decay *000*
Pressure–temperature–time (*P-T-t*) path *340*

Second critical endpoint *343*
Ridge subduction/slab window *345*
High-alumina basalt *346*
Fluxed melting *347*
High-Mg# andesite *347*
Adakite/sanukite/boninite *347*
Melt fluxing *347*
Choke point *349*

Review Questions and Problems

Review Questions and Problems are located on the author's web page at the following address: **http://www.prenhall.com/winter**

Important "First Principle" Concepts

- Subduction zones act as "filters" in that they process mantle and mantle-derived material (such as oceanic crust) and selectively let more silicic material pass through to form arc crust (and eventually continental crust).

- Most of the material that fails to be so filtered returns to the deep mantle. Some has been enriched (compared to primitive mantle) at mid-ocean ridges or elsewhere to produce the (variably altered) oceanic crust, islands/seamounts, and sediments. This material may eventually become part of the enriched deep mantle melt sources tapped by ocean island magmas (Chapter 14).

- H_2O substantially lowers the melting point of silicate rocks, making subduction zone magmatism possible and shifting the composition of mafic melts toward more siliceous compositions.

- LIL elements are more H_2O-soluble than HFS elements. In typical relatively dry mafic and ultramafic melting processes, they behave similarly. When H_2O is present, either as a fluid or a very hydrous melt, LIL elements are partitioned much more strongly into the mobile phase.

- Subduction zone magmatism is a multicomponent, multistage process involving aqueous fluids derived from the subducted slab, melts of several possible sources, and mixing of derived liquids.

- At present, the hydrated mantle wedge above the subducted slab is the preferred source of most island-arc magmas.

- The calc-alkaline series is a hallmark of subduction zone magmatism. It may develop at the stage of source melting or by shallow crystal fractionation under hydrous conditions.

- Shallow crystal fractionation is much more prevalent at island arcs than at mid-ocean ridges and ocean islands, perhaps due to the presence of H_2O in the melts or the low-density arc crust that reduces melt buoyancy.

Suggested Further Readings

Bebout, G. E., D. W. Scholl, S. H. Kirby, and J. P. Platt (eds.). (1996). *Subduction Top to Bottom*. Monograph **96**. American Geophysical Union. Washington, DC.

Davies, J. H., and D. J. Stevenson. (1992). Physical model of source region of subduction zone volcanics. *J. Geophys. Res., 97*, 2037–2070.

Eiler, J. (ed.). (2003). *Inside the Subduction Factory*. Monograph **138**. American Geophysical Union. Washington, DC.

Gill, J. B. (1981). *Orogenic Andesites and Plate Tectonics*. Springer-Verlag. Berlin.

Goldstein, S. L. (ed.). (2007). Chemical and physical processes affecting element mobility from the slab to the surface. *Chemical Geology* **239**, 3–4. A special issue.

Grove, T. L., and R. J. Kinzler. (1986). Petrogenesis of andesites. *Ann. Rev. Earth Planet. Sci., 14*, 417–454.

Hawkesworth, C. J., K. Gallagher, J. M. Hergt, and F. McDermott. (1993). Mantle and slab contributions in arc magmas. *Ann. Rev. Earth Planet. Sci., 21*, 175–204.

Island Arc: Earth Sciences of Convergent Plate Margins and Related Topics. A journal by Blackwell Publishing.

Larter, R. D., and P. T. Leat (eds.). (2003). *Intra-Oceanic Subduction Systems: Tectonic and Magmatic Processes*. Special Publication **219**. Geological Society. London.

Maruyama, S., and S. Santosh (eds.). (2007). Island arcs, past and present. *Gondwana Research* **11**, 3–6. A special issue.

McCulloch, M. T., and J. A. Gamble. (1991). Geochemical and geodynamical constraints on subduction zone magmatism. *Earth Planet. Sci. Lett., 102*, 358–374.

Nixon, G. T., A. D. Johnston, and R. F. Martin (eds.). (1997). Nature and origin of primitive magmas at subduction zones. *Can. Mineral, 35*, 2. (Special volume on subduction zone magmas.)

Pearce, J. A., and D. W. Peate. (1995). Tectonic implications of the composition of volcanic magmas. *Ann. Rev. Earth Planet. Sci., 23*, 251–285.

Pritchard, H. M., T. Alabaster, N. B. W. Harris, and C. R. Neary (eds.). (1993). *Magmatic Processes and Plate Tectonics*. Special Publication **76**, 405–416. Geological Society. London.

Rudnick, R. L. (ed.). (2003). *Treatise on Geochemistry. Volume 3: The Crust*. Elsevier Science. Amsterdam.

Schmidt, M. W., and S. Poli. (2005). Generation of mobile components during subduction of oceanic crust. In: *Treatise on Geochemistry*. (ed. R.L. Rudnick), Chapter 3.17, pp. 567–591. Elsevier. Amsterdam.

Tatsumi, Y. (2005). The subduction factory: How it operates in the evolving earth. *GSA Today, 15*, 4–10.

Tatsumi, Y., and S. Eggins. (1995). *Subduction Zone Magmatism*. Blackwell. Oxford.

Thorpe, R. S. (ed.). (1982). *Andesites. Orogenic Andesites and Related Rocks*. John Wiley & Sons. New York.

van Keken, P. E. (2003). The structure and dynamics of the mantle wedge. *Earth Planet. Sci. Lett., 215*, 323–338.

van Keken, P. E., and S. D. King (eds.). (2004). Thermal structure and dynamics of subduction zones. *Phys. Earth Planet. Inter., 149*, 1–2. A special issue.

Wilson, M. (1989). *Igneous Petrogenesis: A Global Tectonic Approach*. Unwin Hyman. London. Chapter 6.

17

Subduction-Related Igneous Activity, Part II: Continental Arcs

Questions to be Considered in this Chapter:

1. In what principal ways do continental arcs differ from island arcs?
2. What influences does the thick continental crust have on the composition and physical nature of the igneous rocks now exposed?
3. In what way does the subcontinental lithospheric mantle play a distinctive role in arc magmatism?
4. How are the major batholith belts formed, and how do they differ from the volcanic suites?

17.1 INTRODUCTION

When one side of a convergent plate margin is of continental character, the result is called a **continental arc** or an **active continental margin**. Continental crust is too buoyant to be subducted far, so the only stable long-term continental-arc geometry is one in which an oceanic plate subducts beneath the edge of a continent. When a continental land mass arrives at a subduction zone as part of the subducting plate, it eventually plugs the system, and subduction either ceases entirely, or, if the overriding plate is oceanic, the dip of the subduction zone may reverse to a new stable geometry. Continental arcs are geometrically and mechanically similar to island arcs, as summarized in the last chapter. Subduction of oceanic crust and upper mantle is accompanied by dehydration (and perhaps melting) of the subducted crust and sediments, leading to fluid-fluxed melting of the mantle wedge above the subducted slab. The result is a magmatic arc along the edge (called the "leading edge") of the continent. The main difference between continental arcs and island arcs is the presence of continental crust up to 70 km thick overlying the subduction zone. Although the crust is a shallow feature in the context of the subduction zone as a whole, it has some important characteristics that may affect continental-arc magmas in a number of ways. At the onset we might propose the following:

1. Magmas generated in the mantle wedge or subducting slab must traverse the thick layer of sialic and incompatible element-enriched crust before reaching the surface. Because of the contrasting geochemical compositions between primary magmas and this crustal material, the potential for noticeable crustal contamination of rising liquids is great.
2. The low density of the crust may significantly retard the buoyant rise of mafic to intermediate magmas, resulting in more differentiation and/or assimilation in stagnated bodies.
3. The melting point of the continental crust may be low enough to be partially melted by the addition of heat associated with the rise of subduction zone magmas, adding considerable silicic magmatism to the system.
4. Significant amounts of continentally derived sediment may be transported and deposited in the nearby trench, subducted, and thereby enrich the *primary* magmas formed at depth.

In addition, as mentioned at the end of Chapter 15, the **subcontinental lithospheric mantle** (**SCLM**, the mantle portion of the continental plate) may be quite different than in the lithosphere beneath the ocean basins. The SCLM was probably depleted initially when the overlying crust was derived from it. Instead of being subducted and recycled, like oceanic lithosphere, however, the subcontinental lithospheric mantle has been anchored like a keel to the overlying continent since its formation. As we shall see in Chapter 19, xenoliths of subcontinental mantle are brought to the surface in kimberlites, and suggest that the SCLM may have been at least locally enriched during the substantial period (as long as 4 Ga) that it has stagnated beneath the continent since its initial depletion. If such enriched mantle comprises a part of the subduction zone wedge, some primary magmas may also be enriched by this material.

If any or all of the considerations listed above are effective, we would expect continental-arc volcanics, on average, to be more evolved and enriched than corresponding island-arc volcanics. We thus approach our survey of continental-arc magmatism with these initial ideas in mind.

There is a further distinction between continental arcs and island arcs. Along continental arcs, particularly in segments where volcanism has ceased to provide a renewed surface cover of accumulated lavas and pryoclastics, erosion may expose a complex linear chain of plutons. These **batholith belts** are so common to the continental-arc environment that they are widely used as indicators of ancient and extinct active margins, even where those margins are no longer at the edge of a continent due to collision and accretion of land masses.

The thick sialic crust and possibly an enriched and heterogeneous mantle make the continental-arc environment even more complex than island arcs, comprising the most complex igneous/tectonic environment on Earth. These are the great **orogenic belts** where deformation, metamorphism, and volcanism + plutonism combine to create the great mountain chains of the past and present. Their complexity has challenged geologists for centuries. In the ensuing (necessarily limited) survey, we shall attempt to find representative processes, products, and spatial or temporal trends, but in these complex environments with such variety possible, it would be unwise to regard any example as typical.

Let's begin by embarking on a brief look at some continental-arc occurrences in an attempt to evaluate the influence of continental lithosphere on the subduction processes that we have identified in the last chapter. The obvious choice for such a survey is the western coast of the Americas (also called the American Cordillera). This coast extends for over 17,000 km, from the Aleutian Islands in the north to Terra del Fuego in the south (see Figure 16.1), and has been a convergent plate boundary for most of the Phanerozoic, beginning when this was the edge of the supercontinent Pangea.

In keeping with such a complex environment, the tectonic and magmatic style at any one place has not remained constant for the duration of plate interaction, nor is the style now the same along the full length of the orogen. The western edge of the American plate interacts with five different plates (Figures 16.1 and 17.1). From south to north these plates are the Antarctic (southern Chile), Nazca (Chile–Peru), Cocos (Central America), Pacific (California, British Columbia, and Alaska), and Juan de Fuca (N. California, Oregon, Washington, and southern British Columbia, Figure 17.10). Some areas are presently undergoing subduction with active volcanism, whereas others are magmatically quiescent. The dip of the subduction zone varies, as does the horizontal direction of

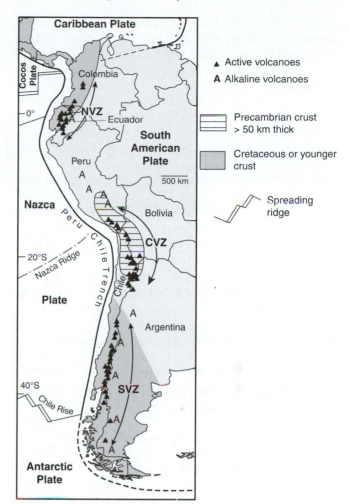

FIGURE 17.1 Map of western South America, showing the plate tectonic framework and the distribution of volcanics and crustal types. NVZ, CVZ, and SVZ are the northern, central, and southern volcanic zones. After Thorpe and Francis (1979), Thorpe et al. (1982), Harmon et al. (1984), Barragan et al. (1998), and Samaniego et al. (2005).

plate incidence, which (presently) ranges from nearly normal to the plate margin in South America to oblique subduction ("transpression") in the north to purely strike-slip motion in much of British Columbia, California, and southernmost Chile. Although most of British Columbia and southern California are not experiencing subduction now, the large batholiths in these areas testify to earlier subduction processes in the Mesozoic and Cenozoic. The geology has been further complicated by the accretion of several **allochthonous** or **exotic terranes**, either offshore island arcs or continental fragments that have collided with North America and/or been transported northward by strike-slip motion (Coney et al., 1980). The only event that apparently has not occurred along the west coast of North and South America is the collision of two major continental landmasses.

17.2 THE VOLCANIC ANDES OF SOUTH AMERICA

Subduction of the Nazca plate along the western edge of South America has produced a broad orogen with a continuous belt of igneous products developed over at least the last

500 Ma. The South American section of the Cordillera is for the most part uncomplicated by collision events, and the term **Andean-type margin** is commonly used to denote an igneous/orogenic belt developed strictly by subduction at a continental edge. At present, active volcanism is restricted to three zones (Figure 17.1) separated by inactive gaps: the northern volcanic zone (**NVZ**, in Colombia and Ecuador from latitude 5°N to 2°S), the central volcanic zone (**CVZ**, largely in southern Peru and northern Chile, 16 to 27°S), and the southern volcanic zone (**SVZ**, largely in southern Chile, 33 to 55°S). Some investigators distinguish an Austral Volcanic Zone from the southern SVZ, separated by a small inactive gap (from 46 to 49°S), but we shall combine the two here. The active zones are separated by magmatically inactive areas. Notice in Figure 16.1 that the subducting Nazca plate is quite young at the northern and southern ends.

The active volcanic zones correspond to steeply dipping (25 to 30°) segments of the subducting slab, whereas the inactive areas correspond to segments with much shallower dips (10 to 15°) of a segmented Nazca plate. The zones of low dip angle correspond roughly to places where thicker and less dense oceanic crust is being subducted. The extra thickness is attributed to the Nazca Ridge (an aseismic seamount chain) and the Chile Rise (Figure 17.1). Among the active zones, the subduction angle is slightly lower in the SVZ, so the depth from the volcanic arc to the subduction zone is only 90 km, compared to the ca. 140 km in the NVZ and CVZ.

Why active volcanism is restricted to the steeper-dipping segments is not clear but may be related to the presence of asthenospheric mantle in these segments, as shown in Figure 17.2. In zones of low subduction dip, the mantle wedge above the slab is much thinner, leaving only the shallow lithospheric mantle in the western portion with the lower, presumably more fertile, asthenospheric component displaced to the east (Barazangi and Isacks, 1979). In a comprehensive geophysical study of subduction zone magmatism, Cross and Pilger (1982) noted that zones with low subduction angles in many island arcs and continental arcs corresponded to areas in which magmatism was either greatly subdued and far behind the trench or absent entirely. The lack of asthenosphere described above is commonly cited as the reason for diminished volcanic activity, but a low subduction angle should merely displace the asthenospheric portion of the wedge further behind the trench where it could still melt. Small dip changes, in fact, are considered a major cause of historic migration of the magmatic front in several arcs. Nonetheless, the dip angle (and distance from the trench) must play an important, though presently enigmatic, role in controlling whether or not magmatism occurs. Perhaps the low angle inhibits mantle recirculation in the wedge, so the wedge gradually becomes depleted and refractory. Slab dip at a particular segment may also vary with time, so many presently non-magmatic zones were once more steeply-dipping and active (see Kay et al., 1991, for example).

The three active zones of the Andes differ in the thickness and type of crust that comprises the continental plate. In the NVZ and SVZ, the crust is dominantly Mesozoic and Cenozoic (shaded in Figure 17.1), at least part of which

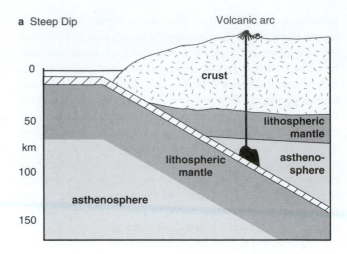

a Steep Dip

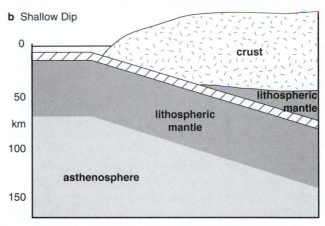

b Shallow Dip

FIGURE 17.2 Schematic diagram illustrating how a shallow dip of the subducting slab can pinch out the asthenosphere from the overlying mantle wedge.

consists of accreted oceanic and island-arc terranes. Crustal thickness in these zones is on the order of 30 to 45 km. The crust is thicker in the CVZ, and outcrops of Precambrian metamorphic rocks along the CVZ coast have led most investigators to conclude that thick Precambrian basement of the Brazilian Shield underlies the central Andes (cross-hatched in Figure 17.1). In fact, thicknesses of up to 75 km make the CVZ the thickest crustal section for any subduction zone on Earth. This thickness may be the result of subduction zone magmas accumulating at depth in the crust (called **magmatic underplating**), but Isacks (1988) suggested that such underplating cannot account for the full thickness, and concluded that shortening by orogenic folding and thrusting of the crustal rocks must also have contributed. If the thick continental crust does indeed play a significant role in continental-arc magmatism, the Andes, with their varying crustal character, should provide some useful information to test this hypothesis.

17.2.1 Petrology and Geochemistry of the Andean Volcanics

All three active zones are dominated by rocks with a typical calc-alkaline trend on AFM diagrams (Figure 17.3). Each zone, however, has its own distinctive petrologic characteristics (Thorpe and Francis, 1979; Thorpe et al., 1982). The **southern SVZ** ($>37°$S) comprises high alumina basalts and

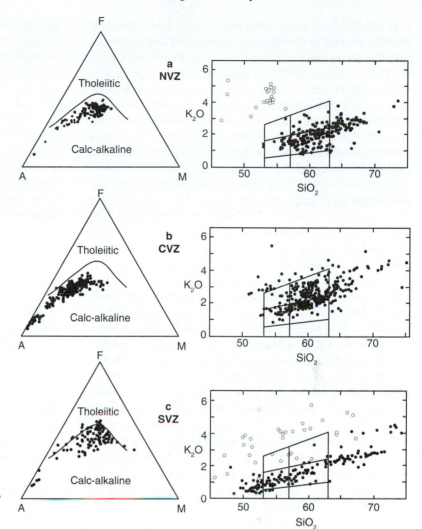

FIGURE 17.3 AFM and K₂O versus SiO₂ diagrams (including Hi-K, Med.-K, and Low-K types of Gill, 1981; see Figures 16.4 and 16.6) for volcanics from the **(a)** northern, **(b)** central, and **(c)** southern volcanic zones of the Andes. Open circles in the NVZ and SVZ are alkaline rocks. Data from Thorpe et al. (1982, 1984), Geist (personal communication), Deruelle (1982), Davidson (personal communication), Hickey et al. (1986), López-Escobar et al. (1981), Hörmann and Pichler (1982).

basaltic andesites on a platform of older andesites. Olivine, two pyroxenes, and plagioclase are the common phenocrysts. SiO_2 is concentrated in the range 50 to 65%, with only a few more silicic varieties. The thin (30 to 35 km) crust has an island-arc-like character, and of the three zones, the chemical composition of the SVZ magmas most resembles island arcs (Harmon et al., 1984). There is considerable variation in K_2O content, however (Figure 17.3c), with some alkaline series extending through hawaiites to highly potassic leucite basanites toward the east. North of 36°S the young arc-intrusive basement becomes thicker (50 to 60 km) and more variable, including Paleozoic and Triassic rocks of more continental affinity (Hildreth and Moorbath, 1988). Here the basalt and basaltic andesite shield volcanoes that are so dominant in the south become subordinate to more silicic hornblende-bearing andesite–dacite stratovolcanoes (Frey et al., 1984; Davidson et al., 1988).

Geochemical data are more limited for the **NVZ** than for the other zones, in spite of significant recent volcanism. The available published data suggest that the NVZ consists of olivine- and two-pyroxene basaltic andesites and andesites as in the SVZ, plus evolved dacites and rhyolites that are more common than in the SVZ. The AFM diagram in Figure 17.3a suggests a rather typical island-arc calc-alkaline

suite on the west, and the K_2O-SiO_2 diagram suggests a more potassic andesite and latite suite to the east, also supported by Barragan et al. (1998). This possible K-*h* relationship (Section 16.4), continues to alkali and shoshonitic rocks about 50 km further east, but these are east of the suture between western Mesozoic and Cenozoic arc crust and eastern Jurassic–Cretaceous metasediments. Bourdon et al. (2002) and Samaniego et al. (2005) noted that some volcanics at two volcanic centers in Ecuador have adakite-like characteristics, which they ascribed to a component of melted young slab near the subducted Carnegie ridge. Garrison and Davidson (2003) disputed the adakite claims, however. They used Sr/Y as a proxy for slab melts because melting of basalt at high p_{H_2O} should melt plagioclase (yielding Sr) and stabilize garnet (holding Y) and found no systematic change in this ratio with latitude across the trace of the extension of the ridge.

The **CVZ** has been studied in most detail and has a more diverse assortment of basaltic to rhyolitic volcanics (Figure 17.3b). On average the volcanics are more silicic than those of the SVZ and NVZ (mostly within the range 56 to 66% SiO_2). Andesites and dacites are the most common CVZ volcanic, whereas basalts are rare. A considerable volume of extensive dacite-rhyolite ignimbrite sheets, some covering over 200,000 km[2], occur in this zone. Plagioclase is the dominant phenocryst in the volcanics, with olivine and pyroxenes in the basaltic andesites. Hornblende is more common than in the NVZ, southern SVZ, and island arcs, and biotite is also common in the more evolved rocks. There is a vague suggestion of a K-h relationship in the central zone, with more low-K_2O volcanics in the west and alkalic shoshonites in the east.

The geochemical composition of nearly 400 samples of Andean volcanics from all three zones is compiled from a number of sources in the Excel file AndesVolc.xls at www.prenhall.com/winter. The major element data from the Andes and other active margins shows that continental-arc magmatism has essentially the same compositional range as that of island arcs, implying that the same general processes of island-arc magmatism described in Chapter 16 apply to continental arcs as well. The distribution of rock types in continental arcs, however, is skewed toward calc-alkaline and alkaline series, with fewer of the tholeiites and low-K types commonly found in island arcs (compare Figures 17.3 and 16.6). The Andes volcanics are also on average more differentiated within a series than their island-arc counterparts. Basalts are rare, particularly the more primitive tholeiitic basalts that occur mostly in immature island arcs where rapid ascent to the surface is easiest. Basaltic andesites are also less common than in island arcs. Andesites and dacites are more common, and dacite-rhyolite ignimbrites occur almost exclusively in continental settings. More evolved Andean magmas are most frequent in the CVZ, where the crust is thickest and oldest. Thus, although the processes of melt generation may be similar for island arcs and continental arcs, the effect of the thick continental crust on producing more evolved and enriched calc-alkaline to alkaline magmas at the surface is evident.

Trace element and isotopic data support the hypothesis that Andean volcanics have an origin similar to those of island arcs, and also shed some light on the processes of crustal interaction and magmatic evolution. Figure 17.4 is a REE

diagram for some of the samples in the file AndesVolc.xls. I attempted to select only the less silica-rich samples in order to minimize the effects of fractional crystallization on REE enrichment. The sparse data on NVZ volcanics, however, required more silicic samples to be included (but the lowest SiO_2 sample provided the highest REE values of that group). The SVZ lavas show a rather tight REE pattern with low slope (Ce/Yb). This is most like the medium-K calc-alkaline island-arc magmas (Figure 16.10), which is in accordance with the thin arc-like SVZ crust. The flat and high HREE pattern suggests a garnet-free source, which reflects the shallower slab dip and lesser depth of magma genesis in the SVZ. (Remember from Chapters 9 and 10 that garnet fractionates the HREE and is stable in the mantle only at depths over about 100 km.) López-Escobar et al. (1977, 1984) were able to model the major- and trace-element characteristics of some parental SVZ magmas on the basis of 10 to 15% partial melting of spinel–peridotite with a subduction zone LIL enrichment, followed by fractional crystallization of about 20% olivine and clinopyroxene.

The NVZ and CVZ patterns, where the subduction angle is somewhat steeper, show more depleted HREE than for the SVZ, suggesting that residual garnet in a deeper source may have fractionated them. The CVZ lavas are also enriched in the LREE, as we might expect if the thick continental crust were somehow involved in their enrichment.

Figure 17.5 is a MORB-normalized spider diagram, in which continental-arc magmas also show the decoupled LIL/HFS pattern and Ta-Nb trough that is now accepted as a characteristic of subduction zone magmas (Figure 16.11a). This supports our postulate that continental-arc magmas are probably created by similar processes as island-arc magmas, including LIL enrichment of the mantle wedge via aqueous fluids derived from dehydration of the altered oceanic crust of the subducting slab and subducted sediments. Enrichment in the NVZ and SVZ lavas is similar to that of island arcs (Figure 16.11a), whereas enrichment in the CVZ is considerably

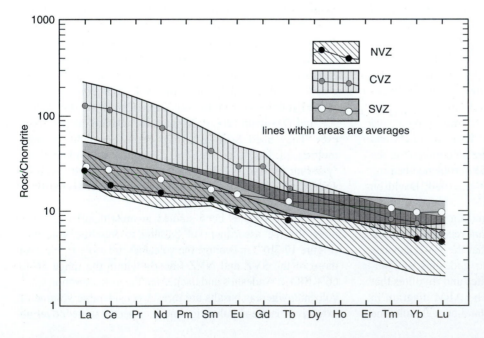

FIGURE 17.4 Chondrite-normalized REE diagram for selected Andean volcanics. NVZ (6 samples, average SiO_2 = 60.7, K_2O = 0.66, data from Thorpe et al. 1984; Geist, personal communication). CVZ (10 samples, average SiO_2 = 54.8, K_2O = 2.77, data from Deruelle, 1982; Davidson, personal communication; Thorpe et al., 1984). SVZ (49 samples, average SiO_2 = 52.1, K_2O = 1.07, data from Hickey et al., 1986; Deruelle, 1982; López-Escobar et al., 1981).

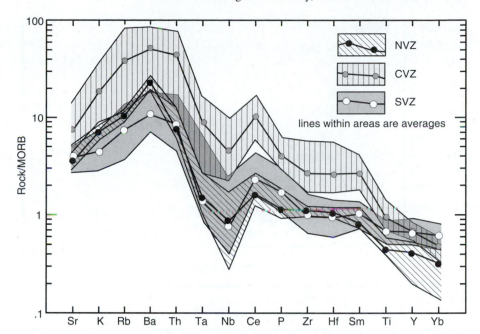

FIGURE 17.5 MORB-normalized multielement (spider) diagram (Pearce, 1983) for selected Andean volcanics. See Figure 17.4 for data sources.

greater. Pearce (1983) attributed the enriched central "hump" shape on MORB-normalized spider diagrams to intraplate enrichment (including both LIL elements and more mobile HFS elements, such as Ce and P), by analogy with intraplate basalts (see Figures 14.5 and 15.9). He suggested that the enriched subcontinental lithospheric mantle (SCLM) was responsible for the enriched pattern in the source of the CVZ magmas, which overlie thicker and older crust. He noted that such a component, containing elevated Ta, Nb, Zr, and Hf, is present in many continental-arc magmas in amounts higher than found in island arcs, and proposed that it may be a good indicator of the involvement of SCLM in magma genesis. Davidson et al. (1990, 1991) and Wörner et al. (1994) argued that the trace element (and isotopic) trends correlate too well with crustal thickness, and concluded that the thick crust is responsible for the enrichment patterns, not the lithospheric mantle. The crustal sources and their geochemical signature can be variable, however, and difficult to interpret (but certainly no more so than SCLM, samples of which we rarely see).

Figure 17.6 shows ^{143}Nd/^{144}Nd versus ^{87}Sr/^{86}Sr for a range of Andean volcanics. Common values for MORB (depleted mantle) are included for reference. Low ^{143}Nd/^{144}Nd and high ^{87}Sr/^{86}Sr you should now recognize as characteristics of an isotopically enriched source such as continental crust. The rather concave shape of the Andean trend is similar to that of the Banda arc and New Zealand toward a Sr-enriched crustal EMII-type reservoir, as shown in Figure 16.12. The NVZ and southern SVZ show only minor isotopic enrichment. In these zones, the slab-derived component, including a contribution from altered oceanic crust and minor sediments, could account for this enrichment, as it does in some island arcs, without requiring any additional crustal assimilation. Because the local basement in these areas is of accreted arc character, we cannot prove that crustal assimilation did or did not take place because it would involve a crustal component that contrasts little with the primitive magmas generated beneath the crust. The CVZ (and northern

SVZ), on the other hand, exhibit substantial isotopic variation that requires a crustal contribution, consistent with the older and thicker crust in these areas.

As with most orogenic magmas, Andean volcanics are enriched in ^{207}Pb/^{206}Pb and ^{208}Pb/^{206}Pb, relative to the NHRL, so they lie above the mantle-MORB field on standard Pb-Pb isotope diagrams (Figure 17.7). Andean enrichments are not much greater than for many enriched OIBs in this figure, and could thus be developed by Pb derived almost solely from a subducted sediment component introduced into the magma source region (as has been proposed for OIBs). ^{207}Pb, however, is enriched above the values for Nazca plate sediments, and thus seems to require an additional crustal component (such as EMI or EMII).

The value of δ^{18}O for mantle rocks is generally between 5 and 6‰, whereas the continental crust is enriched in ^{18}O (δ^{18}O = 10 to 25‰ or even higher) as a result of interaction between the crust and the hydrosphere and/or metamorphic fluids. Elevated δ^{18}O values have thus been used to provide strong *stable* isotopic evidence for crustal involvement in Andean magma genesis. For a comprehensive discussion of the application of oxygen isotopes to Andean volcanism, see Taylor (1986). If we plot the isotopic characteristics of the volcanics as a function of latitude (Figure 17.8) we can readily see that the Pb, Sr, Nd, and O isotopic ratios are enriched principally in the CVZ and northern SVZ (18° to 36°S). Remember that *low* ^{143}Nd/^{144}Nd means enriched. Δ7/4 and Δ8/4 in Figure 17.8 are measures of Pb enrichment expressing the height above the NHRL on these two diagrams (see Section 14.5).

There is certainly lots of scatter in the isotope patterns in Figures 17.6 through 17.8. More detailed analysis of individual suites shows that Sr enrichments may not always correlate with Nd or Pb enrichments, and that some isotopic systems may not have consistent correlations with trace element trends. Such variations almost certainly reflect the complex lithologies of the local crustal basement, as well as variations in the

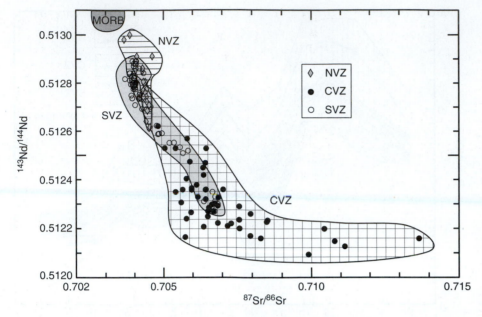

FIGURE 17.6 Sr versus Nd isotopic ratios for the three zones of the Andes. Data from James et al. (1976), Hawkesworth et al. (1979), James (1982), Harmon et al. (1984), Frey et al. (1984), Thorpe et al. (1984), Hickey et al. (1986), Hildreth and Moorbath (1988), Geist (personal communication), Davidson (personal communication), Wörner et al. (1988), Walker et al. (1991), De Silva (1991), Kay et al. (1991), Davidson and deSilva (1992), Bourdon et al. (2003), and Samaniego et al. (2005).

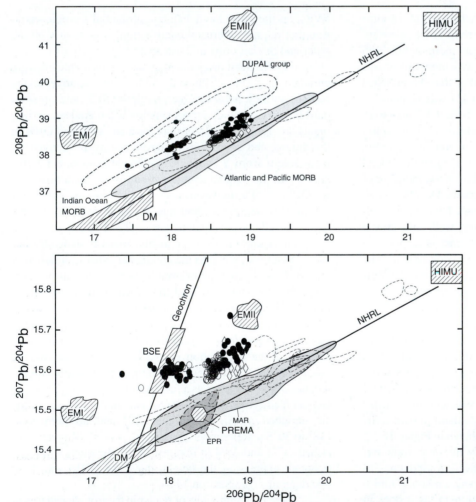

FIGURE 17.7 $^{208}Pb/^{204}Pb$ versus $^{206}Pb/^{204}Pb$ and $^{207}Pb/^{204}Pb$ versus $^{206}Pb/^{204}Pb$ for Andean volcanics plotted over the OIB fields from Figures 14.9 and 14.11. Symbols and sources as in Figure 17.6.

subduction zone source regions below. The lower continental crust generally has low Pb enrichment (in some cases lower than the mantle), whereas the upper crust is enriched in Pb, Sr, Nd, and O ratios, but not in a strictly sympathetic fashion. There are substantial local variations seen in the country rocks

and in various xenoliths. Different depths and styles of assimilation, plus different assimilants, could result in a variety of geochemical trends.

In spite of this "noise," Figures 17.4 through 17.8 clearly indicate that the enriched Andean isotope and trace

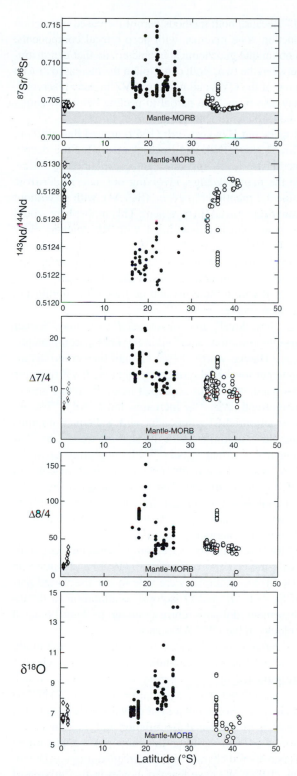

FIGURE 17.8 $^{87}Sr/^{86}Sr$, $^{143}Nd/^{144}Nd$, $\Delta 7/4$, $\Delta 8/4$, and $\delta^{18}O$ (‰) versus Latitude for the Andean volcanics. $\Delta 7/4$ and $\Delta 8/4$ are indices of ^{207}Pb and ^{208}Pb enrichment over the NHRL values of Figure 17.7 (see Rollinson, 1993, p. 240). Shaded areas are estimates for mantle and MORB isotopic ranges from Chapter 10. Symbols and data sources are given in Figure 17.6.

element characteristics correlate well with the major difference between the active zones: the thick Precambrian crust beneath the CVZ and Paleozoic crust beneath the northern SVZ. The Pb enrichments shown in the NVZ also reflect an

increase eastward (toward thicker and older crust), which does not show up on the basis of latitude in Figure 17.8. Similar enrichments away from the trench are also found in some volcanics in the SVZ and CVZ.

17.2.2 Petrogenesis of Andean Volcanic Rocks

Major element and trace element data indicate that Andean magmatism originates from the fluid-fluxed and LIL-enriched mantle wedge above the subducting and dehydrating Nazca plate in the same manner as island-arc magmas described in the previous chapter. Hydrous melts may occasionally be derived from the slab, particularly if the crust is young, such as near hot slab windows associated with subducted ridge segments (see Section 16.8.3). If we accept fluid-fluxed wedge melting as our initial model, we can then focus on how the Andes, and presumably other continental-arc magmas, develop differently subsequent to this initial stage. The good correlation between trace element and isotopic characteristics of the rocks in the three active Andean volcanic zones and the nature of the local crust is strong evidence for the interaction of continental-arc magmas with the crust through which they must pass en route to the surface. As mentioned in the beginning of this chapter, the exact nature of this crustal influence could have many forms, and there is considerable disagreement as to the relative importance of *source* variables (enriched subcontinental lithospheric mantle, depleted MORB-like mantle, variably enriched OIB-like mantle, and subducted components) and subsequent variables such as crustal melting or assimilation, magma mixing, and increased fractional crystallization that may occur in mafic magma chambers that pond at depth as they lose their buoyancy at or near the base of the less dense crust.

The similarity between Andean and island-arc magmas indicates that Andean melts *originate* in the mantle wedge. The arc-like nature of the crust in the NVZ and southern SVZ provides little geochemical or isotopic contrast with the initial melts, so crustal assimilation in these zones should leave little trace. The more evolved and enriched composition of igneous rocks in the CVS and northern SVZ, where the crust is older and thicker, however, must be the result of *crustal* interactions.

Unfortunately, there are simply too many variables to consider, and no unequivocal way of resolving or choosing between them to evaluate the nature of most crustal interaction. Given the number of poorly constrained assimilants, and the complexities within and above the subduction zone system, it is possible to create the same geochemical characteristics in a volcanic suite by more than one process. Considerable debate centers on the relative importance of the processes involved, but most investigators agree upon the field of possibilities. Debates involve rather detailed arguments concerning trends and correlations between trace elements, trace element ratios, and isotopic signatures among the various volcanic sequences. It is certainly beyond the scope or intent of this text to go into the details of models that have not yet been sufficiently universal or compelling to resolve the debate and provide a single model for Andean petrogenesis. I shall therefore (with relief) settle for a brief summary of some of the favored hypotheses of the day. Each of the hypotheses has geochemical support. Perhaps all of the ideas expressed below operate somewhere in continental-arc systems, with any single process dominating locally.

Hildreth and Moorbath (1988) studied a section of the SVZ from 33 to 37°S and concluded that the differences in along-arc enrichments in K_2O, Rb, Cs, Ba, Th, and LREEs reflect the local crustal country rocks. Increasing enrichment was particularly noticeable in the northern part of the zone where the country rocks are older and thicker, as described earlier. Further, they noted that the accompanying changes in $^{87}Sr/^{86}Sr$ and $\delta^{18}O$ were small enough to require a low-Sr, low-$\delta^{18}O$ source for the enrichment, which correlates best with a *deep*, depleted crustal source. Lack of HREE enrichment accompanying the increased LREE also implied that garnet was present in the crustal component, which further supported a deep interaction. Based on these and other geochemical criteria, they proposed that primary subduction zone magmas, which are less dense than the mantle but more dense than the crust, ponded at the mantle–crust boundary, where they incorporated significant quantities (tens of percent) of deep crustal material already at or near its solidus temperature. Incorporation involved a combined process of crustal melting and assimilation, plus storage, and homogenization (which they called **MASH**). The MASH process involved somewhat different crustal lithologies of varying age and depth along the length of the arc. Once the mafic melts became sufficiently enriched and silicic to regain buoyancy with respect to the deep crust, they rose to shallower magma chambers. These MASHed magmas provided the "base level" geochemical signature for the various eruptive centers. The ascending magmas may also have scavenged, or selectively incorporated, mid- to upper-crustal silicic to alkalic melts and wall-rock components during ascent, and have experienced shallow fractional crystallization, AFC, and/or magma mixing in open-system shallow chambers.

Stern (1991a,b) stressed the importance of a lack of any accretionary prism or sediments in the Peru–Chile trench and the truncation of Andean structures at the coast. He suggested that these features resulted from the scraping, removal, and subsequent subduction of the sediment and the upper plate lithosphere along the subduction zone contact by the subducting Nazca plate. This process is commonly known as **subduction erosion** (or **tectonic erosion**). The eastward migration of volcanic activity since the Paleozoic may also be related to the removal of the westernmost crust and accompanying migration of the subduction system to the east with respect to the upper plate. Stern (1991a,b) also noted that the dip angle of subduction decreases in the northern portion of the SVZ, which contacts a longer section of lithosphere along the subduction zone, and thus increases the potential for subduction erosion in this area. Sediment and slivers of continental crust, including deep crust, along the subduction zone may thus be scraped and incorporated into the magma source regions. Hickey-Vargas (1991), obviously a potato lover, referred to this alternative to MASHed hybrid magmas as

"**peeled**." Because both methods permit increased interaction of subduction zone magmas with deep crustal components, there is no unique geochemical characteristic that can distinguish between the two. Both can explain the increased crustal enrichment of the CVZ and northern SVZ in deep reservoirs. Both may also occur. I'm partial to peeling before mashing, but this has little to do with geology.

Rogers and Hawkesworth (1989) argued that shifts in $^{87}Sr/^{86}Sr$, $^{143}Nd/^{144}Nd$, and trace elements for the deeply developed "base-level" magmas can be explained by differences in the mantle wedge, involving old, late Proterozoic subcontinental mantle *lithosphere* (SCLM), with or without significant deep crustal involvement. This lithosphere would also be geographically associated with the thicker, older crust in the CVZ and northern SVZ. It is not possible at present to choose unambiguously from among these models because all of the observed rocks at the surface are further differentiated within the crust and no longer in equilibrium with the deep source.

As "base-level" magmas ascend they may further scavenge and assimilate mid and upper crustal components. Rogers and Hawkesworth (1989) thought they could distinguish an upper level evolutionary trend in the CVZ from the deep MASHed, peeled, or scalloped trend discussed above. In the deep trend, $^{87}Sr/^{86}Sr$ increases and $^{143}Nd/^{144}Nd$ decreases as an enriched component with subequal amounts of Sr and Nd is assimilated. In the shallow trend $^{87}Sr/^{86}Sr$ varies much more than $^{143}Nd/^{144}Nd$, suggesting assimilation of an upper crustal component with a high Sr content. This latter increase in $^{87}Sr/^{86}Sr$ is accompanied by a decrease in total Sr, implying that the assimilation of the high Sr crust was accompanied by Sr loss due to the fractional crystallization of plagioclase (an AFC process).

There is considerable variation in trends between individual volcanic centers. Wörner et al. (1992) and Aitcheson et al. (1995), for example, were able to use the varied patterns of Sr, Nd, Pb, and O isotopic assimilation to distinguish basement domains corresponding to fault-bounded crustal blocks in the CVZ Altiplano.

A few centers can easily be modeled by essentially closed-system fractional crystallization patterns, whereas most are modeled best via open-system behavior in which melts or assimilated material is introduced into a shallow magma chamber. Some open-system centers show both the curved pattern of combined assimilation and fractional crystallization (AFC) on variation diagrams plus some straight patterns considered characteristic of magma mixing (Davidson et al., 1988). We have learned, however, that mixing of mafic and silicic magmas can cause crystallization of the mafic magma, and also produce curved variation patterns. It is thus very difficult to distinguish AFC, in which the foreign material is assimilated, from a similar system in which the foreign material is a melt. The AFC algorithm would work for either situation. The common occurrence of mixed and disequilibrium phenocryst assemblages, reaction rims, and complex zoning/resorption patterns further indicates that magma mixing occurs in many "AFC" magmas. Some centers show little variation in isotopic character with differentiation indices, whereas others show more significant changes that also suggest AFC–magma mixing processes.

The compositional overlap between the dacite-rhyolite ignimbrites, so common in the CVZ and the other calc-alkaline rocks in the zone, led Thorpe and Francis (1979) to conclude that they were both probably of the same origin and the dacites and rhyolites were simply products of more advanced fractional crystallization (with some minor crustal assimilation). De Silva (1989), on the other hand, noted a late Miocene "ignimbrite flare-up" between latitudes 21 and 24°S in the CVZ, and attributed much of the ignimbrite activity to crustal fusion. The timing of the event corresponds to a period of crustal thickening in the area, and the major ignimbrite occurrences correlate to the thickest crustal sections, which would be most susceptible to partial fusion at the deep base of the crustal section. The ignimbrites also show elevated $^{87}Sr/^{86}Sr$ and low $^{143}Nd/^{144}Nd$, suggestive of a substantial crustal contribution.

Alkaline rocks (including shoshonites), on the eastern margin of the Andes resemble intra-continental magmatism (Chapter 19) and are considered to result more from convection in the back-arc region than from the main subduction zone process (Thorpe et al., 1982). The alkaline rocks lack the decoupled LIL/HFS character of subduction zone magmas and show the broad convex upward, or humped, pattern on MORB-normalized spider diagrams typical of intraplate mantle melts unmodified by fluid-transported slab components. The alkaline character may reflect an enriched intraplate subcontinental lithospheric mantle source, decreased partial melting (resulting perhaps from less H_2O behind the slab dehydration zone), or both. Stern et al. (1990) pointed out that there is commonly an eastward gradation from the calc-alkaline magmas to the alkaline types, reflecting perhaps a gradual decrease in the slab contribution (and perhaps an increase in enriched subcontinental lithospheric mantle).

In summary, the geochemical nature of Andean igneous rocks is sufficiently similar to that of island-arc rocks that we can reasonably conclude that they too have their main source in the depleted mantle wedge above the subducting oceanic slab. Significant differences, however, provide a clear indication that many Andean volcanics are modified by some process(es) unique to the continents. On average, they are more evolved and enriched than their island-arc equivalents. Figure 17.9 shows the relative frequency of analyzed volcanics from the Andes and the Southwest Pacific island arcs. Although sampling reflects collecting bias, the trends are probably a reasonable representation of the true distributions. Andesites dominate both areas, but basalts and basaltic andesites are more common in the island arcs, whereas dacites and rhyolites are more common in the Andes. In addition, the Andes are more enriched in LIL elements compared to island arcs, and they have more enriched isotopic signatures. Although these trends could be explained by variably enriched and heterogeneous mantle sources (including subcontinental lithospheric mantle) plus subducted components, the evidence reviewed above suggests that a considerable degree of crustal participation affects the magmas derived from these deep sources. Individual

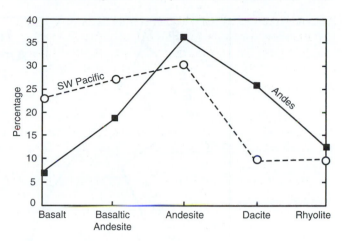

FIGURE 17.9 Relative frequency of rock types in the Andes versus SW Pacific Island arcs. Data from 397 Andean and 1484 southwestern Pacific analyses in Ewart (1982).

magmas can vary, but, collectively, magmas show the influence of local crust in their trace element and isotopic signatures. The sympathetic variation between several geochemical indicators of crustal influence, and the correlation of them with the character and thickness of the underlying country rock, even to the extent that we can infer specific crustal domains, is particularly compelling.

Of course there are many complexities and exceptions to the above generalizations. Every volcanic center in the Andes has unique aspects, and commonly some aspect fails to conform to the rather sweeping generalizations made above. Even whole continental-arc provinces have their own distinctive characteristics. As an example, I'll take you northward for a brief look at our own continental-arc province in the northwestern United States. The Cascade Range has many similarities to the Andes but provides some instructive contrasts as well.

17.3 THE CASCADES OF THE WESTERN UNITED STATES

The Cascade magmatic arc extends for about 1000 km and corresponds to the length of the Juan de Fuca plate system that is subducting obliquely beneath the North American plate (Figure 17.10). The Quaternary chain of large, relatively well-spaced andesitic stratovolcanoes extends from Lassen Peak in northern California to Mt. Meager in southern British Columbia. As I mentioned earlier, the Juan de Fuca and Cocos plates (Figure 17.1) are remnants of the larger Farallon plate that was subducted and produced more extensive igneous activity along a much longer stretch of the North and South American plate during much of the Phanerozoic (Atwater, 1970). To the north and south of the Juan de Fuca plate, the Pacific plate is in direct contact with the North American plate. The relative motion between these latter two plates is right-lateral strike-slip, so igneous activity is now absent along these sections. Basin and Range extension takes place behind the southern portion of the Cascade arc. This may be a back-arc spreading phenomenon, or it

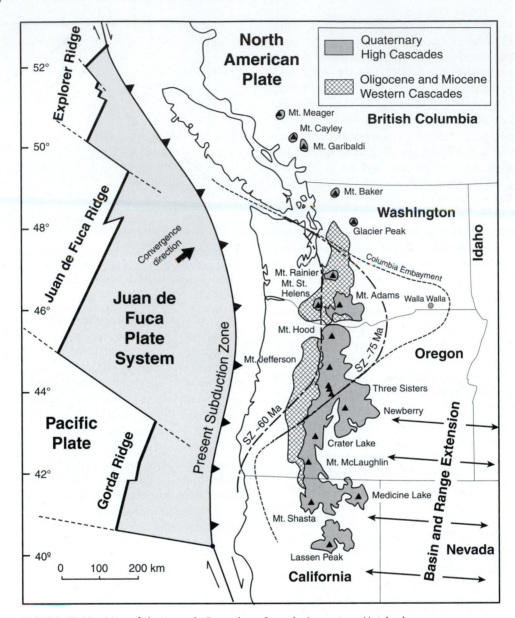

FIGURE 17.10 Map of the Juan de Fuca plate–Cascade Arc system. Hatched area = Oligocene–Miocene Western Cascades and shaded = Quaternary High Cascades (after McBirney and White, 1982). Also shown is the Columbia Embayment (the western margin of pre-Tertiary continental rocks) and approximate locations of the subduction zone as it migrated westward to its present location (after Hughes, 1990 © American Geophysical Union with permission). Due to sparse age constraints and extensive later volcanic cover, the location of the Columbia Embayment is only approximate (particularly along the southern half).

may instead relate to deep thermal instabilities associated with passage of the Pacific–Farallon ridge and/or the Yellowstone hotspot beneath the North American plate.

There is no trench associated with the Juan de Fuca–North American system, and seismicity is weak and only poorly defines a Wadati-Benioff zone. Most investigators attribute the diminished trench and seismicity to shelf sedimentation and/or the proximity of the ridge system, the latter causing the subducting plate to be thinner, warmer, and more ductile than for most subduction systems.

The impressive, dominantly andesitic cones, collectively referred to as the High Cascade Range (Figure 17.10), have existed for less than 1 Ma. The young cones, however,

sit atop a complex base that is mostly volcanic, revealing a long history of broadly subduction-related igneous activity extending back to the Paleozoic. Past activity has not been restricted to the location of the modern chain. The earlier Mesozoic chain is now largely eroded to its plutonic roots, and shows up well as a batholith belt in Figure 17.15a. This belt follows the Columbia Embayment, which defines the western edge of the pre-Tertiary continental crust (including some accreted arc-like terranes). In southern Washington and northern Oregon subduction migrated westward across the embayment from the Cretaceous plate boundary in Idaho to its present location, as shown in Figure 17.10. As a result, a sizable section of oceanic crust and mantle was accreted or trapped behind

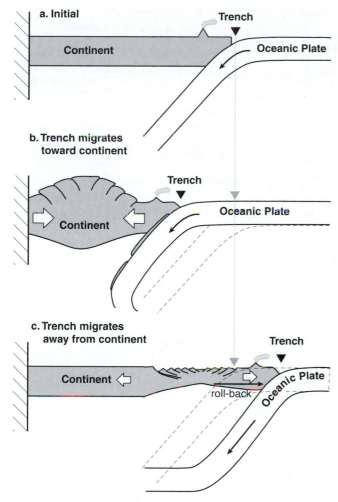

FIGURE 17.11 Schematic cross sections of a volcanic arc showing an initial state **(a)** followed by trench migration toward the continent **(b)**, resulting in a destructive boundary and subduction erosion of the overlying crust. Alternatively, trench migration away from the continent **(c)** results in extension and a constructive boundary. In this case, the extension in **(c)** is accomplished by "roll-back" of the subducting plate. An alternative (although less popular) method involves a jump of the subduction zone away from the continent, leaving a segment of oceanic crust (original dashed) on the left of the new trench.

the migrating subduction zone, as were various oceanic and arc volcanics. This type of continental-arc boundary has been called a **constructive boundary**, as it adds to the area behind the arc. This is in contrast to the situation in the Andes, a **destructive boundary** in which tectonic erosion scrapes away at the leading continental edge (Figure 17.11). Becker et al. (1999b) attempted to model these interactions, concluding that the rate of plate convergence, the density contrasts between the overriding and subducting plate, and the viscosity contrast between the subducting plate and the asthenospheric mantle controlled whether or not the lower plate rolled back or advanced. For example, older subducting plate is denser and should roll back, whereas young plates should compress the arc and erode the crust and wedge. North of Mt. Rainier, the basement is somewhat older, and the margin is not constructive. For the purposes of Cascade–Andean comparison,

we shall concentrate on the southern portion of the arc in Oregon and southern Washington.

Igneous activity in southwestern Washington and western Oregon began in the early Cenozoic. Radiometric data suggest that volcanism in the Western Cascades has been active continuously during the past 40 Ma (Verplanck and Duncan, 1987; Priest, 1990), but there appear to have been several pulses of increased activity. The early volcanism was predominantly oceanic (tholeiitic and island-arc like) in character, probably reflecting an incipient arc with a trapped oceanic section behind it in the Columbia Embayment. Calc-alkaline rocks appeared in the late Eocene, following a strong tectonic event. By early Oligocene times, a broad zone of calc-alkaline volcanoes, over 200 km wide, composed the Western Cascades (Figure 17.10), though some tholeiitic activity occurred in the westernmost portion.

During the late Miocene, 8 to 10 Ma ago, the High Cascades began to form. Activity began with the development of a predominantly mafic platform of coalescent shield volcanoes. The mafic magmatism appears to be related to extension, and a graben developed along the eastern side of the Western Cascades, along which early High Cascade activity concentrated. The large Quaternary stratovolcanoes of more evolved calc-alkaline nature were constructed as localized centers on this mafic platform in a very short period. Few of the cones have reversed magnetic polarity, suggesting that the bulk of the volcanoes were built in the 670,000 years since the last reversal. Radiometric ages suggest that most are less than 200,000 years old.

Erosion and cover by later deposits makes estimating the volume and rate of Cascade volcanism difficult, but most estimates agree in broad terms. Figure 17.12 shows the estimates for Central Oregon by Priest (1990), who related the decrease in volcanism from 35 to 7.5 Ma to a reduced convergence rate. The subsequent increase in volcanism, in spite of further reductions in convergence rate, he related to E-W extension, which faulted the crust of the upper plate, allowing easier surface access for the magma. This extension has been related to various combinations of back-arc spreading, migration of Basin and Range extension into the area, decreased convergence, roll-back of the subducting plate (causing migration of the southern portion of the trench westward, Figure 17.11c), and thermal weakening of the crust by arc volcanism.

Figure 17.12 is based on incomplete and selective sampling, but seems to indicate that the amount of true andesite has been rather minor in the Cascades during the last 35 Ma. Even in the High Cascade volcanoes, where andesites comprise a major proportion of the cones, the proportion of basalts can be as much as 85% when we include the associated mafic shields. The amount of basalt and basaltic andesite is high throughout the Cenozoic history of the Cascade arc. The high mafic component is a major difference between the Cascade and Andean arcs.

Hughes (1990) and Leeman et al. (1990) have analyzed numerous samples of late Cenozoic mafic lavas and found that they vary from light-REE-depleted MORB-like tholeiitic basalts, to variably enriched basalts that chemically resemble

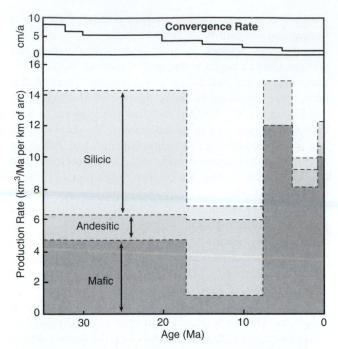

FIGURE 17.12 Time-averaged rates of extrusion of mafic (basalt and basaltic andesite), andesitic, and silicic (dacite and rhyolite) volcanics (Priest, 1990) and Juan de Fuca–North American plate convergence rates (Verplanck and Duncan, 1987) for the past 35 Ma. The volcanics are poorly exposed and sampled, so the timing should be considered tentative.

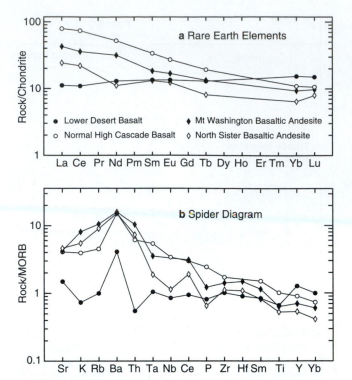

FIGURE 17.13 (a) Rare earth element and (b) multielement (spider) diagrams for mafic platform lavas of the High Cascades. Data from Hughes (1990).

ocean island basalts (OIBs), to basalts and basaltic andesites with high LIL/HFS characteristic of subduction zone fluid alteration (Figure 17.13). Hughes (1990) related the MORB-OIB characteristics to the existence of heterogeneous oceanic mantle (with MORB and OIB parent domains) in the oceanic wedge within the Columbia Embayment, and attributed the variable LIL/HFS pattern results to secondary fluid additions from the subducting slab into these domains. Leeman et al. (1990) noted that the high LIL/HFS types are rare in southern Washington, and attributed this to the young, hot subducting slab that dehydrates before it reaches a depth where the fluids could cause melting of the overlying mantle wedge. Melting in this region may result less from hydration effects than from convective upward mantle flow and decompression melting in the wedge under a thin, extending, overriding crust.

The high proportion of mafic magma in the Cascades may also be related to extension and the resulting normal faults (plus right-lateral strike-slip faults) in the area, which provide conduits for the mafic magmas to reach the surface without ponding at depth where they would fractionate or mix with crustal components. We should note in Figure 17.12, however, that silicic magmas, such as dacites and rhyolites, may be more common than andesites. Such **bimodal** mafic–silicic volcanism (usually basalt–rhyolite) is fairly common in continental volcanic provinces. As mentioned in previous chapters, the basaltic component is attributed to partial melting of the mantle, but the high proportion of silicic magmas and subordinate intermediate ones casts doubt on their origin by fractional crystallization, which should produce successively diminished quantities of more evolved liquids. Rather, the silicic magmas are considered to be *crustal*

melts, created by hot, ponded mafic magmas underplating the base of the continental crust (Huppert and Sparks, 1988a). If the mafic melts are underplated, the fault conduits appear to be less effective.

Most of the Quaternary Cascade volcanoes exhibit a range of compositions from basalts and/or basaltic andesites to dacites, but the high cones are dominated by andesite. Trends on variation diagrams are typically smooth. Remember that our first examples of Harker diagrams (Figure 8.2) and AFM diagrams (Figure 8.3) were for Mt. Mazama/Crater Lake, a Cascade volcano. Many andesites and more evolved magmas can be modeled as being created by fractional crystallization from basalt, but trends are temporally less coherent than a simple progression from mafic to more evolved magmas through time. Multiple magma chambers and recharge of variable mafic components, or mixing of silicic and derivative melts, are generally required. Some cones, such as Lassen Peak (Clynne, 1990; Bullen and Clynne, 1990), Mt. St. Helens (Smith and Leeman, 1993; Gardner et al. 1995), and Medicine Lake (Condie and Hayslip, 1975), are best modeled by mixing of two magmas, one basaltic and the other dacitic (probably a partial melt of metabasaltic sub-arc crust). Trace element concentrations and ratios commonly rule out crystal fractionation or AFC models. Bimodal mafic-silicic volcanism may thus be manifested at single centers, as well as regionally. Magma mixing, either by recharge of more primitive magmas into evolving chambers, exchange between variably evolved chambers from similar parental magmas, or mixing of magmas from distinct sources, seems to be a major process at High Cascade centers, and may be a more common phenomenon in many volcanic arcs than we have realized.

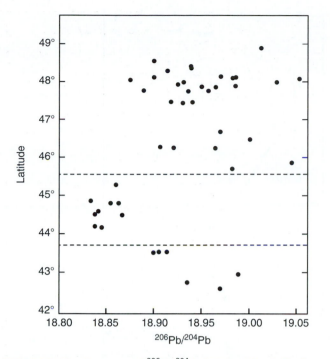

FIGURE 17.14 Summary of $^{206}Pb/^{204}Pb$ from sulfides in Tertiary Cascade intrusives as a function of latitude. After Church et al. (1986). Copyright © with permission from Elsevier Science.

The contrast of crustal types in the upper plate, with accreted oceanic crust in the Columbia Embayment and continental crust to the north and south, should provide an excellent place to further assess the effects of assimilation and crustal contamination of rising mantle-derived melts. Isotopic data are surprisingly sparse for the Cascade volcanoes. The most comprehensive published data come from Pb isotopes from lead sulfide minerals in Tertiary plutons (presumably volcanic equivalents). Figure 17.14 shows enriched Pb isotopic signatures north of 45°30′ and south of 43°40′, with less enriched ratios between these boundaries. This pattern is compatible with Mesozoic and older continental-type crust in Washington and southern Oregon, and younger mafic and intermediate rocks in the embayment. The southern boundary of the isotopically defined embayment in Figure 17.14 agrees with the embayment illustrated in Figure 17.10. The northern boundary in Figure 17.14, however, is closer to the Oregon–Washington boundary, and is considerably farther south than the boundary in Figure 17.10. Because of the extensive volcanic cover in southern and central Washington, it is difficult to locate the exact boundary of the Columbia Embayment. Perhaps Figure 17.14 provides a more accurate location than the classical boundary illustrated in Figure 17.10, and the embayment may be limited to northwestern Oregon.

In summary, the Cascades have many of the characteristics of continental-arc magmatism that we encountered in the Andes. The Cascades differ from the Andes by having a much larger proportion of mafic lavas, and less evidence of assimilation of isotopically enriched crust. We might relate these characteristics to the notion that the Cascades represent a *constructive* margin with accretion of oceanic and arc terranes, whereas the Andes are more of a *destructive* margin where the outer edge of the continental crust is being

scraped away. Thus the crust of the upper plate in the Andes (at least in the CVZ) is Precambrian, whereas the Cascade crust is much younger and more oceanic. There is also more extensional and strike-slip faulting in the Cascades, which could provide more conduits through the crust. The Cascades, then, are transitional to island arcs, but still show many mature continental-arc-type geochemical trends. It is interesting to note that in the southernmost Andes (the Austral Volcanic Zone), which has a less mature crust and more strike-slip component with slower convergence, there is also a significantly greater mafic component to the magmatism.

The volcanics, however, are only part of the story of continental arcs. Much of the subduction zone magma never reaches the surface because the density of the rising magmas may not be significantly less than the thick continental crust through which they must pass. Thus the driving buoyancy forces are reduced and the magmas may crystallize before reaching the surface. Hydrous magmas are also more likely to solidify at depth. Thus are created the large batholith belts. Let's take a brief look at these occurrences in order to complete the picture of orogenic magmatism.

17.4 PLUTONIC BELTS OF CONTINENTAL ARCS

An enormous quantity of mostly Mesozoic and Tertiary intrusive rock is exposed along the American Cordillera (Figure 17.15). The spatial and temporal association of these plutons with volcanics such as those described above strongly implies that both are expressions of the same magmatic and orogenic activity. Indeed, the term **Cordilleran-type batholith** is commonly used to denote any large composite plutons created in mountain belts at a leading plate edge.

These great Cordilleran batholiths are *composite* bodies, consisting of hundreds or even thousands of individual intrusions which may span 10^7 to 10^8 years. The range from basalts to rhyolites in coeval volcanics is matched by the plutonic sequence: gabbro–diorite–tonalite–granodiorite–granite. In the Coastal batholith of Peru, for example, gabbro and diorite comprise 7 to 16% of the overall abundance, tonalite 48 to 60%, granodiorite 20 to 30%, and true granite (as defined in Figure 2.2) 1 to 4% (Pitcher, 1978). Thus the habit of referring to the rocks of these batholiths collectively as "granites" is certainly a misnomer, and I find preferable the term **granitoid**, which may be loosely applied to a spectrum of coarse-grained felsic rocks, from tonalites to syenites.

Their geochemical similarity to the volcanics suggests that the plutonics are simply the exposed plumbing system feeding continental-arc volcanics. Where volcanism is active, there is a net transfer of material to the surface. This deflates the crust, and a carapace of lavas and pyroclastics is continually developed that generally keeps pace with erosional processes. But where volcanism has ceased, erosion and uplift of the thickened crust expose the plutonic components. Plutons are thus most commonly seen in arc segments where volcanism is no longer active. Reasons for the waning of volcanism may be a trench-ward migration of subduction (Idaho batholith), decreased subduction rate, or transfer to

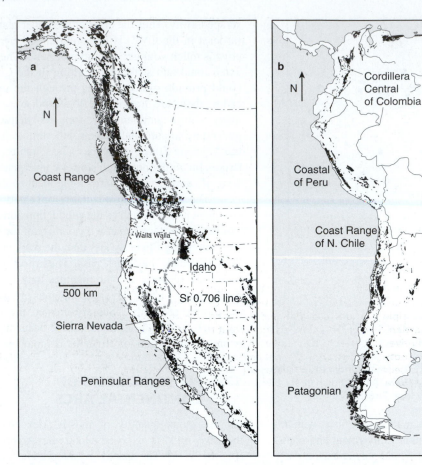

FIGURE 17.15   Major plutons of the North American **(a)** and South American **(b)** cordilleras, the principal segments of a continuous Mesozoic–Tertiary belt from the Aleutians to Antarctica. **(a)** From the Geologic Map of North America, kindly supplied by the Geological Society of America and the United States Geological Survey. Copyright © The Geological Society of America, Inc. **(b)** From Schenk et al. (1999). The Sr 0.706 line in North America is after Kistler (1990), Miller and Barton (1990), and Armstrong (1988). Major composite batholith belts are labeled.

strike-slip motion (Sierra Nevada, Coast, and Peninsular Range batholiths of North America), or decreasing angle of subduction (Peru and Chile batholiths, which occupy the gap between active volcanic CVZ and SVZ segments of South America). Batholiths are a major feature of continental-arc magmatism, and the exposed volume of intrusive rocks in the Cordillera exceeds that of the volcanics by more than an order of magnitude. Field relationships of plutonic bodies, most of which apply to these batholiths, were discussed in Chapter 4.

The Cordilleran-type batholiths are commonly linear arrays of plutons. Pitcher (1993) reviewed the evidence relating the batholiths of South America (particularly the Coastal batholith of Peru) to the development of linear Mesozoic block-faulted marginal basins in a back-arc-like setting some 200 km inland from the trench. He suggested that they were created in a pre-Cretaceous extensional regime that ended with a global increase in sea-floor spreading rates. The thinning of the crust during extension and complementary up-arching of the enriched subcontinental lithospheric mantle presumably led to the decompression melting of the latter. Early dike-like gabbros intruded the sedimentary and mafic volcanic fill. The more mafic character befits the thin crust and extensional regime where numerous faults provided conduits to the surface.

The transition from the mafic-extensional phase to compression brought huge upwellings of silicic magmas to South America, concentrated along the axes of the extensional marginal basins. First came tremendous volumes of tonalites and granodiorites, followed by lesser short-lived pulses of increasing felsic character. Mafic magma was still present, as recorded by numerous syn-plutonic mafic dike swarms. Repeated alternation of extension and compression occurred throughout the Mesozoic, presumably in response to fluctuations in global spreading rates. Extensional periods probably promoted upwelling and extrusion, whereas compressional periods inhibited magmatic rise, providing sufficient crustal residence time for magma ponding and fractionation. Rise of later melts presumably concentrated along the hot, weakened pathway of the extension-related melts, producing the linear composite batholiths.

Extensive detailed mapping of the 1600 km long by 65 km wide Peruvian Coastal batholith showed that there were about 1000 individual plutons. Separate surges of magma commonly produced cross-cutting relationships within plutons (see Figure 4.32). In Peru the constituent rock types of the myriad plutons is limited, and the rocks can be grouped naturally into a relatively small number of **units**, readily recognized in the field on the basis of similar mode and texture.

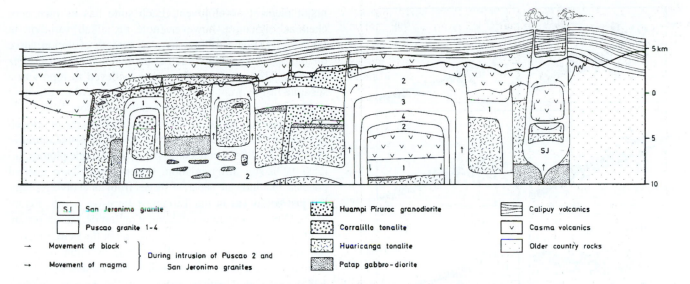

San Jeronimo granite — S.J

Puscao granite 1-4

→ Movement of block
→ Movement of magma
} During intrusion of Puscao 2 and San Jeronimo granites

Huampi Piruroc granodiorite

Corralillo tonalite

Huaricanga tonalite

Patap gabbro-diorite

Calipuy volcanics

Casma volcanics

Older country rocks

FIGURE 17.16   Schematic cross section of the Coastal batholith of Peru. The shallow flat-topped and steep-sided "bell jar"-shaped plutons are stoped into place. Successive pulses may be nested at a single locality. The heavy line is the present erosion surface. From Myers (1975). Copyright © The Geological Society of America, Inc.

Each unit can be recognized as the constituent rock of numerous separate plutons. They are most commonly represented by individual intrusions separated by contacts from other units, and it seems likely that each represents a single magmatic pulse into many rising lobes over a broad area (Pitcher, 1985).

In Peru a unit usually occurs in close temporal and spatial association with a few other units, together forming 12 to 13 well-defined consanguineous **super-units**. Similar associations were noticed in other batholith belts and were called by various authors **suites**, **intrusive suites**, or **sequences**. They show distinctive mineralogical, geochemical, and textural patterns that are evident even over a range of composition. The units within a suite represent a limited span of ages, and the suites are considered to represent single deep magmatic fusion events, typically at intervals of 10 to 20 Ma, that are then differentiated at shallow levels. Investigators in several batholith belts have commented on how easily recognizable these suites were, even in plutons separated by hundreds of kilometers. Can it be that such huge volumes are created at once, or are the sources and generation processes sufficiently similar that nearly identical smaller pulses are created throughout an area? The Peruvian Coastal batholith has also been divided into three to five **segments**, each of which is characterized by a unique assemblage of suites (Cobbing et al., 1977). Perhaps these segments are related to structural and geometric discontinuities in the subducted plate, much like the segments of the present volcanic arc.

Figure 17.16 is a schematic cross section of the Peruvian Coastal batholith, showing how the granitoid magmas rose to, and froze at, a similarly shallow sub-volcanic level in the crust. Sufficient relief now exists to get a reasonable three-dimensional perspective. Individual plutons commonly have a flat-topped and steep-sided ("bell jar") shape. In map view, they are commonly rectilinear, with flat sides parallel to the regional fracture patterns (considered at least partially induced by the rising magma). Wall rocks along the steep sides show little distortion. These features, and the common

occurrence of ring dikes, indicate that the plutons were emplaced at these shallow levels principally by **cauldron subsidence** (stoping of huge fault-bounded crustal blocks), perhaps aided by roof-uplift. The Peruvian and several other batholiths apparently intruded their own associated volcanic ejecta, indicating that they reached the pre-eruptive surface of the Earth and are practically giant viscous flows!

Plutons are commonly emplaced in nested **complexes**, spaced at regular intervals of about 120 km. This spacing is reminiscent of the spacing of volcanic centers discussed previously (Section 16.1) and may reflect the spacing of gravitationally-induced melt diapirs at the source. Deeper erosion in other batholiths exposes the mesozone and catazone where ductility of the country rock increases, intrusive contacts are less distinct, and mechanisms are more diverse (see Section 4.2.5). As the depth of intrusion increases, it becomes more and more difficult to recognize large, discrete plutonic bodies. It is fascinating to speculate as to whether this is the result of less contrast between the igneous and metamorphic rocks at depth, or whether perhaps the bulk of the low-density magma rises to shallow levels, leaving mostly dikes and pods in the deflated crust in its wake.

17.4.1 Geochemistry of the Peruvian Coastal Batholith

We shall address the petrology of granitoid rocks in general in Chapter 18. The geochemistry, however, has bearing on our present endeavor because it addresses both the nature of plutonism and the link between plutonism and volcanism along continental arcs. There is ample opportunity for crystal accumulation in slow-cooled plutons, and many investigators have questioned whether plutonic rocks represent true liquid compositions. The geochemical composition of the plutons in continental arcs, however, corresponds closely to that of the associated volcanics. This and the absence of signs of crystal accumulation within many plutons provide strong

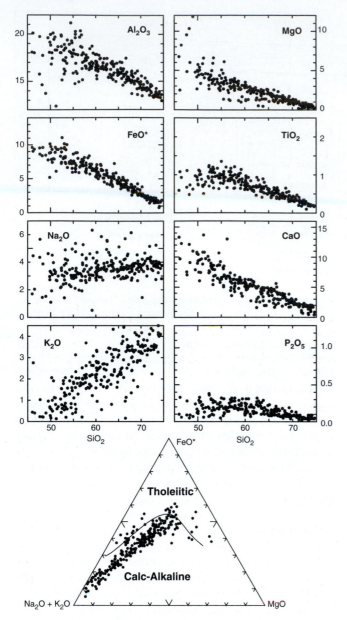

FIGURE 17.17 Harker-type and AFM variation diagrams for the Coastal batholith of Peru. Data span several suites from Pitcher et al. (1985).

major element development. Each suite has its own geochemical signature, however (note the alkali variation in Figure 17.17), and follows its own smooth evolution curve on appropriate diagrams, such as the residual Ab-An-Or-Qtz diagram. Thus, again, each suite is believed to represent a separate fusion event and a mafic-to-silicic rhythm reflecting shallow differentiation (Atherton and Sanderson, 1985).

The Peruvian suite evolutionary rhythms are developed in a wide variety of circumstances in the field. Some intrusive centers show diffuse concentric zonation in single plutons, indicating in situ differentiation. Others show sharp internal contacts that indicate separate surges from an evolving chamber just below (as in the Tuolumne Intrusive Series, Figure 4.32). Other suites show differentiation trends, but pulses are scattered widely among separate plutons in the field, suggesting deeper and larger parental chambers. Magma mixing of various pulses is common as well (Atherton and Sanderson, 1985). Complete chemical analyses of the Andean plutonics from the appendix of Pitcher et al. (1985) are in the Excel spreadsheet AndesPlut.xls at www.prenhall.com/winter. Interested students can compare plutonic and volcanic geochemistry further using the spreadsheet data.

Trace element data for granitoids are complicated by the accessory phases commonly found in felsic rocks. Several such phases act as sinks for a variety of otherwise incompatible trace elements. These accessories may be selective in the trace elements they concentrate, and the bulk D values of the rocks may be substantially affected. In spite of this, the trace element data are similar to the associated volcanics, suggesting that, in most cases, this effect is minor. Figure 17.18 compares the REE data for the Coastal batholith of Peru with the related volcanics in the area, which show how closely correlated some trace element patterns can be. Atherton and Sanderson (1985) modeled both the major element and trace element evolution of the Coastal batholith. They found that the best model was that of an original basaltic magma derived from an enriched mantle source (enriched subcontinental lithospheric mantle or fluid-modified mantle wedge) giving rise to the gabbros that differentiate at depth (fractionating olivine and clinopyroxene) to form the ubiquitous tonalites. Alternatively, tonalites may be derived by partial fusion of ponded gabbros that underplate the crust (Cobbing and Pitcher, 1983). The tonalites then fractionate at higher levels in the crust to produce the super-unit trends discussed above. We shall return to these contrasting ideas shortly.

Isotopic signatures of the plutons also exhibit the same range of values as the volcanics. The Sr data for the Coastal batholith of Peru are shown in Figure 17.19. The Lima segment is intruded into younger and thinner crust, and the initial $^{87}Sr/^{86}Sr$ ratios reflect the mantle-derived parental magmas. The Arequipa and Toquepala segments are intruded into the older cratonal rocks of the Arequipa massif, where a limited number of plutons show elevated initial $^{87}Sr/^{86}Sr$ ratios (although only in a few selective samples), probably caused by assimilation or melting of old, high $^{87}Sr/^{86}Sr$ crust (Figure 17.19a). The Pb data of Mukasa and Tilton (1984) are shown in Figure 17.19b. The Lima segment, (with no old basement exposed) probably reflects the Pb isotopic character of the

circumstantial evidence that the geochemical composition of the rocks approaches that of the intruded liquids. Rather than repeat the volcanic geochemistry, I shall mention only a few additional points relevant to the intrusives.

The chemical composition of the plutons in the Coastal batholith of Peru is a representative example for geochemical purposes. The major element data follows smooth trends on Harker-type variation diagrams (Figure 17.17), that are consistent with fractional crystallization of plagioclase and pyroxene ± magnetite, later giving way to hornblende and biotite, from initial gabbroic, tonalitic, or quartz dioritic parental material. Similarly, the AFM diagram in Figure 17.17 shows a clear calc-alkaline magmatic evolution trend so common in arc magmas. Both trends correspond well with the associated volcanics. Figure 17.17 is a composite for all the Peruvian suites, showing that they all follow a common

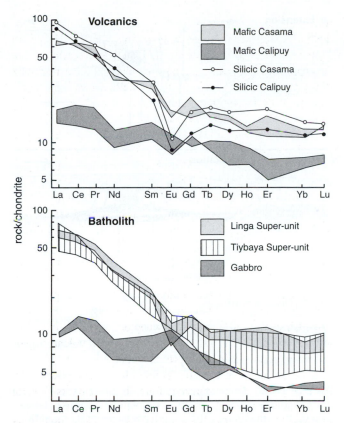

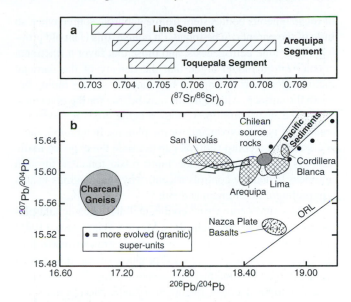

FIGURE 17.18 Chondrite-normalized REE abundances for the Linga and Tiybaya super-units of the Coastal batholith of Peru and associated volcanics. From Atherton et al. (1979).

FIGURE 17.19 (a) Initial $^{87}Sr/^{86}Sr$ ranges for three principal segments of the Coastal batholith of Peru (after Beckinsale et al., 1985). (b) $^{207}Pb/^{204}Pb$ versus $^{206}Pb/^{204}Pb$ data for the plutons (after Mukasa and Tilton, 1984). ORL = ocean regression line for depleted mantle sources (similar to oceanic crust).

mantle source, which, because it lies well above the regression line for depleted mantle sources and Nazca plate basalts, is probably an enriched mantle source such as subcontinental lithospheric mantle. The Sr and Pb isotopes for *Chilean* volcanics and intrusives are uniform and indicate a homogeneous enriched sub-Andean SCLM reservoir (labeled "Chilean source rocks" in Figure 17.19b). The Lima data are close to the Chilean data and suggest a similar source. The slightly higher $^{206}Pb/^{204}Pb$ and lower $^{207}Pb/^{204}Pb$ in the Lima segment are attributed by Mukasa and Tilton (1984) to assimilation of small quantities of shallow and highly radiogenic Jurassic and Cretaceous basin-fill sediments. They also explained the trend of the highly silicic intrusives (black dots in Figure 17.19b) and the Cordillera Blanca granites to assimilation of these sediments as well. The Arequipa and San Nicolás segments require a third component. The Pb data for these segments exhibit a near-linear trend (arrow in Figure 17.19b) toward the values for the lower crustal Charcani gneiss of the Arequipa massif, indicating that assimilation of lower crustal material by the rising magma also took place.

These geochemical examples are but a small portion of the available data. As we have seen, geochemical variation *along* the axis of the intrusives, like the volcanics, can be related to variations in subducted plate geometry, mantle source components, and/or the nature of the upper plate crust. There are also cases in which the chemical composition of the plutons varies systematically *across* the axes of many batholith belts. We shall shortly turn to some examples of such cases because consistent across-arc trends may shed some light on

the geometry and nature of the continental margin and subduction zone components, and how these might affect the generated magmas. But first I digress for a moment to consider whether the geochemical similarity of the volcanics and plutonics necessarily requires that the intrusives are really just the exposed conduits feeding the volcanism.

17.4.2 Volcanic/Plutonic Equivalence

The complete overlap and common correlation of major element, trace element, and isotopic geochemistry for intrusive and extrusive rocks in an area supports the contention that all of the rocks are manifestations of the same plate margin subduction processes, and may simply represent different levels of emplacement and erosion. Is this true?

In Chapter 16, we discussed a model in which the full spectrum of island-arc volcanics could be evolved products of primitive mantle melts. Earlier in this chapter, I concluded that the Andean volcanics indicate that continental-arc volcanics could also be created by partial melting of the mantle wedge above the subducted slab, followed by evolutionary processes that include more advanced fractional crystallization and assimilation or melting of sialic continental crust to produce a spectrum of rocks that is typically more silicic and enriched than in island arcs.

The chemical composition and isotopic signatures of the Peru–Chile intrusives show complete overlap with the volcanics, but are, on average, more evolved and felsic. Pitcher (1993) noted that gabbros and quartz diorites are present in the Coastal batholith of Peru, but these mafic magmas are mostly early phases. The huge volume of tonalite, plus the subordinate granodiorite and granite, occur with only a minor coeval mafic counterpart. One would expect about nine parts gabbroic parent to one part granitoid produced via fractional crystallization, so it is hard to explain how such great quantities of felsic igneous bodies stem from a gabbroic source when so little of the parent can be found. Of course, this is an old problem for granite petrogenesis, and those that favor a fractional crystallization model argue that the density of the parental melts stalled their ascent, and that they are simply hiding beneath the surface. Although this may be true for the volcanics, where the distribution of various magma types is more reasonable, it is not convincing for the plutons. In addition, there is commonly a compositional gap between the exposed mafic gabbros and diorites and the more felsic tonalites through granites. If density controls ascent, why did magmas in this gap preferentially remain at depth?

In a study comparing continental-arc volcanic and plutonic rocks in western Chihuahua, Mexico, Bagby et al. (1981) concluded that the volcanic andesites and intrusive quartz diorites may both be the fractionated products of partially melted mantle peridotites, and that the dacites and rhyolites were produced by more extensive fractional crystallization from the andesites. The granodiorites, however, could not be related to the andesites or the quartz diorites by fractionation of their liquidus phases: pyroxene and plagioclase. Bagby et al. (1981) concluded that the granodiorites must have been products of partial melting of a hornblende-bearing crustal rock.

Now many investigators in batholith belts agree that, although mafic intrusives are still derived from the mantle wedge, the great volume of tonalites are more likely to be the product of fusion of basaltic/gabbroic magmas that pond and crystallize at the base of the crust (underplating). Numerous experimental studies indicate that tonalitic melts can be produced by partial fusion of basaltic rocks under hydrous conditions (Green and Ringwood, 1968; Holloway and Burnham, 1972; Helz, 1973, 1976; Stern and Wyllie, 1978, Johannes and Holtz, 1996; Wyllie et al., 1997). Cobbing and Pitcher (1983) provided a model (Figure 17.20) for the generation of granitoid magmas from a mafic crustal underplate. Figure 17.20a shows the up-arched mantle during the extensional phase of Andean magmatism (Section 17.4) which results in decompression partial melting of the hydrated subcontinental lithospheric mantle. Extension-related faults allow the gabbroic magma to rise into the back-arc basin area. Because of its density, most of the gabbro accumulates at the base of the crust. Here it may undergo the MASH process, and some may solidify. Compression then terminates the normal faulting and thickens the crust, restricting the upward access so that later magma adds to the underplate. During the compression and later relaxation stages, crustal thickening and heat supplied by added underplating magmas remelts some of the earlier underplate to produce tonalites, which are less dense and rise toward the surface to solidify as huge, commonly homogenous tonalitic plutons. Shallow differentiation (fractional crystallization, assimilation of country rock, etc.) produces the lesser amounts of granodiorites and true granites. Whether the underplated mafic magmas are the result of up-arched mantle in a back-arc regime, or more typical subduction zone wedge melting is an open question. One might also question whether large regional underplates are necessary. Perhaps a series of smaller plutons and sill or dike complexes may have the same effect.

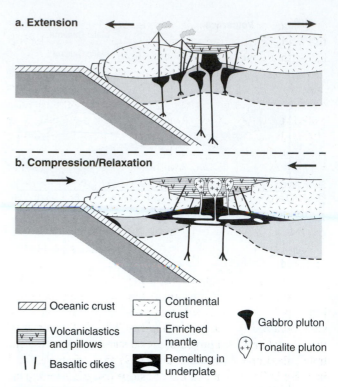

FIGURE 17.20 Schematic diagram illustrating **(a)** the formation of a gabbroic crustal underplate at a continental arc and **(b)** the remelting of the underplate to generate tonalitic plutons. After Cobbing and Pitcher (1983). Copyright © The Geological Society of America, Inc.

If the remelting model is true for most batholith belts, then the plutons (at least the more felsic ones) may *not* be true intrusive equivalents of the prevalent continental-arc volcanics. The source of both is ultimately the same mantle wedge peridotite, but whereas the volcanics have generally been attributed to differentiated *single-stage* mantle melts, the granitoid plutons may result from *two-stage* melting. Of course, it would be silly to assume that *all* of the felsic plutons and *all* of the volcanics strictly followed these separate trends. Certainly there are plutons beneath all volcanoes, and volcanics escape from many plutons. Because the process of crystallization at the base of the crust and remelting are nearly exact opposites, the major and trace element characteristics of the magmas produced by the single- and two-stage processes should be practically indistinguishable, making it hard to substantiate the models on a geochemical basis. Nonetheless, the localization of many batholiths behind the volcanic front, and the arguments based on relative quantities of the plutonic magma types, make the two-stage tonalite model appealing. It is certainly an intriguing proposition, but it is difficult for me to see why processes should not be similar in volcanic and plutonic terranes, and perhaps both single-stage and two-stage melts are generated in each.

17.4.3 Across-Axis Batholith Variations

In the same way that across-axis trends (such as the K-*h* relationship) occur in some volcanic belts, interesting trends also develop in some composite batholiths. Many batholiths show no systematic trends, whereas others show trends other than, or even opposite to, those discussed below. What follows is a brief summary of some of the more well-known batholithic asymmetries.

In an early statement on compositional zoning among intrusives Lindgren (1915) noted that the early Mesozoic plutons of the North American Cordillera were ultramafic to mafic to the west and became more felsic to the east. Buddington (1927) showed that the plutons of the Alaskan Coast Range could be arranged in parallel belts that became progressively richer in SiO_2 and K_2O eastward, away from the trench, and noted the same trends in the Sierra Nevada. Moore (1959) synthesized the available data in the western United States, and distinguished a western zone in which the dominant intrusive rock is a quartz diorite from an eastern zone in which granodiorite and quartz monzonite are more common. He proposed a **quartz diorite line** separating these two zones. Moore et al. (1963) extended the quartz diorite line to and across Alaska.

In Peru there is a decrease in mafic components and increase in SiO_2 between the Coastal batholith and the Cordillera Blanca batholith to the east, but most traverses of the Coastal batholith proper show no such variation. In central Chile, López-Escobar et al. (1979) showed an eastward increase in Sr content and La/Yb (slope of REEs) for the intrusives. The Sierra Nevada batholith is rather complicated, but west-to-east increases in SiO_2, K_2O, $Al_2O_3/(K_2O + Na_2O + CaO)$, U, Th, Be, Rb, oxidation, and $(F/OH)_{biotite}$ and decreases in CaO, FeO, and MgO have been described for various segments (Bateman and Dodge, 1970; Bateman, 1988; Ague and Brimhall, 1988).

Another interesting, though not universal, trend is a *temporal* one. Many batholith belts show a decrease in age inward from the trench, at least during some stages of their history. Figure 17.21a shows the relationship between age and distance from the trench for the Coastal to eastern Cordillera Blanca batholiths of Peru. Likewise the plutonic and volcanic belts of northern and central Chile show an eastward migration during much of the Mesozoic and Cenozoic (Pitcher, 1993, p. 212). In the Peninsular Ranges batholith there is an eastward migration (Figure 17.21b), but it appears to be more of a jump from a static western arc to a static or easterly migrating eastern arc (Walawender et al., 1990).

The Sierra Nevada batholith is more complicated in having been intruded over a considerable time span as a set of superimposed Triassic, Jurassic, and Cretaceous belts with slightly different geographic trends (Armstrong and Ward, 1991). Triassic intrusives are less common, and occur sporadically on the eastern side of the belt. The locus of intrusion jumps westward in the Jurassic and is crossed by the Cretaceous belt at a low angle. There appears to be a more systematic decrease in age eastward within the Cretaceous belt. Similarly, the Coast Range belt of British Columbia appears to exhibit an eastward younging that is repeated in several belts.

A decrease in intrusive ages away from the trench appears to be a common feature of many composite batholiths, but the trend applies better to single belts of shorter duration than to long-term trends for more enduring arcs. Trenchward jumps are not uncommon. Such is the case for our familiar South American example, where an early Paleozoic

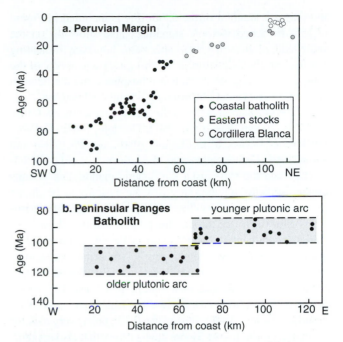

FIGURE 17.21   Isotopic age versus distance across **(a)** the Western Cordillera of Peru (Cobbing and Pitcher, 1983) and **(b)** the Peninsular Ranges batholith of southern California/Baja Mexico (Walawander et al., 1990).

intrusive arc (commonly called "pre-Andean") is exposed to the east of the main Mesozoic Andean arc.

If we address the larger North American Cordillera, we face a peculiar eastward extension of igneous activity some 600 km inland during the Laramide. The eastward shift is commonly attributed to a lower dip of subduction, for which there is some geophysical support (Miller and Barton, 1990). Contemporary igneous activity at the coast, however, suggests that the eastern Laramide activity represents more of a broadening of the belt than a shift. Some investigators have proposed that tectonic erosion and/or underplating may have increased the heat flux from the mantle to the crust, or that crustal thickening had a heating and insulating effect that may have resulted in crustal melting (Chapter 18).

As we might expect for continental-arc magmatism traversing the edge of a continent, isotopic ratios show perhaps the most obvious and consistent variations across the marginal batholith belts. Virtually all belts show correlated isotopic enrichment patterns inland from the trench, including increased initial $^{87}Sr/^{86}Sr$, decreased $^{143}Nd/^{144}Nd$, elevated $\delta^{18}O$, and enriched Pb isotopes. In Chile, initial $^{87}Sr/^{86}Sr$ increases from 0.702 in the west to 0.708 eastward (Aguirre, 1983). A similar increase from 0.704 to 0.708 is observed for the Sierra Nevada batholith (Kistler and Peterman, 1973, 1978). The eastward increase in initial $^{87}Sr/^{86}Sr$ was so consistent and obvious in the Cordillera of North America that isopleths of constant $(^{87}Sr/^{86}Sr)_0$ were drawn to contour the western margin of the continent. As discussed previously, this increase has been interpreted as the result of assimilation of old continental crust by transiting magmas, and the value of 0.706 is a common dividing line between largely unmodified mantle-derived melts, and those that have experienced appreciable cratonal contamination (Section 9.7.2.2). The approximate location of the **0.706 line** is included in Figure 17.15a. Enriched isotopic signatures also show up on the mafic end of the magmatic spectrum, however, indicating that the enriched signature is at least partly a property of the mantle source regions (enriched lithospheric mantle beneath the ancient crust) and not strictly a crustal contribution.

The correlation between the isotopic enrichments for Sr, Nd (DePaolo and Wasserburg, 1977; DePaolo, 1981a; DePaolo and Farmer, 1984), $\delta^{18}O$ (Taylor, 1986), and Pb (Chen and Tilton, 1991) further underscores the influence of old continental crust and the sub-crustal lithospheric mantle, and permits similar isopleths for the other isotopic systems to be drawn across the western Cordillera of North America. These isopleths allow us to see through the plutons that intrude and obscure the continental margin and image the edge of the ancient craton. Thus the 0.706 line in Figure 17.15a is interpreted to be the boundary between Precambrian cratonal rocks on the east and accreted oceanic and arc terranes on the west. The irregularities in the southwest are the result of later strike-slip offsets, including the San Andreas fault. The 0.706 line is very close to the quartz diorite line of Moore et al. (1963), that also approximates the edge of thick continental crust which causes the more dense mafic magmas to pond and evolve beneath it.

In the western portion of the Idaho batholith an abrupt eastward increase in $^{87}Sr/^{86}Sr$ (0.704 → 0.712) and decrease in ε_{Nd} (+5 → −16) presumably marks the suture between the Precambrian crystalline basement of North America and western accreted pre-Cenozoic oceanic arc terranes, a contact now completely obscured by the batholith itself (Fleck, 1990). Chen and Tilton (1991) used $^{208}Pb/^{204}Pb$ and $^{206}Pb/^{204}Pb$ to characterize the crustal contribution to granitoids of the Sierra Nevada batholith as being of two distinct types. One type with high $^{206}Pb/^{204}Pb$ and $^{208}Pb/^{204}Pb$ they attributed to (meta) sediments derived from Precambrian basement, and one (further east) with low $^{206}Pb/^{204}Pb$ they attributed to U-depleted lower crustal metamorphics. Similarly, Kistler (1990) used $^{87}Sr/^{86}Sr$ and $\delta^{18}O$ to identify the boundary between two continental lithospheric types, similar to the two just mentioned, in the Sierra Nevada batholith. An effective technique based on these isotopic ratios is apparently being refined that can not only indicate the continental contribution to plutons and its geographic location, but also partially characterize the nature of the source of the enrichment and estimate its age.

It now seems clear that the western granitoids are of ultimate mantle origin (whether single or two stage), and that an increased continental component (probably of more than one type) is added at some distance inland from the trench. Yet there is still some disagreement on the mechanism(s) by which mixing of the enriched component is added. Some investigators (e.g., DePaolo, 1981a,c) have favored assimilation (AFC) of continental crust (or sediments derived therefrom) as the magmas rise, whereas others (e.g., Doe and Delevaux, 1973; Taylor, 1986; Hill et al. 1988; Pickett and Saleeby, 1994) have argued for mixing of the various components *at the source*, which may include SCLM, underplated magmas, and/or lower crust (melted or MASHed).

In one detailed study of the Peninsular Ranges batholith, Gromet and Silver (1987) derived a model to

explain the transverse variations in geochemical composition across that range. The Peninsular Ranges batholith is simpler than the Sierra Nevada batholith because it represents a single event of shorter duration and is not complicated by superimposed arcs or a tectonic collage of country rock terranes. The whole batholith appears to have been emplaced west of old continental crust into thick pre-Cenozoic metavolcanic and metasedimentary sequences, presumably on oceanic lithosphere. The sedimentary blanket, however, becomes more of an apron of continental clastics toward the east, assimilation of which may explain why $(^{87}Sr/^{86}Sr)_0$ in the plutonic rocks increases from 0.703 to 0.707 eastward, correlated with an increase in $\delta^{18}O$ from +6 to +12‰, more radiogenic Pb, and decreased $^{143}Nd/^{144}Nd$.

There are also subdued eastward increases in Al_2O_3, K_2O, and Na_2O in the Peninsular Ranges batholith. Gromet and Silver (1987) subdivided the batholith into three N-S trending zones: a western, a central, and an eastern belt. In contrast to the more subdued major element trends, there are some sharp variations in trace elements across these zones. Of particular interest are the rare earth elements. Figure 17.22 shows the REE patterns for the three zones. Because magmatic differentiation processes can affect REE distributions, only tonalites are compared, so the differences between the three zones are independent of rock type and large fractionation differences. Note that the western zone shows steep LREE and a negative Eu anomaly, but the mid to heavy REE are essentially unfractionated. As discussed in Section 9.3, this is a typical pattern for melts in equilibrium with plagioclase, which retains Eu and fractionates light REE, but not mid or heavy REE. The light REE of the central zone are comparable to those in the west, but the mid and heavy REE are now fractionated and the Eu anomaly is gone. Garnet is a common mineral that fractionates the mid and heavy REE. The eastern zone shows very steep LREE and similar mid and heavy REE to the central zone.

Accompanying the changes in REE patterns from the western to the central zone is an increase in total Sr content, whereas K_2O and Rb change hardly at all. We cannot explain the changes in REE by fractional crystallization, assimilation, AFC, or magma mixing because all the rocks compared are tonalites, and other major elements and trace elements don't allow it. After this exercise, we are faced with the only reasonable conclusion: the geochemical differences between the three zones must result from differences in the *source* of the magmas.

Because all of the changes from the west to the central zone occur in unison, they must be related. One single parameter that could do this is to have a plagioclase-bearing source (gabbro or amphibolite) beneath the west zone and a garnet-bearing source of the same general composition (eclogite or garnet amphibolite) beneath the central zone. Partial melting of a plagioclase-bearing rock, if some plagioclase remains in the solid residue, will deplete the melt in Sr and Eu (but not K and Rb) and fractionate the LREE. Similarly, garnet will not hold the Sr and Eu, and will fractionate the middle and heavy REE. Gromet and Silver (1987) related the plagioclase- and garnet-bearing sources to depth along the subduction zone. Plagioclase is more stable at low pressure and garnet more stable at high

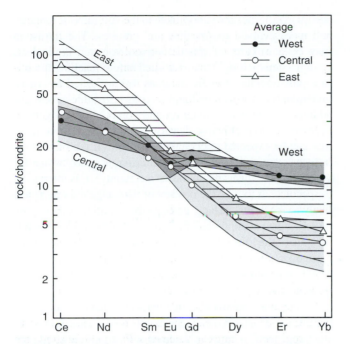

FIGURE 17.22 Range and average chondrite-normalized rare earth element patterns for tonalites from the three zones of the Peninsular Ranges batholith. Data from Gromet and Silver (1987).

pressure in a subcrustal mafic underplate, or perhaps in older slabs of subducted oceanic crust and sediments dismembered and sliced into the mantle wedge during earlier cycles of subduction. Once again, we have a model that considers the tonalites to be the product of melting of a mafic source, rather than of fractional crystallization from an ultramafic one.

The enrichment and fractionation of the LREE associated with the transition from the central to the eastern zone in the Peninsular Ranges batholith is accompanied by elevated values of $(^{87}Sr/^{86}Sr)_o$ and $\delta^{18}O$, which Gromet and Silver (1987) attributed to a *compositional* difference in the source, rather than to a *mineralogical* difference (as in the west → central zones). They suggested a source enriched in LREE, such as altered basalts or an addition of marginal sediments. Taylor (1986) proposed that a mélange of basalt, sediment, and volcaniclastics similar to those in the Franciscan formation of California may be responsible.

In the experimental results of Wolf and Wyllie (1993) partial melting of amphibolite produced a garnet-clinopyroxene residuum at 1 GPa pressure. This implies that the existence of a garnet-bearing residuum is also a function of the degree of partial melting, and may expand the depth at which garnet exists to shallower levels than considered by Gromet and Silver (1987).

The model of Gromet and Silver (1987) is a fine example of the application of geochemistry to petrological problems. Although the model of a geochemically homogenous but mineralogically contrasting source applies well to the west-central Peninsular Ranges batholith, it is unlikely to be representative of all batholith belts, particularly those where ancient continental crust and an old keel of subcontinental lithospheric mantle comprises a thickening wedge toward the inland portion of the orogenic/plutonic belt.

The across-axis variations in geochemical composition and isotopic ratios of various belts are far from uniform, but the common increases inland in SiO_2, K_2O, Rb, Th, U, and other incompatible trace elements, plus enriched isotopes, must reflect some enriched component. The enriched sources may differ from one locality to another and include components of the crust and the subcontinental lithospheric mantle. A variety of processes may also be responsible for the incorporation of the enriched components into the intrusive melts. Individual batholith belts can commonly be modeled in terms of the obvious asymmetries of continental-arc subduction zones in which the depth to the subducted slab, the thickness and nature of the overlying continental crust, and the influence of the subcontinental lithospheric mantle all increase with distance from the trench on the continent side.

Only the temporal inboard drift has no obvious tie to the factors just mentioned. Suggestions are myriad. Some investigators attribute the drift to tectonic erosion and the accompanying migration of the subduction zone toward the continental interior. Others relate the migration to shallowing of the dip of subduction with time, but one must ask why this change should accompany maturity of a continental arc. Perhaps it could be related to the increase in subduction rate or subduction of a ridge system that have been credited with producing a magmatic pulse. Others have suggested that the magmatic underplate (as in Figure 17.20) accumulates and diverts later rising melts toward the interior. Yet another suggestion is that the dense solid residues of fractional crystallization, or of partial melting of the earlier underplated magmas, descends under the initial magmatic zone, diverting later melts toward the interior. Bird (1978, 1979) made a similar proposal, arguing that the dense subcontinental lithosphere delaminates from the crust and sinks (Figure 18.8c2), drawing material inland.

17.5 PETROGENESIS OF CONTINENTAL-ARC MAGMAS

Continental arcs are widely considered to be the most complex tectonic and petrologic systems on Earth. As illustrated in Figure 17.23, continental-arc magmatism is a multisource, multistage process. As with its island-arc counterpart, continental-arc magmatism has its principal origins in the peridotites of the mantle wedge, where melting is induced by the addition of LIL-enriched fluids (and perhaps melts) from the dehydrating subducted plate. The resulting primary magma is probably an olivine tholeiitic basalt, but primitive basaltic magmas only reach the surface in young island arcs where the overlying crust is thin and passage to the surface is not greatly impeded. Where this crust is thick and light, as with continental arcs, primary magmas are likely to be ponded at the base of the crust where they undergo extensive fractional crystallization, assimilation, and probably also melting of the less refractory lower crustal rocks. The term *MASH* has been applied to the various combinations of melting, assimilation, storage, and homogenization that may take place at the base of the crust. Only during extensional episodes are the more primitive magmas likely to reach the surface along faults through a thinning crust in a continental arc. The crust–mantle interface must be an effective density trap, and a great deal of magma should crystallize at this level, adding to the crust in a process called magmatic underplating.

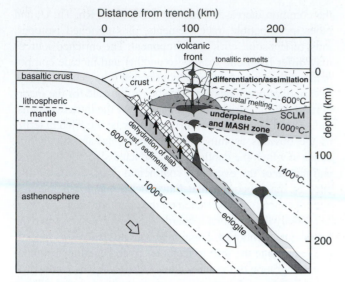

FIGURE 17.23 Schematic cross section of a continental-arc subduction zone (based on Figure 16.18), showing the dehydration of the subducting slab, hydration and melting of a heterogeneous mantle wedge (including enriched subcontinental lithospheric mantle), crustal underplating of the mantle-derived melts where MASH processes may occur, as well as crystallization of the underplates. Remelting of the underplate to produce tonalitic magmas (white bodies) and a possible zone of crustal anatexis are also shown. As magmas pass through the continental crust they may differentiate further and/or assimilate continental crust.

As discussed in Chapter 16, parental calc-alkaline magmas are either primary melts or produced by deep-level fractional crystallization of primary tholeiites, high-alumina basalts, or another of the panoply of proposed primary arc magmas described in Section 16.8.4. Subsequent combined processes produce more evolved and enriched calc-alkaline magmas. The bias of continental arcs toward more siliceous magmas with higher concentrations of K_2O, Rb, Cs, Ba, Th, and LREE and enriched isotopes, and the correlation of these enriched trends with the presence of thick continental crust, clearly indicate the important role of the thick sialic crust in continental-arc petrogenesis. Assimilation of crustal constituents by rising magmas, crustal melting, and magma mixing are all likely processes by which mantle melts can incorporate crustal material. The crustal imprint, in many cases, is clear enough to permit it to be used to delineate deep unexposed crustal domains. The old re-enriched keel of subcontinental lithospheric mantle may also contribute to the enrichment of continental-arc magmas in areas with a thick, ancient craton.

In continental arcs undergoing tectonic erosion, such as the Andes, the scraping and dragging of old crust (and sediment) into the zone of melting may provide a further enriched source. In constructive margins, such as the Cascades, rollback and migration of the trench away from the continent will not carry enriched crust downward. The mantle in the wedge above the subduction zone in such areas is probably trapped oceanic mantle of a depleted nature. All this and the extension behind the retreating trench mean that more mafic magmas should reach the surface in constructive margins.

The plutonic component of continental-arc magmatism is, on average, even more evolved than the volcanics. Gabbros

and quartz diorites are subordinate to the voluminous tonalites and more evolved monzonites and granites. The intrusives may be a mixture of mantle-derived volcanic equivalents ("single-step melting") that solidified before reaching the surface plus remelts of mafic magmas that solidified as crustal underplates ("two-step melting"). As the crust thickens by the addition of solidifying magmas and compressive orogeny, it becomes a more effective density trap, and even more magmas can be expected to crystallize at depth. If the bulk of the intrusives are really produced by a different process than the volcanics (and this is by no means a certainty), it is interesting to speculate as to why the product of two-stage melting is less likely to reach the surface. Perhaps the resulting tonalites are more evolved at depth, and thus more viscous and less mobile (in spite of the lower density this implies). Or perhaps they are more hydrous and therefore these low-T melts hit the negative dP/dT slope of the hydrous solidus as they rise and thus solidify at greater depth (Figure 17.24).

Possible continental-arc sources include subducted oceanic crust and sediments, the mantle wedge (itself probably a heterogeneous mixture of depleted and various enriched peridotites), heterogeneous continental crust, crustal underplates, and the subcontinental lithospheric mantle. Any of these sources may be mixed and/or partially melted.

Slab dehydration accompanies subduction, and the fluids, with their dissolved constituents, may metasomatize higher-level reservoirs and also induce melting. Fractional crystallization may occur in the ponded underplate, during ascent, or at shallow emplacement levels. Mixing of ascending magmas may also occur. Assimilation of heterogeneous crust produces chemical changes that depend upon the composition, temperature, and thickness of the crust, the heat content of the magma, and the magma's residence time. Isotopic changes are greatest when the initial concentration of the

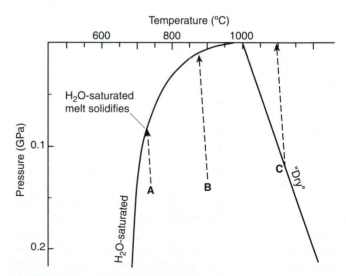

FIGURE 17.24 Pressure–temperature phase diagram showing the solidus curves for H_2O-saturated and dry granite. An H_2O-saturated granitoid just above the solidus at A will intersect the solidus as it rises and will therefore solidify. A hotter, H_2O-undersaturated granitoid at B will rise further before solidifying. Dry melts (C) are much more likely to reach the surface. Note: the pressure axis is inverted to strengthen the analogy with the Earth, so a negative $\partial P/\partial T$ Clapeyron slope will appear positive.

element in question is low in the primary magma and high in the contaminant. In immature crust, or depleted lower crust, the effects of contamination may not be detectable (because the concentration of the contaminant is low). In addition, the tectonic regime may also affect the magma that reaches shallow crustal levels. Compression causes ponding of dense magmas at depth, whereas extension or strike-slip motion can provide conduits for the rapid ascent of magma into arc, or back-arc, areas. All of these sources and processes probably contribute to continental-arc magmas in one place or another.

The formation of arc crust by partial melting of the mantle is commonly considered to be the manner in which much, if not all, of the continental crust forms. In Chapter 1, we discussed the formation of the Earth, and the separation of the mantle and core. At that time, I proposed that the bulk of the *crust* is too young, and did not form at this time. Here, then, appears to be the method by which the crust forms, supplying the final part of that puzzle. As described at the end of Chapter 16, subduction zones act as a sort of *filter*, in the sense that mantle material passes through the zone, and only selected portions of it become incorporated into the arc crust. The thickened and mature arcs then assemble by collisions when one arc attempts to override another. Arc crust is too buoyant to subduct into the deeper mantle, however, so these collisions form small micro-continents that further assemble to form the continents and the thick continental crust.

If this is true, it is important to know the nature of the *primary* mantle melts. Many crustal estimates suggest that the average composition of the continents is close to that of an andesite (e.g., Ellam and Hawkesworth, 1988; Taylor, 1995; Rudnick, 1995; Kelemen et al., 2003a). If primary mantle melts are andesitic, as originally thought with the melting slab model, then the partial melting of the mantle yields the sialic crust directly. In these last two chapters, however, it has become clear that the primary magmas that reach the crust are predominantly *basaltic*. Thus either the continental crust is considerably more mafic than we think, or a completely different flux from the mantle to the crust is required.

Ellam and Hawkesworth (1988) argued that Archean and Proterozoic arc magmatism may have been different than in modern arcs, and that more felsic material was generated in the past when heat flow was higher. The Phanerozoic mantle-to-crust flux, however, has been basaltic, and many investigators suggest that it has been so at least since the Archean. If this is true, then the continental composition does not equal the initial magmas that reach it. In order to produce an average andesitic crust, fractional crystallization of mantle-derived basalt and/or partial melting of a crystallized mafic underplate must leave a very mafic solid residue at depth, probably at the base of the crust, to compensate for the felsic magmas that rise into it. If so, then presumably this dense residuum would delaminate (dare we suggest as the "anti-crust"?) at some point and return to the mantle. As we discussed earlier, this sinking ultramafic mass may be responsible for the migration of igneous activity away from the trench in some continental-arc segments.

Although we have developed a general understanding of subduction zone magmatism, there are still many unanswered questions. To what extent does the oceanic slab melt, or is the slab contribution to arc magmas really mostly via fluids? How heterogeneous is the mantle wedge above the subduction zone, and how much recirculation takes place in it to re-enrich it following depletion by partial melting? What is the nature of the subcontinental lithospheric mantle, and what is the manner and extent of its enrichment? How universal is the high LIL/HFS "subduction zone signature," and is it truly a reliable measure of subduction zone magmatism? To what extent is the geochemical and isotopic variation of island-arc and continental-arc magmas due to mantle/source heterogeneity versus incorporation of crustal or subducted sediments, and how can we adequately determine source versus contamination? How variable is the crustal section, particularly the deep crust? How much remelting takes place?

Of course, there are many more unanswered questions than these. As far as the nature of the deep crust and mantle is concerned, those of us who have worked in deeply eroded high grade metamorphic terranes realize that the lower crust is very heterogeneous, both vertically and laterally. The seismic results of the American Consortium for Continental Reflection Profiling (COCORP) have shown similar heterogeneity on the scale of a few kilometers or less. The COCORP data further suggest that the transition to the mantle is complex beneath the continents, with a gradual transition containing numerous laminations and a much more obscure low-velocity layer than in the ocean basins. It would appear that thrusting and interlayering of oceanic and lower crustal slabs may be an important process in subduction zones, but the laminations may also reflect underplates and cumulate zones as well.

All of the possible variables make work in island-arc and continental-arc environments a tremendous challenge but also great fun. Each occurrence is unique in some way or another, a lesson that probably applies to all of the igneous petrogenesis discussed in Chapters 12 through 20. The challenge is to continue to devise ingenious ways to evaluate the contributions of the various reservoirs and processes, and to explore the petrology, structure, and geophysics of as many occurrences and in as much detail as possible.

Summary

The thick, low-density, sialic continental crust and enriched subcontinental lithospheric mantle affect the diversification and dynamics of primary subduction zone magmas. The principal effects are contamination by an old, geochemically and isotopically enriched crust, and impeding the rise of primary magmas so as to enhance differentiation. Continental-arc magmatism therefore tends to be more calc-alkaline and alkaline (and less tholeiitic/low-K series) than island arcs. Basalts are also subordinate and exposed rocks are skewed toward more evolved types. Silicic hydrous magmas are typically explosive, and rhyolitic ignimbrites are common. Basalts are more buoyant than the mantle, but less so than the crust, so they may be expected to stall and pond at the base of the crust. Here they may interact with the deep crust via some combination of melting, assimilation, crystal fractionation, and magma mixing (all of which may also occur with further ascent). Many magmas never reach the surface, and extensive uplift and erosion of the thickened crust (typically in areas where volcanism ceases) expose large composite batholith belts. The batholith belts are comparatively silicic, and their formation probably involves substantial remelting of mantle-derived gabbros that underplate the crust. The resulting tonalites then fractionate to form other subordinate granitoids. The combination of dehydration of the subducting slab (with or without some melting), fluid-fluxed partial melting of the mantle wedge, stranding of melts at the base of the crust where they differentiate and/or solidify, and partial melting of the solidified bits, is how subduction zones gradually "filter" mantle material, so that selectively more silicic and evolved portions become the continental crust.

Key Terms

Continental-arc/active continental margin *352*
Subcontinental lithospheric mantle (SCLM) *352*
Batholith belts *353*
Orogenic belts *353*
Allochthonous terrane/exotic terrane *353*

Andean-type margin *354*
Himalayan-type margin *000*
Magmatic underplating *354*
MASH *360*
Subduction erosion/tectonic erosion *360*
Constructive/destructive plate boundary and slab roll-back *363*

Bimodal (basalt/rhyolite) volcanism *364*
Cordilleran-type batholith *365*
Granitoid *365*
Cauldron subsidence *367*
Quartz diorite line *371*
0.706 line *372*

Review Questions

Review Questions are located on the author's web page at the following address: **http://www.prenhall.com/winter**

Important "First Principle" Concepts

- The subduction zone beneath continental arcs is very similar to that beneath island arcs.
- Magmatic rocks of continental arcs are, on average, more evolved (silicic) and enriched than their island-arc equivalents. This is probably due to the thick, low-density arc crust, which stalls mafic primary magmas at depth, where they differentiate and incorporate crustal material.
- The subcontinental lithospheric mantle has been attached to the continental crust as a keel for as much as 1 Ga or more and has become a source of enriched components since that time.

- Granitic batholiths probably result from remelting of gabbroic magmatic underplates of the crust ("two-stage" melting) to form tonalites, which then fractionate further within the crust to form granodiorites, granites, etc.
- Hydrous magmas are much more likely than anhydrous magmas (with a high positive $\partial P/\partial T$ solidus slope) to intersect the solidus as they rise, so that they tend to stall as plutons.

Suggested Further Readings

Continental-Arc Volcanics

Harmon, R. S., and B. A. Barreiro (eds.). (1984). *Andean Magmatism: Chemical and Isotopic Constraints*. Shiva. Nantwich, UK.

Hess, P. C. (1989). *Origins of Igneous Rocks*. Harvard. Cambridge, MA. Chapter 11.

Pitcher, W. S, M. P. Atherton, E. J. Cobbing, and R. D. Beckensale (eds.). (1985). *Magmatism at a Plate Edge. The Peruvian Andes*. Blackie. Glasgow.

Thorpe, R. S. (ed.). (1982). *Andesites. Orogenic Andesites and Related Rocks*. John Wiley & Sons. New York.

Wilson, M. (1989). *Igneous Petrogenesis: A Global Tectonic Approach*. Unwin Hyman. London. Chapter 7.

Continental-Arc Intrusives

Anderson, J. L. (1990). *The Nature and Origin of Cordilleran Magmatism*. Memoir, **174**. Geological Society of America. Boulder, CO.

Atherton, M. P., and J. Tarney (eds.). (1979). *Origin of Granite Batholiths: Geochemical Evidence*. Shiva. Kent, UK.

Clarke, D. B. (1992). *Granitoid Rocks*. Chapman Hall. London.

Harmon, R. S., and C. W. Rapela (eds.). (1991). *Andean Magmatism and Its Tectonic Setting*. Special Paper **265**. Geological Society of America. Boulder, CO.

Kay, S. M., and C. W. Rapela (eds.). (1990). *Plutonism from Antarctica to Alaska*. Special Paper **241**. Geological Society of America. Boulder, CO.

Pitcher, W. S. (1993). *The Nature and Origin of Granite*. Blackie. London.

Pitcher, W. S, M. P. Atherton, E. J. Cobbing, and R. D. Beckensale (eds.). (1985). *Magmatism at a Plate Edge. The Peruvian Andes*. Blackie. Glasgow.

Roddick J. A. (ed.). (1983). *Circum-Pacific Plutonic Terranes*. Memoir 159. Geological Society of America. Boulder, CO.

18

Granitoid Rocks

Questions to be Considered in this Chapter:

1. What is a *granitoid*, and why do we use this term when addressing granitic rocks?
2. What variety of granitoids do we find, and how are they classified?
3. What controls the occurrence and distribution of the various granitoid types?
4. What is the nature of crustal and mantle inputs to granitoid petrogenesis?
5. How and when did the continental crust develop?

In this chapter, we shall survey a diverse assortment of silicic quartzo-feldspathic plutonic rocks. As I mentioned in Chapter 17, the term *granite* has been loosely applied to a wide variety of felsic intrusive rocks. Because *granite* (*sensu stricto*) is a term defined by a restricted range of quartz–plagioclase–alkali feldspar proportions in the IUGS classification (Figure 2.2), this dual use can be confusing. Several investigators have adopted the (non-IUGS) term **granitoid** for the broader (*sensu lato*) usage. Granitoids are the most abundant plutonic rocks in the upper continental crust. Such a broad spectrum of rock types will have a broad range of sources and genetic processes, and their genesis is not limited to the subduction zone processes described in the previous chapter. Because of the diversity of granitoids, this chapter breaks from my usual approach of dealing with specific tectonic settings. In this case it is much more efficient to address a spectrum of settings pertaining to these common rocks.

Because of their diversity, granitoid rocks have been the subject of considerable study and controversy for over two hundred years. As evidence, I cite the title of H. H. Read's 1957 classic *The Granite Controversy*, or the following quote from Joseph Jukes, who, in 1863, as director of the Irish Geological Survey, and arbiter of a vigorous 19th century debate on the origin of granite, said, "Granite is not a rock which was simple in its origin but might be produced in more ways than one." Of this we may still be certain. I shall attempt to describe the petrology and geochemistry of common granitoid rocks and to develop a satisfactory classification and a framework for granitoid petrogenesis.

Although this chapter concentrates on granitoids, magmatism at a particular locality is rarely confined strictly to intrusive or extrusive expressions. Thus granitoid emplacement may commonly be associated with cogenetic and chemically equivalent felsic volcanism, such as large rhyolite ash-flow tuffs or alkalic volcanism. This is obvious in modern occurrences, but the volcanic portion is typically eroded away where older intrusives are exposed. The present discussion will focus on the intrusive suites. Also, because subduction zone granitoids were discussed in Chapter 17, less emphasis will be placed on them here.

I begin with a few broad generalizations about granitoid genesis. Although it is the nature of generalizations to have exceptions and dissenters, they serve to give us an initial concept of our subject:

1. Most granitoids of significant volume occur in areas where the continental crust has been thickened by orogeny, either continental-arc subduction or collision of sialic masses. Many granitoids, however, may postdate orogeny by tens of millions of years.

2. Because the crust is solid in its normal state, some thermal disturbance is required to form granitoids.

3. Most investigators are of the opinion that the majority of granitoids are derived by crustal anatexis, but that the mantle commonly is also involved. The mantle may merely supply heat for crustal anatexis, or it may contribute material as well.

18.1 PETROGRAPHY OF GRANITOIDS

Granitoid rocks naturally have a medium to coarse grain size that reflects slow cooling and the presence of volatiles, particularly H_2O, which facilitates mineral growth. Plagioclase, quartz, and alkali feldspar are the predominant phases in most granitoids, although any may be absent in our broadly defined category, which includes tonalites and syenitic varieties. Plagioclase (An_{10} to An_{93}) tends to be an early and dominant phase. Quartz and alkali feldspar typically form later. The proportion of alkali feldspar varies with bulk composition, being predictably higher in more alkaline rocks. Large alkali feldspar megacrysts are common in granites. Hornblende (grading from brown Ti-rich to green Ti-poor varieties) and biotite are the dominant mafic phases, a testament to the elevated pressure and H_2O content of the crystallizing magma. Muscovite is common in aluminous granites, and occurs as both primary igneous crystals and as secondary replacement during pluton cooling. Clinopyroxene is subordinate, present principally in the more mafic rocks of a plutonic suite, but may extend into more evolved granites, particularly in more sodic varieties. Augite is a common early phase in experiments on granitic systems, but in natural rocks it is often partially or completely replaced by reaction with the hydrous residual melt to produce hornblende and/or biotite (Figure 3.20). At greater depth and high H_2O pressure the hydrous mafics may entirely supplant the pyroxene field and be primary liquidus phases. Olivine and orthopyroxene are rare. Fayalitic olivine occurs in some alkaline granitoids, and orthopyroxene occurs in high-temperature anhydrous granitoids called charnockites.

Layering and obvious cumulates are rare in these highly viscous magmas. The reliable phenocryst/matrix textural criteria so useful for modeling fractional crystallization in mafic rocks and volcanics are no longer available. Only by careful petrographic analysis, using textural criteria, can we hope to determine the sequence of crystallization. Even then it may prove difficult. Hypidiomorphic granular (interlocking) textures (Figure 1.1) are common, and are considered to record the mutual accommodation of crystals as they interfere with each other during growth (Chapter 3). Traditional reliance on the principles of idiomorphism (how well formed a mineral is), inclusions, or replacement features has been questioned for granitoids by Flood and Vernon (1988). Nevertheless, textures indicate a general order of crystallization in granitoids, with distinct, though largely overlapping, growth of minerals. Generally first to appear are accessory minerals such as zircon, apatite, pyrite, ilmenite, etc., followed by plagioclase and the mafics (principally hornblende and biotite). Quartz and

alkali feldspar are typically late interstitial phases, but may occur as large plates in true granites, probably due more to poor nucleation than to early formation. Even then they exhibit irregular grain boundaries, indicating that they crystallized along with the earlier phases toward the final stages. This indicates that plagioclase and the mafic minerals were in equilibrium with the felsic liquid over a wide temperature interval. Coarse granophyric or graphic textures, reflecting eutectic quartz–alkali feldspar–plagioclase co-crystallization, are common (Figure 3.9). This sequence may serve as a general guide, but there are many complexities, subject to variations in P, T, f_{H_2O}, f_{O_2}, and bulk composition. H_2O content inhibits nucleation and facilitates crystal growth and the crystallization of hydrous minerals.

Experiments suggest that grain size is *not* always a good indicator of sequence. Feldspar growth rates commonly far exceeds nucleation, so some coarse feldspars (particularly alkali feldspars) may be *late* arrivals. Some experiments show megacrysts and groundmass growing simultaneously.

Low-Ca alkaline granites may have both a sodic and a potassic feldspar, or else a single alkali feldspar of intermediate character. The former are commonly called **subsolvus** granites and are crystallized in systems with high H_2O pressure, where the alkali feldspar solvus intersects the solidus (Figure 6.17c), creating two distinct feldspar fields. Single-feldspar granites are called **hypersolvus** and crystallize in systems with low H_2O pressure and a single feldspar field above the solvus (Figures 6.16 and 6.17a and b). Only a few-percent-An component, however, shifts the feldspar into the ternary solvus (Figure 7.11) so that most granites have two feldspars, regardless of p_{H_2O}. Subsolidus recrystallization and exsolution are common in virtually all slow-cooled hydrous granitoid systems so that many textures may not represent true magmatic crystallization. Perthites (Figure 3.18a) are common, and myrmekite (wormy quartz grains in sodic plagioclase, typically embaying K-feldspar, Figure 3.21) may form by exsolution or small-scale metasomatism.

Minor (or accessory) minerals in granitoids include apatite, zircon, magnetite, ilmenite, monazite, titanite, allanite, tourmaline, pyrite, and fluorite, plus a host of others, depending on the concentration of incompatible trace elements (that generally concentrate in silicic magmas).

Aluminous granites may contain Al-bearing phases that are more typical of metamorphic rocks than igneous. Most of these minerals have high melting temperatures, and may represent, in part at least, minerals in the granite melt source rock that remained solid during melting. Complete melting is a petrologic rarity, and liquids are generally extracted when melting reaches some critical proportion, as described in Section 11.1 and revisited below. The solid refractory material remaining is called **restite**. It may remain as coherent mineral clusters or as disaggregated minerals entrained in the migrating melt. Melting of micaceous rocks consumes SiO_2 from residual quartz and may leave a silica-depleted restite containing garnet, cordierite, an aluminum silicate polymorph, orthopyroxene, zircon, or even corundum. Because the melt was in equilibrium with these phases in the partially melted source prior to extraction, they may

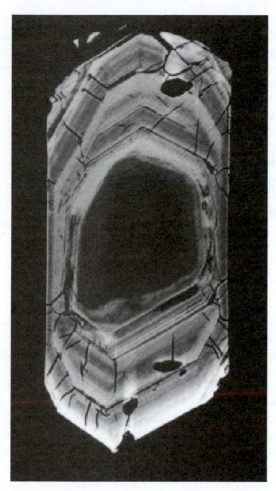

FIGURE 18.1 Backscattered electron image of a zircon from the Strontian Granite, Scotland. The grain has a rounded, un-zoned core (dark) that is an inherited high-temperature non-melted crystal from the pre-granite source. The core is surrounded by a zoned epitaxial igneous overgrowth rim, crystallized from the cooling granite. From Paterson et al. (1992). Reproduced by permission of the Royal Society of Edinburgh.

also crystallize as igneous minerals when the magma cools (depending on pressure). It may be difficult to determine if these phases represent true primary igneous minerals or entrained restite minerals. Careful textural and chemical work is required to make the distinction, when possible at all.

One relatively clear restite example is the polygenetic zircon **xenocrysts** (foreign crystals) found in many granitoids (Figure 18.1). These zircons have been studied well because they concentrate uranium and form the basis for U-Pb dating of many rocks. Zircons are very hard and refractory, and commonly survive both chemical and mechanical weathering. They can thus be eroded from an igneous source, transported, deposited, buried, and metamorphosed without decomposing or being totally recrystallized. Zircons of purely metamorphic origin are generally unzoned. If metamorphic rocks are partially melted, their zircons may be carried off by the magma in the manner proposed above. As a result, many granites **inherit** much of the Pb of their sedimentary precursors in the form of these zircons, resulting in erroneously high age determinations. Some zircons may also be entrained by wall rock melting. Restite zircons provide nuclei for the epitaxial growth of igneous zircon rims, much of which is chemically zoned. Figure 18.1 is a backscattered electron image of a polygenetic zircon xenocryst from the Strontian granite of Scotland, with an unzoned inherited core and a euhedral zoned rim, developed during later granitic crystallization. The rounded nature of the core suggests that it has been either partially melted or abraded during a clastic phase of its tour through the rock cycle. We now have the ability to date single zircons, even portions of single zircons, using the ion microprobe. Under ideal conditions this technique may permit us to determine not only the igneous age (from the rim) but also the age of the precursor rock (from the core).

Xenoliths are commonly observed in intrusive rocks (see Section 4.2.3) and are generally assumed to be pieces of country rock plucked from the walls of the conduit through which a magma rises. More detailed study, however, suggests that there are more than one occurrence of such inclusions, and the term **enclave** has been proposed for the more general case of igneous inclusions. The various types of enclaves and their names are listed in Table 18.1. They include xenoliths of country rock, stirred up cumulates or bits of the fine-grained chill zone, hot mafic magma injected into a felsic chamber and quenched (Figure 11.9), or restites (most easily distinguished when they contrast with the country rock, are refractory, and are micaceous or aluminous because these compositions are exclusively sedimentary).

18.2 GRANITOID GEOCHEMISTRY

The chemical composition of granitoids (granite *sensu lato*) is obviously variable. Table 18.2 illustrates the more granite-like portion of the spectrum with selected average analyses of some granitoid rocks from a number of contrasting settings. I have included oceanic ridge "plagiogranites" (tonalites) plus ocean island, continental rift, and Archean granitoids, plus the S-I-A-M granites (see Section 18.4.1), along with estimates of bulk average, upper, and lower crust for comparison. Regrettably, there is no standard prescribed set of trace elements, resulting in a lack of uniformity from varied literature sources. Elements not analyzed in published averages are thus left as blanks in the table. Note that granitoids are similar in composition to the estimation of the average upper crust by Taylor and McLennan (1985), but the lower crust, by comparison, is depleted in SiO_2, alkalis (especially K_2O), and the incompatible and LIL trace elements.

As with any igneous rock, the chemical composition of a granitoid is controlled by the chemical composition of the source; the pressure, temperature, and degree of partial melting; and the nature and extent of subsequent assimilation and differentiation processes. The variability of source areas, both crustal and mantle, produce considerable variation in granitoid series. Calc-alkaline types dominate over tholeiitic, but there are also alkali-calcic and alkaline varieties (using the criteria of Peacock (1931), Section 8.7). Alumina saturation varies from metaluminous, through peraluminous to peralkaline (using the criteria of Shand (1927), Figures 8.10 and 18.2). Because these characteristics generally hold

TABLE 18.1 The Various Types of Enclaves

Name	Nature	Margin	Shape	Features
Xenolith	Piece of country rocks	Sharp to gradual	Angular to ovoid	Contact metamorphic texture and minerals
Xenocryst	Isolated foreign crystal	Sharp	Angular	Corroded reaction rim
Surmicaceous Enclave	Residue of melting (restite)	Sharp, biotite rim	Lenticular	Metamorphic texture micas, Al-rich minerals
Schlieren	Disrupted enclave	Gradual	Oblate	Coplanar orientation
Felsic Microgranular Enclave	Disrupted fine-grained margin	Sharp to gradual	Ovoid	Fine-grained igneous texture
Mafic Microgranular Enclave	Blob of coeval mafic magma	Mostly sharp	Ovoid	Fine-grained igneous texture
Cumulate Enclave (Autolith)	Disrupted cumulate	Mostly gradual	Ovoid	Coarse-grained cumulate texture

After Didier and Barbarin (1991, p. 20).

for an entire magma series, they must be determined by the nature of the source and its melting.

The notion that granitoids chemically reflect their source forms the basis for many granitoid classifications, but the chemical compositions of granites are more complex than simply reflecting the characteristics of a single uniform source. Hybrid sources are probably common, as is assimilation upon ascent. In the case of mantle-derived magmas, an imprint of crustal assimilation is often isotopically detectable. In the case of crustal melts it is difficult, if not impossible, to assess the extent of assimilation processes because numerous crustal inputs could occur at any point from the source to the level of emplacement.

Any magma containing sufficient normative plagioclase, orthoclase, and quartz (most granitoids) should evolve along a liquid line of descent toward the temperature minimum of the Ab-Or-Qtz system (Figures 18.3 and 11.3: the low-P minimum-temperature composition is commonly referred to as **haplogranite**). The plotting of so many granitoids at or near the low-pressure ternary minimum suggests that either fractional crystallization is rather efficient, or that the granitoids are minimum melts of quartzo-feldspathic crustal sources. Note in Figure 18.3 that not all of our granitoids from Table 18.2 plot near the minimum. This is particularly true for mantle-derived granitoids. The average oceanic plagiogranite is depleted in K_2O, as we might expect for the product of extensive fractional crystallization from MORB (Chapter 13). The S-I-A-M classification of granitoids will be discussed below. Briefly, I-types and M-types are believed to be mantle derived, either as fractionated single-stage mantle partial melts or as remelts of mantle-derived gabbroic crustal underplates or deep crustal plutons (Sections 17.4.2 and 18.4.1). S-types and A-types have more crustal components. The average M-type granitoid and the average I-type also plot toward the low-K_2O portion of the diagram, reflecting their lack of elements favored in the crust. The average Archean granitoid is also depleted in K_2O and may have a fairly direct mantle affiliation (see Section 18.7 below). The other granitoids plot near the

minimum composition, which is not as invariant as we might wish. H_2O pressure shifts the position of the minimum toward the Ab-Or side of the diagram, as does the presence of B and F in the melt. Increasing anorthite component in plagioclase shifts the eutectic toward the Or-Qtz side of the Ab-Or-Qtz diagram (Figure 18.3).

When plotted in a MORB-normalized spider diagram (Figure 18.4; see Figures 9.7 and 16.11a for an explanation), the analyses in Table 18.2 fall into two groups. Figure 18.4a includes those granitoids with a subduction zone, decoupled high LIL/HFS signature. Figure 18.4b shows that plagiogranite has a flat, MORB-like trend and the other granitoids have a lesser LIL/HFS offset and large Ba and Ti troughs reflecting an intraplate nature.

18.3 CRUSTAL MELTING

Granitoids are diverse, with origins typically involving both the mantle and crust. We discussed mantle sources and melting dynamics in previous chapters, and we shall concentrate here on production of granitoids where the crustal component predominates (although some of that crust may not have been resident for long). Under normal circumstances, the crust is solid, so either thickened crust or increased heat is a common requirement for our granite melting pot. What can we expect when we heat the crust to very high temperatures? This question bridges the gap between metamorphic and igneous processes. We shall deal with the metamorphic reactions that lead to melting more specifically in Section 28.4 and concentrate on the melt and its liberation now.

Consider an example of the partial melting of typical aluminous meta-sedimentary (shale or mudstone) source for a crustal-derived granitoid in a thickened orogen (Section 18.4.2.2). At high metamorphic grades of 600°C or better this would probably be a muscovite + biotite + aluminum − silicate gneiss with considerable quartz and feldspar. Simplified representative melting reactions are shown in Figure 18.5a. There are really more than two reactions that involve muscovite and biotite (Vielzeuf and Holloway, 1988, and Section 28.4),

TABLE 18.2   Representative Chemical Analyses of Selected Granitoid Types.

		1 △	2 ▲	3 □	4 ■	5 ◊	6 ♦	7 ○	8 ●	9 +	10 ✳	11 X	12 ⊥
	Oxide	Plagiogr.	Ascen.	Nigeria	M-type	I-type	S-type	A-type	Archean	Modern	AvCrust	U. Crust	L. Crust
Major Elements	SiO_2	68.0	71.6	75.6	67.2	69.5	70.9	73.8	69.8	68.1	57.3	66.0	54.4
	TiO_2	0.7	0.2	0.1	0.5	0.4	0.4	0.3	0.3	0.5	0.9	0.5	1.0
	Al_2O_3	14.1	11.7	13.0	15.2	14.2	14.0	12.4	15.6	15.1	15.9	15.2	16.1
	FeO^*	6.6	4.0	1.3	4.1	3.1	3.0	2.7	2.8	3.9	9.1	4.5	10.6
	MnO	0.1	0.1	0.0	0.1	0.1	0.1	0.1	0.1	0.1	0.6	0.3	0.8
	MgO	1.6	0.2	0.1	1.7	1.4	1.2	0.2	1.2	1.6	5.3	2.2	6.3
	CaO	4.7	0.1	0.5	4.3	3.1	1.9	0.8	3.2	3.1	7.4	4.2	8.5
	Na_2O	3.5	5.5	3.9	4.0	3.2	2.5	4.1	4.9	3.7	3.1	3.9	2.8
	K_2O	0.3	4.7	4.7	1.3	3.5	4.1	4.7	1.8	3.4	1.1	3.4	0.3
	P_2O_5	0.1		0.0	0.1	0.1	0.2	0.0	0.1	0.2			
	Total	99.6	98.1	99.3	98.4	98.5	98.3	98.9	99.7	99.6	100.7	100.2	100.8
CIPW Norm	q	31.9	23.1	31.7	25.5	27.5	33.7	28.6	24.0	22.8	8.2	16.8	5.5
	or	1.8	28.3	28.2	7.8	21.2	25.1	28.3	10.6	20.3	6.5	20.1	1.8
	ab	29.6	36.8	35.6	36.6	29.4	23.2	37.5	44.0	33.5	27.8	35.0	25.1
	an	21.9	0.0	2.5	20.1	14.4	8.4	1.6	15.2	14.2	26.2	13.9	30.5
	cor	0.0	0.0	0.7	0.0	0.0	2.8	0.0	0.0	0.2	0.0	0.0	0.0
	di	0.7	0.4	0.0	0.8	0.6	0.0	1.4	0.0	0.0	8.4	5.5	9.4
	hy	9.4	4.1	0.3	6.0	4.1	3.7	0.0	3.6	5.8	19.2	5.9	23.8
	wo	0.0	0.0	0.0	0.0	0.0	0.0	0.3	0.0	0.0	0.0	0.0	0.0
	ac	0.0	4.8	0.0	0.0	0.0	0.0	0.0	0.0	0.0	0.0	0.0	0.0
	mt	3.2	0.0	0.0	2.1	2.0	2.1	1.9	1.9	2.1	2.5	2.1	2.6
	il	1.3	0.3	0.0	0.7	0.6	0.6	0.4	0.4	0.7	1.3	0.7	1.4
	hem	0.0	0.0	1.0	0.0	0.0	0.0	0.0	0.0	0.0	0.0	0.0	0.0
	ns	0.0	2.2	0.0	0.0	0.0	0.0	0.0	0.0	0.0	0.0	0.0	0.0
Incompatible	Ni	12			3*	8	11	1	14	11	128	44	156
	Co	9			16*	10	10	3			29	17	33
	Cr	8			7*	20	30	2	29	23	185	83	219
	Cu	4		45	8*	9	9	2			75	25	90
	Zn	13		99	31*	48	59	120			80	71	83
	V			3	71*	57	49	6	35	76	230	107	271
Rare Earth	La	4	91	116	9*	31	27	55	32	31	16	30	11
	Ce	11	274	166	22*	66	61	137	56	67	33	64	23
	Nd		122		13*	30	28	67	21	27	16	26	13
	Sm	3	17		3*	6	6	16	3	5	4	5	3
	Eu	1	2		1*	1	1	2	1	1	1	1	1
	Gd	4			4*			14	2	6	3	4	3
	Tb	1	4		1*			2	0	1	1	1	1
	Dy				5*					1	5	4	4
	Yb	5	17		3*	3	3	9	1	3	2	2	2
	Lu	1			1*	1	1	1	0	1	0	0	0
LIL	Rb	4	94	471	30*	164	245	169	55	110	32	112	5
	Ba	38	53	94	232*	519	440	352	690	715	250	550	150
	Sr	124	1	20	154*	235	112	48	454	316	260	350	230
	Pb			42	5	19	27	24			8	20	4
HFS	Zr	97	1089	202	131*	150	157	528	152	171	100	190	70
	Hf	3	42	9				8	5	5	3	6	2
	Th	1	24	52	1	20	19	23	7	12	4	11	2
	Nb	7	168	124	3*	11	13	37	6	12	8	12	7
	Ta	1	16							1	1	1	1
	U	0			0	5	5	5	2	3	1	3	1
	Y	30	92	191	34*	31	32	75	8	26	20	22	19

1: Average of 6 ophiolite plagiogranites from Oman, Troodos (Coleman and Donato, 1979).

2: Granite from Ascension Island (Pearce et al., 1984).

3: Average of 11 Nigerian biotite granites (Bowden et al., 1987).

4: average of 17 M-type granitoids, New Britain arc (Whalen et al., 1987).
  * Trace elements from Saito et al. (2004), central Japan.

5: Average of 1074 I-type granitoids, Lachlan fold belt, Australia (Chappell and White, 1992).

6: Average of 704 S-type granitoids, Lachlan fold belt, Australia (Chappell and White, 1992).

7: Average of 148 A-type granitoids (Whalen et al., 1987; REE from Collins et al., 1982).

8: Average of 355 Archean grey gneisses (Martin, 1994).

9: Average of 250 <200-Ma-old I- and M-type granitoids (Martin, 1994).

10–12: Estimated average, upper, and lower continental crust (Taylor & McLennan, 1985; McLennan et al., 2005).

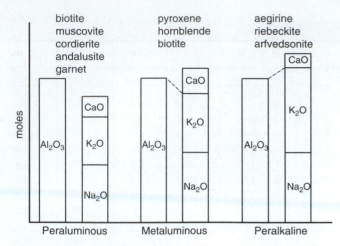

FIGURE 18.2 Alumina saturation classes based on the *molar* proportions of $Al_2O_3/(CaO + Na_2O + K_2O)$ ("A/CNK"), after Shand (1927). Common non-quartzo-feldspathic minerals for each type are indicated. After Clarke (1992). Copyright © by permission of Kluwer Academic Publishers.

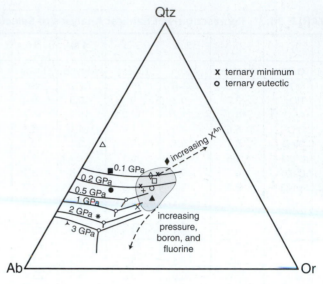

FIGURE 18.3 The Ab-Or-Qtz system with the ternary cotectic curves and eutectic minima from 0.1 to 3 GPa. Included is the locus of most granite compositions from Figure 11.2 (shaded) and the plotted positions of the norms from the analyses in Table 18.2. Symbols are given in Table 18.2. Note the effects of increasing pressure and the An, B, and F contents on the position of the thermal minima.

but the ones shown serve to illustrate the process. The temperature and nature of melting depends upon the rock composition, the H_2O content, and the particular pressure–temperature–time (*P-T-t*) path (Section 16.8) of that portion of the crust. In situations involving continental collision the underthrust crust will probably experience increased pressure first, and then it will heat up as the depressed isotherms relax or rebound, thus following a *clockwise P-T-t* path on a pressure–temperature diagram (Section 25.5). The source will initially follow a low-T geotherm such as 15 to 20°C/km. Crustal rocks being underplated by mantle-derived melts or involved in post-orogenic collapse (discussed below) will follow a higher-T geotherm, and may even follow a *counterclockwise P-T-t* path that is heated first and compressed later. The melting temperature, pressure, and melt fraction will obviously depend upon the *P-T-t* path. For example, in Figure 18.5a we can see that H_2O-*saturated* mica schists or gneisses begin to melt at 0.45 GPa and 640°C (point a) if the area follows a 40°C/km geotherm, or 0.7 GPa and 620°C (point b) along a 25°C/km geotherm, or 1.1 GPa and 610°C (point c) along a 15°C/km geotherm.

The slow heating and temperatures of high-grade metamorphism will in practically all cases drive off most "excess" water (existing as a free fluid phase), so crustal anatexis is predominantly an H_2O-undersaturated (also called "fluid-absent") phenomenon. H_2O present in the lower crust is mainly contained in hydrous minerals, particularly micas and amphiboles. The H_2O-saturated melting curve (solidus) is thus of little importance to the melting of most deep-seated rocks (other than setting a low-temperature limit required for melting), and a dehydration reaction must be reached (above the H_2O-saturated solidus limit) before H_2O can be liberated and directly induce the rock to melt (as discussed in Section 10.2.3).

As an example of what may occur as a micaceous gneiss is heated, let's follow the 40°C/km geotherm in Figure

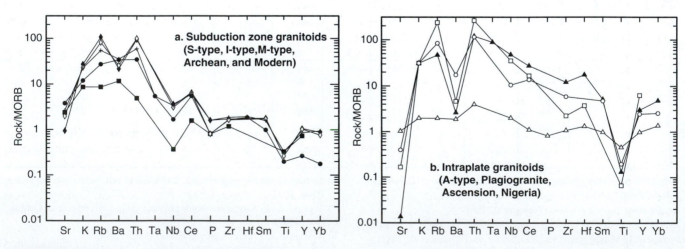

FIGURE 18.4 MORB-normalized multielement (spider) diagrams for the analyses in Table 18.2 (symbols as in Table 18.2).

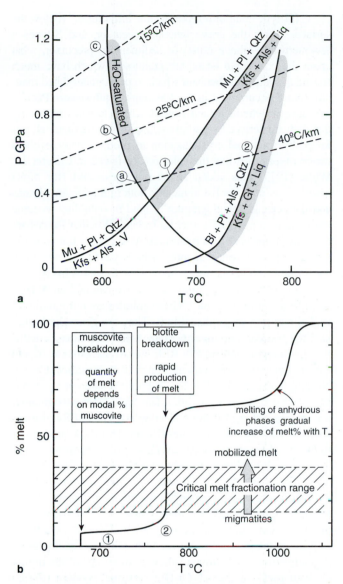

FIGURE 18.5 **(a)** Simplified *P-T* phase diagram and **(b)** quantity of melt generated during the melting of muscovite-biotite-bearing crustal source rocks, after Clarke (1992) and Vielzeuf and Holloway (1988). Shaded areas in **(a)** indicate melt generation.

sufficiently low that less than 10% of the rock will melt at this temperature. The composition of the phases varies little, so the reaction is nearly invariant and runs to completion (all the muscovite is consumed) over a small temperature interval. The vertical line at point 1 in Figure 18.5b represents this incremental production of a few percent melt. The amount of melt may be less than the critical fraction deemed necessary for mobilization, so the melt may remain in the source rock. If so, the result is a **migmatite**, or "mixed rock," with blebs, lenses, or small dikelets of granitic melt intermixed with the gneiss (see Section 28.5).

According to this model, only when biotite begins to break down at 760°C (point 2 in Figure 18.5 a and b) is enough granitic melt believed to be generated to become mobile and rise as true magma. Other potential melt-generating biotite breakdown reactions are considered in Section 28.4. Melt generation at this point is now relatively rapid and up to 60% of the rock may be melted at this temperature (Figure 18.5b). The solid residue (restite) contains K-feldspar, garnet, plagioclase, sillimanite, perhaps quartz, and some refractory accessory phases. The remaining rock is now anhydrous, requiring much higher temperatures to gradually melt these phases. Such high temperatures are not commonly attained, and these phases form the aluminous portion of the deep crustal granulites that we find in deeply eroded terranes.

Partial melting can result from a number of possible reactions (Wyllie, 1977a; Vielzeuf and Holloway, 1988; Thompson and Connolly, 1995; Patiño Douce and Beard, 1995; Johannes and Holtz, 1996; Vigneresse, 2004). At depths appropriate for most melting phenomena, we can expect H_2O-deficient melting, as described above, to be the norm. Melting thus results principally from dehydration of some hydrous phase above the temperature of H_2O-saturated melting for the bulk composition in question. Although only one dehydration reaction for muscovite and one for biotite were illustrated here, other reactions and other hydrous minerals may break down and create melts in different systems (Section 28.4). Dehydration of hornblende, for example, is illustrated in Figures 10.6 and 16.16. Very hydrous melts may not rise far from the point of origin (Figure 17.24), and low melt fractions may also be immobile. Immobile melts are much more important to us in crustal melting situations than in mantle melts. Not only will the critical melt fraction affect the composition of the melt that finally does become mobile, but the source is not very dense, and thus uplift and erosion may expose these migmatitic source areas for our interpretation.

18.5a. At 640°C the *P-T-t* path encounters the H_2O-saturated solidus. Wyllie (1983) contended that a small amount of H_2O-saturated melt may form at this point due to the existence of some minor free water, but the scant melt, if it forms at all, quickly becomes undersaturated as the temperature rises. The amount of melt is well below the critical melt fraction required to become mobile (Section 11.1). At about 680°C, and 0.5 GPa (point 1 in Figures 18.5a and b) muscovite begins to break down, releasing H_2O. Because this dehydration reaction occurs above the H_2O-saturated melting curve, liberated H_2O immediately induces an increment of melting. The reaction is probably: Mu + Plag + Qtz = Kfs + Sil + Melt (Reaction 28.30). The amount of granitic melt generated at this point depends upon the amounts of the reactants (muscovite in particular because quartz and plagioclase are usually more abundant in high grade metasediments). For most common metasedimentary gneisses, the amount of muscovite is

18.4 GRANITOID CLASSIFICATION

There have been numerous attempts to classify the diverse spectrum of granitoid rocks. Of course, a simple mineralogical classification (Figure 2.2) may be readily applied and is useful in naming hand specimens, but tells us little about their nature. The indication that granitoid chemical composition reflects their source and setting suggests that an alternative classification based on genetic criteria may be developed. Field relationships can also refine such an approach. A genetic classification permits us to group and study granitoids in this critical context, so that we may focus on the petrogenetic processes. I shall begin by introducing the S-I-A-M classification scheme, both for historical reasons and because this classification is useful and is often found in the literature. I shall then, however, advocate and develop a different scheme.

18.4.1 The S-I-A-M Classification of Granitoids

In 1974, Bruce Chappell and Allan White described two types of granitic suites from the rather deeply eroded subduction-related early Paleozoic Lachlan Fold Belt of eastern Australia. The suites range from quartz diorites to true granites and have distinctive petrographic, and field characteristics as well as chemical differences that cannot be related to one another by any process such as fractional crystallization or contamination. Only by originating from chemically and isotopically different source rocks could the differences be explained.

One type (for a given value of SiO_2) has higher Na, Ca, Sr, Fe^{3+}/Fe^{2+}, and $^{143}Nd/^{144}Nd$ and lower Cr, Ni, $\delta^{18}O$ and $^{87}Sr/^{86}Sr$. The common oxide is magnetite, and the rocks are hornblende-rich and are either metaluminous or weakly peraluminous (Figure 18.2). They are chemically similar to the continental-arc granitoids discussed in Sections 17.4. These were called **I-type granitoids**, and the chemical composition suggests that they are derived by partial melting a mafic mantle-derived *igneous* source material (probably a subcrustal underplate, but subducted-slab crust or older high-level pluton sources cannot be excluded).

The second suite has the opposite chemical trends to those listed above, and is peraluminous (often strongly so). The common oxide is ilmenite, and the rocks are biotite rich, and normally contain cordierite. They may also contain muscovite, andalusite, sillimanite, and/or garnet. These rocks were called **S-type granitoids**, and the chemical composition suggests that they are produced by partial melting of already peraluminous *sedimentary* source rocks imprinted by weathering at the Earth's surface. These rocks correspond to the crustal melts described in the previous section.

We have dealt mostly with field relationships of I-type granitoids so far, but S-types also occur in the eastern portions of the Andes, the Sierra Nevada batholith, and the Peninsular Ranges batholith, as well as in all of the Idaho batholith except the westernmost portion (west of the suture and 0.706 line, Figure 17.15). The common concentration of S-types inland from the trench correlates well with their sediment-derived nature because we are much more likely to find deeply-buried mature metasedimentary source rocks within the continental crust. In the Andes the "Pre-Andean" Paleozoic granitoids to the east of the Mesozoic and Cenozoic Andean I-type igneous rocks discussed in Section 17.4.1 are also principally S-types. This early (Paleozoic) eastern to later western activity described in Section 17.4.3 runs contrary to the more typical west-to-east trends, and the early S-type to later I-type is also unexpected for a maturing arc, but several authors (see Pankhurst, 1990) suggest that the poorly preserved "Pre-Andean" rocks may represent a different, intraplate form of magmatism associated with continental growth and collision/ accretion (see below) rather than typical continental-arc magmatism.

If I-type granitoids represent an "infracrustal" (below the surface, within the crust) igneous precursor, and S-types a "supracrustal" (above crust) or sedimentary precursor, what about the occasional island-arc granitoids, which have much less thick sialic crust beneath which to pond underplated magmas? Such occurrences in the Aleutians, southwestern Pacific, etc. have geochemical and isotopic characteristics indicating that they are derived by partial melting of mantle material, followed by fractional crystallization to produce more evolved magma types, just like the volcanics reviewed in Chapter 16. White (1979) thus added an **M-type granitoid** (for direct *mantle* source) to our list to include both immature arc plutons and the oceanic "plagiogranites" found in ophiolites–oceanic crust and in eroded OIBs, such as Iceland. Whether I-types are two-stage remelts of underplates and M-types are fractionation products of single-stage mantle melts is yet to be confirmed, and a range of types between these two extremes is likely.

As a final major granitoid type we can add an **A-type granitoid** (for *anorogenic*). These granites are commonly intruded into non-orogenic settings. They are generally higher in SiO_2 (mafic and intermediate types are rare), alkalis, Fe/Mg, halogens (F and Cl), REE, Ga/Al, Zr, Nb, Ga, Y, and Ce and lower in trace elements compatible in mafics (Co, Cr, Ni) and feldspars (Ba, Sr) than I-types (Whalen et al., 1987; Eby, 1990). They have an intraplate trace element signature, commonly lacking the high "decoupled" LIL/HFS ratios associated with subduction zones (Figure 16.11). A-type granitoids are diverse, both chemically and in terms of their genesis, and will be discussed further in Section 18.4.2.4.

The S-I-A-M classification is summarized in Table 18.3. It was an advancement in the study of granitoids because it demonstrated that there can be several sources for granitic rocks, and that the source rocks leave a chemical imprint on the granitoids produced. In other words, the granitoids in some ways "image" their sources. All of the highly evolved rocks in the Ab-Or-Qtz "residua" system (Figure 18.3) look similar: they're granites. The detailed geochemistry, however, shows variations that can be interpreted in terms of sources and processes.

As with most classifications, however, S-I-A-M gives the impression that the types of granitoids are really distinct, however artificial the boundaries may be. In reality, intermediate and hybrid magmas are common. I-type and S-type hybrids are likely to be produced in orogenic belts by mixing of source reservoirs and/or magmas. In fact, few orogenic belts have shown the I-S distinction as well as the original Lachlan example. Crustal assimilation is a process that may readily incorporate an S-type component into I-type melts. M-types and I-types are also likely to be mixed in mature island arcs and continental arcs. The reservoirs are so similar that it may be impossible to geochemically or isotopically distinguish them. Magmas that classify geochemically as A-type are not restricted to the original anorogenic setting. Enriched and alkaline magmas may be created in orogenic belts and hybridize with I- and S-types. In such complex spawning grounds, hybrids (leading to the suggestion of H-types!) are more likely to be the rule than the exception. Because the S-I-A-M classification ignores variations in mantle reservoirs

TABLE 18.3 The S-I-A-M Classification of Granitoids

Type	SiO_2	K_2O/Na_2O	Ca, Sr	A/(C+N+K)*	Fe^{3+}/Fe^{2+}	Cr, Ni	$\delta^{18}O$	$^{87}Sr/^{86}Sr$	Misc	Petrogenesis
M	46–70%	Low	High	Low	Low	Low	<9‰	<0.705	Low Rb, Th, U Low LIL and HFS	Subduction zone or ocean-intraplate Mantle-derived
I	53–76%	Low	High in mafic rocks	Low: metaluminous to peraluminous	Moderate	Low	<9‰	<0.705	High LIL/HFS medium Rb, Th, U hornblende magnetite	Subduction zone Infracrustal Mafic to intermed. igneous source
S	65–74%	High	Low	High metaluminous	Low	High	>9‰	>0.707	Variable LIL/HFS high Rb, Th, U biotite, cordierite Als, Grt, ilmenite	Subduction zone Supracrustal sedimentary source
A	High → 77%	Na_2O high	Low	Variable peralkaline	Variable	Low	Variable	Variable	Low LIL/HFS high Fe/Mg high Ga/Al high REE, Zr high F, Cl	Anorogenic Stable craton Rift zone

*molar $Al_2O_3/(CaO+Na_2O+K_2O)$

Data from White and Chappell (1983), Clarke (1992), and Whalen (1985).

and mixed mantle and crustal inputs, several authors have recently questioned the present usefulness of the "alphabet granitoid" classification (see Clarke, 1992, pp. 13–15, 215–218). In addition to the criticisms mentioned above, S-, I-, and M-types are based on source chemical characteristics, whereas A-types are based on tectonic regime, so the classification basis in not consistent. We might better be served by placing our emphasis on deciding what controls the chemical composition of magmas, and what chemical signatures indicate a particular process. If we are to attempt a classification of granitoids on tectonic setting, we should do so directly, without cryptic letters and mixed basis.

18.4.2 A Classification of Granitoids Based on Tectonic Setting

A number of investigators have suggested that a classification of granitoids based on tectonic setting would be an improvement over one such as the alphabet granitoids based on a mixture of chemical composition and sources. Table 18.4 is such a classification, modified from that of Pitcher (1983, 1993). Although a chemical classification may provide a researcher with some criteria with which she or he may attempt to characterize possible source materials, a classification based on tectonic environment provides the student of granitoids with a conceptual framework to help understand the occurrence of granitoids and their genesis. As you have perhaps realized by now, I am not a great lover of classifications. In spite of having given you several, I think they force the lovely variability and continuity of nature into inappropriately neat compartments,

and they commonly have the same effect on our thought processes. But when we first embark on the investigation of a subject, having a conceptual framework is of great value in providing an initial perspective, and this outweighs the disadvantages. For a comprehensive review of granitoid classifications see Barbarin (1990, 1999).

Table 18.4 reveals that granitoids occur in a wide variety of settings. These can broadly be grouped into orogenic and anorogenic settings. **Orogenic** is rather narrowly defined here as mountain-building resulting from compressive stresses associated with subduction. **Anorogenic** refers to magmatism within a plate or at a spreading plate margin. **Post-orogenic** is rather difficult to classify because it depends on a prior event (rather than the event that created it) to have meaning. It has thus been classified as orogenic by some investigators (who stress the link with orogeny) and anorogenic by others (who stress the tectonic regime at the time of formation). I've called it **transitional**, meaning that it is not really either but has some aspects of each. This is not to say that transitional granitoids must occur between orogenic and anorogenic magmatic events, although this is commonly true.

18.4.2.1 OCEANIC GRANITOIDS Anorogenic granitoid magmatism may be separated into oceanic and continental types. The most common oceanic anorogenic granitoids are probably the **plagiogranites** formed at mid-ocean ridges (Section 13.3 and Figure 13.4), although exposures are rare because of the scarcity of eroded ophiolites and their subordinate role in them. Another example is **ocean island granitoids**. These are simply the intrusive equivalents of the

TABLE 18.4 A Classification of Granitoid Rocks Based on Tectonic Setting After Pitcher (1983, 1993) and Barbarin (1990).

	OROGENIC			TRANSITIONAL	ANOROGENIC	
	Oceanic Island Arc	Continental Arc	Continental Collision	Post-Orogenic Uplift/Collapse	Continental Rifting, Hot Spot	Mid-Ocean Ridge, Ocean Islands
	● = granitoid magma; underplated mantle melts	mantle wedge melting	batch melting / local anatexis	decompression melting	decompression melting / hot spot plume	hot spot plume
Examples	Bougainville, Solomon Islands, Papua New Guinea	Mesozoic Cordilleran batholiths of west Americas Gander Terrane	Manaslu and Lhotse of Nepal, Amorican Massif of Brittany	Late Caledonian Plutons of Britain, Basin and Range, late Variscan, early Northern Proterozoic	Nigerian ring complexes, Oslo rift, British Tertiary Igneous Province, Yellowstone hotspot	Oman and Troodos ophiolites; Iceland, Ascension, and Reunion Island intrusives
Geo-chemistry	calc-alkaline > thol. M-type & I-M hybrid metaluminous	calc-alkaline I-type > S-type metalum. to sl. peral.	calc-alkaline S-type peraluminous	calc-alkaline I-type S-type (A-type) metalum. to peralum	alkaline A-type peralkaline	tholeiitic M-type metaluminous
Rock types	qtz-diorite in mature arcs	tonalite & granodior. > granite or gabbro	migmatites & leucogranite	bimodal granodiorite + diroite-gabbro	granite, syenite + diorite-gabbro.	plagiogranite
Associated Minerals	Hbl > Bt	Hbl, Bt	Bt, Ms, Hbl, Grt, Als, Crd	Hbl > Bt	Hbl, Bt, aegirine fayalite, Rbk, arfved.	Hbl
Opaque Oxides	magnetite series (Ishihara, 1977)		ilmenite series	(Ishihara, 1977)		magnetite series (Ishihara, 1977)
Associated Volcanism	island-arc basalt to andesite	andesite and dacite in great volume	often lacking	basalt and rhyolite	alkali lavas, tuffs, and caldera infill	MORB and ocean island basalt
Classification Barbarin (1990)	T_IA tholeiite island arc	H_CA hybrid calc-alkaline	C_ST C_CA C_CI continental types	H_LO hybrid late orogenic	A alkaline	T_OR tholeiite ocean ridge
Barbarin (1999)	ATG arc tholeiitic granitoids	ACG Amp-bearing (low-K) calc-alk. granitoids	MPG CPG Ms and Crd per-aluminous granitoids	KCG K-rich calc-alk granitoids	PAG peralkaline and alkaline granitoids	RTG mid-ocean ridge tholeiitic granitoids
Pearce et al. (1984)	VAG (volcanic arc granites)		COLG (collision granites)		WPG and ORG (within plate and ocean ridge granites)	
Maniar & Piccoli (1989)	IAG island arc granite	CAG contin. arc granite	CCG cont. collision gran.	POG post-orogenic gran.	RRG CEUG rift & aborted/hotspot	OP ocean plagiogranite
Origin	partial melting of mantle-derived mafic underplate	PM of mantle-derived mafic underplate + crustal contribution	partial melting of recycled crustal material	partial melting of lower crust + mantle and mid-crust contrib.	partial melting of mantle and/or lower crust (anhydrous)	partial melting of mantle and frac-tional crystallization
Melting Mechanism	subduction energy: transfer of fluids and dissolved species from slab to wedge. Melting of wedge, transfer of heat upward		tectonic thickening plus radiogenic crustal heat	crustal heat plus mantle heat (rising asthen. + magmas)	hotspot and/or adiabatic mantle rise	

After Pitcher (1983, 1993), Barbarin (1990, 1999)

evolved phonolites, trachytes, and rhyolites discussed in Chapter 14. Because both ocean ridges and islands occur within the mafic oceanic province, assimilation of felsic crustal material is highly improbable, and both intrusive and extrusive magmas are attributed to formation via fractional crystallization from mantle-generated basaltic magmas or remelting of deep mafic crust. The chemical characteristics of granitoids formed by fractional crystallization can be explained satisfactorily on the basis of relatively simple fractionation models, and the process has been described in previous chapters. The tholeiitic trend is most common in the intraplate oceanic realm, but alkaline types occur in the alkaline provinces, such as Ascension and Réunion islands.

Oceanic *orogenic* granitoids occur where one oceanic plate is subducted beneath another to form an **island arc** (Chapter 16). Immature arcs are dominated by tholeiitic volcanic magmatism, with only a few evolved plutons exposed. Mature arcs have calc-alkaline plutons. As with their volcanic counterparts, these granitoids are largely products of partial melting of the mantle wedge above the subducted oceanic slab, where melting is caused by the ingress of slab-derived hydrous fluids. These are the typical M-type (mantle) granitoids, although some hybridization via interaction with or remelting of arc crust is likely. Because the crust is also M-type this interaction can be difficult to document.

18.4.2.2 CONTINENTAL OROGENIC GRANITOIDS Mature arcs amalgamate to form continents and **continental arcs**, which are the home of the major batholith belts described in Chapter 17. Continental arcs are very complex, and the participation of mature, evolved continental crust and sediments has been documented in many granitoids. The model for continental arcs and some mature arcs with thick crust that is presently in vogue proposes that the batholiths are the result of two-stage melting of the mantle wedge (Figure 17.20). The first stage involves a mantle source and produces basaltic magma (as in the island arcs above). Much of this magma pools at the base of the less dense crust where several processes may take place, including assimilation of crust and solidification of a gabbroic (or slightly hydrous amphibolitic) crustal underplate. Subsequent partial melting of this mafic underplate (now part of the crust) may result from heat carried upward by continued basaltic magmatism. The result is typically a tonalitic magma that is light enough to ascend to a shallow level where further fractionation and solidification take place. These are the typical I-type granites because the underplate is crustal and of magmatic origin. Assimilation of fertile overlying crust by rising magmas must be common, as would be localized crustal melting due to the advective heat transported upward by the melts. The chemical and structural complexity of the crust means that numerous sources may contribute, either separately or premixed, at the point of assimilation (for example by sedimentary layering or thrust stacking). Thus S-types (sedimentary) or hybrid types may also be created. Although metaluminous to weakly peraluminous I-type tonalites and granodiorites are predominant, a wide variety of granitoids occurs at an Andean-type boundary, including strongly peraluminous S-types and even some M-type gabbros and diorites when extension provides conduits.

Our discussion of subduction thus far has been limited to island arcs and continental arcs such as the Andes, where subduction of oceanic crust has been continuous throughout the Mesozoic and Cenozoic. Commonly, however, a segment of continental lithosphere will ride in on the subducting plate, resulting in **continental collision**. The collision of two continents creates an impressive orogeny, with dramatic thrusting, folding, and crustal thickening. This has significant consequences for crustal melting that differ from the types we have discussed so far in this text. We shall thus investigate this situation in a bit more depth than the brief reviews above.

When a thick section of continental crust enters a subduction zone, its buoyancy keeps it from being subducted very far. Eventually the underthrusting of the subducting low-density mass ceases, and the subduction zone terminates. Continued plate convergence may cause a new subduction zone to form behind one of the involved continents, or a global plate tectonic rearrangement may accommodate the change.

The type locality for continental collision is the Himalayas, where the Indian subcontinent was recently partially thrust beneath the southern portion of Asia. Older, but obvious, collisions also occurred in the Appalachian–Caledonian orogen and in the Alps. It is now apparent that collisions of continents and smaller masses (micro-continents

and island arcs) were common throughout the geologic past. The North American Cordillera is a well-documented collage of such micro-continents. Work in the cratons of all continents indicate that they are similar collages, and that continental growth involved both the addition of magmas and sediment aprons at active margins, but also the assembly of such collages.

Ancient continent, micro-continent, or arc collision zones are recognized today by **suture zones**: fault zones separating terranes of contrasting lithology, paleo-fauna, and/or magnetic paleo-latitude. Slivers of oceanic crust and upper mantle (ophiolites) are commonly faulted into the zone as further clear evidence that an ocean basin (or oceanic back-arc basin) once separated the terranes.

Magmatism is a further key to identifying such zones. Because pre-collision continental-arc magmatism must have occurred only on the overriding continental plate, one can use the eroded remnants of the igneous products to determine the **polarity** (direction of dip) of the subduction that led to collision. The collision itself leads to intense later magmatism (mostly granitoid) as well. Figure 18.6 is an analog of Figure 16.18 with the addition of an underthrust section of continental crust, the buoyancy of which caused subduction to shallow and eventually terminate. I have taken liberties in modifying the subduction-related isotherms for such a situation, allowing them to drift upwards with the shallowing plate, but not yet totally rebound. You get the idea from Figure 18.6 that it would be simple to reach the 650 to 700°C minimum melting temperature of H_2O-saturated granites or the 800 to 850°C melting temperature for undersaturated ones within the doubly thickened crust. Relaxation of the depressed isotherms after subduction has ceased and possible heating due to frictional heating or the concentration of heat-producing radioactive elements (notably U-Th-K) may result in even higher temperatures than those illustrated. Melting of the sialic crustal rocks produces syn-orogenic S-type granitoids, or even later transitional post-orogenic A-types.

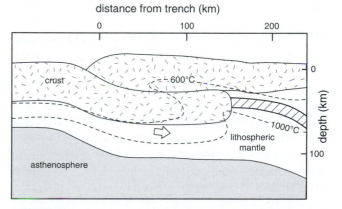

distance from trench (km)

FIGURE 18.6 A simple modification of Figure 16.18, showing the effect of subducting a slab of continental crust, which causes the dip of the subducted plate to shallow as subduction ceases and the isotherms begin to "relax" (return to a steady-state value). Thickened crust, whether created by underthrusting (as shown) or by folding or flow, leads to sialic crust at depths and temperatures sufficient to cause partial melting.

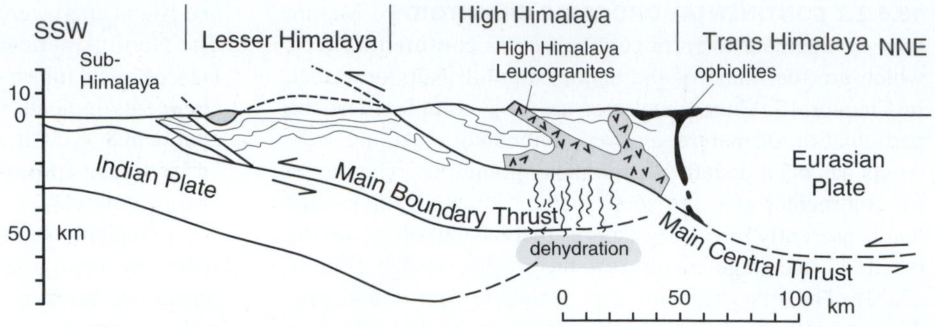

FIGURE 18.7 Schematic cross section of the Himalayas, showing the dehydration and partial melting zones that produced the leucogranites. After France-Lanord and Le Fort (1988).

Le Fort (1988) has shown that there are five separate belts of granitoid intrusions parallel to the axis of the Himalayas, each with its own compositional, temporal, and genetic relationships. One of them, the peraluminous two-mica leucogranites of the High Himalaya, is particularly well exposed in three dimensions, and serves as a good example of collision-generated peraluminous S-type granitoids. The bodies range from 100 m wide sills to plutons 30 km across and 10 km thick (Manaslu leucogranite). They have high Rb, Cs, Ta, and U concentrations and low Ba, Sr, and $(La/Yb)_N$ (=REE slope). $\delta^{18}O = 9 - 14\%$, $(^{87}Sr/^{86}Sr)_o = 0.703 - 0.825(!)$, and $\varepsilon_{Nd} = -11.5$ to -17, indicating a large crustal component. The eastern bodies are intruded between the High Himalaya Sedimentary series and the tectonically underlying Precambrian–Paleozoic High Himalayan migmatitic gneisses (Figure 18.7). The western bodies appear as anatectic melts within the gneisses.

These spatial relations, and the isotopic similarity to the lower gneisses, indicate that the granites originate as localized small-degree partial melts of the High Himalayan gneisses (only some of the melts escaped the source). France-Lanord and Le Fort (1988) proposed that melting of the gneisses was induced by the introduction of hydrous fluids rising from the hydrated upper crustal rocks of the Indian plate that were pushed beneath the Main Boundary Thrust (Figure 18.7). Dehydration of the lower slab due to the structural thickening of the crust may release fluids upward into previously deeper and still hotter crustal material where melting is induced (note the inverted isotherms in Figure 18.6). Radiogenic heating in the U-Th-K-rich crust and the insulating effect of the upper plate may all have contributed to melting in the area (Patiño Douce et al., 1990). Nabelek and Liu (2005) cited several lines of evidence indicating that partial melting was not H_2O saturated, however, requiring temperatures hotter than 700 to 750°C and, therefore, a supplemental source of heat. They suggested that shear strain heating can affect a significant volume of rocks near large crustal thrust shear zones and supply the required excess heat (Zhu and Shi, 1990; England and Molnar, 1993; Harrison et al., 1998).

Most orogenic granitoids are associated with (or follow) an orogenic event in which deformation, metamorphism, and magmatic activity culminate. This event may occur as a result of continental collision or docking of smaller terranes (island arcs, mini-continents, or oceanic plateaus), as described above. Other proposed trigger events include periods of rapid subduction, slab break-off (separation of dense subducting oceanic lithosphere from more buoyant trailing material, e.g., Davies and von Blankenburg, 1995), and ridge subduction (Iwamori, 2000; Iwamori et al., 2007).

18.4.2.3 TRANSITIONAL GRANITOIDS Transitional (post-orogenic) **granitoids** are somewhat enigmatic. They are characteristically generated in an orogenic belt between 10 and 100 Ma, after compressive deformation has ceased. They are typically emplaced during a period of uplift or extensional collapse that follows orogeny. As with the intrusions associated with continental collision they have a predominantly crustal chemical and isotopic signature, and are created in areas of previously overthickened crust. The prevalence of post-orogenic granitoids in such settings suggests that thickening of the continental crust is somehow responsible for both transitional granitoid generation and uplift or collapse.

Compressive orogeny can lead to crustal thickening by some combination of flattening, thrust stacking, folding, and magmatic addition/underplating. The crust in areas of continental collision is dramatically thickened, but continental arcs (without collision) are also unusually thick with respect to average continental crust (Figure 18.8a). Compression and isostatic equilibrium can maintain a thickened and elevated crustal welt as long as compression is maintained (Figure 18.8b), so why are granites generated, and why does extensional collapse so commonly ensue? The thermal models of Patiño Douce et al. (1990), beginning with thrust-stacked crust as in Figure 18.8a1, indicated that heat from the normal mantle flux, **thermal relaxation** (rebound of subduction-depressed isotherms), and radioactive heat from within the U-Th-K enriched crust itself is sufficient to generate partial melts in the crust, provided that a fertile source material is available. They concluded that if the crust is thickened by at least 150%, mobile granitoids can be generated, without requiring any advective heat transfer by shear or rising mantle melts. The "thermal blanketing" effect of thickened crust alone may thus be able to create crustal granitoids in situations such as Figure 18.8a and b, and Figure 18.7.

Other investigators argue that increased mantle input of heat is required for crustal melting (e.g., Brown, 1994). Such is the situation for continental-arc magmatism in which the underplated mafic magmas provide considerable heat by advection. Let's begin with a general case, such as Figure 18.8b (regardless of how arrived at), and suppose that the crust is too cool to melt. Perhaps the simplest way to increase the heat flow is by surface erosion and the responding isostatic rise of the

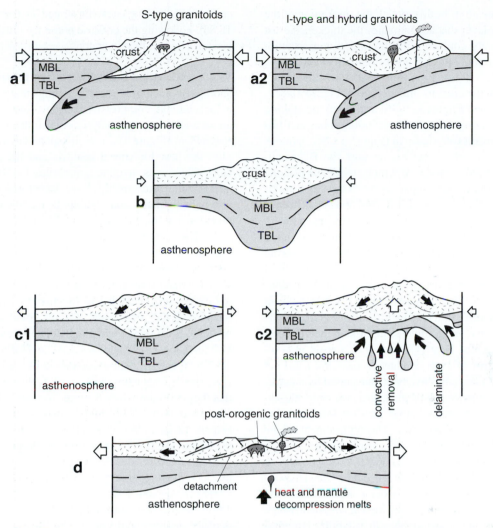

FIGURE 18.8 Schematic models for the uplift and extensional collapse of orogenically thickened continental crust. Subduction leads to thickened crust by either continental collision **(a1)** or compression of the continental arc **(a2)**, each with its characteristic orogenic magmatism. Both mechanisms lead to a thickened crust and probably thickened mechanical and thermal boundary layers (MBL and TBL), as in **(b)**. Following the stable situation in **(b)**, either compression ceases **(c1)** or the thick, dense thermal boundary layer is removed by delamination or convective erosion **(c2)**. The result is extension and collapse of the crust, thinning of the lithosphere, and rise of hot asthenosphere **(d)**. The increased heat flux in **(d)**, plus the decompression melting of the rising asthenosphere, results in bimodal post-orogenic magmatism with both mafic mantle and silicic crustal melts.

crust. If this process is sufficiently rapid, rising crust and mantle may raise the isotherms physically before the temperature can readjust, a case of simple advection. If the rocks at the base of the pile are near their solidus, near-adiabatic decompression partial melting may ensue (Section 10.2.2 and Figure 10.4). Although this process may work in some situations, post-orogenic magmatism is typically associated with extensional collapse of an orogen. Thus we seek a mechanism that can create extension, crustal thinning, and sufficient heat to produce melts in combination. One way may simply be to cease compression (Figure 18.8c1). The thickened crust is then no longer supported laterally, and the welt may collapse and spread (Figure 18.8d), similar to a thick dollop of pancake batter in a skillet.

An alternative model for heating and collapse of a thickened orogen that does not rely on the cessation of compression and timely rebound has been developed by Houseman et al. (1981), England and Houseman (1989), and England (1993). These investigators suggested that the hotter mobile convective asthenosphere cannot directly contact the cooler rigid lithosphere above without some form of transitional layer developing between them. This layer is the horizontal moving part of the convection cell that loses heat upward by conduction. They call the rigid lithosphere the **mechanical boundary layer** and the transitional layer beneath it the **thermal boundary layer**. Both layers are part of the lithospheric mantle. In the oceans the mechanical boundary layer is proposed to be about 80 km thick and the thermal boundary layer about 50 km thick. Their thickness beneath the continents is more speculative. The thermal boundary layer cools and becomes more dense than the warmer asthenosphere beneath it. In this model, the thickening of the thermal boundary layer that accompanies the thickened crust enhances the density contrast with the underlying asthenosphere, so the thermal boundary layer eventually sinks.

Bird (1978, 1979) proposed that the lithosphere **delaminates** from the crust to sink *en masse* (Figure 18.8c2, right). Houseman and England (1986) attributed the sinking to a piecemeal process of smaller descending lobes and bits that they called **convective removal** (Figure 18.8c2, center). Black and Liégeois (1993) and Davies and von Blanckenburg (1995) preferred a model in which the dense oceanic portion of the subducted slab separated from the stuck continental portion and sank, a process referred to as **slab break-off**. Because I chose a genetically neutral starting point in Figure 18.8b, there is no apparent slab, but you can imagine the slabs in Figures 18.8a1 and a2 peeling away, breaking off, and sinking.

Regardless of the exact mechanism, as thick portions of the slab or thermal boundary layer sink, warmer, buoyant asthenospheric material flows upward as a replacement. The rising asthenosphere is subject to decompression melting, producing mafic magmas. Heat flux into the crust is also increased, in the form of both conductive heat from the hot asthenospheric welt and advective heat transported by magma. Either source can generate crustal melts in a post-orogenic setting following the time-lag required for destabilization and loss of the thermal boundary layer. The replacement of the dense boundary layer by the hot asthenosphere (whether by delamination or convective removal) decreases the density of the section, resulting in uplift of the crust. This model has been applied to the delayed uplift of the Tibetan Plateau following the Himalayan orogeny. Uplift places crust at sufficiently high elevations that the compressive stresses that stabilized the situation in Figure 18.8b can no longer provide sufficient lateral support and the orogen extends and collapses (Figure 18.8d) under its own weight (as with that pancake batter).

We thus have some reasonable models that explain why a thickened orogen can lead to a post-orogenic phase dominated by extension, normal and detachment faulting, crustal thinning, and magmatism. The granitoids are principally of crustal origin, but decompression melting of the upwelling asthenosphere under the extending lithosphere can contribute material (basalts) as well as heat. The composition of post-orogenic granitoids ranges from strictly crustal peraluminous S-types, to metaluminous I-types and hybrids, to those with A-type chemical characteristics. The high-K calc-alkaline series (Section 16.2) typically predominates, however, with subordinate shoshonitic rocks (Liégeois, 1998; Liégeois et al., 1998). The mafic mantle magmas have access to the surface via the normal faults that accompany extension, which can explain why post-orogenic magmatism is commonly *bimodal*. Volcanism in the extending Basin and Range of the southwestern United States is characteristically basalt–rhyolite. Extension is typically accompanied by listric fault systems and large, low-angle normal detachment faults. Hot and ductile lower crust may then rise under the extending detachment to form **metamorphic core complexes** (Coney, 1980). More often than not, there is a crustal granitoid intrusion associated with the rising mobile

core of these complexes (illustrated in the center of Figure 18.8d). Following the collapse phase the crust in most orogens becomes thinner than in cratons.

18.4.2.4 CONTINENTAL ANOROGENIC GRANITOIDS Crustal anorogenic granitoids occur in settings that are not genetically associated with compressive orogeny. One such occurrence is above a subcontinental hotspot, such as the Yellowstone–Snake River Plain (Figure 15.5). At hotspots, the mantle plume supplies the heat for crustal anatexis and the generation of locally voluminous rhyolitic ignimbrites (>2000 km^3 of ash at Yellowstone), presumably above deeper granitoids. The plume also partially melts, contributing the basaltic component of another form of bimodal basalt-rhyolite magmatic suite.

The more typical anorogenic peralkaline A-type granitoids occur in association with broad diffuse extension or intracontinental rifts, either failed rifts (aulacogens) or the incipient stages of main-stage rifting that lead to the formation of spreading centers and ocean basins (Section 19.1). As mentioned in Section 18.4.1, anorogenic A-type granitoids are generally higher in SiO$_2$ (mafic and intermediate types are rare), alkalis, Fe/Mg, halogens (F and Cl), Ga/Al, and HFS elements such as Zr, Nb, Y, and Ce than I-type granitoids (Whalen et al., 1987; Eby, 1990). They have an intraplate trace element signature, in that they commonly lack the high "decoupled" LIL/HFS ratios associated with H$_2$O fluid transport in subduction zones (Figure 18.4). Collins et al. (1982) suggested that anorogenic granitoids originate by remelting of a deep-seated halogen-rich granulite facies residue that was depleted and dehydrated by the extraction of orogenic granitoids and left in the lower crust. Others (Anderson, 1983; Whalen et al., 1987; Creaser et al., 1991; Skjerlie and Johnston, 1993) have argued that such a depleted source could not give rise to such alkaline magma, and proposed that the source is a deep, dehydrated orogenic tonalite or granodiorite. The high temperature and H$_2$O-poor/halogen-rich nature of anorogenic granitoids are crucial for their interpretation. A dehydrated source requires fairly high temperatures to generate the critical melt fraction needed to become mobile. Fluorine may flux the source somewhat, and it is believed that the presence of the halides creates selective trace element-halide complexes in the melt. This "counter ion effect" draws Ga and the HFS elements into the melt, and increases their concentration in anorogenic granitoids.

Eby (1990) recognized that granitoids characterized geochemically as anorogenic A-types occur in a number of situations, mostly continental, but including oceanic islands such as Ascension and Réunion. He suggested that anorogenic granitoids may result from a variety of processes, including the aforementioned remelts, but also including fractionation from enriched mantle-derived mafic magmas, alkali–halogen metasomatism, and/or crustal assimilation by mantle magmas. Melting of sedimentary rocks rarely, if ever, produces peralkaline magmas.

The types of anorogenic magmas associated with incipient rifting commonly occur as shallow intrusives and ring-dikes. This probably results from the mobility of such hot melts and the presence of extensional faults. Classic examples include the Oslo Graben of Norway, the Niger–Nigerian alkaline-ring complexes, and granitoids associated with the opening of the Red Sea.

The greatest anorogenic magmatic event(s) occurred in the Proterozoic, when a huge volume of anorogenic rocks were emplaced along a 6000-km-long belt, extending from southern California to Labrador, across Greenland, and into the Baltic Shield (all part of a single supercontinent at the time). A suite of intrusives, commonly called the **anorthosite–mangerite–charnockite–granite**, or **AMCG suite**, is the typical Proterozoic association. The anorthosites form astoundingly large massifs of nearly pure plagioclase (Chapter 20). Many mafic rocks of mantle origin are associated with the AMCG suite, so that this Proterozoic form of magmatism is also bimodal. The felsics are mostly of crustal origin, and range from syenites to granites, with predominant quartz syenites and monzonites. Orthopyroxene is very common in these dry rocks, hence the terms *charnockite* (orthopyroxene granite) and *mangerite* (orthopyroxene monzonite). Charnockites are high-temperature, nearly anhydrous rocks, and can be of either igneous (Kilpatrick and Ellis, 1992; Frost and Frost, 2008) or high-grade metamorphic origin. We usually associate orthopyroxene with mafic rocks, but high-temperature dry granitoids may contain hypersthene, and peralkaline granitoids commonly have Na- and Fe-rich pyroxenes, and some even have fayalitic olivine. Although we usually consider olivine to be unstable with quartz, this is not true at the extreme Fe-rich end of the system (Figure 12.7), and Fe-rich olivine is stable with quartz in some of these granitoids.

The Proterozoic belt of anorogenic magmatism was associated with broad, mild extension (Anderson, 1983; Bickford and Anderson, 1993). The heat required for such extensive crustal melting has been attributed to mantle convection, perhaps resulting from the thermal blanketing effect of the crust subsequent to supercontinent formation near the end of the Archean (similar to the model illustrated in Figure 18.8). We might extend this model to consider a general sequence of subduction, collision, heating, collapse, and finally rifting as heat is carried up toward the extending crust. According to this theory, the late stages of collapse lead to rifting and anorogenic magmatism, which may then lead to continental-arc orogeny on one or both rifted continental edges, hence leading to a *cycle* of tectonic settings. Such a scenario led J. T. Wilson (1966) to propose the cyclic closing and reopening of the Iapetus/Atlantic system. These **Wilson Cycles** (as they are now called) result in the repeated amalgamation and dispersal of supercontinents. The now-famous breakup of Pangea (Deitz and Holden, 1970) is only the latest dispersal episode. The occurrence of such cycles does not necessarily imply that all orogens must lead to rifts because a number of collision zones, such as the Urals, never reopened and the process ceased.

In summary, an area might experience a sequence of tectonic regimes, and the granitoids may record that sequence. Harris et al. (1986) described a sequence of four series of magmas that are found in many collision zones. The first is a typical pre-collision I-type continental-arc series. This is followed by a syn-collision peraluminous leucogranite series such as the High Himalayan series described above. Next is a late- or post-collision calc-alkaline series that

may be mantle derived but undergoes considerable crustal contamination. A final series is post-collision alkaline intrusives with an intraplate signature (lacking high LIL/HFS). Brown et al. (1984) also noted a transition from tholeiitic to calc-alkaline to peraluminous to peralkaline/alkaline chemical composition through time associated with subduction to collision to extension. Although this sequence may be common, it is certainly not the only possibility. These transitions are generally not as distinct as Table 18.4 implies. They may be gradual, and even overlap in many cases.

18.5 GEOCHEMICAL DISCRIMINATION OF TECTONIC GRANITOIDS

In Section 9.6, we discussed attempts to determine the tectonic setting of ancient basalts on the basis of their trace element characteristics. Several investigators have attempted to do the same for granitoid rocks. We have seen above that the tectonic setting can be very broadly correlated with granitoid chemical composition, but that considerable variety in sources and melting processes is inevitable. We are also aware that elements incompatible in basalts are commonly incorporated in the mineral phases of granitoids, so the concentrations of many trace elements may be controlled by the occurrence of accessory phases, even of restites. Nonetheless, the ability to constrain the tectonic setting of ancient granitoids that are no longer recognizably in the setting in which they were generated is certainly a desirable goal.

De la Roche et al. (1980) used a variation diagram based on combined major element parameters to classify volcanic and plutonic rocks. Batchelor and Bowden (1985) used this "R1–R2" diagram [$R1 = 4 Si - 11(Na + K) - 2(Fe + Ti)$ and $R2 = 6 Ca + 2 Mg + Al$] in an attempt to discriminate amongst granitoid rocks from various settings. Although there is considerable overlap, and the fields converge toward values of low R2 and high R1, some degree of separation between mantle plagiogranites, subduction zone-, late-orogenic-, anorogenic-, and crustal melt-granitoids was apparent. Maniar and Piccoli (1989) used a wide variety of major elements to discriminate between the groupings listed in Table 18.4 (see row 11). No single diagram was successful, but by applying several, they could at least partly constrain the origin of some granitoids.

Pearce et al. (1984) used combinations of trace elements, such as Nb versus Y, Ta versus Yb, Rb versus (Y + Nb), and Rb versus (Yb + Ta), to distinguish between granitoids (row 9 in Table 18.4). Harris et al. (1986) used a triangular plot of Rb-Hf-Ta to discriminate between subduction zone, collision, post-tectonic calc-alkaline, and post-tectonic alkaline types. Figure 18.9 shows two examples of the discrimination diagrams used by Pearce et al. (1984). In both diagrams all the granitoids in Table 18.2 plot in the VAG (volcanic arc) area, except Ascension, Nigeria, and A-type, which appropriately plot in the WPG (within plate) field. The plagiogranite does not plot in the ORG (ocean ridge) field, but it's close.

As discussed in Chapter 9, these discrimination diagrams are empirical, and, given the complexity of most igneous (and subsequent metamorphic) processes, attempting to

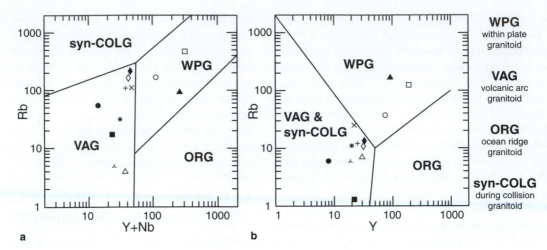

FIGURE 18.9 Examples of granitoid discrimination diagrams used by Pearce et al. (1984), with the granitoids of Table 18.2 plotted (symbols as in Table 18.2).

determine the tectonic setting of igneous and meta-igneous rocks on the basis of chemical parameters is a tenuous proposition. This is particularly true for granitoids, in which minor phases, mixed sources, volatile interaction, and variable partial melting and restite accumulation can have considerable effects on trace element geochemistry. If the rocks are very old, however, and no longer in their original setting, field criteria may be obscure and this approach may be the only available method to reconstruct past geodynamic conditions. If this is the case, I advocate using as many discrimination parameters as possible. If a number of schemes indicate the same setting, one might infer a tectonic setting with slightly more confidence.

Much of my skepticism about the reliability of discrimination among granitoids results from my conviction that the Earth is much more complex than any simple classification, such as those in Table 18.4 and Figure 18.8, can imply. All attempts to classify nature are at least partly artificial. Some granitoids have features of more than one type listed in Table 18.4. Also, several types of granitoids may be associated in the same area. As Barbarin (1990) pointed out, most of the types listed above occur together in the French Massif Central. In the Andes, we can find M-types at the leading edge and then encounter I-types, S-types, and even A-types as we progress eastward. Several of these are coeval. In the western United States, orogenic continental-arc magmatism occurs in the Cascades where the Juan de Fuca plate subducts beneath the North American plate, while at the same time in the Basin and Range province to the south no subduction occurs, but post-orogenic bimodal basalt–rhyolite magmatism and extension does (Figure 17.10). It would be wonderful if discrimination diagrams were able to distinguish these various types in these subsettings, but such fine-tuning is unlikely, and we would be fortunate if we could simply discriminate larger subdivisions.

18.6 THE ROLE OF THE MANTLE IN GRANITOID GENESIS

In the discussion above, we assumed a crustal source for most granitoid melts, but what exactly does the mantle contribute? Does it supply heat, material (melts), or both? In a poll of granitoid petrologists by Jean-Paul Liégeois in the granite listserv discussion group (granite-research@listserv.umd.edu), the responses to the question of the source of granites were:

Crust only	1%
Crust with only minor mantle	21%
Both, but more crust	41%
Both, unknown proportion	15%
Both, but more mantle	7%
Mantle with only minor crust	1%
Mantle only	0%
Other (including "I don't know")	13%

Thus 85% of those with some degree of expertise on the subject think that both the crust and mantle are involved. The nature of the involvement is where most of the controversy lies. The question is important because if the only mantle contribution is heat, then the continental crust is largely recycled, but if mantle material is supplied, then we're looking at crustal *growth*. In the case of oceanic plagiogranites and ocean island granitoids, the mantle appears to supply everything to the granitoid recipe. In the continental crust, the role of mantle material is probably more variable, and is the subject of some debate.

In highly overthickened collision orogens, the mantle may simply supply the "normal" heat flux to the crust, which, as discussed above, may in some situations be sufficient to induce crustal anatexis. In other cases, however, extra heat from the mantle may be required to reach crustal melting temperatures. This heat may result from asthenospheric upwelling or rising mantle melts, typically as they underplate the base of the crust. Finally, the mantle may supply material to the crust in the form of partial melts. M-type granitoids are believed to be fractionated mantle melts. Underplated mantle melts may assimilate crustal components, or they may generate and mix with crustal melts. The MASH hypothesis discussed in Chapter 17 is one such possibility. The two-step model for continental-arc granitoids illustrated in Figure 17.20 is

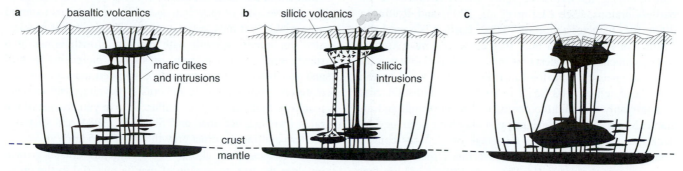

FIGURE 18.10 Evolution of silicic magma chambers formed by the emplacement of mantle-derived basaltic magma into the crust. **(a)** Early stage with colder, brittle crust, when much of the mafic magma reaches the surface via faults. **(b)** Mafic magma heats the crust and causes local crustal anatexis. Denser mafic magmas are trapped in the more ductile crust, but silicic melts rise to form shallow intrusions and feed volcanics. **(c)** When a large region of the crust is close to melting, large silicic magma bodies can be generated, often accompanied by major ignimbrite eruptions and caldera collapse. After Huppert and Sparks (1988a). Reprinted by permission of Oxford University Press.

another, in which the underplate becomes a crustal source to be remelted and form tonalites.

Huppert and Sparks (1988a) suggested a model by which the injection of mantle-derived basalts into the crust generates silicic crustal melts (Figure 18.10). Such a model may best be applied to bimodal magmatism in post-orogenic and anorogenic situations that are dominated by extension and asthenospheric upwelling. In the model, injected basaltic magmas exploit extensional faults in cool, brittle crust to form dikes, sills, and intrusions (Figure 18.10a). As advective heat introduced by these magmas warms the crust, silicic crustal melts are generated (Figure 18.10b), and mafic volcanics give way to silicic ones. Mafic volcanism wanes as the crust warms and fracture-type conduits become scarce. The very hot mafic intrusions melt the crust at the floor, walls, and roof of the chamber. Light crustal melts pond at the roof, forming a stable layer with negligible mixing with the hot dense magma below. Huppert and Sparks (1988) calculated that thick mafic sills can create large volumes of overlying silicic magma in just a few hundred years. In the later stages of warmed crust (Figure 18.10c), large granitoid magma bodies may coalesce and ascend to shallow levels where caldera collapse and ignimbrite fields develop.

The role of the mantle in granitoid generation can thus be variable and complex but appears able to involve both heat and material transfer to the crust. The material contribution may differ from area to area, and even from one magma to the next within an area. The isotopic signature of various reservoirs is commonly distinctive enough to supply important criteria for evaluating the mantle contribution, but the reservoirs are also complex enough in orogenic areas that the job of distinguishing them is seldom as straightforward as we might wish.

18.7 ORIGIN OF THE CONTINENTAL CRUST

The debate over the contribution of the mantle to crustal magmatism leads to uncertainty in the rate of formation of the felsic continental crust because this must be accomplished by the gradual addition of magmatic material from the mantle. At present, this transfer is being accomplished

mostly at subduction zones, as discussed with the petrogenetic model at the end of Chapter 17, but mantle material can be added by post-orogenic and anorogenic processes as well. The debate over the true mantle contribution reflects the difficulty in evaluating the rate of crustal development. Exactly how much of the continental crust was created when, and how, is thus pretty speculative and controversial stuff. For reviews of the various estimates of crustal growth over time, see Ashwal (1989), Rudnick (1995), and several chapters in Brown and Rushmer (2006). Rather than attempt to untangle this web, I'll settle on providing a very general outline of early crustal history based on a summary by Windley (1995). Although there are other viewpoints and plenty of arguments over details, I think this scenario provides a suitable impression of the broad aspects of crustal genesis.

In Chapter 2, I provided a brief summary of the early differentiation of the Earth into a silicic mantle and metallic core, which probably took place within the first 100 Ma of Earth history. But the crust was generated later. A huge gravitational overturn of the magnitude of the mantle-core separation must have released considerable thermal energy. Early radioactive heat production must also have been higher, and Windley (1995) outlined an Archean Earth in which the heat production was two to three times that of today. One ramification of the elevated Archean heat flow is that mantle melting was extensive. For example, **komatiites**, ultramafic lavas with eruption temperatures of about 1650°C, must result from large fractions of mantle melting. They occurred almost exclusively in the Archean.

Most investigators agree that some form of plate tectonics occurred in the Archean, and most of the Earth's high heat flow was probably lost at ocean ridges. Convection must have been vigorous, and production and recycling of oceanic crust (composed of tholeiitic and komatiitic lavas plus intrusives) was correspondingly rapid. Subduction of the warm crust probably resulted in shallow dipping subduction zones with high geotherms. There was thus probably considerable melting of the subducted slab itself, instead of dehydration and fluxing of the mantle wedge as is more common today. The result was the production of tonalites, trondhjemites and granodiorites (the Archean **TTG suite**) that dominate the extensive "gray gneiss terranes" of the earliest cratons. (See Martin et al., 2005, and Rollinson, 2006, for characterization of TTGs.) Early TTGs probably formed in small island arcs that coalesced into Archean mini-continents. The oldest known terrestrial rocks (at present) are gray gneisses from the Slave Province of northwestern Canada, dated at 4.03 Ga. The Archean continental crust was probably thin and the continental nuclei small and unstable. Much of the continental crust may have been recycled back into the mantle during this stage of vigorous convection. In addition to the gray gneiss terranes, there were also numerous **greenstone belts**. These are 100- to 300-km-long by 10- to 25-km-wide tectonic slices of mafic to silicic volcaniclastic, clastic and chemical sediments plus some granitoids that represent tectonic mélanges of continental, oceanic, trench, and arc material generated at early active margins, perhaps as rifted marginal basins. The Pilbara craton in West Australia, illustrated on the front cover, is an example of such a belt.

There was a major jump in the rate of crust formation in the late Archean. Currently popular estimates of crustal growth rates suggest that 50 to 80% of the present crustal volume was created by 2.5 Ga ago, at the beginning of the Proterozoic (see the introduction in Brown and Rushmer, 2006, and Rollinson, 2006, for summaries). The small, transient Archean crustal nuclei were assembled into a few large supercontinents by this time. There was less coastline, and continental-arc orogenic belts became subordinate in the Proterozoic to the anorogenic AMCG type of magmatism, which, along with mafic dikes, was associated with broad rifting of the northern supercontinent as discussed above. Several crustal growth models indicate significant episodes of crustal evolution. For example, McCulloch and Bennett (1994) concluded that Nd and Pb ages of the Australian, North American, and Scandinavian cratons indicated episodic growth with pulses at $3.6 \rightarrow 3.3$, $2.7 \rightarrow 2.5$, and $2.0 \rightarrow 1.8$ Ga. Condie (2004) argued for peaks at 2.7, 1.9, and 1.2 Ga, which he related to superplume events (Chapter 14) and continental amalgamation.

The estimated compositions of the upper and lower continental crust (columns 11 and 12 in Table 18.2) are distinctive. If these estimates are correct, the continental crust has differentiated vertically, with more of the LIL and incompatible elements migrating upward. The chemical transport may be accomplished by a combination of deep crystal fractionation, rising deep crustal melts, and metamorphic fluids, all of which preferentially incorporate the incompatible elements in the upwardly mobile medium. The result is that the upper crust gradually develops a more evolved composition than the deep crust.

During the Proterozoic, the heat flow gradually reduced to nearly that of today, and plate tectonics probably assumed its present character. Archean TTG and Proterozoic AMCG magmatism gave way to Phanerozoic granitoids dominated by the orogenic calc-alkaline suite. Post-Archean continental crustal growth occurred by the processes described in association with Table 18.4, whereas some crustal material was lost and returned to the mantle as

sediment in subduction zones. As I said above, estimates of the relative roles of recycling and addition of crustal material are variable, but most estimates suggest episodes of accelerated continental growth with modern growth either at a steady state (no net growth) or slightly increasing. Scholl and von Huene (2004) recently concluded that new crustal additions from the mantle and crustal recycling back into the mantle at modern arcs are similar. To give you an idea of the rates involved, Reymer and Schubert (1984) calculated a modern continental crust growth rate of 1.65 km^3/yr.

As mentioned above, true continental crustal growth involves the transformation of mantle material to the crust. Granitoids formed by partial melting of crustal rocks is recycling of crust, not generation. Chappell (2005) claimed that most granites result from partial melting within the crust, which seems to emphasize recycling. Johannes and Holtz (1996) were less exclusive, claiming only that nearly all granites are restricted to the continental crust and were associated with its formation. Vigneresse (2004) argued that one-third to three-quarters of granitoids had some mantle component. Patiño Douce (1999) concluded that the SiO_2 content of only peraluminous leucogranites was high enough to correlate with experimental melts of crustal metasediments and could therefore represent purely crustal melts. Other granitoids were not silicic enough, he found, and must therefore represent some form of hybrid of mafic and supracrustal components. Johannes and Holtz (1996) suggested that granitoid formation from the mantle involves two to three steps. The first step is partial melting of the mantle to form basalts, either at mid-ocean ridges, hotspots, or from the mantle wedge in subduction zones. The second step involves partial melting of the basalts, probably hydrated to amphibolite, either as subducted oceanic crust or as mafic underplates or as plutons in the crust, to produce tonalites or trondhjemites (the TT part of TTG). This is the two-step mechanism described earlier. Granodiorites (the G part of TTG) and other silicic granitoids they attributed to partial melting of tonalites and/or metasediments. All of these processes continue to occur, along with crystal fractionation, assimilation, magma mixing, etc., so that hybridization between earlier-step products and later and between liquids and restites and products of weathering and sedimentary reworking are probably common.

Crustal growth is concentrated at subduction zones, but also occurs above subcontinental plumes (mafic underplates and remelting) and even above sub-oceanic plumes (where oceanic plateaus are formed and may be too thick and buoyant to subduct so are accreted to the continents). As mentioned in Chapter 17, the bulk composition of the continental crust is close to that of an *andesite*, but the average product of modern volcanism, both at intraplate hotspots and at subduction zones, is *basaltic*. So how can processes that produce basalt end up producing andesite over time? The secular approach to solving this paradox is to reason that the modern basaltic flux is not representative of earlier times. Perhaps the higher heat flow in Archean and Proterozoic times (when most of the crust was created) resulted in more extensive melting of subducted or underplated mafic rocks, thus producing a net andesitic/tonalitic flux. If, however, the flux from the mantle to the crust has indeed been basaltic over time, there must then be some intra-crustal differentiation mechanism that further refines the product to andesite. A popular mechanism involves crystal fractionation in deep crustal or underplated basaltic magma bodies followed by ascension of evolved liquids and eventual delamination of the mafic to ultramafic residues, which return to the mantle. The delamination model illustrated in Figure 18.8c2 may therefore include material that was once part of the lower crust. Partial melting of underplates and lower crust could produce similar restites. Neither model is exclusive, and both may have occurred. For discussions of these ideas, see Kay and Kay (1986, 1991), Rudnick (1995), and Davidson and Arculus (2005).

Summary

Granitoids comprise a diverse assortment of felsic plutonic rocks ranging from tonalites to syenites with equally diverse origins. The S-I-A-M classification subdivided granitoids on the basis of chemical and isotopic characteristics and showed that granites, despite their similarities, retained a geochemical imprint of different sources and genetic processes. Granitoids, however, might better be classified on the basis of their mode of origin. Orogenic granitoids are the most common, and occur in subduction-related environments, either island-arc, continental-arc, or continental collision types. They may be generated by partial melting of mafic crustal underplates, subducted mafic slabs, or the overlying crust. Transitional (post-orogenic) granitoids follow an orogenic event and occur during a period of uplift or extensional collapse. Anorogenic granitoids are related to rising mantle at rifts or hotspots.

Granitoids include a wide range of compositions, and they occur in a variety of tectonic settings. Some may be derived predominantly from the mantle, but most are largely of crustal origin and occur in areas of thickened continental crust. The mantle generally contributes heat and matter to crustal granitoids, the latter most notably in the form of mafic crustal underplates that may fractionate or be remelted. H_2O plays an important role in crustal melting, as it can lower the melting point of the crust to as low as 600°C in H_2O-saturated rocks. Fluid-absent melting is more common, however, and typical crustal melting temperatures are in the range of 700°C (where muscovite breaks down) to over 950°C (for melting in dehydrated lower crust). H_2O-excess conditions promote peraluminous S-type granitoids, whereas hot, peralkaline magmas occur in H_2O-poor conditions. The critical melt fraction required to create mobile granitoid melts can be anywhere in the range of 5 to 30%, depending on the viscosity of the melt and the state of stress. The role of the mantle and the rate of crustal growth are controversial. Extension and strike-slip motion provide enhanced access of melts to shallow levels.

There is a complex interplay between multiple processes involved in granitoid genesis and the associated tectonic regimes. Characterization and classification of granitoids is correspondingly difficult. This should be done on a wide variety of chemical, mineralogical, textural, and field characteristics. Tidy schemes, such as those presented in Tables 18.3 or 18.4, are likely to result in overly simplistic concepts.

The continental crust probably developed episodically, with the majority created during the Archean and Proterozoic. The present rate of crustal formation is approximately the same as its destruction, resulting in a near steady-state situation. The present flux from the mantle is overwhelmingly basaltic, yet the average crust is more andesitic in composition. Either the crust-forming processes were more andesitic in the past (more mafic slab or underplate melting) or the crust has differentiated from basalt to andesite and the mafic residuum has sunk back into the mantle.

Key Terms

Granitoid *377*
Subsolvus/hypersolvus granites *378*
Restite *378*
Xenocryst/enclave/
   schlieren/autolith *379*
Haplogranite *380*
Suture zone *387*

Subduction polarity *387*
S-I-A-M granitoids *388*
Orogenic/transitional/anorogenic
   granitoids *386–391*
Thermal relaxation *388*
Mechanical/thermal boundary
   layer *389*

Lithospheric delamination/convective
   removal/slab break-off *390*
Metamorphic core complex *390*
Wilson cycle *391*
TTG (tonalite trondhjemites
   granodiorite) suite *394*
Greenstone belts *394*

Review Questions

Review Questions are located on the author's web page at the following address: **http://www.prenhall.com/winter**

Important "First Principle" Concepts

- The subduction zone beneath continental arcs is very similar to that beneath island arcs.
- Most granitoids are of crustal origin but typically require heat and matter derived from the mantle as well.
- Continental crust accumulated after the main mantle–core separation event, mostly through subduction processes, but there may also be a hotspot component.
- The composition of the continental crust is more silicic (andesitic) than the modern flux from the mantle to the crust (basaltic). Either earlier crust-forming processes were more silicic or the crust has differentiated and lost a mafic component back to the mantle.

Suggested Further Readings

Barbarin, B. (1990). Granitoids: Main petrogenetic classification in relation to origin and tectonic setting. *Geol. J.,* **25,** 227–238.

Brown, M. (1994). The generation, segregation, ascent and emplacement of granite magma: The migmatite-to-crustally-derived granite connection in thickened orogens. *Earth Sci. Rev.,* **36,** 83–130.

Brown, M., and T. Rushmer (eds.). (2006). *Evolution and Differentiation of the Continental Crust.* Cambridge University Press. Cambridge, UK.

Castro, A., C. Fernández, and J. L. Vigneresse (eds.). (1999). *Understanding Granites: Integrating New and Classical Techniques.* Special Publication **168.** Geological Society. London.

Clarke, D. B. (1992). *Granitoid Rocks.* Chapman & Hall. London.

Johannes, W., and F. Holtz. (1996). *Petrogenesis and Experimental Petrology of Granitic Rocks.* Springer-Verlag. Berlin.

Lithos. (1998). **45,** 1–4. Special issue on post-collisional magmatism.

Pitcher, W. S. (1993). *The Nature and Origin of Granite.* Blackie. London.

Sial, A. N., W. E. Stephens, and V. P. Ferriera (eds.). (1999). Granites: Crustal evolution and associated mineralization. *Lithos* **46,** 3 (special issue).

Hutton Symposia

Brown, M., and B. W. Chappell. (1992). *The Second Hutton Symposium on the Origin of Granites and Related Rocks.* Special Paper **272.** Geological Society of America. Boulder, CO.

Brown, M., P. A. Candela, D. L. Peck, W. E. Stephens, R. J. Walker, and E-an Zen. (eds.). (1996). *The Third Hutton Symposium on the Origin of Granites and Related Rocks.* Special Paper **315.** Geological Society of America. Boulder, CO.

Barbarin, B., W. E. Stephens, B. Bonin, J.-L. Bouchez, D. B. Clarke, M. Cuney, and H. Martin (eds.). (2001). *Fourth Hutton Symposium. The Origin of Granites and Related Rocks.* Special Paper **350.** Geological Society of America. Boulder, CO.

Ishihara, S., W. E. Stephens, S. L. Harley, M. Arima, and T. Nakajima (eds.). (2005). *Fifth Hutton Symposium. The Origin of Granites and Related Rocks.* Special Paper **389,** 11–22. Geological Society of America. Boulder, CO.

The Sixth Hutton Symposium was held in Stellenbosch, South Africa in 2007. 15 papers will form volume 100 (2009) of the Transactions of the Royal Society of Edinburgh. Expect to see a Geological Society of America Special Paper as well.

19

Continental Alkaline Magmatism

Questions to be Considered in this Chapter:

1. What principal alkaline rock types occur in non-subduction continental settings?

2. How does continental alkaline magmatism differ from oceanic intraplate and subduction zone magmatism?

3. Why are continental alkaline rocks so geochemically diverse, and what is responsible for this diversity?

4. Why are continental rift zones accompanied by alkaline magmatism?

5. What are carbonatites? Are such carbonate-rich rocks truly igneous, and, if so, how might they form?

6. What are lamproites, lamprophyres, and kimberlites? How do they form, and what do they tell us about the nature of the continental lithosphere?

7. How does the subcontinental lithospheric mantle (SCLM) differ from the oceanic lithospheric mantle? What evidence do we have, and what is responsible for the difference?

The title of this chapter is intended to reflect a certain selectivity rather than to imply an exclusive correlation between continental magmatism and alkalinity. Alkaline rocks occur in all tectonic environments, including (as we saw in Chapter 14) the ocean basins. Conversely, Chapters 12, 15, 17, and 18 have shown us that magmatism on the continents can be highly varied, including tholeiitic and calc-alkaline varieties. Now I'd like to focus on the alkaline rocks that compose an extremely diverse spectrum of magmas occurring predominantly in the anorogenic intraplate portions of continental terranes.

Rocks may be considered "alkaline" in various ways. In Chapter 8, we reviewed several chemical classifications of rock suites in which alkaline rocks were distinguished from sub-alkaline ones. Alkaline rocks are generally considered to have more alkalis than can be accommodated by feldspars alone. The excess alkalis then appear in feldspathoids, sodic pyroxenes/amphiboles, or other alkali-rich phases. In the most restricted sense, *alkaline* rocks are deficient in SiO_2 with respect to Na_2O, K_2O, and CaO to the extent that they become "critically undersaturated" in SiO_2, and *nepheline* or *acmite* appears in the norm. On the other hand, some rocks may be deficient in Al_2O_3 (and not necessarily in SiO_2) so that Al_2O_3 may not be able to accommodate the alkalis in normative feldspars. Such rocks are called *peralkaline* (see Figure 18.2) and may be either silica undersaturated or oversaturated.

Attempting to describe alkaline rocks is similar to opening Pandora's box. They are petrologically fascinating but exhaustively diverse. They have attracted the attention of petrologists to an extent out of all proportion to their abundance. As mentioned in Chapter 2, about half of the formal igneous rock names apply to alkaline rocks. When we consider that these rocks constitute less than 1% of the total volume of exposed igneous rocks, we might question the rationality of attempting so complete a description of such a bewildering array at this survey level. I have thus decided to be selective.

TABLE 19.1 Nomenclature of Selected Alkaline Igneous Rocks (Mostly Volcanic/Hypabyssal)

The mildly alkaline series (e.g., Hawaii): **Ankaramite** (alkali picrite)**, Alkali Basalt, Hawaiite, Mugearite, Benmoreite, Trachyte** is discussed in Section 14.3 (see Figure 14.3).

Basanite   Feldspathoid-bearing basalt. Usually contains nepheline, but may have leucite + olivine

Tephrite   Olivine-free basanite.

Leucitite   Volcanic rock that contains leucite + clinopyroxene ± olivine. It typically lacks feldspar.

Nephelinite   Volcanic rock that contains nepheline + clinopyroxene ± olivine. It typically lacks feldspar. Figure 14.3.

Urtite   Plutonic nepheline–pyroxene (aegirine–augite) rock with over 70% nepheline and no feldspar

Ijolite   Plutonic nepheline–pyroxene rock with 30 to 70% nepheline.

Melilitite   A predominantly melilite–clinopyroxene volcanic (if >10% olivine, called olivine melilitite).

Shoshonite   K-rich basalt with K-feldspar ± leucite.

Phonolite   Felsic alkaline volcanic with alkali feldspar + nepheline. See Figure 14.3. (plutonic = **nepheline syenite)**.

Comendite   Peralkaline rhyolite with molar $(Na_2O \pm K_2O)/Al_2O_3$ slightly >1. May contain Na-pyroxene or amphibole.

Pantellerite   Peralkaline rhyolite with molar $(Na_2O + K_2O)/Al_2O_3 = 1.6 - 1.8$. Contains Na-pyroxene or amphibole.

Lamproite   A group of peralkaline, volatile-rich, ultrapotassic, volcanic to hypabyssal rocks. The mineralogy is variable, but most contain phenocrysts of olivine + phlogopite ± leucite ± K-richterite ± clinopyroxene ± sanidine. Table 19.6.

Lamprophyre   A diverse group of dark, porphyritic, mafic to ultramafic hypabyssal (or occasionally volcanic), commonly highly potassic (K > Al) rocks. They are normally rich in alkalis, volatiles, Sr, Ba, and Ti, with biotite-phlogopite and/or amphibole phenocrysts. They typically occur as shallow dikes, sills, plugs, or stocks. Table 19.7.

Kimberlite   A complex group of hybrid volatile-rich (dominantly CO_2), potassic, ultramafic rocks with a fine-grained matrix and macrocrysts of olivine and several of the following: ilmenite, garnet, diopside, phlogopite, enstatite, chromite. Xenocrysts and xenoliths are also common.

Group I kimberlite   Typically CO_2 rich and less potassic than group 2 kimberlite.

Group II kimberlite (orangeite)   Typically H_2O rich and has a mica-rich matrix (also with calcite, diopside, and apatite).

Carbonatite   An igneous rock composed principally of carbonate (typically calcite, ankerite, and/or dolomite), and commonly with clinopyroxene, alkalic amphibole, biotite, apatite, and/or magnetite. The Ca-Mg-rich carbonatites are technically not alkaline, but are commonly associated with, and thus included with, the alkaline rocks. Table 19.3.

For more details, see Sørensen (1974), Streckeisen (1978), Woolley et al. (1996), and Le Maitre et al. (2002).

Before we proceed, let's briefly review some terminology of the alkaline rocks (Table 19.1) because many of the descriptions to follow will use these specialized names. It often seems that every alkaline rock has its own name, so I shall provide only the broadest categories. Table 19.1 concentrates on volcanic and shallow plutonic (hypabyssal) varieties, which are the most common forms of these rocks. Plutonic alkaline rocks certainly occur, and their names can be found in the most recent IUGS classification of Le Maitre et al. (2002). Several aspects of the classification of alkaline rocks are confusing and controversial, reflecting uncertainty about genetic relationships and the tectonic setting of many occurrences. Some investigators attempt to classify on the basis of genetic similarities (like the magmatic series in Chapter 8), whereas others, recognizing the difficulty in assessing the genesis of such diverse rocks in complex continental settings, rely on more easily determinable features, such as chemical, mineralogical, and/or textural similarities. The confusion is most evident in the highly potassic lamprophyre–lamproite–kimberlite group, a diverse array of mafic to ultramafic rocks with high volatile contents. The numerous intertwined petrographic and genetic

similarities and contrasts in this broad group present quite a classification challenge.

Alkalis are separable, of course, and the Na/K ratio in alkaline series varies, particularly in the continental magmas that are the subject of this chapter. In typical oceanic alkaline occurrences, the Na/K range is fairly narrow (concentrated in the shaded region of Figure 19.1), and Na usually exceeds K. There are some rare K-rich oceanic localities, however, such as Tristan da Cunha, that extend beyond the common limits. In continental alkaline suites (Figure 19.1b), the Na/K ratio varies from some Na-rich types (e.g., Hebrides, Otago) to several occurrences of high-K types, all well beyond oceanic values.

In the interest of expediency, I am choosing to simplify an open-ended task and limit this chapter to instances providing the most new information on concepts not covered previously in the text. I confess that some of my choices may reflect simple favoritism, but I think that they illustrate some uniquely continental processes and thus provide insight into the nature of the continental lithosphere and its magmatism.

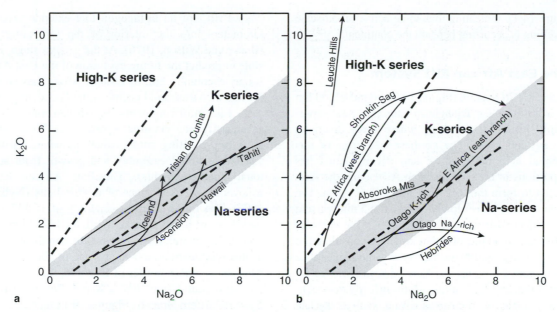

FIGURE 19.1   Variations in alkali ratios for **(a)** oceanic and **(b)** continental alkaline series. The heavy dashed lines distinguish the alkaline magma subdivisions from Figure 8.14, and the shaded area represents the range for the more common oceanic intraplate series. After McBirney (1993). Jones and Bartlett Publishers, Sudbury, MA. Reprinted with permission.

Alkaline rocks generally occur in three principal settings:

1. Continental rifts
2. Continental and oceanic intraplate settings with no clear tectonic control
3. Subduction zones, particularly in back-arc settings or in the waning stages of activity

We discussed the oceanic portion of setting 2 in Chapter 14, and setting 3 in Chapters 16 to 18. This chapter will thus concentrate on setting 1 and the continental portion of setting 2. I will begin with a brief review of continental rift alkaline activity and then proceed to some uniquely continental intraplate occurrences: carbonatites, lamproites, lamprophyres, and kimberlites.

19.1 CONTINENTAL RIFT-ASSOCIATED ALKALINE MAGMATISM

I begin with a caveat that alkaline magmatism is not the only manifestation of continental rifting, and it may not even be the hallmark of such rifting, as has classically been considered. We know from Chapter 15, for example, that plateau basalts of tholeiitic composition occur at several rifts. Continental rifting, however, is commonly associated with alkaline magmatism, and is the setting in which the greatest volumes of continental alkaline rocks are generated. Examples include the active Rhine Graben (Europe), Baikal Rift (Russia), and Rio Grande Rift (southwestern United States); as well as the Permian Oslo Rift (Norway) and Precambrian Gardar Province (Greenland). We shall consider in detail the best-known example, the East African Rift, which has been active through most of the Cenozoic and exhibits nearly the entire range of alkaline igneous rock types.

Rifting may occur in response to incipient plate divergence (eastern Africa), post-collision or post-subduction collapse/extension (Rhine, Rio Grande, Basin and Range), or back-arc extension (Columbia Plateau of the northwestern United States). There may be a broad spectrum of volcanism from the voluminous fissure-fed tholeiitic flood basalts (perhaps associated with vigorous mantle advection of heat) to more quiescent settings with small localized alkaline centers.

Rifting and separation of opposing continental blocks is associated with asthenospheric upwelling and associated high heat flow, but there is some controversy as to the mechanism involved. As mentioned in Chapter 13, some investigators think that asthenospheric upwelling *forces* the lithosphere apart (**active rifting**), whereas others think that the plates are pulled apart and the lithosphere collapses or is stretched, so the asthenosphere rises to fill the gap (**passive rifting**). Active rifting should affect a broader area and lead to earlier magmatism in a rift sequence than would passive rifting. The structure and magmatic patterns of most rifts, however, are sufficiently complex and episodic to obscure such simple evaluation. Active rifting is generally favored for most continental rifts, but there is no reason to limit rift style exclusively to either type. Some recent studies in the East African Rift indicate that an initial period of subsidence preceded the doming. This suggests that the rift began in a passive mode with lithospheric thinning, which may have led to more active rifting as the asthenosphere responded. The real issue for petrologists in active versus passive rifting is the type of mantle involved. If active rifting is driven by vertical mantle motion, it should involve a considerable column of mantle, including deep-seated plumes tapping more enriched OIB-type reservoirs (Figures 14.19 and 15.15). Passive rifting should involve principally the uppermost MORB-source type of asthenosphere (at least in the early stages). Recent seismic imaging beneath Africa has revealed a large "superswell" beneath western

Africa (Figure 14.1), indicating that some active broad-scale mantle upwelling is occurring beneath the continent.

19.1.1 The East African Rift System

The East African Rift System (Figure 19.2) extends 2000 km southward from the Afar Triangle in Ethiopia. Afar represents a **triple junction** of a type where three divergent (spreading) systems meet. The northern two arms of this junction spread at about 2 cm/yr and oceanic crust has formed in portions of the rifts between Africa and the Arabian peninsula to form the Red Sea and Gulf of Aden. The southern arm spreads more slowly (about 1 to 5 mm/yr) and there is no evidence for oceanic crust beneath the volcanic pile. The Red Sea and Gulf of Aden arms are considered to be new parts of the worldwide rift system and true plate boundaries (Figure 13.1), but the East African Rift System has not (yet?) reached that status. Many investigators consider the early stages of continental rifting, such as the late Jurassic incipient separation of Africa and South America, to be associated with a string of hotspot plumes that form triple junctions like Afar (see Figure 15.16). Two arms of adjacent triple junctions join to interconnect a string of adjacent junctions, which then evolves to become a mid-ocean rift. The third arm of each triple junction aborts to become a

"failed rift," or an **aulacogen** that extends several tens of kilometers into the interior of the separating continents (Burke and Wilson, 1976). At the present stage, it is impossible to predict the future evolution of the East African Rift. It may eventually fail, or Eastern Africa may separate from the rest of Africa and become an independent continent.

Like most continents, the African continent comprises several Archean **cratons** separated by Proterozoic mobile belts (representing suture zones associated with Precambrian continental assembly and growth from smaller cratonal microcontinents). The East African Rift System follows the weaker zones of mobile belts (Nyblade et al., 1996). The Eastern Rift is generally 40 to 80 km wide and traverses two broad areas of uplift, the Ethiopian and Kenyan domes. In general, the rifting is oldest in the north and propagates southward at an average rate of about 2.5 to 5 cm/yr. Near the southern border of Ethiopia, the rift splits (around the Tanzanian craton) into a K-rich Western Branch and a Na-rich Eastern Branch (Figures 19.1 and 19.2). At present, the Eastern Branch rift terminates to the south in a diffuse (~300 km) zone of normal faults, some of which fracture the craton. Recent work suggests that the main locus of rifting is progressing along the eastern edge of the cratonal block. The Western Branch extends farther south into Malawi and Mozambique.

Extension at a given locality generally begins with lithospheric thinning and diffuse faulting (the *prerift stage*), followed by listric faulting and the development of half-grabens. Faulting in the opposite sense to the initial listric faults usually occurs slightly later, resulting in the formation of an axial graben system (the *rift stage*).

Recent geophysical work has revealed a low-velocity zone in the upper mantle beneath the Kenya Rift and Eastern and Western branches to the south, extending beneath a thinned crust down to about 160 km (Prodehl et al., 1994; Macdonald et al., 1994; Nyblade et al., 1996). Magnetotellurics reveals high conductivity in the shallow rift mantle that suggests that the mantle beneath the rift is partially molten. These data, combined with the high heat flow and a Bouguer gravity low at the rifts, are consistent with an interpretation of hot, low-density, partially molten asthenosphere having risen to replace the cooler lithospheric mantle that has been laterally displaced by rifting. Beyond the rift proper, the crustal thickness and mantle character are more normal (the crust is >40 km thick adjacent to the rift in Kenya and Ethiopia).

Prerift volcanism began to the north in Ethiopia in the Eocene (~43 Ma) and in Kenya (Eastern Branch) in the early Oligocene (33 to 30 Ma) with the extrusion of extensive flood basalts covering the thinning lithosphere (Kampunzu and Mohr, 1991). Rhyolites and rhyolitic ignimbrites accompanied the later flood basalt activity. Such bimodal basalt–rhyolite volcanism is familiar to us by now, and is believed to represent a combination of mafic mantle and silicic crustal melts. The rift stage began with a half graben about 15 Ma ago, and a full graben at about 4 to 3 Ma. Western Branch activity began in the Miocene (~14 Ma) with transitional basalts giving way to alkaline basalts. The Western

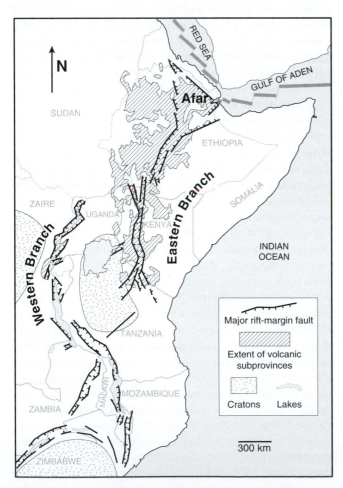

FIGURE 19.2 Map of the East African Rift system (after Kampunzu and Mohr, 1991). Copyright © with permission from Springer-Verlag.

Branch took form about 7 Ma ago. The Western Branch is narrower than the Eastern, with less lithospheric thinning and volcanic activity. About 3 Ma ago, volcanism in the Western branch became more alkaline, more silica-undersaturated, and notably more potassic than in the Eastern Branch (at least in its northern parts).

Since its inception, volcanism in any segment has continued episodically to the present. Volcanic accumulations in the rift may attain thicknesses of over 3 km. Magmatism is not restricted to the rift grabens. Major volcanic episodes have overspilled the rifts and covered the adjacent plateaus. Notable among these episodes are extensive (17 to 8 Ma) flood phonolites and (<4 Ma) trachytic ignimbrites and flows in central Kenya. Numerous volcanic centers also occur outside the axial rift, including Mt. Kenya and Mt. Kilimanjaro.

19.1.2 Magma Series of the East African Rift

The magmas of the East African Rift can be grouped into four main series (Kampunzu and Mohr, 1991):

1. Alkaline
2. Ultra-alkaline and carbonatitic
3. Transitional
4. Tholeiitic

The **alkaline** series spans the range from basanites and alkaline basalts (with >5% normative *ne*) to tephrites, phonolites, and trachytes (see Figure 14.3). In Ethiopia, mid-Miocene shields of alkaline basalt dominate the closing stages of the Tertiary flood basalts (similar to the common tholeiitic to alkaline temporal sequence in Hawaii). Localized flows and small plutons also occur scattered over the plateaus and in the graben to the south.

Ultra-alkaline, highly silica-undersaturated volcanic and hypabyssal rocks occur at the southern end of the Kenya Rift and in the Western Branch where they are K rich. The rocks of this series are rich in feldspathoids (both as phenocrysts and groundmass), most commonly nepheline and leucite. They may also contain one or more of the minerals sodalite, cancrinite, melilite, garnet, perovskite, kalsilite, and monticellite (calcic olivine). Carbonatites (Section 19.2) accompany some of the nephelinites, melilitites, and phonolites in the Southern and Western rifts. In the Kenya Rift, Cenozoic occurrences of the carbonatite–alkaline silicate rock association (Section 19.2) are restricted to areas underlain by the Tanzanian Craton in the south, whereas the transitional rocks of series 3 tend to dominate in areas of thinned crust north of the craton (Macdonald et al., 1994).

Transitional basalts are so named because they have geochemical characteristics near the boundary between alkaline and tholeiitic basalts. These include the bulk of the flood basalts in Afar, Ethiopia, and Kenya that were erupted immediately before (and during) the major graben faulting and flank uplift. They also form a substantial component of the north Western Branch basalts, where they generally lack the associated late-stage silicic volcanics that accompany the mafic lavas in the east. Differentiation of the Quaternary rift transitional basalts leads to a slightly alkaline variation of the typical tholeiite trend, including ferrobasalts, mugearites, alkali trachytes, and peralkaline rhyolites (Barberi et al., 1975). Basalts, trachytes, and phonolites predominate, making the series distinctly bimodal, with a notable "Daly gap" (see Section 8.5.1) between 50 and 60% SiO_2.

The **tholeiitic** series is associated with transitional basalts (to which it is volumetrically subordinate). Pre-rift tholeiites are known from various rift segments, and tholeiites occur at the initial and late stages of rift development. They are similar to MORBs, but are somewhat enriched in the LREE and highly incompatible elements.

Although it is useful to distinguish the four series as we attempt to understand rift magmatism, the mineralogy and geochemistry of the series really form a continuum, and the series should be considered gradational into one another. Such continuity does not require a continuously variable source, however. Hybridized sources, contamination, magma mixing, mixed volatiles, and variable depth and degree of anatexis contribute to create melts of mixed affinities, even in cases with distinct source reservoir types. What really distinguishes the magmatism of the East African Rift is the diversity of geochemical composition expressed by the tholeiitic to ultra-alkaline spectrum over such a limited area. Although most individual volcanic centers produce lavas of a limited compositional range within a single magma series, neighboring centers may differ markedly. Rarely do we see such a variation and range of compositions between centers in such close proximity. Some of the larger centers, such as mounts Kenya, Kilimanjaro, and Olokisalie, are not as uniform as their smaller neighbors and produce highly varied lavas representing more than a single series, usually in seemingly random sequences.

The petrological diversity in the East African Rift is rich in detail and has entertained geologists for decades. The literature is extensive, but still insufficient to describe adequately the diversity of rocks and to constrain detailed petrogenetic models. Trace element data are becoming more comprehensive as XRF and ICP techniques have become automated and labs have proliferated, but isotopic analyses are still expensive, so the isotope data are not abundant. We will only be able to address the broader trends now anyway, and the data are sufficient to provide a reasonable basis for that.

19.1.3 Geochemistry of East African Rift Volcanics

Table 19.2 lists the major element compositions and CIPW norms for some representative lavas from the East African Rift. The Excel spreadsheet EAfr.xls at www.prenhall.com/winter has a more comprehensive set of analyses, including trace elements. The tholeiitic basalt (column 11) and the transitional series (columns 7 to 10) are all silica saturated (*q* in the norm), whereas the two alkaline series are not. The alkaline rocks tend to have normative nepheline and/or leucite. Highly alkaline or calcic rocks also have some weird normative "minerals," such as sodium metasilicate (*ns*) and calcium silicate (*cs*), that were invented by the developers of the norm calculation in order to accommodate

TABLE 19.2 Representative Chemical Analyses of East African Rift Volcanics.

Oxide	Series 1: Alkaline		Series 2: Ultra-Alkaline				Series 3: Transitional Basalt–Rhyolite				Series 4
	1	2	3	4	5	6	7	8	9	10	11
SiO_2	45.6	51.7	46.2	33.1	44.1	55.4	47.6	61.8	70.3	72.5	50.8
TiO_2	2.4	0.9	1.6	2.6	2.8	0.5	2.0	1.0	0.3	0.2	1.4
Al_2O_3	15.6	19.3	18.6	11.3	17.0	20.8	14.8	14.2	7.6	10.3	14.9
FeO*	11.3	5.9	8.9	12.4	10.0	4.6	11.4	6.4	8.4	4.0	10.1
MnO	0.2	0.2	0.2	0.3	0.2	0.2	0.2	0.3	0.3	0.1	0.2
MgO	6.9	1.1	2.3	7.3	3.7	0.5	6.4	0.5	0.0	0.0	6.9
CaO	10.4	4.1	7.3	17.2	8.4	2.9	11.5	1.8	0.4	0.2	9.8
Na_2O	3.2	8.9	9.3	3.2	4.3	9.2	2.7	6.2	7.3	5.9	2.6
K_2O	1.3	4.6	4.2	3.6	7.2	5.5	0.8	5.2	4.3	4.4	0.4
P_2O_5	0.6	0.3	0.5	1.9	1.2	0.1	0.3	0.2	0.0		0.4
Total	97.5	97.0	99.1	92.9	98.9	99.7	97.7	97.6	98.8	97.6	97.4

CIPW NORM

q	0.0	0.0	0.0	0.0	0.0	0.0	2.1	9.0	41.7	35.8	9.1
or	8.9	29.8	27.5	0.0	31.0	34.2	5.5	33.7	28.1	27.8	2.6
ab	31.4	28.6	8.0	0.0	0.0	30.2	26.5	48.3	16.7	30.4	25.2
an	28.3	0.0	0.0	7.3	6.5	0.0	30.0	0.0	0.0	0.0	31.9
lc	0.0	0.0	0.0	20.7	13.2	0.0	0.0	0.0	0.0	0.0	0.0
ne	0.0	28.3	39.1	18.2	22.2	27.1	0.0	0.0	0.0	0.0	0.0
di	14.2	6.5	13.7	15.0	16.7	2.8	20.8	2.9	0.1	0.0	12.7
hy	0.0	0.0	0.0	0.0	0.0	0.0	8.8	0.0	0.0	0.0	13.8
wo	0.0	3.9	5.7	0.0	0.0	4.1	0.0	0.9	0.6	0.0	0.0
ol	9.4	0.0	0.0	10.9	1.8	0.0	0.0	0.0	0.0	0.0	0.0
il	0.5	0.5	0.5	0.8	0.5	0.4	0.5	0.7	0.6	0.1	0.5
ti	4.7	0.0	0.0	0.0	0.0	0.0	5.0	1.8	0.1	0.5	3.3
ap	1.6	0.8	1.3	5.5	3.1	0.2	0.8	0.5	0.1	0.0	1.0
pf	1.0	1.3	2.6	4.8	4.9	0.5	0.0	0.0	0.0	0.0	0.0
ns	0.0	0.4	1.7	0.0	0.0	0.4	0.0	2.1	12.0	5.3	0.0
cs	0.0	0.0	0.0	16.8	0.0	0.0	0.0	0.0	0.0	0.0	0.0

1. Average of 32 alkaline basalts, Kenya (B).   2. Average of phonolite (B).   3. Average of Kenyan nephelinite (B).   4. Melilitite, Western Rift (KM).
5. Leucitite, Western Rift (KM).   6. Average of 55 phonolites, Uganda (B).   7. Average of 31 transitional basalts (B).   8. Average of 40 trachytes (B).
9. Pantellerite (KM).   10. Comendite (KM).   11. Average of 26 tholeiitic basalts (KM).   KM = Kampunzu and Mohr (1991), B = Baker (1987).

alkali and calcium excess over other available oxides (Appendix B).

Rocks of the East African Rift are fairly representative of alkaline and peralkaline rocks from around the world. The tholeiitic, transitional, and alkaline series are similar to their oceanic counterparts (Tables 13.2, 14.2, and 14.3), although the alkalinity can be higher and more variable in the continents, particularly when we consider the ultra-alkaline series 2 above. We could spend ages in an attempt to cover the spectrum of alkaline rock chemistry and mineralogy, which is not my intention. Our short time is better spent addressing the petrogenesis of these rocks, using only the geochemical and petrologic data that are of direct use in that endeavor (mostly isotopic and trace element data).

19.1.4 Isotopic and Trace Element Characteristics of East African Rift Magmas

Figure 19.3 illustrates the isotopic ratios of Sr and Nd for some East African Rift lavas, concentrating on mafic lavas to minimize the effects of crustal contamination (which should produce more evolved compositions) and allow us to focus on source characteristics. The data conform well to the mantle array (Figures 10.16 and 14.8). The transitional, flood, and alkaline basalts from Afar to Kenya are all more enriched than MORB and have characteristics closer to OIB. This is consistent with an intraplate setting in only the incipient stages of rifting (no added rift influence). $^{87}Sr/^{86}Sr$ values in the

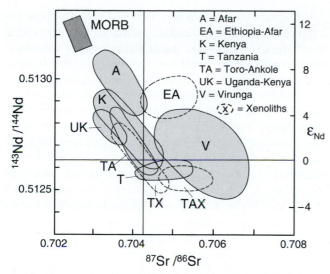

FIGURE 19.3 $^{143}Nd/^{144}Nd$ versus $^{87}Sr/^{86}Sr$ for East African Rift lavas (shaded) and xenoliths (unfilled dashed areas). The "crosshair" intersects at bulk Earth. After Kampunzu and Mohr (1991). Copyright © with permission from Springer-Verlag.

ultra-alkaline series are generally in the range 0.703 to 0.707 (suggesting a mantle source) but may be as high as 0.711. As we shall see, highly potassic magmas and mantle xenoliths along the rift have $^{87}Sr/^{86}Sr$ values up to 0.713, so these high values do not necessarily imply a crustal component. ε_{Nd} values range from −4 to +4 in the ultra-alkaline lavas.

The isotopic data for the more evolved rocks show signatures similar to the mafic rocks. This implies that most magmas evolved by magmatic differentiation with little contamination by assimilated continental crust. This is rather surprising, given the thickness of ancient radiogenic crust in which the evolving magma chambers should have resided. In only a few localities, such as the active Quaternary Boina center near Afar (Barberi et al., 1975), are the trachytes and pantellerites more enriched than the associated basalts, suggesting either a different mantle source or crustal contamination.

Norry et al. (1980) made an interesting argument in which essentially the *same* Nd and Pb isotopic ratios for trachytes and basalts at Emuruangogolak (northern Kenya), combined with different Sr ratios, argue *for* crustal contamination at this center. The authors reason that the very low Sr contents of the silicic lavas leave them susceptible to crustal Sr contamination, whereas the higher Nd and Pb contents are less susceptible. Thus contamination by only a few percent crustal components would dominate the Sr signature and have little effect on Nd and Pb ratios.

Baker (1987) and Kampunzu and Mohr (1991) noted that the Sr and Nd isotopic ratios do not correlate with the parent/daughter Rb/Sr and Nd/Sm *elemental* ratios. The trace element data for the lavas of the East African Rift are generally incompatible-element enriched (as one might expect for an alkaline province). Such enrichment results in high Rb/Sr and Nd/Sm ratios, which should produce elevated $^{87}Sr/^{86}Sr$ and $^{143}Nd/^{144}Nd$ over time. The isotopic ratios, however, are fairly *low* (depleted with respect to bulk Earth, Figure 19.3), suggesting *low* Rb/Sr and Nd/Sm ratios in the ancient source. The authors concluded that the trace element enrichment of

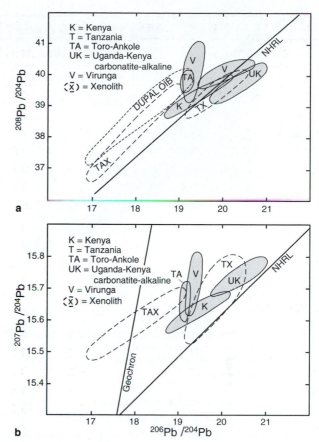

FIGURE 19.4 $^{208}Pb/^{204}Pb$ versus $^{206}Pb/^{204}Pb$ **(a)** and $^{207}Pb/^{204}Pb$ versus $^{206}Pb/^{204}Pb$. Note two distinct Virunga trends reflecting some source heterogeneity. **(b)** Diagrams for some lavas (shaded) and mantle xenoliths from the East African Rift. After Kampunzu and Mohr (1991). Copyright © with permission from Springer-Verlag.

the mantle source *must have happened recently,* following a much earlier depletion event, leaving insufficient time for the element enrichment to produce a difference in the isotopes. Data from ultramafic xenoliths in many of the lavas are also included in Figure 19.3. These data correlate well with the associated lavas, suggesting that the trends reflect *mantle,* rather than magmatic, processes.

Pb isotopic data are scarce, but the lava and xenolith data of Kampunzu and Mohr (1991) are shown in Figure 19.4. The data, like those for Sr and Nd, are comparable to OIBs (Figures 14.9 and 14.10). Kampunzu and Mohr (1991) defined two groups on the basis of Pb systematics. The first group comprises rocks from W. Kenya and E. Uganda, which plot near the NHRL (Figure 14.9) with low radiogenic ^{207}Pb and ^{208}Pb. The second group comprises potassic volcanics of the Western Branch, Tanzanian volcanics, and the Ethiopian flood basalts. Rocks of this group have low ^{206}Pb (like group one) but exhibit a wide variety of ^{207}Pb and ^{208}Pb values (in both lavas and xenoliths). These values, although admittedly sparse, are much like the southern hemisphere OIB values (Figure 14.9) and may indicate that the DUPAL anomaly of the southern oceanic mantle, described in Section 14.5, might extend beneath the African continent as well.

The trace element geochemistry of the rocks of the East African Rift is highly variable. Figure 19.5 shows

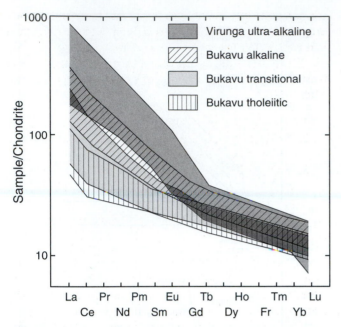

FIGURE 19.5 Chondrite-normalized REE variation diagram for examples of the four magmatic series of the East African Rift. After Kampunzu and Mohr (1991). Copyright © with permission from Springer-Verlag.

chondrite-normalized REE patterns representative of African rift lavas. Note that the slope gets steeper as alkalinity increases from tholeiitic to ultra-alkaline. The steeper slopes could reflect lower degrees of partial melting in the more alkaline magmas, but the trends hold for similar values of Mg#. The variations in incompatible trace element ratios are too large to be explained away by variance in D values, and Kampunzu and Mohr (1991) attributed the variation to different mantle source compositions. Norry et al. (1980) also noted such a correlation between the degree of SiO_2 undersaturation and the abundances of incompatible trace elements (including

LREE enrichment), which they attributed to metasomatic enrichment of previously depleted subcontinental lithospheric mantle by CO_2-rich fluids. Again, the lack of any correlation between trace elements and isotopic ratios was taken to mean that the enrichment occurred shortly before magma generation.

Although the total incompatible trace element concentrations tend to increase as alkalinity increases, the *ratios* of some highly incompatible trace elements tend to be relatively constant at a given location, and evolve along different trends (Figure 19.6). The constant ratios are attributed to fractional crystallization processes that exclude both elements equally. Baker (1987) found similar constant ratios of Zr/Nb and Hf/Ta. Crustal contamination would certainly alter these ratios, so its effect, again, is probably minor. The variation in trace element ratios from one location to another may reflect different degrees of partial melting at the source (Ferrara and Treuil, 1974), as suggested in Figure 19.6a, but there is also some correlation with the degree of rift extension (Figures 19.6a and b). This may reflect the evolution of the mantle sources (loss of incompatible trace elements) as rifting and melt extraction progresses. It may also reflect a change in depth of melting as rifting progresses. Barberi et al. (1975, 1980), for example, attributed at least some of the variation in trace element ratios to progressively deeper tapping of a vertically zoned mantle. Kampunzu and Mohr (1991) attributed the trends in Figure 19.6b to increased involvement of asthenospheric mantle as the rift evolves.

19.1.5 Mantle Enrichment and Heterogeneity Beneath the East African Rift

The occurrence of primitive ultra-alkaline (and alkaline) rocks in the East African Rift, together with the trace element and isotopic data discussed above, suggest that the mantle beneath Africa may be incompatible-element enriched. Experimental

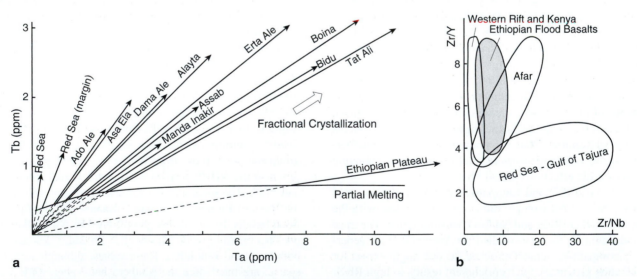

FIGURE 19.6 **(a)** Ta versus Tb for rocks of the Red Sea, Afar, and the Ethiopian Plateau. Rocks from a particular area show nearly constant ratios of the two excluded elements, consistent with fractional crystallization of magmas with distinct Ta/Tb ratios produced either by variable degrees of partial melting of a single source or varied sources (Treuil and Varet, 1973). Copyright © with permission from Société Géol. de France. **(b)** Zr/Y versus Zr/Nb for some East African Rift lavas showing similar constant ratios of excluded trace elements (after Kampunzu and Mohr, 1991). Copyright © with permission from Springer-Verlag.

melting studies suggest that it is difficult, if not impossible, to derive K-rich undersaturated liquids from normal mantle lherzolites. Experiments on the high-K liquids derived by melting the volcanics further suggest that they were never in equilibrium with normal lherzolite. As we shall see in Section 19.4, many of the xenoliths from kimberlites and more alkaline lavas show signs of metasomatic alteration.

As described above, geochemical and petrological heterogeneity in volcanics can be recognized at various scales. Small-scale variations are notable at the larger complex single centers, such as Mt. Kenya and Mt. Kilimanjaro. Medium-scale variations can be seen when we compare trace element ratios among sub-provinces (as in Figure 19.6). On the large scale, Kampunzu and Mohr (1991) pointed out systematic differences in trace element characteristics between the Western Branch–Tanzania–Southern Kenya zones and Northern Kenya–Ethiopia. The former have negative anomalies on normalized multielement (spider) diagrams for Rb, K, and Hf and a positive La anomaly. The negative K anomaly correlates with a residual K-rich phase in the mantle source (phlogopite and alkali feldspar are present in many mantle xenoliths). The northern regions exhibit lower La and higher Rb, and have no carbonatites (which occur in the southern regions). The southern and northern areas also differ in their Pb isotopic signatures, as was pointed out above. The K-rich Western Branch and the Na-rich Eastern Branch in the south is another large-scale contrast. Lloyd et al. (1991) reported higher K, P, Sr, Ba, Nb, La, and volatiles (CO_2, H_2O, F, Cl, S) in the Western Branch.

There are also pronounced large-scale regional differences in mantle xenoliths (most are from ultra-alkaline and alkaline volcanic suites). Xenoliths from the alkaline basalts in the Danakil Block of northeastern Afar include spinel–harzburgite, olivine pyroxenite, and lherzolite, suggestive of equilibration in the spinel peridotite domain (Figure 10.2). Sr isotopic disequilibrium between xenolith whole-rock values and those of the component minerals again attests to *recent* enrichment in radiogenic Sr from circulating fluids or melts. Kampunzu and Mohr (1991) concluded that the Danakil data reflect a zoned upper mantle with lower $^{87}Sr/^{86}Sr$ where alkaline basalts are generated. In the Kenyan and Tanzanian blocks, as well as the Toro–Ankole subprovince of the K-rich Western Branch, the xenoliths are restricted to the ultra-alkaline suite. Many common mantle minerals, such as garnet, spinel, orthopyroxene, and plagioclase, are absent in the Toro-Ankole xenoliths, which are commonly richer in potassium and incompatible trace elements (as are the lavas). The occurrence of calcite and potassium feldspar indicates that associated metasomatism was a shallow event. Experiments on some of the enriched xenoliths indicate that 20 to 30% partial melting of the xenoliths would produce a lava much like the highly alkaline host.

Xenoliths from other areas in the Eastern Branch in Kenya and Ethiopia are more typical peridotites. Many features of the xenoliths correlate well with the host lavas, suggesting that the xenoliths were derived from the mantle lava source. Regional differences in xenoliths appear to reflect variations in the mantle beneath the Eastern and Western branches.

The great geochemical variation in the igneous rocks, combined with the geochemical and petrologic variation of the xenoliths they contain, indicates that the mantle is heterogeneous beneath Africa, probably in both the horizontal and vertical sense and on various scales.

The data suggest that on a regional-scale there are at least three principal mantle reservoirs (Kampunzu and Mohr, 1991):

1. A slightly depleted, DM-like MORB source
2. A HIMU-like source with high radiogenic ^{207}Pb and ^{208}Pb
3. A DUPAL-like EM source (high $^{87}Sr/^{86}Sr$, ^{207}Pb, and ^{208}Pb) south of the equator, a signature also of the lithosphere derived initial flood basalts in the north. Perhaps this is EMII or subducted pelagic sediment

(See Section 14.5 for detailed descriptions of mantle reservoir types.) Reservoirs 2 and 3 imply that the mantle beneath Africa includes recycled oceanic lithosphere. The last subduction in Africa occurred approximately 600 Ma ago, which would allow sufficient time for the isotopic development.

In addition to these older characteristics based on isotopic signatures, the mantle beneath Africa has been variably enriched during a Cenozoic event that imparted the incompatible trace element character to many of the magmas and xenoliths (but was too recent to affect the isotopes). Mineral textures indicate an early stage of silicate liquid metasomatism (with high F and CO_2) followed by a carbonatite injection stage (Kampunzu and Mohr, 1991).

Hart et al. (1989), in a study of Ethiopian transitional basalts, concluded that there are *four* reservoirs beneath Ethiopia:

1. MORB-like DM (northern Afar and Red Sea)
2. OIB-like PREMA (northern and western Central Afar)
3. Enriched heterogeneous subcontinental lithospheric mantle (SCLM)—enriched in $^{143}Nd/^{144}Nd$ at ~constant $^{87}Sr/^{86}Sr$ (called "LoNd" by these authors, it may be EMI or HIMU)
4. Enriched heterogeneous SCLM—enriched in both $^{143}Nd/^{144}Nd$ (low) and $^{87}Sr/^{86}Sr$ (high), perhaps EMII?

Macdonald et al. (1994) noted that most African Rift basalts last equilibrated with the mantle at depths less than 15 km (within the lithosphere), so we can't unequivocally distinguish lithospheric and asthenospheric sources on the basis of magma compositions. Nonetheless, we can still tell something about the mantle. First, it's clearly heterogeneous on various scales. Second, it has experienced at least one Proterozoic depletion event producing a DM-type isotopic mantle reservoir. Later enrichment event(s) then imparted many of the trace element characteristics shortly before Cenozoic magmatism.

Questions still remain as to the exact nature of the lithospheric and asthenospheric components. Is the asthenosphere beneath Africa heterogeneous? Are there both shallow and deeper mantle plumes? These issues are discussed in general terms in Chapter 14. If both questions are eventually answered in the affirmative, we may not need a lithospheric component. Is the asthenosphere the same as for OIB? If so, the East African Rift differs from OIB in several ways, and a lithospheric component is required. As yet, we cannot answer these questions with confidence.

19.1.6 Trends in East African Rift Magmatism

There are two principal types of trends in the geochemical composition of East African Rift magmas: *temporal* and *spatial* trends. Baker (1987), in his summary of the development of the Kenya Rift, noted a general sequence of early pre-rift volcanism dominated by the highly alkaline nephelinite–carbonatite series 2 above, followed by the half-graben stage characterized by alkaline basalts and phonolites of series 1 and later by more transitional series 3 and eventually tholeiitic (series 4) magmas. Baker (1987) and Wendlandt and Morgan (1982) attributed this trend to progressively decreased pressure of partial melting of an upwelling mantle source. Norry et al. (1980) allowed that this may work, or, alternatively, that less CO_2 was available later (Section 10.3). A tempting model is one in which early alkaline magmatism occurs in the continental setting and gradually becomes less alkaline, finally giving way to MORB-like tholeiites as the setting develops a more oceanic character. Kampunzu and Mohr (1991), however, were careful to point out that tholeiitic basalts, although subordinate, are always associated with (and commonly confused with) transitional basalts in the East African Rift system. There may be some broad trends of decreasing alkalinity with time, but tholeiites are present during all stages of rift development, notably including the earliest stages, where they tend to precede transitional basalts. Kampunzu and Mohr (1991) found no compelling reason to accept the proposed temporal trend (other than in the late stages when a rift becomes oceanic and clearly tholeiitic). They concluded that the structural setting exerts a vital (though not always predictable) control on the geochemical composition of the magmas, but that heterogeneous mantle sources are always present and the level of partial melting does not follow any simple progression.

A common *spatial* trend is that the flanks of a rift tend to be more silica-undersaturated and alkaline than the rift itself (similar to volcanism at a mid-ocean ridge, Chapter 13). Kampunzu and Mohr (1991) attributed the increase in alkalinity away from the axis to less vigorous asthenospheric rise, and hence lower degrees of partial melting, at the fringes of the mantle diapir.

Mixed spatial/temporal trends include the southern propagation of activity and the tendency for volcanism to migrate toward the axis as the structural style changes from one of broad doming to an increasingly narrow fault-defined valley (Macdonald et al., 1994).

19.1.7 Magma Evolution in the East African Rift

As mentioned above, most of the evolved magmas in the East African Rift appear to be the result of fractional crystallization from mantle-derived partial melts without significant contamination by ancient continental crust. For example, at the Boina volcanic center in Afar, Barberi et al. (1975) concluded that fractional crystallization at shallow depth of olivine, plagioclase, clinopyroxene, Fe-Ti oxide, and alkali feldspar (in order of appearance) was sufficient to explain the geochemical trends in the transitional basalt-pantellerite sequence there.

Shallow fractional crystallization results in four evolutionary trends, one for each series described in Section 19.1.2. The tholeiite → high-silica sequence of series 4 is familiar by now from Chapters 13 and 14. Boina is an example of the transitional basalt → peralkaline rhyolite (pantellerite) trend of series 3. Series 2 begins with basanites or nephelinites and follows an alkali-rich trend toward phonolites. Series 1 proceeds from alkali basalts to trachytes and then to phonolites (see Figure 14.3).

The various evolution paths in alkaline and transitional alkaline series are best illustrated using the Ne-Ks-SiO_2-H_2O system (Figure 19.7). In this system we can see that there are *two* thermal minima on the liquidus, separated by a thermal barrier extending from albite to orthoclase-leucite. The thermal barrier separates magmatic liquids that evolve to become phonolitic from those that become rhyolitic. Magmas that plot within the Ab-Or-Ks-Ne sub-field are silica undersaturated and evolve by fractional crystallization along liquid lines of descent toward the silica-undersaturated ternary minimum (M_U), resulting in an alkali–feldspar–nepheline–leucite phonolite. Liquids that plot within the Ab-SiO_2-Or sub-field are silica oversaturated and evolve toward the silica oversaturated ternary minimum (M_S), becoming an alkali feldspar-quartz rhyolite. Thus, similar trachytic magmas that plot near (but across) the divide may evolve along different liquid lines of descent to become radically different derivative magmas. Because the barrier is close to the Ab-Or

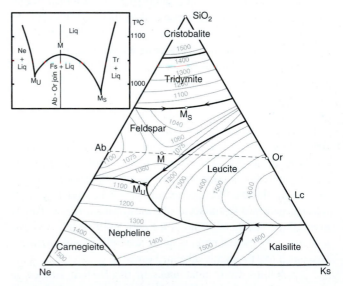

FIGURE 19.7 Phase diagram for the system SiO_2-$NaAlSiO_4$-$KAlSiO_4$-H_2O at 1 atm pressure. Insert shows a T-X section from the silica-undersaturated thermal minimum (M_U) to the silica-oversaturated thermal minimum (M_S) that crosses the lowest point (M) on the binary Ab-Or thermal barrier that separates the undersaturated and oversaturated zones. After Schairer and Bowen (1935 © American Geophysical Union with permission) and Schairer (1950 © the University of Chicago).

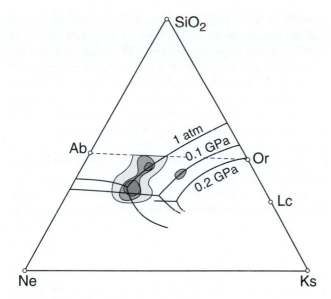

FIGURE 19.8 Part of the Ne-Ks-SiO$_2$-H$_2$O system at 1 atm, 0.1 GPa, and 0.2 GPa, illustrating the reduction in the leucite field with increasing p_{H_2O}. At 0.2 GPa the Lc-liquid field crosses the Ab-Or join, and the system goes from peritectic to eutectic behavior. Also shown are contours for analyses of 122 undersaturated volcanics. After Gittins (1979). Copyright © reprinted by permission of Princeton University Press.

join, fractionation of feldspar exerts a dominant control on the evolution.

At low pressure, the incongruent melting of leucite permits some liquids near Or on Figure 19.7 to descend from the silica-undersaturated field across to the oversaturated field. The leucite field shrinks rapidly with increasing p_{H_2O} (Figure 19.8) so that this crossover behavior ceases when p_{H_2O} is greater than 0.2 GPa (approx. 6 km) and the feldspar join becomes a true barrier. Figure 19.8 also shows the contoured densities of plotted compositions of 122 silica-undersaturated phonolites and trachytes. Their proximity to the thermal minima of the Ne-Ks-SiO$_2$-H$_2$O diagram implies that these rocks are a result of shallow fractional crystallization.

Silicic lavas and pyroclastics constitute a significant proportion of the total volume of East African Rift volcanics. In Ethiopia they compose one-sixth of the volcanic pile, and in Kenya one-half (Williams, 1982). In all subprovinces of the East African Rift system, intermediate lavas (~52 to 57 wt. % SiO$_2$) are subordinate to the mafic and silicic lavas. This bimodality presents a problem for the commonly proposed model of fractional crystallization as the dominant process for the evolution of the silicic lavas because the volume of evolved lavas should progressively drop as fractional crystallization removes material from the magmatic systems. This is particularly true for the vast Kenyan Miocene plateau phonolites. Barberi et al. (1975), in their model for the Boina center, noted that the transition to peralkalinity occurs at $F = 0.2$ [F is the fraction of melt remaining, Equation (9.8)] because of the effect of plagioclase removing so much Ca and Al and rendering the remaining liquid rich in Na and K. They said that the Daly gap and bimodal distribution is due to the sudden drop in f_{O_2} near this

peralkaline transition, resulting in a marked compositional change over a small fractionating interval. Other suggestions for the high proportion of silicic lavas include mantle enrichment that might produce more silicic melts, stratified shallow magma chambers in which the upper silicic portions are tapped (leaving the lower mafic-enriched portion behind), and the sinking of high density intermediate ferrobasalt in some evolutionary sequences. The isotopic data argue against significant crustal assimilation or melting in all but a few cases. I'm rather partial to a model that involves a mafic underplate, similar to that developed in Chapter 17 for granitoids. Much of the dense basaltic mantle melt would stall beneath the lighter continental crust, where it then fractionates until it becomes buoyant again or else crystallizes and becomes remelted by the high heat flux to produce more silicic magmas. Such a model has been proposed for the East African Rift by Hay et al. (1995). It would be difficult to distinguish the products of fractional crystallization of a mantle-derived basalt from those produced by the solidification and partial remelting of that same magma on the basis of isotopic or trace element patterns.

19.1.8 A Model for East African Rift Magmatism

The data presented above provide some constraints for the development of a petrologically reasonable model for magmatism at the East African Rift. Initial rifting may be either active or passive, but the geophysical evidence and the isotopic data suggest some OIB-like asthenospheric mantle is involved. This and the great magma volume for a mere 6 to 10 km of extension and the early onset of flood basalt magmatism suggest that *active* rifting was predominant (i.e., the mantle was rising actively rather than responding to lithospheric separation). Following the very early development of a broad shallow depression, the active rifting stage begins with lithospheric extension over a rising asthenospheric plume or diapir (Figure 19.9a). The rise of less dense asthenosphere and the heat flux causes some uplift and topographic doming. The upper crust becomes faulted, whereas the lower crust and lithospheric mantle are thinned in a ductile fashion. Primary crustal heterogeneity controls the location of rifting, which focuses on the weaker ancient suture areas between amalgamated cratonic blocks. Such a model would be consistent with a major superplume rising beneath Africa and spawning several subsidiary plumes (Figures 14.1 and 14.23).

Ascent of the asthenosphere results in decompression partial melting at a variety of depths. Partial melting at shallow depths (as shallow as 50 km) results in tholeiites (Chapter 10). Transitional basalts are produced at intermediate depths, and alkaline magmas at greater depths. The early rapid ascent favors shallow melting, and hence the tholeiitic and transitional series.

The rift-stage (Figure 19.9b) is characterized by the development of a graben in the stretched crust and the arrival at the surface of alkaline magmas from a deep asthenospheric source because deeper melts should take longer to rise. This trend later reverses as asthenospheric activity wanes, unless

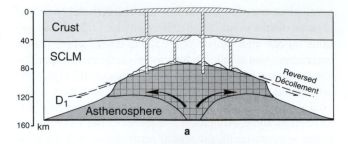

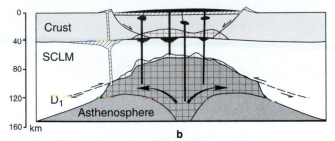

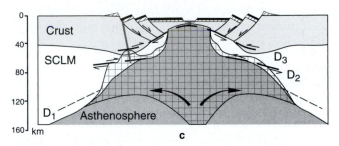

FIGURE 19.9 Hypothetical cross sections (same vertical and horizontal scales) showing a proposed model for the progressive development of the East African Rift System. **(a)** Pre-rift stage, in which an asthenospheric mantle diapir rises (forcefully or passively) into the lithosphere. Decompression melting (crosshatches indicate areas undergoing partial melting) produces variably alkaline melts. Some partial melting of the metasomatized subcontinental lithospheric mantle (SCLM) may also occur. Reversed décollements (D_1) provide room for the diapir. **(b)** Rift stage: development of continental rifting, eruption of alkaline magmas (black) mostly from a deep asthenospheric source. Rise of hot asthenosphere induces some crustal anatexis. Rift valleys accumulate volcanics and volcaniclastic material. **(c)** Afar stage, in which asthenospheric ascent reaches crustal levels. This is transitional to the development of oceanic crust. Successively higher reversed décollements (D_2 and D_3) accommodate space for the rising diapir. After Kampunzu and Mohr (1991) and P. Mohr (personal communication). Copyright © with permission from Springer-Verlag.

oceanic seafloor spreading becomes established (Kampunzu and Mohr, 1991).

In the late stages, such as to the north in Afar (Figure 19.9c), the rising asthenosphere traverses the lithospheric mantle and reaches the crust. Kampunzu and Mohr (1991) proposed that the lithosphere accommodates space for the rising diapir by an upward succession of reverse décollements (underthrust-like crustal detachment faults). Should extension continue beyond this point, oceanic crust would be generated by a system of sea-floor spreading (Chapter 13), as now occurs in the Red Sea and Gulf of Aden.

Melt sources can be highly variable in all stages, including rising asthenosphere, heterogeneous SCLM, and, as heat is added to the shallower portions of the system, crustal

material as well. Density entrapment and underplating of various mantle melts at the base of the crust are to be expected at any stage. Fractionation of primary magmas will thus occur, and heat will be added to the crust to permit some degree of assimilation and crustal anatexis. Crustal assimilation is apparently subordinate in the East African Rift, and more primitive magma is common, both attributable to the many extensional faults permitting fresh mantle melts to reach the surface.

Some melting of enriched subcontinental lithospheric mantle (SCLM) may also produce alkaline magmas (e.g., Menzies, 1987), particularly in cratonal areas and along the margins of the area affected by an asthenospheric diapir. This may explain the trend toward more alkalinity away from the rift axis, but the role of the lithospheric source is far from clear. Kampunzu and Mohr (1991) suggested that African ultra-alkaline activity is typically located at transverse structural zones, especially south of the equator. Macdonald et al. (1994) proposed that the ultra-alkaline carbonatite–nephelinite–kimberlite series is restricted in the Cenozoic to areas of cratonal crust such as the Tanzanian Craton, whereas mainly alkaline basalt–trachyte dominates in areas of thinned crust north of the craton.

The mantle beneath Africa is clearly heterogeneous, including zones depleted during the Precambrian and zones enriched by several events involving various alkaline and carbonatitic melts and fluids. Contrasts such as the DUPAL-like Pb signature in the southern hemisphere and the K-enriched Western Branch versus the Na-enriched Eastern Branch testify to broad regional mantle differences. Marked geochemical and isotopic contrasts between adjacent centers, or even within single centers over time, attest to a considerable degree of small-scale heterogeneity as well. The old model of less silica-undersaturated and alkaline magmatism giving way to progressively more tholeiitic magmatism as the rift changes from continental to oceanic character is clearly too simplistic (except in the MORB-like later stages of truly oceanic development such as the Red Sea–Gulf of Aden). Kampunzu and Mohr (1991) noted that essentially the full compositional range of melts coexist and are available for all of the 33 Ma magmatic history of the East African Rift. They appropriately concluded that the processes that led to East African Rift magmatism are considerably more diverse than has previously been assumed. For an analogy, recall the discussion that accompanied Figure 14.20 of heterogeneous mantle in plumes, possibly containing slivers of subducted crustal and carbonate components.

More evolved magmas appear to develop principally via fractional crystallization, but remelted crustal underplates, assimilation of crustal components, and magma mixing may all occur at various centers. The underplates and metasomatized mantle may also contribute to the anomalously low density and broad domal uplifts that characterize much of the East African Rift system.

Volcanics vastly predominate over plutonics in the East African Rift. Older rift provinces, such as the Proterozoic Gardar province of south Greenland (Macdonald and Upton, 1993) and the Permian Oslo Graben of Norway, however, exhibit considerably more plutonic components, which is almost certainly due to the more extensive erosion in the older areas, which exposes the magmatic roots. Gravity anomalies suggest that the same plutonic infrastructure exists beneath Africa, and that mafic and silicic intrusions may exceed remnant continental crust at depth beneath the rift. If true, this may explain the subdued indicators of crustal contamination in most areas.

There are certainly pronounced differences in structure, chemistry, petrology, and dynamics of the various continental rift provinces, but the East African Rift probably provides a representative example of the general characteristics of such magmatism. Barberi et al. (1982) divided continental rifts into low- and high-volcanicity types. Low-volcanicity rifts (e.g., the Western Branch of the East African Rift, the Rhine Graben, Baikal) are characterized by low eruptive volumes, slow divergence, and slower asthenospheric upwelling. The degree of partial melting is thus lower, and the metasomatized SCLM presumably less attenuated, so that the magmas are generally more strongly alkaline. They are also more geochemically variable. High-volcanicity rifts (e.g., Ethiopia, Afar) are faster, have more voluminous volcanism, and are less alkaline. Low-volcanicity rifts also tend to be more basic and unimodal, whereas high-volcanicity rifts have more silicic products and are commonly bimodal.

As noted earlier, not all continental alkaline rocks are rift-related, and for many occurrences it is difficult to determine the tectonic setting. This may be due in part to the complex geometry and history of the continental crust, but it may also reflect the anorogenic nature of many alkaline provinces. For the student interested in continental alkaline provinces, several good summaries are listed at the end of the chapter. Although the alkaline rocks are commonly considered to be the hallmark of continental rifting, and are certainly the most interesting products, it should be clear from the previous discussion (and Table 8.7) that tholeiites are generated in this setting as well. Although not indicated by Table 8.7, *calc-alkaline* magmas, however rare, are also known in extensional settings (Hooper et al., 1995; Hawkesworth et al., 1995). Such calc-alkaline situations are generally inboard from an active or extinct subduction zone, and the calc-alkaline geochemical signature of the rift zone rocks is probably imparted by hydrous fluids related to a previous episode of dehydration of a descending lithospheric slab beneath it.

19.2 CARBONATITES

There are many unusual and interesting rock types among alkaline rocks of the East African Rift and other continental provinces, but the most unusual type of all must be a variety of carbonate-dominated rocks collectively known as **carbonatites**. The concept of igneous carbonates is so unusual that for a long time geologists refused to believe that they were indeed igneous, and proposed instead that they were remobilized limestones or huge carbonate xenoliths. They consist predominantly of calcite, dolomite, or ankerite, together with varied amounts of clinopyroxene, alkali amphibole, biotite, magnetite, and apatite. This means, of course, that most carbonatites are not alkaline. Alkaline carbonatite is known from only one area in Tanzania so far. I include all carbonatites in this chapter because they typically occur as relatively small intrusions associated with alkaline silicate rocks, suggesting a common genetic link.

19.2.1 Carbonatite Classification and Mineralogy

Carbonatites, by IUGS definition, contain more than 50 modal % carbonate minerals. Table 19.3 shows the terminology that is applied to the more common carbonatites. The first column gives the recommended modern names based on the most abundant carbonate mineral, whereas the other columns give names that are common in the older literature. Sövite is still used for the more abundant coarse-grained calcite–carbonatites. The corresponding term for coarse dolomitic carbonatites, rauhaugite, is much less commonly used. Although a few ferrocarbonatites contain ankerite or siderite, they are typically fine-grained mixtures of calcite and hematite (or hydrated iron oxides). Natrocarbonatite (Na-K-Ca carbonatite) is *very* rare, and known for certain from only one volcanic center.

For carbonate-bearing rocks with 10 to 50% carbonates, the IUGS recommends the use of the modifying terms "calcitic" or "dolomite" preceding the igneous rock name based on the remaining silicate assemblage (e.g., "calcite ijolite"). "Silico-carbonatite" is a term that appears in the literature for rocks with 10 to 50% carbonate. It is not among the terms recommended by the IUGS.

Over 280 minerals are known to occur in various carbonatites, reflecting the exotic diversity of carbonatite compositions. Table 19.4 lists some of the more common or diagnostic minerals. Although carbonates are certainly well represented in the table and comprise the bulk of the bodies, a wide variety of silicates, oxides, and other mineral types are also found.

19.2.2 Carbonatite Occurrences

Of the approximately 350 known carbonatites, over half occur in Africa. Most carbonatites occur in stable continental intraplate settings, although some occur at continental margins and may be linked with orogeny or plate separation. So far, only two carbonatites are known from ocean basins (also intraplate): one in the Cape Verde Islands, and the other in the Canary Islands. The proximity of these islands

TABLE 19.3 Carbonatite Nomenclature

| Name | Alternative | |
	Coarse	Medium–Fine
Calcite–carbonatite	Sövite	Alvikite
Dolomite–carbonatite	Rauhaugite*	Beforsite
Ferrocarbonatite		
Natrocarbonatite		

* Rarely used; *beforsite* may be applied to any grain size.

TABLE 19.4   Some Minerals in Carbonatites

Carbonates	Sulfides
Calcite	Pyrrhotite
Dolomite	Pyrite
Ankerite	Galena
Siderite	Sphalerite
Strontanite	**Oxides-Hydroxides**
Bastnäsite (Ce,La)FCO$_3$	Magnetite
*Nyerereite (Na,K)$_2$Ca(CO$_3$)$_2$	Pyrochlore
*Gregoryite (Na,K)$_2$CO$_3$	Perovskite
Silicates	Hematite
Pyroxene	Ilmenite
Aegirine-augite	Rutile
Diopside	Baddeleyite
Augite	Pyrolusite
Olivine	**Halides**
Monticellite	Fluorite
Alkali amphibole	**Phosphates**
Allanite	Apatite
Andradite	Monazite
Phlogopite	
Zircon	

* Only in natrocarbonatite.

From Heinrich (1966) and Hogarth (1989).

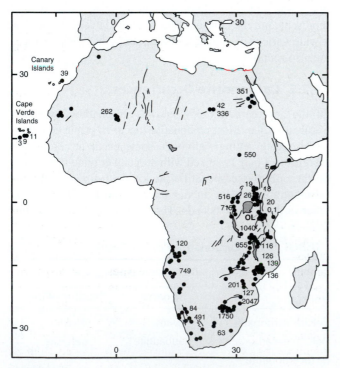

FIGURE 19.10 African carbonatite occurrences and approximate ages in Ma. OL = Oldoinyo Lengai natrocarbonatite volcano. After Woolley (1989).

to the carbonatite-rich African continent (Figure 19.10) suggests that these occurrences might be related to African continental (probably subcrustal) processes. The crust beneath the Canary Islands is suspected to include some continental Hercynian basement (Araña and Ortiz, 1991). There are very few Archean carbonatites, and there is a general increase in the number of carbonatites with time, though they tend to be mysteriously episodic.

The majority of carbonatites occur in relatively small intrusive complexes together with alkaline silicate rocks (typically nephelinites, phonolites, nepheline syenites, ijolites, and urtites). Less common associations include melilitolites, pyroxenites, peridotites, group I kimberlites, or lamprophyres (Bell et al., 1998). Carbonatites that are not associated with coeval silicate rocks tend to be rich in dolomite. So common is the association of carbonatites and alkaline silicate rocks that it is commonly referred to as the **carbonatite-alkaline silicate rock association**. The commonly held view that the spatial association is also genetic (carbonatites are either derived from alkaline silicate magmas or they share a common parent) has been questioned by Gittins (1989) and by Harmer (1999).

Many alkaline silicate rocks and kimberlites contain accessory carbonate (almost exclusively calcite), and many kimberlites have a carbonate-rich matrix. Some kimberlites even have carbonate-rich dikes. Although most of the kimberlite–associated dikes and segregations qualify as carbonatites on the basis of their containing over 50% carbonate, they rarely contain any of the other common carbonatite indicator minerals in Table 19.4 and should not be thought of as carbonatites in the generally accepted usage of the name. Mitchell (1986) suggested that the presence of calcite in both kimberlites and carbonatites is more indicative of their common origin within the same portions of the upper mantle, and not of a true kimberlite–carbonatite genetic relationship. It is increasingly thought that the genesis of kimberlites, carbonatites, and many alkaline silicate magmas involves partial melting of carbonate-bearing peridotite (or at least peridotite with a significant p_{CO_2}). Alkaline silicate melts from such a source should eventually develop a p_{CO_2} sufficient to stabilize carbonate (albeit in very small amounts). We will deal with kimberlites and their carbonates in Section 19.3.3, and will now concentrate on the more common carbonatite–alkaline silicate association.

Figure 19.10 shows the distribution of carbonatites in Africa, one of the world's predominant carbonatite regions (and provides some continuity to the previous discussion of alkaline magmatism associated with the East African Rift). African carbonatites tend to be located individually on crustal domes as little as 1 to 2 km across, or in clusters on domal swells up to 1000 km across. Many carbonatites follow major faults or define lineaments of unknown origin, but others show no obvious structural control. There is rarely an age progression along the faults or lineaments as one would expect if they were produced at hotspots. Major episodes of carbonatite activity occurred in the Proterozoic–Early Cambrian, the Cretaceous, and again in the Cenozoic (the latter was concentrated along the East African Rift).

Carbonatites in Africa, as elsewhere, tend to form in clusters, or provinces, within which carbonatite activity repeats over long time spans, suggesting attached *lithospheric*, as opposed to deeper mobile asthenospheric mantle control. Although the nature of the subcontinental lithospheric mantle (SCLM) may exert a major influence over alkaline and carbonatitic activity, one cannot rule out an ultimate asthenospheric plume contribution, which may have imparted an enriched alkaline character to the lithosphere at an earlier stage of development (as well as heat required for melting). Le Bas (1971) proposed that nephelinite–carbonatite–kimberlite activity marks locations of deep mantle degassing, which modifies the composition of the overlying lithosphere and reduces its density, leading to the broad domal swells that characterize many carbonatite provinces. Whether this may also be plume or superswell related is difficult to evaluate.

19.2.3 Field Characteristics of Carbonatites

Carbonatites can occur as volcanics or intrusive bodies. As noted earlier, carbonatites commonly occur within or satellitic to alkaline intrusive centers, where the carbonatite phase usually comes late in an intrusive series, after the alkaline silicate magmas. Some carbonatites, however, do not have associated silicate rocks. Carbonatite complexes are generally less than 25 km^2 in exposed area and are composite, with multiple intrusions of both silicate and carbonatite magma. Carbonatite typically composes less than 10% of the exposed igneous rocks. Exposed intrusive carbonatites include small plugs, cone sheets, and occasional ring-dikes. Planar dikes or dike swarms of both silicate rocks and carbonatites commonly cut the entire intrusive complex. The wall rocks may have a fractured appearance suggesting a high volatile content of the carbonatite melts.

In a typical sequence, shallow early ijolite and/or nepheline syenite plugs are followed by carbonatites that cut the earlier silicate complex (Figure 19.11). Sövites (typically with over 90% calcite) are the most common type of carbonatite in these complexes and may represent the only carbonatite at a locality. Other common carbonatites contain both calcite and dolomite. Less common are those in which dolomite or ankerite are predominant. The later manifestations of igneous activity in many complexes are the emplacement of dikes or cone sheets of iron-rich carbonatites, collectively called ferrocarbonatite. The most common of these contain fine-grained calcite and hematite, but some are ankeritic, and only a few contain siderite. Finally, pipe-like bodies of Fe and REE-rich carbonatite (some are distinctly radioactive) may be emplaced. The last episodes are typically brecciated and exhibit replacement textures and fluorine addition. They appear to involve late-stage fluids that may be hydrothermal in nature. All of the stages are rarely developed at a single locality.

Dolomitic/ankeritic carbonatites, while less abundant than calcite carbonatites, may have been substantially underestimated in the literature. Harmer and Gittins (1997) suggested that dolomitic/ankeritic carbonatites are more

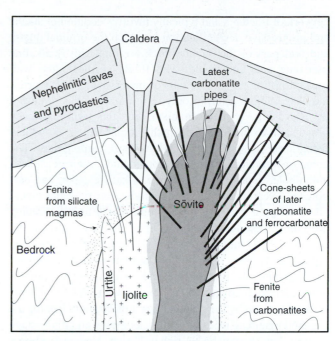

FIGURE 19.11 Idealized cross section of a carbonatite–alkaline silicate complex with early ijolite cut by more evolved urtite. Carbonatite (most commonly calcitic) intrudes the silicate plutons, and is itself cut by later dikes or cone sheets of carbonatite and ferrocarbonatite. The last events in many complexes are late pods of Fe and REE-rich carbonatites. A fenite aureole surrounds the carbonatite phases and perhaps also the alkaline silicate magmas. After Le Bas (1987). Copyright © reprinted by permission of John Wiley & Sons, Inc.

common than is generally appreciated, particularly in the Precambrian cratons, such as the Zimbabwe (Figure 19.2) and Kaapvaal (South Africa) cratons, and in the Archean parts of the Canadian Shield. The relationships between successive stages of carbonatites are variable and complex. Replacement of calcite–carbonatites by dolomitic and later ferrocarbonatite has been described in the literature, but this subject is controversial at present.

Rocks intermediate between carbonatites and alkaline silicate rocks are less common than either extreme and usually appear to be the result of carbonatite magma having intruded and mingled with previously consolidated or semicrystalline silicate rocks, rather than being the product of progressive fractionation. Estimated temperatures of emplacement, calculated from mineral geothermometry (Section 27.4), range from 550°C to over 1000°C.

An almost universal characteristic of carbonatite complexes is the presence of a distinctive metasomatic aureole in which the wall rocks (most commonly quartzo-feldspathic gneiss) have been converted to aegirine-rich and alkali amphibole-rich rocks, and in some cases to K-feldspar-rich rocks. The metasomatic rocks are commonly called **fenites** (Figure 19.11), and the process **fenitization**, after the Fen alkaline complex in southern Norway. Replacement textures indicate that hot, reactive, alkaline fluids are involved, but opinion varies as to whether they are derived solely from the carbonatitic magma or from the alkaline silicate magmas as well (note that the interpretation in Figure 19.11 includes silicate-generated fenites). Fenitization begins along a network of fractures and typically involves addition of alkalis and progressive desilicification in which the original quartz and feldspars of the country rock are replaced by alkaline pyroxene and amphibole. The fluids permeate the wall rocks, forming an aureole tens of meters wide. In the original usage, fenite referred only to the mafic rocks produced by K-Na-Mg-Fe metasomatism, but the term has been broadened to include feldspathization, which is widely recognized in some very shallow-level carbonatites. The common products of fenitization include aegirine, alkali-amphibole, nepheline, phlogopite, alkali feldspar, and carbonates. Near the contact the production of replacement alkali feldspar may be so thorough that a pseudo-syenite, difficult to distinguish from an igneous rock, is produced. Kramm (1994) and Kramm and Sindern (1998) have even argued on the basis of Sr and Nd isotopic ratios that some of the magmatic alkaline silicate rocks accompanying carbonatites are the products of melting of wall rocks that were highly fenitized by the carbonatites. Fenitization, like the associated magmatism, is typically a complex and multi-stage process. Fenites may be Na-rich or K-rich, producing albite ± sodic amphibole ± aegirine or alkali feldspar, respectively. Fenites are not associated with the ultramafic/ kimberlitic associated carbonatites.

Volcanic carbonatites are far less common than intrusive ones, although many exposed intrusive carbonatites may once have had volcanic edifices. Most volcanic carbonatites are calcitic, with flows (both aa and pahoehoe) and pyroclastic material (including ash, lapilli, and bombs).

19.2.4 Carbonatite Geochemistry

Table 19.5 lists typical values of the major- and trace-element constituents of the four types of carbonatite listed in Table 19.3. The very low SiO_2, very high REE (highest of any igneous rock), and volatile (CO_2, F, Cl, S) contents characterize carbonatites as completely unique. They are further characterized by unusually high Ba, Sr, LREE, P, and Nb. All are characteristics that cannot realistically be produced by fractional crystallization of typical mantle melts and must be inherently high in the primary carbonatite magma sources. Carbonatites are the principal economic sources of Nb and REE, and increasingly of phosphate. Some are mined for Ba, fluorite, Sr, V, Th, U, Zr, and Cu (see Mariano, 1989, for details). As we shall see, carbonatites may largely be cumulates, so the compositions given do not necessarily represent the liquids from which they crystallized. We will discuss other aspects of carbonatite geochemistry, particularly the isotope data, in conjunction with carbonatite origin in the next section.

19.2.5 The Origin of Carbonatites

The high melting point of calcite (>1340°C at 0.1 GPa) and the high p_{CO_2} required to prevent its dissociation at this pressure, together with the absence of high-temperature metamorphic aureoles around carbonatite complexes, deterred early investigators from considering carbonatites as

TABLE 19.5 Representative Carbonatite Compositions

%	Calcite–Carbonatite	Dolomite–Carbonatite	Ferro–Carbonatite	Natro–Carbonatite
SiO_2	2.72	3.63	4.7	0.16
TiO_2	0.15	0.33	0.42	0.02
Al_2O_3	1.06	0.99	1.46	0.01
Fe_2O_3	2.25	2.41	7.44	0.05
FeO	1.01	3.93	5.28	0.23
MnO	0.52	0.96	1.65	0.38
MgO	1.80	15.06	6.05	0.38
CaO	49.1	30.1	32.8	14.0
Na_2O	0.29	0.29	0.39	32.2
K_2O	0.26	0.28	0.39	8.38
P_2O_5	2.10	1.90	1.97	0.85
H_2O+	0.76	1.20	1.25	0.56
CO_2	36.6	36.8	30.7	31.6
BaO	0.34	0.64	3.25	1.66
SrO	0.86	0.69	0.88	1.42
F	0.29	0.31	0.45	2.50
Cl	0.08	0.07	0.02	3.40
S	0.41	0.35	0.96	–
SO_3	0.88	1.08	4.14	3.72
ppm				
Li	0.1	–	10	–
Be	2	<5	12	–
Sc	7	14	10	–
V	80	89	191	116
Cr	13	55	62	0
Co	11	17	26	–
Ni	18	33	26	1
Cu	24	27	16	–
Zn	188	251	606	88
Ga	<5	5	12	<20
Rb	14	31		178
Y	119	61	204	7
Zr	189	165	127	0
Nb	1204	569*	1292	28
Mo	–	12	71	125
Ag	–	3	3	–
Cs	20	1	1	6
Hf	–	3	–	0
Ta	5	21	1	0
W	–	10	20	49
Au	–	–	12	–
Pb	56	89	217	–
Th	52	93	276	4
U	9	13	7	11
La	608	764	2666	545
Ce	1687	2183	5125	645
Pr	219	560	550	–
Nd	883	634	1618	102
Sm	130	45	128	8
Eu	39	12	34	2
Gd	105	–	130	–
Tb	9	5	16	–
Dy	34	–	52	2
Ho	6	–	6	–
Er	4	–	17	–
Tm	1	–	2	–
Yb	5	10	16	–
Lu	1	0		0

* One excluded analysis contained 16,780 ppm Nb.

From Woolley and Kempe (1989); natrocarb. from Keller and Spettel (1995).

magmatic. For many years carbonatites were thought to be remobilized limestone, marble xenoliths, or precipitates from hydrothermal solutions. The first seeds of doubt were planted by Lt. Col. W. Campbell Smith of the British Museum, who studied specimens routinely sent to London by geologists of the Colonial Geological Surveys mapping in various parts of Malawi, Kenya, Uganda, and Tanzania during the 1930s. From careful petrographic observations he pointed out the presence of typically igneous silicates, which prompted the re-examination of these rocks in the field, revealing sharp cross-cutting contacts and chilled margins. In the late 1950s, Tuttle and Wyllie (1958) and Wyllie and Tuttle (1960) demonstrated that calcite begins to melt at much lower temperatures (600 to 700°C) at $p_{H_2O} = 0.1$ GPa in the presence of a dense H_2O-CO_2 vapor. Finally, geologists observed alkaline carbonatite lava flows in action at the 3000-m tall volcano Oldoinyo Lengai, northern Tanzania, in 1958 (Dawson, 1962, 1966). For the location, see "OL" in Figure 19.10. This is still the only known active carbonatite volcano, and because the recent carbonatitic lavas there represent one of only two known natrocarbonatites, they cannot be related directly to the common calcitic or dolomitic carbonatites. Diehards still resisted, however, and proposed that the Oldoinyo Lengai flows were the result of melted trona beds, found nearby. Nonetheless, the eruption of Oldoinyo Lengai, when combined with the experimental melting of calcite, and the by now voluminous record of field relations and the distinctive mineralogy and chemistry of carbonatites, served to convince even the most skeptical petrologists that carbonatites are magmatic igneous rocks.

Oldoinyo Lengai also provided a unique opportunity to determine the properties of carbonatite magma (at least of the unusual natrocarbonatites that occur there). Krafft and Keller (1989) measured temperatures in natrocarbonatite flows and a lava lake in the range 491 to 544°C. Pinkerton et al. (1995) determined that the natrocarbonatite flows were extruded at temperatures between 576 to 593°C. Carbonate-rich liquids are thought to be largely non-polymerized ionic liquids with very low viscosity: 0.15 Pa·s for non-vesicular lavas, nearly like water (Norton and Pinkerton, 1997), and low density: 1.15–2.17 g/cm³ (Pinkerton et al., 1995; Treiman and Schedl, 1983).

During the burst of geological mapping and geochemical study of carbonatites in Africa during the 1950s and 1960s, it became abundantly clear that carbonatites are entirely unlike sedimentary or metasedimentary carbonate rocks, and are much more like a complex igneous rock. During the same period, carbonatites were found to have $^{87}Sr/^{86}Sr$ values typical of mantle material and far lower than crustal sedimentary carbonate rocks. Later, Nd isotopic studies yielded similar results. The high Sr and Nd content of carbonatites (and some of the alkaline silicate rocks) makes the $^{87}Sr/^{86}Sr$ and $^{143}Nd/^{144}Nd$ ratios less susceptible to change with contamination, making them fairly reliable indicators of their mantle sources.

The carbonatites of eastern Canada have initial $^{87}Sr/^{86}Sr$ and $^{143}Nd/^{144}Nd$ ratios that cluster sufficiently close

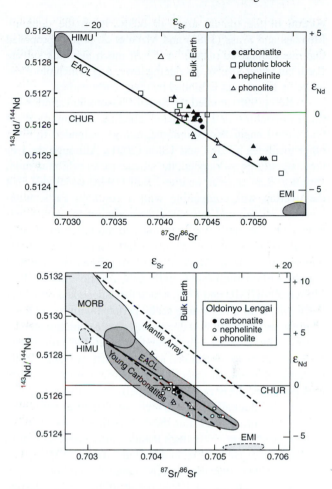

FIGURE 19.12 Initial $^{143}Nd/^{144}Nd$ versus $^{87}Sr/^{86}Sr$ diagram for young carbonatites (dark shaded), and the East African Carbonatite Line (EACL), plus the HIMU and EMI mantle reservoirs (from Bell and Blenkinsop, 1987, 1989). Also included are the data for Oldoinyo Lengai natrocarbonatites and alkali silicate rocks (from Bell and Dawson, 1995). Copyright © with permission from Springer-Verlag. MORB values and the mantle array are from Figure 10.16.

together that Bell and Blenkinsop (1989) explained them as derived from a single reservoir, which they characterized as depleted mantle (DM) at least 3 Ga old. They found that the East African carbonatites (Figure 19.12) describe a linear trend (which they called the "East African Carbonatite Line"). Such a linear array implies mixing involving two long-lived and isotopically distinct reservoirs. A slightly depleted lithospheric and a more enriched asthenospheric pair, similar to HIMU and EMI, respectively, provide the best match to the overall African isotope characteristics. The Sr-Nd data for numerous young carbonatites are included in Figure 19.12 (shaded). These carbonatites also plot along a trend similar to the East African Carbonatite Line. The natrocarbonatites of the active Oldoinyo Lengai volcano plot on the line as well, as do the nephelinites and phonolites at the volcano. The similarity in isotopic signatures between carbonatites and the associated alkaline silicate rocks suggests that the two suites at Oldoinyo Lengai are genetically related. Figure 19.12 also includes the Sr-Nd field of MORBs and the mantle array from Figure 10.16 for reference. Carbonatites plot either in the low portion of the mantle array or below it on the diagram. Menzies and Wass (1983) and Meen et al. (1989) proposed that metasomatized mantle has Rb:Sr much like bulk silicate Earth but lower Sm:Nd, producing lower ε_{Nd}, which plots low in the Sr-Nd plot, as exhibited by many carbonatites.

Bell (1998) studied the Sr-Nd-Pb isotopic variations in the carbonatites and associated alkaline silicate rocks at Oldoinyo Lengai and Shombole, a work extended to eight other localities by Bell and Tilton (2001). Although the isotopic signatures overlapped, the silicate rocks exhibited more variation than the carbonatites. Bell (1998) concluded that the overlap was compatible with a common carbonatite-silicate heritage and attributed the variation to mantle heterogeneity for the nephelinites, and to assimilation of deep crustal granulites for the phonolites (after separation of a carbonatite fraction). Bell and Tilton (2001) similarly concluded that the carbonatite isotopes reflect the mantle end-members (principally HIMU and EMI, which they suggested were spatially mixed on a fairly fine scale, perhaps within a plume), whereas variations in the silicate rocks were contaminated by various other sources. Harmer and Gittins (1998) found that carbonatites in some cratonal areas on Earth are isotopically more variable and enriched than the African ones described thus far, particularly in EMI and EMII. They also found that the associated alkaline silicate rocks show greater variability and greater enrichments than the carbonatites (some showing crustal enrichment, and others mantle enrichment, similar to the findings of Bell, 1998, and Bell and Tilton, 2001). They noted, however, that the isotopic range for carbonatites is much less than for ocean island basalts (Chapter 14), despite the greater thickness and variable nature of the continental crust that they must traverse.

The trace element and mantle-like isotope geochemistry finally convinced even the most resistant pro-sedimentary champions that carbonatites are indeed igneous and mantle derived, and not remobilized or assimilated limestone. The exact nature of the mantle source, however, is not always clear. Bell and Blenkinsop (1987, 1989), on the basis of their Nd-Sr values, could not unequivocally distinguish between asthenospheric and lithospheric sources. Nelson et al. (1988) considered the Nd-Sr-Pb characteristics of carbonatites to be similar to ocean island basalts, and thus favor an asthenospheric plume source. Kwon et al. (1989), in their overview of Pb isotopes in carbonatite-alkaline complexes, also liked an asthenospheric component. Bell and Simonetti (1996) found broadly linear relationships in $^{87}Sr/^{86}Sr$ versus $^{143}Nd/^{144}Nd$ and $^{208}Pb/^{204}Pb$ or $^{207}Pb/^{204}Pb$ versus $^{206}Pb/^{204}Pb$ in carbonatites, as well as in their alkaline silicate rock associates and xenoliths. They proposed a two-stage model in which the release of metasomatic agents from a HIMU-like asthenospheric plume enriches the subcontinental lithospheric mantle (SCLM), which has EMI characteristics. Ensuing variable degrees of discrete partial melting of the resulting heterogeneous and metasomatized lithosphere might then produce the array of melts and isotopic signatures. Our experience with plumes in Chapter 14 suggests that plumes may be heterogeneous and thus carry pods of HIMU and EM components.

By now there is complete agreement that carbonatites are clearly igneous, but there is still controversy over how carbonatite magmas are formed and how the various types of carbonatite and associated alkaline silicate magmas are related. Current debates center around these principal issues:

1. Are carbonatite and silicate magmas developed separately by partial melting of carbonated mantle peridotite, or are carbonatites derived from a carbonated parental silicate magma by differentiation in the mantle or crust?
2. If carbonatite magma is derived from parental silicate magma, is it by fractional crystallization or by liquid immiscibility, and at what depth?
3. If carbonatite magma is not derived via differentiation of a silicate magma, what is the nature of the parental carbonatite liquid? Is it calcic, dolomitic, or sodic? Is the source lithospheric, asthenospheric, or both?

Field, experimental, and isotopic data have been advanced to support all of the possibilities, but nothing is yet completely definitive. Experimental data have been reviewed by Wyllie (1989), Wyllie et al. (1990), and Lee and Wyllie (1994). Bell et al. (1998) also provided a brief summary of our understanding of carbonatite problems.

19.2.5.1 CARBONATITES AS PRIMARY MAGMAS   In his early work on the Alnö Complex of Sweden, Von Eckermann (1948) proposed a parental carbonate liquid. Development of carbonatite magmas by partial melting of carbonate-bearing peridotite was advocated by Wallace and Green (1988). Earlier advocates of a mantle origin of carbonatitic liquids, before the existence of carbonate in the mantle was widely acknowledged, include Dawson (1966), Cooper et al. (1975), and Koster van Groos (1975). The composition of volatile-rich magmas derived from lherzolite vary with pressure, volatile content (and hence types of volatile-bearing minerals), and the temperature excess above the solidus (Wendlandt and Eggler, 1980a, b; Wyllie, 1987; Eggler, 1989; Wyllie and Lee, 1998). As we discussed in Chapter 10, melting of volatile-free peridotite–lherzolite, at various pressures and melt fractions, produces liquids ranging from komatiites to picrites to tholeiites and alkaline basalts (Eggler, 1989). More exotic compositions require volatiles in the melt source or metasomatically altered mantle material, either SCLM or enriched asthenosphere. The nature of the vapor phase in the mantle is poorly constrained due to the mobility of the volatile constituents and uncertainty about the redox state of the mantle. There is general agreement that the composition of the volatile phase, both in the crust and mantle, is dominated by the system C-O-H, but halogens and S may also be important local constituents. Phases in the system C-O-H may exist as CO_2-H_2O fluid mixtures, if sufficiently oxidized, but may also include CH_4, H_2, and graphite/diamond under more reducing conditions. For now it is instructive to assume oxidizing conditions and hence a CO_2-H_2O fluid. Dasgupta and Hirschmann (2007) estimated that the CO_2 content of the mantle beneath ocean island basalt occurrences was 0.1 to 0.4 wt. %.

When present, H_2O-CO_2 fluids, depending on the proportions of each, will generate mantle lherzolite containing minor amounts of amphibole, phlogopite, dolomite, and/or magnesite, any of which can significantly affect the composition of melts produced at the lherzolite solidus. Uncertainty still remains, however, about the position of the solidus curve in the system lherzolite-C-O-H because it varies with the rock and fluid composition (for a discussion, see Wyllie, 1979). Figure 19.13 is the solidus (from Wyllie, 1989) for X_{CO_2} = $CO_2/(CO_2 + H_2O)$ = 0.8 based on experiments by Olafsson and Eggler (1983). At low pressure, the solidus assemblage is composed of melt plus an amphibole-spinel lherzolite and a CO_2-rich vapor. Amphibole controls (buffers) the fluid composition at X_{CO_2} = 0.8 (hence the choice for the overall system in Figure 19.13). The position of the solidus lies between the dry and the H_2O-saturated solidi, as expected for a mixed-volatile system (see, for example, Figure 7.24).

In the vicinity of 70-km depth, the presence of CO_2 begins to convert silicates to carbonates, and dolomite joins amphibole as a CO_2-bearing subsolidus phase (consider this reaction running right-to-left):

$$CaMg(CO_3)_2 + 4\,MgSiO_3 \qquad (19.1)$$
$$\text{Dol} \qquad\qquad \text{Opx}$$
$$= CaMgSi_2O_6 + 2\,Mg_2SiO_4 + 2\,CO_2$$
$$\text{Cpx} \qquad\quad \text{Ol} \qquad\quad \textit{fluid}$$

and the solidus temperature drops dramatically toward point Q in Figure 19.13, forming a nearly isobaric "ledge" in the solidus (in the words of Wyllie, 1989). The temperature drop

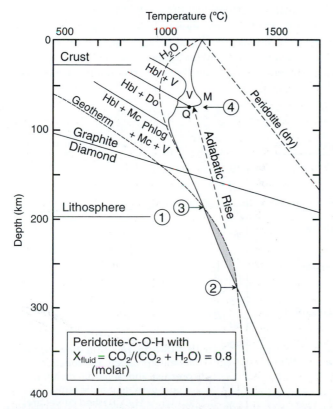

FIGURE 19.13 Solidus curve (heavy solid curve) for lherzolite-CO_2-H_2O with a defined ratio of CO_2:H_2O = 0.8. Dashed curves = H_2O-saturated and volatile-free peridotite solidi and the approximate shield geotherm. After Wyllie (1989).

is caused by the increase in the $MgCO_3$ component in the melt when dolomite becomes a mantle phase, and as much as 45 wt. % CO_2 may dissolve in the liquid, stabilizing it to lower temperatures. At higher pressures yet, amphibole reaches its high-pressure stability and a phlogopite–carbonate–lherzolite is the solidus rock assemblage (at X_{CO_2} = 0.8). This again deflects the solidus, and the carbonate-buffered solidus essentially follows the H_2O-saturated normal lherzolite solidus to higher pressures from there.

The composition of near-solidus liquids (low % partial melting) at pressures greater than point Q in Figure 19.13 (the first appearance of a carbonate at the solidus with increasing pressure) reflects the increased solubility of carbonate in the melt. Wyllie (1977b, 1987, 1989) concluded that the liquid under these conditions corresponds to a *dolomitic carbonatite magma* with Ca/Mg > 1, enriched in alkali carbonates, but with no more than 10 to 15% dissolved silicates (Wyllie et al., 1990). At temperatures above the solidus (increased percentage of partial melting) melt compositions become more silica-rich, and include melilitites and nephelinites (Wendlandt and Eggler, 1980a; Eggler, 1989).

In Figure 19.13, it is assumed that there is no free volatile phase (in excess of that required to make amphibole or phlogopite) at pressures greater than point Q. In my opinion this seems like a safe bet for the mantle, which should become progressively depleted in volatiles, but it is not assured. Koster van Groos (1975) found that a CO_2-bearing excess vapor can react with sodic or calcic minerals in the mantle and produce droplets of Ca-Mg-Na carbonatitic melts at temperatures as low as 650°C.

The effect of H_2O on the melting of carbonates has been neglected in Figure 19.13. The presence of H_2O can bring the melting point of pure calcite down from over 1300°C to as low as 600°C. Wyllie (1989) concluded that excess H_2O in any calcite-, dolomite-, and/or magnesite-bearing mantle assemblage will produce a trace of melt down to about 650°C. Whether these traces can actually separate and accumulate as carbonatites is not known.

In summary, the introduction of oxidized volatiles (H_2O and CO_2) can produce dolomitic or carbonated phlogopite–peridotites in the mantle at depths greater than 70 km. Primary carbonatite melts can then be generated near the solidus at these depths.

The predominance of calcite–carbonatite, and its occurrence as the first carbonatite in localities where a sequence of carbonatite liquids are emplaced, suggests that *calcite* carbonatite may be the initial carbonatite liquid and parental to the other types of carbonatite. *Dolomite* carbonatites, however, appear to be the types formed by experimental partial melting of typically Mg-rich mantle lherzolites. Wyllie and Lee (1994) concluded that calcite–carbonatites, which are not in equilibrium with orthopyroxene at these depths (see below), can only be created by partial melting of wehrlites, which are not considered to be a common constituent of typical, or even residual depleted mantle (harzburgite or dunite). Dalton and Wood (1993) circumvented this problem by proposing that calcite carbonatites could be developed by low-pressure wall rock interaction between primary dolomitic carbonatites and mantle lherzolites to consume orthopyroxene and produce calcite–carbonatites and wehrlite-lined conduits.

Given reasonable estimates of the mantle source composition, carbonatite melts result only from very low degrees of partial melting, which consumes the solid carbonate phase and thus forms very small quantities of volatile-rich melts. In Section 14.7, we introduced the idea that plumes may also contain disrupted and stretched slivers of subducted oceanic crust, converted to eclogite and perhaps containing some remnant carbonate or hydrous phases. As described in that section, Yaxley and Brey (2004) and Dasgupta et al. (2004, 2006) found that carbonate lowered the solidus temperature of eclogite and that carbonatite liquids may form in carbonate-eclogite pods as the first components to melt in rising asthenosphere. Due to their very low viscosity and dihedral angles with adjacent silicates, such carbonatite melts, from either peridotite or eclogite, may become mobile with as little as 0.01% partial melting (e.g., Bell et al., 1998). As we will see below, as they rise these melts are likely to devolatilize and crystallize or react with the overlying peridotites and thus be consumed (and, as a result, refertilize their hosts). Wyllie (1989), Wyllie et al. (1990), and Lee and Wyllie (1994) argued that this method of carbonatite genesis may apply better to the kimberlite-associated carbonates (which will be left to Section 19.3.3, on kimberlites). The carbonatites we address here (those that occur by themselves or in the carbonatite–alkaline silicate rock association) are different, they propose. The intimate association of these carbonatites and alkaline silicate magmas, as well as many geochemical similarities between the two magma types, suggests that such carbonatites form in conjunction with the alkaline silicate rocks. The late injection and subordinate volume of the carbonatites furthermore indicate that many carbonatites are more likely to be derived from a silicate parent, either by fractional crystallization or liquid immiscibility (Barker, 1989).

19.2.5.2 CARBONATITES AS DIFFERENTIATION PRODUCTS OF PRIMARY ALKALINE SILICATE MELTS

Although carbonatites can be produced by partial melting in the lherzolite-C-O-H system, as demonstrated above, alkaline silicate magmas are common associates, including basanites, nephelinites, and melilitites as near-solidus melts of amphibole-bearing peridotites. In hydrous peridotite most H_2O probably goes into alkali-bearing amphiboles, which melt early and release the alkalis in the depth range of 65 to 80 km (Eggler, 1989). Figure 19.14 shows the grid of Green (1970) that illustrates the progressive melting of model peridotite (their "pyrolite," Chapter 10) with only 0.1 wt. % H_2O as a function of pressure. Although most of the diagram reinforces the discussions of Chapter 10 on basalt genesis, highly alkaline nephelinites and melilitites are the primary melt products at depths >60 km with less than 10% partial melting. The addition of CO_2 and/or an enriched-metasomatized mantle source would make the generation of alkaline melts even more likely over a broader range of conditions. For reviews of alkaline petrogenesis, see Green (1970), Basaltic Volcanism Study Project (1981), Edgar (1987), Green et al. (1987), Eggler (1989), Hirose (1997), and Green and Falloon (2005).

Given the relatively high ductility of the asthenospheric mantle, and the low viscosity of the melts, extraction of melt fractions as low as 0.2% may be plausible (McKenzie, 1985, argues that it is *inevitable* under these conditions). For a low degree of partial melting in the presence of CO_2-H_2O, however, rising melts should intersect the ledge in the solidus in Figure 19.13 and solidify, as will be discussed below. In order to avoid the ledge and reach the upper crust, the temperature, and thus the melt fraction, must be higher,

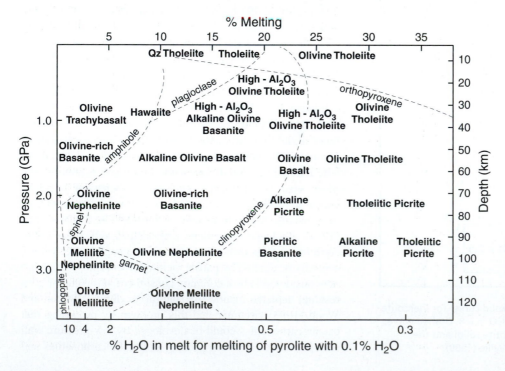

FIGURE 19.14 Grid showing the melting products as a function of pressure and % partial melting of model pyrolite mantle with 0.1 wt. % H_2O. Dashed curves are the stability limits of the minerals indicated. After Green (1970) © American Geophysical Union with permission.

implying a silicate melt rather than a carbonatite [unless the carbonatite is isolated from the host peridotite by a reaction zone of wehrlite as proposed by Dalton and Wood (1993) above and by Moore and Wood (1998)].

In carbonatite–alkaline silicate rock complexes, the volumetric predominance, early emplacement, and more primitive character of the alkaline silicate magmas suggest that they would be parental to the later carbonatites. Some alkaline silicate melts can dissolve 10 to 20 wt. % carbonate (Wyllie, 1978, 1989), and fractional crystallization of non-carbonate phases from a carbonated Ne-normative magma could concentrate CO_2 in the remaining liquid to the point of carbonate saturation, resulting in a residual carbonatitic liquid (Lee and Wyllie, 1994).

It is unlikely, however, that this fractional crystallization path leads to the common carbonatites we see today. Gittins (1989) argued that fractional crystallization won't produce the very high values of Nb, REE, and other incompatible trace elements characteristic of carbonatites, nor do the parent nephelinites have enough CO_2 to produce the volume of associated carbonatites found in most localities. Although there seems to be a strong link between the silicate and carbonatite magmas, it now appears unlikely that a carbonatite liquid is derived from an alkaline silicate parent by fractional crystallization.

19.2.5.3 CARBONATITES AS PRODUCTS OF LIQUID IMMISCIBILITY

Several lines of field evidence suggest that alkaline silicate and carbonatite liquids may be immiscible. For example, ijolites in carbonatite–alkaline silicate rock complexes form discrete intrusions, distinct from the carbonatite (Figure 19.11). Gradations between the two are rare and usually reflect mixing, or brecciation, and injection of one component into the other. Carbonatite globules are commonly found in alkaline silicate glasses and in apatites from ijolite in silicate xenoliths. The common overlap in isotope characteristics also indicates that the two liquids are closely related.

Derivation of a carbonatite liquid via liquid immiscibility became an attractive possibility with the discovery by Koster van Groos and Wyllie (1963, 1966) of a wide immiscibility gap between silicate-rich and carbonate-rich liquids in the system ($NaAlSi_3O_8$-Na_2CO_3-CO_2) above 870°C at 0.1 GPa. This gap persisted in the expanded $NaAlSi_3O_8$-Na_2CO_3-K_2O-H_2O-CO_2 system, even with the addition of minor amounts of Fe_2O_3 and MgO up to at least 1 GPa (Koster van Groos, 1975). Indeed, the studies showed the presence of *three* immiscible non-solid phases coexisting at equilibrium in these systems: an alkaline silicate liquid, a sodic carbonate liquid, and a volatile alkaline fluid, the latter being an attractive candidate for the fenitizing solutions known to have been present in most carbonatite–alkaline silicate rock complexes.

Since these initial studies, there have been several experimental investigations on synthetic and natural carbonated alkaline systems (reviewed by Wyllie et al., 1990; Lee and Wyllie, 1994, 1996). Once again, sövites are usually the earliest and dominant carbonatite, so *calcite* carbonatite should be the initial immiscible liquid and parental to the other types

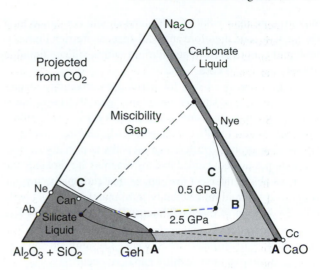

FIGURE 19.15 Silicate–carbonate liquid immiscibility in the system Na_2O-CaO-SiO_2-Al_2O_3-CO_2 (modified by Freestone and Hamilton, 1980, to incorporate K_2O, MgO, FeO, and TiO_2). The system is projected from CO_2 for CO_2-saturated conditions. The dark shaded liquids enclose the miscibility gap of Kjarsgaard and Hamilton (1988, 1989) at 0.5 GPa, which extends to the alkali-free side (A-A). The lighter shaded liquids enclose the smaller gap (B) of Lee and Wyllie (1994) at 2.5 GPa. C-C is the revised gap of Kjarsgaard and Hamilton. Dashed tie-lines connect some of the conjugate silicate-carbonate liquid pairs found to coexist in the system Nye = nyerereite, Can = cancrinite, Geh = gehlenite. After Lee and Wyllie (1996). Reprinted by permission of Oxford University Press.

of carbonatite. Of critical importance for the origin of carbonatites, then, is the system Na_2O-CaO-Al_2O_3-SiO_2-CO_2. The triangular diagram Na_2O-CaO-(Al_2O_3 + SiO_2), projected from CO_2, used by Freestone and Hamilton (1980) is shown in Figure 19.15 to illustrate the experimentally determined immiscibility gaps in this system. Kjarsgaard and Hamilton (1988, 1989) projected silicate–carbonatite immiscibility in alkaline systems to the alkali-free baseline (the dark-shaded curve A in Figure 19.15) at 0.2 and 0.5 GPa. Lee and Wyllie (1994), on the other hand, found that the miscibility gap at 2.5 GPa did *not* extend to the alkali-free base (curve B in Figure 19.15). Because the miscibility gap expands at higher pressure, these results are at odds with one another. This is an important difference because if the immiscibility gap extends to the alkali-free side, then calcite–carbonatites can coexist with a silicate magma and be a parental liquid to the carbonatite suite. If the solvus does not extend to the alkali-free base, then the common sövites are not primary melts, but are somehow derivative from more alkaline carbonatite liquids that are on the true solvus, but are exceedingly rare in nature. Kjarsgaard and Hamilton (1988, 1989) later reinterpreted some of their textures, deciding that the rounded calcite grains in several of their charges did not represent immiscible liquids. With this correction, true immiscible liquid compositions defined curve C in Figure 19.15, and not curve A. Given the pressure difference, curve C is in good agreement with the data of Lee and Wyllie (1994). Lee and Wyllie (1996) reviewed the experimental results and experimented further to constrain the miscibility gap liquidus surface. They concluded that no immiscible calcite-carbonatite liquid can contain more than ~80% $CaCO_3$ and should contain at least 5% Na_2CO_3, and therefore an alkali-free calcite-rich immiscible liquid cannot exist in equilibrium with a silicate liquid at pressures up to 2.5 GPa.

The addition of Mg for peridotite–carbonatite assemblages results in a smaller miscibility gap at 1 GPa, but it expands at lower pressures (Lee and Wyllie, 1994, 1996). Baker and Wyllie (1990) compared the Mg-bearing miscibility gap with the probable differentiation paths of primary carbonatite magmas and concluded that the magmas are probably too low in alkalis to intersect the gap, so that carbonatite magma formation by liquid immiscibility from mantle melts is highly unlikely. A point of dispute remains: The experiments of Lee and Wyllie were all undersaturated with respect to CO_2, whereas those of Brooker (1998) were CO_2 saturated and suggest that a suitable immiscibility gap does occur.

Lee and Wyllie (1996) further noted that the carbonate liquid in equilibrium with the silicate liquid in these systems does not begin to crystallize carbonates until after a long period of silicate crystallization. The carbonate liquid must first separate from the carbonated silicate parent (above 1100°C), and fractionate down a steep liquidus to temperatures in the vicinity of 600°C (which is closer to the eruptive temperatures at Oldoinyo Lengai and the calculated intrusion temperatures for many carbonatites). Even then the liquids are fairly alkaline. Calcite, however, becomes a liquidus phase at these temperatures. Given the extremely low viscosity of carbonatite liquids, fractional crystallization and crystal settling should be effective in them. Lee and Wyllie (1996) endorsed the widely held view that most calcite carbonatites are *cumulates*, settling out from somewhat more alkaline carbonatite liquids. Some others proposed that calcite-rich liquids evolved from more alkaline ones by alkali loss to a fluid phase.

19.2.5.4 CARBONATITE PETROGENESIS.
From the discussion above, we can conclude that carbonatites are genuine igneous rocks and are ultimately produced by melting of the mantle. All three methods, direct partial melting of hydrous-carbonated lherzolite, fractional crystallization, and liquid immiscibility (the latter two from a parental alkaline silicate melt), are capable of producing carbonate-rich melts. Perhaps any one of the three methods works in some locality or another, although the production of "typical" carbonatites via fractional crystallization from an alkaline silicate liquid may be unlikely.

Peter Wyllie proposed a model for the genesis of carbonatites (and kimberlites, covered in Section 19.3.3) based on the *P-T* lherzolite-C-O-H phase diagram depicted in Figure 19.13 (Wyllie, 1989; Wyllie et al., 1990). The HIMU–EMI isotopic character of carbonatites suggests that mantle plumes are involved in their origin (Bell and Tilton, 2001). If there is sufficient CO_2 and H_2O available in an asthenospheric mantle plume rising along the geotherm, melting will occur as the rising lherzolite exceeds the solidus temperature at level 2 in Figures 19.13 and 19.16 (or somewhat deeper if the source is entrained eclogite in the plume). With reasonable mantle amounts of CO_2 and H_2O, only small quantities of primary carbonatitic (and kimberlitic)

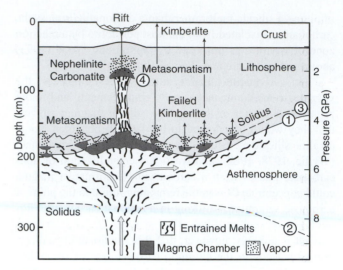

FIGURE 19.16 Schematic cross section of an asthenospheric mantle plume beneath a continental rift environment, and the genesis of nephelinite-carbonatites and kimberlite–carbonatites. Numbers correspond to Figure 19.13. After Wyllie (1989) and Wyllie et al. (1990).

melt will be generated in this manner. In the unlikely situation of excess volatile constituents, carbonatites may be generated at much lower temperatures, as discussed above. If the melt pockets are small, they may become entrained in the flowing mantle (Figure 19.16). If they accumulate and erupt, the high volatile content is more likely to give rise to explosive kimberlite diatremes (Section 9.3.3.2) than carbonatite–alkaline silicate rock complexes.

Primary carbonatite melts might accumulate at the lithosphere-asthenosphere boundary (level 1 in Figures 19.13 and 19.16), where they meet some mechanical resistance due to the more rigid nature of the lithosphere. Others may work their way up into the lithospheric mantle, following the geothermal gradient, until they encounter the solidus again at level 3 and begin to solidify at subsolidus temperatures. In either situation, crystallization of the volatile-saturated melts releases volatiles (and their dissolved constituents), resulting in metasomatism of the adjacent mantle wall rocks. Vapors may crack the country rock, allowing the melt to rise further before crystallizing fully. Should the asthenosphere rise beneath a rift zone (for which Figure 19.16 was designed), the rising asthenospheric plume may result in higher degrees of partial melting and produce alkaline silicate melts, such as nephelinites or melilitites, in addition to carbonatites and kimberlites. If the melts rise adiabatically, they may reach level 4 in both figures at about 75-km depth, where they will encounter the solidus ledge in Figure 19.13. At this point they ought to crystallize, again releasing volatiles and metasomatizing and fracturing the wall rocks. Hotter magmas may miss the ledge and reach shallow crustal levels. Dalton and Wood (1993) proposed that the rising dolomitic melts may react with orthopyroxene in the lherzolitic mantle at this point via Reaction (19.1) and create localized zones of wehrlite (orthopyroxene-free peridotite). If sufficient carbonatitic magma can be produced to form a wehrlite-lined conduit to the crust, a primary carbonatite may reach the surface. Harmer and Gittins (1997) proposed such a model for primary carbonatites in which the first batches of dolomitic carbonatite froze in the lithosphere at and above the ledge as they produced wehrlites plus localized metasomatic alteration. Subsequent melt batches reacted with the wehrlites via the reaction:

$$3 \, Dol + Di = 2 \, Fo + 4 \, Cal + 2 \, CO_2 \qquad (19.2)$$

which results in more Ca-rich calcite-carbonatites. Later primary dolomitic carbonatites could reach the surface, they proposed, by following the wehrlite-lined plumbing system. Whether this is a common mechanism for the common calcite–carbonatites is questionable, but it may well explain the dolomitic carbonatites, particularly those that occur in the cratons, in many instances without associated silicate rocks.

Melts that rise from level 4 in Figures 19.13 and 19.16, and most of those that reach the crust, probably form fairly large bodies capable of differentiating. Primary alkaline silicate magmas are not of a composition that would be in equilibrium with carbonatite. As they fractionate, however, the fractionation path may encounter the carbonatite–silicate solvus, resulting in immiscible carbonatite liquids, or (less likely) miss the solvus and produce residual carbonatite liquids. An alkaline volatile-rich phase is also produced that acts as the fenitizing agent accompanying both the ijolites and the carbonatites. Wyllie proposed that the carbonatites probably then separate from the silicate liquid at temperatures above 1100°C and cool to 550 to 650°C, where crystallization begins. There is evidence suggesting that some carbonatites may be intruded at temperatures as high as 1000°C, but the Wyllie model does not preclude crystallization at higher temperatures. Although calcite–carbonatites are generally the earliest carbonatite emplaced in a given locality, the primary carbonatite magma is probably a little more alkaline, and many coarse sövites are believed to form by crystal settling and accumulation of calcite. Diversification within the carbonatite system depends on the composition of the parent magma, the pressure and temperature at which liquid immiscibility and fractional crystallization take place, wall-rock exchange, the nature of volatile constituents, and the loss of alkalis to fenitization.

In my opinion, carbonatites derived from each of the above processes probably reach the surface at one location or another. Regardless of the exact process, the asthenosphere appears to be the predominant contributor to carbonatite magmas, and hotspot or volatile-driven plumes may play a major role in their genesis. If the depleted isotopic component in Figure 19.12 is indeed lithospheric, some lithosphere must also be incorporated into the rising melts. Whether this is via direct partial melting of the lithosphere or by assimilation is an open question. Haggerty (1989), for example, proposed that kimberlite originates from a deep (75 to 100 km) K-rich layer, similar to the present model, and that carbonatite stems from metasomatized horizons in the lithosphere. Lithospheric thinning at cratonal rifts and edges, he stated, results in thicker metasomes due to enhanced activity, more lithospheric underplates, and aborted melts that provide the fluids.

The relationship between carbonatites and the commonly associated alkaline silicate magmas is complex and still not completely understood. Whether both melts are fractionation products of the same parental magma or whether they are more independent and follow the same conduits to the surface is the subject of considerable debate. The common association and isotopic similarity of both rock types suggest a common source. Although the isotopic signatures of several African carbonatites are similar to the more primitive melilitites and olivine nephelinites, they are less enriched than many of the more evolved silicate rocks. Harmer and Gittins (1998) argued on this basis that the carbonatites, if they are products of immiscible liquids, must have separated early in the evolution of the silicate rocks (when they had primitive characteristics), and not late. They concluded that the carbonatites must have existed as discrete magmas in the mantle and are not derived from the associated silicate rocks at crustal levels. They proposed that some alkali silicate magmas may be the product of partial melting of the subcontinental lithospheric mantle induced by the metasomatic effects created by the influx of carbonatites. Just as carbonatites probably arise from more than one process and source, so may the accompanying silicate rocks.

19.2.6 The Natrocarbonatite Problem

As I mentioned above, the only historically active carbonatite volcano is at Oldoinyo Lengai in Tanzania. The flows were observed in action in 1958 to 1966, and 1983 to the present, and are also the only known occurrence of natrocarbonatite (except for a tephra at nearby Keramasi volcano). Dawson et al. (1995) and Pinkerton et al. (1995) described active aa and pahoehoe flows of jet-black mobile lava. The lavas are hygroscopic, so they soon absorb atmospheric moisture, imparting a white surface. Natrocarbonatite is unique, and is composed of phenocrysts of nyerereite: $(Na_{0.82}K_{0.18})_2Ca(CO_3)_2$ and gregoryite: $(Na_{0.78}K_{0.05})_2\,Ca_{0.17}$ CO_3 in a matrix consisting of nyerereite, gregoryite, fluorite, and sylvite, together with a number of accessory minerals. Gregoryite exhibits moderate compositional variation, mainly in the form of more sodic rims. A typical natrocarbonatite composition is listed in Table 19.5. Natrocarbonatite is only a minor component of the volcanic edifice at Oldoinyo Lengai, which consists principally of nephelinite and subordinate phonolite.

Deans and Roberts (1984) interpreted tabular calcite crystals in some carbonatite lavas (principally in Kenya) as pseudomorphic replacement of nyerereite. This view was based upon the perceived impossibility of calcite crystallizing at atmospheric pressure. If this interpretation is true, although they are not natrocarbonatites today, these carbonatites may have been so upon eruption, and were subsequently altered to produce more Ca-rich mineralogy. Dawson et al. (1987) described what they believed to be progressive stages of "calcitized" nyerereite in samples of lavas from Oldoinyo Lengai. Because (1) modern lavas are natrocarbonatite, (2) calcite–carbonatites are not primary immiscible carbonatite liquids, and (3) carbonatites are commonly surrounded by *alkaline* fenite aureoles, some investigators have proposed that natro-carbonatites may be parental carbonatites (derived from nephelinitic liquids), and that the more common calcite- and dolomite–carbonatites are derivatives, generated perhaps as alkalis are lost to the fenitizing fluids (Dawson, 1964; Cooper et al., 1975, Gittins et al., 1975; Le Bas, 1981, 1987). Such an interpretation presents a problem in that the proposed parental natrocarbonatite magma is exceedingly rare, whereas the evolved products are dominant, and commonly the earliest type of carbonatite in complexes. Parental natrocarbonatite is not a view that has found popular support and has been shown by Wyllic to be untenable from an experimental viewpoint.

A contrasted view of the origin of natrocarbonatite was taken by Twyman and Gittins (1987), who argued that it could fractionate from mildly alkaline carbonatitic liquids. In the "normal" case, fractional crystallization causes the alkalinity and H_2O content of carbonatite magma to increase until the magma is H_2O saturated. The alkalis are then lost to the fluid and eventually become fenites, whereas calcite–dolomite crystallizes to form the common calcite and dolomite carbonatites. An alkalic (Oldoinyo Lengai) trend, they propose, may be caused by low H_2O and high F and Cl content. The apparent difficulty of the nyerereite composition representing a thermal barrier in the system Na_2CO_3-K_2CO_3-$CaCO_3$ (Cooper et al., 1975) was shown by Jago and Gittins (1991) to disappear when fluorine is added to the system (note the high F and Cl content of natrocarbonatite in Table 19.5). Jago and Gittins argued that crystallization of nyerereite and gregoryite (neither of which contains significant halogens) must concentrate F and Cl sufficiently to allow an alkaline magma to develop in which fluorite and sylvite are eventually stabilized, and crystallize in the matrix. The halogens, they argue, preclude H_2O saturation and the production of a hydrous fluid to remove the alkalis. Gittins and Jago (1991) showed that calcite can crystallize from an F-rich calcitic liquid at atmospheric pressure, so there is no need to explain the tabular calcite in the lavas discussed above as nyerereite pseudomorphs.

The results of a symposium session dedicated solely to Oldoinyo Lengai were published by Bell and Keller (1995). There seems to be agreement among the participants that the natrocarbonatites of Oldoinyo Lengai are the result of extreme fractional crystallization of a Na-rich olivine melilitite, resulting in a strongly peralkaline melt with $(Na + K)/Al \geq 1.5$. This liquid intersected the silicate carbonatite miscibility gap with wollastonite nephelinite + natrocarbonatite as conjugate liquids (Peterson and Kjarsgaard, 1995; Kjarsgaard et al., 1995, Dawson et al., 1996). The authors concluded that this natrocarbonatite is exotic and evolved from an unusually alkaline parent derived by very low degrees of partial melting of volatile-enriched subcontinental lithospheric mantle source.

Natrocarbonatite remains a fascinating and enigmatic rock type, probably unique, and with little applicability to the broader problems of carbonatite magma origin and evolution. Field interpretations are complicated by its transitory existence, as it is very soluble in rainwater. The briefly held hypothesis of carbonatites being remobilized trona is now reversed, and alkalis are now believed to be leached by rain water from the natrocarbonatite lavas into the groundwater, eventually precipitating as trona in nearby Lake Natron.

19.2.7 The Source of Volatiles in the Mantle

The C-O-H volatiles that so many investigators have relied upon to help generate alkaline and carbonatite melts may occur in either of two ways. First, they may be derived from deeper primordial mantle material (hydrous and carbonate phases). Wyllie (1989) proposed that dense hydrous magnesium silicates may rise in a plume and dehydrate to produce olivine + vapor at about 300 km depth. Second, we learned in Section 16.8.4 that some of the volatile content of a subducting lithospheric slab (principally H_2O and CO_2) may escape subduction magmatism and be returned to the mantle, and that Kerrick and Connolly (2001b) estimated that about $2.5 - 3.5 \times 10^{12}$ moles of CO_2 are carried into the deeper mantle each year. If present in the mantle, calcite would react with peridotite to form dolomite or magnesite and could then remain in long-term mantle storage. Hydrous phases may also survive subduction as dense high-Mg silicates. Schreyer et al. (1987) made similar arguments that subducted crustal rocks may release K-Mg-H_2O rich fluids that interact with mantle rocks to convert olivine and clinopyroxene to phlogopite and K-richterite.

19.3 HIGHLY POTASSIC ROCKS

As discussed in Section 8.7, and in conjunction with Figure 19.1, alkaline rocks may belong to either a sodic or a potassic series. Potassic rocks are defined as those with $K_2O > Na_2O$, but whether that is on a molar or weight basis has been rather loosely handled. Because sodic rocks are more common than potassic, the IUGS (Le Bas et al., 1986; Le Maitre et al., 2002) has recommended that the term **sodic** be applied when $(Na_2O - 4) > K_2O$ and **potassic** when $K_2O > Na_2O$ (both by wt. %). Potassic rocks may further be subdivided into potassic and ultrapotassic varieties. **Ultrapotassic** rocks are conventionally defined as those in which the *molar* K/Na ratio is greater than 3 (Mitchell and Bergman, 1991). Because the K/Na ratio in a magma series increases with differentiation, mafic rocks with a high K/Na ratio are in a sense more potassic than a silicic rock with the same ratio. Foley et al. (1987), who focused on mafic–potassic rocks, defined ultrapotassic rocks as those with $K_2O > 3$ wt. %, $K_2O/Na_2O > 2$, and MgO > 3 wt. %.

Potassic rocks are mineralogically, geochemically, and texturally diverse and have suffered from an overabundance of classification schemes and localized, and commonly redundant, rock names. The tectonic or genetic environment of these rocks is normally obscure, and classifications have tended to rely on mineralogical, chemical, or textural similarities, which may result in rocks of dissimilar origin being lumped together. Such a method works well for the common igneous rocks (Chapter 2) because we have come to realize that familiar rock types, such as basalts or granites, may originate in more than one way. For the potassic rocks, however, the great diversity leads only to a plethora of names and little comprehension. The most recent effort to establish order in this realm by the IUGS (Woolley et al., 1996; Le Maitre et al., 2002) is hierarchical and based on modal mineralogy (as is the general IUGS igneous rock classification). Attempts to simplify the situation include works by Sahama (1974), Barton (1979), Dawson (1987), Bergman (1987), Rock (1987, 1991), Foley et al. (1987), and Mitchell (1994b), yet the basic problem of how to categorize rocks of uncertain genesis still persists. Woolley et al. (1996) stressed that their IUGS system, although useful for naming a rock, is not definitive, has little genetic content, and is provisional.

The latest attempt to organize the potassic rocks on a genetic basis is that of Mitchell and Bergman (1991), who distinguished the following groups:

- Lamproites
- Kimberlites
- Potassic lamprophyres
- Subduction-related leucite, plagioclase, and alkali feldspar–bearing volcanic rocks of the Roman Province (e.g., Mt. Vesuvius), Greece, Indonesia, and the Philippines
- Mafic kalsilite-, melilite-, and leucite-bearing lavas, such as those of the K-rich Western Branch of the East African Rift ("kamafugites" of Sahama, 1974)
- Shoshonites: subduction-related alkalic rocks, typically formed well behind an arc, and presumably resulting from increasing K content with depth to the Wadati-Benioff zone (K-h relationship). See Morrison (1980) for a review
- Potassic to ultrapotassic leucitites and leucite basalts such as those found in central New South Wales
- K-rich intrusive rocks. A diverse group of variable origins, including shonkinite, yakutite, etc.

Mitchell and Bergman (1991) excluded in this scheme potassic variants of the common (generally evolved) rock types, such as granites, trachytes, etc.

Because the potassic rock spectrum is genetically and petrologically so diverse I have opted to avoid protracted discussions of some of the more rare types, many of which are fascinating, but many also appear to be hybrids. The potassic rocks have an overall mafic to ultramafic character and high compatible element concentrations, both of which strongly imply that they are primitive mantle derivatives. At the same time they have high incompatible element concentrations which, as we shall see below, are also generally believed to reflect the nature of the mantle source. These rocks thus provide us with a unique opportunity to evaluate at least some aspects of the mantle beneath the continents. Toward this end, we shall briefly discuss the lamproites and then the lamprophyres (mostly to clarify what is meant by this confusing group) and then focus on kimberlites because they offer a complementary view of the subcontinental lithospheric mantle, due largely to their rock and mineral inclusions.

19.3.1 Lamproites

Beginning with Sahama (1974), lamproites (his "orendites") have come to be recognized as a distinct group of rocks with some common genetic characteristics, yet diverse chemistry and mineralogy. The diversity makes it difficult to confidently recognize a lamproite, but the work of Scott-Smith and Skinner (1984a, b), Jacques et al. (1984), Mitchell (1985), Bergman (1987), and Mitchell and Bergman (1991) have provided a set of geochemical and mineralogical characteristics which serve to define them. Lamproites, group II kimberlites, and minette lamprophyres (all defined below) are about the only rock types that are both ultrapotassic (molar K/Na > 3) and perpotassic (molar K/Al > 1.0). Lamproites are also peralkaline ([K + Na]/Al commonly >1.0) and have high Mg# (usually >70) as well as high concentrations of the compatible trace elements Ni and Cr. At the same time they are highly enriched in incompatible elements, such as K, Ti, Rb, Zr, Sr, Ba, and F. They are depleted in Ca, Na, and Al, which indicates that the mantle source was depleted in these elements by earlier episodes of partial melting. Lamproite REE patterns, illustrated in Figure 19.17, are extremely LREE enriched, much more so than other common igneous rocks, and even more than the continental crust! Because many lamproite enrichments are greater than the crust we can infer that that assimilation of

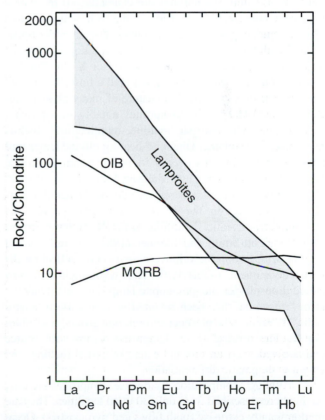

FIGURE 19.17 Chondrite-normalized rare earth element diagram showing the range of patterns for olivine-, phlogopite-, and madupitic-lamproites from Mitchell and Bergman (1991). Copyright © by permission Kluwer Academic Publishers. Typical MORB and OIB from Figure 10.14 for comparison.

TABLE 19.6 Lamproite Nomenclature

Old Nomenclature	Recommended by IUGS
wyomingite	diopside–leucite–phlogopite lamproite
orendite	diopside–sanidine–phlogopite lamproite
madupite	diopside madupidic lamproite
cedricite	diopside–leucite lamproite
mamilite	leucite–richterite lamproite
wolgidite	diopside–leucite–richterite madupidic lamproite
fitzroyite	leucite–phlogopite lamproite
verite	hyalo–olivine–diopside–phlogopite lamproite
jumillite	olivine diopside–richterite madupidic lamproite
fortunite	hyalo–enstatite–phlogopite lamproite
cancalite	enstatite–sanidine–phlogopite lamproite

From Mitchell and Bergman (1991).

crustal material cannot cause the enrichments, and thus some other process must be involved, presumably at the source.

The mineralogy of lamproites reflects their peralkaline-perpotassic nature. Lamproites are characterized by widely varying amounts (0 to 90%) of the following primary phases: phenocryst and groundmass Ti-rich phlogopite, Ti- and K-rich richteritic amphibole, olivine, diopside, leucite, and sanidine. They may also have lesser K-Ba titanites, K-Zr-Ti silicates, perovskite, spinel, and a host of other accessory minerals. The hydrous nature of many phases indicates high H_2O content. Lamproites notably *lack* primary plagioclase, sodic feldspar, melilite, monticellite, kalsilite, nepheline, and sodalite. Diamond-bearing olivine lamproites have recently been discovered in northwestern Australia; Prairie Creek, Arkansas; and Mahgaw, India.

Once recognized, a lamproite may be further classified into a subgroup on the basis of petrography. The old parochial type-locality lamproite terminology has been revised by Scott-Smith and Skinner (1984a,b) and Mitchell (1985), with a more descriptive classification reflecting the rock's constituents (Table 19.6). Thus wyomingites are now called diopside–leucite–phlogopite lamproites, etc. Only the term "madupite" has been retained as a modifier ("madupidic") to signify the presence of poikilitic groundmass phlogopite. The mineralogy of lamproites varies with source characteristics, conditions and extent of partial melting, and extent of magmatic differentiation.

Although compositionally diverse, lamproites are rare, having been described from only 30 to 40 localities. They are predominantly extrusive (both flows and pyroclastics). Occasional intrusive forms are generally hypabyssal (shallow) dikes, sills, and vent pipes. Lamproites are produced in a short magmatic episode (<3 to 10 Ma) and show few effects of differentiation. They occur strictly in continental-intraplate

areas with thick crust (>40 to 55 km) and thick lithosphere (>150 to 200 km). Only a few lamproites occur within ancient cratons, but most concentrate at cratonic margins in areas that have experienced one or more stages of compressive orogeny, aborted rifting, and/or post-collisional collapse. That *virtually all lamproites occur in areas that overlie extinct subduction zones* must have genetic significance, and the hydrous, incompatible element-enriched fluids released above these subduction zones (Chapters 16 and 17) are likely to play an important role in developing the unique geochemical composition and mineralogy of these rocks. The low Nb and Ta contents of lamproites are reminiscent of the decoupled HFS-LIL signature of subduction-related magmas (Figure 16.11). The few cratonic lamproites may tap similarly modified lithospheric mantle.

Figure 19.18 shows the Sr-Nd and ^{207}Pb-^{206}Pb isotope systematics for lamproites. By now you should be familiar enough with the Nd-Sr diagram to recognize that lamproites are all enriched with respect to bulk Earth (low $^{143}Nd/^{144}Nd$ and high $^{87}Sr/^{86}Sr$). Because of the high REE and Sr contents of lamproites, these ratios should not be susceptible to change resulting from contamination by crustal components, and are thus believed to reflect *mantle* values at the source of lamproite melts. Many of the continental alkaline magmas suggest that much of the SCLM may have a very enriched isotopic signature (note the scale of this diagram), and that we should be wary of simply using high values of $^{87}Sr/^{86}Sr$ to universally indicate crustal contamination in continental areas!

There appear to be two trends in the Nd-Sr data in Figure 19.18. The West Kimberly and Spanish lamproites follow a shallow trend reflective of mixing between a depleted or unenriched component and a component with both enriched (high) $^{87}Sr/^{86}Sr$ and (low) $^{143}Nd/^{144}Nd$. The Sisimiut, Leucite Hills, and Smoky Butte lamproites follow a steeper mixing line between a depleted or unenriched component and a component with an enriched REE (low $^{143}Nd/^{144}Nd$) signature, but without a corresponding enrichment in Rb/Sr. The enriched components are widely considered to be variations within enriched SCLM, but the unenriched component may be either DM-type asthenosphere (perhaps in a plume) or another component within the heterogeneous lithosphere. In either case, the components must have behaved as isolated systems for at least 1 to 2 Ga in order to develop their isotopic character from the geochemical differences.

Figure 19.18b shows the variations in $^{207}Pb/^{204}Pb$ versus $^{206}Pb/^{204}Pb$ for lamproites. Each locality is unique on this diagram, suggesting a distinct U/Pb ratio for each source. All but the Spanish samples fall to the left of the geochron, indicating that they cannot have had a single-stage evolution in which $^{238}U \rightarrow {}^{206}Pb$ and $^{235}U \rightarrow {}^{207}Pb$ in unison. Their sources must have experienced ancient U-Pb fractionation events that reduced their U/Pb relative to MORB and OIB. Because Pb is a LIL element, and is thus generally more incompatible than the HFS element U (particularly in a hydrous fluid), this fractionation was probably an enrichment event (Pb added). The variation in Pb ratios, then, could reflect mixing between a high U/Pb (MORB-like) reservoir and an enriched low U/Pb reservoir (perhaps

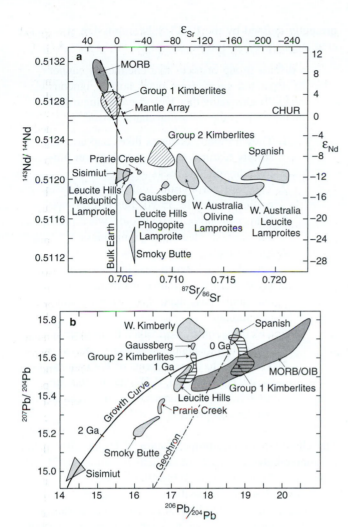

FIGURE 19.18 **(a)** Initial $^{87}Sr/^{86}Sr$ versus $^{143}Nd/^{144}Nd$ for lamproites (shaded) and kimberlites (hatched). MORB and the mantle array are included for reference. **(b)** $^{207}Pb/^{204}Pb$ versus $^{206}Pb/^{204}Pb$ for lamproites and kimberlites. After Mitchell and Bergman (1991). Copyright © by permission Kluwer Academic Publishers.

hydrated SCLM). The lamproites that plot above the mantle growth curve (which represents simultaneous evolution of $^{238}U \rightarrow ^{206}Pb$ and $^{235}U \rightarrow ^{207}Pb$ for primitive mantle U concentrations) require more complex models. Fraser et al. (1985) and Nelson et al. (1986) suggested a three-stage model that could accomplish this and argued that at least part of the enriched lithospheric component was developed with such three-stage events over 1.8 Ga ago.

The isotopic systems require mixing of two to three distinct sources, even within most provinces. The low Na-Al-Ca contents and high Ni and Cr suggest that a refractory harzburgite, from which clinopyroxene and an aluminous phase were preferentially melted, may be an appropriate source, provided that it was subsequently re-enriched with K, Ti, Rb, Zr, Sr, Ba, and F. Mitchell and Bergman (1991) suggested the following model for the generation of lamproites:

1. A depleted harzburgite is created, either by melt extraction within a rising asthenospheric plume or by long-term depletion of the subcontinental lithospheric mantle

(SCLM), perhaps associated with crustal genesis. The isotopes indicate that this event must have occurred at least 1 to 2 Ga ago.

2. Later enrichment adds incompatible elements to the harzburgite. This may occur in the form of subduction zone fluids rising from the dehydrating slab into the overlying SCLM or via melt infiltration (including carbonatites), underplating, stalled and crystallizing hydrous melts in rift or subduction zones, escaping juvenile fluids, or a combination of these factors. Enriched aqueous fluids will produce phlogopite, and perhaps K-richterite, as the principal mantle metasomatic phases, which act as incompatible element repositories. Other than the introduction of K, the enrichment affects trace elements far more than major elements. Enrichment processes may occur in several stages and affect only portions of the lithosphere, resulting in a heterogeneous subcontinental mantle with variably fertile pockets.

3. The depleted then enriched heterogeneous SCLM source is partially melted. This may be triggered by a new plume that supplies thermal energy and/or a sudden volatile influx, or it may result from collapse of an orogen and decompression melting of the rising asthenospheric welt (Section 18.4.2.4). Given the complex and speculative nature of the source, it is impossible to constrain the degree of partial melting from the geochemistry of the lamproites. 1 to 10% partial melting probably occurs under H_2O- and F-rich conditions at a single eutectic point, resulting in a primitive phlogopite-lamproite magma with 52 to 55 wt. % SiO_2 and a limited compositional range.

Given their concentration in areas above extinct subduction zones, the lithospheric source and subduction zone fluid enrichment is an attractive combination for the source of lamprophyres. Mitchell and Bergman (1991) proposed that all other members of the lamproite clan are derived by fractionation and hybridization from the primitive phlogopite-lamproite that appears to be the parental magma in a number of provinces. They suggested that the olivine–lamproites, which have lower silica contents and thus appear more primitive, do not represent primary eutectic melts, but rather are hybrids of phlogopite lamproites that have assimilated olivine-rich mantle material.

19.3.2 Lamprophyres

During most of my petrological career, I never understood lamprophyres well at all. Through research and discussions to prepare this chapter, I discovered that I am in good and plentiful company. The concept began clearly enough when Von Gümbel (1874) coined the term to describe the lustrous porphyritic biotite-rich rocks of the Fichtelgebirge in Germany. *Lampros porphyros* is Greek for "glistening porphry," making "lamprophyre" one of the more appropriate and descriptive terms in the igneous lexicon. Since then, the meaning has been broadened and degraded, as the term was applied to a wide variety of hypabyssal rocks containing mafic phenocrysts. The group eventually became a repository for any difficult-to-characterize porphyritic rock. The unfortunate practice of type-locality nomenclature led to the introduction of a "legion of obscure rock types named after equally obscure European villages" (Rock, 1991). Lamprophyres thus became an extremely diverse group of rocks with unintelligible (and commonly unpronounceable) names that had little in common other than their porphyritic character. No wonder I had such a problem. As with lamproites, however, many of the lamprophyres were composed of hydrous mineral assemblages dominated by mafic minerals (mica and amphibole), both as phenocryst and groundmass phases, and had a similar hybrid nature, being rich in both compatible and incompatible elements. They thus provided important information bearing on the nature of the subcontinental lithospheric mantle.

Rock (1987, 1991) has provided us with modern summaries of the lamprophyre group. He proposed a "lamprophyre clan," which included lamprophyres, lamproites, and kimberlites. His use of the term *clan* was meant to refer to a group of rocks that appear similar, are commonly associated in the field, and have a number of common characteristics, such as porphyritic texture, high volatile content, and hypabyssal occurrence. Most petrologists use the term *clan* as synonymous with *series*, to refer to a group of rocks derived from a particular parental magma type (tholeiitic, calc-alkaline, etc. as first presented in Chapter 8), and Mitchell (1994a, b) vigorously opposed Rock's adaptation of the term. He also opposed Rock's inclusion of lamproites and kimberlites in a lamprophyre clan on the basis that only some lamproites and kimberlites qualified, and thus lamprophyres were only being further diluted and confused by the addition of inappropriate members. The IUGS Subcommission (on which both Rock and Mitchell served) agreed that the extended "lamprophyre clan" should not be endorsed, and that lamproites and kimberlites are best considered separately.

The great diversity of the lamprophyre group precludes an extensive discussion of their petrogenesis here. Because many of the members of the group are volatile-rich variants on other series and associations that we have covered already, much of the discussion would overlap with the petrogenetic models presented earlier. Rather, I prefer to simply provide some clarification as to what this group of rocks comprises.

If we exclude lamproites, kimberlites, and other improperly grouped rock types, lamprophyres *sensu stricto* are presently defined by the IUGS Subcommission (Le Maitre et al., 2002, p.19) as follows:

a diverse group of rocks that chemically cannot be separated easily from other normal igneous rocks. Traditionally they have been distinguished on the basis of the following characteristics:

1. They normally occur as dikes and are not simply textural varieties of common plutonic or volcanic rocks
2. They are porphyritic, with M' (modal % mafics) typically 35-90, but rarely >90
3. Feldspars and/or feldspathoids, when present, are restricted to the groundmass
4. They usually contain essential biotite and/or amphibole and sometimes clinopyroxene
5. Hydrothermal alteration of olivine, pyroxene, biotite, and plagioclase (when present), is common
6. Calcite, zeolites, and other hydrothermal minerals may appear as primary phases
7. They tend to have contents of K_2O and/or Na_2O, H_2O, CO_2, S, P_2O_5, and Ba that are relatively high compared to other rocks of similar composition

The classification and nomenclature of lamprophyres, and the mineral characteristics of each, as recommended by the IUGS in 2002, are given in Table 19.7.

An earlier IUSG classification (Le Maitre et al., 1989) classified lamprophyres into three broad categories: calc-alkaline lamprophyres, alkaline lamprophyres, and melilitic lamprophyres. Rock (1987, 1991) preferred to call the melilitic lamprophyres "ultramafic lamprophyres" instead, arguing that melilite-free and melilite-rich varieties commonly coexist. Not all melilitic lamprophyres were ultramafic, however, making either choice a compromise. The latest attempts by the IUGS Subcommission to classify the melilitic rocks (Woolley et al., 1996, Le Maitre et al., 2002) considered them as varieties of a separate melilitic rock group, and not lamprophyres at all.

Calc-alkaline lamprophyres occur in subduction zone environments, generally in association with calc-alkaline

---

TABLE 19.7 Lamprophyre Classification and Nomenclature Based on Mineralogy

Light-Colored Constituents		Predominant Mafic Minerals		
Feldspar	Foid	Biotite >Hornblende, ± Diopsidic Augite, (±olivine)	Hornblende, Diopsidic Augite, ± Olivine	Brown Amphibole, Ti-Augite, Olivine, Biotite
or > pl	—	minette	vogesite	
pl > or	—	kersantite	spessartite	
or > pl	feld >foid			
pl > or	feld >foid			sannaite
—	glass or foid			camptonite
				monchiquite

After Le Maitre et al. (2002), Table 2.9, p. 19.

granitoid suites or with the more alkaline shoshonites inboard from the volcanic front. Alkaline lamprophyres and melilitic rocks typically occur in intraplate and rift environments, accompanying other alkaline magmas, from mildly alkaline gabbros to highly alkaline carbonatite-alkaline silicate rock complexes.

Even when limited to the IUGS-approved types, lamprophyres encompass a wide range of compositions, from ultramafic to silicic and with variable alkalinity and K/Na ratios (although generally ultrapotassic). Thus the term "lamprophyre" is best used only as a field term, and the more specialized names in Table 19.7 are preferred when the petrography is known. As a group they tend to have high (but variable) H_2O, CO_2, F, Cl, SO_3, K, Na, Sr, Th, P, Ba, and LREE, along with basaltic levels of Y, Ti, HREE, and Sc. Petrographically, they are characterized by euhedral to subhedral phenocrysts (panidiomorphic) of mica and/or amphibole with lesser clinopyroxene in a (usually light colored) groundmass which may contain glass, plagioclase, alkali feldspar, feldspathoid, mica, amphibole, olivine, carbonate, monticellite, perovskite, and/or an Fe-Ti oxide. Like the lamproites, they tend to be intriguing hybrids with mafic character, yet enriched in incompatible elements.

Lamprophyres typically occur as minor hypabyssal intrusions (sills, dikes, stocks, pipes, or volcanic necks), but there are now several well-documented examples of extrusive and larger plutonic lamprophyres as well. Perhaps the more deeply eroded plutonic lamprophyres fed into more typical shallow hypabyssal types. Calc-alkaline lamprophyres commonly accompany larger granitoid intrusions as components of multiple or composite plutonic bodies. In such cases they occur either as dike swarms, enclaves, globules (presumably immiscible liquids), or late-stage dikes. They commonly exhibit many characteristics of the associated granitoids and are thus not simply associated by coincidence.

The high volatile content (particularly H_2O) and the resulting abundant mica–amphibole phenocrysts are the predominant uniting characteristics of the lamprophyre group. The implication is that lamprophyres develop as a consequence of volatile retention via crystallization at high pressure, or by prolonged normal differentiation processes (Mitchell, 1994a). If so, many lamprophyres may be nothing more than the hydrous crystallization products of common magma types that occur under unusually H_2O-rich conditions. They are still worth considering, however, because they record part of the long-term evolution of some magma chambers, and the mechanisms by which such hydrous variants are accomplished are both interesting and probably diverse (depending on the initial magma type and the conditions under which it crystallizes).

Recognizing that lamprophyres are polygenetic and may be variations on familiar themes, Mitchell (1994a) proposed adopting the *facies* concept from our sedimentary and metamorphic colleagues for use in the sense that rocks in a given facies have some similar conditions in common (in this case, crystallization under volatile-rich conditions), regardless of other genetic differences. Thus the micaceous kimberlites might constitute a lamprophyric facies of the

kimberlite group, or camptonites may be a lamprophyric facies of the alkali basalt series. This idea has merit, but, as of this writing, it has not been fully accepted or integrated into common usage. Mitchell and Bergman (1991) also suggested that grouping rocks of such diverse character and origin under the lamprophyre banner "serves no rational petrogenetic purpose." They further stated that the practice of extending the lamprophyre group to include even more diverse rocks, such as lamproites and kimberlites, is "regressive, and that petrologists, instead of creating more lamprophyres, should be working toward the complete elimination of the term." Whether this suggestion is appropriate depends upon the relative efficacy of considering lamprophyres as a group or not. Do we gain more by thinking in terms of a diverse group of rocks, all of which evolve under very hydrous conditions, or in terms of how many of the magma series with we are now familiar could have a hydrous variant? Only time will tell.

19.3.3 Kimberlites

Kimberlites are fascinating for a number of reasons. They are K-rich, typically ultramafic hybrid rocks that occur in ancient cratons. They are volatile rich, tend to rise from mantle depths rapidly and are emplaced violently if they reach the near-surface environment. Many contain diamond and coesite, which indicate a fairly deep mantle origin. They also contain a plethora of crystal and rock fragments, mostly gathered from the wall rocks as they rose, which provide an unparalleled collection of reasonably fresh mantle materials from beneath the continents. No rock type has contributed more to our knowledge of the subcontinental mantle and its volatiles.

Kimberlites are currently divided into two groups. Group 1 kimberlites are the archetypal ultramafic kimberlites, first described from Kimberly, South Africa, but known to occur on all continents. Group 2 kimberlites are micaceous kimberlites, the occurrence of which is presently limited to South Africa, where they are older (100 to 200 Ma) than the group 1 kimberlites in the same area (<100 Ma).

19.3.3.1 PETROGRAPHY OF KIMBERLITES Group 1 kimberlites are volatile-rich (principally CO_2) potassic ultramafic rocks. In addition to their xenolith content, they commonly exhibit a distinctively inequigranular texture caused by the presence of rounded, anhedral, and fragmented **macrocrysts** (a non-genetic term for 0.5 to 10 mm diameter crystals) and, in some cases, **megacrysts** (similar, yet larger, generally 1 to 20 cm diameter crystals) set in a fine-grained matrix. Some of these crystals are certainly *xenocrysts* (probably disaggregated constituents of xenoliths of lherzolite, harzburgite, eclogite, and metasomatized peridotite), whereas others may be of cognate origin (phenocrysts or disaggregated cumulates). Olivine is generally predominant, but may be accompanied by ilmenite, pyrope, diopside, phlogopite, enstatite, and chromite. Large crystals of subhedral to euhedral habit are considered to be true phenocrysts, and are so named when properly identified. The matrix typically contains a second generation of fine euhedral to subhedral olivine, plus one or more of the primary minerals: monticellite, phlogopite, perovskite, spinel, and apatite. Carbonate and serpentine typically constitute a late groundmass. Many group 1 kimberlites contain a late poikilitic Ba-rich phlogopite. Nickeliferous sulfides and rutile are common accessory phases. Many kimberlites exhibit 1- to 10-mm-sized rounded globular to amoeboid-shaped segregation masses of calcite + serpentine. Whether these form as immiscible liquids, vesicle-gas condensates, or later vesicle fills is still in question. Replacement of early primary minerals by deuteric serpentine and calcite may occur during cooling in these volatile-rich systems.

Group 2 kimberlites, although texturally similar to those of group 1, are distinctive mineralogically and geochemically. They are ultrapotassic, peralkaline, and H_2O-rich. Phlogopite is the dominant macrocryst and groundmass phase. Olivine is also common, although essentially a xenocryst. Other characteristic primary phases include diopside (commonly rimmed by aegirine), spinel, perovskite, apatite, REE-rich phosphates, K-Ba-titanites, rutile, and ilmenite. The fine groundmass may contain calcite, dolomite, REE-carbonates, witherite, norsethite, zirconium silicates, and/or serpentine. A distinct group of evolved types may contain sanidine and richterite, and even quartz.

Group 1 and 2 kimberlites are distinctive isotopically (see Figure 19.18). Mitchell (1994b, 1995) and Mitchell and Bergman (1991) concluded that the two groups must represent different magma types. They suggested that the differences are substantial enough that group 2 kimberlites should be separated from kimberlites proper (meaning group 1 kimberlites) and be renamed **orangeites** (after early descriptions by Wagner, 1928, and named for the Orange Free State of South Africa to which they are essentially restricted). Mineralogically, group 2 kimberlites are similar to lamproites, but have sufficient petrological differences to warrant that they be considered separately from these rocks as well. The arguments are compelling, and I will henceforth use the term *kimberlite* to mean group 1 kimberlites, and address group 2 kimberlites as *orangeites*.

Diamonds occur principally in kimberlites, orangeites, and some lamproites. Rare occurrences of diamond, or graphite pseudomorphs after diamond, are also known from some alkali basalts, lamprophyres, alpine peridotites, and even in some *crustal* rocks in ultra-high pressure metamorphic terranes (Section 28.6). Kimberlite diamonds are very minor phases and occur only in a few kimberlites. A rich economic deposit may have a diamond concentration of 1 to 1.4 grams per ton of rock, but their economic value accentuates their importance. Probably because of their common euhedral octahedron shape, they were initially believed to be *phenocrysts*. After Sr-Nd and U-Pb dating of diamond inclusions showed that they are older than their host rock, they are now recognized as *xenocrysts*. Diamond-bearing xenoliths and inclusions in diamonds indicate that the original diamond host rocks were garnet harzburgites and eclogites. The latter may represent either subducted basalts or basaltic magmas trapped at the lithosphere-asthenosphere boundary

(underplates). Harzburgite and eclogite mineral inclusions do not coexist in the same diamond (Meyer, 1985). $^{13}C/^{12}C$ in most diamonds is similar to diamonds found in meteorites, suggesting that the carbon is primordial, but some eclogitic diamonds appear to contain marine carbon (suggesting a subduction origin for them). Cosmochemical arguments suggest that the upper mantle may contain ~700 ppm carbon, and that the lower mantle may contain 1000 to 3700 ppm C (Pearson et al., 2007). The diamond-bearing inclusions are largely disaggregated during transport to the surface in the gas-charged kimberlite matrix. We can attribute the euhedral shape of the diamonds to their great hardness and resistance to abrasion, although slow ascent and high f_{O_2} in some kimberlite magmas may result in complete resorption of original diamonds into the melt. From Figure 19.13 we can see that the graphite–diamond stability curve intersects the continental geotherm at about 140 km, indicating a *minimum* depth for diamond stability. Diamond suites are variable, both within and between kimberlites, suggesting a combination of sources. Diamonds, we can conclude, are *not* genetically associated with kimberlite magmas. Kimberlites merely entrain them from wall rocks, should they encounter them, and transport them to the surface. For more information on diamonds, see the reviews by Meyer (1985), Haggerty (1999), Navon (1999), and Pearson et al. (2007).

19.3.3.2 KIMBERLITE–ORANGEITE FIELD RELATIONSHIPS

Kimberlites and orangeites are products of continental intraplate magmatism and are concentrated within ancient cratons. Some also occur in younger (mostly Proterozoic) accreted terranes, but the diamond-bearing ones are essentially restricted to terranes underlain by rocks older than 2.5 Ga (Clifford, 1966; Janse, 1991). Perhaps the higher Archean heat flow and copious komatiite production led to a deeper depleted SCLM root zone beneath Archean cratons (roots that extended to the diamond stability range). In the Proterozoic, on the other hand, the lower heat flow limited melting, resulting in a less deep root (Groves et al., 1987). Kimberlites and orangeites generally occur in clusters within larger provinces, where, like carbonatites, they tend to occur in repeated cycles. Shallow fault zones and lineaments control local kimberlite–orangeite emplacement, whereas deeper plumes or metasomatic processes may initiate magma formation. Kimberlites and orangeites are not associated with transform fault extensions, subduction zones, obvious hotspots, or major continental rifts.

Kimberlites and orangeites can occur as hypabyssal dikes or sills, diatremes (see below), crater-fill, or pyroclastics, depending largely on the depth of erosion and exposure (Figure 19.19). The dikes are generally 1 to 3 m thick and generally occur in swarms, where they tend to bifurcate into anastomosing stringers. Most dikes tend to pinch out toward the surface and thicken with depth. Sills are less common, and they may be up to several hundred meters thick. Some dikes expand locally near the top into lenticular enlargements called "blows" (Figure 19.19), which may be up to 10 to 20 times the dike width and 100 m long. Blows may feed into the root zones of diatremes.

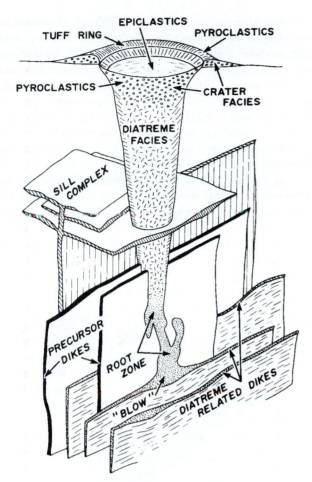

FIGURE 19.19   Model of an idealized kimberlite system, illustrating the hypabyssal dike-sill complex leading to a diatreme and tuff ring explosive crater. This model is *not to scale*; the diatreme portion is expanded to illustrate it better. From Mitchell (1986). Copyright © by permission Kluwer Academic Publishers.

Diatremes (Figure 19.19) are 1- to 2-km deep carrot-shaped bodies with circular-to-elliptical cross sections, vertical axes, and steeply dipping sides (80 to 85°). They taper downward and terminate in a "root zone," an irregularly shaped multiphase intrusion zone, transitional into hypabyssal kimberlites–orangeites. The nature of the volcanic processes that produce diatremes is still the subject of much debate. The diatreme represents the expansion of the volatiles in the magma as it approaches the surface and the confining pressure is lowered. In the model of Clement (1979), multiple batches of magma exsolve CO_2 because of pressure reduction, shattering the wall rocks to form subsurface breccias. An upwardly progressing sequence of stalled "buds" form in this fashion until they reach approximately 300 to 400 m depth, where hydrovolcanic interaction with groundwater produces gas with sufficient violence to break through to the surface. At this point, either rapid degassing and vapor exsolution in response to progressive pressure release resulting from unroofing or increased groundwater flow into the crater and pipe results in a downward migrating zone of violent brecciation and mixing to form the diatreme (Mitchell, 1986). Diatreme facies kimberlites–orangeites, at least near the surface, are

more fragmented than their hypabyssal equivalents and take on a volcaniclastic appearance. Breccias containing abundant country rock inclusions and subordinate earlier hypabyssal kimberlite–orangeite solid fragments, from a few centimeters to microscopic size, are the most common rock type. Megacrysts and macrocrysts are also common. The fragmental nature grades downward into non-brecciated kimberlite.

The *surface* expression of kimberlites is only preserved in a few areas where erosion has been negligible. Only one possible kimberlite lava flow has been described, at Igwisi Hills, Tanzania. The other occurrences are tuffs and tuff rings or cones. Poorly sorted tuffs and tuff breccias are overlain by well-stratified tuffs with ash to lapilli-sized particles. For a more extensive description of the rocks and textures at various levels, see Mitchell (1986).

19.3.3.3 KIMBERLITE–ORANGEITE GEOCHEMISTRY   Kimberlites and orangeites are complex hybrids in which an undeterminable quantity of foreign and cumulate material has been integrated, disaggregated, and variably absorbed into the liquid. It is thus difficult, if not impossible, to determine what constituents are components of the original melt, and what has been incorporated en route to the surface. The nature of the primitive liquids is thus largely unknown and interpreting the geochemistry is obviously a problem. Table 19.8 lists some rough averages of kimberlite, orangeite, and lamproite analyses. They are averages of samples judged to be less contaminated by xenolithic material, but represent highly variable rock types and are unlikely to have much geochemical significance. They are intended for crude comparative purposes only.

Differentiation processes also lead to the concentration of macrocryst + phenocryst phases and evolved liquids, eventually resulting in the evolution of carbonate-rich residua. Major element concentrations thus vary widely as a result of contamination, accumulation, and fractionation processes. When shown on typical variation diagrams (Harker, AFM, etc.) the data are usually interpreted as liquid lines of descent, but this is seldom appropriate in the present case, where they principally represent varying proportions of olivine and calcite in kimberlites and olivine, phlogopite, and carbonate in orangeites. Variations in phenocryst compositions suggest that most kimberlites and orangeites are mixtures that result from the coalescence of smaller magma batches as they rise. Late stages of intrusion typically take the form of a crystal–liquid slurry in which the relative proportions of the constituent minerals might easily vary. The wide variations in SiO_2, CaO, MgO, CO_2, and H_2O in Table 19.8 should thus not be surprising. Low concentrations of Al_2O_3 and Na_2O, however, appear to be a consistent attribute of kimberlite–orangeite magmatism.

The high levels of *compatible* trace elements (Ni, Cr, Sc, V, Co, Cu, Zn) reflect the mantle source of kimberlite and orangeite magmas as well as the incorporation of mantle xenoliths and olivine macrocrysts. Table 19.8 shows that the concentration of these elements is fairly similar for both orangeites and kimberlites, except for Ni and Cr, which reflect higher proportions of macrocrystal olivine and chromite in

TABLE 19.8 **Average Analyses and Compositional Ranges of Kimberlites, Orangeites, and Lamproites**

	Kimberlite		Orangeite		Lamproite*
SiO$_2$	33.0	27.8–37.5	35.0	27.6–41.9	45.5
TiO$_2$	1.3	0.4–2.8	1.1	0.4–2.5	2.3
Al$_2$O$_3$	2.0	1.0–5.1	2.9	0.9–6.0	8.9
FeO*	7.6	5.9–12.2	7.1	4.6–9.3	6.0
MnO	0.14	0.1–0.17	0.19	0.1–0.6	–
MgO	34.0	17.0–38.6	27.0	10.4–39.8	11.2
CaO	6.7	2.1–21.3	7.5	2.9–24.5	11.8
Na$_2$O	0.12	0.03–0.48	0.17	0.01–0.7	0.8
K$_2$O	0.8	0.4–2.1	3.0	0.5–6.7	7.8
P$_2$O$_5$	1.3	0.5–1.9	1.0	0.1–3.3	2.1
LOI	10.9	7.4–13.9	11.7	5.2–21.5	3.5
Sc	14		20		19
V	100		95		66
Cr	893		1722		430
Ni	965		1227		152
Co	65		77		41
Cu	93		28		–
Zn	69		65		–
Ba	885		3164		9831
Sr	847		1263		3960
Zr	263		268		1302
Hf	5		7		42
Nb	171		120		99
Ta	12		9		6
Th	20		28		37
U	4		5		9
La	150		186		297
Yb	1		1		1

* Leucite Hills madupidic lamproite.
Data from Mitchell (1995) and Mitchell and Bergman (1991).

kimberlites, on average. The range of Ni and Cr in kimberlites and orangeites, however, overlaps considerably.

The *incompatible* trace elements in these very mafic rocks are largely introduced into the source regions by metasomatic and/or melt additions. Figure 19.20 shows REE and spider diagrams for kimberlites, unevolved orangeites, and primary phlogopite lamproites. Many of the mineral phases in kimberlites and orangeites have high concentrations of traditional incompatible elements (thus the term *incompatible* is not really appropriate for these rocks). Variations in traditionally incompatible element concentrations in the mica-rich orangeites are generally linked to the modal amounts of primary phlogopite, apatite, and carbonate present. Models by Fraser and Hawkesworth (1992) showed that the incompatible trace element concentrations of orangeites, at least, cannot be produced from primitive mantle compositions, even by extremely small (<0.5%) degrees of partial melting.

The REE patterns (Figure 19.20a) are all similar in slope, but the lamproites show the greatest LREE enrichment, perhaps associated with the proposed subduction-zone enrichment of their source areas. Kimberlites and orangeites are not distinguishable on the basis of the REE, but the kimberlites do show a wider range of values. The differentiation within the HREE of all three types implies that garnet was a residual phase in their source regions (suggesting that they are generated in the garnet–lherzolite region of the mantle, as shown in Figure 10.2).

In the chondrite-normalized spider diagrams (Figure 19.20b), the elements are arranged along the abscissa in the approximate order of decreasing incompatibility (right to left), so that normal mantle partial melts display a fairly smooth trend with a negative slope (such as OIB). Spikes and troughs in trends (positive and negative anomalies, respectively) may

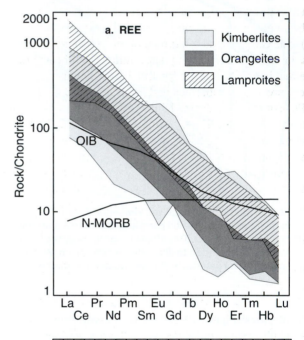

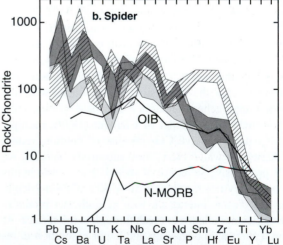

FIGURE 19.20 Chondrite-normalized REE **(a)** and multielement (spider) **(b)** diagrams for kimberlites, unevolved orangeites, and phlogopite lamproites (with typical OIB and MORB). After Mitchell and Bergman (1991) and Mitchell (1995). Copyright © by permission Kluwer Academic Publishers.

reflect nonstandard mantle sources or a residual phase that is left behind and retains an incompatible element (negative anomaly only). Both kimberlites and orangeites have negative K, Rb, and Sr anomalies. The K and Rb anomalies are compatible with residual phlogopite in the source. Sr may be caused by residual phosphate or the earlier removal of clinopyroxene. The phlogopite lamproites have a different pattern, with pronounced negative anomalies for Ta-Nb (a common subduction zone signature) and U-Th. This suggests that kimberlites and orangeites have a different mantle source than have lamproites.

The isotopic patterns in Figure 19.18 provide the strongest evidence for different sources of kimberlites (group 1) and orangeites (group 2). Kimberlites have a Sr-Nd signature that is unenriched or slightly depleted with respect to bulk Earth. Although interpretations vary, the source is generally accepted to be the convectively stirred and partly homogenized asthenospheric mantle. Orangeites have a Sr-Nd isotopic signature that is enriched and intermediate between the two lamproite enrichment trends (Figure 19.18). The conventional interpretation of this source is that it lies in the non-convecting subcontinental lithospheric mantle that has been isolated for 1 to 2 Ga following an enrichment process. Kimberlites and orangeites also differ in their Pb isotopic signatures. Kimberlites follow a mixing line between EMI and HIMU and overlap with MORB/OIB, supporting an asthenospheric source. Janney et al. (2002) suggested that the EMI signature may represent heterogeneous SCLM, and the HIMU component may be either asthenospheric peridotite or pods of recycled oceanic crust in a plume (the African superplume may explain why a HIMU-like isotopic signature is so widespread across Africa). Orangeites are enriched, probably in more than a single stage, in a fashion similar to that discussed previously in conjunction with lamproites. The most likely enriched mantle source is lithospheric.

Thus all three of the coherent high potassic groups that we have discussed in this section have different sources that can be distinguished on the basis of the trace element and isotopic characteristics. The most straightforward interpretation is that archetypal kimberlites are derived from the asthenosphere, whereas orangeites stem from enriched cratonic lithosphere, and lamproites from continental lithosphere that has been modified by earlier subduction zone processes.

19.3.3.4 PETROGENESIS OF KIMBERLITES AND ORANGEITES

Given the meager state of our knowledge concerning primary kimberlite and orangeite magmas and the nature of mantle sources, there are few constraints on petrologic creativity in models for the generation and development of these rocks. For excellent summaries and critiques of the various petrogenetic theories, see Mitchell (1986, 1995). The model I summarize below conforms to the data at present and appears to provide the simplest explanation for kimberlite and orangeite genesis. I have little doubt, however, that it will be revised as we learn more about these unusual rocks.

Earlier in this chapter, we concluded that highly alkaline and volatile magmas could not be derived from normal four-phase lherzolites, but rather from carbonated and hydrous lherzolites. Experiments in the system lherzolite-C-O-H show that small degrees of partial melting of a garnet lherzolite containing either phlogopite or richterite and magnesite or dolomite could produce broadly kimberlitic magmas at depths on the order of 100 to 300 km and 1000 to 1300°C (Canil and Scarfe, 1990; Wendlandt and Eggler, 1980a,b; Wyllie et al., 1990). Although the *slopes* of the REE patterns could be produced this way, the *absolute abundances* of the REE, P, Sr, Th, U, Zr, and Ta are still difficult to attain without adding an REE-rich phosphate or titanite to the source. Thus some form of metasomatic enrichment of primordial mantle peridotites appears to be a necessary precursor to the generation of kimberlite–orangeite melts.

Our model begins with that of Wyllie (1980, 1989; Wyllie et al., 1990), discussed in Section 19.2 in conjunction with carbonatite genesis and Figures 19.13 and 19.16. In this model, introduction of CO_2 and H_2O lowers the solidus of the lherzolite to that shown in Figure 19.13. A rising plume of hydrous-carbonated asthenosphere, if it follows the geotherm, intersects the solidus and a small quantity of kimberlitic melt is produced at point 2 (260 to 270 km). Entrained pods of eclogite in the plume may also contribute in a similar fashion. Primary carbonatites may also be produced in this way, but, as described in Section 19.2, these are not the common carbonatites associated with alkaline complexes. The partially melted mantle diapir continues to rise either along the geotherm or in a near-adiabatic fashion. The kimberlite magmas may stall upon meeting some mechanical resistance at the lithosphere–asthenosphere boundary (level 1 in Figure 19.13), or they may continue to rise into the lithosphere until they encounter the "ledge" in the solidus (level 4), where they should crystallize to a phlogopite–dolomite peridotite and release vapor, fracturing and metasomatizing the wall rocks. Haggerty (1989) and Thibault et al. (1992) proposed that a complex sequence of metasomatic layers, or a stockwork of veins, may form in this fashion over a depth range of 60 to 100 km. The depleted lithospheric harzburgites in these layers or veins react with the carbonated-hydrous melts to produce enriched phlogopite, K-richterite, and/or carbonate-rich wehrlites and dunites with incompatible-element-rich accessory phases. The plume component may supply the HIMU isotopic signature and SCLM may supply the EMI signature.

Anderson (1979) proposed a variation in which the release of vapor in rising volatile-rich plumes may enhance crack propagation, if the overlying lithosphere is in tension, and provide channelways to the surface. Kimberlitic magma might then separate and rise independently up the established conduit, bypassing the ledge in the solidus for equilibrium lherzolite + melt behavior. Wyllie (1980) proposed that cracks to the surface will be followed by the explosive eruption of vesiculating magma directly from the separation depth of 90 km. This idea runs contrary to the extensive hypabyssal dike-sill complexes and a diatreme expression that is restricted to the shallowest levels. Also, magmas that separate at 90 km will not be diamond-bearing, as the rapid rise begins shallower than the stability range of diamonds

(Figure 19.13). To produce diamonds, Wyllie (1980) proposed that once a conduit is established, further escape of volatiles triggers diapiric ascent from successively deeper levels, ultimately reaching the diamond stability range. Although the initial aspects of the Wyllie model, including kimberlite magma genesis and metasomatism of the subcontinental lithospheric mantle, are attractive, these latter details are less compelling.

Returning to Wyllie's broader model, the rising plume of volatile-enriched asthenosphere initiates melting. The melts, in turn, release fluids that metasomatize the overlying mantle. An alternative melting mechanism is one in which metasomatism reduces the melting point of a refractory peridotite to the extent that it begins to melt at ambient conditions (or conditions only slightly disturbed from ambient). In the former case, melting leads to metasomatism, whereas in the latter, metasomatism leads to melting. Of course, these two processes are not mutually exclusive; and multiple processes of magmatism, crystallization, metasomatism, upwelling, and heating may produce several generations of alkaline melts and metasomatism at progressively shallower levels. Such a veined and metasomatically enriched lithosphere is an attractive source for orangeites, as proposed by Skinner (1989), who, contrary to the popular interpretation of the isotopic data, also attempted to derive kimberlites from the same lithospheric source.

Our favored model, then, is one in which kimberlites (group 1) are generated from volatile-enriched garnet lherzolites in upwelling portions of the asthenosphere. The similarity of kimberlites all over the world makes a convecting (homogenizing) asthenospheric (plume) source most likely. (See Section 14.7 and Figure 14.20 for a general plume melting model.) Kimberlite source rocks are probably a magnesite-phlogopite-garnet lherzolite with some added Ti, K, and Ba. Deeper mantle phases may include K-Ti-richterite, K-diopside, garnet, etc. Some theories place the depth of origin of kimberlitic melts as deep as the core-mantle boundary, but the peridotite solidus appears to get progressively farther from the geotherm below point 2 in Figure 19.13, so we shall assume that they originate at depths less than 300 km. Carbonatites, melilitites, and other alkaline magmas may also be created in these rising cells, depending on initial mineralogy, added components, and depth and degree of partial melting. The alkaline silicate magmas that require more extensive asthenospheric melting may only develop in areas experiencing substantial lithospheric thinning and rifting (more active plumes or more sustained adiabatic rise). Less buoyant plumes may terminate and experience modest partial melting, resulting in kimberlite magmatism in unrifted cratons. The more rapid rise of asthenospheric plumes in oceanic settings results in more extensive partial melting, which precludes kimberlite magmatism.

The ubiquitous, predominantly olivine, megacrysts in kimberlites exhibit compositional and textural variations, even within kimberlite suites. Textures suggest that most are phenocrysts. Mitchell (1986, 1995) preferred the interpretation that most are cumulates from several batches of kimberlite magma deposited at high pressures (probably trapped at the base of the lithosphere). As discussed above, the hypabyssal kimberlites show evidence for magma mixing and interaction of numerous kimberlite magma batches. Later batches disaggregate the cumulates from earlier ones, resulting in complex hybrids of mixed magmas and a spectrum of partially resorbed cumulates. Other macrocrysts and megacrysts are truly xenocrystal, resulting from disaggregated xenoliths. The lack of orthopyroxene macrocrysts implies that this material may be derived largely from garnet dunites (including diamond-bearing varieties), located primarily near the base of the lithosphere. The alternative: complete resorption of orthopyroxene, requires high degrees of xenolith assimilation. Olivine and minor phlogopite crystallize during kimberlite ascent and the hybrid assemblage of melt, xenocrysts, and phenocrysts is emplaced in the upper crust as the dike-sill complex and eventually diatreme as described above in Section 19.3.3.2. Rates of kimberlite ascent have been estimated on the basis of comparing experimentally determined versus actual decompression rates of garnet dissolution (Canil and Fedortchouk, 1999) and argon diffusion in phlogopite rims (combined with Ar dating, Wartho and Kelley, 2003). Durations of ascent from ~ 400 km depth are 0.9 to 6.9 days for diamond-bearing South African kimberlites and 2 to 15 hours for Siberian kimberlites (Wartho and Kelley, 2003). These yield average rates of ascent of 2.4 to 18 km/hr (0.7 to 5.1 m/sec) for the African bodies and up to 200 km/hr (55 m/sec) for the Siberian one! McGetchin and Ullrich (1973) estimated an ascent rate of 70 km/hr.

Although CO_2-rich kimberlites and carbonatites may be produced by partial melting of carbonated lherzolite, H_2O-rich melts, such as orangeites and lamproites, are probably not (Mitchell, 1995). The isotopic signatures suggest a *lithospheric* origin, and the models for metasomatically layered or net-veined lithosphere provide an attractive enriched source for these melts. The best geochemical models for orangeites suggest a depleted harzburgite source that is subsequently enriched (by asthenospheric melts/fluids) in K, Pb, Rb, Ba, LREE, SiO_2, and H_2O. Lack of orthopyroxene xenocrysts indicates that the source may also be a dunite. Either source becomes enriched in phlogopite, K-richterite, apatite, and carbonate. If the enriched minerals occur as veins in a dunite substrate, melting of the combined vein–wall–rock assemblage may produce orangeitic melts.

Foley (1992) proposed a mechanism of combined vein–wallrock melting in which fusion is initiated in the enriched veins, which have a lower solidus temperature than the wall rocks. According to the model, the strongly alkaline vein melt incorporates wallrock components due to solid–solution exchange between minerals that are the same in both the vein and the wall rock, as well as by dissolving some of the wallrock minerals that are not in equilibrium with the vein–melt assemblage. Foley (1992) applied the model directly to the development of orangeites, showing how it could lead to a hybridized melt with both incompatible and refractory constituents. Xenolith and diamond inclusion studies suggest that orangeites are developed near the base of the lithosphere within the diamond stability field at depths of 150 to 200 km. Lamproites may also be generated in a similar vein-plus-wallrock fashion, but with an enriched source dominated by K-Ti richterite, diopside, and K-Ba titanites that presumably has a subduction-zone-related enrichment process.

Figure 19.21 is a schematic cross section of an Archean craton with a marginal Proterozoic mobile belt and a modern rift. A remnant of subducted eclogite slab remains beneath the orogenic edge, and some eclogitized basaltic lithosphere underplates are also shown. Due to the lower geothermal gradient in the poorly radioactive cratonal areas, the diamond–graphite transition is elevated into the deep cratonal lithospheric mantle root (see Figure 19.13). Diamonds thus concentrate in the lherzolites, depleted harzburgites–dunites, and eclogites of those roots, and not beneath the rifts or mobile belts. Only melts generated in or beneath these roots can entrain and disaggregate diamond-bearing xenoliths. Kimberlites may pass through various rock types, picking up harzburgite, eclogite, or lherzolite xenoliths. Lithospheric orangeites may also traverse diamond-bearing levels. Lamproites occur more commonly in the old mobile belts, where they develop the high LIL/HFS subduction signature, but some may be generated in diamond-bearing material at the cratonal edge. Diamond inclusions are generally eclogitic in diamondiferous lamproites, and the source may be the remnant subducted slab.

Figure 19.21 suggests that *any* melt that traverses the deep diamond-bearing horizons of the cratonal roots may incorporate diamonds. Mitchell (1995) mentioned that even some rare alkali basalts and melilitites do so. Most melts rise sufficiently slowly and are oxidized enough to destroy any diamonds that they may encounter.

Nephelinites and most other alkaline magmas require greater degrees of partial melting at shallower depths than do

kimberlites, orangeites, and lamproites. These requirements are commensurate with asthenospheric rise and decompression melting beneath rift areas, as discussed in Section 19.1.

Mitchell (1995) noted that there are distinct differences between the potassic magmas of each craton. We know that orangeites occur only in the Kaapvaal craton of South Africa (where lamproites are absent). In the Wyoming craton of the United States, lamproites occur with a wide variety of potassic rocks of the shonkinite suite. In the Aldan craton of Russia, lamproites and lamproite-like rocks occur with a host of extremely silica-undersaturated rocks of the kalsilite–leucite–biotite–orthoclase suite. Although some form of mantle metasomatism–underplating was effective in developing the sources of all these rocks, the nature, duration, extent, and depth of the process are apparently unique for each craton.

19.4 MANTLE METASOMATISM AND MANTLE XENOLITHS

Several times in Chapters 13 through 19, we have turned to mantle metasomatism to explain the production of enriched magmas from otherwise unenriched mantle sources. In particular, metasomatism of the subcontinental lithospheric mantle is indicated as an important contributor to the elevated incompatible element enrichment of highly alkaline continental magmas. Possible agents of metasomatism include:

1. Volatile primary magmas, such as kimberlites and carbonatites that stall at the resistant lithosphere or the solidus ledge where they solidify and release an enriched volatile phase into the wall rocks
2. Supercritical volatile fluids from the deep asthenosphere or subduction zones
3. Silicate melts

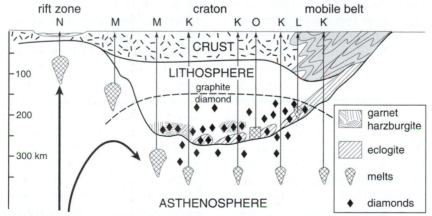

FIGURE 19.21 Hypothetical cross section of an Archean craton with an extinct ancient mobile belt (once associated with subduction) and a young rift. The low cratonal geotherm causes the graphite-diamond transition to rise in the central portion. Lithospheric diamonds therefore occur only in the peridotites and eclogites of the deep cratonal root, where they are then incorporated by rising magmas (mostly kimberlitic—"K"). Lithospheric orangeites ("O") and some lamproites ("L") may also scavenge diamonds. Melilitites ("M") are generated by more extensive partial melting of the asthenosphere. Depending on the depth of segregation they may contain diamonds. Nephelinites ("N") and associated carbonatites develop from extensive partial melting at shallow depths in rift areas. After Mitchell (1995). Copyright © by permission Kluwer Academic Publishers.

Numerous authors include melt addition (dikes, lithospheric underplates, and other stalled bodies) under the general term *metasomatism*. Such usage is technically incorrect, and this metamorphic term should be restricted to rock alteration resulting from diffusion or pervasively infiltrating fluids (metasomatism will be defined and explored more fully in Chapters 21 and 30). Melts may nonetheless be an important agent for enriching the upper mantle. They may add material by injection, and, if they are volatile rich, they may alter adjacent rocks by (true) metasomatic fluid emanations. On a small scale (such as a vein network), the melt/fluid distinction becomes difficult. Distinction at any scale is complicated further at mantle depths, where melts and fluids become similar and indistinguishable beyond the "second critical endpoint" (Section 16.8.2). Problematic terminology is understandable here, and perhaps we should broaden the scope of metasomatism when we consider mantle processes. Either fluids or melts, or some combination, may produce enriched zones or veins in mantle peridotites, and it is not a matter of finding some universally correct process.

Some investigators have distinguished two types of mantle metasomatism (Dawson, 1984; Meen et al., 1989; Jones, 1989). The first is called **patent metasomatism**, or **modal metasomatism**, in which a melt or fluid produces secondary replacement or veining. In patent metasomatism one can see textural recrystallization and new mineral growth in the veins or matrix, and the major and trace element chemical composition is clearly modified. The second type of metasomatism is called **cryptic metasomatism**, in which only certain minor and trace elements are enriched by substitution in existing phases without textural changes or new mineral growth.

Mantle metasomatism has attracted considerable interest in the past two decades. We have seen several instances in which a metasomatized source was required in order to model adequately the high incompatible trace element concentrations or isotopic enrichment of diverse alkaline rocks that are clearly of mantle origin. Minerals such as phlogopite, K-richterite, diopside, LIL-titanites, calcite, apatite, and zircon added to previously depleted harzburgite are required to generate lamproites, orangeites, etc. We have also discussed experiments on hydrated and carbonated lherzolites that show how highly alkaline melts can be derived from such volatile-enhanced sources. Other experiments, such as those discussed by Meen et al. (1989), show how volatile alkaline melts can interact with peridotites, for example beneath carbonatite-alkaline silicate rock complexes, dissolving olivine and orthopyroxene and precipitating Ca-rich clinopyroxene, perhaps accompanied by carbonate or amphibole in vein networks of clinopyroxenite or wehrlite. Numerous models for mantle metasomatism have been proposed (Hawkesworth et al., 1983, 1990; Wilshire, 1987; Haggerty, 1989; McDonough, 1990; Thibault et al., 1992; Ionov et al., 2002) to explain the enrichment process.

All of the approaches above have the air of speculation without some hard evidence, such as a sample of metasomatized mantle. Although a variety of magma types are known to bring ultramafic xenoliths to the surface, most are cognate (typically cumulates) or refractory residua. Kimberlites and orangeites, on the other hand, perhaps due to their more rapid rise, have delivered the most varied and extensive collection of crustal and mantle xenoliths of any magma type. The deep-seated inclusions include garnet peridotite, spinel peridotite, dunite, harzburgite, websterite, and eclogite, as we might expect. Many appear to be pristine and relatively unaltered during transport to the surface. Some, however, contain minor amounts of amphibole or mica. These mantle samples have provided experimental petrologists with valuable natural mantle specimens for stability and melting studies that allow us to create model systems of the mantle and its derivative magmas to test against the natural phenomena. We have reviewed and used many of those models in the preceding sections of this text.

In addition to "typical" mantle samples, kimberlites and orangeites contain a host of enriched and metasomatized ultramafic samples, including phlogopite clinopyroxenites, "glimmerites" (phlogopite-rich rocks), apatite–amphibole pyroxenites, and the "MARID" suite (mica–amphibole–rutile–ilmenite–diopside rocks; Dawson and Smith, 1977). See Frost (2006) and Luth (2007) for reviews of metasomatized mantle samples. Some patent-metasomatized xenoliths in alkaline volcanics and kimberlites also exhibit veins with amphibole, mica, diopside, carbonates, and/or REE-rich titanite pegmatites. In some peridotite xenoliths, magmatic dikelets of K-rich magma have added K, Fe, Ti, OH, and REE metasomatically to previously depleted wall rocks (Dawson and Smith, 1988a). Rudnick et al. (1993) and Ionov (1998), among others, reported evidence for carbonatitic melts enriching depleted mantle xenoliths. The evidence included secondary carbonate, amphibole, clinopyroxene, spinel, apatite, and silicate glass with trace element concentrations indicating equilibration with carbonate-rich melts. These are certainly not your typical primordial mantle lherzolite or depleted harzburgite, dunite, etc. They are variably enriched in H_2O, CO_2, F, Na_2O, Al_2O_3, K_2O, CaO, TiO_2, Fe, P_2O_5, S, Cl, Rb, Y, Zr, Nb, Ba, and REE. Some of these, like the MARID xenolith suite, may be created by crystallization of kimberlite or lamproite magma stalled in the subcontinental lithospheric mantle. Others are excellent prospects for the type of enriched mantle that could give rise to the alkaline magmas that we have just been discussing. Several works, including Pilet et al. (2002), Pearson et al. (2002, 2006), Beyer et al. (2006), and the extensive reviews by Pearson et al. (2007) and Luth (2007), cite evidence for early depletion of cratonic lithospheric keels followed by subsequent re-enrichment. There is now ample evidence indicating that the lithospheric keels beneath continents are of similar age to the crust immediately above, supporting the proposal that such keels have been anchored within the lithosphere since its creation. One could reasonably conclude that the depletion event was of similar age and associated with crustal genesis (related to Precambrian subduction and/or plume-LIP-underplate processes). Enrichment takes many forms, suggesting multiple events of differing character (agents include kimberlitic and other silicate melts, carbonatites, and infiltrating fluids). Here, finally, we find evidence for all of my vague (but enriching!) earlier references to enriched subcontinental lithospheric mantle.

As a minor postscript, among the more interesting xenoliths and diamond inclusions are a very few that appear to be brought up from *very* deep sources. Haggerty and Sautter (1990) and Sautter et al. (1991) described garnet–clinopyroxene xenoliths from diamondiferous South African kimberlites containing clinopyroxene exsolved from grains of garnet. The original solution of such clinopyroxene in garnet is predicted to occur at depths of 300 to 400 km, suggesting that these xenoliths come from eclogites (presumably subducted) that reached this depth. McCammon (2001) reported occurrences of diamond with very high-pressure inclusions from at least 12 localities, representing eight cratons. The presence of ferropericlase, plus enstatite and calcium silicate in some of these diamonds, interpreted as the low-pressure polymorphs or exsolved equivalents of high-pressure precursors ($MgSiO_3$ and $CaSiO_3$ perovskites), strongly suggests an origin in the *lower* mantle. Corundum associated with $MgSiO_3$-perovskite and ferropericlase indicates an origin even deeper in the lower mantle and that free aluminum may exist there. One inclusion of extremely iron-rich ferropericlase has been interpreted to indicate an origin near the core-mantle boundary. If interpreted correctly, these inclusions support the idea that mantle plumes originate at and below the 600-km transition zone, that very deep-seated plumes can penetrate the transition zone (a question addressed in Sections 1.7.2 and 14.6), and that kimberlites are produced in such plumes.

Summary

Igneous rocks in continental settings are highly variable in composition, extending to highly alkaline and carbonatitic compositions. The geochemical data indicate that high alkalinity is present even at the mafic (parental) end of the spectrum, which may also exhibit highly enriched isotopic ratios. This suggests that the alkaline character in many igneous rocks is related to mantle processes and may be attributed to deeper melting or to re-enrichment of subcontinental lithospheric mantle during the eons since it was first depleted and attached to the early continental crust. Continental rifting, particularly in the incipient stages, may cause such enriched mantle to rise and experience decompression partial melting. Mantle plumes may also contain slivers of enriched subducted material that melt during rise. Rising melts may also assimilate alkalis and other incompatible elements from the subcontinental lithospheric mantle or from the continental crust itself. The alkaline silicate magmas produced may be K-rich or Na-rich, range from ultramafic to felsic, and are generally silica undersaturated.

Carbonatites are igneous rocks with more than 50% carbonate in the mode (as strictly defined). The carbonate mineral may be calcite, dolomite, or, more rarely, siderite, ankerite, or a Na-K-bearing carbonate. Most are not alkaline, therefore, but they typically occur as small intrusive complexes together with alkaline silicate rocks, such as nepheline syenites, ijolites, and/or urtites. Carbonatites may be generated as primary melts of carbonated peridotite or eclogite in ascending plumes, by liquid immiscibility from a cooling and crystallizing alkaline silicate magma, or by fractionation from such a silicate liquid. Melt volumes of primary carbonatites, however, are probably small, and most melts doubtfully reach crustal levels before being consumed by reaction with enveloping/overlying peridotites with which they are no longer in equilibrium shallower than 70 to 80 km. These melts may thus be an important agent that enriches the subcontinental lithospheric mantle, which then becomes an attractive source for many of the alkaline and carbonatite rocks discussed in this chapter.

Lamproites are highly alkaline, typically porphyritic mafic rocks. They have high H_2O and incompatible element concentrations in addition to an abundance of compatible elements. They occur in areas previously affected by subduction and are probably created by partial melting of subcontinental lithospheric mantle that has been enriched by rising subduction zone fluids and/or melts. Lamprophyres are an (exceedingly) diverse group of hydrous, strongly porphyritic rocks with mafic phenocrysts. Most are hybrids, and their origins may be as diverse as their members. They may represent hydrous or otherwise enriched variations on other well-known petrogenetic processes, such as alkaline or calc-alkaline evolution.

Kimberlites are potassic, volatile-rich, typically ultramafic rocks that rise rapidly through the continental lithosphere and tend to erupt explosively as diatremes. Due to their rapid rise they transport a host of xenoliths and xenocrysts from depth, which inform us about the nature of the mantle beneath continents. Kimberlites are also the principal source of diamonds, which are included in entrained peridotite or eclogite xenoliths or become scattered in the kimberlite matrix as the xenoliths disaggregate. Kimberlites are probably generated by low degrees of partial melting of hydrated-carbonated peridotite in rising asthenospheric plumes.

The subcontinental lithospheric mantle (SCLM), as interpreted from xenoliths in kimberlites and other lavas, is broadly similar to its sub-oceanic counterpart but has several distinct differences. The age of the subcontinental lithospheric mantle is essentially the same as that of the immediately overlying crust, indicating that it has been anchored to that crust since its creation and not subducted like oceanic lithosphere. The lithospheric mantle beneath cratons can thus be several billion years old. There are several geochemical and mineralogical indicators of early depletion of the SCLM (presumably associated with crustal extraction) followed by re-enrichment. Enrichment may take many forms, suggesting multiple events of differing character. Agents include kimberlitic and other silicate melts, carbonatites, and infiltrating fluids.

Key Terms

Active/passive rifting *399*
Aulacogen *400*
Craton *400*
Carbonatite (nomenclature in
Table 19.3) *409*

Carbonatite–alkaline silicate rock
association *410*
Fenite/fenitization *411*
Lamproite *421*
Lamprophyre *423*

Kimberlite/orangeite *426*
Diatreme *427*
Patent, or modal/cryptic,
metasomatism *432*
Glimmerites and MARID xenoliths *432*

Review Questions

Review Questions are located on the author's web page at the following address: **http://www.prenhall.com/winter**

Important "First Principle" Concepts

- The subcontinental lithospheric mantle is heterogeneous and apparently much different from the sub-oceanic lithospheric mantle, having been attached to the continents since their formation and re-enriched over substantial periods of time.

- Protracted enrichment following its initial depletion probably makes the subcontinental lithospheric mantle a significant source of alkalinity in many continental intraplate magmas.

Suggested Further Readings

General Alkaline Rocks

Fitton, J. G., and B. G. J. Upton (eds.). (1987). *Alkaline Igneous Rocks*. Blackwell Scientific. Oxford, UK.

Foley, S. F., G. Venturelli, D. H. Green, and L. Toscani. (1987). The ultrapotassic rocks: Characteristics, classification, and constraints for petrogenetic models. *Earth Sci. Rev.,* **24**, 81–134.

Kampunzu, A. B., and R. T. Lubala (eds.). (1991). *Magmatism in Extensional Settings, the Phanerozoic African Plate*. Springer-Verlag, Berlin.

Le Maitre et al. (eds.) (2002). Igneous Rocks: A Classification and Glossary of Terms. Sections 2.3 to 2.9. Cambridge University Press. Cambridge, UK.

Lithos. (1992). **28**, 1. Special section on potassic and ultrapotassic magmas.

Mitchell, R. H. (ed.). (1996). *Undersaturated Alkaline Rocks: Mineralogy, Petrogenesis, and Economic Potential*. Short Course **24**. Mineralogical Association of Canada.

Mitchell, R. H., G. N. Eby, and R. F. Martin (eds.). (1996). *Alkaline Rocks: Petrology and Mineralogy. Can. Mineral.,* **34**, 2. Special issue on alkaline rocks.

Morris, E. M., and J. D. Pasteris (eds.). (1987). *Mantle Metasomatism and Alkaline Magmatism*. Special Paper **215**. Geological Society of America. Boulder, CO.

Sørensen, H. (ed.). (1974). *The Alkaline Rocks*. John Wiley & Sons. New York.

Woolley, A. R., S. C. Bergman, A. D. Edgar, M. J. Le Bas, R. H. Mitchell, N. M. S. Rock, and B. S. Smith. (1996). Classification of lamprophyres, lamproites, kimberlites, and the kalsilitic, melilitic, and leucitic rocks. *Can. Mineralogist,* **34**, 175–196.

Mantle Samples and the Subcontinental Lithospheric Mantle

Carlson, R. W., D. G. Pearson, and D. E. James. (2005). Physical, chemical, and chronological characteristics of continental mantle. *Rev. Geophys.,* **43**, 1–24.

Fei, Y., C. M. Bertka, and B. O. Mysen (eds.). (1999). *Mantle Petrology: Field Observations and High-Pressure Experimentation: A Tribute to Francis R. (Joe) Boyd*. Geochemical Society. Houston. Volume 6.

Frost, D. J. (2006). The stability of hydrous mantle phases. In: *Water in Nominally Anhydrous Minerals (eds. H. Keppler and J. Smyth)*. Reviews in Mineralogy and Geochemistry, *Mineral. Soc. Amer./Geochem. Soc.,* **62**, 243–271.

Luth, R.W. (2007). Mantle volatiles—Distribution and consequences. In: *Treatise on Geochemistry*, Volume 2 (ed. R. W. Carlson). *The Mantle and Core*. Chapter 2.07. pp. 319–361.

Menzies, M. A., and C. J. Hawkesworth (eds.). (1987). *Mantle Metasomatism*. Academic Press. London.

Nixon, P. H. (1987). *Mantle Xenoliths*. Wiley. Chichester, UK.

Pearson, D. G., Canil, D., and S. Shirey. (2007). Mantle samples in volcanic rocks: Xenoliths and diamonds. In: *Treatise on Geochemistry*, Volume 2 (ed. R. W. Carlson). *The Mantle and Core*. Chapter 2.05. pp. 171–275.

Pearson, D. G., and G. M. Nowell. (2002). The continental lithospheric mantle: characteristics and significance as a mantle reservoir. *Phil. Trans. R. Soc. Lond.,* **360**, 2383–2410.

Carbonatites

Bailey, D. K. (1993). Carbonatite magmas. *J. Geol. Soc. London,* **150**, 637–651.

Bell, K. (ed.). (1989). *Carbonatites: Genesis and Evolution*. Unwin Hyman, London.

Bell, K., and J. Keller (eds.). (1995). *Carbonatite Volcanism: Oldoinyo Lengai and the Petrogenesis of Natrocarbonatites*. Springer-Verlag. Berlin.

Heinrich, E. W. (1966). *The Geology of Carbonatites*. Rand McNally, Chicago.

Journal of African Earth Sciences. (1997). **25**, 1. Special issue on carbonatites.

Journal of Petrology. (1998). **39**, 11–12. Special issue on carbonatites.

Le Bas, M. J. (1977). *Carbonatite–Nephelinite Volcanism.* Wiley. New York.

Le Bas, M. J. (1987). Nephelinites and carbonatites. In: *Alkaline Igneous Rocks* (eds. J. G. Fitton and B. G. J. Upton). Blackwell Scientific. Oxford, UK. pp. 53–83.

Tuttle, O. F., and J. Gittins (eds.). (1966). *Carbonatites.* Wiley-Interscience. New York.

South African Journal of Geology. (1993). **96**, 3. Special issue on carbonatites.

Lamproites and Lamprophyres

Bergman, S. C. (1987). Lamproites and other potassium-rich igneous rocks: A review of their occurrence, mineralogy, and geochemistry. In: *Alkaline Igneous Rocks* (eds. J. G. Fitton and B. G. J. Upton). Blackwell Scientific. Oxford, UK. pp. 103–190.

Mitchell, R. H., and S. C. Bergman. (1991). *Petrology of Lamproites.* Plenum. New York.

Rock, N. M. S. (1987). The nature and origin of lamprophyres: an overview. In: *Alkaline Igneous Rocks* (eds. J. G. Fitton and B. G. J. Upton). Blackwell Scientific. Oxford, UK. pp. 191–226.

Rock, N. M. S. (1991). *Lamprophyres.* Blackie. Glasgow.

Kimberlites

Dawson, J. B. (1980). *Kimberlites and Their Xenoliths.* Springer-Verlag. Berlin.

Mitchell, R. H. (1986). *Kimberlites: Mineralogy, Geochemistry, and Petrology.* Plenum. New York.

Mitchell, R. H. (1995). *Kimberlites, Orangeites, and Related Rocks.* Plenum. New York.

Proceedings of the Kimberlite Conferences

MacGregor, I.D. (1974). First Kimberlite Conference, Republic of South Africa. Geology, **2**, 151-152.

Boyd, F. R., and H. O. A. Meyer (eds.). (1979). *Kimberlites, Diatremes, and Diamonds: Their Geology, Petrology, and Geochemistry.* (Second International Kimberlite Conference) Volume 1. American Geophysical Union, Washington, D.C.

Kornprobst, J. (ed.). (1984). *Kimberlites I: Kimberlites and related rocks.* Proceedings of the Third International Kimberlite Conference. Elsevier. Amsterdam.

Kornprobst, J. (ed.). (1984). *Kimberlites II: The mantle and crust–mantle relationships.* Proceedings of the Third International Kimberlite Conference. Elsevier. Amsterdam.

Ross, J., A. L. Jacques, J. Ferguson, D. H. Green, S. Y. O'Reilly, R. V. Danchin, and A. J. A. Janse (eds.). (1989). *Kimberlites and Related Rocks.* Proc. Fourth International Kimberlite Conference. Australia Special Publication. Volume 14.

Meyer H. O. A., and O. H. Leonardos (eds.). (1994). *Kimberlites, related rocks and mantle xenoliths.* Proceedings of the Fifth International Kimberlite Conference. CPRM. Rio de Janeiro.

Sobolev, N. V., and R. H. Mitchell (eds). (1997). Proceedings of the Sixth International Kimberlite Conference. *Russian Journal of Geology and Geophysics,* **38**. Also published by Allerton Press, New York.

Gurney, J. J., J. L. Gurney, M. D. Pascoe, and S. H. Richardson (eds.). (1999). Proceedings of the Seventh International Kimberlite Conference. P. H. Nixon Volume. Red Roof Design, Cape Town, South Africa.

Mitchell, R. H., H. S. Grütter, L. M. Heaman, B. H. Scott Smith, and T. Stachel. (2005). Proceedings of the Eighth International Kimberlite Conference. Elsevier, Amsterdam. Also *Selected Papers. Lithos,* Volumes 76–77.

The ninth conference was held in Frankfurt, Germany, in August 2008.

20

Anorthosites

Questions to be Considered in this Chapter:

1. What are the principal anorthosite types, and how do they differ from one another?

2. What processes can generate significant (often enormous) volumes of nearly pure plagioclase?

3. Why are the most common and voluminous terrestrial anorthosites of Precambrian age, and how and why are Archean anorthosites different from Proterozoic ones?

4. What is ultimately responsible for generation of the parent magmas and their differentiation to produce anorthosites?

Anorthosites are defined in Chapter 2 as plutonic rocks with over 90% plagioclase (there are no known volcanic equivalents). Their highly felsic nature and their location in continental areas are characteristics they share with granitoid rocks. The felsic mineral, however, is a calcic plagioclase, which, along with subordinate associated high-temperature mafic minerals, suggests a stronger similarity to basaltic rocks. The mafic similarity is supported by our previous encounter with anorthosites: as layers in layered mafic intrusions (Chapter 12). Hence this short final igneous chapter that briefly addresses these relatively uncommon and difficult-to-categorize rocks.

Ashwal (1993) listed six major types or anorthosite occurrences:

1. Archean anorthosite plutons
2. Proterozoic "massif-type" anorthosite plutons
3. 1-cm- to 100-m-thick layers in layered mafic intrusions
4. Thin cumulate layers in ophiolites/oceanic crust
5. Small inclusions in other rock types (xenoliths and cognate inclusions)
6. Lunar highland anorthosites

We will concentrate here on the first two types because they are the classic examples. Type 3 was discussed in Chapter 12. Type 4 is created by cumulate processes at mid-ocean ridge magma chambers similar to those discussed in Chapters 12 and 14. Type 5 is a very minor occurrence and may represent fragments of material derived from the other types. Although type 6 is extraterrestrial, it presents an interesting opportunity to compare igneous processes on Earth with those on other bodies, so I will describe them briefly at the conclusion of the chapter. If you are interested in investigating any of these or other occurrences further, see Ashwal (1993) for a more comprehensive review.

20.1 ARCHEAN ANORTHOSITES

Most Archean anorthosites cluster in the age range 3.2 to 2.8 Ga. They typically occur as kilometer-scale lenses in Archean high-grade metamorphic gneiss terranes. Some bodies are hundreds of kilometers long, but most are tectonically disrupted and metamorphosed. They are generally less than 1 km thick and appear to be sheet-like, conformable

FIGURE 20.1 **(a)** "Snowflake" clusters of plagioclase crystals. **(b)** Typical texture of Archean anorthosite. Both from the Fiskenæsset complex, western Greenland. **(a)** Myers, 1985. **(b)** Photo courtesy John Myers. Copyright Geological Survey of Denmark and Greenland.

sills. The extensive deformation typical of Archean terranes makes the original thickness of the anorthosites difficult to determine.

Archean anorthosites are associated with gabbroic rocks and tend to be internally layered (as in layered mafic intrusions; Chapter 12). Other associated rock types range from very plagioclase-rich leuco-gabbros to ultramafic rocks. Archean anorthosites are similar to layered mafic intrusions, but plagioclase is much more prominent.

Many Archean anorthosite bodies are emplaced as shallow sills into "**supracrustal**" rocks (a common term among Precambrian geologists, used to indicate rocks deposited at the Earth's surface: on top of the crust, not within it). The original shallow-water supracrustal sediments are now pelitic schists, quartzites, and marbles. These may be associated with basaltic volcanics (now amphibolites). This package is a common constituent of Archean *greenstone belts* (Section 18.7). There are some plagioclase-rich basaltic flows associated with a few Archean anorthosites. The flows are always more mafic than true anorthosite. Later granitoids generally obscure many primary contacts and structures. See Ashwal (1993) for a comprehensive list and map of individual occurrences, as well as descriptions of some specific examples.

20.1.1 Petrology and Geochemistry of Archean Anorthosites

The plagioclase crystals in Archean anorthosites are subhedral to euhedral megacrysts ranging in size from 0.5 to 30 cm in diameter (most are 1 to 5 cm). The megacrysts are relatively equidimensional, a strange shape for plagioclase, which usually forms elongate laths. Phinney et al. (1988) proposed that the equant plagioclase may result from an initial quench texture that produces "snowflake" plagioclase clusters (Figure 20.1a) that are later filled in and recrystallized. The megacrysts are unusually homogeneous and calcic in composition (An_{80-95}) and are surrounded by a finer-grained mafic matrix (ranging from 0 to 50% of an outcrop). Because of their highly calcic plagioclases, several investigators have called these anorthosites *labradorite type*. The large white plagioclase megacrysts set in a dark matrix (Figure 20.1) have invited several colloquial names, including "leopard rock" and "football" or "baseball" anorthosite. The mafic matrix is typically dominated by metamorphic amphibole, but some primary pyroxene or olivine may occasionally remain, as may oxides, such as chromite or magnetite. The original mineralogy suggests that the magma was initially dry and was hydrated during subsequent metamorphism.

Cumulate texture is well developed in undeformed bodies, where layering is also more obvious. There is also significant adcumulate growth (see Figure 3.14), so that the interstitial liquid has been at least partly expelled and presumably exchanged with a larger magma reservoir associated with the anorthosite cumulates. The major element concentrations are controlled by the percentage of accumulated plagioclase, and the bulk chemical composition of anorthosites, therefore, is *not* that of the initial liquid.

The interstitial mafics are not the same throughout most Archean anorthosite bodies. The Mg# of the mafic minerals, for example, typically decreases upward. Thus the mafic material is not representative of a single residual liquid, as its interstitial texture might otherwise suggest. Variation diagrams are of little value in the classical sense because they reflect the cumulate processes and don't represent a liquid line of descent.

The REE concentrations of Archean anorthosites (Figure 20.2) are relatively primitive, with flat HREE near chondritic values and the LREE slightly enriched ($3\times$ to $10\times$ chondrite). Because of the high accumulative plagioclase concentrations, all anorthosite REE diagrams show a pronounced *positive* Eu anomaly. Initial Sr and Nd isotopic ratios indicate a range of depleted to enriched contributions, although depleted types dominate. Most investigators suspect a depleted mantle (DM) source with some crustal contamination. In several cases, however, an enriched mantle source cannot be eliminated.

20.1.2 The Parent Liquid of Archean Anorthosites

The predominance of cumulate material, combined with the absence of any volcanic equivalents or modern anorthosite occurrences, makes it difficult to determine the parent magma of Archean anorthosites. Attempts to formulate a

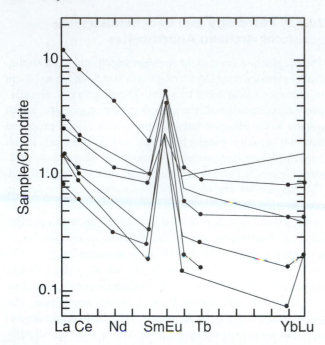

FIGURE 20.2 Chondrite-normalized rare earth element diagram for some typical Archean anorthosites from the Bad Vermillion (Ontario) and Fiskenæsset (Greenland) bodies. Data from Seifert et al. (1977), Simmons and Hanson (1978), and Ashwal et al. (1989).

parental composition (as attempted for layered mafic intrusions in Chapter 12) have followed one of three methods. One way is to determine the compositions of the various layers and sub-units and then sum them for the whole layered complex on the basis of estimated relative volumes of each constituent in the exposed body. Results vary somewhat, but the most common average bulk composition is a normative-plagioclase-rich and Fe-rich leuco-gabbro with high Al_2O_3 (23 to 25 wt. %) and CaO (13 to 14 wt. %). Another approach is to look at marginal areas and flows, dikes, and sills, all of which are finer and less differentiated than the main anorthosite bodies. These rocks are variable, but generally basaltic, with a similar high Al_2O_3 and plagioclase content that may constitute up to 80% of the rock. The final method is to use the minerals present in the anorthosite and the crystal-liquid partition coefficients to infer the composition of the liquid that would be in equilibrium with the solid assemblage. Because the partition coefficients are not known with accuracy, the liquids so determined are somewhat variable. Phinney and Morrison (1990) used this method to determine a parent that was a relatively fractionated Fe-rich tholeiitic basalt (Mg# = 50−60). The basalt is still very plagioclase rich (69 to 74% normative plagioclase), with correspondingly high Al_2O_3 and CaO.

Although there is some variation, the general consensus is that the parental magma for Archean anorthosites is a tholeiitic basalt that is rich in Fe, Al, and Ca (Ashwal, 1993; Ashwal and Myers, 1994). The parent is enriched in plagioclase components beyond any parental basalt that we have yet encountered in our survey of igneous petrogenesis, including layered mafic intrusions. The high Fe content of the parental

basalt indicates that it cannot be a primary magma in equilibrium with the mantle, and thus it must be differentiated at depth from a more primitive magma, either a basalt, a picrite, or even a komatiite (the latter is also uniquely Archean).

20.1.3 Petrogenesis of Archean Anorthosites

Anorthosites are strange beasts and have inspired a wealth of speculation as to their origins. Archean anorthosites have been interpreted in various ways, including as metamorphosed layered mafic intrusions (but most lack cryptic compositional variations and the parent is too plagioclase rich), tectonic slices of lower crust (but many are clearly intruded into supracrustals), pieces of primordial crust (but they are too young for that), and layers from Archean ophiolites (but they are much thicker than modern anorthosite layers in ophiolites).

The original settings of Archean anorthosites are difficult to assess because there are no modern analogs, and the Archean terranes are now badly deformed. The common association of Archean anorthosites with the mafic lavas (including pillow lavas) of greenstone belts has led several investigators to conclude that most of these bodies are oceanic and consanguineous with mafic magmatism (Ashwal, 1993). Their location in continental cratons would then be due to their being caught up in the later amalgamation of Archean arcs and microcontinents. If so, they represent old suture zones that were later metamorphosed at mid-crustal levels.

Windley (1995) noted that the Canadian examples occur in rocks that have more of an island-arc than an oceanic affinity. He thus preferred an island arc or continental arc setting for Archean anorthosites. He cited the Peruvian Andes (Sections 17.2 and 17.4) as a modern analog. Early in the development of the Andean arc, ensialic marginal basins formed by extensional processes filling with pillow basalts, andesites, tuffs, etc. (see Section 17.4 and Atherton et al., 1985). These Andean marginal basins were intruded initially by mafic gabbro-anorthosite layered cumulate complexes (Regan, 1985), followed by the main-stage tonalite batholiths. Windley (1995) proposed that the Archean anorthosites may well be analogous mafic intrusions into marginal basins filled with greenstone-belt-like volcanics and sediments. Like the Andean analogs, they were then intruded by granitoid rocks and metamorphosed to high grades. Their occurrence in cratonal settings is thus a natural consequence of their initial setting.

Both of these models have merit, and either may produce one or another anorthosite body. Phinney et al. (1988) proposed that the parental Fe-rich tholeiite must have been derived from a more primitive tholeiite at depth, via crystallization of olivine and pyroxene in a magma chamber stalled by its density at the crust–mantle boundary. The implication is that these parental melts are *underplated* beneath low-density continental crust, and, if true, this would favor the arc-related model. A model for fractionation in crustal underplate magma chambers will be developed further in conjunction with Proterozoic anorthosites, so we will postpone a more detailed description of this process until then.

20.2 PROTEROZOIC ANORTHOSITES

Proterozoic anorthosites are generally referred to as **massif-type** anorthosites. The term is borrowed from the French and was applied to this type of anorthosite to indicate a plutonic mass of large size. Proterozoic anorthosites differ from their Archean counterparts in several ways. They are larger and less sill-like, the plagioclase crystals are shaped in the common tabular form and are less anorthitic, they contain less mafic matrix or mafic cumulates, and they are associated more with granitoids and not greenstone belts/supracrustals.

The shape of Proterozoic anorthosite bodies is highly varied, ranging from funnel-shaped, to lopoliths, to large sheets. They range in age from 1.7 to 0.9 Ga. The sheets are slabs 2 to 14 km thick, and some cover areas up to 17,000 km^2. On a trip to do some fieldwork in Labrador, I recall thinking that I had been flying over almost pure plagioclase for over an hour (okay, it was a slow plane, but that's a lot of concentrated feldspar). Most bodies are composite, with multiple intrusions accumulated over a relatively brief 20 to 30 Ma span. Many anorthosites are domal, presumably because of late solid-state isostatic rise.

Proterozoic anorthosites are almost always associated with nearly anhydrous pyroxene-bearing granitoid rocks (**charnockites**), as well as with Fe-rich and K-rich diorites, monzonites, and other K-rich granitoids. Ages of the associated plutons scatter about the anorthosite ages, and many of these are slightly younger and intrude the anorthosites. Some investigators have referred to the association as **AMCG complexes** (for anorthosite–mangerite–charnockite–granite).

The tectonic setting is characteristically anorogenic, and the massifs are intruded into thick, stable cratonic crust: high-grade gneiss terranes of Proterozoic or Archean age. Some investigators propose that the mafic parental magmas are intruded during a period of incipient rifting and continental breakup or post-orogenic collapse (Section 18.4.2.3). The linearity of many massifs and the bimodal granitoid-mafic nature of the igneous assemblage argue in favor of such a setting, but it is hard to conceive of incipient rifting lasting nearly 800 Ma.

Herz (1969) first pointed out that the Proterozoic anorthosites defined two belts in reconstructed Pangea. A northern hemisphere belt (in Laurasia) extends from the Ukraine, through Fennoscandia and Greenland into North America. A southern hemisphere belt (in Gondwanaland) extends from India through Madagascar into Africa. Proterozoic anorogenic granitoids (Section 18.4.2.4) follow the northern belt and extend into the southwestern United States. This belt is complex and long-lived, containing a variety of rock types and evidence for both extensional and compressional (Grenville) events. Both the northern and southern belts are intriguing, but their significance remains controversial.

Many investigators have speculated that the anorthosites were created during some special Proterozoic event. Although their development during a single 1.7 to 0.9 Ga time interval suggests that some special conditions existed unique to the Proterozoic, one could hardly call it an "event," having lasted longer than all of Phanerozoic history.

20.2.1 Petrology and Geochemistry of Proterozoic Anorthosites

Proterozoic anorthosites are dominated by massive to weakly layered plutons containing 75 to 95% plagioclase. The plagioclase crystals are tabular, or lath-shaped, and commonly 1 to 10 cm long, but may reach 1 m. The composition is typically in the range An$_{40-65}$. (They are thus referred to by some as *andesine-type* anorthosites, as compared to the Archean *labradorite-type* anorthosites.) The more sodic plagioclase composition may be due to deeper crystallization (high pressure favors lower An content) or a more sodic continental environment than the calcic oceanic or island-arc terranes that may typify the Archean. The composition of the plagioclase is in the vicinity of one of the three known solvi in the Ab-An system, so that some of the slowly cooled crystals develop very fine exsolution lamellae, fine enough to internally reflect light producing the well-known colorful chatoyance of some forms of labradorite.

Like their Archean counterparts, Proterozoic anorthosites exhibit cumulate textures. The prominent adcumulates result in highly plagioclase-rich masses in which the intercumulate liquid largely escaped from the anorthosite proper. True anorthosites are dominant, but in some areas, the mafic mineral content exceeds the 10% limit that defines anorthosite *sensu stricto*. The rocks are then called leuco-norite, leuco-gabbro, or leuco-troctolite (anorthositic rocks with over 10% orthopyroxene, clinopyroxene, or olivine, respectively). The mafic content probably correlates with the percentage of trapped intercumulate liquid. The mafic-mineral Mg# is low (40 to 70). There is a minor quantity of more mafic rocks, such as norite, gabbro, and troctolite, and there are also some ilmenite–magnetite layers or pods. Ultramafic rocks are notably rare or absent. Al-rich orthopyroxene megacrysts are widely distributed in massif-type anorthosites (e.g., Emslie, 1975). Although a minor constituent, they are important in that they yield crystallization pressures of 1.0 to 1.3 GPa (35–45 km) via orthopyroxene geobarometry (Section 27.4).

The geochemistry of Proterozoic anorthosites suffers from the same problems as the Archean ones. Because they are cumulates, the major element composition does not represent that of a parental or derivative liquid, but rather an accumulation of plagioclase crystals with an interstitial liquid in open-system exchange with another magma reservoir. The distribution of mafics is irregular and may reflect diffusion of late liquid over considerable distances and significant temperature ranges. Large poikilitic pyroxenes may grow late and drive the interstitial liquid elsewhere. As a result, chemical variations between analyzed samples of a single body may reflect the size of the sample and local redistribution of solid and liquid components rather than a liquid evolutionary trend.

Trace element concentrations of Proterozoic anorthosites are similar to the Archean anorthosites, with near-chondritic REE (slightly enriched in LREE) and positive Eu anomalies. ϵ_{Nd} and ($^{87}Sr/^{86}Sr$)$_i$ values extend from depleted mantle values (indicating an ultimate mantle source) but extend toward local crustal values, suggesting that crustal contamination is influential. In several bodies, the isotopic enrichment trends correlate with local gneissic crustal values, and in some cases, where massifs span major tectonic boundaries, local trends correlate with tectonic province (Ashwal and Wooden, 1983; Emslie et al., 1994). Scoates and Frost (1996) modeled Nd and Sr trends in the Laramie anorthosite, finding they could be explained by progressively mixing up to 10% of the local Archean meta-pelites and orthogneisses with a parental basaltic magma having a slightly depleted mantle character. Further evidence for crustal interaction comes in the form of numerous crustal xenoliths and enriched trace element and isotopic signatures in the marginal zones of the bodies.

20.2.2 The Parent Liquid of Proterozoic Anorthosites

As with their Archean cousins, the nature of the parent liquid of Proterozoic anorthosites is not easily determined. The parent liquid is like the holy grail of anorthosite petrology. The quest is even more difficult for the huge volumes of nearly pure plagioclase that constitute some of the Proterozoic occurrences. Historically, estimates of the parent have ranged from picrites to granites.

Before investigators could really address the question of a parent liquid for Proterozoic anorthosites, it was necessary to first determine whether the nearly ubiquitous associated granitoids represent a fractionate from the anorthosite system. This was a difficult early problem, but modern geochemical approaches have come to our aid. The trace element characteristics of the anorthosites and granitoids are not compatible as anorthosite–liquid counterparts, and the enriched isotopic signature of the granitoids indicates a *crustal* source that contrasts sharply with the more depleted anorthosites. Thus we can eliminate the granitoids as a co-magmatic component of the anorthosite massifs.

We are left with the anorthosite system itself as a magmatic product. We might then begin with the simplest explanation: that of anorthosite melts (nearly pure melted plagioclase). The idea goes back a long way, and Bowen himself argued strongly against it. As we saw in Chapters 6 and 7, additional components usually tend to lower the melting point of chemically simpler systems. Consider, for example, the melting point of pure anorthite in Figure 7.2, indicating that anorthosite liquids should crystallize in the range of 1450 to 1550°C, which is too high for upper mantle processes (even in the Precambrian). When diopside and olivine components are added, however, the eutectic liquid coexists with anorthite at 1270°C. The cumulus textures and lack of a volcanic equivalent also suggest that differentiation, not straight equilibrium crystallization, was an important process in the formation of anorthosites.

The mineralogy of anorthosites is typically basaltic (labradorite, olivine, augite, magnetite, and apatite); only the relative proportions differ. The same techniques employed to estimate the parent of Archean anorthosites apply equally well

to Proterozoic anorthosites, and investigators again come up with a fractionated tholeiitic basalt parent with low Mg# and high Al_2O_3 (17 to 18 wt. %): an unusual and very plagioclase-normative type of mafic magma. There is still contention as to whether such parental magmas result from partial melting of typical mantle material, anomalous Fe-Al-rich mantle (Olsen and Morse, 1990), or mafic crustal sources (e.g., Taylor et al., 1984; Duchesne et al., 1999). The phase equilibrium calculations of Longhi et al. (1999) and Longhi (2005) failed to match the compositional array of anorthosites if they began with melts of normal or Al-Fe-rich mantle, either by fractional crystallization or AFC (fractional crystallization plus crustal assimilation) models, but they could do so with mafic crustal sources. An Fe-Al-rich mantle would also have unusual TiO_2, K_2O, and P_2O_5 (in addition to FeO and Al_2O_3) contents for producing a suitable parent. On the other hand, crustal melting would have to be extensive (75% or more). Emslie (1985) and Emslie et al. (1994) contended that a sufficiently aluminous basaltic parent may be created by melting of normal mantle in the spinel or plagioclase peridotite stability fields (Figure 10.2), but not in the deeper garnet peridotite field. Early plagioclase saturation is required to fractionate anorthosite from a parent, and parental basaltic magmas from any of the proposed sources could equilibrate relatively early with plagioclase. Plagioclase content and production would also be enhanced by assimilation of plagioclase-rich lower crust. But production of the extensive quantities represented by Proterozoic anorthosites is still not likely, Emslie et al. (1994) concluded, in a closed system. If anorthosite massifs are derived from such Al-rich basaltic parents in an open system by the accumulation of plagioclase and separation from the mafic residue, then we would expect to find the complementary associated ultramafic cumulate masses somewhere in the area. They are not found. Ultramafics are notably absent in the vicinity of Proterozoic anorthosite massifs, and gravity surveys do not indicate that they exist in the crust immediately beneath the exposed anorthosites.

20.2.3 Petrogenesis of Proterozoic Anorthosite Massifs

The genesis of massif-type anorthosites has inspired a wide range of speculation. Models include proposals that anorthosites are sedimentary (Hunt, 1862; Logan et al., 1863), metasomatized sediments (Hietanen, 1963; Gresens, 1978), residues left after the extraction of crustal partial melts (de Waard, 1967), or crystallized from anorthosite melts (Buddington, 1939). We are now aware that anorthosites are cumulates derived from basaltic parent liquids, but how and where this is accomplished, and the fate of the mafic counterparts, is the subject of considerable debate (see Wiebe, 1992; and Ashwal, 1993, for reviews).

Figure 20.3 illustrates Lew Ashwal's (1993) model for the genesis of plagioclase-rich magmas in general and Proterozoic anorthosites in particular. Note that the term *magma* can mean anything from a pure liquid melt to a crystal mush, rich in suspended minerals. The model is a summary of modern opinion, including suggestions of Emslie (1985),

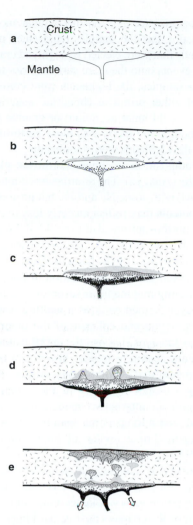

FIGURE 20.3 Model for the generation of massif-type anorthosites. **(a)** Mantle-derived magma underplates the crust as it becomes density equilibrated. **(b)** Crystallization of mafic phases (which sink) and partial melting of the crust above the ponded magma. The melt becomes enriched in Al and Fe/Mg. **(c)** Plagioclase forms when the melt is sufficiently enriched. Plagioclase rises to the top of the chamber, whereas mafics sink. **(d)** Plagioclase accumulations become less dense than the crust above and rise as crystal mush plutons. **(e)** Plagioclase plutons coalesce to form massif anorthosite, whereas granitoid crustal melts rise to shallow levels as well. Mafic cumulates remain at depth or detach and sink into the mantle. From Ashwal (1993). Copyright © with permission from Springer-Verlag.

Phinney et al. (1988), and others. It proposes that anorthosite genesis begins with a high-Al basaltic magma created by partial melting of a depleted mantle source (although enriched mantle may contribute to some occurrences). Although not specified in Ashwal's model, such localized mantle melting suggests a *plume* origin, and the enormous size of massif anorthosites indicates a large igneous province (LIP), generally attributed to the surfacing of a newly-initiated plume head, as described in Chapter 14. As the plume rises and begins to melt in the spinel- or plagioclase-peridotite stability field, the aluminous magma so generated rises through the mantle but is denser than the crust, so it ponds at the base of the continental crust as a liquid underplate (Figure 20.3a). Here olivine and Al-rich pyroxenes crystallize and sink,

accumulating at the bottom of the chamber. The heat released by the crystallization induces partial melting of the crust at the chamber roof (Figure 20.3b). This situation is identical to the underplates and MASH processes proposed for the generation of magmas at active continental margins in Chapter 17, but was probably more extensive due to the high Proterozoic heat flow. Assimilation or partial melting of plagioclase-rich mafic lower crust is considered an important prerequisite for creating liquids capable of crystallizing large quantities of plagioclase. It is presently contentious whether the magma at this stage is dominated by a mantle-derived melt with significant crustal assimilation or an extensive crustal melt heated by a large crystallizing subcrustal underplate.

Either way, some combination of crystal fractionation, partial melting, and assimilation at or near the base of the continental crust (30 to 35 km, as indicated by the orthopyroxene megacrysts) causes the evolved melt to increase in Al_2O_3, Fe/Mg, and LREE until the liquid reaches the plagioclase cotectic and andesine also crystallizes. The residual melt is now approximately an Fe-rich high-Al tholeiite. Large compositionally homogeneous plagioclase crystals grow due to slow cooling and perhaps recharge of more primitive magma into the chamber. Recharge permits extensive crystallization of plagioclase while still maintaining a high concentration of plagioclase components in the liquid. Plagioclase is buoyant in basaltic magmas (especially in dense Fe-rich ones) at these depths. Its buoyancy decreases at shallow depths because the liquid expands more than the plagioclase as pressure is reduced. Plagioclase crystals float and accumulate at the top of the chamber (Figure 20.3c).

A low-density, plagioclase-rich crystal mush gradually builds up at the chamber top. The mush becomes less dense than lower continental crust, at which point it rises to shallow crustal levels as a series of plagioclase–liquid mush diapirs, coalescing there to form thick sheet-like composite anorthosite intrusions (Figure 20.3d and e). Further accumulation of plagioclase and adcumulus liquid expulsion (probably due to compaction) may occur in the shallow chambers. The ultramafic cumulates are left at the base of the crust, where they either remain or delaminate and sink back into the mantle. This would explain why they are not detected in the crust near the anorthosite massifs.

If we consider the Fo-An-Qtz system (Figure 7.4), a decrease in pressure causes the orthopyroxene-anorthite cotectic curve to shift away from the An apex. As a result, the composition of a high-pressure cotectic melt becomes located in the An field as the cotectic shifts away. In order to return to the shallow equilibrium cotectic, a lot of plagioclase must form. This effect may contribute to the unusually high plagioclase content of the bodies.

Emslie et al. (1994) extended the anorthosite model to include the entire AMCG suite for the Nain plutonic suite in Labrador. The Nain granites are typically K rich and orthopyroxene bearing, ranging from monzonites (mangerites) to quartz monzonites to granites/charnockites, and have Nd and Sr isotopic ratios that correspond much more closely to crustal gneisses than to the anorthosites, indicating that they are crustal melts (generated by heat from the mafic under-plates described above). These partial melts leave behind hot plagioclase–pyroxene granulite residues, which may readily be assimilated by the parental basaltic liquids residing at the crust/mantle boundary. The subordinate ferrodiorites have isotopic ratios and REE patterns suggesting equilibrium with the anorthosites and an origin as dense intercumulus residual liquids, expelled after plagioclase, olivine, and pyroxene re-moval. Their high density would cause them to collect at lower levels, explaining why they represent a minor compo-nent at the exposed structural level of the anorthosites.

Why was the process illustrated in Figure 20.3 re-stricted to the Proterozoic? As discussed in Section 18.7, the Archean was dominated by small arcs and microcontinents that finally amalgamated to produce large continents in the Proterozoic. Archean plumes (assuming they existed) there-fore probably rose into oceanic basins or island-arc-like crustal fragments, where they cooled relatively rapidly, forming smaller Archean mafic intrusions and anorthosites. Perhaps the thermal blanketing effect of the large new conti-nents, accompanied by relatively high Precambrian heat flow, warmed the sub-continental mantle (Hoffman, 1989). Convective removal of the thermal boundary layer (Figure 18.8) may also result in increased anorogenic magmatism, producing anorthosites and the Proterozoic anorogenic gran-itoids. Thus large massif-type anorthosites could not be pro-duced earlier because the continents were not extensive enough to provide the necessary insulation, and they could not be produced later due to secular cooling of the mantle.

20.2.4 Lunar Anorthosites

Samples returned by the *Apollo 11* landing on the moon in-cluded some brecciated anorthosites. The landing area was on *Mare Tranquilitatis*, and the maria are known to be basaltic.

The surprising presence of fragments of lunar anorthosites led to the suggestion that the samples represented pieces of the lunar highlands that were ejected by highland cratering mete-orite impact events onto the mare surface. Prior to the collec-tion of these samples, the highlands were considered to be comprised of either primitive chondritic material (by those who favored a cold lunar accretion) or granitic material (by those who thought the moon was more Earth-like). Nobody thought that the highlands were anorthositic.

The composition of lunar anorthosite plagioclase is *very* anorthitic (An_{94-99}). The anorthosites contain abundant Si, Ca, and Al, with some Na and Fe, but little else. The low Na and K contents may reflect an early loss of alkalis in the moon. The anorthosites are also *very* old: 4.4 Ga.

There are two principal theories regarding the origin of lunar anorthosites. The favored theory is that they formed by crystallization and flotation of plagioclase from a moon-encircling magma layer several hundred kilometers thick. The layer formed early as a melting response to ac-cretion and gravitational collapse of the moon (hence the age). If so, as the ages suggest, the anorthosites spelled the end of the cold accretion model. Several investigators ex-pressed a concern that a suitable heat source for such mas-sive melting was not available in the primordial moon. They proposed an alternative model: that the highlands were created over a longer time span by intermittent or se-rial magmatism. For a review of lunar anorthosites, see Ashwal (1993).

Our survey of igneous petrogenesis is now at an end. My apologies for omissions, but it is impossible to cover every rock type or theory. Classes may find it difficult to cover several of the chapters as it is, but I hope material left uncovered will at least make a good reference if you decide to explore more later. Now we turn to Part II: Metamorphic Petrology.

Summary

Anorthosites are large (occasionally enormous) plutonic bodies of nearly pure plagioclase. They are thus as felsic as any granite, but their mineralogy (plagioclase + pyroxene ± olivine) conforms more to mafic rocks. The two classic types are Archean and Proterozoic (massif-type) anortho-sites. Equivalent liquids would be prohibitively hot and dif-ficult to produce, which, along with the typical cumulate texture, indicates that anorthosites are the products of pla-gioclase accumulation from a mafic liquid parent. Archean anorthosites typically form as sheet-like, internally layered sills associated with mafic rocks in high-grade gneiss ter-ranes. Most have large rounded plagioclase megacrysts (An_{80-95}) in a mafic matrix. The parent is apparently a Fe-Al-Ca-rich basaltic liquid that has evolved from a more primitive melt. They were probably emplaced in Archean oceanic crust or volcanic arc environments.

Proterozoic, or massif-type, anorthosites are larger and less sill-like compared to their Archean counterparts. Plagioclase crystals are more tabular and are less anorthitic,

and there is less associated mafic matrix or cumulates. They are typically intruded into stable (anorogenic) high-grade continental gneiss terranes or into areas undergoing post-orogenic rifting and collapse. They are generally associated with granites (mangerites and charnockites), collectively called AMCG complexes, plus subordinate ferrodiorites.

The favored model for (terrestrial) anorthosite pet-rogenesis involves a mantle plume (head?) that induces peridotite melting in the spinel– or plagioclase–herzolite stability field. The resulting aluminous basaltic liquid rises and ponds at the base of the crust. The heated crust melts to produce granitic liquids, leaving a hot plagioclase–pyroxene–rich granulite residue that is readily assimi-lated by the basaltic liquids. Whether mantle or crustal melts predominate at this stage is debated. Crystal frac-tionation produces olivine and pyroxene, which sink, and plagioclase, which floats. The upper plagioclase-rich crystal–liquid mush rises in several pulses to shallower levels, and the dense Fe-rich interstitial liquid is expelled downward, leaving adcumulus masses of anorthosite. Anorthosites also constitute the lunar highlands, where they rose and crystallized either from a thick magma layer that encircled the entire moon or did so from a series of intermittent melting events.

Key Terms

Massif-type anorthosite *436*   Supracrustal rocks *437*   AMCG complex *439*

Review Questions

Review Questions are located on the author's web page at the following address: **http://www.prenhall.com/winter**

Suggested Further Readings

Ashwal, L. D. (1993). *Anorthosites*. Springer-Verlag. Berlin.

Ashwal, L. D., and J. S. Myers (1994). Archean anorthosites. In K. C. Condie (ed.). *Archean Crustal Evolution*. Elsevier. Amsterdam. pp. 315–355.

Emslie, R. F. (1985). Proterozoic anorthosite massifs. In: *The Deep Proterozoic Crust in the North Atlantic Provinces* (eds. A. C. Tobi and J. L. R. Touret). D. Reidel. Dordrecht. pp. 39–60.

Wiebe, R. A. (1992). Proterozoic anorthosite complexes. In: *Proterozoic Crustal Evolution* (ed. K. C. Condie). Elsevier. Amsterdam. pp. 215–261.

Windley, B. F. (1995). *The Evolving Continents*. John Wiley & Sons. New York.

PART

II

Metamorphic Petrology

21

An Introduction to Metamorphism

Questions to be Considered in this Chapter:

1. What is metamorphism, and what changes may accompany it?

2. What are the defining limits of metamorphism, and how is metamorphism distinguished from diagenetic and igneous processes?

3. What are the principal controlling agents of metamorphic change? What affect might changes in their relative magnitudes have on the style of metamorphism and the resulting rocks?

4. What types of strain may metamorphic rocks manifest, and how can strain affect the structures and textures of metamorphic rocks?

5. What are the principal types of metamorphism, in what field settings does each occur, and what rock types and textures are produced in each?

6. How might we recognize the precursor rocks (protolith) to metamorphism, and what is a reasonable scheme for categorizing them?

7. What metamorphic field gradients occur in metamorphic terranes, and how can we map and interpret the mineral changes that reflect those temperature, pressure, and X_{fluid} gradients?

This chapter serves as a general survey of metamorphism, introducing aspects of metamorphic classification, processes, rock types produced, and field settings. Most of the concepts introduced here will be dealt with in greater detail in ensuing chapters.

For a moment, let's return to the concept of a chemical system developed in Chapter 5. Any natural chemical system (such as a rock) at equilibrium will manifest itself as a particular assemblage of coexisting phases in accordance with the concepts of thermodynamic equilibrium and the phase rule. A basaltic melt is thus stable as a liquid within some range of temperature and pressure. If the melt rises toward the surface and is cooled to temperatures below this *P-T* range, it will eventually crystallize to an aggregate of plagioclase, pyroxene, and perhaps olivine, ilmenite, and some accessory minerals (a basalt or gabbro). As the physical conditions changed from those at which the initial melt was stable, the system thus responded to the new set of conditions by transforming to a different form (an assemblage of minerals). Between these two forms, the system maintained equilibrium with the changing conditions by progressively transforming itself to the final mineral assemblage by crystallizing along some liquid line of descent, as described in Chapters 6 and 7.

In our dynamic Earth, the mafic rock generated above may be created at or brought to the surface by uplift and erosion. Basalts or gabbros at the Earth's surface are then weathered and broken down to become a sediment or soil. This weathering process is another example of a chemical system being exposed to a set of physical conditions different than those at which it formed (it crystallized at high temperatures). The feldspars and mafic minerals thus become unstable and react with surface- or groundwater to become clays, oxides, etc.

In general, when rocks or melts are transported or exposed to conditions unlike those under which they initially formed, they react in response to those new conditions (unless kinetic factors preclude it). The relatively simple examples above occur at temperatures more than 1000°C apart. The broad range of conditions between these extremes (representing igneous crystallization and surficial weathering) is the realm of *metamorphism*. If our basalt had been exposed to some intermediate conditions of pressure and temperature for a sufficient time it might have equilibrated to a different assemblage of phases that reflect equilibration to those conditions.

Just as an igneous rock may become exposed to lower temperatures and pressures, a sediment, such as a shale, may become buried and heated at temperatures and pressures *higher than* those at which it initially formed. The clays and fine pieces of quartz may become heated to the point that they recrystallize to become coarser grains, and new minerals may also form. The adjustments of the basalt or shale to the new physical conditions fall under the area of study known as metamorphic petrology.

The term *metamorphism* comes from the Greek μετα μορφη (*meta morph*), meaning "change of form." In petrology, metamorphism refers to changes in a rock's mineralogy, texture, and/or composition that occur predominantly in the solid state under conditions between those of diagenesis and large-scale melting. Given that weathering and diagenesis occur only in the thin uppermost veneer of sediments and that melting is an exceptional process with respect to normal geotherms (Chapter 10), we may expect metamorphism to be the dominant process taking place throughout most of the Earth's crust and mantle. As a historical note, the Scottish physician and farmer James Hutton (1726–1797) of Edinburgh was the first to propose that some crystalline rocks were originally sedimentary in nature and had been subsequently transformed by subterranean heat. British geologist Charles Lyell (1797–1855) elucidated Hutton's ideas in the first edition of his *Principles of Geology* (1833) and proposed the term *metamorphic* for the altered strata.

Similar to the mandate for igneous rocks discussed in Chapter 2, the International Union of Geological Sciences (IUGS) in 1985 formed the Subcommission on the Systematics of Metamorphic Rocks (SCMR), with the objective of providing a scheme for naming and describing metamorphic rocks. It also attempted to organize much of the vocabulary concerning metamorphic processes. The SCMR recommendations were recently published (Fettes and Desmons, 2007), and I shall attempt to adhere to their recommendations in the chapters that follow.

21.1 THE LIMITS OF METAMORPHISM

At first, it may seem pretty simple and straightforward to constrain the limits of metamorphism. We understand the processes of weathering, cementation, and diagenesis as *sedimentary* processes that occur only near the Earth's surface. **Weathering** is the alteration of rocks at or near the Earth's surface by atmospheric agents. It involves the physical disintegration or chemical decomposition of rocks, and produces a mantle of loose waste or soil. **Diagenesis** comprises all of the chemical, physical, and biological changes that a sediment undergoes during and after lithification in the near-surface environment. We also know a melt when we see one. On closer inspection, however, the limits of metamorphism are somewhat arbitrary. At the low end, the processes involved in weathering and diagenesis are largely the same as those that occur in metamorphism. In each, solid and (usually) fluid phases plus dissolved constituents react to produce recrystallized or different solids and altered fluids. To a metamorphic petrologist the simplest distinction of metamorphism is to limit her or his attention only to those products that are not produced in the zones of weathering or diagenesis.

Regrettably, the products and zones are far from distinct. A variety of minerals, notably zeolites, are generated in very-low-temperature, diagenetic environments as well as in rocks that are clearly recrystallized and metamorphic by any reasonable standard. Even such characteristically high-temperature minerals as alkali feldspar, and pyroxene, can be created in diagenetic environments if the composition of the fluid is appropriate. In addition, the temperature at which recrystallization or new mineral formation takes place depends strongly on the initial material (called **protolith**). Glass, volcanic ash, organic matter, or evaporites may alter at much lower temperatures than silicate or carbonate minerals.

Because mineral tranformations may occur at practically any point following deposition and burial, we must decide on some type of standard, however arbitrary, to distinguish metamorphism from diagenesis. There is a general consensus that metamorphism begins in the range of 100 to 150°C for the more unstable types of protolith, and may be marked by the formation of minerals such as laumontite, analcime, heulandite, carpholite, paragonite, prehnite, pumpellyite, lawsonite, glaucophane, or stilpnomelane. Some zeolites have been considered diagenetic and others metamorphic. Perhaps you can now see why the distinction is somewhat arbitrary. For further details on the problems of distinguishing diagenesis and low-grade metamorphism, see Frey and Kisch (1989) or Chapter 2.5 of the IUGS/SCMR recommendations (Fettes and Desmons, 2007).

At the high-temperature end, we encounter similar problems in distinguishing metamorphic and igneous processes. Although we may all recognize a melt, we may not be so adept at recognizing the solid products crystallized from one. During my studies in high-grade metamorphic terranes in Greenland and Labrador, I often encountered small, elongate, fairly coarse-grained segregations of granitoid material in the gneisses. Whether these were thin pockets of locally derived melt, precipitates from fluids, or fluid-enhanced recrystallization along fluid-filled fractures was a problem for us to distinguish, particularly when the contacts were in many cases gradational into the surrounding gneisses. At very high temperatures and pressures the distinction between the precipitated products of a silicate-saturated aqueous fluid and a fluid-saturated silicate melt may not always be as clear as we might like. In addition, you learned in Part I that crystallization and melting processes take place over a considerable range of temperature. For granitoid rocks this melting range may begin as low as 600°C but depends on the H_2O content of the system (Figure 18.5). Over the melting range, solids and liquids coexist. If we heat a metamorphic rock until it melts, at what point in the melting process does it become igneous? We can agree that the melted portion is igneous, but the solid portion ought to be considered metamorphic. Xenoliths, restites, and other enclaves (see Table 18.1) are considered part of the igneous realm only because the melt that accompanies them is so dominant, but we can imagine a gradation in melting between initial melt segregations and final scattered restites. Any distinction between igneous and metamorphic realms near the transition becomes vague and disputable. There is a common rock, known as **migmatite** ("mixed rock"), that occurs where high-grade metamorphics grade into igneous crustal melts. We shall take a closer look at migmatites and partial melts in Section 28.5.

Perhaps we should acquiesce and leave both the high and low boundaries of metamorphism as vague as they are in nature. I see no reason why metamorphic petrologists cannot work in conjunction with those who study diagenesis or with igneous petrologists where their areas of expertise overlap.

The **pressure** limits of metamorphism are also fairly broad. At low pressures, an abnormally high geothermal gradient may be required to heat rocks sufficiently to initiate metamorphism. Substantially metamorphosed rocks can be generated near the contact of shallow intrusions very near the Earth's surface. At the high-pressure end, solid rocks extend through the mantle and occur again in the solid inner core. Due to their high density and the distances involved, few, if any, of the truly deep rocks make it back to the surface, and to consider them metamorphic is more of an academic question because we shall never see them. Mantle xenoliths from kimberlites (Section 19.4) record pressures up to 4 GPa (>120 km) or more and are generally regarded as metamorphic (yet are generally given igneous names, such as lherzolite, presumably to best convey their mineralogy). Extensive mantle samples are exposed in ophiolites (Chapter 13) and comprise the uppermost mantle beneath thin oceanic crust (initially perhaps 10 to 20 km deep). Nonetheless, the vast majority of metamorphic rocks that we see at the surface and study are *crustal* rocks. The practical limits of pressure for the study of metamorphics thus rarely exceeds 3 GPa for continental crust. In Section 28.6, we shall see some recently discovered spectacular examples of continental rocks that have returned from depths of at least 90 km, but these seem to be very rare. Oceanic crust, however, is more readily subducted to great depth and samples of eclogite are delivered back to the surface in many localities.

With the above reservations in mind, the IUGS/SCMR proposed the following definition of metamorphism (Fettes and Desmons, 2007):

> Metamorphism: *a process involving changes in the mineral content/composition and/or microstructure of a rock, dominantly in the solid state. This process is mainly due to an adjustment of the rock to physical conditions that differ from those under which the rock originally formed and that also differ from the physical conditions normally occurring at the surface of the Earth and in the zone of diagenesis. The process may coexist with partial melting and may also involve changes in the bulk chemical composition of the rock.*

Other conventional boundaries of metamorphic petrology are that it does not generally include the study of coal, petroleum, or ore deposits. These fields are left to specialists even though the processes involved in their evolution are typically of a metamorphic nature.

21.2 METAMORPHIC AGENTS AND CHANGES

Because metamorphism takes place when a rock is exposed to a physical or chemical environment that is significantly different from that in which it initially formed, any parameter that can effect such a change in environment is one that could cause metamorphism. These include temperature, pressure, the nature of the fluid phase, and the state of stress. The reacting phases in a rock are solids (minerals and amorphous solids such as glass, organics, etc.) and commonly a pore fluid, including the dissolved material in that fluid.

21.2.1 Temperature

Changes in **temperature** are probably the most common cause of metamorphism. Typical oceanic and continental geotherms are shown in Figure 1.11. Note that in the 10- to 40-km depth range of the typical mid- to lower continental crust the continental geotherm is significantly lower than the oceanic one (due largely to the thicker continental lithosphere, which stalls convective mantle rise at a deeper level, as illustrated in Figure 1.9). Increasing temperature has several effects on sedimentary or volcanic rocks.

First of all, increasing temperature promotes recrystallization, which generally results in increased grain size. This effect is particularly true for fine-grained rocks, especially in a static environment because shear stresses typically act to reduce grain size (as discussed in Section 23.1). In Chapter 3 we discussed the excess energy (instability) of grain surfaces that lack a fully surrounding crystal lattice to maintain electrostatic site stability. In very small mineral grains, as the surface/volume ratio gets quite large, the less stable near-surface zone represents a large proportion of the crystal, significantly lowering the overall stability. Clays, tuffs, fine-grained clastic sediments, and some chemical precipitates are composed of very small grains. Increasing temperature will eventually overcome kinetic barriers to recrystallization, and these fine aggregates tend to coalesce into larger grains. Theoretically a single huge grain of each mineral present is the most stable configuration for a rock, but there are limits to the extent to which constituents can migrate by diffusion to growing grains. Even in monomineralic rocks, however, such as quartzites or carbonates, the size of grains tends to increase with temperature, suggesting that there are also limits to the extent to which atoms in the lattices of neighboring grains are able to rearrange themselves as grains coalesce and grow. The process of grain coalescence and growth is essentially that of **annealing**, observed in metals and ceramics when maintained at high temperatures (Section 23.1).

Second, rocks being heated may eventually reach a temperature at which a particular mineral is no longer stable or a group of minerals is no longer stable together. When the physical conditions are outside the stability range of some mineral assemblage, a **reaction** will take place that consumes unstable mineral(s) and produces new minerals that are stable under the newly achieved conditions (as long as there are no kinetic factors preventing it). An analogous situation in igneous petrology is that of a melt being replaced by some assemblage of minerals as it cools below the stability range of the melt.

To summarize, if we were to walk across a metamorphosed area (now uplifted, eroded, and exposed) such that we traversed in the direction of increasing metamorphic temperature, we should expect to observe two principal changes that we could attribute largely to the temperature change. First, the average grain size of the rocks generally increases. Second, we should observe a succession of mineral types that reflect the temperature–pressure conditions that existed during the metamorphic event at any point along our traverse.

A number of different types of reactions involving minerals may occur with increasing temperature (as will be discussed in Chapter 26). Among the most common are **devolatilization** reactions (usually **dehydration** or **decarbonation** reactions). As a general rule, volatile-bearing minerals (such as hydrous minerals and carbonates) tend to lose their volatiles as temperature rises. And the more volatiles a mineral contains, the more susceptible it is to thermal decomposition. Very hydrous minerals, such as clay minerals, zeolites, chlorite, or serpentine, thus characterize diagenesis or low grades of metamorphism, and they are typically the first to dehydrate as temperature rises. All volatile-bearing minerals have an upper temperature stability limit, and very high grades of metamorphism are typically characterized by a volatile-free mineral assemblage. That is not to say, however, that only volatile-bearing minerals become unstable with increasing temperature. Many volatile-free minerals may also succumb to increased metamorphism. Many will also become hydrated or carbonated at lower metamorphic grades.

Crystallization includes the formation of new minerals and the **recrystallization** of existing ones. Some authors use the term **neocrystallization** to clearly distinguish the crystallization of new minerals.

A third effect of increased temperature is that it overcomes kinetic barriers that might otherwise preclude the attainment of equilibrium. At low temperatures (in the areas of diagenesis and very low-grade metamorphism), disequilibrium may thus be common, and we may find metastable materials or associations of minerals that would otherwise be unstable together. At higher temperatures, however, reaction and diffusion rates increase to the extent that equilibrium is much more likely.

21.2.2 Pressure

Rocks are generally metamorphosed at depth within the Earth where temperatures are high. This cannot happen, of course, unless **pressure** increases also. Remember from Section 1.7.1 that the pressure increase with depth is due to the weight of overlying rocks, and is called **lithostatic pressure** (also called **confining pressure**). The relationship between depth and temperature is the **geothermal gradient** (Section 1.7.2). Although the geothermal gradients in Figure 1.11 are probably good estimates of long-term, or steady-state, gradients, these can be perturbed in a number of ways. We discovered in Chapter 16, for example, that the geothermal gradient in subduction zones is very low (Figure 16.15). Higher-than-average gradients may be caused by igneous intrusions, plumes/hotspots, crustal extension, lithospheric mantle delamination, ridge subduction, etc. Most of these perturbations are transient phenomena, and an area will eventually return to the steady-state values when the disturbance (subduction, upwelling, intrusion, rifting, etc.) ceases and its thermal effects slowly dissipate. Metamorphism is typically associated with some of these disturbing events, however, and may therefore retain an enduring record of such temporary variations in the geothermal gradient.

Figure 21.1 illustrates several estimates of metamorphic temperature–pressure relationships from ancient orogenic belts. These estimates are based on *P-T* estimates for rocks exposed at the surface in these areas along a traverse from lowest to highest metamorphic conditions. Such traverses typically move toward increasingly uplifted areas of previously more deeply buried portions of a metamorphic belt at the time of metamorphism, but rarely represent originally *vertical* sections (except in fortuitous cases) and hence are not equivalent to true geothermal gradients (temperature versus depth). Nonetheless, because pressure correlates with depth, the variations in these **metamorphic field gradients** (also called "metamorphic trajectories" or "metamorphic arrays") are probably a reasonable reflection of similar variations in true geothermal gradients at the time of metamorphism. But the geothermal gradient can be expected to vary from place to place across an orogen, generally increasing toward the central axis where heat and plutonism are concentrated. Because postmetamorphic uplift and erosion vary as well, metamorphic field gradients generally span the gradual transition from low *T/P* gradients at the margins of an exposed orogen to higher *T/P* gradients toward the core. Figure 21.1 indicates that metamorphism can accompany a wide range of *P/T* gradients.

Metamorphic grade is a convenient term that is commonly used to express the general increase in degree of metamorphism without specifying the exact relationship between temperature and pressure. The term should concentrate on temperature, but not constrain pressure. We may thus refer to "high-grade" rocks or "low-grade" rocks from any area, such as those depicted in Figure 21.1.

There are pressure limits to the stability of minerals and mineral associations, just as there are temperature limits. Rocks experiencing changes in metamorphic grade along a high-pressure *P-T* path, such as the Franciscan path in Figure 21.1, can thus be expected to have different

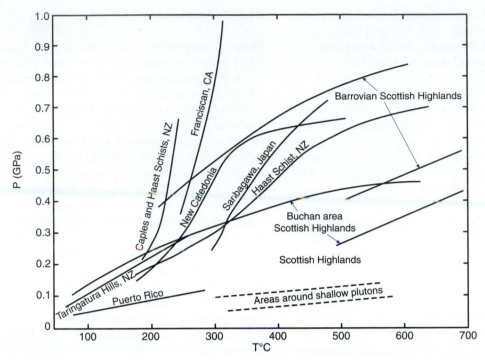

FIGURE 21.1 Metamorphic field gradients (estimated *P-T* conditions along surface traverses directly up metamorphic grade) for several metamorphic areas. After Turner (1981).

metamorphic mineral assemblages than rocks that follow a low-pressure *P-T* path, similar to the Scottish ones. If we accept the idea that temperature is the most important metamorphic agent in most cases, a somewhat artificial, yet useful, approach may be to consider pressure as a modifier, in the sense that temperature can increase along any number of pressure-varied paths. Along some of these paths pressure may be low, favoring the formation of low-density metamorphic minerals as temperature rises. Alternatively, pressure may be high, and dense minerals tend to occur instead.

Lithostatic pressure is generally considered to be equal in all directions (**hydrostatic**), similar to pressure in deep water. We assume this to be the case for many metamorphic environments. If not, and the pressure in one direction were significantly greater than in another direction, the rock would yield until the motion offset the pressure difference. Such deformation occurs when the pressure differential exceeds a material's strength, which we can expect to occur beneath a relatively shallow zone of low pressure with relatively cool and brittle rocks. Rocks under lithostatic conditions, regardless of the pressure, will not change *shape* (i.e., deform). Pressure may cause a volume loss, but it will be uniform in all directions (just as the exerted pressure). This volume loss is facilitated by the formation of low-volume (high-density) minerals, which is why high-pressure metamorphism favors dense minerals, as mentioned above.

21.2.3 Deviatoric Stress

Only when the pressure is unequal in various directions will a rock be deformed. Unequal pressure is usually called **deviatoric stress** (whereas lithostatic pressure is **uniform**

stress). We can envision deviatoric stress as being resolvable into three mutually perpendicular stress (σ) components: σ_1 is the maximum principal stress, σ_2 is an intermediate principal stress, and σ_3 is the minimum principal stress. In hydrostatic situations, all three are equal. Deviatoric stress may be maintained as long as the application of the differential continues to be applied and keeps pace with any tendency of the rock to yield. This occurs most commonly in orogenic belts, extending rifts, or in shear zones (i.e., generally at or near plate boundaries). The yielding of the rock is **deformation**, or **strain**. *Stress*, then, is an applied force acting on a rock (over a particular cross-sectional area), and *strain* is the response of the rock to an applied stress. A recent advertising campaign mistakenly recommended that you use their pain reliever to relieve *stress*. Pain relievers, however, can only relieve *strain* (the response to stress). To relieve stress, you probably need a vacation, graduation, or a different job.

Deviatoric stress affects the textures and structures in rocks but not the equilibrium mineral assemblage. The addition of some strain energy in deformed rocks may also provide the impetus to overcome kinetic barriers to reactions that would otherwise occur had not the barriers been effective. Deformation may thus have a catalytic effect and eliminate metastable mineral associations in favor of stable ones. Deformation cannot, however, change the nature of the stable state itself. An early school of thought proposed that some minerals were stabilized or favored in the presence of deviatoric stress. All of these so-called **stress minerals** (Harker, 1932) have since been experimentally synthesized under lithostatic conditions, and the idea has lost favor.

Deviatoric stresses can be lumped into three principal conceptual types: tension, compression, and shear. In **tension**

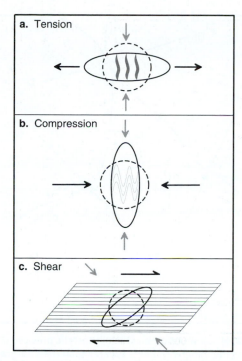

FIGURE 21.2 The three main types of deviatoric stress with an example of possible resulting structures. **(a)** Tension, in which one stress in negative. "Tension fractures" may open normal to the extension direction and become filled with mineral precipitates. **(b)** Compression, causing flattening or folding. **(c)** Shear, causing slip along parallel planes and rotation. The dashed circle in each figure represents the outline of an initial circle (a sphere in 3D), whereas the solid ellipse represents the deformed result (a "strain ellipsoid"). The light vertical arrows indicate the lithostatic (confining) component of pressure.

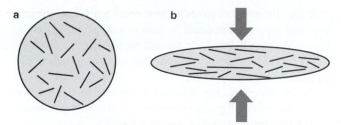

FIGURE 21.3 Flattening of a ductile homogeneous sphere **(a)** containing randomly oriented flat disks or flakes. In **(b)**, the matrix flows with progressive flattening, and the flakes are rotated toward parallelism normal to the predominant stress.

(Figure 21.2a), σ_3 is negative, and the resulting *strain* is *extension*, or pulling apart. Tension can occur only at shallow depths, and the response is largely brittle faulting. A common result is the development of **tension fractures**, which generally fill with fluids that precipitate minerals in the extending fractures.

In **compression** (Figure 21.2b) one stress direction (σ_1) is dominant, which may cause folding or a more homogeneous deformation called **flattening** (similar to stepping on a rubber ball). Existing minerals with a platy or elongated shape may be rotated during either folding or flattening. For example, imagine a spherical lump of cookie dough with a number of randomly oriented flat disks (imagine sequins) in it (Figure 21.3a). If you place your hand on the lump and flatten it against a table top (you supply a vertical σ_1), the dough will flow to a pancake-like ellipsoid (a **strain ellipsoid**), whereas the rigid disks will be physically rotated toward parallelism with the table top in the ductile medium (Figure 21.3b). When we sample a rock that has been similarly deformed, we can rarely see directly the change in overall shape because we usually have no idea what the original rock shape was, but we can easily see the parallel alignment of minerals such as micas with platy habits. Alternatively, if new metamorphic micas *grow* during the compression, they will tend to grow perpendicular to σ_1. This way the maximum directions of mineral growth do not

extend against the maximum compressive stress. In either case, rotation or growth, the platy minerals tend to become aligned normal to the principal compression direction. The general term for a planar texture or structure is called **foliation**. The term has no genetic implications and may include sedimentary bedding or igneous layering, etc. Metamorphic foliations, including **cleavage**, **schistosity**, and **gneissose structure**, shall be described more fully in the next two chapters.

If the dough ball that we flatten contains elongated elements, such as toothpicks, they too will rotate so that their elongation is in the plane normal to the maximum compression direction. *Within* that plane, however, there is no preference (because $\sigma_2 = \sigma_3$), so they may be randomly oriented in it (just as if you had dropped a box of toothpicks on the floor). A foliation can thus be defined by an array of linear elements.

If you compress the dough ball vertically at the same time that a friend compresses it the same amount laterally, then $\sigma_1 = \sigma_2 > \sigma_3$ and the resulting shape, or strain ellipsoid, looks like a salami. Any elongated minerals, such as amphiboles, in a rock experiencing this type of deformation will either rotate or grow so that their maximum elongation is parallel to the longest axis of the deformed ellipsoid. **Lineation** is the non-genetic term that refers to such a parallel alignment of elongated features. A pure lineation without a foliation will occur if an object is stretched and elongated, as described above with $\sigma_1 = \sigma_2 > \sigma_3$. If you compress the dough ball vertically, and your friend compresses it horizontally again, but less so than you, then $\sigma_1 > \sigma_2 > \sigma_3$, and the dough will deform such that one horizontal direction stretches out more than the other into a shape rather like an ellipsoidal bar of soap. In this case, if the rock has both platy and elongated minerals, it may exhibit *both* foliation *and* lineation simultaneously. The interpretation of these textural elements and their relationship to metamorphic mineral growth supply valuable information concerning the thermal and deformational evolution of orogenic belts, and will be discussed more fully in Chapter 23.

Shear (Figure 21.2c) is an alternative response to compression in which motion occurs along a set of planes at an angle to σ_1, like pushing the top of a deck of cards. The strain ellipsoid resulting from shear may be identical to that resulting from flattening. Structural geologists refer to flattening as "pure shear" and shear as defined by Figure 21.2c as "simple shear." Distinguishing between the two in a deformed rock after the event requires some careful analysis of the textures and is covered in detail in most structural geology texts. This study goes beyond our present purposes, but as an indication of the technique, compare the relationships between the strain ellipsoid and the foliation in Figures 21.2c and 21.3b. In flattening, the foliation is normal to the shortest axis of the strain ellipsoid, whereas it is at a non-orthogonal angle to the shortest axis in (simple) shear.

21.2.4 Metamorphic Fluids

Petrologists agree that most metamorphic rocks, at least during metamorphism, contain an intergranular fluid phase. We use the term **fluid** to avoid specifying the exact physical nature of the phase. At low pressures, the fluid is either a liquid or a gas, but at pressures and temperatures above the **critical point** of water there is no difference between liquid and gas. The critical point of pure H_2O is at 374°C and 21.8 MPa (Figure 6.7) but extends to higher temperatures and pressures when the water contains dissolved electrolytes or other fluid species. Under conditions beyond the critical point (realized in most metamorphic regions) the non-solid phase is called a **supercritical fluid**. Direct evidence for such a phase is difficult to obtain because the fluids escape as the rocks are uplifted and exposed by erosion. By the time we examine the rocks they are essentially dry. Some direct evidence comes from fluid inclusions (Figure 7.17). Some of these inclusions cluster in planar arrays, suggesting that they are late, post-metamorphic ("secondary") fluids that penetrated along fractures and got trapped by annealing of the crack. Physical and chemical evidence in other inclusions, however, suggests that they formed *during* a metamorphic event ("primary fluids") and were in equilibrium with the metamorphic mineral assemblage before being trapped by mineral growth (Roedder, 1972; Touret, 1977). Other evidence for metamorphic intergranular fluids comes from theoretical considerations, such as the need for some H_2O or CO_2 pressure in order to stabilize observed hydrous and/or carbonate minerals in metamorphic rocks at the temperatures of metamorphism. Without such a fluid, these minerals would quickly devolatilize and disappear. Of course, some devolatilization does take place, but the fluids liberated by the dehydration or decarbonation reactions contribute to the metamorphic intergranular fluids until equilibrium is attained. As we shall see in future discussions, the volatile-bearing minerals in many rocks, and the reactions that involve them, occur at metamorphic grades that require the existence of a fluid in equilibrium with the solid phases.

In shallow porous rocks the fluid forms a continuous network extending to the Earth's surface. The lithostatic pressure is exerted by the weight of the overlying minerals in mutual contact, and is equal to $\rho_{minerals}gh$ [see Equation (1.1)]. The intergranular fluid (essentially groundwater in these shallow systems), on the other hand, is independently open to the surface, so the hydrostatic **fluid pressure** at the same depth is $\rho_{water}gh$. Because water is less dense that the minerals $P_{fluid} < P_{lith}$. In some cases, an impervious sedimentary cap may exert an additional pressure on the fluid, but P_{fluid} will

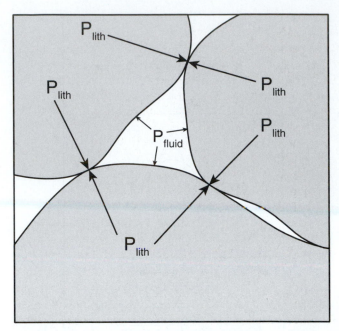

FIGURE 21.4   A situation in which lithostatic pressure (P_{lith}) exerted by the mineral grains is greater than the intergranular fluid pressure (P_{fluid}). At a depth around 10 km (or T around 300°C), minerals begin to yield or dissolve at the contact points and shift toward or precipitate in the fluid-filled areas, allowing the rock to compress. The decreased volume of the pore spaces will raise P_{fluid} until it equals P_{lith}.

still be less than P_{lith}. At depths greater than about 10 km, the minerals cannot maintain an independently supporting network. At the points of mineral contact (Figure 21.4), the stress may be much higher than in the pores. As a result, the mineral strength may not be sufficient to maintain the differential stress, and the mineral grains will yield, compressing the fluid-filled pore space until $P_{fluid} = P_{lith}$. Another common process in such situations is **pressure solution**. In this case, the free energy of the mineral at the stressed mutual contacts is higher than it is adjacent to the pore spaces. The overall free energy of the system can thus be lowered by dissolving the mineral at the contacts and re-precipitating it in the pores (Figure 23.1). This also reduces the volume of the pore spaces and raises P_{fluid} until it reaches P_{lith}. Whether deformation or pressure solution dominates depends on the minerals, depth, temperature, fluids, etc. The $P_{fluid} = P_{lith}$ condition is probably maintained as metamorphism proceeds, but once the peak of a metamorphic event is reached and temperature drops, fluids that have not escaped may be quickly reabsorbed into previously dehydrated minerals. Yardley and Valley (1997) argued that P_{fluid} in stable continental lower crust (where most rocks are past any orogenic peak of metamorphism) is probably one to three orders of magnitude less than P_{lith}.

Intergranular metamorphic fluids are usually dominated by H_2O, but CO_2 may also be present in some rocks, as may CH_4. Minor fluid components may include S and N_2, as well as dissolved species, notably alkalis and halides. When the fluid is composed of several volatile components, P_{fluid} indicates the total fluid pressure, which is the sum of the **partial pressures** of each volatile component $(P_{fluid} = p_{H_2O} + p_{CO_2} + \cdots)$. Another common treatment is to consider the **mole fractions** of the components, which must sum to 1.0 ($X_{H_2O} + X_{CO_2} + ... = 1.0$). The partial pressure of a fluid component is then equal to the mole fraction times the total fluid pressure ($p_{H_2O} = X_{H_2O} \cdot P_{fluid}$). Although there is some consensus that fluids are important constituents of most metamorphic systems, there is less agreement on the nature and source of the fluids, or even whether they are present at some higher metamorphic grades. Fluids can come from meteoric sources, juvenile magmatic sources, subducted material, trapped sedimentary brines, dehydrating metamorphics, or degassing of the mantle.

The motion of fluids may transport various chemical species over considerable distances. This is particularly true for fluids released by crystallizing plutons into the adjacent country rocks (Section 11.2.2). If the physical or chemical nature of the rocks through which fluids pass differ markedly from that of the fluid entering, the fluid may exchange material with the new host rocks. When substantial chemical change accompanies metamorphism the process is called **metasomatism**. As we shall see, metasomatism may involve fluid transport, as described here, or diffusion of constituents through minerals or intergranular fluids. The scale of the latter is generally limited to a few centimeters or less. Metasomatism should exclude added melts, but those melts may release metasomatizing fluids. A further problem confronts mantle petrologists, where melts and fluids become similar, even indistinguishable beyond the "second critical endpoint" (Section 16.8.2). Chemical changes in the mantle (described periodically in Chapters 13 through 19), even if in the form of stranded melts, are generally considered on a broad scale and collectively referred to as "mantle metasomatism." This is a rather special exception to the accepted definition of metasomatism.

Chemical analyses of a wide variety of metamorphic rocks are now available. In general, the chemical composition of metamorphic rocks correlates well with common igneous and sedimentary protoliths. We conclude that metamorphism commonly approximates an **isochemical** process, meaning that little is added, transported, or removed during metamorphism (except for volatiles like H_2O and CO_2, much of which escapes from heated and compressed sediments). Because it is impossible to measure the amount of fluid that was once present in rocks (or has infiltrated through), and because we expect the fluid phase to be relatively mobile, we ignore the volatile components when we say that a metamorphic rock was produced "isochemically." Because isochemically metamorphosed rocks involve virtually no nonvolatile chemical change, and metasomatic rocks involve substantial change, there is obviously a gap between these categories with no adequate terminology, and the boundary is left (conveniently) vague. Also, because isochemical metamorphism is more the norm, petrologists rarely specify that a metamorphic rock is isochemical unless they want to emphasize the point.

Keep in mind that metamorphism is a response to changes in external parameters, and that in nature there are **gradients** in temperature, pressure, and fluid composition.

As a result, we can expect there to be a **zonation** in the mineral assemblages constituting the rocks that equilibrate across an expanse of these gradients. We should thus be able to walk a multi-km-scale traverse in an eroded metamorphic area and cross from non-metamorphosed rocks through zones of progressively higher metamorphic grade or through centimeter-to-meter-scale zones adjacent to a pluton that reflect metasomatic composition gradients. We shall discuss fluids and metasomatism more fully in Chapter 30.

21.3 THE TYPES OF METAMORPHISM

There are several approaches to classifying metamorphic processes, and geologists also disagree on the categories pertaining to any single approach. One approach is to classify metamorphism on the basis of the principal agent or process involved. Thus **thermal metamorphism** results when heat transfer is the dominant agent (such as near plutons). **Dynamic metamorphism** occurs when deviatoric stress results in deformation and recrystallization. **Dynamo-thermal metamorphism** results when temperature and stresses are combined, as in orogenic belts. To these three classic types, one might add **metasomatism** because fluid-enhanced infiltration and alteration is a process distinct from the above three. Although this approach has the merit of concentrating on the immediate process and avoiding any bias as to field setting, I think it is a bit antiseptic for our needs in an introduction to metamorphism. At this stage, I prefer to keep the typical field settings in mind because they keep us aware of real Earth systems. I thus propose that we stick with a more traditional classification based mainly on field setting.

Following the IUGS/SCMR recommendations (with some modifications), we shall use the following classification. The indented types are subsets of the major ones:

> Contact Metamorphism
> > Pyrometamorphism
>
> Regional Metamorphism
> > Orogenic Metamorphism
> > Burial Metamorphism
> > Ocean Floor Metamorphism
>
> Hydrothermal Metamorphism
>
> Fault-Zone Metamorphism
>
> Impact or Shock Metamorphism

The IUGS/SCMR recommendations also include Hot-slab, Combustion, and Lightning Metamorphism, all of which occur in rare and localized occurrences (Fettes and Desmons, 2007).

21.3.1 Contact Metamorphism

Contact metamorphism occurs adjacent to igneous intrusions, principally as a result of the thermal (and possibly metasomatic) effects of hot magma intruding cooler shallow rocks (see Section 4.2.3). The rocks surrounding a pluton are typically called **country rocks**, or **host rocks** (wall-, floor-, and roof-rocks may also be specified locally, depending on geometry). Contact metamorphism can occur wherever igneous activity does, and, although probably most common at plate boundaries, it is not restricted to any particular setting. Because plutons can rise and transmit heat to even the shallow crust (the epizone in Section 4.2.5), this type of metamorphism may occur over a wide range of pressures, extending nearly to the surface. A **contact aureole** of metamorphosed rock typically forms in the country rocks surrounding a pluton. Contact metamorphic effects are generally most dramatic when plutons intrude to shallow epizonal levels due to the substantial thermal contrast between the melt and the shallow country rocks. In the intermediate-depth mesozone, plutons cool more slowly and thus maintain metamorphic temperatures for a longer time, so the contact aureole may be wider. The country rocks are probably already metamorphic, however, so the contact effects may not be as easy to distinguish. In the deep catazone, the temperature of the country rocks may not differ much from that of the melts, and contact effects may be minor to insignificant.

The thermal effects associated with intrusion of hot magma into cooler country rocks are fairly well understood, and can be analyzed using heat-flow models (Jaeger, 1968; Ghiorso, 1991; Spear, 1993). Figure 21.5 shows the results of one of Jaeger's calculations for a vertical 1-km-thick basaltic dike intruded at 1200°C into country rocks at 0°C. According to the calculations, the temperature 1 km from the contact with this very hot magma is raised by about 200°C, but this takes several thousand years. Here the temperature remains near this value for at least a million years. Temperatures only a few hundred meters from the contact never approach the temperature of the initial magma. The magnitude of the thermal gradients, and their duration, depend upon the initial thermal contrast and the size, shape, and orientation of the igneous body. Small dikes may have millimeter-sized contact zones, whereas batholiths may have aureoles extending for several kilometers.

If the country rocks are permeable and sufficient fluid is available, convection of the fluid (driven by thermal gradients) will help cool the magma body (Figure 4.38) but will also transfer heat and matter farther from the contact, extending the aureole. Contact metamorphic rocks are commonly affected

by substantial metasomatism associated with these fluids. Oxygen isotopic data indicate that hydrothermal fluids circulating above plutons are typically dominated by meteoric water (groundwater), but some are also juvenile and expelled by the cooling magma. Metasomatism is most evident in situations in which the chemical composition of the country rock differs considerably from that of the melt. This is particularly evident in carbonate metasediments. As hot, acidic, silica-rich waters are driven from the pluton into the country rocks they react with the carbonates, producing a variety of calc–silicate minerals in a rock type called a **skarn** or **tactite** (Chapter 30).

As mentioned above, contact metamorphism is most evident (and most commonly studied) in low-pressure environments. Shallow rocks have relatively high yield strength, and metamorphism typically occurs under conditions of low deviatoric stress. The resulting metamorphic rocks thus typically display nearly random textural fabrics and may generally be called a **granofels** (or **hornfels** if hard, compact, and displaying conchoidal fracture). Textures inherited from the parent rocks are commonly preserved because there is little deformation to destroy them.

Plutonism is generally associated with contemporaneous orogeny, as described with respect to the Andes in Section 17.4. In these situations, contact metamorphic effects occur in conjunction with deformation and orogenic metamorphism. It is not unusual, however, to see the effects of contact metamorphism overprinting the effects of orogenic metamorphism in a polymetamorphic sequence. A common rock of this type is a "spotted phyllite" (Figure 23.14). The foliated phyllite formed during a regional event and the later ovoid "spots" are typically low-pressure minerals that grew during a later contact event. The contact overprint on regional rocks may reflect a "lag time" involved between the creation of the pluton, presumably at depth during the thermal maximum of a metamorphic terrane, and its migration and final emplacement in the lower grade rocks above. Alternatively, late plutonism may reflect post-orogenic collapse magmatism (Section 18.4.2.3). **Polymetamorphism** involves the overprint of one metamorphic event on one or more older events, and need not be restricted to any particular types of metamorphism.

Pyrometamorphism is a minor type of contact metamorphism characterized by very high temperatures at very low pressures, generated by a volcanic or sub-volcanic body. It is most typically developed in xenoliths enclosed in such bodies, but may also occur at country rock contacts. Pyrometamorphism is typically accompanied by varying degrees of partial melting. Critical minerals are: spurrite, tilleyite, rankinite, larnite, and/or merwinite in low-SiO_2 carbonate rocks; mullite and glass in aluminous rocks; or tridymite and glass in high-SiO_2 rocks.

21.3.2 Regional Metamorphism

Regional metamorphism, in the most general sense, is any metamorphism that affects a large body of rock, and thus covers a great lateral extent (typically tens of kilometers or

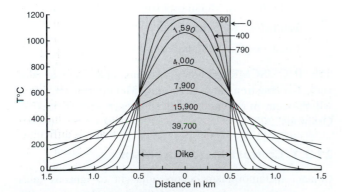

FIGURE 21.5 Temperature distribution within a 1-km-thick vertical dike and in the country rocks (initially at 0°C) as a function of time. Curves are labeled in years. The model assumes an initial intrusion temperature of 1200°C and cooling by conduction only. After Jaeger (1968).

more). Using this definition, regional metamorphism can be of three principal types: orogenic metamorphism, burial metamorphism, or ocean-floor metamorphism.

Orogenic metamorphism is the type of metamorphism associated with **convergent plate margins**. It thus occurs during the development of island arcs (Chapter 16), continental arcs (Chapter 17), and continental collision zones (Section 18.4.2.2). Most studies of metamorphism have focused on orogenic belts, and many petrologists consider the term "regional metamorphism" to be synonymous with "orogenic metamorphism." Orogenic metamorphism is dynamo-thermal, involving one or more episodes of orogeny with combined elevated geothermal gradients and deformation (deviatoric stress). Most affected rocks therefore display a definite foliation (slates, phyllites, schists, gneisses, etc.). Orogenic–metamorphic episodes may be due to variations in plate motion (accelerated subduction rates or changes in plate direction), continental (or arc) collision, ridge subduction, or post-orogenic collapse (Chapters 16 to 18).

Figure 21.6 illustrates a model for the sequential development of an orogenic belt at an active continental margin without a collision event. Part (a) represents the incipient stages of subduction. In part (b), an "orogenic welt" is created by crustal thickening due to compression, thrust stacking of oceanic slices, and/or addition of magmatic material from below. Underthrusting of oceanic lithosphere in the forearc area may migrate trenchward, adding successive ophiolite slabs to the base of the outer welt. This process had been called **tectonic underplating**, not to be confused with the **magmatic underplating** that occurs when mantle-derived melts stall and accumulate at the base of the crust (Section 17.2). Heat may be added to the growing welt by

rising plutons, underplated magmas, increased radioactive heat generation in the thickened enriched crust, and induced mantle convection in the mantle wedge above the subducted slab. Temperature increases both downward and toward the axial portion of the welt where plutons are concentrated, and metamorphism is widespread. Deep axial rocks may become heated to the point of melting. Whether or not the deep rocks melt, they may become sufficiently ductile to be mobilized, rising into the extending welt as **metamorphic core complexes** or gneiss domes. The Lepontine Alps and Adirondack Mountains are classic examples of such thermal domes. Coney (1980) described the metamorphic core complexes of the North American Cordillera.

Uplift of the thickening welt in Figure 21.6b, and ensuing erosion, result in *advective* heat transfer upward and exposure of the metamorphic rocks. Because heat dissipates slowly, metamorphism generally continues after major deformation (such as the thrusting observed in the "foreland" area on the left of Figure 21.6b and c) ceases. When this occurs, the metamorphic pattern is simpler than the structural one. In such cases, the folding and thrusting may be complex, but the metamorphism may exhibit a simple domal pattern, centering on the metamorphic/igneous core where heat input (as well as thickening and uplift) were the greatest. Subsequent erosion will be nearly proportional to uplift, leaving an exposed surface pattern of increasing metamorphic grade from both directions toward the core area. Naturally this is a very simplified example, and most orogenic belts have several episodes of deformation and metamorphism, thus creating a more complex polymetamorphic pattern (e.g., Figure 23.49).

Orogeny involving continental collision involves the interaction of an active continental margin such as Figure 21.6 with a continental mass having a "passive" margin and an apron of sediments extending from the continental shelf. Such collisions usually produce even more complex structural, magmatic, and metamorphic patterns (based largely on the themes developed in Figure 21.6). A modern example is the Himalayas.

Although batholiths are usually present in the highest grade areas of regional terranes, the metamorphism described above isn't considered to be contact metamorphism because it develops regionally, and the pattern of metamorphic grade does not relate directly to the proximity of the igneous contacts. In other words, the metamorphism in these situations is not *caused by* the intrusions. Rather, both the metamorphism and the intrusions are produced by a large-scale thermal and tectonic disturbance (subduction and orogeny). Of course, contact metamorphism may develop locally within regional terranes. In many cases intrusive rocks are plentiful and closely spaced, so that it is difficult or impossible to distinguish regional metamorphism from overlapping contact aureoles. Spear (1993) called such situations **regional contact metamorphism**.

Burial metamorphism is a term coined by Coombs (1961) for low-grade metamorphism that occurs in sedimentary basins due to burial by successive layers. Coombs worked in the Southland Syncline in southernmost New Zealand, where a thick pile (>10 km) of Triassic and Jurassic

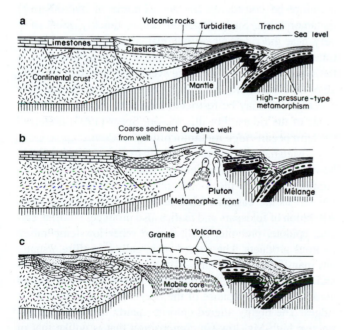

FIGURE 21.6 Schematic model for the sequential (a → c) development of a "Cordilleran-type" or continental arc orogen. The dashed and black layers on the right represent the basaltic and gabbroic layers of the oceanic crust. From Dewey and Bird (1970) and Miyashiro et al. (1979).

material, almost entirely volcaniclastics, had accumulated. Deformation was mild, and no igneous intrusions have been discovered. The volcaniclastics contain numerous fine-grained, high-temperature phases, including glassy ash, the relative instability of which makes them very susceptible to metamorphic alteration. The metamorphic effects are attributed to increased pressure and temperature related to burial, and range from diagenesis to the formation of zeolites, prehnite, pumpellyite, laumontite, and at deeper levels, minerals found in the lower grades of some exposed orogenic belts. The metamorphic minerals are commonly restricted to veins, which, combined with the occurrence of opal, suggests that silica-saturated hydrous fluids were important agents of metamorphism. Many areas of burial metamorphism are also hydrothermal fields, suggesting an elevated geothermal gradient and fluids may be important elements in at least some cases of burial metamorphism. Coombs (1961) also proposed a type of metamorphism called **hydrothermal metamorphism**, caused by hot H_2O-rich fluids and usually involving metasomatism. Hydrothermal metamorphism is a difficult type of metamorphism to constrain because hydrothermal processes generally play some role in most of the other types of metamorphism.

Many early concepts of orogeny attributed the metamorphic effects to the normal geothermal gradient, implying that burial metamorphism and orogenic metamorphism are the same, and rocks such as those described from New Zealand are simply the shallow manifestations of orogenic metamorphism. Although orogenic areas may grade at the low-T end into mineral assemblages characteristic of burial metamorphism, the settings are now considered to be different. Burial metamorphism, as defined here, occurs in areas that have not experienced significant deformation or orogeny. It is thus restricted to large, relatively undisturbed sedimentary piles away from active plate margins. The Gulf of Mexico, fed by the Mississippi River, represents a modern example of such a pile. Another is the Bengal Fan, fed by the Ganges and Brahmaputra rivers, which has the form of a sedimentary wedge accumulating along a passive continental margin. According to the interpretation of Curray (1991), the seismic data indicate a sedimentary pile in the Bay of Bengal in excess of 22 km. A typical geothermal gradient suggests temperatures of 250 to 300°C at the base, where the pressure would be about 0.6 GPa. These conditions are well into the metamorphic range, and the weight of the overlying sediments may cause sufficient compression to impart a foliation to the metamorphic rocks forming at depth. It may be impossible to distinguish a hand specimen of a rock retrieved from these depths from one picked up in the lower grade regions of an orogenic belt.

I should also emphasize that passive continental margins become active continental margins with the initiation of subduction or by continental collision. Areas of burial metamorphism may thus become areas of orogenic metamorphism. Usually, the orogenic event(s) obliterate the features of any earlier burial metamorphism, but the foreland areas, such as on the left of Figure 21.6c, may not suffer greatly from the orogenic metamorphism, and thus may be of transitional character. It is easy to recognize classic examples of burial metamorphism and orogenic metamorphism, but, as in practically all classifications, the boundaries are generally transitional, rather than abrupt.

Ocean-floor metamorphism was coined by Miyashiro et al. (1971) to describe the type of metamorphism affecting the oceanic crust near ocean ridge spreading centers (Chapter 13). This form of metamorphism was discovered only by post-war exploration of the ocean basins. At first dredges, and then drilling and submersibles, retrieved numerous metamorphosed basalts and gabbros in addition to the typical MORB igneous suite.

Humphris and Thompson (1978) described a variety of metamorphic minerals in ocean-floor rocks, representing a wide range of temperatures (at relatively low pressure). The metamorphic rocks exhibit considerable metasomatic alteration, notably loss of Ca and Si and gain of Mg and Na in most cases. These changes can be correlated with exchange between basalt and hot seawater. The intensity of metamorphism varies extensively on the local scale, and probably relates to the distribution of pervasive fractures that act as fluid conduits (Mottl, 1983). Direct evidence for such a process came when submersibles encountered the hot springs and "black smokers" at mid-ocean ridges (Chapter 13) with their unique biological communities. These observations, plus drill cores, have documented fracture systems ranging from the major fracture zones (spaced kilometers apart), through meter-spaced fractures, to centimeter-spaced cracks attributed to cooling. Seawater penetrates down these fracture systems, where it becomes heated and leaches metals and silica from the hot basalts. The hot water circulates convectively back upward, exchanging components with the rocks with which it comes in contact. Ocean-floor metamorphism may therefore be considered another example of hydrothermal metamorphism. Such alteration occurs quickly, most of it very near the ridge where magmatism and heat is concentrated. If so, this type of metamorphism, although regional in the sense that affected rocks are eventually spread to virtually all of the oceanic crust, is actually more localized because the process itself may be restricted largely to the near-axial regions of the ridges. For this reason, Spear (1993) preferred the term **ocean-*ridge* metamorphism**.

Metamorphic alteration varies from incipiently altered basalts to highly altered chlorite–quartz rocks. Incipient alteration causes the plagioclase of the basalts to become "albitized" as the basalts exchange Ca for Na in the seawater. Alteration of feldspars and mafics also produces chlorite, calcite, epidote, prehnite, zeolites, and other low-temperature hydrous products. The altered rock, called a **spilite**, usually has many inherited textures of the basalt, including vesicles and pillow structures. The incipient stages may be produced across a broader expanse of the ocean basin than at just the ridges. The highly altered chlorite–quartz rocks have a distinctive high-Mg, low-Ca composition that is unlike that of any other known sedimentary or igneous rock. These rocks are a prime candidate for the protolith of the unusual cordierite–anthophyllite metamorphic rocks found at higher grades of metamorphism in some orogenic areas.

21.3.3 Fault-Zone and Impact Metamorphism

Fault-zone and impact metamorphism occur in areas experiencing relatively high rates of deformation and subsequent strain with only minor thermal recrystallization effects. Fault-zone metamorphism occurs in areas of high shear stress. The term *fault* is to be interpreted broadly in this context, and includes zones of distributed shear that can be up to several kilometers across (see below). The IUGS/SCMR uses the term **dislocation metamorphism**, and others have used **shear-zone metamorphism** instead. **Impact metamorphism** (also called **shock metamorphism**) occurs at meteorite (or other bolide) impact craters. Both fault-zone and impact metamorphism correlate with **dynamic metamorphism**, or Spear's (1993) **high-stress metamorphism**. Although the latter terms are based on the metamorphic agent (discussed above) instead of the setting, they are commonly used as synonyms for fault-zone metamorphism (because impact metamorphism is so rare).

As will be discussed in more detail in Chapter 23, strain of the lattice in a mineral grain raises the energy of that grain, and promotes recrystallization back to an unstrained lattice state. If the strain rate is high enough, and the temperature low enough, minerals may be broken, bent, or crushed without much accompanying recrystallization. This process is known as **cataclasis**, and occurs at impacts and in the very shallow portions of fault zones where rocks behave in a brittle fashion (Figure 21.7a). Common products in shallow fault zones are **fault breccia** (a broken and crushed filling in fault zones) and **fault gouge** (a clayey alteration of breccia resulting from interaction with groundwater that permeates down along the porous fault plane). With increased depth, faults gradually change from brittle fractures to wider shear zones involving a combination of cataclasis and recrystallization (Figure 21.7b). Intense localized shear produces a fine-grained foliated flint-like rock called **mylonite**. At deeper levels yet, shear movement is distributed more evenly throughout the zone, and rocks are almost entirely ductile. The change in deformation style reflects the temperature increase that accompanies depth, so that recrystallization accompanies the deformation. Under a particular set of *P-T*-stress conditions, different minerals respond to strain in different ways. Thus at shallow levels quartz may deform in a brittle fashion, whereas associated calcite is ductile. We shall return to the subject of strain and recrystallization in the following two chapters, when we attempt to classify high-strain rocks and interpret their textures.

21.4 THE PROGRESSIVE NATURE OF METAMORPHISM

The term **prograde** refers to an increase in metamorphic grade with time as a rock is subjected to gradually more severe metamorphic conditions. **Prograde metamorphism** refers to the changes in a rock that accompany increasing metamorphic grade. **Retrograde** refers to decreasing grade as a body of rock cools and recovers from a metamorphic or igneous event, and **retrograde metamorphism** describes any accompanying changes.

Although chemical equilibrium may not be attained in rocks at the lowest metamorphic grades, most workers contend that, beyond the incipient grades, the mineral assemblage in metamorphic rocks maintains equilibrium as grade increases. **Progressive metamorphism** expresses the idealized view that a rock at a high metamorphic grade *progressed* through a sequence of mineral assemblages as it passed through all of the mineral changes necessary to maintain equilibrium with increasing temperature and pressure, rather than hopping directly from an unmetamorphosed rock to the metamorphic rock that we find today. Whether the *temporal* change in mineralogy at a single place in a metamorphic terrane is similar to the *spatial* change in mineralogy (for rocks of similar composition) as one approaches the same point walking up the metamorphic gradient will be discussed in Sections 25.4 and 27.4.4.

Strong evidence for the progressive nature of metamorphic rocks comes from textural studies in which metastable relics of lower-grade minerals are found only partly reacted to the higher-grade mineral assemblage. Thus the prograde reaction did not run to completion, perhaps for kinetic reasons. Although such relics are far from ubiquitous, they are common and indicate that temporally progressive metamorphism does occur. It is an article of faith among petrologists that metamorphic rocks passed through a more complete progressive sequence, but there seems little reason to doubt it (unless at the lowest grades where equilibrium may be impeded).

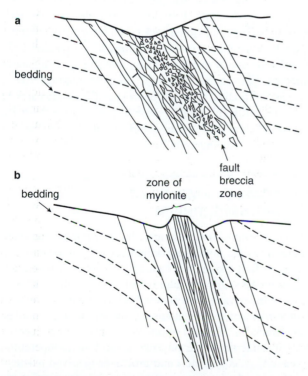

FIGURE 21.7 Schematic cross section across fault zones. **(a)** A shallow fault zone with fault breccia. **(b)** A slightly deeper fault zone (exposed by erosion) with some ductile flow and fault mylonite. After Mason (1978).

Of course, all of the rocks that we now find must also have cooled to surface conditions. If a metamorphosed sedimentary rock experienced a cycle of increasing metamorphic grade, followed by decreasing grade, at what point on this cyclic *P-T-t* **path** (pressure–temperature–time path, see Sections 16.8 and 25.4) did its present mineral assemblage last equilibrate?

The zonal distribution of metamorphic rock types preserved in a geographic sequence of increased metamorphic grade suggests that each rock preserves the conditions of the *maximum metamorphic grade (temperature)* experienced by that rock during metamorphism. It follows that retrograde metamorphism is of only minor significance, and is usually detectable by observing textures, such as the incipient replacement of high-grade minerals by low-grade ones at their rims. This proposal is also an article of faith to some extent, but again, I can find no compelling reason to doubt it in most cases. Prograde metamorphic reactions are generally endothermic (they consume heat), and the heat supplied to progressive metamorphic rocks should quickly drive the reactions, particularly the common dehydration and decarbonation reactions that have large volume and enthalpy changes. Guiraud et al. (2001) proposed that equilibrium is much more likely to be maintained during prograde metamorphism while a fluid is present (due to the devolatilization reactions) but that equilibrium ceases when that fluid is used up during the earliest stages of retrogression. Either way, rocks should readily maintain equilibrium during prograde metamorphism. Retrograde reactions are exothermic, and there is little force to drive them as the rocks cool, nor is a fluid available to facilitate the requisite elemental redistribution. Rehydration and re-carbonation requires infiltration of metamorphic fluids back into rocks from which they have been released. This is not as easy as driving them out in the first place. As mentioned above, the composition of metamorphic rocks correlates well with various types of protolith, except for a loss of volatiles.

Although the mineralogy and texture of metamorphic rocks typically reflects the maximum grade attained, the *composition* of the minerals may not always do so. The technique of **geothermobarometry** uses the temperature (and in some cases pressure) dependence of metamorphic reactions between coexisting minerals (Section 27.4) to estimate the T and P conditions of metamorphism. Geothermobarometry is predicated on the idea that the chemical composition of the minerals reflects the maximum metamorphic conditions (or compositional zoning, if present, reflects the *P-T-t* path of metamorphism). Although the results of many studies suggest that this is commonly the case, exchange reactions, such as Fe-Mg exchange between two mafic minerals in contact, may not require much driving energy to keep abreast with slowly falling temperatures in metamorphic rocks, at least for awhile, until the thermal activity drops to the point that kinetic factors impede further equilibration. Different minerals may re-equilibrate to different extents, and only a thin zone at the contacting rims may be involved. We shall return to this concept again in Chapter 23.

21.5 TYPES OF PROTOLITH

The initial chemical composition of a rock profoundly affects the mineralogy of its metamorphic offspring. When we study metamorphic rocks, it is important to keep the "parental" rock type in mind. From a metamorphic point of view, the chemical composition of the protolith is the most important clue toward deducing the parent. We can lump the common types of sedimentary and igneous rocks into six broad compositionally based groups:

1. **Ultramafic** rocks. Mantle rocks, komatiites, and cumulates. Very high Mg, Fe, Ni, and Cr.
2. **Mafic** rocks. Basalts, gabbros, and some graywackes. High Fe, Mg, and Ca.
3. **Shales** and **mudstones** (or **pelitic** rocks). The most common sediment. Fine grained clastic clays, muds, and silts deposited in stable platforms or offshore wedges. High Al, K, and Si.
4. **Carbonates** (or **calcareous** rocks). Mostly sedimentary limestones and dolostones. High Ca, Mg, and CO_2. Impure carbonates (**marls**) may contain sand or shale components.
5. **Quartz** rocks. Cherts are oceanic, and sands are moderately high-energy continental clastics. Nearly pure SiO_2.
6. **Quartzo-feldspathic** rocks. Arkose or granitoid and rhyolitic rocks. High Si, Na, K, and Al.

Of course, these six categories are to a degree gradational, and they cannot possibly include the full range of possible parental rocks, but they do cover most types and provide an easy frame of reference from the metamorphic perspective. One gradational rock type that is fairly common is a sand–shale mixture (called **psammitic**). Important rocks not included above are evaporites, ironstones, manganese sediments, phosphates, laterites, alkaline igneous rocks, coal, and ore bodies. One can be as specific as one cares in naming protolith (if suitable evidence exists to support it), using such names as "meta-conglomerate," "metabasalt," "meta-arkose," "metagranite," etc.

21.6 SOME EXAMPLES OF METAMORPHISM

The classification, textures, and details of metamorphism that are presented in the following chapters will be easier to grasp if we first take a brief look at some examples of metamorphism. We shall only address a few contrasting types of metamorphic terranes at a survey level in order to provide a broad context of how metamorphism affects rock bodies. Bear in mind that the goal of practicing metamorphic petrology is to understand the physical conditions (temperature, pressure, X_{rock}, X_{fluid}, etc.) and processes involved in metamorphism, including recrystallization, formation of metamorphic minerals, deformation, and metasomatism. Such investigations are directed toward interpreting the conditions and evolution of metamorphic bodies, mountain belts, and ultimately the evolution of the Earth's crust. Metamorphic rocks may retain enough inherited information from their protolith to allow us to interpret much of the pre-metamorphic history as well.

21.6.1 Orogenic Regional Metamorphism of the Scottish Highlands

In what is now a classic work on orogenic regional metamorphism, George Barrow (1893, 1912) made one of the first systematic studies of the variation in rock types and mineral assemblages with progressive metamorphism (in the circled area in Figure 21.8). Metamorphism and deformation in the southeastern Highlands of Scotland occurred during the Caledonian orogeny, which reached its maximum intensity about 500 Ma ago. Deformation in the Highlands was intense, and the rocks were folded and thrust into a series of nappes. Numerous large granites were also intruded toward the end of the orogeny, after the main episode of regional metamorphism. Later plate tectonic rifting split up Laurasia and the orogenic belt, and fragments are now found in Scandinavia, Greenland, and North America (the Appalachians). The Scottish rocks in Barrow's study belong to the latest Precambrian to Cambrian Dalradian Supergroup, comprising some 13 km thickness of conglomerates, sandstones, shales, limestones, and mafic lavas.

Although sandstones show little change across the area, Barrow noted significant and systematic mineralogical changes in the *pelitic* rocks (originally shales). He found that he could subdivide the area into a series of **metamorphic zones** (Figure 21.8), each based on the appearance of a new mineral in the metamorphosed pelitic rocks as metamorphic grade increased (which he could correlate to increased grain size). The new mineral that characterizes any particular zone is termed an **index mineral**. The sequence of zones now recognized in the Highlands, and the rocks and typical metamorphic mineral assemblage in each, are:

- *Chlorite zone.* Pelitic rocks are slates or phyllites and typically contain chlorite, muscovite, quartz, and albite.
- *Biotite zone.* Slates give way to phyllites and schists, with biotite, chlorite, muscovite, quartz, and albite.
- *Garnet zone.* Schists with conspicuous red almandine garnet, usually with biotite, chlorite, muscovite, quartz, and albite or oligoclase.
- *Staurolite zone.* Schists with staurolite, biotite, muscovite, quartz, garnet, and plagioclase. Some chlorite may persist.
- *Kyanite zone.* Schists with kyanite, biotite, muscovite, quartz, plagioclase, and usually garnet and staurolite.
- *Sillimanite zone.* Schists and gneisses with sillimanite, biotite, muscovite, quartz, plagioclase, garnet, and perhaps staurolite. Some kyanite may persist (although kyanite and sillimanite are both polymorphs of Al_2SiO_5).

This sequence of mineral zones has been recognized in other orogenic belts in the world, and is now so well established in the literature that the zones are commonly referred to as the **Barrovian zones**. The *P-T* conditions represented (see the range of estimated metamorphic field gradients in Figure 21.1) are also referred to as **Barrovian-type** (or style) metamorphism, which is fairly typical of many orogenic metamorphic belts. C. E. Tilley (1925) and W. Q. Kennedy

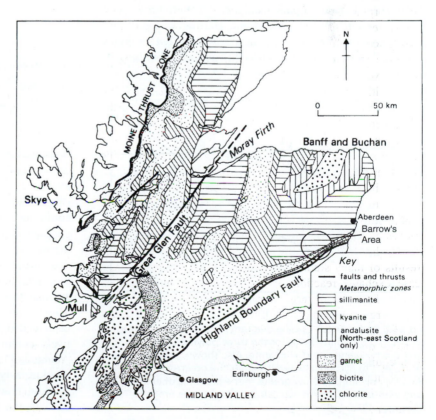

FIGURE 21.8 Regional metamorphic map of the Scottish Highlands, showing the zones of minerals that develop with increasing metamorphic grade. From Gillen (1982).

(1949), among others, have confirmed Barrow's zones and extended them over a much larger area of the Highlands (Figure 21.8). Tilley coined the term **isograd** for the boundary that separates zones. An isograd, then, was intended to indicate a line in the field of *constant* metamorphic *grade*. Isograds are perhaps best considered to be irregular curved surfaces in space, and the mapped isograds thus represent the intersection of the isogradic surface with the Earth's surface.

To summarize, an *isograd* (in this classical sense) represents the first appearance of a particular metamorphic *index mineral* in the field as one progresses up metamorphic grade. When one crosses an isograd, such as the biotite isograd, one enters the biotite *zone*. Zones thus have the same name as the isograd that forms its low-grade boundary. Because classic isograds are based on the first appearance of a mineral, and not its disappearance, an index mineral may still be stable in higher grade zones. Chlorite, for example, is still stable in the biotite zone (and even up into the staurolite zone in some situations). If equilibrium is maintained, however, the transition from the kyanite zone to the sillimanite zone crosses the boundary of a polymorphic transformation (Figure 21.9) and should eliminate kyanite at the sillimanite isograd. That this rarely happens so sharply is due to the small volume and enthalpy changes associated with the kyanite → sillimanite reaction, so that kyanite may remain metastably in the sillimanite zone.

Barrow (1893) attributed the regional metamorphism in his area to the thermal effects of the "Older Granite" that, although only locally exposed, he assumed was largely still hidden beneath the Dalradian schists. He thus considered the area to be an example of what some would now call regional contact metamorphism. Harker (1932) was the first to ascribe both the metamorphism and the granites to the thermal effects of a major orogenic event. Barrow was probably wrong about the origin of the metamorphism, but his zones live on. Scientists don't have to be right on everything to make a lasting contribution. Good *observations* tend to endure longer than *interpretations*.

As I mentioned above, the Barrovian sequence of zones has been recognized in orogenic terranes worldwide. It serves as a good way of comparing the metamorphic grade from one area to another. We may thus talk of biotite-zone rocks in the Alps, or the Appalachians, and have some concept of correlative metamorphic conditions between the areas. As we shall see in Chapter 25, subdividing metamorphic assemblages into broader categories called **metamorphic facies** is another way to do this. We must not, however, allow our thinking and observation to become confined by these zones. When mapping a metamorphic terrane one should keep an eye open for the appearance of *any* new minerals that may prove to be systematically related to metamorphic grade, and not simply search for biotite, garnet, etc. Differences may result from variations in geothermal gradient (pressure effects) or different rock compositions. For example, the pelitic rocks in Barrow's area represent a rather narrow compositional range. More iron and aluminum-rich shales are common in the Appalachians, and a regional *chloritoid* isograd can be mapped in many lower grade rocks as a result. Why a mineral occurs in some pelites and not in others at the same grade must depend on the bulk composition, but it is hard, at this point, to visualize why. Bulk composition may affect the grade at which a particular mineral first appears, or whether it occurs at all. We shall develop ways to visualize and understand these effects in the ensuing chapters.

Another difference occurs in the area just to the north of Barrow's, in the Banff and Buchan district (Figure 21.8). Here the pelitic compositions are similar to those in Barrow's area, but the sequence of isograds is:

chlorite

biotite

cordierite

andalusite

sillimanite

From the phase diagram for the Al_2SiO_5 system (Figure 21.9), we can see that the stability field of andalusite occurs at pressures less than 0.37 GPa (ca. 10 km), whereas kyanite can only give way to sillimanite at the sillimanite isograd above this pressure. Also the molar volume of cordierite is relatively large, indicating that it too is a low-pressure mineral. From this we can conclude that the geothermal gradient (T/P) in this northern district was higher than in Barrow's area, and rocks at any equivalent temperature must have been at a lower pressure (see the range of estimated metamorphic field gradients in Figure 21.1). This lower P/T variation on the Barrovian-type theme has been called **Buchan-type** metamorphism. It too is relatively common. Miyashiro (1961), from his work in the Abukuma Plateau of Japan, called a similar low P/T variant **Abukuma-type** metamorphism. Both terms, Buchan and Abukuma, are common in the literature, and they mean essentially the same thing.

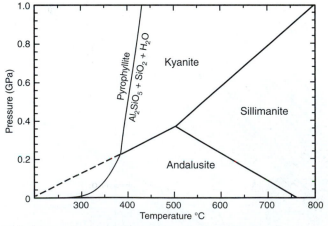

FIGURE 21.9 The *P-T* phase diagram for the system Al_2SiO_5, showing the stability fields for the three polymorphs: andalusite, kyanite, and sillimanite. Also shown is the hydration of Al_2SiO_5 to pyrophyllite, which limits the occurrence of an Al_2SiO_5 polymorph at low grades in the presence of excess silica and water. The diagram was calculated using the program TWQ (Berman, 1988, 1990, 1991).

21.6.2 Regional Burial Metamorphism, Otago, New Zealand

Much of New Zealand (Figure 21.10) was the site of voluminous Permian through Jurassic sedimentation and intermittent volcanism, depositing graywackes, tuffs, and some volcanics in a deep trough that was metamorphosed during the Cretaceous. The fine grain size and immature nature of the material makes it highly susceptible to metamorphic alteration, even at low grades, and, thanks to the work of D. S. Coombs (Coombs, 1954, 1961; Coombs et al., 1959), this area is the "type locality" of burial metamorphism. Coombs and colleagues showed that these low-grade rocks could be treated in the same systematic zonal fashion as other regional rocks and are not the result of random acts of senseless alteration. Although the metamorphic grade reached that of the medium Barrovian zones to the north, it is the lowest grades that are of interest here, as they add to the knowledge that we already have from our Scottish examples above. Deformation and plutons are inconsequential in the area, and the degree of metamorphism is related directly to depth of burial.

The isograds mapped at the lower grades, listed below, are well represented in the Haast River section (Figure 21.11):

1. Zeolite
2. Prehnite-Pumpellyite
3. Pumpellyite (-actinolite)
4. Chlorite (-clinozoisite)
5. Biotite
6. Almandine (garnet)
7. Oligoclase (albite at lower grades is replaced by a more calcic plagioclase)

The zeolite zone is developed in the southern tip of the island (Figure 21.10). At these lowest grades, metamorphic and diagenetic minerals develop only as joint fillings and in the very fine-grained and glassy matrix. Volcanic glass is altered to the zeolites heulandite or analcite, plus some secondary quartz and fine low-temperature phyllosilicates. Original *minerals*, however, even high-temperature igneous minerals that are far from stable, remain relatively unaltered. With depth heulandite gives way to laumontite and then prehnite and pumpellyite. Plagioclase becomes albitized because more calcic compositions are unstable at low temperature. At higher grades the rocks become more fully reconstituted and recrystallized, developing a good metamorphic schistosity. The rocks grade into the zones characteristic of regional orogenic metamorphism.

At the lower grades, reaction rates are slow, and metastability is common, so that the tenet of progressive metamorphism does not necessarily apply to such incipient types of metamorphism. Original igneous textures and minerals are commonly preserved, and it appears that igneous rocks can react to host zeolite zone minerals, or transform directly to mineral assemblages characteristic of any zone up to the chlorite or biotite zones, probably depending on the penetration of hot water. Nonetheless, the zonation becomes clear if we concentrate on the spatial development of certain mineral assemblages with the understanding that they may occur only sporadically.

The low-grade metamorphic rocks found in southern New Zealand represent the transition from diagenesis to metamorphism, and therefore the very beginnings of metamorphism. We might thus expect to find mineral assemblages

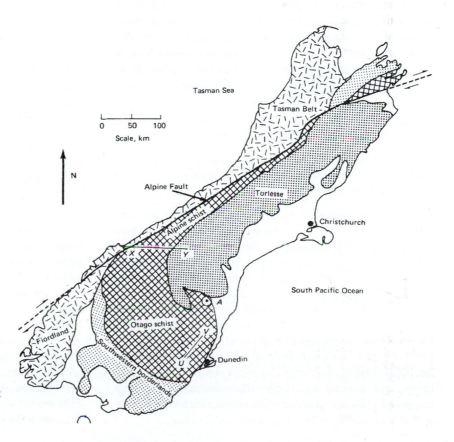

FIGURE 21.10 Geologic sketch map of the South Island of New Zealand, showing the Mesozoic metamorphic rocks east of the older Tasman Belt and the Alpine Fault. The Torlese Group is metamorphosed predominantly in the prehnite–pumpellyite zone, and the Otago Schist in higher grade zones. X-Y is the Haast River Section of Figure 21.11. From Turner (1981).

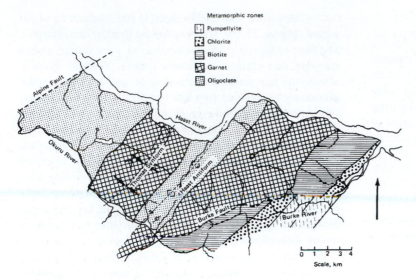

FIGURE 21.11 Metamorphic zones of the Haast Group (along section X-Y in Figure 21.10). After Cooper and Lovering (1970) and Turner (1981). Copyright © with permission from Springer-Verlag.

characteristic of the zeolite, prehnite–pumpellyite, and pumpellyite-actinolite zones to occur also at the low-grade end of most orogenic belts. Worldwide searches have found them in only a few cases. Igneous and sedimentary rocks generally proceed directly to the chlorite or biotite zones. Why do these low-grade rocks occur in some areas and not in others? The development of the very low-grade zones in New Zealand may reflect the highly unstable nature of the tuffs and graywackes, and the availability of hot water; whereas typical pelitic sediments may not react until higher grades. Zen (1961) also suggested that low-temperature hydrous Ca-bearing minerals (such as laumontite, prehnite, and pumpellyite) that are characteristic of burial metamorphism are stable in H_2O-rich fluids. If CO_2 is present, he proposed, Ca might be sequestered in calcite, inhibiting the formation of these characteristic low-temperature minerals. Some confirmation of this idea comes from the Salton Sea geothermal field in southernmost California, where the hydrothermal fluids do contain CO_2 and zeolites are absent, but epidote and chlorite are common (Muffler and White, 1969). Thompson (1971) showed experimentally that only small quantities of CO_2 are required to suppress the formation of laumontite and prehnite.

21.6.3 Paired Orogenic Metamorphic Belts of Japan

As a final example of regional metamorphism, we turn briefly to the islands of Shikoku and Honshu in Japan, where a pair of parallel metamorphic belts are exposed along a northeastern–southwestern axis parallel to the active subduction zone (Figure 21.12). These belts have different metamorphic signatures but are of the same age, suggesting that they developed together. The northwestern belt (called the "inner" belt, in the sense that it is landward, or away from the trench) is the Ryoke (or Abukuma) Belt. As mentioned above, this is a low P/T type of regional orogenic metamorphism (similar to the Buchan type). The dominant rocks are meta-pelitic sediments, and isograds up to the sillimanite zone have been mapped. The Ryoke–Abukuma belt is thus a high-temperature/low-pressure belt, and granitic plutons are common.

Of greater interest to us now is the outer belt, called the Sanbagawa Belt. This belt is composed of late Paleozoic volcanic/sedimentary filling with the metamorphic grade increasing toward the northwest. It is of a high-pressure/low-temperature nature compared to the Ryoke belt. Only the garnet zone is reached in the pelitic rocks. Basic rocks are more common than in the Ryoke belt, however, and in these, glaucophane is developed (giving way to hornblende at higher grades). The presence of glaucophane is characteristic of most high-pressure/low-temperature metamorphics and imparts a distinct blue color to the rocks. As a result, the rocks are commonly called **blueschists**.

The two belts are separated along their whole length by a major fault zone called the Median Line. The lithologies in the Ryoke–Abukuma belt are similar to sediments that we might expect to be derived from a relatively mature volcanic arc (Chapter 16), whereas those in the Sanbagawa belt are more akin to the oceanward accretionary wedge where distal arc-derived sediments and volcanics mix with oceanic crust and marine sediment.

The thermal model of a subduction zone system in Figure 16.15 suggests that the 600°C isotherm, for example, could be as deep as 100 km in the trench-subduction zone area and as shallow as 20 km beneath the volcanic arc. Miyashiro (1961, 1973) noted the paired nature of the Ryoke–Sanbagawa belts and suggested that the occurrence of coeval metamorphic belts, an outer, high-P/T belt, and an inner, lower-P/T belt ought to be a common occurrence in a number of subduction zones, either contemporary or ancient (Figure 21.6). He called these **paired metamorphic belts** and proposed several examples in addition to Japan, both island arcs and continental arcs, mostly in the circum-Pacific area (Figure 21.13). Paired belts may be separated by 100 to 200 km of less-metamorphosed and less-deformed material (the "arc–trench gap") or closely juxtaposed, like the Ryoke–Sanbagawa example. In the latter cases, the contact is typically a major fault and may show considerable strike-slip, as well as dip-slip, offset. Most of these paired belts are complex, and in several, the high-P/T and lower-P/T belts are not coeval. Nonetheless, the idea of such paired belts in an attractive concept.

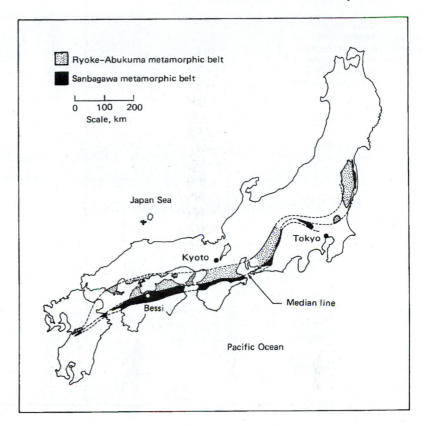

FIGURE 21.12 The Sanbagawa and Ryoke metamorphic belts of Japan. From Turner (1981) and Miyashiro (1994).

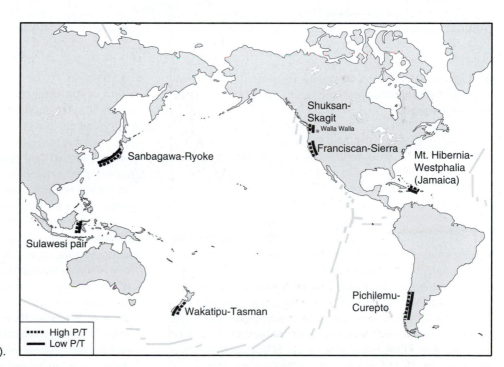

FIGURE 21.13 Some of the paired metamorphic belts in the circum-Pacific region. From Miyashiro (1994).

21.6.4 Contact Metamorphism of Pelitic Rocks in the Skiddaw Aureole, United Kingdom

The Ordovician Skiddaw Slates in the English Lake District are intruded by several granite and granodiorite bodies. The intrusions are shallow, and, following a common pattern, the contact effects are overprinted on an earlier phase of low-grade regional orogenic metamorphism, during which the rocks had been metamorphosed to chlorite-zone slates. Outside the aureole, the slates are tightly folded and typically contain muscovite (or, at these low grades, sericite–phengite), quartz, chlorite, chloritoid, and opaques (Fe-oxides, sulfides, and graphite) ± biotite. The aureole around the Skiddaw granite

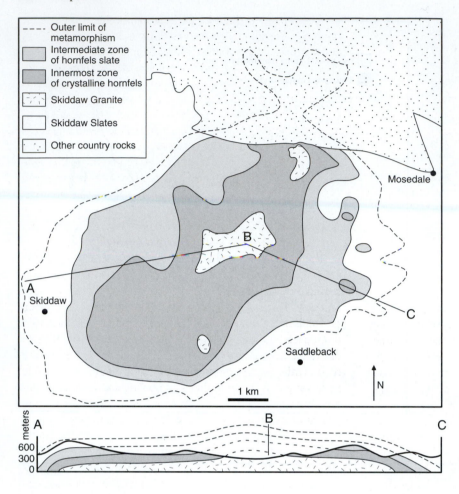

Outer limit of metamorphism (- - -)
Intermediate zone of hornfels slate
Innermost zone of crystalline hornfels
Skiddaw Granite
Skiddaw Slates
Other country rocks

Mosedale

A
Skiddaw
B
C
Saddleback

1 km

N

A   B   C
meters
600
300
0

FIGURE 21.14 Geologic map and cross section of the area around the Skiddaw Granite. Lake District, UK. From Eastwood et al. (1968). With permission British Geological Survey.

(Figure 21.14) was first described by Rastall (1910) and has been subdivided into three zones, principally on the basis of textures:

Increasing grade →

Unaltered slates
Outer zone of spotted slates
Middle zone of andalusite slates
Inner zone of hornfels
Skiddaw granite

The aureole is as much as 2 km wide. The great width suggests that the exposed igneous contact dips outward at a gentle angle. At the outer reaches, the first contact effects are noticed in the form of 0.2-mm- to 2.0-mm-sized black ovoid "spots" in the slates. At the same time, recrystallization of the fine slate minerals results in a slight coarsening of the grain size and degradation of the pronounced slaty cleavage. The black color of the spots is caused by opaque material that seems to concentrate around the centers of crystallization. In this section, the matrix of the spots shows essentially the same mineralogy as areas outside the spots, although the amount of muscovite is greater within them. Something must have grown here to form the spots. Rastall (1910) proposed that some chemical components concentrated in the spot zones preceding the formation of some spot-forming mineral, but this is not a very satisfactory explanation. Usually such localized small-scale chemical concentration is the result of diffusion induced by the growth of some mineral and is not likely to anticipate such growth. It seems more likely that the spots were once single larger minerals that grew and enveloped smaller grains. The spots were probably cordierite or andalusite, but they have since rehydrated and retrograded back to fine aggregates dominated by muscovite. Both cordierite and andalusite are found at higher grades (including the inner portion of the outer spotted slate zone, where they are also partly retrograded), but they have not been found farther out. If this theory is true, the spots that we now see in most of the spotted slates are **pseudomorphs**, representing replacement of one mineral by others, which retain the shape of the original mineral.

As one crosses to the middle zone, the slates are more thoroughly recrystallized, and typically contain biotite + muscovite + cordierite + andalusite + quartz + opaques. Cordierite characteristically forms equidimensional crystals with irregular outlines and numerous inclusions, in this case of biotite, muscovite, and opaques (Figure 21.15). The biotite and muscovite inclusions typically retain the orientation of the slaty cleavage outside the cordierites. This indicates that the growing cordierite crystals enveloped aligned micas that grew during the earlier regional event. This is excellent textural evidence for the overprint of a later contact metamorphism on an earlier regional one (Section 23.4.5). Mica outside the cordierite is generally larger and more randomly oriented, suggesting that it formed or recrystallized during

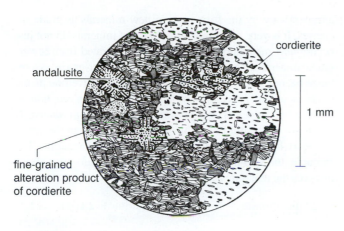

FIGURE 21.15 Cordierite–andalusite slate from the middle zone of the Skiddaw aureole. From Mason (1978).

the later thermal event when it was not subject to the stress of the earlier orogenic metamorphism that produced the foliated slates. The andalusite crystals have fewer inclusions than does cordierite, and many show the cruciform pattern of fine opaque inclusions known as **chiastolite** (Section 23.5; see the incipient form in Figure 21.15 and more advanced development in Figure 21.16). As discussed above, both andalusite and cordierite are minerals characteristic of low-pressure metamorphism, which is certainly the case in the Skiddaw aureole, where heat is carried up into the shallow crust by the magmas.

The rocks of the inner zone at Skiddaw are characterized by coarser and more thoroughly recrystallized textures. They contain the same mineral assemblage as the middle zone. Cordierite is free of mica but still contains trains of opaques that record the original orogenic foliation. Some rocks are schistose, but in the innermost portions the rock fabric loses the foliation, and the rocks are typical granofelses. The texture in the quartz-rich areas of Figure 21.16 exhibits a tendency for equidimensional crystals that meet in 120° triple junctions and is typical of near-hydrostatic metamorphism.

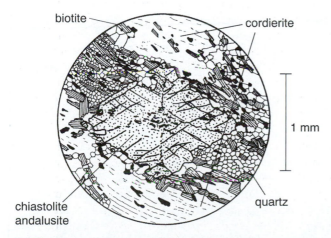

FIGURE 21.16 Andalusite–cordierite schist from the inner zone of the Skiddaw aureole. Note the chiastolite cross in andalusite (see also Figure 22.49). From Mason (1978).

As mentioned above, the zones at Skiddaw were determined by Rastall (1910) on a *textural* basis. A more modern and appropriate approach would be to conform to the practice used in the regional example above and use the sequential appearance of minerals and isograds to define the zones. This is now the common approach for all types of regional and contact metamorphism. At Skiddaw, however, the sequence of mineral development with grade is difficult to determine accurately. Most of the minerals at Skiddaw first appear in the outer zone and retrogression of andalusite and cordierite in this zone is practically complete. The first new contact metamorphic mineral in most slates is biotite, followed by the approximatcly simultaneous development of cordierite and andalusite. A textural zonation appears to be more useful at Skiddaw and perhaps in some other areas of contact metamorphism.

In the inner aureole at Comrie (a diorite intruded into the Dalradian schists up north in Scotland), the intrusion was hotter and the rocks were metamorphosed to higher grades than at Skiddaw. Tilley (1924) described coarse-grained non-foliated granofelses containing very high-temperature mineral associations, such as orthopyroxene + K-feldspar, that have formed due to the dehydration of biotite and muscovite in the country rocks. Orthopyroxene occurs in pelitic and quartzo-feldspathic rocks only at very high grades of contact and regional metamorphism, grades that may not be reached prior to melting in many instances. A typical mineral assemblage in these rocks is hypersthene + cordierite + orthoclase + biotite + opaques. More silica-rich rocks may contain the Ca-free amphibole cummingtonite, along with quartz, plagioclase ($\sim An_{38}$), biotite, and opaques. Some very interesting silica-undersaturated rocks also occur in the inner aureole. These contain such non-silicate high-temperature phases as corundum and Fe-Mg spinel. Tilley noted that the low-silica rocks occur only in the inner aureole and attributed their origin to loss of SiO_2 into the diorite. Perhaps a better alternative explanation is that SiO_2 (and H_2O) were concentrated into, and removed by, partial melts produced in the sediments adjacent to the contact with the very hot diorite.

21.6.5 Contact Metamorphism and Skarn Formation at Crestmore, California

As a final example in our brief survey, we turn to a much different protolith and some unusual metamorphic conditions. At the Crestmore quarry in the Los Angeles basin, a quartz monzonite porphyry of unknown age intruded Mg-bearing carbonates (either late Paleozoic or Triassic). As in the previous examples, the country rocks had experienced an earlier regional event and are now mostly brucite-bearing calcite marbles beyond the contact aureole. Crestmore is unusual in two ways. The temperature attained in the inner aureole is exceedingly high for such low pressures: it is an example of pyrometamorphism. These conditions are documented in only a few places on Earth. Second, the Crestmore quarry is operated by a cement company, and virtually all of the aureole has been removed in order to supply cement for building in Los Angeles. As a result, the data discussed below document an occurrence that cannot be observed and studied further today.

Fortunately, the study of Burnham (1959) was extensive (and samples of the unusual rocks from Crestmore can be found in the collections of many geology departments in the United States and abroad).

Burnham (1959) mapped the following zones and the mineral assemblages in each (numbers list assemblages in order of increasing grade):

Forsterite Zone:
1. calcite + brucite + clinohumite + spinel
2. calcite + clinohumite + forsterite + spinel
3. calcite + forsterite + spinel + clintonite

Monticellite Zone:
4. calcite + forsterite + monticellite + clintonite
5. calcite + monticellite + melilite + clintonite
6. calcite + monticellite + spurrite (or tilleyite) + clintonite
7. monticellite + spurrite + merwinite + melilite

Vesuvianite Zone:
8. vesuvianite + monticellite + spurrite + merwinite + melilite
9. vesuvianite + monticellite + diopside + wollastonite

Garnet Zone:
10. grossular + diopside + wollastonite

In this progression of zones, we can again see the sequential development of index minerals toward the igneous contact, such as clinohumite, followed by forsterite, clintonite, monticellite, melilite, spurrite/tillyite, merwinite, vesuvianite, diopside, wollastonite, and finally grossular garnet (don't worry about the avalanche of mineral names). Figure 21.17 is an idealized cross section through the aureole. The list of zones is at first confusing, and again serves to illustrate a common problem faced by petrologists (and probably all scientists). We can collect quality data but can become overwhelmed by the quantity at times, and it may be difficult to recognize meaningful patterns.

Two approaches are helpful in this case. First, the mineral associations in adjacent zones (in this and all metamorphic terranes) vary by the formation of new minerals as grade increases. It is very important to realize that minerals do not just occur. New metamorphic minerals are generated from lower-grade mineral assemblages because the lower-grade assemblage becomes less stable than the new alternative one as the grade increases. *This can only occur by a chemical reaction, in which some minerals are consumed and others produced.* If this is so, one ought to be able to relate minerals in the lower and next higher zone by a balanced chemical reaction. For example, the step from the first to the second sub-zone above involves the reaction:

$$2\,\text{Clinohumite} + \text{SiO}_2 \rightarrow 9\,\text{Forsterite} + 2\,\text{H}_2\text{O} \quad (21.1)$$

And the formation of the vesuvianite zone involves the reaction:

$$\text{Monticellite} + 2\,\text{Spurrite} + 3\,\text{Merwinite} + 4\,\text{Melilite}$$
$$+ 15\,\text{SiO}_2 + 12\,\text{H}_2\text{O} \rightarrow 6\,\text{Vesuvianite} + 2\,\text{CO}_2 \quad (21.2)$$

It may not be easy, however, to deduce the reaction (or even the mineralogy) in the field. Later in the lab, a researcher can look at thin sections and mineral formulas in an attempt to determine the responsible reaction. When we address isograds as reactions, we can then turn to what variables are involved. In the present case, for example, we discover that the majority of these prograde reactions consume SiO_2. Because quartz is not found in the Crestmore aureole, the SiO_2 must have been added in the form of dissolved silica in hydrothermal fluids. We can thus conclude that infiltration of silica from the monzonite into the carbonate country rocks must have played a critical role in the aureole development. It has become accepted practice to treat all isograds as reactions, whenever possible. We shall develop this notion throughout the remainder of this text. Classical isograds, based simply on the first appearance of an index mineral in the field, should be clearly indicated as such. Various names have been used, such as **mineral-in isograds** or even **provisional isograds**.

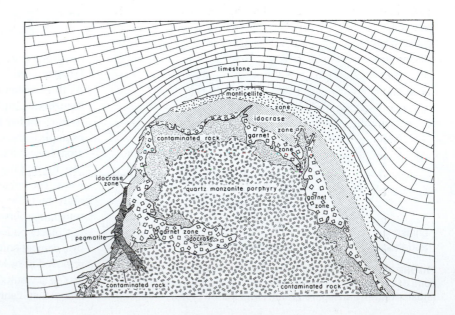

FIGURE 21.17 Idealized north–south cross section (not to scale) through the quartz monzonite and the aureole at Crestmore, CA. From Burnham (1959). Copyright © The Geological Society of America, Inc.

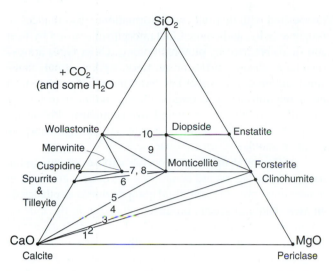

FIGURE 21.18 CaO-MgO-SiO$_2$ diagram at a fixed pressure and temperature showing the compositional relationships among the minerals and zones at Crestmore. Numbers correspond to zones listed in the text. After Burnham (1959) and Best (1982).

A second important approach to treating accumulated data is to find a way to *display* it all in simple, useful ways. Addressing the list of mineral assemblages in the zones at Crestmore can be bewildering. If we think of the aureole as a chemical system, we note that most of the minerals consist of the components CaO-MgO-SiO$_2$-CO$_2$-H$_2$O (with minor Al$_2$O$_3$). If we graphically plot the minerals on a triangular CaO-MgO-SiO$_2$ diagram (Figure 21.18), we get a fairly simple view of how the minerals are related chemically. As we shall see, these **chemographic compatibility**, or **composition-paragenesis, diagrams** are very useful in helping us understand mineral assemblages in metamorphic zones.

The theory and construction of chemographic compatibility diagrams will be developed in Section 24.3. For now, I'll simply cover a few basic points. A researcher must first figure out which chemical system is best. For Mg-bearing calcite marbles with silica added, this is easy. As we shall see later, we can effectively "ignore" volatile components, such as CO$_2$ and H$_2$O, in the context of most of the diagrams. We then plot the minerals on an appropriate diagram, connecting coexisting minerals by tie-lines so that the coexisting mineral assemblages form sub-triangles. Any point within the CaO-MgO-SiO$_2$ triangle represents a bulk-rock composition with a fixed proportion of the three components (disregarding volatiles and Al$_2$O$_3$ in this case). The ideal mineral assemblage that correlates to any such point should then consist of the three minerals at the corners of the sub-triangle in which the point falls.

If the point happens to fall on a line, the mineral assemblage consists of the two minerals that plot at the ends of the line. In the unlikely case that a chosen point coincides with a mineral composition, a rock with the composition corresponding to the point should consist of that single mineral only.

When we do this for the Crestmore data, an interesting pattern emerges (Figure 21.18). The sequence of prograde zones listed above at Crestmore (labeled 1 → 10 in the figure) corresponds to a simple progression of increasing silica content (as we also deduced from the reactions). The zones at Crestmore could thus result from progressive infiltration of SiO$_2$ from fluids released by the porphyry into the country rock, resulting in a gradient of SiO$_2$ content that is highest at the contact and decreases outward.

Remember also that H$_2$O lowers the melting point of a magma, and thus the loss of H$_2$O as a magma rises will rapidly raise the melting point, causing the magma to suddenly become undercooled (even at a constant temperature). Many porphyries are created in this fashion. The phenocrysts form while the magma rises, where it is nearly H$_2$O saturated and only slightly undercooled. The finer groundmass forms when the magma is emplaced in a shallow environment and H$_2$O escapes due to the pressure drop, so that the melting point rises and the magma becomes undercooled. The porphyritic nature of the pluton at Crestmore thus supports the idea of fluid release. The escaping silica-saturated water, in this case, permeates the silica-free marbles and a gradient in silica content results due to the diffusion of SiO$_2$. The silica reacts with the carbonates to produce skarns consisting of Ca-Mg silicates, whereas CO$_2$ is liberated by the reactions.

According to Figure 21.18, the zones at Crestmore could have formed at constant temperature and reflect an infiltration gradient in SiO$_2$ only. This is probably not the case, however, because temperature should also increase toward the pluton. Only by knowing the pressure–temperature stability ranges of the minerals, and the pressure–temperature dependence of the reactions relating them, could we fully understand the processes at Crestmore. We shall learn more about these techniques in later chapters.

So ends our broad survey. You now have a general grasp of metamorphism and a vocabulary to help explore the concepts in more detail in the chapters to follow. As you may be noticing by now, metamorphic rocks contain a wide variety of minerals. In the interest of saving space, it will occasionally help to abbreviate the names of minerals. The generally accepted mineral abbreviations are listed inside the back cover so you can look up abbreviations with which you are not familiar.

Summary

This chapter is a survey of metamorphism: a classification and an investigation of the processes, rocks produced, and field settings. Metamorphism involves the solid-state textural and/or mineralogical adjustment of a rock to physical conditions different from those under which it originally formed. It involves processes such as recrystallization of existing minerals, reaction among minerals and between minerals and fluids (to consume some and produce other new ones), and deformation. Metamorphism grades into diagenesis and weathering near the Earth's surface and into melting and igneous processes at high temperatures. The principal agent of metamorphism is temperature, the main factor controlling metamorphic grade. Increasing temperature tends to promote recrystallization (increasing grain size) and drives a sequence of mineral reactions as a rock progressively equilibrates to increasing metamorphic grade. Metamorphic temperatures (grades) may increase along trajectories ranging from high-pressure to low-pressure (high-P/T to low-P/T paths), depending on local tectonic and igneous regimes. Most metamorphic rocks have been deformed, although some low-pressure metamorphic rocks may not be. Some sort of foliation (planar texture) and/or lineation (linear texture) is thus a characteristic of most metamorphic rocks.

Metamorphic fluids, when present, are principally aqueous and supercritical (although CO_2 may predominate in some situations). Prograde metamorphism refers to the changes occurring in a rock in response to increasing metamorphic grade and typically releases fluids from lower-temperature H_2O- or CO_2-bearing minerals. Metamorphism is generally considered to be progressive, in that rocks progressively re-equilibrate in response to increasing metamorphic grade, gradually transforming through a series of lower-temperature to higher-temperature mineral assemblages. Retrograde metamorphism is rarely extensive because it is typically impeded by kinetic factors in cooling rocks and by the lack of sufficient fluids. Metamorphic rocks are thus thought to record the conditions of the maximum temperature (grade) reached.

There are many types of metamorphism. The two predominant types are regional (orogenic) metamorphism (associated with thermal and deformational disturbances in orogenic belts) and contact metamorphism (caused by heat and fluids released by plutonic bodies). Other types are pyrometamorphism, burial, ocean-floor, hydrothermal, fault-zone, and impact (or shock) metamorphism. Metamorphism involving substantial chemical change (other than loss or gain of fluid species) is called metasomatism. Polymetamorphic rocks exhibit an overprint of one metamorphic event on another.

Differential uplift and erosion exposes metamorphic rocks, where they record conditions and natural gradients in temperature, pressure, and fluid composition at depth across an area. Those gradients produce a zonation in the mineral assemblages constituting the rocks that equilibrate across the area. For example, we might walk a traverse in an eroded orogenic belt and cross from non-metamorphosed rocks through zones of progressively higher metamorphic grade or through centimeter-to-meter-scale zones, reflecting metasomatic composition gradients adjacent to a pluton. When mapping such metamorphic rock sequences, it is convenient to locate isograds, representing the first appearance of a particular metamorphic index mineral in the field as one traverses up metamorphic grade. Metamorphic zones occupy the area between isograds and have the same name as the isograd (and index mineral) that forms its low-grade boundary. It is now preferable, when possible, to identify the reaction responsible for isograd changes.

Key Terms

Metamorphism *447*
Weathering/diagenesis *447*
Devolatilization (dehydration, decarbonation) *449*
Crystallization (recrystallization, neocrystallization) *449*
Lithostatic (confining) pressure *449*
Metamorphic field gradient *449*
Metamorphic grade *449*
Hydrostatic pressure *450*
Deviatoric stress *450*
Strain (deformation) *450*
Tension/extension *450*
Compression/flattening *451*
Strain ellipsoid *451*
Foliation *451*
Lineation *451*
Shear *451*

Supercritical fluid *452*
Fluid pressure *452*
Pressure solution *452*
Partial pressure *452*
Metasomatism/isochemical metamorphism *453*
Thermal/dynamic/dynamo-thermal metamorphism *453*
Contact metamorphism *453*
Contact aureole *454*
Polymetamorphism *454*
Pyrometamorphism *454*
Regional metamorphism *454*
Orogenic metamorphism *455*
Burial metamorphism *455*
Ocean-floor metamorphism *456*
Hydrothermal metamorphism *456*
Fault-zone metamorphism *457*
Impact (shock) metamorphism *457*

Tectonic versus magmatic underplating *455*
Metamorphic core complex *455*
Prograde/retrograde metamorphism *457*
Progressive metamorphism *457*
P-T-t path *458*
Geothermobarometry *458*
Protolith (and types) *458*
Metamorphic zone *459*
Index mineral *459*
Barrovian-type metamorphism *459*
Isograd *460*
Buchan-type and Abukuma-type metamorphism *460*
Paired metamorphic belts *462*
Chemographic compatibility, or composition-paragenesis, diagrams *467*

Review Questions

Review Questions are located on the author's web page at the following address: **http://www.prenhall.com/winter**

Important "First Principle" Concepts

- Geologists are attempting to piece together the history of the Earth's development. Metamorphic petrologists contribute by using the mineralogy, textures, and geometry of metamorphic rocks and areas to interpret the development of past orogenic belts, contact aureoles, shear zones, etc. The protolith also yields information about the nature of the pre-metamorphic conditions.
- Metamorphism is largely a solid-state phenomenon, involving mineralogical or textural readjustment to changing conditions of temperature, pressure, composition, and/or stress.
- Metamorphic rocks seem to readily equilibrate to prograde metamorphism, but not well to retrograde. They thus generally record the maximum grade (temperature) attained during metamorphism.

- Differential uplift and erosion expose metamorphosed rocks spanning a range of grades. The metamorphic rocks can be mapped using isograds (defining reactions or the first appearance of a new "index" mineral), which separate metamorphic zones, each having characteristic mineral assemblages.
- Barrovian-type metamorphism represents a common P/T type of regional orogenic metamorphism, with the sequence of zones chlorite–biotite–garnet–staurolite–kyanite–sillimanite. Other P/T trajectories may produce higher-pressure or lower-pressure minerals in other orogenic belts. Isograds and zones may also be used in contact or other types of metamorphism.
- Mineral changes are an important hallmark of metamorphism. Such changes involve chemical reactions, consuming reactant minerals ± fluids and producing other minerals ± fluids.

Suggested Further Readings

Bucher, K., and M. Frey. (2002). *Petrogenesis of Metamorphic Rocks*, 7th ed. Springer-Verlag. Heidelberg.

Fettes, D. and J. Desmons (eds.) (2007). *Metamorphic Rocks: A Classification and Glossary of Terms. Recommendations of the International Union of Geological Sciences Subcommission on the Systematics of Metamorphic Rocks, p. 256.* Cambridge University Press. Cambridge.

Miyashiro, A. (1994). *Metamorphic Petrology.* Oxford University Press. Oxford, UK.

Treolar, P. J., and P. J. O'Brien (eds.). (1998). *What Drives Metamorphism and Metamorphic Reactions?* Special Publication, **138**. Geological Society. London.

Vernon, R. H. and G. L. Clarke (2008). *Principles of Metamorphic Petrology.* Cambridge University Press. Cambridge, U.K.

Yardley, B. W. D. (1989). *An Introduction to Metamorphic Petrology.* Longman. Essex, UK.

22

A Classification of Metamorphic Rocks

Questions to be Considered in this Chapter:

1. On what basis are metamorphic rocks classified and named?

2. What general terms are used to describe the basic textural and compositional parameters?

Like other rocks, metamorphic rocks are classified on the basis of *texture* and *composition* (either mineralogical or chemical). Although we may consider deriving a proper name for a rock as an exercise in finding the right pigeonhole, we are much better served by considering a good name as a concise way of imparting information about a rock. Unlike igneous rocks, which have been plagued by a proliferation of local and specific names, metamorphic rock names are surprisingly simple and flexible. I appreciate the flexibility in how we approach naming a metamorphic rock. In addition to a "proper" (or "root") name from the scheme that follows, we may choose some prefix-type modifiers to attach to names if we care to stress some textural or mineralogical aspects that we deem important or unusual.

The classification and textural description of metamorphic rocks are intimately related, and a proper job requires good thin section and hand specimen work. Metamorphic textures are quite varied, capable of recording a complex interplay between deformation and crystallization. The description and interpretation of metamorphic textures will be left to the following chapter. In the present chapter we shall concentrate on hand-specimen description and classification, using only the textural features necessary to do so.

The IUGS-SCMR had recommended a flowchart and glossary for naming a metamorphic rock (Fettes and Desmons, 2007). Those desiring a more formalized and detailed approach are referred to that work. In what follows, I shall take a slightly simplified approach that should work for all but the most unusual specimen. Rocks that are only slightly metamorphosed and dominated by the original igneous or sedimentary textures are typically named for the protolith with the prefix "meta-" (e.g., meta-siltstone, meta-basalt). In the simplest and most general sense, for a more clearly metamorphosed rock, we approach naming it by first deciding if it is a high-strain rock or not. The decision may require thin section and/or field relationships to help us decide. High-strain rocks have their own classification scheme (Section 22.5). If a rock is not high strain, we must next observe whether the rock is foliated/lineated. If the rock is not high strain but is foliated, we determine the nature of the foliation and call the rock a slate, phyllite, schist, or gneiss. If the rock is non-foliated, we call it a granofels (or hornfels; the distinction is explained further below).

The foregoing simple classification would cover virtually all metamorphic rocks, but the categories are much too broad to provide a satisfactory subdivision that imparts much useful information. Knowing that a rock is a schist does not really give you an adequate picture of the nature of the rock other than that it is foliated and fairly fine grained. To improve this situation, we may want to add a few mineral and/or textural modifiers (such as "staurolite–kyanite schist," or "andalusite spotted hornfels") to convey some further information about a rock's composition or texture. When we do so, it is good to have some general guidelines. For example, is there more staurolite or more kyanite in a staurolite–kyanite schist? In what order do we place them? There must be other minerals in such a schist, why not name them all? What textural information should one include? Some of these decisions are left to our own judgment and the information that we want to convey; others have been more formalized. In addition, there are a number of metamorphic rock types that are so common that we have specific names for them. These names usually take precedence over our simple foliated versus non-foliated scheme above. For example, metamorphosed quartz sandstone and limestone are rarely foliated and are called quartzite and marble, respectively, instead of quartz granofels or calcite granofels. We would probably call them the same thing even if foliated (although we might add a modifier, such as "schistose marble"). We shall have to become familiar with these special names if we are to develop a comprehensive system for naming metamorphic rocks that conforms with common practice.

22.1 FOLIATED AND LINEATED ROCKS

As introduced in Section 21.2.3 and discussed more fully in Section 23.4, **foliation** and **lineation** refer to planar and linear fabric elements, respectively, in a rock and have no genetic connotations. Some high-strain rocks may also be foliated, but, as mentioned above, these are treated separately. In the next chapter we shall discuss the textures in more detail, including mechanisms by which they may have been generated. For now we shall discuss the textures only to the extent that we can identify them and use them to aid in classification. Rocks with multiple foliations and/or lineations are also possible.

Minerals may exhibit a preferred orientation in two ways. **Dimensional preferred orientation (DPO)** is the more obvious, meaning that some minerals are either platy or elongated, and aligned such that one can easily see the parallelism of shapes. **Lattice preferred orientation (LPO)** is a preferred orientation of crystallographic elements (most commonly crystallographic axes). Of course when phyllosilicates are aligned so that the plates are coplanar, their c-axes are also parallel, so LPO usually accompanies DPO. Some minerals, however, rarely exhibit good crystal shapes in metamorphic rocks, yet may still have a lattice preferred orientation of crystallographic elements. Quartz and olivine are two notable examples of minerals that typically appear granular, thus lacking a dimensional orientation, yet may have a lattice-preferred orientation. Determining such a crystallographic alignment requires some tedious labor using a petrographic microscope and a universal stage. We shall restrict our present discussion to DPO and things that we can see in hand specimen.

In general, foliations in non-high-strain rocks are caused by orogeny and regional metamorphism, and the type of foliation varies with metamorphic grade. In order of increasing grade, they are:

Cleavage. Traditionally: the property of a rock to split along a regular set of subparallel, closely spaced planes. A more general concept adopted by some geologists is to consider cleavage to be any type of foliation in which the aligned platy phyllosilicates are too fine grained to see individually with the unaided eye.

Schistosity. A preferred orientation (DPO) of inequant mineral grains or grain aggregates produced by metamorphic processes. Aligned minerals are coarse grained enough to see with the unaided eye. The orientation is generally planar, but linear orientations are not excluded.

Gneissose structure. Either a poorly developed schistosity or segregation into layers by metamorphic processes. Gneissose rocks are generally coarse grained.

The rock names that follow from these textures are given below. Again, these names are listed in a sequence that generally corresponds with increasing grade:

Slate. Figure 22.1a. A compact, very fine-grained, metamorphic rock with a well-developed cleavage. Freshly cleaved surfaces are dull. Slates look like shales, but have a more ceramic ring when struck with a hammer.

Phyllite. Figure 22.1b. A rock with a schistosity in which very fine phyllosilicates (sericite/phengite and/or chlorite), although rarely coarse enough to see unaided, impart a silky sheen to the foliation surface. Phyllites with both a foliation and lineation (typically crenulated fold axes) are very common.

Schist. Figure 22.1c. A metamorphic rock exhibiting a schistosity. By this definition, schist is a broad term, and slates and phyllites are also types of schists. The more specific terms, however, are preferable. In common usage, schists are restricted to those metamorphic rocks in which the foliated minerals are coarse enough to see easily in hand specimen.

Gneiss. Figure 22.1d. A metamorphic rock displaying gneissose structure. Gneisses are typically layered (also called banded), generally with alternating felsic and darker mineral layers. Gneisses may also be lineated, but must also show segregations of felsic-mineral-rich and dark-mineral-rich concentrations. Gneissic layers or concentrations need not be laterally continuous.

22.2 NON-FOLIATED AND NON-LINEATED ROCKS

This category is simpler than the previous one. Again, this discussion and classification applies only to rocks that are not produced by high-strain metamorphism. A comprehensive term for any isotropic rock (a rock with no preferred orientation) is a **granofels**. **Granofels(ic) texture** is then a texture characterized by a lack of preferred orientation. An outdated alternative is *granulite*, but this term is now used to denote very high-grade rocks (whether foliated or not) and is not endorsed here as a synonym for granofels. A **hornfels** is a type of granofels that is typically very fine grained and compact, and it occurs in contact aureoles. Hornfelses are tough and tend to splinter or display conchoidal fracture when broken. (The SCMR allows that hornfelses may be of any grain size.)

As we shall see in the next chapter, many metamorphosed rocks experience more than one deformational and/or metamorphic event. Such **polymetamorphosed** rocks typically exhibit the overprinted effects of the successive events.

FIGURE 22.1 Examples of foliated metamorphic rocks.
(a) Phyllite. (b) Slate. Note the difference in reflectance on the foliation surfaces between a and b: phyllite is characterized by a satiny sheen. (c) Garnet muscovite schist. Muscovite crystals are visible and silvery, garnets occur as large dark porphyroblasts.
(d) Quartzo-feldspathic gneiss with obvious layering.

22.3 SPECIFIC METAMORPHIC ROCK TYPES

As mentioned above, some rock types are sufficiently common that they have been given special names, typically based on a common and specific protolith (Section 21.5),

but many also imply a specific range of metamorphic grade. It is also proper to name a metamorphic rock by adding the prefix meta- to a term that indicates the protolith, such as meta-pelite, meta-ironstone, etc.

The commonly used names that specify a particular rock type are listed below. As a rule, these names take precedence over the textural names described above:

Marble. A metamorphic rock composed predominantly of calcite or dolomite. The protolith is typically limestone or dolostone.

Quartzite. A metamorphic rock composed predominantly of quartz. The protolith is typically sandstone. Some confusion may result from the use of this term in sedimentary petrology for a pure quartz sandstone.

Greenschist/greenstone. A low-grade metamorphic rock that typically contains chlorite, actinolite, epidote, and albite. Note that the first three minerals are green, which imparts the color to the rock. Such a rock is called greenschist if foliated and greenstone if not. The protolith is either a mafic igneous rock or graywacke.

Amphibolite. A metamorphic rock dominated by hornblende + plagioclase. Amphibolites may be foliated (gneissose) or non-foliated. The protolith is either a mafic igneous rock or graywacke.

Serpentinite. An ultramafic rock metamorphosed at low grade, so that it contains mostly serpentine.

Blueschist. A blue-amphibole-bearing metamorphosed mafic igneous rock or mafic graywacke. This term is so commonly applied to such rocks that it is even applied to non-schistose rocks. Glaucophane is the most common blue amphibole, and *glaucophane schist* has been commonly applied to rocks known to contain it.

Eclogite. A green and red metamorphic rock that contains clinopyroxene and garnet (omphacite + pyrope). The protolith is typically basaltic. Eclogites contain no plagioclase.

Calc-silicate rock (granofels or schist). A rock composed of various Ca-Mg-Fe-Al silicate minerals, such as grossular, epidote, tremolite, vesuvianite, etc. The protolith is typically a limestone or dolostone with silica either originally present as clastic grains or introduced metasomatically.

Skarn. A calc-silicate rock (see immediately above) formed by contact metamorphism and silica metasomatism from a pluton into an adjacent carbonate rock. **Tactite** is a synonym.

Granulite. A high-grade rock of pelitic, mafic, or quartzo-feldspathic parentage that is predominantly composed of OH-free minerals. Muscovite is absent, and plagioclase and orthopyroxene are common.

Migmatite. A composite silicate rock that is heterogeneous on the 1- to 10-cm scale, commonly having a dark gneissic matrix (melanosome) and lighter felsic portions (leucosome). Migmatites may appear layered, or the leucosomes may occur as pods or form a network of cross-cutting veins (see Section 28.5).

For a more comprehensive description of these rock types and their textures, see Chapter 2 of Shelley (1993) or several chapters and the glossary in Fettes and Desmons (2007). Any of these terms, or a term indicating the protolith, may be combined with the textural classification terms if it helps describe a rock more fully. One may choose to use either term as a modifier (e.g., *schistose amphibolite, amphibolitic schist, pelitic schist, phyllitic meta-tuff*, etc.).

22.4 ADDITIONAL MODIFYING TERMS

Remembering that the main purpose of naming a rock is to impart information about the nature of the rock to others, you can add modifying terms to the above list as you see fit. Aspects of a rock to consider include the mineralogy, structure, protolith, chemical composition, and metamorphic grade or conditions. For example, if you want to emphasize some structural aspect of the rock, you may add terms such as lineated, layered, banded, or folded. **Porphyroblastic** means that a metamorphic rock has one or more metamorphic minerals that grew much larger than the others. Each individual crystal is a **porphyroblast**. If you want to call attention to this texture in a sample, you may use a name such as *kyanite porphyroblast schist*. Some porphyroblasts, particularly in low-grade contact metamorphism, occur as ovoid **spots** (Figure 23.14). If such spots occur in a hornfels or a phyllite (typically as a contact metamorphic overprint over a regionally developed phyllite), the term **spotted hornfels** or **spotted phyllite** would be appropriate. Some gneisses have large eye-shaped grains (commonly feldspar) that are derived from pre-existing large crystals by shear (as described in Section 23.1). Individual grains of this sort are called **auge** (German for *eye*), and the (German) plural is **augen**. An **augen gneiss** is a gneiss with augen structure (Figure 23.18). We will investigate these and other textural terms more fully in Chapter 23.

Other modifying terms that we may want to add as a means of emphasizing some aspect of a rock may concern such features as grain size, color, and chemical aspects (aluminous, calcareous, mafic, felsic, etc.). As a general rule, we use these when the aspect is unusual. Obviously a *calcareous marble* or *mafic greenschist* is redundant, as is a *fine-grained slate*.

Two common prefixes that pertain to protolith are the terms ortho- and para-. **Ortho-** indicates an igneous parent, and **para-** indicates a sedimentary parent. The terms are used only when they serve to dispel doubt. For example, many quartzo-feldspathic gneisses could easily be derived from either an impure arkose or a granitoid rock. If some mineralogical, chemical, or field-derived clue permits the distinction, terms such as *orthogneiss, paragneiss,* or *orthoamphibolite* may be useful.

Although textural criteria are commonly emphasized in metamorphic rock names, it may be desirable to include some information about the mineralogy of a metamorphic rock in the name as well. The mineralogical classification of metamorphic rocks is much less formally defined than that of igneous rocks (Chapter 2). One would include mineral-content modifiers when they are significant in terms of content or as indicators of grade (such as an index mineral). The number of minerals listed depends on the intent and emphasis of the project. If, for example, an investigator wants to distinguish metamorphosed pelitic sediments on the basis of composition, she may choose a longer list of minerals in the names. If the objective is to emphasize metamorphic grade, however, she may elect to include only the index minerals. The SCMR recommends listing all major minerals in a rock. By convention, minerals are listed in the order of *increasing modal abundance* in a rock.

22.5 HIGH-STRAIN ROCKS

Table 22.1 shows a simple scheme for classifying dynamically metamorphosed rocks associated with fault zones (including more ductile shear zones). For the moment, we shall consider the terms as purely descriptive, so that we can use Table 22.1 to readily identify high-strain rocks based on observed textural criteria in hand specimen and thin section. We shall discuss the processes involved in more detail in the next chapter. To name a high-strain rock, one must determine whether the rock is cohesive or whether it falls apart and then estimate the relative proportions of large clasts versus fine matrix (presumably derived mechanically from the once-larger clasts). A rock without cohesion is either a **fault breccia** or **fault gouge** (left column of Table 22.1). Such cohesionless rocks are typically restricted to very shallow crustal levels. Gouge is usually altered by groundwater to a clay-rich matrix.

Cohesive rocks are further distinguished by being either foliated or non-foliated, both of which are usually more fine-grained than the non-cohesive varieties. Non-foliated

TABLE 22.1	**Classification of High-Strain Fault Zone Rocks**				
% fine matrix	Rocks without primary cohesion	Rocks with primary cohesion			
		Non-foliated	Foliated		Glass in matrix
50	Fault breccia	Microbreccia	Protomylonite	Blastomylonite (if significantly recrystallized)	Pseudotachylite
70			Mylonite		
90	Fault gouge	Cataclasite	Ultramylonite		

After Higgins (1971)

cohesive rocks are either **microbreccias** (<70% clasts) or **cataclasites** (>70% clasts). Foliated cohesive rocks are **mylonites** and are subdivided as shown in the third column of Table 22.1. Some geologists use the term **phyllonite** to indicate a mica-rich mylonite. As with practically any other classification, all the categories in Table 22.1 are gradational.

Virtually all high-strain processes involve grain size reduction. The degree of recrystallization in the matrix is not a factor unless it is very advanced (perhaps aided by a later metamorphic event). The prefix **blasto-** is then added, meaning that the mylonitic texture is still apparent but largely inherited from an earlier high-strain event. In high-strain rocks, the prefix blasto- is usually restricted to foliated rocks because a recrystallized cataclasite may be difficult to recognize without the foliation as a clue. In extreme cases of high-strain deformation, thin, commonly anastomosing seams of glassy rock, known as **pseudotachylite** (Shand, 1916), are generated (Figure 23.16b). The glass is attributed to melting due to frictional heat.

Figure 22.2 shows the common distribution of fault-related rock types with depth. In shallow areas, the fault zone is narrower and involves more brittle behavior, resulting in anastomosing faults containing rocks that commonly lack cohesion (breccia and gouge). Below this zone, the confining pressure forces the walls of the fault together and deformation becomes more thorough and pervasive so that the constituents become finer. The rocks thus grade downward into cataclasites and microbreccias, occasionally with pseudotachylites. Deeper yet, the rocks take on a foliated character, and recrystallization accompanies deformation, producing mylonites. The processes will be discussed in

more detail in the next chapter. The width of the shear zone increases with depth and becomes both more ductile and more evenly distributed throughout the rock matrix. At the greatest depths, the shear is distributed over a wide area and is not very intense at any single place. The rocks are ductile, and gneisses generally develop that are indistinguishable from gneisses of regional orogenic metamorphism.

The terminology for high-strain rocks is complicated by recent advances in our understanding of the deformation processes involved in their generation. The classification in Table 22.1 was developed before our modern understanding of the importance of recrystallization and the ductile processes that commonly accompany shear-zone deformation. It suffers from using terms that have genetic implications of purely brittle behavior. The term **cataclasis**, as originally conceived, refers to a process of mechanical crushing and granulation of a rock and its mineral constituents (with no accompanying recrystallization). This was once generally accepted as the predominant process operating in most fault and shear zones. **Cataclastic** is an adjective form used to describe the processes and textures resulting from cataclasis. Lapworth (1885) first used the term **mylonite** to describe a fine-grained laminated rock from the Moine Thrust of Scotland. The term was derived from the Greek word *mule*, for mill, meaning "to grind." Although cataclasis is clearly the dominant process acting in shallow brittle fault zones, resulting in broken up fault gouges and breccias, this is usually not the case at even a modest depth.

Figure 22.2 indicates that purely mechanical grinding in fault zones quickly gives way with depth to processes that combine brittle deformation with recrystallization. As we

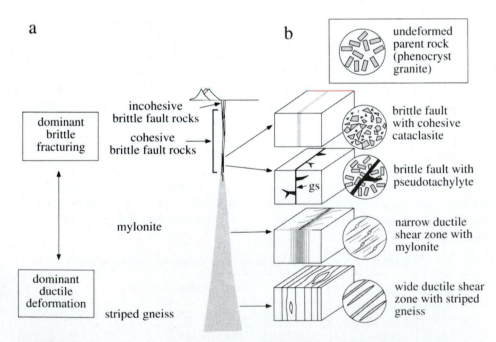

FIGURE 22.2 Schematic cross section through a shear zone, showing the vertical distribution of fault-related rock types, ranging from non-cohesive gouge and breccia near the surface through progressively more cohesive and foliated rocks. Note that the width of the shear zone increases with depth as the shear is distributed over a larger area and becomes more ductile. Circles on the right represent microscopic views or textures. From Passchier and Trouw (2005). Copyright © with permission from Springer-Verlag.

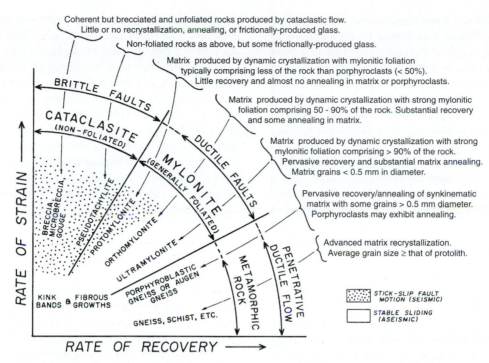

FIGURE 22.3 Terminology for high-strain shear-zone related rocks proposed by Wise et al. (1984). Copyright © The Geological Society of America, Inc.

shall see in the next chapter, deformation increases the energy of strained crystal lattices and enhances recrystallization. Only a small amount of heat may be sufficient to permit recrystallization of the small strained grains produced by high degrees of shear. Our understanding of mylonites advanced significantly due to the work of Bell and Etheridge (1973), who demonstrated that the processes involved in the production of mylonites were ductile processes (rapid strain recovery and recrystallization), not mere brittle crushing. Although our understanding improved, our terminology did not because the roots of the terms cataclasite and mylonite had direct implications of purely cataclastic processes.

Attempts to remedy this situation and avoid using terms in ways contrary to their original intent have thus far traded one problem for another. Figure 22.3 is a classification proposed by Wise et al. (1984). These workers clearly recognized that deformation in shear zones commonly involves coeval deformation and recovery–recrystallization processes, and they attempt to relate these processes in a clear fashion by comparing the rate of strain versus the rate of recovery. Although I can appreciate the attempt to understand the processes involved and relate them to the products, it is very difficult to infer the rates of strain versus recovery from a rock in hand specimen. We have thus traded a classification that we can easily use for one that we can't easily use, but is based on a better understanding of the processes and involves less misuse of terms. I prefer to have a classification that is easy to use, so I choose Table 22.1 for naming a shear-related rock and refer to Figure 22.3 for an understanding of how the processes of deformation and recovery relate to the continuous spectrum of rock types from truly cataclastic fault breccias to well-recrystallized schists. Several of the terms used in the figure are discussed in the next chapter, so it may be a

more useful reference for you after you have read further and return to when addressing hand specimens.

Perhaps we can use Table 22.1 and discuss the rock types and processes if we use the terms *cataclasite* and *mylonite* in a purely descriptive sense. Bell and Etheridge (1973), among others, have advocated that we redefine the term *mylonite* to strip it of any genetic implications. Remembering the root of the term, this makes it a bit of a

FIGURE 22.4 Shatter cones in limestone from the Haughton Structure, Northwest Territories. Sample is ~9 cm across. Photograph courtesy of Richard Grieve © Natural Resources Canada.

misnomer, but I think it is a fair compromise. Sixty-one participants at a 1981 Penrose Conference on mylonitic rocks generally agreed, although not unanimously, that mylonites should occur in a relatively narrow zone (< a few tens of kilometers wide), involve grain-size reduction, and have a distinct foliation and/or lineation (Tullis et al., 1982). Purely brittle behavior was no longer required. Likewise, a cataclasite may have some accompanying recrystallization and may thus not be purely cataclastic.

Impact rocks are a separate category of high-strain rocks and do not really lend themselves to classification using Table 22.1. Impacts are catastrophic events, producing shock waves and raising the temperature at impact to thousands of degrees, melting (and even vaporizing) some minerals. The rocks can all be called **impactites**. All impactites are breccias and have characteristic field settings (impact crater and ejecta blanket, if preserved) and textural features, such as shocked crystals (highly deformed lattices, commonly with multiple sets of planar deformation features), and amorphous glassy phases. High-pressure silica polymorphs, such as coesite and stishovite, may be present. Impact areas may have macroscopic nested cone-like structures called shatter cones (Figure 22.4). See Hibbard (1995) and IUGS-SCMR (Fettes and Desmons, 2007, which had a subcommittee working on a classification of impact-related rocks) for descriptions.

Summary

Assigning a name to a metamorphic rock involves the following steps:

1. Is the rock only slightly metamorphosed and dominated by the original igneous or sedimentary textures? If so, name if after the protolith, with the prefix "meta-" (e.g., meta-conglomerate). If not, proceed.
2. Is the rock a high-strain rock? If so, use Table 22.1. If not, proceed.
3. Is the rock foliated? If so, name it a slate, phyllite, schist, or gneiss, based on the nature of the foliation.

Add appropriate mineral constituents (in order of increasing abundance). If not, proceed.

4. Non-foliated rocks are granofelses (or hornfelses), also with appropriate mineral names.
5. Would one of the more specific terms in Section 22.3 be more appropriate or informative?
6. Is there any other information that would be useful or that you would like to emphasize, based on mineralogy, structure, protolith, chemical composition, and metamorphic grade or conditions?

Key Terms

Metamorphic rock names *470*
Dimensional preferred orientation (DPO) *471*
Lattice preferred orientation (LPO) *471*

Cleavage *471*
Schistosity *471*
Gneissose structure *471*
Porphyroblast/spot *473*

Augen *473*
Ortho-/para- *473*

Review Questions

Review Questions are located on the author's web page at the following address: **http://www.prenhall.com/winter**

Important "First Principle" Concept

■ Metamorphic rock names are not as rigidly prescribed as are igneous names. Remember, you are trying to convey information about the rock, so a good name is one that describes the rock well.

Suggested Further Readings

Fettes, D., and J. Desmons (eds.). (2007). *Metamorphic Rocks: A Classification and Glossary of Terms. Recommendations of the International Union of Geological Sciences Subcommission on the Systematics of Metamorphic Rocks.* Cambridge University Press. Cambridge, UK.

Hibbard, M. J. (1995). *Petrography to Petrogenesis.* Prentice Hall. Englewood Cliffs, NJ. Chapter 23 and Chapter 24.

Higgins, M. W. (1971). *Cataclastic Rocks.* USGS Prof. Paper **687**.

Shelley, D. (1993). *Igneous and Metamorphic Rocks Under the Microscope.* Chapman & Hall. London.

Tullis, J., A. W. Snoke, and V. R. Todd. (1982). *Penrose Conference Report. Significance and petrogenesis of mylonitic rocks. Geology,* **10**, 227–230.

Yardley, B. W. D. (1989). *An Introduction to Metamorphic Petrology.* Longmans. Essex, UK.

23

Structures and Textures of Metamorphic Rocks

Questions to be Considered in this Chapter:

1. What textures develop during the various types of metamorphism?

2. How do solid-state metamorphic textures differ from igneous (and sedimentary) textures, and how can we use those textures to interpret the style of and conditions accompanying metamorphism?

3. What processes occur within crystals and at grain boundaries during solid-state deformation, recrystallization, and/or reaction?

4. How can metamorphic textures be used to interpret the metamorphic, orogenic, and even pre-metamorphic history of rocks?

5. How can geochronology be linked more specifically to textural criteria to determine the age of specific (even multiple) events and the rates of metamorphic or tectonic processes?

I use the term **texture** to refer to small-scale features in a rock that are **penetrative**, meaning that the texture occurs in virtually all of the rock body at the microscopic scale. **Structures** are larger-scale features that occur on the hand sample, outcrop, or regional scale. Materials scientists, who typically work with metals or ceramics, use "texture" to refer strictly to preferred orientations. Several geologists (including the IUGS-SCMR) advocate conforming to the material science literature and using the term **micro-structure**, instead of texture, to describe any small-scale features of rocks. This relatively new usage, however, restricts the meaning of *texture*, as applied elsewhere in geologic literature, and it remains to see if it will be widely adopted. **Fabric** is also used by many geologists as a synonym for texture. I'm old fashioned, I suppose, and will stick with the definition of texture as first defined above.

This chapter concentrates on textures observable in thin section, although many may also be seen in hand specimen. As I said in Chapter 20, geology generally behaves like fractals because the structures seen on a regional scale (such as folding) are similar in style and orientation to those seen in the outcrop, hand specimen, and thin section. Petrographic data can thus be coordinated with petrology and structural geology as part of the ultimate goal of regional (and eventually global) synthesis of geologic history.

As in igneous rocks (Chapter 3), the textures of metamorphic rocks reflect the combined processes of crystal nucleation, crystal growth, and diffusion of matter (Daniel and Spear, 1999). These processes, however, are generally more complex in solid-state metamorphic systems. For example, the growth of a new metamorphic mineral typically results from the reaction of other minerals that have become unstable. Some overstepping of the actual *P-T* boundary conditions is usually necessary before any reaction is really effective, particularly under lower temperature metamorphic conditions. Except for some cases of polymorphic transformations, metamorphic mineral growth generally involves:

1. Detachment of ions from the surface of reacting minerals (presumably dissolving into a small amount of intergranular fluid in most situations)
2. Nucleation of the new mineral(s)

3. Diffusion of material to the sites of new mineral growth

4. Growth of the new mineral(s), incorporating components carried to the surface and transportation of excess waste constituents away

The step that is slowest under any set of conditions is the *rate-determining* step for the process.

The combined effects of deformation and recovery–recrystallization, as well as variations in the timing of deformation versus crystallization, make for more textural variability in metamorphic rocks than is common in igneous ones. The majority of these effects leave clear textural imprints, but the interaction of several processes and the repetition of some make it more challenging to interpret metamorphic histories on the basis of textural criteria.

In most cases, the principles behind textural interpretation are clear and relatively simple, but the application of the principles may at times be ambiguous. In this chapter, we shall attempt to review the principles involved so that we can interpret the processes responsible for the development of a metamorphic rock by careful observation of the products.

First, let's define a few basic textural terms that will make the ensuing discussion easier. The suffix **-blast** or **-blastic** indicates that a feature is of metamorphic origin. Thus **porphyroblastic** means a porphyritic-like texture (large grains in a finer matrix) that is of metamorphic origin. Regrettably, the *prefix* **blasto-** (meaning that a feature is *not* of metamorphic origin but is inherited from the parent rock) is too easily confused with the *suffix* form meaning just the opposite. For example, *blastoporphyritic* indicates an igneous porphyritic texture that survived metamorphism to the extent that it can still be recognized. The similarity between *blastoporphyritic* and *porphyroblastic*, and numerous other such pairs, is unfortunate. The term **relict**, like the prefix blasto-, indicates that a feature is inherited from the protolith. We can thus speak of relict bedding in metasediments, or relict porphyritic texture, or even of individual relict minerals. Because of the blasted confusion associated with the prefix and suffix forms of *blast*, I prefer to use the term *relict* rather than the prefix *blasto-*.

Perfectly straightforward terms such as euhedral, subhedral, and anhedral (indicating progressively less-well-shaped crystals) have been replaced in some of the metamorphic literature with **idioblastic**, **hybidioblastic**, and **xenoblastic**, respectively. The terms stem from somewhat dated general ones (idiomorphic, hypidiomorphic, etc.), but I doubt that a clear metamorphic distinction justifies the inflation of our lexicon in this case, and I prefer the simple universal terms listed first. The metamorphic terms are well established, however, and useful in some contexts.

The IUGS-SCMR recommended use of *phaneritic* and *aphanitic* as grain size terms, following the usage advocated in Chapter 2 for igneous rocks (Fettes and Desmons, 2007). The IUGS-SCMR recommended that other grain size terms not be associated with any absolute values because there are no universal standards. For those stubbornly requiring absolute values, however, the IUGS-SCMR suggested the following scheme (numeric values in parentheses represent the limiting diameters): ultra-fine grained (0.01 mm), very fine grained (0.1 mm), fine grained (1 mm), medium grained (4 mm), coarse grained (16 mm), and very coarse grained. That these limits do not correlate to those recommended for igneous rocks in Chapter 2 seems to strengthen the IUGS-SCMR's point.

In what follows, I shall attempt to summarize the principal mechanisms of solid-state deformation and recrystallization, followed by a review of the textures characteristic of the three principal types of metamorphism: contact, regional/orogenic, and fault zone. For a more detailed analysis, see the excellent summaries of metamorphic textures and deformation mechanisms by Passchier and Trouw (2005) and Vernon (2004).

23.1 THE PROCESSES OF DEFORMATION, RECOVERY, AND RECRYSTALLIZATION

Deformation of crystalline solids involves a number of processes, several of which may work in unison. The dominant processes at any particular time depend on both intrinsic rock factors (mineralogy; grain size and orientation; and the presence, composition, and mobility of intergranular fluids) and externally imposed factors (temperature, pressure, deviatoric stress, fluid pressure, and strain rate). The principal deformation mechanisms are listed below in the general order of increasing temperature and/or decreasing strain-rate, following the approach of Passchier and Trouw (2005).

1. Cataclastic flow is the mechanical fragmentation of a rock and the sliding and rotation of the fragments. These brittle processes include frictional grain-boundary sliding and fracture (Vernon, 2004). The products are fault gouge, breccia, or cataclasite.

2. Solution transfer (Figures 21.4 and 23.1), also called **pressure solution**, requires intergranular fluid to be effective. Grain contacts at a high angle to σ_1 (the shortening direction) become highly strained and have higher energy. The material at these contacts dissolves more readily as a result of the higher energy and produces higher concentrations of dissolved species at these locations. This, in turn, sets up an activity gradient so that dissolved species migrate from high activity to low activity places (which are also low-strain places) where the material precipitates.

3. Intracrystalline deformation of a plastic type involves no loss of cohesion in the rock. Several processes may be involved, often simultaneously. Simple torquing and bending of bonds in a crystal is elastic and quickly recoverable. *Permanent* deformation requires more significant changes in the position of atoms/ions, typically involving the breaking of chemical bonds. In intracrystalline deformation, this happens most easily by movement of lattice **defects**. (Defects are described in most mineralogy texts and can be imaged and studied by electron microscopy techniques.) **Point defects** include *vacancy* and *interstitial* types, and **line defects** include *edge* and *screw dislocations*. There is also a variety of combined defects.

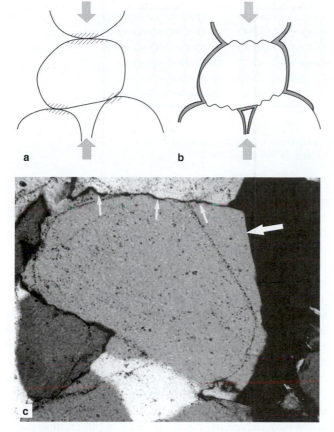

FIGURE 23.1 Pressure solution in grains affected by a vertical maximum stress and surrounded by a pore fluid. **(a)** Highest strain in areas near grain contacts (hatch pattern). **(b)** High-strain areas dissolve and material precipitates in adjacent low-strain areas (shaded). The process is accompanied by vertical shortening. **(c)** Pressure solution of a quartz crystal in a deformed quartzite (σ_1 is vertical). Pressure solution results in a serrated solution surface in high-strain areas (small arrows) and precipitation in low-strain areas (large arrow). ~0.5 mm across. The faint line within the grain is a hematite stain along the original clast surface. After Hibbard (1995).

Figure 23.2 illustrates lattice deformation via a process of vacancy migration to accommodate stress by shortening (flattening). Figure 23.3 shows accommodation of shear by edge-dislocation migration. Defect migration in either case is crystallographically controlled and can happen only along certain **slip directions** in certain **slip planes**; the combination is called a **slip system**. Both processes in the two figures allow deformation to occur by displacing only a tiny part of a crystal at a time, thus permitting strain to occur with much less deviatoric stress than if the whole lattice had to yield along a fracture at once. Defects, in other words, weaken rocks. Point defect migration also enhances *diffusion* through a lattice because the motion of a vacancy in one direction is equivalent to migration of matter in the opposite direction.

Dislocation glide involves the migration of dislocations along a slip system (as in Figure 23.3). Most plastically or ductilely deformable crystals (quartz, carbonates, ice) have several potential slip systems with different orientations, which enhances their capacity to accommodate stress. When glide along different active slip systems intersects,

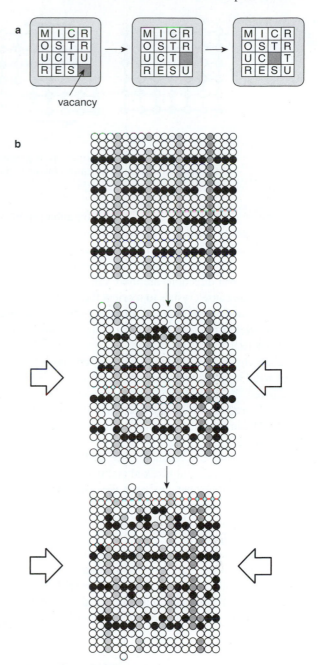

FIGURE 23.2 **(a)** Migration of a vacancy in a familiar game. **(b)** Plastic horizontal shortening of a crystal by vacancy migration. From Passchier and Trouw (2005). Copyright © with permission from Springer-Verlag.

migrating dislocations can interfere, forming dislocation "tangles," inhibiting further motion on all slip systems. A greater degree of differential stress is then required to overcome the tangle and permit slip. Because more stress is required for deformation, material with numerous tangles is effectively stronger. This is called **strain hardening**. Obstructions can be overcome by **dislocation creep**, in which vacancies migrate to the dislocations and allow them to "climb" over a blocked site.

As dislocations form and migrate, portions of a crystal's lattice become reoriented. An easily observed result of this is **undulose extinction** (Figure 23.4), because the extinction position of a mineral under the polarizing

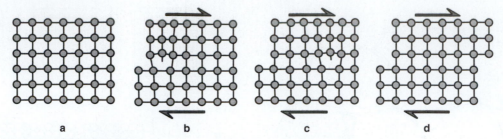

FIGURE 23.3 Plastic deformation of a crystal lattice (experiencing dextral shear) by the migration of an edge dislocation (as viewed down the axis of the dislocation).

microscope is dependent upon its crystallographic orientation. **Deformation twinning** (Figure 3.19) can also accommodate a limited amount of strain in specific crystallographic directions. It is common in calcite and plagioclase. Because some slip and twin directions are more effective than others, deformation can cause crystals to reorient themselves to more easily accommodate an imposed stress, resulting in **lattice preferred orientation (LPO)** of many of the crystals in a deformed rock (Section 22.1).

4. Recovery. Permanent strain in crystals depends largely on defects. The magnitude of the strain, unless somehow relieved, is proportional to the density of defects in a crystal. Stored strain energy decreases the stability of a mineral and can be lowered by migration of defects in the following ways:

> Migration of vacancies to dislocation tangles straightens out blocked and tangled areas.
>
> Dislocations that are bent can straighten out by their migration.
>
> Dislocations can migrate and arrange themselves into networks.
>
> Dislocations of opposite sense that migrate and meet will annihilate one another.

These mechanisms constitute *recovery*. During deformation, the formation and disorderly migration of defects competes with the recovery processes, so that the stored strain energy reaches some constant value dependent on strain rate, temperature, etc. When deformation wanes, recovery predominates (if the temperature is sufficiently high), and the dislocation density and strain energy are greatly reduced.

Figure 23.5 illustrates the migration of scattered dislocations to planar arrays, or "dislocation walls." This takes a strained crystal and creates two distinct unstrained **subgrains**. Subgrains are portions of a grain in which the lattices differ by a small angle (the misorientation angle is exaggerated in Figure 23.5b). *Separate* grains differ in that they have high-angle misorientations. Figure 23.4b shows elongate subgrains in a single larger quartz grain. Note that the number of dislocations in Figure 23.5 may be the same (although some annihilation of dislocations of opposite sense may reduce it), but the *density* of subgrain dislocations (if you consider the subgrains separately) is much smaller, as is the total lattice strain. Note that subgrains also produce undulose extinction, but more abruptly at the boundaries of the extinction domains, which are much sharper and correspond to "dislocation walls."

5. Recrystallization is another way to reduce stored lattice strain energy. Recrystallization involves the movement of grain boundaries or the development of new boundaries, both of which produce a different configuration of *grains* (with high-angle orientation differences from neighbors), not *subgrains*.

FIGURE 23.4 **(a)** Undulose extinction in quartz. 0.2 mm across. **(b)** Elongate subgrains in deformed quartz, also undulose. 0.7 mm across.

a strained grain with undulose extinction

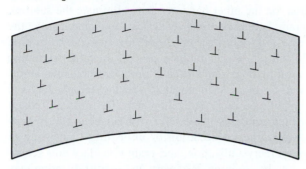

b recovery produces two strain-free subgrains

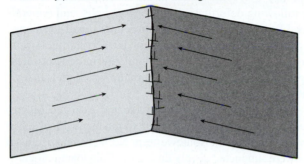

FIGURE 23.5   Illustration of a recovery process in which dislocations migrate to form a subgrain boundary.

a grain boundary migration

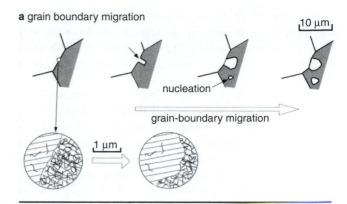

b sub-grain rotation

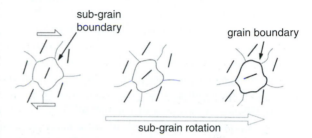

FIGURE 23.6   Recrystallization by **(a)** grain-boundary migration (including nucleation) and **(b)** subgrain rotation. From Passchier and Trouw (2005). Copyright © with permission from Springer-Verlag.

Recrystallization occurs by grain boundary migration or subgrain rotation (Figure 23.6). In **grain boundary migration** (Figures 23.6a and 23.7a), the atoms in a grain with higher strain energy migrate by diffusion and add to a neighboring grain with lower strain energy (also called **bulging**). Less commonly, an isolated new grain may nucleate in a highly

strained area of a deformed grain. **Subgrain rotation** (Figure 23.6b) occurs when dislocations are free to creep from the lattice plane of one subgrain to that of another and add (almost continuously) to a neighboring subgrain. The added dislocations cause the receiving subgrain lattice to rotate until it is sufficiently different from its neighbors to qualify as a separate

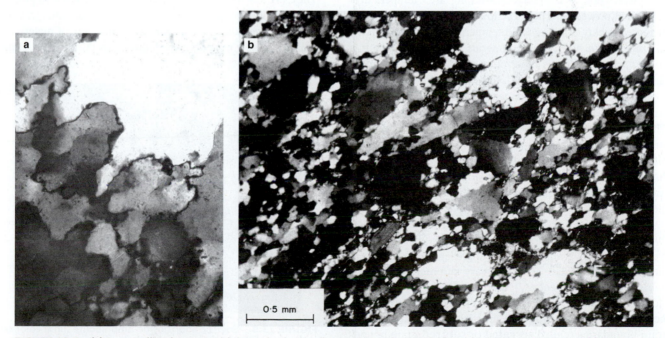

FIGURE 23.7   **(a)** Recrystallized quartz with irregular (sutured) boundaries, formed by grain boundary migration. Width 0.2 mm. **(b)** Deformed polycrystalline quartz with a foliation attributed to rotation of grains plus intracrystalline slip. Small grains are probably formed by subgrain rotation recrystallization. From Borradaile et al. (1982). Copyright © with permission from Springer-Verlag.

grain. Figure 23.7b shows irregular quartz grains with both low-angle subgrains and high-angle grains, the latter probably created by subgrain rotation. These two forms of recrystallization may occur during deformation as well as after deformation ceases.

At higher temperatures, crystals can deform solely by the migration of vacancies through the lattice (Figure 23.2). This process is called **solid-state diffusion creep**. **Crystalplastic deformation** is a loose term used to describe deformation by a combination of dislocation and diffusion creep. In fine-grained aggregates, crystals may also slide past one another by **grain boundary sliding**. Voids between sliding crystals are prevented by solid-state diffusion creep or solution-precipitation involving a fluid. At such high temperatures the rocks are weak, and deformation may be rapid.

A final recrystallization process is **grain boundary area reduction**. As discussed in Chapter 3, grain boundaries are surfaces where lattices end, and bonds are unsatisfied. The high energy of these boundaries constitutes a smaller percentage of the volume of a rock when crystals become coarser (a smaller surface-area-to-volume ratio) and/or when boundaries become straighter. Both of these processes constitute grain boundary area reduction and reduce the overall energy of the rock. Ostwald ripening (Figure 3.17) is a form of grain boundary area reduction in which grain boundaries migrate in the direction of curvature, thereby eliminating smaller grains and enlarging others as grain boundaries are straightened. Reduction in the free energy of a rock system by grain boundary area reduction is generally much less than by other recovery or recrystallization processes. It cannot keep pace with deformation, and its effects are much more obvious and dominant after deformation ceases, especially at high temperatures.

Recovery and recrystallization are temperature-dependent processes. They are slow at low temperatures but increase quickly at a threshold temperature *that depends on the minerals involved*. Some minerals may thus behave in a brittle fashion, whereas other coexisting minerals are more plastic. For example, in a deforming granite at shallow depths quartz is usually stronger than feldspar, but both are brittle and deform by fracture. At slightly greater depths where temperatures reach 200 to 300°C, quartz becomes ductile and loses strength, whereas feldspars remain brittle. The resistant feldspars typically form augen in the more ductile matrix of quartz and micas (Figure 23.18). At higher temperatures both quartz and feldspar behave in a more ductile fashion.

There is an important interrelationship between deformation and recovery–recrystallization. Deformation creates strain in grains (raising lattice energy), which, in turn, drives the processes of recovery and recrystallization (which lower lattice energy). Recovery has a lower threshold energy than recrystallization, so it occurs more readily. Recovery typically produces irregular and interlocking grain boundaries (called **serrated** or **sutured** grain boundaries, shown in Figures 23.7a and 23.15b), as highly strained material at grain boundaries is incorporated onto either of the adjacent larger crystals by grain boundary

migration and subgrain rotation. More advanced recrystallization, on the other hand, tends to eliminate the suturing by grain boundary area reduction. In some cases, the processes of deformation and recovery–recrystallization act in unison. The combined process is called **dynamic recrystallization**, which usually produces elongated grains and well-developed schistosity. Under low-temperature and dry conditions, recrystallization may be impeded, but at higher temperatures (especially with aqueous fluids present), **static recrystallization** or **annealing** (consisting of recovery, recrystallization, and grain boundary area reduction) may be common. Recovery and recrystallization may completely overcome lattice strain if temperature remains high following deformation in orogenic belts, and minerals aligned by dynamic recrystallization into a well-developed schistosity may retain little or no intracrystalline strain. Recrystallized textures also predominate in contact metamorphic aureoles where the temperature was high and deviatoric stress was low.

Deformation may also provide the extra energy required to overcome kinetic barriers to reactions. Figure 23.8 shows an outcrop in southwestern Norway where crustal rocks were subducted deeply during the Caledonian orogeny. Reaction from unstable (plagioclase–hornblende) amphibolites to stable (clinopyroxene–garnet) eclogites at the pressures involved proceeded only where the strain was high within a shear zone. Without the extra instability of strain-induced lattice energy (plus perhaps introduction of fluids along the shear zone, which may also enhance reactivity), the amphibolites outside the shear zone resisted reacting. Baxter and DePaolo (2004) noted that reaction rates in natural settings are generally orders of magnitude faster than in the lab and just slower than associated strain rates (Section 23.7). They argued that strain-induced grain boundary migration and diffusion creep expose grain interiors to the intergranular transport medium, thereby enhancing diffusion of material to the sites of new mineral growth during reactions. If so, strain

FIGURE 23.8 Gneissic anorthositic–amphibolite (light color on right) reacts to become eclogite (darker on left) as left-lateral shear transposes the gneissosity and facilitates the amphibolite-to-eclogite reaction. Bergen area, Norway. Two-foot scale. Courtesy of David Bridgwater.

may commonly control reaction progress and explain why reaction rates are usually slightly slower than strain rates and faster in nature than in static lab experiments. Although strain may permit reactions to proceed toward equilibrium, it cannot change the nature of the equilibrium mineralogy, as was once proposed for "stress minerals" that were mistakenly believed to form only under conditions of deviatoric stress.

23.2 TEXTURES OF CONTACT METAMORPHISM

As discussed in Section 21.3.1, contact metamorphism occurs in aureoles around intrusive bodies and is a response of cooler country rocks to the thermal $\pm$ metasomatic effects of the intrusion. The textures of contact metamorphism are typically developed at low pressures and thus under conditions of low deviatoric stress. The thermal maximum in a contact aureole should also occur much later than any stress imparted by forceful intrusion. Crystallization (including recrystallization) may therefore occur in a near-static environment, and contact metamorphism is typically characterized by a lack of significant preferred mineral orientation. Many minerals are equidimensional, rather than elongated, and the elongated minerals that do form are orientated randomly. Relict textures are also common in contact metamorphic rocks because there is little accompanying shear to destroy them.

Static recrystallization occurs after deformation ceases at elevated temperatures or when a thermal disturbance occurs in a low-stress environment. As described above, the process acts to reduce both lattice strain energy and overall surface energy. The textures that result typically depend on the minerals involved. In "structurally anisotropic" minerals, the surface energy of a grain boundary depends strongly on the lattice orientation of the boundary in question, whereas in "structurally isotropic" minerals the grain boundary energy is about the same for any surface. If the orientation dependence is low (as in quartz or calcite), no particular faces are preferentially developed. Thus in *monomineralic* aggregates of structurally isotropic minerals (quartzites or marbles), grain boundary area reduction leads to an equilibrium texture in which grains meet along straight boundaries (resulting in low surface area for each grain). The texture is called **granoblastic polygonal** (or **polygonal mosaic**), and grains appear in two-dimensional thin sections as equidimensional polygons with grain boundaries that meet in triple junctions with approximately 120° between them (Figure 23.9a). The size of the polygonal grains depends mainly on temperature and the presence of fluids (higher temperatures and aqueous fluids promote larger crystals).

Structurally anisotropic minerals, such as micas and amphiboles, have some crystallographic surfaces with much lower energy than others. This affects the shape of the grains, and thus the equilibrium recrystallization texture. High surface energy boundaries grow faster, so that low-energy

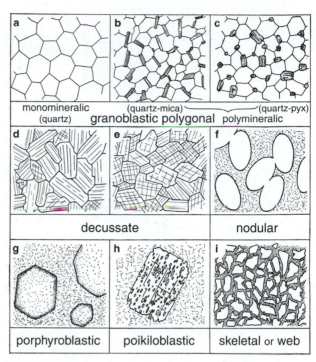

FIGURE 23.9 Typical textures of contact metamorphism. From Spry (1969).

surfaces become larger. To understand this, imagine a cube on a table. If the four vertical faces are of higher energy (less stable) and grow more quickly than the top and bottom faces, the cube will expand laterally faster than it does vertically, and it gradually becomes more plate-like. Low-energy surfaces in such minerals (e.g., the top and bottom of the plate) predominate in the final static recrystallization texture, even in monomineralic rocks, so that simple regular polygons are no longer abundant. The result is called **decussate** texture (Figures 23.9d and e).

In *polymineralic* rocks, the grain boundary energy also depends on whether the boundary is between like minerals or unlike minerals. In general, same-mineral (A-A) grain boundaries have higher energy than different-mineral (A-B) boundaries, so that the final static equilibrium texture will tend to minimize the total area of A-A boundaries and increase the area of A-B boundaries (thereby resulting in lower total energy). This brings us back to the concept of **dihedral angles** (as in Figure 11.1, only substituting a different mineral for the melt). In Figure 23.10, if the dihedral angle, θ, becomes smaller, the total area of A-B grain boundaries increases with respect to A-A boundaries. Polymineralic rocks thus develop a modified type of granoblastic texture in which 120° angles occur only at A-A-A or B-B-B triple junctions, and the other junctions depend on the relative grain-boundary energies of the different minerals involved (Figure 23.10c). Figure 23.9b and c also illustrate modified polygonal textures of polymineralic aggregates of quartz–mica and quartz–pyroxene, respectively.

In highly structurally anisotropic minerals in which the energy of one face is unusually low, such as {001} in mica, that face will predominate and will not be altered

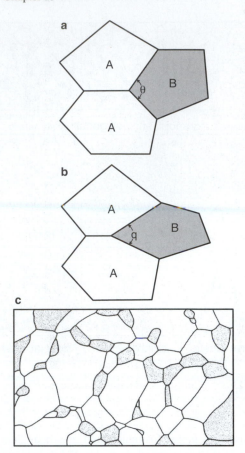

FIGURE 23.10 **(a)** Dihedral angle between two mineral types. When the A-A grain boundary energy is greater than for A-B, the angle θ will decrease **(b)** so as to increase the relative area of A-B boundaries. **(c)** Sketch of a plagioclase (light)-clinopyroxene (dark) hornfels showing lower dihedral angles in clinopyroxene at most cpx-plag-plag boundaries. (c) from Vernon (1976).

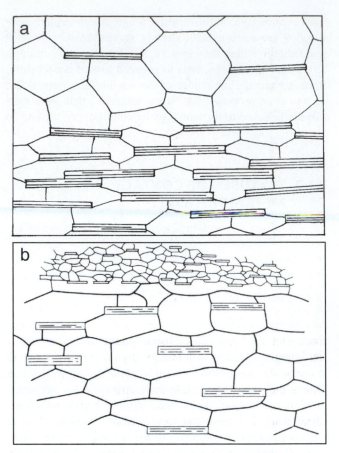

FIGURE 23.11 Drawings of quartz–mica schists. **(a)** Closer spacing of micas in the lower half causes quartz grains to passively elongate in order for quartz–quartz boundaries to meet mica (001) faces at 90°. From Shelley (1993). **(b)** Layered rock in which the growth of quartz has been retarded by grain boundary "pinning" by finer micas in the upper layer. From Vernon (1976).

much by the interfaces of more isotropic minerals such as quartz (Figure 23.9b). As a result, quartz–quartz interfaces usually intersect {001} faces of mica at right angles or join to the edges of mica flakes (Figure 23.11). In situations where micas are foliated and closely spaced, this tendency for right-angle intersections may cause the quartz grains to be elongate (Figure 23.11a, lower half), contrary to their usual tendency. This dimensional elongation is not caused by deformation of the quartz crystals but is an artifact of quartz conforming to the mica foliation.

Another common feature of polymineralic rocks is that the occurrence of a minor phase, such as graphite, may retard grain boundary migration of a more abundant phase and "pin" (or anchor) the boundaries, resulting in a finer grain size than might otherwise have occurred (Figure 23.11b).

Although not restricted to contact metamorphic rocks, **porphyroblasts** (Figure 23.9g) are very common. In the examples of contact metamorphism discussed in Chapter 21, the rocks were typically characterized by large porphyroblasts of biotite, andalusite, or cordierite.

Poikiloblasts are porphyroblasts that incorporate numerous inclusions (equivalent to poikilitic texture in igneous

rocks). This texture is common in garnet, staurolite, cordierite, and hornblende. Poikiloblastic texture (Figure 23.9h) is a high-energy texture because it represents a high surface area situation. It probably occurs as a result of poor nucleation and rapid porphyroblast growth, which envelops neighboring grains. In some cases, a high grain boundary energy between inclusion minerals may be reduced by the formation of a layer of another (poikiloblast) mineral between them. The surface energy of an included mineral is also lowered by rounding the edges, which is common among poikiloblast inclusions. Some inclusions may be inert phases that were passed up by the advancing front of the poikiloblast growth. Most, however, are believed to be a co-product of the poikiloblast-forming reaction, or a reactant that wasn't completely consumed. Poikiloblast thus typically grow at the same time as their inclusions, although at a faster rate (Vernon, 2004). The inclusion-free rims commonly observed surrounding poikiloblastic garnet cores may result from a late stage of slow growth or perhaps from a different garnet-forming reaction than the one that formed the cores.

Skeletal texture (also called **web**, or **spongy**) is an extreme example of poikiloblast formation (Figures 23.9i and 23.12), in which the inclusions form the bulk of the

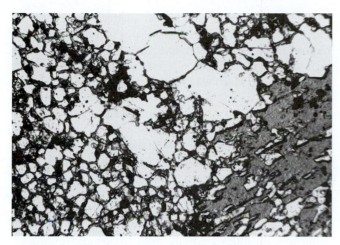

FIGURE 23.12   Skeletal or web texture of staurolite in a quartzite. The gray intergranular material, and the mass in the lower right, are all part of a single large staurolite crystal. Pateca, New Mexico. Width of view ~5 mm.

FIGURE 23.13   Light-colored depletion haloes around cm-sized garnets in amphibolite. Fe and Mg were less plentiful, so that hornblende was consumed to a greater extent than was plagioclase as the garnets grew, leaving hornblende-depleted zones. Sample courtesy of Peter Misch.

rock and the enclosing mineral phase occurs almost as an intergranular, crystallographically continuous, network. It may develop at the margins of some porphyroblasts where they are growing into the matrix, but the occurrence is more pervasive in some rocks. Skeletal texture may result from rapid poikiloblast growth or from the introduction of poikiloblast-building or reactive components via an intergranular fluid.

Porphyroblasts typically form in minerals for which nucleation is impeded. They are larger than the groundmass minerals and are separated by greater distances. Crystallizing groundmass minerals typically grow among preexisting grains and thus have little nucleation problem, whereas porphyroblasts must nucleate from scratch. Porphyroblast growth requires that components diffuse farther in order to add onto the growing porphyroblast surface. Perhaps poikiloblastic and skeletal texture are metamorphic equivalents of igneous dendritic quench textures (Section 3.1.1) in which the growing grain surface reaches out toward a source of components in cases where diffusion is slow. In many cases the distance that a critical component must travel may control the size of porphyroblasts. Once the distance becomes too great, porphyroblast growth ceases. Diffusion of components that are scarce in the area surrounding a growing porphyroblast may create a zone surrounding the porphyroblast that is depleted in those components. Such areas are called **depletion haloes** (Figure 23.13). Porphyroblasts tend to be larger at higher grade. Perhaps we can attribute this to less overstepping of the true reaction conditions at elevated temperature (hence subdued nucleation, Figure 3.1) combined with more effective diffusion.

Cordierite, biotite, and some other minerals commonly form ovoid porphyroblasts in contact aureoles, particularly when the matrix is very fine grained. The texture is called **nodular** (Figures 23.9f and 23.14a). A field term for rocks that in hand specimen contain small porphyroblasts (usually ovoid, but not necessarily so) in a fine

matrix is **spotted**. If the matrix is non-foliated the rock is commonly called a **spotted hornfels**. As discussed in Chapter 21, contact metamorphism overprinting regional metamorphism is common (reflecting either post-orogenic magmatism or the time required for magmas to rise to shallow regions following an orogenic metamorphic-plutonic event). The result for low-grade regional metamorphic rocks is **spotted slates** or **spotted phyllites** (Figure 23.14b).

You have probably noticed in the lab or in the field that some metamorphic minerals tend to be more euhedral than others. In contrast to igneous rocks, this capacity is no longer determined by which minerals grew earliest (early igneous minerals are surrounded by melt, so growth is unencumbered by contact with other minerals). Because all metamorphic minerals grow in contact with others, the tendency for a mineral to be more euhedral must then be a property of the mineral itself. Garnet and staurolite, for example, are typically euhedral, whereas quartz and carbonates tend to be anhedral. Crystals that develop good crystal faces have anisotropic surface energy that is high for some faces, but low for others, so that, when developed, the low-energy faces reduce the overall surface energy of the mineral. Rather than attempt a rigorous theoretical treatment, it is far simpler to create an empirical list of the common metamorphic minerals in terms of their tendency to form euhedral (idioblastic) crystals. Such a list has been available since Becke (1913) first proposed it as the "crystalloblastic series."

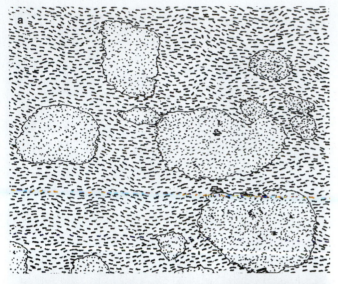

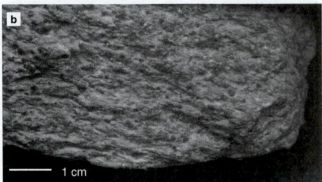

FIGURE 23.14 Overprint of contact metamorphism on regional. **(a)** Nodular texture of cordierite porphyroblasts developed during a thermal overprinting of previous regional metamorphism (note the foliation in the opaques). From Bard (1986), approximately 1.5 × 2mm. Copyright © by permission Kluwer Academic Publishers. **(b)** Spotted phyllite in which small porphyroblasts of cordierite (and one larger pyrite) develop in a preexisting phyllite.

THE CRYSTALLOBLASTIC SERIES

Most Euhedral

> *Titanite, rutile, pyrite, spinel*
> *Garnet, sillimanite, staurolite, tourmaline*
> *Epidote, magnetite, ilmenite*
> *Andalusite, pyroxene, amphibole*
> *Mica, chlorite, dolomite, kyanite*
> *Calcite, vesuvianite, scapolite*
> *Feldspar, quartz, cordierite*

Least Euhedral

Porphyroblasts, even of the same mineral, may exhibit euhedral outlines in one rock and anhedral ones in another. As with igneous crystals, euhedral shapes are generally attributed to growth into a liquid rather than interference with adjacent solids. In metamorphic rocks, this may involve only a thin surface film of aqueous fluid, perhaps released by a prograde dehydration reaction in which the porphyroblast is one product (Vernon, 2004). Euhedral porphyroblasts are rare in dry high-grade metamorphic rocks, which supports this idea.

Cashman and Ferry (1988) plotted the number of grains as a function of grain size (called **crystal size distribution [CSD]** plots) for olivine, pyroxene, and magnetite in high-temperature hornfelses (~1000°C) from Skye. They found that CSDs for the hornfelses plot in a log-linear fashion with ever-increasing numbers of grains as crystal size decreases. This is the same pattern developed in volcanics, which Cashman and Ferry (1988) interpreted to reflect both continuous nucleation and growth that quickly cease due to cooling and kinetic constraints. As we shall see, the CSD patterns in many areas of *regional* metamorphism do not conform to this log-linear trend.

As metamorphic grade (mostly temperature) increases, recrystallization becomes more dominant. Thus we might summarize the following as the most pronounced *textural* effects of increasing metamorphic grade in contact aureoles:

1. Fewer relict textures and a more fully recrystallized metamorphic texture
2. Increased grain size
3. Straighter grain boundaries and less evidence of strain

23.3 HIGH-STRAIN METAMORPHIC TEXTURES

Because impacts are rare, we will concentrate here on fault/shear zones. Refer again to Figure 21.7 for a cross section of a fault/shear zone and how it varies with depth. In shallow fault zones where the rocks are cooler and behavior is more brittle, the cataclastic processes described in the beginning of Section 23.1 dominate over the recovery and recrystallization processes. Broken, crushed, rotated, and bent grains dominate in fault breccias.

Below this shallowest zone, confining pressure increases, and deformation is more pervasive. Cataclastic processes continue to be dominant at intermediate depths, and affect the rock on a finer scale. The rocks are more coherent microbreccias and cataclasites. Phyllosilicates slip readily along (001) and may become disrupted by slip. The thin plates produced by slip easily bend or break, which produces a more frayed or **shredded** appearance at high strain rates. **Undulose extinction** is ubiquitous. Undulose extinction of deformed lattices in plagioclase may be distinguished from compositional zoning by the generally concentric extinction pattern of the latter. Remnants of broken larger pre-deformational grains are called **clasts** (from cata*clast*ic). Larger initial grains (sedimentary grains or phenocrysts) or more resistant minerals may remain larger in a matrix of crushed material. These larger fragments are called **porphyro*clasts*** (to distinguish them from porphyro*blasts* that *grew* larger). Some porphyroclasts may be surrounded by a matrix of fine crushed material that is derived from them as they are rotated and ground down, a texture called **mortar** texture (Figure 23.15b). **Pseudotachylite** is produced by localized rapid fragmentation and melting due to shear heating, generally attributed to earthquake shock energy in dry rocks (Sibson, 1975; Vernon, 2004). Figure 23.16b shows the typical outcrop structure of pseudotachylite, with dark sharply

(1) Hand Specimen   (2) Thin Section

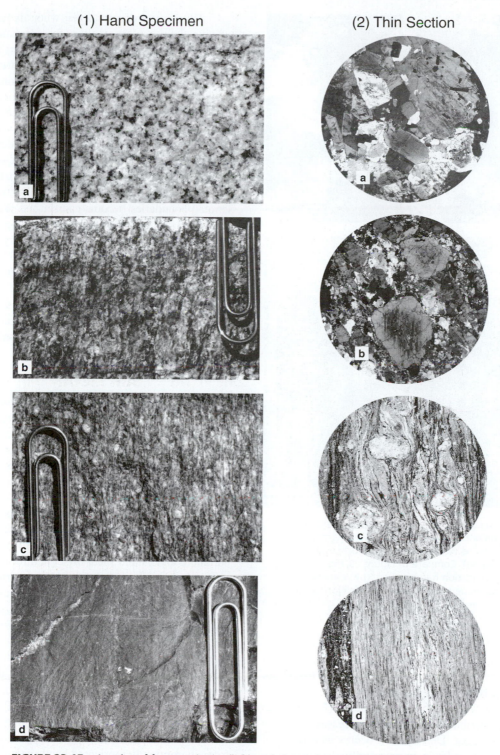

FIGURE 23.15 A series of four specimens (left) and photomicrographs (right, 10×) showing the progressive mylonitization of granitic rocks in the southeastern San Gabriel Mountains, California. **(a)** Undeformed granite. **(b)** Mortar texture in which large porphyroclasts of quartz are surrounded by a crushed matrix derived by cataclasis and recrystallization. **(c)** Mylonitization imparts a distinct foliation as shear and recrystallization causes elongated elements. **(d)** Ultramylonite, in which very little of the original grains remain. Ribbon texture predominates. © Shelton © (1966).

bounded irregular dendritic veins containing deformed grains suspended in a glassy matrix. Rapid cooling of the narrow veins preserves both the glass and the deformed textures. The name comes from tachylite (basaltic glass). Curiously, the melt composition is typically mafic, even when formed in more silicic rocks. Camacho et al. (1995), for example, reported the occurrence of pseudotachylite with a mafic glass composition that developed in felsic granulites (orthopyroxene meta-granitoids). Apparently, pyroxene and plagioclase structures are more easily disturbed by the shock than associated

FIGURE 23.16   **(a)** Large polygonized quartz crystals with undulose extinction and subgrains that show sutured grain boundaries caused by recrystallization. From Urai et al. (1986 © AGU with permission). Compare to Figure 23.15b, in which little, if any, recrystallization has occurred. **(b)** Vein-like pseudotachylite developed in gneisses, Hebron Fjord area, northern Labrador, Canada.

Recovery processes, even when dynamic and accompanying deformation, result in the formation of subgrains. Traditional descriptions of the formation of smaller reduced-strain subgrains and grains from larger deformed grains refer to the collective process as **polygonization**. Figures 23.16a and 23.4b are photomicrographs of polygonized quartz.

Recovery and recrystallization become progressively more important as deformation wanes. Highly deformed rocks have high strain energy and tend to recrystallize readily. Grain boundary migration and subgrain rotation result in serrated-sutured boundaries (Figure 23.16a). If the temperature is high enough, recrystallization may progress further, and serrated boundaries become coarser and less sharply curved (Figure 23.7a), and the shapes of the larger crystals appear more **amoeboid**. In cases of advanced recrystallization, grain boundary area reduction straightens the boundaries further, eventually producing a granoblastic polygonal texture that is completely recrystallized (annealed). The collective process by which larger grains form from subgrains or by the addition of smaller grains by grain boundary migration recrystallization has been traditionally referred to as **coalescence**.

23.3.1 Shear Sense Indicators

In order to properly interpret deformation, it is useful to know the sense of shear between opposing blocks across a fault zone that is no longer active. Plate tectonics has increased our interest in knowing the sense of shear, as it may help us to determine the relative motions of crustal blocks involved in orogeny. The *direction* of motion in a shear zone is in the shear zone plane and assumed to be parallel to fault striations or mineral elongation. The *sense* of shear (which of the two opposite possible motions associated with a given direction), however, is more difficult to determine. It is important to understand shear offset in three dimensions, as the sense of offset on a two-dimensional surface may be misleading if the direction of motion has not clearly been determined. Orientation of samples is critical, and it will be assumed in the following discussion that we are observing textures in a direction that is both *in the shear plane and normal to the direction of motion*. In the figures that follow, the shear plane will thus be normal to the page and the direction of shear will be right–left (and the sense will consistently be dextral (clockwise). Proper orientation may be difficult to manage on two-dimensional outcrops in the field but can be accomplished by cutting thin sections from appropriately reoriented hand specimens.

If a planar or linear structure is cut by the shear plane, we can use the offset or drag-curvature features to easily determine the sense of offset (as in the dark dikelet in Figure 23.17a, top, or the foliation in Figure 23.8). Many shear zones, however, do not show foliations that curve into them, at least not on the outcrop or smaller scale. When offset or curved markers are absent, careful textural analysis and interpretation are required.

Oblique foliations, such as those developed in the schematic quartz grains of Figure 23.17a, cut across an S-foliation developed as shear offsets. To understand the two

quartz and K-feldspar so that the mafic component melted. This is an interesting example of high energy dynamics that produces partial melting trends that contrast sharply to equilibrium melting experiments we generally rely upon.

Shear zones typically broaden at depth (Figure 22.2) so that shear is distributed and ductile processes gradually become more prevalent. Deformation occurs as a result of combined cataclastic, plastic intracrystalline deformation, and recovery processes. Twinned and ductilely elongated grains are common. Foliated **mylonites** are the predominant type of rock. Quartz may have the form of highly elongate **ribbons** (Figure 23.15c and d).

Foliations are important in the interpretation of mylonites, especially in determining the sense of shear motion in the shear zone (see below). Foliations occur in shear zones and in orogenic belts of regional extent. They are more varied in the latter, and I shall postpone discussing the types and generation of foliations until we address orogenic metamorphic textures in Section 23.4.

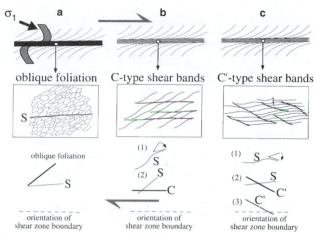

oblique foliation   C-type shear bands   C'-type shear bands

FIGURE 23.17 Some features that permit the determination of sense of shear. All examples involve dextral shear. σ_1 is oriented as shown. **(a)** Passive planar marker unit (shaded) and foliation oblique to shear planes. **(b)** S-C foliations. **(c)** S-C' foliations. σ_1 is the direction of maximum principal stress (Section 21.2.3). After Passchier and Trouw (2005). Copyright © with permission from Springer-Verlag.

sets of foliations, consider a deck of cards (Figure 21.2c). If we shear the deck (slip between the individual cards), any material in the volume being sheared (such as a line or circle drawn on the side of the deck) will elongate and rotate to define a foliation (the long axis of the ellipse in Figure 21.2c). The shear planes (between cards) define another foliation. With progressive shear, the two foliations approach one another, but can never become truly parallel. The foliation labeled S in Figure 23.17a is the card-parallel shear foliation and the oblique foliation is the stretching (long axis) of the material elements. The quadrants with the *acute* angle between the S-foliation and the oblique foliation indicate the shear sense direction.

Shear bands (labeled C in Figure 23.17b) are spaced cleavages that transect a well-developed simultaneous mineral foliation (S in Figure 23.17b) at a small angle. The combined texture is called **shear band cleavage** or **S-C texture**. Mylonites showing shear band cleavage are called **S-C mylonites** (Berthé et al., 1979; Lister and Snoke, 1984). The C-type surfaces are parallel to the main shear zone offset, and the sense of shear is determined by the acute angle between the S and C foliations. A **C'-type** of cleavage may also form in some instances, usually later than the S- and C-types (Figure 23.17c). C' is oriented in a *conjugate* sense to the C-type cleavage (σ_1 bisects the acute angle between them), so it is oblique to the shear zone boundaries (by 15 to 35°). C'-type cleavages are usually more weakly developed, shorter, and more wavy than C-types.

Porphyroclasts developed in sheared mylonite typically develop tapered rims of fine-grained material. If the rims have the same mineralogy as the porphyroclast, they are assumed to be derived from the porphyroclast by grinding and are called **mantles**. **Mantled porphyroclasts** typically develop from more resistant feldspars in a matrix of quartz, feldspar, and mica in sheared granites, or from dolomite in sheared calcite–dolomite marbles. Mantles are interpreted to form by ductile crystal deformation and

storage of dislocation tangles in the rim of porphyroclasts in response to flow in the matrix. The rim then recrystallizes to form the mantle. Coarse **augen** texture in mylonites and more ductile sheared granitic gneisses (Figure 23.18) are types of mantled porphyroclasts. The mantle is finer grained than the porphyroclast core and can be further deformed by shear to form **tails** that extend from the porphyroclast in both directions into the mylonitic foliation (Vernon, 2004).

The shape of the deformed tails on *some* mantled porphyroclasts can be used as shear-sense indicators. Mantled porphyroclasts can be subdivided into five types (Simpson and Schmidt, 1983; Simpson, 1985; Passchier and Simpson, 1986; Passchier and Trouw, 2005). As illustrated in Figure 23.19, **θ-type** mantles have no tails, and **φ-type** mantles have symmetrical tails. Neither can be used as sense-of-shear indicators. **σ-type** mantled porphyroclasts are asymmetrical and have wide tails, with a nearly straight outer side. The inner side is usually concave toward the median plane (the plane parallel to the shear zone and bisecting the porphyroclast). The shape has been described as **stair-step** because the two outer sides offset in two directions, just like the rise and run of a stair. The tail shape is believed to form as the foliation drags the softer mantle. The stair-step can be used to infer the sense of shear as illustrated for the σ-type in Figure 23.19.

In **δ-type** mantled porphyroclasts, both sides of the mantle are curved, and an embayment is formed on the inner side. δ-types are believed to begin as σ-types, and the curvature probably forms as the core rotates during further shear. The shape is related to the median plane and sense of shear as illustrated in Figure 23.19. Care must be taken when interpreting sense of shear from δ-type and σ-type mantled porphyroclasts. Note that if you mentally straighten the curvature of the δ-type tails in Figure 23.19, it looks like an σ-type, *but if improperly interpreted as σ-type, the sense of*

FIGURE 23.18 Augen gneiss with eye-shaped porphyroclasts (augen) of K-feldspar in a sheared granitic matrix.

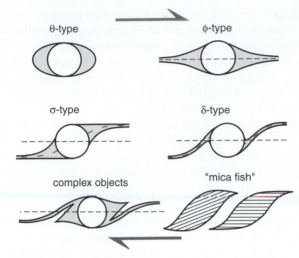

FIGURE 23.19   Mantled porphyroclasts and "mica fish" as sense-of-shear indicators. After Passchier and Simpson (1986). Copyright © with permission from Elsevier Science.

shear it then indicates is opposite the true sense. You must therefore be certain of the type of porphyroclast that you are observing before you can interpret it correctly.

Complex object mantled porphyroclasts are generated by further rotation of δ-types, which again stretches out the mantle in a renewed σ-type fashion. By using the shape in Figure 23.19, the sense of shear can be determined.

"Mica-fish" are single mica crystals (not porphyroclasts) that are shaped much like σ-type mantled porphyroclasts (Figure 23.19). They are most common in mica–quartz mylonites and ultramylonites. Like σ-type porphyroclasts, their long axis is oriented in the direction of extension (see below). This and the "stair-step" form can be used to indicate sense of shear. The mica {001} cleavages may be parallel to the elongation direction, or they may be oriented parallel to the slip direction of the shear zone (as shown in the two examples in Figure 23.19). Trails of mica fragments typically extend from the tips well into the matrix. Lister and Snoke (1984) proposed that mica-fish form by a combination of slip on {001}, rotation, boudinage (see below), and recrystallization at the edges.

Other sense-of-shear indicators (Figure 23.20) include **quarter structures** that form on unmantled porphyroclasts. Quarter structures are so named because the four quadrants defined by the foliation and its normal are not symmetrical. In Figure 23.20, for example, the northwestern and southeastern quadrants experience shortening, and the northeastern and southwestern ones experience extension. In **quarter folds** (Figure 23.20a), shear extension drags small folds in the foliation around porphyroclasts. **Quarter mats** (Figure 23.20b) are concentrations of mica that result from dissolution of quartz in the shortened quadrants and precipitation in the extending quadrants. The vergence of **asymmetric folds** (Figure 23.20c) is also a good shear-sense indicator, as are pairs of pre-shear dikelets if one is in the quadrants of extension and the other in the quadrants of compression (Figure 23.20d). Other shear-sense indicators may also be used, but most are either more cumbersome, more specialized, or inconsistent.

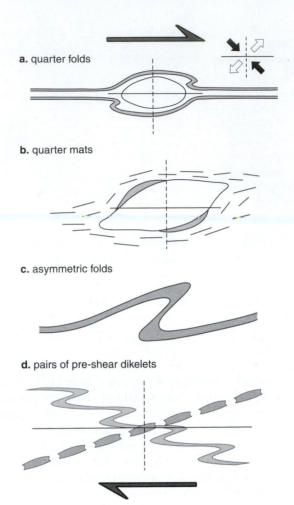

a. quarter folds

b. quarter mats

c. asymmetric folds

d. pairs of pre-shear dikelets

FIGURE 23.20   Other methods to determine sense of shear.

See Passchier and Trouw (2005) and Vernon (2004) for detailed discussions of shear-sense indicators.

23.4 REGIONAL OROGENIC METAMORPHIC TEXTURES

The textures discussed in this section are appropriately called **dynamothermal** textures because they occur in any situation where deformation and heat are combined. Such situations range from deep shear zones to strained contact aureoles, but the majority of dynamothermal rocks are found in ancient orogenic belts, so we will concentrate on this setting.

As we learned in Chapters 16 to 18, orogenic belts are complex tectonic environments where plate convergence produces a number of deformational and thermal patterns. We can envision an **orogeny** as a long-term mountain building process, such as the Appalachian Orogeny or the Alpine Orogeny. Orogenies are not continuous, however, and an orogeny may comprise more than one **tectonic event**, believed to result most commonly from short-term changes in plate motion, such as accelerated subduction rate or (micro)continent accretion. Tectonic events, in turn, may consist of more than one **deformation phase**. A deformation phase, according to Passchier and Trouw (2005), is a distinct period of active deformation with a specific style and orientation.

Deformation phases may be separated by periods of reduced or absent deformation, during which the orientation of the stress field may change. Thus one or more deformation phases may constitute a tectonic event, and one or more tectonic events may constitute an orogeny.

Metamorphism accompanies many of these deformational processes but not necessarily in a direct one-to-one manner. We might best consider metamorphism separately as occurring in one or more **metamorphic reaction events** that in turn occur in a larger **metamorphic cycle** (in which rocks are buried, heated, metamorphosed, and then brought back to the surface by uplift and erosion). There may also conceivably be more than one metamorphic cycle in an orogeny. Metamorphic events typically last 1 to 10 Ma, but multiple events in an orogen may span over 1 Ga. Metamorphism may accompany only some deformational events, and the style of deformation and grade of metamorphism may vary in both time and space within a single orogen. A single metamorphic event may even have more than one **phase** of heating within it. The IUGS-SCMR (Fettes and Desmons, 2007) proposed the term **polyphase** for a broad *P-T-t* cycle (event) with more than one temperature climax (phases). **Monophase** metamorphic events have only one climax. **Polymetamorphic** (introduced in Chapter 22) is a term that describes more than one event.

Deformation tends to break minerals down to smaller grains and subgrains, whereas the heat of metamorphism tends to build them back up again. Such a complex set of processes allows for myriad interactions and overprints between metamorphic mineral growth and deformation, making the study of textures in orogenic rocks a challenge and often leading to controversy over textural interpretations. Of course, whenever things get complex, there is more useful information to be gathered, so complexity becomes a benefit when the features can be interpreted properly. We can do a respectable job on the basis of a few guiding principles. In the following summary, I review the useful principles upon which geologists tend to agree and discuss some of the remaining controversial aspects of the textural interpretation of orogenically metamorphosed rocks.

As I alluded to above, the crystal size distribution (CSD) curves for regional metamorphism differ from those of contact metamorphism. In contrast to the log-linear plots for contact metamorphism (and volcanics), Cashman and Ferry (1988) found that regional CSDs have a bell-shaped pattern. The grain size of the maximum of the CSD curves increases with grade, as we would expect. Cashman and Ferry (1988) attributed the bell shape to initial continuous nucleation and crystal growth (as in contact rocks) followed by a period of annealing in which nucleation ceases and larger grains grow at the expense of smaller ones (as in Ostwald ripening). The minimum grain size in regional metamorphism thus appears to reflect continued growth after nucleation ends.

23.4.1 Tectonites, Foliations, and Lineations

A **tectonite** is a deformed rock with a texture that records the deformation by developing a preferred mineral orientation of some sort. The fabric of a tectonite is the complete spatial

and geometrical configuration of its textural and structural elements. As discussed in Sections 21.3.2 and 22.1, **foliation** is a general term for any planar textural element in a rock, whereas **lineation** similarly applies to linear elements. Foliations and lineations can be subdivided into **primary** (pre-deformational) ones, such as bedding, and **secondary** (deformational) ones. Minerals may be oriented by either dimensional preferred orientation (DPO) or lattice preferred orientation (LPO), or both (Section 21.1). Although they are treated separately, there is probably a complete natural gradation from pure foliations through combined foliations and lineations to pure lineations.

23.4.1.1 FOLIATIONS. A number of features can define a secondary foliation (Figure 23.21), including platy minerals, linear minerals, layers, fractures, and flattened elements.

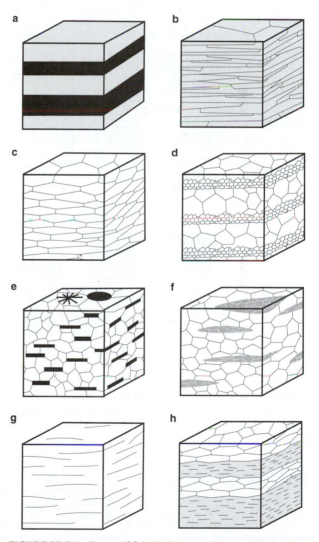

FIGURE 23.21 Types of fabric elements that may define a foliation. **(a)** Compositional layering. **(b)** Preferred orientation of platy minerals. **(c)** Shape of deformed grains. **(d)** Grain size variation. **(e)** Preferred orientation of platy minerals in a matrix without preferred orientation. Note that linear minerals may also define a foliation if randomly oriented in a plane. **(f)** Preferred orientation of lenticular mineral aggregates. **(g)** Preferred orientation of fractures. **(h)** Combinations of the above. From Turner and Weiss (1963) and Passchier and Trouw (2005).

Metamorphic foliations are divided into cleavages (fine penetrative foliations), schistosity (coarser penetrative foliations), and gneissose structure (poorly developed coarse foliations or segregated layers). The nomenclature of **cleavages** suffers from an overabundance of terms that were used in different ways by different investigators and often confused the purely descriptive from the genetic aspects. A Penrose Conference was held in 1976 to reach a consensus on the description and origins of rock cleavages (Platt, 1976). Although consensus was not immediately forthcoming, Powell (1979) followed it up and proposed a classification of cleavages (including schistosity) that was purely descriptive, thereby allowing us a

universal nomenclature unencumbered by inferences about cleavage origins.

Geological processes are usually slow, and the majority of petrological phenomena that we see today have remained unchanged since scientists first observed them. Interpretations as to the origin of the features, on the other hand, change often as new ideas and data come to light. Thus it is always a good idea to separate our observations from our interpretations, so that we may speculate freely after we have agreed upon what we see. Powell's (1979) classification has thus met with approval from the majority of investigators in the field and is presented in Figure 23.22,

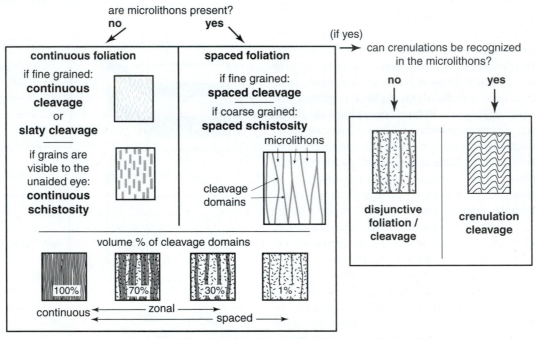

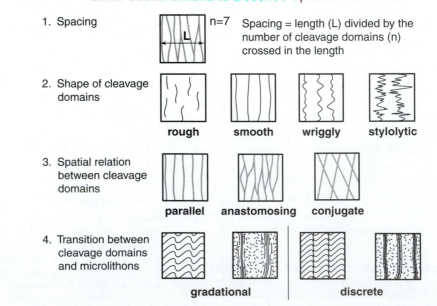

FIGURE 23.22 A morphological (non-genetic) classification of foliations. After Powell (1979), Borradaile et al. (1982), and Passchier and Trouw (2005).

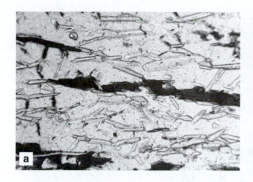

FIGURE 23.23 Continuous schistosity developed by dynamic recrystallization of biotite, muscovite, and quartz. **(a)** Plane-polarized light, width of field 1 mm. **(b)** Another sample with crossed-polars, width of field 2 mm. Although there is a definite foliation in both samples, the minerals are entirely strain free.

as modified slightly by Borradaile et al. (1982), Passchier and Trouw (2005), and me. The description of cleavages in some cases may be done in hand specimen, but it is best done with the aid of thin sections. Figure 23.22 is thus based largely on thin section–scale features.

The first step in characterizing a cleavage is to decide whether the foliation elements are *continuous* or *spaced*. In a **continuous foliation**, the aligned minerals define a foliation that does not vary across the area of the thin section (Figure 23.23). In a **spaced foliation**, the fabric of the rock is separated into unfoliated **microlithons** separated by **cleavage domains** (fractures or concentrations of platy minerals). Of course, there is a continuum between these idealized end-members, and continuous cleavage grades through zonal to spaced cleavage (Figure 23.22). Fine-grained continuous cleavage may also be called **slaty cleavage** because it typifies slates. As Vernon (1998, 2004) pointed out, however, slates may develop some degree of spacing into anastomosing lenticular P-rich and Q-rich domains (Section 23.4.3), even at very low grades. Most petrologists still refer to such finely spaced cleavages as slaty cleavage. Phyllites have a slightly coarser continuous cleavage, and, when the individual aligned crystals become large enough to see with the unaided eye, the foliation is called a **schistosity**.

Spaced foliations are further subdivided and described on the basis of the spacing and shape of the cleavage domains, the presence of crenulations, the spatial relations between the domains, and whether the domains grade into the microlithons. These characteristics are also illustrated in Figure 23.22. Terms that describe these aspects may be combined when appropriate (e.g., a "smooth, anastomosing, discrete cleavage," etc.).

A **crenulation cleavage** is actually *two* cleavages. The first cleavage may be a slaty cleavage or schistosity that

becomes microfolded. The fold axial planes typically form at a high angle to σ_1 of the second compressional phase. The folds may be *symmetrical* (Figure 23.24a) or *asymmetrical* (Figure 23.24b). Quartz grains tend to dissolve by pressure solution from the fold limbs (or just the steeper limb of asymmetrical folds) either to precipitate in the hinge areas or be transported further away (Figure 23.24). If the metamorphic temperature is high enough, new micas may also grow normal to σ_1 during the second phase. Pressure solution and the growth of new minerals may both enhance the second foliation and obscure the first, as can be seen in the sequence $1 \rightarrow 5$ in Figure 23.25. In such cases, the first foliation is most evident in the microlithons between the second cleavage (see also Figures 23.24a, 23.42, and 23.48b).

Compositional **layering** is a common type of foliation that can be either primary or secondary. Primary layering, such as bedding, cumulate layering, or igneous flow features, is inherited from the pre-metamorphic rock. A common form of secondary layering is **gneissose structure**. We shall discuss metamorphic layering in more detail below.

23.4.1.2 LINEATIONS. Foliations generally occur when $\sigma_1 > \sigma_2 \approx \sigma_3$ and lineations generally occur when $\sigma_1 \approx \sigma_2 > \sigma_3$, or when shear smears out an object. As shown in Figure 23.26, there are also several types of lineations. They usually result from the elongation of minerals or mineral aggregates (**stretching lineations**). Stretched pebbles in deformed conglomerates is a common example. Lineations may also result from parallel **growth of elongate minerals**, **fold axes**, or **intersecting planar elements** (cleavage and bedding, or two cleavages). Note in Figure 23.26c that, under some circumstances, *planar* minerals can create a *linear* fabric, just as *linear* minerals can create a *planar* fabric in Figure 23.21e.

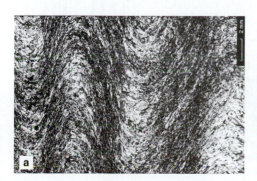
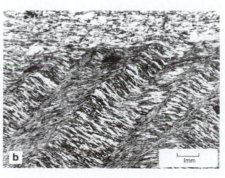

FIGURE 23.24 Crenulation cleavages. **(a)** (top) Symmetrical crenulation cleavages in amphibole–quartz-rich schist. Note concentration of quartz in hinge areas. **(b)** Asymmetric crenulation cleavages in mica–quartz-rich schist. Note horizontal compositional layering (relict bedding) and preferential dissolution of quartz from one limb of the folds. From Borradaile et al. (1982). Copyright © with permission from Springer-Verlag.

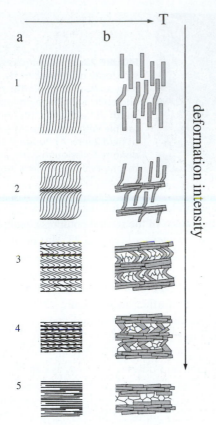

FIGURE 23.25 Stages in the development of crenulation cleavage as a function of temperature and intensity of the second deformation. From Passchier and Trouw (2005). Copyright © with permission from Springer-Verlag.

23.4.2 Mechanisms of Tectonite Development

Secondary (metamorphic) fabric elements in regionally metamorphosed rocks develop in response to deformation ± mineral growth. We shall concentrate on foliations here, but lineations are created in a similar fashion, depending largely on the relative magnitudes of σ_1, σ_2, and σ_3. Figure 23.27 illustrates the principal ways that we presently think a foliation may develop in a rock. How a rock or a mineral grain responds depends upon the type of mineral, pressure, temperature, magnitude and orientation of the stress field, and composition and pressure of the fluid. For reviews of the mechanisms by which foliations develop, see Shelley (1993) or Passchier and Trouw (2005) and the references therein.

Mechanical rotation (Figure 23.27a) was described in conjunction with Figure 21.3. The minerals behave as fairly rigid objects, although some minerals may deform internally by cataclasis during rotation. This mechanism occurs in low-temperature areas, such as shallow shear zones and low-temperature regional metamorphics.

Oriented new mineral growth involves either nucleation or preferred growth of existing minerals in advantageous orientations (Figure 23.27b and c). Many minerals grow most readily in directions in which the compression is least, and thus platy minerals typically grow normal to σ_1 and columnar or accicular minerals grow elongated parallel to σ_3. Existing minerals that are elongated in the direction of

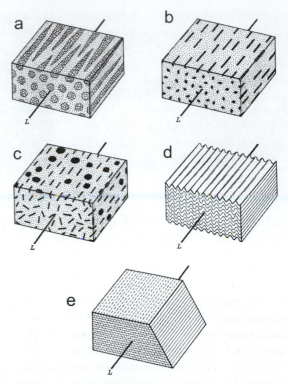

FIGURE 23.26 Types of fabric elements that define a lineation. **(a)** Preferred orientation of elongated mineral aggregates. **(b)** Preferred orientation of elongate minerals. **(c)** Lineation defined by platy minerals. **(d)** Fold axes (especially of crenulations). **(e)** Intersecting planar elements. From Turner and Weiss (1963). Copyright © with permission of the McGraw Hill Companies.

σ_1 may not grow at all or may even dissolve and become shorter (Figure 23.27c). Favorably aligned minerals may thus grow at the expense of unfavorably aligned ones. This process has been called **competitive growth**.

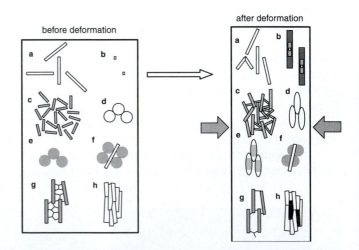

FIGURE 23.27 Proposed mechanisms for the development of foliations. **(a)** Mechanical rotation. **(b)** Preferred growth normal to compression. **(c)** Grains with advantageous orientation grow whereas those with poor orientation do not (or dissolve). **(d)** Minerals change shape by ductile deformation. **(e)** Solution transfer. **(f)** A combination of (a) and (e). **(g)** Constrained growth between platy minerals. **(h)** Mimetic growth following an existing foliation. After Passchier and Trouw (2005). Copyright © with permission from Springer-Verlag.

Crystal-plastic deformation and recrystallization may also result in elongated minerals during shear or flattening (Figures 23.27d and 21.2). Figure 23.28 illustrates the development of continuous mineral foliations by both processes. Flattening causes the foliation to be elongated normal to σ_1 which parallels the shortest axis of the strain ellipsoid. Shear, as discussed in conjunction with Figures 21.2 and 23.17, rotates a preexisting foliation toward the direction normal to the shortest axis of the strain ellipsoid, but it never reaches it. Note in Figure 23.28c that, if the initial state is already foliated as shown, flattening will enhance the foliation in its original orientation, and shear will *reorient* it. The reorientation of a foliation by folding or shear is called **transposition**.

Minerals such as quartz, which tends to be equidimensional in metamorphic rocks, even under conditions of regional metamorphism, may deform to elongated (Figure 23.29) or even **ribbon** shapes by crystal plastic mechanisms in mylonites (Figure 23.15d).

Solution transfer (Figures 23.27e and f), discussed earlier in conjunction with Figure 23.1, can produce elongated mineral shapes or enhance other foliations by dissolving minerals from high-pressure areas (surfaces at a high angle to σ_1). Loss of quartz from the limbs of folds (Figure 23.24) makes the mica foliation much more evident, as does removing carbonate or quartz, to leave an insoluble residue that defines a cleavage domain. Solution all but requires the presence of a fluid.

Mineral elongation and parallel growth may also occur in a passive sense. For example (as described earlier), quartz may become elongated due to constraints imposed by neighboring oriented mica grains. Thus in Figures 23.11 and 23.27g the increase in grain size of recrystallizing quartz is confined by the aligned micas and the stability of the mica {001} faces. Thus the quartz crystals develop a dimensional

FIGURE 23.29 Deformed quartzite with elongated quartz crystals following shear, recovery, and recrystallization. Note the broad and rounded suturing due to coalescence. Field width ~1 cm. From Spry (1969).

preferred orientation (DPO) as they conform to the micas. **Mimetic** mineral growth is also a common phenomenon that can produce oriented minerals even under nearly lithostatic stress conditions. If, for example, dynamic crystallization produces a preferred orientation of mica in a schist, and that schist is recrystallized during a later non-deformational thermal event, static mica growth during the annealing event ought to be random. Recrystallization, however, may be easier by conforming to the preexisting micas. Other minerals may also conform, either by epitaxial nucleation of new minerals or due to restrictions on growth imposed by neighboring mica grains. Thermodynamic (chemical and mineralogical) equilibrium is usually established before textural equilibrium. Thus textural equilibrium is particularly dependent upon the length of time that a rock is maintained at high temperature.

23.4.3 Gneissose Structure and Layers

As defined in Chapter 21, gneissose structure is either a secondary layering in a metamorphic rock or a poorly developed schistosity in which the platy minerals are dispersed. Gneissose structure can range from strongly planar to strongly linear. Fabric elements range from nearly continuous layers to discrete lensoidal shapes.

Several metamorphic rock types exhibit layers or lenses on the centimeter to several millimeter scale, and in gneisses it is practically characteristic. Such layering in fine-grained, low-grade rocks is generally relict bedding or igneous layering because secondary separation into contrasting layers requires diffusion, which is most effective at elevated temperature. Shear separation onto P-rich and Q-rich domains (see below) in slates and phyllites is an exception (although typically much finer). Also the layers in some schists may be small locally derived veins, typically quartz or calcite. Shear may transpose these veins so that they become nearly parallel to the schistosity.

Layers in higher-grade schists and gneisses appear to develop in initially unlayered rocks, and the layering is

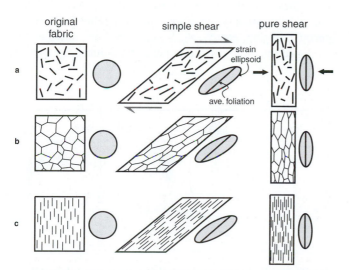

FIGURE 23.28 Development of foliation by simple shear and pure shear (flattening). **(a)** Beginning with randomly oriented planar or linear elements. **(b)** Beginning with equidimensional crystals. **(c)** Beginning with preexisting foliation. Shaded figures represent an initial sphere and the resulting strain ellipsoid. After Passchier and Trouw (2005). Copyright © with permission from Springer-Verlag.

attributed to a poorly defined processes collectively called **metamorphic differentiation** (Stillwell, 1918). Metamorphic differentiation runs contrary to the concept of increasing entropy with temperature, and processes that cause initially dispersed constituents to segregate into contrasting layers of more uniform composition and mineralogy are therefore interesting. Numerous mechanisms have been proposed, but it is still uncertain what is responsible. Proposals include solution transfer and redeposition via an aqueous phase (governed perhaps by local pressure gradients), local melt segregation, diffusion-controlled mineral development, and segregation of minerals as a result of their response to shear or stress differences (Stephens et al., 1979). Schmidt (1932) first proposed that micas segregate into high-shear areas due to their ability to slip readily on {001} cleavages, leaving quartzo-feldspathic layers between the high-shear domains. Robin (1979) proposed a similar mechanism in which a layered muscovite-quartz ("M-Q") rock would develop an unequal distribution of differential stress $(\sigma_1 - \sigma_3)$, which would be higher in the less competent "M-rich" layers (now popularly called **P-rich**, for phyllosilicate), and that this would create localized pressure gradients. The pressure gradients cause unequal pressure solution and migration of matter to the low-stress (**"Q-rich"**) layers (solution transfer). SiO_2 is more mobile than Al_2O_3, so quartz–muscovite segregation becomes enhanced with time. The process, Robin argued, can be initiated by relatively small initial inhomogeneities. As mentioned earlier, segregation into fine P-rich and Q-rich domains may even occur at very low metamorphic grades in slates. Perhaps all the mechanisms proposed above are possible, and each occurs in one rock or another. For summaries of metamorphic differentiation, see Hyndman (1985), Shelley (1993), Kretz (1994), or Vernon (2004).

23.4.4 Other Regional Metamorphic Textures

Other textural and structural elements that may develop in deformed rocks and minerals are **folds** and **kink bands** (Figure 23.30). **Boudinage** is a process in which elements such as dikes or elongate minerals that are less ductile than their surroundings, stretch and separate into tablets or sausage shapes (called **boudins**, French for "sausages") as the surroundings deform by ductile flow (Figures 23.20d, 23.34e, and 23.36).

Proper study of tectonites usually requires that the hand specimen and thin section can be spatially related to the outcrop. This way, we can use our textural observations to estimate the local or regional stresses. **Well-oriented samples** are thus required. Techniques for collecting and recording oriented samples are described by Passchier and Trouw (2005). Laboratory work for lattice-preferred orientations (LPO) requires a **universal stage** that permits a petrographer to orient a crystal in a thin section in three dimensions and to record the spatial orientation of the crystal axes. See Turner and Weiss (1963), Emmons (1943), or Passchier and Trouw (2005) for universal stage techniques. The orientation of a sound statistical sampling of crystal axes can then be plotted on an equal-area stereonet, and the density distribution of points can be contoured to produce **petrofabric diagrams**. The two petrofabric diagrams in Figure 23.31 will be familiar to students of structural geology, who use similar diagrams for fold axes, etc. See Turner and Weiss (1963) or Passchier and Trouw (2005) for further information on petrofabric diagrams.

23.4.5 Deformation Versus Metamorphic Mineral Growth

Following the lead of Sander (1930, 1950) and Turner and Weiss (1963), foliated rocks are called **S-tectonites**, and we can refer to the foliations as **S-surfaces** (as a shorthand notation: **S**). Linear elements are **L-tectonites** (shorthand: **L**). If two or more geometric elements are present, we can add a numeric subscript to denote the *chronological* sequence in which they were developed and superimposed. S_0 and L_0 are reserved for primary structures, such as relict bedding or igneous layering, etc. S_1, S_2, S_3, etc. are then subsequently developed foliations, whenever present. Similarly, L_1, L_2, and

FIGURE 23.30 Kink bands involving cleavage in deformed chlorite. Inclusions are quartz (white), and epidote (lower right). Field of view ~1 mm.

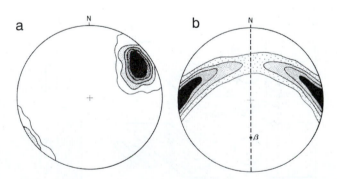

FIGURE 23.31 Examples of petrofabric diagrams. **(a)** Crystal c-axes cluster in a shallow inclination to the northeast. **(b)** Crystal axes form a girdle of maxima that represents folding of an earlier LPO. Poles cluster as normal to fold limbs. β represents the fold axis. The dashed line represents the axial plane and suggests that σ_1 was approximately east-west and horizontal. From Turner and Weiss (1963). Copyright © with permission of the McGraw-Hill Companies.

L_3 are successive secondary linear elements. By using this notation we can conveniently describe the sequential development of fabric elements in a tectonite. D_1, D_2, D_3, etc. can be used to refer to deformational phases, and M_1, M_2, M_3, etc. to metamorphic events so that we can then relate fabric elements and the minerals involved to the events that created them, provided of course that we can determine the relationships on the basis of textures. We cannot always figure out every detail of a rock's deformational and metamorphic history because some rather special textures are required to have developed, and later events may obscure evidence of earlier ones. Nonetheless, careful observation can often reveal a number of important textural clues. Also note that only the *relative* sequence of events is recorded petrographically, and *absolute* ages for multiple events require detailed isotopic work combined with structural–sedimentological–paleontological studies (Section 23.7).

As an example of the petrographic technique, suppose we begin with a pelitic sediment that has fine laminations of more sand-rich and clay-rich bedding (left and right halves of Figure 23.32). We conventionally call this primary foliation S_0. Suppose the sediment then becomes buried and metamorphosed (M_1) at some later date, and deformation (D_1) during the metamorphic event produces an excellent continuous cleavage in the metamorphic rock, which we then call S_1 (the sub-vertical foliation in Figure 23.32). S_1 is commonly subparallel to S_0 because the weight of the overlying strata causes a considerable vertical stress component on the originally horizontal bedding. Pelitic sedimentary rocks that have become compressed but not strongly deformed or metamorphosed may develop a **bedding-plane foliation** in which original clays or micas rotate in response to compaction. But horizontal compression is also a common facet of orogenic metamorphic events, and this adds a component of stress that reorients σ_1 a little from true vertical. This and later folding of S_0 causes S_1 to form most commonly at an angle to S_0 in the average thin section (although the angle may be quite small, as in Figure 23.32). This S_1

tends to obliterate any transient bedding-plane foliation. A later D_2 event then affected the rock, folding S_1 into a spaced crenulation cleavage (sub-horizontal in Figure 23.32), which we can call S_2. Metamorphism may or may not accompany D_2 and the folding of S_1. If it does, new M_2 mica growth will develop parallel to S_2.

S_1 is most obvious in the microlithons between the spaced S_2 cleavages in the micaceous layer in Figure 23.32 because micas are platy and define cleavages well. S_1 and S_2 are also visible in the slightly more mica-rich layer within the sandy bed. Crenulation cleavages become more distinctive when the fold limbs get tighter and asymmetrical (Figure 23.25) and quartz grains are removed from the limbs by solution transfer (Figure 23.24). If D_2 is accompanied by a metamorphic event (M_2), growth of new micas along the steep limbs or axial planes will further define S_2 (Figures 23.24 and 23.25). If D_2 is intense or recrystallization accompanying M_2 is sufficiently thorough, S_1 textures may become overprinted to the extent that they are no longer visible. In Figure 23.32, M_2 metamorphism did accompany S_2 development because quartz has been removed from the S_2 cleavages, leaving a micaceous and opaque graphite residue (horizontal black bands).

Although these observations may appear to be tiny details seen at the microscopic scale, let me remind you that large-scale and dramatic events, such as the collision of India and Asia, may produce some awesome geologic and topographic features, but erosion will one day remove the grandeur of such spectacles. All that will remain is the exposed deformed metamorphic rocks, which hold the only lasting clues to those ancient events. Structural geologists and metamorphic petrologists unravel the history of those events by the types of observations that I have just described. The Taconic and Acadian orogenies in New England, for example, produced some mountains that once probably looked much like the Himalayas. Today we might see an Acadian S_2 crenulation cleavage overprint a Taconic S_1 cleavage, that may be the only local record of that history in some places.

In order to interpret the metamorphic and deformational history of a rock, it is useful to be able to distinguish metamorphic mineral growth from deformational phases. Deformation and metamorphism generally occur in unison as dynamic recrystallization in orogenic metamorphism, but this is not always the case. Metamorphic mineral growth can be constrained on the basis of the timing of growth with respect to deformation. Mineral growth may thus be characterized as **pre-kinematic**, **syn-kinematic**, or **post-kinematic**. Many investigators use the term "tectonic" in place of "kinematic," but this has genetic implications (however broad), so I prefer the less encumbered term that simply indicates motion. A fourth category of porphyroblasts, **inter-kinematic**, implying mineral growth that is post-kinematic to one deformational event and pre-kinematic to the next, is less commonly used, but the term may be useful when the distinction is clear. It may be difficult to clearly assign the growth of a mineral to one of the categories. For example, minerals that define a good crystallization schistosity (as in Figure 23.23) may have

FIGURE 23.32 Pelitic schist with three s-surfaces. S_0 is the compositional layering (bedding) evident as the quartz-rich (left) half and mica-rich (right) half. S_1 (subvertical) is a continuous slaty cleavage. S_2 (subhorizontal) is a later crenulation cleavage. Field width ~4 mm. From Passchier and Trouw (2005). Copyright © with permission from Springer-Verlag.

grown syn-kinematically, but growth may have outlasted deformation and continued post-kinematically.

Porphyroblasts are among the most useful tools for interpreting metamorphic-deformational histories for several reasons. First, porphyroblasts, being larger than the matrix around them, are mechanically more resistant to deformation, and can thus become porphyroclasts during later shearing, and be used as sense-of-shear indicators. Second, porphyroblasts may envelop and include some finer grains as they grow, thus becoming **poikiloblasts**. The nature and pattern of the inclusions may be very useful in interpreting deformation-mineral growth histories.

If the matrix was foliated before poikiloblast growth, the inclusions within the porphyroblast may record that earlier foliation as an **internal S (S_i)**, as seen in two of the bands in Figure 23.33. Note that a foliated S_i within a poikiloblast cannot be formed by deformation of a poikiloblast with random inclusions (unless the poikiloblast is unusually weak and severely deformed). Internal S foliations must have been developed *prior to* porphyroblast growth and then enveloped by the growing porphyroblast. If a foliation becomes partially or completely obliterated in a rock matrix due to strong later deformation ($\pm$ metamorphism), the S_i in the poikiloblast may be the only surviving record of the earlier S-forming event.

A porphyroblast may also include a mineral that becomes a reactant in a later metamorphic reaction. Because diffusion within a crystal is slow, the porphyroblast may

protect the inclusions, which may then be the only remnant of that mineral if it is completely consumed in the matrix. Such protected inclusions are called **armored relics**. If the inclusion is an important index mineral or an indicator of metamorphic grade, it is particularly valuable in the interpretation of the history of the rock.

A careful petrographer may be able to infer several useful clues about the history of metamorphism, deformation, and growth of metamorphic minerals by observing inclusions and their orientation with respect to the foliation outside a porphyroblast (Zwart, 1960, 1962). When S_i is compared to an external foliation, the external one is typically referred to as the **external S**, or S_e, regardless of whether S_e is S_1 or S_2, etc.

As described earlier, porphyroblasts grow larger than matrix minerals because their nucleation rate is much slower than their growth rate. Nucleation from a melt or a glass is considered *homogeneous* because the matrix in which they grow is uniform. Nucleation in metamorphic rocks is *heterogeneous*, however, because there is usually a variety of preexisting surfaces on which a new mineral may nucleate. Various crystal surfaces in different lattice orientations, different mineral-pair grain boundaries, crack and fracture surfaces, fluids, impurities, irregularities and deformed areas all make for a plethora of possible nucleation sites in rocks. Some investigators use this variety to argue that nucleation must be easy, and that porphyroblasts must then result from rapid growth rates for certain minerals (perhaps the introduction of some critical component in a fluid). But there are clearly fewer nuclei formed, so others argue that a variety of nucleation sites does not guarantee that all minerals will nucleate equally easily, and that porphyroblasts require either larger initial clusters to be stable or very special sites. One thing is clear: phyroblasts involve more extensive growth on fewer nuclei than the other minerals in a rock. Common porphyroblast-forming minerals are garnet, staurolite, andalusite, kyanite, cordierite, and albite, but a number of other minerals may form porphyroblasts under favorable circumstances.

As I said above, it may not always be possible to determine unequivocally the relationship between metamorphic mineral growth and deformation, but the following is intended to be a useful guide.

Pre-kinematic crystals (Figure 23.34), when clearly recognizable, show the usual characteristics of minerals affected by later deformation, most of which are described above for high-strain rocks. These include undulose extinction, cracked and broken crystals, deformation bands and twins, kink bands, pressure shadows, porphyroclasts with mortar texture or sheared mantles, etc. **Pressure shadows** (Figure 23.34c) occur when solution transfer dissolves a mineral (usually a matrix mineral) from high-stress areas (generally in the σ_1 direction) and re-precipitates it in low-stress areas adjacent to a porphyroblast. In addition, the external foliation may be **wrapped**, or compressed about a pre-kinematic porphyroblast, a result of flattening in the matrix. S_i is of little use in determining the relationship between a porphyroblast and *later* deformational events.

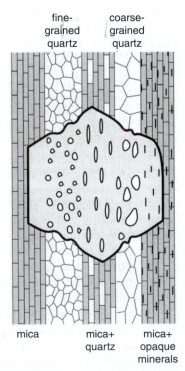

fine-grained quartz — coarse-grained quartz

mica — mica+quartz — mica+opaque minerals

FIGURE 23.33 Illustration of an Al_2SiO_5 poikiloblast that consumes more muscovite than quartz, thus inheriting quartz (and opaque) inclusions. The nature of the quartz inclusions can be related directly to individual bedding substructures. Note that some quartz is consumed by the reaction, and that quartz grains are invariably rounded. From Passchier and Trouw (2005). Copyright © with permission from Springer-Verlag.

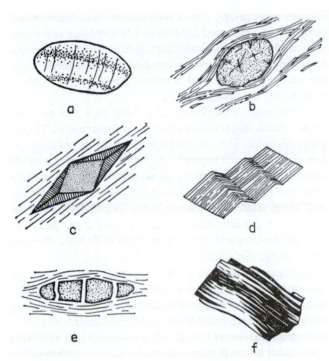

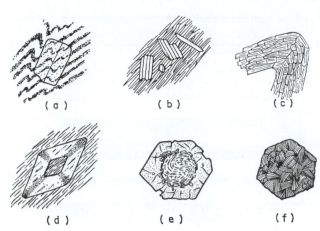

FIGURE 23.35 Typical textures of post-kinematic crystals.
(a) Helicitic folds in S_i. **(b)** Randomly oriented elongate crystals.
(c) Polygonal arc of straight crystal segments in a fold.
(d) Chiastolite with central S_i concordant with S_e. **(e)** Late,
inclusion-free rim on a poikiloblast. The rim *may* be post-
kinematic. **(f)** Random aggregate of grains of one mineral
pseudomorphing a single crystal of another (chlorite after
garnet). The chlorite is post-kinematic. From Spry (1969).

FIGURE 23.34 Typical textures of pre-kinematic crystals.
(a) Bent crystal with undulose extinction. **(b)** Foliation wrapped
around a porphyroblast. **(c)** Pressure shadow or fringe. **(d)** Kink
bands or folds. **(e)** Fragmented and stretched crystals
(microboudinage). **(f)** Deformation twins. From Spry (1969).

Post-kinematic crystallization either outlasted deformation or occurred in a distinct later thermal or contact
event. When clearly distinguishable, both the previous deformation and the later mineral growth or recrystallization
must be apparent. Post-kinematic growth results in unstrained crystals. In many cases, the crystals are randomly
oriented and cut across an earlier foliation (Figure 23.35b),
although mimetic growth may cause a large proportion of
later crystals to grow parallel to the foliation anyway.

Pseudomorphs (Figure 23.35f), particularly those in
which a crystal is replaced by an aggregate of random smaller
crystals, suggest that the replacement was post-kinematic.
Bent crystals may polygonize during later static growth to
smaller unstrained crystals. A dramatic example of this is the
polygonal arc (Figure 23.35c), in which folded elongate minerals polygonize to an arcuate pattern consisting of smaller
straight crystals. If post-kinematic recrystallization is extensive, a granoblastic or decussate texture may be developed,
and an earlier foliation may be difficult or impossible to discern. When an internal S is visible in poikiloblasts, the pattern
of S_i may be useful. For example, **helicitic folds** (Figure
23.35a) mean that S_i is folded. Because a poikiloblast with a
straight S_i cannot be deformed so that the S_i becomes folded,
helicitic folds indicate that the porphyroblast grew after both
S_1 and S_2 (the axial surfaces of the folds) had formed. Helicitic folds are thus an indicator of post-kinematic porphyroblast growth.

A poikiloblast with a straight S_i that is parallel to,
and continuous with, S_e (Figure 23.35d, Figure 23.37-3a)
is also *probably* post-kinematic because it is most easily

explained by the growth of a porphyroblast in a static situation over an existing schistosity. A later deformation may rotate an earlier porphyroblast or transpose S_e. Thus if S_i is not
parallel to, or continuous with, S_e the porphyroblast is either
pre-kinematic to D_2, or syn-kinematic and deformation outlasted porphyroblast growth. It may be impossible to distinguish between these two possibilities.

Syn-kinematic mineral growth is probably the most
common type in orogenic metamorphism because metamorphism and deformation are believed to occur generally in
unison. It is also the most difficult to demonstrate unequivocally. A continuous schistosity, as illustrated in Figure
23.23, was probably generated by a process of dynamic
(syn-kinematic) recrystallization. Aligned grains that are a
mixture of bent crystals with undulose extinction and
straight recrystallized grains support this conclusion more
strongly, although they could also be a less likely combination of pre-kinematic and post-kinematic crystallization. It
is probable, but largely a matter of faith, then, that syntec-
tonic crystallization occurs during the development of a
schistosity.

Misch (1969) described one clear indicator of defini-
tively syn-kinematic mineral growth. Figure 23.36 is an
example of "syn-crystallization micro-boudinage." Although
boudinage of individual crystals has been described from
several localities, it is usually an indication of pre-kinematic
crystals that have been stretched after their growth. In the
case described by Misch, however, the crystals are clearly
zoned, as seen in thin section by a gradation from colorless
amphibole cores to blue rims. The concentric pattern of zoning occurs *between* the separated tablets, as well as around
them, reflecting growth added *during* separation.

It is also widely believed that syn-kinematic porphyroblasts are the most common type of porphyroblast, probably due to the catalyzing effect that deformation has on
nucleation, reaction, and diffusion rates (Bell, 1981; Bell

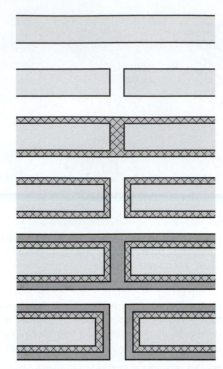

FIGURE 23.36 Syn-crystallization micro-boudinage. Syn-kinematic crystal growth can be demonstrated by the color zoning that grows and progressively fills the gap between the separating fragments. After Misch (1969). Reprinted by permission of the *American Journal of Science*.

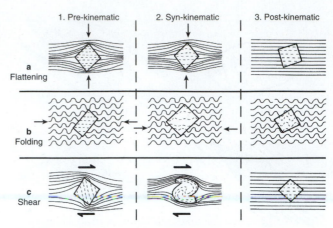

FIGURE 23.37 S_i characteristics of clearly pre-, syn-, and post-kinematic crystals (columns 1–3: left-to-right) as proposed by Zwart. **(a)** Top row: flattening. **(b)** Middle row: folding. **(c)** Bottom row: shear/rotation. The pre-, syn-, and post- prefixes refer to the flattening, folding, or shear event, not the earlier development of S_1. After Zwart (1962).

and Hayward, 1991; Baxter and DePaolo, 2004), but proof of syn-kinematic growth is rare. In certain instances, however, the pattern of S_i inclusions in a porphyroblast may provide unequivocal evidence for syn-kinematic porphyroblast growth (Zwart, 1960, 1962). Figure 23.37 illustrates porphyroblast growth incorporating an S_i, associated with (a) flattening, (b) folding, and (c) shear. Let's begin with row (a)—*flattening*. Imagine a deformation (D_1) that produced a foliation (S_1), followed by *post-kinematic* poikiloblast growth that then incorporated the *uniform* S_1. This would be the case for column 3 (right side) of row (a), or 3(a), where the poikiloblast must have grown entirely after the foliation was developed. Next imagine that S_1 becomes flattened. In box 1(a) the poikiloblast grew after S_1 but is *pre-kinematic* to flattening, during which the foliation was compressed around the poikiloblast. This may be a good example of *inter-kinematic* porphyroblast growth, but the illustration is attempting to address the flattening, which the bending around the porphyroblast suggests was produced by a second D_2 event. The S_i pattern in the center figure of row (a), box 2(a) is (finally) an example of demonstrable syn-kinematic porphyroblast growth because it shows a systematic variation in the spacing of S_i from core to rim. The wider spacing of S_i at the core indicates that the poikiloblast grew initially over a less flattened S_1, and successive layers of poikiloblast growth incorporated a progressively more closely spaced foliation that was progressively flattened as the poikiloblast grew. The poikiloblast is thus demonstrably syn-kinematic to flattening. The foliation and the flattening in this case may reflect a single D_1 event.

Row (b) of Figure 23.37 illustrates a similar scenario, in which S_1 is *folded* and the folds in the center *syn-kinematic* example become progressively more intensely folded as the porphyroblast grows and incorporates them. 1(b) illustrates poikiloblast growth following D_1-S_1, but *pre-kinematic* to folding. Poikiloblast growth in 3(b) is entirely *post-kinematic* to both S_1 development and folding.

The bottom row in Figure 23.37 corresponds to *shear*. I'm sure you can now interpret 1(c) and 3(c) as poikiloblast growth that is pre-shear and post-shear. 2(c) is *syn-kinematic* and shows a **spiral** pattern of the S_i inclusions that is not found in the matrix foliation. The traditional interpretation of this spiral pattern is that the porphyroblast rotated as it grew (Figure 23.38), progressively incorporating the external foliation, like rolling a snowball. This spiral S_i texture is particularly common in garnets. Some call the spiral texture **rotated**, or, if the rotation is extensive, **snowball** (Figure 23.39). Note that the spiral creates a fold pattern, but the spiral is a special case of a fold and has rotational symmetry. It is important to distinguish this spiral pattern as separate from helicitic folds (which indicate post-kinematic poikiloblast growth).

In all three situations in the left column of Figure 23.37, pre-kinematic porphyroblasts have a straight, parallel S_i (= S_1), and the S_e has either rotated, wrapped, or folded after porphyroblast growth, presumably during a separate later deformational event (D_2). Barker (1990) argued, however, that porphyroblasts can grow relatively quickly, and may envelop a straight S_i during the early stages of a deformation event and still be rotated, flattened, or folded during the same event. Syn-kinematic porphyroblasts (center column in Figure 23.37) show progressively spiraled, flattened, or folded S_i. Post-kinematic porphyroblasts (right column in Figure 23.37) overgrow the sheared, compressed, or folded S_1 without a discontinuity between S_i and S_e. When S_i is not continuous with S_e it means that either a separate deformation affected it, or that porphyroblast growth ceased before deformation did, and a single deformation created the S_i and then the discontinuity.

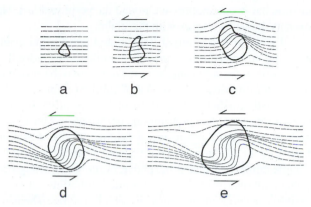

FIGURE 23.38 Traditional interpretation of spiral S_i train in which a porphyroblast is rotated by shear as it grows. From Spry (1969).

The traditional interpretation of rotated porphyroblasts has been challenged by Ramsay (1962), Bell (1985), and others. Ramsay (1962) first suggested that rigid porphyroblasts (at least equidimensional ones such as garnet) may remain spatially fixed during deformation, and that the *foliation is rotated* about the porphyroblast instead. Some works (Fyson, 1975, 1980; Johnson, 1990; de Wit, 1976; Bell et al., 1992) have demonstrated consistent porphyroblast orientations on a regional scale, or even on opposite sides of folds (where the sense of shear should be opposite), which support the notion that the foliation was rotated, and not the porphyroblast. It may be hard to imagine rotation of a regional foliation more than 90° or so, and extreme examples of spiral inclusion trails (Figure 23.39) may support rotated porphyroblasts, but for the much more common occurrence of <90° rotations, it is still controversial as to what mechanism occurred. There is little disagreement that elongate porphyroblasts may be rotated, and the controversy is over equidimensional ones, such as garnets. For discussions of this controversy, see Bell et al. (1992); Passchier et al. (1992); Williams and Jiang (1999), Vernon (2004), and Passchier and Trouw (2005).

FIGURE 23.39 "Snowball garnet" with highly rotated spiral S_i. Porphyroblast is ~5 mm in diameter. From Yardley et al. (1990). Copyright © reprinted by permission of John Wiley & Sons, Inc.

Why all the worry if there is agreement that the spiral represents syn-kinematic growth? Note in Figure 23.38 that the sense of shear in the rotating porphyroblast is *sinistral* (anticlockwise). If, on the other hand, the external foliation is rotated and the porphyroblast remains stationary, the same pattern would require *dextral* shear. Thus the shear sense that we deduce depends on our interpretation of what rotates. Until this controversy is resolved, it would be wise to avoid depending upon spiral inclusion trails alone to determine sense of shear. Similarly, mechanism-biased terms, such as *snowball* and *rotated* porphyroblasts, are best avoided in favor of the simple geometric term *spiral*.

Bell (1985) and Bell et al. (1986) suggested that deformation in a rock body experiencing shear is partitioned unequally as a result of primary or secondary mechanical inhomogeneities (e.g., the front cover of this text). They proposed that high-shear domains (S_2 in this situation, and they are white in Figure 23.40) are separated from lozenge-shaped zones of little or no strain (dark shaded) by transitional zones of mostly flattening or shortening (light shaded), depending on the style of deformation. As mentioned previously, phyllosilicates can accommodate shear better than most other minerals. High shear, they proposed, causes the other minerals to dissolve (a form of solution transfer) and migrate to low-shear areas where, they believe, porphyroblasts tend to form. This results in *metamorphic differentiation* into mica-rich and mica-poor/porphyroblast-rich domains (P and Q, as described previously).

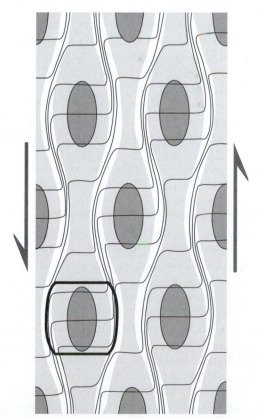

FIGURE 23.40 Non-uniform distribution of shear strain as proposed by Bell et al. (1986). Blank areas represent high shear strain and dark areas are low strain. Lines represent initially horizontal inert markers (S_1).

Note that a porphyroblast that grows in a low-shear area but eventually extends into the marginal area of transition, such as the one outlined in Figure 23.40, can develop an inclusion trail that is straight in the center (from an earlier S_1) and spiraled toward the rims. This is a common shape for S_i in poikiloblasts, which Bell et al. (1986, 1992) and Bell and Johnson (1989) interpreted to be *unrotated* because the porphyroblast simply grows over a folded foliation. The straight to spiraled S_i in the porphyroblast was attributed above to pre-kinematic (to S_2) core growth, followed by a syn-kinematic rim, but the porphyroblast in question here is entirely late syn-kinematic to S_2 if this interpretation is correct. Bell et al. (1992) also proposed that the sheared zones (white areas in Figure 23.40) may expand with time into the low-shear areas and the porphyroblasts, causing porphyroblasts to dissolve and the S_i to become truncated. They suggested that this common pattern of straight S_i in porphyroblast cores with spiraled rims occurs because porphyroblasts do not tend to grow during the first deformation (D_1) of pelitic sediments (the one that produced the initial S_1 in Figure 23.40) because poor lithification results in more evenly distributed strain and more compression as the sediments lose pore fluids. Only during D_2, when the rock is more competent, does the above scenario take place and S_1 is found in the porphyroblast cores.

The deformational history of porphyroblasts without an internal S is difficult to interpret with confidence. Undeflected S_e at porphyroblast margins is traditionally taken to indicate post-kinematic porphyroblast growth, which does not disturb a preexisting foliation, but deflection of S_e is less certain. Ferguson and Harte (1975) suggested that deformed and deflected S_e around a porphyroblast could reflect pre-, syn-, or post-kinematic porphyroblast growth and warned against using any interpretation of S_e at porphyroblast margins when S_i is missing. When pressure shadows are present, or the external foliation is deflected, the porphyroblast is probably pre-kinematic, but there could be several kinematic events.

A long-standing controversy has focused on foliations that wrap around porphyroblasts (Figure 23.34b). Some investigators (Ramberg, 1947; Misch, 1971, 1972a, b) have argued that growing crystals exert a "force of crystallization" that can physically bow out an earlier foliation, similar to ice-wedging during weathering. Others (Rast, 1965; Spry, 1969, 1972; Shelley, 1972) have argued that metamorphic porphyroblasts grow by infiltration along grain boundaries and replacement of preexisting minerals and cannot exert enough force to displace neighboring grains. Yet others (Yardley, 1974; Ferguson, 1980) argued that the foliation may migrate and appear bowed by pressure-solution associated with a growing porphyroblast. I think most petrologists lean toward the notion that it is simpler for a foliation to deform around an existing porphyroblast than for the porphyroblast to affect the foliation. In several situations, this has been amply demonstrated to be the case, whereas the opposite is less certain. Wrapping of foliation about a porphyroblast, particularly if it is well developed or associated with a discontinuity between S_i and S_e is thus generally taken to indicate that the foliation has been compressed around the porphyroblast. No warranty, however, comes with this claim.

Passchier and Trouw (2005) noted two further complications in interpreting mineral growth–deformation relationships in metamorphic rocks. First, it may be possible for two deformation phases to be of similar orientation and thus indistinguishable texturally or structurally. Second, the opposite may also be true: one deformation phase may produce both a foliation and a folding of that foliation (Figure 23.41). In this case both S_1 and S_2 are produced by a single consistent and progressive event, even though S_2 appears to overprint S_1. Such folded foliations are observed in several shear zones, in which a single shearing event is all but assured. I doubt that the development of a penetrative crenulation cleavage can be developed over a broad area in this fashion, however, and suspect that single-stage folded foliations develop more locally, initiated perhaps by inhomogeneities. Even in the shear zone examples, the secondary folds typically occur as local folds that do not develop in all layers. Although we must understand that our interpretations may miss some historical event, we can only assume the simplest and most straightforward explanations of the textures that we observe. We thus interpret foliations to represent a single phase of deformation, and folds to represent a separate phase for the structures that they affect. Only when intensive and well-distributed

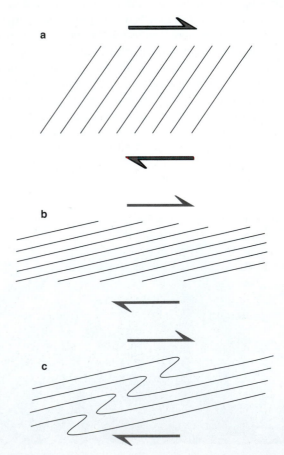

FIGURE 23.41 **(a–b)** Initial shear strain causes transposition of foliation. **(c)** Continued strain during the same phase causes folding of the foliation.

sampling indicates otherwise should we propose scenarios such as those illustrated in Figure 23.41b to c.

23.4.6 Analysis of Polydeformed and Polymetamorphosed Rocks

Let's apply what we have learned to the analysis of some metamorphic textures. A good approach, following careful observation of several thin sections for each sample, is to graphically portray our interpretations, showing the growth of each major mineral in a rock versus time, and to infer deformation and metamorphic events on the same time plot. Of course, it may not always be possible to determine with certainty the full metamorphic and deformational history of a rock. The last episode, particularly if deformation was intense and/or metamorphism was of high grade, may have obliterated any evidence of earlier events. Careful observation, however, will generally produce some textural clues.

To illustrate this approach, consider the texture shown in Figure 23.42. The fabric of the rock exhibits three secondary foliations, which we infer to represent three phases of deformation, D_1, D_2, and D_3. D_1 produced a very good continuous cleavage (S_1—sub-horizontal in Figure 23.42), which is penetrative, affecting the entire volume of the rock at the smallest scale. From this we might infer that D_1 was more intense than D_2 or D_3, both of which are spaced cleavages. The micas are well aligned in S_1, and the quartz is elongated as well.

Because the minerals are so well aligned, we assume that metamorphism M_1 accompanied D_1 and that mineral growth kept pace with deformation. A second deformation (D_2) produced the crenulation cleavage (S_2—nearly vertical in Figure 23.42). Quartz is missing from the steeper limbs of the asymmetric folds, suggesting that quartz dissolution and recrystallization (metamorphic differentiation) accompanied D_2. We want to look closely at the micas in the fold hinges to see if they are bent, or recrystallized to form polygonal arcs. Let's suppose that many of the micas are straight, but a few are bent, suggesting that mica crystallization accompanied D_2 as well, but D_2 just barely outlasted mica crystallization so that some bending remains. Finally, D_3 folded S_2 and bent all of the micas so that little or no recrystallization is apparent. Our graphical representation might look like Figure 23.43. The size of the deformation "humps" reflects our interpretation of the intensity of the deformation, and the horizontal lines represent the metamorphic crystallization of each mineral (dashed where less certain). Timing on such a diagram is strictly relative, and the length of deformation and metamorphic events, and the intervals between them, are unknown. In this case, the deformational features of the rock are the most obvious ones to analyze.

As a second example, suppose we observe the texture illustrated in Figure 23.44. The quartz-rich and mica-rich layers may be relict bedding (S_0) or represent metamorphic differentiation (S_1). The preferred orientation of micas is certainly S_1. S_1 is perfectly parallel to the layering and the grade of metamorphism is in the garnet zone (medium grade), both of which suggest that metamorphic differentiation *may* be a

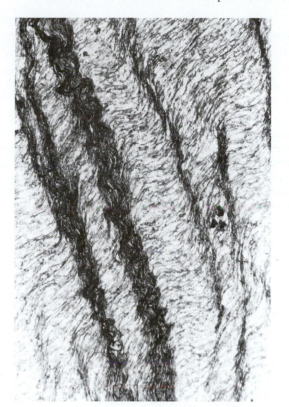

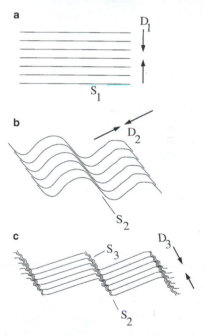

FIGURE 23.42   (top) Asymmetric crenulation cleavage (S_2) developed over S_1 cleavage. S_2 is folded, as can be seen in the dark sub-vertical S_2 bands. Field width ~2 mm. (bottom) Sequential analysis of the development of the textures. From Passchier and Trouw (2005). Copyright © with permission from Springer-Verlag.

more likely origin than bedding. We might look at the hand specimen to see if the layers are continuous or lensoidal. Lensoidal layers are more likely to result from metamorphic differentiation. The garnets have a good spiraled S_i of elongated quartz and opaque inclusions. The spiral suggests rotation of ~170°. Given only what we see in the figure, it is not

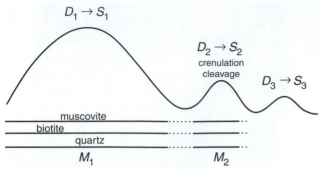

FIGURE 23.43 Graphical analysis of the relationships between deformation (*D*), metamorphism (*M*), mineral growth, and textures in the rock illustrated in Figure 23.42.

possible to say with certainty whether the matrix rotated 170° dextrally (clockwise) or the garnet rotated 170° sinistrally (anticlockwise). The slight asymmetry in the upper quartz-rich layer, stair-stepping upward to the left, vaguely suggests sinistral rotation (Figure 23.19), but unless other shear-sense indicators are found elsewhere in the rock, any conclusion based strictly on the S_i in garnet should be considered provisional. Note also that the thickness of the quartz-rich layers is greater at the garnets. The simplest explanation for this effect is that the layering was compressed about the garnets by the same deformation responsible for the rotation.

Note also that quartz has a well-developed granoblastic polygonal texture in the quartz-rich layers but is elongated in the mica-rich layers. The good polygonal texture suggests that crystallization outlasted deformation, which is also reflected in the generally straight micas and continuity between S_i and S_e of the garnets. The elongation of quartz in the micaceous layers is probably controlled by the micas (as discussed above: Figure 23.11). The quartz inclusions in the garnet are also elongated, and these inclusions may have been part of originally quartz-rich layers. If so, this suggests that perhaps the quartz crystals were elongated in both the mica-rich and mica-poor of layers when the garnet was growing, and the granoblastic texture in the quartz-rich

layers occurred later. Alternatively, perhaps the garnet began to grow prior to metamorphic differentiation and completely replaced the micas (scavenging the Al, and perhaps Fe and Mg) that controlled quartz elongation. Metamorphic differentiation then separated the "Q-rich" and "M-rich" layers at a later point, perhaps due to more advanced shear or at the metamorphic peak temperature.

A final factor is the number of deformational phases involved in Figure 23.44. Clearly, there was an S_1 prior to garnet growth. The simplest explanation for the growth of the garnet and the rotation that created the spiraling is that the garnet grew late during S_1, and the same deformation caused the rotation and the compression of the layering. This is consistent with the notion that garnet occurs at a higher metamorphic grade than quartz recrystallization and muscovite + biotite growth. A graphical interpretation would then look like Figure 23.45. Had S_i been discontinuous with ($S_i \neq S_e$) then the deformation curve would have extended past the lines representing crystallization. Although a second minor deformation event cannot be ruled out, a single event is a simpler explanation.

As a third example, consider Figure 23.46. In this case, there is a distinct S_1 in muscovite that has been folded to develop a good set of S_2 axial planes. Large andalusite porphyroblasts then formed in a random orientation with good helicitic folds in S_i. Andalusite growth is interpreted as the result of a later contact metamorphic overprint (presumably supported by field relationships). Note that the S_i pattern could almost be a spiral, but the fact that $S_i = S_e$ in identical external folds shows that the inclusion texture must be helcitic. The well-developed polygonal arcs could indicate that muscovite crystallization outlasted D_2 deformation or that the contact event caused muscovite recrystallization that was largely mimetic to the preexisting folded muscovite. Figure 23.47 is a graphical interpretation. Whether D_1 and D_2 were deformation phases associated with a single M_1 or two metamorphic events is purely speculative.

It is imperative to remember that this type of analysis can contribute valuable data to a structural synthesis of an area, but to be done properly it must be integrated with good structural field work. A simple straight S_1 in thin section, for example, may be parallel to the axial planes of folds too large to be seen in a thin section. Note also that polydeformed and polymetamorphosed rocks are relatively common, and

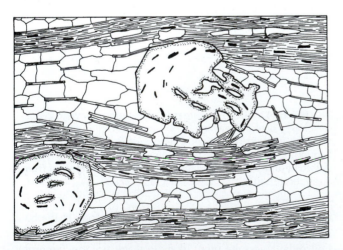

FIGURE 23.44 Composite sketch of some common textures in Pikikiruna Schist, New Zealand. Garnet diameter is ~1.5 mm From Shelley (1993). Copyright © by permission Kluwer Academic Publishers.

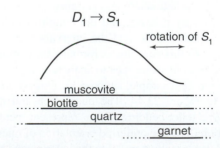

FIGURE 23.45 Graphical analysis of the relationships between deformation (*D*), metamorphism (*M*), mineral growth, and textures in the rock illustrated in Figure 23.44.

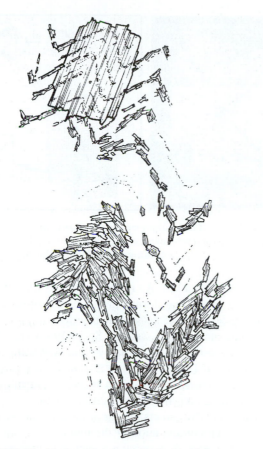

FIGURE 23.46 Textures in a hypothetical andalusite porphyryoblast–mica schist. After Bard (1986). Copyright © by permission Kluwer Academic Publishers.

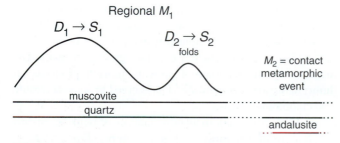

FIGURE 23.47 Graphical analysis of the relationships between deformation (D), metamorphism (M), mineral growth, and textures in the rock illustrated in Figure 23.46.

that the classification of crystallization into pre-, syn-, and post-kinematic types may often be oversimplified. In polymetamorphic rocks, one should be clear as to which event or textural element a prefix is related (e.g., pre-kinematic to S_2).

As a more complex example of a polymetamorphic rock, consider Figure 23.48, which shows three stages in Spry's (1969) interpretation of the final rock texture (Figure 23.28c). The matrix minerals are muscovite and quartz, whereas the porphyroblasts are garnet (G) and albite (Ab). Notice how S_2 mica growth has become sufficiently pronounced to obliterate most of the evidence for S_1, which is only discernable in the microlithons and S_i of the poikiloblasts. S_0 is only preserved as an S_i as well.

As a final note on orogenic metamorphism and deformation in an orogenic belt, consider Figure 23.49, a hypothetical scenario for the development of an orogen. In this scenario, an island arc terrane develops offshore from a continental mass. At stage I, deformation (D_1) occurs in the forearc subduction complex (A) associated with the arc (B). In detail, this deformation typically migrates away from the volcanic arc and toward the trench (to the left in the figure). Thus D_1 deformation is not concurrent at all points within A and does not occur within B or C at all. By stage II, D_2 deformation overprints D_1 within A in the form of sub-horizontal folding and back-thrusting, which develops as the right side of the forearc package is pushed against the resistant igneous arc crust. D_1 is still occurring on the left side of A at the same time. Subduction is initiated at C in stage II, and D_1 deformation (as locally defined) begins here (also migrating left toward the trench). Subduction-related igneous activity develops behind C as the arc matures. The same development of D_2 deformation against the continental crust may develop in area C as occurred during stage I in area A. In stage III, the offshore terrane is accreted to the continent, thereby creating a **suture zone**, and terminating subduction in area C. Because of the intensity of the collision, local D_1 deformation may penetrate the resistant sialic crust of both the arc and the continent. Within area B, the initial D_1–S_1 foliation has been overprinted by S_2 folds during stage III, perhaps as parts of a single event (as in Figure 23.41). Renewed (D_2) thrusting cuts the outermost foliation in the outer portions of the wedge in area C. D_2 deformation may also be associated with a later phase of collapse of the orogen.

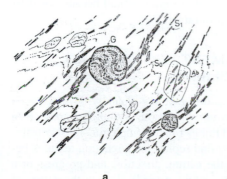
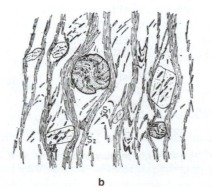

a   **b**   **c**

FIGURE 23.48 Interpreted sequential development of a polymetamorphic rock. From Spry (1969).

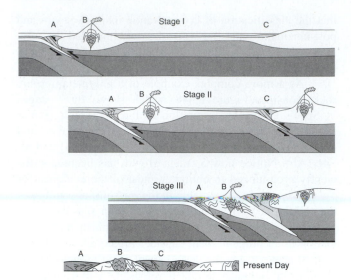

FIGURE 23.49 Hypothetical development of an orogenic belt involving development and eventual accretion of a volcanic island arc terrane. See the text for discussion. After Passchier and Trouw (2005). Copyright © with permission from Springer-Verlag.

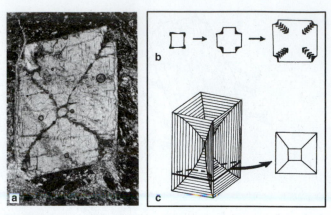

FIGURE 23.50 Chiastolite cross in andalusite **(a)** and Frondel's (1934) theory for its development **(b** and **c)**. From Shelley (1993). Copyright © by permission Kluwer Academic Publishers.

The scenario depicted above is hypothetical but consistent with the types of processes that we believe occur during the plate tectonic development of an orogenic belt. Note that deformation is neither of the same style nor is any phase simultaneous throughout the resulting belt. D_1, at one point, may be a different deformation than D_1 elsewhere, and may occur at the same time as D_2 or D_3 in another part of the orogen. Most mountain belts are an amalgamation of assembled terranes, each with its own unique history. Similarly, deformation may change style in a single locality, and deformation of a single style may migrate, so as not to be synchronous, even in adjacent rock packages. One can only unravel the history of such a complex orogen by careful integration of petrographic, structural, isotopic, sedimentary, and igneous data collected throughout the orogen.

Analysis of deformational and metamorphic histories based on textures and structural data is a difficult task and readily subject to oversimplification and interpretive errors. This has been only a cursory introduction to the techniques. Students are advised to read Jamieson (1988), Passchier and Trouw (2005), and Vernon (2004) for guidance and warnings before attempting independent work. In Section 23.7 we will see how we can combine textures with geochronology to determine the absolute timing of some events. Among other benefits this approach may eventually solve some of the textural controversies described above.

23.5 CRYSTALLOGRAPHICALLY CONTROLLED INCLUSIONS

Not all inclusion patterns reflect passive envelopment of a preexisting matrix texture into a growing porphyroblast. Some patterns are controlled by the lattice or growth surfaces of the porphyroblast itself. For example, inclusions may adsorb preferentially onto a face or some other part of a euhedral porphyroblast. A common example is the feathery **chiastolite cross** that forms in some andalusite crystals (Figure 23.50). There are several theories on the origin of chiastolite, summarized well by Spry (1969) and Shelley (1993). The best explanation, in my opinion, is that proposed by Frondel (1934), who proposed that the cross forms by the selective attachment of impurities (typically graphite) at the rapidly growing corners of the andalusite (see Figure 3.3 for an explanation of why corners may grow more rapidly than crystal faces). The concentration of the impurity (Figure 23.50) retards corner growth, forming a re-entrant, where the impurity gradually becomes incorporated by further porphyroblast growth. Repetition of the growth-retardation-growth scenario results in the feather-like appearance along four radiating arms. Similar patterns in other minerals, such as garnet, have been reported, but are much less common. Another crystallographically controlled inclusion pattern is **sector zoning**, in which inclusions are preferentially incorporated on some faces of a growing crystal (typically the faster-growing faces). If this occurs only on one pair of growing faces it produces an **hourglass** pattern (Figure 23.51), observed in some staurolite and chloritoid crystals.

These oriented inclusion patterns should be distinguished from **exsolution** of a dissolved phase, such as perthite, etc., which can also produce crystallographically oriented lamellae of one mineral in another (see Section 3.2.3). The crystallographically controlled needles of rutile in some biotite or hornblende crystals, for example, are exsolution phenomena, and are not oriented because of their envelopment by the host mineral.

23.6 REPLACEMENT TEXTURES AND REACTION RIMS

Replacement and reaction textures typically develop when reactions do not run to completion. Although it may seem more tidy to have all things run to completion, most petrologists are delighted to find replacement/reaction textures because they indicate the nature, direction, and progress of a reaction. These textures may provide clues to the nature of the protolith or the history of rocks sampled.

FIGURE 23.51 Euhedral chloritoid porphyroblasts in a fine-grained schist with hourglass zoning and rhythmic layer zoning, both of which are due to differing concentrations of inclusions. The hourglass zoning is crystallographically controlled, whereas the layer zoning must be related to varying conditions of growth. Field width ~8 mm. From Passchier and Trouw (2005). Copyright © with permission from Springer-Verlag.

Replacement occurs when the reaction products replace a reacting mineral. Several replacement textures were discussed in Section 3.2.4. in conjunction with "autometamorphic" processes associated with cooling igneous rocks (actually a form of retrograde metamorphism). Some degree of retrograde metamorphism typically occurs during the cooling of plutons because the rocks are maintained at metamorphic temperatures for extended periods. A wide range of retrograde reactions are possible in both igneous and metamorphic rocks. Perhaps the most common are hydration reactions, such as those that produce chlorite rims on mafic minerals. As discussed in Section 21.4, retrograde metamorphism is less likely than prograde to reach equilibrium, commonly resulting in replacement textures.

In many replacement reactions, a **pseudomorph** may develop, in which the reaction products retain the shape of the original mineral. Even if the pseudomorphing reaction runs to completion, the texture might still be used to interpret the reaction if the original mineral can be recognized by the shape of the pseudomorph. Some reactions produce intimate, typically wormy-looking, intergrowths of two or more minerals, a texture called **symplectite**. Pseudomorphs may thus be either monomineralic or a symplectitic intergrowth. Some replacement styles can be specific to certain mineral types. For example, in serpentinitized ultramafics the original olivines are typically replaced by serpentine along cracks in a net-vein-like **mesh** pattern (Figure 23.52a). Small relict olivine islands may remain in the network, forming an array of subgrains that all go extinct in unison, testifying to their once having been part of a larger single crystal. Orthopyroxene, if present in the same ultramafic protolith, is usually completely pseudomorphed by more uniform areas of fibrous serpentine, called **bastite** (Figure 23.52b).

Reaction rims involve reaction between minerals where they meet at grain boundaries, resulting in the partial

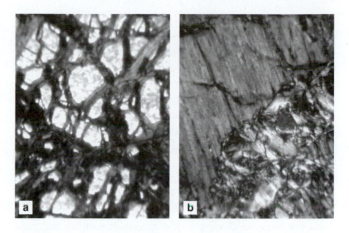

FIGURE 23.52 (a) Mesh texture in which serpentine (dark) replaces a single olivine crystal (light) along irregular cracks. (b) Serpentine pseudomorphs orthopyroxene to form **bastite** in the upper portion of the photograph, giving way to mesh olivine below. Field of view ca. 0.1 mm. Fidalgo serpentinite, Washington.

replacement of either or both minerals adjacent to their contact. If the reaction product forms a complete rim around a mineral it is called a **corona**. Coronas can be either monomineralic or polymineralic. Polymineralic intergrowths of small elongate grains are called **symplectitic coronas**.

Reaction rims occur when reactions did not reach completion, and are thus frozen records of reactions caused by changes in metamorphic conditions. They may reflect retrograde alteration, polymorphic transformations, diffusion of material along grain boundaries, or solid-state reactions. The latter process forms the most dramatic **coronites** (rocks with prominent corona textures), which are most common in high-grade dry rocks where slow cooling permits retrograde reactions and the lack of water keeps the reactions from running more readily to completion.

Figure 23.53 shows two minerals, A and B, that react to form either a new mineral C or two minerals C and D. If diffusion is limited and/or reaction time is short, the reaction product(s) may be limited to a zone at the A-B contact.

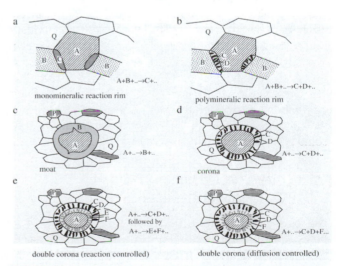

FIGURE 23.53 Reaction rims and coronas. From Passchier and Trouw (2005). Copyright © with permission from Springer-Verlag.

If one reactant is plentiful, the product(s) may form a continuous corona around the more limited reactant (Figure 23.53b). Monomineralic coronas (Figure 23.53c) are called **moats** by some investigators. Once a continuous rim is created between the reactants, they are separated, and the reaction consuming them can only continue by diffusion through the corona (Section 30.2.2). As the corona becomes thicker the extent of diffusion must be greater, which may eventually stall the reaction, particularly in a cooling dry rock.

Coronas may be single rims (Figure 23.53c and d) or complex multiple concentric layers (Figure 23.53e and f). Double or multiple corona rims may be created as conditions continue to change to the extent that either the core mineral or the matrix mineral is no longer stable adjacent to the first corona, thereby reacting with it and forming another layer between them (Figure 23.53e). Alternatively, gradients in diffusing components may generate multiple rims at the same time (Figure 23.53f), a concept developed further in Chapter 30. Spectacular multiple coronites are developed between olivine and plagioclase in some deep-seated gabbros or anorthosites (Figure 23.54).

Corona textures can be combined with mineral stability analysis (Chapter 27) to provide detailed reconstruction of *P-T-t* histories. For example, Griffin (1971) interpreted the sequence in Figure 23.54 as developing in several stages during a cycle of deep burial and subsequent uplift and erosion of the Jötun Nappe. Harley et al. (1990) described Al-rich granulites that are retrogressed in stages from 1.1 GPa and 1000°C to 0.45 GPa and 750°C. The first stage of uplift involved reactions between sillimanite and garnet or orthopyroxene that produce rims of cordierite (a low-pressure mineral) on garnet (a high-pressure mineral) or orthopyroxene, and symplectites of cordierite and sapphirine replacing sillimanite. Further reaction consumed K-feldspar, pyroxene, and sapphirine to produce biotite, and later still

biotite reacted with plagioclase and quartz to develop symplectitic rims of second-generation orthopyroxene and cordierite. Such detailed reconstructions are possible because the reactions are incomplete and the products occur as coronas at original grain boundaries. Coronas are common in scenarios that involve decompression. Perhaps decompression provides enough driving energy to initiate the reaction, but the reaction is still easily stalled by the diffusion required for the reactants to communicate through the rims.

23.7 TEXTURAL GEOCHRONOLOGY

The radiometric dating techniques discussed in Chapter 9 have traditionally required mineral separation and concentration from a significant mass of rock in order to analyze the isotopic ratios of phases present in only accessory amounts. Such dates typically represented average ages from homogenized mineral concentrates. Because these minerals may consist of zones representing several growth episodes (e.g., Figure 18.1), such averaged ages may be nearly meaningless. Recent advances in analytical technology are now allowing us to extract age information from very small volumes of a single mineral as observed in thin section. This is opening up some exciting avenues of research in which the timing of individual tectonic/metamorphic events may be determined and pressure–temperature–time (*P-T-t*) paths may finally have real meaning in terms of time. But even these more specific ages can be a challenge to relate with confidence to other critical petrologic information, such as temperature, pressure, or specific orogenic and metamorphic events. The challenge is to know exactly what is being dated. I shall outline some of the critical new techniques in the present section and provide a few examples in which geochronology has been linked to specific tectonic/metamorphic events, emphasizing works relying on textures to make the link. In later

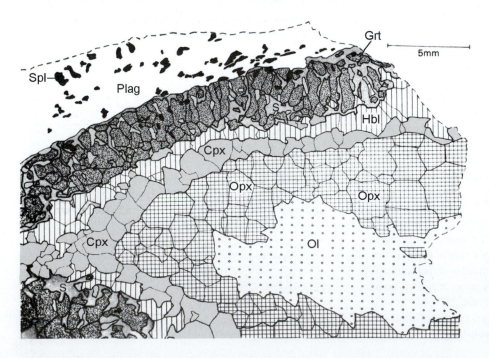

FIGURE 23.54 Portion of a multiple coronite developed as concentric rims due to reaction at what was initially the contact between an olivine megacryst and surrounding plagioclase in anorthosites of the upper Jötun Nappe, western Norway (from Griffin, 1971). The orthopyroxene–clinopyroxene coronas and subsequent garnet developed as the anorthosite cooled during burial and compression due to thrust stacking and nappe development. The amphibole corona developed during uplift as water became available, and later rapid uplift and decompression at relatively high temperatures resulted in the decomposition of garnet to produce a symplectitic intergrowth (S) of fine-grained orthopyroxene + plagioclase ± spinel ± clinopyroxene.

chapters we will address linking geochronology to specific reactions and/or geothermobarometry, thereby placing some time constraints on *P-T-t* paths.

23.7.1 Analytical Techniques and Suitable Minerals

Techniques for age determination of minute samples are rather specialized. The list below acts as an addendum to the more general analytical techniques discussed in Chapters 8 and 9. For detailed reviews, see Parrish (1990), Harrison et al. (2002), Müller (2003), Vance et al. (2003), and Williams et al. (2007).

Thermal ionization mass spectrometry (TIMS) involves ionization of very small samples (subnanogram) on a filament and mass spectrometry of the ionized isotopes. The method is good for Sr and Pb isotopes of mineral separates. A finely pointed tiny diamond-tipped drill mounted on a microscope can extract (*microsample*) pieces of single mineral grains a few tens of microns across for analysis. Mineral samples can thus be observed in thin section and microsampled from specific textural situations.

Laser-ablation inductively coupled plasma mass spectrometry (LA-ICPMS). A high-power laser ablates a sample in situ, and particles are fed as a dry aerosol into an argon or helium plasma and passed into a mass spectrometer. Ultraviolet (UV) laser ablation has a spatial resolution $\leq 10\ \mu m$, so it can determine ages within zoned minerals and inclusions in porphyroblasts as they are observed microscopically.

Ion microprobe (IMP, SHRIMP), also called **secondary-ion mass spectrometry (SIMS)**, uses an ion beam (typically Cs or O) to sputter ions from a sample surface while observed in thin section and feed them into a mass spectrometer. Resolutions down to $30\ \mu m$ are possible.

Electron microprobe (EMP) was described in Chapter 8. Some new probes are now optimized for trace element analysis and geochronology. See the discussion of monazite below for more details. Goncalves et al. (2005) have developed a monazite age-mapping software program for use with electron microprobes. Such maps are capable of providing useful information for unraveling metamorphic and tectonic histories and assisting in gathering and interpreting results using other geochronological techniques.

Desirable properties for a mineral to yield useful age information on specific petrogenetic events are listed below:

1. Highly variable composition that reflects *P-T-X* of the host rock at the time of the mineral's growth.
2. Variations in composition during mineral growth, which are preserved as distinct domains representing successive generations.
3. Slow diffusion, so zoning is stable. Also a very high "blocking temperature" so the mineral can hold

chemical and isotopic signatures during metamorphism or even melting without significant diffusion.
4. Stability in a wide variety of igneous and metamorphic rocks. If a mineral is also resistant to weathering, it may be a useful constituent in clastic sediments.

Some minerals with these properties include zircon (Figure 18.1), monazite, xenotime, allanite, titanite/sphene, rutile, apatite, baddeleyite (all for U-Th-Pb dating), garnet (Sm/Nd, Lu/Hf), mica, and K-feldspar (Rb/Sr, K/Ar, $^{40}Ar/^{39}Ar$).

With the exception of the K-Ar system, geochronometry involves parent elements present in trace quantities (Rb, Sm, U, Th, etc.), which typically concentrate in accessory phases. Because accessory minerals are not necessarily controlled by major elements that reflect the *P-T* history of a rock, it may not be obvious when they crystallized. The ability to analyze a specific area of a mineral in thin section, either extracted by microdrill or in situ, is of greatest advantage when the development of that accessory mineral (or a zone within it) can be directly linked to an inherited, orogenic, or metamorphic event. Methods for doing so include:

1. Getting ages from major phases instead: K/Ar and Rb/Sr in micas, feldspars, or amphiboles; Sm/Nd and Lu/Hf in garnet.
2. Directly linking tectonic or metamorphic events to the development of datable accessory minerals using textures.
3. Linking an accessory phase to a specific reaction so that you can date a metamorphic mineral growth event (typically an isograd). We will explore examples of this technique in Chapter 28.
4. Chemically partitioning with a major phase to estimate development of datable mineral by its relationship to the major phase (e.g., Rubatto, 2002; Rubatto and Hermann, 2007). Partitioning can even lead to new accessory phase geothermometers (Pyle and Spear, 2000; Pyle et al., 2001).
5. Combining with geothermobarometry to add real ages (*t*) to *P-T-t* paths (Chapter 27).

23.7.2 Examples of Textural Geochronology

Geochronology on minute samples of single minerals that have been linked texturally to a specific structural or metamorphic element is a rapidly proliferating field. Most examples below rely largely on the second technique listed above, which is most closely related to the subject of this chapter.

Christensen et al. (1989) used TIMS to measure $^{87}Sr/^{86}Sr$ in single garnets from southeastern Vermont (an example of technique 1 above). ^{87}Rb in K-rich matrix minerals (such as biotite) decays over time to ^{87}Sr, which was then incorporated into growing garnet (which accepts Ca, hence Sr, but not Rb). The garnets grew during the Acadian orogeny ($\sim$380 Ma). Christensen et al. (1989) managed to separately determine core and rim ages for three garnets, which allowed them to calculate the average duration of garnet growth to be 10.5 ± 4.2 Ma. Then, by measuring garnet radii, they calculated the average growth rate: 1.4 mm/Ma.

One garnet with spiral inclusions also yielded a rotational shear strain rate of $7.6 \times 10^{-7} \, a^{-1}$ (0.76 per Ma). Strain is dimensionless per time interval because it represents a ratio: the change in length of a strain ellipsoid axis divided by unit initial length. A strain rate of 0.76, simply put, means the long axis of the strain ellipsoid stretched by 76% in a million years. (More precise results require integrating the strain over shorter time intervals.) Similar isotopically based estimates of garnet growth and strain rates, as well as reaction rates, are listed by Baxter (2003). So here we see how detailed geochronology can permit estimation of the rates of some metamorphic and deformational processes.

Simpson et al. (2000) determined TIMS isotopic ratios on monazite separates from metapelites from the Everest region of Nepal. The collision of India with Asia occurred $\sim$54 to 50 Ma, and leucogranites were emplaced 24 to 17 Ma. Previous Sr-Nd ages on garnet and U-Pb ages on bulk monazite yielded 33 to 28 and 37 to 29 Ma, respectively. Of particular importance in this study was the careful observation of textures in two rocks from which monazite was extracted. Older euhedral monazites in one sample contain oriented sillimanite inclusions parallel to S_1 in the matrix (Figure 23.55a), which were therefore interpreted as having grown either syn- or post-sillimanite and post S_1. Monazite separates from this rock yielded ages of 32.2 ± 0.4 Ma. Monazite in this sample is often armored within plagioclase (Figure 23.55b) or other minerals, which kept them from reacting in a later event and thereby preserved this early post-collision sillimanite-zone metamorphic age. Relict kyanite indicates high-pressure Barrovian-style metamorphism. Later lower-pressure metamorphism produced cordierite and irregular-shaped monazite (Figure 23.55c). Although there is no clear textural link, the shape and lack of armoring suggest a separate event (probably associated with the obvious low-P metamorphism), and Pb isotopes yielded 22.7 ± 0.2 Ma. The Everest granite was dated using monazite and xenotime at 21.3 to 20.5 Ma. These results place far better constraints than earlier traditional works on post-collision metamorphism, with early Barrovian style metamorphism peaking at 32 Ma and a second low-P event at 22.7 Ma (post-orogenic collapse?) and granites (probably related to collapse) emplaced at 20 to 21 Ma.

Möller et al. (2003) performed in situ SHRIMP analysis of U-Th-Pb isotopic ratios in zircons in the contact aureole around the Rogaland anorthosite/norite complex in southeastern Norway. Inherited zircons and zircon cores exhibited oscillatory zoning, which is characteristic of igneous zircon growth (see the igneous rims in Figure 18.1). They yielded an age of $\sim$1035 Ma (the age of the initial igneous complex). A 600 to 700°C metamorphic event (M_1) followed at $\sim$1.00 Ga, producing unzoned zircon rims with variable Th, U, and REE. A second very-high-temperature M_2 metamorphic event produced orthopyroxene, magnetite, spinel, and osumilite. Zircons occurring intergrown with, or as inclusions in, those minerals suggest syn-M_2 growth and yield SHRIMP Pb ages of 927 ± 7 Ma. Finally, zircons found outside M_2 minerals or overgrown by later garnet coronas or garnet-quartz or garnet-orthopyroxene symplectites

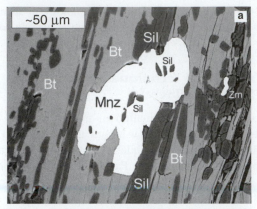

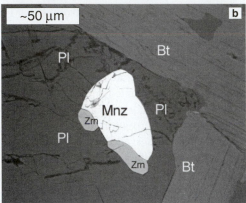

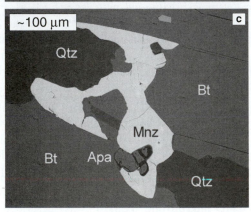

FIGURE 23.55   Backscattered SEM images of textural relationships of monazites from the Everest region of Nepal. **(a)** Well-developed M_1 monazite enveloping sillimanite inclusions aligned subparallel to external S_1 foliation. **(b)** Nearly euhedral M_1 monazite in the same sample surrounded by armoring plagioclase. **(c)** Irregularly shaped M_2 monazite in a nearby sample. From Simpson et al. (2000).

yield ages of 908 ± 9 Ma for a retrograde M_3 event. Earlier TIMS isotopic studies on zircon separates provided sparse evidence for the highest-grade M_2 event. Möller et al (2003) attributed this to the traditional magnetic separation of zircons, which selectively removed M_2 zircons from those analyzed because they were intimately associated with magnetite.

Cliff and Meffan-Main (2003) used the microdrill technique to extract white micas from sheared samples from the Sonnenblick Dome of the Austrian Pennine Alps. Using TIMS Rb-Sr ages, they determined a main schistosity age of 27.3 ± 0.8 Ma, which they distinguished from recrystallized micas in a later crenulation cleavage at 25.5 ± 0.3 Ma.

Whole-rock Rb-Sr dating would have sampled both mica populations and yielded an intermediate age, but careful textural and microsampling not only avoided Sr isotopic heterogeneity but took advantage of it to date individual tectono-metamorphic events.

Müller et al. (2000) microsampled carbonate and quartz–chlorite from incrementally developed strain fringes of σ-type mantles (Figure 23.19) on pyrite porphyroclasts from a shear zone in the northern Pyrenees. The fringes developed during two distinct phases of shear (D_2 and D_3), following an earlier period of crustal shortening (D_1, which created a foliation preserved as straight inclusion trails within the pyrites). The pyrites thus grew as post-D_1 porphyroblasts but were deformed during D_2 and D_3. Figure 23.56 shows two examples of the porphyroclasts and fringes and the TIMS Rb/Sr dates of four fringe increments. Notice that successive increments developed between the porphyroclast and receding earlier fringe, not at the ends of the fringe tails. Figure 23.57 indicates the ages versus strain (ε as a percent), showing a relatively slow D_2 period (strain rate $\sim 3.5 \times 10^{-8}$ a^{-1}) lasting from $\sim$ 87 to 66 Ma, followed by a period of increasing D_3 strain rate ($\sim 2.4 \times 10^{-7}$a^{-1}) for about 4 Ma, correlated with an abrupt change in fiber growth direction (and interpreted as a stress field transformation from D_2 gravitational collapse to renewed D_3 crustal shortening). Compressive strain then waned to earlier D_2 rates until $\sim$50 Ma ago.

UV-LA-ICPMS has been used to address a variety of problems. For example, Mulch et al. (2002) found that undeformed muscovite porphyroblasts in an S-C mylonite from the Italian Alps yielded a ^{40}Ar/^{39}Ar age of 182.0 ± 1.6 Ma (interpreted as the age of greenschist facies metamorphism, deformation and associated uplift). In situ UV-LA-ICPMS ^{40}Ar/^{39}Ar dating and furnace step heating (Section 9.7.2.1) within strongly deformed mica grains from the mylonite displayed a range of systematically younger apparent ages. Mulch et al. (2002) then modeled protracted cooling through argon closure temperatures with argon loss via microstructural defect-controlled intragranular diffusion pathways. From this they estimated a cooling rate of about 2°C/Ma

over 50 to 60 Ma for post-182 Ma uplift of the Ivrea-Verbano Zone. Sherlock et al. (2003) dated slaty cleavage development in Welsh slates at $\sim$396 Ma by laser step-heating and ^{40}Ar/^{39}Ar dating of white micas developed synkinematically in strain fringe around rigid pyritized graptolites. Accurate dating of low-grade slaty cleavage was impossible by conventional isotopic methods. Wartho and Kelley (2003) used in situ UV-LA-ICPMS and ^{40}Ar/^{39}Ar to date phlogopite core-rim traverses in mantle xenoliths in kimberlites. The cores yielded ages of formation of the magmatic or metasomatic events creating the parent rock, presumably in the SCLM. They then modeled the Ar loss profiles toward the rims, fit to experimentally determined Ar diffusion rates in phlogopite, to estimate the rate of ascent of the kimberlites (reported in Section 19.3.3.4).

Monazite U-Th-Pb dating using the electron microprobe is a rapidly proliferating technique with a wealth of applications spanning a broad range of igneous, metamorphic, sedimentary, and hydrothermal processes (see reviews by Harrison et al., 2002; and Williams et al., 2007). I shall mention only a few investigations, concentrating on ones that link ages to textures in metamorphic rocks that have experienced multiple deformation events. Monazite is a REE-phosphate mineral, which has all of the useful properties of a suitable mineral for textural geochronology listed above. Its development in metamorphic rocks is typically associated with garnet breakdown because garnet is the principal REE-bearing major silicate mineral. Monazite picks up U and Th but virtually zero Pb, so any Pb detected is derived over time from U or Th decay. We can thus use (U or Th)/Pb from *chemical* analysis (EMP) to yield an age. This enables many more labs with EM facilities but no mass spectrometer to determine ages. Because an EMP cannot distinguish isotopic ratios for direct age determination, the technique assumes that all Pb in monazite is radiogenic and that the parental U isotopes occur in average crustal proportions. If we determine total U using the microprobe, we can then estimate isotopic concentration and then Pb isotopic concentrations from each U isotope's decay rate. Both assumptions seem reasonable and justified by the good

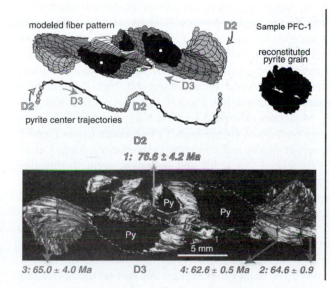

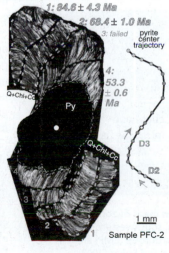

FIGURE 23.56 **(a)** Broken pyrite porphyroblast with sigmoidal fibrous carbonate–quartz–chlorite strain fringe and kinematic reconstruction above. Areas generated during D_2 and D_3 events are outlined with dashed lines in the photomicrograph (with arrows indicating the Rb-Sr ages) and shaded in the reconstruction (with arrows indicating the direction of fiber growth). **(b)** Photomicrograph of an unbroken pyrite porphyroblast and strain fringe with outlined growth zones and Rb-Sr ages. After Müller et al. (2000).

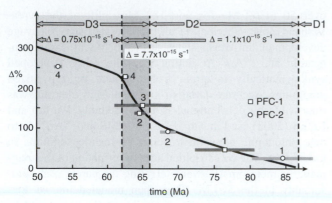

FIGURE 23.57 Strain in percent versus time for the two strain fringes analyzed in Figure 23.56. Note accelerated strain rate associated with the transition from D_2 to D_3. From Müller et al. (2000).

correlation between EMP and isotopic ages from studies addressing both. The technique is best if sufficient Pb has accumulated (i.e., early Paleozoic and older monazites). The blocking temperatures for diffusion in monazite are in excess of 800°C, so monazite can be used to date high-grade metamorphic and even many igneous events.

Pyle and Spear (2003) and Pyle et al. (2005) detected four distinct generations of monazite in migmatites sampled from the Chesham Pond Nappe of southwestern New Hampshire. The first generation occurs as high-yttrium cores in zoned monazites (bright in Figure 23.58). In situ EMP U-Th-Pb dating yielded an age of 410 ± 10 Ma for domain 1 cores. Pyle et al. (2005) speculated that these cores represent inherited pre-metamorphic monazites. Domain 2 monazite occurs as rims on domain 1 cores and as inclusions associated with xenotime in garnet and yield an age of 381 ± 8 Ma. Pyle and Spear (2003) attributed domain 2 development to a prograde metamorphic reaction: Chl + Qtz + Bt + Plg + Xno = Grt + Ms + Mnz + Ap + H_2O. Xenotime supplies the Y and REE necessary for monazite growth, and monazite reveals the age of this reaction and the development of garnet. Domain 3 monazite (372 ± 6 Ma) grew in the absence of xenotime and is thus is low in yttrium (dark in Figure 23.58). It surrounds and/or embays domain 2

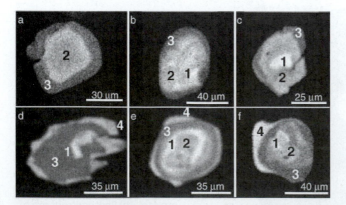

FIGURE 23.58 Yttrium (Y) distribution maps of zoned monazite crystals from the Chesham Pond Nappe, southwestern New Hampshire, determined by electron microprobe (EMP) analysis. Brighter areas are higher in Y. From Pyle and Spear (2003).

monazite or occurs as single grains in the matrix. It is associated with the breakdown of garnet and the development of sillimanite: Grt + Ms + Ap = Sill + Bt + Mnz. Domain 4 monazite (352 ± 14 Ma) occurs as thin discontinuous rims on earlier monazite and has very high Y content. Pyle and Spear (2003) attributed this stage to cooling and crystallization of local partial melts and leucosome development in the migmatites via the reaction: Crd + Kfs + Grt + Melt = Sil + Bt + Qtz + Plg + Mnz ± Xno. The ability to distinguish texturally separate growth stages of accessory minerals and determine ages of micron-scale domains from EMP chemical analysis is a valuable new tool. Relating those stages and domains to specific events and/or mineral reactions during prograde or retrograde metamorphism is an important step in relating these observations to the petrogenetic history of the rocks and area. Pyle and Spear (2003) used the reactions and geothermobarometry (Section 27.4) to estimate the temperatures and pressures of the dated stages of petrogenesis. They concluded that stages 2, 3, and 4 occurred along a nearly isobaric prograde path of metamorphism at about 0.3 GPa from ~500°C to melting just over 700°C. Pyle et al. (2005) related pre-metamorphic (domain 1) monazites to the local New Hampshire Granite Series of Acadian age (~390 to 410 Ma). The domains 2–4 regional metamorphism were attributed to a later heating event, ascribed to lithospheric mantle delamination (Figure 18.8c2) and related asthenospheric upwelling. Cooling to crystallize domain 4 monazite was probably associated with overthrusting of the Chesham Pond Nappe, which may then be constrained to have begun roughly 355 Ma ago.

Mahan et al. (2006) used EMP-based geochronology to date events in high-P-T Precambrian granulites associated with the ductile Legs Lake shear zone in the Lake Athabasca region, Snowbird Tectonic Zone of the Canadian Shield. They also found multiple growth episodes (five) in zoned monazites. Figure 23.59 summarizes their findings. Monazite events 1 (2570 ± 11 Ma) and 2 (2544 to 2486 Ma) are high-Y and occur as inclusions in garnet. They appear to have grown during high-P-T granulite facies metamorphism prior to or coeval with garnet growth. Monazite 3 is lower in yttrium (suggesting ample garnet was present and sequestered much Y) and occurs principally in the matrix or in garnet cracks. A wide range of ages (2529 to 2160 Ma) derived from event 3 monazites suggests episodic growth with unclear significance. Event 4 monazite (1937 to 1884 Ma) was interpreted as developed during a second high-P-T granulite metamorphic event. It is also low in Y and coexists with garnet. Monazites of event 5 (~1850 Ma) were correlated with garnet breakdown (hence high-Y) to produce lower T and P retrograde biotite and cordierite (+ monazite). Mahan et al. (2006) related this uplift and hydration event to thrusting along the Legs Lake shear zone. Hydration, they speculated, was aided by loading and dehydration of the footwall metasediments with fluid channeled up the shear zone.

Finally, Dahl et al. (2005) used EMP monazite ages to constrain the timing of three Proterozoic fabric-forming events in metapelites of the Wyoming craton in South Dakota. Figure 23.60 shows the textures. An east-northeast-trending

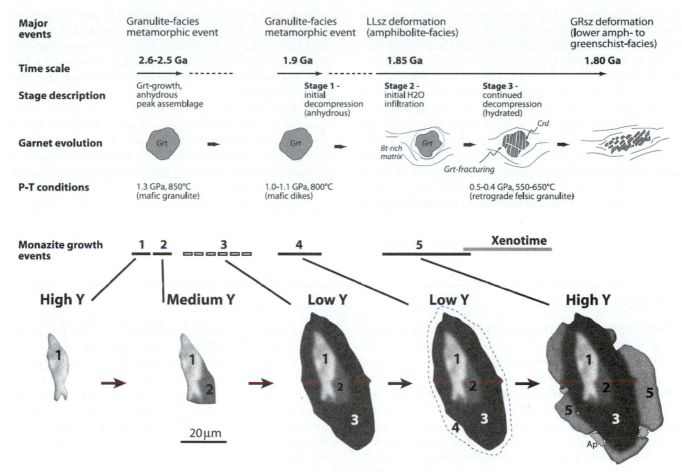

FIGURE 23.59 Summary model for the evolution of felsic granulites in retrograde shear zones, Snowbird Tectonic Zone, Saskatchewan, Canada. Bottom images are yttrium element maps of a zoned monazite crystal from which age determinations for the events have been derived. (Brighter areas are higher in Y.) Possible intermediate periods of resorption are not shown. LLsz = Legs Lake shear zone. After Mahan et al. (2006).

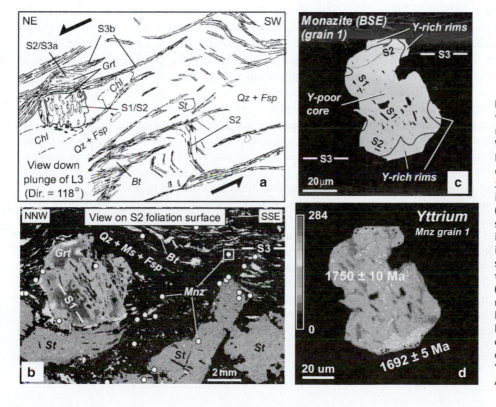

FIGURE 23.60 Polydeformed textures and monazite in a metapelite from the eastern Wyoming craton, Black Hills, South Dakota. **(a)** Sketch of textures viewed down plunge of L_3 lineation showing S_1/S_2 in garnet porphyroblasts, and S_2 in microlithons between S_3 overprint. **(b)** Photomicrograph parallel to S_2 surface showing S_1 in garnet and S_3 in matrix. (S_2 cannot be seen because it is in the plane of the section). The square in the upper right surrounds the monazite in c and d. **(c)** Backscattered SEM image of monazite with internal S_1 in core leading into spiral S_2 toward rim and later Y-rich overgrowth. **(d)** Yttrium element map of the same monazite crystal (brighter areas are higher in Y) showing spots analyzed for ages. After Dahl et al. (2005).

S_1 is attributed to a north-verging nappe/thrusting accretion event and is preserved only in garnet and staurolite porphyroblasts. S_2 (attributed to folding associated with ~east-west collision of continental fragments) is a rotational extension of S_1 in some porphyroblasts and dominates many microlithons between the dominant foliation, S_3, which overprinted and transposed S_2 and has been ascribed to doming associated with intrusion of a local granite. The consistent east-northeast orientation of S_1 in several porphyroblasts suggests they didn't rotate during growth (in spite of the spiral S_2 tails). EMP ages from a complex monazite (in the small square in Figure 23.60b and magnified in the BSE image Figure 23.60c), and a Y element map (Figure 23.60d) yielded ages of 1750 ± 10 Ma for the core with internal S_1 and S_2 (providing a minimum age for the foliation events) and 1692 ± 5 Ma for the high-Y rim, which truncates S_1/S_2 and is associated with resorption textures of garnet (supplying Y to monazite). This later event is best correlated with intrusion of the nearby Harney Creek granite.

Hopefully, this section has given you an idea of the many applications of analyzing individual minerals, or even mineral zones, to determine the ages of specific metamorphic, tectonic, or igneous events to which the minerals can be correlated. I'm sure many more creative uses will be developed and techniques refined.

Summary

Metamorphic textures develop in the solid state as mineral grains interact with their neighbors during deformation, recrystallization, and/or growth. Applied stress commonly accompanies metamorphism, whereby it causes deformation (strain) in solids. Crystalline deformation at low temperatures is dominated by cataclastic flow, involving mechanical breaking or crushing of crystals. Plastic deformation takes over at higher grades where no cohesion is lost. Defect and dislocation migration greatly facilitate plastic deformation. Recovery involves migration of defects to grain boundaries or new subgrain boundaries, leaving behind regions of reduced lattice strain. Static recrystallization involves a more advanced migration of grain or subgrain boundaries to produce a mosaic of low-strain grains. Deformation and recovery/recrystallization generally interact because deformation adds energy to crystal lattices. Higher energy states are less stable, so processes that reduce it, such as recrystallization, are encouraged. A strained crystal is therefore more apt to recrystallize than a neighboring unstrained one. Dynamic recrystallization is the simultaneous action of deformation and recovery/recrystallization. It typically produces a good crystallization schistosity. If recrystallization lasts longer than deformation or occurs in a static environment, such as a contact aureole, it is called annealing.

The hallmark texture of contact metamorphism in a low-pressure relatively static environment is granoblastic polygonal. It develops best in monomineralic quartz or feldspar regions and is characterized by polygonal grains with straight boundaries, typically meeting at three-grain boundaries in thin section at 120° angles. Minerals that tend more toward euhedral or subhedral shapes develop decussate texture. Porphyroblasts of larger metamorphic minerals in a finer groundmass are common in (although certainly not restricted to) contact metamorphism, as are inclusion-bearing poikiloblasts.

Cataclastic processes dominate over recovery and recrystallization in fault/shear zones, particularly at shallow depths, where broken, crushed, and bent pre-deformational grains (called *clasts*) are common. Shear zones typically broaden with depth, where shear is distributed while recovery processes join cataclasis as dynamic recrystallization. Mylonitic foliations are generally produced. Determining the sense of shear in zones no longer active can be challenging, and a number of shear-sense indicators have been proposed.

Textures of regional (orogenic) metamorphism can be very complex (and thus very informative). Orogeny may involve several cycles and events of deformation and metamorphism, each leaving textural and mineralogical traces. Deciphering multiple foliations, lineations, and periods of mineral growth requires careful microscopic observation and practice. The pattern of inclusions within poikiloblasts (S_i), when developed, may be particularly informative, as they may record events obliterated in the matrix by later deformation.

Replacement and reaction textures typically develop when reactions fail to run to completion. Such textures capture metamorphic reactions texturally and indicate the nature, direction, and progress of a reaction. We can use these textures to understand the nature of the protolith and the changing conditions experienced by a sample.

Technological advances are now allowing us to extract age information from very small volumes of a mineral, even from grains observed in thin section. This is opening up some exciting avenues of research in which the timing of individual tectonic/metamorphic, igneous, or sedimentary events may be determined. To be meaningful, these ages must be related to other critical petrologic information, such as textures, mineral reactions, and/or thermodynamic models that estimate pressures and temperatures of evolving rocks.

TABLE 23.1   Textures of Metamorphic Rocks

General Textural Terms:

Relict	A remnant texture inherited from the parent rock and not yet overprinted or obscured by metamorphism.
Blasto-	A prefix indicating a relict texture (e.g., "blastophitic").
-blast	A suffix indicating a texture of true metamorphic origin.
Idioblastic	Euhedral
Hypidioblastic (subidioblastic)	Subhedral
Xenoblastic	Anhedral
Porphyroblast	A larger metamorphic mineral in a matrix of smaller ones.
Poikiloblast	A porphyroblast with numerous inclusions.
Skeletal, web	An extreme variation of a poikiloblast in which the including mineral occurs as a thin, optically continuous film surrounding the inclusions.
Nodular	Ovoid porphyroblasts.
Spotted	Used to describe the appearance of small (commonly ovoid) porphyroblasts in hand specimen.
Depletion haloes	Zones surrounding porphyroblasts that have had one or more minerals removed by reaction and diffusion to the growing porphyroblast.

Non-Foliated Texture:

Granoblastic Polygonal (polygonal mosaic)	Equidimensional generally anhedral crystals of approximately equal size with regular polygonal shapes (usually 5- or 6-sided) and fairly straight boundaries meeting at triple junctions. Best displayed in monomineralic rocks (marble, quartzite). Poymineralic rocks approach this, but are modified by the differences in degree of idioblastic form.
Decussate	Randomly oriented interlocking subhedral prisms. Tends to form in minerals with strong anisotropy of surface energy, such as mica, pyroxene, or hornblende.
Symplectite	An intimate intergrowth of two or more minerals that grew simultaneously.
Corona	A rim of one or more minerals surrounding another mineral.
Moat	A coronitic rim that completely surrounds a mineral.

Textures of Dynamic Metamorphism:

Cataclastic	A general texture resulting from the predominantly brittle deformation of minerals (it typically involves grain-size reduction).
Mylonitic	Foliated cataclastic/plastic texture.
Pseudotachylite	A dynamically metamorphosed rock containing glass generated by localized shear melting.
Undulose extinction	Irregular extinction resulting from deformation of the lattice of a crystal.
Cracked	Crystals showing obvious cracks.
Crushed	Crystals are ground by shear and exhibit fine-grained margins derived from the original larger crystals. The proportion of crush material can vary considerably and even comprise 90% of the rock.
Deformation band	Dimensionally elongated bands within a crystal resulting from cracking or slip.
Porphyroclast	A large, relict crystal in a fine crush matrix.
Mortar	Rounded porphyroclasts rimmed by a crush matrix.
Ribbon	Very elongate crystals (usually quartz) resulting from cataclasis and intense plastic deformation.
Shredded	The intense breakup of minerals (usually phyllosilicates) along cleavages to produce very fine locally crushed margins.
Kink band	Zones bounded by parallel planes in which some feature, usually cleavages, have a different orientation.
Deformation twin	Twinning or lamellae produced by deformation. Common in carbonates and recognized in feldspars when the twins/lamellae are bent or dominantly wedge shaped.

Textures of Dynamic Metamorphism:

Polygonized	Incipient recrystallization where larger deformed crystals break down into smaller, undeformed subgrains. The outlines of the larger crystals are still distinguishable.
Sutured	Incipient recrystallization in which the larger crystals differentially incorporate the marginal crush matrix to produce an interdigitation of grain boundaries.

(Continued)

TABLE 23.1   Textures of Metamorphic Rocks (*Continued*)

Textures of Dynamic Metamorphism:

Oblique foliation	Foliations that cut across the foliation developed due to shear offsets.
Shear band cleavage or S-C texture	A texture that has shear bands (or C foliations), which are spaced cleavages that transect well-developed mineral foliation (S foliation) at a small angle.
Mantled porphyroclast	Resistant porphyroclast with a rim that has the same mineralogy as the porphyroclast. The mantles are assumed to be derived from the porphyroclast by grinding.
Augen, flaser	Eye-shaped mantled porphyroclasts.
Quarter structure	A structure in which the four quadrants defined by the foliation and its normal are not symmetric.
Quarter fold	Small fold in the foliation around porphyroclasts due to drag.
Quarter mat	Concentration of mica resulting from dissolution of quartz in the shortened quadrants of porphyroclast.

Texture of Regional Metamorphism:

Foliation	Any planar fabric element.
Lineation	Any linear fabric element.
Spaced foliation	A foliation developed in zones separated by non-foliated "microlithons."
Continuous foliation	A foliation that is not spaced, but occurs continuously.
Cleavage	Any type of foliation in which the aligned platy phyllosilicates are too fine to see with the unaided eye.
Slaty cleavage	Fine-grained continuous cleavage.
Crenulation cleavage	A cleavage or schistosity that becomes microfolded.
Schistosity	A planar orientation of elongated mineral grains or grain aggregates produced by metamorphic processes. Aligned minerals are coarse enough to see with the unaided eye.
Gneissose structure	Either layered by metamorphic processes or a poorly developed schistosity.
Layering, banding	A foliation consisting of alternating layers of different composition.
Prekinematic	Refers to crystals which were present prior to deformation/metamorphism and were deformed.
Synkinematic	Refers to crystals which grew during deformation/metamorphism.
Postkinematic	Refers to crystals which grew after deformation.
Inter-kinematic	Mineral growth that is post-kinematic to one event and pre-kinematic to another.
S-surface	A foliation or planar fabric in a deformed rock. Commonly numbered (S_1, S_2, S_3, etc.) if there are successively developed foliations. S_0 is reserved for pre-metamorphic foliations, such as bedding, flow banding, cumulate layering, etc.
S_i	Internal S, the alignment of inclusions within a poikiloblast.
S_e	External S, the foliation of the matrix outside a poikiloblast.
L	A lineation (can also have L_1, L_2, etc.).
Pressure shadow	Concentration of a matrix mineral in low-stress areas adjacent to a porphyroblast due to pressure solution (solution transfer) and re-precipitation.
Wrapped	A foliation compressed about a pre-kinematic porphyroblast (or possibly bowed out by the growing porphyroblast).
Helicitic	Folded internal S, indicates a post-kinematic porphyroblast.
Spiral (rotational, snowball)	Showing a texture, such as a spiraled inclusion trail, that indicates rotation.
Polygonal arc	A post-kinematic recrystallization of bent crystals into an arcuate pattern of smaller straight crystals. A form of polygonization.

See also the glossary in Vernon, 2004.

Key Terms

See Table 23.1 for definitions of many textural terms.

Texture (micro-structure)/fabric, structure 477

Penetrative features 477

-blast, blasto- 478

Cataclastic flow 478

Solution transfer 478

Point/line defects 478

Slip system 479

Dislocation glide 479

Strain hardening 479

Dislocation creep 479

Undulose extinction 479

Deformation twinning 480

Recovery 480

Grains/subgrains 480

Recrystallization 480

Review Questions and Problems

Review Questions and Problems are located on the author's web page at the following address: **http://www.prenhall.com/winter**

Important "First Principle" Concepts

- Deformation adds strain energy to the lattices of crystals in the form of bond bending and defects. As we learned in Chapter 5, systems tend naturally toward states of lowest energy, so strained lattices are less stable than unstrained ones. Recovery and recrystallization are ways to reduce that stored energy, so are assisted by deformation.

- Orogenic structures and textures are "fractal" in the sense that deformational features, orientations, and styles are typically

evident and correlative from the scale of the mountain to the outcrop to the hand specimen and thin section.

- Orogenic deformation and metamorphism may involve several events that are neither synchronous nor universal across the orogen. Careful petrographic work, combined with structural analysis and good field work, is required to unravel such complex histories.

Suggested Further Readings

Barker, A. J. (1990). *Introduction to Metamorphic Textures and Microstructures*. Blackie. Glasgow.

Jamieson, R. A. (1988). Textures, sequences of events, and assemblages in metamorphic rocks. In: *Short Course on Heat, Metamorphism, and Tectonics* (eds. E. G. Nisbet and C. M. R. Fowler). Short Course **14**, 213–257. Mineralogical Association of Canada.

Müller, W. (2003). Strengthening the link between geochronology, textures, and petrology. *Earth Planet. Sci. Lett.,* **206**, 237–251.

Parrish, R. R. (1990). U-Pb dating of monazite and its application to geological problems. *Can. J. Earth Sci.,* **27**, 1431–1450.

Passchier, C. W., and R. A. J. Trouw. (2005). *Mictotectonics*. Springer-Verlag. Berlin.

Platt, L. (1976). A Penrose Conference on cleavage in rocks. *Geotimes,* **21**, 19–20.

Spry, A. (1969). *Metamorphic Textures.* Pergamon. Oxford, UK.

Vance, D., W. Müller, and I. M. Villa (eds.). (2003). *Geochronology: Linking the Isotopic Record with Petrology and Textures.* Special Publication **220**. The Geological Society. London.

Vernon, R. H. (2004). *A Practical Guide to Rock Microstructure.* Cambridge University Press. Cambridge, UK.

Williams, M. L., M. J. Jercinovic, and C. J. Hetherington. (2007). Microprobe monazite geochronology: Understanding geologic processes by integrating composition and chronology. *Ann. Rev. Earth Planet. Sci.,* **35**, 137–175.

Yardley, B. W. D., W. S. McKenzie, and C. Guilford. (1990). *Atlas of Metamorphic Rocks and Their Textures.* Longman. Essex, UK.

24

Stable Mineral Assemblages in Metamorphic Rocks

Questions to be Considered in this Chapter:

1. How can the phase rule from Chapter 6 be applied to metamorphic systems and what implications may we draw from analyzing metamorphic mineral assemblages with the phase rule in mind?

2. How might metamorphic mineral assemblages be displayed graphically in order to understand them better?

3. What chemographic diagrams have been found to be particularly useful so far, and how might we choose the best one for any particular rock type or purpose?

24.1 EQUILIBRIUM MINERAL ASSEMBLAGES

At thermodynamic equilibrium under specific conditions of temperature (T), pressure (P), and composition (X), a rock system manifests itself as a stable assemblage of minerals (perhaps with a fluid and/or melt). Thus, *regardless of the path to it* (cooling, heating, burial, change in X, etc.), the equilibrium mineralogy and the composition of each mineral will be entirely fixed by the T, P, and X of the system. Petrologists use the term **mineral paragenesis** to refer to such an equilibrium mineral assemblage. Many petrologists, myself included, simply use the term **mineral assemblage**, conventionally assuming equilibrium for the term. Relict pre-metamorphic minerals or later alteration products are thereby excluded from consideration unless specifically stated. Incidentally, the term *petrogenesis* is not to be confused with *paragenesis*. The former is a description of the process by which a rock was created, which may be complicated and span several million years.

We work with metamorphic rocks that are now at Earth's surface, and it is impossible to demonstrate unequivocally that a mineral assemblage represents thermodynamic (chemical) equilibrium under prior metamorphic conditions. Indirect support for such a conclusion comes from rocks that display the following:

- There are no obvious reaction or disequilibrium textures, such as replacement textures, coronas, or compositional zoning. Reaction textures, when developed, may represent equilibrium between the reactants and products, or arrested equilibrium due to slow reaction rates. It may be impossible to distinguish between these alternatives texturally. In the case of zoning, the outermost zone *may* be in equilibrium with the rest of the mineral assemblage.

- Each mineral is in physical contact with every other mineral somewhere in a thin section and in contact with the grain-boundary network of the rock as well. Minerals that occur only as inclusions in a poikiloblast, or in the core of a corona, are probably not in equilibrium with other minerals in the rock.

- Non-layered rocks are more likely to be in equilibrium than layered ones. Diffusion can be limited, even over centimeter scales, so that the minerals within one layer may not be in equilibrium with the minerals of another. Equilibrium may be maintained *within* each layer, however, so that layered rocks may be treated as a series of mineral assemblages, each in *local equilibrium*.

- Textural equilibrium is more difficult to achieve than chemical equilibrium. Metamorphic rocks in which the original igneous or sedimentary texture is obliterated and replaced by a metamorphic texture are generally considered to be in chemical equilibrium.

- The rock conforms to the phase rule (Sections 6.2 and 24.2). Rocks in which the number of phases is greater than the number of components, or in which two or more polymorphs are found together, are suspected to be out of equilibrium. As we shall soon see, application of this criterion requires practice and careful consideration.
- Microprobe analysis of minerals (or rims of chemically zoned minerals) demonstrates that the composition of each mineral is nearly constant.

If a rock passes the above tests, it is generally considered to represent chemical equilibrium. Innocent until proven guilty! Failure to meet some criterion, such as a reaction texture, may allow us to eliminate one mineral (e.g., retrograde chlorite around garnet) but still treat the other minerals as an equilibrium assemblage. Whether equilibrium represents peak metamorphic conditions is another question, of course, as discussed in Section 21.4.

Field criteria for suites of metamorphic rocks provide even stronger support for thermodynamic equilibrium. For example, if all rocks of similar composition in a given area have the same mineral assemblage, this implies that they all equilibrated to the same set of conditions. If such rocks demonstrate *systematic* changes in mineral assemblage over larger distances, this implies that the metamorphic conditions varied along a gradient, and the rocks equilibrated locally to conditions along the gradient.

24.2 THE PHASE RULE IN METAMORPHIC SYSTEMS

Recall from Chapter 6 the Gibbs phase rule, as applied to systems *at equilibrium*:

$$F = C - \phi + 2 \quad \text{(Gibbs phase rule)} \quad (6.1)$$

where: ϕ = number of phases in the system

C = number of components: the *minimum* number of chemical constituents required to specify every phase in the system

F = number of degrees of freedom: the number of *intensive* parameters of state (such as temperature, pressure, the composition of each phase, etc.) that may be varied independently without changing ϕ

You may want to review Section 6.2 to refresh your memory on the phase rule and these definitions.

Imagine that you grab a typical sample from some outcrop in a regional metamorphic terrane. The odds are strong that you will select a sample from within a zone, and not from a location right on an isograd. Alternatively, close your eyes and stab your pencil point anywhere on a phase diagram, such as Figures 6.8 or 21.9. In nearly every case, your pencil point will land within a divariant field and not right on a univariant curve or invariant point. In either case above, the most common situation is divariant ($F = 2$), meaning that P and T are independently variable (over a

limited range) without affecting the number of minerals present (ϕ). In complex natural systems, there may be one or more compositional variables as well, so that F may be greater than two. As Goldschmidt (1912b) pointed out, the common occurrence of certain metamorphic mineral assemblages worldwide supports this contention that $F \geq 2$ because such assemblages are much more likely to represent variable *P-T-X* conditions (areas in the field) than more restricted situations (on isograds).

If $F \geq 2$ is the most common situation, then the phase rule may be adjusted accordingly:

$$F = C - \phi + 2 \geq 2$$
$$\phi \leq C \qquad (24.1)$$

The above simplified phase rule states that, in the most common situation for a rock at equilibrium, the number of phases is equal to or less than the number of components. It has been called **Goldschmidt's mineralogical phase rule**, or simply the **mineralogical phase rule**. It is useful in evaluating whether or not a rock is at equilibrium, as noted in the criteria listed in the previous section. In order to do this, of course, we must be able to determine the appropriate number of components, as discussed shortly.

Suppose we have determined the number of components, C, for a rock. Consider the following three scenarios:

$\phi = C$. This is accepted as the standard divariant situation in metamorphic rocks. The rock probably represents an equilibrium mineral assemblage from within a metamorphic zone. To some authors the mineralogical phase rule is $\phi = C$.

$\phi < C$. This is common with mineral systems exhibiting solid solution. For example, in the plagioclase system (Figure 6.8) or the olivine system (Figure 6.9), we can see that under metamorphic conditions (below the solidus) these two-*C* systems consist of a single mineral phase. As we shall soon see, we can combine components that substitute for each other into a single mixed component. This has the effect of reducing C, so that $\phi = C$ is again the common rule. Alternatively, C may erroneously include "mobile" components, as we shall investigate in a moment.

$\phi > C$. This is a more interesting situation, and at least one of the three situations listed below must be responsible:

1. $F < 2$. In other words, the sample is collected from a location right on a univariant reaction curve (isograd) or invariant point. The Gibbs phase rule [Equation (6.1)] indicates that for every decrease in F, ϕ increases by one. In the simple one-component system Al_2SiO_5 (Figure 21.9), we expect only one of the three possible polymorphs to occur in common rocks. Only under special univariant conditions (on an isograd) would we find, for example, stable coexisting andalusite + sillimanite, or sillimanite + kyanite. All three polymorphs coexist stably only at the specific *P-T* conditions of the invariant point (ca. 0.37 GPa and 500°C).

2. *Equilibrium has not been attained.* The phase rule applies only to systems at equilibrium. There could be any number of minerals coexisting if equilibrium is not attained. Consider, for example, a graywacke, containing fragments of a number of different rocks in a matrix of grains derived from a rapidly eroding source. Dozens of minerals may be present, but they are not in equilibrium with one another. A number of igneous and metamorphic rocks also have retrograde reactions that begin, but do not run to completion. In your labs you have probably seen several examples of chlorite replacing biotite or pyroxene, or of sericite replacing feldspar. We can avoid this particular $\phi > C$ situation by simply ignoring the replacement phases and addressing only the main mineral assemblage that reflects equilibrium at elevated metamorphic conditions.

In several localities, two coexisting Al_2SiO_5 polymorphs have been noted over areas too large to correspond to an isograd. Coexistence of all three polymorphs has also been described more often than one would expect for such specific *P-T* conditions. Given the small ΔV and ΔH of the polymorphic transformations, it is difficult to drive these reactions to completion by changing pressure or temperature. It is generally believed that the majority of these two-polymorph occurrences represent disequilibrium in which some of the reactant phase remains metastably in the stability field of the product. For example, all the andalusite in a rock may not be consumed by the reaction to sillimanite during prograde metamorphism, and some may linger as relict crystals.

How can you tell situations 1 and 2 apart? It may be difficult because both represent arrested reactions. Textures are generally most helpful, such as obvious alteration, or perhaps you can observe prograde rims rather than retrograde ones. In some cases you may be able to estimate grade from the dominant mineralogy, and know that one mineral represents a much different grade and is thus a later alteration product. In the field, you may notice that $\phi > C$ assemblages occur only at a certain metamorphic grade (along a mappable isograd).

3. We didn't choose the number of components correctly. This may sound silly, but it is a common problem and can be difficult to evaluate. It is not sufficient to chemically analyze a rock and simply assume that each constituent is a component. This would lead to dozens of components, including trace elements. In order to manageably treat equilibrium mineral assemblages, it is helpful to limit the number of components and phases with which we must deal to those that are essential. An appropriate choice of *C* is generally to select those components that play an important role in influencing the equilibrium mineral assemblage. Some guidelines for the choice follow.

If we begin with a one-component system, such as $CaAl_2Si_2O_8$ (anorthite), there are three common types of major/minor components that we can add:

a. *Components that generate a new phase.* As we discovered in Chapter 6, adding a component such as $CaMgSi_2O_6$ (diopside) results in an additional phase. In Figure 6.10, for example, diopside coexists with anorthite below the solidus. If a component is a major constituent of only a single phase (e.g.,

$P_2O_5 \rightarrow$ apatite, or $TiO_2 \rightarrow$ ilmenite), we can simplify our analysis by ignoring both the component and the phase because the phase rule is unaffected by reducing ϕ and *C* by the same amount. As a result, many minor elements, such as P_2O_5 and TiO_2, are commonly ignored along with the corresponding accessory phase. Otherwise, components that generate additional phases must be considered.

b. *Components that substitute for other components.* Alternatively, we could add a component such as $NaAlSi_3O_8$ (albite) to the one-*C* anorthite system. This type of component would dissolve in the anorthite structure, resulting in a single solid-solution mineral (plagioclase) below the solidus. Fe and Mn commonly substitute for Mg. Al may also substitute for Si in some minerals, or Na for K. These components may just dissolve in phases (particularly if they are not present in great quantities). If so, we may decide to combine the components that substitute for one another into a single, mixed component such as (Fe + Mg + Mn). As a rule, we don't include trace elements in the phase rule, as they behave as substitutes only and are not present in quantities sufficient to affect the number and nature of the mineral phases present.

Although there is no need to ignore or combine major components such as this, providing that we understand the nature of the mineral solutions in our rocks, we shall see that it is a handy device to limit the number of components when we wish to construct diagrams to display chemical–mineralogical data. We can mathematically manage eight or more components using the phase rule, but it is hard to depict more than three graphically on paper.

c. *"Perfectly mobile" components.* Mobile components are either a freely mobile fluid component (H_2O, CO_2, etc.) or a component that dissolves readily in a fluid phase and can be transported easily (Na^+, K^+, Cl^-, etc.). The chemical activity of such components is controlled by factors external to the local rock system. If so, they are commonly ignored in deriving *C* when applying the mineralogical phase rule to metamorphic systems. In order to understand why, consider the very simple metamorphic system, MgO-H_2O. The possible natural phases in this system are periclase (MgO), aqueous fluid (H_2O), and brucite ($Mg(OH)_2$). How we treat this system and deal with H_2O depends upon whether water is perfectly mobile.

First suppose that H_2O is perfectly mobile and that we follow the guideline above and ignore it as a component. If we don't treat a constituent as a component, we cannot treat it as a phase either, so we ignore the aqueous pore fluid in this case as well (the fluid phase is usually gone by the time we look at the rock anyway). A reaction can occur in this system: MgO + H_2O $\rightarrow$ $Mg(OH)_2$ as written is a retrograde reaction that would occur as the rock cools and hydrates (Figure 24.1). If we begin with periclase at a temperature above the reaction equilibrium curve, $\phi = 1$ (water doesn't count) and *C* = 1 (MgO), so the mineralogical phase rule holds and $\phi = C$. As we cool to the temperature of the univariant reaction curve, periclase reacts with water to form brucite. Under the univariant conditions (at the pressure and temperature of the reaction), we have coexisting

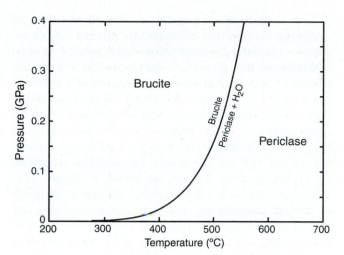

FIGURE 24.1 *P-T* phase diagram illustrating the reaction brucite = periclase + water, calculated using the program *TWQ* of Berman (1988, 1990, 1991).

periclase and brucite, so $\phi > C$, and $F = 1$ (the second reason to violate the $\phi = C$ norm stated above). At temperatures below the curve, all the periclase will be consumed by the reaction, and brucite will be the solitary phase. In this situation $\phi = 1$ and $C = 1$ again. Note: If H_2O is perfectly mobile and we "erroneously" include it in C, then ϕ (if it includes only mineral phases found in a hand sample) will typically be less than C ($\phi = 1$ and $C = 2$). Counting mobile components in C may thus be an alternative explanation for $\phi < C$.

By "perfectly mobile," I mean that H_2O behaves such that it can be added as it is needed, and excess amounts will leave as it is produced. Pore water in such a situation is simply an available reservoir: available in sufficient quantity that it enables the immobile magnesium to exist as either periclase or brucite, depending on which is stable under the given hydrous *P-T* conditions.

We can imagine the diagram in Figure 24.1 to be a field area and the reaction an isograd. For any *common* sample in the area, $\phi = 1$ (either periclase or brucite because we ignore H_2O and don't really see the fluid in the rock), and thus $\phi = C$ in accordance with the mineralogical phase rule. Only an *uncommon* sample, collected exactly on the isograd, will have $\phi > C$ (periclase + brucite).

If H_2O is *not* perfectly mobile, the system behaves differently. In this case, water is not a freely permeating phase, and it may thus become a limiting factor in the reaction. Imagine cooling periclase in Figure 24.1 under limited-H_2O conditions until it intersects the reaction curve. Periclase can react only with the quantity of water that is available or can diffuse into the system.

At this point, I should stress an obvious but commonly overlooked aspect of chemical reactions:

A reaction involving more than one reactant can proceed only until any one of the reactants is consumed.

In the present situation, periclase + H_2O react to form brucite. As the system cools at equilibrium (on the univariant reaction curve), the reaction will proceed until *either* periclase *or* water is consumed. If either one runs out, the other has nothing more to react with. Both reactants will be consumed only under the very special circumstances in which they occur in equal molar quantities (because they react in 1:1 proportions). If water is not perfectly mobile and is limited in quantity, it will be consumed first. Once the water is gone, the *excess periclase remains stable as conditions change into the brucite stability field* because there is no water with which to react.

Please note the following, then, as a general rule: Reactions such as the one in Figure 24.1 represent the absolute stability boundary of a phase such as brucite when it is the *only* reactant in the prograde reaction (brucite $\rightarrow$ periclase + H_2O). The reaction, however, is *not* the absolute stability boundary of periclase, or of H_2O, because either can be stable across the boundary if the other reactant is absent. Of course, this is true for any reaction involving multiple phases. It is amazing how many people will look at a reaction on a phase diagram such as the one illustrated in Figure 24.1 and conclude that periclase is stable only on the high-temperature side of the reaction equilibrium curve.

We can now conclude that *periclase can be stable anywhere on the whole diagram, if H_2O is present in insufficient quantities to permit the reaction to brucite to go to completion*. In H_2O-deficient systems, then, we are likely to find brucite and periclase coexisting under the conditions on the low-T side of the univariant curve in Figure 24.1. In the case of insufficiently mobile water, we find that $\phi = 2$ over a range of metamorphic grades and thus over a significant portion of the field outcrops as well. $C = 2$ as well now, however, because water is not perfectly mobile, and it becomes a limiting factor, controlling the equilibrium mineral assemblage, and must therefore be counted. $\phi = C$, therefore, still holds. At temperatures above the reaction curve, we appear to have only one phase (periclase), but if we count H_2O as a component, we must also include water as a phase: the fluid phase. Thus, at any point on the diagram (other than on the univariant curve itself), we would expect to find *two* phases, not one: $\phi =$ brucite + periclase below the reaction curve (if water is limited) or periclase + water above the curve.

How do you know which treatment is correct? *The rocks should tell you.* Remember, the phase rule (including the mineralogical phase rule) is not to be used as a *predictive* tool and does not tell the rocks how to behave. It is an *interpretive* tool, used to understand how the rocks do behave. If in the field you see only low-ϕ assemblages (as we found periclase *or* brucite in the simple MgO-H_2O system), then some components may be mobile. Mobility is common for fluid components. If, on the other hand, you often observe assemblages that have many phases in an area (corresponding to periclase + brucite), it is unlikely that so much of the area is exactly on a univariant curve and may require the number of components to include otherwise mobile components, such as H_2O or CO_2, in order to apply the phase rule correctly.

In metasomatic rocks (Chapter 30), the mobile components may increase to include components that ordinarily do not diffuse far. Fluids permeating rocks can carry several dissolved silicate species, including virtually every major element, occasionally leaving the minimum number of immobile components (one). Even if *every* component is mobile, we must consider one immobile because only $C - 1$ components are independently variable, and the final component must be determined by the requirement that all the components must sum to 100% (and if C were reduced to zero, then $\phi = 0$, which is impossible in a rock).

Examples of highly metasomatic alteration are discussed in Chapter 30 and include the "blackwall" zones that form around some metasomatized ultramafic bodies, in which one or more monomineralic marginal zones of chlorite, phlogopite, tremolite, talc, anthophyllite, etc. form (e.g., Figure 30.17). In these cases, we have lots of rock with $\phi = 1$. It would be unlikely that the original composition of the rock within any zone exactly equaled that of a single mineral, and we are left to conclude that $C = 1$, and all the other components were mobile. Components present in excess of the amount contained in the single mineral phase present in a zone have moved beyond that particular zone. The mineralogical phase rule thus leads us to the important conclusion that monomineralic zonal rocks were witness to strong metasomatic processes.

The guidelines above will help you to apply the Gibbs phase rule and the mineralogical phase rule to metamorphic mineral assemblages. I'm sure that it is a bit confusing at this point, but proper application of the phase rule requires practice before you can expect to master the art. We shall apply the phase rule to several systems in the rest of this chapter.

24.3 CHEMOGRAPHIC DIAGRAMS

Chemographics refers to the graphical representation of the chemistry of mineral assemblages. As a simple example, let's look at the olivine system as a linear $C = 2$ plot:

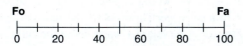

Olivine may be represented as a binary system of forsterite–fayalite solid solution along a line. The numbers along the line = %Fa ($100 \cdot Fe/(Fe + Mg)$). In metamorphic petrology, the elemental proportions are generally expressed as *molar* quantities and not on a weight basis, as normally used by igneous petrologists. The diagram, however, can be used in either way. Any intermediate olivine composition is simply plotted an appropriate distance along the line. Fo_{50} would plot in the center, Fo_{25} would plot one-quarter of the way from Fa to Fo, etc. This should be familiar to you from the abscissa of the numerous binary phase diagrams in Chapter 6.

A three-component system can be illustrated on an equilateral triangle. The techniques for plotting data on triangles are discussed in conjunction with Figure 2.1. In order to understand how these diagrams are applied to metamorphic rocks, imagine that we have a small area of a metamorphic terrane in which the rocks correspond to a hypothetical three-component system with variable proportions of the components x, y, and z. Suppose the rocks in the area are found to contain six minerals with the fixed compositions x, y, z, xy, xyz, and x_2z. The mineral compositions are plotted on a chemographic diagram, as shown in Figure 24.2. When applying such a chemographic diagram to the study of mineral assemblages in a particular area, phases that coexist at equilibrium in a rock are connected by *tie-lines*. For example, suppose that the rocks in our area have one of the following five mineral assemblages:

x-xy-x_2z

xy-xyz-x_2z

xy-xyz-y

xyz-z-x_2z

y-z-xyz

Minerals that coexist in any of these five assemblages are connected by tie-lines in Figure 24.2. Note that this subdivides the chemographic diagram into five sub-triangles, labeled (A)−(E). A diagram such as this is a type of phase diagram commonly employed by metamorphic petrologists. It corresponds to an isothermal (and isobaric) slice through a three-component phase diagram such as the one illustrated in Figure 7.9 (although it may be applicable over a limited P-T range rather than a single P-T point).

If you were to close your eyes and stab your pencil point anywhere within the diagram, the point would represent

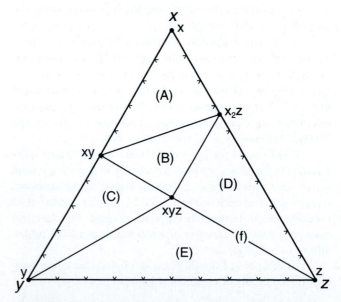

FIGURE 24.2 Hypothetical three-component chemographic compatibility diagram illustrating the positions of various stable minerals. Minerals that coexist compatibly under the range of P-T conditions specific to the diagram are connected by tie-lines. After Best (1982).

a specific bulk rock composition. From that bulk composition, you can use the diagram to determine the corresponding mineral assemblage that develops at equilibrium under the conditions for the diagram. For example, should your pencil point fall within the sub-triangle (E) in Figure 24.2, the corresponding mineral assemblage corresponds to the corners of the (E) sub-triangle, or y-z-xyz. *Any* rock with a bulk composition plotting within triangle (E) should develop that same mineral assemblage (although in different proportions). The common mineral assemblage for any point where your pencil is likely to land contains three phases; thus $\phi = C$, in accordance with the standard situation for the mineralogical phase rule. If desired, the lever principle can be used to determine the relative proportions of the minerals, as described in conjunction with Figure 7.3. If your bulk composition is in sub-triangle (E) and very close to the point xyz, for example, then the rock will have more of this mineral and less of minerals y and z.

If you randomly place your pencil point in the diagram several times, it will nearly always fall in one or another of the sub-triangles, each corresponding to one of the five commonly observed mineral assemblages listed above. Such diagrams explain clearly why *different rocks, even though equilibrated at the same metamorphic grade, may develop different mineral assemblages.* For example, if we shift the bulk composition, even only slightly, from sub-triangle (E) across the line into sub-triangle (D), the rock will then contain the mineral x_2z (along with z and xyz) and no longer the mineral y that we had before in (E).

Only rocks with bulk compositions corresponding to sub-triangle (A), for example, develop the mineral x because they are rich in the x component. Suppose some mineral, such as mineral xyz, corresponds to an important index mineral, perhaps garnet. The diagram can readily explain why all pelitic rocks need not contain garnet, even if they are all equilibrated within the garnet zone. Any rock with a bulk composition in sub-triangle (A), for example, will be garnet free, and not necessarily too low grade to develop garnet. When mapping an isograd in the field, then, we must not expect every rock to develop an appropriate index mineral. We must examine a variety of rocks and constrain the first appearance of the index mineral as best we can on the basis of a broad survey.

Figure 24.2 also demonstrates why some mineral pairs *cannot coexist.* Mineral x, for example, cannot coexist at equilibrium with y, xyz, or z under the conditions represented by Figure 24.2 because tie-lines separate these pairs. Likewise, mineral z cannot coexist with x or xz. These minerals are incompatible under the *P-T* conditions for which Figure 24.2 was created.

What happens if you pick a composition that falls directly on a tie-line, such as point (*f*) in Figure 24.2? In this case, the mineral assemblage consists of xyz and z only (the ends of the tie-line) because by adding these two phases together in the proper proportion you can produce the bulk composition (*f*). In such a situation $\phi = 2$, but, because C is defined as the *minimum* number of components required

to characterize all the phases in a system (the rock in question), in this special case $C = 2$ (redefining the *components* as y and xyz). In the unlikely event that the bulk composition equals that of a single mineral, such as xyz, then $\phi = 1$, but $C = 1$ as well. Such special situations, requiring fewer components than normal, have been described by the intriguing term **compositionally degenerate**.

Any valid chemographic phase diagram, if it is to have petrological significance when applied to metamorphic rocks, *must be referenced to a specific range of P-T conditions, such as a zone in some metamorphic terrane*, because the stability of the minerals and their groupings will vary as P and T vary. Figure 24.2 refers to a *P-T* range in which the fictitious minerals x, y, z, xy, xyz, and x_2z are all stable and occur in the groups shown. Such chemographic diagrams, when applied to a group of coexisting mineral assemblages under a limited range of *P-T* conditions, are called **compatibility diagrams** (or **composition–paragenesis diagrams**, Miyashiro, 1994).

Figure 24.3 is a three-component diagram in which many of the minerals exhibit limited *solid solution*. Only minerals x and y occur as pure phases. Minerals $x(y, z)$ and $x_2(y, z)$ show limited solid solution of components y and z on one type of lattice site. Mineral $x(y, z)$ was found to allow more y in the lattice than does mineral $x_2(y, z)$. Minerals $(xyz)_{ss}$ and z_{ss} (the subscript denotes solid solution) show limited exchange of all three components.

Suppose we poke our pencil point (choose a bulk rock composition) in the shaded field of the mineral $(xyz)_{ss}$ in Figure 24.3. In this situation, $\phi = 1$ but the system is *not* degenerate. Due to the variable nature of the composition of the phase, C must still equal 3, and the (full) Gibbs phase

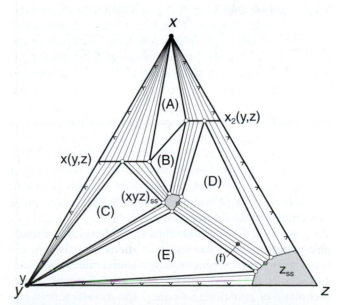

FIGURE 24.3 Hypothetical three-component chemographic compatibility diagram illustrating the positions of various stable minerals, many of which exhibit solid solution. Mineral compositions that coexist compatibly under the *P-T* conditions specific to the diagram are connected by tie-lines. After Best (1982).

rule tells us that $F = C - \phi + 2 = 4$. Thus P, T, and any two of the three components in the phase are independently variable (the final component must be fixed because all three components must total 100%). The shaded area representing the composition of mineral $(xyz)_{ss}$ is thus a two-dimensional *area* (compositionally divariant), and our single-phase rock can have any composition within the shaded solid solution limits. $\phi < C$ in this case because of the solid solution and compositional variance (F is not the standard value of 2). This is the reason that the mineralogical phase rule in Equation (24.1) includes this $\phi < C$ possibility.

What if we pick a composition such as (f), represented by the dark spot on a tie-line in Figure 24.3? As in the previous example, there are two coexisting phases: $(xyz)_{ss}$ and z_{ss}. Because $\phi = 2$ and C is still 3, the Gibbs phase rule tells us that $F = 3 - 2 + 2 = 3$. Because P and T are independently variable, the composition of each phase is univariant and must vary along the lines where the bundles of tie-lines end. Although the composition of $(xyz)_{ss}$ can vary anywhere within the shaded area, the composition of $(xyz)_{ss}$ *that coexists with z_{ss}* is constrained to the edge of the area facing z. A degree of freedom is thus lost as a phase is gained. Likewise, the composition of z_{ss} that coexists with $(xyz)_{ss}$ is constrained to a portion of the edge of the shaded z_{ss} area. The composition of the two minerals that correspond to bulk rock composition (f) are indicated by the two shaded dots at the ends of the tie-line through (f). Such a two-phase situation occurs for any rock composition in the tie-line striped area between $(xyz)_{ss}$ and z_{ss}. Only a few of the infinite number of possible tie-lines are illustrated. The same is true for any of the tie-line-dominated two-phase fields.

Any bulk composition that falls within one of the three-phase triangles $(A)-(E)$ acts as in Figure 24.2. In such situations $\phi = C = 3$, as predicted by the mineralogical phase rule. Because $F = 3 - 3 + 2 = 2$ and corresponds to P and T, the phase rule tells us that all of the compositional variables for each phase are fixed. Indeed, for any rock composition in sub-triangle (E), for example, the compositions of the minerals z_{ss} and $(xyz)_{ss}$ are fixed and equal to the corners of the sub-triangle (small open dots). In other words, the only composition of the mineral z_{ss} that can coexist with y and $(xyz)_{ss}$ (for the P and T represented by the diagram) is that of the dot where the (E) sub-triangle touches the z_{ss} area. The composition of any solid-solution mineral that coexists with any two other minerals is fixed in the same way by phase rule constraints, and each is indicated in Figure 24.3 as an open dot.

Because tie-lines link coexisting phases on compatibility diagrams, and the nature of the phases that coexist necessarily varies with changing metamorphic grade, any single compatibility diagram can be applied to only a limited range of grade (e.g., a zone). *The transition from one zone to another involves reactions between the minerals present, which, in turn, involve the loss or gain of minerals on the diagrams and reconfiguring the tie-lines.* We shall return to this notion in Chapter 26, when we discuss metamorphic reactions.

24.3.1 Common Chemographic Diagrams for Metamorphic Rocks

Because graphical compatibility diagrams offer powerful insights into the stable mineral assemblages and the controls of metamorphism, they are routinely used in studies of metamorphic terranes. Which diagram is most suitable depends on the rocks to be studied. Most common natural rocks contain the major oxides, SiO_2, Al_2O_3, K_2O, CaO, Na_2O, FeO, MgO, MnO, and H_2O, such that $C = 9$. This is clearly too complex, and we must simplify if we want to display the system in a convenient graphical way. As we learned in Chapter 7, three components is the maximum number that we can easily depict graphically in two dimensions (as a triangle). When we try to illustrate four components, we must use three-dimensional tetrahedra, and we lose even a semi-quantitative sense of depth in the diagram (see Figure 7.12 for an example). Systems with more than four components are much too complex to depict graphically, and algebraic techniques must be used to analyze these systems.

What is the "right" choice of components if we want to reduce the number to three? We turn to the following simplifying methods:

1. Simply "ignore" components. As discussed above, this works well for trace elements, elements that enter only a single phase (we can drop both the component and the phase without violating the phase rule), or perfectly mobile components. Some minerals are stable over a large range of metamorphic conditions and thus are not useful as indicators of metamorphic grade. Even if present in significant quantities, they may be eliminated from consideration without impairing the analysis, particularly if they are the only phase containing a particular component. This is commonly the case for albite (Na_2O), K-feldspar (K_2O), magnetite (Fe^{3+}), ilmenite (TiO_2), titanite (TiO_2), and apatite (P_2O_5).

2. Combine components, such as those that substitute for one another in a solid solution [e.g., $(Fe + Mg)$].

3. Limit the types of rocks to be shown. Deal with only a subset of rock types for which a simplified system works.

4. Use projections. (I'll explain this shortly.)

The phase rule works well, and compatibility diagrams are rigorously correct, when they fully represent the systems under study. Thus triangular diagrams, such as those illustrated in Figures 24.2 and 24.3, apply rigorously only to true three-component systems (which are rare in nature). When we begin to drop components and phases, or to combine components, or project from phases, we face the same dilemma we did when using phase diagrams based on simplified systems in order to understand igneous processes in Chapters 6 and 7. We gain considerably by being able to graphically display the simplified system, and many aspects of the system's behavior become apparent. At the same time, we lose a rigorous correlation between the behavior of the simplified system and reality. As we shall see, it is nearly always advantageous to be able to see what is going on by adopting an appropriate and easily visualized system, even if simplified. When we do this, however, the diagrams may at times exhibit some inconsistencies (as we shall see below). If we choose our simplified system well, inconsistencies will be minimal. When an inconsistency occurs, careful analysis using a lesser degree of simplification usually reveals the reason for it.

Several diagrams have been proposed to analyze various chemical rock types. In all but a few simple natural systems, they are heroic attempts at reducing the true number of components from eight or nine to an accessibly graphical three. Let's look at a few diagrams in common use and see how they are produced, what they can do for us, and their limitations; we will make extensive use of them later.

24.3.1.1 THE *ACF* DIAGRAM.
Eskola (1915) proposed the *ACF* diagram as a way to illustrate metamorphic mineral assemblages on a simplified three-component triangular diagram. He concentrated only on the minerals that appeared or disappeared during metamorphism, thus acting as indicators of metamorphic grade. Figure 24.4 illustrates the positions of several common metamorphic minerals on the *ACF* diagram. Note: Figure 24.4 is presented only to show you where a number of important phases plot. It isn't referenced to a *P-T* range and has no compatible mineral tie-lines, and it is therefore *not* a true compatibility diagram. You may use it to help plot minerals on *ACF* diagrams, but it has no petrological significance.

The three pseudo-components *A*, *C*, and *F* are all calculated on a *molecular* basis and have rather odd definitions:

$$A = Al_2O + Fe_2O_3 - Na_2O - K_2O$$
$$C = CaO - 3.3 P_2O_5$$
$$F = FeO + MgO + MnO$$

To calculate *A* for a mineral, you must combine the *molecular* proportions of Al_2O_3 and Fe_2O_3 (Fe^{3+}) in the mineral formula and then subtract Na_2O and K_2O. If you

begin with an ideal mineral formula, Fe^{3+} is rare, except in a few minerals. In ferromagnesian minerals, Fe is probably in both valence states but generally is dominated by Fe^{2+}. Because Fe^{3+} typically substitutes for Al^{3+}, rather than joining Fe^{2+}, it is included in *A*, not F.

Why the subtraction? It is assumed that Na and K in the average mafic rock are combined with Al to produce K-feldspar and albite, respectively (or combined as alkali feldspar). Eskola (1915) was interested only in the amount of Al_2O_3 beyond that required for the formation of K-feldspar and albite. Perhaps you remember this from your *norm* calculation. If not, take a peek at the norm calculation procedure in Appendix B. After allocation of Al_2O_3 to K_2O (step 4a) and Na_2O (step 4c), the remaining Al_2O_3, if any, is combined with other chemical constituents to create other normative minerals. In the *ACF* diagram, we are interested only in those other metamorphic minerals, and thus only in the amount of Al_2O_3 that occurs in excess of that combined with Na_2O and K_2O. Because the ratio of Al_2O_3 to Na_2O or K_2O in feldspars is 1:1, we subtract from Al_2O_3 an amount equivalent to Na_2O and K_2O in the same 1:1 ratio. Although it appears as though we are subtracting sodium and potassium from aluminum, we are really subtracting from Al an amount *of Al* equivalent to the amounts of Na + K and then dealing with the Al that is left over. As we'll see in Section 24.3.3, this subtraction is more appropriately considered as a projection from K-feldspar and albite, thereby eliminating K_2O and Na_2O as components.

Combining Al and Fe^{3+} in this fashion is not very rigorous, and projecting from K-feldspar is justified only when this phase is present (which is rare in mafic rocks). The amounts of Fe^{3+} and K_2O are usually minor, however, and this is seldom a major problem. Projecting from Ab is more reliable because this component is present in the more common plagioclase feldspar.

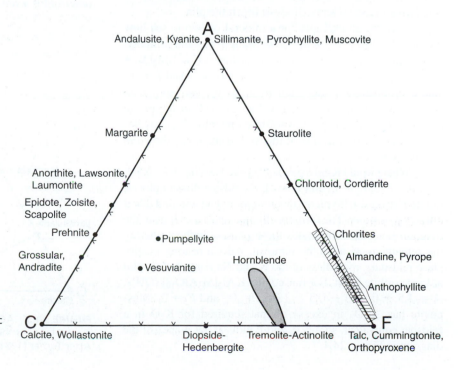

FIGURE 24.4 Chemographics of the ACF diagram showing the location of the ideal compositions of various common metamorphic minerals. After Ehlers and Blatt (1982) and Miyashiro (1994).

C is formulated in a fashion similar to A. All the P_2O_5 in most rocks is combined with CaO (in the ratio 1:3.3) to create apatite. Apatite is a ubiquitous accessory mineral, and in the ACF diagram, we are interested only in the CaO that exists in excess of that captured by P_2O_5 to create apatite. Thus we subtract from CaO an amount equal to 3.3 times the amount of P_2O_5 to eliminate P_2O_5 as a component and apatite as a phase without altering CaO in the rest of the system.

F is a combined pseudo-component based on the common exchangeability of Fe, Mg, and Mn in solid solution in most mafic minerals. If they behave in such a similar fashion, it makes some sense to treat them as a single component.

By creating these three pseudo-components, Eskola (1915) reduced the number of components from eight to three, suitable for a triangular diagram. H_2O is omitted, under the assumption that it is perfectly mobile. Note that SiO_2 is simply ignored. We will also see in Section 24.3.3 that this is equivalent to projecting from quartz. I'll make one point about projections now: In order for a projected phase diagram to be truly valid, the phase from which it is projected must be present in the mineral assemblages represented. Thus, to be valid, the ACF diagram must have quartz, alkali feldspar, and plagioclase present. An ACF compatibility diagram may still work when these phases are lacking, but the result may also violate the mineralogical phase rule on occasion.

To create an ACF diagram, you first calculate A, C, and F from the formulas of the minerals in the rocks found in an area. Because triangular diagrams work only if the sum of the three components equals 1.0, we must then normalize the preliminary values of A, C, and F to 1.0 before finally plotting. This is done by multiplying each of the initial values by 1.0 divided by the sum of the preliminary values $[1.0/(A + C + F)]$. When dealing with equilibrium mineral assemblages in a metamorphic area, you then connect coexisting phases with tie-lines to finish the plot.

Let's try an example for a mineral or two to see how the calculations work. Suppose we want to plot the composition of anorthite. The formula for anorthite is $CaAl_2Si_2O_8$. A is thus $1 + 0 - 0 - 0 = 1$, $C = 1 - 0 = 1$, and $F = 0$. These provisional values sum to 2, so we can normalize to 1.0 by multiplying each value by 1/2, resulting in $A = 0.5$, $C = 0.5$, $F = 0$. Anorthite thus plots halfway between A and C on the side of the ACF triangle, as shown in Figure 24.4.

Where would K-feldspar plot? For $KAlSi_3O_8$, $A = 0.5 + 0 - 0.5 = 0$, $C = 0$, and $F = 0$. K-feldspar doesn't plot on the ACF diagram. If you try this for albite, you will find that it doesn't plot either. The formula eliminates Na and K from the diagram (and thus eliminates these phases) without altering the remaining Al after the removal. All Ca-bearing plagioclase feldspars, regardless of the K and Na content, thus plot at the anorthite point. For muscovite, $KAl_2[Si_3AlO_{10}](OH)_2$, $A = 1.5 + 0 - 0 - 0.5 = 1.0$; $C = 0$, and $F = 0$. Muscovite has Al_2O_3 in excess of that required for K_2O in a feldspar equivalent, so $A > 0$, and it plots at the A apex of the diagram in Figure 24.4.

Figure 24.5 is a typical ACF compatibility diagram, referring to a specific range of P and T (the kyanite zone in the Scottish Highlands). The compositions of most mafic rocks fall in the hornblende–plagioclase field or the hornblende–plagioclase–garnet triangle, and thus most metabasaltic rocks occur as amphibolites or garnet amphibolites in this zone. More aluminous rocks develop kyanite and/or muscovite and not hornblende. More calcic rocks lose Ca-free garnet and contain diopside, grossular, or even calcite (if CO_2 is available). Again, you can see how the diagram allows us to interpret the relationship between the chemical composition of a rock and the equilibrium mineral assemblage that develops. Bulk rock compositions, when available, can also be plotted after conversion from wt. % oxides to molecular proportions (as step 1 does in the norm calculation).

Because the ACF diagram of Eskola lumps several chemical species and represents a projection into a reduced number of components, it is not a rigorous representation of equilibrium mineral assemblages in terms of the phase rule as applied to all components. Some authors have used alternative projections, claiming that they are equally valid in terms of reduced ACF components and either they are easier to calculate or the shifted mineral positions expand important fields to provide better visualization. Spear (1993), for example, used:

$$A = AlO_{3/2}$$
$$C = CaO$$
$$F = FeO + MgO$$

Using this formulation, most minerals shift closer to the A apex. Plagioclase, for example, would plot at $A = 2$, $C = 1$, and the range of Al-contents of hornblende and biotite expands. Half of the diagram in Figure 24.5 would no longer be occupied by the kyanite–plagioclase–almandine sub-triangle, and details within the expanded other fields would appear less cramped.

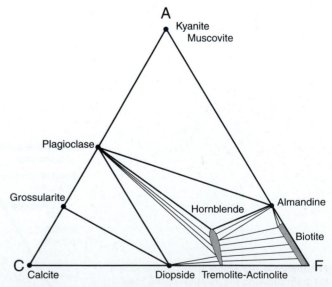

FIGURE 24.5 ACF compatibility diagram for quartz-bearing assemblages of the kyanite zone of the Scottish Highlands. After Turner (1981).

24.3.1.2 THE *AKF* DIAGRAM. Because pelitic sediments are high in Al_2O_3 and K_2O, and low in CaO, Eskola (1915) proposed a different diagram that included K_2O to depict the mineral assemblages that develop in them. In his ***AKF* diagram**, the pseudo-components are:

$$A = Al_2O_3 + Fe_2O_3 - Na_2O - K_2O - CaO$$
$$K = K_2O$$
$$F = FeO + MgO + MnO$$

Figure 24.6 illustrates the positions of several common metamorphic minerals on the *AKF* diagram. Note that CaO is now subtracted from Al_2O_3 in calculating *A*. As with Na_2O and K_2O, and their corresponding feldspars, this eliminates CaO and plagioclase from the diagram also (equivalent to projecting from anorthite). We are now interested only in the Al that occurs in excess of that combined with K, Na, and Ca to make any feldspar. As a result of this formulation, no plagioclase plots on the *AKF* diagram. Only one atom of Ca is subtracted from two of Al because the Al:Ca ratio in anorthite is 2:1. Although $A = 0$ in K-feldspar, it plots on the *AKF* diagram nonetheless because of its K_2O content.

Figure 24.7 shows an *AKF* compatibility diagram used by Eskola (1915) to illustrate the paragenesis of pelitic hornfelses in the Orijärvi region of Finland. Al-poor rocks there contain biotite and may contain an amphibole if sufficiently rich in Mg and Fe, or microcline if not. Rocks richer in Al contain andalusite and cordierite. Notice that three of the most common minerals in the area, andalusite, muscovite, and microcline, all plot as distinct points in the *AKF* diagram. Andalusite and muscovite plot as the same point in the *ACF* diagram, and microcline doesn't plot at all, making the *ACF* diagram much less useful for pelitic rocks that are rich in K and Al. The *AKF* diagram is also useful for metamorphosed granitoid rocks and some meta-graywackes.

For the same reasons mentioned above for *ACF* diagrams, several authors have used alternative formulations for the *AKF* diagram. Spear (1993), among others, used:

$$A = \frac{1}{2}\left[AlO_{3/2} - KO_{1/2}\right]$$
$$K = KO_{1/2}$$
$$F = FeO + MgO$$

AKF diagrams using this formulation are shown in Figures 28.1, 28.3, and 28.4.

24.3.2 Projections in Chemographic Diagrams

In this section, we will explore the methods of chemographic projection, explaining why we ignored SiO_2 in the *ACF* and *AKF* diagrams and what that subtraction was all about in calculating *A* and *C*. It will also help you better understand the *AFM* diagram in the next section and some of the shortcomings of projected metamorphic phase diagrams. For detailed discussions, see J. B. Thompson (1982a, b).

24.3.2.1 PROJECTION FROM APICAL PHASES. As an example of projection, let's begin with the ternary system: $CaO-MgO-SiO_2$ (the *CMS* system). This system is almost truly ternary in natural metamorphosed siliceous dolomites (although with CO_2) and some ultramafic rocks (with some Fe and other relatively minor components). The diagram is straightforward: $C = CaO$, $M = MgO$, and $S = SiO_2$ . . . none of that fancy subtracting business! Let's plot the following minerals (abbreviations are defined inside the back cover):

$$Fo - Mg_2SiO_4 \qquad Per - MgO$$
$$En - MgSiO_3 \qquad Qtz - SiO_2$$
$$Di - CaMgSi_2O_6 \quad Cal - CaCO_3$$

Forsterite has $M = 2$, $S = 1$, and $C = 0$. Normalized, that becomes $M = 0.67, S = 0.33, C = 0$. Similarly, En

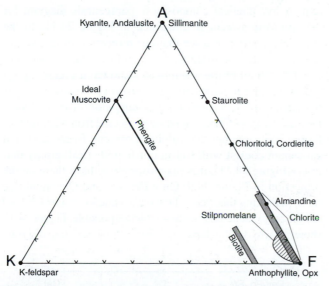

FIGURE 24.6 Chemographics of the *AKF* diagram showing the location of the ideal compositions of various common metamorphic minerals. After Ehlers and Blatt (1982).

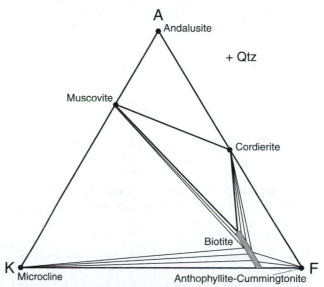

FIGURE 24.7 *AKF* compatibility diagram for the silica-saturated low-pressure metamorphic rocks of the Orijärvi region, Finland. After Eskola (1915) and Turner (1981).

yields $M = 0.5, S = 0.5, C = 0$, and Di yields $M = 0.25$, $S = 0.5, C = 0.25$, etc. The results are plotted as black dots in Figure 24.8 (ignoring CO_2 as perfectly mobile).

In order to understand projections in multiple dimensions, it is easiest to begin with a simple example that we can visualize readily. Suppose that three components were too difficult for us, and we decided to project the three-C system to only two components. For the purpose of illustration, let's eliminate CaO by projecting all phases from Cal (at the CaO apex of the triangle) to the Mg-Si side. Because the other minerals are already along the MgO-SiO_2 side, only Di needs to be projected. To do this, we draw a line from Cal through the Di point to the *M-S* side of the triangle (dashed in Figure 24.8). The line intersects the *M-S* side at a point equivalent to 33% MgO and 67% SiO_2. Note that *any* point on the dashed line from *C* through Di to the *M-S* side, including Di itself, has a constant Mg:Si ratio of 1:2.

By this projection, we can simplify the diagram from a ternary one to a pseudo-binary Mg-Si diagram in which Di is projected to a point at 33% Mg to 66% Si, as illustrated in Figure 24.9. By performing the projection, we have managed to reduce the ternary system to a pseudo-binary one, thus making it simpler. Note that the *geometric* projection of diopside from Cal or CaO is *mathematically* equivalent to ignoring CaO in the formula of the projected phase. The ratio of MgO to SiO_2 in Di is 1:2 or 33:67, so that we could calculate and plot Di in the pseudo-binary system directly without having to go through the complicated procedure of plotting the three-C system, drawing the (dashed) projection line, and then redrawing the result.

In a similar fashion, we could project Di from SiO_2 to the MgO-CaO pseudo-binary and get $C = 0.5, M = 0.5$. Di would plot at the midpoint of the Ca-Mg side, as shown in Figure 24.10. We get the same result mathematically by simply ignoring SiO_2 in the formula of the projected phase. Note that periclase, forsterite, and enstatite all stack up on the same point in this projection, limiting its usefulness.

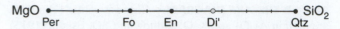

FIGURE 24.9 The pseudo-MgO-SiO_2 system in which the composition of diopside has been projected from CaO.

Suppose next that we were to treat Figure 24.9 as a compatibility diagram for some rocks belonging to the *CMS* system. If we pick a point along the MgO-CaO line (equivalent to specifying a MgO/CaO bulk composition), the mineral assemblage developed should be the immediate endpoints of the line segment we picked. For example, if our chosen point plotted between Per and Fo, the rock should contain periclase + forsterite. We note that, in accordance with the mineralogical phase rule (in which ϕ commonly equals *C*), we could have any of the following two-phase mineral assemblages in our two-component system:

$$Per + Fo$$
$$Po + En$$
$$En + Di$$
$$Di + Q$$

As I mentioned earlier, if we project from any phase, that phase must be present in all rocks to which we plan to apply the projected diagram. Because Figure 24.9 is projected from Cal-CaO, calcite (the apical phase) must be present in each of the above assemblages. If we violate this stipulation, our projected diagram may not work to characterize the mineral assemblages as intended. It appears from Figure 24.9, for example, that quartz cannot coexist with enstatite, nor can diopside coexist with forsterite. Imagine our surprise as we check back with our field notes and discover that enstatite and quartz commonly coexist, as do diopside and forsterite. These are the types of "inconsistencies" to which I referred earlier that might arise in combined or projected phase diagrams.

What's wrong? In order to find out, we need to go back to the more rigorously correct three-component system. A hypothetical composition–paragenesis diagram for the complete system is illustrated in Figure 24.11. In the complete diagram, we see that our projection from Cal collapsed several important tie-lines and that calcite does not coexist with all of the assemblages shown in Figure 24.9. I repeat the caveat made earlier: *to project from some point is to assume that a phase plotting at that point coexists with the projected phases.* Thus a diagram projected from Cal, such as Figure 24.9, assumes that calcite is present. Because Fo and En cannot coexist with Cal in our hypothetical ternary diagram (Figure 24.11), it is inappropriate to have them on the projection in Figure 24.9. Only Per, Di, and Qtz should be on it, indicating that Per + Di (+ Cal) and Di + Qtz (+ Cal) are the only calcite-bearing assemblages possible. Figure 24.11 shows us why we can have Di + En + Fo (coexisting diopside

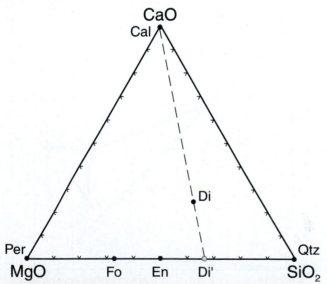

FIGURE 24.8 Chemographic relationships in a portion of the CaO-MgO-SiO_2 system. Di is projected from CaO to the MgO-SiO_2 base.

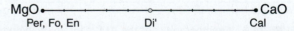

FIGURE 24.10 The pseudo-MgO-CaO system in which the composition of diopside has been projected from SiO_2.

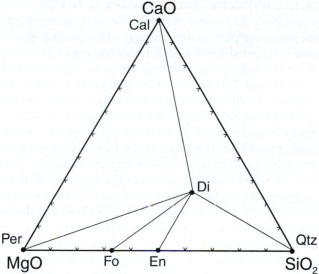

FIGURE 24.11 Hypothetical compatibility diagram based on the CaO-MgO-SiO$_2$ system. Several minerals that occur in nature are left out for simplification.

and forsterite) and Di + En + Q (coexisting enstatite and quartz), each of which is a calcite-free three-phase assemblage. By referring to the "real" system (Figure 24.11), we understand why the projected system gave us some spurious results. The *apparent* incompatibility of En + Q and Di + Fo in Figure 24.9 is an artifact of the projection. It's easy to project any phase, but we must remember that by doing so, we assume that calcite is the third phase in each projected assemblage!

Diagrams such as *ACF* and *AKF* eliminate SiO$_2$ by ignoring SiO$_2$ in the formulas of the projected phases. This is equivalent to projecting from quartz. The mathematical process is easy: projecting from an **apex** phase is equivalent to ignoring the apex component. We thus ignored CaO in the Di formula in Figure 24.9, SiO$_2$ in the Di formula in Figure 24.10, and SiO$_2$ in the mathematical calculations of the projected minerals in the *ACF* and *AKF* diagrams. A potential shortcoming of these projections is that they compress the true relationships as a dimension is lost. To be properly applicable, quartz must be present in every mineral assemblage on the *ACF* and *AKF* diagrams (or SiO$_2$ must at least be present to the point of saturation). Otherwise, we may get minerals that appear to coexist in projection, but they are not at the corners of the same sub-triangle (or polygon) with the projection phase in the unprojected system.

You may be wondering why we would want to create projected pseudo-systems when they can lead to such inconsistencies. Reducing a ternary system to a binary one, as we just did, does not provide much of a benefit because three-component and two-component systems are both easily illustrated in two-dimensional diagrams. The detriments are thus more obvious than the benefits, but systems with more than three components are no longer so easily visualized. The ability to reduce the number of components from eight or nine to three provides many a petrologist with the motivation to combine components, such as FeO, MgO, and MnO; or to project from SiO$_2$ in order to find a useful

pseudo-three-component system. As you have seen above in the *ACF* and *AKF* diagrams, the ability to visualize the geometric relationships, and how the bulk rock composition affects the resulting equilibrium mineral assemblage, is usually an advantage that far outweighs the shortcomings.

To better understand how this applies to projecting from SiO$_2$ to triangular *ACF* or *AKF* diagrams, consider the model four-component system *a-b-c-q*, with the equivalent phases *A*, *B*, *C*, and *Q* illustrated in Figure 24.12. Two compounds, *X* (formula *abcq*) and *Y* (formula *a$_2$b$_2$cq*), plot within the *abcq* compositional tetrahedron. If minerals coexisting within a metamorphic zone are connected by tie-lines, the tetrahedron becomes a compatibility diagram, subdivided into triangular-sided sub-polyhedra: *AYQX*, *BYQX*, etc. A bulk rock composition that falls within any one polyhedral volume will consist of the four-phase equilibrium assemblage represented by the four corners of that sub-polyhedron.

Next, visualize a projection from phase *Q* (at the apex) to the *abc* plane. Point *X* is in the center of the tetrahedron, and point *Y* is within the *ABQX* volume. The projection process works the same way as in the ternary example above. Projecting from phase *Q* projects points *X* and *Y* to the *abc* base as points *X'* and *Y'*, respectively, in Figure 24.12. Along any single projection line, such as *Q-X-X'* or *Q-Y-Y'*, the ratio of *a:b:c* is constant. We can mathematically perform the projection by simply ignoring the *q* component in the formula of the projected phase and normalizing the remaining *a + b + c* to 100. For *X* the result is *a:b:c* = 1:1:1 = 33:33:33 so that *X* plots as *X'* in the center of the projected pseudo-ternary diagram. *Y* also plots as *Y'* with *a:b:c* = 2:2:1 = 40:40:20, resulting in the pseudo-ternary *abc* compatibility diagram illustrated in Figure 24.13.

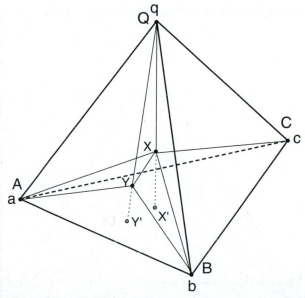

FIGURE 24.12 Spatial relationships in the hypothetical four-component system *abcq*. Two four-component compounds, *X* and *Y*, are projected from the interior of the *abcq* tetrahedron to the *abc* base.

The resulting *abc* projection in Figure 24.13 is simplified with respect to the quaternary system, and we need no longer attempt to estimate the depth of points *X* and *Y*, as in Figure 24.12. But the projection suffers from the same pitfalls that we saw above. If we remember our projection point, we conclude from Figure 24.13 that the following assemblages are possible:

$$(Q) - B - X - C$$
$$(Q) - A - X - Y$$
$$(Q) - B - X - Y$$
$$(Q) - A - B - Y$$
$$(Q) - A - X - C$$

The assemblage *A + B + C* appears to be impossible in Figure 24.13. When you refer to Figure 24.12, however, you can discern a sub-polyhedron *ABCX*, indicating that *A*, *B*, and *C* can indeed coexist. Phases *A*, *B*, and *C* are truly on the base, but *X* (and *Y*) are above it. Again, the diagram in Figure 24.13 is not truly ternary, but a projection, and we must remember that! The pseudo-ternary triangle is really part of a quaternary system, and because it is projected from *Q*, *it should be clearly referenced to coexisting phase Q!* Although *A*, *B*, and *C* are separated by several tie-lines in Figure 24.13 and appear to be incompatible together, they can coexist, but with *X* and *without Q*! So, we're not simply ignoring the component *q* and phase *Q* when we create the projection in Figure 24.13. Phase *Q* is implicitly part of the projection, and the sub-triangles in the compatibility diagram are valid groupings only when phase *Q* is included as a part of each mineral assemblage. To do this properly, conscientious investigators include a message such as "+Qtz" in the corner of diagrams that are projected from quartz (*ACF*, *AKF*, etc.). Multiple projections are possible, each successive one reducing the number of remaining components. The *ACF* diagram, for example, is projected from SiO_2 and then from alkali feldspar. In such cases all projection phases should be noted in the diagrams.

24.3.2.2 PROJECTING FROM NON-APICAL PHASES.

Projecting from a non-apex point is as simple geometrically, but more complex mathematically. The ensuing discussion is intended for the advanced student who is curious about the subtraction of components in some formulations, such as K_2O and Na_2O from Al_2O_3 in *A* of the *ACF* diagram. It will also help you to better understand the *AFM* diagram discussed in the next section. It is not necessary if all you wish to do is use the diagrams and are willing to put up with some of the inconsistencies. If so, you may choose to skip ahead to the next section.

If you're still here, consider the hypothetical three-component *abc* system in Figure 24.14, with three possible mineral compounds: *ac*, a_2c, and abc_2. Suppose that we desire to eliminate the *a* component of the diagram and deal only with a projection to *b-c*. As we've learned, projecting from an apex phase, such as *A*, is simple. Minerals *ac* and a_2c project to point *c*, and mineral abc_2 projects to the *b-c* base at a point equivalent to bc_2. The projection can be performed mathematically by simply ignoring *a* in each formula, as discussed in the previous section. Ignoring *a* in the formula of abc_2, for example, results in $b{:}c = 33{:}67$, and thus the point plots as shown in the projected pseudo-binary *b-c* system below the three-component triangle in Figure 24.14.

But suppose no mineral corresponding to *A* occurs in the rocks, and projecting from *A* crosses tie-lines involving *ac* or a_2c. As we saw above, this can lead to some spurious apparent groupings. In such a situation, the pseudo-binary projection would be more accurate if we projected from a phase that contains *a* and actually occurs in the rocks.

What if we project from phase *ac*? The graphical method is still simple (at least for three-component to two-component projections). Just draw the line from phase *ac* through abc_2 to the *b-c* base (Figure 24.15). Note that the

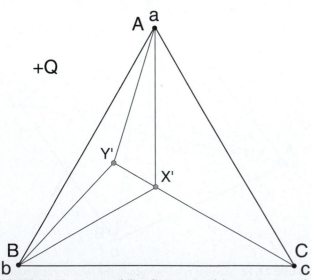

FIGURE 24.13 Compatibility diagram involving projection of compounds, *X* and *Y*, from the *Q* apex of the *abcq* tetrahedron (Figure 24.12) onto the *abc* base.

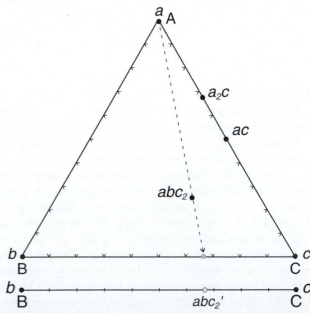

FIGURE 24.14 Spatial relationships in the hypothetical three-component system *abc*. The phase abc_2 is projected from the *a* apex to the *bc* base.

projected position of abc_2 is in a different position in this projection to b-c than if we had projected from the a apex. How do we do this *mathematically*? The projection source (ac) has an a:c ratio of 1:1. Thus, every atom of a combines with, or "consumes," a single c atom. The projection is a compensation for that ratio, or the loss of an atom of c for each atom of a in the projection source phase. The formula of our projected mineral is abc_2, so to plot it mathematically on the b-c side of the triangle we ignore the a content (because we're projecting to the "a-free" side of the triangle), b is unaffected (because there is no b in the projection source), and we then *subtract one c atom for each a atom in the formula of the phase to be projected*. In other words, our resulting b-c projection is really a pseudo-binary b-c' projection, in which:

$$b = b$$
$$c' = c - a$$

So to mathematically project abc_2 in this system, we calculate the components in this compound according to the formulas above. Thus $b = 1$ and $c' = 2 - 1 = 1$, and therefore abc_2 should plot at the midpoint of the b-c' diagram. This agrees with the graphical method shown in Figure 24.15. Although the formula for c' looks as though we are subtracting a from c, it really means that we are subtracting from c an amount *of c* equivalent to the amount of a in the compound. (Once again, the 1:1 ratio reflects the ratio in the ac projection source phase.)

Now recall those funny formulas for A in the ACF and AKF diagrams. The ACF diagram was a projection from alkali feldspar and quartz. We first project from the SiO_2 apex by ignoring SiO_2 in the formulas of the projected minerals. Projecting from K-feldspar requires that the calculation of A involves subtraction from Al an amount equivalent to K, (Al_2O_3-K_2O), and similarly subtract an amount equivalent to Na for the Ab projection (Al_2O_3-Na_2O). The ratio of Al:K and Al:Na is 1:1 because that is the ratio of these components

in the minerals from which the projection is made. The result is $A = Al_2O_3 - K_2O - Na_2O$, which reflects the combined projection first from quartz, then from K-feldspar, then from albite. Fe_2O_3 is added because of the Al-Fe^{3+} solid solution effect, similar to Fe-Mg-Mn.

We can project geometrically, as in the Figures 24.14 and 24.15, or mathematically by the method of redefining c'. The result is the same. If there are fewer than four components, the graphical method is easier to visualize, but the mathematical method, once understood, is much easier, and can be applied to any number of initial components.

If you are catching on to the method, we can test it by trying to project from a_2c instead of ac. We do it graphically as in Figure 24.16, and we see that abc_2 projects to the b-c side of the triangle at a point 60% of the distance from b to c. To derive a mathematical formula for the new projected c component, c'', we must again consider the a:c ratio in the projection source mineral a_2c, which is obviously 2:1. Thus each a atom in the source "consumes" only half a c atom. Thus our new formulas for the b-c projection, which we may refer to as b-c'', is:

$$b = b$$
$$c'' = c - \frac{1}{2}a$$

We thus subtract from c an amount of c equivalent to half of the a atoms in the projected mineral. For abc_2, then, $b = 1$ and $c'' = 2 - 0.5 = 1.5$. Normalizing to 1.0 by multiplying each by 1/2.5 (the sum of b and c'') gives us $b = 0.4$ and $c'' = 0.6$, and 40:60 agrees with the graphical method in Figure 24.16.

24.3.3 J. B. Thompson's *AKFM* Diagram

With perhaps a better understanding of projections, we turn to a particularly useful diagram for pelites. J. B. Thompson (1957) proposed the **AKFM diagram** (commonly called simply the **AFM projection**) as an alternative to the AKF diagram

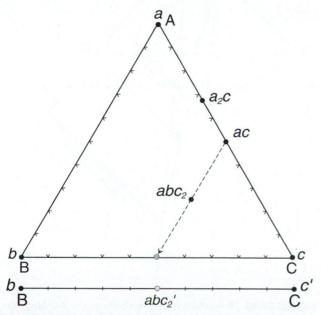

FIGURE 24.15 Projection of the phase abc_2 from the phase ac to the b-c base in the abc system.

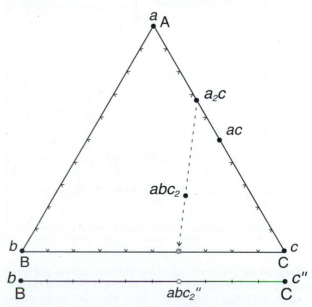

FIGURE 24.16 Projection of the phase abc_2 from the phase a_2c to the b-c base in the abc system.

for metamorphosed pelitic rocks. Pelites are sensitive to temperature and pressure, and undergo many mineralogical changes as grade increases. It is thus good to have a comprehensive diagram to display these changes. Although the *AKF* is useful in this capacity, Thompson (1957) noted that Fe and Mg do not partition themselves equally between the various mafic minerals in most rocks. The unequal distribution of elements in solid solution was evident to us in our discussion of distribution coefficients in Chapter 9 and will be treated again, quantitatively, in Chapter 27. Figure 24.17 confirms that Fe and Mg are not evenly distributed among the minerals common to metamorphosed *ultramafic* rocks. Talc is always richer in Mg than coexisting diopside, diopside more so than tremolite, etc. All but anthophyllite are more Mg rich than olivine. The same type of partitioning occurs in virtually all mafic minerals, including those in pelitic rocks. Garnet, staurolite, and chloritiod have low Mg/Fe ratios, whereas cordierite has a higher ratio. Thompson (1957) concluded that it would be advantageous in many situations to account for this unequal distribution because it could play a role in the relative stabilities of the ferromagnesian minerals. The Fe/Mg ratio of a rock, Thompson argued, could thus exert control on the stable mineral assemblage that develops at a particular metamorphic grade. To address this he developed the *AFM* projection, in which Fe and Mg are no longer combined as a single component.

Thompson (1957) neglected minor components in pelitic rocks (including CaO and Na_2O) and considered H_2O as perfectly mobile. He eliminated SiO_2 by projecting from quartz. Because quartz is nearly always present in metapelites, this requirement of the projection is generally met. Four principal components thus remain, and Figure 24.18

shows the *AKFM* tetrahedron ($A = Al_2O_3$, $K = K_2O$, $F = FeO$, $M = MgO$). To avoid dealing with a three-dimensional tetrahedron, Thompson projected the phases in the system to the *AFM* face, thereby eliminating K_2O as an explicit component.

Thompson (1957) recognized that projecting from the K_2O apex of the tetrahedron would not work because no phase corresponds to this apical point, and projections would cross important tie-lines. A realistic diagram must be projected from a phase that is present in the mineral assemblages to be studied. Because muscovite is the most widespread K-rich phase in metapelites, he decided to project from muscovite (Ms) to the *AFM* base, as shown in Figure 24.18.

Projecting from muscovite can lead to some strange looking *AFM* projections. Note that Ms is still rather K poor, and only mineral phases in the volume A-F-M-Ms in Figure 24.18 will be projected to points within the *AFM* face of the *AKFM* tetrahedron. Phases in the volume Ms-*x*-F-M-*y* project to points on the K-free plane beyond the *F-M* end of the *AFM* triangle. Biotite, for example, is such a phase. The broad gray band within the tetrahedron corresponds to a full Fe-Mg range for biotites. As you can see, projecting from *Ms* causes biotite to plot as a band outside the *AFM* triangle.

Muscovite may be absent in some pelitic rocks, particularly at higher grades when it dehydrates, giving way to K-feldspar as the common high-K phase. When this is the case, the *AFM* projection should be projected from

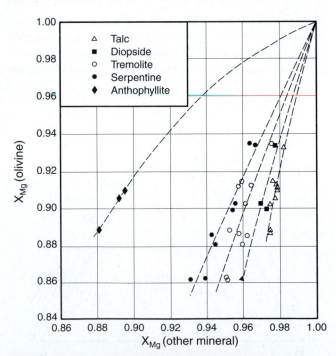

FIGURE 24.17 Partitioning of Mg between olivine and other minerals in metamorphosed ultramafic rocks from the Bergell Aureole, Italy. $X_{Mg} = Mg/(Mg + Fe + Mn + Ni)$ on an atomic basis. From Trommsdorff and Evans (1972). Reprinted by permission of the *American Journal of Science*.

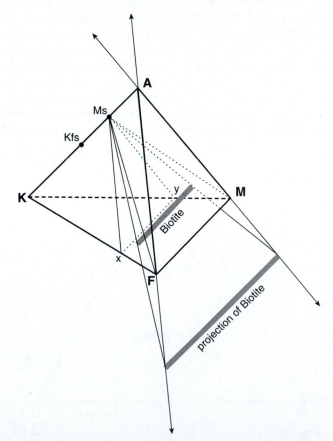

FIGURE 24.18 Projecting from muscovite in the *AKFM* system. After Thompson (1957). Copyright © the Mineralogical Society of America.

K-feldspar if the mineral assemblages of the diagram are to have significance. When projected from Kfs, biotite projects within the *AFM* triangle. If both Ms and Kfs are present, we generally project from Ms.

We avoid having to create diagrams such as Figure 24.18 and attempting three-dimensional projections by calculating the projected components of the *AFM* projection mathematically, using the following formulas:

$$A = Al_2O_3 - 3K_2O \text{ (if projected from Muscovite)}$$
$$= Al_2O_3 - K_2O \text{ (if projected from K-feldspar)}$$
$$F = FeO$$
$$M = MgO$$

We can then plot any mineral composition as a point directly on the *AFM* projection. When *A*, *F*, and *M* all have positive values, plotting a point is straightforward. *A* may be negative, however, as with biotite, and the negative-*A* process requires an extra step. Suppose we have biotite with the composition $KMg_2FeSi_3AlO_{10}(OH)_2$. When projecting from *Ms*, for example, we calculate the components by using the formulas above:

$$A = 0.5 - 3(0.5) = -1$$
$$F = 1$$
$$M = 2$$

To normalize, we multiply each value by $1.0/(2 + 1 - 1) = 1.0/2 = 0.5$. Thus $A = -0.5$, $F = 0.5$, and $M = 1$, which sum to the required 1.0. To plot the point, we extend a line from *A* at a constant *M/F* ratio. Because $M/(F + M) = 0.67$, we use the dashed line extending from A to 0.67 on the $M/(F + M)$ scale on the bottom of the diagram in Figure 24.19. Because $A = -0.5$, we next extend a *vertical* line from *A* to a distance equal to half the distance from *A* to the *F-M* base, but we extend this distance *beyond* the base because *A* is negative. It thus extends to $A/(A + F + M) = -0.5$. This line is also illustrated in

Figure 24.19, where it corresponds to the bottom boundary of the biotite field. Where a horizontal line at this value of *A* intersects the first line of constant *F:M* is the location of our biotite on the *AFM* projection (point X). The broad biotite field in Figure 24.19 is caused by both Fe-Mg exchange (right–left) and coupled substitution of (Al + Al) for (Fe or Mg) + Si (up–down) in biotite (making some biotite more aluminous than the ideal biotite we just plotted).

From Figure 24.18 you can see that K-feldspar, when projected from muscovite, projects away from the *AFM* projection. Projecting K-feldspar from Ms in the *AFM* places it at either the *A*-apex or at negative infinity (the latter is equivalent to the arrow to Kfs at the bottom of Figure 24.19).

Figure 24.20 is an example of an *AFM* compatibility diagram, applied to the mineral assemblages developed in some metapelitic rocks in New Hampshire. Due to extensive Mg-Fe solid solution in biotite and garnet, much of the area is dominated by two-phase fields with tie-lines connecting coexisting mineral compositions. These fields are really four-phase fields when we include the Qtz and Ms projection phases.

Although we can easily plot ideal mineral formulas on the *ACF* and *AKF* diagrams, for a real mafic phase to be plotted on an *AFM* projection, we must know Mg/(Fe + Mg), which can only be determined accurately by chemical analysis of the minerals, generally performed using the electron microprobe. If analyses are unavailable, we can approximate the correct positions on the basis of typical relative Mg/(Fe + Mg), based on our knowledge of numerous analyses of these minerals available in the literature. From these we know that Mg-enrichment occurs typically in the order: cordierite > chlorite > biotite > chloritoid > staurolite > garnet. We can do this more accurately if we know the Fe/Mg distribution coefficients for coexisting mafic phases.

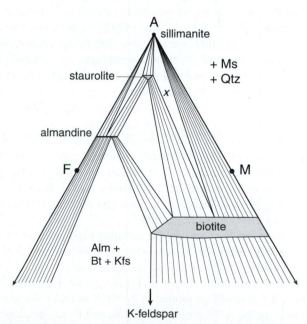

FIGURE 24.20 *AFM* compatibility diagram projected from muscovite, showing the mineral assemblages observed in metapelitic rocks in the lower sillimanite zone of west-central New Hampshire. After J. B. Thompson (1957). Copyright © the Mineralogical Society of America.

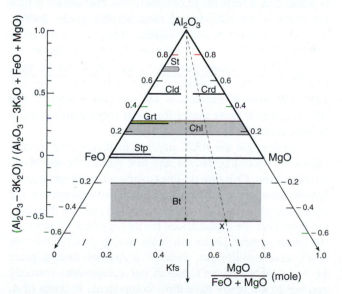

FIGURE 24.19 The *AFM* projection from muscovite. After J. B. Thompson (1957). Copyright © the Mineralogical Society of America.

24.3.4 Choosing the Appropriate Compatibility Diagram

The *ACF*, *AKF*, and *AFM* diagrams are the most commonly used compatibility diagrams, but several others have been employed for rocks of different compositions. In Figure 21.18, the *CMS* ($CaO - MgO - SiO_2$) diagram was applied to dolomitic marbles (we will use it extensively in Chapter 29). The *CFM* projection, in which $C = CaO + Na_2O + K_2O - Al_2O_3$, $F = FeO - Fe_2O_3$, and $M = MgO$, was developed by Abbott (1982, 1984) for use in metabasaltic rocks. Thompson and Thompson (1976) showed that assemblages of Fe-Mg-free minerals in metapelites could be successfully analyzed using an $Al_2O_3 - NaAlO_2 - KAlO_2$ diagram. Several other diagrams have also appeared in the literature (see Miyashiro, 1994, for a review). You can also make up your own if you like. The choice of components and projection point depends on the system and mineral assemblages under study.

For example, suppose we have a series of pelitic rocks in an area. The pelitic system consists of the nine principal components: SiO_2, Al_2O_3, FeO, MgO, MnO, CaO, Na_2O, K_2O, and H_2O. How do we lump those nine components to get a meaningful and useful diagram? A reasonable choice of simplification steps is listed below that permits the development of a pseudo-three-component diagram. Each step simplifies the resulting system, however, and may overlook some aspect of the rocks in question.

Although the steps below have proved useful for most pelitic studies, some productive research may arise from studies of the components otherwise overlooked. For example, MnO is commonly lumped with $FeO + MgO$, or else ignored, as it usually occurs in low concentrations and enters solid solutions along with FeO and MgO. This does not mean that MnO is an unimportant constituent in metamorphic rocks. Symmes and Ferry (1992) have shown that a small amount of MnO is essential in stabilizing many of the commonly observed mineral assemblages in metapelites (particularly those involving garnet, Figure 28.15). Once the assemblages are developed, however, they can be dealt with graphically without necessarily having to consider MnO any longer as a geometrically distinct component.

In metapelites, Na_2O is usually significant only in plagioclase, so we may commonly ignore it also, or project from albite. As a rule, H_2O is sufficiently mobile to be ignored as well. We are left with SiO_2, Al_2O_3, K_2O, CaO, FeO, and MgO. Which combination and diagram is best depends on the rocks to be studied.

Suppose we consider the common high-grade mineral assemblage Sil-St-Ms-Bt-Qtz-Plag. A bulk rock composition corresponding to this assemblage is marked with an *x* in the *AFM* projection shown in Figure 24.20. Figure 24.21 shows the assemblage plotted in the *ACF* and *AKF* diagrams as well. In the *AFM* projection Sil, St, and Bt can be plotted (with Ms + Qtz as associated phases, required by the projection). In the *ACF* diagram, Sil, St, and Plag are the phases than can be plotted. In both the *ACF* and *AFM* diagrams, $C = \phi = 3$ on the diagrams, so we have "valid" divariant

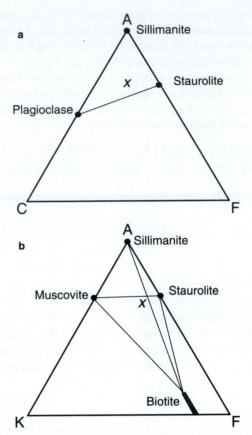

FIGURE 24.21 The mineral assemblage sillimanite–staurolite–muscovite–biotite–quartz plagioclase plotted on **(a)** the *ACF* diagram and **(b)** the *AKF* diagram. Refer to Figure 24.20 for the *AFM* diagram The bulk rock composition is marked as *x*. After Ehlers and Blatt (1982).

assemblages as far as the mineralogical phase rule and plotted phases are concerned.

The *ACF* diagram in Figure 24.21a is less useful for pelites because plagioclase is the only Ca-bearing phase and is stable over a wide range of conditions. Plagioclase is thus not much of an indicator of metamorphic grade. Biotite, which is an important pelitic mineral that in many instances indicates both grade and composition, does not appear. Although our assemblage plots as a three-phase sub-triangle, important pelitic assemblages are constrained to the *A-F* edge, and most of the area of the diagram is wasted. The *ACF* diagram is more useful for mafic rocks with more Ca-bearing minerals.

Notice that the *AKF* diagram in Figure 24.21b contains four coexisting phases and that the Sil-Bt and Ms-St tie-lines cross. Crossing tie-lines can be a troublesome problem because we cannot choose a single sub-triangle for our rock composition *x*. Remember from Section 24.2 that there are three possible explanations for this apparent violation of the mineralogical phase rule now that $\phi > C$: Either we don't have equilibrium, there is a reaction taking place ($F = 1$), or we haven't chosen our components correctly and we do not really have three components in terms of *A*, *K*, and *F*. This is a common assemblage, which argues against the first two possibilities. If we compare the *AKF* and *AFM* diagrams, we see that Fe/Mg partitioning between Bt and St is relatively strong and should not be ignored. Consider the *AKF* diagram in Figure 24.21b as the base of an *AKFM* tetrahedron. Because biotite is more Mg rich than staurolite, its composition actually plots higher into the tetrahedron than St from the *AKF* base. As a result, the four phases appearing in the *AKF* diagram are not really coplanar in a more comprehensive four-component *AKFM* tetrahedral diagram. Thus the Sil-Bt and Ms-St tie-lines do not really intersect in space (and Sil-Bt is not compositionally equivalent to Ms-St).

We conclude that the *AFM* diagram works best for the assemblage in question. It may not be the best for other assemblages, however, and the *AFM* and *AKF* diagrams are commonly used in combination to depict the relationships between rock composition, mineral assemblage, and metamorphic grade in metamorphosed pelitic rock sequences (as we shall do in Chapter 28).

Numerous compatibility diagrams have been proposed to analyze paragenetic relationships in various metamorphic rock types. Most simplify complex systems to pseudo-ternary ones, such as the *ACF*, *AKF*, and *AFM* projections. Such simplifications may produce anomalies in which ϕ becomes greater than *C*, typically involving crossed tie-lines in compatibility diagrams. Reference to more comprehensive and realistic systems, including more components, will generally reveal the cause for the inconsistency, either graphically or by using the phase rule. Analytical techniques to resolve these anomalies, based on linear algebra, have been developed by Greenwood (1967b), Perry (1967), and Braun and Stout (1975).

Variations in the mineral assemblage that develops in metamorphic rocks result from (1) differences in bulk composition or (2) differences in other intensive variables, such as *T*, P_{H_2O}, etc. (metamorphic grade). A good compatibility diagram permits easy visualization of the first situation, providing a graphical portrayal of how the equilibrium mineral assemblage depends on X_{bulk} *within a particular metamorphic zone*. The dependence of mineral assemblage on metamorphic grade can be determined by a balanced reaction (representing an isograd) in which one rock's mineral assemblage contains the reactants and another contains the products. The differences from one zone to the next can effectively be visualized by comparing separate compatibility diagrams, one for each grade. We will explore these techniques in the ensuing chapters.

Summary

The phase rule states that $F = C - \phi + 2$, and the variance of a typical metamorphic rock sample is generally equal to two. Therefore the number of phases (ϕ) in a sample (if equilibrium is attained) is typically equal to the number of components (*C*), or $\phi = C$ (sometimes called Goldschmidt's mineralogical phase rule). Situations in which $\phi < C$, can usually be explained by solid solution effects. When $\phi > C$ any of three possibilities may be responsible: (1) equilibrium has not been attained and some phases coexist with others incompatibly, (2) the sample was collected on an isograd and an incomplete reaction has been preserved ($F < 2$), or (3) *C* was not determined correctly. Perfectly mobile components (generally fluid species such as H_2O or CO_2) are typically excluded from *C* in the application of the mineralogical phase rule, because, when so mobile, they do not control the presence of a phase. If we exclude them as components, we must also ignore the fluid phase in ϕ (subtracting one from both *C* and ϕ does not affect the $C = \phi$ equivalence).

Chemographic diagrams are graphical depictions of phase compositions and coexisting phase relationships. Most are triangular diagrams that address a particular metamorphic grade or a limited *P-T* range, such as within a metamorphic zone (and are then called compatibility diagrams). Coexisting phases are connected by tie-lines that subdivide the overall triangular diagram into several smaller three-phase sub-triangles. The mineralogical phase rule thus generally holds, because $C = \phi = 3$ in most situations. *C* is reduced to the requisite value of three by a combination of ignoring components, combining components, and projecting. Well-conceived triangular chemographic diagrams reveal how the bulk composition (position on the diagram) controls the phase assemblage developed (corners of the sub-triangle within which X_{bulk} plots) and rarely have situations in which $\phi > 3$ (typically manifested by crossing tie-lines). Solid solution, when present in any phase and for which the compositional range can be analytically determined, should be indicated in chemographic diagrams whenever possible. The most commonly used chemographic diagrams are the ACF, *AKF*, and AFM diagrams. When a diagram reveals an anomaly, such as crossing tie-lines, the reason can generally be found by turning to a more comprehensive system ($C > 3$) and the original Gibbs phase rule.

Key Terms

Review Questions and Problems

Review Questions and Problems are located on the author's web page at the following address: **http://www.prenhall.com/winter**

Important "First Principle" Concepts

- At equilibrium, the stable mineral assemblage (mineral paragenesis) of a rock (including the composition of each phase) is determined by pressure (P), temperature (T), and composition (X).
- A reaction involving more than one reactant can proceed only until any *one* of the reactants is consumed. Further reaction is impossible, and all the remaining reactants can coexist stably with the products.
- The number of phases in "typical" metamorphic rocks is equal to the number of components: $\phi = C$.
- If $\phi > C$, we might suspect that solid solution is involved with some of our components or that some components are perfectly mobile and should not be included in C.
- If $\phi > C$, there is either disequilibrium or a reaction is taking place among the phases.
- Compatibility diagrams provide an excellent way to analyze how variations in X_{bulk} affect the mineral paragenesis that develops in a rock at a particular grade (typically within a zone).
- Compatibility diagrams typically reduce the number of displayed components to three by ignoring, combining, and/or projecting from components. Because such diagrams are not really ternary, some may appear to violate the $\phi = C$ standard by having more than three plotted phases and hence crossed tie-lines when coexisting phases are connected. We can usually figure out the reason by referring to the full phase rule with fewer component simplifications. Rocks at equilibrium should *always* conform to the phase rule.
- Changing from one zone to the next involves a reaction (isograd) that affects the phases and/or tie-line geometry so that analysis of mineral paragenesis in the next zone requires a new compatibility diagram.

Suggested Further Readings

Miyashiro, A. (1994). *Metamorphic Petrology*. Oxford University Press. New York.

Thompson, J. B., Jr. (1957). The graphical analysis of mineral assemblages in pelitic schists. *Amer. J. Sci.,* **42**, 842–858.

Thompson, J. B., Jr. (1982a). Composition space: An algebraic and geometric approach. In: *Characterization of Metamorphism Through Mineral Equilibria* (ed. J. M. Ferry). *Min. Soc. Amer. Reviews in Mineralogy* **10**, 1–31.

Thompson, J. B., Jr. (1982b). Reaction space: An algebraic and geometric approach. In: *Characterization of Metamorphism Through Mineral Equilibria* (ed. J. M. Ferry). *Min. Soc. Amer. Reviews in Mineralogy* **10**, 33–52.

25

Metamorphic Facies
and Metamorphosed Mafic Rocks

Questions to be Considered in this Chapter:

1. How do we use the metamorphic facies concept to characterize the grade of metamorphism affecting a rock or an area?

2. What are the commonly recognized metamorphic facies and what mineral parageneses and *P-T* conditions characterize them?

3. How can we combine metamorphic field gradients and the facies concept to explore the sequence of facies that develop across a broader area?

4. The facies subdivisions are based on mineral assemblages developed in metamorphosed mafic rocks. What are these assemblages, and how are they developed?

5. What happens at the extreme pressure and temperature limits of crustal metamorphism, and how do we interpret rocks that have been subjected to these conditions?

6. What types of pressure–temperature–time (*P-T-t*) paths may metamorphic rocks follow during burial, metamorphism, uplift, and erosion? How do *P-T-t* paths relate to metamorphic field gradients?

25.1 METAMORPHIC FACIES

The concept of metamorphic facies is fundamental to our understanding and characterization of metamorphic rocks. Although Pentii Eskola coined the term and initially developed the concept, we should step back a few more years to the work of V. M. Goldschmidt to understand more fully its background.

Goldschmidt (1911, 1912a) studied a series of contact-metamorphosed pelitic, calcareous, and psammitic hornfelses of Paleozoic age in the Oslo region of southern Norway. Although the chemical composition of the rocks varied considerably, Goldschmidt found relatively simple mineral assemblages of fewer than six major minerals in the inner zones of the aureoles around granitoid intrusives (recall his mineralogical phase rule in Section 24.2). Goldschmidt was the first to formally note that the equilibrium mineral assemblage of a metamorphic rock (at a particular metamorphic grade) could be related directly to its bulk composition. He noticed that the aluminous pelites contained Al-rich minerals, such as cordierite, plagioclase, garnet, and/or an Al_2SiO_5 polymorph (andalusite or sillimanite). The calcareous rocks, on the other hand, contained Ca-rich and Al-poor minerals, such as diopside, wollastonite, and/or amphibole. From the correlation between a rock's chemical and mineralogical composition, he concluded that the rocks in the inner aureoles had achieved chemical equilibrium.

As support for the equilibrium concept, Goldschmidt noted that certain mineral pairs, such as anorthite + hypersthene, were consistently present in rocks of the appropriate composition, whereas the *compositionally equivalent* pair, diopside + andalusite, was not. If two alternative assemblages are compositionally equivalent, we must be able to relate them by a *reaction*. In this case, the reaction is simple:

$$MgSiO_3 + CaAl_2Si_2O_8 = CaMgSi_2O_6 + Al_2SiO_5 \quad (25.1)$$
$$En \qquad\quad An \qquad\qquad Di \qquad\quad And$$

Goldschmidt thus concluded that the left side of the reaction was always stable under the metamorphic conditions existing at Oslo. If equilibrium were not attained, he reasoned, perhaps either pair or all four minerals may have been found.

Eskola (1914, 1915) worked in similar hornfelses in the Orijärvi region of southern Finland, where he confirmed the ideas of Goldschmidt (1911, 1912a) by noting a consistent relationship between the mineral assemblage and the chemical composition of a wide range of rock types. Many of the assemblages were the same as at Oslo. Some rocks, however, although chemically equivalent to some at Oslo, contained a different mineral assemblage. For example, rocks that contained K-feldspar and cordierite at Oslo contained the compositionally equivalent pair biotite + muscovite at Orijärvi. Eskola concluded that the difference must reflect differing physical conditions between the regions. Using strictly qualitative thermodynamic reasoning (as we learned to do in Chapter 5), he correctly deduced that his Finnish rocks (with a more hydrous nature and lower volume assemblage) equilibrated at lower temperatures and higher pressures than Goldschmidt's Norwegian ones.

On the basis of the predictable relationship between rock composition and mineral assemblage, and the worldwide occurrence of virtually identical mineral assemblages, Eskola (1915) developed the concept of **metamorphic facies**. In his words,

> In any rock or metamorphic formation which has arrived at a chemical equilibrium through metamorphism at constant temperature and pressure conditions, the mineral composition is controlled only by the chemical composition. We are led to a general conception which the writer proposes to call metamorphic facies.

At the time Eskola made this proposal, the only ways to categorize metamorphic conditions were to use either the metamorphic zones based on isograds or the "depth zones" of Grubenmann (1904). As described in Section 4.2.5, Grubenmann's epizone, mesozone, and catazone were very broad and poorly defined. The isograd-based metamorphic zones were restricted to pelitic compositions at the time, and were too narrow for easy correlation from one locality to another, particularly if different rock types were involved. Eskola's facies were initially based on metamorphosed mafic rocks. Because basaltic rocks occur in practically all orogenic belts, and the mineral changes in them define broader *T-P* ranges than those in pelites, facies provided a convenient way to compare metamorphic areas around the world.

There is a dual basis for the facies concept. First is the purely *descriptive* basis: the relationship between the composition of a rock and its mineralogy. Eskola based his proposed facies on the mineral assemblages developed in metamorphosed mafic rocks, and this descriptive aspect was a fundamental feature of the concept. By this definition, a metamorphic facies is a set of repeatedly associated metamorphic mineral assemblages. If we find a specified assemblage in the field (or, better yet, a sequence of compatible assemblages covering a range of compositions), then a certain facies may be assigned to the area.

The second basis is *interpretive*: the range of temperature and pressure conditions represented by each facies. Although in the beginning of the century, the application of thermodynamics to geological systems was in its infancy, and the facies concept was largely descriptive, Eskola was certainly aware of the temperature–pressure implications of the concept (as is clear in the quotation above), and he correctly deduced the *relative* temperatures and pressures represented by the different facies that he proposed. He did so on the basis of the densities and volatile content of the various mineral assemblages, as described above. Since that time, advances in experimental techniques and the accumulation of experimental and thermodynamic data have allowed us to assign relatively accurate temperature and pressure limits to individual facies.

Eskola (1920) proposed five original metamorphic facies, which he named **greenschist**, **amphibolite**, **hornfels**, **sanidinite**, and **eclogite** facies. Each of these facies was easily defined on the basis of distinctive mineral assemblages that develop in mafic rocks. *The mafic assemblages are clearly reflected in the facies names, most of which correspond to characteristic metamorphic mafic rock types.* In his final account, Eskola (1939) added the **granulite**, **epidote–amphibolite**, and **glaucophane–schist** facies, and he changed the name of the hornfels facies to the **pyroxene hornfels** facies. His facies, and his estimate of their relative temperature–pressure relationships, are shown in Figure 25.1. Since then, several additional facies types have been proposed. Most notable are the **zeolite** and **prehnite–pumpellyite** facies, resulting from the work of Coombs (1954, 1960, 1961) in the "burial metamorphic" terranes of New Zealand (Section 21.6.2). Fyfe et al. (1958) also proposed the **albite–epidote hornfels** and

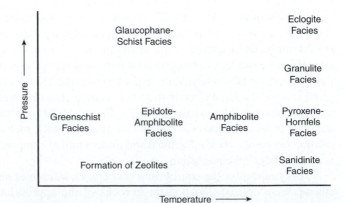

FIGURE 25.1 The metamorphic facies proposed by Eskola and their relative temperature–pressure relationships. After Eskola (1939).

hornblende hornfels facies. Numerous other facies, and even sub-facies, have been proposed, but most have been dropped in favor of simplicity and ease of correlation. When more detail is required for in-depth studies, one can still refer to more detailed subdivisions (e.g., Turner and Verhoogen, 1960; Turner, 1981; Liou and Zhang, 2002).

Now that we can assign approximate T-P limits to the facies types, we face a fundamental dilemma resulting from the dual basis behind the facies concept as we interpret facies today. Do we remain faithful to Eskola's original method of defining facies as a set of mineral assemblages and restrict our choice of facies types only to those defined by unique parageneses in mafic rocks? Or do we rely on our modern ability to assign temperature–pressure limits to facies and stress the importance of these regimes? The controversy lasts to this day and affects the number and types of facies presented in introductory petrology texts such as this one, particularly over the proposed facies of low-pressure metamorphism. The mineral assemblages that develop in mafic rocks at low pressure do not differ significantly from those of medium to high pressure. Do we thus reject the albite–epidote hornfels and hornblende hornfels facies because the mineralogy is essentially the same as in the greenschist and amphibolite facies, respectively, so they cannot be defined on the basis of unique mineral assemblages? Or do we accept them as representing important low-pressure environments?

Texts appear to fall in about equal numbers on both sides of this argument. I have decided to include the hornfels facies types (although the IUGS-SCMR does not [Fettes and Desmons, 2007], and I have left the boundaries as dashed lines). Although there may not be a significant difference in the *mafic* mineral assemblages between low-pressure and

medium-pressure facies types, in *pelitic* rocks garnet is rare in the hornfels facies types, whereas cordierite and andalusite are common. Furthermore, I see no reason why we can't modify the original concept of facies in light of modern advances. Eskola was certainly aware of the T-P implications of the various facies.

The metamorphic facies used in this text, and their generally accepted temperature and pressure limits, are shown in Figure 25.2. The boundaries between metamorphic facies represent T-P conditions in which key minerals in mafic rocks are either introduced or lost, thus changing the mineral assemblages observed (e.g., Figure 26.19). Facies are thus separated by *mineral reaction isograds*. The facies limits are approximate and gradational because the reactions vary with rock composition and the nature and composition of the intergranular fluid phase (as we shall investigate further in the ensuing chapters). The definitive mineral assemblages that characterize each facies (for mafic rocks) are listed in Table 25.1. Details and variations will be discussed in Section 25.3.

The "typical" continental geothermal gradient (Figure 1.11) represents an average gradient for stable continental interiors. The local gradient can be either higher or lower, depending on the tectonic environment. See Spear (1993) for a discussion of various geotherm models.

As Yardley (1989) pointed out, it is convenient to consider metamorphic facies in four groups:

1. Facies of high pressure. The blueschist and eclogite facies are characterized by the development of low molar volume phases and assemblages under conditions of high pressure. The lower-temperature blueschist facies occurs in areas of unusually low T/P gradients, characteristically

FIGURE 25.2 Temperature–pressure diagram showing the generally accepted limits of the various facies used in this text. Boundaries are approximate and gradational. The "typical" or average continental geotherm is from Brown and Mussett (1993). The 30°C/km geothermal gradient is an example of an elevated orogenic geothermal gradient. The stability ranges of the three Al$_2$SiO$_5$ polymorphs are from Figure 21.9, and the minimum melting curve for H$_2$O-saturated granite is from Figure 18.7. The "forbidden zone" is bounded by the 5°C/km(150°C/GPa) gradient beyond which conditions are generally considered unrealized in nature (e.g., Liou et al., 2000).

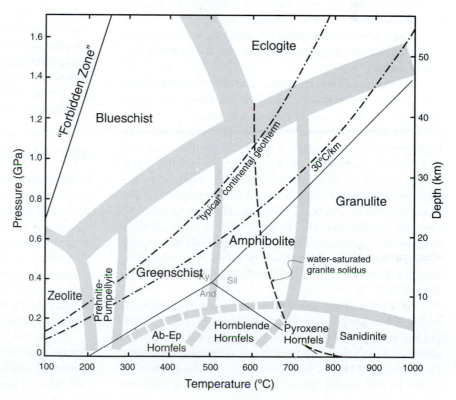

TABLE 25.1   Definitive Mineral Assemblages of Metamorphic Facies

Facies	Definitive Mineral Assemblage in Mafic Rocks
Zeolite	zeolites: especially laumontite, wairakite, analcime (in place of other Ca-Al silicates such as prehnite, pumpellyite, and epidote).
Prehnite–pumpellyite	prehnite + pumpellyite (+ chlorite + albite)
Greenschist	chlorite + albite + epidote (or zoisite) + actinolite ± quartz
Amphibolite	hornblende + plagioclase (oligoclase, andesine) ± garnet
Granulite	orthopyroxene + clinopyroxene + plagioclase ± garnet
Blueschist	glaucophane + lawsonite or epidote/zoisite (± albite ± chlorite ± garnet)
Eclogite	pyralspite garnet + omphacitic pyroxene (± kyanite ± quartz). No plagioclase.
Contact facies	mineral assemblages in mafic rocks of the facies of contact metamorphism do not differ substantially from those of the corresponding regional facies at higher pressure.

After Spear (1993), Fettes and Desmons (2007).

developed in subduction zones (see Figure 16.15). Although the term *glaucophane schist facies* has precedence over *blueschist facies*, having been used earlier, the latter term is more common. This is due to the blue-gray color of most mafic blueschists, imparted by sodic amphiboles. Eclogites occur in subduction zones, and, because they are also stable under normal geothermal conditions, they also develop wherever mafic rocks occur in the deep crust or mantle (stranded crustal magma chambers and dikes, or sub-crustal magmatic underplates). We will explore some recent discoveries of "ultra-high pressure" metamorphism in Section 25.3.4.

2. Facies of medium pressure. Most metamorphic rocks now exposed at the surface of the Earth represent metamorphism in the greenschist, amphibolite, or granulite facies. As you can see in Figure 25.2, the greenschist and amphibolite facies conform to the "typical" continental geothermal gradient. All three facies also conform to elevated geotherms typical of most orogenic regions, such as the 30°C/km geotherm illustrated. Granulite facies rocks occur predominantly in deeply eroded continental cratons of Precambrian age.

3. Facies of low pressure. The albite–epidote hornfels, hornblende hornfels, and pyroxene hornfels facies are typically developed in contact metamorphic terranes, although they can also occur in regional terranes with very high geothermal gradients. The sanidinite facies is rare and limited to xenoliths in basic magmas and the innermost portions of some contact aureoles adjacent to hot basic or anorthosite intrusives (pyrometamorphism). Because intrusions can stall at practically any depth, and orogenic geotherms are diverse, there is considerable overlap and gradation in terms of genesis and mineralogy between the low pressure facies types and the corresponding medium-pressure facies immediately above them on Figure 25.2.

4. Facies of low grades. As discussed in Chapter 21, rocks commonly fail to recrystallize thoroughly at very low grades, and equilibrium mineral assemblages do not always develop. As a result, the zeolite and prehnite–pumpellyite facies may not always be represented, and the greenschist facies is the lowest grade developed in many regional

terranes. The zeolite and prehnite–pumpellyite facies are thus treated separately. These facies are best developed where the protolith is immature and susceptible to metamorphism and where there are a high geothermal gradient and abundant hydrous fluids. As a result, they are most common in areas of burial or hydrothermal metamorphism affecting immature volcanics. Due to the poorly developed mineral assemblages and the complex dependency of these two facies types on mineral compositions, many petrologists have given up trying to distinguish between them and have taken to using the term **sub-greenschist facies** to refer to them collectively.

When describing or mapping metamorphic rocks, we tend to combine the concept of facies with those of isograds and zones, discussed in Section 21.6. For example, we may speak of the "chlorite zone of the greenschist facies," the "staurolite zone of the amphibolite facies," the "cordierite zone of the hornblende hornfels facies," etc. Maps of metamorphic terranes generally include isograds that define zones and ones that define facies boundaries. Rarely can one characterize a facies or zone on the basis of a single rock, and it is most reliably done when several rocks of varying composition and mineralogy are available.

25.2 FACIES SERIES

A traverse across a typical metamorphic terrane typically reveals a sequence of zones and facies that developed in response to gradients in temperature and pressure. This led Miyashiro (1961, 1973) to extend the facies concept to encompass progressive sequences and propose **facies series**. According to the facies series concept, any large-scale traverse up grade through a metamorphic terrane should follow one of several possible metamorphic field gradients (Figure 21.1), and, if extensive enough, cross through a sequence of facies.

The contrast between Miyashiro's higher-*T*-lower-*P* Ryoke-Abukuma belt (Section 21.6.3) and the classical Barrovian sequence also led him to suggest that more than one type of facies series was probable. Miyashiro (1961) initially proposed five facies series, most of them named for a specific representative "type locality." The series were:

1. Contact facies series (very low P)
2. Buchan or Abukuma facies series (low P regional)
3. Barrovian facies series (medium P regional)
4. Sanbagawa facies series (high P, moderate T)
5. Franciscan facies series (high P, low T)

As indicated in Figure 21.1, metamorphic field gradients are highly variable, even in the same orogenic belt, and transitional series abound. The specific nature of the series proposed above could thus lead to a proliferation of series types, and Miyashiro (1973, 1994) chose to limit the number instead to the three broad major types of facies series (also called "baric types") illustrated in Figure 25.3. Although the boundaries and gaps between them are somewhat artificial, the three series represent common types of tectonic environments.

The **high P/T series**, for example, typically occurs in subduction zones where "normal" isotherms are depressed by the subduction of cool lithosphere faster than it can equilibrate thermally (see Figures 16.15 and 16.16). The facies sequence here is (zeolite facies)–(prehnite–pumpellyite facies)–blueschist facies–eclogite facies. Facies in parentheses may not be developed. The **medium P/T series** is characteristic of common orogenic belts (Barrovian type), where the sequence is (zeolite facies)–(prehnite–pumpellyite facies)–greenschist facies–amphibolite facies–(granulite facies). As we learned in Section 18.4, crustal melting under H_2O-saturated conditions occurs in the upper amphibolite facies (the H_2O-saturated granite solidus is indicated in Figure 25.2). The granulite facies, therefore, occurs only in H_2O-deficient rocks, either dehydrated lower crust or areas with high X_{CO_2} in the intergranular fluid. The **low P/T series** is characteristic of high heat-flow orogenic belts (Buchan or Ryoke-Abukuma type), rift areas, or contact metamorphism. The sequence of facies may be a low-pressure version of the medium P/T series described above (but with cordierite and/or andalusite in aluminous rocks) or the sequence (zeolite facies)–albite–epidote hornfels facies–hornblende hornfels facies–pyroxene hornfels facies. Sanidinite facies rocks are rare and generally localized, requiring the introduction of great heat to very shallow levels.

Zones, facies, and facies series provide a convenient way to describe and compare metamorphic rocks from different areas and to categorize metamorphic belts. We can thus speak of greenschist facies rocks in general, or greenschist facies areas in the Appalachians, Alps, Caledonides, etc. Or we can compare Barrovian-type belts to Buchan-type belts or to high P/T blueschist facies belts, and so on.

Figure 25.4 is an attempt by Ernst (1976) to relate facies and facies series to plate tectonics in a typical subduction-related orogenic arc setting. Notice the development of blueschist and eclogite facies rocks in the high P/T subduction zone complex and the higher-temperature inboard belt where amphibolite and granulite facies rocks are generated. Recognizing this, Ernst (1971) argued, the dip of ancient subduction zones should be in the direction of increasing

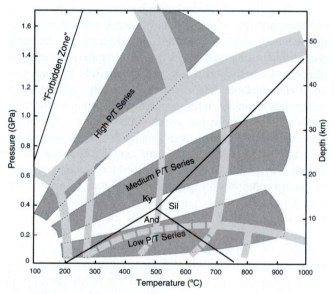

FIGURE 25.3 Temperature–pressure diagram showing the three major types of metamorphic facies series proposed by Miyashiro (1973, 1994). Included are the facies boundaries from Figure 25.2 and the Al_2SiO_5 phase diagram from Figure 21.9. Each series is representative of a different type of tectonic environment. The high P/T series is characteristic of subduction zones, medium P/T of continental regional metamorphism, and low P/T of high-heat-flow orogens, rift areas, and contact metamorphism.

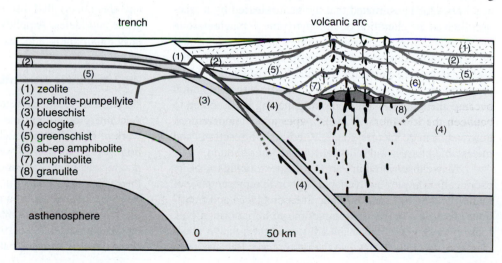

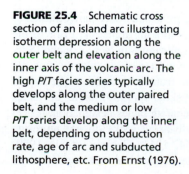

FIGURE 25.4 Schematic cross section of an island arc illustrating isotherm depression along the outer belt and elevation along the inner axis of the volcanic arc. The high P/T facies series typically develops along the outer paired belt, and the medium or low P/T series develop along the inner belt, depending on subduction rate, age of arc and subducted lithosphere, etc. From Ernst (1976).

(1) zeolite
(2) prehnite-pumpellyite
(3) blueschist
(4) eclogite
(5) greenschist
(6) ab-ep amphibolite
(7) amphibolite
(8) granulite

metamorphic grade. Such a generalized diagram provides a useful framework for conceptualizing the development of metamorphic facies in one common setting, but nature is far more complex and variable, and metamorphic rocks develop in a wider variety of situations (as described in Chapter 21). Variations in subduction rate, age of subducted crust, subduction of LIPs or ridges, etc. may have profound influences on thermal regime and facies development. In addition, the facies ranges in Figure 25.4 indicate where the conditions are appropriate for the development of the particular facies, but not necessarily where rocks of a facies may still be found metastably. For example, the lower oceanic crust may be metamorphosed to amphibolite facies near mid-ocean ridges and retain such mineral assemblages to the subduction zone, even if in the stability area of greenschist facies during orogeny (as indicated in Figure 25.4).

25.3 METAMORPHISM OF MAFIC ROCKS

Because Eskola defined metamorphic facies on the basis of metamorphosed mafic rocks (also called "metabasites"), we will focus first on these rock types as we look in more depth at the types of mineral changes and associations that develop with increasing metamorphic grade along *T-P* gradients characteristic of the three facies series shown in Figure 25.3. We will look at other rock types, such as pelites, calcareous rocks, and ultramafics, in Chapters 28 and 29. In the present context, mafic rocks include volcanics from basaltic to andesitic composition, gabbros and diorites, and immature mafic graywackes. Metamorphosed volcanics are by far the most common, and the volcanic parent rocks occur in a wide variety of tectonic settings, from oceanic crust to volcanic arcs to rifts.

Hydrous minerals are not common in high-temperature mafic igneous rocks, so *hydration* is a prerequisite for the development of the metamorphic mineral assemblages that characterize most facies. Unless H_2O is available, mafic igneous rocks will remain largely unaffected in metamorphic terranes, even as associated sediments are completely re-equilibrated. Coarse-grained intrusives are the least permeable and thus most likely to resist metamorphic changes, whereas tuffs and graywackes are the most susceptible.

Hydration reactions are strongly exothermic. Bucher and Frey (2002) estimated that the heat released by a reaction such as the alteration of clinopyroxene + plagioclase in basalt to prehnite + chlorite + zeolites at low grades could raise the temperature of a metamorphic rock by a further 100°C, if it were perfectly insulated. If H_2O is available, these low-grade types of reactions tend to run to completion because the reactants are metastable, having been cooled to well below their stability ranges. At very low temperatures, however, kinetic factors generally retard the reactions and metastable phases, and relict igneous textures abound.

As we discovered in Part I of this text, mafic rocks are chemically complex and variable. Although representative of mafic rocks in general, the mineral assemblages and transitions discussed below are susceptible to variations in bulk rock composition, fluid availability, and the relative abundances of H_2O versus CO_2 in the fluid phase. Because most

reactions are continuous reactions (Section 26.5), the grade at which a particular mineral appears or disappears during prograde metamorphism may vary from one rock or location to another. Many of the reactions that govern petrogenesis of metamorphosed mafic rocks are illustrated in Figure 26.19 (in simplified form). You may want to refer to this figure from time to time in the following discussion. Bucher and Frey (2002) have a more detailed analysis of the reactions governing metamorphism of mafic rocks in their Chapter 9.

Most of the common minerals in mafic rocks exhibit extensive solid solution. Because these minerals can accommodate a lot of chemical diversity, metabasites tend to have fewer phases than pelites. This results in fewer reactions and isograds. This is one of the reasons that Eskola chose to define his facies on these rocks because it resulted in broader subdivisions and made it easier to categorize and to correlate different areas.

The principal mineral changes in metabasalts are due to the breakdown of the two most common basalt minerals: plagioclase and clinopyroxene. As temperature is lowered, the more Ca-rich plagioclases (the high-temperature igneous plagioclases in the Ab-An system, Figure 6.8) become progressively unstable. There is thus a general correlation between temperature and the maximum An-content of the stable plagioclase. At low metamorphic grades, only albite (An_{0-3}) is stable. In the upper-greenschist facies, oligoclase becomes stable. The "peristerite" solvus spans approximately An_{7-17}, and few plagioclase compositions are found within this composition gap. The An content of plagioclase thus jumps from An_{1-7} to An_{17-20} (oligoclase) as grade increases. Andesine and more calcic plagioclases become stable in the upper amphibolite and granulite facies. The composition of plagioclase can be determined using the microprobe, or it can be estimated fairly accurately petrographically using the extinction angle to the (010) cleavage in crystals having their *a*-axis oriented normal to the thin section (see Hibbard, 1995). The excess Ca and Al released when a more An-rich igneous plagioclase breaks down to albite or oligoclase may produce calcite, an epidote mineral, titanite, or amphibole, etc., depending on the grade and rock composition. Clinopyroxene breaks down to a number of mafic minerals, depending on grade. These minerals include chlorite, amphibole, epidote, a metamorphic pyroxene, etc., and the one(s) that form are generally diagnostic of the grade and facies. Amphibole compositions may be particularly important as a facies indicator for some grades.

25.3.1 Mafic Assemblages at Low Grades

Very-low-grade metabasites are usually only partly altered to very-fine-grained and messy-looking minerals. Prior to the work of D. S. Coombs and colleagues in New Zealand (Section 21.6.2), such rocks were written off as "altered" and did not receive much attention from petrologists. The words *formation of zeolites* in Eskola's (1939) table of facies (Figure 25.1), in lieu of a separate facies designation, reflect his impression that zeolite-bearing rocks do not represent equilibrium. Coombs (1954, 1961; Coombs et al., 1959) recognized systematic patterns in these rocks and showed that they could be studied in the same way as other metamorphic rocks. This led to the acceptance of the zeolite and prehnite–pumpellyite facies. The rocks in New Zealand are predominantly volcanic tuffs and graywackes and hence susceptible to metamorphism, even at low grade. The area is only slightly deformed, and the metamorphic grade is related to depth of burial (leading Coombs to coin the term *burial metamorphism*, as described in Section 21.3.2). Boles and Coombs (1975) showed that metamorphism of the tuffs was accompanied by substantial changes in bulk composition due to circulating fluids, and that these fluids played an important role in determining which metamorphic minerals were stable. Thus the classic area of burial metamorphism has a strong component of hydrothermal metamorphism as well.

In New Zealand, at the lowest grades of the **zeolite facies**, volcanic glass is altered to the zeolites heulandite or stilbite (less commonly to analcime), along with phyllosilicates such as celadonite, smectite, kaolinite, or montmorillonite, plus secondary quartz and carbonate. Crystalline igneous minerals remain essentially intact. At slightly greater depths, albite replaces more calcic detrital igneous plagioclase, chlorite appears, and heulandite is replaced by laumontite (and analcime by albite). Because stilbite, heulandites, and analcime are stable under diagenetic conditions, some authors consider the stabilization of laumontite to mark the low T limit of metamorphism. Wairakite is another zeolite that may occur, generally becoming stable at a slightly higher grade than laumontite. The first appearance of laumontite (the laumontite isograd) occurs before heulandite or analcime are totally lost, and these zeolites may coexist over an interval.

Although prehnite and pumpellyite are stable in the upper zeolite facies, eventually laumontite is lost with increasing grade, and prehnite + pumpellyite + quartz becomes stable (generally along with albite, chlorite, phengite, and titanite). This transition marks the beginning of the **prehnite–pumpellyite facies** at a temperature of about 200°C and a depth of 3 to 13 km (depending on composition and geothermal gradient). Minor epidote is also found. Other minerals in the prehnite–pumpellyite facies include lawsonite (above ~0.3 GPa) and stilpnomelane. Metamorphic minerals in both the zeolite and prehnite–pumpellyite facies are generally concentrated in veins, amygdules, cavity fillings, and the fine groundmass (see Plates 39, 40, and 57 of Yardley et al., 1990).

Prehnite and pumpellyite disappear at higher grades. Prehnite generally disappears first, giving way to actinolite.

Some investigators (Hashimoto, 1966; Turner, 1981) have separated a **pumpellyite–actinolite facies** as a result, but most regard this as part of the prehnite–pumpellyite facies. Several investigators in low-grade terranes around the world have recognized and defined various zones within the zeolite and prehnite–pumpellyite facies (e.g., Seki et al., 1969). These zones generally differ from one area to the next, due to compositional or physical variations, and are based on either texture or mineralogy (usually the type of zeolite or phyllosilicate). For example, the zeolite facies in New Zealand can be subdivided into a shallower heulandite– analcime zone (or sub-facies) and a deeper laumontite–albite zone. Other zeolites, such as stilbite, mordenite, wairakite, etc., are reported elsewhere. Such zones represent a more detailed degree of subdivision than is appropriate for the present survey.

ACF diagrams with representative mineral assemblages that develop in metamorphosed mafic rocks in the zeolite and prehnite–pumpellyite facies are shown in Figure 25.5. For example, most mafic rock compositions (shaded area) in the zeolite facies develop a mineral assemblage such as chlorite + heulandite (or laumontite) + calcite + quartz + albite. Kaolinite develops in more aluminous rocks. The transition from zeolite to prehnite–pumpellyite facies is marked by the loss of huelandite and laumontite, and the development of prehnite, pumpellyite, and (at a higher grade) actinolite. The crossing tie-lines in Figure 25.5b are a result of combining both the lower-grade prehnite-bearing zone with the higher-grade pumpellyite–actinolite zone on the same diagram. Thus we might expect chlorite + prehnite + calcite in the lower prehnite–pumpellyite facies, and pumpellyite + actinolite + chlorite in the upper part.

Pumpellyite disappears as rocks enter the greenschist facies, where it is replaced by actinolite + epidote at ~270 to 300°C (see Figure 26.19). As discussed in Section 21.6.2, the low-grade (sub-greenschist) facies are not always realized in many occurrences of metabasites, where rocks proceed directly to the greenschist facies, with no lower-grade minerals developing. This may be due to lack of fluids (or of a circulating hydrothermal system), poor attainment of equilibrium at low grades, or, as Zen (1961) suggested, the presence of CO_2 (see Section 21.6.2).

Figure 25.6 summarizes the mineral changes that typically occur in metabasites in the zeolite, prehnite–pumpellyite, and lowest greenschist facies. Different mineral stability ranges characterize the rocks throughout various hydrothermal fields

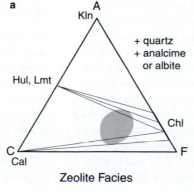

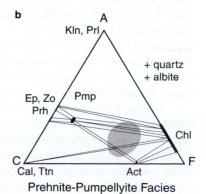

FIGURE 25.5 *ACF* compatibility diagrams illustrating representative mineral assemblages for metabasites in the **(a) zeolite** and **(b) prehnite–pumpellyite facies.** Actinolite is stable only in the *upper* prehnite–pumpellyite facies. The composition range of common mafic rocks is shaded. See inside cover for mineral abbreviations.

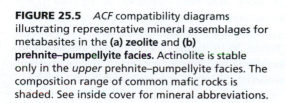

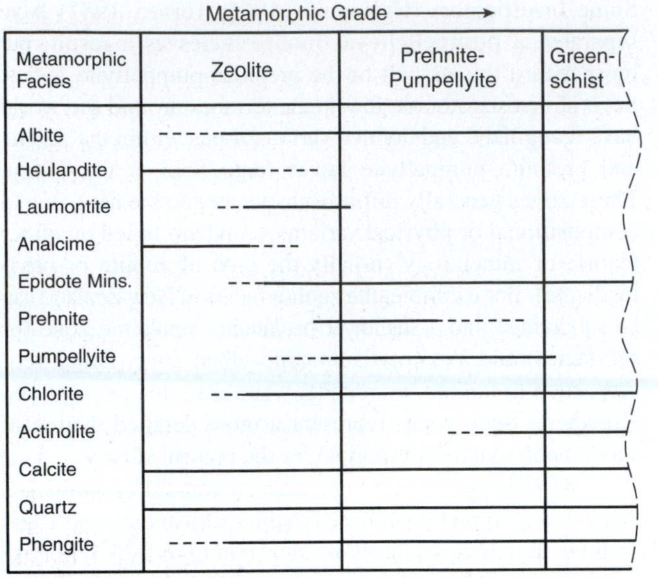

Metamorphic Grade ⟶

Metamorphic Facies	Zeolite	Prehnite–Pumpellyite	Green-schist
Albite	– – – –		
Heulandite	– – – –		
Laumontite	– –		
Analcime	– – –		
Epidote Mins.	– –		
Prehnite		– – – –	
Pumpellyite		– – – –	
Chlorite	– –		
Actinolite		– – –	
Calcite			
Quartz			
Phengite	– – – –		

FIGURE 25.6 Typical mineral changes that take place in metabasic rocks during progressive metamorphism in the zeolite, prehnite–pumpellyite, and incipient greenschist facies.

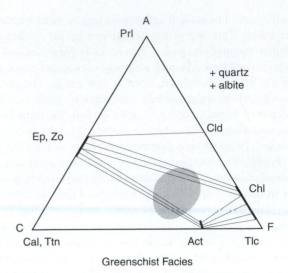

Greenschist Facies

FIGURE 25.7 *ACF* compatibility diagram illustrating representative mineral assemblages for metabasites in the **greenschist facies.** The composition range of common mafic rocks is shaded.

and burial metamorphic terranes as a result of variations in composition, fluids, and local geothermal gradient. See Chapter 9 in Bucher and Frey (2002) for some other examples.

25.3.2 Mafic Assemblages of the Medium *P/T* Series: Greenschist, Amphibolite, and Granulite Facies

The greenschist, amphibolite and granulite facies constitute the most common facies series of regional metamorphism. Both the classical Barrovian series of pelitic zones and the lower-pressure Buchan-Ryoke-Abukuma series are *P/T* variations on this trend (see Figure 25.3). We return to the Barrovian area of the Scottish Highlands (described in Section 21.6.1) and discuss the progressive mineral changes that occur in mafic rocks across the terrane.

The zeolite and prehnite–pumpellyite facies are not represented in the Scottish Highlands, and metamorphism of mafic rocks is first evident in the **greenschist facies**, which correlates with the chlorite and biotite zones of the associated pelitic rocks. Although relict igneous textures may be preserved, the mineralogy reflects re-equilibration to metamorphic conditions. Typical metabasic minerals include chlorite, albite, actinolite, epidote, quartz, and, in some cases, calcite, phengite or biotite, or stilpnomelane. The predominance of chlorite, actinolite, and epidote impart the green color from which the mafic rocks and the facies get their name. Although some schists that are not of greenschist facies may be green, the greenschist (and greenstone) rock names are most appropriately applied to rocks in this facies.

Figure 25.7 is an *ACF* diagram representing typical metamorphic mineral assemblages in metabasites of the greenschist facies. The most characteristic mineral assemblage is chlorite + albite + epidote + actinolite ± quartz. Chloritoid, you may remember from Section 21.6.1, is not recognized in the classical Barrovian sequence of the Scottish

Highlands but is widespread in the Appalachians, where Fe-Al rich pelitic rocks are more common. A sericitic or phengitic white mica may be present in K-bearing mafics at lower grades in the greenschist facies, giving way to biotite toward higher grades.

The transition from greenschist to amphibolite facies in mafic rocks involves *two* major mineralogical changes. The first is the transition from albite to oligoclase (increased Ca content of stable plagioclase with temperature across the peristerite gap). The second is the transition from actinolite to hornblende as amphibole becomes able to accept increasing amounts of aluminum and alkalis at higher temperatures. Both of these transitions occur at approximately the same grade (near 500°C). The reactions that generate calcic plagioclase and hornblende are complex and continuous, involving the breakdown of epidote and chlorite to supply the Ca and Al for the anorthite and Al-hornblende components (see Cooper, 1972; Liou et al., 1974; Maruyama et al., 1983). One such generalized reaction proposed by Liou et al. (1974) is:

$$Ab + Ep + Chl + Qtz = oligoclase \qquad (25.2)$$
$$+ \; tschermakite + Mt + H_2O$$

Tschermakite, ideally $Ca_2(Mg)_3Al_2Si_6Al_2O_{22}(OH)_2$, is an Al-rich component in the amphibole and not a separate phase, representing an $(Al^{3+})^{VI} + (Al^{3+})^{IV} \Longleftrightarrow (Mg^{2+}, Fe^{2+})^{VI} + (Si^{4+})^{IV}$ substitution. (Roman numerals indicate coordination numbers.)

The transition to more calcic plagioclase consumes epidote, and, as pressure increases, the temperature of the calcic plagioclase reaction(s) appears to increase more than the temperature of the reaction(s) responsible for the actinolite–hornblende transition (note the steeper slope of the former in Figure 26.19. In the higher-pressure Barrovian sequence, therefore, hornblende appears before oligoclase, and a transitional **(albite-) epidote-amphibolite facies**—proposed by several investigators, including Eskola,

Figure 25.1, and recommended by the IUGS-SCMR (Fettes and Desmons, 2007)—may be useful in higher-*P/T* terranes. In low *P/T* terranes and in contact aureoles, the plagioclase transition occurs first, resulting in a transitional zone with co-existing actinolite + calcic plagioclase (as calcic as labradorite). The transition to amphibolite facies usually occurs in the garnet zone of associated pelitic rocks (occasionally in the upper biotite zone).

Metabasites in the **amphibolite facies** are characterized by the assemblage hornblende + plagioclase ($An_{>17}$), with lesser amounts of garnet, clinopyroxene, quartz, and/or biotite (perhaps with some chlorite or epidote in the lower-grade portions of the facies). The amphibolite facies corresponds generally to the upper garnet, staurolite, kyanite, and lower sillimanite zones in associated pelitic rocks. Figure 25.8 is an *ACF* diagram illustrating the typical assemblages. Because of hornblende solid solution, much of the compositional area for mafic rocks lies in the two-phase Hbl-Plag field. Most amphibolites in hand specimen are thus predominantly black rocks with variable amounts of white plagioclase. An astute student might notice that the shaded field of common mafic compositions is close to hornblende, and that applying the lever principle suggests that the ratio of Hbl to Plg should range from 1.0 to about 0.7. Amphibolites may have more plagioclase than this because *ACF* diagrams, such as Figure 25.8, are projected from albite (a component of plagioclase). If the albite component is appreciable, the amount of plagioclase would be underrepresented on the diagrams.

In many cases, hornblende becomes more brown and less green in thin section as grade increases. The brown color is attributed to higher Ti content. Pyralspite garnet occurs in the more Al-Fe-rich and Ca-poor mafic rocks and clinopyroxene in the Al-poor/Ca-rich ones, adding bits of burgundy red or green to the hand specimen. Cummingtonite also occurs in Ca-Al-poor metabasites.

Some distinctive and unusual cordierite–anthophyllite rocks may be present in amphibolite facies metamorphism. Their high-Mg/low-Ca composition is unlike any other known sedimentary or igneous rock In Section 21.3.2, we

attributed the composition to "ocean-floor metamorphism," in which the basalts of the oceanic crust were hydrothermally altered by hot recirculating brines in the vicinity of mid-ocean ridges (Vallance, 1967). Although cordierite is absent in Figure 25.8, it occurs in many amphibolite facies rocks, especially at lower pressure and higher temperature.

The transition from amphibolite to granulite facies occurs in the range 650 to 850°C. In the presence of an aqueous fluid, associated pelitic and quartzo-feldspathic rocks (including granitoids) begin to melt in this range at low to medium pressures (see Section 18.3 and the solidus in Figure 25.2). Migmatites may form and the melts may become mobilized. As a result, not all pelites and quartzo-feldspathic rocks reach the granulite facies prior to melting. H_2O is typically removed by the partial melts, however, and the remaining rocks may become depleted in H_2O, developing granulite facies mineral assemblages. Mafic rocks generally melt at somewhat higher temperature, and granulite facies mineral assemblages are common at high metamorphic grades. Hornblende, typically the last remaining major hydrous phase, decomposes and orthopyroxene + clinopyroxene appear. This reaction occurs over a temperature interval of at least 100°C (Figure 26.19).

The **granulite facies** is characterized by the presence of a largely anhydrous mineral assemblage. In metabasites the critical mineral assemblage is orthopyroxene + clinopyroxene + plagioclase + quartz. Garnet is also common, and minor hornblende and/or biotite may be present. In pelitic and quartzo-feldspathic rocks the granulite facies is characterized by the dehydration and absence of muscovite, and the presence of sillimanite, cordierite, garnet, quartz, K-feldspar, and orthopyroxene (Chapter 28). Hornblende may remain stable in metabasites in the lower granulite facies, but it dehydrates in the upper granulite facies. Note that a basalt that has been progressively metamorphosed from the greenschist to the granulite facies has returned to a mineralogy dominated by plagioclase and pyroxene, just as in the original basalt. The *texture*, however, is generally gneissic and grain shapes are granular or polygonal, thus bearing no resemblance to the original basalt. Figure 25.9 is an *ACF* diagram for typical mineralogy in granulite facies metabasites.

Miyashiro (1994) noted that, although single-pyroxene metabasites can occur in the granulite facies, they are not diagnostic and cannot be used to identify the facies. Clinopyroxene is common in low-Al/high-Ca mafic rocks in the amphibolite facies, as discussed above (see Figure 25.8). In high Mg-Fe metabasites, cummingtonite may break down to orthopyroxene at lower temperatures than hornblende breaks down, resulting in orthopyroxene-bearing amphibolites.

The origin of granulite facies rocks is complex and controversial. There is general agreement, however, on two points:

1. Granulites represent unusually hot crustal conditions. Granulite facies rocks represent temperatures in excess of 700°C, and geothermometry by several investigators, including myself, has yielded very high temperatures, some in excess of 1000°C. Average geotherm temperatures for granulite facies depths should be in

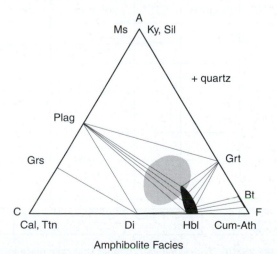

FIGURE 25.8 *ACF* compatibility diagram illustrating representative mineral assemblages for metabasites in the **amphibolite facies**. The composition range of common mafic rocks is shaded.

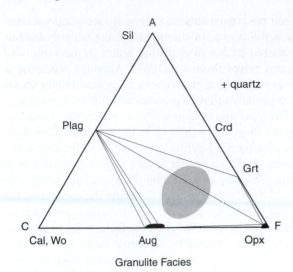

FIGURE 25.9 *ACF* compatibility diagram illustrating representative mineral assemblages for metabasites in the **granulite facies.** The composition range of common mafic rocks is shaded.

the vicinity of 500°C, suggesting that granulites are the products of crustal thickening and excess heating (Section 25.4).

2. Granulites are dry. The only reason these rocks didn't melt on a large scale was the lack of available H_2O.

The granulite facies encompasses a wide range of pressure, giving way to the eclogite facies when the pressure reaches the point that plagioclase is no longer stable. Metabasites in low pressure granulites lack garnet, whereas very-high-pressure granulites may lose orthopyroxene by reaction with plagioclase such as:

$$4(Mg, Fe)SiO_3 + CaAl_2Si_2O_8 \qquad (25.3)$$
$$\text{Opx} \qquad\qquad \text{Plag}$$
$$= (Mg, Fe)_3Al_2Si_3O_{12} + Ca(Mg, Fe)Si_2O_6 + SiO_2$$
$$\text{Grt} \qquad\qquad \text{Cpx} \qquad \text{Qtz}$$

This reaction involves only solid phases and has a large ΔV associated with it. It is therefore pressure sensitive [low slope on a *P-T* phase diagram, as indicated by the Clapeyron equation, Equation (5.15)]. The reactant side is thus stable at low pressure, and the product side at high pressure. The reaction occurs at pressures higher than those in Figure 26.19 (~1.3 GPa at 800°C).

It is a long-held tenet that granulite facies terranes represent the deeply buried and dehydrated roots of the deep continental crust. Indeed, most exposed granulite facies rocks are found in the deeply eroded areas of the Precambrian continental shields. In his extensive review, however, Harley (1989) demonstrated that granulites are highly diverse, and Bohlen and Metzger (1989) showed that many have mid-crustal, not deep-crustal, origins. Touret (1971, 1985) showed that the fluid inclusions in granulite facies rocks of southern Norway are CO_2 rich, whereas those in the associated amphibolite facies rocks are more H_2O rich. This may imply that dehydration of at least some granulite facies terranes may be due to the infiltration of CO_2 replacing H_2O, rather than the

elimination of fluids altogether (Santosh and Omori, 2008). We will discuss this idea further in Section 29.1.3. Alternatively, H_2O may have been extracted by hydrous partial melts.

The bulk composition of typical deep-crustal granulite facies rocks differs from that of corresponding amphibolite facies as well. Granulite facies rocks tend to be depleted in LILs and other incompatible elements. Whether this is caused by depletion by fluids, melts, or initial compositional differences is the subject of spirited debate (see Sørensen and Winter, 1989, for a brief review).

Figure 25.10 summarizes the typical mineralogical changes occurring in metabasites in medium-pressure greenschist to granulite facies series. The positions of the pelitic zones of Barrovian metamorphism are included for comparison. Once again, the point at which a given mineral transformation takes place is subject to chemical and fluid variations, so the ranges of the minerals (and the zones) should be considered approximate. We shall address these considerations more fully in Chapters 28 and 29.

25.3.3 Mafic Assemblages of the Low *P/T* Series: Albite–Epidote Hornfels, Hornblende Hornfels, Pyroxene Hornfels, and Sanidinite Facies

As mentioned above, the mineralogy of metabasites in the low-pressure facies is not appreciably different from that in the medium-pressure facies series just addressed. As Figure 25.2 shows, the albite–epidote hornfels facies correlates with the greenschist facies into which it grades with increasing pressure. Similarly, the hornblende hornfels facies correlates with the amphibolite facies, and the pyroxene hornfels correlates with the granulite facies (although olivine is stable with plagioclase in the pyroxene hornfels facies but not in the granulite facies). The facies of contact metamorphism can be distinguished from those of medium-pressure regional metamorphism much better in metapelites (in which andalusite and cordierite develop).

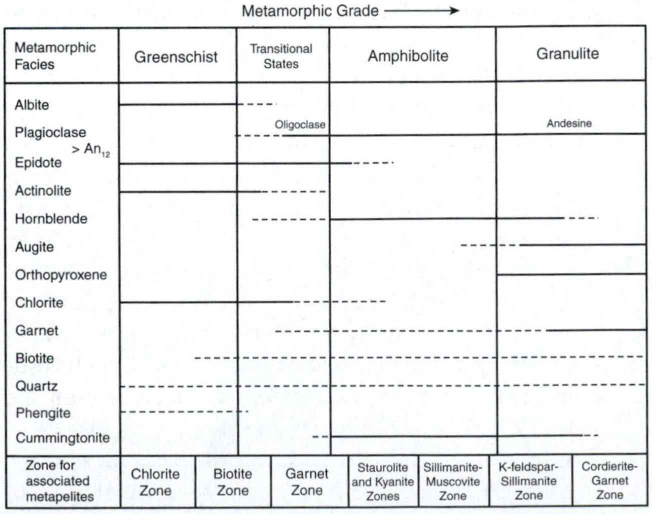

	Metamorphic Grade ⟶			
Metamorphic Facies	Greenschist	Transitional States	Amphibolite	Granulite
Albite		- - - -		
Plagioclase > An₁₂		Oligoclase		Andesine
Epidote			- - - - -	
Actinolite		- - - - - - -		
Hornblende		- - - - - - - -		- - - - - -
Augite			- - - -	- -
Orthopyroxene				
Chlorite			- - - -	
Garnet			- - - - - -	
Biotite	- - - - - - - - - - - - -			- - - - - -
Quartz	- -			- - - - - -
Phengite	- - - - - - - - -			
Cummingtonite		- -		
Zone for associated metapelites	Chlorite Zone Biotite Zone Garnet Zone		Staurolite and Kyanite Zones Sillimanite-Muscovite Zone	K-feldspar-Sillimanite Zone Cordierite-Garnet Zone

FIGURE 25.10 Typical mineral changes that take place in metabasic rocks during progressive metamorphism in the medium *P/T* facies series. The approximate location of the pelitic zones of Barrovian metamorphism are included for comparison.

At low pressure, the albite–oligoclase transition occurs before the actinolite–hornblende transition (Figure 26.19), and it marks the end of the albite–epidote hornfels facies. A transitional zone (the "actinolite–calcic plagioclase zone") typically separates it from the hornblende hornfels facies in contact aureoles. Pyralspite garnet, being a dense phase and thus favored at higher pressures, is rare or absent in the hornfels facies. Ca-poor amphiboles, such as cummingtonite, are more widespread at lower pressure. The innermost zone of granitic aureoles rarely reaches the pyroxene hornfels facies. When the intrusion is hot enough and dry enough, however, a narrow zone develops in which amphiboles break down to orthopyroxene + clinopyroxene + plagioclase + quartz (without garnet), characterizing this facies. Sanidinite facies is not evident in basic rocks and will be discussed in Chapters 28 and 29, when we cover other rock types.

25.3.4 Mafic Assemblages of the High *P/T* Series: Blueschist and Eclogite Facies

Metabasites contribute greatly to our understanding of metamorphism at high pressure, particularly of high *P/T* subduction zone metamorphism (Chapter 16). In contrast to low *P/T* metamorphism discussed in the previous section, it is the mafic rocks, and not so much the pelites, that develop conspicuous and definitive mineral assemblages under high *P/T* conditions. Most mafic blueschists are easily recognizable by their color, and are useful indicators of ancient subduction zones. Precambrian blueschists are rare. This has been interpreted to indicate that either the Precambrian geothermal gradient was too high or that most have been overprinted by later metamorphism or eroded away. The great density of eclogites suggests that subducted basaltic oceanic crust becomes more dense than the surrounding mantle. The implications of this on subduction processes were discussed in Chapter 16, and on mantle dynamics and isotopes in Chapter 14.

In some areas the high *P/T* facies series begins with rocks in the zeolite facies, whereas in others the lowest grade observed is in the blueschist facies. **Zeolite facies** rocks in high *P/T* series are not significantly different from those in the medium *P/T* series because the pressure is not significantly higher. (The *P-T* paths of the series types must all converge toward the surface.) Between the zeolite facies and blueschist facies a transitional zone may occur, characterized by the high-pressure phase lawsonite, although the grade is not yet high enough to produce glaucophane or jadeite (Coombs, 1960). You can see in Figure 26.19 that the prograde lawsonite-forming reactions are all pressure sensitive between 0.3 to 0.6 GPa and below 400°C. Some investigators have proposed a separate facies for this transition, the **lawsonite–albite–chlorite facies** (see Miyashiro, 1994).

Alternative paths to the blueschist facies depend on the local geothermal gradient and may follow paths such as zeolite facies → prehnite–pumpellyite facies → blueschist facies, or (zeolite facies) → (prehnite–pumpellyite facies) → greenschist facies → blueschist facies. You can imagine appropriate geotherms for each of these series on Figure 25.2.

The **blueschist facies** is characterized in metabasites by the presence of a sodic blue amphibole (notably high in glaucophane component), stable only at high pressures. The association glaucophane + lawsonite is diagnostic. Several reactions can generate glaucophane and define the transition to the blueschist facies. Examples are:

$$Tr + Chl + Ab = Gln + Ep + H_2O \text{ or} \quad (25.4)$$
$$Tr + Chl + Ab = Gln + Lws$$

and:

$$Pmp + Chl + Ab = Gln + Ep + H_2O \quad (25.5)$$

All are pressure sensitive and are shown in Figure 26.19.

Albite breaks down at high pressure by reaction to jadeitic pyroxene + quartz:

$$\underset{\text{Ab}}{NaAlSi_3O_8} = \underset{\text{Jd}}{NaAlSi_2O_6} + \underset{\text{Qtz}}{SiO_2} \quad (25.6)$$

The assemblage jadeite + quartz indicates high-pressure blueschist facies. If we apply the principle we learned recently about reactions, we note that jadeite *without quartz* is stable into the albite field at lower pressure. A low-pressure blueschist facies can be distinguished by the presence of jadeite without quartz, but this assemblage is more typical of ultramafic than of mafic rocks. Low-pressure metabasites are distinguished by glaucophane + epidote without jadeite + quartz. The Sanbagawa belt of Japan (Section 21.6.3) represents the low-pressure type of blueschist facies, whereas the high-pressure type occurs in the Franciscan Formation of western California, New Caledonia, Sifnos and Syros (Greece), and Sesia (western Alps).

Lawsonite is stable down to 0.3 GPa and 200°C (into the zeolite and prehnite–pumpellyite facies) but remains stable throughout most of the blueschist facies and even into the eclogite facies (e.g., Schmidt, 1995; Okamoto and Maruyama, 1999). Aragonite also replaces calcite as the stable polymorph of $CaCO_3$ (Figure 26.1). Paragonite is common, even in some metabasites. Chlorite, titanite, stilpnomelane, quartz, albite, sericite, and pumpellyite are also common. Among the less common minerals, carpholite and chloritoid occurs in pelitic rocks of the blueschist facies, and the hydrous iron silicates deerite, howieite, and zussmanite occur in Fe-rich metasediments. Figure 25.11 is an *ACF* diagram representing the blueschist facies. See Section 16.8.4 for a discussion of the stability of hydrous phases to great depths in subducted oceanic crust.

The transition to the **eclogite facies** is marked by the development of the characteristic mafic assemblage omphacitic pyroxene (an augite–jadeite solution) + almandine–pyrope–grossular garnet, creating dense, beautiful green and red rocks. Figure 25.12 is an ACF diagram showing the mineral assemblages characteristic of the eclogite facies. The eclogite facies represents pressures at which plagioclase is no longer stable. The albite component of plagioclase breaks down in the blueschist facies via glaucophane-producing reactions or Reaction (25.6), so that Na is stored in sodic amphiboles or pyroxenes, and not in plagioclase.

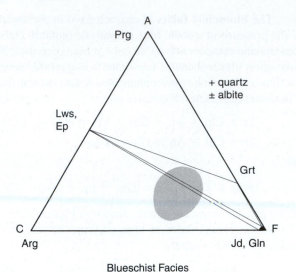

Blueschist Facies

FIGURE 25.11 *ACF* compatibility diagram illustrating representative mineral assemblages for metabasites in the **blueschist facies.** The composition range of common mafic rocks is shaded.

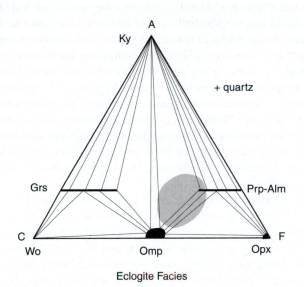

Eclogite Facies

FIGURE 25.12 *ACF* compatibility diagram illustrating representative mineral assemblages for metabasites in the **eclogite facies.** The composition range of common mafic rocks is shaded.

The blueschist facies to eclogite facies transition involves glaucophane and paragonite reacting to form the garnet–omphacite assemblage:

$$Gln + Pg = Prp + Jd + Qtz + H_2O \qquad (25.7)$$

at pressures greater than 1.2 GPa (Figure 26.19), but glaucophane alone may remain stable into the eclogite facies.

Along higher geothermal gradients, the amphibolite facies, or even the granulite facies, may give way at high pressure to the eclogite facies as well. At higher temperature, the anorthitic component must be considered, and plagioclase reacts with amphibole, orthopyroxene, clinopyroxene, and/or olivine via a number of possible reactions to form omphacitic clinopyroxene + pyrope–grossular garnet (and in some cases quartz and/or kyanite) as pressure increases. Equation (25.3), for example, characterizes the granulite–eclogite facies transition.

Ultimately, anorthite breaks down in the absence of other mineral phases via the reactions:

$$An + H_2O = Zo + Ky + Qtz \qquad (25.8)$$

and

$$An = Grs + Ky + Qtz \text{ or } An = Ca\text{-}Ts + Qtz \qquad (25.9)$$

defining the point at which the rocks least susceptible to the effects of high pressure finally succumb. Any of the plagioclase breakdown reactions mark the transition from amphibolite or granulite facies to eclogite facies (see Bucher and Frey, 2002; and Miyashiro, 1994, for reviews). Na, Ca, and Al, rather than residing in plagioclase, are then ensconced in the pyroxene and garnet in eclogites.

Coleman et al. (1965) subdivided eclogites on the basis of their mode of occurrence into:

Group A: xenoliths in kimberlites and basalts (see Chapters 19 and 14)

Group B: bands or lenses in migmatitic gneisses

Group C: bands or lenses associated with blueschists

Carswell (1990) subdivided eclogites on the basis of equilibration temperature into low-temperature (~450 to 550°C), medium-temperature (~550 to 900°C), and high-temperature (~900 to 1600°C) groups.

Group C eclogites of Coleman et al. (1965) correlate with Carswell's low-temperature subdivision, although higher-temperature and -pressure eclogites in the western Alps are also mixed in. These are typical eclogite–blueschist associations from subduction zones that are usually tectonically mixed and disrupted following metamorphism. Epidote, zoisite, and quartz, plus hydrous phases such as amphibole, phengite, and paragonite, are common, and the garnets contain <30% pyrope component. Bearth (1959) described an unusual example from near Zermatt, Switzerland, in which deformed (yet recognizable) basaltic pillows exhibit eclogite cores and blueschist rims. Such an occurrence indicates that the protolith was clearly a surface lava and not mantle rock, yet reached very high pressure. The close association of blueschist and eclogite facies rocks has spurred some debate as to whether one mineral assemblage replaces another as prograde or retrograde overprints, or whether they developed under the same (transitional) *T-P* conditions, but with more H_2O available to stabilize the hydrous blueschists. See Yardley (1989) for a discussion.

Group B eclogites occur as lenses in high-grade gneiss terranes. Most are of the medium-temperature type of Carswell and may contain quartz, kyanite, zoisite, paragonite, or Ca-amphibole. The pyrope content of garnet is 30 to 55%. Surrounding rocks are mostly in the amphibolite facies, but there is local evidence for granulite facies and even some blueschist facies rocks. An interesting hypothesis for many of these eclogites is that thrust stacking of crustal units subjected the base of the crust to pressures in excess of 2 GPa and metamorphism in the eclogite facies (see the next section).

Group A eclogites are high-temperature eclogites and represent subducted oceanic crust (Chapters 14 and 16) or delaminated mafic crustal underplates (Figure 18.8) that have been reincorporated into the mantle and picked up by rising magmas. A variety of (mostly anhydrous) minerals occur in them, including kyanite, coesite, and diamond. Pyrope content of the garnet is >50%. Another scheme for subdividing the eclogite facies on the basis of hydrous mineral stabilities is presented in the next section.

25.4 UHP AND UHT METAMORPHISM: THE EXTREMES OF CRUSTAL METAMORPHISM

In recent years, the pressure and temperature domain of crustal metamorphism has vastly expanded, with new extremes of ultra-high-pressure and ultra-high-temperature crustal metamorphism recognized in the geological record. Ultra-high-pressure metamorphism has important implications for how the continental crust behaves during subduction, continental collision, and crustal thickening. Ultra-high-temperature metamorphism may revolutionize ideas on crustal thermal behavior and rheology under extreme conditions previously considered impossible so far beyond the beginning of melting of many rock types.

Ultra-high-pressure (UHP) metamorphism is defined as the mineralogical and structural modification of predominantly continental and minor associated oceanic *crustal* protoliths and associated mafic-ultramafic rocks at mantle pressures generally in the range of 2.7 to 5.0 GPa (90 to 170 km) and temperatures about 700 to 950°C. Chopin (1984) first described **coesite** as inclusions within garnet in garnet–kyanite-bearing "whiteschists" from the Dora-Maira

massif of the western Alps of Italy. Very soon thereafter, Smith (1984) described coesite from crustal eclogites in western Norway. Coesite is stable only at pressures in excess of 2.5 GPa at 600°C (Figure 6.6), equivalent to about 90-km depth. Since Chopin's and Smith's initial discoveries, similar "ultra-high-pressure" crustal metamorphic rocks have been described from more than 20 localities (Figure 25.13) where crustal rocks have reached previously unsuspected depths and then been exhumed. (Recent summaries of UHP metamorphism are listed in the Suggested Readings at the end of this chapter.)

Studies of UHP metamorphism concentrate on discontinuous belts several hundred kilometers long and 10 to 50 km wide associated with subduction and continental collision where the continental crust is partly subducted and overthickened by stacking and compression. New mineralogical discoveries continue to increase the maximum depths to which crustal rocks are found to have been subjected (presently on the order of 150 to 200 km). This extends by almost an order of magnitude the depth traditionally ascribed to the metamorphic cycling of continental crust and demonstrates that, in spite of its thickness and low density, such crust can be subducted to considerable depths (before finally clogging and suturing the subduction zone). This may also explain the curiously selective definition of UHP metamorphism: it specifies crustal rocks, particularly with continental affinity, excludes eclogites created in the mantle from stalled mafic melts, and largely ignores eclogites created by normal subduction of mafic oceanic crust. For sediments to reach depths in excess of 150 km (without melting) is impressive enough. For them to resurface (with a record of the trip) is really astonishing. Exhumation must have been rapid, occurring before thermal relaxation could raise the temperature of the cool subducted material sufficiently to induce melting (see Section 25.5).

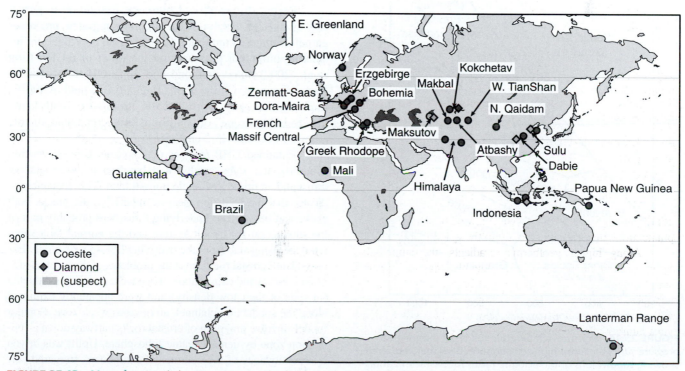

FIGURE 25.13   Map of presently known occurrences of ultra-high pressure (UHP) metamorphic rocks (after Liou et al., 2005).

Figure 25.14 shows how our facies concept has been extended by investigators in UHP (and UHT) terranes. The eclogite facies in particular has been subdivided to allow more specific characterization of the conditions to which these unusual rocks have been subjected. Notice the occurrence (and stability limits) of the hydrous phases lawsonite, amphibole, and epidote (actually zoisite) in the eclogite facies (also discussed in regard to re-enriching the mantle with fluids in Section 16.8.4). Amphibole eclogites, for example, occur at pressures greater than the breakdown of plagioclase, but less than the breakdown of amphibole (Schmidt and Poli, 2003). They are thus eclogites and not amphibolites. High-pressure granulites (e.g., O'Brien and Rötzler, 2003) are characterized by the key associations garnet + clinopyroxene + plagioclase + quartz (in basic rocks) and kyanite + K-feldspar (in metapelites and felsic rocks) and are typically orthopyroxene-free in both basic and felsic bulk compositions [see Reaction (25.3)]. These rocks are thus transitional between granulites (with orthopyroxene) and eclogites (without plagioclase). The facies subdivisions at extreme pressures and temperatures in Figure 25.14 are rather specialized and not commonly used (hence left out of Figure 25.2).

Indicators of the crustal nature of UHP rocks include interlayered quartzite, marble, mica schist, and quartzofeldspathic gneiss, reflecting sedimentary sequences of sandstone, limestone, shale, and graywacke, respectively. Such sequences are typical of a continental forearc accretionary wedge. Concordant layers of eclogite indicate mafic lava flows or sills. The spectacular talc + kyanite ± garnet ± phengite whiteschists are probably sheared and hydrated

granites. Low $\delta^{18}O$ and $^2H/^1H$ in most of these rocks indicate the influence of meteoric water, which strengthens the case for initial shallow (continental) crustal provenance. The high SiO_2 concentrations in many Chinese UHP eclogites indicates the protolith basalts were of continental affinity, supported by high $^{87}Sr/^{86}Sr$ (0.706 to 0.713) and low ε_{Nd} (−6 to −20). Liou et al. (2004) distinguished intracontinental "collision-type" from (typically smaller) continental-arc "Pacific-type" UHP terranes (subducting oceanic lithosphere). The crustal origin of the Pacific-type is recorded in the accretionary wedge (including chert and trench turbidite) components and tonalite–trondhjemite–granitiod (TTG) lithologies. Jahn et al. (2003) found that the geochemical and isotopic signature of several eclogites in UHP localities had a distinct continental affinity.

The principal UHP mineral indicators are the high-pressure polymorphs of silica (coesite) and carbon (microdiamond) that escaped back-transformation during rise to the surface, typically as micron-scale inclusions in garnet, clinopyroxene, and zircon in volumetrically subordinate (but widespread) mafic eclogite and ultramafic peridotite pods and slabs. The lattices of the host minerals are generally strong enough to maintain pressure on the inclusions following uplift and retard their inversion to lower-pressure polymorphs (which involves a volume increase). Note the stability of coesite and diamond in Figure 25.14. The quartzo-feldspathic gneisses and other rocks of continental affinity typically re-equilibrate upon exhumation to lower pressure assemblages. Coesite and microdiamond inclusions in zircon, garnet, and clinopyroxene in some quartzo-feldspathic gneisses are an important link to direct subduction of the continental crustal lithologies, offsetting speculation that they were merely interleaved later by faulting with UHP eclogites and not themselves exposed to UHP conditions.

Recent discoveries of pyroxene exsolution in garnet and of coesite exsolution in titanite suggest a precursor garnet or titanite containing Si in six-fold coordination, indicating minimum pressures on the order of 6 GPa. Liou et al. (2000) noted diamond and topaz-OH in the Dabie-Sulu terrane of China, indicating ~6 GPa and <900°C, implying that crust there actually ventured (slightly) into the "forbidden zone" of conditions considered unattainable (shown in Figures 25.2 and 25.14).

Exhumed UHP terranes are typically exposed as thin sub-horizontal slabs sandwiched between a lower shallow thrust and an upper low-angle normal fault. This sequence indicates that the UHP terrane rose relative to the blocks both above and below. The underlying thrust was probably part of the subduction zone fault system and the normal fault separated the rising slab from the overlying lithospheric block. Exposed UHP crustal rocks pose the problems of the mechanisms of their burial and exhumation. They were certainly subducted (in spite of their low density) and were apparently exhumed along the subduction channel, an obvious weak zone. Popular models involve dragging of crustal rocks partway down a subduction zone by dense oceanic lithosphere. Uplift was driven by the buoyancy of the continental component, triggered perhaps by "slab break-off" (separation and sinking of leading

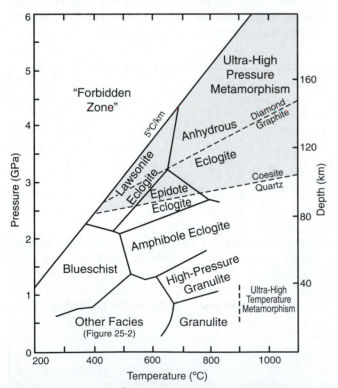

FIGURE 25.14 Pressure–temperature ranges of ultra-high-pressure (UHP) and ultra-high-temperature (UHT) metamorphism, showing subdivisions of the eclogite facies for characterization of UHP rocks. After Liou and Zhang (2002).

oceanic lithosphere) and maybe enhanced by delamination of subcontinental lithospheric mantle (Figure 18.8). The UHP terrane may thus have popped up with respect to blocks above and below, in some cases quite rapidly (see Section 25.5). It's impossible to tell if some crustal material of continental character was carried and recycled back into the mantle, but the possibility carries some interesting ramifications for enriched components in the mantle (Section 14.6).

Widespread Barrovian-style metamorphism in most collisional orogens typically overprints the UHP signature of any such slabs present, masking their earlier deep history. Some investigators suspect that coesite-bearing UHP rocks were generated in the majority of exhumed collision orogens, but their identification requires careful searching for the tiny relict inclusions. Chopin (2003, p.10) concluded that continental subduction to UHP depths "must be considered a normal process, inherent to continental-plates convergence, collision, and orogeny—at least since the Proterozoic." Who knows what Figure 25.13 will look like in a few decades?

Remember, these occurrences are of extremely deep burial with emphasis on *continental crustal* material. They should not be confused with those rare kimberlite xenoliths described in Section 19.4 in which mineral inclusions in diamond and exsolution phenomena indicate that garnet pyroxenites were carried into the *lower* mantle. Those latter are neither coherent packages nor can they be shown to be crustal with confidence. I realize this distinction is rather artificial because the kimberlite xenoliths are certainly metamorphic under extreme pressure conditions and very interesting. But the interest of those investigating UHP rocks as defined here is in crustal terranes that have been dragged to great depths as semi-coherent packages and subsequently returned to the surface.

Ultra-high temperature (UHT) metamorphism was defined by Harley (1998, 2004) as a division of medium-pressure granulite facies in which peak temperatures reach 900 to 1100°C at 0.7 to 1.3 GPa. There are more than 40 known UHT occurrences (see Table 1 in Harley, 1998, and the map in Kelsey, 2008). Brown (2007a) attributed most UHT metamorphism to continental back-arc settings where crustal thinning and mantle rise adds significant heat. Detachment of subcontinental lithospheric mantle may also be responsible for added heat. UHT metamorphism may have been particularly prevalent in the Precambrian when heat flow was greater, but examples are known throughout Earth history.

Geothermobarometry (Section 27.4) typically fails to record UHT temperatures because Fe/Mg exchange tends to reset with cooling to 700 to 850°C, the typical "closure temperature" for such exchange (Section 26.6). It is therefore preferable to use mineral parageneses rather than geothermometry. Indicator mineral assemblages for UHT conditions include sapphirine + quartz, orthopyroxene + sillimanite, and osumilite; or wollastonite + scapolite + quartz + grossular in calc-silicates (Chapter 29). Al_2O_3 content greater than 8 to 2 wt. % in orthopyroxene coexisting with garnet, sillimanite, or sapphirine also indicates UHT metamorphism. UHT terranes cool/decompress along a variety of *P-T* paths, ranging from isobaric cooling to isothermal decompression.

25.5 PRESSURE–TEMPERATURE–TIME (*P-T-t*) PATHS

The concept of metamorphic facies series (Section 25.2) suggests that a traverse upgrade through a metamorphic terrane following a metamorphic field gradient may cross through a sequence of facies. This is supported by a walk through most metamorphic terranes. But does progressive metamorphism of a rock in the upper amphibolite facies, for example, require that it pass through exactly the same sequence of mineral assemblages encountered via a traverse upgrade to that rock through greenschist facies, etc.? In other words, are the *temporal* and *spatial* mineralogical changes in an area the same?

Let's consider the complete set of *T-P* conditions that a rock now at the surface may experience during a metamorphic cycle from burial to metamorphism (and orogeny) through uplift and erosion. Such a cycle, called a **pressure–temperature–time path**, or *P-T-t* **path**, was first introduced in Section 16.8, where we attempted to assess the conditions experienced by progressively subducted crust and convecting mantle wedge materials as they moved through the subduction zone complex.

Metamorphic *P-T-t* paths may be addressed in several ways. For example, we might observe partial overprints of one mineral assemblage upon another (Jamieson, 1988; Ghent et al., 1988). The relict minerals may indicate a portion of either the prograde or retrograde path (or both) depending upon when they were created. We learned in Section 21.4 that thermodynamic equilibrium is generally maintained during prograde metamorphism, so that relict minerals are more common during the retrograde path, although both types have been observed. Alternatively, we might be able to apply geothermometers and geobarometers (Section 27.4) to the core versus rim compositions of chemically zoned minerals to document the changing *P-T* conditions experienced by a rock during mineral growth (Spear et al., 1984; Selverstone and Spear, 1985; Spear, 1989). Other analytical techniques (summarized in Ghent et al., 1988) include oxygen isotope geothermometry, fluid inclusion data, and isotopic cooling ages (including the Ar^{40}/Ar^{39} method for estimating uplift rates discussed in Section 9.7.2.1). These methods are typical "inverse" methods because they begin with the product (the rock) and estimate the process that created it. Even under the best of circumstances, these methods can usually document only a small portion of the *P-T-t* path to which a rock was subjected. We thus rely more on "forward" heat-flow models for various tectonic regimes to model more complete *P-T-t* paths and evaluate them by comparison with the results of the inverse methods.

As classically conceived, regional metamorphism was thought to occur as a result of deep burial or by intrusion of hot magma. Since the advent of plate tectonics, we now realize that regional metamorphism is rarely produced by simple burial but is a result of crustal thickening and increased crustal radioactive heating, plus heat input during orogeny at convergent plate boundaries (see Section 21.3.2). Heat-flow models have been developed for various regimes, including burial (Fowler and Nisbet, 1982), progressive thrust stacking as might be expected in subduction zones and illustrated in Figure 21.6 (Oxburgh and Turcotte, 1971, 1974; Bickle et al., 1975), thrust stacking expanded to include crustal doubling by continental collision (England and Richardson, 1977; England and Thompson, 1984), and the effects of crustal anatexis and magma migration (Lux et al., 1986; DeYoreo et al., 1989). For reviews, see Nisbet and Fowler (1988) and Peacock (1989). Nearly all models indicate that heat flow higher than the normal continental geotherm is required for typical greenschist–amphibolite medium P/T facies series, and that uplift and erosion have a fundamental effect on the geotherm and must be considered in any complete model of metamorphism.

Figure 25.15 illustrates some examples of modeled P-T-t paths representing common types of metamorphism. The paths illustrated are schematic, and numerous variations are possible, depending upon the style of deformation and the rates of thickening, heat transfer, magmatism, and erosion, etc. that are assumed. Path (a) in Figure 25.15 is considered to be a typical P-T-t path for an orogenic belt experiencing crustal thickening. During the thickening stage the pressure increases much more rapidly than the temperature because of the time lag required for heat transfer (pressure equilibrates nearly instantaneously, but heat conducts very slowly through rocks). A rock in the thickened crustal block thus quickly approaches its maximum depth (P_{max}) while remaining relatively cool.

The increased thickness of crust is rich in radioactive LIL elements, so the heat flux increases. Subduction zone magmatism may also enhance the heating effect, as noted in several published models that accommodate the process. So the rocks at depth gradually heat up, and path (a) curves toward higher temperature. The new geotherm is higher than

the steady state geotherm of the initial stable continent, but is a transient phenomenon, and lasts only as long as the thickened crust and subduction related heat generation lasts. Erosion soon affects the thickened crust and the pressure at depth usually begins to decrease before the rocks can equilibrate with the higher orogenic geotherm. The temperature is thus still increasing due to the slow heat transfer (most models address heat transfer by conduction only), so that the P-T-t path has a negative slope after reaching P_{max} on Figure 25.15. Uplift and erosive pressure release, plus increased heat input, may also result from delamination of a thermal boundary layer of dense lithospheric mantle immediately below the crust and its replacement by hotter, lighter, asthenosphere during post-orogenic collapse, as illustrated in Figure 18.5.

A rock reaches T_{max} when the cooling effect of uplift and erosion catches up to the increased geotherm, so that the thermal perturbation of crustal thickening is dampened and begins to fade. From this point the P-T-t path follows a positive slope as both temperature and pressure fall while the rock moves toward the surface (due to uplift and erosion of the overthickened crust), and the geotherm gradually returns to a normal continental geotherm. Only when the thickened crust is stable for sufficient time will T_{max} occur on a static elevated geotherm. It is more likely that the elevated geotherm will begin to relax due to erosional thinning so that T_{max} occurs during a waning thermal system.

Although the exact shape, size, and position of an orogenic P-T-t path such as path (a) may vary with the constraints of the model, most examples of crustal thickening have the same general looping shape, whether the model assumes homogeneous thickening or thrusting of large masses, conductive heat transfer or additional magmatic rise. Two cooling-uplift variations (to be discussed shortly) are labeled a_1 and a_2 in Figure 25.15. Paths such as (a) are called **"clockwise" P-T-t paths** in the literature and are considered to be the norm for regional metamorphism.

Figure 25.16b illustrates a P-T-t path based on the chemical zonation of garnet (Figure 25.16a) coexisting with hornblende, kyanite, staurolite, biotite, chlorite, and quartz in a schist from the Tauern Window (southwestern Alps). The method assumes that the core → rim zonation is due to successive growth layers added during changing P-T conditions and employs geothermobarometry (Section 27.4) following the method developed by Selverstone et al. (1984). P-T calculations from the core to the rim also produce a "clockwise" P-T-t path, as shown in Figure 25.16b. Most orogenic regional metamorphics, when they retain a record of metamorphic history, show such paths (although only segments of the overall path), supporting the results of the common crustal thickening heat-flow models.

Path (b) in Figure 25.15 represents a different situation in which a rock is heated and cooled at virtually constant pressure by magmatic intrusion at shallow levels. This may be an appropriate P-T-t path for contact metamorphism. Depending upon the extent of magmatic activity and its contribution to the crustal mass, any number of paths transitional between (a) and (b) can be imagined, representing a gradation from high-pressure blueschist metamorphism to (Barrovian) regional

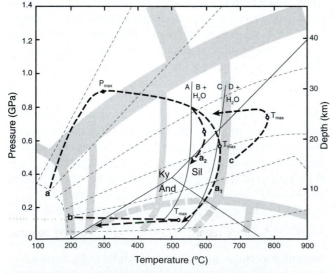

FIGURE 25.15 Schematic pressure–temperature–time paths based on heat-flow models for **(a)** crustal thickening, **(b)** shallow magmatism, and **(c)** some types of granulite facies metamorphism. The Al$_2$SiO$_5$ phase diagram and two hypothetical dehydration curves are included. Open circles represent maximum temperatures attained. Facies boundaries and facies series from Figures 25.2 and 25.3.

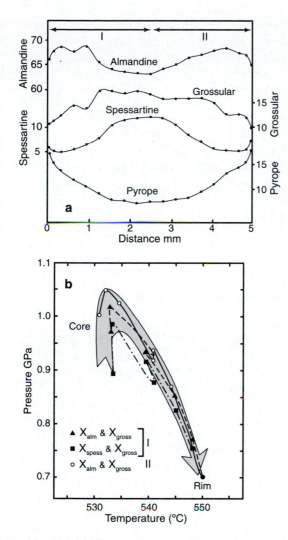

FIGURE 25.16 (a) Chemical zoning profiles across a garnet from the Tauern Window. (b) *P-T* diagram showing three modeled "clockwise" *P-T-t* paths computed from the profiles using the method of Selverstone et al. (1984) and Spear (1989). After Spear (1989).

metamorphism to "regional contact metamorphism" with numerous plutons to local contact metamorphism.

Path (c) is an example of what is conventionally called a **"counterclockwise" *P-T-t* path** ("anticlockwise" elsewhere in the English-speaking world). This commonly occurs in high-grade gneisses and granulite-facies terranes (Bohlen, 1987, 1991) and is believed to result from the intrusion of relatively large quantities of (usually mafic) magma into the lower and middle crust. The rapid introduction of magmatic heat and mass causes both the pressure and temperature to increase in unison *below* the intrusions. This is followed by nearly isobaric cooling because the high density of the mafic magma does not lead to crustal buoyancy, so that uplift and erosion are limited.

Not all granulite facies metamorphism exhibits this "counterclockwise" behavior, and two paths are commonly recognized (Bohlen, 1987; Harley, 1989). One type exhibits near-isobaric cooling, as in path (c) above. The other exhibits near-isothermal decompression in the range 550 to 950°C. The latter path may represent an uplift or collapse stage

following crustal thickening as in path (a). Frost and Chacko (1989) pointed out that inferred *P-T-t* paths based on thermobarometry in granulite facies rocks may re-equilibrate during cooling because they are held at very high temperatures for extended periods. Selverstone and Chamberlain (1990) suggested that some isobaric cooling paths appear to be artifacts of the core-rim thermobarometric technique, and do not agree with the record preserved in the progression of overprinted mineral assemblages. The *P-T-t* paths for granulites based on geothermobarometry should thus be interpreted with caution, and may not exhibit the exact path that the rock followed. More robust methods for reconstructing *P-T-t* paths are based on partial replacement textures (coronites/reaction rims) to reconstruct a sequence of mineral parageneses related to *P-T* conditions using petrogenetic grids (Section 26.11). The isobaric cooling and isothermal decompression trends are commonly derived for granulite facies rocks using either method, causing most investigators to conclude that these two processes are common in the late evolution of deep crustal domains.

"Counterclockwise" *P-T-t* paths are not restricted to granulite terranes. The behavior has been shown to occur in other settings, usually associated with high magmatic heat flow. Spear (1993) discussed the Acadian terrane in the northeastern United States, in which the western zone has a high-pressure (up to 1 GPa) Barrovian clockwise *P-T-t* path and the eastern zone a low-pressure (less than 0.5 GPa) Buchan counterclockwise *P-T-t* path. The contrast is attributed to difference in tectonic setting. The clockwise path in the west is typical for crustal thickening followed by erosion and cooling. The eastern zone is riddled with numerous synmetamorphic plutons, perhaps because of crustal extension and upwelling mantle supplying sufficient heat to partially melt the lower crust.

Although the paths that rocks are believed to follow during metamorphism are called pressure–temperature–time paths, time is not quantitatively illustrated on the *P-T* diagrams produced. Until recently (see below) there was no indication of the *rate* of pressure and temperature change, so that distance along a loop need not be directly proportional to time. In general, the increase in pressure to P_{max} in path (a) is generally considered the most rapid, and erosional unroofing should be much slower. On the basis of their models, England and Thompson (1984) suggested that temperature remains within 50°C of T_{max} for nearly one-third of the burial–uplift cycle. Several investigators refer to these diagrams as simply "*P-T* paths," which may be more accurate but fails to emphasize the (qualitative) temporal component.

Given the broad agreement between the forward and inverse techniques regarding *P-T-t* paths, we may assume that the general form of a path such as (a) in Figure 25.15 represents the path followed by a typical rock during orogeny and regional metamorphism (at least during a single event). If this assumption is correct, the shape of the typical *P-T-t* path has some instructive ramifications:

1. Contrary to the classical treatment of metamorphism, temperature and pressure do not both increase in unison as a single unified "metamorphic grade." Their relative magnitudes vary considerably during virtually any process of metamorphism.

2. P_{max} and T_{max} do not generally occur at the same time. In the usual case of "clockwise" P-T-t paths, such as path (a), P_{max} occurs much earlier than T_{max}. T_{max} should represent the maximum grade at which chemical equilibrium is "frozen in" and the equilibrium metamorphic mineral assemblage is developed. This occurs at a pressure well below P_{max}. P_{max} is uncertain because a mineral geobarometer should record the pressure of T_{max}. "Metamorphic grade" should refer to the temperature and pressure at T_{max} because the grade is determined via reference to the equilibrium mineral assemblage.

3. Some variations on the cooling–uplift portion of the "clockwise" path (a) in Figure 25.15 indicate some surprising circumstances. For example, the kyanite $\rightarrow$ sillimanite transition is generally considered a prograde transition (as in path a_1), but path a_2 crosses the kyanite fi sillimanite transition as temperature is *decreasing*. This may result in only minor replacement of kyanite by sillimanite during such a retrograde process. If the P-T-t path is steeper than a dehydration reaction curve, it is also possible that a dehydration reaction can occur with decreasing temperature (although this is only likely at low pressures where the dehydration curve slope is low).

4. Now that we have some concept of the P-T-t path that a rock follows we can return to the question of whether the temporal mineralogical changes associated with progressive metamorphism are the same as the spatial changes exhibited across the exposed terrane. In Section 21.2.2, we defined the metamorphic field gradient as the change in temperature and pressure of metamorphism along a traverse at the Earth's surface headed directly up metamorphic grade, and distinguished this from the geothermal gradient during peak metamorphism, which is the variation in temperature with depth. It was clear in Section 21.2.2 that a spatial progression of mineral assemblages followed the traverse up the metamorphic field gradient, but we avoided speculating whether the temporal progressive changes in mineralogy of a high grade rock mimicked the spatial changes. If the P-T-t path models are correct the answer is now available and is definitely negative. Figure 25.17 illustrates the difference between the spatial changes along the metamorphic field gradient and the temporal changes (P-T-t paths) for several rocks along the gradient. Note that every rock follows a path involving considerably higher pressures and lower temperatures than the final locus of T_{max} versus (P at T_{max}) points along the metamorphic field gradient suggests.

5. To expand upon this spatial-temporal difference further, in Figure 25.15 it appears that typical "clockwise" P-T-t paths for Barrovian-type regional metamorphism initially follow the high P/T metamorphic facies series toward blueschist facies and finally equilibrate to a lower pressure facies at a later time. Blueschist metamorphism may thus be considerably more common than we thought, and not necessarily restricted to subduction zones proper or the outer member of paired metamorphic belts (Section 21.6.3). In

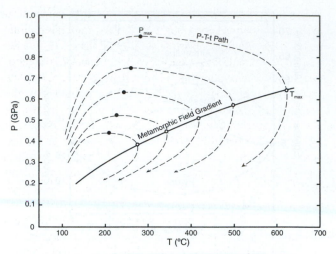

FIGURE 25.17 A typical Barrovian-type metamorphic field gradient and a series of metamorphic P-T-t paths for rocks found along that gradient in the field.

addition to this implication based on simple crustal thickening models, Oxburgh and Turcotte (1971) proposed that progressive underthrusting in the subduction zone would tectonically underplate successive slabs beneath the accretionary wedge, so that, over time, the subduction zone would migrate away from the arc, and rocks previously in the subduction zone would move into the high heat flow regime of the inner belt, thereby providing a second mechanism for early high P/T metamorphism followed by medium or low P/T metamorphism. Ernst (1988) listed several blueschist facies terranes that display a later greenschist–amphibolite overprint, including the Alps, Turkey, Greece, the Klamath Mountains of California, the Sanbagawa belt, the Haast Schists, New Caledonia, and several areas in northern Asia and China (including the UHP terranes discussed above). Richardson (1970) demonstrated that not only rapid burial but also rapid *uplift* is required to preserve blueschist facies conditions and expose blueschists before they evolve into the higher temperature portion of the P-T-t path. Perhaps blueschists would be much more common if uplift were not delayed.

In spite of the implication that blueschists may occur in the early stages of medium P/T facies series, known blueschists still appear to correlate well with subduction-zone metamorphism. This suggests that either (1) the initial pressure increase of most medium P/T regional P-T-t paths may not be as steep as path (a) in Figure 25.15 and the paths in Figure 25.17, or (2) early and rapid uplift may also be a characteristic of subduction zones, so that blueschist *preservation*, and not simply generation, is a common subduction zone process. Tectonic underplating by underthrust imbrication beneath the accretionary wedge may be a mechanism by which this is accomplished, so that preserved blueschists would still correlate with subduction zones.

The rapidly developing application of "textural geochronology" (Section 23.7), whereby small accessory minerals may be accurately dated (even *in situ* in thin section) and related texturally to specific tectonic-metamorphic events is now enabling us to quantify the time aspect of at least a few points along some pressure–temperature–time paths. For example, Burton and O'Nions (1991) used Sm/Nd, U/Pb, and Rb/Sr ages of zoned garnets in amphibolite facies metasediments in northern Norway to date Barrovian-style prograde metamorphism there. Using the ages and estimated temperatures and pressures from geothermobarometry, they estimated average rates of heating at 8.6°C/Ma and burial at 0.8 km/Ma. Using these estimates, burial from point a to P_{max} in Figure 25.15 should involve a (dP/dT) slope about half as steep in Norway and take about 16 Ma. They then noted continued Rb and Sr exchange between coexisting silicates for another 37 Ma, from which they extracted a cooling rate of about 4°C/Ma. Cooling (and presumably uplift) was thus much slower than heating and burial, as generalized a few paragraphs previously. Foster et al. (2004) used monazite inclusions in garnets for dating and garnet zoning for P-T estimates for a sample from the Himalayas and two from the Canadian Cordillera. They determined that the Himalayan sample went from about 550°C at ~80 Ma to about 620°C at ~55 Ma (heating rate approximately 2.4 ± 1.2°C/Ma), and a Canadian sample went from about 570°C at ~75 Ma to about 670°C at ~55 Ma (heating rate approximately 5 ± 2°C/Ma), both along clockwise P-T-t paths. Cooling rates were more difficult to constrain as retrograde reactions were minor.

In closing, two studies of P-T-t paths of UHP rocks relate the last two sections of this chapter. Rubatto and Hermann (2001) determined ion-microprobe U/Pb ages of zoned titanites from the original Alpine Dora Maira UHP locality of Chopin (1984), combined with P and T estimates using geothermobarometry. They found peak metamorphic titanites in calc-silicates yielding ages of 35.1 ± 0.9 Ma at a pressure of ~3.5 GPa. Titanite formed during two decompression metamorphic stages at 32.9 ± 0.9 Ma and 1 ± 0.15 GPa, and at 31.8 ± 0.5 Ma and ~0.4 to 0.5 GPa. From these data they derive mean exhumation rates of 3.4 and 1.6 cm/yr! Liou et al. (2007) reported ion-microprobe U/Pb ages of zoned zircons from the Dabie-Sulu UHP locality of China. The zircon cores yielded a protolith age 680 Ma. An intermediate zircon overgrowth at 231 ± 4 Ma contained coesite and was correlated to an estimated 750 to 950°C at 4.0 to 6.7 GPa (reaching into the "forbidden zone"). Finally, zircon rims at 211 ± 4 Ma contain lower-pressure quartz and plagioclase and formed during retrograde metamorphism at much lower pressures, providing exhumation rates of about 5 km/Ma (0.5 cm/yr). These exhumation rates are incredibly rapid and may explain the preservation of the UHP imprint. Such rates are faster than isostatic uplift and erosion could possibly be and support the contention that tectonic processes (including crustal buoyancy and normal faulting) were important in transporting these rocks to the surface.

Summary

Metamorphic facies subdivisions are based on the mineral assemblages that develop in metamorphosed mafic rocks. Because basaltic rocks occur in practically all orogenic belts and the mineral changes in them define broader T-P ranges than in pelites, the facies concept provided a convenient way to compare metamorphic areas around the world. Although the facies concept was introduced on a descriptive basis (the relationship between a rock's composition and the mineral assemblage developed at some particular grade of metamorphism), the interpretive aspect (the temperature and pressure conditions represented) was originally and is increasingly well understood. Figure 25.2 summarizes the P-T limits of the various facies used in this text. We can combine the concepts of facies and metamorphic zones (as indicated by Figure 25.10) so that we can speak of the chlorite zone of the greenschist facies, etc. When considered on a broad scale, we may notice a sequence of several facies occurring transitionally along a metamorphic field gradient. Miyashiro called such sequences *facies series* and eventually proposed three broad types, reflecting high, medium, or low P/T ratios of the gradient. The facies and facies series that develop in a particular metamorphic terrane can reveal a lot about the dynamics leading to metamorphism (subduction zone, orogenic belt, back-arc area, etc.).

Mafic igneous rocks were originally assemblages of plagioclase, pyroxene, and perhaps olivine, and ilmenite or magnetite. The metamorphic minerals that replace them include zeolites and prehnite or pumpellyite at low grades, giving way to albite, chlorite, actinolite, and/or epidote in the greenschist facies, and hornblende plus plagioclase in the amphibolite facies. If the rocks become sufficiently dehydrated, partial melting may be avoided and pyroxene may develop in the granulite facies. At high pressures, glaucophane is the hallmark of the blueschist facies (and may be joined by lawsonite), and pyrope-rich garnet plus omphacitic pyroxene define the eclogite facies (surpassing the high-pressure stability limits of plagioclase).

New extremes of ultra-high-pressure (UHP) and ultra-high-temperature (UHT) crustal metamorphism have recently been recognized in the geological record. UHP localities indicate that continental crust can be carried to depths previously unrecognized and then exhumed relatively rapidly. UHT metamorphism also indicates that some crustal rocks have been subjected to temperatures previously considered impossible so far beyond the beginning of melting of many rock types.

A pressure–temperature–time (P-T-t) path represents the progressive series of T-P conditions that a rock experienced during a metamorphic cycle from burial to metamorphism (and orogeny) through uplift and erosion to its present position at the surface. We can document parts of this path using chemical zoning and geothermometry, as well as partial replacement and reaction textures. We can even place actual time constraints on some P-T-t paths using texturally constrained geochronometry techniques when suitable minerals are available. We have discovered that the *temporal* (*P-T-t* path) sequence of metamorphic mineral changes experienced by a single rock mass are not the same as the *spatial* sequence of mineral parageneses seen in the field by following a metamorphic field gradient.

Key Terms

Metamorphic facies *538*

Facies series (high *P/T*, medium *P/T*, and low *P/T*) *541*

Low-grade facies: zeolite and prehnite–pumpellyite facies *543*

Medium *P/T* facies: greenschist, amphibolite and granulite facies *544*

Low *P/T* facies: albite–epidote hornfels, hornblende hornfels, pyroxene hornfels, and sanidinite facies *544*

High *P/T* facies: blueschist and eclogite facies *547*

Ultra-high-pressure (UHP) metamorphism *549*

Ultra-high-temperature (UHT) metamorphism *551*

Pressure–temperature–time (*P-T-t*) paths *551*

Review Questions

Review Questions are located on the author's web page at the following address: **http://www.prenhall.com/winter**

Important "First Principle" Concepts

■ At any particular grade of metamorphism, the mineral paragenesis that develops in a rock (under equilibrium conditions) depends only on the bulk composition. Rocks metamorphosed at the same grade with the same mineral assemblage have similar compositions, and rocks with different mineral assemblages have different compositions. This can readily be seen in any chemographic mineral paragenesis (compatibility) diagram.

■ The transition from one facies to another involves one or more mineral reactions (Chapter 26), so the facies are thus separated by mineral reaction isograds. As we shall soon see, this involves a change in the minerals and/or tie-lines on chemographic mineral paragenesis diagrams from one facies to another.

Suggested Further Readings

Metamorphic Facies and Facies Series

Coombs, D. S. (1960). Lower grade mineral facies in New Zealand. *21st Int'l. Geol. Congress Rept.,* Part **13**, 339–351.

Coombs, D. S. (1961). Some recent work on the lower grade metamorphism. *Australian J. Sci.,* **24**, 203–215.

Eskola, P. (1915). On the relations between the chemical and mineralogical composition in the metamorphic rocks of the Orijärvi region. *Bull. Comm. Geol. Finlande,* **44**.

Eskola, P. (1920). The mineral facies of rocks. *Norsk. Geol. Tidsskr.,* **6**, 143–194.

Fettes, D., and J. Desmons (eds.). (2007). *Metamorphic Rocks: A Classification and Glossary of Terms. Recommendations of the International Union of Geological Sciences Subcommission on the Systematics of Metamorphic Rocks.* Cambridge University Press. Cambridge, UK. See especially Section 2.2 and Section 2.8.

Miyashiro, A. (1961). Evolution of metamorphic belts. *J. Petrol.,* **2**, 277–311.

Metamorphism of Mafic Rocks

Bucher, K., and M. Frey. (2002). *Petrogenesis of Metamorphic Rocks.* Springer-Verlag. Berlin. Chapter 9.

Carswell, D. A. (ed.). (1990). *Eclogite Facies Rocks.* Blackie. Glasgow.

Cooper, A. F. (1972). Progressive metamorphism of metabasic rocks from the Haast Schist group of southern New Zealand. *J. Petrol.,* **13**, 457–492.

Poli, S., and M. W. Schmidt. (2002). Petrology of subducted slabs. *Ann. Rev. Earth Planet. Sci.,* **30**, 207–235.

Spear, F. S. (1993). *Metamorphic Phase Equilibria and Pressure–Temperature–Time Paths.* Monograph **1**. Mineralogical Society of America, Washington, DC. Chapter 11.

Yardley, B. W. D. (1989). *An Introduction to Metamorphic Petrology.* Longman. Essex, UK. Chapter 4.

UHP and UHT Metamorphism

Journal of Metamorphic Geology, vol. **21** #6 (August, 2003). Special issue: Petrochemical and Tectonic Processes of UHP/HP Terranes. vol. **25**, #2 (February, 2007) Special issue: Multidisciplinary approaches to ultrahigh-pressure metamorphism: A celebration of the career contribution of Juhn G. Liou.

Lithos vol. **52** (2000) #1–4 is a special issue on ultra-high pressure rocks.

Chopin C. (2003): Ultrahigh-pressure metamorphism: tracing continental crust into the mantle. *Earth Planet. Sci. Lett.,* **212**, 1–14.

Coleman, R. G., and X. Wang (eds.). (1995b). *Ultrahigh Pressure Metamorphism.* Cambridge University Press. Cambridge, UK.

Ernst, W. G., and J. G. Liou (eds.). (2000). *Ultrahigh-Pressure Metamorphism and Geodynamics in Collision-Type Orogenic Belts.* International Book Series, **4**. Geological Society of America. Boulder, CO, and Bellwether Publishing Ltd, Columbia, MD.

Hacker, B. R. (2006). Pressures and temperatures of ultrahigh-pressure metamorphism: Implications for UHP tectonics and H_2O in subducting slabs. *Int. Geol. Rev.,* **48**, 1053–1066.

Hacker, B. R., W. C. McClelland, and J. G. Liou (eds.). (2006). *Ultrahigh-Pressure Metamorphism: Deep Continental Subduction.* Special Paper **403**. Geological Society of America. Boulder, CO.

Harley, S. L. (1998). On the occurrence and characterization of ultrahigh-temperature metamorphism. In: *What Drives Metamorphism and Metamorphic Reactions?* (eds. P. J. Treloar and P. J. O'Brien). Special Publication **138**. Geological Society. London. pp. 81–107.

Harley, S. L. (2004). Extending our understanding of ultrahigh temperature crustal metamorphism. *J. Min. Petrol. Sci.,* **99**, 140–158.

Kelsey, D. E. (2008). On ultrahigh-temperature crustal metamorphism. *Gondwana Res.,* **13**, 1–29.

Liou, J. G., and R.-Y. Zhang. (2002). Ultrahigh-pressure metamorphic rocks. *Encycl. of Phys. Sci. and Tech.,* **17**, 227–244.

Liou J. G., T. Tsujimori, R.-Y. Zhang, I. Katayama, and S. Maruyama. (2004). Global UHP metamorphism and continental subduction/collision: The Himalayan model. *Int. Geol. Rev.,* **46**, 1–27.

Liou J. G., T. Tsujimori, I. Katayama, and S. Maruyama. (2005). UHP metamorphism and continental subduction/collision.

In: *Metamorphism and Crustal Evolution* (ed. H. Thomas). Atlantic Publishers and Distributors, pp. 285–313.

Rumble, D., J. G. Liou, and B. M. Jahn. (2003). Continental crust subduction and ultrahigh pressure metamorphism. *Treatise on Geochemistry, vol. 3, The Crust* (ed. R. L. Rudnick). Chapter 3.09, pp. 293–319. Elsevier. Amsterdam.

P-T-t Paths

Ghent, E. D., M. Z. Stout, and R. L. Parrish. (1988). Determination of metamorphic pressure–temperature–time (*P-T-t*) paths. In: *Short Course on Heat, Metamorphism, and Tectonics* (eds. E. G. Nisbet and C. M. R. Fowler). Short Course **14**, 155–212. Mineralogical Association of Canada.

Spear, F. S. (1993). *Metamorphic Phase Equilibria and Pressure-Temperature-Time Paths.* Monograph **1**. Mineralogical Society of America, Washington, DC.

Spear, F. S., and S. M. Peacock. (1989). *Metamorphic Pressure–Temperature–Time Paths.* Short Course in Geology **7**. American Geophysical Union.

Spear, F. S., J. Selverstone, D. Hickmott, P. Crowley, and K. V. Hodges. (1984). *P-T* paths from garnet zoning: A new technique for deciphering tectonic processes in crystalline terranes. *Geology,* **12**, 87–90.

26

Metamorphic Reactions

Questions to be Considered in this Chapter:

1. What advantages are gained when we consider changes in mineral parageneses (such as isograds) as reactions?

2. What various types of reactions occur in metamorphic rocks, and what principal intensive variables control the progress of each?

3. What are the effects of variable fluid and mineral compositions on the grade and temperature interval over which a reaction takes place?

4. How can we deal with multiple reactions represented on a single phase diagram, such as a pressure–temperature diagram? What happens when some reactions have several phases in common, and their intersection reduces phase rule variance?

5. In what ways might actual reaction processes differ from the reactions as typically written?

Although metamorphism may involve a number of changes, including recrystallization of preexisting phases and diffusion, the most dramatic and useful changes involve metamorphic reactions that generate new mineral phases or modify the composition of existing ones. In several instances in the preceding chapters, we have run across mineral changes that involve reactions (whether explicitly stated or not). In this chapter, we address these issues more formally.

The classic notion of an **isograd** is that it is a line in the field that demarcates the first appearance of a new mineral phase as one progresses up metamorphic grade (Section 21.6.1). Such an isograd is useful in the field because a worker need only be able to recognize new minerals in a hand specimen. More recently, we have recognized that an isograd can also demarcate the *disappearance* of a mineral as grade increases. Such isograds are commonly called **mineral-out** isograds (e.g., "muscovite-out" isograd, Reaction (26.5)), to distinguish them from the traditional (**mineral-in**) isograds. If the "in" or "out" is not stated, an isograd is accepted to be a traditional mineral-in type.

When we realize that reactions are always responsible for introducing or consuming mineral phases during metamorphism, we gain considerably in our understanding of metamorphic processes and isograds. If we treat isograds as *reactions*, we can then understand what physical and/or chemical variables affect the location of a particular isograd. Some investigators have advocated that we distinguish simple field-based mineral-in and mineral-out isograds (without a specified reaction) from reaction-based isograds. Miyashiro (1994), for example, referred to simple isograds as "tentative isograds," implying that more detailed petrographic and laboratory work on rocks below, at, and above the isograd would reveal the nature of the reaction responsible. Although every isograd must represent some sort of reaction, the exact reaction is not determined in every case, so many isograds are still characterized by, and named for, the index mineral for which they mark the appearance or disappearance.

A student may wonder why, after over a century of research in metamorphic terranes, we can't simply look up which reaction is responsible for the introduction of any particular mineral. The problem is that more than one reaction can produce any single mineral. In 10 minutes of searching my office bookshelf, I found 23 reactions that could produce biotite as one of the reaction products. I'm sure there are many more. Some of the reactions were prograde and others retrograde. Because biotite occurs in a variety of igneous rocks as well, we also know that it can be generated by reactions involving the crystallization of melts. Which reaction is responsible for the "biotite isograd" depends upon the minerals present below the isograd (the potential reactants), which, in turn, are determined by the rock composition and the metamorphic grade (*P* and *T*). Care must therefore be exercised when addressing isograds based simply on the appearance or disappearance of a mineral, and an attempt should always be made to determine the responsible reaction or reactions.

If we understand the nature of the reactions that produce metamorphic minerals, the physical conditions under which any particular reaction occurs, and what physical variables affect a reaction and how, we can use this knowledge to understand metamorphic processes better. If we have good experimental and thermodynamic data on minerals and reactions we can locate a reaction in *P-T-X* space and constrain the conditions under which a particular metamorphic rock formed (Chapters 27–29). In this chapter we will review the various types of metamorphic reactions and discuss what affects them and how.

26.1 POLYMORPHIC TRANSFORMATIONS

Single-component polymorphic transformations, such as among the polymorphs of SiO_2 (Figure 6.6) or Al_2SiO_5 (Figure 21.9), graphite–diamond (Figure 25.14), or calcite–aragonite (Figure 26.1), etc. are in many ways the simplest

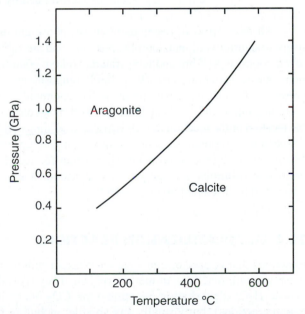

FIGURE 26.1 A portion of the equilibrium boundary for the calcite–aragonite phase transformation in the $CaCO_3$ system. After Johannes and Puhan (1971). Copyright © with permission from Springer-Verlag.

to deal with. Because the alternative phases are of essentially fixed and identical composition, the transformations depend on temperature and pressure only and are minimally affected by variations in the composition of the system in which one or another polymorph is found. For example, as long as pure $CaCO_3$ is stable in a rock system, it should occur as calcite at pressures below the equilibrium curve in Figure 26.1, and as aragonite at pressures above the curve. This explains why aragonite is the stable $CaCO_3$ polymorph typically found in blueschist facies terranes. Similar arguments hold for essentially all polymorphic transformations, and, provided that the boundary curves have been located accurately by experiments and that the mineralogy reflects equilibrium conditions, the presence of one or another polymorph may conveniently be used to set limits on the temperature and pressure conditions under which a rock formed. In Chapter 21, for example, we used the presence of andalusite to indicate low-pressure metamorphic conditions, and by referring to Figure 21.9, we can effectively limit the pressure of andalusite-bearing rocks to values below ~0.38 GPa.

The presence of two coexisting polymorphs in a single rock has generally been taken to indicate that the metamorphic peak corresponded to equilibrium conditions along the univariant boundary curve separating the pair. If an independent estimate of either pressure or temperature is available, the other parameter may then be estimated from the location of the equilibrium curve. For example, if kyanite and sillimanite were to be observed together, and the pressure were estimated via geobarometry to be 0.5 GPa, then the temperature of equilibration could be determined from Figure 21.9 to be approximately 560°C. *If* all three Al_2SiO_5 polymorphs were to be found in *stable* coexistence, the assemblage would indicate conditions at the invariant point (ca. 500°C and 0.38 GPa).

Because of the small changes in entropy and volume for most polymorphic transformations, the difference in Gibbs free energy (Chapter 5) between two alternative polymorphs may be small, even at temperatures or pressures relatively far from the equilibrium boundary. There is thus little driving force for the reaction to proceed, and crystals of one polymorph may remain as *metastable* relics in the stability field of another. Coexisting polymorphs may therefore represent non-equilibrium states, reflecting overstepped equilibrium curves or polymetamorphic overprints. By carefully observing the textures, one may be able to distinguish partial replacement and metastable coexistence from true stable equilibrium grain boundaries. For example, Hietanen (1956) reported the coexistence of all three Al_2SiO_5 polymorphs in northern Idaho and proposed that the complex sequence of regional and contact events in the area occurred near the invariant point. Others have proposed that kyanite is partially replaced by sillimanite during a prograde event near the kyanite–sillimanite boundary and that andalusite partially replaces kyanite during a later event at lower pressure. See Kerrick (1990) and Kretz (1994) for other examples of coexisting Al_2SiO_5 polymorphs and how they have been interpreted.

Another complication results from (usually minor) variations in the compositions of polymorphs from the pure phases. For example, several authors have noted the presence of Fe^{3+} in some Al_2SiO_5 polymorphs, and microprobe analyses indicate that andalusite tends to admit about twice as much Fe_2O_3 (up to $\sim$2.6 wt.%) as either kyanite or sillimanite (Kerrick, 1990). If we consider this in light of Le Châtelier's Principle, the addition of Fe^{3+} to the Al_2SiO_5 system in which andalusite and sillimanite, for example, were at equilibrium would be offset by the formation of extra andalusite. The effect of adding Fe to the system would thus be to enlarge the stability field of andalusite at the expense of the fields of kyanite and sillimanite. The concentrations of most impurities, however, including Fe, are generally considered to be too low to cause a significant displacement in the position of the equilibrium curves separating the Al_2SiO_5 polymorphs in Figure 21.9. Even in a few cases in which hematite is abundant, Kretz (1994) calculated that the shift in the andalusite–sillimanite equilibrium curve caused by Fe^{3+} would be on the order of 30°C, which is too small to account for the broad andalusite $\rightarrow$ sillimanite transition observed in some field areas. The coexistence of andalusite and sillimanite in such cases is commonly attributed to the very low ΔG of the andalusite $\rightarrow$ sillimanite reaction so that the two phases coexist metastably as the transition is overstepped with rising T.

26.2 EXSOLUTION REACTIONS

Exsolution reactions occur in solid-solution mineral series when a **solvus** is encountered, typically upon cooling (Section 6.5.4) or decompression. The process involves the unmixing of the solid solution within the composition range covered by the solvus. Typical examples include the alkali feldspar solvus (Figures 6.16 and 6.17), the peristerite gap in Na-rich plagioclase (Section 25.3), and the orthopyroxene– clinopyroxene and calcite–dolomite solvi. Exsolution need not involve minerals of the same family. For example, high-pressure pyroxenes in deep-seated high-grade rocks may dissolve quite a bit of Al_2O_3. When uplifted, the Al-rich pyroxenes may exsolve garnet or plagioclase. We discussed some other decompression exsolution phenomena in Sections 19.4 and 25.4. Exsolution may result in the development of crystallographically oriented rods or lamellae in the exsolving host or, if diffusion is favorable, distinct separate grains.

26.3 SOLID–SOLID NET-TRANSFER REACTIONS

Solid–solid net-transfer reactions involve solids only and differ from polymorphic transformations in that they involve solids of different composition. Matter must therefore be transferred from one site in the rock to another in order for the reaction to proceed. The four "general" steps involved in metamorphic reactions listed at the very beginning of Chapter 23 characterize this type of reaction. Net-transfer reactions differ from solid–solid ion exchange reactions (Section 26.6) in that progress of net-transfer reactions results in a change in

the modal amounts of the phases involved. Several examples were encountered in the previous chapter, including:

$$NaAlSi_2O_6 + SiO_2 = NaAlSi_3O_8 \quad (26.1)$$
$$\text{Jd} \qquad \text{Qtz} \qquad \text{Ab}$$

$$MgSiO_3 + CaAl_2Si_2O_8 = CaMgSi_2O_6 + Al_2SiO_5 \quad (26.2)$$
$$\text{En} \qquad \text{An} \qquad \text{Di} \qquad \text{Sil}$$

$$4(Mg,Fe)SiO_3 + CaAl_2Si_2O_8 \quad (26.3)$$
$$\text{Opx} \qquad \text{Plag}$$
$$= (Mg,Fe)_3Al_2Si_3O_{12} + Ca(Mg,Fe)Si_2O_6 + SiO_2$$
$$\text{Grt} \qquad \text{Cpx} \qquad \text{Qtz}$$

When diffusion becomes a limiting factor (generally in rocks with very little pore fluid), the reactions may become arrested, as in the corona-forming reactions discussed in Section 23.6. Minerals involved in solid–solid reactions may contain some volatiles, but the volatiles are conserved in the reaction so that no fluid is generated or consumed. For example, the reaction:

$$Mg_3Si_4O_{10}(OH)_2 + 4\,MgSiO_3 = Mg_7Si_8O_{22}(OH)_2 \quad (26.4)$$
$$\text{Tlc} \qquad \text{En} \qquad \text{Ath}$$

involves hydrous phases but conserves H_2O. It may therefore be treated as a solid–solid net-transfer reaction.

When solid–solution is limited, solid–solid net-transfer reactions are **discontinuous** reactions in the sense described in Section 6.5.2 and later. Discontinuous reactions are univariant and tend to run to completion at a single metamorphic grade (where the P-T-t path for which pressure and temperature are related crosses the univariant reaction curve). There is thus an abrupt (discontinuous) change from the reactant assemblage to the product assemblage at the reaction isograd. (The reaction behaves as though invariant in such situations, because P and T are not independent, but constrained by the P-T-t path, so $F = C - \phi + 1$.)

All three types of reactions discussed above are relatively straightforward metamorphic reactions and are subject to variations in pressure and temperature, without complications due to variations in rock or fluid compositions. The presence of reactants versus products has commonly been used, in conjunction with experimental work that constrains the location of the reaction in P-T-X space, to set limits on the temperature and pressure conditions of a metamorphic event. When solid-solution is pronounced, net-transfer reactions become **continuous** and subject to compositional effects, which will be discussed in Section 26.5.

26.4 DEVOLATILIZATION REACTIONS

Reactions that release or consume volatiles are among the most common reactions in metamorphism. They typically evolve H_2O (dehydration reactions) or CO_2 (decarbonation reactions), but virtually any volatile, including O_2, H_2, CH_4, F, Cl, SO_2, etc. may be involved under appropriate circumstances. The present discussion will concentrate on H_2O-CO_2 volatile systems, but the principles involved may be applied to any reaction releasing volatiles.

Because a volatile species is involved, the reactions are dependent not only upon temperature and pressure but also upon the **partial pressure** of the volatile components. For example, the location on a P-T phase diagram of the dehydration reaction:

$$KAl_2Si_3AlO_{10}(OH)_2 + SiO_2 \qquad (26.5)$$
$$\text{Ms} \qquad\qquad \text{Qtz}$$

$$= KAlSi_3O_8 + Al_2SiO_5 + H_2O$$
$$\text{Kfs} \quad \text{Al–silicate} \quad \text{fluid}$$

depends upon the partial pressure of H_2O (p_{H_2O}). This dependence is easily demonstrated by applying Le Châtelier's Principle to the reaction at equilibrium. Reaction (26.5), incidentally, marks the disappearance of muscovite in typical (quartz-dominant) metapelites and thus represents the muscovite-out isograd. Al_2SiO_5 and K-feldspar are both stable at lower grades (although K-feldspar is rarely encountered for most pelite compositions), so this may not make a good mineral-in isograd.

Figure 26.2 is a P-T phase diagram that shows the equilibrium reaction curve for Reaction (26.5). The heavy equilibrium curve on the right represents equilibrium between the reactants and products under H_2O-saturated conditions ($p_{H_2O} = P_{lithostatic}$). This is a common assumption, and the heavy curve represents the typical shape of equilibrium curves for dehydration reactions reported in the literature. The hydrous assemblage is nearly always on the low-temperature side of the curve, and the evolved fluid phase is liberated as temperature increases. The concave upward shape is characteristic of all devolatilization equilibrium curves at low pressure because the slope, as determined by the Clapeyron equation:

$$\frac{dP}{dT} = \frac{\Delta S}{\Delta V} \qquad (5.15)$$

is low at low pressures due to the high volume of the fluid phase but steepens quickly at higher pressures because the fluid is most easily compressed. Thus ΔV decreases much more than ΔS with increasing pressure.

As can be seen in Figure 16.16, at very high pressures, the vapor becomes so compressed that many devolatilization curves bend back upon themselves to attain a negative slope. Theoretically, complete devolatilization curves form a closed loop and bend back again at high pressure and low temperature to a positive slope again (Figure 26.3), but the full loop is rarely, if ever, stable for any reaction. Some greenschist → blueschist reactions, for example, are unusual in that they have the low-T/high-P portion of the loop stable and are distinguished by having the hydrous phase on the high-temperature side, so that "retrograde dehydration" becomes possible. If we deal only with normal crustal pressures and geothermal gradients, however, devolatilization curves have the shape in Figure 26.2, and we shall consider this shape to be typical.

Suppose H_2O is withdrawn from the system at some point on the H_2O-saturated equilibrium curve in Figure 26.2, so that $p_{H_2O} < P_{lithostatic}$. According to Le Châtelier's Principle, removing H_2O at equilibrium will be compensated by Reaction (26.5) running to the right, thereby producing more H_2O to compensate for the loss. This stabilizes the right side of the reaction at the expense of the left side. In other words, as H_2O is withdrawn, the Kfs + Al_2SiO_5 + H_2O field expands at the expense of the Ms + Qtz field, and the reaction curve shifts toward lower temperature in Figure 26.2. I have

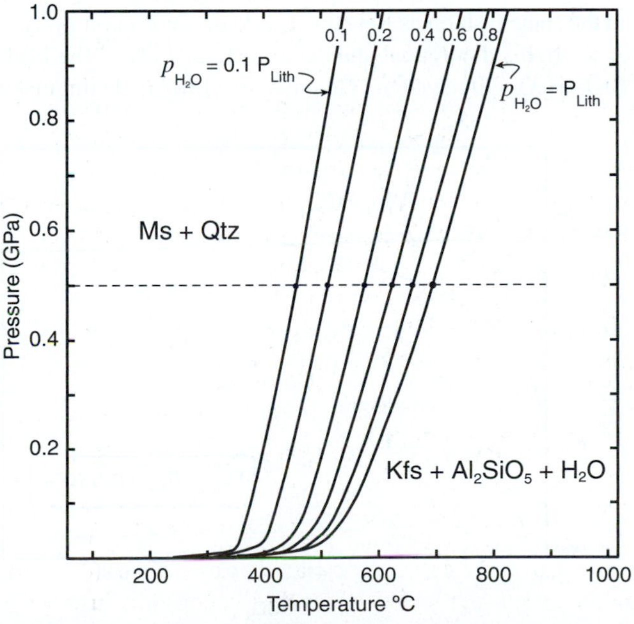

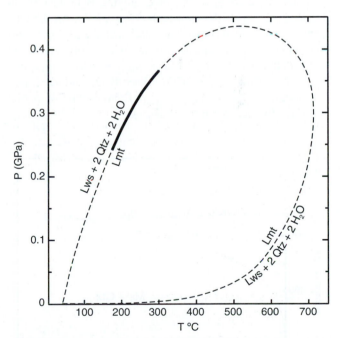

FIGURE 26.3 Calculated P-T equilibrium reaction curve for a dehydration reaction illustrating the full loop that is theoretically possible. Rarely, if ever, is the loop completely stable. In the present case, only the upper portion of the Lmt = Lws + 2 Qtz + H_2O reaction is stable. The experimentally determined equilibrium curve is shown in Figure 26.19, where the stable section is located between approximately 180 and 300°C at 0.3 GPa (solid portion). Dehydration occurs with *decreasing* temperature for this stable segment of the reaction curve (an unusual situation).

FIGURE 26.2 P-T phase diagram for the reaction Ms + Qtz = Kfs + Al_2SiO_5 + H_2O, showing the shift in equilibrium conditions as pH_2O varies (assuming ideal H_2O-CO_2 mixing). Calculated using the program TWQ by Berman (1988, 1990, 1991).

calculated the curve shift for values of $p_{H_2O} = 0.8, 0.6, 0.4,$ 0.2, and 0.1 times P_{lith} and included the curves in the figure.

p_{H_2O} can become less than P_{lith} in either (or both) of two ways. First, P_{fluid} can become less than P_{lith} by simply drying out the rock and reducing the fluid content ($p_{H_2O} = P_{fluid} < P_{lith}$). Second, P_{fluid} can remain equal to P_{lith}, but the H_2O in the fluid can become diluted by adding another fluid component, such as CO_2 or some other volatile phase ($p_{H_2O} < P_{fluid} = P_{lith}$). I calculated the curves in Figure 26.2 for the latter case, assuming ideal dilution of H_2O at $P_{fluid} = P_{lith}$.

An important point arising from Figure 26.2 is that *the temperature of an isograd based on a devolatilization reaction is sensitive to the partial pressure of the volatile species involved.* An alternative way to show this is to use a ***T-X*$_{fluid}$ phase diagram** (Greenwood, 1967a). Because H_2O and CO_2 are by far the most common metamorphic volatiles, the X in most *T-X* diagrams is the mole fraction of CO_2 (or H_2O) in H_2O-CO_2 mixtures. Thus $X_{CO_2} = n_{CO_2}/(n_{CO_2} + n_{H_2O})$, where n is the number of moles or molecules of each species in the fluid mixture. *T-X*$_{fluid}$ diagrams, however, can be created for any volatile mixtures desired. In common *T-X*$_{fluid}$ diagrams, temperature is the ordinate (y-axis), and X_{CO_2} or X_{H_2O} is the abscissa (x-axis). Because pressure is also an important variable, a *T-X*$_{fluid}$ diagram must be created for a specified pressure.

Figure 26.4 is a *T-X*$_{H_2O}$ diagram for Reaction (26.5) in which $P_{lith} = 0.5$ GPa, calculated with the same assumptions as Figure 26.2 ($P_{fluid} = P_{lith}$, and diluting H_2O ideally). In Figure 26.2, I have drawn a dashed isobaric line at 0.5 GPa and shown the intersection of this line with the series of equilibrium curves as a sequence of dots. These dots correspond to the dots in Figure 26.4 because $p_{H_2O} + p_{CO_2} = P_{lith}$ in a binary H_2O-CO_2 mixture (CO_2 dilutes H_2O), and ideal mixing assumes $p_{H_2O} = X_{H_2O} \cdot P_{fluid}$ (Dalton's law of partial pressures for ideal mixtures of gases).

Once the equilibrium curve is plotted in Figure 26.4, it is easy to label the fields by remembering that the hydrous mineral assemblage is stable at low temperature, and the volatile phase is liberated as temperature increases. Note also that the maximum stability temperature of the hydrous assemblage is for pure H_2O ($X_{H_2O} = 1.0$), because this is the maximum p_{H_2O} possible at the pressure specified, and we can imagine optimal H_2O being forced into the hydrous muscovite, enhancing the mineral's stability. At very low p_{H_2O} there is little H_2O pressure, so muscovite breaks down. A hydrous phase is not stable in an absolutely H_2O-free environment, so the equilibrium curve never really reaches $X_{H_2O} = 0$, but becomes asymptotic to it at low temperature.

The shape of *all* dehydration curves on *T-X*$_{fluid}$ diagrams is similar to the curve in Figure 26.4. They have a maximum temperature at the pure H_2O end and a slope that is gentle at high X_{H_2O}, but increasingly steep toward low X_{H_2O}, becoming nearly vertical at very low X_{H_2O}. The temperature of the reaction can thus be practically any temperature below the maximum representing $p_{H_2O} = P_{lith}$. For most of the X_{H_2O} range in Figure 26.4, however, the reaction temperature varies by less than 200°C. Nonetheless, *one should take great care to constrain the fluid composition, if at all possible, before using a devolatilization reaction to indicate metamorphic grade.*

Decarbonation reactions may be treated in an identical fashion. For example, the reaction:

$$CaCO_3 + SiO_2 = CaSiO_3 + CO_2 \qquad (26.6)$$
$$\text{Cal} \qquad \text{Qtz} \qquad \text{Wo}$$

can be shown on a *T-X*$_{CO_2}$ diagram and has the same form as Reaction (26.5), only the maximum thermal stability of the assemblage containing the carbonate mineral occurs at $X_{CO_2} = 1.0$ (Figure 26.5). The temperature of a wollastonite-in isograd based on this reaction obviously depends upon p_{CO_2} in the same fashion as Reaction (26.5) depends upon p_{H_2O}.

In his theoretical and experimental study of the MgO-SiO_2-H_2O-CO_2 system, Greenwood (1967a) distinguished

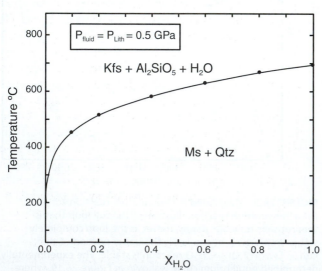

FIGURE 26.4 *T-X*$_{H_2O}$ phase diagram for the reaction Ms + Qtz = Kfs + Sil + H_2O at 0.5 GPa, assuming ideal H_2O-CO_2 mixing, calculated using the program TWQ by Berman (1988, 1990, 1991). The dots correspond to those in Figure 26.2.

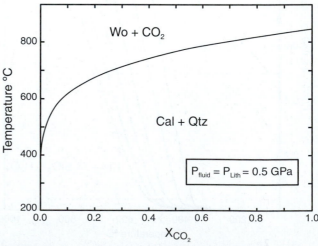

FIGURE 26.5 *T-X*$_{CO_2}$ phase diagram for the reaction Cal + Qtz = Wo + CO_2 at 0.5 GPa, assuming ideal H_2O-CO_2 mixing, calculated using the program TWQ by Berman (1988, 1990, 1991).

five principal types of equilibria involving CO_2-H_2O fluids [other than those that conserve the volatile species, such as Reaction (26.4)]. Each type has its own characteristic shape on T-X_{fluid} diagrams. The five types are based on which volatile component is consumed or liberated *as temperature increases*:

1. Dehydration reactions, such as Reaction (26.5).
2. Decarbonation reactions, such as Reaction (26.6).
3. Combined dehydration–decarbonation reactions:

$$5\,MgCO_3 + Mg_3Si_4O_{10}(OH)_2 \quad\quad (26.7)$$
$$\text{Mgs} \quad\quad\quad \text{Tlc}$$
$$= 4\,Mg_2SiO_4 + 5\,CO_2 + H_2O$$
$$\text{Fo}$$

4. Prograde reactions that consume H_2O and liberate CO_2:

$$3\,MgCO_3 + 4\,SiO_2 + H_2O \quad\quad (26.8)$$
$$\text{Mgs} \quad\quad \text{Qtz}$$
$$= Mg_3Si_4O_{10}(OH)_2 + CO_2$$
$$\text{Tlc}$$

5. Prograde reactions that consume CO_2 and liberate H_2O:

$$2\,Ca_2Al_3Si_3O_{12}(OH) + CO_2 \quad\quad (26.9)$$
$$\text{Zo}$$
$$= 3\,CaAl_2Si_2O_8 + CaCO_3 + H_2O$$
$$\text{An} \quad\quad\quad \text{Cal}$$

The typical shapes of these reaction types on T-X_{fluid} diagrams are illustrated in Figure 26.6. The shapes make some

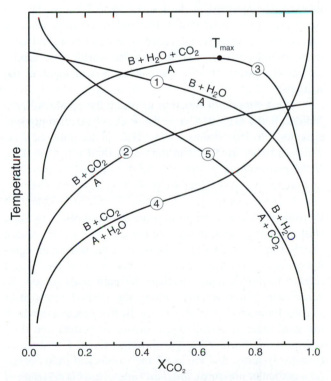

FIGURE 26.6 Schematic T-X_{CO_2} phase diagram, illustrating the general shapes of the five types of reactions involving CO_2 and H_2O fluids. *A* and *B* represent one or more solid phases, and reaction stoichiometry is ignored. After Greenwood (1967a). Copyright © reprinted by permission of John Wiley & Sons, Inc.

intuitive sense when considered in light of hydrous phases breaking down at very low X_{H_2O} and carbonate phases breaking down at very low X_{CO_2}. For example, in type 4 reactions, the mineral assemblage (A) in Figure 26.6 must include a carbonate mineral (because CO_2 is on the opposite side of the reaction). The carbonate-bearing assemblage breaks down along an isothermal traverse in Figure 26.6 toward the H_2O-rich end because there is little CO_2 to stabilize the carbonate. Assemblage (B), on the other hand, includes a hydrous mineral, which breaks down toward the CO_2-rich end. In type 5 reactions, (A) includes a hydrous phase, which breaks down toward low X_{H_2O}, and (B) includes a carbonate, which breaks down toward low X_{CO_2}. In reactions of type 3, assemblage (A) includes both a hydrous phase and a carbonate phase. The hydrous phase breaks down toward high X_{CO_2}, and the associated carbonate is consumed by the reaction even though it would otherwise be stable. Similarly, the carbonate breaks down at high X_{H_2O}, consuming the associated hydrous phase. There must be a thermal maximum to type 3 reactions in a T-X_{fluid} diagram. Any reaction that does not liberate or consume H_2O or CO_2 will not be affected by X_{CO_2} and will thus form a horizontal line on a T-X_{fluid} diagram.

Greenwood (1967a) demonstrated theoretically that the location along the X-axis of the thermal maximum (T_{max}) of type 3 reactions is determined by the ratio of the stoichiometric coefficients of CO_2 and H_2O in the reaction (at least for ideal mixing behavior). If equal molar quantities of CO_2 and H_2O are liberated, T_{max} occurs at $X_{CO_2} = 0.5$. For Reaction (26.7), which liberates 5 moles of CO_2 for each mole of H_2O, T_{max} is located at $X_{CO_2} = 5/(5 + 1) = 0.83$ (assuming ideal CO_2-H_2O mixing). T_{max} for the reaction illustrated by the hypothetical curve (type 5) in Figure 26.6 is at $X_{CO_2} = 0.67$, indicating a reaction of the type $A = B + 2CO_2 + H_2O$.

Devolatilization reactions involve a potentially mobile fluid, and thus the pressure, temperature, and progress of these reactions depend on such physical rock properties as porosity and permeability. Recall the discussion in Section 24.2 in which we distinguished perfectly mobile versus less mobile fluid components. When a system is permeable, the fluids are free to migrate and are thus highly mobile. In such cases, fluids released from greater depths may pass through shallower rocks, which may be treated as **open systems** with respect to those fluids. If the external fluid reservoir is large with respect to the shallower system, and fluids pass through readily, the fluid composition may be controlled by the larger external reservoir and thus affect the mineral assemblage of the more shallow one. When permeability is low, on the other hand, fluids remain local and are more likely to equilibrate with the minerals present, so that the mineral assemblage of the shallow system controls the fluid composition and not vice versa. As an example of open-system behavior, imagine the nominal effect of your breathing on the composition of the air around you in a breezy open field. For a closed system, imagine the same effect if you were shut in a small, air-tight container.

For an example of how this concept might work during metamorphism, consider Figure 26.7, which illustrates two devolatilization reactions. The low-temperature reaction

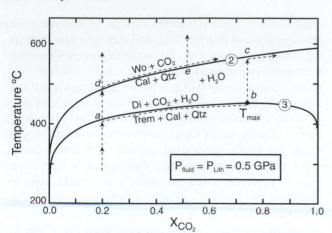

FIGURE 26.7 T-X_{CO_2} phase diagram in the CaO-MgO-SiO$_2$-H$_2$O-CO$_2$ system, for the reactions Cal + Qtz = Wo + CO$_2$ and and Tr + 3 Cal + 2 Qtz = 5 Di + 3 CO$_2$ + H$_2$O at 0.5 GPa, assuming ideal mixing of non-ideal gases, calculated using the program TWQ by Berman (1988, 1990, 1991).

is Reaction (26.5), a decarbonation reaction of type 2 above, and the other is:

$$Ca_2Mg_5Si_8O_{22}(OH)_2 + 3\,CaCO_3 + 2\,SiO_2 \qquad (26.10)$$
$$\text{Tr} \qquad\qquad \text{Cal} \qquad \text{Qtz}$$
$$= 5\,CaMgSi_2O_6 + 3\,CO_2 + H_2O$$
$$\text{Di}$$

a type 3 reaction that liberates both H$_2$O and CO$_2$. Because the molar ratio of CO$_2$:H$_2$O liberated in Reaction (26.10) is 3:1, T_{max} occurs at $X_{CO_2} = 3/(3 + 1) = 0.75$ in Figure 26.7. Other reactions occur in the CaO-MgO-SiO$_2$-H$_2$O-CO$_2$ system, as shall be discussed in Section 29.1.1.1, and may affect the reactions considered here and play an important role in meta-carbonate petrogenesis. They are ignored for the sake of simplicity in the present discussion so that we can focus on fluid behavior associated with some representative reactions rather than what mineral assemblage actually develops.

Suppose we begin with a marble that contains calcite + quartz + tremolite at low temperature (already above the tremolite isograd that has been left out of the diagram) and that the initial intergranular fluid phase is composed of 20 mol % CO$_2$ and 80 mol % H$_2$O. If we heat this mixture, the fluid follows the vertical arrow at $X_{CO_2} = 0.2$. At about 410°C, the system reaches the equilibrium curve for Reaction (26.10) (point a), and Tr + Cal + Qtz react to form diopside, liberating 3 moles of CO$_2$ and 1 mole of H$_2$O for each mole of tremolite consumed. First, let's assume that the rock is porous and permeable (*open-system* behavior), so these volatile species are free to escape and leave the system without altering the composition of the intergranular fluid phase that passes through the rock, which would then remain at $X_{CO_2} = 0.2$ if it is controlled at that composition by some large external reservoir. If this is the case, Reaction (26.10) will behave as an invariant (discontinuous) reaction. According to the phase rule, $F = C - \phi + 2 = 5 - 5 + 2 = 2$, but, because P and X_{CO_2} are externally fixed, F is reduced by 2, to 0. Alternatively, we could consider CO$_2$ and H$_2$O

as perfectly mobile, so $C = 3$ and $\phi = 4$ (you can't count the fluid phase if you don't count CO$_2$ and H$_2$O as components). Thus $F = 3 - 4 + 1$ (isobaric) $= 0$. Because the reaction is invariant, the system will remain at 410°C (and $X_{CO_2} = 0.2$) until the reaction runs to completion (defined as the point at which *one of* the reactant phases is consumed; remember that a reaction can proceed only as long as *all of the* reactants are present). Any heat added while $F = 0$ is consumed by the endothermic reaction at a constant temperature.

Once a reactant is consumed, the reaction is considered complete, and temperature may rise again along the fixed X_{CO_2} fluid path. If tremolite is the limiting phase and is consumed first, the system may be heated to point d, where the remaining calcite and quartz will react to produce wollastonite via Reaction (26.5) and liberate only CO$_2$. Again, if the fluid is perfectly mobile and controlled by an external reservoir at $X_{CO_2} = 0.2$, then this too will act as a discontinuous invariant reaction (in the CaO-SiO$_2$-CO$_2$ system) and remain at 487°C until either calcite or quartz is consumed before the temperature can rise again with Wo + Qtz or Wo + Cal present.

Alternatively, if the initial temperature were around 350°C and $X_{CO_2} = 0.2$, but nearly pure H$_2$O fluids were released into the shallow system from a crystallizing granite below, X_{CO_2} would then decrease to approximately zero, causing both "prograde" reactions to occur, producing first diopside and then wollastonite *under isothermal conditions*. Such a process may be responsible for the occurrence of very wollastonite-rich rocks associated with numerous marbles. Tracy et al. (1983) described a situation in Connecticut where quartz veins inject marble lenses and created calc-silicate assemblages at the boundaries. They interpreted the veins as representing externally controlled SiO$_2$-H$_2$O-rich fluid conduits, and the addition of H$_2$O caused the marble to decarbonate locally, producing the calc-silicate assemblages at the margins.

Next consider a situation in which the permeability is limited. If the original Cal + Tr + Q is heated to point a, Reaction (26.10) produces CO$_2$:H$_2$O in the ratio of 3:1, which is richer in CO$_2$ than the initial fluid ($X_{CO_2} = 0.2$). If the permeability is low, the fluid produced by the reaction mixes with the original fluid, shifting X_{CO_2} to higher values. Because the reaction tremolite + calcite + quartz → diopside has already begun in order to accomplish this fluid shift, all four mineral phases are present, so the fluid cannot simply shift isothermally to the right of the univariant equilibrium curve on the T-X_{fluid} diagram because this would require losing Di again. Rather, the path must *follow the equilibrium reaction curve* along the dashed line (offset slightly for clarity). This is a classic **buffer** process, in which the solid mineral assemblage controls, or buffers, the fluid composition. Because X_{fluid} is no longer fixed, $F = 1$ (in an *isobaric* system). As long as all four solids and fluid coexist (at a particular pressure), *the fluid composition is determined by the temperature*, along the equilibrium curve in Figure 26.7. If we continue to heat the system, the composition of the fluid will be buffered toward higher X_{CO_2} as temperature rises and the reaction proceeds, consuming Tr, Cal, and Q and producing Di and CO_2-enriched fluid over that interval. The reaction is thus a univariant one and occurs over a temperature interval across which the proportions of the reactants and products vary.

At what point will a buffering reaction cease? It depends upon the quantities of the reactants and fluid and the permeability of the rock. The porosity is probably very small in most regionally metamorphosed rocks and restricted to an intergranular film (less than 1 to 2% of the rock volume). In such a case, growth of only a few volume % of diopside would release over 10% fluid, which would greatly modify (and perhaps flush out) the small amount of fluid initially present. The prograde reaction will thus modify the fluid toward $X_{CO_2} = 0.75$ as the temperature rises toward T_{max} at 455°C (point b on Figure 26.7). Suppose such a system evolves to point b. At this point, the fluid produced by the reaction is equal to the existing pore fluid, and neither the temperature nor the fluid composition can change as long as all four minerals are present. Thus the reaction will run to completion at 455°C, until one reactant is consumed. Greenwood (1975) argued that this should be a common situation and that even though a reaction path such as a → b in Figure 26.7 spans approximately 45°C as it progresses along the Di-producing reaction, only a minor amount of Di is created until X_{CO_2} of the fluid is stabilized at 0.75. If Greenwood is correct, the reaction will produce and maintain a reservoir of fluid with $X_{CO_2} = 0.75$ at point b during most of the reaction process. This, in turn, is capable of "externally buffering" the fluid for other rocks (generally situated above this system because fluids tend to rise). A similar external buffer may have been responsible for maintaining $X_{CO_2} = 0.2$ in the open system described above.

Once a reactant is consumed, the temperature of the system may rise from the equilibrium curve for Reaction (26.10). Again, if tremolite is consumed first, calcite + quartz will react to wollastonite at point c, where pure CO_2 is produced at 570°C. In the case of low permeability, the Cal + Qtz + Wo assemblage will then buffer the fluid with rising temperature toward $X_{CO_2} = 1.0$. Again, if Greenwood (1975) is correct, the generation of very little Wo may be required to raise X_{CO_2} of the fluid to nearly 1.0, and most of the Wo will be created when the fluid composition becomes stabilized at that composition at around 590°C.

If one of the reactants is present in very small quantities, or if the rock is semi-permeable, either reaction may run to completion before the composition of the pore fluid becomes stabilized at the value of the fluid produced by the reaction. Is such situations, the reaction (and its buffering ability) will cease at some point along the reaction-buffering path before reaching T_{max}. For example, Reaction (26.10) may cease when tremolite (or another reactant) is consumed before X_{CO_2} of the fluid reaches 0.75 in Figure 26.7. The T-X_{fluid} path of the process may thus leave a reaction curve at any point between the point it first encounters the equilibrium curve and the ultimate stabilized fluid composition. The path a → d → e and upward from e in Figure 26.7 illustrates a fluid being buffered along Reaction (25.5) when either calcite or quartz is consumed at an arbitrary point e, and the buffer assemblage is lost so temperature may increase from there with $X_{CO_2} = 0.52$. In Section 30.2.1.2, we shall discuss methods by which modal mineralogical changes can be used to estimate fluid fluxes.

It is generally impossible to predict at what T-X_{fluid} point a devolatilization reaction will run to completion. The open-system and buffer-to-T_{max} scenarios represent the two extreme situations. Even in low-porosity, well-buffered situations, such as the path a → b in Figure 26.7, as the composition of the intergranular fluid approaches point b during the progress of Reaction (26.10), more reaction progress is required to modify it because the fluid being produced is approaching that already existing in the pore spaces. Greenwood (1975) may be correct in that only a small degree of reaction progress is required to buffer the fluid composition quickly to the *vicinity of* T_{max}. The reaction, however, may still run to completion before it finally gets there.

So the temperature interval over which a buffered reaction takes place is wider than for unbuffered open systems. Rather than having the reactants and products coexist only at a single temperature (open system, discontinuous) they can coexist over a temperature range, in many cases in excess of 100°C. As in our phase rule discussion in Section 24.2 with regard to periclase, brucite, and H_2O, the fluid phase is not perfectly mobile in buffered situations, and thus H_2O and CO_2 must be considered as components, and the fluid considered as a phase, thereby explaining why so many phases (reactants plus products) coexist over a temperature interval (and perhaps across a modest spatial zone). Rather than being discontinuous, the reactions become continuous, as the composition of at least one phase (in this case, the fluid) changes over the temperature interval in which the reaction proceeds. Such metamorphic continuous reactions are analogous to the continuous melting reactions discussed in Chapters 6 and 7. We shall develop this concept further in the next section, when we generalize to include solid-solutions as well.

26.5 CONTINUOUS REACTIONS

Imagine an idealized field area of steeply dipping metamorphosed, mostly pelitic sediments that strike directly up metamorphic grade (Figure 26.8). The bulk composition of each unit is homogeneous but differs somewhat from the other units in the area. One can thus compare how each composition behaves as the grade of metamorphism increases by walking within each unit along strike toward higher grade. The garnet isograd (based on the first appearance of garnet) is shown in Figure 26.8 as a dashed line. Note that the isograd is not located at the same grade in each unit. This may occur for one of two reasons (assuming that the rocks represent equilibrium mineral assemblages):

1. The rocks may be of such contrasting compositions that garnet is produced by different reactions. For example, in some rocks, garnet may be created by the (unbalanced) reaction:

$$Chl + Ms + Qtz \rightarrow Grt + Bt + H_2O \quad (26.11)$$

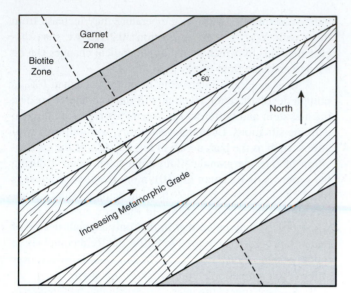

FIGURE 26.8 Geologic map of a hypothetical field area in which metamorphosed pelitic sediments strike directly up metamorphic grade. The isograd based on the first appearance of garnet is shown as a dashed line. The offsets of this isograd at unit contacts are discussed in the text.

In more Fe-rich and K-poor pelites, garnet might be generated by an (unbalanced) reaction involving chloritoid:

$$\text{Chl} + \text{Cld} + \text{Qtz} \rightarrow \text{Grt} + \text{H}_2\text{O} \qquad (26.12)$$

Offsets in a particular isograd, when based on different reactions, may be relatively large. In the blank unit in the center of the map, garnet isn't created at all. This could be a Mg-rich sandy pelite or even a quartzite or marble.

2. The reaction on which the isograd is based is the same in each unit, but it is a **continuous reaction**, and its location is sensitive to the composition of the solutions (either solid or fluid) involved. The offsets this creates in an isograd are usually more subtle than for reason #1, but in some cases they can be substantial.

Isograd offsets due to the first reason are relatively obvious and will be dealt with in more detail when we look at pelitic rocks in Chapter 28. For now, let's concentrate on the second reason.

To understand the effect of solutions on the temperature of an isograd, let's go back to a familiar system: the Fo-Fa solid solution that comprises olivine in Figure 6.10. Notice that the temperature at which melt first occurs depends on the Mg/Fe ratio of the bulk composition that we plan to melt. If we begin with a rock that has 30 wt. % Fo, for example, melt of a composition having about 8 wt. % Fo is first generated at ~1320°C. If we begin with a rock that has 70 wt. % Fo, on the other hand, melt with a composition of about 35 wt. % Fo will first be generated at ~1585°C. The temperature at which melt is first generated (consider it a "melt-in" isograd) between these two examples varies by about 265°C!

While we are referring to Figure 6.10, it may be a good time to recall the difference between discontinuous and continuous reactions. A *discontinuous* reaction occurs at a constant temperature when there are no degrees of freedom

($F = 0$) according to the phase rule. As an example, consider pure forsterite at a constant pressure in Figure 6.10. In the pure system, the number of components, C, is 1. When forsterite begins to melt at 1890°C, $\phi = 2$, so that $F = C - \phi + 1(isobaric) = 1 - 2 + 1 = 0$. Thus the temperature remains constant until all of the forsterite is consumed by the melting reaction. This is discontinuous because a pronounced change occurs in the system at a single temperature (solid below the temperature, liquid above it).

In the two-component system, if we similarly heat Fo_{30}, melting begins at 1320°C but is not completed until 1535°C, at which time the liquid composition becomes equal to the bulk composition (as discussed in Chapter 6). In this case, the melting reaction is a *continuous* reaction because $C = 2$ and $F = 2 - 2 + 1 = 1$. Because there is a degree of freedom, temperature is free to vary, and the composition of the liquid and solid varies along the liquidus and solidus, respectively, as melting progresses.

So, discontinuous reactions occur at a constant temperature (or metamorphic grade). Discontinuous reactions are actually *univariant* ($F = 1$) on *P-T* phase diagrams. Because pressure and temperature are not independent during metamorphism, however, but constrained to follow a geothermal gradient or *P-T-t* path, the *P-T* path thus crosses the reaction at a single *P-T* point. In other words, because T depends upon P (as determined by the local geothermal gradient), a degree of freedom is lost. Suppose Reaction (26.11) were a discontinuous reaction (let's say it occurred for pure Mg end-members) and is responsible for the formation of garnet in the map area of Figure 26.8. We would then expect the reaction to run to completion (when *one* of the reactants was consumed) at a single metamorphic grade. Continuous reactions occur when $F \geq 1$, and the reactants and products coexist over a temperature (or grade) interval. If Reaction (26.11) were a *continuous* reaction, then we would expect to find chlorite, muscovite, quartz, biotite, and garnet all together in the same rock over an interval of metamorphic grade above the garnet isograd. The composition of one or more solution phases (usually all of them) will then vary across the interval, and the proportions of the minerals will change correspondingly until *one of* the reactants disappears with increasing grade.

A continuous metamorphic reaction is illustrated in Figure 26.9, a schematic *isobaric T-X*$_{\text{Mg}}$ representation of Reaction (26.11), simplified to eliminate K_2O and approximate a two-component system analogous to the Fo-Fa melting reaction in Figure 6.10. The reaction is discontinuous in the pure Fe and pure Mg systems, but continuous in the mixed Fe-Mg system (C increases by one, so F does also). Just as in Figure 6.10, the absolute temperature of the first appearance of the new phase (garnet), as well as the width of the temperature interval within which the reactants and products coexist, varies with the Fe/Mg ratio of the bulk rock composition. For the pure-Fe and pure-Mg systems, the reaction occurs at a single temperature (T_{Fe} or T_{Mg}), and the reaction curves on a *P-T* diagram (dashed in Figure 26.9b) are sharp (discontinuous), univariant reactions crossed at a single grade. For a mixed Fe/Mg system, however, the

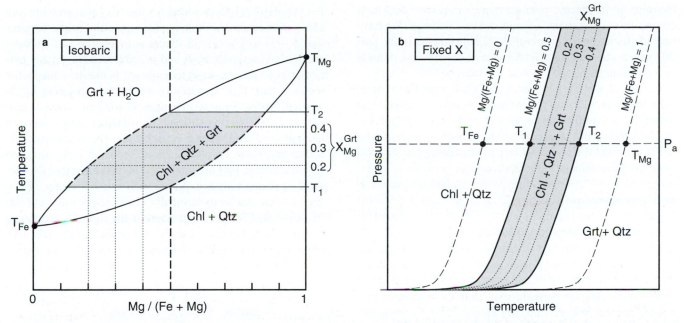

FIGURE 26.9 (a) Schematic isobaric T-X_{Mg} "pseudosection" representing the simplified metamorphic reaction Chl + Qtz → Grt + H$_2$O. (b) Schematic P-T "pseudosection" for a specific bulk composition [in this case for Mg/(Mg + Fe) = 0.5 and quartz-excess], showing the stability fields of Chl + Qtz, Chl + Qtz, and Grt + Qtz. Note the continuous nature of the reaction when all solid phases are present (shaded area). Note also that one can contour the shaded divariant field in (b) for specific compositions of either garnet (as has been done) or chlorite. The boundaries and contours would change for a different X_{bulk} (i.e., different X_{Mg}), and even the field assemblages might change: for example, the higher-temperature garnet + quartz field would be garnet + chlorite if the SiO$_2$ content were so low that quartz were consumed before chlorite by the reaction.

reaction becomes continuous. Suppose, for example, a rock with Mg/(Fe + Mg) = 0.5 is heated (vertical heavy dashed line at the bottom of Figure 26.9a). The garnet-in isograd for such a rock occurs at T_1. Chlorite (or quartz in less common SiO$_2$-poor pelites) is not consumed until T_2 is reached. Between T_1 and T_2, the Chl + Qtz (+ Ms) → Grt + H$_2$O (+ Bt) reaction is continuous, all five minerals are present, and chlorite is gradually consumed as more garnet is produced while the composition of each becomes progressively more Mg-rich (although X_{Mg} at any given temperature is not the same in garnet as it is in biotite or chlorite). On a P-T phase diagram (Figure 26.9b), the reaction for X_{Mg} = 0.5 is shown as the shaded band. If isothermally heated, T_1 is the garnet-in isograd, and chlorite is consumed at T_2. A more likely situation would involve a P-T-t path with positive slope rather than being isothermal, but T_1 and T_2 would nonetheless be fixed by the intersection of the path and the reaction boundaries. Figures 26.9a and b constitute rudimentary T-X and P-T **pseudosections**: grids designed to show reactions *for specific rock compositions*. A hallmark of pseudosections is the depiction of fields of stability of particular mineral assemblages, rather than showing the univariant, typically pure-Mg or pure-Fe, end-member reactions so common on P-T grids. We will encounter more realistic and comprehensive pseudosections in Section 28.2.5. The isograd offsets mapped in Figure 26.8 may thus reflect differences in bulk X_{Mg} from one unit to another (imagine any other bulk composition in Figure 26.9a), and the continuous reaction proceeds across a variable width within the garnet zone immediately above each isograd segment.

When petrologists use the term *continuous reaction*, they are referring to reactions such as the one illustrated in

Figure 26.9, involving *solid* solutions. Devolatilization reactions may be continuous as well, in a sense, because the *fluid* composition may change over a buffered reaction temperature interval. Only when a devolatilization reaction is *also* affected by continuous *solid*-solution behavior would the reaction be considered "continuous" in common usage (as is the case for the reaction in Figure 26.9). We will develop these concepts in more detail in Chapter 28.

26.6 ION EXCHANGE REACTIONS

Ion exchange reactions involve the reciprocal exchange of components between two or more minerals. Although the exchange can involve anions or anionic complexes, such as Cl-F-OH exchange among amphiboles and micas, petrologists have concentrated more on exchange reactions involving cations. Typical examples include:

Fe-Mg exchange between Opx and Cpx:

$$En + Hd = Fs + Di$$

Fe-Mg exchange between garnet and biotite:

$$Annite + Pyrope = Phlogopite + Almandine$$

Note that ion-exchange reactions are conveniently expressed in terms of the opposing pure end-member components ($A_{Fe} + B_{Mg} = A_{Mg} + B_{Fe}$), and an equilibrium constant [K in Equation (27.20)] is then fitted to actual compositions.

Although continuous reactions may also involve shifting Fe/Mg ratios in reacting phases, the modal proportions of the phases also change as a continuous reaction progresses. Ion exchange reactions differ in that *the modal amounts of the phases involved remain constant*; only their composition changes as a result of the exchange. We have become increasingly aware of these adjustments in the past 30 years because of the relative ease of determining mineral compositions using the electron microprobe.

As we saw in Chapter 9 and shall investigate more fully in Chapter 27, the partitioning of cations between two minerals in equilibrium is significantly temperature dependent. Many of the exchanges have been experimentally calibrated, and the compositions of coexisting phases have been used to estimate the temperature or pressure of metamorphic equilibration, leading to the technique known as *geothermobarometry* (Section 27.4). The fixed distribution of Fe and Mg, for example, on an *AFM* diagram representing a given metamorphic grade (such as Figure 24.20) requires that the *tie-lines connecting coexisting phases cannot cross*. Exchange reactions occur as grade changes and cause the tie-lines to rotate as Fe and Mg ratios vary. The tie-lines representing one temperature may thus cross those representing another temperature (see Figure 27.6). Crossing tie-lines, when observed between coexisting solid solutions in a single rock, are good indicators of non-equilibrium, such as partial readjustments to changing grade.

Exchange reactions can occur readily, even during retrograde metamorphism, a process that can significantly upset geothermobarometry. A common example is biotite in high-temperature rocks that cool slowly. When a rock is in equilibrium, the composition of any particular mineral should be the same throughout the rock. If a mineral is zoned, the rim, at least, should be in equilibrium with the rims of the other minerals. Biotite rim compositions in numerous high-grade gneisses vary somewhat, depending on the mafic mineral with which the biotite crystal is in contact. This suggests that some limited exchange occurred during cooling. The temperature determined by a geothermometer involving biotite may thus record a **blocking** or **closure temperature** (the temperature below which kinetic factors impede the retrograde exchange reaction), rather than the metamorphic peak temperature. For cases in which the biotite compositions vary, it may be better to analyze a biotite surrounded by felsic minerals where retrograde Fe-Mg exchange with an adjacent mafic neighbor cannot readily occur.

26.7 OXIDATION/REDUCTION REACTIONS

Oxidation/reduction, or **redox**, reactions involve changes in the oxidation state of ions or ionic complexes that naturally occur in more than one state. Fe^{2+}-Fe^{3+} is probably the most common multi-valent ion of geological interest, but other examples include Cu^+-Cu^{2+}, Mn^{2+}-Mn^{3+}, O^0-O^{2-}, S^0-S^{2-}, C^0-C^{4+}, etc. A simple redox reaction that relates hematite and magnetite is:

$$6\,Fe_2O_3 = 4\,Fe_3O_4 + O_2 \qquad (26.13)$$

In this reaction, two-thirds of the Fe^{3+} ions in hematite are reduced to Fe^{2+} ions in magnetite, and half as many O^{2-} ions are oxidized to O^0 to compensate and maintain electrical neutrality. Notice that, in the three-component system

FeO-O_2-H_2O (H_2O is added so that oxygen pressure can vary), the occurrence of three phases at equilibrium requires that $F = 3 - 3 + 2 = 2$. Because magnetite and hematite are pure phases, only P, T, and p_{O_2} are variable. At any particular pressure, the solid assemblage hematite + magnetite requires that $F = 1$, and the assemblage behaves as an **oxygen buffer**: imposed changes in oxygen concentration are compensated by shifting the relative proportions of hematite and magnetite at equilibrium. The equilibrium reaction curve can be plotted on an isobaric T-p_{O_2} diagram (Figure 26.10). This diagram uses oxygen *fugacity* (f_{O_2}) instead of partial pressure. As we shall see in Chapter 27, fugacity is the thermodynamically effective pressure and can be substituted for partial pressure if the difference between real and ideal behavior can be determined in the lab. Two other oxygen-buffering reactions are included in Figure 26.10. These are:

$$2\,Fe_3O_4 + 3\,SiO_2 = 3\,Fe_2SiO_4 + O_2 \qquad (26.14)$$
magnetite + quartz = fayalite ("FMQ")

$$Fe_2SiO_4 = 2\,Fe^0 + SiO_2 + O_2 \qquad (26.15)$$
fayalite = iron + quartz ("QIF")

Although these buffers are effective at controlling oxygen fugacity in experimental runs, natural rocks are more complex and in many cases control f_{O_2} by equilibria involving several silicate and/or oxide phases. Only in ironstones does the hematite–magnetite (HM) buffer operate approximately as advertised. Nevertheless, the buffers in Figure 26.10 bracket natural f_{O_2} values in metamorphic and igneous rocks because Fe in silicates is seldom as oxidized as in hematite, and it is rarely found in the native iron state. Notice that this limits f_{O_2} values in metamorphic rocks (shaded area in Figure 26.10) to the range of 10^{-10} to 10^{-50} MPa. This may be as little as a few molecules of free O_2 per cubic cm of rock! When present in such minute amounts, oxygen may not be able to diffuse easily through the rock, and values of f_{O_2} may thus vary on a local

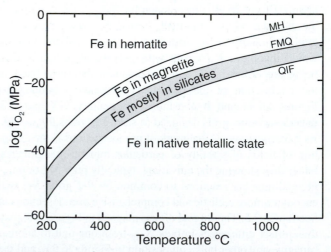

FIGURE 26.10 Isobaric T-f_{O_2} diagram showing the location of Reactions (26.13) to (26.15) used to buffer oxygen in experimental systems. Note the range of f_{O_2} in which iron occurs principally in silicate minerals (i.e., most natural rocks). After Frost (1991). Copyright © the Mineralogical Society of America.

basis. This would explain, for example, why banded iron formations have sub-centimeter-scale layers of hematite and magnetite between which f_{O_2} varies considerably.

Redox reactions are common in a number of metamorphic, igneous, hydrothermal, and near-surface aqueous systems. The presence of carbonaceous material in many dark shales is a powerful reducing agent, resulting in some reduced meta-pelite mineral compositions. For a summary of redox reactions, see a geochemistry text, such as Faure (1998), and for a more specific review of redox in metamorphism, see Eugster (1959).

26.8 REACTIONS INVOLVING DISSOLVED SPECIES

Fluids are capable of dissolving a variety of chemical components as they equilibrate with the minerals in a rock. These components are garnered by reactions between the fluids and minerals, and the dissolved species are generally maintained in equilibrium with the solids, unless the flow rate is too fast and/or the fluids are buffered by some larger external reservoir. The approach to equilibrium between a fluid and the minerals of a rock is generally referred to as **fluid–rock interaction**. This interaction is fundamental to the process of metasomatism (Chapter 30) and is also important during regional metamorphism, weathering, ore deposition, hydrothermal alteration, ocean-floor metamorphism, etc.

A well-used example of this type of reaction is the **hydrolysis** weathering reaction of K-feldspar given in practically every introductory geology text:

$$2\,KAlSi_3O_8 + 2\,H^+ + H_2O \qquad (26.16)$$
$$\text{Kfs} \qquad \text{aqueous species}$$
$$= Al_2Si_2O_5(OH)_4 + SiO_2 + 2K^+$$
$$\text{kaolinite} \qquad \text{aqueous species}$$

By this reaction, orthoclase reacts with somewhat acidic H_2O to form a common clay mineral plus dissolved SiO_2 and K^+ ions. As another example, the albitization of plagioclase during ocean-floor metamorphism involves the substitution of Na^+ dissolved in seawater for Ca^{2+} in the anorthite component of the plagioclase (plus other mineral products; see Section 21.3.2).

Many metamorphic reactions involve such aqueous constituents. The reactions may be understood in terms of the phase rule and the appropriate intensive variables: typically T, P, and the concentrations of the aqueous species. Isobaric T-concentration diagrams, isobaric–isothermal concentration–concentration diagrams (e.g., C_{H^+} versus C_{K^+} for Equation 26.16), or plots of T versus concentration ratios (e.g., T versus C_{K^+}/C_{H^+}) can be used in the same way as Figures 26.4 through 26.10. Activities (thermodynamically effective concentrations) are typically substituted for concentrations, resulting in T-activity or activity–activity diagrams (see Figure 30.12 for an example). As we shall see in Section 26.11, even mineral reactions that do not appear to involve dissolved species in the overall reaction may have them as intermediate states. The "net transfer" in solid–solid

net transfer reactions implies migration of such dissolved material. We shall postpone further discussion of reactions that involve dissolved species until Chapter 30.

26.9 REACTIONS AND CHEMOGRAPHICS: A GEOMETRIC APPROACH

As we learned in Chapter 24, chemographic diagrams can be excellent geometric aids that help conceptualize the relationship between rock chemical composition and stable equilibrium mineral assemblages. An appropriate chemographic compatibility diagram should be limited to a small range of P-T conditions, such as a metamorphic zone. The reason for this restriction is that zones are separated by isograd reactions, and reactions change the geometry of compatibility diagrams. Diagrams representing any two zones will thus appear different, having either different phases, different tie-lines connecting coexisting phases, or both. A change in tie-lines between adjacent metamorphic zones thus means that different mineral groupings result from the reaction that separates them. We typically refer to the overall geometry of phase compositions and tie-lines on a compatibility diagram with the general term **topology**.

To understand how reactions can be dealt with in terms of chemographic geometries, it is probably easiest to begin with a simple binary system. You may remember from Chapter 6 that three collinear phases on a chemographic phase diagram imply a possible reaction because the central phase can be generated by combining the phases at the ends of the line.

For example, consider the binary chemographic diagram in the MgO-SiO_2 system illustrated in Figure 26.11, which makes a good binary compatibility diagram. Specify a rock bulk composition by picking any point along the line, and the mineral assemblage that develops equals the two phases that bracket the point. SiO_2-rich rocks will thus develop En + Qtz, less SiO_2-rich rocks will develop Fo + En, and MgO-rich rocks will develop Per + Fo. We can combine any two phases on the diagram to create a phase between them, such as Per + En = Fo, or Fo + Qtz = En. The geometry indicates that these reactions are possible but does not indicate whether a reaction must occur in nature, nor the conditions under which it might occur (something requiring experimental or thermodynamic data). If we write down the formulas, we find that we can balance any reaction indicated geometrically. For example:

$$Mg_2SiO_4 + SiO_2 = 2\,MgSiO_3 \qquad (26.17)$$
$$\text{Fo} \qquad \text{Qtz} \qquad \text{En}$$

We could get the ratio of Fo:Q for the reaction directly from the chemographic diagram by using the lever principle, if necessary, but it's usually easier to balance the

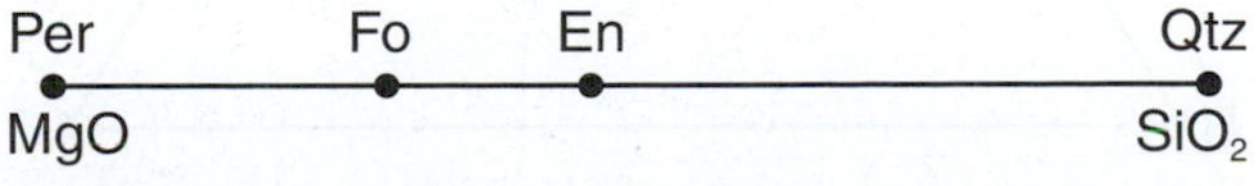

FIGURE 26.11 An example of a possible chemographic diagram in the binary MgO-SiO_2 system, showing the positions of periclase, forsterite, enstatite, and quartz.

reaction mathematically. The simple geometric rule that two minerals can be combined to create a mineral that plots directly between them tells you whether a reaction *can be* balanced; the actual balancing is then a simple matter of bookkeeping.

So how would a compatibility diagram change from one zone to another via such a reaction? Suppose Reaction (26.17) is an "enstatite isograd" for some hypothetical field area. Below the isograd, the linear chemographic diagram might look like Figure 26.11, but it will have only periclase, forsterite, and quartz, whereas above the isograd Figure 26.11 would apply because enstatite is generated and appears on the line between Fo and Qtz. This example is purely hypothetical, of course, because forsterite and quartz never coexist at equilibrium in nature, contrary to the implication of the Fo-Qtz tie-line in the chemographics below our hypothetical "isograd." Alternatively, a reaction may signify the *disappearance* of a phase with increasing metamorphic grade (a "mineral-out" isograd), in which case the phase would appear in the lower-grade diagram and not in the one representing the higher grade.

A similar situation for three components is illustrated in Figure 26.12. From the geometry shown, we could infer that $A + B + C$ could be combined in some proportions to produce composition X because X is coplanar with A, B and C and lies within the A-B-C chemographic triangle. The triangle need not be equilateral and could be any sub-triangle within a larger three-component system. *As long as a phase lies within a triangle, the apices of which are three other phases, the apical phases could be combined to create the phase within.* Again, balancing the reaction is a matter of bookkeeping. The formula of X in Figure 26.12 is A_2BC, so the reaction would be $2A + B + C = X$.

The geometric implication of a point within a triangle for a possible reaction is true whether the triangle is a true ternary diagram or a **projection**. If the chemographic triangle is part of a projection, balancing it will require other phases and/or components (ones involved in the projection). For example, suppose the A-B-C triangle is a projection in some four-component system. If we "un-project" it and view it in three dimensions, it may appear as in Figure 26.13. Note that point X can be construed as either a combination of phases A, B, and C or as a linear combination of phases D and E. We can thus geometrically infer the (unbalanced) reaction: $A + B + C = D + E$. Such a reaction is possible because some combination of phases on both sides of the reaction can be equivalent to X. This is true for any line that pierces a triangle in chemographic space because the point of intersection can be resolved into either the corners of the coplanar triangle or the ends of the line.

A **"tie-line flip"** is perhaps the most common situation in which a reaction causes a change in topology for a chemographic diagram. Consider Figure 26.14, in which a hypothetical three-component system has six phases, A, B, C, D, X, and Y. Suppose Figure 26.14a represents a lower-grade metamorphic zone and Figure 26.14c represents an immediately higher grade zone. The possible mineral assemblages in part (a) are:

$$X + A + D \quad A + D + B \quad A + B + C \quad B + D + Y$$

But the assemblages in part (c) are:

$$X + A + D \quad A + D + C \quad C + D + B \quad B + D + Y$$

Notice that *the minerals in both diagrams are the same, but the groupings (tie-lines) differ*. Figure 26.14b shows the transitional situation at the isograd itself. Remember that coexisting phases are connected by tie-lines. The **crossed tie-lines** indicate the isograd reaction geometrically. Because the point of intersection of the two alternative tie-lines is an intermediate point on either line, it could be composed of either $A + B$ or $C + D$. This implies that the contrasting

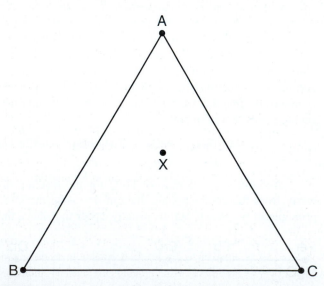

FIGURE 26.12 A hypothetical three-component system, *A-B-C*, having a compound *X* within the triangle. *X* can be generated by a reaction involving *A + B + C*.

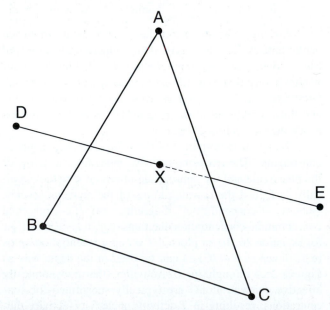

FIGURE 26.13 Three-dimensional perspective view of the *A-B-C* triangle in Figure 26.14, contained in a hypothetical four-component system. *X* can be generated by a combination of *A + B + C* or of *D + E*, implying a possible reaction: *A + B + C = D + E*.

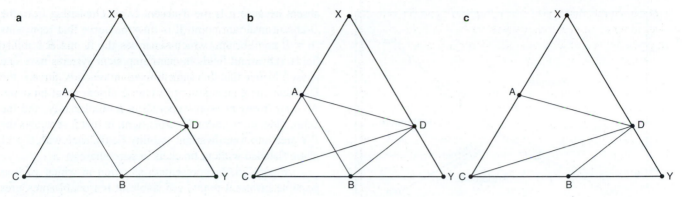

FIGURE 26.14 The sequence of chemographic diagrams for a hypothetical three-component system illustrating a "tie-line flip" resulting from a reaction *A* + *B* = *C* + *D*. **(a)** The topology for the lower grade. **(b)** The topology for the transition. **(c)** The topology for the higher grade.

pairs, if combined in the appropriate proportions, are compositionally equivalent, suggesting the (unbalanced) reaction:

$$A + B = C + D$$

The chemographic diagrams for the two zones (Figures 26.14a and c) are thus related by a "flip" of the *A-B* tie-line to become the *C-D* tie-line. When combined, the tie-lines cross, and this can be used to derive the reaction that relates the two zones. Once again, this is rigorously true only if the four phases at the ends of the tie-lines are indeed coplanar and not a projection. If they are projected, the reaction would be:

$$A + B + \ldots = C + D + \ldots$$

where the ellipses indicate that other phases, composed of the projected components, may be needed to produce a final mathematically balanced reaction. Again, the geometry of crossing tie-lines shows that a balanced reaction is possible, but it can't do the balancing for you.

More complex geometric reaction constructions are possible, including intersecting planes, etc., but the added complexity of visualizing in three or more dimensions soon compromises the advantages of the approach. The simple geometries discussed above are the most useful. For more complex reactions, an approach based on linear algebra is preferable for balancing. The program CSpace and the brief tutorial available via www.prenhall.com/winter may be used to balance any reaction that can be balanced.

26.10 PHASE DIAGRAMS FOR MULTICOMPONENT SYSTEMS THAT INVOLVE SEVERAL REACTIONS

Metamorphic reactions do not occur singly, in isolation from other reactions. For example, if we were to treat only the andalusite = kyanite reaction in Figure 21.9, we might assume that the reaction extrapolated as a straight line across the whole width of the diagram. Only when we consider all of the Al_2SiO_5 minerals (and pyrophyllite, when H_2O and silica are present in sufficient quantities) do we discover more accurately the stability of andalusite and kyanite in the Al_2SiO_5-SiO_2-H_2O system. As systems become more complex, containing four or more components, the phase diagrams

become more complex as well. Pressure–temperature phase diagrams that show a host of reactions in a system are called *petrogenetic grids* (Section 26.11).

Before we look into examples of petrogenetic grids, let's first address a few simple rules that govern the way reactions behave when they intersect on phase diagrams. "Topology" may also be used to describe such geometric patterns of reaction curves. The rules follow the strictly geometric treatment of **Schreinemakers** (see Zen, 1966, for a complete description of the method, or Powell, 1991, for some practical examples). Although initially applied to *T-P* phase diagrams, the method is general and can be applied to *T-X* diagrams, or activity–activity diagrams, or any other diagram involving two intensive variables. We shall use *T-P* diagrams for our examples.

Suppose we begin with a *C* component system. When there are *C* + 2 phases (φ) present at equilibrium, $F = C - \phi + 2 = 0$. This is an invariant situation in which all intensive variables, including *T* and *P*, are fixed, and is represented by an invariant point on a *T-P* phase diagram. If one phase were absent, we would have a univariant situation, represented by a curve on a *T-P* diagram. Because there are *C* + 2 possible phases at the invariant point, any one may be absent, and it follows that there must be *C* + 2 possible univariant curves emanating from the invariant point, each with a different phase absent.

For example, Figure 26.15 shows the one-component Al_2SiO_5 system from Figure 21.9 (if we ignore SiO_2, H_2O, and the formation of pyrophyllite). All three polymorphs (*C* + 2) can coexist at the invariant point. From that point, three (*C* + 2) univariant curves emanate, each missing one phase (*C* + 1 phases are present). The phase that is *absent* serves to identify the reaction represented by any univariant curve. Thus the Ky-And reaction is the Sil-absent reaction, etc. The absent phase (in parentheses) may thus be used as a unique label for each curve, as has been done in Figure 26.15. The short dashed lines in Figure 26.15 are **metastable extensions** of each stable curve, where it extends beyond the invariant point into the field in which the absent phase is stable.

Systems with more than *C* + 2 possible phases are called **multisystems** and contain more than one invariant point. For simple examples, refer to Figures 6.6 for H_2O

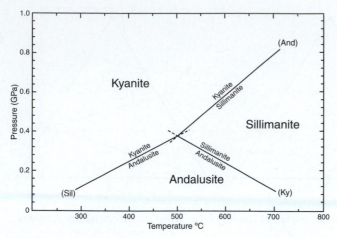

FIGURE 26.15 The Al$_2$SiO$_5$ *T-P* phase diagram from Figure 21.9 (without H$_2$O).

and 6.7 for SiO$_2$. A situation for the general case of a one-component system with four possible phases (A, B, D, and E) is illustrated in Figure 26.16. Because there can be only $C + 2 = 3$ phases at an invariant point, and there are four phases in the system, we can readily deduce that four invariant points are possible, each of which is missing one phase. The four invariant points have been labeled in Figure 26.16 with the absent phase in square brackets. From each invariant point there emanate three univariant curves, each with $C + 1 = 2$ phases at equilibrium along the curve. Univariant curves have been labeled in Figure 26.16, with the absent phase in parentheses. For example, invariant point [D] in the upper right of the diagram has univariant curves (A), (B), and (E) emanating from it. Curve (E), for example, as it emanates from invariant point [D] must have both phase E and phase D

absent because it is the E-absent curve emanating from the D-absent invariant point. It is thus the curve that represents $A = B$ equilibrium, which separates the A and B stability fields. (Divariant fields in one-component systems have one phase.) Notice that this same univariant reaction curve is the D-absent curve emanating from the E-absent invariant point. All four invariant points operate in a similar way, and the entire diagram is internally consistent, in that it separates the P-T area into four divariant stability fields, each with $C = 1$ phase (labeled without brackets or parentheses).

Figure 26.16 represents a situation in which all four possible invariant points are stable at reasonable pressures and temperatures. It can be seen that such a diagram requires substantial curvature of many reaction curves in order to satisfy the Schreinemakers topologies. In most phase diagrams, such as the SiO$_2$ diagram in Figure 6.6, one or more of the possible invariant points may not be stable because they occur at unrealistically high or low P-T conditions, or they are intrinsically metastable (the intersection of metastable extensions, not of stable curves). Figure 26.17 illustrates the concept with a portion of the SiO$_2$ phase diagram, showing the situation for a four-phase, one-component subset of the system. Only two of the possible invariant points, [Liq] and [Trd], are present in Figure 6.6. The [β-Qtz] invariant point occurs at negative pressure, and the [Crs] invariant point is not stable because it occurs at the intersection of metastable extensions, not of stable univariant curves. This diagram also satisfies all the Schreinemakers criteria.

Figure 21.9 is another relatively simple example of a phase diagram with more than one invariant point. The And-Sil-Ky invariant point is the pyrophyllite-absent invariant point [similar to (B) in Figure 26.16], and the Prl-And-Ky invariant point is the Sil-absent invariant point [similar to (E) in Figure 26.16]. The other possible invariant points are not stable. The And-absent invariant point, for example, requires the Ky-Sil reaction to curve back and intersect the Prl-Ky curve, which does not happen at any realistic pressure.

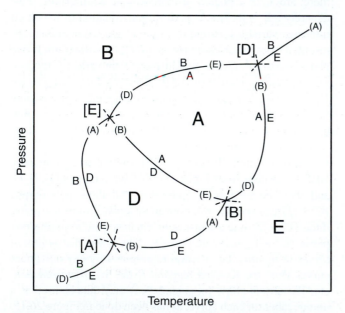

FIGURE 26.16 Schematic one-component *T-P* phase diagram, showing the topology of a four-phase multisystem in which all invariant points are stable. Because only three phases ($C + 2$) coexist at an invariant point, a complete system should have four invariant points, each with one phase absent. Phases absent at invariant points are in square brackets, phases absent for univariant reactions are in parentheses.

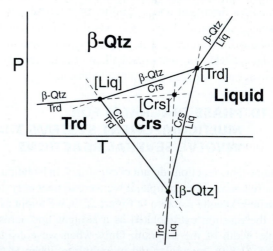

FIGURE 26.17 A portion of the *P-T* phase diagram for SiO$_2$ (Figure 6.6), showing two stable invariant points [Trd] and [Liq] and two metastable ones. [β-Qtz] occurs at negative pressure, and [Crs] is truly metastable in that it is the intersection of metastable extensions. From Spear (1993).

Some important geometric rules for dealing with systems of C components that stem from both the phase rule and the method of Schreinemakers include:

1. When $\phi = C + 2$, an invariant point results.
2. $C + 2$ univariant curves must emanate from each invariant point.
3. If two reactions have fewer than C phases in common, they cross in an **indifferent** fashion: they are independent of each other and do not require an invariant point at the intersection. For example, the high-quartz/low-quartz reaction may cross the And-Sil reaction, and neither is affected by the other.
4. If, on the other hand, two intersecting univariant reactions have C phases in common, their intersection cannot be indifferent, and the intersection generates an invariant point, plus other univariant curves emanating from it, as described in rules 1 and 2 above. This is so because each univariant curve must have $C + 1$ phases at equilibrium, and one of the phases cannot be the same for each intersecting curve (otherwise, they would be the same curve). Thus $C + 2$ phases coexist at the intersection of the curves, and this implies an invariant situation.
5. Consider a univariant reaction such as $D + E = F$ in a two-component system in Figure 26.18a. The reaction is univariant in a two-C system because it has $C + 1 = 3$ phases at equilibrium. If it intersects another univariant reaction curve at an invariant point, *the D-absent and E-absent curves must both occur on the side of the initial reaction opposite to the field in which D and E are stable* (see Figure 26.18b for an example using the D-absent reaction). Similarly, the F-absent reaction must lie on the opposite side of the initial reaction, the side opposite the field in which F is stable. *The metastable extensions of each of these reactions will thus lie on the side of the first reaction in which the absent-phase is stable.* For example, the metastable extension of the D-absent reaction is on the side of the first reaction in which phase D is stable. This is a fundamental axiom of the Schreinemakers approach and must apply to all reactions that meet at an invariant point.
6. It follows from rule 5 that if two reactions that meet at an invariant point are known and can be located and oriented fairly accurately on a phase diagram, the full topology of the invariant point can be deduced.
7. It also follows from rule 5 that any divariant field cannot occupy a sector $>180°$ about any invariant point. Otherwise, the metastable extension would extend into the field in which the phase is unstable.

Figure 26.18 illustrates the Schreinemakers rules for a two-component, four-phase system. Suppose we begin with the reaction $D + E = F$ (Figure 26.18a) and add another well-constrained reaction, such as $F = G + E$ (Figure 26.18b). Because these two reactions have C phases in common (E and F), their intersection requires an invariant point with $C + 2 = 4$ phases in equilibrium and $C + 2 = 4$ reactions emanating from it. The initial reaction must be G-absent, the second reaction must be D-absent, and we can

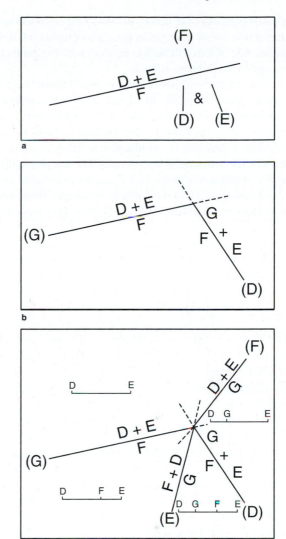

FIGURE 26.18 **(a)** Hypothetical reaction $D + E = F$ in a two-component phase diagram. Note that the D-absent and E-absent curves must both lie on the side of the initial univariant curve opposite the field in which $D + E$ is stable. Likewise, the F-absent curve must lie on the side opposite to the field in which F is stable. **(b)** A second hypothetical univariant curve (D-absent) is added. **(c)** The complete topology of the invariant point can then be derived from the two initial reactions in **(b)**. The chemographics may then be added to each divariant field.

apply rule 5 above to both reactions simultaneously. This implies that reaction (E) must be on the lower side of reaction (G) *and* on the left side of reaction (D) (Figure 26.18c). Reaction (F) must be on the upper side of reaction (G) *and* on the right side of reaction (D). We can then determine what phases occur on which side of each of the remaining two reactions, so that rule 5 applies to all four reactions simultaneously. From that, we have the full topology of the invariant point and the reactions in Figure 26.18c (rule 6). Affirm for yourself that rule 5 applies to all the reactions, and the metastable extension of any reaction (x-absent) is on the side of the other three reactions where x is stable. Note also that all two-phase divariant assemblages occupy sectors $<180°$ about the invariant point (rule 7). The binary chemographic compatibility diagrams have also been added to each divariant field in Figure 26.18c such that each field has a unique compatibility diagram. As an exercise, demonstrate to your satisfaction that each reaction causes changes in the chemographic topologies on either side of it in a fashion consistent with the geometries discussed in the previous section.

Note that rules 5 and 7 also apply to Figure 26.15 and any other phase diagram. In Figure 26.15, the andalusite-absent reaction is on the side of the sillimanite-absent reaction that is opposite to the andalusite field, and the metastable extension is within the andalusite field, etc. Similarly, the crossing of the pyrophyllite reaction with the And-Ky reaction in Figure 21.9 is *not* an indifferent crossing, and thus the kink in the curve as it changes from And + Qtz + H_2O = Prl to Ky + Qtz + H_2O = Prl is required by rules 5 and 7 above.

The geometric approach of Schreinemakers is useful and is used in conjunction with experimental and theoretical data to develop more comprehensive phase diagrams and petrogenetic grids. Although numerous univariant equilibrium curves have been studied experimentally, rarely has every curve that emanates from any given invariant point been located accurately. Just as in Figure 26.18, the method of Schreinemakers permits investigators to infer the general location of the other reactions about an invariant point if two are known from experiments. If some thermodynamic data are available on the phases involved in the reactions, we could calculate the slope from the Clapeyron equation [Equation (5.15)] and further refine the geometries. Even a simple knowledge of relative molar volumes of the minerals and the characteristic shape of devolatilization reaction curves is useful in refining the arrangement of univariant curves about an invariant point.

The geometric method has three shortcomings that cannot be resolved without good experimental and/or thermodynamic data. First, the *location* of any invariant point cannot be determined using the Schreinemakers method; only the topology can be determined. Second, the exact slope or location of any univariant reaction curve likewise cannot be known. Finally, for any topology such as that in Figure 26.18c, an alternative enantiomorphic topology (a mirror image) is also compatible with the Schreinemakers rules. Which choice is real cannot be known from the geometric approach alone. All of these shortcomings can be resolved if sufficient physical–thermodynamic data are available.

There remains one final point to mention. The ideal topologies, such as those illustrated in Figures 26.16 and 26.18, hold only in the general case of non-degenerate systems. **Degenerate** systems are those in which the composition of some of the phases is related in a particular way. For example, in a system of two or more components, the composition of some phases may be identical. This is the case in Figure 21.9. The system illustrated is really the three-component Al_2O_3-SiO_2-H_2O system, and three of the phases all have the composition Al_2SiO_5. Thus the Prl-absent curve that emanates from the five-phase Prl-Qtz-And-Ky-H_2O invariant point is also the *Qtz*-absent and the H_2O-absent curve as well. Thus there are $C - 1$, and not $C + 2$, curves emanating from the invariant point because of the degeneracy. Another type of degeneracy results when three phases plot along a single line in a $C > 3$ system (as discussed in Section 24.3),

or when four phases are coplanar in a $C > 4$ system, etc. In all degenerate cases, there may be fewer than $C + 2$ univariant curves emanating from an invariant point.

26.11 PETROGENETIC GRIDS

A **petrogenetic grid** is a *P-T-X* (or *P-T*, *T-X*, *P-X*) phase diagram for a particular chemical system or bulk composition that illustrates several important reactions for that system. When Bowen (1940) proposed the term, he had the vision of a complete characterization of all divariant mineral assemblages in nature but realized that the magnitude of undertaking the necessary experiments was a huge task that would not be finished for a long time. Modern petrogenetic grids are thus only partially complete (although advances in thermodynamic databases and computer programs allow us to create increasingly more accurate multisystems). Choosing an appropriate chemical system and the reactions to use requires a combination of good lab work on a suite of rocks (and/or good theoretical data) and good judgment. When making the choice, we face the same problem we faced in Chapters 6 and 7 as we attempted to model igneous crystallization and melting behavior. Simple chemical systems generate simple phase diagrams that may allow us to understand some fundamental features, but most are drastic oversimplifications of natural rock systems. Complex chemical systems may be more realistic, but the petrogenetic grid may become so cluttered that it becomes difficult to recognize the reactions that are most important in the petrogenesis of some rock types.

A petrogenetic grid may be as simple as Figure 21.9, or it may be a full multisystem of five to eight components containing dozens of invariant points and a hundred or more reactions. Most good petrogenetic grids are a judicious compromise of three to six components and a selection of the most definitive reactions and invariant points. These reactions provide the best quantitative *P-T-X* constraints on the conditions of metamorphism that produce the common mineral assemblages of a class of metamorphic rocks. Because only selected reactions are included, many grids do not have the complete Schreinemakers topologies for all invariant points.

Figure 26.19 is an example of a petrogenetic grid for metamorphosed mafic rocks (discussed in Chapter 24). It illustrates many of the important reactions that govern the development of mafic mineral assemblages from the zeolite to the granulite facies, as well as the blueschist and eclogite facies at higher pressures. For a more complete version, see Chapter 9 of Bücher and Frey (2002), and for higher pressures see Guiraud et al. (1990), Poli and Schmidt (2002), or some of the Suggested Readings in Chapter 25. The grid illustrates several important reactions, including those that govern the transitions from one facies to the next. As you study the diagram, you may notice the reactions that govern the development of zeolites, prehnite and pumpellyite, the formation of glaucophane or the decomposition of plagioclase at high pressure, the development of epidote and actinolite of the greenschist facies, hornblende and plagioclase of the amphibolite facies, and the dehydration of hornblende to form pyroxene-bearing granulite facies assemblages. Notice

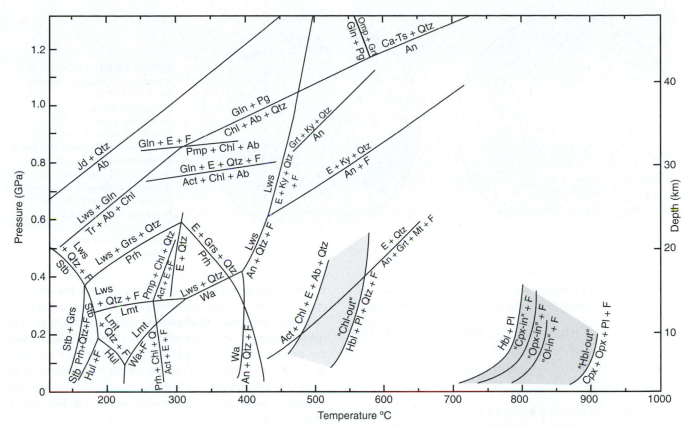

FIGURE 26.19 Simplified petrogenetic grid for metamorphosed mafic rocks, showing the location of several determined univariant reactions in the CaO-MgO-Al$_2$O$_3$-SiO$_2$-H$_2$O-(Na$_2$O) system ("C(N)MASH"). Mineral abbreviations are listed in the inside cover (F = aqueous fluid, E = an epidote mineral: epidote, zoisite, or clinozoisite). Important continuous reactions are shaded (although most reactions are continuous with respect to Fe-Mg, Ca-Na, or some other components). Data from Liou (1970, 1971a, b, c), Liou et al. (1974, 1985, 1987), Cho et al. (1987), Maruyama and Liou (1988), Maruyama et al. (1986), Newton and Smith (1967), Newton (1986), Newton and Kennedy (1963), Evans (1990), Massone (1989), Spear (1981), and Bucher and Frey (2002).

also that many of these important reaction isograds span a limited *P-T* range and would be intersected by only certain facies series or *P-T-t* paths. Other reaction isograds cross (indifferently in many cases). For example, the epidote → anorthite reaction in the center of the grid below 0.6 GPa has a shallower slope than does the actinolite → hornblende reaction. Because fewer than *C* phases are in common between the two reactions, the crossing is indifferent, and an invariant point does not result. This crossing explains why the albite → plagioclase transition occurs before actinolite → hornblende for low-pressure metamorphism, and the opposite is true for higher *P-T-t* paths. We shall use other petrogenetic grids in Chapters 28 and 29, when we evaluate the metamorphism of pelitic, carbonate, and ultramafic rocks.

26.12 REACTION MECHANISMS

The discussion of metamorphic reactions in the preceding sections of this chapter has dealt with relatively simple balanced reactions and the location of the equilibrium reaction boundaries on phase diagrams. Little attention has been paid to the actual *mechanism* by which a reaction might take place in the rocks themselves. Dugald Carmichael (1969) pointed out that the textures seen in thin section may not coincide with the reaction that may be deduced by simply comparing the reactant and product mineral assemblages.

Although there is certainly a gradual reduction in reactant minerals across any reaction isograd, and this is compensated by growth of the product assemblage, the products in many cases do not grow in direct contact with the reactants. Rather, the products and reactants are found in different domains of a thin section, separated by "inert" minerals that are stable on both sides of the reaction. In other words, the most obvious and simple balanced reaction may not reflect the true mechanism by which the overall reaction was accomplished.

For example, the simple reaction of kyanite → sillimanite at the sillimanite isograd in pelites rarely results in the sillimanite directly replacing the kyanite as pseudomorphs. Rather, small sillimanite crystals (commonly as fibrolite) occur embedded in muscovite, biotite, or quartz. A common texture is illustrated in Figure 26.20a, in which several blebs of kyanite and quartz occur in a larger host muscovite crystal. The kyanite and quartz grains appear to be remnants of larger crystals, as neighboring grains of each mineral type are in optical continuity (they go extinct in unison), and kyanite cleavages have the same orientation. Fine sillimanite crystals appear to have grown nearby in the muscovite.

Carmichael (1969) noted that such a texture does not support the simple Ky → Sil reaction occurring as such and used the textures to infer an alternative mechanism involving dissolution and transport on a localized scale. Assuming that

a
b

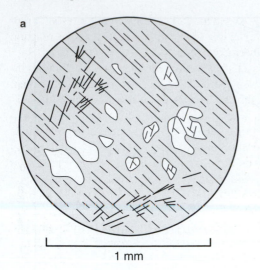
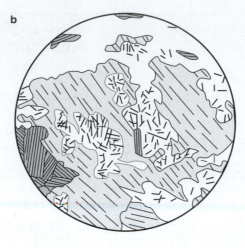

1 mm

FIGURE 26.20 **(a)** Sketch from a photomicrograph, showing small crystals of kyanite (with cleavages) and quartz in a larger muscovite grain (shaded with NW–SE cleavages). Small crystals of fibrolitic sillimanite also occur in the muscovite. Glen Cova, Scotland. **(b)** Sillimanite needles in quartz embaying muscovite. Darker crystals are biotite. Donegal, Ireland. After Carmichael (1969). Copyright © with permission from Springer-Verlag.

Al_2O_3 (a notably insoluble component in most pelites) is conserved, he proposed that kyanite broke down by the reaction:

$$3 Al_2SiO_5 + 3 SiO_2 + 2 K^+ + 3 H_2O \qquad (26.18)$$
$$\text{Ky} \qquad \text{Qtz}$$
$$\rightarrow 2 KAl_2Si_3AlO_{10}(OH)_2 + 2H^+$$
$$\text{Ms}$$

Rims of muscovite appear to grow at the expense of Ky and Qtz, and they gradually encroach on both as the reaction advances, leaving isolated patches of the reactants, as observed in Figure 26.20a. Such a reaction consumes K^+ and H_2O, and releases H^+, all of which involve exchange of ions between minerals and a fluid phase. For the reaction to proceed, there must be a source of K^+ and a receptor for H^+. This may involve an external source (such as a nearby granite) and sink, but other local solution–precipitation reactions may occur in the same rock to fulfill these roles, alleviating the need for an external reservoir. Carmichael (1969) noted another common texture in many pelites in which needles of sillimanite occur in quartz embayments into ragged muscovite crystals (Figure 26.20b). A reaction capable of producing this texture that conserves Al_2O_3 is:

$$2 KAl_2Si_3AlO_{10}(OH)_2 + 2 H^+ \qquad (26.19)$$
$$\text{Ms}$$
$$\rightarrow 3 Al_2SiO_5 + 3 SiO_2 + 2 K^+ + 3 H_2O$$
$$\text{Sil} \qquad \text{Qtz}$$

This reaction consumes muscovite and produces sillimanite and quartz, which explains the embayed muscovite and quartz–sillimanite filling the embayments. If Reactions (26.18) and (26.19) occur simultaneously, then each serves as both source and sink for the fluid and solute species of the other, muscovite is conserved, and the *net reaction* is Ky → Sil.

Figure 26.21 shows that the overall reaction can be considered the sum of two local reactions that operate similarly to the "half-cell" reactions used to demonstrate the electromotive series in chemistry labs.

Carmichael (1969) proposed a number of other such "half-cell" type reactions that may substitute for isograd

reactions and provide better explanations for the observed textures. Any isograd may have a number of alternative combined reactions that involve ionic species that contribute to the net isograd reaction. An alternative mechanism for the Ky → Sil isograd that explains the very common occurrence in some pelites of sillimanite (or fibrolite) in biotite, rather than replacing kyanite, consists of three separate exchange subreactions. The first is Reaction (26.18), and the other two are:

$$K(Mg, Fe)_3Si_3AlO_{10}(OH)_2 + Na^+ + 6 H^+ \qquad (26.20)$$
$$\text{Bt}$$
$$\rightarrow NaAlSi_3O_8 + K^+ + 3(Mg, Fe)^{2+} + 4 H_2O$$
$$\text{Ab}$$

$$2 KAl_2Si_3AlO_{10}(OH)_2 + NaAlSi_3O_8 + 3 (Mg, Fe)^{2+}$$
$$\text{Ms} \qquad\qquad \text{Ab} \qquad (26.21)$$
$$+ 4 H_2O \rightarrow K(Mg, Fe)_3Si_3AlO_{10}(OH)_2 + 3 Al_2SiO_5$$
$$\text{Bt} \qquad\qquad \text{Sil}$$
$$+ 3 SiO_2 + K^+ + Na^+ + 4 H^+$$
$$\text{Qtz}$$

Notice that Reaction (26.21) produces biotite and sillimanite, which are commonly found in association, while consuming albite. Reaction (26.20) replaces the albite, which produces biotite crystals embayed by plagioclase. Reactions

$2 K^+ + 3 H_2O$

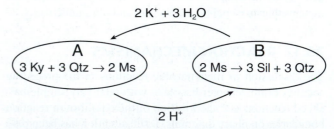

A
$3 Ky + 3 Qtz \rightarrow 2 Ms$

B
$2 Ms \rightarrow 3 Sil + 3 Qtz$

$2 H^+$

FIGURE 26.21 A possible mechanism by which the Ky → Sil reaction can be accomplished while producing the textures illustrated in Figure 26.20a and b. The exchange of ions shown between the two local zones is required if the reactions are to occur. After Carmichael (1969). Copyright © with permission from Springer-Verlag.

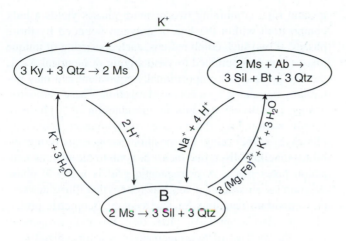

FIGURE 26.22 An alternative mechanism by which the Ky → Sil reaction can be accomplished while producing sillimanite needles associated with biotite, whereas plagioclase occupies embayments in the biotite. The exchange of ions shown between the two local zones is required if the reactions are to occur. After Carmichael (1969). Copyright © with permission from Springer-Verlag.

(26.18), (26.20), and (26.21) can be combined to eliminate the fluid and solute species and produce the Ky → Sil net reaction. Figure 26.22 illustrates the overall process and local exchanges required by such a net reaction. This alternative to Figure 26.21 appears to occur in rocks where sillimanite is associated with biotite instead of with muscovite.

Two requirements must be met if the coupled exchange reactions are to work. First, there must be sufficient fluid and ionic mobility for the exchange to occur between the two zones in the rock. Mass transfer for such reactions may be on the scale of millimeters to (rarely) several meters. Second, the Ky → Sil univariant reaction curve in Figure 26.15 must be overstepped, so that sillimanite, rather than kyanite, is produced by Reaction (26.19). Thus the net reaction of Ky → Sil is still a proper *overall* reaction for the sillimanite isograd. Although the mechanism and textures

differ, the isograd and the *P-T* stability conditions that it represents are still valid. Thus we can still treat the reaction equilibrium–thermodynamically as Ky → Sil.

Carmichael's (1969) proposal suggests that reactions involving dissolved species may be much more common than we previously thought, and it also explains why fluids can catalyze solid–solid reactions such as polymorphic transformations, which do not appear, at first glance, to involve fluids at all. Several investigators have carried this idea beyond Carmichael's initial work. Fisher (1973, 1977), Frantz and Mao (1976), Loomis (1976), Foster (1981, 1983), Likhanov and Reverdatto (2002), and others have developed methods (most based on non-equilibrium irreversible thermodynamics) to model mineral dissolution and diffusion along chemical potential gradients to add a quantitative aspect to such reaction mechanisms. In his summary, Putnis (2002) suggested that mineral replacement reactions (including isochemical replacement and pseudomorphs) occur *primarily* by dissolution–reprecipitation processes involving exchange of dissolved species, typically along a sharp replacement front. Ferry (2000) noted that pseudomorphs are much more common results of retrograde processes than of prograde metamorphism. He noted that prograde reactions are typically devolatilization reactions with positive d*P*/d*T* slopes and positive Δ V of reaction and he suggested that the added volume of the mineral(s) produced when a mineral breaks down exerts a local "force of crystallization" at the breakdown surface. The local pressure rise at the mineral surface would increase strain and inhibit a pseudomorphing process. The products therefore develop better at some site away from the expanding mineral breakdown surface, presumably by a dissolution–migration–precipitation process such as Carmichael and others have described. Fluid release during prograde devolatilization should also greatly enhance such processes. Retrograde reactions, on the other hand, involve a negative Δ V and, according to this theory, would make the interface a more favorable site for the retrograde reaction, which, in combination with reduced fluid availability, makes a pseudomorph more likely.

Summary

Changes in the mineral assemblages in rocks must involve reactions in which one or more reactant minerals are consumed and other product minerals are produced. Reactions can be of various types. Polymorphic transformations involve the change of one mineral to another of the same composition. Exsolution reactions involve the unmixing of solid solution minerals accompanying cooling or, less commonly, decompression. Solid–solid net transfer reactions are a common type of reaction that involves multiple solids, requiring that matter must migrate from reactant minerals as they break down to the sites of product mineral growth. When solid solution is limited, these reactions are discontinuous (they occur abruptly at a single metamorphic grade). These three types of reactions are relatively easy to constrain, as they are dependent only on temperature and pressure (and the attainment of equilibrium, of course).

Devolatilization reactions involve the breakdown of one or more volatile-bearing minerals and release of a volatile species (typically H_2O or CO_2, but any volatile may be possible). Such reactions depend on the partial pressures of the volatiles in the system as well as on temperature and pressure. Such reactions occur at the maximum grade possible when the partial pressure of the volatile species released equals the total pressure. When $P_{fluid} < P_{lithostatic}$ or the fluid is diluted with another volatile species, the reaction occurs at lower grades. If a rock system is open to the passage of fluids, the fluid composition remains constant, and devolatilization reactions occur at a single grade. If fluid motion is limited, a devolatilization reaction will span a range of grades and buffer the fluid composition across the range.

Continuous reactions involve solid solutions and take place over a range of temperatures (grades). The grade at which a product mineral first appears (an isograd) with such a reaction depends on the bulk composition. The reaction spans a range of grades as the reactants are gradually consumed and products generated, and the composition of each varies systematically over the range. Ion exchange reactions involve the reciprocal exchange of components (most commonly Fe versus Mg) between two or more minerals (without changing the modal amounts of any). Oxidation/reduction (redox) reactions involve changes in the ionic charge of atoms or complexes (e.g., Fe^{2+} to Fe^{3+}), typically involving a compensating change in the charge of oxygen. Many reactions involve the release, migration, and capture of dissolved species in order to maintain equilibrium between the reacting phases.

Many reactions can readily be visualized and understood by some simple geometric patterns of phase locations in appropriately developed chemographic diagrams. As a general rule, combining two or more phases yields a bulk composition within the line or polygon bounded by those phases. If two such combinations, each consisting of unique phases, can be combined to yield a point of identical X_{bulk}, the two groups are compositionally equivalent and may potentially be related by a balanced reaction. These geometric changes, involving the loss or introduction of a phase or simply rearranging the tie-lines that connect coexisting phases, are what relate composition paragenesis (compatibility) diagrams from one facies or zone to the next as a reaction takes place. A petrogenetic grid is a *P-T-X* phase diagram for a particular chemical system that illustrates several important reactions for that system. Geometric principles may also be applied to phase diagrams and petrogenetic grids: the method of Schreinemakers is a formalized geometric procedure for organizing the topology of reaction curves about invariant points on such diagrams.

Key Terms

Isograds: mineral-in, mineral-out *558*
Polymorphic transformation reactions *559*
Exsolution reactions *560*
Solid–solid net transfer reactions *560*
Devolatilization reactions *560*
Partial pressure (p_{H_2O}, p_{CO_2}) *561*
T-X_{fluid} phase diagram *562*

Open-system and closed system volatile behavior *563*
Buffer *564*
Continuous reaction *566*
Pseudosection *567*
Ion-exchange reaction *567*
Blocking or closure temperature *568*
Oxidation/reduction (redox) reaction *568*

Oxygen buffer *568*
Dissolved species and hydrolysis *569*
Topology *569*
Tie-line flip *570*
Schreinemakers method *571*
Metastable extension *571*
Multisystem *571*
Indifferent reaction intersection *573*
Petrogenetic grid *574*

Review Questions and Problems

Review Questions and Problems are located on the author's web page at the following address: **http://www.prenhall.com/winter**

Important "First Principle" Concepts

- Because the temperature of a devolatilization reaction depends on the partial pressure of the volatile species involved, one should take great care to constrain the fluid composition, if at all possible, before using a devolatilization reaction to indicate metamorphic grade.

- The metamorphic grade at which a continuous reaction occurs depends on the composition of the phases, which, in turn, is dependent upon X_{bulk}.

- Continuous reactions span an interval of grade, across which the reactants are progressively consumed, products are generated, and the compositions of each vary systematically.

- Reactions involve changes in topology of compatibility diagrams: loss or introduction of a phase or rearranging of tie-lines that connect coexisting phases. Therefore, each facies or zone may be represented by a compatibility diagram that differs from the neighboring zone by a reaction and a very specific geometric change in topology that is related to that reaction.

Suggested Further Readings

Bücher, K., and M. Frey. (2002). *Petrogenesis of Metamorphic Rocks*. Springer-Verlag. Berlin. Chapter 3.

Carmichael, D. M. (1969). On the mechanism of prograde metamorphic reactions in quartz-bearing pelitic rocks. *Contrib. Mineral. Petrol.,* **20,** 244–267.

Kretz, R. (1994). *Metamorphic Crystallization*. John Wiley & Sons. New York. Chapter 3.

Zen, E-an. (1966). Construction of pressure–temperature diagrams for multi-component systems after the method of Schreinemakers—A geometric approach. *U.S. Geol. Survey Bulletin* **1225**.

27

Thermodynamics
of Metamorphic Reactions

Questions to be Considered in this Chapter:

1. How can we apply the thermodynamic principles introduced in Chapter 5 to metamorphic reactions and mineral parageneses to better understand and constrain the physical and chemical conditions of metamorphism?

2. How can we calculate the position of the equilibrium curve for a solid-solid net-transfer reaction on a P-T diagram?

3. How are gas phases, such as H_2O or CO_2, addressed in order to similarly plot the equilibrium curves of devolatilization reactions?

4. How can variable composition be treated in equilibrium calculations, and how can that then be used in reverse to use the composition of natural mineral assemblages to estimate the temperature and pressure of metamorphic or igneous equilibration?

In this chapter, we will proceed from the fundamental principles of thermodynamics developed in Chapter 5 and develop the methods to calculate the equilibrium position of reactions among minerals and fluid phases that comprise metamorphic rocks. A brief summary of the main points from Chapter 5 follows. If, as you read the summary, you are not comfortable with your memory of these concepts, I urge you to review Chapter 5 before proceeding because you will need to be familiar with the background behind the equations in that chapter before continuing and applying them to reactions.

- **Gibbs free energy** is a measure of the energy content of chemical systems. The Gibbs free energy of a *phase* at a specified pressure and temperature can be defined mathematically as:

$$G = H - TS \tag{5.1}$$

G can be calculated for a phase by determining H and S by calorimetry, assuming the third law of thermodynamics. An alternative method, not mentioned in Chapter 5, is to extract G directly from high-T-P experiments on reactions at equilibrium.

- ΔG for a *reaction* at any pressure and temperature can be computed from the G of formation for the phases participating in the reaction by the equation:

$$\Delta G = \Sigma (n_{products}G_{products} - n_{reactants}G_{reactants}) \tag{5.2}$$

- Once ΔG is known at a particular temperature and pressure, ΔG can be found at any other temperature and pressure by integrating the equation:

$$d\Delta G = \Delta V dP - \Delta S dT \tag{5.11}$$

over specific increments of P and T. Remember, Δ refers to the difference in a property (G, S, V, etc.) between the reactants and products in a reaction, and d refers to changes in P, T, and ΔG.

- The stable form of a chemical system is that with the *minimum* possible Gibbs free energy for the given *P-T-X* conditions. G is thus a measure of the stability of a phase and ΔG is a measure of the relative stability of the reactants versus products of a reaction.

27.1 CALCULATING THE LOCATION OF A REACTION EQUILIBRIUM CURVE ON A PHASE DIAGRAM

If a system has two (or more) alternative forms (e.g., SiO_2 may occur as low-quartz, high-quartz, tridymite, cristobalite, a melt, etc.), we can compare the values of G for each of the possible forms at a given pressure and temperature to determine which is the lowest (hence the most stable) under those conditions. Alternatively, we can treat the difference between the free energies of two competing forms directly. The transition from one form to another is a *reaction*, so we will then be dealing with the changes in G involved during such a reaction (ΔG). Reactions can be simply $A = B$ (e.g., low quartz = high quartz) or more heterogeneous, such as $A + B = C + D + E$.

We shall now use Equation (5.11) to find the equilibrium conditions for a reaction and then use the results to plot the equilibrium curve on a pressure–temperature phase diagram.

EXAMPLE PROBLEM: Calculating the Equilibrium Curve for the Reaction Jadeite + Quartz = Albite

This exercise will familiarize you with the methods by which one uses Equation (5.11) and thermodynamic data to determine mineral stabilities and locate univariant reaction curves on phase diagrams. In the exercise, you will be asked to do a large proportion of the work, which is best done with a spreadsheet. I have picked a simple solid–solid net-transfer reaction: jadeite + quartz = albite. You may remember this reaction from Chapters 25 and 26 as the high-pressure stability limit of albitic plagioclase in the eclogite and blueschist facies.

We will use the reference state (298 K and 0.1 MPa) data for the minerals involved in the reaction, use Equation (5.2) to get ΔG, ΔV, and ΔS for the reaction, and use Equation (5.11) to solve for various equilibrium conditions (for which $\Delta G = 0$). By this method, we can calculate the relative stabilities of minerals or mineral assemblages and thereby create pressure–temperature phase diagrams. Such diagrams tell us what minerals would be stable under a given set of conditions attainable at depth. We can then use this information to evaluate mineral assemblages in terms of the pressure and temperature (and compositional) variables in the Earth.

Once again, the reaction we are interested in is:

$$1 \text{ NaAlSi}_2O_6 + 1 \text{ SiO}_2 = 1 \text{ NaAlSi}_3O_8 \qquad (26.1)$$
$$\text{Jd} \qquad\qquad \text{Qtz} \qquad\quad \text{Ab}$$

Table 27.1 lists the reference-state thermodynamic data for the reactants and products, taken from the computer program

TABLE 27.1 Thermodynamic Data (molar) at 298 K and 0.1 MPa, from SUPCRT Database (Helgeson et al., 1978)

Mineral	S (J/K mol)	G (J/mol)	V (cm³/mol)
Low albite	207.25	−3,710,085	100.07
Jadeite	133.53	−2,844,157	60.04
Quartz	41.36	−856,648	22.688

SUPCRT (Helgeson et al., 1978: see Section 27.5). Other sources of thermodynamic data are listed in Section 27.5.

Equation (5.11) tells us that ΔG for the reaction above equals $G_{\text{albite}} - G_{\text{jadeite}} - G_{\text{quartz}}$ (because all stoichiometric coefficients are 1). Similarly, you can determine ΔV (in cm³/mol) and ΔS (in J/K mol). The only equation we need to construct the *P-T* phase diagram is Equation (5.11):

$$d\Delta G = \Delta V dp - \Delta S dT \qquad (5.11)$$

which provides us with the relationship between the change in free energy of a reaction (a measure of relative stability of the reactants versus the products) and pressure and temperature. Conceptually, the most straightforward way to determine the *P-T* conditions of equilibrium is to calculate ΔG for a nearly infinite number of pressures and temperatures, and then see for what *P-T* conditions $\Delta G = 0$. A curve connecting all these *P-T* points would be the equilibrium curve (illustrated in Figure 27.1). But we would like to finish this problem before we get too old, so we must find a more efficient method.

1. Let's begin with the 298 K data. Use a spreadsheet to recreate Table 27.1, and calculate ΔV, ΔS, and ΔG for the jadeite + quartz = albite reaction at 298 K and 0.1 MPa. Consider these questions:

- How does your value for ΔG tell you whether albite or jadeite + quartz is stable at 298 K and 0.1 MPa?
- Why can you find *all three* of these minerals in your mineralogy lab at 298 K and 0.1 MPa?

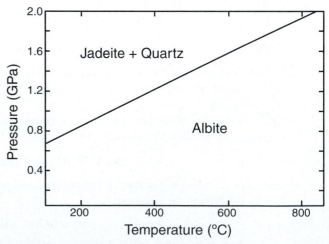

FIGURE 27.1 Temperature–pressure phase diagram for the reaction albite = jadeite + quartz, calculated using the program TWQ (Berman, 1988, 1990, 1991).

2. You now have ΔG for the reaction at 298 K and 0.1 MPa. We next want to locate the equilibrium curve. Albite, jadeite, and quartz are at equilibrium when both sides of the reaction have the same free energy ($\Delta G = 0$). So our problem is: *If we know the value of* ΔG *at one set of* P-T *conditions, and we have an equation that expresses the change in* ΔG *with changes in* P *and* T, *how can we use this equation to find some* P-T *conditions for which* ΔG *reduces to 0?* Once again, we want to calculate the change in ΔG of reaction ($d\Delta G$). If we integrate Equation (5.11) at constant T and assume ΔV *is constant*, we get the algebraic equivalent of Equation (5.6):

$$\Delta G_{P_2} - \Delta G_{P_1} = \Delta V(P_2 - P_1) \quad \text{(constant } T) \qquad (27.1)$$

We can then use Equation (27.1) to calculate the equilibrium pressure at 298 K. If P_1 is 0.1 MPa, then ΔG_{P_1} is the value of ΔG that you just calculated at 0.1 MPa. ΔG_{P_2} is 0 because we're looking for the equilibrium pressure (P_2). The only remaining variable is P_2, for which you can now solve directly, knowing ΔV. *Be careful of your units here.* You must use V in m^3/mol and P in Pa (or use the cm^3-MPa shortcut that we used in the example problem in Chapter 5). The resulting P_2 is the location of the equilibrium curve that separates the field of albite stability from that of jadeite + quartz at 298 K. Don't worry if the value of P_2 is a bit strange. Using thermodynamics, we can calculate unrealistic or unattainable values for pressure or temperature. The equations permit this because they simply don't know what is possible. If we ask a silly question, we get a silly answer. It is up to us to interpret the results. In this case, we can extrapolate from an unrealistic pressure to determine the location of the equilibrium curve at realistic ones.

3. We can next integrate Equation (5.11) at constant pressure (assuming ΔS is constant) to yield:

$$\Delta G_{T_2} - \Delta G_{T_1} = -\Delta S(T_2 - T_1) \qquad (27.2)$$

which is the reaction equivalent to Equation (5.8). We can apply Equation (27.2) to the initial ΔG (298, 0.1) to calculate the approximate ΔG at 0.1 MPa and 600 K and then at 900 K.

4. Repeat step 2 to calculate the equilibrium pressure (P_2) at 600 K and 900 K. You will then have three *P-T* points along the equilibrium curve.

5. Plot the points on a P (GPa) versus T (Celsius) diagram. Be sure that you have converted the units correctly. Use the standard orientation of P as the y-axis and T as the x-axis. Connect the points to generate the equilibrium curve, truncate the curve, if necessary, to consider only a positive and realistic P and T range, and label the fields of stability Albite and Jadeite + Quartz.

6. Compare your results with the curve in Figure 27.1, which is determined by a computer program that treats the volume and enthalpy changes in a rigorous manner, and discuss the appropriateness of your assumptions of constant ΔS and constant ΔV for metamorphic reactions that involve only solids.

The results of this example problem should show you that our assumptions of constant ΔV and ΔS are reasonable approximations. In fact, ΔV and ΔS change less than do V and S for each mineral in the reaction. This is because all of these solids respond in a similar (but not identical) fashion to changes in pressure and temperature, the changes in the products thus partially offset those of the reactants. For example, if the molar volume of the products is twice that of the reactants at low pressure, and all phases compress by 50% at higher pressure, the products will still be half the size of the reactants. We can thus use the integrated algebraic form of Equation (5.11) to get fairly reliable results (at least for reactions involving only solids). Once a gaseous phase is involved in a reaction, however, the changes in ΔV and ΔS become much greater.

27.2 GAS PHASES

The relatively simple method outlined above works well for reactions that involve only solids. When gas/fluid phases are included in the reaction, however, Equation (27.1) fails. Why? To answer this, return to Equation (5.5):

$$G_{P_2} - G_{P_1} = \int_{P_1}^{P_2} VdP \quad \text{(constant } T) \qquad (5.5)$$

When we assessed this equation the first time, we were forced to come up with a relationship between V and P so that we could perform the integration. Not being dummies, we came up with the easiest possible relationship: V is a constant and has no relationship to P whatsoever. By doing that we could avoid integrating a volume expression, and we came up with Equation (27.1), which worked reasonably well for solid phases. For a gas, the volume changes considerably as pressure varies, so we need to find another relationship between P and T. If the gas behaves ideally, we can use the **ideal gas equation ($PV = nRT$)**, where n is the number of moles ($n = 1$ for molar properties) and R is the "gas constant," an empirically derived constant equal to 8.3144 J/mol K. The ideal gas equation is a way of quantifying the pressure versus volume relationship (at a given temperature) in the piston-cylinder apparatus illustrated in Figure 5.5. It works reasonably well for gases with few molecular interactions, particularly at low pressures. If we substitute the ideal gas equation value of RT/P for V in Equation (5.5) (for a single mole of the gas), we get:

$$G_{P_2} - G_{P_1} = \int_{P_1}^{P_2} \frac{RT}{P} dP \qquad (27.3)$$

and, because R and T are definitely independent of changes in P, this equation becomes:

$$G_{P_2} - G_{P_1} = RT \int_{P_1}^{P_2} \frac{1}{P} dP \qquad (27.4)$$

Integrating Equation (27.4) between P_1 and P_2 gives us:

$$G_{P_2} - G_{P_1} = RT \ell n P_2 - RT \ell n P_1 = RT \ell n(P_2/P_1) \qquad (27.5)$$

because $\int \frac{1}{x} dx = \ell nx$. Rearranging, we can express Equation (27.5) for the free energy of a gas phase at a given pressure and temperature, which, when referred to an atmospheric pressure reference state ($P = 0.1$ MPa), becomes:

$$G_{P,T} = G_T^o + RT\ell n(P/P^o) \qquad (27.6)$$

which can be read to mean that the free energy of some gas phase at a specific pressure and temperature (denoted by the subscripts P, T) is equal to the free energy at some reference state, plus the remaining pressure term. The superscript o refers to the reference state, the pressure of which, in this case, is 0.1 MPa. The reference temperature, however, need not be 298 K. Rather, it is the same temperature for which we are determining G. Equation (27.6) is a constant temperature calculation, so the subscript T indicates the singular temperature for G on both sides of the equation. $P^o = 1$ if bars are used for pressure, so that P^o drops from Equation (27.6) altogether and makes the form more convenient. With P expressed in Pa, this is not possible, and $P^o = 10^5$ Pa must be substituted. I think the equation using Pa is more reliable because P^o is there as a reminder of the reference state.

If the gas does not behave as an ideal gas, and most gases at geologically realistic pressures do not, we can substitute the term f for P, such that:

$$f = P\gamma \qquad (27.7)$$

f is called the **fugacity**, and it is the real-gas (or fluid) equivalent of pressure. γ is called the **fugacity coefficient**, and it is simply a conversion factor that is determined experimentally and adjusts for non-ideality. Tables of fugacities and fugacity coefficients are available for a number of gas species. For any particular gas species, γ varies with T, P, and the proportions of other gas species with which it may be mixed. For H_2O and CO_2 mixtures (the most common species in geological fluids), values of γ are tabulated in Kerrick and Jacobs (1981), among other sources. Virtually any gas would behave ideally at a pressure as low as 0.1 MPa, so $f^o = P_o$ and the fugacity equivalent of Equation (27.6) needs no fugacity coefficient in the denominator.

If a reaction involves both solid mineral phases and a real gas, Equation (5.2) can be integrated for the solids and vapor separately to yield:

$$\Delta G_{P,T} = \Delta G_T^o + \Delta V_s(P - P^o) + RT\ell n(f/P^o) \quad (27.8)$$

where $\Delta G_{P,T}$ is the ΔG of the reaction (including the gas) at some pressure and temperature in question (and is equal to 0 at equilibrium), ΔG_T^o is the ΔG for the same reaction at the same temperature and the reference pressure (0.1 MPa or 1 bar), and ΔV_s is the reaction volume change *for the solids only*. Thus the third term in Equation (27.8) is the pressure correction for the solids, and the last term is the pressure correction for the fluid. If the fluid is an ideal gas, P can be substituted for f.

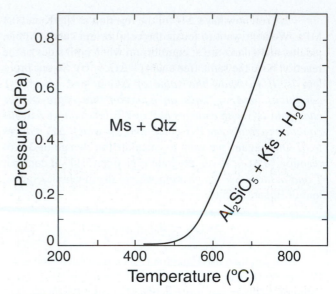

FIGURE 27.2 Pressure–temperature phase diagram for the reaction muscovite + quartz = Al_2SiO_5 + K-feldspar + H_2O, calculated using SUPCRT (Helgeson et al., 1978), assuming $P_{H_2O} = P_{lith}$.

Figure 27.2 is a pressure–temperature phase diagram for a metamorphic reaction involving a fluid phase. The reaction:

$$KAl_2Si_3AlO_{10}(OH)_2 + SiO_2 \qquad (26.5)$$
$$\text{Ms} \qquad\qquad \text{Qtz}$$
$$= KAlSi_3O_8 + Al_2SiO_5 + H_2O$$
$$\text{Kfs} \qquad \text{Al-silicate} \quad \text{fluid}$$

is a typical dehydration reaction in which a hydrous phase (muscovite) reaches the limits of its thermal stability while coexisting with quartz, and dehydrates by reacting with quartz to form an anhydrous solid assemblage and liberating H_2O (Section 26.4).

At the risk of being repetitious, the shape of this dehydration curve is characteristic of all devolatilization reactions (at least under typical crustal conditions), asymptotic to the T axis at low pressure and with a much higher slope at high pressure. We can understand this variation in slope by applying the Clapeyron equation [Equation (5.15)]. Because there is a free volatile phase involved, ΔV for the reaction will be very large at low pressure, and the large denominator results in a low dP/dT slope. As pressure increases, the volume of the gas decreases significantly more than the solids, so that ΔV also decreases and the slope increases. Of course the volatile-bearing mineral phase (in this case muscovite) is typically on the low-temperature side of the reaction because increasing temperature liberates the volatile.

27.3 COMPOSITIONAL VARIATION

So far we have looked only at the pressure and temperature effects on phases of fixed composition. When the composition of a phase varies, as in most minerals, melts, and fluids, the free energy of the phase must vary as well. Therefore, we must find a way to include compositionally dependent parameters in Equations (5.3) and (5.11). We can add compositional terms for the dependence of G upon changes in the number of moles of components i, j, k, etc., and modify Equation (5.3) to the form:

$$dG = VdP - SdT + \sum_i \mu_i \, dn_i \quad (27.9)$$

where n_i is the number of moles of component i in the system, and μ_i is the **chemical potential** of component i. There must be a $\mu_i dn_i$ term for each component, hence the summation in Equation (27.9). μ_i can be defined analogously with Equations (5.12) and (5.13) as:

$$\mu_i = \left(\frac{\partial G}{\partial n_i}\right)_{P,T,n_{j \neq i}} \quad (27.10)$$

where the subscript $n_{j \neq i}$ means that all components other than i are held constant (as are T and P). μ_i *thus expresses the manner in which the free energy of a phase changes with the number of moles of component i in the phase, all other things being constant.*

The chemical potential is an important property of most natural chemical systems because the systems generally include several minerals that exhibit solid-solution behavior. An alternative definition of the Gibbs free energy (at any specified T and P) is that it is the sum of the chemical potentials contributed by each of the components:

$$G = \sum_i n_i \mu_i \quad (27.11)$$

If we have olivine with a composition intermediate between forsterite (Fo = Mg_2SiO_4) and fayalite (Fa = Fe_2SiO_4), we can express the olivine composition as some mixture of Fo and Fa components. At a constant temperature and pressure, the free energy of such an olivine can be determined, from Equation (27.11), as $G^{ol} = n_{Fo}\mu_{Fo} + n_{Fa}\mu_{Fa}$. If the olivine were pure Fo, then $n_{Fa} = 0$, and $G^{ol} = G^{Fo} = n_{Fo}\mu_{Fo}$, and $\mu_{Fo} = G^{Fo}/n_{Fo}$, which is, by definition, G^{Fo}, or *the molar Gibbs free energy of pure forsterite*. In other words, the chemical potential of a pure phase is equal to its molar Gibbs free energy, and the chemical potential continues to reflect the free energy of that component as it becomes diluted by other components.

Another important point: *At equilibrium, the chemical potential of a given component must be the same in every coexisting phase that contains it.* This follows from the fundamental notion that the stable state is the state with the minimum possible free energy. If we specify that the system is at equilibrium, then that must be the minimum energy state. If the chemical potential of a component is lower in one phase than in another, the free energy of the system could be lowered by migration of the component from a phase with higher value of μ_i and adding it to a phase with lower μ_i. This is because Equation (27.9) would lower the G of the high μ phase (by $-\Delta n_i$ times a larger μ_i) by more than it would raise the G of the low μ phase (by $+\Delta n_i$ times a smaller μ_i). In fact, the difference in the chemical potentials of various components is what drives the diffusion process associated with metasomatism and alteration at places such as

veins or igneous contacts (Chapter 30). In such cases, species migrate from places with high chemical potential toward places with lower chemical potential.

We now come to the final useful compositional term, the activity. The **activity** of component i in phase A (a_i^A) describes the difference between the chemical potential of i in A under some specified temperature and pressure conditions, and the value of the chemical potential in the reference state. That relationship takes the form:

$$\mu_i^A = \mu_i^o + RT \ell n a_i^A \quad (27.12)$$

The similarity of the form of Equation (27.12) with that of Equation (27.6) for pressure is not coincidental. From Equations (27.6) and (27.12), we can define the activity of a gas as:

$$a_i^A = P_i/P_i^o \quad \text{(for an ideal gas)} \quad (27.13)$$

or:

$$a_i^A = f_i/f_i^o \quad \text{(for a real gas)}$$

If the gas is in the reference state, then, by definition, $\mu_i^A = \mu_i^o$ and $\ell n a_i^A = 0$, or $a_i^A = 1$. The same applies to solid phases, such as olivine above. The usual reference state, as defined long ago by chemists, refers to pure substances at 298 K and 0.1 MPa. Because geologists are not bound to the laboratory, as is your poor average chemist, and our study areas enjoyed conditions far removed from 298 K and 0.1 MPa, we can choose any reference state we please. Petrologists generally choose as a reference state the pure phase at the temperature and pressure of interest, thus μ_{Fo}^{ol} for pure forsterite $= a_{Fo}^{ol}$ and $\mu_{Fo}^A = 1$ at any P and T we choose. In simple words, the activity of a pure phase in the reference state equals 1. We can adjust the reference state for our pure phase from some tabulated value, such as 298 K and 0.1 MPa, to any temperature and pressure of interest using Equation (5.3), as we did for quartz in the example problem in Chapter 5.

If phase A is not pure, but the reference state is chosen as above, then μ_i^A and μ_i^o are at the same pressure and temperature, so the only difference between them is based on composition, which is then embodied in the expression $RT \ell n \, a_i^A$ in Equation (27.12).

To understand this better as applied to solid solutions, we can turn to a reaction, such as the jadeite + quartz = albite reaction dealt with in the example problem above, and address variable compositions. When the phases were pure, we calculated the location of the reaction curve from the relationship $\Delta G = G_{Ab} - G_{Jd} - G_{Qtz} = 0$. If the phases are not pure (e.g., the plagioclase may be An_{10}, and the jadeitic clinopyroxene Jd_{95}), we can say that, at equilibrium, the following condition is required:

$$\mu_{Ab}^{Pl} - \mu_{Jd}^{Cpx} - \mu_{SiO_2}^{Qtz} = \Delta G = 0 \quad (27.14)$$

If this were not true, the free energy of the system could be lowered by having the reaction progressing in one direction toward the phases with the lower chemical potential. We can then substitute Equation (27.12) into (27.14):

$$\mu_{Ab}^o + RT \ell n \, a_{Ab}^{Pl} - \mu_{Jd}^o - RT \ell n a_{Jd}^{Cpx} \quad (27.15)$$

$$- \mu_{SiO_2}^o - RT \ell n \, a_{SiO_2}^{Qtz} = 0$$

Rearranging, we get:

$$\mu^o_{Ab} - \mu^o_{Jd} - \mu^o_{SiO_2} = \Delta G^o$$

$$= -RT\ell n\left(\frac{a^{Pl}_{Ab}}{a^{Cpx}_{Jd} \cdot a^{Qtz}_{SiO_2}}\right) \qquad (27.16)$$

Because $\mu^o_{Ab} = \overline{G}^{Ab}$, etc., ΔG^o is the calculated ΔG of the reaction in the reference state (pure phases, and preferably at the pressure and temperature of interest) and can be readily calculated from standard thermodynamic data, as we did in the example problem above.

The stoichiometric coefficients in the given reaction all happen to equal 1. If, for some reaction, a coefficient is unequal to unity (e.g., 3 moles of quartz participate), it must act as a *multiplier* to the appropriate μ^o_i terms ($3\,\mu^0_{SiO_2}$) and must appear as an *exponent* on the appropriate activity terms $[(a^{Qtz}_{SiO_2})^3]$ because of the log function. The combined activity product in the bracket in Equation (27.16) expresses the difference between equilibrium for the pure phases and the actual mineral compositions. At equilibrium this term is a constant and is referred to as *K*, or the **equilibrium constant**. In its simplest form, Equation (27.16), at any given *P* and *T*, becomes:

$$\Delta G = 0 = \Delta G^o + RT\ell nK \qquad (27.17)$$

or: $\Delta G^o = -RT\ell nK$

Now this looks like a pretty simple equation (after all the mathematics it took to get to it). All we need to do is add a few activity terms to the equation we used for pure phases, and we can calculate the stabilities of mineral equilibria at any pressure and temperature, as before, but now at any composition as well.

As we learned in Chapter 8, it is a relatively easy matter to determine the composition of a phase that is not pure by using the electron microprobe. Unfortunately, we cannot directly measure or determine in a single analysis μ^A_i, and hence a^A_i, of a component in a solid-solution phase. Calculating the Gibbs free energy of a *pure* phase is relatively easy, as we have found in our example problem. Determining μ^A_i, and hence the relationship between activity and composition of a multicomponent phase, on the other hand, can be quite vexing, particularly for *solid* solutions. The relation between the activity of a component *i* in a phase *A* and the measurable mole fraction of *i* in *A* may not be a simple linear function. Further complications arise when more than one component is mixed because the energetics of mixing components *i* and *j* may depend upon the amounts of *k*, *l*, etc. in the solution.

A considerable proportion of modern experimental study is being devoted to determining the **activity–composition (a-X) relationships** in the more common silicate minerals. Analyzing the distribution of two or more components between two coexisting phases at equilibrium from experiments at a known pressure and temperature (and over a sufficient composition range) can permit us to model the relationship between activity and composition. Models can be as simple as an **ideal solution**, or more complex **real solutions**, such as "regular" solutions, "asymmetric," or "quasiregular" solutions, etc. (see Saxena, 1973, for a review). As

the solution model becomes progressively more complex, and more components are considered, more empirically based "mixing," or "excess," terms are typically required (and more experimental data are needed to constrain those terms). The simplest model that fits the data and adequately describes the behavior of a particular mixture, be it solid, melt, or fluid, is the appropriate choice. Only when reliable data justify and require a more complex model with extra fitting parameters should such a model be used.

As a simple example of a mixing model, we will take a look at the ideal solution, in which the activity–composition relationship is:

$$a^A_i = (X^A_i)^y \quad \text{(ideal solution)} \qquad (27.18)$$

where X^A_i is the **mole fraction** of mixture component *i* in phase *A*. $X^A_i = n_i/(n_i + n_j + n_k \dots)$, where n_i is the number of moles (or molecules) of component *i* in a solution of *i*, *j*, *k*, etc. The exponent *y* is the number of crystallographic sites on which the mixing takes place. Please recognize that an ideal *solution* is not the same as an ideal *gas*. An ideal gas is one which behaves according to the ideal gas law ($PV = nRT$), whereas an ideal solution behaves according to Equation (27.18). It is therefore possible to consider ideal mixtures of non-ideal gases, or vice versa. In our calculations, we will deal only with this ideal mixing model because the mathematics are simple, and it is an appropriate model for many mineral solutions. More complex models derive formulations for **activity coefficients** (γ_i), which may vary with *X*, and are applied by the equation:

$$a^A_i = (\gamma_i X^A_i)^y \qquad (27.19)$$

Note that the activity is essentially a thermodynamically effective analog of concentration, just as the fugacity is the thermodynamically effective pressure. Both are adjusted for non-ideal behavior with the activity or fugacity coefficient. Ionic species are also treated effectively using activities. We will rarely deal with ions in this text, but the simplest assumption is to equate the activity with the concentration (*C*), usually expressed as molality, the number of moles of the ionic species (solute) per kg of H_2O. More sophisticated treatment uses an activity coefficient such that $a_i = \gamma_i C_i$. Numerical values for γ can be calculated from general ionic theories (such as the Debye-Hückel theory) that model the ionic strength of solutions (see Faure, 1998, for further explanation).

Figure 27.3 is an example of the activity–composition relationship for the Fe-Mg exchange in orthopyroxene. The straight-line model is the ideal solution model. The curves, representing a type of regular solution model called the *simple mixture model*, are a much better fit to the experimental data (open circles). The models are generated by regressing appropriate activity coefficients in solution model expressions to fit curves (such as those shown in Figure 27.3) to the experimental data. Note that the orthopyroxene solution approaches an ideal mixture with increasing temperature. The extra thermal vibrational energy makes the M1 and M2 sites in pyroxene less selective as temperature increases, so that mixing of Fe and Mg becomes more evenly

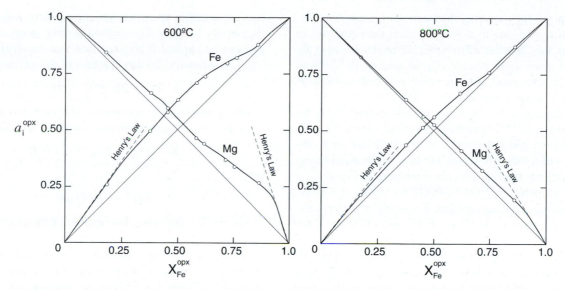

FIGURE 27.3 Activity–composition relationships for the enstatite–ferrosilite mixture in orthopyroxene at 600°C and 800°C. Circles are data from Saxena and Ghose (1971); curves are model for sites as simple mixtures (from Saxena, 1973). Copyright © with permission from Springer-Verlag.

distributed (ideal). Most real solutions approach the ideal solution at high temperatures or when sufficiently concentrated. Even at 600°C, the real curves in Figure 27.3 approach the ideal line above $X = 0.85$ or so.

At the other extreme (very dilute solutions), the relationship between activity and composition becomes approximately linear. This linearity is the essence of **Henry's law**, and the dilute region is generally called the "Henry's law region." According to Henry's law, the activity is proportional to the concentration (although the proportionality ratio need not be 1.0), so the activity coefficient must be a constant over that range. Although the solution is not ideal, the constant activity coefficient makes it easier to predict solution behavior in the dilute region.

In a moment, we will use the ideal solution model to estimate the effect of compositional changes on the jadeite + quartz = albite equilibrium calculated in the example problem above. But first we must understand the model. The composition of a given mineral can be determined by using an electron microprobe. From the composition we can derive X_i^A, and we can then estimate a_i^A using an appropriate a-X model (such as the ideal solution model). In order to do this, we need to know what components are competing with i in the solution and something about the mechanism of the solution.

Let's begin with a simple solution: Mg_2SiO_4 and Fe_2SiO_4 in olivine. The mixing of the Fe and Mg ions involves a simple exchange, but we must remember that there are 2 moles of Fe or Mg ions per mole of olivine. From your mineralogy course, you know that these ions occupy two slightly different sites (M1 and M2) in olivine. On each site, $X_{Mg} = n_{Mg}/(n_{Mg} + n_{Fe})$ for a simple binary Mg-Fe system. If there are more components, $X_{Mg} = n_{Mg}/(n_{Mg} + n_{Fe} + n_{Mn} + n_{Mn} + n_{Ni} + \dots)$, including whatever other species share the M sites with Mg included in the denominator. The situation is similar to having 2 moles of $MgSi_{0.5}O_2$ in a reaction involving olivine, which would put the reaction on a 1-cation mixing basis. If we did this,

we would have to double the μ_o term and square the exponent on the olivine activity term in Equation (27.16). For the μ_o term, this is of little importance because $2\mu_{MgSi_{0.5}O_2}^o = \mu_{Mg_2SiO_4}^o$. For the exponent, however, it makes a significant difference. Because we routinely treat forsterite as Mg_2SiO_4 and use 1 mole of Mg_2SiO_4 in our calculations, we should adjust the activity term appropriately to compensate for the loss of the squared term we would have used in the single-cation formula. This is the reason for the exponent in the ideal solution model, Equation (27.18), and a proper ideal model for olivine would thus be $a_{Fo}^{ol} = (X_{Fo}^{ol})^2$. Even with our simple ideal solution, we must make allowances for multiple cation sites.

Although SiO_2 is also a component of olivine, it does not participate in the Fo-Fa mixture. It is thus not included in X_{Mg} or the solution model. Only components that actually substitute for Mg are included.

For **coupled substitutions**, such as Ca-Al for Na-Si in plagioclase, we do not need to express the solution for both Na $\leftrightarrow$ Ca and Al $\leftrightarrow$ Si separately because each Na $\leftrightarrow$ Ca exchange *requires* an Si $\leftrightarrow$ Al exchange. To use both in the model would be like counting the same exchange twice. Thus the ideal solution model for Ab-An in plagioclase is $a_{Ab}^{Plag} = X_{Ab}^{Plag}$, where X_{Ab}^{Plag} is simply $n_{Ab}/(n_{Ab} + n_{An})$, which is the same as $n_{Na}/(n_{Na} + n_{Ca})$.

Note from Equation (27.17) that K *is temperature dependent* (and, to a lesser degree, pressure dependent because ΔG^o is determined at a given P as well as T). Adding other components will thus shift the pressure and temperature of the equilibrium curve for a reaction. For example, in our earlier example problem for the reaction Jd + Q = Ab, we calculated ΔG^o from standard tabulated thermodynamic data. If we change the equilibrium by adding Ca and Mg, for example, the equilibrium conditions shift as a function of the distribution of An versus Ab in plagioclase and Di versus Jd in clinopyroxene. (Quartz accepts very few impurities and remains essentially pure.)

Suppose, for example, that some Di component dissolves in the pyroxene to a greater extent than An does in plagioclase. Le Châtelier's Principle tells us that because the impurity is more stable dissolved in Jd than in Ab, the field of Jd will be expanded, and the curve will shift to lower P and higher T—*but by how much?*

We can calculate the reaction shift quantitatively by using Equation (27.17) and the appropriate values for ΔG^o (at the pressure and temperature of interest) and for K. By doing so, we can determine the location of the equilibrium curve for any value of K. This has been done in Figure 27.4 using the computer program SUPCRT for values of K in increments of 0.2. You can familiarize yourself with the mathematical process to accomplish this by doing Problem 2 for this chapter.

Notice that the Jd + Qtz = Ab reaction, when treated as pure Na phases (the heavy curve in Figure 27.4), is a *discontinuous solid–solid net-transfer reaction* (Section 26.3). A metamorphic field gradient, following some curved P-T path (Figure 21.1) in Figure 27.4, would cross the reaction isograd at a single point, so that, if equilibrium were perfectly maintained, the reaction should run to completion *at* the isograd, and an abrupt discontinuity in mineral assemblages would occur as albite breaks down and is replaced by jadeite + quartz. In natural rocks, some of the reactants may remain as metastable relics for a short distance above the isograd because of local disequilibrium effects, but these are usually easy to identify texturally.

If, on the other hand, the phases involve Na-Al-Ca-Mg solution, the reaction becomes *continuous*. The temperature and pressure of the isograd now depend on the Na/Ca ratio of the phases involved and may occur at different grades as this ratio changes from one rock type to the next, as described in Section 26.5. The isograd is no longer an abrupt change from reactants to products but becomes a zone. Below the zone the reactants are stable, and above the zone the products are stable, but within the zone *both* are stable, and the composition of certain phases varies across the zone as the continuous reaction progressively consumes the reactants in favor of the products (and K varies with grade). The Jd + Qtz = Ab reaction typically involves relatively pure phases, and the example above is somewhat artificial. I use it

because it builds upon the example problem with which we are already familiar. The method is quite general, however, and can be applied to any reaction that involves solutions.

In summary, for any reaction of the generalized type:

$$2A + 3B = C + 4D$$

in which the composition of one or more phases is variable, we can solve for the equilibrium curve using the equation:

$$\Delta G = 0 = \Delta G^o + RT \ell nK \qquad (27.17)$$

or:

$$\Delta G^o = -RT \ell nK$$

where ΔG^o is the Gibbs free energy of the reaction for the pure phases *at the pressure and temperature of interest*, and K is the equilibrium constant, or the product of the activities raised to the power of their stoichiometric coefficients for the reaction. Thus, for our hypothetical reaction:

$$K = \frac{a_c \cdot a_D^4}{a_A^2 \cdot a_B^3} \qquad (27.20)$$

ΔG^o can be calculated using Equations (5.1), (5.2), and (5.11). We can then substitute the appropriate values for the activities. For minerals we substitute $a_i = (\gamma_i X_i)^y$, and for gases we substitute the fugacity, where $a_{gas} = f_{gas}/P^o$ and $f_{gas} = \gamma_{gas}p_{gas}$ [Equation (27.13)]. For ionic species we use $a_i = \gamma_i C_i$, where C_i is the concentration of component i in the solution. The fugacity/activity coefficient, γ, is either empirically or theoretically determined. For minerals the activity of a component i can be difficult to determine over the full compositional range, and activity–composition (a-X) relationships may require very sophisticated empirically based models. We have limited our present calculations to an ideal solution model in which $a_i^A = (X_i^A)^y$ and $\gamma_i = 1$. This version of the ideal model accounts for mixing on individual sites because the exponent y encompasses both the number of moles of the species in the reaction and the number of sites in each mineral on which mixing takes place.

In Section 9.1, we discussed an *empirical* constant, K_D, called the **distribution constant**, which was used to describe the distribution of a component between phases. The distribution constant is empirical because it is based on

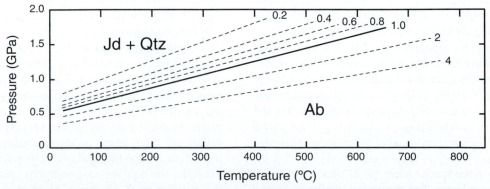

FIGURE 27.4 P-T phase diagram for the reaction jadeite + quartz = albite for various values of K. The equilibrium curve for $K = 1.0$ is the reaction for pure end-member minerals (Figure 27.1). Data from SUPCRT (Helgeson et al., 1978).

measured values of X_i. It is related to the true equilibrium constant, K, by the following equation:

$$K = \frac{a_c \cdot a_D^4}{a_A^2 \cdot a_B^3} = \frac{X_C \gamma_c \cdot X_D^4 \gamma_D^4}{X_A^2 \gamma_A^2 \cdot X_B^3 \gamma_B^3} \qquad (27.21)$$

$$= \frac{X_C \cdot X_D^4}{X_A^2 \cdot X_B^3} \cdot \frac{\gamma_C \cdot \gamma_D^4}{\gamma_A^2 \cdot \gamma_B^3} = K_D \cdot K_\gamma$$

For the simple reaction in Chapter 9 involving the melting of a crystal to a liquid, the distribution occurred as the result of an *exchange* reaction (Section 26.6) of the type:

$$i_{(liquid)} \longleftrightarrow i_{(solid)} \qquad (9.1)$$

and the equilibrium constant for this reaction is:

$$K = \frac{a_i^{solid}}{a_i^{liquid}} \qquad (27.22)$$

In Chapter 9, we dealt with trace elements. A trace element (which is necessarily very dilute) should obey Henry's law (Figure 27.3), and γ^i is thus a constant in the equation $a_i^A = \gamma_i C_i^A$, relating the activity of component i in phase A to its concentration, C. We can substitute the γC terms for the a terms in Equation (27.22) and then incorporate all three constants into the left side of the equation (into K), and by defining:

$$K_D = K \cdot \frac{\gamma_i^{liquid}}{\gamma_i^{solid}} \quad (= K/K_\gamma) \qquad (27.23)$$

we get:

$$K_D = \frac{C_S}{C_L} \qquad (9.3)$$

This is the equation used in Chapter 9, where C_S and C_L are the concentrations of a trace element in the solid and liquid, respectively. Although the expression for K_D in equation (27.23) looks complex, the activity coefficient ratio acts to undo the same ratio already included in K, so that K_D is simply the ratio of the measured concentration of a component in two coexisting phases. K_D remains a constant when applied to sufficiently dilute solutions. As you may remember, when referring to trace elements, K_D is commonly expressed simply as D.

27.4 GEOTHERMOBAROMETRY

As part of determining the history of rocks and regions, geologists have for a long time attempted to estimate the temperature and/or pressure of formation of igneous and metamorphic rocks. Early approaches were mostly qualitative. The classic pelitic index minerals, for example, were qualitative indicators of metamorphic grade (mostly temperature) and were used to estimate the *relative* conditions of metamorphism across a terrane or to compare one terrane to another. Certain minerals, such as andalusite and cordierite, also indicate low pressures, whereas other minerals, such as jadeite and glaucophane, indicate high pressures. Estimation of the stratigraphic overburden above shallow plutons at the time of emplacement was also used to approximate the pressure of igneous crystallization and contact metamorphism.

Experimental petrology, in which investigators model natural geological systems at elevated temperatures and pressures in the lab, has permitted more quantitative approaches to addressing the conditions of magmatic and metamorphic processes. In earlier chapters, we reviewed techniques such as estimating the depth of separation of a magma based on the pressure at which the liquidus was multiply saturated in phenocryst phases (Section 10.4 and Figure 10.13) and using REE patterns to identify garnet in the mantle source region of primitive basalts as an indicator of the depth of partial melting. We have also seen in this chapter how the mineral reactions in a petrogenetic grid can be used to limit the P-T conditions under which a particular mineral assemblage is stable, a technique we will use more fully in the next two chapters.

Another powerful technique based on the work of experimental petrology is the calculation of equilibrium temperatures and pressures from the measured distribution of elements between coexisting phases. So far in our studies of metamorphic mineral assemblages, we have concentrated on the types of minerals present and the reactions that separate them, but we have become increasingly aware that the compositions of virtually every solid-solution phase in a rock, whether metamorphic or igneous, varies with T and P because of exchange or continuous reactions. The compositions of the minerals involved in many of these reactions can be sensitive indicators of the P-T conditions. If we can model and replicate some of the equilibria in the lab, we can relate the distribution of exchanged components (K or K_D) in natural assemblages to the laboratory T-P conditions. This general technique is called **geothermobarometry** and can be separated into its two common components: **geothermometry**, the evaluation of the temperature at which a rock formed, and **geobarometry**, the evaluation of the pressure.

It would be an impossibly long job to reproduce the full P-T-X spectrum of metamorphic and igneous rocks in the laboratory, and we are thus forced to settle for exploring only a relatively small subset of natural conditions and compositions. We can quantify our estimates of P and T from values of K_D and extend the range of application somewhat beyond the experimental base by extracting thermodynamic data from the experiments, and by fitting appropriate activity–composition (a-X) models to the mineral solutions. This is the approach we just took to model the shift in the Jd + Qtz = Ab equilibrium with variations in the compositions of the phases. We must be careful not to extrapolate too far, however. It is possible to use thermodynamic equations to calculate far beyond the P-T-X conditions for which we have reliable data, and the further we go beyond this, the greater the possible error. We can get erroneous results from injudicious use of this powerful tool.

In order to understand how the distribution of components between two or more phases at equilibrium can be used to estimate temperatures and/or pressures of equilibration, let's return to Equation (27.17). The temperature (and pressure) dependence of K is more directly obvious from a rearrangement of this equation to yield:

$$\ell nK = -\Delta G^o/RT \qquad (27.24)$$

which expresses ℓnK (a compositional variable) as a function of T (and P because ΔG^o, for reactions involving solids, represents the ΔG of reaction for pure phases at a specific P and T of interest). As we saw earlier in this chapter, ΔG^o can be determined by calculating ΔG at 0.1 MPa and 298 K, from tabulated thermodynamic data, and then adjusting to higher P and T, using Equation (5.11). In this section we will continue to adopt our simplification that ΔV, ΔS, and ΔH do not vary with T or P in our calculations, so they will be easier. For more rigorous treatments, we would need to integrate the heat capacities and compressibilities to compute ΔH, ΔS, and/or ΔV at elevated pressures and temperatures. This adds more terms and complexity to the equations, but can readily be done (see Wood and Fraser, 1977; or Spear, 1993, for the relevant equations).

Equation (5.1), when applied to a reaction at constant pressure, yields:

$$\Delta G = \Delta H - T\Delta S \qquad (27.25)$$

If we combine Equation (27.25) with Equations (27.24) and (27.21), and use $\Delta V dP$ to correct ΔG [as in Equation (5.11), using our simplifying assumptions], we come up with:

$$\ell nK = \ell nK_D + \ell nK_\gamma \qquad (27.26)$$
$$= -\Delta H^o/RT + \Delta S^o/R - (\Delta V/RT)dP$$

Equation (27.26) is the basis for calculating the T and P of equilibration from values of the easily measured distribution constant, K_D (if ΔH, ΔS, and ΔV are reasonably constant). It is best to explore the potential of this equation by briefly analyzing a commonly used geothermometer and geobarometer. In Chapter 15 of his excellent monograph, Frank Spear (1993) explained the basis for the garnet–biotite exchange geothermometer and the *GASP* geobarometer. I'll simply follow his lead.

27.4.1 The Garnet–Biotite Exchange Geothermometer

Ferry and Spear (1978) performed a series of experiments at 0.207 GPa and temperatures from 500 to 800°C on the Fe-Mg exchange reaction between biotite and Ca-free garnet:

$$Fe_3Al_2Si_3O_{12} + KMg_3Si_3AlO_{10}(OH)_2 \qquad (27.27)$$
$$= Mg_3Al_2Si_3O_{12} + KFe_3Si_3AlO_{10}(OH)_2$$

They began at each temperature with garnet of composition $Alm_{90}Prp_{10}$ (X_{Fe} = molar Fe/(Fe + Mg) = 0.9) and two different biotite compositions known to bracket the equilibrium Fe-Mg distribution value with that garnet composition. Garnet is notoriously sluggish to equilibrate at metamorphic temperatures, so Ferry and Spear (1978) used charges in which garnet and biotite were mixed in molar proportions of 98:2. The biotite composition would thus have to change far more to accommodate the garnet composition in attaining the equilibrium K_D value (as we shall see below). Two biotite compositions were used in each starting mixture so that equilibrium was approached from both directions, thereby providing a bracket to the equilibrium value. Such experiments are called *reversed* because equilibrium is approached from both directions. If equilibrium were approached from only one direction, it would be impossible to be certain that equilibrium was really attained, and the value of K_D at the time the experiment ended might simply reflect an interrupted approach toward the true equilibrium value. The experiments of Ferry and Spear (1978) lasted from 13 to 56 days in order to give equilibration a good chance to be attained.

Table 27.2 is a spreadsheet that shows the results of Ferry and Spear's experimental runs. The first three columns are the experimental conditions and results. Notice in the first two runs at 799°C, for example, that an initial garnet composition of $X_{Fe} = 0.9$ was combined with a biotite of $X_{Fe} = 1.0$, and another was combined with $X_{Fe} = 0.5$. When the runs were stopped and quickly cooled, the final biotite in the first run had a composition of $X_{Fe} = 0.750$, and in the second run $X_{Fe} = 0.710$. This represents a good reversed (or bracketed) experiment in which both biotites adjusted toward the equilibrium value, which must then lie between $X_{Fe} = 0.750$ and 0.710. The composition of the garnet in each run must then have changed by an amount equal to 2/98 times the change in the corresponding biotite to create column 5. For run 1, this is:

$$\Delta X_{Fe}^{Grt} = \frac{2}{98}\Delta X_{Fe}^{Bt} = \frac{2}{98}(1.0 - 0.750) = 0.005 \quad (27.28)$$

TABLE 27.2 Experimental Results of Ferry and Spear (1978) on a Garnet–Biotite Geothermometer

T°C	Initial X_{Bt}^{Fe}	Final X_{Bt}^{Fe}	Final X_{Grt}^{Fe}	Final $(Mg/Fe)_{Grt}$	Final $(Mg/Fe)_{Bt}$	K	T kelvins	1/T kelvins	ℓnK
799	1.00	0.750	0.905	0.105	0.333	0.315	1072	0.00093	−1.155
799	0.50	0.710	0.896	0.116	0.408	0.284	1072	0.00093	−1.258
749	0.50	0.695	0.896	0.116	0.439	0.264	1022	0.00098	−1.330
738	1.00	0.730	0.906	0.104	0.370	0.281	1011	0.00099	−1.271
698	0.75	0.704	0.901	0.110	0.420	0.261	971	0.00103	−1.342
698	0.50	0.690	0.896	0.116	0.449	0.258	971	0.00103	−1.353
651	0.75	0.679	0.901	0.110	0.473	0.232	924	0.00108	−1.459
651	0.50	0.661	0.897	0.115	0.513	0.224	924	0.00108	−1.497
599	0.75	0.645	0.902	0.109	0.550	0.197	872	0.00115	−1.623
599	0.50	0.610	0.898	0.114	0.639	0.178	872	0.00115	−1.728
550	0.75	0.620	0.903	0.107	0.613	0.175	823	0.00122	−1.741
550	0.50	0.590	0.898	0.114	0.695	0.163	823	0.00122	−1.811

Thus the final garnet composition is $0.90 + 0.005 = 0.905$, as shown in Table 27.2. Notice that the garnet composition changed much less than that of the biotite due to the relative amounts in the experimental charge, as was the intent. Notice also that Fe/Mg is much higher in garnet than in coexisting biotite, as has been mentioned previously.

If we divide Reaction (27.27) by 3 to put it on a single-cation basis and assume an ideal solution model [all $\gamma = 1$ so that $\ell nK_\gamma = 0$ in Equation (27.21)], the equilibrium constant is:

$$K = K_D = \left(\frac{X_{Fe}^{Bt} X_{Mg}^{Grt}}{X_{Mg}^{Bt} X_{Fe}^{Grt}} \right) \quad (27.29)$$

We can rearrange Equation (27.29) as follows:

$$K = K_D = \frac{(X_{Mg}/X_{Fe})^{Grt}}{(X_{Mg}/X_{Fe})^{Bt}} = \frac{(Mg/Fe)^{Grt}}{(Mg/Fe)^{Bt}} \quad (27.30)$$

K_D can readily be determined from microprobe analyses of coexisting garnets and biotites. The final column in Table 27.2 is the natural log of K_D, which may then be plotted against $1/T$ in kelvins to produce the graph in Figure 27.5. Because the experimental data are for a single pressure, Equation (27.6) reduces to:

$$\ell nK_D = -\Delta H^o/R \cdot (1/T) + \Delta S^o/R \quad (27.31)$$

which is the equation for a straight line of versus $1/T$ in kelvins. The data in Figure 27.5 conform well to a straight line suggesting that the ideal solution model is an adequate model for Fe-Mg exchange between garnet and biotite. The results of a linear regression for the data are included in Figure 27.5. Ferry and Spear (1978) got essentially the same results:

$$\ell nK_D = -2109/T(K) + 0.782 \quad (27.32)$$

Notice that this equation relates temperature to measurable K_D and is thus a quantitative geothermometer. This geothermometer is valid at pressure $= 0.207$ GPa, but Spear (1993) argued that it is relatively insensitive to pressure changes and should apply to garnet–biotite pairs extending to mid-crustal levels (see more below).

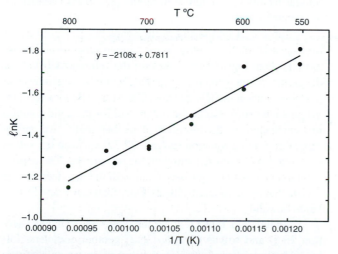

FIGURE 27.5 Graph of *lnK* versus *1/T* (in kelvins) for the Ferry and Spear (1978) garnet–biotite exchange equilibrium at 0.2 GPa from Table 27.2.

We can extract thermodynamic data from these experimental results. Because Table 27.2 and Figure 27.5 refer to one-third of Reaction (27.27) we must multiply everything by 3 to get appropriate values for the reaction as written. If ΔV is constant, $\int_{0.1}^{P} \Delta V dP$ is simply $\Delta V (P - 0.1) = P\Delta V$ (because P is usually much greater than 0.1, we can ignore 0.1 in our calculations). Equation (27.26) then becomes:

$$\Delta G_{P,T} = \Delta H_{0.1, 298} - T\Delta S_{0.1, 298} \quad (27.33)$$
$$+ P\Delta V + 3RT\ell nK_D = 0$$

Rearranging, we get:

$$\ell nK_D = \frac{-\Delta H - P\Delta V}{3R}\left(\frac{1}{T}\right) + \frac{\Delta S}{3R} \quad (27.34)$$

Comparing this with Equation (27.32) shows us that the slope of the regression line in Figure 27.5 is $(-\Delta H - P\Delta V)/3R$, and the y-intercept is $\Delta S/3R$. We can thus extract these fundamental thermodynamic parameters for the reaction from the experimental data. This is an important alternative to thermodynamic data derived from calorimetry because it produces high temperature–pressure data and is less subject to errors in extrapolating 298 K-one-atmosphere calorimetric data to geologic conditions. The linear fit also implies that the assumption of constant ΔH and ΔS as temperature varies is reasonable. Using the value of $R = 8.3144 \, J \cdot mol^{-1} \cdot K^{-1}$ from the inside cover, we derive:

$$\Delta S = 3 \cdot 8.3144 \cdot 0.782 = 19.506 \, J/K \, mol$$

ΔV in Robie and Hemingway (1995) is 2.494 J/MPa, and, because the experiments were run at 207 MPa:

$$\Delta H = 3 \cdot 8.3144 \cdot 2109 - 207 \cdot 2.494$$
$$= 52.09 \, kJ/mol$$

The garnet–biotite geothermometer assumes ideal Fe-Mg solution and constant values of ΔH, ΔS, and ΔV, all of which are supported by the linearity of Figure 27.5. If the solution is really non-ideal but linear and modeled as ideal, the activity coefficients become incorporated into the ΔH and ΔS terms by the regression process. As a result, the geothermometer will still work, but the regressed values of ΔH and ΔS will be in error (usually only slightly if the fit is still linear).

Figure 27.6 illustrates a pair of *AFM* diagrams for coexisting garnet and biotite at approximately 500°C and 800°C based on the K_D values of the thermometer. Notice again that garnet tends to incorporate Fe preferentially with respect to biotite and that the distribution is less equal at 500°C than at 800°C. At higher temperatures, the greater thermal vibration in the crystal lattices make the minerals less selective about the size of the ions that occupy the sites, and the distribution becomes more even ($K_D \rightarrow 1.0$). The tie-lines are thus closer to radii emanating from the Al_2O_3 apex at 800°C. Such radii represent constant Fe/Mg, and hence equal Fe/Mg in both phases. As discussed in Section 26.6, changing metamorphic grade causes the tie-lines between coexisting phases to rotate on an *AFM* diagram. We can now see that this occurs because K_D varies with temperature. When tie-lines such as these are observed to be crossed in a single rock, more than one K_D,

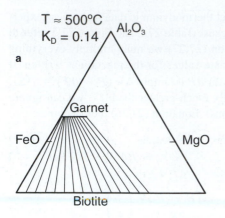
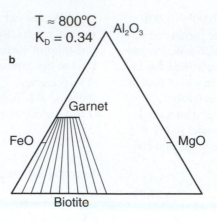

FIGURE 27.6 *AFM* projections, showing the relative distribution of Fe and Mg in garnet versus biotite at approximately 500°C **(a)** and 800°C **(b)**. From Spear (1993).

and thus more than one equilibrium temperature, is indicated. This is a clear indication of disequilibrium in the rock and may represent chemical zoning, retrograde alteration or multiple metamorphic events.

Ferry and Spear substituted the values for ΔH and ΔS back into Equation (27.33) to give:

$$52{,}090 - 19.506T(K) + 2.494P(\text{MPa}) \quad (27.35)$$
$$+ \ 3 \cdot 8.3144 \cdot T(K)\ell nK_D = 0$$

Their regression values differed very slightly from mine, and I've kept mine in the equation above for uniformity. Rearranging gives:

$$T°C = \frac{52{,}090 + 2.494 \ P(\text{MPa})}{19.506 - 24.943 \ \ell nK_D} - 273 \quad (27.36)$$

From this, plus an estimate of P, we can calculate a temperature of equilibration from measured K_D of coexisting garnets and biotites in natural rocks. Application to pressures differing greatly from the 207 MPa of the experiments is risky, however, particularly because $\Delta H = -3 \cdot R \cdot 2109 - P\Delta V$, which is not really constant as pressure varies [as was assumed in Equation (27.36)].

Figure 27.7 is a *P-T* diagram showing curves for various values of K_D, called **isopleths**, calculated by Spear (1993) on the basis of his version of Equation (27.36). These isopleths are like those calculated for the Jd + Qtz = Ab equilibrium in Figure 27.4. Note the very steep slope of the isopleths. This indicates that the reaction is much more sensitive to temperature than to pressure, the hallmark of a good geothermometer. Remember that a small ΔV in the denominator of Clapeyron equation [Equation (5.15)] makes for a steep slope on a *P-T* diagram. ΔV for Reaction (27.27) is small because the reactants and products are so similar (garnet + biotite constitute both). This also implies that a pressure estimate need only be approximate in Equation (27.36) to determine a reasonable temperature.

The garnet–biotite geothermometer formulation of Ferry and Spear (1978) was based on Ca-free experiments and appears to work well for low-Ca garnets in the greenschist and amphibolite facies. When the garnet has significantly more calcium (commonly at higher pressures), Equation (27.26) is less reliable. Mixing of Ca in garnet is non-ideal, and therefore a correction for the effect of Ca on Fe-Mg exchange, typically

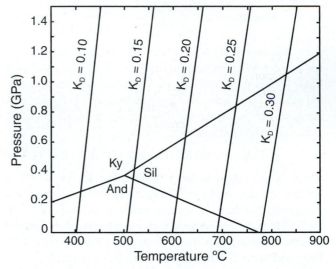

FIGURE 27.7 Pressure–temperature diagram similar to Figure 27.4, showing lines of constant K_D for the garnet–biotite exchange reaction. The Al_2SiO_5 phase diagram is added. From Spear (1993).

based on a complex "regular solution" activity–composition (*a-X*) model, must be included (see Ganguly and Saxena, 1984; Berman, 1990; Holdaway et al., 1997; and Holdaway, 2000). Similar factors for Al and Ti in biotite have been included by several investigators.

Table 27.4 (at the end of the chapter) is a partial list of many geothermobarometers in use among igneous and metamorphic petrologists today. Among the references listed for the garnet–biotite geothermometer are a number of studies that propose solution models that take Ca, Al, and Ti into account. When we wish to change the *a-X* model for the solid solution in a published geothermometer of geobarometer, we cannot simply correct an equation such as (27.32), or keep the values of ΔS and ΔH and use them with the new model. The experimental data must be regressed again with the new formula for K (which may no longer be ideal). Only then can a new formulation be valid.

Included in Table 27.4 are the common exchange (Section 26.6) and solvus (Section 26.2) geothermometers, followed by a number of other geothermobarometers in common use. Many of these latter are continuous types of solid–solid net-transfer reactions (Sections 26.5 and 26.3), which involve variations in the relative proportions of minerals as well as their composition.

27.4.2 The *GASP* Continuous Net-Transfer Geobarometer

A good geobarometer has $\Delta V \gg \Delta S$ and thus a low slope on a *P-T* diagram. This makes it more sensitive to pressure than to temperature. As an example of a very common geobarometer, we turn to the garnet–aluminosilicate–silica–plagioclase geobarometer, affectionately called *GASP*. It is a well-constrained geobarometer that may be applied to many high-grade pelitic schists containing plagioclase, garnet, quartz, and an Al_2SiO_5 mineral. It is based on the reaction:

$$3\, CaAl_2Si_2O_8 = Ca_3Al_2Si_3O_{12} + 2\, Al_2SiO_5 + SiO_2$$
$$\quad An \qquad\qquad Grs \qquad\qquad Ky \qquad Qtz$$
(27.37)

You may remember this as the high-pressure stability limit of anorthite [Reaction (25.9)], and the reaction curve for the pure end-member reaction is shown at the top-center of the mafic petrogenetic grid in Figure 26.19. The relationship between $\Delta G°$ and the equilibrium constant K for the reaction is:

$$\Delta G° = -RT \ln K$$
(27.38)

$$= -RT \ln \left(\frac{a_{SiO_2}^{Qtz} (a_{Al_2SiO_5}^{Ky})^2 a_{Grs}^{Grt}}{(a_{An}^{Plag})^3} \right)$$

Because quartz and kyanite are essentially pure phases in nature, the activity of each is equal to 1.0, and their activity terms drop out of the K expression.

Figure 27.8 is a *P-T* diagram showing the location of the univariant equilibrium curve for Reaction (27.38), based on the experiments of Koziol and Newton (1988). Each sample ("charge") in the experiments initially included all four phases, finely ground and mixed. The open triangles in Figure 27.8 represent runs in which the amount of anorthite increased during the experiment; the solid triangles represent runs in which the amount of grossular, kyanite, and quartz increased; and the half-filled triangles represent runs in which no significant change occurred. The runs are thus also reversed, in that the reaction was bracketed by experiments in which it ran in either direction. The shaded area is Koziol and Newton's (1988) estimate of the uncertainty associated with the location of the equilibrium curve. They estimated that the curve is accurate to within 0.1 GPa (or about 3.5 km).

The best-line fit of Koziol and Newton (1988) to their data yields:

$$P(MPa) = 2.28\, T(°C) - 109.3$$
(27.39)
$$= 2.28\, T(K) - 731.7$$

If we make our customary simplifying assumptions that ΔS, ΔH, and ΔV are constant and look up ΔV ($=-66.08$ J/MPa), we can again use Equation (27.33) in combination with Equation (27.39) to extract ΔH and ΔS for Reaction (27.37). In the present case, however, $K = 1$, because Figure 27.8 is for the pure Ca-end-member reaction. Spear (1993) did this and determined that $\Delta H =$

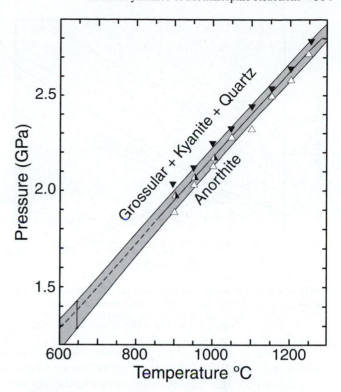

FIGURE 27.8 *P-T* phase diagram, showing the experimental results of Koziol and Newton (1988), and the equilibrium curve for Reaction (27.37). Open triangles indicate runs in which An grew, closed triangles indicate runs in which Grs + Ky + Qtz grew, and half-filled triangles indicate no significant reaction. The univariant equilibrium curve is a best-fit regression of the data brackets. The line at 650°C is Koziol and Newton's estimate of the reaction location based on reactions involving zoisite. The shaded area is the uncertainty envelope. After Koziol and Newton (1988). Copyright © The Mineralogical Society of America.

$-48,375$ J/mol and $\Delta S = -150.66$ J/mol K. He then inserted these values back into Equation (27.33) to derive an equation for the temperature and pressure dependence of K for the *GASP* reaction:

$$\Delta G = 0 = -48,357 + 150.66\, T(K)$$
(27.40)
$$-66.08\, P(MPa) + RT \ln K$$

where: $K = \dfrac{a_{Grs}^{Grt}}{(a_{An}^{Plag})^3} = \dfrac{(X_{Grs}^{Grt})^3}{(X_{An}^{Plag})^3}$ (if ideal)

X_{Grs}^{Grt} is cubed because mixing occurs on three sites. Spear (1993) then used this equation to create a *P-T* diagram for the *GASP* geobarometer, contoured for various isopleths of K (Figure 27.9). Note that slopes of the reaction curves are shallow compared to the garnet–biotite exchange equilibria in Figure 27.7. The *GASP* reaction is thus a much better geobarometer than a geothermometer. The curves are still inclined, however, and application of the *GASP* geobarometer requires an independent estimate of temperature (presumably from an exchange or solvus geothermometer) in Equation (27.40) in order to derive a reasonable pressure estimate. As with the garnet-biotite geothermometer, the assumption that ΔH does not vary with pressure is a weakness of the formulation, but may be minor if not extrapolated too far from the experimental pressures.

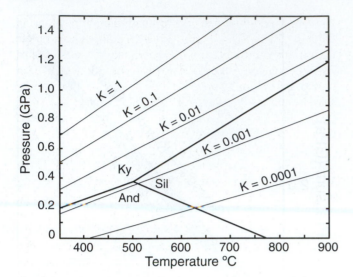

FIGURE 27.9 *P-T* diagram contoured for equilibrium curves of various values of *K* for the *GASP* geobarometer reaction 3 An = Grs + 2 Ky + Qtz. From Spear (1993).

Neither grossular–pyrope–almandine solutions in garnet nor anorthite–albite solutions in plagioclase are ideal (as one could easily infer from the solvi in both series). As a result, most recent calibrations of the *GASP* geobarometer use non-ideal models for the activity–composition (*a-X*) relationships in the expression for *K* before regressing against experimental *P* and *T*.

27.4.3 Application of Geothermobarometry to Rocks

Although Table 27.4 lists a number of potential geothermometers and geobarometers, it would be impossible to review them adequately in this introduction. In fact, it is impossible to keep up with the burgeoning number and refinements of these equilibria, and the table, although useful as an initial reference, can never be complete. For more details see some of the reviews in the literature listed at the end of the chapter. We will settle for a single example.

One approach to estimating the temperature and/or pressure of a rock (whether igneous or metamorphic) is to select an appropriate geothermometer or geobarometer from those available on the basis of the mineral assemblages available in the field. Some of the geothermobarometers can be applied to pelitic rocks, others to mafic rocks, and others to granites, etc. The next step is to survey the calibrations available for each thermometer or barometer of interest, and to judge which may best be applied. For example, if you are studying pelites with low-Ca garnet and biotite low in Al and Ti, the Ferry and Spear (1978) calibration may be appropriate, and its simplicity is attractive. If you are dealing with high-pressure rocks with high-Ca garnets, or high-temperature rocks with Ti and F in biotite, a more complex garnet–biotite geothermometer may be preferable.

For any given mineral assemblage, there may be several reactions that we can write to relate various subsets of minerals, and more than one such reaction may have been calibrated as a geothermometer or geobarometer. Thus we can

use not only several calibrations of a single equilibrium but several simultaneous equilibria as well. Different equilibria and different calibrations of the same equilibrium will invariably yield a range of temperatures and/or pressures. Knowing which values to use requires a good knowledge of the geothermobarometer used and careful interpretation of the results.

As an example, Hodges and Spear (1982) applied several thermobarometer calibrations to some pelitic schists from Mt. Moosilauke, New Hampshire. On petrographic grounds, and the regional occurrence of all three Al₂SiO₅ polymorphs, the area is believed to have equilibrated near the Al₂SiO₅ invariant point. Spear (1993) chose a single sample from the Mt. Moosilauke suite to illustrate the approach: sample 90A, containing quartz, muscovite, biotite, plagioclase, garnet, sillimanite, staurolite, ilmenite, and graphite. The composition and mineral formulas of the solid-solution minerals in the sample are given in Table 27.3.

Both the garnet–biotite geothermometer and the *GASP* geobarometer can be applied to this mineral assemblage. Application of geothermobarometers requires that we calculate mineral formulas and allocate the cations to the proper sites in order to deal with the values of *X* properly. Even the simplest case of ideal mixing requires that we treat independently the mixing of cations on each site where mixing takes place [see

TABLE 27.3 Mineral Compositions, Formulas, and End-Members for Sample 90A from Mt. Moosilauke, New Hampshire

Wt. % Oxides	Garnet	Biotite	Muscovite	Plagioclase
SiO₂	37.26	34.22	44.50	64.93
Al₂O₃	21.03	18.97	34.50	22.59
TiO₂		1.23	0.40	
FeO	32.45	17.50	0.70	
MgO	2.46	9.98	0.46	
MnO	6.08	0.12	0.02	
CaO	1.03	0.01	0.03	2.90
Na₂O		0.27	1.64	9.36
K₂O		7.79	8.05	0.45
Total	100.31	90.09	90.30	100.23
		Cations		
Si	3.00	5.43	6.17	2.84
Al^{IV}	2.00	2.57	1.83	1.17
Al^{VI}		0.98	3.81	
Ti		0.15	0.04	
Fe	2.19	2.32	0.08	
Mg	0.30	2.36	0.10	
Mn	0.42	0.02	0.00	
Ca	0.09		0.00	0.14
Na		0.08	0.44	0.83
K		1.58	1.42	0.03
Fe/(Fe + Mg)	0.88	0.50	0.46	
End members	Prp 10			An 14
	Alm 73			Ab 83
	Sps 14			Or 3
	Grs 3			

From Hodges and Spear (1982) and Spear (1993).

Equation (27.18)]. For example, a_{An}^{Plag} for an ideal solution equals Ca/(Ca + Na + K) on a single site (on an atomic basis), and we need not deal with the substitution of Al $\leftrightarrow$ Si because it is coupled to the Ca $\leftrightarrow$ Na-K substitution. $a_{Phl}^{Bt} = (Mg/(Fe + Mg))^3$ for ideal phlogopite–annite substitution on the three octahedral sites in biotite. Garnet solutions deal with essentially independent mixing of (Mg-Fe^{2+}-Ca-Mn) on the three divalent sites and (Al-Fe^{3+}-Cr) on the two trivalent sites. Thus the ideal solution model for $a_{Prp}^{Grt} = (Mg/(Mg + Fe^{2+} + Ca + Mn))^3 \cdot (Al/(Al + Fe^{3+} + Cr))^2$. See Powell (1978) for a more detailed explanation of ideal mixing on multiple sites.

Of the three examples above, plagioclase and garnet are significantly non-ideal, and the ideal formulations will not work for them, particularly if the grossular content of the garnet is significant. Fortunately, the Mt. Moosilauke garnet is very low in Ca, and the ideal model is a close fit. Non-ideal solutions are well beyond the scope of this introductory review and require more sophisticated site occupancy allocations and activity parameters. Some references for typical a-X models can be found in the Suggested Readings at the end of the chapter. The sources for each geothermobarometer listed in Table 27.4 will generally describe the activity–composition model used in their regression. The model used when applying any particular geothermobarometer *must* be the same as that used in its formulation. This can lead to a mathematical headache when applying geothermobarometers, and several computer programs are available to handle the calculations, beginning with simple chemical analyses of coexisting minerals. The most common programs are listed in Section 27.5 (and the a-X models they employ are generally listed in the references that accompany the programs).

Using the program THERMOBAROMETRY, by Spear and Kohn (1999), on the analyses in Table 27.3 produced curves for each calibration of the garnet–biotite

and *GASP* equilibria included in the program, resulting in the P-T plot shown in Figure 27.10. The differences reflect the various approaches to formulating the a-X relationships for the mineral phases, the experiments on which each model is based, and the degree to which a model is extrapolated beyond its P-T-X experimental base. Hodges and Spear (1982) and Spear (1993) discussed the differences between the various formulations in more detail.

The appropriate choice of geothermobarometer is difficult without knowing the P-T answer beforehand. If the Mt. Moosilauke samples occur near the Al_2SiO_5 invariant point, as proposed by some, this gives us an independent check on the results. Note from Figure 27.10 that the shaded area bracketed by the estimates of pressure and temperature for sample 90A has a pressure–temperature range of ~0.2 GPa and ~150°C and that the bracket includes the Al_2SiO_5 invariant point. The bracket is fairly broad, and we commonly find that the various models seldom agree to within a few degrees and megapascals. Also, even if a certain thermometer or barometer provides a good temperature or pressure estimate on one rock, it might not work as well on another.

The problems involved in applying a plethora of largely independent thermometers and barometers based on a variety of a-X models are simply the price we have to pay for advances in accuracy and an ever-increasing amount of experimental data leading to "new and improved" formulations. Geothermobarometry may have taken a stride forward with the development of **internally consistent** databases of thermodynamic data and computer programs that deal with the calculations of phase diagrams. Application of geothermobarometry has evolved over the past three or four decades from calculating a few relevant calibrations by hand, to programs and spreadsheets that handle microprobe data on mineral compositions and apply several geothermometers

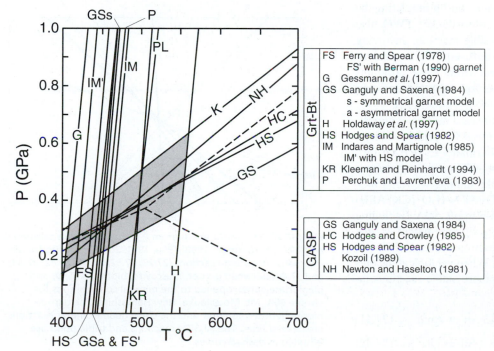

FIGURE 27.10 P-T diagram showing the results of garnet–biotite geothermometry (steep lines) and *GASP* barometry (shallow lines) for sample 90A of Mt. Moosilauke (Table 27.3). Each curve represents a different calibration, calculated using the program THERMOBAROMETRY, by Spear and Kohn (1999). The shaded area represents the bracketed estimate of the P-T conditions for the sample. The Al_2SiO_5 invariant point (dashed) also lies within the shaded area.

and geobarometers of choice, to using modern programs based on such internally consistent databases.

Internally consistent data have been compared with all the other data in the database and adjusted for mutual conformity. For example, suppose that three independent labs each studied one of the And = Sil, Sil = Ky, and And = Ky reactions and then extracted thermodynamic data for the two phases in their selected reaction. If we were to use the independent data from each lab to reconstruct the equilibrium curves on a *P-T* phase diagram, it is unlikely, given the errors inherent in the experimental methods, that the three independent curves would intersect exactly at a single invariant point (as they must). Clearly, the free energy of sillimanite at the pressure and temperature of the invariant point is the same, whether it coexists with andalusite or with kyanite. If we compared the thermodynamic data for all three phases and adjusted the values within the limits of experimental uncertainty so that they agreed and intersected at a single point, the data would be internally consistent and probably more accurate because the data for any single phase are consistent with several experimentally calibrated reactions and not with just one. For large modern data sets with several components and phases, the data are cross-referenced in this fashion with all of the other data in the set, a relatively simple job for modern computers.

Several computer programs combine internally consistent thermodynamic databases for minerals and fluids with the ability to calculate the location of reaction curves in *P-T-X* space (using more rigorous *P* and *T* corrections and sophisticated activity models for many common solid solution minerals). I use several programs and describe them briefly in Section 27.5.

Both TWQ and THERMOCALC are capable of accepting mineral composition data and calculating equilibrium curves based on a consistent set of calibrations and activity–composition mineral solution models. As an example, we shall use Berman's (1988, 1990, 1991, 2007) TWQ 2.32 program to calculate the relevant equilibria relating the phases in sample 90A from Mt. Moosilauke. TWQ also searches for and computes all possible reactions involving the input phases, a process called **multi-equilibrium calculations** by Berman (1991). Output from these programs yields a single equilibrium curve for each reaction and should produce a tighter bracket of *P-T-X* conditions.

To use TWQ, we must first create a mineral composition file with the wt. % oxides in each variable mineral (from Table 27.3). We then run Berman's CMP.EXE program to calculate mineral formulae and cation site occupancies. Finally, for the Mt. Moosilauke sample, we run TWQ. EXE for the $CaO-K_2O-FeO-Al_2O_3-SiO_2-H_2O$ (CKFASH) system and choose the phases in Sample 90A (including muscovite) from the appropriate program menu. At that point, TWQ calculates the *P-T* reaction curves (which we can see by running WPLOT.EXE). When we look at the output, we see that TWQ found four more stable reactions involving the phases:

$$Alm + Grs + Ms = 3\,An + Ann \qquad (27.41)$$

$$Alm + Ms = Ann + 2\,Sil + Qtz \qquad (27.42)$$

$$Prp + Ms = Phl + 2\,Sil + Qtz \qquad (27.43)$$

$$3\,An + Phl = Prp + Ms + Grs \qquad (27.44)$$

Figure 27.11 shows the TWQ output (without removing the metastable extensions and with my addition of the Al_2SiO_5 reactions for reference). Only four of the six TWQ reactions are independent because (any) three of the reactions with muscovite (above) are simply linear combinations of the independent ones. Notice the relatively tight intersection of the (three independent) curves in Figure 27.11, centered around approximately 550°C and 0.48 GPa. As Berman (1991) suggested, such a tight intersection provides support for equilibrium having been attained, and the more rigorous calculations and non-ideal mixing models in TWQ *should* lead to a more reliable temperature and pressure estimate than by using independent equilibria.

The THERMOCALC program (Holland and Powell, 1985, 1990, 1998) is also based on an internally consistent data set and produces similar results, which Powell and Holland (1994) called "optimal thermobarometry" using the Average PT (AvePT) module. THERMOCALC accepts mineral compositions as input and, like TWQ, searches for other reactions involving the stable mineral assemblage. But it also considers activities of each of the end-members of the phases to be variable within the uncertainty of each activity model, defining bands for each reaction within that uncertainty. It then calculates an optimal *P-T* point within the correlated uncertainty of all relevant reactions via least squares and estimates the overall activity model uncertainty (see more on uncertainty below). Figure 27.12a is a *P-T* phase diagram of the garnet–biotite exchange and *GASP* net-transfer reactions for a pelite, the analysis of which was used in several papers by Powell and colleagues (e.g., Powell, 1985; Powell and Holland, 1988, 1994; Holland and Powell, 1990;

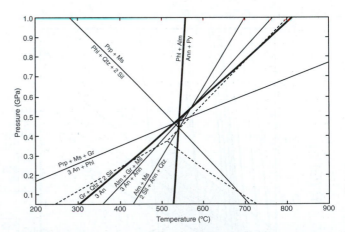

FIGURE 27.11 Multi-equilibrium *P-T* phase diagram calculated by TQW 2.32 (Berman, 1988, 1990, 1991, 2007), showing the six internally consistent reactions (27.27, 27.32, 27.41, 27.42, 27.43, and 27.44) between garnet, muscovite, biotite, Al_2SiO_5 and plagioclase, when applied to the mineral compositions for sample 90A, Mt. Moosilauke, New Hampshire. The garnet–biotite geothermometer and the *GASP* geobarometer reactions are heavier lines. The Al_2SiO_5 reactions and triple point are included as dashed curves.

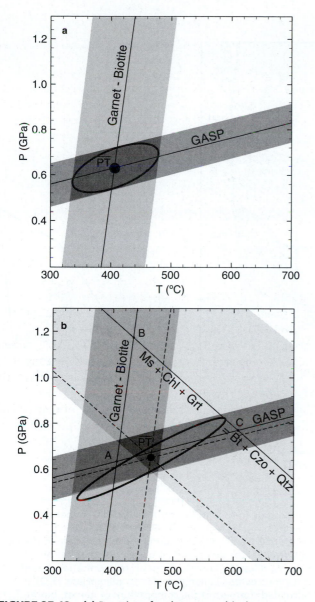

FIGURE 27.12 **(a)** Reactions for the garnet–biotite geothermometer and *GASP* geobarometer, calculated using THERMOCALC and its internally consistent thermodynamic database with the mineral compositions from sample PR13 of Powell (1985). Uncertainty estimates for each reaction (shaded), a *P-T* uncertainty ellipse, and the "optimal" AvePT ($\overline{PT}$) were calculated from correlated uncertainties using the approach of Powell and Holland (1994). **(b)** Addition of a third independent reaction generates three intersections (A, B, and C). The calculated AvePT lies within the consistent band of overlap of individual reaction uncertainties (yet outside the ABC triangle). The large uncertainty associated with the third reaction is to actually increase the overall uncertainty of the correlated ellipse. After Powell and Holland (1994).

Powell et al., 1998; Worley and Powell, 2000), as determined by THERMOCALC, along with the estimated uncertainties. The *P* and *T* uncertainties for the Grt-Bt and *GASP* equilibria are about ±0.1 GPa and 75°C, respectively. A third independent reaction was also found involving the phases present (Figure 27.12b). Notice how the uncertainty *increases* when the third reaction is included, due to the effect of the larger uncertainty for this reaction on the

correlated overall uncertainty. The average *P-T* value ($\overline{PT}$) is higher due to the third reaction and *may* be considered more reliable when based on all three. See Powell and Holland (1994) for a complete description of the methodology.

The pressure and temperature of the intersection in Figures 27.11 and 27.12 are *probably* more reliable than a slew of independent calibrations, but systematic calibration errors are still possible in an internally consistent data set. Powell and Holland (2008) suggested that thermobarometry may best be practiced using the pseudosection approach (Sections 26.5 and 28.2.5) of THERMOCALC (or Perple_X), in which a particular whole-rock bulk composition is defined and the mineral reactions delimit a certain *P-T* range of equilibration for the mineral assemblage present. Comparing calculated mineral compositions with those analyzed in the rock may then provide additional refinement and evaluation of the results. Figure 27.13 is an example of this approach. It takes a bulk composition for an average Permian Plattengneiss from the eastern Alps, near the Austrian–Italian border (Habler and Thöni, 2001), as the basis for the pseudosection. The *P-T* range containing the peak metamorphic mineral assemblage, garnet + muscovite + biotite + sillimanite + quartz + plagioclase + H_2O, is shaded most darkly (and is considerably smaller than the uncertainty ellipse determined by the AvePT approach). The calculated compositions of garnet, biotite, and plagioclase within the shaded area are also contoured (inset). They compare favorably with the reported mineral compositions of Habler and Thöni (2001) and can further constrain the equilibrium *P* and *T*.

Many investigators are turning to the integrated approach to geothermobarometry, using programs such as TWQ, THERMOCALC, or Perple_X and their periodically updated, internally consistent thermodynamic databases (with increasingly sophisticated mineral solution activity models). Internal consistency does not guarantee accuracy, however; well-calibrated individual geothermobarometers from Table 27.4 are still quite useful, and they can be applied simply, effectively, and directly.

27.4.4 Calculating *P-T-t* Paths from Zoned Crystals

As we shall see in the next two sections, a number of factors can lead to errors in geothermobarometry. One major problem results from compositional variation in minerals. If one or more of the minerals involved in a geothermometer or geobarometer is compositionally heterogeneous, the calculated temperature and/or pressure will necessarily vary with the composition, and therefore with the area of the crystal that is microprobed. Careful work, however, can in some cases change such a potential problem into a benefit. Consider the case in which a mineral does not fully re-equilibrate during growth but is compositionally variable, either because of poor internal diffusion (resulting in a zoned crystal) or because it becomes an inclusion in another mineral and is thus isolated from the rest of the rock. In such situations, it may be possible to use the variation in mineral composition if the minerals involved constitute a geothermometer or

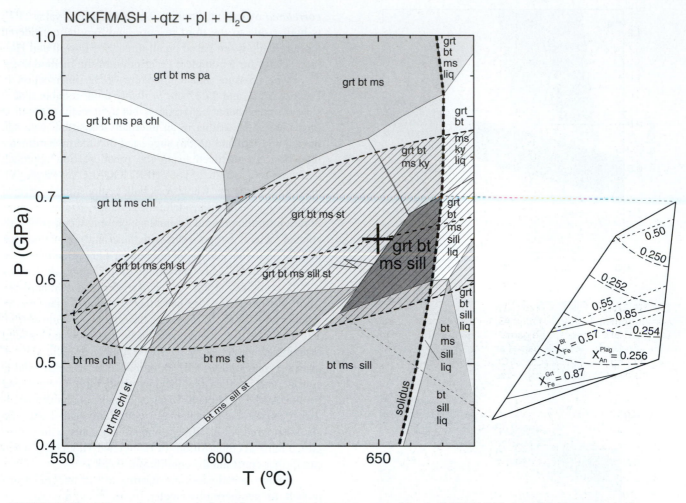

NCKFMASH +qtz + pl + H$_2$O

FIGURE 27.13 *P-T* pseudosection calculated for a computed average composition in NCKFMASH for a pelitic Plattengneiss from the Austrian eastern Alps (molar proportions: SiO$_2$ = 79.63, Al$_2$O$_3$ = 10.16, CaO = 1.47, MgO = 2.87, FeO = 4.50, K$_2$O = 1.94, Na$_2$O = 2.14, with H$_2$O in excess). The large + is the calculated Average PT (= 650°C and 0.65 GPa), using the mineral data of Habler and Thöni (2001) and THERMOCALC v.3.23. The heavy curve going through AvePT is the average P calculated from a series of temperatures (Powell and Holland, 1994). The shaded ellipse is the AvePT error ellipse (R. Powell, personal communication, based on Powell and Holland, 1994). The peak mineral assemblage, garnet–biotite–muscovite–sillimanite (+quartz, plagioclase, H$_2$O), is constrained to the darkest shaded polygon. Inset shows THERMOCALC calculated contours of the compositions of garnet, biotite, and plagioclase within the polygon, which can be used to place further constraints on peak *P* and *T*. After Tenczer et al. (2006). We will discuss the nature of pseudosections in Section 28.2.5.

geobarometer to document the changes in temperature and/or pressure that occurred during the portion of the *P-T-t* path to which the zoning is related (see also Section 25.4). St-Onge (1987) provided a careful and well-documented example of such an approach.

In a study of the Proterozoic Wopmay Orogen of the Northwest Territories, Canada, St-Onge (1987) used zonation of garnet and plagioclase in pelitic schists to document the variation in *P-T* conditions that occurred during metamorphic garnet growth. His sample #3 provides a good example of the type of techniques employed. The sample is a garnet–biotite–muscovite–sillimanite–plagioclase–quartz schist from the sillimanite zone of the upper amphibolite facies. Figure 27.14a illustrates the chemical zoning in poikiloblastic garnets from the sample with typical Mn-Ca-rich cores and more Fe-Mg-rich rims. This strong zoning across the entire grain suggests that the zonation is the result of partial equilibrium during growth and was not modified by later diffusion.

The garnets in sample #3 contain inclusions of plagioclase, biotite, quartz, and an Al$_2$SiO$_5$ polymorph. The plagioclase inclusions in garnet also show a systematic compositional variation with distance from the garnet core, a variation that corresponds to the reverse zonation (increasing An-content outwards) in larger plagioclase grains in the matrix (Figure 27.14b). The zoning patterns observed in garnet and plagioclase support the hypothesis that both phases grew simultaneously, and that the composition of a plagioclase inclusion, and of the garnet adjacent to it, represent the equilibrium compositions at a time when the plagioclase was growing in the matrix and the adjacent area of the garnet was at the rim of the growing porphyroblast as it was beginning to envelop the plagioclase. Garnet and plagioclase are typically sluggish in re-equilibrating during growth (at least at temperatures of the amphibolite facies and below) so that the compositions of early growth stages became frozen in place to produce the observed zoning.

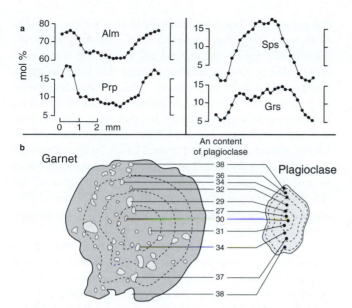

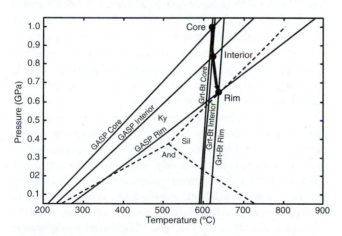

FIGURE 27.14 Chemically zoned plagioclase and poikiloblastic garnet from meta-pelitic sample #3, Wopmay Orogen, Canada. **(a)** Chemical profiles across a garnet (rim → rim). **(b)** An-content of plagioclase inclusions in garnet and corresponding zonation in neighboring plagioclase. After St-Onge (1987). Reprinted by permission of Oxford University Press.

The mineral assemblage in the sample permits calculation of temperatures via the garnet–biotite geothermometer and of pressures using the *GASP* geobarometer. Biotite was also zoned, so that the garnet–biotite geothermometer could be applied to paired Grt-Bt cores, interiors, and rims in order to record the change in temperature during mineral growth. Similarly, the *GASP* geobarometer could be applied to paired plagioclase inclusions and adjacent garnet compositions to record the accompanying pressure changes. Figure 27.15 shows the results of such calculations for the core area, the rim area, and a selected intermediate area. The intersection of

FIGURE 27.15 The results of applying the garnet–biotite geothermometer and the *GASP* geobarometer reactions using TWQ 2.32 (Berman, 1988, 1990, 1991, 2007) to the core, interior, and rim composition data of St-Onge (1987). The three intersection points yield *P-T* estimates that define a *P-T-t* path for the growing minerals showing near-isothermal decompression.

the Grt-Bt isopleth with the *GASP* isopleth for each area provides a *P-T* estimate for the sample at the corresponding period of mineral growth. The three points define a portion of the *P-T-t* path during which this growth occurred. In this case, the *P-T-t* path shows nearly isothermal decompression, suggesting rapid uplift and erosion following orogenic crustal thickening, consistent with several geophysical models (see Section 18.2). Even if the calibration of either reaction is imprecise, and the absolute *P-T* conditions are in error, the *relative* positions of the points would still be credible because each is calculated using the same technique. Notice also that, although the sample is currently in the sillimanite zone, most of the *P-T-t* path is in the kyanite stability field. This is supported by the preservation of relict kyanite in many of the samples, both in the matrix and as inclusions in garnet. For more details on the method of calculating *P-T-t* paths from zoned crystals, see Spear (1989).

Recent advances in textural geochronology (Section 23.7) have, in some cases, allowed age estimates for some points along a *P-T-t* path, finally placing the "*t*" term in "*P-T-t*" on a similar quantitative basis as *P* and *T*. For example, Foster et al. (2004) modeled temperature and pressure evolution of two amphibolite facies metapelites from the Canadian Cordillera and one from the Pakistan Himalaya. They used a combination of AvePT in THERMOCALC, conventional garnet–biotite geothermometry on zoned garnets, and monazite–xenotime (Heinrich et al., 1997; Gratz and Heinrich, 1997) and monazite–garnet geothermometry (Pyle et al., 2001) on zoned monazites and garnets. Three to four stages of monazite growth were recognized texturally in the samples and dated on the basis of U-Pb isotopes in Monazite analyzed by LA-ICPMS. Figure 27.16 shows the

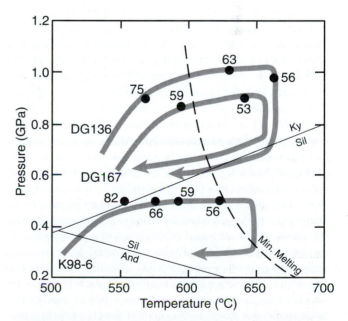

FIGURE 27.16 Clockwise *P-T-t* paths for samples D136 and D167 from the Canadian Cordillera and K98-6 from the Pakistan Himalaya. Monazite U-Pb ages of black dots are in Ma. Small-dashed lines are Al$_2$SiO$_5$ polymorph reactions and the dashed curve represents the H$_2$O-saturated minimum melting conditions. After Foster et al. (2004).

P-T-t paths for the three samples and the ages of monazite growth events. Foster et al. (2004) used the *P-T-t* paths to constrain the timing of thrusting (pressure increase) along the Monashee décollement in Canada (it ceased about 58 Ma b.p.), followed by exhumation beginning about 54 Ma. They also were able to estimate a heating rate of $5 \pm 2°C/Ma$ for sample DG136. The Himalayan sample records periods of monazite formation during garnet growth at 82 Ma, followed by later monazite growth during uplift and garnet breakdown at 56 Ma, and a melting event during subsequent decompression. These temporally-constrained data, when combined with field recognition of structural features, can not only elucidate the metamorphic and tectonic history of an area but can also place constraints on kinematic and thermal models of orogeny.

27.4.5 Sources of Error in Geothermobarometry

Geothermobarometry is relatively easy to apply and can yield useful results when applied judiciously. It is all too easy, however, to apply the technique in a haphazard fashion, yielding nearly meaningless results. Proper application of geothermobarometry requires careful observation and interpretation of the textures and compositional heterogeneity (if any) of the minerals and an understanding of the phase equilibria, chemical evolution, and reaction history of the host rock. Geothermobarometry is based on several assumptions, and there are many ways to err in its application. Among the most important assumptions and pitfalls are the following:

1. Geothermobarometry assumes that the minerals represent an **equilibrium** assemblage and that their compositions also represent equilibrium distribution of the components. Although a number of textural and chemical characteristics discussed above and in Section 24.1 may indicate that equilibrium was *not* attained, it is impossible to prove that it *was*. The absence of disequilibrium textures and consistent distribution of elements between phases (including tight clustering of reaction intersections in TWQ multi-equilibrium calculations [Figure 27.11] and small error ellipses in Figure 27.12) support an equilibrium assumption, but we can never be certain.

2. Particular care must be exercised when minerals are **chemically zoned**. Comparing the core of one mineral with the rim of another in a geothermobarometer usually leads to spurious results because the two compositions were never in equilibrium with each other. It is commonly assumed that the rim compositions of zoned minerals are in equilibrium with the intergranular fluid and with each other. Mineral cores were once rims of growing crystals, of course, but may not represent earlier simultaneous equilibrium states because one mineral may have nucleated and grown before another became stable. For example, the order of the classic pelitic isograds indicates that biotite usually forms before garnet.

3. For reasons discussed in Section 21.4, it is assumed that most mineral assemblages represent peak metamorphic temperatures. **Retrograde effects**, however, commonly upset the system. This is particularly true for rocks that get hot and then cool relatively slowly, such as granulite facies rocks and plutonic igneous rocks. Obviously, analyzing a retrograde rim of one mineral and comparing it to a non-retrograded portion of another mineral will produce meaningless temperatures or pressures. When retrograde textures, such as exsolution lamellae or replacement and altered rims, are obvious, we can avoid them. In other cases, we may have to analyze numerous points to detect retrogression (such as the rim zoning reversals in Figure 27.14a). In situations with exsolution, we may be able to analyze numerous points within an exsolved grain and mathematically "rehomogenize" the mineral by combining the analyses. Bohlen and Essene (1977), for example, did this for two-feldspar and Fe-Ti oxide geothermometry in granulite facies rocks from the Adirondacks.

In Section 26.6, I described how the composition of biotite in some granulite facies gneisses showed the tendency to vary as a function of the type of adjacent mafic mineral, presumably due to retrograde Fe-Mg exchange reactions. Many of us found that selection of biotites surrounded only by felsic minerals produced much more consistent analyses, which in turn yielded better temperatures and pressures. Nonetheless, net-transfer exchange reactions can be susceptible to retrograde exchange. This is particularly notable for isotopic exchanges, and the temperatures derived may represent **blocking temperatures**, the temperatures at which kinetic factors finally inhibit further exchange during cooling. Retrograde effects can invalidate the assumption discussed above that mineral rims are reliable recorders of peak metamorphic conditions, so great care must be exercised to recognize and understand the patterns and basis for compositional heterogeneities. Kohn and Spear (2000) described a method based on microprobe element maps that attempts to correct for retrograde net-transfer and locate or approximate peak element distributions, a method they called "retrograde net-transfer reaction insurance."

4. A good geothermobarometer must be *well calibrated* and based on reversed experiments and good thermodynamic data. The activity–composition (*a-X*) model for the minerals should be appropriate. The minerals in the experiments must also have the same **structural state** as those in the natural assemblage. If the experimental synthetic minerals are disordered or are a high-temperature polymorph, and the metamorphic minerals are ordered or are a low-temperature polymorph, the geothermobarometer may produce errors.

5. Most geothermobarometers are based on experiments using simple mineral systems. For example, the garnet–biotite thermometer of Ferry and Spear (1978) was performed on synthetic minerals in the Fe-Mg series, whereas many calibrations of the *GASP* geobarometer are based on synthetic minerals in the $CaO-MgO-Al_2O_3-SiO_2$ system. In natural rocks, however, the minerals are usually complex mixtures, containing Fe, Mn, Ti, Na, etc. **Additional components**, when present in sufficient concentrations, commonly affect the ratio of other components in a mineral and thereby the distribution of those components between coexisting minerals. Ca in garnet, for example, has a strong and non-ideal effect on Fe/Mg. Non-ideal *a-X* relationships can be complex, and the most complex models lead to computational hardships, so many investigators avoid them. Well-constrained *a-X* models also require that experiments be performed over a broad spectrum of compositions as well as temperatures and pressures, and only a small part of that job is presently complete. Perhaps the most common and significant errors resulting from the application of geothermobarometry are related to applying a system beyond the compositional range over which it has been calibrated. Modern programs, such as TWQ, THERMOCALC, and Perple_X, periodically update the *a*-X models they use and also ease the computational hardships of calculating activities by hand. But this ease of use, simply plugging in compositional data without carefully reading the documentation, is probably what most commonly has users erroneously applying equilibria beyond the calibrated limits.

6. Many geothermobarometers are based on experiments at high temperatures because equilibrium can be difficult to attain at lower metamorphic temperatures in a reasonable laboratory time period. For example, notice in Figure 27.8 that the direct reversed experiments of Koziol and Newton (1988), on which the GASP geobarometer was calibrated above, were all at temperatures in excess of 900°C, whereas the barometer itself is typically applied to metamorphic rocks in the 500 to 800°C range (e.g., see Figures 27.9, 27.10, and 27.15). The longer the **extrapolation in temperature or pressure** from the experiments to the application, the larger the possible error. Even a small uncertainty, when extrapolated far enough, can lead to large errors.

7. Because the microprobe is not capable of distinguishing oxidation states, the **Fe^{3+}/Fe^{2+} ratio** for an analyzed mineral must be calculated on the basis of charge balance. For relatively simple minerals, in which most octahedral cations are divalent except for Fe, this is considered to be reliable, but it still depends on factors such as Si^{4+}/Al^{3+}. For more complex minerals, such as amphiboles and phyllosilicates, with mixing of ions representing several valences, as well as variable H_2O-content and cation site vacancies, charge-balance calculations are suspect at best. Because the Fe^{3+}/Fe^{2+} ratio affects the amount of Fe^{2+}, this can profoundly influence the Fe^{2+}/Mg^{2+} ratio, which is the basis for many exchange geothermobarometers. Many investigators are forced to assume that all Fe is Fe^{2+}, or that Fe^{2+}/Fe^{3+} is fixed at some common ratio, but if the ratio is substantially different from that of the experimental calibration, errors may result. Accurate determination of Fe^{2+}/Fe^{3+} by wet-chemical analysis or Mössbauer spectroscopy is desirable but time-consuming and expensive.

8. Some reactions are **sensitive only over a restricted temperature or pressure range**. For example, solvus geothermometers have shallow slopes at the top of the solvus and progressively steeper limbs at lower temperatures (see Figure 6.16). Mineral composition will thus be a sensitive measure of temperature at the shallow (high-temperature) portion of the temperature range but not at the steep (low-*T*) portion. Isotopic geothermometers, on the other hand, such as fractionation of $^{18}O/^{16}O$ or $^{13}C/^{12}C$ between minerals, are more sensitive at low temperatures because the distribution becomes more even at temperatures over 800°C.

9. Polymetamorphism may affect both the mineralogy and the mineral composition. Great care must be taken to properly interpret textures when applying a geothermobarometer to rocks that have been metamorphosed more than once.

In order to avoid most of these pitfalls, we should carefully examine the textures and mineralogy (including accessory minerals) of rocks to be studied. Element mapping using microprobe or SEM imaging can provide a detailed look at compositional heterogeneity (see the color images of zoned garnet on the back cover). Textural studies should be combined with structural evidence to assess multiple metamorphic episodes. Zoned minerals should be treated carefully. As many different thermometers and barometers as possible should be evaluated, especially those with internally consistent data sets. Varying Fe^{2+}/Fe^{3+} for each will give an idea of the uncertainties involved with this variable. If the results from several geothermobarometers give consistent results, the chances improve that the calculated temperatures and/or pressures are accurate.

27.4.6 Precision and Accuracy in Geothermobarometry

Simply plugging numbers into a computer or calculator and plotting the results may provide an estimate of temperature and/or pressure, but without some concept of the uncertainties involved, the results may not be as accurate or reliable as we wish. Several investigators have stressed the need to understand the propagation of uncertainties when applying geothermobarometers to rocks (e.g., Demerest and Haselton, 1981; Hodges and McKenna, 1987; McKenna and Hodges, 1988; Kohn and Spear, 1991a, b; Spear, 1993; Powell and Holland, 1994; Worley and Powell, 2000; Ashworth et al., 2004). Details are available in these works, which have developed the mathematics and applied them to the application of geothermobarometry. It suits our present purposes to simply discuss the various potential sources of error and illustrate the magnitude of typical uncertainties involved.

The term **precision** is used to describe the reproducibility of a technique with randomly distributed errors. For example, because microprobe analyses are based on counts per second of emitted fluorescent x-rays, there will always be fluctuations in the results based on counting statistics. If the fluctuations are relatively small and the results don't vary much when a mineral is analyzed repeatedly, the results are considered to be precise. Reproducible results, however, do not mean that they are *correct*. If the standards are not correctly analyzed, for example, there will be a *systematic error* in the probe results because the software is calibrating the data on the basis of the wrong value. The term **accuracy** is used to describe how close the results are to the correct value. Imagine a target at a shooting range (Figure 27.17). If a series of shots are tightly clustered, but not at the bulls-eye, the gun and shooter are precise (reproducible), but not accurate (Figure 27.17a). If the shots are all over the target, but average out to the bulls-eye, the gun and

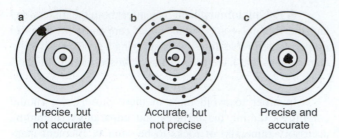

FIGURE 27.17 An illustration of precision versus accuracy. **(a)** The shots are precise because successive shots hit near the same place (reproducibility). However, they are not accurate because they do not hit the bulls-eye. **(b)** The shots are not precise because of the large scatter, but they are accurate because the average of the shots is near the bulls-eye. **(c)** The shots are both precise and accurate.

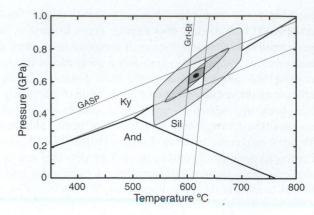

FIGURE 27.18 *P-T* diagram illustrating the calculated uncertainties from various sources in the application of the garnet–biotite geothermometer and the *GASP* geobarometer to a pelitic schist from southern Chile. After Kohn and Spear (1991b) and Spear (1993). Copyright © The Mineralogical Society of America. See text for details.

shooter are accurate, but not precise (Figure 27.17b). Only when the shots are tightly clustered in the bulls-eye are the results both precise and accurate (Figure 27.17c).

The common sources of uncertainty in geothermobarometry include:

1. Analytical imprecision of microprobe analyses
2. Uncertainty and inaccuracy of microprobe standards and correction software
3. Accuracy of the calibration of experiments on which a geothermobarometer is based, such as the *P-T* bracket of the reversals, the accuracy of the regression, and the values of ΔH and ΔS extracted
4. Inaccuracy in ΔV of a reaction, which is usually determined independently of the calibration experiments
5. Effects of errors in temperature estimates on barometers and pressure estimates on thermometers because both are to some extent dependent on both T and P
6. Uncertainties in activity–composition models of minerals
7. Uncertainty arising from compositional heterogeneity in natural minerals

All of the above contribute to the final degree of uncertainty. Figure 27.18 shows the results of calculations of Kohn and Spear (1991b) that illustrate their evaluation of the magnitude of the uncertainties from the various sources as they contribute to the *GASP* geobarometer with temperature constrained by the garnet–biotite geothermometer as applied to a schist from southern Chile. The small black dot in the center is the uncertainty (1σ) associated with the analytical precision of the microprobe. For a well-tuned probe the error is relatively small, on the order of ±0.015 GPa. The long, thin ellipse is the total uncertainty that can be treated statistically (uncertainty in experimental calibration and ΔV, microprobe analysis, and cross-correlation of *T-P* estimates). This is on the order of ±0.07 GPa and 125°C.

Non-statistical errors are illustrated as parallelograms in Figure 27.18. The small shaded parallelogram around the dot represents uncertainty associated with natural compositional heterogeneity, which is about ±0.05 GPa in the Chilean sam-

ple. If there is no chemical zoning, this value would reduce to zero. The larger light sharp parallelogram represents uncertainty in the *a-X* models for the minerals, which Kohn and Spear (1991b) determined by using a range of *a-X* models for garnet and plagioclase in the *GASP* barometer. This is on the order of ±0.1 GPa. This is the principal uncertainty addressed in the ellipse of Figure 27.12. Finally, the largest shaded region is a combination of all these uncertainties and provides an estimate of the total uncertainty of the technique, a range of ±0.3 GPa (1σ) and 150°C. The pressure uncertainty corresponds to a depth uncertainty of about ±10 km, or about 15% of the total crustal thickness of an orogenic belt. This value is dominated by uncertainty in the mineral activity models and should improve as better-constrained models are developed. The calculated uncertainty varies with the sample to which a geothermobarometer is applied, and from one geothermobarometer to another, but the example illustrated in Figure 27.18 gives us an idea of the magnitudes with which we are dealing.

As mentioned above, although such uncertainties are associated with *absolute* values of P and T calculated from any particular geothermobarometer, the *change* in pressure or temperature from one sample to the next, when a single barometer or thermometer is used repeatedly, should be more reliable because many of the larger errors (such as the *a-X* model) apply to all calculations and thus cancel out. See Ashworth et al. (2004) and Powell and Holland (2008) for further discussions of uncertainties.

27.5 SOURCES OF DATA AND PROGRAMS

Calculations such as those for reaction curves require thermodynamic data on minerals, fluids, etc. The data set by Robie and Hemingway (1995) is available from the United States Geological Survey. "Internally consistent" databases are ones in which the data for each phase have been coordinated with all the other phases in the set.

Programs and web sites for calculating phase diagrams and geothermobarometry (generally with documentation) include the following:

TWQ by Rob Berman (1988, 1990, 1991, 2007) calculates and can plot P-T and T-X diagrams for several reactions at once in a chosen system. TWQ is available for download from Rob Berman's Geological Survey of Canada web site (http://gsc.nrcan.gc.ca/ sw/twq_e.php).

THERMOCALC by Holland and Powell (1985, 1990, 1998) is probably the most popular program for creating petrogenetic grids, and particularly for constructing pseudosections (Section 28.2.5). It is available for the Mac or Windows/DOS; see www. earthsci. unimelb. edu. au/tpg/thermocalc/ or http://rock.esc.cam.ac.uk/astaff/holland/index.html. Documentation and course notes for a THERMOCALC workshop are available from the first of these sites. The mineral database is large and the learning curve is pretty steep.

Perple_X is James Connolly's (1990, 2005) set of programs that can calculate phase diagrams based on reactions and an internally consistent database, available at www .perplex.ethz.ch. There are also modules for creating pseudosections for specified bulk compositions (Connolly and Petrini, 2002). The algorithm for calculating pseudosections uses a free-energy minimization approach, which is much easier to use than THERMOCALC, but the output needs some "cleaning up" in a drawing program in order to look good.

A version of **SUPCRT** (Helgeson et al., 1978) that I have modified to work in a DOS window on PCs can be downloaded from my web site (www.prenhall. com/winter/). A newer version, **SUPCRT92** (Johnson et al., 1992) with a somewhat more opaque front end is also available.

GPT by Reche and Martinez (1996) is one of several geothermobarometry spreadsheets, and it is available from the authors. A web search of keywords *geothermometry spreadsheet* turns up others.

Gibbs by Frank Spear is another thermodynamic database and calculation routine. **GTB** (GeoThermoBarometry), by Matt Kohn and Frank Spear, calculates pressures and/or temperatures of equilibrium from mineral compositions. Both are for the Mac and are available from Frank Spear's web site, at http://ees2. geo.rpi.edu/MetaPetaRen/Software/Software.html.

You can access all these programs through my web site, www.prenhall.com/winter/.

In addition to the programs, many petrologists have designed tutorials and class exercises using some of them. My favorite source of education-related materials is the Science Education Resource Center (SERC) site, presently hosted at Carleton College: http://serc.carleton.edu. This is an excellent site that is worth browsing. Specifically searching the web for thermodynamics, petrology, geothermometry, or one of the programs above will help find tutorials.

Summary

We can use tables of thermodynamic data for minerals and gas/fluid species to determine ΔG of a reaction over a range of metamorphic temperatures and pressures, and then we can find values of P, T, and X for which $\Delta G = 0$, yielding the conditions of equilibrium and allowing us to plot the equilibrium curve for the reaction on a P-T-X phase diagram. The volume of gas species is certainly not constant as pressure varies, so we must integrate the gas VdP term in Equation (5.5) resulting in the $RT\ell n(f/f^{o})$ term for dG/dP, where f is the fugacity (equal to pressure times a fugacity coefficient: $P\gamma$) at any particular pressure. Compositional variation in minerals is treated in a similar fashion with chemical potential (μ) or activity (a) terms in the free energy equation. In the general case, then, at any chosen P and T, we can calculate ΔG^{o} for the pure phases using standard state G-V-S data and an equation of the form of Equation (5.11), which is then equal to $-RT\ell nK$ [Equation (27.17)], where K is equal to the products of the appropriate activity terms [as in Equation (27.20), all to the power of the stoichiometric coefficients of the reaction]. We then substitute appropriate values for the activities: $a_i = (\gamma_i X_i)^y$ for minerals and $a_{gas} = f_{gas}/P^o$ and $f_{gas} = \gamma_{gas}p_{gas}$. For ionic species, we use $a_i = \gamma_i C_i$, where C_i is the concentration of component i in the solution. The fugacity/activity coefficient, γ, is either empirically or theoretically determined (from equilibrium values in experiments). In other words, we must be able to determine the composition of each phase involved in the equilibrium and then apply a coefficient of non-ideality based on some activity-composition model.

Geothermobarometry involves the calculation of temperatures and/or pressures of equilibration (typically peak metamorphic grade) from the measured distribution of elements between coexisting phases and experimental knowledge of how such distributions depend on temperature and pressure. Good geothermometers are based on equilibria, such as Fe-Mg exchange, that are not pressure sensitive. Likewise, good geobarometers are not as temperature sensitive. Modern thermodynamic software packages using internally consistent databases, rigorous P and T corrections for calculating ΔG^o, and sophisticated activity models for many common solid solution minerals can now be used to estimate pressure and temperature of equilibrium based on the intersection of reactions based on input tables of mineral analyses. This is the preferred modern approach to geothermobarometry. P-T-t paths can now be modeled using microprobe analyses of zoned minerals, and advances in textural geochronology have recently led to quantifying the timing of such paths as well.

Key Terms

Gibbs free energy, enthalpy, entropy
  579
Ideal gas equation *581*
Fugacity (*f*)/fugacity coefficient
  (γ) *582*
Chemical potential (μ) *583*
Activity (*a*)/activity coefficient (γ) *583*

Equilibrium constant (*K*) *584*
Activity–composition (*a-X*)
  relationship *584*
Ideal/real solutions *584*
Henry's law *585*
Geothermobarometry/geothermometry/
  geobarometry *587*

Reversed experiments *588*
Internally consistent thermodynamic
  data *597*
Multi-equilibrium calculations *598*
Blocking temperatures *602*
Precision/accuracy *603*

Review Questions and Problems

Review Questions and Problems are located on the author's web page at the following address: **http://www.prenhall.com/winter**

Important "First Principle" Concepts

- When comparing alternative states of a system, such as reactants versus products for a reaction, the state with the lowest Gibbs free energy (*G*) at the conditions of interest is most stable.
- Δ*G* for a reaction is zero at equilibrium because the reactants and products are equally stable.
- At equilibrium, the chemical potential (μ_i^A) of any given component (*i*) must be the same in every phase (*A*) that contains it.
- We can use tables of thermodynamic data for minerals and fluid species (with activity models for solutions) to calculate condi-

tions of equilibrium for many metamorphic reactions. This allows us to plot equilibrium reaction curves on *P-T* diagrams or calculate temperatures and/or pressures of equilibration.
- Experimentally determined phase equilibria provide the basis for nearly all geothermobarometry. Equilibrium thermodynamics then provides equations of an appropriate shape to fit the experimental data so that they can be extrapolated to *P*, *T*, and *X* values other than those of the experiments.

Suggested Further Readings

General Thermodynamics

Fletcher, P. (1993). *Chemical Thermodynamics for Earth Scientists*. Longman Scientific. Essex, UK.
Powell, R. (1978). *Equilibrium Thermodynamics in Petrology*. Harper & Row. London.
Saxena, S. (1992). *Thermodynamic Data: Systematics and Estimation*. Springer-Verlag. New York.
Wood, B. J., and D. G. Fraser (1976). *Elementary Thermodynamics for Geologists*. Oxford University Press. Oxford, UK.

Geochemistry (Especially Ionic Species)

Faure, G. (1998). *Principles and Applications of Geochemistry*. Prentice Hall. Englewood Cliffs, NJ.

Geothermobarometry

Berman, R. G. (1991). Thermobarometry using multi-equilibrium calculations: a new technique, with petrological applications. Can. Mineral, **29,** 833–855.
Essene, E. J. (1982). Geologic thermometry and barometry. In: *Characterization of Metamorphism Through Mineral Equilibria* (ed. J. M. Ferry). *Reviews in Mineralogy,* **10,** 153–206. Mineralogical Society of America. Washington, DC.
Essene, E. J. (1989). The current status of thermobarometry in metamorphic rocks. In: *Evolution of Metamorphic*

Belts (eds. J. S. Daly, R. A. Cliff, and B. W. D. Yardley). Special Publication **43.** Geological Society. London. pp. 1–44.
Powell, R. (1985). Geothermometry and geobarometry: A discussion. J. Geol. Soc. Lond., **142,** 29–38.
Powell, R. and T. J. B. Holland (1994). Optimal geothermometry and geobarometry. *Amer. Mineral.,* **79,** 120–133.
Powell, R. and T. J. B. Holland (2008). On thermobarometry. J. Met. Geol., **26,** 155–179.
Spear, F. S. (1993). *Metamorphic Phase Equilibria and Pressure–Temperature–Time Paths.* Monograph **1**. Mineralogical Society of America. Washington, DC. Chapter 15.

Solution Models for Minerals

Chatterjee, N. D. (1991). *Applied Mineralogical Thermodynamics*. Springer-Verlag. New York.
Ganguly, J., and S. Saxena. (1987). *Mixtures and Mineral Reactions*. Springer-Verlag. New York.
Newton, R. C., A. Navrotsky, and B. J. Wood. (1981). *Thermodynamics of Minerals and Melts*. Springer-Verlag. New York.
Saxena, S. (1973). *Thermodynamics of Rock-Forming Crystalline Solutions*. Springer-Verlag. New York.

TABLE 27.4a   Some Useful Geothermobarometers in Petrology: Exchange and Solvus Equilibria

Exchange Thermometers

Garnet–biotite	Fe-Mg exchange	Thompson (1976), Goldman and Albee (1977), Ferry and Spear (1978), Perchuk and Lavrent'eva (1981), Hodges and Spear (1982), Pigage and Greenwood (1982), Ganguly and Saxena (1984), Indares and Martignole (1985), Chipera and Perkins (1988), Berman (1990), Perchuk (1991), Bhattacharya et al. (1992), Patiño Douce et al. (1993), Kleemann and Reinhardt (1994), Kullerud (1995), Alcock (1996), Holdaway et al. (1997), Gessmann et al. (1997), Holdaway (2000, 2004), Kaneko and Miyano (2003)
Garnet–cordierite	Fe-Mg exchange	Currie (1971), Hensen and Green (1973), Thompson (1976), Holdaway and Lee (1977), Perchuk and Lavrent'eva (1981), Martignole and Sisi (1981), Bhattacharya et al. (1988), Perchuk (1991), Kaneko and Miyano (2003)
Garnet–clinopyroxene	Fe-Mg exchange	Råheim and Green (1974), Mori and Green (1978), Ellis and Green (1979), Saxena (1979), Ganguly (1979), Dahl (1980), Powell (1985), Pattison and Newton (1989), Krogh (1988), Carswell and Harley (1989), Brey and Köhler (1990), Perchuk (1991), Green and Adam (1991), Ai (1994), Nikitina and Ivanov (1995), Berman et al. (1995), Aranovich and Pattison (1995), Ganguly et al. (1996), Ravna (2000a)
Garnet–orthopyroxene	Fe-Mg exchange	Mori and Green (1978), Harley (1984), Sen and Bhattacharya (1984), Lee and Ganguly (1988), Carswell and Harley (1989), Brey and Köhler (1990), Perchuk (1991), Bhattacharya et al. (1991), Lal (1993), Ganguly et al. (1996), Carson and Powell (1997), Aranovich and Berman (1997), Pattison et al. (2003)
Garnet–hornblende	Fe-Mg exchange	Graham and Powell (1984), Perchuk (1991), Himmelberg et al. (1994), Ravna (2000b)
Garnet–chlorite	Fe-Mg exchange	Dickenson and Hewitt (1986), Laird (1988), Grambling (1990), Perchuk (1991)
Garnet–staurolite	Fe-Mg exchange	Perchuk (1991)
Chloritiod–(garnet or chlorite or biotite)	Fe-Mg exchange	Perchuk (1991), Vidal et al. (1999)
Garnet–epidote	Fe-Mg exchange	Perchuk (1991)
Garnet–olivine	Fe-Mg exchange	Kawasaki and Matsui (1977), O'Neil and Wood (1979, 1980), Smith and Wilson (1985), Carswell and Harley (1989), Brey and Köhler (1990)
Garnet–olivine	Ni-(Fe-Mg) exchange "Ni-in-garnet"	Griffin et al. (1989, 1996), Canil (1994)
Garnet–ilmenite	Fe-Mg and Fe-Mn exchange	Pownceby et al. (1987a, b), Feenstra and Engi (1998)
Garnet–phengite	Fe-Mg exchange	Krogh and Råheim (1978), Green and Hellman (1982), Hynes and Forest (1988), Carswell and Harley (1989), Coggon and Holland (2002)
Olivine–spinel	Fe-Mg exchange	Fabriès (1979), Roeder et al. (1979), Ozawa (1982), Engi (1983)
Olivine–Orthopyroxene	Fe-Mg exchange	Docka et al. (1986), Carswell and Harley (1989), Brey and Köhler (1990), Kock-Mueller et al. (1992)
Olivine–clinopyroxene	Fe-Mg exchange	Powell and Powell (1974), Sengupta et al. (1989), Köhler and Brey (1990), Brey and Köhler (1990), Perkins and Vielzeuf (1992), Loucks (1996)
Orthopyroxene–spinel	Fe-Mg exchange	Liermann and Ganguly (2003, 2007)
Orthopyroxene–ilmenite	Fe-Mg exchange	Docka et al. (1986), Carswell and Harley (1989)
Clinopyroxene–ilmenite	Fe-Mg exchange	Docka et al. (1986), Carswell and Harley (1989)
Olivine–ilmenite	Fe-Mg exchange	Docka et al. (1986), Carswell and Harley (1989)
Orthopyroxene–biotite	Fe-Mg exchange	Fonarev and Konilov (1986), Sengupta et al. (1990), Wu et al. (1999)
Orthopyroxene–sapphirine	Fe-Mg exchange	Kawasaki and Sato (2002)
Cordierite–spinel	Fe-Mg exchange	Vielzeuf (1983)
Cordierite–orthopyroxene	Fe-Mg exchange	Sakai and Kawasaki (1997)
Nepheline–feldspar	Na-K exchange	Perchuk et al. (1991)
Oxygen isotope thermometry	O^{18}-O^{16} exchange	O'Neil and Clayton (1964), Bottinga and Javoy (1973, 1975, 1987), Javoy (1977), Clayton (1981), Valley et al. (1986), O'Neil and Pinkthorn (1988), Chiba et al. (1989), Bechtel and Hoernes (1990), Zheng and Simon (1991), Agrinier (1991), Clayton and Kieffer (1991), Eiler et al. (1993), Sharp et al. (1993), Farquahar et al. (1993, 1996), Matthews (1994), Krylov and Mineev (1994), Hoffbauer et al. (1994), Ghent and Valley (1998), Moecher and Sharp (1999), Valley (2001, 2003)
Carbon isotope thermometry	C^{13}-C^{12} exchange	Valley and O'Neil (1980), Wada and Suzuki (1983), Valley et al. (1986), Dunn and Valley (1992), Morrison and Barth (1993), Eiler et al. (1993), Kitchen and Valley (1995), Satish-Kumar (2000), Valley (2001)

(continued)

TABLE 27.4a	Some Useful Geothermobarometers in Petrology: Exchange and Solvus Equilibria (*Continued*)

Solvus Thermometers

Two-pyroxene	Ca-Mg-Fe exchange	David and Boyd (1966), Wood and Banno (1973), Warner and Luth (1974), Ross and Huebner (1975), Saxena and Nehru (1975), Saxena (1976), Nehru (1976), Lindsley and Dixon (1976), Wells (1977), Mori and Green (1975, 1976, 1978), Sachtleben and Seck (1981), Kretz (1982), Lindsley (1983), Lindsley and Andersen (1983), Finnerty and Boyd (1984), Nickel and Brey (1984), Nickel and Green (1985), Nickel et al. (1985), Carswell and Gibb (1987), Brey and Kohler (1990), Fonarev and Graphchikov (1991)
Two-feldspar	Na-K exchange	Stormer (1975), Whitney and Stormer (1977), Powell and Powell (1977a), Brown and Parsons (1981, 1985), Ghiorso (1984), Green and Udansky (1986), Fuhrman and Lindsley (1988), Elkins and Grove (1990), Perchuk et al. (1991), Kroll et al. (1993), Voll et al. (1994)
Calcite–dolomite	Ca-Mg exchange	Goldsmith and Heard (1961), Goldsmith and Newton (1969), Anovitz and Essene (1987a), Powell et al. (1984)
Muscovite–paragonite	Na-K exchange	Eugster et al. (1972), Chatterjee and Flux (1986)
Monazite–xenotime	Ce-Y exchange in phosphates	Heinrich et al. (1997), Gratz and Heinrich (1997)

TABLE 27.4b	Some Useful Geothermobarometers in Petrology: Continuous Net-Transfer and Miscellaneous Equilibria

Continuous Net-Transfer Equilibria

Garnet-Al$_2$SiO$_5$-quartz-plagioclase (*GASP*)	3 An = Grs + 2 Al$_2$SiO$_5$ + Qtz	Ghent (1976), Ghent et al. (1979), Newton and Haselton (1981), Hodges and Spear (1982), Ganguly and Saxena (1984), Hodges and Royden (1984), Powell and Holland (1988), McKenna and Hodges (1988), Koziol and Newton (1988), Koziol (1989), Dasgupta et al. (1991), Ganguly et al. (1996), Holdaway (2001, 2004)
Garnet-rutile-al$_2$sio$_5$-ilmenite-quartz (*GRAIL*)	Several reactions	Bohlen et al. (1983b), Ghent and Stout (1984, 1994), Bohlen and Liotta (1986), Essene and Bohlen (1985)
Garnet–rutile–ilmenite–plagioclase–quartz (*GRIPS*)	Grs + 2 Alm + 6 Rut = 6 Ilm + 3 An + 3 Qtz	Bohlen and Liotta (1986), Anovitz and Essene (1987b)
Garnet–plagioclase–muscovite–biotite	Bt + Grs + Mu = 3 An + Bt	Ghent and Stout (1981), Hodges and Crowley (1985), Hoisch (1991), Powell and Holland (1988)
Garnet–plagioclase–muscovite–quartz	Prp + Grs + 3 (Al-Al) Mu + 6 Qtz = 6 An + 3(Fe-Si)Mu	Hodges and Crowley (1985), Hoisch (1991)
Garnet-muscovite-quartz-Al$_2$SiO$_5$	Prp + 3 (Al-Al) Mu + 4 Qtz = 3 (Fe-Si)Mu + 4 Al$_2$Si$_2$O$_5$	Hodges and Crowley (1985), Hoisch (1991)
Garnet-muscovite-biotite-quartz-Al$_2$SiO$_5$	Prp + Mu = Bt + 2 Al$_2$SiO$_5$ + Qtz	Hodges and Crowley (1985), Holdaway et al. (1988), Holdaway (2004), Hoisch (1991)
Garnet–biotite–plagioclase–quartz	>1	Wu et al. (2004)
Garnet–plagioclase–hornblende–quartz	Complex reactions involving Si-Al exchange in Hbl and Plag + Fe-Mg exchange	Kohn and Spear (1989, 1990), Dale et al. (2000)
Garnet–plagioclase–olivine	3 Fo + 3 An = Grs + 2 Prp	Wood (1975), Johnson and Essene (1982), Bohlen et al. (1983a, c)
Garnet–plagioclase–orthopyroxene–quartz (*GAES-GAFS*)	3 Opx + 3 An = 2 Prp-Alm + Grs + 3 Qtz	Wood (1975), Newton and Perkins (1982), Bohlen et al. (1983a), Perkins and Chipera (1985), Powell and Holland (1988), Bhattacharya et al. (1991), Eckert et al. (1991), Faulhaber and Raith (1991), Lal (1993)

(continued)

TABLE 27.4b   Some Useful Geothermobarometers in Petrology: Continuous Net-Transfer and Miscellaneous Equilibria (Continued)

Garnet–plagioclase–clinopyroxene–quartz (GADS-GAHS)	3 Cpx + 3 An = 2 Alm-Pry + 2 Grs + 3 Qtz	Newton and Perkins (1982), Perkins (1987), Powell and Holland (1988), Moecher et al. (1988), Eckert et al. (1991)
Garnet–plagioclase–orthopyroxene–clinopyroxene–quartz	Prp-Alm + Di-Hd + Qtz = En-Fs + An	Paria et al. (1988)
Garnet–cordierite–sillimanite–quartz	3 Cord = 2 Prp-Alm + 4 Al$_2$SiO$_5$ + 5 Qtz	Currie (1971), Hensen and Green (1973), Weisbrod (1973), Thompson (1976), Tracy et al. (1976), Hensen (1977), Holdaway and Lee (1977), Newton and Wood (1979), Martignole and Sisi (1981), Lonker (1981), Aranovich and Podlesskii (1983), Perchuk (1991)
Biotite–cordierite–Al$_2$SiO$_5$–quartz Garnet–biotite–Al$_2$SiO$_5$–quartz garnet–cordierite–biotite–quartz	Several reactions	Patiño Douce et al. (1993)
Garnet–clinopyroxene	2 Grs + Prp = 3 Di + 3 CaTs	Mukhopadhyay (1991)
Garnet–monazite	Y-Al-Grt + OH-Ap + Qtz = Grs + An + YPO$_4$-Mnz + H$_2$O	Pyle and Spear (2000), Pyle et al. (2001)
Garnet–carbonate	3 Dol + Grs = 6 Cal + Pyr	Yaxley and Brey (2004)
Pyroxene–plagioclase–quartz	Jd + Qtz = Ab	Johannes et al. (1971), Holland (1980), Hemingway et al. (1981), Carswell and Harley (1989), Perchuk (1994), Meyre et al. (1997)
Pyroxene–plagioclase–quartz	CaTs + Qtz = An	Gasparik and Lindsley (1980), Newton (1983), Gasparik (1984), McCarthy and Patiño Douce (1998)
Pyroxene–olivine–quartz	2 Fs = Fo + Qtz	Bohlen et al. (1980), Bohlen and Boettcher (1981), Newton (1983), Gasparik (1984), Carswell and Harley (1989)
Amphibole–plagioclase	Several equilibria	Spear (1980, 1981), Plyusnina (1982), Blundy and Holland (1990), Fershtater (1991), Holland and Blundy (1994), Bhadra and Bhattacharya (2007)
Amphibole–plagioclase–clinopyroxene–quartz	Ts + Di + Qtz = Tr + An	Liogys and Jenkins (2000)
Wollastonite–plagioclase–garnet–quartz (WAGS)	Grs + Qtz = An + 2 Wo	Huckenholz et al. (1981)
Garnet–clinopyroxene–phengite–kyanite–quartz	Several	Waters and Martin (1993), Ravna and Terry (2004)
Garnet–spinel–sillimanite–quartz	Alm + 2 Sill = 3 Hc + 5 Qtz	Bohlen et al. (1986)
Garnet–spinel–sillimanite–corundum	Alm + 5 Crn = 3 Hc + 3 Sill	Shulters and Bohlen (1989)
Orthopyroxene–spinel–olivine	En + Sp = MgTs + Fo	Fujii (1976), Gasparik and Newton (1984), Witt-Eickschen and Seck (1991), Ballhaus et al. (1991)
Spinel barometers	Several equilibria	Perchuk (1991)
Talc–kyanite–phengite–coesite Garnet–zoisite–kyanite–coesite Garnet–omphacite–kyanite–coesite	Several equilibria	Massone and Schreyer (1989), Okay (1995), Nakamura and Banno (1997)
Titanite–kyanite–plagioclase–rutile	Ttn + Ky = An + Rt	Manning and Bohlen (1991), Tropper and Manning (2008).

(continued)

TABLE 27.4b   Some Useful Geothermobarometers in Petrology: Continuous Net-Transfer and Miscellaneous Equilibria (*Continued*)

Other Geothermobarometers

Phengite barometry	Phengite content of white mica	Powell and Evans (1983), Massone and Schreyer (1987), Bucher-Nurminen (1987)
Stilpnomelane–chlorite–phengite	Several equilibria	Currie and Van Staal (1999).
Sphalerite–pyrrhotite–pyrite barometry	Fe content of sphalerite	Scott (1973, 1976a, b), Lusk and Ford (1978), Jamieson and Craw (1987), Bryndzia et al. (1988, 1990), Toulmin (1991)
Magnetite–ilmenite thermometry	$4\ Mt + O_2 = 6\ Hem$ $Fe_2 TiO_4 + Fe_2O_3 = Fe_3O_4 + FeTiO_3$	Buddington and Lindsley (1964), Powell and Powell (1977b), Spencer and Lindsley (1981), Ghiorso and Sack (1991), Lindsley and Frost (1992), Anderson et al. (1993), Sauerzapf et al. (2008)
Al-in-hornblende thermometry	Al-content of hornblende in some igneous rocks	Hammarstrom and Zen (1986), Hollister et al. (1987), Johnson and Rutherford (1989), Vyhnal et al. (1991), Schmidt (1992), Anderson and Smith (1995)
Na-in-cordierite thermometry	Na-content of cordierite coexisting with albite and NaOH	Kalt et al. (1998)
Chlorite thermometry	Al^{IV} in chlorite	Stoessell (1984), Cathelineau and Nieva (1985), Walshe (1986), Caritat et al. (1993), Xie et al. (1997), Vidal et al. (2001, 2005, 2006)
Ti-in-zircon and Zr in rutile	$Rt + Zr = Ti\text{-}Zr + Qtz$ and other equilibria	Zack et al. (2004), Zack and Luvizotto (2006), Watson et al. (2006), Ferry and Watson (2007), Tompkins et al. (2007), Fu et al. (2008)
Ti-in-biotite		Henry et al. (2005)
Ti-in-quartz		Wark and Watson (2006)
Magmatic Epidote		Naney (1983), Vyhnal et al. (1991)
Fluid Inclusion thermometry		Hollister and Crawford (1981), Vityk et al. (1994)
Plagioclase–liquid		Mathez (1973), Loomis (1979), Glazner (1984), Marsh et al. (1990), Ariskin and Barmina (1990), Housh and Luhr (1991), Sugawara (2001), Putirka (2005)
Basaltic glass	Wt. % MgO and CaO	Helz and Thorber (1987), Helz et al. (1995)
Solid inclusion piezothermobarometry	Elastic effects around inclusions	Rosenfeld and Chase (1961), Harris et al. (1970), Rosenfeld (1969), Adams et al. (1975a, b), Cohen and Rosenfeld (1979), Graham and Cybriwsky (1981), Van der Molen and Van Roermund (1986), Izraeli et al. (2000), Guiraud and Powell (2006)

28

Metamorphism of Pelitic Sediments

Questions to be Considered in this Chapter:

1. What mineral assemblages develop during progressive metamorphism of pelitic protoliths along various *P-T-t* paths?

2. How can we keep track of the equilibrium mineral assemblages and isograd reactions for a variety of bulk compositions by addressing both discontinuous and continuous reactions?

3. How does progressive metamorphism of pelites differ at high *P/T*, medium *P/T*, and low *P/T*?

4. What occurs at very high grades involving partial melting and migmatites?

Pelitic sediments are very fine grained (commonly <2 μm) mature clastic sediments derived from weathering and erosion of continental crust. Sedimentary petrologists refer to them as **mudstones** and **shales**. They characteristically accumulate in the distal portions of a wedge of sediment off the continental shelf/slope of both active and passive continental margins. Pelites grade into coarser graywackes and sandy sediments toward the continental source. True pelites are much less common in the fore-arc wedges off island arcs, where immature graywackes are characteristic derivatives of the more primitive arc crust. Although they begin as humble mud, clay, or shale, metapelites represent a classic and distinguished family of metamorphic rocks because this range of bulk compositions is sensitive to variations in temperature and pressure, undergoing extensive changes in mineralogy during progressive metamorphism. They also tend to form beautiful mica schists with porphyroblasts of garnet, staurolite, kyanite, etc. Metapelites are the rocks of the classic studies in the Scottish Highlands in which the Barrovian sequence of isograds was developed. It might be wise to briefly review the classic sequence of isograds (chlorite → biotite → garnet → staurolite → kyanite → sillimanite; see Section 21.6.1) if you have forgotten them.

The mineralogy of pelitic sediments is dominated by fine Al-K-rich phyllosilicates, such as clays (montmorillonite, kaolinite, or smectite), fine white micas (sericite or phengite), plus chlorite, all of which may occur as detrital or authigenic grains. Phyllosilicates may compose more than 50% of the original sediment. Fine quartz constitutes another 10 to 30%. Other common constituents include feldspars (albite and K-feldspar), iron oxides and hydroxides, zeolites, carbonates, sulfides, and organic matter. Table 28.1 lists some average compositions of shales (analyses 1–3) and metapelites (analyses 4 and 5). Muds and shales may contain over 30% H_2O and CO_2, most of which is driven off during diagenesis and metamorphism. I have therefore recalculated all the analyses on an anhydrous basis by ignoring volatiles and normalizing the analyses to 100% to aid comparison between the sediments and the metamorphics. Other than volatiles, the composition of a typical metapelite is not significantly different than that of the shale precursor, indicating that (other than loss of volatiles) metamorphism is largely isochemical.

The geochemical characteristics that distinguish pelites from other common rocks are high Al_2O_3 and K_2O (and usually SiO_2) and low CaO. These characteristics reflect the high clay and mica content of the original sediment and lead to the dominance of muscovite and quartz throughout most of the range of metamorphism. The high proportion of micas in

TABLE 28.1 Chemical Compositions* of Shales and Metapelites

	1	2	3	4	5
SiO_2	64.7	64.0	61.5	65.9	56.3
TiO_2	0.80	0.81	0.87	0.92	1.05
Al_2O_3	17.0	18.1	18.6	9.1	20.2
MgO	2.82	2.85	3.81	2.30	3.23
FeO*	5.69	7.03	10.0	6.86	8.38
MnO	0.25	0.10			0.18
CaO	3.50	1.54	0.81	0.17	1.59
Na_2O	1.13	1.64	1.46	0.85	1.86
K_2O	3.96	3.86	3.02	3.88	4.15
P_2O_5	0.15	0.15			
Total	100.00	100.08	100.07	99.98	96.94

*Reported on a volatile-free basis (normalized to 100%) to aid comparison.

1. "North American Shale Composite." Gromet et al. (1984). **2.** Average of ~100 published shale and slate analyses (Ague, 1991). **3.** Ave. pelite–pelagic clay (Carmichael, 1989). **4.** Ave. of low-grade pelitic rocks, Littleton Formation New Hampshire (Shaw, 1956). **5.** Ave. of ~150 amphibolite–facies pelitic rocks (Ague, 1991).

metapelites results in the typical development of *foliated* rocks, such as slates, phyllites, and mica schists. Pelite compositions range from high-Al types to low-Al types and grade through semi-pelites (more sandy) to lithic sandstones with a more granitic composition. We shall restrict our attention to pelites with a predominance of clays ("true pelites").

The chemical composition of pelites can be represented by the system K_2O-FeO-MgO-Al_2O_3-SiO_2-H_2O (*KFMASH*). If we treat H_2O as mobile, the petrogenesis of pelitic systems is represented well in *AKF* and *A(K)FM* diagrams (Sections 24.3.1.2 and 24.3.4). In the discussions that follow we will often be discussing the variance of mineral assemblages. According to the phase rule, $F = C - \phi + 2$. The *KFMASH* system has six components, but we will typically be dealing with triangular diagrams that represent projections down three "components." It's easier to conform to the geometry of these diagrams by addressing the three-component systems, although the full six-component systems are more rigorous. The variance is unaffected, however, if the phases from which the diagrams are projected (typically muscovite and quartz), as well as an aqueous fluid phase, are present. In other words, a three-phase assemblage, such as garnet + staurolite + chlorite, is divariant in the three-component *AFM* or *AKF* system (which we shall call the *reduced* system), and in the six-component *KFMASH* *full* system, if we recall that the assemblage is actually garnet + staurolite + chlorite + muscovite + quartz + fluid. The additional three phases and three components compensate for each other in applying the phase rule. So, for simplicity, I will generally refer to the reduced system and forgo the disclaimer (of the other three necessary phases) each and every time. Before going further you may also want to quickly review the geometric approach to reactions and chemographics in Section 26.9.

28.1 DIAGENESIS AND LOW-GRADE METAMORPHISM OF PELITES

Diagenesis and low-grade metamorphism are gradational across a rather broad and arbitrarily defined zone in the range of approximately 200°C and 0.15 GPa, but it is impossible to place a distinct boundary on the onset of metamorphism proper. Frey (1987), Frey and D. Robinson (1999), and Merriman and Peacor (1999) have provided excellent reviews of low-grade processes in pelites. During diagenesis, the original clays are gradually replaced by chlorite and the more thermally stable clay, illite (a precursor to K-Al-rich "white micas" of the muscovite family). As temperature increases, the layers of the clay minerals gradually become less mixed and the lattices more ordered. As a result, the x-ray peaks of illite become progressively more sharply defined as diagenesis and low-grade metamorphism advance. This "illite crystallinity" has been quantified for use as a type of geothermometer (Kisch, 1966, 1980), measuring the degree of recrystallization as an estimate of metamorphic grade. Carbonaceous matter (responsible for the black color of many shales) is also progressively converted from complex hydrocarbons to graphite, so that the reflectance of the material on a polished surface (called "vitrinite reflectance") has also been used as a grade indicator to estimate the degree to which this process has advanced. The conversion of clays and carbonaceous matter is accompanied by coalescence of fine quartz and feldspar grains to larger grain sizes, while recrystallization and compaction under the weight of overlying sediments expels intergranular water and imparts a characteristic foliation to the rock. The common product of diagenesis and lowest-grade metamorphism of pelitic sediments is an argillite (if massive) or slate (if foliated), typically composed of illite, sericite–phengite, quartz, chlorite, albite, K-feldspar, and minor carbonates, sulfides, organic material, and hematite.

White micas can be compositionally diverse. As mentioned above, sericite and phengite occur at the lowest grades. Ideal muscovite has the formula $KAl_2^{VI}Si_3Al^{IV}O_{10}(OH)_2$. *Sericite* may have high contents of SiO_2, MgO, Na_2O, and H_2O. Sericite may also contain some of the hydrous Al-silicate, *pyrophyllite*. *Phengite* and *Al-celadonite* are components involving an increasing extent of the coupled substitution of $(Si^{4+})^{IV}$-$[(Fe, Mg)^{2+}]^{VI}$ on the tetrahedral (IV) and octahedral (VI) sites, respectively, for $(Al^{+3})^{IV}$-$(Al^{+3})^{VI}$ in ideal muscovite. *Celadonite* involves substitution of Fe^{3+} for Al^{VI}. *Muscovite, paragonite*, and *margarite* represent the K-rich, Na-rich, and Ca-rich white micas, respectively, all of which may occur in metapelites and are separated by immiscibility gaps. The composition of the white mica in a metapelite is controlled by both bulk composition and metamorphic grade. Paragonite occurs in more aluminous pelites, and its compositional range is greater at higher pressures. Margarite occurs in Ca-rich metapelites and marls, as well as in some metamorphosed mafic rocks. Muscovite is the most common of the three in metapelites. At low metamorphic grades muscovite typically contains more of the phengite–celadonite component (Si, Mg, and Fe). At higher grades, muscovite gradually becomes less abundant and more pure.

The celadonite content also increases markedly as pressure increases. White micas are stable well into the amphibolite facies. See Guidotti and Sassi (1976, 1998) and Guidotti (1984) for more extensive reviews of the behavior of metamorphic white micas. I shall limit the following discussion to the occurrence of the most common K-rich muscovites.

28.2 MEDIUM *P/T* METAMORPHISM OF PELITES: THE BARROVIAN SEQUENCE

Metamorphism of pelites provides us with an excellent set of rocks with which to integrate the principles discussed in Chapters 24–27. We shall concentrate on stable mineral assemblages and the reactions responsible for the transitions between them (isograds). Pelites span a considerable range of compositions, and a great variety of mineral assemblages is possible, more than we can adequately cover here. The following discussion will concentrate on some of the more common parageneses among metapelites to illustrate the principles involved. Students wishing to explore the many other parageneses for research or laboratory purposes may then apply the principles to other conditions, bulk rock compositions, or equilibria.

Beginning students generally think of the first appearance, or the final disappearance, of a particular mineral during progressive metamorphism as reflecting the absolute stability limits of that mineral. We expect sillimanite, for example, to be unstable below the sillimanite isograd. In pelites, however, this is rarely the case, and the appearance or disappearance of a mineral is typically due to reactions that occur *within* the broader stability range of that mineral. The grade at which a mineral appears or disappears is usually governed by the composition of the rock (typically the Al content or the Fe/Mg ratio) and may reflect either a discontinuous or a continuous reaction. These concepts will become clearer below, as we explore some examples.

28.2.1 The Chlorite Zone

Pelites in the chlorite zone, the lowest of the classical Barrovian zones, are typically slates with the mineralogy described above. Figure 28.1 illustrates representative *AKF* and *AFM* diagrams for mineral assemblages in the chlorite zone. Equilibrium is difficult to attain in experiments at these low temperatures, so that *P-T* conditions are not certain, but they are believed to be in the 350 to 450°C range. Chlorite, K-feldspar, and muscovite all plot on the *AKF* diagram (as does pyrophyllite, although it occurs only in more Al-rich pelites). The *AFM* diagram in Figure 28.1b is not as useful as at higher grades because only chlorite and pyrophyllite plot on it, and the composition of muscovite, from which the projection is made, contains significant amounts of Fe and Mg (phengite component), so that the projection used (from pure muscovite) distorts the true relationships somewhat. Included in the diagrams are the shaded ranges of typical pelites and granitoid rocks. Pelites extend from low-Al to high-Al types (with chlorite the approximate boundary). The analyses from Table 28.1 are plotted as dots within the pelite field. Most of the analyzed pelite compositions contain quartz, plus chlorite

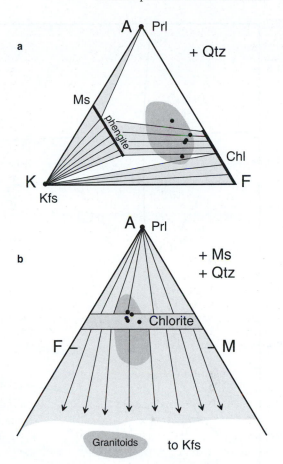

FIGURE 28.1 (a) *AKF* (using the Spear, 1993, formulation) and (b) *AFM* (projected from Ms) compatibility diagrams for pelitic rocks in the **chlorite zone** of the lower **greenschist facies**. Dark shaded areas represent the common range of pelite and granitoid rock compositions. Small black dots are the analyses from Table 28.1. In this and most following compatibility diagrams single- and 2-phase (reduced system) areas are lightly shaded whereas 3-phase areas are blank.

and phengitic muscovite in the two-phase field in Figure 28.1a. Chl-Ms tie-lines connect coexisting chlorite and mica compositions that vary in $(Al + Al) \leftrightarrow ((Fe, Mg) + Si)$ substitution, as a function of the Al-content of the rock. Only Al-poor pelites have K-feldspar. Remember, K-feldspar, when projected from muscovite in the *AFM* diagram (Figure 28.1b), plots at infinity toward the bottom or top of the diagram.

Figure 28.2 is a petrogenetic grid for pelites (largely above the chlorite zone). There are a number of published grids available, and I have selected the one by Spear and Cheney (1989), modified later by Spear (1999) and by Spear, Cheney, and Pattison (personal communication). For more comprehensive grids, see Powell and Holland (1990), Xu et al. (1994), and Wei et al. (2004). We shall refer to Figure 28.2 often as we discuss the reactions and mineral changes that occur as pelitic sediments are progressively metamorphosed. The grid assumes that $p_{H_2O} = P_{total}$, so the dehydration reactions on it represent *maximum* temperatures. If $P_{fluid} < P_{total}$, or the fluid phase is diluted by CO_2 or CH_4, etc., these reactions will occur at somewhat lower grades (as explained in Section 26.4). The grid is a useful guide for the *P-T* conditions of the reactions, but it has some weaknesses (discussed

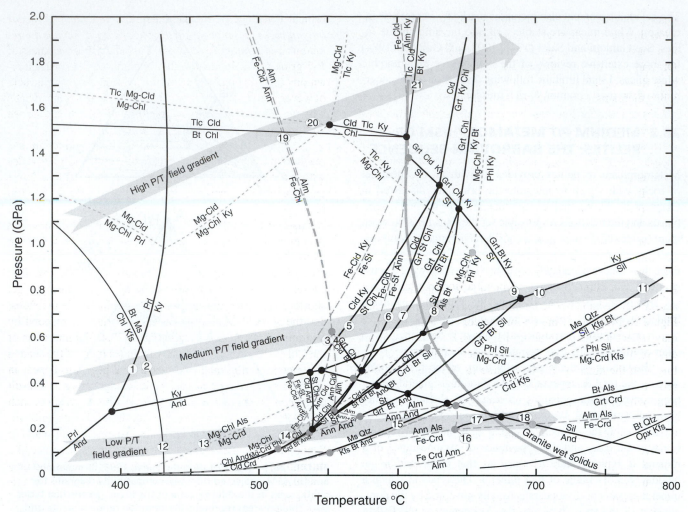

FIGURE 28.2 Petrogenetic grid for the system *KFMASH* at $p_{H_2O} = P_{total}$. Gray dashed curves represent the system *KFASH* and gray small-dashed curves represent the system *KMASH*. Reactions are not balanced and commonly leave out quartz, muscovite, and H_2O, which are considered to be present in excess. Typical high, medium, and low *P/T* metamorphic field gradients are represented by broad shaded arrows. After Spear and Cheney (1989), Spear (1999), and Frank Spear and Dave Pattison (personal communication).

shortly), and the processes and controls governing mineral development are best understood by studying the grid in conjunction with appropriate *AFM* and *AKF* diagrams. Trajectories corresponding to the metamorphic field gradients of the high, low, and medium *P/T* facies series (Figure 25.3) have been added to Figure 28.2 as broad arrows. Remember, these simply represent three alternative gradients among a broad spectrum of possible *P/T* relationships (e.g., Figure 21.1). In the present section, we are following the medium *P/T* gradient.

The reaction crossed at point 1, just above 400°C along the medium *P/T* metamorphic field gradient in the lower-left portion of Figure 28.2, shows that pyrophyllite breaks down to produce kyanite via:

$$Prl = Ky + Qtz + H_2O \qquad (28.1)$$

The classical kyanite isograd in the Scottish Highlands (and typically elsewhere) occurs at much higher grades, as we shall see, but kyanite *can* occur in unusually high-Al pelites (ones in which pyrophyllite occurs as a distinct phase, and not merely as a component in muscovite) at these lower grades. This reaffirms two critical points that have become

increasingly apparent in the past few chapters: (1) When we understand the reactions responsible for isograds, and illustrate the compositional controls on mineral development with the use of appropriate compatibility diagrams, we are able to understand the nature of the mineralogical changes associated with metamorphism. (2) The grade at which a mineral first appears in the field can thus vary, and need not conform to the classical sequence in the Scottish Highlands.

28.2.2 The Biotite Zone

Many reactions can produce biotite. One biotite isograd reaction that can affect pelites is encountered at point 2 in Figure 28.2, in which chlorite reacts with K-feldspar to produce biotite and phengite-rich muscovite:

$$Chl + Kfs = Bt + Ms (+ Qtz + H_2O) \qquad (28.2)$$

Because chlorite is more plentiful than K-feldspar in most pelites, Reaction (28.2) typically marks the loss of Kfs toward higher grades. As with most reactions in this chapter, Reaction (28.2) is *not balanced*. To balance it properly requires knowing the exact compositions of the solid-solution phases. In reduced system diagrams, such as the *AKF* and *AFM* diagrams, additional phases from the full *KFMASH* system may also be required to balance a reaction. In this case Qtz and H_2O (in parentheses) are present and needed to balance Reaction (28.2), but do not plot on the projected compatibility diagrams. Occasionally I shall ignore such non-diagram phases in reactions in order to concentrate on the relationship between reactions and tie-line geometry in the diagrams, as discussed in Section 26.9. If a reaction conforms to these geometric "rules" we shall generally content ourselves with the knowledge that it can be balanced, without formally doing so.

Reaction (28.2) can be demonstrated geometrically by resorting to the *AKF* projection (Figure 28.3). At grades below this biotite–isograd reaction, Chl and Kfs are stable together, as indicated in Figure 28.3a by the tie-lines connecting them. Above the isograd (Figure 28.3b), the new Bt - phengitic Ms tie-line separates Chl and Kfs, so that Bt + Ms are stable together and Chl + Kfs are not, as Reaction (28.2) indicates. Remember from Section 26.9 and Figure 26.14 that crossing tie-lines in compatibility diagrams may reflect a reaction relating the minerals that plot at the ends of the crossed lines. The replacement of the Chl-Kfs tie-line with the crossing Bt-Ms tie-line (a "tie-line flip") thus reflects Reaction (28.2). Note from the full reaction above: if muscovite or quartz is absent, or H_2O is scarce, biotite would be stable at lower temperatures (biotite plots on Figure 28.3a, but natural pelite compositions are generally too Al rich for it to develop). Quartz and muscovite are abundant in pelites, however, and H_2O is plentiful at low grades. These phases are thus considered **in excess**, meaning that they are not generally consumed by a reaction (so their elimination should not limit a reaction's progress). The lowest dot representing our analyzed pelites from Table 28.1 is in the Chl-Kfs-Ms sub-triangle in Figure 28.3a (no biotite) and is in the Chl-Bt-Ms sub-triangle in Figure 28.3b. Thus only Kfs is lost in this rock (not chlorite), and biotite appears at the isograd. The other analyzed rocks, however, will not manifest the reaction (so will not develop biotite) because they do not plot in the

critical Chl-Bt-Kfs-Ms quadrilateral. *A reaction will occur only in rocks that plot within the quadrilateral having the reactants and products as corners*. This is a major weakness of *P-T* grids. Good ones show all of the important univariant reactions for all compositions in the system. Any single rock, however, will manifest but a few of these reactions. That's why we compare grids to compatibility diagrams and, as we will see in a moment, pseudosections.

Although Reaction (28.2) appears to be a *discontinuous* reaction in Figure 28.3b, the temperature of the isograd varies with the Fe/Mg ratio (Fe-rich biotite becomes stable at lower temperatures). At metamorphic grades above the tie-line flip reaction in Figure 28.3, the composition of the white mica that coexists with biotite and chlorite gradually becomes less phengitic via *continuous* reactions by which the most Al-poor white mica breaks down, and the amounts of chlorite and biotite increase (as they become more Al- and Mg-rich). This causes the Ms-phengite solid-solution range (line) to get shorter in the *AKF* diagram as metamorphic grade increases, and the Ms-Chl-Bt and Ms-Bt-Kfs sub-triangles respond by migrating toward the A apex in Figure 28.4. The migration of the Chl-Bt-Ms triangle gradually includes more of the shaded pelite composition range, so that biotite begins to appear in progressively more aluminous pelite compositions as a result of the *continuous* reactions, not a *discontinuous* reaction, such as (28.2), which we tend to associate in our minds with isograds. This is another way of illustrating the point made in Section 26.5 that an isograd based on the first appearance of a mineral may depend on the composition of the rocks and may not occur in all rocks at the same grade. We will discuss this effect more thoroughly later, when we have a little more experience with reactions and isograds on *AFM* diagrams. For now, notice that *continuous* reactions on *AFM* diagrams generally involve the migration of three-phase (divariant) triangles and *discontinuous* reactions involve rearrangement of tie-lines in four-phase (univariant) situations. The variance of each reaction type is reduced from the phase rule value by one because *P* and *T* are related by the *P-T-t* path, so

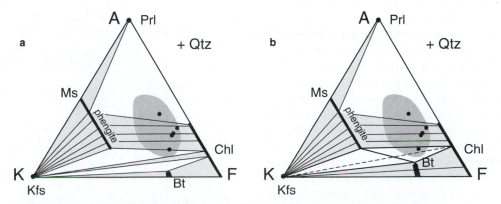

FIGURE 28.3 Greenschist facies *AKF* compatibility diagrams (using the Spear, 1993, formulation) showing the **biotite-in isograd** reaction as a "tie-line flip." In **(a)** below the isograd, the tie-lines connecting chlorite and K-feldspar show that the mineral pair is stable. As grade increases the Chl-Kfs field shrinks to a single tie-line. In **(b)** above the isograd, biotite + phengite is now stable, and chlorite + K-feldspar are separated by the new biotite-phengite tie-line, so they are no longer stable together. Only the most Al-poor portion of the shaded natural pelite range is affected by this reaction.

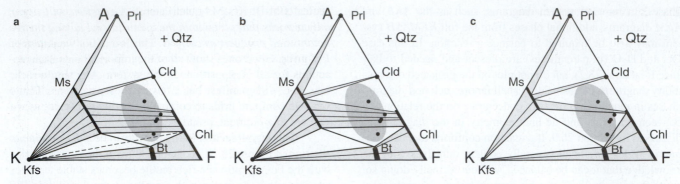

FIGURE 28.4 A series of *AKF* compatibility diagrams (using the Spear, 1993, formulation) illustrating the migration of the Ms-Bt-Chl and Ms-Kfs-Bt sub-triangles to more Al-rich compositions via continuous reactions in the **biotite zone** of the **greenschist facies** above the biotite isograd.

univariant reactions begin and run to completion at a single grade (hence the invariant-like discontinuity for univariant reactions).

Chloritoid may first be introduced into some Fe-rich pelites by the reaction:

$$Chl + Prl = Cld \; (+ Qtz + H_2O) \qquad (28.3)$$

The reaction occurs at temperatures below $250°C$ in the Mg-free (*KFASH*) system (below the grid in Figure 28.2). Because chloritoid lies between chlorite and pyrophyllite along the *A-F* side of Figure 28.4, Reaction (28.3) is indicated by a geometry in which the ends of a line from Chl to Prl combine to form something between them (Cld). Although chloritoid partitions Fe > Mg strongly, natural chloritoids contain significant Mg. Because Mg^{2+} is smaller than Fe^{2+}, it has a higher charge density and thus forms stronger bonds. The Fe end-member, as a general rule, therefore breaks down during prograde metamorphism before the Mg-end member for most mafic minerals (this is not true for *all* mafic minerals). The temperature of Reaction (28.3) in natural pelites should therefore be at a higher temperature than 250°C based on the reaction involving Fe-chlorite.

Most solid reaction curves in Figure 28.2 are continuous reactions in terms of Fe-Mg exchange, and each could be contoured on the *P-T* grid with the Fe-end-member reaction typically defining the low-temperature limit and progressively more Mg-rich reactions at higher temperatures (see Figure 26.9 for an example). The exact temperature at which chloritoid forms in natural pelites via Reaction (28.3) thus varies with bulk Fe/Mg, and cannot be known exactly from Figure 26.2, but it may be in the upper chlorite zone or the lower biotite zone. I have included chloritoid in Figure 28.4 for the biotite zone, but (somewhat arbitrarily) have left it out in Figures 28.1 and 28.3.

An alternative chloritoid-forming reaction in the Mg-free *KFASH* system is encountered at point 3 in Figure 28.2 along the medium *P/T* trajectory (at about 520°C). In this reaction:

$$Fe\text{-}Chl \; (+Ms) = Ann + Fe\text{-}Cld \; (+Qtz + H_2O) \qquad (28.4)$$

Fe-chlorite breaks down by this reaction to form Fe-biotite (annite), and biotite is thus stable in Fe-rich pelites and K-rich granitoids at metamorphic grades below that of Reaction (28.2) in what is generally considered the chlorite zone for

typical pelites (Figure 28.3a). This reaction rarely affects normal pelites. Although Fe-rich chlorites become unstable at these low grades, Mg-chlorite is stable to much higher temperatures (notice the Mg-Chl breakdown reactions in Figure 28.2).

Figure 28.5 illustrates an *AFM* projection for the biotite zone in which chloritoid (and kyanite) are stable. Note from Figures 28.4 and 28.5 that high-Al and a high Fe/Mg ratio favor chloritoid over chlorite. As discussed in Section 21.6.1, the composition range of the pelites in most of the classical Barrovian area of the Scottish Highlands is dominated by low-Al and fairly low-Fe types, so chloritoid was not commonly developed, and poor chloritoid missed out on having a classical isograd and its share of fame. Chloritoid is common in pelites elsewhere. Reaction (28.4) can be understood in regard to Figure 28.5 by imagining the disappearance of Fe-Chl along the A-F side of the triangle and the formation of Cld and Bt (between which Chl plots). Notice how the chlorite Fe/Mg solid solution range begins to shorten in Figure 28.5 as compared to Figure 28.1b.

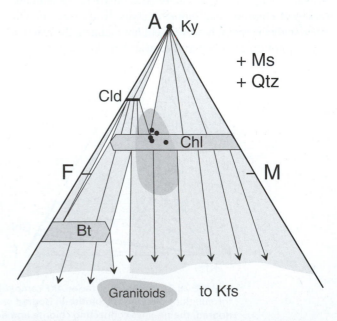

FIGURE 28.5 *AFM* compatibility diagram (projected from Ms) for the **biotite zone, greenschist facies**, above the **chloritoid isograd**. The compositional ranges of common pelites and granitoids are darkly shaded.

28.2.3 The Garnet Zone

Garnet may become stable in the Mg-free *KFASH* system due to any of several reactions. The reaction for the breakdown of Fe-chlorite to Fe-garnet (almandine):

$$\text{Fe-Chl (+Qtz)} = \text{Alm (+H}_2\text{O)} \qquad (28.5)$$

which is intersected near point 3 at pressures above the invariant point on the Reaction (28.4) curve (just above the medium *P/T* trajectory on Figure 28.2). Note that Reaction (28.5) must occur at a pressure higher than that of Reaction (28.4), otherwise Fe-Chl would not be available, having already reacted to chloritoid. As a result of Reaction (28.5), garnet appears on the *AFM* diagram where the Chl solid-solution band would extend to the *A-F* side of Figure 28.5 (see Figure 28.6 for the location). Alternatively, if Fe-Chlorite is less aluminous it may break down to almandine plus a small amount of Fe-biotite (annite):

$$\text{Fe-(low-Al)-Chl (+ Ms + Qtz)}$$
$$= \text{Alm + Ann (+ H}_2\text{O)} \qquad (28.6)$$

If chloritoid is present, almandine may be created by the reaction:

$$\text{Fe-Cld + Ann (+Qtz)} = \text{Alm (+ Ms + H}_2\text{O)} \quad (28.7a)$$

which is nearly coincident with Reaction (28.5) on Figure 28.2. Reaction (28.7a) produces almandine garnet along the *A-F* side of the *AFM* triangle between the Cld and Bt points. These three almandine-producing reactions introduce garnet along the Mg-free edge of the *AFM* diagram, but do not produce garnet in natural pelites if the Cld-Bt tie-line separates garnet from the shaded pelite field (Figure 28.6). Therefore, they rarely represent the usual garnet isograd. When Mg is added, Reaction (28.7a) becomes a *continuous* reaction in the *KFMASH* system:

$$\text{Cld + Bt (+Qtz)} = \text{Grt (+ Ms + H}_2\text{O)} \quad (28.7b)$$

and this causes the collinear assemblage Cld + Bt + Grt in *KFASH* to become a sub-triangle in the *AFM* diagram (because

biotite contains more Mg than coexisting chloritoid or garnet), which continues to migrate toward the right in Figure 28.6 as a continuous reaction in which all three coexisting mafic phases become more Mg-rich with increasing grade. Figure 28.6 represents the *AFM* diagram for the upper biotite zone once the Cld-Bt-Grt sub-triangle has expanded and shifted slightly to the right via this continuous reaction, allowing the range of garnet compositions to extend from the Al-Fe edge.

Garnet may also develop in some natural pelites due to the reaction:

$$\text{Cld + Bt (+ Qtz + H}_2\text{O)} = \text{Grt + Chl (+Ms)} \quad (28.8)$$

encountered at point 4 in Figure 28.2, marking the garnet isograd for some Fe-rich natural pelites in which chloritoid is present. The stoichiometry makes Reaction (28.8) an unusual *prograde hydration* reaction. It results in a tie-line flip, in which the Cld-Bt tie-line of Figure 28.6 gives way to the crossing Grt-Chl tie-line in Figure 28.7. The Grt-Cld-Chl and Grt-Bt-Chl sub-triangles in Figure 28.7 now overlap the Fe-rich portions of the shaded natural pelite range, meaning that garnet occurs in these rocks at this grade.

Let me review an important theoretical point, mentioned earlier, and implicit in a number of reactions discussed since: The first appearance (or the final disappearance) of a mineral in metapelites rarely reflects the *absolute* stability limits of the mineral, but typically occurs within its overall stability range. Recall a discussion from an earlier chapter: The stability of any mineral A typically extends beyond the limits of the association of A plus any other mineral B. For example, if A + B react to form C + D, then, *if A is present in sufficient quantity*, the reaction will cease when B is consumed, and A will remain stable with C and D beyond the reaction curve that limits the stability of A + B combined. The same argument applies to minerals B, C, and D. Garnet must therefore be intrinsically stable at temperatures below that of Reaction (28.8). If muscovite and chlorite are subordinate to garnet, we might expect

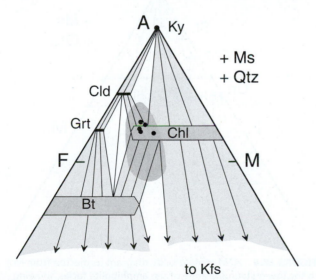

FIGURE 28.6 *AFM* compatibility diagram (projected from Ms) for the **upper biotite zone, greenschist facies**. Although garnet is stable, it is limited to unusually Fe-rich compositions and does not occur in natural pelites (shaded).

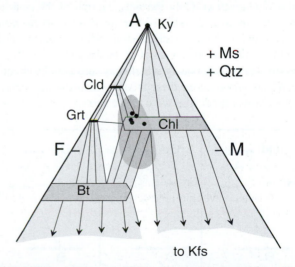

FIGURE 28.7 *AFM* compatibility diagram (projected from Ms) for the **garnet zone, transitional to the amphibolite facies,** showing the tie-line flip associated with Reaction (28.8) (compare to Figure 28.6), which introduces garnet into the more Fe-rich types of common (shaded) pelites. After Spear (1993).

to find garnet at lower grades. This is rarely the case in nature because muscovite and chlorite are typically plentiful at this grade. For the same reasons, garnet + chlorite (+ muscovite) is stable over a more limited range than just garnet + muscovite, explaining why Reaction (28.7a) occurs at a lower temperature than Reaction (28.8) in Figure 28.2.

As an extension of this argument, Reaction (28.8) does not add any new phases to the *AFM* diagram of Figure 28.7 that were not already present in Figure 28.6. The tie-line flip associated with Reaction (28.8) merely *changes the mineral associations*. This is made clearer in the expanded view in Figure 28.8. Figure 28.8a corresponds to the topology of Figure 28.6 prior to Reaction (28.8) when Cld + Bt are stable together (the Cld-Bt tie-line connects these coexisting phases). At the isograd itself (Figure 28.8b) all four phases are in equilibrium (the system is invariant when *P* and *T* are related by the *P-T-t* path), and the Cld-Bt and Chl-Grt tie-lines both exist and cross. Above the isograd (Figure 28.8c) Cld + Bt is no longer stable and Grt + Chl is stable. There are four areas within the Grt-Cld-Chl-Bt quadrilateral affected by the reaction. Any rock with a bulk composition that plots in area 1 (Figure 28.8b) will contain Cld + Grt + Bt (+ Ms + Qtz) below the isograd and Cld + Grt + Chl (+ Ms + Qtz) above the isograd. Such a rock would, in a reversal of the classical isograd sequence, lose biotite and gain chlorite at the isograd. In area 2, rocks would lose biotite and develop small porphyroblasts of garnet. In area 3, rocks would lose chloritoid and gain chlorite. And in area 4, rocks would lose chloritoid and gain garnet. *The minerals that are developed or lost in a rock at the isograd thus depend upon the bulk composition, and do not reflect the absolute stability limit of any mineral involved.* From Figures 28.6 and 28.7 we can see that natural pelites overlap into areas 2 and 4 in Figure 28.8. Garnet is therefore introduced into them by Reaction (28.8), well after garnet first becomes theoretically stable. Most typical pelites in the shaded area do not fall within the Cld-Chl-Bt-Grt quadrilateral, and thus do not change at all due to this reaction. Only the most Fe-rich of the common pelites fall in areas 2 and 4, so that this is the garnet isograd for them only. The above discussion addresses garnet in the *KFMASH* system only. Garnet is unusually susceptible to non-*KFMASH* components, which may substantially affect its stability. Notable among these is Mn, which stabilizes garnet to much lower grades, as we will see shortly.

28.2.4 The Staurolite Zone

Fe-staurolite becomes stable on the *AFM* diagram by a reaction between Fe-chloritoid and kyanite in the Fe (*KFASH*) system (encountered just above point 4 in Figure 28.2). Staurolite therefore appears between Cld and Ky along the *A-F* side of the *AFM* diagram. Because of the relatively low positive *dP/dT* slope of this reaction, at lower pressure St can occur before Reaction (28.8), or even before Reaction (28.4) in the classic biotite zone. Fe-St could thus be added to Figures 28.6 or 28.7 for lower *P/T* terranes. The point is academic, however, because this only affects the ultimate stability of St on the chemographic diagrams and not its occurrence in metapelites, because sub-triangles with St at a corner at this grade are too Al-Fe-rich to overlap the shaded pelite area in those figures. In *KFMASH*, staurolite is produced by the next reaction that is crossed along the medium *P/T* trajectory in the petrogenetic grid (encountered at point 5 in Figure 28.2, at about 570°C), which is:

$$Cld + Ky = St + Chl \ (+ \ Qtz + H_2O) \quad (28.9)$$

Geometrically this implies a tie-line flip on the *AFM* diagram from the Cld-Ky tie-line to the crossing St-Chl tie-line (Figure 28.9). The new St-bearing sub-triangles in Figure 28.9 only overlap a portion of higher-Al pelite compositions, so Reaction (28.9) only represents the staurolite isograd for these rocks. Although the staurolite isograd and zone may be mapped in the field on the basis of such high-Al rocks, Figure 28.9 reminds us that only a limited range of metapelites would be staurolite-bearing. Other assemblages for the shaded pelite area include Ky + Chl, Chl + Bt, Chl + Bt + Kfs, Grt + Chl + Bt, Grt + Cld + Chl, and Cld + Chl (all include Ms + Qtz). If no Al-rich pelites are found in the field, however, staurolite would not yet be developed, and then the area would still technically be in the garnet zone. This is why

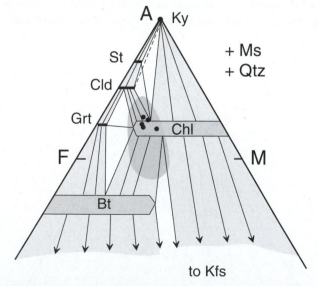

FIGURE 28.9 *AFM* compatibility diagram (projected from Ms) in the **lower staurolite zone** of the **amphibolite facies**, showing the change in topology associated with Reaction (28.9) in which the lower-grade Cld-Ky tie-line (dashed) is lost and replaced by the St-Chl tie-line. This reaction introduces staurolite to only a small range of Al-rich metapelites. After Spear (1993).

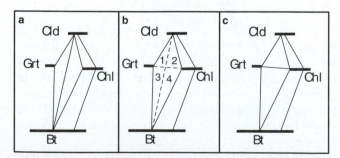

FIGURE 28.8 An expanded sketch of the Grt-Cld-Chl-Bt quadrilateral from Figures 28.6 and 28.7 illustrating the tie-line flip of Reaction (28.8). **(a)** Before flip. **(b)** During flip (at the isograd). **(c)** After flip (above the isograd).

petrogenetic grids must be used in conjunction with appropriate compatibility diagrams to interpret field occurrences adequately.

Above this grade we approach the upper stability limits of chloritoid. Again, the Fe-rich varieties are the first to break down, and the range of Cld compositions shown in Figure 28.9 gradually shrinks as a Cld-Grt-St sub-triangle forms and migrates to the right in the *AFM* projection because of a continuous reaction of the form:

$$Cld = Grt + St \ (+ \ Qtz + H_2O) \qquad (28.10)$$

Eventually the Cld composition range shrinks to a point, and is then lost because of the reaction:

$$Cld \ (+ \ Qtz) = Grt + Chl + St + H_2O \qquad (28.11)$$

at point 6 along the medium *P/T* trajectory in Figure 28.2 (~590°C). Because Cld lies within the Grt-Chl-St triangle, chloritoid simply disappears from the (quartz-bearing) *AFM* diagram, and its components are redistributed into the phases defining the triangle that surrounds it (Figure 28.10). Reactions that lose a phase entirely from the compatibility diagrams, rather than rearrange (flip) tie-lines, are referred to as **terminal** reactions. Reaction (28.11) is thus a terminal reaction for chloritoid. As mentioned previously, some investigators refer to prograde terminal reactions as "-out" isograds, an interesting adaptation of the isograd concept. We may thus refer to Reaction (28.11) as the **chloritoid-out** isograd. Examination of Reaction (28.11) tells us that quartz must be present in excess of chloritoid for it to be really terminal, but quartz is generally plentiful in metapelites.

Although staurolite is stable in some pelites at lower temperatures, as described above, it is introduced into most common pelites by the reaction:

$$Grt + Chl \ (+ \ Ms) = St + Bt \ (+ \ Qtz + H_2O) \qquad (28.12)$$

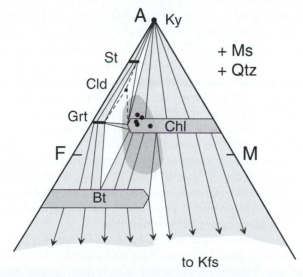

FIGURE 28.10 *AFM* compatibility diagram (projected from Ms) in the **staurolite zone** of the **amphibolite facies**, showing the change in topology associated with the terminal Reaction (28.11), in which chloritoid is lost (lost tie-lines are dashed), yielding to the Grt-St-Chl sub-triangle that surrounds it.

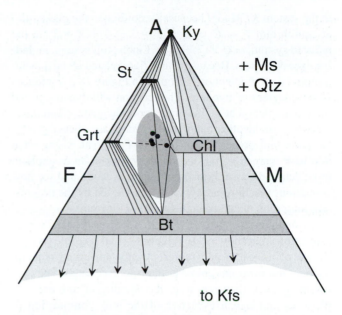

FIGURE 28.11 *AFM* compatibility diagram (projected from Ms) for the **staurolite zone, amphibolite facies,** showing the tie-line flip associated with Reaction (28.12), which introduces staurolite into many low-Al common pelites (dark shaded).

encountered at point 7 in Figure 28.2 (~610°C). This reaction produces a tie-line flip in which the Grt-Chl tie-line is lost on the *AFM* diagram and a St-Bt tie-line is created (Figure 28.11). Mineral assemblages that include staurolite now extend to a much wider range of natural pelite compositions, including many on the low-Al side of the Grt-Chl tie-line. Garnet may also be lost in some metapelites via this reaction (pelites to the right of the new St-Bt tie-line). Because this isograd reaction marks only the introduction of staurolite into low-Al pelites (the common Barrovian type), Carmichael (1970) proposed that the isograd and zone be called "staurolite-biotite," because these two minerals are not stable *together* (in quartz-bearing rocks) below the isograd, but are stable together above it, whereas either mineral alone is stable over a larger *T-P* range. This reasoning has considerable merit, and could be used to clarify virtually all isograds. It would require renaming the classical isograds, however, and has not yet had the widespread application it (richly) deserves.

28.2.5 Pseudosections and High-Variance Reactions in *KFMASH*

Before proceeding further up metamorphic grade, let's review for a moment and develop the concept of continuous reactions more fully as they affect phase equilibria in pelites. I will also introduce pseudosections, which address high-variance reactions much better than *P-T* grids. These concepts will help us as we continue to investigate progressive metamorphism of pelitic rocks.

By now we are becoming familiar with the nature of discontinuous reactions, and how a tie-line flip or the terminal loss of a phase affects mineral parageneses and compatibility diagrams. Reaction (28.12) above is a discontinuous reaction in the system *KFMASH* because (according to the phase rule) at equilibrium $F = C - \phi + 2 = 3 - 4 + 2 = 1$ (in the reduced system, or $6 - 7 + 2 = 1$ in the full system, including excess phases). Because P and T are related along a metamorphic field gradient or *P-T-t* path in Figure 28.2, F reduces to zero degrees of freedom. The reaction will thus occur and run to completion at a single metamorphic grade. *Continuous* reactions, on the other hand, have fewer phases, hence greater variance, and proceed spanning a *P-T* interval. Using what you have learned so far, we can now look at an example in more detail, discover how pseudosections work, and see how continuous reactions are typically manifested by the progressive migration of sub-triangles on *AFM* diagrams.

As mentioned above, *P-T* grids such as Figure 28.2 are projections that show phase relations for all possible compositions of a system and represent only univariant reaction curves (and invariant points at their intersections). They are useful because they constrain the stability of any possible phase assemblage irrespective of the bulk composition of the system. The weakness of such grids is that several different mineral assemblages may be stable within any given region of the diagram, depending on X_{bulk} (as indicated by compatibility diagrams), and only a very small subset of the many reactions depicted on a grid may be relevant for any specified composition. Moreover, because most phases in natural systems are solutions, as we have recently seen, continuous reactions of high variance (not included in *P-T* grids) generally limit phase stabilities in any rock. Phase diagram sections computed for a specified bulk composition, called **pseudosections**, offer an alternative to *P-T* grid projections and avoid many of these complexities. Hensen (1971) first demonstrated their utility, but the term was coined by Roger Powell in the mid-1980s and their use has flourished due to the ability to construct them using THERMOCALC and Perple_X software (described in Section 27.5). In contrast to grid projections, there is only one *P-T-X* state represented by any point within a pseudosection, so that at each point the composition and proportions of the phases and the thermodynamic properties of the system are uniquely determined. Pseudosections are typically *P-T*, *T-X*, or *P-X* diagrams and, because they depict the phase assemblage at any point on them, they can reflect both continuous and discontinuous reactions and compliment *P-T* grids and compatibility diagrams very nicely.

Figure 28.12 is a portion of a *T-X*$_{Mg}$ pseudosection for a fixed bulk composition (except for Fe/Mg) extending over a limited *P-T-X* range (for simplicity). It is part of a larger pseudosection created by Powell et al. (1998) using their program THERMOCALC. This type of pseudosection can either be isobaric or may conform to a fixed *P/T* path (equilibrium at each temperature is determined for a different pressure on the path). I have modified the temperatures from the original isobaric diagram (at 0.6 GPa) to fit our medium *P/T* path in Figure 28.2. Each field is labeled by the (reduced) mineral assemblage that is stable within it (excess phases are assumed present). By convention, higher variance fields are shaded darker in pseudosections. The nature of continuous *KFMASH* reactions may become more

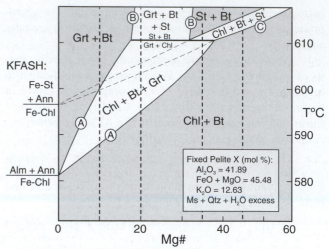

FIGURE 28.12 *T-X*$_{Mg}$ pseudosection diagram in the system *KFMASH* of variable Mg/Fe for a "common pelite" of specified composition: Mg#=100*Mg/(Mg + Fe) (molar). I have modified the temperatures of the original isobaric diagram to conform with the specified medium *P/T* trajectory in Figure 28.2.

clear when we recognize that Figure 28.12 consists of a combination of three divariant loops (A, B, and C, all surrounding unshaded three-phase fields; six phases if you include excess Ms, Qtz, and H_2O). Each loop is similar to the binary Fo-Fa divariant (continuous) melting reaction loop (Figure 6.10) or the schematic chlorite $\rightarrow$ garnet continuous reaction and pseudosection illustrated in Figure 26.9. Only a portion of each loop is stable, however, in this multicomponent system. I have added a dashed metastable extension of the St-Bt-Chl loop (C) to make the overall shape more obvious.

Pure Fe or Mg end-member reactions are *discontinuous* because the variance in *KFASH* (including the fluid phase) is $F = C - \phi + 2 = 5 - 6 + 2 = 1$, which, if isobaric or limited by a specified *P/T* relationship (as in Figure 28.12), reduces to zero. The reaction, if stable, would therefore occur at a single temperature along the *KFASH* (left) side of the diagram, where I have labeled them (ignoring excess Ms, Qtz, and H_2O). Within the *KFMASH* interior of the diagram, however, the reaction becomes continuous because C increases by one while ϕ remains the same as Fe and Mg mix in the mafic phases. Hence the range of temperatures within loops A, B, and C away from the Fe side. The loops (or their metastable extensions) close again to a discontinuous reaction as $X_{Mg} \rightarrow 1.0$ in *KMASH* (not shown). The analogy with the Fo-Fa (olivine) and An-Ab (plagioclase) systems in Chapter 6 should be clear to you. As is typical of most (although not all) continuous reactions, the Fe-end-member reaction (*KFASH* system) occurs at a lower temperature than the Mg-end-member (*KMASH*) reaction.

As an example of how to interpret Figure 28.12, let's return down grade for a moment and consider what happens on Figure 28.12 when we heat a pelite with Mg/(Mg + Fe) = 0.10 (Mg# =10) and follow the vertical dashed line. With the bulk composition now fully fixed, the low-*T* mineral assemblage is Chl + Bt (+ Ms + Qtz + H_2O). The bulk composition would thus plot somewhere between the Chl and Bt fields in a compatibility diagram similar to Figure 28.10, which is shown schematically in Figure 28.13a as the dot labeled "10." Upon heating, this system eventually intersects loop A in Figure 28.12. In the pure-Fe (*KFASH*) system this reaction is the discontinuous Reaction (28.6), but in *KFMASH* it is the *continuous* reaction:

$$\text{Chl} (+ \text{Ms} + \text{Qtz}) = \text{Grt} + \text{Bt} (+ \text{H}_2\text{O}) \quad (28.13)$$

When heating a single bulk composition (any vertical line in Figure 28.12), the temperature of an isograd (the point at which any loop in Figure 28.12 is intersected and a new phase is introduced) is variable, depending upon X_{Mg} of the rock in question. This follows the discussion in Section 26.5. Reaction (28.13) introduces garnet in our Mg# 10 rock at about 585°C. The reaction progresses over a temperature interval as Chl is gradually consumed and Grt and Bt are produced. Although the *bulk rock* composition must conform to the plane of the pseudosection, the compositions of the *phases* generally do not: Al : (Fe + Mg) : K is not the same for chlorite, biotite, garnet, muscovite, quartz, etc., but they must, in appropriate proportions, sum to that of X_{bulk}. This is not a true section (such as the binary Fo-Fa system), therefore, because the phases must be projected to it. Hence the term *pseudo*section. This does not, however, detract from the ability of a pseudosection to accurately depict the proper phase assemblage at any P, T, and X_{bulk}. Although the phase compositions cannot directly be determined from Figure 28.12 (nor can their proportions be determined by the lever principle as we could in Chapter 6), the position of chlorite to the right of loop A and of garnet to the left, with biotite intermediate, indicates that chlorite is more Mg-rich than the associated biotite, which, in turn, is more Mg-rich than garnet when all five phases of Reaction (28.13) are in equilibrium. This is consistent with the chemographics in Figure 28.13.

The slope of loop A also indicates that all three mafic phases become more Mg-enriched as the continuous Reaction (28.13) progresses up the temperature range. This is illustrated in Figure 28.13 by the migration of the three-phase triangle Chl-Bt-Grt (see also Figure 28.10) toward the right (white block arrow) as the shaded compositional range of Chl continues to contract. *This shift in a three-phase triangle due to a continuous reaction can result in the formation or disappearance of minerals in metamorphosed rocks.* For example, our Mg# 10 dot is eventually overrun by the migrating triangle in Figure 28.13, at which point garnet is introduced into the rock (corresponding to the point at which the dashed Mg# 10 curve in Figure 28.12 intersects loop A). *Univariant continuous reactions typically involve the migration of a three-phase triangle in compatibility diagrams.* We can think of these moving triangles as sweeping up tie-lines before them (like a bulldozer or snowplow) and leaving a trail of new tie-lines in their wake. *Minerals then develop or are lost in specific bulk compositions when the migrating triangles first encompass or abandon (respectively) corresponding X_{bulk} points.*

You may wonder how all three phases in the Grt-Chl-Bt sub-triangle can become richer in Mg as the continuous reaction progresses and the sub-triangle migrates toward the right. Where did the necessary Mg come from? Remember, the compositional shift is accommodated by a *change in the proportions of the phases* due to the reaction. Because chlorite is more Mg-rich than either garnet or biotite, as it breaks down it releases Fe and Mg in a high Mg/Fe ratio. This is rich enough in Mg to cause the garnet and biotite that form to be more Mg-rich than they were previously.

Note from both Figures 28.12 and 28.13b that the Mg# 10 rock at 590°C has Grt + Bt + Chl, whereas the other two compositions have only Bt + Chl at this temperature. With further heating the Mg# 10 assemblage eventually reaches the upper limit of loop A in Figure 28.12, at which point chlorite is finally lost. This is illustrated in Figure 28.13c, where the Mg# 10 dot is left in the wake of the migrating Grt-Chl-Bt triangle and is now in the two-phase Grt-Bt field. This example shows that discontinuous (univariant) reactions

All fields + Ms + Qtz + H₂O

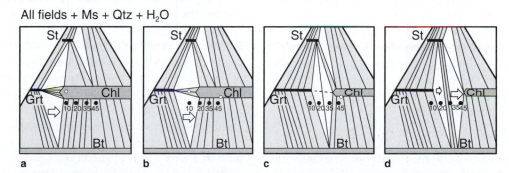

a b c d

FIGURE 28.13 A schematic expanded view of the Grt-St-Chl-Bt quadrilateral from Figure 28.11 illustrating the progressive metamorphism of compositions with 100*Mg/(Mg + Fe), or Mg#, of 10, 20, 35, and 45 from Figure 28.12. **(a)** At a grade below 585°C at which all four compositions contain chlorite + biotite (+Ms + Qtz). **(b)** As Reaction (28.12) proceeds, the most Fe-rich chlorite breaks down, and the Chl-Grt-Bt sub-triangle shifts to the right (arrow). **(c)** Further shift of the Chl-Grt-Bt sub-triangle due to Reaction (28.12) encompasses Mg# 20 and 35 and leaves Mg# 10. The Grt-Chl field shrinks to a single tie-line, then disappears as Reaction (28.12) causes a tie-line flip to St-Bt. Composition Mg# 20 thereby loses Chl and gains St as composition Mg# 35 loses Grt and gains St. **(d)** Migration of the new Chl-St-Bt sub-triangle (arrow) due to Reaction (28.14) encompasses Mg# 45 (which develops St) and leaves Mg# 35 (which loses Chl).

that we usually associate with isograds are not the only reactions that can introduce a mineral into a rock. *Continuous reactions can cause a mineral, such as biotite, garnet, or staurolite, to appear (in appropriate rocks) at grades above that at which the corresponding discontinuous reaction would take place.* This also holds true for a mineral's disappearance. Notice that this Mg# 10 rock developed new garnet and later lost chlorite over a temperature interval of ~585 → 600°C, *all due to continuous reactions.*

Let's briefly look at the Mg# 20 and 35 compositions in Figures 28.12 and 28.13. Heating both encounters loop A and develops garnet via Reaction (28.13) at a grade slightly below 595°C for Mg# 20 and slightly over 605°C for Mg# 35. This occurs at some condition just after Figure 28.13b for Mg# 20 and just before Figure 28.13c for Mg# 35 when the migrating Grt-Chl-Bt triangle first overruns and includes their respective points.

At about 610°C the *discontinuous* Reaction (28.12) is reached, resulting in the Grt-Chl → Bt-St tie-line flip (as discussed previously and illustrated in Figure 28.11). This marks the transition from loop A to loops B and C in Figure 28.12 and the introduction of staurolite in these low-Al pelites if Mg# is appropriate. Because this reaction is invariant (at fixed P or P/T) the reaction curve is horizontal in Figure 28.12, occurring entirely at a single grade. This is the only reaction curve in Figure 28.12 that is univariant on a P-T grid and therefore shown in Figure 28.2 and thus common to both figures. Any X_{bulk} between Mg# 17 and 32 will lose chlorite before garnet as a result of this reaction and any composition between Mg# 32 and 38 will lose garnet first. For example, the Mg# 20 line in Figure 28.12 crosses from the Chl + Bt + Grt field to the St + Bt + Grt field, and likewise plots in the Chl + Bt + Grt sub-triangle below the isograd in Figure 28.13c and in the St + Bt + Grt field above it. You should be able to confirm the corresponding sequence for Mg# 35. A rock with Mg# 10 (or less) will be unaffected by this reaction (because it lacks chlorite as a reactant, and thus plots outside the critical Grt-St-Chl-Bt quadrilateral in Figure 28.13), as would a rock with Mg# 38 or greater (which lacks garnet).

Once the discontinuous Reaction (28.12) is complete (within the new "staurolite-biotite zone"), loops B and C occur simultaneously (although they cannot simultaneously affect the same rocks). For example, a rock with $X_{bulk} =$ Mg# 35 is affected by loop C in Figure 28.12, which corresponds to the *continuous* reaction:

$$Chl\,(+Ms) = St + Bt\,(+Qtz + H_2O)\quad(28.14a)$$

by which the most Fe-rich chlorite (which now coexists with St + Bt) continues to break down and the Chl compositional range shrinks further as the Chl-St-Bt triangle migrates to the right in Figure 28.13d (block arrow). As it does so, the St-Bt tie-line created by Reaction (28.12) expands to a two-phase field (with new parallel tie-lines added progressively to the right side) and the St-Chl field eventually shrinks to a single tie-line as tie-lines on the left are lost. The triangle eventually abandons Mg# 35 (chlorite is lost and only biotite and staurolite remain [+ Ms + Qtz, of course]).

A rock with Mg# 45, in contrast, will develop new staurolite via this same continuous reaction as the migrating triangle envelops it.

Loop B causes the St-Grt-Bt sub-triangle in Figure 28.13d to migrate toward the right above 610°C also, due to the reaction:

$$Bt + St\,(+Qtz) = Grt\,(+Ms + H_2O)\quad(28.14b)$$

which affects more Fe-rich rocks, introducing garnet or eliminating staurolite in biotite-rich rocks, depending on X_{Mg}. The steeper limbs of the loop in Figure 28.12 mean that the St-Grt-Bt sub-triangle migrates more slowly than the St-Chl-Bt sub-triangle as grade increases, explaining why the St-Bt 2-phase field expands (toward the right) from Figure 28.13c to 28.13d.

Multivariate continuous reactions such as Reactions (28.13) and (28.14a and b) occur in virtually all metamorphic zones between the discontinuous (univariant) reactions. The importance of continuous reactions in metamorphic petrology was overlooked for some time. We are now beginning to realize that most phase changes in metamorphic rocks probably occur as a result of *continuous*, rather than *discontinuous*, reactions.

Perhaps we can now recognize the wisdom of mapping an isograd in the field simply as the first appearance of an index mineral, although it should still be related to a specific reaction whenever possible. Although rocks Mg# 10 and Mg# 35 did not develop staurolite at the grade of Figure 28.13c, this does not mean that they were below the grade at which staurolite became stable via Reaction (28.12). Rather it means that they were not of a suitable composition (Mg# 10 was too Fe-rich and Mg# 35 was too Mg-rich) for the reaction to have been manifested in them. If staurolite does occur in rock Mg# 20, it would be proper to consider neighboring rocks to also be of the same grade, i.e., above the staurolite isograd, which is based on the first appearance of staurolite in a regional sense, even if it does not appear in all rocks. These arguments apply to nearly all of the tie-line flip isograds that we discuss.

Figure 28.14 is a (perhaps more common) *P-T* **pseudosection**. Because X_{bulk} is entirely fixed in *KFMASH* by specifying Al_2O_3, FeO, MgO, and K_2O and saturating (buffering) in Qtz (SiO_2) and H_2O, P and T are both permitted as independent variables and the state of the system (and the phases present, including their composition) is fully determined at any P-T point. X_{bulk} in this figure is the same as for Figure 28.12, but with Mg# = 40, so it relates to it at 0.4 on the abscissa. Pseudosections typically show mineral assemblage fields only. The reactions separating them are generally not labeled and must be determined or inferred from the P-T grids or compatibility diagrams (or output from the computer programs used to construct them). I have labeled the true univariant reactions (shown on Figure 28.2) in Figures 28.12 and 28.13, and made those curves thicker for easier recognition. They separate adjacent three-phase fields (unshaded) in both figures (so that four phases, again ignoring excess Ms, Qtz, and fluid, are present and thus $F = 3 - 4 + 2 = 1$).

As a means of introduction to *P-T* pseudosections, consider the sequence of mineral parageneses that might

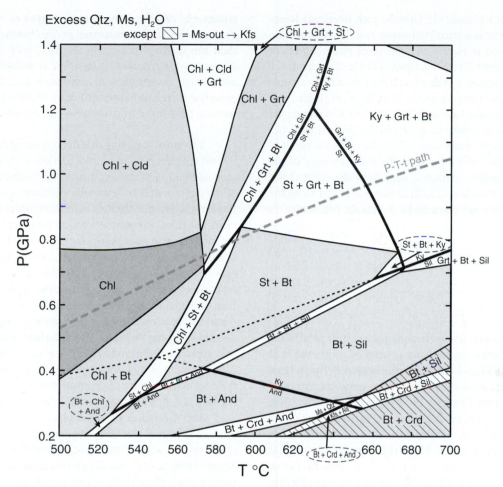

FIGURE 28.14   *P-T* pseudosection in *KFMASH* for mol. % SiO$_2$ = 76.14, Al$_2$O$_3$ = 11.25, MgO = 4.89, FeO = 7.33, and K$_2$O = 3.39. This composition has Qtz and Ms in excess, and H$_2$O was set to saturated. Calculated using both THERMOCALC (a somewhat time-consuming task) and Perple_X (a simpler, essentially automated, process that uses a free energy minimization approach; see Connolly and Petrini, 2002) and the November 2003 Holland-Powell internally consistent thermodynamic database, with quite similar results. Based on Powell et al. (1998). Extensions of Al$_2$SiO$_5$ polymorph reactions shown as dashed curves for clarity.

develop following the *P-T-t* path in Figure 28.14. This path is near the medium-*P* path in Figure 28.2, but I raised the pressure a bit to provide a more "interesting" and illustrative petrogenetic sequence. The path begins in the Chl-only field, which is common at low grades in many *P-T* pseudosections that I have seen. This large one-phase field (+ Ms + Qtz + H$_2$O) reflects a correspondingly broad chlorite compositional range. As grade rises, the range shrinks toward more stoichiometric chlorite compositions (similar to the behavior of muscovite discussed in Section 28.1). Along the *P-T-t* path indicated garnet joins chlorite at about 570°C. This is a high-variance situation (for a fixed *P/T*, *F* = 3 within the Chl field and *F* = 2 beyond that) so the reactions are poorly constrained. Because *X*$_{bulk}$ is entirely fixed, the transition must be related to the instability of the Fe-richest chlorite and associated encroachment of the Grt-Chl field to the right on *AFM* diagrams until it encompasses this *X*$_{bulk}$ (as illustrated by the two corresponding white dots in Figures 28.13a and b). Alternatively, below about 0.7 GPa an up-grade path would enter the Bt + Chl field, indicating that, at lower pressures, the most Al-poor

chlorite is less stable with increasing grade and the Chl-band becomes narrower (as illustrated in Figure 28.13) by reacting with associated Ms + Qtz to eventually recede past *X*$_{bulk}$, which is then encompassed by the Chl + Bt field (illustrated by the white dots in Figures 28.13c and d). At some point between the Chl → Chl + Bt sequence at low grades and the Chl → Chl + Grt at higher temperature and pressure, Reaction (28.8) was surpassed and garnet became stable. Because this *X*$_{bulk}$ is not in the proper compositional quadrilateral (see Figures 28.6 and 28.7), this reaction is not manifested and therefore not shown in Figure 28.14. Thermodynamically, however, the products in Reaction (28.8) became more stable than the reactants, so garnet became a potentially stable phase, created in the present case by a later continuous reaction. As grade increases further along the dashed *P-T-t* path in Figure 28.14, biotite joins Chl + Grt (for the same reason it joined chlorite in the low-*P* example). The sequence for this particular rock and *P-T-t* path is thus Chl (F = 4) → Chl + Grt (F = 3) → Chl + Grt + Bt (F = 2), which doesn't quite conform to the classical Barrovian sequence.

At about 585°C and 0.75 GPa the path intersects Reaction (28.12), which, as a "true" univariant reaction, is therefore also on the *P-T* grid in Figure 28.2 and takes place entirely at this single grade for a fixed *P-T-t* path. For the X_{bulk} and *P-T* path in question, garnet is consumed before chlorite as staurolite is produced and our path enters the Chl + St + Bt field. Sophisticated programs such as THERMOCALC compute the stable mineral compositions at any *P* and *T*, which, if you consider it for a moment, must vary (this is the basis for geothermobarometry). This, in turn, affects the stoichiometry of the reaction. For example, extracted data from the THERMO-CALC output for Reaction (28.12) at two *P-T* points along the stable reaction curve in Figure 28.14 reveals the following:

P(GPa)	T(C)	x(Chl)	x(Bt)	x(St)	x(Grt)	x(Ms)
				where x = molar Fe/(Fe + Mg)		
0.70	576.9	0.5885	0.6470	0.9144	0.9078	0.5836
	27 Chl + 24 Grt + 63 Ms = 63 Bt + 10 St + 100 Qtz + 87 H₂O					
1.20	633.2	0.301	0.3439	0.7501	0.7237	0.3177
	29 Chl + 20 Grt + 61 Ms = 61 Bt + 10 St + 91 Qtz + 95 H₂O					

The proportions of reacting garnet:chlorite are 0.89 at 0.7 GPa and 0.69 at 1.2 GPa (qualitatively: garnet is more stable at higher pressures so less is consumed). Had the *P-T* path been at higher pressure, then chlorite would be consumed first for the X_{bulk} in question, resulting in St + Grt + Bt rather than St + Chl + Bt (as shown in Figure 28.14). Returning to the dashed path, further heating enters the St + Bt field because Reaction (12.14a) causes the Chl-St-Bt triangle to move past X_{bulk} (just as Mg# 35 entered the St + Bt area in Figure 28.13d). Garnet (interestingly) reappears as Reaction (28.14b) causes the system to enter the Grt-St-Bt field in Figure 28.13.

We can conclude that the evolution of progressively metamorphosed pelites (or any rocks, for that matter) is governed by a combination of discontinuous and continuous reactions. Continuous reactions affect the compositional ranges of solid-solution minerals and occur over a range of temperature-pressure conditions, causing virtually all of the sub-triangles in AFM diagrams to migrate (usually toward the right as grade increases). As mentioned earlier, as a sub-triangle migrates, it consumes the tie-lines of contracting two-phase fields ahead of it (ignoring the phases Ms, Qtz, and fluid) and generates new tie-lines in expanding two-phase fields behind it (as in Figure 28.13). Once a two-phase field shrinks to a single tie-line, the next step is for that line to disappear and be replaced by a new tie-line that crosses it (a tie-line flip) reflecting a discontinuous reaction. The new tie-line then typically expands to a two-phase field as a new continuous reaction takes over. If the compositional range of any single solid-solution mineral shrinks to a point (as did chloritoid in Figure 28.10), the next step is for it to disappear entirely in a "terminal" discontinuous reaction, resolving to three phases defining the sub-triangle that surrounds it. Discontinuous reactions control the ultimate (intrinsic) stability of minerals. If bulk compositions are suitably diverse, discontinuous reactions will be the major regional isograd-forming reactions, and continuous reactions operate within the zones. As we have seen, however, continuous reactions can also generate new minerals in rocks of varying compositions at different grades and are the most common types of reactions. Their predominance is demonstrated in the literature, where, other than the relatively common Al_2SiO_5 polymorph reactions, discontinuous reactions contribute a minority of the phase assemblage boundaries in published pseudosections (e.g., only four such boundaries in Figure 28.14, and many pseudosections have none whatsoever—e.g., Figure 28.24).

You are discovering that three types of diagrams are useful and are generally used in conjunction. Petrogenetic grids, such as Figure 28.2, are complex, but useful in keeping track of the major reactions (typically discontinuous ones), but any single rock generally fails to manifest more than a few of these many reactions. Pseudosections, on the other hand, show all of the mineral changes a particular rock should experience, governed by much fewer pertinent reactions, but are limited to only a single bulk composition (or linear range on *T-X* or *P-X* types). Compatibility diagrams such as *AFM*, *AKF*, etc. illustrate the variations in mineral assemblages for a variety of rock types, but only for a limited range of metamorphic grade. Each diagram thus has different strengths and weaknesses, so the best approach seems to keep a *P-T* grid as reference and use a set of pseudosections for important bulk compositions and a series of compatibility diagrams to see how reactions of differing variance affect a variety of different rock compositions.

The migration of three-phase sub-triangles and creation of new ones due to continuous and discontinuous reactions, respectively, can be fascinating. Powell et al. (1998), Spear (1999), and Connolly (1990) have designed their thermodynamic algorithms (THERMOCALC, GIBBS, and Perple_X, respectively) to generate *AFM* diagrams at any *P* and *T* on their petrogenetic grids (or pseudosections). Closely spaced diagrams along a *P/T* trajectory can then be saved and combined to create wonderful (as well as informative) compatibility diagram animations, complete with migrating sub-triangles and tie-line reorganizations at discontinuous isograds. You can download some example animations by Roger Powell from his THERMOCALC web site (www.earthsci.unimelb.edu.au/tpg/thermocalc/). I keep copies of some on my web site as well (www.prenhall.com/winter/).

Now that we have an understanding of these processes and graphical aids, let's continue along our medium *P/T* path and examine the remaining important isograd reactions of the Barrovian series.

28.2.6 The Kyanite Zone

As we follow the medium *P-T* field gradient in Figures 28.2, the next reaction encountered (at point 8, ~630°C) is:

$$St + Chl\ (+\ Ms + Qtz) = Ky + Bt\ (+H_2O) \quad (28.15)$$

In this discontinuous reaction the two-phase St-Chl field (see Figure 28.11), now shrunk to a single tie-line by Reaction (28.14a) with increasing grade in Figure 28.13d, disappears as it flips with the crossing tie-line Ky-Bt (Figure 28.15). Although kyanite is theoretically stable in very Al-rich pelites at much lower grades because of the dehydration of pyrophyllite (as we have seen), the *AFM* diagram in Figure 28.14 shows how Reaction (28.15) introduces kyanite into many common pelites (shaded).

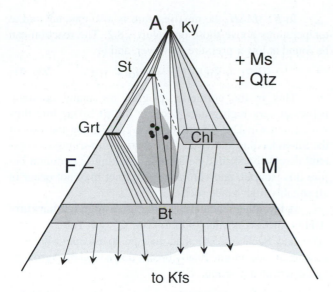

FIGURE 28.15 *AFM* compatibility diagram (projected from Ms) for the **kyanite zone, amphibolite facies**, showing the tie-line flip associated with Reaction (28.15) that introduces kyanite into many low-Al common pelites (dark shaded). After Carmichael (1970).

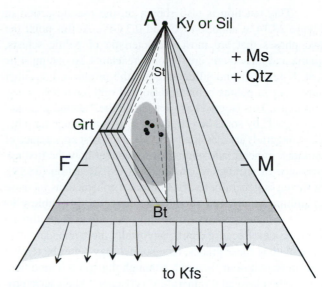

FIGURE 28.16 *AFM* compatibility diagram (projected from Ms) above the sillimanite and staurolite-out isograds, **sillimanite zone, upper amphibolite facies**.

The *solidus curve* for H_2O-saturated granitic compositions is also encountered in Figure 28.2 at about this metamorphic grade. This means that rocks containing quartz, plagioclase, and K-feldspar, if excess H_2O is available, will begin to melt. Many pelites are of a suitable composition for melting to occur, but H_2O may be bound in the micas and not available in excess as a free intergranular fluid phase. Melting may still be possible in fluid-absent rocks above the solidus temperature, but the amount of melt generated depends on the amount of H_2O liberated by dehydration reactions and made available to enter the melt (as discussed in Sections 7.5.1, 10.2.3, and 18.3). The free H_2O content of metamorphic rocks can vary from H_2O-saturated to nearly dry. The latter is much more common according to Yardley and Valley (1997), so melting rarely occurs at this point in pelites because of a lack of excess H_2O. We shall return to melting of pelites again shortly.

Above the kyanite isograd, the Chl-Ky-Bt triangle continues to migrate toward higher Mg/Fe caused by a continuous reaction:

$$Chl\ (+\ Ms\ +\ Qtz) = Ky + Bt\ (+H_2O) \quad (28.16)$$

At these grades, the compositional range of chlorite shrinks dramatically with increasing temperature toward the pure Mg end-member, and then Mg-rich chlorite disappears in the *KMASH* system at the Mg-Chl = Phl + Ky reaction just above point 8 in Figure 28.2. Chl thus finally disappears from the *AFM* diagram, yielding biotite and kyanite (which it plots between).

The staurolite-out isograd in common pelites is due to the reaction:

$$St\ (+\ Ms\ +\ Qtz) = Grt + Bt + Als + H_2O \quad (28.17)$$

which occurs at point 9 in Figure 28.2. The *P-T-t* path for the pelite modeled in the Figure 28.14 pseudosection intersects this reaction at about 660°C and 1 GPa. This is another terminal reaction, and St simply disappears from the *AFM* diagram at

this isograd because it plots within the Grt-Bt-(Ky or Sil) subtriangle (Figure 28.16) and is typically subordinate to muscovite and quartz in metapelites. This may introduce an Al_2SiO_5 polymorph into a range of low-Al pelites.

28.2.7 The Sillimanite Zone

The sillimanite isograd is encountered along the *P-T-t* path at point 10 in Figure 28.2 ($\sim$690°C), which is the polymorphic transformation:

$$Ky = Sil \quad (28.18)$$

As discussed in Section 26.11, the actual mechanism for this reaction is not as simple as the reaction itself, and sillimanite rarely replaces kyanite directly as the reaction implies. More commonly sillimanite nucleates as tiny *fibrolite* needles on micas, and the reaction may occur as a series of interrelated reactions involving (limited) mass transfer. Similar arguments may be made for several reactions. Thermodynamically, however, regardless of the mechanism, sillimanite ought to occur at the grade of the reaction indicated in Figure 28.2. The only change in the *AFM* diagram caused by Reaction (28.18) is the replacement of Ky by Sil at the A apex.

Because the curve for Reaction (28.18) crosses that of Reaction (28.17) near the chosen medium *P/T* trajectory in Figure 28.2, either curve could be encountered first (depending on the pressure), and the Al_2SiO_5 mineral generated by Reaction (28.17) can therefore be either sillimanite (as for the medium *P/T* path in Figure 28.2) or kyanite (as for the slightly higher *P/T* path in Figure 28.14).

28.2.8 Changes Above the Kyanite → Sillimanite Isograd

At point 11 in Figure 28.2 ($\sim$790°C) the medium *P/T* gradient intersects the important high-grade pelitic reaction (see also Figure 28.2):

$$Ms + Qtz = Kfs + Sil + H_2O \quad (28.19)$$

This reaction is also shown on the pseudosection in Figure 28.14 at about 640°C and 0.3 GPa. At this point the two phases that are most characteristic of pelitic schists, quartz and muscovite, are the only reactants, so one must be lost. K-feldspar and sillimanite are the products, but either can occur in pelites below the isograd (although Kfs is less common). This isograd has been called the "second sillimanite isograd" by some investigators. This is a strange term because isograds are classically based on the first appearance of an index mineral, and one is left to wonder how the first appearance of sillimanite can occur twice! Reaction (28.19), however, is generally responsible for a conspicuous increase in sillimanite and may produce the first readily visible sillimanite in hand specimen (fibrolite is too tiny to see without a microscope). Because quartz is typically more abundant than muscovite at these grades, this isograd may better be termed the "muscovite-out" isograd, although this too is a modification of the classical ("mineral-in") concept. Others have proposed the "K-feldspar + sillimanite isograd," which I like.

So above this isograd muscovite is typically lost, sillimanite is more plentiful, and K-feldspar is common in pelitic rocks. H_2O released by the reaction may induce partial melting because the H_2O liberated at this point is above the wet granite solidus. This may well be the case in the crosshatched area in the lower-left corner of Figure 28.14, where the loss of muscovite via Reaction (28.19) (which occurs at lower grades as pressure drops: Figure 28.2) produces K-feldspar and probably melt. If H_2O escapes (by streaming along fractures, being flushed by more deeply derived CO_2, or being withdrawn into melts) the remaining dry rocks may continue to be heated to higher metamorphic grades. Some petrologists consider Reaction (28.19) to be the transition from the amphibolite facies to the **granulite facies** in pelitic rocks. The granulite facies is developed principally in ancient deep- to mid-level continental crust that has been substantially dehydrated (and probably melt depleted, White and Powell, 2002). Another popular approach is to define the granulite facies threshold as the first appearance of orthopyroxene in quartz-bearing rocks (see below).

The loss of muscovite in pelites via Reaction (28.19) has two effects. First, the loss of mica and development of abundant feldspar causes the rocks to become less schistose and appear more granular or gneissic. Second, *AFM* diagrams above this grade must be projected from K-feldspar and not muscovite (see Section 24.3.4), causing the positions of many phases (most noticeably biotite) to shift.

At higher temperatures (and typically somewhat lower pressures) **cordierite** appears in pelites. Cordierite is Mg-rich, and in the *KMASH* system, the reaction:

$$Phl + Sil (+ Qtz) = Mg\text{-}Crd (+ Kfs + H_2O) \quad (28.20)$$

causes cordierite to appear on the *A-M* side of the *AFM* diagram between phlogopite and sillimanite. The curve for Reaction (28.20) is dashed in Figure 28.2 (no Fe) and is intersected by the medium *P/T* field gradient at very high temperatures (beyond the temperature limit of the figure) and typically at lower pressures. Cordierite occurs only in unusually magnesian pelites by this reaction, however, and is separated from common pelites by the sillimanite–biotite field.

In *KFMASH*, the next reaction is also encountered at temperatures above those on Figure 28.2. The reaction can be found at lower pressure, however, and is:

$$Bt + Sil = Grt + Crd + H_2O \quad (28.21)$$

This is the cordierite isograd for many common pelites at very high grades because the Bt-Sil tie-line flips with a new Grt-Crd tie-line (Figure 28.17), and the loss of the Sil-Bt tie-line allows the Crd-bearing sub-triangles to extend across the shaded pelite area. Melting is common before this isograd, and cordierite + garnet may not occur in all suitably high-grade rocks.

A few other reactions occur in **ultra-high temperature (UHT)** metasediments (900 to 1100°C, see Harley, 1998, 2004 and Section 25.4) that manage to escape wholesale melting. In one, biotite finally breaks down to **orthopyroxene** by the terminal reaction:

$$Bt + Qtz = Opx + Kfs + H_2O \quad (28.22)$$

This reaction occurs more typically in less aluminous quartzo–feldspathic rocks (meta-arkoses or granitoids) than in pelites, but may affect pelites, and I have included the equilibrium curve on Figure 28.2 (hi-*T*/low-*P* corner). Orthopyroxene in quartzo-feldspathic rocks is a common indicator of the granulite facies, where the rocks typically take on a greenish hue. Granulite facies rocks are generally reduced in LIL elements compared to their amphibolite facies equivalents (LILs are probably drawn into the departing fluid phase and/or melts). Mineral assemblages that occur in metasediments at even higher grades include *sapphirine* + quartz, *spinel* + quartz, and *osumilite* + garnet. These unusual rocks are typically developed in Precambrian terranes where considerable heat was carried by dry igneous bodies (such as anorthosite complexes, Chapter 20) from the mantle to the lower and middle continental crust. For petrogenetic grids and pseudosections that address these phases see Kelsey

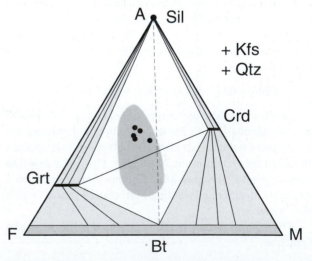

FIGURE 28.17 *AFM* compatibility diagram (projected from K-feldspar) above the cordierite-in isograds, **granulite facies**. Cordierite forms first by Reaction (29.14), and then the dashed Sil-Bt tie-line is lost and the Grt-Crd tie-line forms as a result of Reaction (28.17).

et al. (2004), and White et al. (2007). We will return to the very high temperature changes that occur in pelites in Section 28.5 when we discuss partial melting in more detail.

28.2.9 The Effects of Non-*KFMASH* Components

Although SiO_2, Al_2O_3, K_2O, FeO, MgO, and H_2O (*KFMASH*) predominantly control mineral parageneses in metapelites, other components may play significant roles. CaO, Na_2O, and MnO are probably the most important in this regard, and *P-T* grids and pseudosections for *MnNCKFMASH* can be found in the literature (e.g., Vance and Mahar, 1998; Tinkham et al., 2001; Zuluaga et al., 2005). Na_2O will enter and stabilize plagioclase and paragonitic white mica, and CaO will likewise affect garnet, plagioclase, epidote/zoisite, and margaritic white mica. In their theoretical treatment of the *NCKFMASH* system, Worley and Powell (1998) found that margarite and paragonite occurred only in unusually Mg-rich compositions (for the *P-T-X* range they addressed) and that the presence of plagioclase did not radically alter the stability of the *KFMASH* phases. Detailed systematic studies of natural mineral stabilities and compositions are presently lacking. From a theoretical perspective, both Na_2O and CaO bond with Al_2O_3 and SiO_2 in the phases they enter, which reduces the availability of these elements to *KFMASH*, thereby shifting

X_{bulk} on *KFMASH* compatibility diagrams away from the high-Al apex. To compensate for plagioclase, suitable compatibility diagrams (and input to thermodynamic computer programs) can be created in *KFMASH* by projecting X_{bulk} from anorthite and albite (subtracting from Al_2O_3 a molar amount equivalent to that of CaO and/or Na_2O) as done for Na_2O by Tinkham et al. (2001). Accounting for CaO in other phases is more challenging. Because CaO and Na_2O occur in relatively minor concentrations (Table 28.1) the effects are generally ones of detail rather than substantive. White et al. (2000) also calculated the effects of TiO_2 and Fe_2O_3 in pelites in the greenschist and amphibolite facies, finding that small to moderate amounts of these components had little effect on silicate mineral equilibria at these grades (and we get to add *KFMASHTO* to our acronym list!).

Manganese, on the other hand, can have a significant effect on the stability of garnet, particularly as it extends to lower grades (Symmes and Ferry, 1992; Mahar et al., 1997). Figure 28.18 is a pseudosection created by Tinkham et al. (2001) that illustrates this effect. The pseudosection is for the *KFMASH* system (molar Mg# 47), which indicates that garnet becomes stable at about 580°C and over 0.8 GPa, whereas they calculated garnet stability in *MnKFMASH* as low as 420°C at 0.95 GPa and 520°C down to 1.5 GPa (stippled overlay on Figure 28.18). Clearly, we must address Mn when considering garnet stability in metapelites.

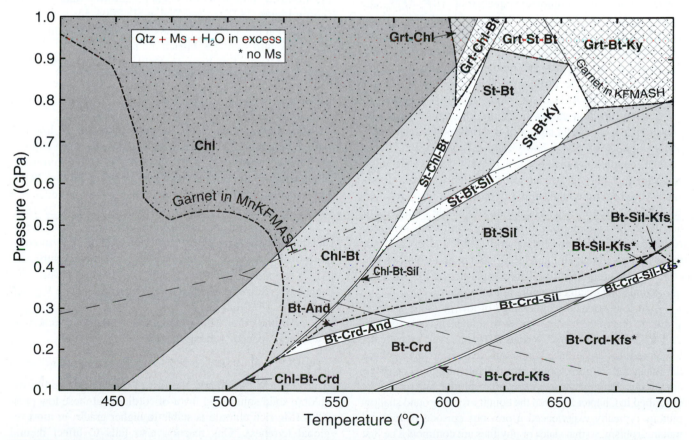

FIGURE 28.18 *P-T* pseudosection in *KFMASH* for X_{bulk}: Al_2O_3 = 45.80, FeO = 21.93, MgO = 19.59, and K_2O = 9.01 (in mol. %), calculated using the program THERMOCALC by Tinkham et al. (2001). The crosshatched area in the upper right is the stability range of garnet in *KFMASH*. The dashed curve is an overlay of the stability limit of garnet in *MnKFMASH* (after Tinkham et al., 2001).

28.3 LOW *P/T* METAMORPHISM OF PELITES

Recall from previous chapters that, as pressure lowers, *P/T* field gradients can range continuously from the medium *P/T* Barrovian sequence, to lower *P/T* Buchan or Abukuma-type regional metamorphism, to "regional contact" metamorphism, to contact aureoles around shallow plutons. Metamorphism along the low *P/T* trajectory in Figure 28.2, taken from the middle of the low *P/T* facies series in Figure 25.3, provides an example of the type of metamorphism that characteristically occurs in contact aureoles, or in regions where high heat flow accompanies copious plutonic activity, and is presented as a low-*P* contrast to the previous discussion. A rigorous exploration of every conceivable variation in *P/T* is well beyond the scope of our present endeavor. The facies series developed (Figures 25.2 and 25.3) along such a low *P/T* gradient is typically (zeolite → prehnite–pumpellyite) → albite–epidote hornfels → hornblende hornfels → pyroxene hornfels (rarely extending to sanidinite or ultra-high-temperature metamorphism).

Unlike metamorphosed mafic rocks, the mineralogy of pelites is relatively pressure sensitive and can be used effectively to distinguish a variety of lower-pressure metamorphic gradients (e.g., Pattison and Tracy, 1991). We analyzed the medium *P/T* sequence in considerable detail above to illustrate the ways that minerals can appear and disappear in metapelites due to discontinuous and continuous reactions. Rather than be repetitious, we shall concentrate on significant differences between the low *P/T* and medium *P/T* parageneses. For a more complete sequence of *AKF*, *AFM*, and *AKM* diagrams for the medium, low, and high *P/T* gradients, see Chapter 13 of Spear (1993), or Chapter 7 of Bucher and Frey (2002), or generate your own sequence of *AFM* diagrams for virtually any *P/T* gradient using Frank Spear's program GIBBS (see Problem 1 for this chapter), Roger Powell's THERMOCALC, or Jamie Connolly's Perple_X.

Many reaction curves in Figure 28.2 converge at low pressures to a confusing cluster of invariant points (especially in the vicinity of 520 to 540°C and 0.1 to 0.2 GPa). Reactions at low pressure thus occur in closely spaced succession and vary in nature and sequence with slight pressure differences. The proximity and similarity in slope of many reaction curves leaves their relative positions vulnerable to uncertainties in the thermodynamic properties of the minerals involved. Neither the Spear and Cheney grid (1989 and update) nor those by Powell and Holland (1990), Xu et al. (1994), or Wei et al. (2004) are perfectly reliable at predicting the sequence of isograds that occurs in nature at low pressures.

At low grades the mineralogy of low-pressure pelites is similar to that of the medium-pressure equivalents (see Sections 28.1 to 28.2.2). Of course, various *P/T* paths must converge at low grades toward surface conditions, so the pressure variance diminishes. Low-grade pelites are typically argillites or slates, containing phengitic muscovite, quartz, chlorite, and biotite. As described in Chapter 21, even the country rocks around shallow plutons typically experienced a previous episode of regional metamorphism, so that slates or phyllites are common. The first recognizable isograd encountered in many pelitic contact aureoles is the formation of **biotite** and muscovite via Reaction

(28.2) at about 440°C in the albite–epidote hornfels facies (point 12 in Figure 28.2 at ~430°C). These new minerals are generally coarser and more randomly oriented than the fine phyllosilicates found outside the aureole. **Chloritoid** may form at slightly higher grades in Fe-rich pelites caused by Reaction (28.3). Also, Fe-chlorite breaks down to Bt + Cld just above 500°C in Figure 28.2.

The major mineralogical differences that distinguish low-pressure metapelites from higher-pressure types are the occurrences of andalusite and/or cordierite. The compositional range of the low-temperature phase, chloritoid, shrinks as the most Mg-rich end of the range breaks down to And + Chl due to a continuous reaction (the resulting migration of the Cld-And-Chl triangle is indicated by a block arrow in Figure 28.19a). **Andalusite** may be introduced into some Al-Fe-rich pelites by this reaction. Although andalusite is stable at lower grades as a result of the breakdown of pyrophyllite [Reaction (28.1) at low pressure], it is rarely developed in natural rocks at that stage.

The Chl-Bt-Cld sub-triangle also opens and migrates toward more Mg-rich compositions, indicated by the right-pointing block arrow in Figure 28.19a. Eventually, the Chl + Cld field of tie-lines shrinks to a single tie-line and then disappears as:

$$\text{Chl} + \text{Cld} \, (+ \, \text{Ms}) \rightarrow \text{Bt} + \text{And} \, (+ \, \text{Qtz} + H_2O) \quad (28.23)$$

(The dashed tie-line in Figure 28.19b flips to the solid crossing tie-line.) This reaction occurs shallower than 0.2 GPa (not in Figure 28.2, but see Wei et al., 2004, for a more detailed grid at low pressure). This reaction may also introduce andalusite into some Fe-Al-rich metapelites.

Cordierite first becomes stable in the *KMASH* system due to the reaction:

$$\text{Mg-Chl} + Al_2SiO_5 \, (+ \, \text{Qtz}) = \text{Mg-Crd} \, (+ \, H_2O) \quad (28.24a)$$

which occurs at about 450°C at point **13** along the low *P/T* path in Figure 28.2. Cordierite thus appears between chlorite and andalusite along the *AM* side of the *AFM* diagram in the albite–epidote hornfels facies (not shown in Figure 28.19, but it first establishes Crd as a point along the A-M edge at a grade above that of Figure 28.19a and below that of 28.19b).

At this grade andalusite and cordierite are both intrinsically stable, but rarely present in metapelites due to compositional restrictions. Natural pelites typically lack sufficient Al_2O_3 for andalusite at this grade, and they are never sufficiently Mg rich to be affected by cordierite (along the Mg-rich end of the *AFM* diagram).

Around point 14 in Figure 28.2, a number of reactions are closely spaced. Among them is:

$$\text{Mg-Chl} \, (+ \, \text{Ms} + \text{Qtz}) = \text{Mg-Crd} + \text{Phl} \, (+ \, H_2O) \quad (28.24b)$$

which (in *KMASH*) marks the loss of the Mg-rich end of the chlorite solid solution in favor of cordierite at these low pressures (Mg-rich chlorite is stable to higher grades in most regional terranes). This reaction also fails to affect natural pelites, but, as grade increases the compositional range of cordierite then expands toward more Fe-rich compositions as

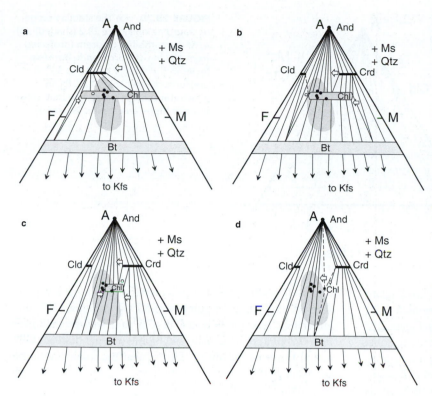

FIGURE 28.19 Schematic *AFM* compatibility diagrams (projected from Ms) for low *P/T* metamorphism of pelites. **(a)** Cordierite forms between andalusite and chlorite along the Mg-rich side of the diagram via Reaction (28.24b) in the **albite–epidote hornfels facies**. Chloritoid has formed earlier [via Reaction (28.4)] and the Chl-Cld-Bt sub-triangle migrates toward the right (block arrow) while the Chl-Cld-And sub-triangle migrates toward the left. **(b)** Chl migrates off the A-M edge to form a Chl + Bt + Crd sub-triangle via the continuous version of the same Reaction (28.24b) in *KFMASH*. The compositional range of chloritoid and chlorite are reduced and that of cordierite expands as the And-Chl-Crd and Chl-Crd-Bt sub-triangles all migrate toward more Fe-rich compositions. The Chl + Ctd area shrinks to a single-tie-line (dashed), which then "flips" to the crossing And + Bt tie-line. Andalusite and cordierite may be introduced into the shaded region of pelite compositions by these combined processes. **(c)** Migration of the Chl-And-Bt sub-triangle to the left (arrow) results from the discontinuous reaction Chl (+ Ms + Qtz) → And + Bt in the lower to mid **hornblende hornfels facies**. **(d)** Chlorite is lost in Ms-bearing pelites as a result of Reaction (28.25). Partially created using the program GIBBS (Spear, 1999).

both Reactions (28.24a) and (28.24b) become continuous in *KFMASH*, and the And-Chl-Crd and Chl-Bt-Crd sub-triangles first open and then migrate toward the left in *AFM* compatibility projections (arrows in Figure 28.19b). Reaction (28.24a) involves the leading edge of Cld extension to more Fe-rich compositions, and Reaction (28.24b) follows the receding stability of Mg-richer chlorite. Both reactions have the effect of introducing cordierite into progressively more Fe-rich pelites.

Notice how sub-triangle migration in Figure 28.19b, c, and d converges on the shaded range of pelite bulk compositions, introducing andalusite or cordierite into them. The continuous migration of the And-Chl-Crd sub-triangle due to Reaction (28.24a) indicates that the grade of the first appearance of andalusite (an *Fe-Mg-free* mineral) may commonly (and ironically) be controlled by the bulk-rock Fe/Mg ratio.

At low pressure the compositional range of chlorite shrinks dramatically over a short temperature range, and, at about 20°C over the grade of Reaction (28.24b), it shrinks to a point, and the terminal reaction for chlorite (in rocks with abundant Ms and Qtz) is encountered:

$$Chl\ (+\ Ms + Qtz) = Crd + And + Bt\ (+H_2O)\quad (28.25)$$

Because chlorite falls in the triangle formed by the reaction products, it disappears entirely from the *AFM* diagram in Figure 28.19d. Chlorite is thus lost in low-*P* (typically quartz-muscovite-excess) metapelites in about the middle of the hornblende hornfels facies. The loss of Chl on the *AFM* diagram indicates that the entire shaded area of common pelite compositions corresponds to either And-Bt, And-Bt-Crd, or Crd-Bt, in order of increasing Mg/(Mg + Fe).

The *cordierite* and *andalusite* isograds in the (more common) lower-Al pelite types can thus occur as a result of

multiple continuous or discontinuous reactions, depending on the bulk composition. For the composition represented by the pseudosection in Figure 28.14, note that at 0.2 to 0.3 GPa, heating causes andalusite to join the initial Chl + Bt assemblage around 520°C before chlorite is lost less than 10°C hotter (can you figure out why from Figure 28.19?). Then cordierite joins Bt + And. This requires that Reaction (28.25) takes place *without affecting this rock bulk composition*, so it is not shown in the pseudosection, but then the new Chl + Bt + And sub-triangle in Figure 28.19d migrates to the left until it overtakes our X_{bulk}. In many metamorphic aureoles (e.g., at Skiddaw, Section 21.6.4), andalusite and cordierite form as conspicuous ovoid porphyroblasts (spots) in slates and phyllites at about the same distance from the contact, presumably as a result of the close spacing of Reactions (28.23) to (28.25). Figure 28.19 indicates how the relative appearance of andalusite versus cordierite depends on Mg/Fe as well as upon Al/(Mg + Fe).

Cordierite has a relatively high molar volume, typically placing it at the low-*P*–high-*T* side of reactions with shallow *dP/dT* Clapeyron slopes. At low pressures, cordierite thus becomes stable at much lower temperatures (450 to 500°C) than in Barrovian-type metamorphic terranes (over 800°C) where rocks may melt first. This explains why cordierite is more common in low-pressure terranes.

At pressures greater than about 0.2 = GPa, a low *P/T* trajectory may enter the low-*P* tip of the range of **staurolite** stability (Figure 28.20a) via reaction:

$$Cld + And = St + Chl\ (+\ Qtz + H_2O)\quad (28.26)$$

Deeper than 0.2 GPa, this reaction (because of its steeper *P/T* slope) occurs at a lower grade than Reaction

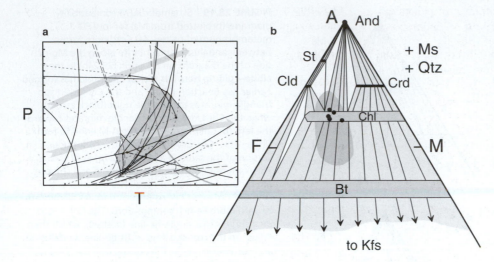

FIGURE 28.20 (a) The stability range of staurolite on Figure 28.2 (shaded). **(b)** *AFM* compatibility diagram (projected from Ms) in the **hornblende hornfels facies** in the vicinity of 530 to 560°C at pressures greater than 0.2 GPa, in which staurolite is stable and may occur in some high-Fe-Al pelites (shaded).

(28.24b). Staurolite would thus appear on the *AFM* diagram between Cld and And in Figure 28.19c if following a *P-T-t* path along the upper edge of the low *P/T* traverse in Figure 28.2 (see Figure 28.20b). Staurolite may thus occur in some high-Fe-Al pelites over a narrow range of temperatures. Staurolite is stable to ~700°C along the medium *P/T* trajectory, and to only ~560°C along the low *P/T* trajectory, and is no longer stable shallower than 0.1 GPa.

As with staurolite, some low-*P/T* trajectories may traverse the low-*P* tip of the *garnet* stability range (greater than ~0.15 GPa at these temperatures), and garnet may thus be generated by the breakdown of either chlorite or chloritoid (as described earlier for the medium-pressure traverse). The *AFM* projection for garnet-bearing rocks at this grade is similar to Figure 28.6 (but with cordierite present). Garnet is more stable at high pressures and is rare in the lower-grade zones of most low-pressure regional terranes and contact aureoles, even with the stabilizing effects of Mn (Figure 28.18) and Ca.

At these intermediate grades, the rocks become thoroughly recrystallized hornfelses, exhibiting decussate or granoblastic polygonal textures. Cordierite begins to break down, beginning with the Fe-richer varieties, at higher grades (~650°C) approaching the pyroxene hornfels facies. As the compositional range of Crd shortens in Figure 28.21a, the Crd-And-Bt sub-triangle encroaches onto the shaded pelite field. Muscovite then breaks down via Reaction (28.19) at point 15 in Figure 28.2, although the Al_2SiO_5 polymorph is andalusite, not sillimanite, at low pressures.

Almandine garnet occurs more commonly at low pressures in the vicinity of point 16 (650°C) in Figure 28.2, due either to a reaction between Fe-Crd and annite (shallower than 0.2 GPa) or andalusite and biotite (deeper than 0.2 GPa and at lower temperature). This is near the transition to the pyroxene hornfels facies.

At point 17 the reaction:

$$Bt + Als (+ Qtz) = Grt + Crd (+ Ms + H_2O) \quad (28.27)$$

results in the tie-line flip illustrated in Figure 28.21b and may introduce garnet into the pelites of some inner contact aureoles.

Above this temperature, the wet granite solidus is encountered and local **partial melts** may be generated. Because of the difference in the characteristic curvature of H_2O-saturated melting and dehydration reactions (Figure 28.2), the temperature of wet granite melting rises as pressure drops, whereas the temperature of the dehydration reactions falls. The muscovite and biotite breakdown reactions may thus occur before melting, so that orthopyroxene, spinel, etc. tend to be more common at low pressures. We must bear in mind, however, that the heat source for low *P/T* metamorphism is typically a rising plutonic body. Granite batholiths usually intrude at 700 to 800°C, so that the maximum temperature in the aureole may be limited. In the case of large basic intrusions, the temperature in the aureole may exceed 800 to 900°C. The slope of the And = Sill polymorphic transition is shallower than the granite solidus, so sillimanite may not be generated (point 18 in Figure 28.2) prior to melting in H_2O-saturated rocks.

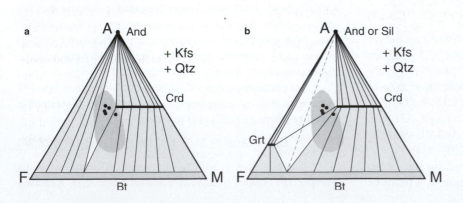

FIGURE 28.21 *AFM* compatibility diagrams (projected from Kfs) in the lowermost **pyroxene hornfels facies**. **(a)** The compositional range of cordierite is reduced as the Crd-And-Bt sub-triangle migrates toward more Mg-rich compositions. Andalusite may be introduced into Al-rich pelites. **(b)** Garnet is introduced to many Al-rich pelites via Reaction (28.27).

To summarize, low *P/T* metamorphism of pelites differs from medium *P/T* metamorphism in the following ways:

1. Reactions on the *P-T* grid are closely spaced and many intersect (particularly at low pressures), forming several invariant points. As a result new-mineral isograds may be tightly grouped and complex, varying significantly with *P*, *T*, and *X*.
2. Cordierite is a common mineral and occurs at low to intermediate grades.
3. Andalusite is the stable Al_2SiO_5 polymorph at low grades, giving way to sillimanite at very high grades.
4. Chlorite breaks down at lower grades.
5. Staurolite and garnet are less common. Staurolite does not occur at all below $\sim$0.2 GPa.
6. High-grade assemblages (orthopyroxene, spinel, etc.) are more likely to be developed before melting occurs.

A common sequence of isograds in a contact aureole of pelitic rocks is:

- The biotite isograd [Reaction (28.2)].
- The andalusite isograd [Reaction (28.23), or, more likely, the continuous reaction Chl $\rightarrow$ And + Bt].
- The cordierite isograd [perhaps most commonly due to the continuous version of Reaction (28.24b)].
- The andalusite-Kfs isograd [Reaction (28.19)].

And if the temperature is sufficiently high:

- The sillimanite isograd (And = Sil).
- The orthopyroxene isograd [Reaction (28.16)].

Localized anatexis may accompany either of the last two isograds, producing granitoid liquids (Figure 28.22). Partial melting of pelitic rocks can occur at temperatures as low as 650°C, if excess H_2O is available. Because these temperatures are readily attainable during high-grade metamorphism, we should investigate the equilibria involved. For a more detailed description of low-*P* metamorphism of pelites,

FIGURE 28.22 Veins developed in pelitic hornfels within a few meters of the contact with diorite. The vein composition contrasts with that of the diorite and suggests that the veins result from localized partial melting of the hornfels. Onawa aureole, Maine.

including a variety of low *P/T* gradients, see the reviews by Pattison and Tracy (1991) and Pattison and Vogl (2005).

28.4 PARTIAL MELTING OF PELITES

Spear et al. (1999) proposed a simplified petrogenetic grid for high-grade pelites in the *NKFMASH* system that addresses anatexis (Figure 28.23). This diagram is a high-*T* extension of Figure 28.2, with the addition of sufficient Na to stabilize plagioclase (albite) and a few simplifications (it ignores staurolite and several sub-solidus equilibria in order to concentrate on melting reactions). Spear et al. (1999) made three assumptions as to the behavior of typical high-temperature pelites that affect their melting:

1. The only H_2O available for melting is that derived from dehydration reactions (none is added externally).
2. In the subsolidus region a minute amount of vapor is nearly continuously evolved due to dehydration reactions, but most fluid leaves the rock (in spite of low porosity). Initial melting may thus occur at H_2O-saturated conditions, but the extent of such vapor-saturated melting is very limited.
3. Once melting begins, any H_2O evolved by dehydration reactions immediately dissolves into the melt phase and remains in the rock as long as liquid is not removed.

These assumptions reduce the numerous possible melting reactions to only a few because the only vapor-saturated melting reaction that is stable becomes the one at lowest temperature (the heavy curve in Figure 28.23). The fluid released by this reaction induces ("fluid-fluxed") melting and enters the melt, rendering subsequent reactions vapor-absent. The activity of H_2O (a_{H_2O}) is thus not externally controlled, but is buffered by equilibrium with hydrous silicates (cordierite and biotite).

Several experimental studies have investigated the melting of rocks with bulk compositions in the *NKFMASH* system. As Carrington and Harley (1995) pointed out, melting reactions, like many reactions we've discussed, are high-variance continuous reactions, and most are specific to a particular bulk composition. Spear et al. (1999) suggested that the reactions in Figure 28.23 are the predominant univariant reactions that limit the high-variance ones. Each of the univariant reactions in Figure 28.23 thus bounds a *P-T* field in which higher variance reactions occur. Spear et al. (1999) described in detail variations in Fe/Mg of the mafic minerals involved in these continuous reactions, and the implications this has for compositional zoning, particularly among retrograde garnets, where it might be preserved and used as an indicator of the *P-T-t* path of anatexis and cooling. We, however, shall restrict our attention to the melting reactions themselves.

Along the medium *P/T* metamorphic field gradient in Figure 28.23 (or at any pressure greater than invariant point 1), Spear et al. (1999) proposed that melting should occur only as a result of the vapor-excess muscovite breakdown:

$$Ms + Ab + Qtz + H_2O = Al_2SiO_5 + L \quad (28.28)$$

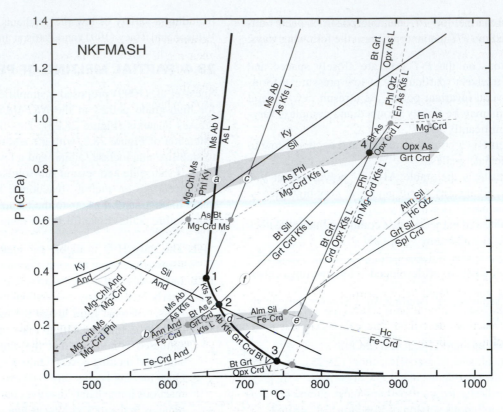

FIGURE 28.23 Simplified high-temperature petrogenetic grid, showing the location of selected melting and dehydration equilibria in the Na$_2$O-K$_2$O-FeO-MgO-Al$_2$O$_3$-SiO$_2$-H$_2$O (*NKFMASH*) system, with sufficient sodium to stabilize albite. Also shown are some equilibria in the *KFASH* (dotted) and *KMASH* (dashed) systems. Reactions are not balanced and commonly leave out Ms and Qtz, which are considered to be present in excess. The medium and low *P/T* metamorphic field gradients from Figure 28.2 (broad shaded arrows) are included for comparative purposes. The Al$_2$SiO$_5$ triple point is shifted, as shown, to 550°C and 0.45 GPa following the arguments of Pattison (1992), allowing for the coexistence of andalusite and liquid. *V* = H$_2$O-rich vapor, when present in fluid-saturated rocks. After Spear et al. (1999).

at point *a* in the *NKFMASH* system of Figure 28.23 (Ab is absent from the reaction in the *KFMASH* system and invariant point 1 rises to ~0.6 GPa and 730°C). Because of assumption (2) above, however, very little free aqueous fluid is available, so little melt will be generated at this reaction curve. At pressures less than invariant point 1 (such as point *b* on the low *P/T* gradient in Figure 28.23) muscovite-bearing rocks will dehydrate below the solidus (without melting) to produce alkali feldspar and Al$_2$SiO$_5$ via:

$$Ms + Ab + Qtz = Al_2SiO_5 + Kfs + H_2O \quad (28.29)$$

which is a Na-bearing analog to Reaction (28.15). Rocks heated at pressures greater than invariant point 1 will thus have alkali feldspar in small melt segregations, whereas those heated at pressures less than invariant point 1 will have alkali feldspar porphyroblasts.

Heating a rock along the medium *P/T* path will continue beyond point *a* with little or no melt (because of the limited vapor content). Continuous reactions involving muscovite, quartz, biotite, Al$_2$SiO$_5$, garnet, and K-feldspar may evolve minor liquid. Not until point *c*, however, will the first major melting reaction be encountered:

$$Ms + Ab + Qtz = Al_2SiO_5 + Kfs + L \quad (28.30)$$

Spear et al. (1999) calculated that the amount of melt generated by Reaction (28.30) is equal to about 70% of the volume of muscovite consumed. Typical pelites with 10 to 30 modal % Ms (at this grade) should thus develop 7 to 20% melt, corresponding to the first appearance of significant melt segregations in garnet–biotite–sillimanite–K-feldspar–plagioclase gneisses. Further liquid, along with orthopyroxene and cordierite, may be generated by the dehydration melting of biotite (due to several possible reactions) at ~850°C, well into the granulite facies. The experiments of Carrington and Harley (1995) constrained invariant point 4 in Figure 28.23 in the *KFMASH* system. The temperature at which biotite dehydration melting occurs is shifted toward significantly higher temperature by increased fluorine and titanium in biotite.

Phase relationships beyond point *b* along the *low P/T* gradient are very pressure specific. Cordierite may be generated by univariant Reactions (28.24a, 28.24b, or 28.25), all of which are below the minimum melting temperature (heavy curve) in Figure 28.23. Vapor-saturated melting occurs at point *d* due to the reaction:

$$Kfs + Ab + Grt + Crd + Bt + Qtz + H_2O = L \quad (28.31)$$

but, again, little melt is produced because of the low vapor content. Not until point *e* is significant melt produced by biotite breakdown via the reaction:

$$Bt + Grt + Qtz = Crd + Opx + Kfs + L \qquad (28.32)$$

At slightly higher pressure, between invariant points 1 and 2, initial limited vapor-saturated melting may occur by the reaction:

$$Kfs + Ab + Sil + Qtz + H_2O = L \qquad (28.33)$$

but significant melting is not generated until point *f* because of biotite breakdown via the reaction:

$$Bt + Sil + Qtz = Grt + Crd + Kfs + L \qquad (28.34)$$

Remember that because of the three assumptions described above, H_2O is limited, and its activity is buffered by the hydrous silicates present. The plethora of melting reactions at intermediate (externally controlled) H_2O activities and the continuous dehydration reactions are not shown, so the melting reactions described here can be considered as *minimum* melting equilibria in the *NKFMASH* system.

White et al. (2001, 2007) also modeled partial melting reactions in pelites using THERMOCALC. In the broadest sense, their results agree with those of Spear et al. (1999), but they cite other important dehydration/fluid-fluxed melt-generating reactions. Figure 28.24 is a pseudosection in *NCFMASH* from White et al. (2001) illustrating the mineral assemblages developed and the melt mode contours (volume fraction of melt generated, which is strongly influenced by the 20.4 mol. %H_2O they specified: equivalent to less than 3 wt. %). *P-T* pseudosections are simpler than grids in the sense that they portray the equilibrium phase assemblage at any *P* and *T* (and of course the boundaries that mark the transition from one assemblage to another). We can thus quickly recognize, for example, the minimum temperature of liquid (the vapor-saturated solidus), the upper-pressure stability limits of cordierite, and the lower-pressure stability limits of garnet. Figure 28.24 also reveals the predominance of high-variance (shaded) mineral assemblages and the complete lack of any univariant reactions affecting this X_{bulk} (other than the polymorphic Ky/Sill transformation). Continuous reactions determine ALL other mineralogical changes (and melt generation). Melt contours (dashed) are very

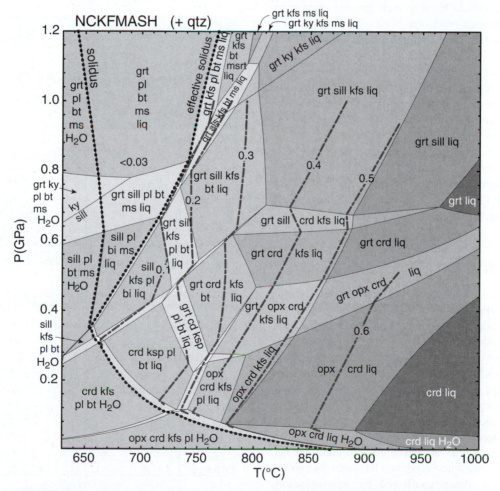

FIGURE 28.24 *P-T* pseudosection for the average pelite composition of Powell et al. (1998) with "representative" Na_2O and CaO added and just sufficient H_2O to saturate immediately subsolidus at 0.6 GPa. mol. % Al_2O_3 = 30.66, FeO = 23.74, MgO = 12.47, CaO = 0.97, Na_2O = 1.94, K_2O = 9.83, and H_2O = 20.39 (and quartz in excess). Solidus and melt mode (volume fraction of melt produced) overlain as contours. "Effective solidus" is when the melt fraction exceeds a few hundredths of a percent. After White et al. (2001).

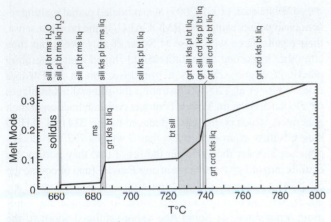

FIGURE 28.25 Melt mode produced (molar on a one-oxygen basis) upon heating at 0.5 GPa of the same X_{bulk} as in Figure 28.24. After White et al. (2001).

steeply sloped, reinforcing the idea that temperature is much more important than pressure in generating more melt. The contours tend to go through mineral assemblage fields, not along boundaries, indicating that melt production results largely from continuous reactions. Figure 28.25 shows the calculated melt fraction produced by isobaric heating of the same X_{bulk} at 0.5 GPa. Here, too, only a small amount of melt is produced at the vapor-saturated solidus (too little free fluid) and melt then increases progressively across the entire interval above the solidus (as H_2O is liberated by various continuous reactions). Several temperature intervals, however, are responsible for enhanced melt production. The first is involved with muscovite breakdown (~685°C). Three other melt-producing dehydration reactions occur over a small temperature range between 725−740°C [including Reaction (28.34)], collectively producing about 15% partial melting for this particular X_{bulk}.

28.5 MIGMATITES

Only dehydrated rocks and restites become dry and depleted granulite facies rocks. Many other rocks at high grades develop felsic segregations and are called **migmatites**. The term was coined by Sederholm (1907) and means "mixed rock," referring to a mixture of apparently igneous and metamorphic components. The rocks do indeed appear mixed, having a dark schistose component (the **melanosome**) that is intimately associated with light colored, coarser-grained, centimeter-scale layers, veins, or pods of poorly schistose material (the **leucosome**). The term **paleosome** (or **mesosome**) refers to material in migmatites that is intermediate in character between melanosome and leucosome, and has been interpreted to be original rock zones unaffected by migmatization. Migmatites appear to represent the culmination of high-grade metamorphism under more hydrous conditions than characterize the granulite facies. Migmatites are best developed in metapelites, but also occur in metamorphosed sandy and arkosic sediments, mafic rocks, and granitoids.

Migmatites are classified structurally on the basis of the relationship between the leucosome and melanosome. The most common types are:

Vein-type migmatites. The leucosome forms a fairly random network of distinct veins that separate irregular blocks of melanosome (Figure 28.26d).

Stromatic migmatites. The most common type, in which the leucosome forms concordant layers that commonly parallel the schistosity of the melanosome. The layers are rarely continuous, however, and typically die out and/or crosscut the melanosome at some point along their length (Figure 28.26e).

Nebulite. The leucosome occurs as irregularly shaped patches that grade into the melanosome (Figure 28.26h).

Agmatite. Numerous blocks of paleosome are surrounded by subordinate and relatively narrow veins of leucosome (Figure 28.26a). Agmatites are typically enclave-rich zones marginal to granitoid intrusions.

Other common structures are also illustrated in Figure 28.26. These include **net-like** structures similar to veins, but accompanied by shear (Figure 28.26b); **rafts** of material that are

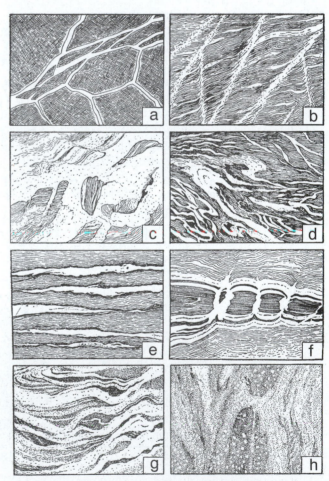

FIGURE 28.26 Some textures of **migmatites**. **(a)** Breccia structure in agmatite. **(b)** Net-like structure. **(c)** Raft-like structure. **(d)** Vein structure. **(e)** Stromatic, or layered, structure. **(f)** Dilation structure in a boudinaged layer. **(g)** Schleiren structure. **(h)** Nebulitic structure. From Mehnert (1968). Copyright © with permission from Elsevier Science.

similar to agmatite, but smaller, more rounded, and commonly sheared (Figure 28.26c); **dilation** structure, in which the leucosome fills openings in stretched competent layers (Figure 28.26f); and **schleiren**, stretched or sheared irregular streaks of melanosome that taper at the ends (Figure 28.26g). Schleiren may be found in several types of migmatites, most commonly in stromatic ones. For a more detailed classification, and a discussion of migmatite structures and textures, see Mehnert (1968) or Ashworth (1985), or see Fettes and Desmons (2007) for the IUGS-SCMR opinion.

The origin of migmatites has been controversial since the inception of the term. There are three principal theories:

1. Migmatites form by **injection** of granitic leucosome into dark high-grade schistose rocks.
2. Migmatites form by localized partial melting (**anatexis**). The first melts are granitoids, which compose the leucosome. The melanosome is generally considered to be the **restite**, or the somewhat refractory residuum from which the melts were extracted.
3. Migmatites are created by **metamorphic differentiation** or metasomatic growth of the leucosome, and melts are not involved.

The melanosome is generally more mafic than a typical pelite, and the leucosome is much more felsic. The combination, however, is similar to the paleosome and results in a broadly pelitic composition. In light of this finding, the injection hypothesis lost favor because it could not explain the more mafic character of the melanosome. Both of the other processes derived the leucosome from the melanosome, and they could not be distinguished on the basis of this criterion.

The debate over the igneous versus metamorphic origin of migmatites became part of a major feud (the "granite controversy") in the 1940s and 1950s concerning the origin of granitoids in general. Some champions of the metamorphic origin of the leucosome proposed that the process could occur on a scale large enough to produce metasomatic granite bodies of batholithic size, and that many granites were of metamorphic, not igneous, origin. When experimental petrologists demonstrated that granitic liquids could be generated by melting of H_2O-saturated pelitic sediments at temperatures as low as 650°C, the controversy was ended in favor of the magmatic side.

Although petrologists now ascribe an igneous origin to granites, the origin of the leucosome in migmatites is still controversial. The composition of the leucosome is generally more tonalitic than granitic, and thus does not always correspond to minimum-melts. Crystals of feldspar in nebulites are typically isolated and have the appearance of porphyroblasts, not of crystals formed from a melt. Many migmatites that I have observed have a complex structure that appears to have resulted from multiple events and processes, generally with several generations of crosscutting leucosomes (Figure 28.27). As can be seen in Figure 28.26, the textures and structures of migmatites are highly variable. The leucosome in some migmatites may represent partial melts, whereas it may be the product of metamorphic segregation in others. Some migmatites may contain both. Melts are likely suspects when the leucosome

FIGURE 28.27 Complex migmatite textures including multiple generations of concordant bands and crosscutting veins. Angmagssalik area, eastern Greenland. Outcrop width ca. 10 m.

approaches the minimum-melt composition in the Ab-Or-Qtz system (Figure 11.3), whereas metamorphic growth is more likely when the leucosome is dominated by plagioclase or alkali feldspar. In either case, migmatites represent high-grade metamorphic rocks in which fluids, and perhaps melts, played a substantial role. White et al. (2001) used a thermodynamic model for melts (in THERMOCALC) to analyze partial melting in pelites and concluded that leucosome compositions in many natural occurrences are not those expected of partial melts. They suggested that melt loss (including *from leucosomes*) is a common occurrence and may well be responsible for the lack of expected retrogression on cooling and the preservation of granulite facies rocks. For a recent review of melting and melt extraction processes, see Brown (2007b).

28.6 HIGH *P/T* METAMORPHISM OF PELITES

High-pressure metamorphism is characteristic of subduction zones and most commonly affects basic igneous rocks and fore-arc sediments such as graywackes. High *P/T* metamorphism of true pelites is much less common, and studies of such rocks from the Alps, Greece, and Indonesia have only recently been published.

At low temperatures, high *P/T* metamorphism of pelites is characterized by the presence of **talc**, the expanded stability of **phengite**-rich and **celadonite**-rich muscovite (at the expense of biotite), and **paragonite** (Na-rich white mica), plus the minerals **sudoite** (an Fe-Mg-poor, Al-rich chlorite) and **carpholite** (Fe, Mg, Mn) $Al_2Si_2O_6(OH)_4$. Phengite is considered by many to be the characteristic mica of high-*P*-low-*T* metapelites. **Garnet**, **chloritoid**, and **kyanite** are also common.

Sudoite occurs at very low grades and breaks down with increasing grade to form carpholite below 350°C and less than 0.8 GPa (Bucher and Frey, 2002). Neither is considered in Figure 28.2. Carpholite is a low-grade near-equivalent of chloritoid and breaks down to chloritoid + quartz + H_2O at modest grades.

Talc is introduced into Mg-rich pelites in the *KMASH* system by the reaction:

$$\text{Mg-Chl} + \text{Phl} = \text{Tlc} + \text{Pheng-Ms} + H_2O \quad (28.35)$$

The reaction (dashed in Figure 28.2) has a low dP/dT slope, so it is pressure dependent. Although the Mg-end member reaction suggests that talc + phengitic-muscovite is stable at pressures as low as 0.3 GPa in Figure 28.2, the slope and position of the reaction curve are sensitive to compositional changes (particularly the Al-content of chlorite) and occur at higher pressures for natural compositions.

Reaction (28.35) results in the appearance of talc on the *AFM* diagram between Mg-rich chlorite and biotite in Figure 28.28a. Talc appears in more typical pelitic bulk compositions at higher grade as the result of the reaction:

$$Bt + Chl = Tlc + Cld + H_2O \qquad (28.36)$$

which occurs at pressures greater than 1.5 GPa in Figure 28.2 (around point 20 in the eclogite facies along the high P/T metamorphic field gradient). As a result of this reaction, the Chl-Bt tie-line disappears as the Tlc-Cld tie-line forms (Figure 28.28b). Below this isograd the Chl-Bt tie-line separates Tlc from the shaded field of common pelite compositions, but after this reaction, the Tlc-Bt-Cld triangle extends across much of the shaded pelite field. Reaction (28.36) is thus responsible for the introduction of talc into many common pelites at high pressures.

Talc + phengite, talc + kyanite, talc + pyrope, and talc + Mg-rich chloritoid assemblages have been reported from several high-pressure blueschist and eclogite facies terranes (Abraham and Schreyer, 1976; Schreyer, 1977; Chopin, 1981). The predominance of talc + phengite imparts a distinctly white color to the rocks, which Schreyer (1977) called **whiteschists** (many are now interpreted as metasomatically altered granitic protolith).

Garnet forms prior to Reaction (28.29) in the *KFASH* system (point 19 in Figure 28.2) by either Reaction (28.7a) or the breakdown of Fe-Chl, which is why Grt already exists along the A-F side of the *AFM* diagram in Figure 28.28b, even though it does not occur in many metapelites at that grade. As the Grt-Cld-Bt triangle moves to the right in Figure 28.28b via a continuous reaction, all phases get richer in Mg and the triangle envelops more of the common pelite shaded area so that garnet appears in them. The blueschist–eclogite facies transition in *mafic* rocks occurs approximately between points 19 and 20 in Figure 28.2.

Chlorite is lost to talc + kyanite ($\pm$chloritoid) via several high-pressure reactions shown in Figure 28.2. Kyanite is thereby introduced, but only into very high-Al pelites, not into most common pelites, because the Cld-Ky-Tlc triangle does not overlap the shaded pelite area (Figure 28.28c). Kyanite probably occurs in most high-pressure metapelites when the Tlc-Cld tie-line gives way to the Bt-Ky tie-line via the reaction:

$$Tlc + Cld\,(+\,Ms) = Bt + Ky\,(+\,Qtz + H_2O) \quad (28.37)$$

at point 21 in Figure 28.2. At this point the Ky-Bt-Cld and Ky-Tlc-Bt sub-triangles form in Figure 28.28c, which encompass much of the shaded pelite region. The H_2O-saturated granite solidus indicates that melting may also occur near these grades if sufficient H_2O is present. The major change above this grade is the loss of chloritoid and the stabilization of garnet + biotite + kyanite above ~1.8 GPa.

At pressures over 2.0 GPa (well into the eclogite facies) the assemblage talc + phengite + pyrope + kyanite + a silica polymorph develops in metapelites. As described in Section 25.4, Chopin (1984) first described **coesite** as inclusions in pyrope in such high-pressure pyrope-kyanite-bearing whiteschists from the Dora Maira massif of the western Alps of Italy, leading to several more discoveries of **ultra-high-pressure (UHP)** crustal metamorphic rocks (Figure 25.13 is a map of presently known occurrences). Diamond inclusions in garnet from crustal gneisses from Kokchetav were then described by Sobolev and Shatsky (1990), indicating even higher pressures (>4 GPa, or 150 km). These ultra-high-pressure rocks are *crustal* rocks, including pelites, quartzites, marbles, gneisses, and granitoids. The impact of these discoveries was substantial because they provided a record of crustal rocks being subducted to considerable depths and returning to the surface in orogenic belts. For sediments to reach depths in excess of 100 km (without melting) is impressive and they must have taken the round-trip rapidly in order to remain cool and escape melting. Estimates for the time lapse between peak metamorphism and exposure range from 40 to 100 Ma (Coleman and Wang, 1995). For more complete high-pressure pelitic grids and pseudosections, see Wei and Powell (2003) and Proyer (2003). See Section 25.4 for a more extensive general discussion of UHP rocks.

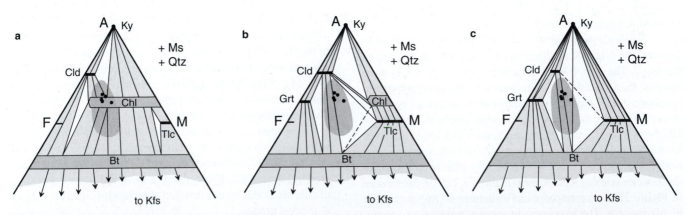

FIGURE 28.28 *AFM* compatibility diagrams (projected from muscovite) for the **eclogite facies** of high P/T metamorphism of pelites. **(a)** Talc forms between biotite and chlorite along the Mg-rich side of the diagram via Reaction (28.35). **(b)** At a higher grade, the Chl-Bt tie-line flips to the Tlc-Cld tie-line via Reaction (28.36). **(c)** After chlorite breaks down, the kyanite forms in many metapelites via Reaction (28.36). After Spear (1993).

Summary

Pelitic protoliths are mudstones and/or shales that typically develop as forearc wedge deposits off both active and passive continental margins. Because of the high Al_2O_3, K_2O, and SiO_2 contents of these sediments, muscovite and quartz predominate over most of the common crustal P-T range. They thus occur *in excess*, so their elimination would never limit the progress of any reaction. The stable mineral assemblage in meta-pelites is quite sensitive to temperature and pressure so that the mineralogy of metapelites varies considerably with the P-T conditions. This variety includes all the classic Barrovian index minerals, plus several others, depending on P, T, and X. The petrogenetic grid in the *KFMASH* system is thus rather complex.

Metapelites at low grades are typically argillites, slates, or fine-grained schists consisting of phengite, quartz, and chlorite. Along medium P/T paths, the principal changes as grade increases conform broadly to the Barrovian sequence of index minerals (plus perhaps chloritoid). Along lower P/T paths, cordierite and/or andalusite may become more prevalent than at medium pressures, whereas staurolite and garnet are less common. At higher pressures, talc, phengite, and several other white micas become more stable. Garnet, chloritoid, and kyanite are also common.

The mineral assemblage at any particular grade (and hence the sequence as grade changes) is quite sensitive to X_{bulk}. This is best understood when we look beyond P-T petrogenetic grids (Figure 28.2), which comprehensively address the full spectrum of possible univariant reactions capable of affecting metapelites, to *AFM* or *AKF* compatibility diagrams and pseudosections. Compatibility diagrams show the dependence of mineral assemblage on X_{bulk}, but are limited a specific grade. P-T pseudosections show the mineral assemblages that can be expected to develop in a pelitic rock at any P and T, but can do so only for a single specified X_{bulk}. A full understanding of progressive metamorphism of pelites (or any rock type, for that matter) is best accomplished by integrating all three types of diagrams. T-X and P-X pseudosections may also be generated; they allow a single compositional variable (Fe/Mg, Al/(Fe + Mg), Al/K etc.) while sacrificing an intensive physical variable (P or T).

The evolution of progressively metamorphosed rocks is governed by a combination of discontinuous and continuous reactions. Continuous reactions affect the compositional ranges of solid-solution minerals and occur over a range of temperature-pressure conditions, causing virtually all of the sub-triangles in *AFM* diagrams to migrate, consuming the tie-lines of contracting two-phase fields ahead of them and generating new tie-lines in expanding two-phase fields behind them. Minerals may develop in a specific bulk composition when a migrating sub-triangle first encompasses the X_{bulk} point. Likewise, minerals are lost when a migrating sub-triangle abandons the corresponding X_{bulk}. Once sub-triangle migration causes a two-phase field to shrink to a single tie-line, the next step is for that line to disappear, and it is replaced by a new tie-line that crosses it (a tie-line flip), reflecting a discontinuous reaction. The new tie-line then typically expands to a two-phase field as a new continuous reaction takes over. If the compositional range of any single solid-solution mineral shrinks to a point, it will then disappear entirely in a "terminal" discontinuous reaction, resolving to three phases defining the sub-triangle that surrounds it.

Most rocks tend to dehydrate progressively and a free aqueous fluid is greatly diminished or absent, particularly at the highest grades. The last bits of hydrous minerals (biotite, muscovite, or cordierite) are finally consumed, releasing some fluid, which, because temperatures are generally above the H_2O-saturated solidus at this point, immediately fluxes the host to produce melts. The melt amounts are usually small and migmatites result. Partial melts scavenge fluids, SiO_2, alkalis, and LIL elements. If melts are mobile and accumulate, they produce granites (particularly S-type), and reduced granulites may be left behind.

Key Terms

KFMASH
In-excess (phases) *611*
Terminal reaction *615*
Mineral-in versus mineral-out
  isograds *615*

Pseudosection *616*
Migmatite *629*
Melanosome/leucosome/paleosome
  (mesosome) *629*

Vein-type, stromatic, nebulite,
  agmatite *629*

Review Questions and Problems

Review Questions and Problems are located on the author's web page at the following address: **http://www.prenhall.com/winter**

Important "First Principle" Concepts

- The appearance or disappearance of any mineral with increasing metamorphic grade is typically due to reactions that occur within the broader stability range of that mineral. The grade at which a mineral appears or disappears in metapelites is usually governed by the composition of the rock (typically the Al-content or the Fe/Mg ratio) and may reflect either a discontinuous or a continuous reaction (but more commonly the latter).

- Continuous reactions, at least divariant ones, can be thought of as migrating three-phase triangles on *AFM* or *AKF* diagrams. Mineral changes thus occur for any specific X_{bulk} when such a migrating triangle overruns and initially encompasses (or alternatively abandons) that composition.

- Discontinuous reactions involve either the absolute disappearance of a phase or a tie-line flip. They affect only bulk compositions that plot on compatibility diagrams within the sub-polygon bounded by the phases involved.

- *P-T* petrogenetic grids (such as Figure 28.2) do a good job of representing the variety of reactions that govern mineral petrogenesis in a system, but they are complex and only display discontinuous reactions, most of which affect different small subsets of pelitic rock bulk compositions. Pseudosections do a much better job of addressing only the reactions, either continuous or discontinuous, that affect a particular rock but are limited by the compositions that they address.

- The facies and isograd concepts may impart the wrong impression that reactions affecting mineral parageneses are typically discontinuous and occur only at specific intervals. But reactions of high variance are happening continuously, so that equilibrium partitioning of elements, mineral modes, and tie-lines on compatibility diagrams are continually shifting. These shifts affect mineral changes in metapelites more commonly than the discontinuous reactions customarily associated with isograds.

Suggested Further Readings

Ashworth, J. R. (1985). *Migmatites*. Blackie. Glasgow.

Bucher, K., and M. Frey. (2002). *Petrogenesis of Metamorphic Rocks*. Springer-Verlag. Berlin. Chapter 7.

Frey, M. (1987). Very low-grade metamorphism of clastic sedimentary rocks. In: *Low Temperature Metamorphism* (ed. M. Frey). Blackie. Glasgow. pp. 9–58.

Frey, M., and D. Robinson (eds.). (1999). *Low-Grade Metamorphism*. Blackwell. Oxford, UK.

Mehnert, K. R. (1968). *Migmatites and the Origin of Granitic Rocks*. Elsevier. Amsterdam.

Pattison, D. R. M., and R. J. Tracy. (1991). Phase equilibria and thermobarometry of metapleites. In: *Contact Metamorphism* (ed. D. M. Kerrick). *Reviews in Mineralogy*, **26**. Mineralogical Society of America. Washington, DC. pp. 105–206.

Pattison, D. R. M., and J. J. Vogl. (2005). Contrasting sequences of metapelitic mineral-assemblages in the aureole of the tilted Nelson batholith, British Columbia: Implications for phase equilibria and pressure determination in andalusite-sillimanite-type settings. *Can. Mineral.*, **43**, 51–88.

Powell, R., and T. Holland. (1990). Calculated mineral equilibria in the pelite system, KFMASH (K_2O-FeO-MgO-Al_2O_3-SiO_2-H_2O). *Amer. Mineral.*, **75**, 367–380.

Proyer, A. (2003). Metamorphism of pelites in NKFMASH—A new petrogenetic grid with implications for the preservation of high-pressure mineral assemblages during exhumation. *J. Metam. Geol.*, **21**, 493–509.

Spear, F. S. (1993). *Metamorphic Phase Equilibria and Pressure-Temperature-Time Paths*. Monograph **1**. Mineralogical Society of America. Washington, DC. Chapter 10.

Spear, F. S., and J. T. Cheney. (1989). A petrogenetic grid for pelitic schists in the system SiO_2-Al_2O_3-FeO-MgO-K_2O-H_2O. *Contrib. Mineral. Petrol.*, **101**, 149–164.

Tinkham, D. K., C. A. Zuluaga, and H. H. Stowell. (2001). Metapelite phase equilibria modeling in MnNCKFMASH: The effect of variable Al_2O_3 and MgO/(MgO + FeO) on mineral stability. *Geol. Materials Res.*, **3**, 1–42.

Wei, C., and R. Powell. (2003). Phase relations in high-pressure metapelites in the system KFMASH (K_2O-FeO-MgO-Al_2O_3-SiO_2-H_2O) with application to natural rocks. *Contrib. Mineral. Petrol.*, **145**, 301–315.

Wei, C., R. Powell, and G. L. Clarke. (2004). Calculated phase equilibria for low- and medium-pressure metapelites in the KFMASH and KMnFMASH systems. *J. Metam. Geol.*, **22**, 495–508.

White, R. W., R. Powell, and T. J. B. Holland. (2001). Calculation of partial melting equilibria in the system Na_2O-CaO-K_2O-FeO-MgO-Al_2O_3-SiO_2-H_2O (NCKFMASH). *J. Metam. Geol.*, **19**, 139–153.

Wimmenauer, W., and I. Bryhni. (2007). Migmatites and related rocks. In: Met*amorphic Rocks: A Classification and Glossary of Terms* (eds. D. Fettes and J. Desmons). Cambridge University Press. Cambridge, UK. pp. 43–45.

Xu, G., T. Will, and R. Powell. (1994). A calculated petrogenetic grid for the system K_2O-FeO-MgO-Al_2O_3-SiO_2-H_2O. *J. Metam. Geol.*, **12**, 99–119.

Yardley, B. W. D. (1989). *An Introduction to Metamorphic Petrology*. Longman. Essex, UK. Chapter 3.

29

Metamorphism of Calcareous and Ultramafic Rocks

Questions to be Considered in this Chapter:

1. What mineral assemblages develop during progressive metamorphism of siliceous carbonates in contact aureoles and due to regional metamorphism?

2. How do fluids evolve in CO_2-H_2O mixtures when calcareous rocks are metamorphosed, and how does fluid release affect metamorphism in systems either open or closed to fluid flow?

3. What mineral assemblages develop during progressive metamorphism of ultramafic rocks at crustal levels?

29.1 METAMORPHISM OF CALCAREOUS ROCKS

Calcareous rocks are predominantly carbonate rocks, usually limestone (Ca carbonate) or dolostone (Ca-Mg carbonate). They typically form in a stable continental shelf environment along a passive margin, such as the present Bahama Banks off the southeastern United States. They may be pure carbonate, or they may contain variable amounts of other precipitates (e.g., chert or hematite) or detrital material (sand, clays, etc.). As the quantity of impurities increases, carbonates grade into calcareous clastic sediments (marls) and then into pelites or arenites. The spectrum from pure carbonate to purely clastic rocks is essentially complete. Although there is no definite cutoff, calcareous rocks include that part of the spectrum with a significant proportion of carbonate material. Carbonates typically become metamorphosed when the passive margin becomes part of an orogenic belt, either when it is transformed into an active margin by the development of a subduction zone inclined beneath the continent or when the margin enters a subduction complex as part of the subducting plate.

Metacarbonates are metamorphosed calcareous rocks in which the carbonate component is predominant. **Marbles** are metamorphic rocks that are nearly pure carbonate. When carbonate is subordinate, the metamorphic rock may be composed of Ca-Mg-Fe-Al silicate minerals, such as diopside, grossular, Ca-amphiboles, vesuvianite, epidote, wollastonite, etc. Metamorphosed rocks composed of these minerals are called **calc-silicate rocks**. A **skarn** is a type of calc-silicate rock formed by metasomatic interactions between carbonates and silicate-rich rocks or fluids. This may occur at the contact between sedimentary layers but is developed most spectacularly at the contact between carbonate country rocks and a hot, hydrous, silicate intrusion, such as a granite. See Rosen et al. (2007) for the IUGS-SCMR classification of calc-silicate rocks.

Metacarbonates compose only a small proportion of metamorphic rocks. Pure limestones can be awfully boring (though lovely) when metamorphosed, because calcite is stable over a wide range of conditions. Over the common range of medium and low *P/T* metamorphism, the only change in limestone marbles is an increase in grain size with grade (or in zones of fluid-enhanced recrystallization), and perhaps the development of a preferred orientation in some deformed rock bodies. They become more interesting only when metamorphosed at unusually high temperatures, when calcite breaks down to spurrite or larnite $+ CO_2$, or at high pressures, when it is transformed to aragonite.

Metamorphism of impure limestones and dolostones, on the other hand, is far more interesting and informative. A number of Ca-Mg silicates can form in metamorphosed siliceous dolostones under commonly attainable metamorphic conditions. As a result, they have received more attention than their limited occurrence would otherwise justify. At the incipient stages of metamorphism, impure carbonates consist of a mixture of carbonate minerals (calcite, dolomite, and ankerite are most common), perhaps some chert, and detrital material such as fine quartz, feldspar, and hydrous minerals (usually clays). Metamorphism of these rocks is interesting because it involves several devolatilization reactions (typically both decarbonation and dehydration reactions) dependent not only on P, T, and the rock composition, but also on the composition of the associated pore fluid. To understand the metamorphism of these rocks we will have to adopt an approach that deals with these fluids as H_2O-CO_2 mixtures (see Section 26.4).

In this chapter we will concentrate on impure carbonates in which the major initial constituents are dolomite, calcite, and silica (chert, detrital quartz, or SiO_2 in solution). Such rocks correspond nicely to the simple five-component system: CaO-MgO-SiO_2-H_2O-CO_2 (**CMS-HC**). When projected from H_2O and CO_2, the triangular **CMS** chemographic system depicts all the major minerals. Other components may also be present, typically FeO, Al_2O_3, K_2O, and Na_2O, but in most carbonate rocks they are present in only minor amounts and do not significantly alter the parageneses.

Figure 29.1 illustrates the **CMS** system and the minerals that typically develop in metamorphosed siliceous carbonates. A comprehensive analysis of metamorphism of these rocks requires that we consider the reactions between these phases as a function of metamorphic grade on a petrogenetic grid, and the changes that the reactions have on the topology of **CMS-HC** compatibility diagrams. This is the approach that proved effective when studying pelitic rocks in the previous chapter. Because the composition of the fluid phase is so important in metacarbonates, however, we shall

adopt a T-X_{CO_2} grid in favor of a T-P grid. To account for the effects of pressure, we can compare T-X_{CO_2} diagrams at different pressures.

We begin our analysis with low-pressure contact metamorphism of relatively simple siliceous dolostones. We will then model regional metamorphism by studying how increased pressure affects the system. As the amount of other constituents increases, the chemical composition and mineralogy becomes more complex and variable. We will attempt only a brief survey of some classic studies of calc-silicates as examples of how the approaches we adopt in the early sections can be extended to more complex systems.

29.1.1 Contact Metamorphism of Siliceous Dolostones

Eskola (1922) first described the sequence of mineral zones that typically develop in metamorphosed dolomitic marbles. In his classic paper on the subject, Bowen (1940) provided a more rigorous theoretical interpretation of the zonation. Although there are some differences between regional and contact metamorphism, the simplest sequence of zones that typically develops (in order of increasing grade) is:

- Talc zone
- Tremolite zone
- Diopside and/or forsterite zones

Figure 29.2 illustrates the isograds mapped by Moore and Kerrick (1976) in the dolostones of the Alta aureole, Utah. Diopside is rare at Alta, so a diopside isograd was not readily located. An inner periclase isograd occurs within a few tens of meters of the contact with the granodiorite of the Alta Stock. Other minerals, such as wollastonite, monticellite, merwinite, etc., may also form in the innermost portions of very hot aureoles.

In order to understand the development of these zones, we turn to the results of experiments in the **CMS-HC** system.

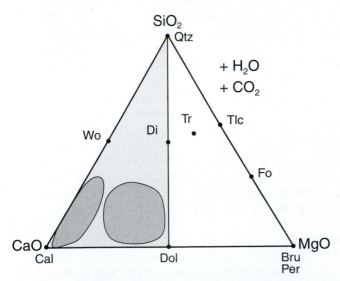

FIGURE 29.1 Chemographics in the CaO-MgO-SiO_2-CO_2-H_2O system, projected from CO_2 and H_2O. The dark shaded areas represent the typical composition range of limestones and dolostones. Because of the solvus between calcite and dolomite, both minerals can coexist in carbonate rocks. The shaded left half of the triangle is the area of interest for metacarbonates. Carbonated ultramafics occupy the right half of the triangle.

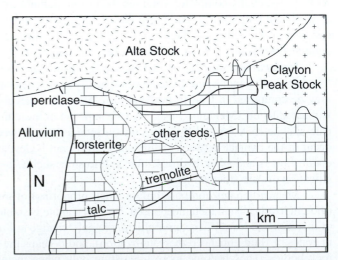

FIGURE 29.2 A portion of the Alta aureole in Little Cottonwood Canyon, southeast of Salt Lake City, Utah, where talc, tremolite, forsterite, and periclase isograds were mapped in metacarbonates by Moore and Kerrick (1976). Reprinted by permission of the *American Journal of Science*.

Figure 29.3 is an isobaric $T\text{-}X_{CO_2}$ phase diagram calculated at 0.1 GPa using Rob Berman's TWQ program (see Section 27.5), based on an internally consistent set of thermodynamic data extracted from experiments at high temperature and pressure. 0.1 GPa is equivalent to a depth of approximately 3 km, suitable for an example of contact metamorphism.

The assemblage Do + Cal + Qtz is stable in siliceous carbonates at temperatures ranging from sediments to the lowest metamorphic grades in Figure 29.3. As shown in the corresponding compatibility diagram (Figure 29.4a), this mineral assemblage develops in any bulk composition in the shaded (carbonate) half of the triangle (the right half covers ultramafic rocks). The first metamorphic reaction that occurs (and the temperature at which it takes place) depends on X_{CO_2} of the pore fluid. $X_{CO_2} < 0.7$ in most situations (the value of invariant point A in Figure 29.3), so the first isograd is typically:

$$3\,\text{Dol} + 4\,\text{Qtz} + H_2O = \text{Tlc} + 3\,\text{Cal} + 3\,CO_2 \quad (29.1)$$

This reaction is of type 4 in Figure 26.6, consuming H_2O and liberating CO_2 as temperature increases. It involves a tie-line flip (Section 26.8) in the *CMS* system whereby the low-temperature Dol-Qtz tie-line is replaced by the higher-temperature Tlc-Cal tie-line (compare Figures 29.4a and b). This reaction introduces **talc** into many metacarbonate rocks. Note in Figure 29.4b that virtually all dolomitic marbles above the isograd fall into the Dol-Tlc-Cal sub-triangle. Very Ca- and Si-rich carbonates fall in the Cal-Tlc-Qtz sub-triangle. Let's assume that we are dealing with a common siliceous dolostone, in which dolomite is the dominant phase, followed by calcite and finally quartz. How Reaction (29.1) behaves and what happens next depend on the physical dynamics of the pore fluid, whether open-system or closed-system (buffered), as described in Section 26.4.

29.1.1.1 OPEN-SYSTEM BEHAVIOR

If the system is *open* (porous and permeable), the fluid composition may be controlled externally and remain essentially constant, so a traverse up metamorphic grade (toward the igneous contact) follows a vertical path on the $T\text{-}X_{CO_2}$ diagram. If we assume that X_{CO_2} of the intergranular fluid phase is initially equal to 0.5, the path is illustrated as the "externally controlled fluid" arrow rising along $X_{CO_2} = 0.5$ in Figure 29.3. In such a situation, Reaction (29.1) is isobarically invariant, because there are $C = 3$ inert components (H_2O and CO_2 are mobile) and $\phi = 4$ coexisting phases, so, according to the phase rule at constant P, $F = 3 - 4 + 1 = 0$. As a result, the temperature remains at that of the isograd, as heat is consumed by the reaction, until one reacting phase is used up. Because we are addressing a dolomitic carbonate with silica impurity, Dol $\gg$ Qtz, so quartz will be consumed first. Once the quartz is gone, the Dol + Cal + Tlc assemblage is univariant and the temperature can rise again. In this situation the temperature will rise until the system intersects the curve for the reaction:

$$5\,\text{Tlc} + 4\,\text{Qtz} + 6\,\text{Cal} = 3\,\text{Tr} + 6\,CO_2 + 2\,H_2O \quad (29.2)$$

This reaction involves the appearance of **tremolite** inside the Tlc-Qtz-Cal sub-triangle in the *CMS* system (Figure 29.4c). This is the "tremolite-in" isograd for many siliceous limestones (shown in Figure 29.2) and the more SiO_2-rich dolostones, but most of the shaded area of dolomitic bulk compositions is still in the Dol-Tlc-Cal field, a field in which Reaction (29.2) does not occur because one reactant (quartz) is missing.

When Reaction (29.2) runs to completion in appropriate metacarbonates, either talc or quartz will be consumed. Noting where most dolomitic marbles plot in Figure 29.4, quartz is most likely to disappear first, and the temperature of the system will then rise again. These rocks, and other

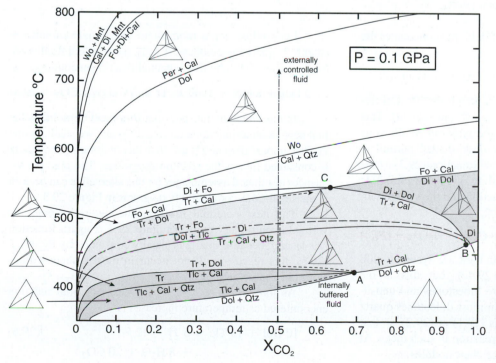

FIGURE 29.3 $T\text{-}X_{CO_2}$ phase diagram for siliceous carbonates at $P = 0.1$ GPa. Calculated using the program TWQ of Berman (1988, 1990, 1991). The light shaded area is the field in which tremolite is stable, and the darker shaded areas are the fields in which dolomite and either talc or diopside is stable. Compatibility diagrams, similar to those in Figure 29.4, show the mineral assemblages in each field.

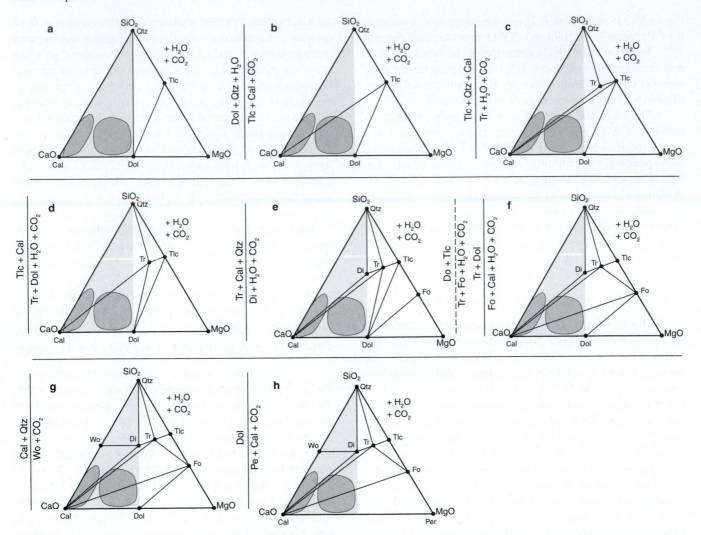

FIGURE 29.4 The sequence of CaO-MgO-SiO$_2$-H$_2$O-CO$_2$ compatibility diagrams for metamorphosed siliceous carbonates (shaded half) along an open-system (vertical) path up metamorphic grade for $X_{CO_2} < 0.63$ in Figure 29.3. The dashed isograd requires that tremolite is more abundant than either calcite or quartz, which is rare in siliceous carbonates. Dark shaded areas are typical ranges of limestone and dolostone compositions. After Spear (1993).

meta-dolostones in the Dol-Tlc-Cal field, next encounter the curve for the reaction:

$$2\,Tlc + 3\,Cal = Dol + Tr + CO_2 + H_2O \quad (29.3)$$

This involves a tie-line flip in which the Tlc-Cal tie-line is replaced by the Dol-Tr tie-line (Figure 29.4d). This reaction introduces tremolite into the majority of siliceous dolostones, as can be seen by the dark shaded portion in Figure 29.4d. For this reason, I chose Reaction (29.3) as the talc–tremolite transition in the shading of Figure 29.3. Talc is likely to be consumed by this reaction and the system will then rise to encounter the reaction:

$$Tr + 2\,Qtz + 3\,Cal = 5\,Di + 3\,CO_2 + H_2O \quad (29.4)$$

By this reaction, Di appears in the *CMS* system within the Tr-Cal-Qtz triangle (Figure 29.4e), but this is the **diopside-in** isograd only in impure limestones and unusually siliceous impure dolostones. In most dolostones quartz was consumed earlier by Reactions (29.1) or (29.2), so that diopside cannot be formed by this reaction in such rocks. At Alta, diopside is found around some chert nodules.

Let's return for a moment to our initial dolomite + calcite + quartz assemblage. If X_{CO_2} of the initial fluid is between 0.7 and 0.97 the first potential reaction encountered is:

$$5\,Dol + 8\,Qtz + H_2O = Tr + 3\,Cal + 7\,CO_2 \quad (29.5)$$

In this situation, talc does not form, and tremolite is the first new metamorphic mineral found. Again, we shall assume that the amount of quartz is less than that of dolomite, so quartz is consumed first as the reaction progresses. The possible sequences of mineral assemblages for this alternative can be seen in the small *CMS* compatibility diagrams in Figure 29.3.

The next potential reaction encountered along this vertical evolutionary path in Figure 29.3 is again Reaction (29.4), and, because quartz is usually consumed by Reaction (29.5), this reaction will not occur in most siliceous dolostones in contact aureoles.

As long as $X_{CO_2} < 0.97$, the next potential reaction encountered as temperature increases is:

$$10\,Dol + 13\,Tlc = 5\,Tr + 12\,Fo \quad (29.6)$$
$$+ 8\,H_2O + 20\,CO_2$$

The reaction replaces the Dol-Tlc tie-line (Figure 28.4e) with the Tr-Fo tie-line and *would be* the forsterite-in isograd. Because the Dol-Tr-Tlc-Fo quadrilateral in Figure 29.4e is the only compositional area affected by this reaction, and this quadrilateral is entirely in the Mg-rich (unshaded) half of the *CMS* triangle, this reaction occurs only in carbonated ultramafic rocks and not in impure marbles. The reaction is thus dashed in Figures 29.3 and 29.4.

The next potential reaction along a high X_{CO_2} (>0.7) path is:

$$\text{Cal} + \text{Tr} = \text{Dol} + 4\,\text{Di} + H_2O + CO_2 \quad (29.7)$$

This reaction may introduce diopside into common impure dolostones, providing that X_{CO_2} in the fluid is greater than 0.7. Note in Figure 29.3 that diopside can also form in very CO_2-rich fluids ($X_{CO_2} > 0.97$) directly from the initial dolomite and quartz via the reaction:

$$\text{Dol} + 2\,\text{Qtz} = \text{Di} + 2\,CO_2 \quad (29.8)$$

But such CO_2-rich fluids are rare at low metamorphic grades. Because the amount of SiO_2 is small, and X_{CO_2} is typically less than 0.7, diopside does not occur in many contact metamorphosed dolostones (at least those in open fluid systems).

If $X_{CO_2} < 0.63$, **forsterite** may be introduced into many impure dolostones via the reaction:

$$11\,\text{Dol} + \text{Tr} = 8\,\text{Fo} + 13\,\text{Cal} + H_2O + 9\,CO_2 \quad (29.9)$$

as shown in Figure 29.4f, which is separated from Figure 29.4e by both Reactions (29.6) and (29.9). If $X_{CO_2} > 0.63$, and diopside is present, forsterite is produced by the reaction:

$$\text{Di} + 3\,\text{Dol} = 2\,\text{Fo} + 4\,\text{Cal} + 2\,CO_2 \quad (29.10)$$

In the inner portions of contact aureoles, **wollastonite** is created in siliceous limestones by the now familiar reaction:

$$\text{Cal} + \text{Qtz} = \text{Wo} + CO_2 \quad (29.11)$$

The reaction occurs at temperatures over 550°C at $P = 0.1$ GPa if $X_{CO_2} > 0.2$. Note in Figure 29.4g that Wo occurs only in limestones and very siliceous dolostones and is separated from virtually all of the dark shaded common dolostone area by the Cal-Di and Cal-Tr tie-lines. In most dolostones, quartz is consumed by early reaction with dolomite to create these tie-lines.

In even hotter aureoles ($T > 700$°C unless the fluids are very H_2O-rich), dolomite finally breaks down to **periclase** + calcite via the reaction:

$$\text{Do} = \text{Per} + \text{Cal} + CO_2 \quad (29.12)$$

By this reaction, periclase is introduced into the more MgO-rich and SiO_2-poor marbles (Figure 29.4h). This accounts for the periclase zone at Alta (Figure 29.2).

Aureoles that attain temperatures above this grade are much less common and require large, hot, intermediate to mafic or anorthosite intrusions. **Monticellite** ($CaMgSiO_4$) may form via one of the reactions below (listed in order of increasing temperature):

$$\text{Fo} + \text{Di} + 2\,\text{Cal} = 3\,\text{Mtc} + 2\,CO_2 \quad (29.13)$$

$$\text{Cal} + \text{Di} = \text{Wo} + \text{Mtc} + CO_2 \quad (29.14)$$

$$\text{Cal} + \text{Fo} = \text{Per} + \text{Mtc} + CO_2 \quad (29.15)$$

depending on whether diopside or forsterite is present. These reactions generally occur at temperatures above 800°C but may take place at temperatures on the order of 600 to 650°C if the fluid is nearly pure H_2O.

Åkermanite ($Ca_2MgSi_2O_7$), **merwinite** ($Ca_3Mg Si_2O_8$), **spurrite** ($Ca_4Si_2O_8CaCO_3$), and **larnite** (Ca_2SiO_4) occur in rarely attained high-*T*/low-*P* **sanidinite facies** rocks, usually at gabbro–carbonate contacts. Åkermanite and merwinite occur in several areas of sanidinite facies metamorphism, including Crestmore, California (Section 21.6.5). Spurrite and larnite occur at even higher temperatures and are known only in a few localities, such as the basalt–chalk contact at Scawt Hill, Northern Ireland (Tilley, 1951a), the alkali gabbro contact in the Christmas Mountains, Texas (Joesten, 1976), the anorthosite contact at Cascade Slide in the Adirondacks, New York (Bohlen et al., 1992; Valley and Essene, 1980), or in limestone xenoliths in mafic rocks.

29.1.1.2 CLOSED-SYSTEM BEHAVIOR

Marbles at depths greater than 2 to 3 km are likely to be much less porous and permeable, and typically behave more like **closed systems**, in that they do not allow free movement of the fluid phase and rapid exchange with an exterior fluid reservoir. As a result, the fluids consumed and produced by dehydration and decarbonation reactions are likely to overwhelm the initial traces of pore fluid and **internally buffer** the fluid composition (Section 26.4).

Let's return to our initial dolomite + calcite + quartz marble at low temperature and again begin to heat it in a contact aureole at 3-km depth with initial $X_{CO_2} = 0.5$ (Figure 29.3) under closed-system conditions. The first reaction encountered is Reaction (29.1), in which talc (+ calcite) is produced from a reaction between dolomite and quartz. At equilibrium $C = 5$, because now the fluid components are no longer perfectly mobile, and $\phi = 5$ (four solids and a fluid) as Reaction (29.1) takes place at equilibrium. F then equals 1 (at a fixed pressure), and the reaction is a *continuous* reaction, in which the composition of the fluid is buffered along the reaction curve in Figure 29.3 toward higher values of X_{CO_2}, as temperature rises and the reaction progresses, consuming H_2O and liberating CO_2 (recall the example discussed in conjunction with Figure 26.6).

How far will the fluid be buffered along the reaction curve as Reaction (29.1) proceeds? This depends on two principal variables. First, the mineral assemblage can buffer the fluid only as long as all four solid phases in the reaction coexist at equilibrium. As soon as a reactant is consumed, the buffering capacity is lost. The distance that the system may proceed along the curve thus depends upon the quantity of quartz present (again assuming that quartz is the least abundant phase in the marble). Second, we must know what effect the progress of the reaction has on the composition of the intergranular fluid. If we assume closed-system behavior, the extent to which a reaction will shift the composition of the fluid depends on the amount of fluid initially present, and how close the composition of that fluid is to the composition of the fluid being generated by the reaction (the stoichiometric coefficients of CO_2 and H_2O in the reaction). If there is only a small volume of initial fluid present and it is far from the composition of the fluid produced, the composition will change quickly. Greenwood (1975), Rice and Ferry (1982), and Spear (1993) described mathematical treatments that relate reaction progress to the change in fluid composition. In one example, using reasonable estimates of porosity, Spear concluded that less than 3% (by volume) of a mineral product need be produced in order to shift the fluid composition from an initial value of $X_{CO_2} = 0.4$ to a final value of 0.89.

Spear's result suggests that the fluid composition in marbles acting as a system closed to fluids will be altered considerably by the reaction. In the present case, this implies that the fluid will be efficiently buffered along the isobarically univariant equilibrium curve for Reaction (29.1) along the dashed arrow in Figure 29.3. Because the reaction consumes H_2O and produces CO_2, the fluid should be buffered toward $X_{CO_2} = 1.0$ but first the system encounters invariant point A. If X_{CO_2} of the initial fluid is 0.5, the system would evolve to point A about 20°C above the first appearance of talc, and, if the calculations of Spear are a reliable indicator, only a little talc may have been produced in the interval. If only a minor amount of quartz is present, however, it may be consumed before reaching point A, in which case the system would leave the univariant curve and rise vertically from the point at which quartz was consumed. Depending on the amount of quartz and the permeability, then, the system could leave the univariant curve at any point between the point at which the reaction curve was first intersected and the invariant point.

Suppose the system reaches invariant point A. At this temperature, tremolite appears along with dolomite, calcite, quartz, and talc. Because this is an isobarically invariant situation, the system must remain at point A until a phase is consumed. Because heat is still being added to the aureole from the pluton, the reaction proceeds, producing more tremolite and consuming dolomite and quartz. What happens next depends on the relative proportions of Dol, Cal, Tlc, and Qtz in the rock (i.e., the bulk rock composition). Whichever phase is consumed, the fluid will be buffered along the reaction emanating from A in which that phase is absent. Remember from the Schreinemakers treatment (Section 26.9), that each reaction curve emanating from an invariant point is missing one of the phases that coexist at the point. Quartz is typically the first phase to disappear in a dolomitic marble, and, if so, the system will consist of Tr + Dol + Cal + Tlc. Then the system will continue to be buffered *back to lower* X_{CO_2} along the univariant curve for Reaction (29.3) as tremolite is produced at the expense of talc and calcite (dashed curves).

The system will leave the curve for Reaction (29.3) and rise again when a reactant is consumed (probably talc). This may occur at any point along the curve between invariant point A and the thermal maximum of the curve. The temperature

maximum along this reaction is at $X_{CO_2} = 0.5$, when the fluid composition is equal to the fluid produced by the reaction. [Equal molar quantities of CO_2 and H_2O are produced by Reaction (29.3).] If the system reaches this point, according to Figure 29.3, it will remain at $T = 430$°C and $X_{CO_2} = 0.5$ until the remaining talc is consumed by the reaction. Then the system will rise in temperature until forsterite is produced by Reaction (29.9). The diopside reaction is less likely to occur because quartz is generally absent at this point, as discussed in the previous section. The fluid will then be buffered along the curve for Reaction (29.9) toward invariant point C in Figure 29.3, where diopside will be created. Diopside is more likely to form in closed systems than in open systems because of this buffering to CO_2-rich fluids.

Which curve is followed from invariant point C again depends upon the bulk rock composition. If tremolite is consumed first, the fluid will be buffered along Reaction (29.10) to very CO_2-rich fluids as diopside + dolomite → forsterite + calcite + CO_2. If dolomite disappears first (unlikely if we begin with a siliceous dolostone), the fluid will follow the curve for the reaction:

$$3\,Tr + 5\,Cal = 11\,Di + 2\,Fo + 3\,H_2O + 5\,CO_2 \quad (29.16)$$

The thermal maximum for this reaction is at $X_{CO_2} = 5/(5 + 3) = 0.625$, which, coincidentally, is essentially the same value as invariant point C at 0.1 GPa. The fluids produced by Reaction (29.16) will therefore not noticeably affect X_{CO_2} of the fluid. Once either of the last two reactions is complete, the system, if hot enough and if quartz is available, will reach the temperature of Reaction (29.11). Because X_{CO_2} in closed systems is equal to 0.62 or greater, the wollastonite isograd, when it occurs at all, corresponds to temperatures above 750°C.

The reactions at higher grades are the same as in open systems, producing monticellite, merwinite, larnite, and/or spurrite. Because all these reactions are strictly decarbonation reactions, they will have the same shape on a T-X_{CO_2} diagram as Reaction (29.11), so that the temperature of the reactions will be high in closed systems in which X_{CO_2} has been buffered to high values. These phases are therefore even less likely to occur than in open systems.

Alternatively, talc (and not quartz) may be consumed first at invariant point A. If this is the case, the fluid will be buffered toward higher X_{CO_2} along reaction curve (29.5) from point A toward invariant point B. If the system reaches point B, diopside will form at about 415°C as X_{CO_2} of the fluid reaches 0.96. If quartz is then consumed at B, the fluid will be buffered back to lower X_{CO_2} along reaction curve (29.7) toward invariant point C.

Closed-system behavior differs from open-system behavior in the following principal ways:

- Closed systems are characterized by univariant mineral assemblages (four minerals coexisting in *CMS*), which occur over broader temperature (and thus spatial) intervals.
- Invariant situations are more common when internal buffering is efficient, because the fluids are readily

buffered to invariant points with little reaction progress. Rice and Ferry (1982) argued that gradual and minor mineralogical changes occur along the univariant curves in such efficient systems, with abrupt changes occurring at the invariant points. In contrast to open systems, which are characterized by zones of divariant assemblages (three minerals coexisting *CMS*) separated by univariant reaction isograds, efficiently buffered closed systems ought to have zones of univariant mineral assemblages separated by invariant discontinuities.

- Fluids may be buffered to high X_{CO_2} so that diopside is more common. Fluids may also be buffered to high X_{H_2O} in situations in which reactions have a negative slope on T-X_{CO_2} diagrams. This is true for some impure metacarbonates with reactions involving zoisite, vesuvianite, and grossular.
- The wollastonite and monticellite isograds occur at higher temperatures because of the high X_{CO_2} produced by the forsterite-producing reactions. This effect may be overcome in inner aureoles by infiltration of H_2O expelled from the pluton.
- Rice and Ferry (1982) pointed out that efficiently buffered systems should evolve along paths that are determined entirely by the grade and initial bulk composition, because the bulk composition determines which phase is consumed first at an invariant point and thus which curve is followed from there. These paths, they noted, involve only selected univariant reaction curves and bypass others that would have been effective reactions in open systems. This feature, they suggest, should cause interlayered rocks of different bulk composition to develop independently, with different mineral assemblages and different fluid compositions in each layer.

Open-system and closed-system behavior are both known to occur, and there is probably a complete spectrum in nature between the two extremes. In a review of the literature, Rice and Ferry (1982) concluded that internal buffering is far more common (although it is difficult to evaluate to what extent it may have been accompanied by some degree of open-system behavior). A particularly good example of closed-system behavior is the Marysville contact aureole in Montana (Rice, 1977), where univariant zones are typically separated by invariant isograds. Rice concluded that little or no externally derived fluids interacted with the carbonates in the bulk of the aureole. Only near the granodiorite contact is there evidence for the introduction of H_2O-rich fluids, and these were restricted largely to vein-like pathways by which the hydrous fluids were transported away without reacting with the main volume of aureole rocks.

As shall be discussed in Chapter 30, there is evidence for infiltration and fluid flow in many contact and regional metamorphic terranes. To cite just a few examples, Cook and Bowman (1994) attempted to model cooling in the Alta aureole and found that simple conductive cooling could not produce the thermal gradients necessary to produce the isograds in Figure 29.2 or the temperatures calculated by calcite–dolomite geothermometry. They concluded that the temperature regime in the southern portion of the Alta aureole was affected significantly by heat transfer involving fluids moving sub-horizontally away from the stock. Cook et al. (1997) attributed the predominantly lateral flow to the existence of sub-horizontal impermeable barriers above the present level of exposure. Bowman et al. (1985) found that the stable isotopic ratios of $\delta^{18}O$ and $\delta^{13}C$ in skarn calcites of the aureole at Elkhorn, Montana, could not result from simple decarbonation of the original marbles. The correlated increase in $\delta^{18}O$ of the minerals and X_{CO_2} of the fluid outwards from the contact was most easily modeled by infiltration of low-$\delta^{18}O$ H_2O-rich fluids from the stock into the country rock. Nabelek (2002) found unshifted O and C isotopic ratios in the bedding-scale patterns in the Notch Peak, Utah, aureole and concluded that little external fluid infiltration had occurred there outside the inner wollastonite zone. Convective heat-fluid transport models for contact aureoles have been proposed by Cathles (1977), Norton and Knight (1977), Parmentier and Schedl (1981), and Furlong et al. (1991). Lüttge et al. (2004) proposed a kinetic model based on competing rates of reaction, fluid migration, and heating. This model deviates from equilibrium in many open-system situations in which reactions cannot keep up with fluid infiltration, implying that metastable reactions may interplay with the stable equilibrium reactions discussed in this chapter. Ague (2000) developed a two-dimensional model of coupled mass transfer, chemical reactions, and heat transport involving cross-layer transport of CO_2 from decarbonating metacarbonates layers into neighboring dehydrating metapelites and of H_2O in the opposite direction. His model indicated that cross-layer mass transfer typically forced rock decarbonation while fluids ascend, dominating the effects of cooling (which would otherwise cause carbonate precipitation and CO_2 depletion in the fluid). Consequently, prograde metamorphism of carbonate-bearing sedimentary sequences containing significant amounts of pelitic rock will generally release CO_2 to regionally migrating fluids, ultimately to the atmosphere and oceans.

29.1.2 Regional Metamorphism of Siliceous Dolostones

Figure 29.5 is a map of the mineral zones developed in regionally metamorphosed dolomitic rocks of the central Alps, spanning the Swiss–Italian border region (Trommsdorff, 1966, 1972). In this area, a regional diopside zone is developed above the talc and tremolite zones. The metamorphic zonation is truncated by later faulting associated with the Insubric Line, so that higher grade zones, if they developed, are missing. Talc is not always developed at low grades in regional metamorphism, and diopside is more common.

Analysis of regional metamorphism of siliceous dolostones requires a T-X_{CO_2} petrogenetic grid for higher pressures than for Figure 29.3. Figure 29.6 is a T-X_{CO_2} phase diagram similar to Figure 29.3, but calculated for $P = 0.5\,GPa$. The pressure is more appropriate for regional metamorphism along a medium P/T metamorphic

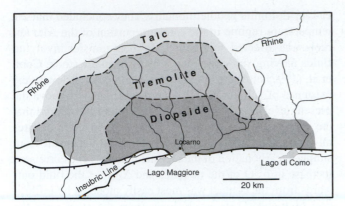

FIGURE 29.5 Metamorphic zones developed in regionally metamorphosed dolomitic rocks of the Lepontine Alps, along the Swiss–Italian border. After Trommsdorff (1966, 1972).

field gradient. The increase in pressure has two principal effects. Because devolatilization reactions at crustal pressures have a characteristic curved shape on a *P-T* phase diagram (see Figure 26.2) in which reaction temperature increases with increasing pressure, the reactions in Figure 29.6 are at higher temperature than in Figure 29.3. Second, if we compare invariant points A, B, and C between the two diagrams, we realize that the talc field is much smaller and the diopside field is much larger at higher pressure.

The reactions involved are the same as at low pressure, and rather than reiterate all the scenarios of the preceding sections, I shall concentrate on the ones most likely to occur, and on the principal differences between medium-pressure regional metamorphism and the low-pressure equivalent that we have already discussed.

If we begin with low-temperature siliceous dolostones, the first obvious difference in regional versus low-*P* contact metamorphism is that X_{CO_2} of the fluid must be less than 0.1 for talc to form by Reaction (29.1). For the

largest range of fluid compositions ($0.1 < X_{CO_2} < 0.87$), *tremolite* will be the first new metamorphic mineral to form [from dolomite + quartz via Reaction (29.5)]. From this point the sequence of reactions for open-system behavior is the same as for contact metamorphism. Closed-system behavior is highly favored at depths below 10 km, however, because the marbles will be compressed and permeability will be limited. Most reactions may thus be expected to internally buffer the fluids.

As an example of a typical situation, consider a path rising in temperature along the dashed curve at $X_{CO_2} = 0.5$ in Figure 29.6. When **tremolite** begins to form, the fluid will probably be buffered along the curve for Reaction (29.5) (dashed curve). The fluid composition becomes enriched in CO_2 as the reaction progresses, consuming H_2O and liberating CO_2. If quartz is present in minor amounts, it may be consumed before the system reaches invariant point B, in which case the system would progress vertically from the curve at whatever temperature quartz disappears. If the example of Spear (1993) discussed above is correct, however, it is more likely that only a small degree of reaction progress is necessary for the fluid to be buffered to point B. At this point **diopside** will form along with dolomite, tremolite, quartz, and calcite in an invariant assemblage.

Which curve is followed from the invariant point again depends upon the relative proportions of the reacting phases (i.e., on the bulk composition). If tremolite and calcite are lost, the system may proceed along the Dol + Qtz = Di + CO_2 curve (Reaction 29.8) toward $X_{CO_2} = 1.0$. It is far more likely that quartz will be consumed first, however, and the system will follow the curve for Reaction (29.7), consuming tremolite and calcite to produce diopside + dolomite.

If buffering is effective in a closed system, the fluid will continue to be buffered along the curve for Reaction (29.7) to near the temperature maximum at $X_{CO_2} = 0.5$. As

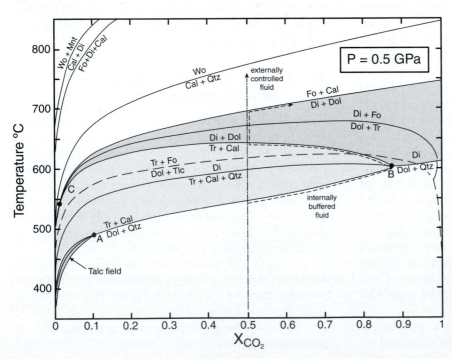

FIGURE 29.6 T-X_{CO_2} phase diagram for siliceous carbonates at $P = 0.5$ GPa, calculated using the program TWQ of Berman (1988, 1990, 1991). The light-shaded area is the field in which tremolite is stable, and the darker shaded areas are the fields in which talc or diopside are stable.

the composition of the intergranular fluid approaches that of the fluid generated by the reaction, a greater degree of reaction progress is required to modify X_{CO_2} of the fluid. As a result, the temperature and fluid composition change less as the temperature maximum is approached. Once a reactant phase is consumed (either tremolite or calcite is possible), the buffer is lost and the temperature rises again until **forsterite** forms via Reaction (29.10). The typical sequence in regionally metamorphosed siliceous dolomites is thus:

1. (Talc if X_{CO_2} is very low)
2. Tremolite
3. Diopside
4. Forsterite

If the system is internally buffered and low in SiO_2, we can estimate the temperature of the isograds at 0.5 GPa from Figure 29.6. The appearance of tremolite depends on the initial X_{CO_2} of the fluid. If talc occurs first, tremolite should appear at invariant point A (~480°C). Diopside would appear at invariant point B (~600°C), and forsterite would appear at ~690°C. Of course, pressure is not constant in metamorphic terranes, and it will affect these temperatures. Raising the pressure from 0.1 to 0.5 GPa causes the Fo isograd to rise from approximately 530°C to 690°C. If the metamorphic field gradient for medium P/T metamorphism in Figure 28.3 is appropriate, the system would reach 700°C at a pressure closer to 0.8 GPa, where the temperature of the Fo-in isograd is approximately 780°C. One could create a series of T-X_{CO_2} diagrams at several pressures to determine the isograd temperatures, or one could calculate a T/P versus X_{CO_2} diagram in which a specific T/P gradient is substituted as the ordinate.

The controlling effect of bulk rock composition on the mineral assemblage can be determined from Figure 29.4.

Forsterite is favored in high-Mg/Fe-low-SiO_2 carbonates, whereas diopside is favored in high-SiO_2-low-Mg/Fe carbonates.

Wollastonite may form in the upper amphibolite facies or in the granulite facies. Because of the high temperatures required, it is not common in regional metamorphic rocks. The higher temperature phases (monticellite, merwinite, etc.) require unusually high temperatures and melting usually occurs first.

29.1.3 Fluid Infiltration in Calcic Rocks

Fluid infiltration refers to open-system behavior in which a fluid permeates into a rock body, generally along lithological layers, vein-like passageways, and/or the interconnected grain-boundary network. In the present context, we shall deal with the effects of the fluid only, and not any dissolved constituents. Dissolution and transport-deposition of dissolved material constitutes *metasomatic* alteration, which we shall discuss in the final chapter.

Carmichael (1970) was probably the first to combine field and theoretical data to demonstrate the effects of fluid infiltration on the petrogenesis of metamorphic rocks. In his study of metamorphism in the Whetstone Lake area, midway between Ottawa and Toronto in Ontario, he identified and mapped five isograds. These reactions and isograds are shown in a T-X_{H_2O} diagram and map in Figure 29.7a. Carmichael's isograds 1 to 4 occur in metapelites and indicate a regional increase in metamorphic grade to the northwest. You may recognize some of these isograds from the previous chapter. Reactions 1, 2, and 4 are dehydration reactions, all of which increase in temperature on Figure 29.7a toward pure H_2O. To facilitate comparison to the map area in Figure 29.7b, the T-X diagram in Figure 27.7a has pure H_2O on the *right*, which is opposite to our customary usage. Reaction 4 is the familiar

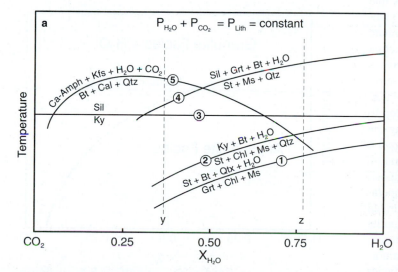
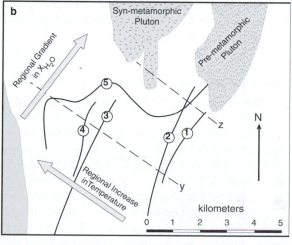

FIGURE 29.7 **(a)** T-X_{H_2O} diagram illustrating the shapes and relative locations of the reactions for the isograds mapped in the Whetstone Lake area. Reactions 1, 2, and 4 are dehydration reactions, and Reaction 3 is the Ky = Sil transition, all in metapelites. Reaction 5 is a dehydration–decarbonation in calcic rocks with a temperature maximum at $X_{H_2O} = 0.25$. **(b)** Isograds mapped in the field. Note that isograd 5 crosses the others in a manner similar to that in part (a). This behavior is attributed to infiltration of H_2O from the syn-metamorphic pluton in the area, creating a gradient in X_{H_2O} across the area at a high angle to the regional temperature gradient, equivalent to the T-X diagram. After Carmichael (1970). Reprinted by permission of Oxford University Press.

kyanite → sillimanite phase transformation, which is not affected by the composition of the fluid.

The fifth isograd occurs in calcareous units in the sedimentary package and involves the reaction:

$$5 \, Bt + 6 \, Cal + 24 \, Qtz = 3 \, Ca\text{-}Amphibole + \quad (29.17)$$
$$5 \, Kfs + 6 \, CO_2 + 2 \, H_2O$$

This reaction liberates both H_2O and CO_2, which, according to the principles discussed in Section 26.4, must have a shape with a thermal maximum at $X_{H_2O} = 2/(2 + 6) = 0.25$. Because of its distinctive shape, this reaction crosses the other reactions on a T-X phase diagram. Note that reactions 3 and 4 also cross at low X_{H_2O}.

Figure 29.7b is Carmichael's (1970) field map of the isograds based on the reactions in Figure 29.7a, with metamorphic grade increasing to the northwest. Note that isograd 5 also crosses isograds 2, 3, and 4, in a fashion similar to that in the T-X diagram. Thus a traverse up metamorphic grade, such as (y) in Figure 29.7b, will cross the isograds in the order 1–2–3–4–5. Traverse (z) crosses the isograds in the order 1–5–2–3–4 (if 3 and 4 are extrapolated), which corresponds to a higher value of X_{H_2O} in Figure 29.7a. The most reasonable explanation for the difference in isograd sequence is a regional gradient in X_{H_2O} at a high angle to the thermal gradient in the area. As a result, the map area behaves as a large T-X_{H_2O} diagram. Carmichael (1970) attributed the X_{H_2O} gradient to infiltration of H_2O expelled from the syn-metamorphic pluton in the northern-central portion of the map. The pluton on the east has been overprinted by the metamorphism and is interpreted as pre-metamorphic. It thus has no effect on the isograds. Since Carmichael's study, several other investigators have documented the effects of infiltrating fluids on the mineral assemblages that developed (see Ferry, 1991).

As I mentioned in Section 25.3.2, the development of the granulite facies in mafic or pelitic rocks may in some cases be the result of CO_2 infiltration. The reactions that represent the transition from the amphibolite facies to the granulite facies in mafic rocks are shown at the high-temperature end of the petrogenetic grid in Figure 26.19. Reaction (28.22) is a typical pelitic reaction in which biotite breaks down to orthopyroxene + K-feldspar + H_2O. All these reactions are dehydration reactions, and the traditional interpretation is that granulite facies terranes represent deep to mid-crustal rocks that have been dehydrated by having the fluids driven off over time at high temperature. Dehydration reactions have T-X_{CO_2} shapes with a maximum temperature at $X_{H_2O} = 1.0$ (see Figure 26.4). Shifting the fluid composition from H_2O-rich to CO_2-rich, even at a constant temperature, may therefore also drive the reaction to the dehydrated state (Figure 29.8). Local examples of amphibolite-granulite transitions are relatively common in the literature. For example, Glassley and Sørensen (1980) described metamorphosed mafic dikes in Greenland with granulite facies centers and amphibolite facies margins that were both generated at the same P-T conditions but different f_{H_2O}. Todd and Evans (1994) attributed local dehydration of amphibolite facies gneisses to granulite facies mineralogy to infiltration of CO_2 from an underlying marble layer. On a *regional*

scale, Touret (1971, 1985) showed that the fluid inclusions in granulite facies rocks of southern Norway are much more CO_2 rich than those in the nearby amphibolite facies rocks. This has led several investigators to propose that some granulite facies terranes are produced as a result of the infiltration of CO_2 into the deep crust from mantle sources (Newton et al., 1980; Newton, 1987; Santosh and Omori, 2008), which displaces H_2O and dehydrates the rock while possibly maintaining a nearly constant fluid content. If true, this process is one of crustal-scale infiltration.

29.1.4 Metamorphism of Calc-Silicate Rocks

The calcic rocks in which Reaction (29.17) occurs in the Whetstone Lake area are calc-silicate rocks, not the carbonate-dominated calcareous rocks discussed in Sections 29.1.1 and 29.1.2. Calc-silicate rocks are rich in Ca, Mg, Fe, Si, and Al. They are generally more common in metamorphic terranes than siliceous limestone and dolostone marbles. Many calc-silicate rocks begin as carbonate-bearing pelitic sediments (sometimes called calcareous pelites, or marls). The clastic component may include quartz, feldspar, clays, and other silicates or oxides. The proportions of clastic and carbonate minerals can vary widely.

When carbonate is subordinate to other clastic material, it may be largely consumed by decarbonation reactions, and the metamorphic rocks, particularly at medium and high grades, are then composed solely of Ca-Mg-Fe-Al-Na-K silicate minerals (**calc-silicate** minerals), such as diopside, the epidote group minerals, grossular, Ca-amphiboles, calcic plagioclase, vesuvianite, titanite, margarite, and/or scapolite. Muscovite, biotite, and chlorite are also common. Some calc-silicate minerals (e.g., zoisite, margarite, vesuvianite, and grossular) are stable in H_2O-rich fluids, whereas others (e.g., anorthite) are stable in CO_2-rich fluids.

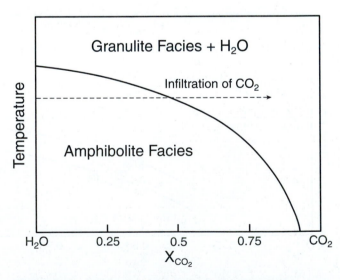

FIGURE 29.8 Schematic T-X_{CO_2} diagram illustrating the characteristic shape of typical dehydration reactions, such as those that generate orthopyroxene from hornblende or biotite. Notice that the amphibolite facies to granulite facies can be accomplished by either an increase in temperature or infiltration of CO_2 at a constant temperature.

Metamorphosed calc-silicate rocks were described in the classic works of Goldschmidt (1911, 1912a and b, 1921) and Eskola (1922, 1939), in which they developed the facies concept (Section 25.1). Kennedy (1949) studied calc-silicate rocks in the Scottish Highlands, where he recognized a series of isograds and zones that he could correlate to the Barrovian zones based on pelite mineral assemblages (Table 29.1).

The range of calc-silicate bulk rock compositions overlaps with that of mafic to intermediate volcanics (and volcaniclastics), so that the metamorphic rocks may not always be distinguishable petrographically. The common practice of inferring protolith from metamorphic rock bulk composition, and interpreting the environment of deposition, must be used with care in such situations, because a warm-water shallow offshore bank differs greatly from a volcanic arc. Although the major element profile for the two rock types may be similar, the trace element and isotopic signatures are distinct and should be determined before attempting any interpretation based on the type of parent.

Because of the complex compositions and variability of calc-silicate rocks, a comprehensive approach to calc-silicate petrogenesis is a formidable task. Even the addition of a single component, such as Al_2O_3, requires the introduction of several new phases and reactions to the petrogenetic grids of Figures 29.3 and 29.6, and projection from that component onto the *CMS-HC* chemographic triangle. If we add FeO, Na_2O, and K_2O we find ourselves dealing with a complex chemical system, requiring either a formidable grid and multiple projections or a series of simplified subsystems. Numerous studies may be found in the literature that deal with one subsystem or another. Kerrick (1974) provided an early review, and Bucher and Frey (2002, Chapter 8) developed several subsystems. We will content ourselves with a single example: the now-classic studies by John Ferry on the calc-silicates of the Vassalboro Formation in the Augusta-Waterville area of south-central Maine.

Figure 29.9 is a metamorphic map of the Augusta-Waterville area, adapted from Ferry (1976a and b, 1983a and b, 1988). Note that isograds can be mapped in both of the major Silurian formations in the area: the dominantly pelitic Waterville Formation and the Vassalboro Formation, which contains numerous calcareous interbeds. Metamorphism in the area was of the lower pressure (0.2 to 0.4 GPa) Buchan type and occurred during the late Paleozoic Acadian

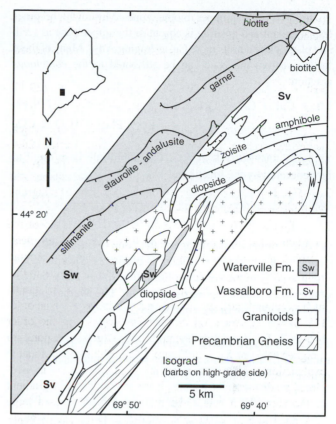

FIGURE 29.9   Map of isograds in the pelitic Waterville and calcareous Vassalboro formations of south-central Maine. After Ferry (1983b). Reprinted by permission of Oxford University Press.

Orogeny. The granitic plutons in the area were intruded at the time of metamorphism, raising the temperature of the shallow rocks and affecting the geometry of the isograds.

Ferry described the following five zones in the calc-silicate rocks. The reactions that define the isograds and the continuous reactions within the zones are complex, and several equilibria may occur simultaneously. The reactions listed below are only approximations, intended as simplified representations.

29.1.4.1 ANKERITE ZONE   Rocks of the lowest grade, observed in the northern end of the area, contain the assemblage ankerite + quartz + albite + muscovite ± calcite ± chlorite. Ankeritic carbonate is a dolomite with over 20% (Fe^{2+} + Mn) replacing Mg and is stable in sedimentary environments. Muscovite, chlorite, and calcite are probably created by metamorphic reaction between the initial ankerite and clays.

29.1.4.2 BIOTITE ZONE   The biotite isograd is related to the inferred reaction:

$$Ms + Qtz + Ank + H_2O = \qquad\qquad (29.18)$$
$$Cal + Chl + Bt + CO_2$$

It occurs at about the same grade as the biotite isograd in the local pelites. Calcareous rocks in the lower biotite zone have the mineral assemblage biotite + ankerite + quartz + albite + muscovite + calcite + chlorite. Biotite is generally less than 3% of the mode in the lower zone, but increases to

TABLE 29.1   Calcareous Mineral Assemblages in the Scottish Highlands

Pelitic Zone	Calcareous Mineral Assemblage
Chlorite	Albite–zoisite–calcite–biotite–hornblende
Biotite	
Garnet	Andesine–zoisite–calcite–biotite–hornblende
Staurolite	Anorthite–hornblende–garnet
Kyanite	
Sillimanite	Anorthite–diopside–garnet

After Kennedy (1949) and Tanner (1976).

16 to 30% in the upper biotite zone. Muscovite becomes less abundant and ankerite is absent in the upper zone as well, and plagioclase admits more Ca, averaging An_{29}. Many of these features across the zone can be attributed to the *continuous* reaction:

$$Ms + Cal + Chl + Qtz + Ab = \qquad (29.19)$$
$$Bt + Plag + H_2O + CO_2$$

29.1.4.3 AMPHIBOLE ZONE Ca-amphibole is inferred to be generated by the reaction:

$$Chl + Cal + Qtz + Ab = \qquad (29.20)$$
$$Act + An + H_2O + CO_2$$

A typical mineral assemblage in the zone is Ca-amphibole + quartz + plagioclase + calcite + biotite ± chlorite. The amphibole is tremolite-actinolite in the lower zone, but becomes more aluminous across the zone. The average plagioclase composition in the zone is An_{70}.

29.1.4.4 ZOISITE ZONE Zoisite (a nearly Fe-free epidote mineral) first appears as reaction rims on plagioclase at its contact with calcite, suggesting the reaction:

$$An + Cal + H_2O = Zo + CO_2 \qquad (29.21)$$

The typical mineral assemblage is zoisite + Ca-amphibole + quartz + plagioclase (average An_{74}) + calcite ± biotite ± microcline. Microcline probably occurs via Reaction (29.17) of Carmichael (1970) given above.

Reaction (29.21) is nearly vertical on a T-X_{CO_2} diagram, with zoisite stable only if X_{H_2O} of the fluid is > 0.95. At suitable temperatures the reaction runs to the right and produces zoisite only if H_2O is infiltrated into the calcareous rocks, not because temperature rises. Most prograde reactions in the calcareous rocks produce more CO_2 than H_2O, so we might expect them to buffer the fluids to high values of X_{CO_2}. Ferry attributes the high X_{H_2O} in zoisite-bearing rocks to infiltration of fluids derived from the local granitic plutons, similar to the findings of Carmichael (1970) in the previous section. High H_2O contents have been noted for the development of vesuvianite and grossular garnet in other studies of calc-silicate rocks. Plutons are an attractive immediate source for H_2O-rich fluids, which may be concentrated along certain local pathways or conduits. An alternative source of H_2O-rich fluids in some cases may be underlying pelitic rocks undergoing dehydration during metamorphism or even CO_2-H_2O immiscibility at low pressures (Labotka, 1991; Trommsdorff and Skippen, 1986). Ferry also notes that the alkali content of the rocks drops considerably, beginning in the amphibole zone. This may result from preferential solution (scavenging) of alkalis by the same H_2O-rich infiltrating fluids.

29.1.4.5 DIOPSIDE ZONE The diopside isograd in the calcareous rocks occurs at about the grade of the sillimanite isograd in the pelites. Diopside growth results from dehydration of amphibole at high grade caused by the reaction:

$$Tr + Cal + Qtz = Di + H_2O + CO_2 \qquad (29.4)$$

The typical mineral assemblage is diopside + zoisite + Ca-amphibole + calcite + quartz + plagioclase (ave. An_{79}) ± biotite ± microcline. Ca-amphibole continues to coexist with diopside in the zone because of the continuous nature of Reaction (29.4) in rocks with Fe, Mg, and Al. Zoisite continues to form via Reaction (29.21) throughout the diopside zone, consistent with the continued introduction of H_2O into the rocks closest to the plutons.

The five zones observed by Ferry are common, but not universal, in calc-silicate rocks. Differences in P, T, bulk composition, and fluid composition may result in the development of several other mineral zones. Zonation similar to that developed on a regional scale in the Augusta-Waterville and other areas is also common on a local scale in many calcareous rocks. The scale may be as small as a centimeter or less where chert nodules or quartz veins occur in carbonates. In such situations, the zonation is typically ascribed to localized variations in the fluid composition. The situation can be particularly clear in the case of quartz veins, which represent aqueous silica-rich fluid conduits along fractures in carbonate rocks (Chapter 30). Diffusion is the major factor that controls the zonation in the small-scale features, because thermal gradients cannot be that steep.

Infiltration, and the chemical changes that accompany it, may be widespread in metamorphic rocks, especially at the higher grades. It has been documented in many calc-silicate rocks, where the effects are most easily recognized. I suspect that fluid infiltration is far more common than many of us realize, but there are fewer clear indicators of the passage of fluids in most other common rock types. Veins, dikes, and pods are generally overlooked as petrologists concentrate on the more prevalent country rocks.

One interesting example of possible fluid infiltration into pelitic rocks concerns the small anomalously high-grade metamorphic areas in New Hampshire described by Chamberlain and Lyons (1983) and Chamberlain and Rumble (1988). The areas are irregularly shaped and a few kilometers across, with mineral assemblages that reflect higher grades than the surrounding regionally metamorphosed rocks. Garnet–biotite geothermometry indicates that temperature increases from $\sim 500°C$ outside to $\sim 700°C$ in the anomalous areas. Chamberlain and Rumble (1988) noted that graphite–quartz veins are common within the areas and rare outside them. They attribute the areas to the flow of high-temperature fluids that were focused into narrow zones with high fracture permeability (now the graphite–quartz veins), which acted as channels for fluids rising from deeper dehydrating terranes.

I have seen several instances of retrogressed granulite facies rocks in Greenland where amphibolite facies mineralogy developed in small irregular areas or along apparent fractures. We attributed these to the introduction of post-granulite facies hydrous fluids along areas similar to those described by Chamberlain and Rumble (1988), but probably more diffuse. In one occurrence in the Angmagssalik area of eastern Greenland, we even noticed the *prograde* overprint of granulite facies mineralogy on amphibolite facies rocks along vein/fracture systems above a large charnockite body. These may reflect the introduction of CO_2-rich fluids from the charnockite into overlying more hydrous rocks at high temperatures, causing them to dehydrate.

29.2 METAMORPHISM OF ULTRAMAFIC ROCKS

As we discussed in Chapters 1, 10, and 13, ultramafic rocks compose the Earth's mantle. Although some ultramafics may originate as cumulates in crustal mafic magma chambers, or even from unusual komatiitic volcanics, the majority of ultramafic rocks found in the crust are mantle derived and occur in orogenic belts as slivers, lenses and pods of purely ultramafic rock or in conjunction with gabbros, basalts (commonly pillow basalts), and deepwater pelagic sediments plus chert. These ultramafic rocks are casually called **alpine peridotites** or **orogenic peridotites** (even when they are metamorphosed and lack the characteristic olivine–pyroxene mineralogy of a true peridotite). The ultramafic–mafic–sediment association was a curiosity prior to the 1960s, but we now recognize it as the typical constituents of the oceanic crust, or **ophiolite suite** (Section 13.3).

Alpine peridotites most likely represent the uppermost mantle that forms the base of slivers of oceanic lithosphere that become incorporated into the continental crust along subduction zones. These slivers are dismembered portions of ophiolites: pieces of oceanic crust and mantle that either separate from the subducting slab and become incorporated into the accretionary wedge of the subduction zone, or get trapped between two terranes during an accretion event. Strings of ultramafic bodies in orogens have been found to follow major fault zones separating contrasting rock bodies. These strings are interpreted as remnants of oceanic crust + mantle that once separated collisional terranes, and thus mark the **suture zone**: the ancient subduction zone that existed prior to the collision. The common association of blueschist facies rocks with the ultramafics, mafics, and sediments further strengthens the argument for a subduction-related origin. The ophiolite may represent either typical oceanic crust between colliding continents or a marginal sea behind an offshore island arc (similar to present-day Japan) in cases where the arc gets pushed back toward the continent behind it.

The rocks that now constitute alpine peridotites were originally mantle lherzolites or more refractory harzburgites or dunites, the latter two representing either cumulates or depleted residues after MORB-type melts were extracted. The initial high-T-P peridotite mineralogy is typically retrograded to serpentine before or during the collisional event. Later orogeny results in prograde regional metamorphism of the serpentinite to higher grade mineral assemblages. In the typical situation, the ultramafic bodies are metamorphosed to the same grade as the surrounding rocks. Evans (1977) called this type of meta-ultramafic occurrence "isofacial" with the country rocks. In some cases, an ultramafic body may be emplaced hot and be only partially retrogressed. Such a body would have a mineral assemblage reflecting a higher grade than the surrounding rocks (the "allofacial" type of Evans, 1977).

Like carbonates, typical metamorphosed ultramafic rocks conform well to the simple CaO-MgO-SiO$_2$-H$_2$O

(CMS-H) system. Other chemical components are relatively minor. Only FeO and Al$_2$O$_3$ occur in amounts that typically exceed a few weight percent, and they have only a minor effect on the petrogenetic relationships. Ultramafics are thus represented by the right half of the CMS triangle (Figure 29.1) and complement the calcareous rocks already discussed in this chapter. Because isofacial ultramafics are more common, and are easier to treat because they represent prograde dehydration reactions that equilibrate more readily, we shall restrict our discussion to these rocks. We could spend considerable time and effort addressing numerous subtle variations, but I'd prefer to address only the most typical assemblages of prograde regional metamorphism corresponding to a medium P/T field gradient or facies series (see Figures 21.1 and 25.3). For a more complete treatment, including a discussion of allofacial rocks, see Evans (1977). We will also concentrate on metamorphism under *crustal* conditions, as appropriate for a general survey of metamorphism. For studies of subducted ultramafics returning to mantle pressures and the capacity of newly recognized high-pressure dense hydrous magnesium silicates to transport H$_2$O into the mantle see Section 16.8.4 or the reviews by Angel et al. (2001), Iwamori (2004), Fumagalli and Poli (2005), Frost (2006), and Kawamoto (2006).

29.2.1 Regional Metamorphism of Ultramafic Rocks in the *CMS-H* System

Figure 29.10 is a P-T petrogenetic grid for ultramafic rocks in the CaO-MgO-SiO$_2$-H$_2$O system, calculated using the TWQ software of Berman (1988). The reaction curves assume that $p_{H_2O} = P_{total}$, and thus represent *maximum* temperatures for the dehydration equilibria. This is a reasonable assumption for completely hydrated serpentinites undergoing prograde metamorphism, because the H$_2$O liberated by the reactions should quickly saturate the limited pore space. The assumption would not apply as well to partly retrograded/hydrated (allofacial) ultramafics.

At low metamorphic grades meta-ultramafic rocks are dominated by the **serpentine** minerals: lizardite, chrysotile, and antigorite. Although nearly polymorphs, slight chemical differences distinguish them. The iron content of serpentine is low. When mantle peridotites are exposed to hydrous conditions below 500°C, olivine and pyroxene become serpentinized. The excess iron from the high-T phases that could not be accommodated in serpentine is incorporated into magnetite. Some serpentinites are beautiful dark green rocks when cut and polished. They are popular decorative building stones, commonly called "verd antique," or "green marble" (a misnomer to petrologists).

Both lizardite and chrysotile occur in low-grade serpentinites. Chrysotile is the asbestosform type of serpentine and is usually the least abundant of the three serpentine minerals. According to Deer et al. (1992), a typical sequence with increasing grade is lizardite → lizardite + chrysotile → chrysotile + antigorite → antigorite. The relationship between lizardite and chrysotile is probably more complex than this and has yet to be fully understood. Some combination of

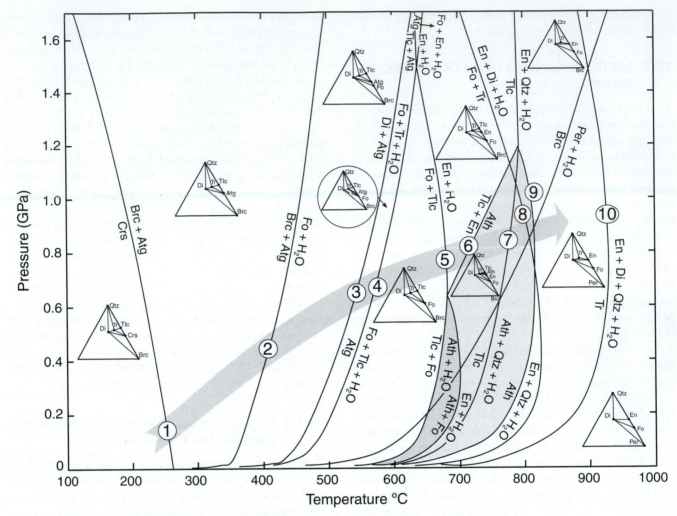

FIGURE 29.10 Petrogenetic grid for water-saturated ultramafic rocks in the system CaO-MgO-SiO₂-H₂O produced using the TWQ software of Berman (1988, 1990, 1991). The shaded arrow represents a typical medium *P/T* metamorphic field gradient. The dark shaded area represents the stability range of anthophyllite in "normal" ultramafic compositions. The lighter shaded area represents the overall stability range of anthophyllite, including more siliceous ultramafic rocks. After Spear (1993).

lizardite and chrysotile generally displays the mesh-type replacement of olivine and bastite pseudomorphs of orthopyroxene illustrated in Figure 23.52 (see Mellini, 1998, for an example). Antigorite does tend to occur at higher grades, and to have a coarser more well-developed flaky habit. Only the chrysotile → antigorite reaction is shown in Figure 29.10 as reaction curve 1, encountered at approximately 250°C. Antigorite is generally less Mg-rich than chrysotile, so the balanced idealized reaction is:

$$17 \, \text{Chr} = 3 \, \text{Brc} + \text{Atg} \qquad (29.22)$$

The compatibility diagrams in the *CMS-H* system (projected from H₂O) are also shown in most fields in Figure 29.10. You can easily derive the topologies in the smaller fields by considering the reactions that separate them from a labeled field. Spear (1993) concentrated on the reactions along a medium *P/T* traverse (corresponding to the broad darkened arrow in Figure 29.10). The enlarged compatibility diagrams of his sequence are reproduced in Figure 29.11, along with the reactions that separate the divariant fields. We will discuss the isograds and changes in mineral assemblage along that traverse. A typical peridotite

contains olivine, orthopyroxene, and clinopyroxene in the approximate proportions 6:3:1. The bulk composition of such a peridotite (see Table 10.2) in the *CMS* system is indicated by the star in Figure 29.11.

A typical peridotite metamorphosed in the greenschist facies would be nearly pure **serpentine**, with some **brucite** and **diopside** (plus magnetite), corresponding to the two lower grade fields of Figure 29.10 and Figure 29.11a. Note that diopside is stable at low grades in the presence of pure H₂O fluids (it requires higher grades in calcareous rocks). Notice also in Figure 29.11a that tremolite and/or talc may be present in more SiO₂-rich serpentinites.

Antigorite breaks down progressively via several reactions in Figure 29.10. If brucite is present, it is less abundant than serpentine and will be consumed by reacting with antigorite at point 2 in Figure 29.10 (~400°C) by the reaction:

$$20 \, \text{Brc} + \text{Atg} = 34 \, \text{Fo} + 51 \, \text{H}_2\text{O} \qquad (29.23)$$

This marks the high-*T* stability limit of brucite in serpentinites and the low-*T* stability limit of **olivine**. A typical serpentinite should then contain antigorite + diopside + forsterite (Figure 29.11b). The olivine that forms from the prograde

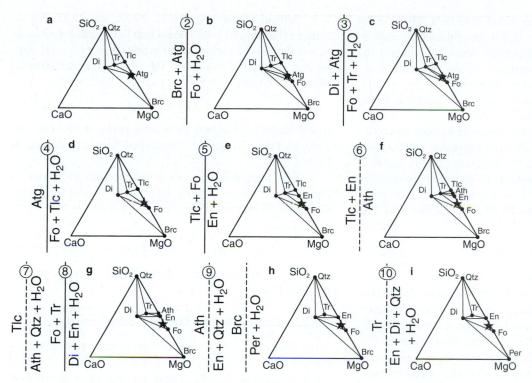

FIGURE 29.11
Chemographics of ultramafic rocks in the *CMS-H* system (projected from H_2O), showing the stable mineral assemblages (in the presence of excess H_2O) and changes in topology caused by reactions along the medium *P/T* metamorphic field gradient illustrated in Figure 29.10. The star represents the composition of a typical mantle lherzolite. Dashed reactions represent those that do not occur in typical ultramafic rocks but rather in unusually SiO_2-rich or SiO_2-poor varieties. After Spear (1993).

dehydration of serpentine tends to be more Mg-rich than the original peridotitic olivine, because of the oxidation of Fe to magnetite upon initial serpentinization.

At point 3 in Figure 29.10 (~530°C), antigorite reacts with diopside via the reaction:

$$8\ Di\ +\ Atg\ =\ 18\ Fo\ +\ 4\ Tr\ +\ 27\ H_2O \quad (29.24)$$

As you can see by comparing Figures 29.11b and 29.11c, the reaction involves a tie-line flip in which the low-*T* Di-Atg tie-line is replaced by the high-*T* Tr-Fo tie-line. The Di-Atg tie-line separated the typical peridotite composition (star) from tremolite at lower grades and the new Tr-Fo tie-line separates the star from diopside at higher grades. Although the reaction does not represent the absolute stability limits of either phase, it results in the loss of diopside and the introduction of **tremolite** in a typical (low-SiO_2) ultramafic. As the chemographics indicate, a more silica-rich ultramafic may contain both diopside and tremolite. I made similar arguments about isograds that introduce a phase without representing its true stability limits in regard to pelitic rocks in Chapter 28, and such situations are common in ultramafic (and calcareous) rocks as well.

Serpentine is finally lost as the stability limit of antigorite alone is reached at point 4 in Figure 29.10 (~570°C) via the reaction:

$$Atg\ =\ 18\ Fo\ +\ 4\ Tlc\ +\ 27\ H_2O \quad (29.25)$$

This also marks the first appearance of **talc** in typical (olivine-bearing) ultramafics (Figure 29.11d). The olivine crystals produced by this reaction can in some cases be very large, elongate or bladed crystals, set in a talc-rich matrix. The texture resembles quenched *spinifex* olivines (Section 3.1.1), but the mineral assemblage is clearly metamorphic.

At pressures above about 1.5 GPa (~55 km) antigorite breaks down via the reactions:

$$14\ Tlc\ +\ Atg\ =\ 45\ En\ +\ 45\ H_2O \quad (29.26)$$

$$Atg\ =\ 14\ Fo\ +\ 10\ En\ +\ 31\ H_2O \quad (29.27)$$

The reaction curves are shown in Figure 29.10, but neither of these reactions occurs with medium *P/T* gradients, and they will not be discussed further.

At point 5 in Figure 29.10 (~650°C), talc is lost and **enstatite** appears, caused by the reaction:

$$2\ Tlc\ +\ 2\ Fo\ =\ 5\ En\ +\ 2\ H_2O \quad (29.28)$$

This is the true low-*T* stability limit of enstatite in H_2O-saturated ultramafic rocks, as can be readily ascertained by noting that enstatite is the only solid phase on the right side of Reaction (29.28). Above this reaction a typical ultramafic rock contains forsterite + enstatite + tremolite (Figure 29.11e). Enstatite-bearing ultramafic rocks are associated in the field with amphibolite facies rocks, and the orthopyroxene in this case does not signify granulite facies conditions, as it typically does in mafic and quartzo-feldspathic rocks.

At lower pressures, **anthophyllite** is introduced into ultramafics due to the reaction:

$$9\ Tlc\ +\ 4\ Fo\ =\ 5\ Ath\ +\ 4\ H_2O \quad (29.29)$$

Anthophyllite has a limited stability field in typical (low-SiO_2) metamorphosed ultramafic rocks (the dark shaded area in Figure 29.10). The upper-temperature limit of anthophyllite in typical ultramafic compositions is limited by the reaction:

$$2\ Ath\ +\ 2\ Fo\ =\ 9\ En\ +\ 2\ H_2O \quad (29.30)$$

because forsterite is more abundant than anthophyllite (see Figure 29.11f). The anthophyllite field is about 40°C across and is constrained to pressures below 0.65 GPa. The grid in Figure 29.10 is simplified in that it recognizes only antho-phyllite and not Mg-cummingtonite, which is common in meta-ultramafic rocks. Petrogenesis of the amphiboles is complex and beyond the scope of our review. For $X_{Fe} = 0.1$, Evans (personal communication) places the transition from anthophyllite to cummingtonite at ~650°C.

Talc is more abundant than forsterite in more siliceous ultramafics (e.g., see Figure 29.11e). Forsterite will thus be consumed first in such rocks via Reaction (29.28), and the assemblage will be talc + enstatite + tremolite above it. In such silica-rich rocks, talc + enstatite will react to form an-thophyllite at point 6 in Figure 29.10. The reaction does not involve H_2O, and is:

$$Tlc + 2\,En = Ath \qquad (29.31)$$

Because this reaction does not occur in typical (olivine-rich) ultramafics, it is dashed in Figure 29.11. The anthophyllite stability field in SiO_2-rich ultramafics is larger than in typical ultramafics and is lightly shaded in Figure 29.10. Although theoretically possible, field evidence for such SiO_2-rich ultramafics is meager. In these speculative SiO_2-rich ultramafics, talc would also remain stable to over 750°C (point 7 in Figure 29.10), where it would finally suc-cumb via the terminal reaction:

$$7\,Tlc = 3\,Ath + 4\,Qtz + 4\,H_2O \qquad (29.32)$$

At higher pressure, the talc-out terminal reaction in SiO_2-rich ultramafic rocks is:

$$2\,Tlc = 2\,Qtz + 3\,En + 2\,H_2O \qquad (29.33)$$

At point 8 in Figure 29.10 (~800°C) **diopside** makes a *reappearance* in typical ultramafic rock compositions due to the reaction:

$$2\,Fo + 2\,Tr = 5\,En + 4\,Di + 2\,H_2O \qquad (29.34)$$

Now we lose the Fo-Tr tie-line that was created in Reaction (29.24), so that the starred composition in Fig-ure 29.11g is once again within a sub-triangle having Di at an apex. This time the diopside-bearing sub-triangle contains diopside + forsterite + enstatite, however, and not diopside + tremolite + serpentine. The loss of tremolite and generation of diopside at point 8 is a change nearly the opposite to that associated with Reaction (29.24).

The loss and reappearance of diopside with increasing grade is unusual. It happens because diopside is stable throughout the full temperature range, and is lost and re-gained due only to changes in the topology of the tie-lines.

By this grade the mineralogy of a meta-ultramafic is similar to the original peridotite (olivine + orthopyroxene + clinopyroxene). Orthopyroxene + clinopyroxene are indica-tive of the granulite facies in ultramafic rocks. Few further changes occur with increasing grade.

Anthophyllite would still be stable in any SiO_2-rich ultramafic rocks, however, and would break down at point 9 (~800°C) via the reaction:

$$2\,Ath = 2\,Qtz + 7\,En + 2\,H_2O \qquad (29.35)$$

Although Reactions (29.31, 29.32, and 29.35) are in-cluded in Figure 29.11, we should bear in mind that they occur only in ultramafics that are richer in SiO_2 than the typical com-position represented by the star. The reactions are dashed to in-dicate this. Similarly, just above point 9 is the reaction:

$$Brc = Per + H_2O \qquad (29.36)$$

which only occurs in very SiO_2-*poor* rocks. Because it, too, does not occur in typical ultramafics, it is also dashed in Figure 29.11.

The final reaction along the medium P/T path in Figure 29.10 occurs at point 10 (> 900°C) and represents the upper thermal stability limit of tremolite *in SiO_2-rich ultramafics*:

$$2\,Tr = 3\,En + 4\,Di + 2\,Qtz + 2\,H_2O \qquad (29.37)$$

At this grade, however, the amphibole is really an Mg-rich hornblende.

29.2.2 The Effect of Other Components

From Table 10.2 we can see that the average peridotite con-tains 1.5 to 2 wt. % Al_2O_3 and 7 to 8 wt. % FeO_T (recalcu-lated to include Fe_2O_3). These are the only nonvolatile components other than CaO-MgO-SiO_2 that occur in amounts greater than 1% (with the occasional exception of Cr_2O_3).

The principal effects of iron are (1) the formation of Fe-oxides or spinels and (2) Fe-Mg solution in mafic silicate minerals. The effect of the latter is to cause the univariant re-actions above to become divariant and continuous. Remem-ber that continuous reactions occur over an interval of grade and are not sharp discontinuous isograds. Products and reac-tants of continuous reactions will overlap in the field to some extent. Analyses of mafic minerals in ultramafic rocks (Figure 24.17) show them to be Mg-rich. Mg/(Mg + Fe) tends to be greater than 0.93, with the exception of antho-phyllite (which can be as low as 0.88). Given this limited Fe-Mg compositional range, the overlap effect of continuous reactions will probably be minor.

The effect of aluminum is to stabilize an Al-rich phase. Chlorite is common in metamorphosed ultramafic rocks in the greenschist, blueschist, and amphibolite facies. It is much more stable in ultramafic rocks (where it may be stable up to 700 to 750°C) than in quartz-bearing rocks, such as pelites (where it typically gives way to garnet at ~450°C). In the upper amphibolite facies and granulite facies, spinel is usually the Al-rich phase at higher pressures, whereas plagioclase is more common at lower pressures. Garnet is the aluminous phase in the eclogite facies. Al_2O_3 is also an important com-ponent in amphiboles and will increase their stability range in ultramafic rocks. Na_2O can have a similar effect. The other minor nonvolatile chemical species will either mix in one of the major phases or stabilize an accessory phase.

29.2.3 The Effect of CO_2

If you remember seeing serpentine in building facades or shop windows, you may recall that it commonly contains white veins of carbonate. Carbonates are relatively common in metamorphosed ultramafic rocks. Sometimes the carbonate is of mantle origin, but most commonly it occurs as secondary minerals. Although simplified, a brief look at the effects of CO_2 on the *CMS-H* system will help us to understand ultramafic paragenesis a bit more fully. If we add CO_2 as a component and dolomite + magnesite as phases, we must consider several reactions in addition to those discussed above. Calcite also occurs, but is generally restricted to more Ca-rich compositions than in typical ultramafic rocks. Figure 29.12 is a simplified $T\text{-}X_{CO_2}$ diagram (at 0.5 GPa) that illustrates most of the stable reactions in the $CaO\text{-}MgO\text{-}SiO_2\text{-}H_2O\text{-}CO_2$ (*CMS-HC*) system. I have listed the additional reactions at the end of the chapter.

The most obvious effect of adding CO_2 is the development of the carbonate minerals dolomite and magnesite. Because these are vapor-bearing minerals, they tend to dominate the low-grade portions of Figure 29.12. Their presence provides a low-temperature limit for tremolite and diopside, both of which were stable at the lowest temperatures considered in the *CMS-H* system (Figure 29.10). Although they must break down to hydrous phases at some point, the carbonates constrain tremolite and diopside to $X_{CO_2} < 0.05$ at temperatures below 450°C.

Increasing X_{CO_2} progressively limits the stability of the hydrous minerals in the *CMS* system. This is particularly true for serpentine and brucite. The thirteen reactions that involve antigorite (listed at the end of the chapter) all occur in the narrow dark-shaded region below 530°C at $X_{CO_2} < 0.03$. The addition of relatively small quantities of CO_2 thus causes antigorite to react, eventually to talc + magnesite. Brucite (not shown) is even more limited (to $X_{CO_2} < 0.01$). The carbonate veins in some decorative serpentinites suggest that the serpentine-carbonate equilibrium is internally buffered, or that the veins represent localized infiltration that is not in equilibrium with the bulk of the serpentinite.

The stability of anthophyllite is also limited as CO_2 is added. The darker shaded area of anthophyllite stability in Figure 29.12 corresponds to the area of typical (low-SiO_2) ultramafic rocks shaded darkly in Figure 29.10. The stability of anthophyllite in CO_2-bearing fluids is limited by the dilution of H_2O in the same dehydration reactions involved in Figure 29.10. Tremolite is also limited at high temperature by CO_2 dilution of the reaction 8 in Figure 29.11, and at low temperature by reaction to carbonates and talc. Although the field of tremolite narrows as X_{CO_2} increases, it is stable up to values of ~0.95.

The evolution of ultramafic rocks with increasing grade and CO_2-bearing fluids is similar to that of carbonates. Diopside, forsterite, and/or enstatite may develop at higher grades, depending on the bulk rock and fluid compositions. We will eschew a detailed analysis of Figure 29.12, similar to the approach we took with Figures 29.3 and 29.6. The discussion would be too repetitive once we understand the approach and the possibilities of internal buffering in closed systems versus externally controlled fluids in open systems. If the carbonate is primary, it is probably minor, and insufficient to buffer and increase X_{CO_2} of the fluid far before being exhausted. In highly carbonated ultramafics the carbonates are probably secondary. For further discussions of metamorphism of ultramafic rocks in general, see the references at the end of the chapter.

Because metamorphosed ultramafic rocks are commonly hydrated or carbonated, and their composition generally contrasts sharply with those that they contact (generally across a thrust fault), **metasomatism** is common at the margin of the bodies. Metasomatism is also common in contact metamorphosed calcareous rocks where hot silica-rich and acidic H_2O-rich plutons are emplaced into carbonates. Metasomatism involves the infiltration of fluids and/or the diffusion of material through the fluid and solid phases. In ultramafic bodies we can commonly observe spectacular, nearly monomineralic marginal zones of chlorite, phlogopite, actinolite, talc, etc. up to several meters in thickness. **Rodingites** are also calc-silicate-rich rocks that

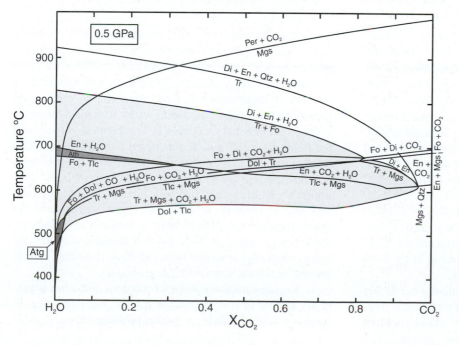

FIGURE 29.12 Simplified $T\text{-}X_{CO_2}$ phase diagram for the system $CaO\text{-}MgO\text{-}SiO_2\text{-}H_2O\text{-}CO_2$ at 0.5 GPa, calculated using the program TWQ of Berman (1988, 1990, 1991). The diagram focuses on ultramafic–carbonate rocks and omits reactions involving quartz. The shaded fields represent the stability ranges of, from darkest to lightest, serpentine (antigorite), anthophyllite in typical low-SiO_2 ultramafics, and tremolite in low-SiO_2 ultramafics.

form in association with serpentinites, usually in mafic to intermediate dike rocks within a serpentinized body. The diffusion of calcium and alkalis necessary to produce rodingites is believed to result from the liberation of these elements as the original peridotite is converted to serpentine. Even aluminum, which is generally regarded as poorly mobile, may migrate during metamorphism in ultramafic systems. **Skarns** of calc-silicate minerals form as a result of diffusion into carbonates. We will address the nature of metamorphic fluids and metasomatism in the next and final chapter of the text.

ADDITIONAL REACTIONS IN THE CaO-MgO-SiO₂-H₂O-CO₂ SYSTEM

$$Atg + 5\,Dol = 24\,Fo + 5\,Di + 31\,H_2O + 10\,CO_2$$

$$Atg + 20\,Dol = 34\,Fo + 20\,Cal + 31\,H_2O + 20\,CO_2$$

$$Atg + 47\,Di + 30\,CO_2 = 16\,Tr + 15\,Dol + 15\,H_2O$$

$$4\,Atg + 141\,Cal + 73\,CO_2$$
$$= 107\,Dol + 17\,Tr + 107\,H_2O$$

$$Atg + 20\,Mgs = 34\,Fo + 31\,H_2O + 20\,CO_2$$

$$48\,Cal + Atg + 14\,CO_2 = 17\,Di + 31\,Dol + 31\,H_2O$$

$$3\,Atg + 20\,Cal = 62\,Fo + 20\,Di + 93\,H_2O + 20\,CO_2$$

$$45\,Cal + 31\,Tr = 107\,Di + Atg + 45\,CO_2$$

$$14\,Cal + 48\,Di + 31\,H_2O = 62\,Wo + Atg + 14\,CO_2$$

$$45\,Mgs + 17\,Tlc + 45\,H_2O = 2\,Atg + 45\,CO_2$$

$$13\,Atg + 40\,Dol = 20\,Tr + 282\,Fo$$
$$+ 383\,H_2O + 80\,CO_2$$

$$34\,Dol + 4\,Atg + 73\,CO_2 = 141\,Mgs$$
$$+ 17\,Tr + 107\,H_2O$$

$$30\,Dol + 47\,Tlc + 30\,H_2O = 15\,Tr + 2\,Atg + 60\,CO_2$$

$$3\,Dol + 4\,Qtz + H_2O = Tlc + 3\,Cal + 3\,CO_2$$

$$5\,Dol + 8\,Qtz + H_2O = Tr + 3\,Cal + 7\,CO_2$$

$$3\,Cal + 2\,Qtz + Tr = 5\,Di + H_2O + 3\,CO_2$$

$$Cal + Qtz = Wo + CO_2$$

$$2\,Ath = 7\,En + 2\,Qtz + 2\,H_2O$$

$$7\,Tlc = 4\,Qtz + 3\,Ath + 4\,H_2O$$

$$Tlc + 2\,En = Ath$$

$$2\,Tlc = 3\,En + 2\,Qtz + 2\,H_2O$$

$$5\,Tlc + 6\,Cal + 4\,Qtz = 3\,Tr + 2\,H_2O + 6\,CO_2$$

$$2\,Dol + 4\,Qtz + Tlc = Tr + 4\,CO_2$$

$$3\,Cal + 2\,Tlc = Tr + Dol + H_2O + CO_2$$

$$2\,Fo + 2\,Tlc = 5\,En + H_2O$$

$$Mgs + 4\,Qtz + H_2O = Tlc + 3\,CO_2$$

$$Tr + 3\,Cal = 4\,Di + Dol + CO_2 + H_2O$$

$$2\,Dol + 3\,Mgs + 8\,Qtz + H_2O = Tr + 7\,CO_2$$

$$Tr + 2\,Dol = 4\,Di + 3\,Mgs + CO_2 + H_2O$$

$$Tr + 5\,Mgs = 2\,Dol + 4\,En + CO_2 + H_2O$$

$$Tr + 3\,CO_2 = 2\,Di + 3\,Mgs + 4\,Qtz + H_2O$$

$$10\,Fo + 2\,Tr + 8\,CO_2 = 13\,En + 4\,Dol + 2\,H_2O$$

$$Dol + 2\,Tr = 3\,En + 5\,Di + 2\,H_2O + 2\,CO_2$$

$$Dol + 2\,Qtz = Di + 2\,CO_2$$

$$3\,Dol + Di = 4\,Cal + 2\,Fo + 2\,CO_2$$

$$2\,Fo + Di + 2\,CO_2 = Dol + 2\,En$$

$$2\,Mgs + Di = Dol + En$$

$$2\,Mgs + En = 2\,Fo + 2\,CO_2$$

$$Cal + Fo = Wo + Per + CO_2$$

$$Cal + 2\,Di = 3\,Wo + Fo + CO_2$$

$$Dol = Per + Cal + CO_2$$

Summary

Calcareous protoliths are limestones or dolostones (perhaps containing chert or detrital silicates) that are typically deposited in stable continental shelf environments. Because devolatilization reactions in carbonates produce CO_2, which mixes with the nearly ubiquitous H_2O in crustal fluids, evolving fluid composition may exert an important control on the mineral assemblages that develop during metamorphism. T-X_{CO_2} diagrams (at a particular P) are particularly useful in analyzing mineral stabilities and fluid composition in the CaO-MgO-SiO₂-H₂O-CO₂ (*CMS-HC*) system. The sequence of zones that typically develops in progressively metamorphosed siliceous dolomites is talc → tremolite → diopside and/or forsterite → wollastonite, but the sequence of isograds, and grade at which each develops, depends on X_{CO_2}. Periclase and more exotic minerals may develop in the inner aureoles of hotter plutons (e.g., anorthosites).

Open-system behavior occurs in relatively permeable rocks when fluids readily move through the rock matrix (typically exiting the system as devolatilization proceeds). Fluid composition in open systems is usually controlled externally and remains essentially constant, so a traverse up metamorphic grade (e.g., toward the igneous contact) follows a vertical path on a T-X_{CO_2} diagram. Mineral assemblages are thus typically divariant (*three* minerals coexisting in *CMS*), with univariant assemblages occurring only at marked isograds. Closed-system behavior occurs in less permeable rocks and the fluid composition is thus buffered internally (controlled by the local minerals and reaction progress). Closed systems are characterized by univariant mineral assemblages (*four* minerals coexisting in *CMS*) over broader temperature (and thus spatial) intervals, in many cases separated by invariant mineral assemblages. Fluids may be buffered in closed systems to high values of X_{CO_2} (so that diopside may develop more commonly) or to low values of X_{CO_2} (resulting, if Al is present, in zoisite, vesuvianite, or grossular).

Regional metamorphism of siliceous dolomites is generally similar to contact metamorphism, but talc has a diminished field of stability at higher pressure ($X_{CO_2} < 0.15$)

and diopside has a larger one (extending to nearly the full range of X_{CO_2} values), so the talc zone is suppressed and the diopside zone is more prevalent. Calc-silicate rocks are more complex, involving additional components (notably Al_2O_3, Na_2O, and K_2O) and a correspondingly larger range of minerals (including plagioclase, zoisite, biotite, hornblende, and garnet).

Ultramafic rocks are usually mantle-derived protoliths that typically retrogress in the upper crust to serpentinite before metamorphosing to higher grades. Metamorphism at crustal levels typically involves the sequential development of serpentine, diopside, forsterite, tremolite, talc, anthophyllite, and enstatite.

Key Terms

Metacarbonate *635*
Marble *635*
Calc-silicate *635*
Skarn *635*

CMS-HC *636*
Open-system behavior *637*
Closed-system behavior *639*
Alpine/orogenic peridotite *647*

Ophiolite suite *647*
Suture zone *647*
Rodingite *651*

Review Questions and Problems

Review Questions and Problems are located on the author's web page at the following address: **http://www.prenhall.com/winter**

Important "First Principle" Concept

■ Fluid flow and external buffering of fluids (open system) versus internal buffering (closed system) can exert a strong control over the mineral assemblages developed during metamorphism.

Suggested Further Readings

Calcareous Rocks

Bowen, N. L. (1940). Progressive metamorphism of siliceous limestone and dolomite. *J. Geol.,* **48**, 225–274.

Bowman, J. R. (1998). Basic aspects and applications of phase equilibria in the analysis of metasomatic Ca-Mg-Al-Fe-Si skarns. In: *Mineralized Intrusion-Related Skarn Systems* (ed. D. R. Lentz). *Mineralogical Association of Canada Short Course Notes,* **26**, 1–49.

Bucher, K., and M. Frey. (2002). *Petrogenesis of Metamorphic Rocks.* Springer. Heidelberg. Chapters 6 and 8.

Eskola, P. (1922). On contact phenomena between gneiss and limestone in western Massachusetts. *J. Geol.,* **30**, 265–294.

Ferry, J. M. (1976). Metamorphism of calcareous sediments in the Waterville-Vassalboro area, south-central Maine: mineral reactions and graphical analysis. *Amer. J. Sci.,* **276**, 841–882.

Greenwood, H. J. (1975). The buffering of pore fluids by metamorphic reactions. *Amer. J. Sci.,* **275**, 573–593.

Kerrick, D. M. (1974). Review of metamorphic mixed-volatile (H_2O-CO_2) equilibria. *Am. Mineral.,* **59**, 729–762.

Lentz, D. R. (ed.). (1998). *Mineralized Intrusion-Related Skarn Systems. Mineralogical Association of Canada Short Course Notes,* **26**.

Spear, F. S. (1993). *Metamorphic Phase Equilibria and Pressure-Temperature-Time Paths.* Monograph **1**. Mineralogical Society of America. Washington, DC. Chapter 12.

Tracy, R. J., and B. R. Frost. (1991). Phase equilibria and thermobarometry of calcareous, ultramafic and mafic rocks, and iron formations. In: *Contact Metamorphism* (ed. D. M. Kerrick). *Mineralogical Society of America Reviews in Mineralogy,* **26**, 207–290.

Yardley, B. W. D. (1989). *An Introduction to Metamorphic Petrology.* Longman. Essex, UK. Chapter 5.

Ultramafic Rocks

Bucher, K., and M. Frey. (2002). *Petrogenesis of Metamorphic Rocks.* Springer. Heidelberg. Chapter 5.

Evans, B. W. (1977). Metamorphism of alpine peridotite and serpentinite. *Ann. Rev. Earth Planet. Sci.,* **5**, 397–447.

Frost, B. R. (1975). Contact metamorphism of serpentinite, chloritic blackwall and rodingite at Paddy-go-easy Pass, central Cascades, Washington. *J. Petrol.,* **16**, 272–313.

Frost, D. J. (2006). The stability of hydrous mantle phases. In: *Water in Nominally Anhydrous Minerals* (eds. H. Keppler and J. Smyth). *Reviews in Mineralogy and Geochemistry,* **62**, 243–271.

Greenwood, H. J. (1967). Mineral equilibria in the system MgO-SiO_2-H_2O-CO_2. In: *Researches in Geochemistry* (ed. P. H. Abelson). John Wiley. New York. pp. 542–567.

Iwamori, H. (2004). Phase relations of peridotites under H_2O-saturated conditions and ability of subducting plates for transportation of H_2O. *Earth Planet. Sci. Lett.,* **227**, 57–71.

Spear, F. S. (1993). *Metamorphic Phase Equilibria and Pressure-Temperature-Time Paths.* Monograph **1**. Mineralogical Society of America, Washington, DC. Chapter 13.

Tracy, R. J., and B. R. Frost. (1991). Phase equilibria and thermobarometry of calcareous, ultramafic and mafic rocks, and iron formations. In: *Contact Metamorphism* (ed. D. M. Kerrick). *Mineralogical Society of America Reviews in Mineralogy,* **26**, 207–290.

Trommsdorff, V., and B. W. Evans. (1972). Progressive metamorphism of antigorite schist in the Bergell tonalite aureole (Italy). *Amer. J. Sci.,* **272**, 423–437.

Trommsdorff, V., and B. W. Evans. (1974). Alpine metamorphism of peridotitic rocks. *Schweiz. Mineral. Petrol. Mitt.,* **54**, 333–352.

30

Metamorphic Fluids, Mass Transport, and Metasomatism

Questions to be Considered in this Chapter:

1. What is the nature of fluids that accompany metamorphism?

2. What volumes of fluid pass through crustal rocks during metamorphism?

3. What occurs when various elements become mobilized during metasomatism?

luids play an influential role in many igneous and metamorphic processes, where they may partake in various types of reactions or act as a transporting medium for the movement of soluble material. We have often encountered processes influenced by fluids and material transport in the preceding pages of this text. We noted the effects of introduced fluids on the melting of rocks and the effects of released fluids from crystallizing igneous bodies. We discussed the role of fluids in seafloor and hydrothermal metamorphism (Chapter 21), in the development of fenites associated with carbonatites, and in metasomatism of the mantle (Chapter 19), in the amphibolite facies-granulite facies transition (Chapters 25 and 28), and in infiltration processes (Chapter 29). Fluids are essential components in pegmatite and ore genesis as well. Fluids can change the transport properties of a rock by dissolution or precipitation of material, or by creating fractures due to fluid pressure ("hydrofracturing"). Fluids can also affect the mechanical properties of rocks, and therefore the nature and extent of deformation in areas under stress. In this final chapter, I summarize what we know about crustal fluids and then provide a brief introduction to mass transport and metasomatic processes.

30.1 METAMORPHIC FLUIDS

As discussed in Section 21.2.4, fluids are believed to be nearly ubiquitous during metamorphism (at least at low and medium grades), but there is little direct remaining evidence concerning their nature in the rocks that we now collect at the Earth's surface. Virtually all of the intergranular high-temperature/high-pressure fluids that were in equilibrium with the metamorphic mineral assemblage during peak metamorphic conditions escaped as pressure was released upon uplift and erosion. The only direct evidence for metamorphic fluids that remains is found in the small high-density fluid inclusions (Figure 7.17) trapped in many metamorphic rocks (see the summaries by Hollister and Crawford, 1981; Roedder, 1984; and Crawford and Hollister, 1986). The existence of fluids trapped in a rock, however, is not an indicator of the *quantity* of fluid that existed during metamorphism. Nor is there any guarantee that the fluids coexisted with the peak equilibrium mineral assemblage and were not trapped during a later event following uplift. Studies of freezing and CO_2-H_2O homogenization temperatures for complex fluid inclusions using a heating-freezing microscope stage, however, suggest that many (but not all) inclusions were trapped under metamorphic conditions. The nature of the fluid inclusions commonly correlates with the metamorphic mineral assemblage as well, further strengthening the argument that some fluid exists as a discrete phase during metamorphism. The heating-freezing technique has even been used as a geothermobarometer to estimate the conditions of metamorphism (e.g., Touret, 1981). Recovery of fluids from geothermal wells provides further direct evidence for the participation of fluids in metamorphism (e.g., White et al., 1963; Muffler and White, 1969; Bird et al., 1984; Schiffman et al., 1984; Yardley et al., 1993).

Other evidence for metamorphic fluids is less direct but also compelling. The existence of hydrous and carbonate minerals under peak metamorphic conditions all but requires that a fluid exists to maintain sufficient volatile pressure to stabilize the volatiles in the minerals. Experiments on mineral–fluid equilibria and thermodynamic treatment support this contention and commonly permit us to characterize the type of fluid present. In Chapter 29, for example, we were able to constrain the temperature of metamorphism and X_{CO_2} of CO_2-H_2O fluid mixtures on the basis of calcareous or ultramafic mineral assemblages and the location of reaction equilibria on T-X_{CO_2} diagrams. We could even speculate as to whether the fluids were infiltrating or internally buffered. In addition, numerous stable isotopic studies indicate that the characteristics of many metamorphic rocks can be explained only if the original isotopic composition (usually $\delta^{18}O$) of the rocks interacted and exchanged with fluids with which they were initially out of isotopic equilibrium (e.g., Garlick and Epstein, 1967; Rye et al., 1976; Rumble et al., 1982; Nabelek et al., 1984; Bowman et al., 1994; Lewis et al., 1998).

When mud or shaly sediment is first deposited, it may have an initial porosity in excess of 50%, the pore space being occupied by meteoric water or seawater brines. Compaction, cementation, and recrystallization usually reduces the initial porosity to vanishingly small values at depths of burial corresponding to the beginning of regional metamorphism. Rice and Ferry (1982) estimated that high-grade rocks probably have less than 0.1% porosity. Initial pore fluids, therefore, are probably driven out early and contribute little to the fluid phase attending subsequent metamorphism. Volatile components bound in hydrous or carbonate minerals, however, are typically released during metamorphism. Average slates contain about 4.5 wt. % bound H_2O and 2.3 wt. % bound CO_2 (Shaw, 1956), whereas high-grade schists contain about 2.4 wt. % H_2O and 0.2 wt. % CO_2 (and granulites contain hardly any volatile components at all). Release of these bound volatiles (by devolatilization reactions) between the slate and schist stages amounts to approximately 1.5 moles of fluid (mostly H_2O and CO_2) per kilogram of rock. Walther and Orville (1980) calculated that, if this fluid were released at once at 500°C and 0.5 GPa, it would occupy about 12% of the rock volume.

Fluids may also be introduced into rocks during metamorphism. Metamorphism of a basalt to a greenschist typically involves the addition of 5 to 10 wt. % H_2O from the surroundings. Fluids added during metamorphism may be meteoric, magmatic, or metamorphic (released by devolatilization reactions at greater depth). Stable isotopic data suggest that meteoric and magmatic waters have interacted with the rocks in numerous metamorphic terranes, particularly those with plentiful shallow-level plutonism and hydrothermal circulation (hydrothermal and "regional contact" metamorphism). A. B. Thompson (1983), however, warned us that, although fluid generation via devolatilization is a

consequence of most forms of metamorphism, fluids may not be present at all times and in all places during any metamorphic event.

30.1.1 The Nature of Metamorphic Fluids

At the high temperatures and pressures of regional metamorphism, a free "vapor" phase is usually a **supercritical fluid** because the conditions are above that of the critical point of CO_2-H_2O mixtures (see Figure 6.7 and Section 6.4). There is no distinction between a gas and liquid, and the density of supercritical fluids at high pressure and temperature is close to that of water at 25°C and 1 atm (Ferry and Burt, 1982).

30.1.1.1 VOLATILE SPECIES   For the moment, let's consider just the volatile species in supercritical metamorphic fluids (see Ferry and Burt, 1982; Labotka, 1991; and Huizenga, 2001, for reviews). These are composed predominantly of O, H, and C, with lesser amounts of S, N, etc. Figure 30.1 illustrates the volatile species that coexist in C-O-H-S fluids at 1100°C and 0.5 GPa. Because a single C-O-H-S fluid phase contains four components, the phase rule tells us that $F = 4 - 1 + 2 = 5$ (or three at fixed P and T). There is thus considerable variance possible in such fluids unless some species are buffered by the solids. In Figure 30.1, sulfur fugacity is controlled by pyrrhotite, and the resulting fluid species are projected from sulfur to the C-O-H face of the compositional space. The resulting fluids in equilibrium with such buffered f_{S_2} are thus only divariant.

Figure 30.1 is divided into regions of bulk fluid composition labeled with the predominant species in each region (other species are always present but occur in very low concentrations). Fluids between the H_2O-CO_2 join and the O corner are relatively oxidized and consist of mixtures of CO_2, H_2O, and O_2. Remember from Section 26.7 that free oxygen is generally buffered to very low concentrations, and thus fluids in the CO_2-H_2O-O_2 triangle (light shaded) probably do not occur during metamorphism. Fluids in the H_2O-CH_4-H triangle are more reduced and are dominantly $H_2O + CH_4 + H_2S$. In all cases, of course, the proportions vary with the bulk composition (X_{bulk}) of the fluid mixtures. Increasing carbon concentration eventually stabilizes graphite, which then buffers the carbon content of the fluids

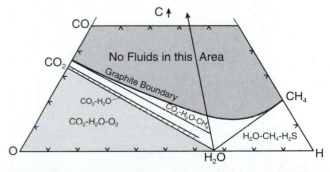

FIGURE 30.1   Fluid speciation in the C-O-H-S system at 1100°C and 0.5 GPa projected from S to the C-O-H triangle (mole proportions). f_{S_2} is determined by pyrrhotite with the composition $Fe_{0.905}S$. After Holloway (1981).

and prevents fluids from developing more C than the buffered assemblage. Fluids in the dark shaded C-rich portion of Figure 30.1 thus cannot ordinarily occur.

The presence of graphite in many metamorphosed pelites and carbonates substantially limits the compositional variation of metamorphic fluids. (They must occur along the graphite boundary in Figure 30.1.) If f_{S_2} and f_{O_2} are buffered and graphite is present, the variance of C-O-H-S fluids is reduced to 2. We can then calculate the exact speciation at any given P and T, using the equilibrium constants for the reactions that interrelate them (H_2 + ½O_2 = H_2O, C + O_2 = CO_2, etc.). Figure 30.2 is a plot of the relative concentrations of the major species in C-O-H-S fluids in equilibrium with graphite at 0.2 GPa, if f_{S_2} is buffered as in Figure 30.1 and f_{O_2} is buffered by quartz–fayalite–magnetite (see Figure 26.10). Note in Figure 30.2 that the common assumption that metamorphic fluids are predominantly CO_2-H_2O mixtures is true for graphite-bearing assemblages at very high temperature, but CH_4 is dominant below 700°C. When graphite is absent, however, fluids may be more oxidized, and methane less common. H_2O, CO_2, and CH_4 are the major species in metamorphic C-O-H-S fluids. H_2S is less abundant in most metamorphic fluids than indicated by Figure 30.2 because the amount of sulfur is usually much more limited than the pyrrhotite buffer.

In Chapter 29, we used several calcareous and ultramafic mineral assemblages to estimate the composition of H_2O-CO_2 fluids, buffered either along univariant T-X_{CO_2} reactions or at invariant points. Ferry and Burt (1982) presented tables of calculated fluid species from carbonates, pelitic schists and granulite facies gneisses.

30.1.1.2 NONVOLATILE SOLUTES Although we commonly treat metamorphic mineral equilibria as solid–solid or solid–volatile species reactions (in Chapters 28 and 29, for example),

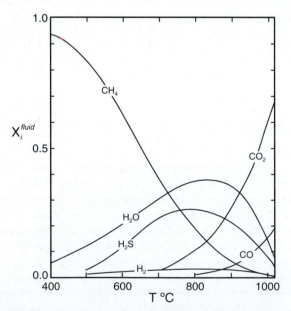

FIGURE 30.2 Speciation in C-O-H-S fluids coexisting with graphite at 0.2 GPa, with f_{O_2} buffered by quartz–fayalite–magnetite and f_{S_2} controlled as in Figure 30.1. x_i^{fluid} is the mole fraction of each species in the fluid. From Holloway (1981).

when a fluid species is present, each solid phase will also progress toward equilibrium with that fluid. Just as in mineral-melt equilibria (Chapters 18 and 19), chemical components will be distributed (partitioned) between every mineral and the fluid in some equilibrium proportionality. We are familiar with this concept for very soluble phases, such as salt in water. Salt will dissolve and Na^+ and Cl^- will be distributed between water and halite for saturated solutions in equilibrium. Consider the reaction:

$$NaCl_{(s)} = Na^+_{(aq)} + Cl^-_{(aq)} \tag{30.1}$$

where the subscript (s) indicates a solid and (aq) indicates an aqueous species. The distribution can be expressed as a constant, K_{SP}, the **solubility product** constant:

$$K_{SP} = a_{Na^+} \cdot a_{Cl^-} \tag{30.2}$$

where a indicates solute activities. The activity of solid NaCl at the pressure and temperature of interest is always 1.0, and is therefore dropped from the denominator of the expression on the right side of Equation (30.2) under saturated conditions.

Of course, most silicates are far less soluble than halite, even at high temperatures, but we can still explore the solubility of minerals in equilibrium with C-O-H fluids experimentally, and we can use the results to model the composition of metamorphic fluids in equilibrium with various mineral assemblages at elevated temperatures and pressures. Helgeson and colleagues have pioneered the quantitative thermodynamic modeling of metamorphic fluids and aqueous species at elevated temperatures and pressures (see Helgeson et al., 1978, 1981). From a combination of thermodynamics and experimental data on mineral solubilities, we can in many cases estimate the composition of metamorphic fluids from the associated mineral assemblages. I will provide some examples of how we can use the relationship between the mineral assemblage and the fluid species activities to model metasomatism shortly. For now, I shall briefly review what we presently know (or think we know) about the dominant solute species in metamorphic fluids.

The minerals in which igneous and metamorphic petrologists are most interested are silicates, and the predominant anion is oxygen. Because of this, and because very little oxygen dissolves in the fluid as O^{2-}, we concern ourselves mostly with the exchange and transport of *cations* with the fluid. Unfortunately, this does not necessarily simplify our job because electrical neutrality requires that an equivalent number of *anions* (or anionic complexes, such as SO_4^{2-}) must also be present in the fluid solution. Petrologists agree that the dominant anion is chloride, but it is generally not possible to estimate the total dissolved chloride in metamorphic fluids (which are now mostly gone). As a result, even if we know the solubility of a mineral in chloride brines, the amount of a cation in solution may be considerably less if little Cl^- or other anionic ligands are available. One way to partially circumvent this problem is to calculate the *ratios* of dissolved cations, rather than the absolute concentration of each (Helgeson, 1967; Eugster, 1986). Many mineral reactions involve the exchange of several species with the fluid (Section 26.8), so that the equilibrium conditions for the reaction also depend on *ratios* of the species. For example, the equilibrium constant for Reaction (26.16) is given below:

$$2\ KAlSi_3O_8 + 2\ H^+ + H_2O \qquad (26.16)$$
$$\text{Kfs} \qquad \text{aqueous species}$$
$$= Al_2Si_2O_5(OH)_4 + 4\ SiO_2 + 2\ K^+$$
$$\text{kaolinite} \qquad \text{aqueous species}$$

$$K_{SP} = \frac{(a_{SiO_{2(aq)}})^4 \cdot a_{K^+}^2}{a_{H^+}^2} \qquad (30.3)$$

To define the K-feldspar/kaolinite equilibrium, we thus need know only the ratios of the activities of the aqueous species, and not the activity of each individual one. Regardless of the convenience of solute ratios, it would be nice at this point to know what silicate mineral constituents are most soluble in high-*T-P* aqueous solutions and to what extent.

Although we are used to considering dissolved solutes in aqueous solutions as ionic species, the dissolved constituents are much more associated in most metamorphic fluids (Helgeson et al., 1981; Eugster, 1986; Labotka, 1991). The **dielectric constant (ε)** of water expresses its ability to shield charged dissolved ligands by *solvation* (surrounding them with a layer of charged bipolar H_2O molecules). ε decreases with increasing temperature and decreasing pressure. When ε is large, as in near-surface waters, solutes are readily shielded and tend to occur as charged ionic species. When ε is small, charged species are suppressed.

We may consider the dissociation of an electrolyte as a reaction of the sort:

$$NaCl_{(aq)}^{\circ} = Na_{(aq)}^+ + Cl_{(aq)}^- \qquad (30.4)$$

where $NaCl_{(aq)}^{\circ}$ is the concentration of associated NaCl molecules in solution [not solid NaCl, as in Reaction (30.1)]. From Reaction (30.4) we can define the **dissociation constant** for NaCl° as:

$$K_{NaCl^{\circ}} = \frac{a_{Na^+} \cdot a_{Cl^-}}{a_{NaCl^{\circ}}} \qquad (30.5)$$

Figure 30.3 shows how the dissociation of NaCl in aqueous solutions varies with temperature and pressure. We can see that NaCl is highly dissociated ($K_{NaCl^{\circ}} > 10$) in saturated surface waters, but the dissociation constant decreases significantly with increasing temperature and decreasing pressure. Under conditions of contact metamorphism $K_{NaCl^{\circ}}$ may be 10^{-2} or less, so that the associated NaCl° molecules outnumber the dissociated Na^+ and Cl^- ions in the fluid by over two orders of magnitude. Under regional metamorphic conditions the associated species are still 5 to 10 times as prevalent as dissociated. Chlorides (NaCl, KCl, $CaCl_2$, $MgCl_2$) are the most common associated species in natural metamorphic fluids, but hydroxides [$Al(OH)_3$, $KAl(OH)_4$], acids (H_4SiO_4), sulfates, carbonates, fluorides, etc. may also be present.

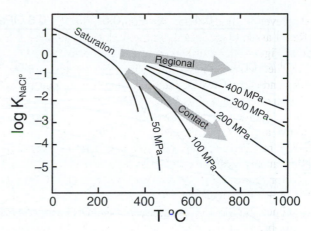

FIGURE 30.3 Variation in the dissociation constant of NaCl in aqueous solutions with temperature and pressure. Shaded arrows indicate regional and contact metamorphic *P-T* paths. After Sverjensky (1987). Copyright © The Mineralogical Society of America.

Natural observations and experiments addressing dissolved species show that their concentrations depend upon temperature, pressure, and the composition of the rock and fluid (including the volatile species and the concentration of other solutes). Although exceptions abound, the following broad generalizations seem justified at our current state of knowledge.

In addition to halides, alkalis, particularly Na, are usually the most soluble constituents. On the basis of fluid inclusions, Crawford (1981) concluded that most fluids are chloride brines in which NaCl dominates over KCl and $CaCl_2$ with only minor $MgCl_2$. Other species are only slightly soluble. Crawford (1981) noted that CO_3^{2-}, HCO_3^-, SO_4^{2-}, and Br^- are also present in many inclusions.

Walther and Helgeson (1977) determined the solubility of quartz (or its polymorphs) in aqueous solutions at elevated temperatures. The solubility ranges from 2 ppm SiO_2 at 25°C to about 10,000 ppm (1 wt. %) at 600°C and 0.5 GPa. Ragnarsdottir and Walther (1985) found that corundum is much less soluble and that the Al content of metamorphic fluids in equilibrium with corundum cannot be much more than a few dozen parts per million.

Eugster and Baumgartner (1987), in their review of mineral solubilities and speciation in metamorphic fluids, used mineral solubility data, extrapolated to metamorphic temperatures and pressures, to estimate the concentrations of solute species. They noted that available data on fluids are limited, so that uncertainty is relatively large, but they were nonetheless able to semi-quantitatively estimate speciation, at least in some simple aqueous supercritical fluids in equilibrium with equally simple mineral assemblages. Figure 30.4 shows the results of Eugster and Baumgartner's calculations of aqueous–chloride fluid speciation associated with ultramafic rocks with a bulk composition in the Tr-Tlc-Atg sub-triangle of the CaO-MgO-SiO$_2$ system (Figure 29.11) from 350 to 650°C. The abrupt breaks in slope for several species between 400 and 500°C correspond to changes in mineralogy associated with Reactions (29.24) and (29.25). Note that dissociated ionic species

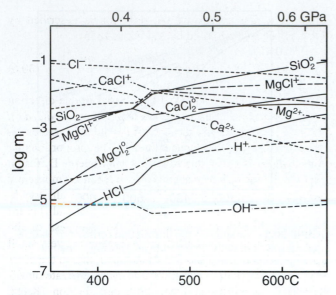

FIGURE 30.4 Speciation in aqueous–chloride fluids calculated for an ultramafic bulk composition, assuming a geothermal gradient of 0.1°C/bar. m_i is the molality of species i in the fluid. Solid curves represent neutral associated species. After Eugster and Baumgartner (1987). Copyright © The Mineralogical Society of America.

(dashed curves for Cl^-, Mg^{2+}, Ca^{2+}, etc.) tend to decrease with increasing grade (H^+ is a notable exception), whereas associated ions ($MgCl^+$ and $CaCl^+$) and uncharged species (SiO_2°, $CaCl_2^\circ$ and $MgCl_2^\circ$) all increase. Na and K, although usually quite soluble, are not considered in Figure 30.4 because of their low abundance in ultramafic systems. Similarly, Fe is not considered in Figure 30.4, although Eugster and Gunter (1981) found that Fe is generally more soluble than Mg.

Walther and Helgeson (1980) found that Ca^{2+} is a thousand times more soluble than Mg^{2+} in metamorphic CO_2-H_2O fluids associated with calcite-bearing calcareous rocks, whereas Mg^{2+} may be more soluble in carbonated metamorphosed ultramafics (which is apparent at high temperatures in Figure 30.4). Helgeson et al. (1981), in their weighty monograph, summarized aqueous solutions as ranging from nearly pure H_2O-CO_2 fluids to concentrated ore-forming solutions, composed predominantly of NaCl, KCl, $CaCl_2$, and $MgCl_2$, with lesser concentrations of bicarbonates and sulfates–bisulfates, together with minor H_2S, SiO_2, and chlorides of Al, Fe, Zn, Pb, Cu, Ag, etc.

Characterization of metamorphic fluids is still in the early stages of development, and the speciation obviously varies, but we are now aware that metamorphic fluids may contain a host of volatile and dissolved species. Dissolved constituents are present mostly in low concentrations, and the relative abundances are commonly Na and K > Si, Ca, Fe, and Mg > Al. In addition to Al, the high field strength (HFS) minor and trace elements such as Ti, Cr, Zr, Y, and Ni are generally insoluble in aqueous fluids, hence their use as indicators of the paleo-tectonic environment of metamorphosed igneous rocks in discrimination diagrams (Section 9.6). The high solubility of large-ion lithophile elements (i.e.,

Rb, Cs, Ba, Pb, Sr, U) explains the "decoupling" of soluble LIL and insoluble HFS elements due to the interaction of hydrous fluids with the mantle source of subduction-related magmas (Chapters 16 and 17).

30.1.2 The Role of Fluids in Metamorphism

We are becoming increasingly aware that fluids are a critical participant in metamorphic and igneous processes. Aqueous fluids reduce the melting point of rocks and enhance melting. They are also released by rising and cooling magmas where they generate pegmatites and ores. Mixtures of exhaled and meteoric fluids feed hydrothermal systems above plutons and in shallow permeable areas of regional metamorphism. As discussed above, fluids released over time by devolatilization reactions may be equivalent to over 10% of the volume of a rock. Fluids can dissolve material, transport heat and solutes, precipitate minerals, exchange components as they react with minerals, and catalyze deformation processes by weakening rocks. Contact metamorphic aureoles are largest where fluids are available to transport heat and matter. Fluid buffering versus open-system behavior (Chapter 29) may control the progress of metamorphism and mineral reactions in many situations. Release and flow of fluids absorb and transfer large quantities of heat and matter. The absorption and transfer by fluids of heat alone may have profound effects upon the temperature distribution in a contact aureole or even throughout an orogenic belt, and hence upon the geologic evolution of an area (Bickle and MacKenzie, 1987). Metamorphism is thus far more complex than simple re-equilibration of minerals to increased temperature and/or pressure upon burial or proximity to an intrusion, as it is commonly perceived.

Fluid transport through rocks has a profound effect on the processes mentioned in the preceding paragraph. Although flow through a porous medium is certainly the most efficient mechanism for fluid transport, rocks are highly porous only in the shallow crust (Section 21.2.4). Below 10 km, a continuous network of large open pore spaces is closed by compaction and recrystallization. Motion of a fluid along fractures or of an intergranular fluid along grain boundaries are the only ways that fluids can migrate in the deeper crust. We will return to this concept in more detail shortly.

30.2 METAMORPHISM

Metasomatism is defined as metamorphism accompanied by changes in whole-rock composition. Because volatiles are so readily released and mobilized during metamorphism, changes in the volatile content of rocks are generally excluded, and the chemical changes that constitute metasomatism (*sensu-stricto*) are usually restricted to the redistribution of nonvolatile species. In Section 21.6.5, we discussed the changes in mineralogy in the contact aureole at Crestmore, California, and found that the mineral assemblages were readily explained by a progressive increase in the amount of SiO_2 in the country rock marbles as the igneous contact was approached (Figure 21.18). SiO_2 was presumably derived from the quartz monzonite porphyry and transported as dissolved silica in aqueous fluids circulating away from the body. Other examples of metasomatism discussed previously include Na metasomatism associated with ocean floor metamorphism (Section 21.3.2), alkali metasomatism associated with carbonatites that produce fenites (Section 19.2.3), mantle metasomatism (Section 19.4), infiltration (Section 29.1.3), and depletion of granulites (Section 25.3.2). We also discussed small-scale migration of dissolved constituents during metamorphic reactions in Section 26.12 and in metamorphic differentiation in Section 23.4.3.

Metasomatism is most dramatically developed in situations where rocks of highly contrasting composition are juxtaposed and elements move easily. Common examples include:

> Shallow plutons, particularly where siliceous magmas contact calcareous or ultramafic rocks and fluids are circulated through open fractures.

> Layers, lenses, and pods with contrasting composition in rocks undergoing metamorphism: ultramafic pods in pelites or carbonates, interbedded carbonate and pelitic sediments.

> Veins where fluids equilibrate with one rock type and then migrate into a contrasting rock type along a fracture.

For a detailed review of the types of metasomatism and their general characteristics, see Barton et al. (1991a). Their main categories are (1) alkaline and alkali earth metasomatism, (2) hydrolytic alteration (hydrogen metasomatism), (3) volatile additions, (4) fenitization, (5) carbonate-hosted skarn formation, and (6) alteration of ultramafics.

30.2.1 Metasomatic Processes

The Russians have an extensive literature on metasomatism, largely because of the pioneering work of D. S. Korzhinskii (1959 and 1970 translations in English). Korzhinskii first distinguished the two principal processes of mass transfer in rocks: diffusion and infiltration. Although conceptually distinguishable, these two processes are probably end-members of a continuous spectrum of combined mechanisms.

30.2.1.1 DIFFUSION **Diffusion** is the process by which components move *through* another medium, either a solid (lattice diffusion) or a stationary fluid. Diffusion is driven by chemical potential (μ) gradients. If the chemical potential of a species is higher in one area than in another, migration of that species from the area of higher μ to that of lower μ will lower the free energy of the whole system. Matter will thus tend to migrate down chemical potential gradients if there is nothing to inhibit such migration. **Bimetasomatism** refers to the diffusion of components in opposite directions in a reciprocal or exchange fashion. The migration of MgO in one direction and SiO_2 in the opposite direction in the hypothetical column of J. B. Thompson (1959), discussed below, is an example of bimetasomatism.

At a first approximation, diffusion is described by **Fick's first law**, proposed by A. Fick in 1855:

$$J_x\,(g\,cm^{-2}\,s^{-1}) = -D(cm^{-2}\,s^{-1})\frac{dC\,(g\,cm^{-3})}{dx\,(cm)} \quad (30.6)$$

which states that the flux (J_x) of material in one direction (x) is proportional to the concentration gradient ($-dC/dx$). We now recognize that the chemical potential gradient is more appropriate than concentration. The proportionality constant (D) is called the **diffusion coefficient**. Diffusion coefficients for various materials are determined empirically and increase with temperature ($\log D$ decreases almost linearly with $1/T$ in kelvin). As a result, the material flux increases with both temperature and the steepness of the concentration gradient. Diffusion coefficients are very small for most silicate minerals. Typical D values are on the order of $10^{-15}\,cm^2/sec$ or less at 1000°C. We can estimate the effectiveness of diffusion from the approximation (Kretz, 1994, p. 283):

$$\overline{x} \approx (Dt)^{1/2} \quad (30.7)$$

where $\overline{x}$ is the mean displacement of material in time t (in seconds). From this equation we can calculate that, if $D = 10^{-15}$, a component will diffuse about 1 cm through a mineral in about 100 million years under typical metamorphic temperatures. This is dreadfully slow and suggests that diffusion of material through minerals is an ineffective metasomatic process. The preservation of fine compositional zoning in plagioclase and garnet, and of coronites and reaction rims in high-grade (dry) metamorphic rocks, supports this conclusion. Of course, geothermobarometry relies upon the sluggishness of diffusive re-equilibration upon cooling.

Diffusion is much more effective through a fluid phase, in which diffusion coefficients are on the order of 10^{-4} or greater. This would allow mean displacements on the order of a few meters per year. If diffusion is to occur over such distances a fluid must be present, of course, and it must be sufficiently interconnected as a grain boundary network. Whether it is interconnected or not depends upon the quantity of fluid present and the dihedral angle (θ) of fluid–mineral triple points (see Figure 11.1). Even in rocks of very low porosity, an intergranular fluid can form a continuous grain boundary network if the dihedral angle is sufficiently low. For the fluid to form a surface film that wets the surfaces of the grains requires a dihedral angle of only a few degrees. Such a film is unnecessary, however, for an interconnected fluid network. It is sufficient only that the fluid forms a network of interconnected tubes at grain edges (Figure 30.5a and c). For fluids occupying only 1 to 2% of a rock volume, this is possible when the dihedral angle of the fluid is less than 60°. When θ is greater than 60°, the tubes pinch off and the fluid beads up as isolated pockets at grain–edge intersections (Figure 30.5b).

Brenan (1991) summarized most of the experimental data on dihedral angles of fluids in mineral aggregates. Experiments with quartz in H_2O-CO_2 mixtures yield dihedral angles that increase from 55° in pure H_2O to over 90° if $X_{CO_2} > 0.86$. Addition of about 10% NaCl, KCl, or CaF_2

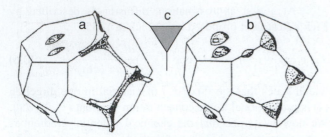

FIGURE 30.5 Three-dimensional distribution of fluid about a single grain at θ < 60° (left) and θ > 60° (right). In the center is a cross section through a fluid tube at the intersection of three mineral grains for which θ = 60°. After Brenan (1991). Copyright © the Mineralogical Society of America.

reduced θ by 10 to 15°. Decreasing pressure from 1.0 → 0.6 GPa also tends to raise θ by 10 to 20°. Initial data from natural rocks seem to agree with the experimental findings that θ is generally >60°, indicating that a continuous grain-surface film is rare, and that an interconnected tube-like network of fluid is unlikely at middle to upper crustal levels, unless the fluids are very saline and H_2O-rich. Although fluids can allow much more rapid diffusion, it appears that fluids, at least under hydrostatic conditions in which volatiles are not being released by reactions, usually exist as isolated pockets. Diffusion through fluid pockets and along dry grain boundaries, however, is still much more rapid than through a crystal lattice.

30.2.1.2 INFILTRATION

Infiltration refers to passive mass transfer of solute species carried in a moving fluid medium. Whereas diffusion occurs in response to internal chemical potential gradients, infiltration is driven by external fluid pressure gradients that cause the fluids to move. Pervasive fluid flow through a porous-permeable medium is described by **Darcy's law**. The Darcy flux increases with increased permeability, fluid pressure gradients, and reduced fluid viscosity. Darcy's law is more complex than we need deal with here. If you are interested, see equation 1 in Ague (2003) and the ensuing discussion. Fluids in motion permit much more extensive transport of matter than does diffusion. Regional-scale mass transport is not possible via diffusion alone and can only occur with the aid of infiltrating fluids. Evidence for fluid flow is widespread, including the progressive devolatilization of rocks with increasing metamorphic grade, the existence of geothermal springs and fluids encountered in deep wells, nearly ubiquitous veins filled with deposited minerals, hydrous alteration along shear zones and thrusts, metasomatic replacement textures, and variations in stable isotopes that are best explained by exchange between host rocks and introduced fluids.

One must wonder how such pervasive fluid flow is possible in light of the evidence that an interconnected network of fluid channels is not stable with typical rock porosities under hydrostatic conditions. Permeability may be enhanced, however, in several possible ways (see the reviews by Walther, 1990; Brenan, 1991; Ferry, 1991, 1994; Rumble, 1994; and Ague, 2003). Among the more popular ideas that have been proposed are the following.

Microfractures in crystalline rocks were first recognized by Adams and Williamson (1923). Microfractures are believed to be disc-shaped slits and are known to affect field-measured bulk transport properties, such as electrical resistivity and permeability, and must therefore form an interconnected network (Ferry, 1994). How pervasive these small cracks are in nature is not yet known. Many microfractures are now healed (recognizable as planar arrays of trapped fluid inclusions), suggesting that they may be transient.

During devolatilization reactions the fluid pressure increases to the point that it becomes greater than the tensile strength of the rocks, resulting in fractures. The process is called **hydraulic fracturing**. Fluids, being less dense than rocks, may move in these fractures as a result of their own buoyancy, promoting upwardly directed fluid-filled hydrofractures. This buoyancy mechanism may be responsible for significant upward metamorphic fluid flow. John Walther developed a model by which devolatilization reactions and expansion of pore fluids during active metamorphism resulted in fluid flow toward the surface along fracture networks (Walther and Orville, 1982; Walther and Wood, 1984; Wood and Walther, 1986; Walther, 1990). Calculations suggest that an isolated fluid-filled fracture can propagate upward at rates approaching 1 km/hr (Rumble, 1994). This rapid propagation of fluid-filled cracks can result in significant fluid fluxes. To the extent that devolatilization reactions are active during discrete intervals of prograde metamorphism (at discontinuous isograds), and because fracture propagation eventually dissipates the fluid pressure, this type of enhanced permeability may be largely episodic. To the extent that such reactions are continuous or buffered, fluids may also be released in lesser quantities on a more prolonged basis.

Many devolatilization reactions produce a smaller volume of solid products than was occupied by the mineral reactants. The volume reduction has been called **reaction-enhanced permeability** (Rumble, 1994). Marchildon and Dipple (1998) and Dipple and Gerdes (1998) discussed the positive feedback in which fluid flow (e.g., flow in a contact aureole related to pluton emplacement) induces mineral reactions, which increase permeability, which in turn enhance fluid flow, etc. This process, which they called **flow focusing**, results in enhanced flow in some areas and irregularly shaped reaction zones. For the same reasons stated above, this type of enhanced permeability may be episodic below the shallowest depths. Russian researchers have also recognized a process known as **thermal decompaction**, which is the creation of voids in rocks upon heating due to the anisotropic expansion of the mineral grains and their variable orientations in rocks (see Zaraisky and Balashov, 1995).

If a rock is subjected to non-isostatic stress during metamorphism, permeability may be enhanced by **dilatancy pumping** (Sibson et al., 1975). Rocks dilate due to rapidly increasing numbers of microfractures just prior to failure under increasing stress. These fractures dilate and draw fluid into the stressed rock. Stress is released upon failure and the fractures collapse, expelling the fluid. The process can be repeated during applied stress, but is self-limiting because the fluid weakens the rock, and a rock with saturated pores will dilate less before failure (Murrell, 1985). More elaborate models for fluid pumping have been proposed by Etheridge et al. (1984) and Oliver et al. (1990).

Although direct evidence for an interconnected network of cracks in rocks of moderate or greater depth is presently lacking, the models above suggest that relatively large fluid fluxes are at least possible. Contact metamorphism is particularly likely to have large, although local, fluid fluxes. Steep temperature and compositional gradients exist near the igneous contact. At low pressure, the porosity of the country rocks is likely to be high, and fracture systems associated with intrusion will usually be extensive, so that expelled magmatic fluids and available meteoric water will be convectively recirculated through the aureole. Flow, however, is probably localized along fractures and in permeable lithologies (see models and reviews by Norton and Knight, 1977; Barton et al., 1991b; Ferry, 1994; and Hanson, 1995).

Several investigators have attempted to estimate the quantity of fluid to pass through a given volume of rock over time, usually expressed as **time-integrated fluid:rock ratios**. Such ratios are not to be confused with porosity, which is the instantaneous ratio of fluid to rock. Connolly (1997) estimated porosity during metamorphism to be 0.1% to 0.001% of rock volume. Time-integrated fluid:rock ratios, however, are intended to reflect the extent of infiltration and flow over the duration of a metamorphic cycle. For example, Rumble et al. (1982) used the progress of the calcite + quartz $\rightarrow$ wollastonite + CO_2 reaction to estimate the quantity of fluid evolved during metamorphism of a bed containing silicified brachiopods at the Beaverbrook fossil locality in New Hampshire. The amount of wollastonite in some of the rocks was about 70%, far more than could be produced if the fluid were internally buffered (Section 29.1.1.2). They calculated the amount of infiltrated fluid required to produce the "excess" wollastonite and derived a volume fluid:rock ratio of 4.6/1, so rocks containing 70% Wo had a volume of at least 4.6 times as much fluid pass through as there was rock. These estimates assume that the fluid is pure H_2O. If CO_2 were present, or if more fluid passed through without reaction, the fluid:rock ratios would have to be greater, so that all fluid:rock ratios are really *minimum* estimates. Wood and Walther (1986) cited estimates of fluid:rock ratios, based on reaction progress of isotopic exchange, ranging from 1 to 5, and calculated that the deposition of 1% quartz in metamorphosed pelites requires a fluid:rock ratio of 6/1.

Calculations of mass balance along a flow path, integrated over time, yield estimates of fluid flux as a volume of fluid passing through a specified rock area (the **time-integrated fluid flux**, or q_{TI}). Ague (2003) reviewed several mathematical approximations for q_{TI} (see Problem 1 for an example). Figure 30.6 summarizes several estimates of q_{TI} from Ague's (2003) review. Ague (1994a) estimated an average pervasive flow during regional metamorphism of $10^{2.7} \pm 0.5$ m^3 of fluid passed through each m^2 cross section of rock (shaded area in Figure 30.6). Methods of estimating such fluxes vary. For example, the work summarized by

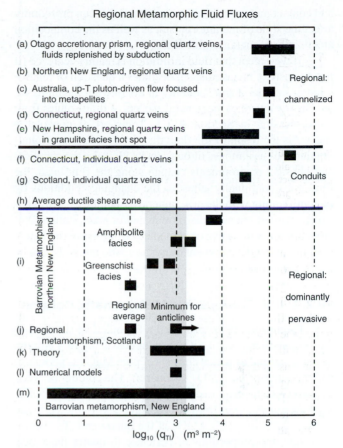

FIGURE 30.6 Selection of time-integrated fluid fluxes from the literature. Grey area is the average regional pervasive flow-dominated flux estimated by Ague (1994a). Sources: (a) Breeding and Ague (2002), (b) Ferry (1992), (c) Ague (1994b), (d) Oliver et al. (1998), (e) Chamberlain and Rumble (1989). Range for (e) computed by Ague (2003) using average flux of 1.5×10 m^3 m^{-2} s^{-1} for 10^5 and 10^6 yr. (f) Ague (1994b), (g) Ague (1997), (h) Dipple and Ferry (1992), (i) Ferry (1992) and Léger and Ferry (1993), (j) Skelton et al. (1995), (k) Walther and Orville (1982) and Walther (1990). Range for (k) computed using total timescales of fluid flow of 10^6 yr and 10^7 yr. by Ague (2003), (l) Hanson (1997), (m) Evans and Bickle (1999). After Ague (2003).

Ferry (1994) is based on mass balance using stable isotope exchange reactions or mineral-fluid reactions, yielding highly variable results. Time-integrated fluxes up to 10^6 m^3 of fluid/m^2 of rock have been estimated for pervasive flow through some contact aureoles and mid-crustal greenschist and amphibolite facies regional terranes. Values up to 10^9 cm^3 of fluid/cm^2 rock are reported for some metamorphic quartz veins. Walther and Orville (1982) calculated that fluid fluxes in a regional metamorphic terrane may reach 10^{-10} to 10^{-9} g/cm^2 sec, which is equivalent to 3 to 30 kg of fluid passing through each cm of crust above the 400°C isotherm during each million years of prograde metamorphism.

It seems fair to conclude that fluid infiltration is extensive in the Earth's crust, even in crystalline rocks with little obvious porosity. We are beginning to realize that the amount of fluid that passes through the crust is much greater than we had previously imagined. The ability of fluids to redistribute both heat and dissolved constituents in regional and contact metamorphism means that fluids can profoundly affect the style of metamorphism, as well as the composition and texture of metamorphic rocks.

If we accept that fluid infiltration is readily achieved (if not pervasive), metasomatism becomes a question of relative mineral saturation and solubility in the flowing medium. We are already aware of the general aspects of fluid solutes from the discussion in Section 30.1.1. Transport of nonvolatile species is the result of the fluid dissolving and/or precipitating minerals as it moves through rocks. Because the chemical and physical environments change along the fluid path, the fluid must continually adjust to the changing conditions by re-equilibrating with the new mineral assemblages. This generally involves renewed dissolution and/or precipitation. Precipitation of vein quartz or carbonates from rising fluids in a fracture, or the development of metasomatic minerals, can be explained in this fashion.

30.2.2 J. B. Thompson's Metasomatic Column

J. B. Thompson (1959) proposed a hypothetical situation in the simple binary MgO-SiO$_2$ system consisting of a column of rock, one end of which consists of pure MgO (as periclase) and the other of pure SiO$_2$ (as quartz). The column between the two ends, he proposed, should have a *continuously varying bulk composition* from 100% MgO next to the periclase end to 100% SiO$_2$ (0% MgO) at the quartz end. To accomplish this, the proportions of periclase to quartz must vary progressively across the column.

Suppose next that this column is held at a temperature and pressure at which both forsterite and enstatite are stable in addition to periclase and quartz. Because the original MgO and SiO$_2$ are not stable together, they will react within the interior of the column to form enstatite and/or forsterite. At any particular point in the interior of the column, the stable mineral assemblage will reflect the initial MgO/SiO$_2$ ratio. As a result, three *interior* zones should develop (Figure 30.7):

1. A zone with periclase + forsterite, with the ratio of Per/Fo decreasing steadily from the pure-periclase rock (0% SiO$_2$) until a surface *F* is reached at which the proportion of periclase drops to zero and forsterite is 100%. Surface *F* marks the beginning of:
2. A zone with forsterite + enstatite. Across this zone Fo/En decreases steadily from zero En at surface *F* to 100% En at surface *E*, which marks the beginning of:

3. A zone of enstatite + quartz. Across this zone En/Qtz decreases steadily from zero % Qtz to 100% Qtz, from which point the rest of the column is pure quartz.

Note that the percentage of SiO$_2$ still increases steadily (and MgO decreases steadily) across the interior zones of Figure 30.7.

The concept of **local equilibrium** (Korzhinskii, 1959; J. B. Thompson, 1959) can be applied to this system. Note that the system as a whole is certainly *not* in equilibrium because it contains such incompatible phases as periclase and enstatite, as well as forsterite and quartz. Within any zone, however, the mineral assemblages are in *local* equilibrium because only compatible phases coexist in mutual contact. We can thus treat the rock column as a sequence of subsystems, each in local equilibrium as long as no incompatible phases are in direct contact.

Our column is not in local equilibrium, however, at surfaces *F* and *E*. Periclase in zone 1 is in contact with enstatite in zone 2 across surface *F*, and forsterite of zone 2 is in contact with quartz of zone 3 across surface *E*. As a result periclase + enstatite should react across surface *F* to form a thin monomineralic layer of forsterite. Similarly forsterite + quartz will react across surface E to form a thin layer of enstatite (Figure 30.8).

Once very thin layers of forsterite and enstatite form, local equilibrium reigns across the column because incompatible phases are no longer in direct contact anywhere. If diffusion permits, however, either through the crystals or some intergranular fluid, forsterite + quartz can still react across the monomineralic enstatite layer. This situation is analogous to the formation and growth of reaction rims and coronas discussed in Section 23.6.

As discussed above, diffusion is driven by chemical potential gradients across a medium. We can see in Figure 30.8 that the *content* of SiO$_2$ (and therefore also of MgO) is constant across the enstatite layer (pure enstatite is 59.8 wt. % SiO$_2$ and 40.1 wt. % MgO). The *chemical potential* of SiO$_2$ and of MgO, however, can vary in enstatite. To understand this, we refer to the phase rule. In a two-component system, such as in our rock column in Figure 30.8, a one-phase zone has $F = C - \phi + 2 = 2 - 1 + 2 = 3$ degrees of freedom. At a fixed temperature and pressure for the hypothetical example, $F = 1$. We can thus vary μ_{SiO_2} (or μ_{MgO}, but not both because they are related in enstatite at fixed P and T). In the two-phase assemblages, Fo + En

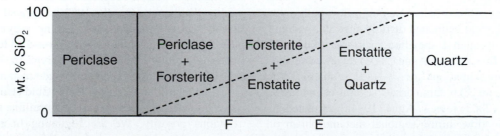

FIGURE 30.7   A hypothetical column of rock proposed by J. B. Thompson (1959). The left end is pure periclase and the right end pure quartz. Between these ends, the bulk composition varies continuously so that the wt. % SiO$_2$ increases linearly from left to right (dashed line).

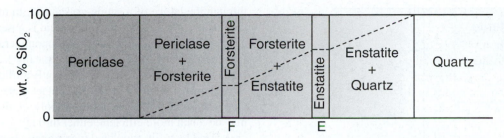

FIGURE 30.8 The hypothetical column of rock of J. B. Thompson (1959) after reactions create monomineralic forsterite and enstatite zones at F and E. The dashed line shows the variation in wt. % SiO_2 across the column. After Thompson (1959).

and En + Q on either side of the enstatite zone, the isothermal isobaric variance is zero ($F = 2 - 2 + 0 = 0$), so that both μ_{SiO_2} and μ_{MgO} are fixed.

Perhaps the situation is more clear on a *G-X* diagram for a fixed *P* and *T* (Figure 30.9). This diagram is only schematic, but shows the general positions of the phases on such diagrams. For pure SiO_2 in a single-component system ($X_{MgO} = 0$) because quartz is in its standard state (defined as the pure phase at the *P* and *T* of interest), μ_{SiO_2} is fixed at $\mu^\circ_{SiO_2}$, which is also G°_{Qtz} [see Section 27.3, especially Equation (27.12)]. Next, suppose MgO is mobile and progressively added to a quartzite. Migration of MgO into quartz gradually increases μ_{MgO}. Because quartz is stable, $\mu^\circ_{SiO_2}$, remains constant, but some MgO is present, presumably introduced along grain boundaries. When μ_{MgO} becomes great enough, a Mg-bearing solid phase will form. In the present case, the added MgO would react with some of the quartz to form enstatite.

It is a fundamental tenet of thermodynamics that, *at equilibrium, the chemical potential of any component is the same in all coexisting phases*. If this were not true (as mentioned earlier), the total free energy of the system could be

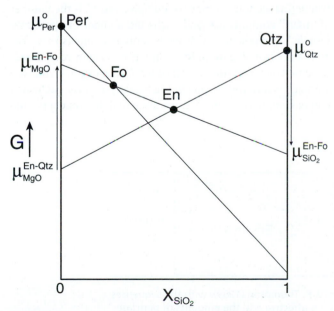

FIGURE 30.9 Schematic G-X_{SiO_2} diagram for the SiO_2-MgO system at fixed temperature and pressure. See text for discussion.

lowered by having some component migrate from a phase with higher μ to a phase with lower μ until it reached the same value in both. Thus μ_{MgO} is the same in both quartz and enstatite (as is μ_{SiO_2}) once enstatite has formed and coexists with quartz. Their values can be determined by extrapolating the line connecting Qtz and En (Figure 30.9.) μ_{SiO_2} in both En and Qtz is equal to the intercept of the line at $X_{SiO_2} = 1$ ($=\mu^\circ_{Qtz}$), and μ_{MgO} in each ($\mu^{En\text{-}Qtz}_{Mgo}$) is equal to the intercept at $X_{SiO_2} = 0$. It may seem strange to speak of μ_{MgO} in quartz, but it does have a finite value. Because quartz admits a vanishingly small amount of MgO, μ_{MgO} increases *very* rapidly in quartz as X_{MgO} is increased in it. In other words, the energy of adding MgO to the quartz crystals is not favored, so that it takes only a minute amount of MgO in the presence of quartz to effectively saturate it, so that any excess MgO will go elsewhere (in this case, by reacting with some quartz to create enstatite). Also, the absence of a component in the solid rock does not mean that it may not have been abundant in a pore fluid phase.

Figure 30.9 agrees with our phase rule conclusion. At a particular value of *P* and *T* in the two-component MgO-SiO_2 system, coexisting En + Qtz is invariant and fixes (or buffers) both μ_{MgO} and μ_{SiO_2}. Suppose we continue to diffuse more MgO into our system. Quartz will continue to react with the added MgO until it is consumed. Although the amounts of quartz and enstatite vary during this process, μ_{MgO} and μ_{SiO_2} (the intensive variables) remain fixed. Once quartz is consumed, only En is present and μ_{MgO} and μ_{SiO_2} are free to vary. In this case, μ_{SiO_2} will now drop (because no quartz is left to fix it) and μ_{MgO} will rise until enstatite becomes effectively saturated in MgO and forsterite forms as any additional MgO reacts with the enstatite. Again we have two phases, and μ_{MgO} and μ_{SiO_2} are again fixed (or buffered) at $\mu^{En\text{-}Fo}_{MgO}$ and $\mu^{En\text{-}Fo}_{SiO_2}$ in Figure 30.9. Further addition of MgO will consume enstatite, after which μ_{MgO} can again rise in forsterite until periclase forms. A completely analogous argument can be made if SiO_2 is mobile and added to pure periclase at the opposite end of Figure 30.9 until forsterite, enstatite, and eventually quartz are formed.

Now let's return to our column in Figure 30.8 and address one of the incipient monomineralic zones, shown expanded in Figure 30.10. Because the proportion of Fo/En decreases toward the right across the two-phase Fo + En zone, X_{SiO_2} must correspondingly increase (and X_{MgO} must decrease). The same is true in the other two-phase zone as

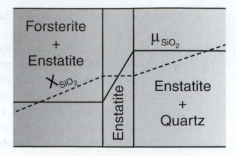

FIGURE 30.10 Expanded view of the monomineralic enstatite zone in Figure 30.8, showing the profiles of X_{SiO_2} and μ_{SiO_2}.

En/Qtz decreases. Within any one-phase zone both X_{SiO_2} and X_{MgO} are constant. The chemical potentials, however, behave differently. Because both two-phase assemblages buffer MgO and SiO_2, the chemical potentials are constant across these zones but are variable across the one-phase zone. The variation in X_{SiO_2} and μ_{SiO_2} is illustrated across Figure 30.10.

The difference in μ_{SiO_2} and μ_{MgO} across the monomineralic zone has the potential (so to speak) to drive diffusion across it. To the extent that diffusion is effective in the enstatite rock, SiO_2 and/or MgO can diffuse through it down their respective chemical potential gradients (only the SiO_2 gradient is illustrated in Figure 30.10). If only SiO_2 is mobile, it will diffuse through the En zone and react with forsterite at the far side to form enstatite, and the En zone will grow at the expense of the Fo + En zone. If only MgO is mobile, it will migrate through the En zone and react with quartz on the far side to form enstatite, and the En zone will grow at the expense of the En + Qtz zone. If both components are mobile, they will both diffuse through En in opposite directions, and the En zone will encroach upon both of the bimineralic zones on each side. One side may grow faster than the other, depending on the differences in diffusion coefficients for SiO_2 and MgO.

Because μ_{SiO_2} and μ_{MgO} are fixed on either side, as the width of the monomineralic zones expand, the gradients in μ become progressively less steep, and the driving force for diffusion drops correspondingly. This decrease in gradient may limit the growth of the zones. In mineral reaction rims and coronites (Section 23.6) diffusion commonly takes place

through the lattice of a single-crystal rim. In many dry granulite facies rocks, an intergranular fluid is absent. In both situations diffusion is slow, and the monomineralic zones that form between reacting phases are typically less than a centimeter in thickness, which is apparently sufficient to inhibit further diffusion and growth. In contact metamorphic aureoles and rocks with a generous supply of fluids, diffusion may be much more efficient, either through a stationary fluid or assisted by infiltration, and metasomatic zones can be several meters thick.

Suppose diffusion in Thompson's hypothetical column is very efficient. The En and Fo zones may eventually grow to the extent that they completely replace the bimineralic zones. The column will then look like the one in Figure 30.11. The two-phase zones and smooth gradients in rock composition (expressed as wt. % SiO_2) in Figure 30.8 have been replaced by one-phase zones, smooth gradients in chemical potential, and abrupt changes in composition. *The reduction in the number of coexisting phases and the accompanying discontinuities in bulk composition (even where none existed before) are the hallmarks of metasomatic processes.* If the amount of periclase or quartz is limited, and metasomatism highly efficient, growth of the enstatite and forsterite zones may replace either phase entirely, resulting in a system composed of En + Fo, En + Qtz, or Fo + Per that is then in complete equilibrium (not just locally) because no incompatible phases occur anywhere in the system.

The reduction in the number of phases as components become mobile is what led Korzhinskii (1936, 1949, 1959, 1970) and Thompson (1959, 1970) to reformulate Gibb's phase rule to be applied to local equilibrium situations and to account for what Korzhinsikii called **"perfectly mobile" components**:

$$F = C - C_m - \phi + 2 = C_i - \phi + 2 \qquad (30.8)$$

where C is the total number of components, C_i is the number of "inert" components, and C_m is the number of "perfectly mobile" components. For each mobile component μ becomes controlled externally to the *local* system considered, in accordance with the notion of local equilibrium. Thus the variance is reduced by one in the local system for each such component. In Section 24.2, we discussed this concept in the

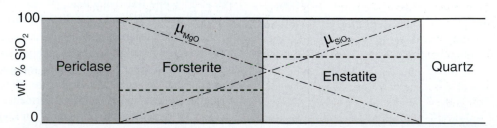

FIGURE 30.11 The hypothetical column of rock of J. B. Thompson (1959), with the sequences of mineral assemblages expected to form if diffusion is effective and the amounts of periclase and quartz prove inexhaustible. The dashed line shows the variation in wt. % SiO_2 across the column and the lighter dot-dashed lines show the variation in μ_{SiO_2} and μ_{MgO}. After Thompson (1959).

context of fluid components, such as mobile H_2O or CO_2. Now we recognize that it may be applied to any mobilized component. The "mineralogical phase rule" [Equation (24.1)] can be similarly modified if P and T are externally controlled:

$$\phi \leq C - C_m \text{ or } \phi \leq C_i \qquad (30.9)$$

This states that the number of minerals is reduced by one for each component that becomes mobilized, which is what we observed in J. B. Thompson's (1959) column above. If neither SiO_2 nor MgO is mobile, bimineralic zones remain. If either becomes mobile, monomineralic zones result. In the event that all components are mobile, we cannot have $\phi = 0$ (a void), so ϕ must equal at least one. Indeed, if everything is moving, it becomes impossible to prove that any one component did not move. This speaks to the problem of *reference frames*, discussed below. The local systems in Figure 30.11 are then generally monomineralic and $C_i = 1$ (the minimum number). At zone contacts μ_{SiO_2} and μ_{MgO} are buffered by the two phases in contact across the interface and thus internally controlled. On a local level C_m at a zone contact then equals zero and $C_i = 2$, so $\phi = 2$. As far as complete equilibrium is concerned, Figure 30.11 is not in equilibrium because the four phases, quartz, enstatite, forsterite, and periclase, are all present, and Per + En, Per + Qtz, and Fo + Qtz are all unstable associations. Only if the En and Fo zones expand to the point that Per and Quartz are exhausted will the column be in complete equilibrium. The concept of local equilibrium has the important attribute that we can apply equilibrium thermodynamics to any part of a larger disequilibrium system if the smaller part is itself in equilibrium. The larger system can then be treated as a **mosaic** of local equilibrium domains.

Vidale and Hewitt (1973) noted that "mobile" components need be able to move only to the extent that μ is controlled externally to the local system under consideration. Such components, they emphasized, may actually *move* less than some "inert" components. For example, if quartz is present in all the mineral assemblages of a system in an amount sufficient to internally control μ_{SiO_2}, then SiO_2 behaves as an inert component, even if it is soluble and easily moved. A particular component may behave as a mobile component in one occurrence and as an inert component in another, or even at different stages in the history of a single occurrence. To avoid inexorably relating external control invariably to high mobility, J. B. Thompson (1970) referred to Korzhinskii's "perfectly mobile" components as "*K*-components" and the "inert" components as "*J* components." Others refer to "perfectly mobile" components as "externally controlled" or "externally buffered" or "boundary value" components. Although some petrologists (e.g., Weill and Fyfe, 1964, 1967) justifiably criticize the Korzhinskii/Thompson phase rule [Equations (30.8) and (30.9)], the approach is effective in analyzing the broad characteristics of metasomatic rocks, and we shall adopt it in the discussions that follow. I shall also refer to externally controlled components as "mobile" because this

is their usual and simplest behavior. We should be aware, however, of the term's limitations.

Because μ of mobile components are externally controlled intensive parameters of state, similar to pressure and temperature, their use as variables in phase diagrams can be very informative. In other words, because variations in μ for mobile phases, exerted from beyond the system, can exert some control on stable mineral assemblages, one can readily substitute μ as a variable for P or T in typical phase diagrams. Thus, instead of the *P-T* or *T-X_{CO_2}* diagrams with which you are by now so familiar, we could alternatively create *T-μ* diagrams, *P-μ* diagrams, or *μ-μ* diagrams (Korzhinskii, 1959). These differ from the *T-X* or *P-X* pseudosections in Chapter 28 in that μ is an *intensive* variable and rigorously true for any phase or component in the plane of the diagram, whereas X is an *extensive* variable and the compositions of each phase must be projected onto the pseudosection.

Because, for any component i in phase A:

$$\mu_i^A = \mu_i^o + RT \ell n \, a_i^A \qquad (27.12)$$

and because μ_i^o is a constant at any T and P, we can also substitute the *activity*, a_i (or preferably the log of a_i to maintain direct proportionality), for the chemical potential of any component in phase diagrams.

Figure 30.12 is a log a_{SiO_2} vs. log a_{H_2O} diagram for the MgO-SiO_2-H_2O system calculated at 600°C and 0.2 GPa using the TQW thermodynamic equilibrium program of Berman (1988, 1990, 1991; described in Section 27.5). In the figure, I chose to add H_2O as an extra variable to determine its effect on the periclase–quartz column discussed above. Although the system is a three-component system, by making a_{SiO_2} and a_{H_2O} independent intensive variables, they may be treated as externally controlled and behave as though "mobile." The diagram is thus dominated by one-phase fields and not three-phase fields. The horizontal gray arrow in Figure 30.12 corresponds to a traverse across J. B. Thompson's column in Figure 30.11. At aqueous silica activities below $10^{-3.5}$ periclase is stable. As a_{SiO_2} increases to $10^{-3.5}$, $SiO_{2\,(aq)}$ reacts with periclase to generate forsterite. When a_{SiO_2} reaches $10^{-0.3}$, forsterite reacts with $SiO_{2(aq)}$ to generate enstatite. And quartz precipitates when a_{SiO_2} reaches $10^{0.0}$, or 1.0, meaning that the solution is now saturated in silica. Of course a_{SiO_2} cannot become larger than 1.0 because the fluid is saturated and any additional SiO_2 simply precipitates as more quartz.

If a_{H_2O} is raised, periclase gives way to brucite (Per + H_2O = Bru) at low a_{SiO_2}. Raising a_{SiO_2} at high H_2O activities (just below water saturation) then traverses the forsterite field (2 Bru + $SiO_{2\,(aq)}$ = Fo + 2 H_2O) and the talc field (3 Fo + 5 $SiO_{2\,(aq)}$ + 2 H_2O = 2 Tlc), until quartz precipitates. Increasing a_{SiO_2} at $a_{H_2O} = 10^{-0.5}$ in Figure 30.12 results in the sequence Per → Fo → En → Tlc → Qtz. The En-Tlc boundary corresponds to the reaction 3 En + 2 $SiO_{2(aq)}$ + 2 H_2O = 2 Tlc. Figure 30.12 not only quantifies the Thompson column approach, but introduces alternative columns as well.

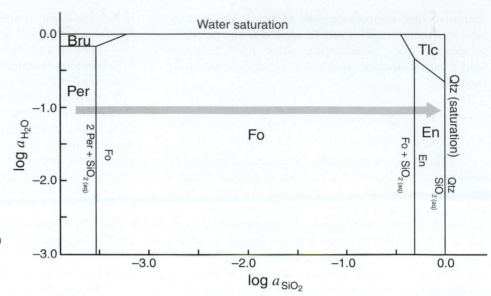

FIGURE 30.12 a_{SiO_2}– a_{H_2O} diagram for fluids in the $MgO-SiO_2-H_2O$ system at 600°C and 0.2 GPa calculated using the TQW program (Berman, 1988, 1990, 1991).

μ-μ or log *a*-log *a* diagrams are superior to the common chemographic diagrams when treating systems with several mobile components, as discussed by Brady (1977). A compatibility diagram, such as a triangular $SiO_2-MgO-H_2O$ diagram, assumes that all components are immobile and that most common compositions on the diagram are represented by three-phase assemblages. Even if H_2O is mobile, a SiO_2-MgO compatibility diagram (Figure 30.13) suggests that two-phase assemblages are the norm. One-phase assemblages only occur when the composition is equal to that of one of the phases (and gives the impression that this is a rather fortuitous situation). Figure 30.9 shows that one-phase assemblages may be expected when SiO_2 and/or MgO become mobile, which agrees with Figure 30.12.

In light of the previous discussion involving metasomatism in a hypothetical column of rock, let's summarize how diffusion and infiltration might operate during mass transfer and effect changes in the composition of rocks. In purely diffusional metasomatism the components diffuse through the rock, either through crystals or between them, most effectively through a stationary intergranular fluid medium. In purely infiltrational metasomatism, a solution flows into a rock and permeates the rock along a network of pores. If the solution is not in equilibrium with the new host rock, it reacts with the rock, exchanging components with the solid phases until local equilibrium is attained (or at least approached). As the fluid moves, it carries solute species and continuously exchanges them with successive rock volumes that it encounters, commonly producing different reactions at different zones of the flow.

Korzhinskii (1970) derived transport equations for both the diffusion and infiltration models and used them to analyze qualitatively the development of metasomatic zones. On the basis of his models he proposed that purely diffusive and infiltrative metasomatism are similar in a number of respects. For example, both produce a sequence of zones with a reduced number of phases, and the zones are typically separated by sharp metasomatic "fronts" (if local equilibrium prevails). On the other hand, diffusion and infiltration may potentially be distinguished on the basis of the following distinctive features (Korzhinskii, 1970; Hofmann, 1972):

- Infiltration can produce much wider zones than diffusion alone. The diffusion zones are commonly measured on the centimeter scale or less, whereas infiltration can produce zones several meters across.
- Diffusion can lead to reactions between rock components and diffusing components, but direct precipitation of a mineral cannot occur by diffusion in one direction because components migrate to areas of reduced activity. Infiltration can result in precipitation.
- Minerals within a zone can vary in composition as a result of diffusion because μ_i can vary steadily across a zone. μ_i is more constant within a zone during infiltration, so that the composition of resulting minerals that exhibit solid solution tends to be nearly constant.
- In infiltration, all components migrate in the same direction. In diffusion, different components can migrate in opposite directions (as J. B. Thompson's bimetasomatic column above). Bimetasomatic diffusion can result in the precipitation of new minerals.

Remember that diffusion and infiltration are simply end-member processes and that metasomatism in nature may consist of varying proportions of fluid motion (infiltration) and diffusion of solute species through that fluid, or diffusion from the fluid into the rocks through which it flows. "Wall–rock metasomatism" is a typical example of the combined process in which solutions percolate along a fracture (infiltration), and dissolved constituents diffuse from the fracture into the wall rocks through a largely stationary intergranular pore fluid.

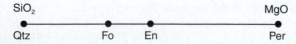

FIGURE 30.13 SiO_2-MgO chemographic diagram assuming only Qtz, Fo, En, and Per are stable.

30.2.3 Changes Associated with Metasomatism

Once we begin to treat metamorphism and metamorphic reactions as involving fluids and potential dissolved species, it becomes possible to balance mineral reactions in several possible ways. For example, in addition to the tremolite $\rightarrow$ diopside reaction discussed in Chapter 29:

$$Tr + 2\,Qtz + 3\,Cal = 5\,Di + 3\,CO_2 + H_2O \qquad (29.4)$$

We could balance the reaction in, among other ways, the following:

$$Tr + 6\,H^+ = 2\,Di + 3\,Mg^{2+}$$
$$+ 4\,SiO_{2(aq)} + 4\,H_2O \qquad (30.10)$$
$$Tr + 2\,Ca^{2+} = 4\,Di + Mg^{2+} + 2\,H^+ \qquad (30.11)$$
$$Tr + 2\,SiO_{2(aq)} + 3\,Ca^{2+} + 2\,H_2O$$
$$= 5\,Di + 6\,H^+ \qquad (30.12)$$

etc.

The reaction that occurs depends on the physical conditions, as well as the nature of the fluid and its solute content. In infiltration, the latter may depend on the nature of the rock through which the fluid most recently passed. In addition to P and T, then, the reaction may be dependent on pH and the activities of Ca^{2+}, Mg^{2+}, and SiO_2, etc. in the fluid. Determining which reaction occurred in situations with component mobility requires good petrography, including careful examination and interpretation of textures, accurate modes across isograds or metasomatic fronts, and an adequate reference frame.

The **reference frame problem** can be illustrated by returning to the metasomatic column of J. B. Thompson (1959). Remember from Figure 30.10 that the monomineralic enstatite zone can grow at the expense of the Fo + En zone on the left if SiO_2 is mobile, or it can grow at the expense of the Q + En zone if MgO is mobile. Alternatively, the En zone can expand in both directions at different rates if MgO and SiO_2 are both mobile, but to different extents. In natural situations of metasomatism that occur across ultramafic–quartzo-feldspathic contacts, for example, it may be difficult to determine which way a metasomatic front moved and what rock was replaced by a particular zone. Brady (1975) discussed the problem of reference frames in diffusional processes. The usual choice is a tacit assumption that diffusion takes place in reference to a single fixed point, which is really equivalent to a reference frame consisting of several points, all in fixed positions relative to each other. Because volume changes are common in metasomatic processes, such a choice of reference frame is impossible, and the extent of diffusion for various components, if so referenced, cannot be realistic. Brady (1975) reviewed several alternative reference frames. A common approach in experimental investigations and quantitative models of metasomatism is to choose some weighted average of the determined velocities of all chemical components, such as the motion of the center of mass. Alternative reference frames of this type may include assuming a fixed mean number of moles of all components, a fixed mean volume or mass, or that some individual component is fixed or immobile.

The reference frame problem is closely related to the problem of determining exactly what chemical changes occurred to produce a metasomatic rock in the field. Suppose we have an ideal situation in which we can determine the original contact (perhaps because some immobile residue was left, such as a layer of graphite, or rutile grains), and we thus know which rock was replaced. Furthermore, suppose we have an essentially unaltered sample of the replaced rock some distance from the metasomatized contact zone (similar to the periclase or quartz zones at the ends of our hypothetical column above). If we compare the chemical composition of the initial and altered rock, we may be tempted to conclude that any constituents that decrease from the initial analysis to the final one were removed, and those that increase were added. This has an implicit assumption, however, that the total mass of the sample hasn't changed. Recall that the meaning of "percent," as in "wt. %," implies that 100 g of the sample is essentially constant, and that the parts that make up those 100 g are simply exchanged through the walls surrounding the 100-g sample, picogram-for-picogram, so that the total remains the same. When matter is in flux, however, several alternative scenarios are possible. Perhaps material was only added during metasomatism, which diluted all of the non-added components. Comparing analyses on a wt. % basis would give the false impression that the diluted components *decreased* because they would have smaller percentages in the analysis of the altered rock than in the original. Alternatively, material may only have been removed, giving the false impression that the conserved constituents increased on a percentage basis, even though they never really moved. This problem is another aspect of the "closure problem" discussed in Section 8.5.1.

Gresens (1967) provided an approach for comparing two chemical analyses that attempts to solve the closure problem in certain cases. He provided a set of equations that allows calculation in terms of initial and final rock analyses and assumptions concerning either the volume change or the gain or loss of any one component. Gresens' (1967) equations have the form:

$$d_i = \left(f_v \frac{g_B}{g_A} c_i^B - c_i^A \right) \qquad (30.13)$$

where d_i is the amount of component i gained or lost, A is the original unaltered rock and B is the altered equivalent, c_i is the concentration of i (wt. %) in rock A or B, and g is the specific gravity or density. f_v is the volume factor, the number by which the total volume of the initial rock should be multiplied to result in the volume of the final rock. Although intended for volume changes, the equation is entirely general. If we wish to retain a weight basis, rather than a volume basis, f_v may be replaced by f_m, a mass factor, and the specific gravities may be ignored. In that case we would be dealing with changes in mass during metasomatism and not changes of volume.

As an example of his approach Gresens (1967) compared two rocks, from the Stavanger region of Norway, the analyses of which are given in Table 30.1. Rock A is a garnet phyllite, and rock B is an albite porphyroblast schist, interpreted as being derived from A by metasomatism. Figure 30.14 is a diagram of the net gains or losses of the oxides (d_i) versus the volume factor (f_v) using Equation (30.13) and specific gravities of 2.838 for the original phyllite and 2.777 for the altered schist.

A vertical line at $f_v = 1.0$ in Figure 30.14 is equivalent to assuming that volume is constant and intersects the oxide curves at values that are equivalent to simply subtracting the wt. % oxide values in column B of Table 30.1 from those in column A (with a volume weighting factor equal to the specific gravity ratio applied). Thus SiO_2, Na_2O, and CaO are added, and the other major oxides are lost during metasomatism if it is considered a constant volume process. Figure 30.14 also shows comparisons using alternative volume change interpretations. For example, if $f_v < 0.6$ (the volume loss is such that the final rock B is less than 60% of the volume of the original rock A), all of the constituents could have been removed in different (but specific) proportions and still produce rock B from rock A. If $f_v > 1.6$, on the other hand, all of the constituents could have been added. There is no textural basis for making a choice among these alternatives because the rocks are deformed, but the usual "choice" of constant mass (equivalent to direct comparison of the analyses) now seems as arbitrary as any alternative. A choice of constant volume may seem the most natural because we may not expect substantial volume changes in a rock at depth, but if matter is in flux, and the rocks are being deformed, there may be no compelling argument to support this contention. If rock A is neatly pseudomorphed during alteration, one may make a reasonable case for a constant volume process, but otherwise not.

TABLE 30.1 Analyses of Two Rocks from the Stavanger Region, Norway

Oxide	A	B
SiO_2	60.6	64.7
TiO_2	0.80	0.59
Al_2O_3	18.5	15.45
FeO*	6.84	5.41
MgO	2.42	1.48
MnO		
CaO	1.66	2.92
Na_2O	1.81	3.09
K_2O	4.02	3.46

A good alternative to assuming constant volume, however, may be indicated in Figure 30.14. Note that the Al_2O_3 and TiO_2 curves both intersect the $d_i = 0$ value (no gain or loss) at about the same value of f_v (about 1.2). Al and Ti are considered to be among the least mobile elements during metamorphism. Perhaps their mutual intersection with $d_i = 0$ indicates an appropriate choice for a volume change consistent with Al-Ti immobility. Such a choice of $f_v \approx 1.2$ (about 20% volume increase) also correlates well with minimal gain or loss of Fe and K. If we choose such a volume change, a vertical line at $f_v = 1.2$ yields corresponding gains or losses of all the oxides. Using this choice we see that SiO_2, Na_2O, and CaO are again the only oxides gained, but FeO* and MgO are all that is lost. The gain in SiO_2 is substantial, but correlates with the high solubility of SiO_2 in aqueous solutions.

Grant (1986) developed an alternative approach to the Gresens (1967) diagram, in which he rearranged Gresens' equations into a linear relationship between the concentration of a component in the altered and original rocks. In a plot of the concentrations of each component in the initial unaltered rock versus the concentration of those same components in the altered rock (Figure 30.15), any line that connects the origin with a point representing the concentration of any single component in both rocks can be called an **isocon** (constant concentration). An isocon for constant Al_2O_3 has been drawn in Figure 30.15. Such a line represents all points for which the concentration of a component in the altered rock (C^A) equals 0.835 times the concentration of the same component in the original unaltered rock (C_o). Note that the points for K_2O, FeO, MgO, and TiO_2 all fall near this line, correlating to the similar intersection of these oxide lines in Figure 30.14. The isocon with slope of 1.0 corresponds to $C^A = C_o$, and hence no mass gain or loss. Points above this are components gained during constant mass alteration, and those below the line are lost (again in agreement with Figure 30.14). Grant's (1986) isocon diagrams are easier to construct and solve, but they are still based on the same criteria developed by Gresens (1967).

The approaches of Gresens (1967) and Grant (1986) underscore the closure problem and the arbitrary nature of our common practice of comparing directly the analyses of altered rocks with their presumed unaltered parent rock. In

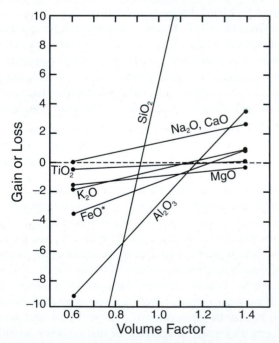

FIGURE 30.14 "Gresens-type" variation diagram showing the gains or losses (in grams per 100 grams of original rock A) as a function of the volume factor, f_v, in Equation (30.13). Rock A is a garnet phyllite from Stavanger, Norway, and rock B is a metasomatized albite schist, supposedly derived from (A). After Gresens (1967). Copyright © with permission from Elsevier Science.

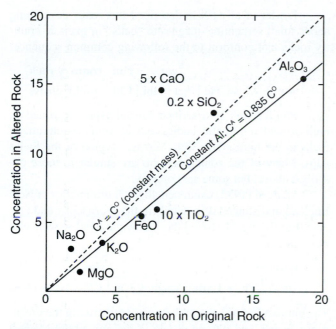

FIGURE 30.15 Isocon diagram of Grant (1986) for the data from Table 30.1. Some oxides have been scaled to provide a better distribution of data points.

some cases these approaches may indicate a preferred alternative, and the changes may be normalized to values consistent with minimal changes in elements regarded as immobile during metamorphism. The Gresens–Grant approaches may be strengthened if trace elements are added, and typically immobile elements such as Zr and Y, etc. also intersect the $d_i = 0$ curve close to the point at which Al and Ti do. The mobility of an element commonly varies with the conditions, rocks, and fluids involved. Al, for example, is generally regarded as an element of low solubility in metamorphic fluids, and of limited mobility under most metamorphic conditions. In low-to-medium grade metamorphosed ultramafic rocks, however, few Al-bearing minerals are stable, and Al may be quite mobile. If I were to apply the Gresens/Grant approach, I would keep an open mind and critically evaluate the patterns, with obvious emphasis on which elements intersect $d_i = 0$, and where they do so. The approach should complement good petrographic work, however, and not be a substitute for it.

Further discussion of metasomatism and the formation of metasomatic zones from a theoretical perspective becomes somewhat abstract. I think that it is far better to discuss the methodology by using some real examples. We shall thus address the two most common metasomatized

rock types: ultramafics and calcareous skarns. Other types certainly occur, including fenites, greisens (fluorine and hydrothermal alteration of granitoids), rodingites (Ca metasomatism of mafic rocks associated with ultramafics), etc., but the two following examples will suffice to illustrate the approaches and processes involved.

30.2.4 Examples of Metasomatism: Ultramafics

Metasomatic zones occur commonly at the margins of pod-like ultramafic bodies (Sections 10.1 and 29.2) in regional metamorphic terranes. Because of sharp initial chemical contrasts across the contact between the ultramafic and the (commonly) pelitic or quartzo-feldspathic country rock, sharp chemical potential (or activity) gradients in SiO_2, MgO, FeO, etc. are common and diffusion or infiltration may result. Descriptions of metasomatic ultramafic bodies date at least back to Gillson (1927), Read (1934), and Phillips and Hess (1936).

The number of low-ϕ zones that develop at metasomatized ultramafic margins, as well as their thickness and mineralogy, depends upon the stability of the various minerals and the ability of the components to migrate. These, in turn, depend upon the P-T conditions, the composition of the juxtaposed rocks, and the nature and mobility of the fluid phase. Because ultramafic rocks correspond closely to the MgO-FeO-SiO_2 system, the mineralogy of the ultramafic is typically relatively simple (≤ 3 principal phases; see Section 29.2). When any of these components become "mobile," according to Equation (30.9) the number of phases (ϕ) is reduced by the number of "mobilized" components. As a result, monomineralic and bimineralic zones are common.

Read (1934) described several types of mineral zonation around small (a few centimeters to 7 m in length) ultramafic pods in pelites in the Shetland Islands that were metamorphosed first in the biotite to kyanite zones and then affected by later chlorite zone retrograde metamorphism. He described an "ideal" zonation in the area as having a serpentine core with successive concentric zones of nearly monomineralic talc, actinolite, chlorite, and biotite as illustrated in Figure 30.16. Carbonate and Fe-Ti oxides are common minor phases. The zones are typically a few cm in width. The dark chlorite and micaceous zones were called **blackwall** zones by early talc miners and quarrymen in the northeastern United States, and the term is now commonly used more loosely by petrologists to refer to nearly any zone that forms around metasomatized ultramafic bodies.

Variations on Read's ideal sequence are consistent with more advanced chemical migration (and/or smaller bodies), which causes some zones to grow at the expense of other

FIGURE 30.16 "Ideal" mineral zonation due to metasomatism in < 3-m-long ultramafic pods in low-grade regionally metamorphosed pelites at Unst, Shetland Islands. After Read (1934).

zones, eventually consuming them. This process was described above in conjunction with J. B. Thompson's hypothetical column, in which the forsterite and the enstatite zones grew to the extent that they consumed adjacent zones. Read (1934) described at least the following variations:

core →   rim

Atg | Tlc | Act | Bt

Tlc | Act | Bt

Tlc | Act | Chl | Bt

Tlc | Chl | Bt

Act | Bt

but several monomineralic pods in the area, such as biotite, chlorite, or actinolite clots, may also represent metasomatized ultramafics.

Phillips and Hess (1936) described similar zoned ultramafic bodies from the Appalachians, some extending several hundred meters in length. They also noted variations in zonation, similar to those of Read (1934), and concluded that two main types of ideal zonation occur there, a low-temperature zonation:

core →   rim   country rock

Atg | Tlc + Mgs | Tlc | Chl | quartz-mica schist

and a high-temperature zonation:

core →   rim   country rock

Atg | Tlc + Mgs | Act | Bt | pelitic schist

The authors noted that chlorite occurs at the Act-Bt boundary in several high-temperature bodies and attributed it to retrograde alteration. Chlorite, however, is stable up to 600 to 700°C, so it may be part of the high-temperature equilibrium assemblage, as it was considered by Read (1934).

Carswell et al. (1974) described symmetrically zoned *veins* in peridotite in the high-temperature gneisses of southern Norway. They noted the sequence:

country rock →   vein center

Peridotite | En | Ath | Tr | Chl

This type of occurrence is distinctive in being not only higher grade but associated with the introduction of components via alkali–halide-rich aqueous fluids along fractures in the host peridotite (~90% olivine, 10% enstatite). (I reversed the center-margin sequence above to conform left → right to the petrology of the other sequences.)

Matthews (1967) found pods in the Lewisian gneisses on the Isle of Skye that are similar to those of Read (1934). The ultramafic core has been entirely replaced by talc-carbonate, which is rimmed by an actinolite and a biotite zone. The absence of a chlorite zone is probably due to higher metamorphic grade at Skye. Sharpe (1980) described the following zonation in an ultramafic rock in the quartzo-feldspathic gneisses of southwestern Greenland:

core →   rim   country rock

Atg | Tlc | Act | Hbl | Chl | q-feld. gneiss

Fowler et al. (1981) described the zones developing about small serpentine–magnesite "balls" in gneissic country rocks that conform to the following common sequence:

core → rim   country rock

Atg + Mgs | Tlc | Act | Hbl | Chl | q-feld. gneiss

Pfeiffer (1987) described several types of zonation, both around ultramafic bodies and as veins in ultramafic rocks in the high-grade Valle Verzasca region of the Swiss Alps. Many of the zoning patterns are similar to those described above, but some are complex.

Sanford (1982) examined 40 to 50 ultramafics in New England and studied 4 in detail (in the greenschist and amphibolite facies). He found the following zonal sequence to be common:

core →   rim

(um) | Tlc + Carb | Amph + Chl | Chl | Trans | CR

The ultramafic (um) assemblage was antigorite in the greenschist facies, and olivine + talc in the amphibolite facies. The country rock (CR) was generally a biotite-bearing mafic rock, more pelitic in some cases. The transitional-to-CR zone was commonly a biotite + amphibole ± plagioclase ± epidote zone.

Zone boundaries in most metasomatically zoned ultramafics are fairly abrupt, as the mineralogy changes quickly, but there is typically some overlap between phases. For example, Read (1934) described the transition between the talc and actinolite zones as one in which the amount of actinolite increased and the amount of talc decreased across a short distance. Similarly, the chlorite content increased within the actinolite zone as the chlorite zone was approached.

Sanford (1982) supplied sufficiently detailed descriptions to construct modes along a traverse. His traverse at the Grafton talc quarry in Vermont is shown in Figure 30.17. The zones at Grafton are:

A | B | C | D | E

Tlc + Ath | Tlc | Act + Chl | Trans | CR

Zone A is the main ultramafic body, and zone E is the quartzo-feldspathic country rock (CR). Note in Figure 30.17 that none of the zones is perfectly monomineralic. Curtis and Brown (1969) attributed the overlap of mineral types to kinetic effects, in that the reaction rate is unable to keep pace with diffusion or infiltration. In addition, the number of phases typically decreases progressively as the number of components that are mobilized increases. It has been noted in several experimental columns in which metasomatic fluids have been introduced that the number of phases decreases from the unaltered rock toward the source of infiltrating fluids, reflecting more elemental mobility in the fluid-rich domains. In cases of bimetasomatism, there is typically a single phase at the initial contact, where μ gradients are steepest, and increasing numbers of phases in both directions from that point. These trends conform to Korzhinskii's (1959, 1970) theory of metasomatic zoning. Regardless of the overlap of minerals in the observed traverses, the trends

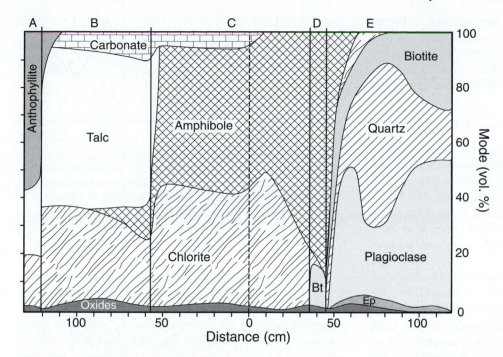

FIGURE 30.17 Variation in mineral proportions across the zones between the ultramafic and quartzo-feldspathic gneiss contact at Grafton, Vermont, after Sanford (1982). Zone letters at the top correspond to the zones listed in the text. Zone letters at the top are: A = Tlc + Ath, B = Tlc, C = Act + Chl, D = transitional, E = quartzo-feldspathic country rock. The vertical dashed line represents the estimated initial contact. After Sanford (1982). Reprinted by permission of the *American Journal of Science*.

toward developing a sequence of zones, each dominated by one or two principal phases, is still clear.

As Brady (1977) pointed out, diffusion in simple two-component systems, such as in the J. B. Thompson column above, can be uniquely modeled in terms of the ideal sequence of mineral zones that form because the variation in composition is a one-dimensional function of the ratio of the two components. Systems with three components or more, however, even if only one component is mobile, can have a number of possible zonal sequences. The mineral assemblages that form, and their sequence, depends not only on the amount of mobile component(s) added or lost, but also upon their ratio and the ratio of "immobile" components at any point along the column. It is thus impossible to predict a single zonal sequence. As is so commonly the case, a theoretical approach is best used to analyze what *has* happened at a given locality, rather than to attempt to predict what *should* happen.

H_2O and CO_2 are clearly mobile in most ultramafic systems and are involved in the mineralogy of several zones. H_2O is typically present in sufficient quantity that it establishes a constant value of μ_{H_2O} within any zone and is rarely buffered by the local mineral assemblages. CO_2 may be buffered by serpentine + talc + magnesite, but carbonate only occurs in an inner talc + magnesite zone. We can avoid complexity by considering the chemical potentials of H_2O and CO_2 to be constant and use variations in μ_{MgO} and μ_{SiO_2} (ignoring carbonates) as a first approximation in attempting to develop a model for the formation of ultramafic metasomatic zones that reproduces the types observed above.

At a constant temperature, pressure, μ_{H_2O}, and μ_{CO_2}, the minerals present in the low-temperature types of zones, such as that of Read (1934), can be represented on an *AMS* diagram, in which $A = Al_2O_3$, $M = MgO + FeO$, and $S = SiO_2$ (Figure 30.18). Points X and Y in Figure 30.18 indicate the bulk composition of the serpentinite and the pelitic country rocks,

respectively. Again, because more than one component is involved, there is no single path between X and Y that is universally applicable and that would provide a unique solution to the development of a sequence of mineral zones. For example, if SiO_2 is the only diffusing component, then the sequence might be:

(1) Atg | Tlc | Ms + Chl | CR (Qtz-Ms schist)

In this case, shown as the two vectors labeled 1 in Figure 30.18, composition Y loses SiO_2 as composition X gains SiO_2. The loss of SiO_2 from Y (Ms + Chl + Qtz) results in Ms + Chl in the adjacent zone, and the gain in SiO_2 to X (Atg) results in the formation of Tlc. Note that because one component is mobile in the pseudo-three-component *AMS* diagram shown, three-phase assemblages do not occur (except at zone boundaries). Thus only two-phase assemblages (represented by successively encountered tie-lines) and one-phase assemblages (points) are to be found within

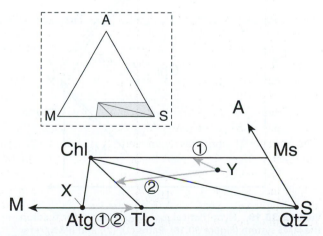

FIGURE 30.18 *AMS* diagram ($A = Al_2O_3$, $M = MgO + FeO$, and $S = SiO_2$), projected from K_2O, for ideal lower-temperature metasomatic zones around ultramafic bodies. After Brady (1977).

the metasomatic zones. Note that the sequence in Figure 30.18 is actually Atg → Atg + Tlc (at the contact or across a short interval) → Tlc.

Alternatively, M may be mobile, so that X loses M, and Y gains it (vectors 2 in Figure 30.18), resulting in this possible sequence:

(2) Atg | Tlc | Tlc + Chl | Chl + Qtz | CR (Qtz-Ms schist)

Figure 30.19 is a qualitative $\mu_M - \mu_{SiO_2}$ diagram (M = MgO + FeO) showing the two sequences above and the sequence observed at low temperatures by Phillips and Hess (1936). Path 2, for example, follows the Atg + Chl boundary, to the Tlc + Chl boundary, to the Qtz + Chl boundary, to the Chl + Ms + Qtz point (pelitic country rock). Note that the Atg, Tlc, and Qtz saturation surfaces buffer the fluid composition, so that μ_M and μ_{SiO_2} must vary in sympathy, even though only one species may be "mobile." For the observed sequence of Phillips and Hess (1936), the fluid composition leaves the saturation surfaces in the monomineralic chlorite zone and cuts through the Chl field, requiring that both M and SiO_2 be diffusing components.

By projecting from K_2O in Figure 30.18, we have ignored its importance as a diffusing component. It may play a role, but the major mineral zones can be derived without addressing it. Aluminum is also assumed to be immobile in balancing the reactions to create this $\mu-\mu$ diagram (Figure 30.19). At higher temperatures, actinolite and biotite are important phases. The role of biotite is similar to that of chlorite in the treatment just described. Actinolite requires CaO, which is presumably derived from the country rocks and must diffuse through the biotite and actinolite zones in order for actinolite to form.

Figure 30.20 shows the positions of analyzed rocks (large dots) from the Shetland Island zones of Read (1934), reported by Curtis and Brown (1969). The analyses match

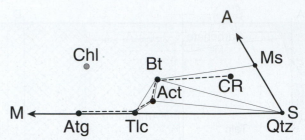

FIGURE 30.20 The same portion of the *AMS* diagram as in Figure 30.18, projected from K_2O and CaO, with the locations of analyzed rocks (large dots) from the metasomatized zones of Read (1934; see Figure 30.15), reported by Curtis and Brown (1969). The dashed curve represents a path through the zonal sequence. After Brady (1977). Copyright © with permission from Elsevier Science.

closely the ideal mineral compositions for monomineralic zones. The dashed curve represents a pathway corresponding to the sequence of zones in Figure 30.16. If chlorite is not a retrograde product in the Shetland rocks, the path must divert drastically from the actinolite zone rocks back to the chlorite zone and back again to the biotite zone. Remember, however, that the process is driven largely by chemical potential or activity gradients of species in the fluid, and these may not necessarily be directly proportional to the bulk rock compositions. It is thus possible to have reversals in the composition of components in the bulk rock along a traverse. It is also possible to have peaks or reversals in some, but not all, activity gradients in a system with several diffusing species. We will see some more well-constrained examples of these features in the following section on calc-silicate skarns.

In his detailed study in New England, Sanford (1982) noted that the composition of several phases varied significantly across zones (noted earlier as more compatible with diffusion than with infiltration). He also placed the initial contact at or near the actinolite–chlorite boundary because both Cr and Ni (typically immobile and concentrated in ultramafic rocks) decrease drastically from this point toward the quartzofeldspathics. Assuming that Ni and Cr are immobile, he calculated estimates for the volume change at three localities, concluding that the altered ultramafic zones increased in volume (from 1.4 to as much as 11 times, but remember that they are thin) and that the country rocks decreased in volume. Of course these factors change if Cr or Ni moved. On the basis of reaction textures and the mineral modes, combined with inferred displacement of zone contacts, Sanford also estimated the reactions and component fluxes associated with metasomatism of the ultramafics and country rocks. Figure 30.21 illustrates the reactions and fluxes estimated for the Grafton quarry, corresponding to the mineralogy shown in Figure 30.17. Note from the proposed fluxes at the top of the figure that every major oxide was to some extent mobile. The major fluxes were of Si from the country rocks into the ultramafic and of Mg from the ultramafic to the country rocks. The other fluxes were of lesser magnitude (with the possible exception of CO_2). Note that Al, commonly regarded as relatively insoluble and immobile, was mobile at Grafton (as in most metasomatized ultramafics).

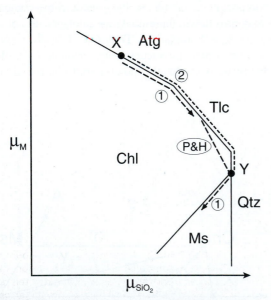

FIGURE 30.19 Hypothetical $\mu_M - \mu_{SiO_2}$ diagram for fluids in the *AMS* system (Figure 30.18). Paths (1), (2), and (P&H) refer to the theoretical paths in Figure 30.18, and the observed sequence of Phillips and Hess (1936). After Brady (1977). Copyright © with permission from Elsevier Science.

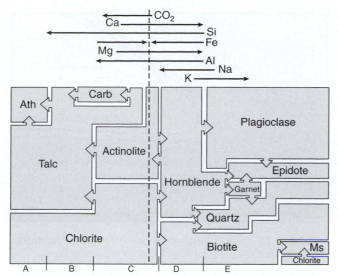

FIGURE 30.21 Schematic representation of major silicate mineral reactions and component fluxes associated with metasomatism of the ultramafic body at Grafton, Vermont. Elemental fluxes across various zones are indicated by the arrows at the top. Arrows between mineral boxes (somewhat distorted from the true modes in Figure 30.16) indicate reactions. When horizontal, these arrows involve metasomatic reactions; when vertical, these arrows are approximately isochemical. The zones listed at the bottom correspond to those in Figure 30.16, and the heavy dashed line is the estimated original contact. After Sanford (1982). Reprinted by permission of the *American Journal of Science*.

Sanford (1982) was able to calculate the chemical potentials of most major elements at several points along the traverses at four of his zoned ultramafic bodies. This is possible at points where sufficient phases were present to buffer μ of a particular component at a given temperature and pressure. For example, remember from J. B. Thompson's column illustrated in Figure 30.10 that μ_{SiO_2} is buffered by either two-phase assemblage (Fo + En or En + Qtz), but is variable across one-phase zones. Thus Sanford (1982) could constrain the μ of several species when the number of phases was

high and the variance therefore low: typically at zone boundaries. Figure 30.22 shows the results of his calculations for the Grafton locality. Any migrating component should move from zones of high μ toward areas of lower μ. Again, note the locally steep gradients in SiO_2 and MgO. Na_2O migrated from the quartzo-feldspathic country rock toward the ultramafic. FeO migrated from both rock types toward the contact zone at Grafton, but only into the ultramafic at the other three localities studied by Sanford. CaO has a high μ in the actinolite zone and decreases toward the country rocks (it is a minor component and unconstrained in the ultramafic body). CaO thus migrated from the ultramafic body to the country rocks. Al_2O_3 shows some interesting peaks and troughs. Assuming that the migration of all species reflects a coherent flux between only two initial contrasting rock types, the peaks and troughs in the profiles of Al_2O_3 and FeO indicate that these components migrated *up* their μ gradients at several points on a local basis. Such "uphill" migration has been discussed by Cooper (1974) and is attributed to interactions between diffusing components. The flux of some major component may stabilize a phase that acts as a "sink" for a less mobile or less abundant component, so that the lesser component becomes concentrated. Conversely, a phase that acts as a "sink" for a lesser component may not be stable due to the fluxes of more dominant components, resulting in local depletion of the lesser component. It follows that "uphill" migration is limited to local effects on less plentiful or mobile elements, and it cannot occur in simple binary systems, but requires several components.

The approach of studying metasomatic rocks by a combination of standard chemographic diagrams linked with chemical potential or activity profiles and μ-μ or log a-log a diagrams has proved fruitful and could be applied to the higher-temperature ultramafic reaction zones as well. Rather than risk overdoing it, we will move on and show how the technique can be applied to another common metasomatic rock type, calcareous skarns.

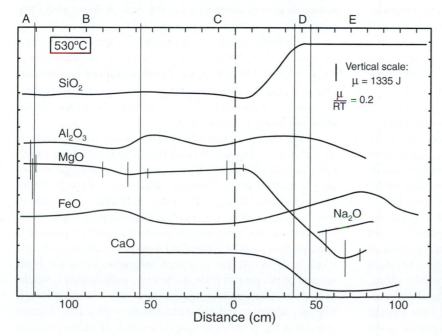

FIGURE 30.22 Variation in chemical potentials of major components across the metasomatic zones at Grafton, Vermont. Estimated temperature = 530°C. Typical data points and error bars are illustrated for the MgO profile. Lettered zones at the top correspond to those in Figure 30.16. The dashed vertical line is the estimated original contact. After Sanford (1982). Reprinted by permission of the *American Journal of Science*.

30.2.5 Examples of Metasomatism: Calcareous Skarns

A **skarn** (or **tactite**) is a rock dominated by Ca-Fe-Mg-rich calc-silicate minerals, usually formed by replacement of carbonate-bearing rocks during either regional or contact metamorphism. Although recrystallization of sufficiently impure carbonate rocks can produce a skarn, most are created by metasomatism, either bimetasomatism across the contact between unlike lithologies or infiltration of silica-rich fluids into carbonate rocks. Some investigators prefer to restrict skarns to include only the replacement types, but this makes it hard to identify a skarn without correctly interpreting its genesis. Others use the term more generally, encompassing a number of replacement rocks, including ultramafic blackwalls, but most investigators prefer restricting it to calc-silicate rocks that replace carbonates. Einaudi et al. (1981), in their extensive review, further subdivided skarns into magnesian skarns (that replace dolomite), calcic skarns (that replace calcite), and various ore skarns (particularly common are magnetite and tungsten–scheelite ore skarns).

Alternatively, skarns can be subdivided on the basis of their geologic setting (Kerrick, 1977), as illustrated in Figure 30.23. **Magmatic skarns** are found at the contact between igneous rocks (usually granitoid plutons) and marbles. **Vein skarns** form along fractures or small dikes in marbles. Infiltration is a dominant process in skarn formation of these two types. **Metamorphic skarns** form at the contact between carbonate and silicate lithologies, usually representing original sedimentary layers, but also including chert nodules in marbles, etc. Diffusion (usually bimetasomatism) directly between the contrasting rock types is usually the dominant process involved in the generation of metamorphic skarns. Vein skarns and magmatic skarns usually involve infiltration, but it may be accompanied by diffusion. These skarns form by metasomatism and are usually only a few centimeters thick. All three of these occurrence types are also possible in ultramafic and other metasomatic rocks. The examples of ultramafic blackwalls above were predominantly metamorphic, but some vein types were also described. Magmatic types are more dramatic for skarns than for other rock types because of the radical compositional contrast between a silicate magma, commonly releasing acidic aqueous solutions, and a carbonate country rock.

Skarns are typically zoned, and the zones have a reduced number of phases, as we would expect when components make the transition from "inert" to "mobile" states. As in zoned ultramafics, the mineralogy, thickness, and number of zones depends upon the P-T conditions, the composition of the juxtaposed rocks, and/or the nature and mobility of the fluid phase. Table 30.2 illustrates some types of zonation observed in selected examples of skarns described in the literature. Magmatic and metamorphic types are ordered toward the marble on the right; vein types are listed from the center of the vein to the marble margin (again on the right), having a symmetric other half not shown. The minerals within a zone, when reported, are listed in order of decreasing abundance. The list is by no means comprehensive, and I have left out many variations in the interest of brevity, tending to select the

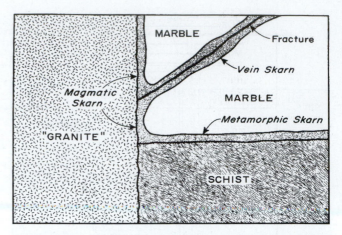

FIGURE 30.23 The three principal types of skarns. From Kerrick (1977). Reprinted by permission of Oxford University Press.

more complex zonation when more than one type was described in any particular work. Higher degrees of mobility tend to simplify the zonation and replace adjacent zones, imparting less information. Table 30.2 merely illustrates a spectrum of the possible types of zoning. Magmatic skarn zones are not to be confused with contact metamorphic isograds and the zones between them. The zones in Table 30.2 are due to metasomatic addition of material by diffusion and/or infiltration, and are much smaller than most contact metamorphic aureoles, which are largely thermal effects (and may or may not have some accompanying metasomatism).

Note in some magmatic examples in Table 30.2 that an **endoskarn** develops within the plutonic rock. Although metasomatism associated with plutons is usually dominated by the infiltration of the carbonate country rocks by fluids released by the cooling pluton, an endoskarn may form when CaO and/or CO_2 migrate back into the igneous body, causing hornblende or biotite to dehydrate and alter to pyroxene. Endoskarns are readily recognized in the field by the gradual transformation of the black igneous mafics to green pyroxene. When endoskarns are present, the term **exoskarn** is commonly used to distinguish the more typical skarns that form outside the contact by replacement of the country rocks.

It is well beyond our scope here to attempt to explain the development of every zonal sequence in skarns, even those listed in Table 30.2. It will suffice for our present purposes to model a typical example or two, so that we might see how our approach can be used to understand calc-silicate mineral zonation in general.

The simplest chemical system in calc-silicate skarns is an initial calcite-quartz contact (CaO-SiO_2-CO_2), an excellent example of which is Joesten's metasomatized chert nodules in marbles in the Christmas Mountains of southwestern Texas (Joesten, 1974, 1991; Joesten and Fisher, 1988), one of which is shown in Figure 30.24. At a given μ_{CO_2}, metasomatism between the chert and carbonate is a matter of diffusion in a binary system of either CaO into the nodule or SiO_2 outward, or both. At the very high temperatures in the inner aureole of the gabbro body in the area, wollastonite is stable, as are the unusual high-temperature calc-silicate minerals tilleyite, spurrite, and rankinite (Section 29.1.1.1). Joesten noted that the nodules developed a series of monomineralic

TABLE 30.2   Some Examples of Metasomatic Zones Developed in Calc-Silicate Skarns

Zones						Type	Width	Reference
Diorite (Hbl+Plag)	Endoskarn (Cpx+Plag)	Pyroxene (Cpx)	Garnet (Grt+Cpx)	Marble (Cal)		Magmatic	3 cm	Kerrick (1977)
Dior.?	Endoskarn (Plag+Grt+Ep)	Garnet (Grt+Cpx+Ep)	Pyroxene (Cpx+Grt+Ep)	Marble (Cal)		Magmatic	6 cm	Kerrick (1977)
"	Endoskarn (Plag+Ep+Cte+Di)	Garnet (Grt+Cpx+Qtz)	Wollastonite (Wo)	Marble (Cal+Cpx+Ep+Grt)		Magmatic	5 cm	Kerrick (1977)
		Garnet (Grt+Qtz)	Wollastonite (Wo+Qtz)	Marble (Cal)		Vein	2 cm	Kerrick (1977)
Amphibolite (Plag+Hbl)	Ep-Trem (Ep+Tr+Plag)	Pyroxene (Cpx+Ep)	Garnet (Grt+Cpx)	Marble (Cal)		Metamorphic	1.5 cm	Kerrick (1977)
Hornfels (Bt+Kfs+Plag)	Pyx-Plag (Cpx + Plag)	Pyx-Ep (Cpx+Ep+Plag)	Garnet (Grt + Cpx)	Woll. (Wo)	Marble (Cal+Qtz+Grt)	Metamorphic	2 cm	Kerrick (1977)
Pelite (Bt+Ms+Qtz)	Amphibole (Qtz+Plag+Amph+Kfs)	Pyroxene (Cpx+Plag)	Garnet (Grt+Cpx)	Marble (Cal+Cpx+Wo)		Metamorphic	7 cm	Thompson (1975)
Pelite (Bt+Kfs+Plag+Qtz)	Amphibole (Amph+Kfs+Plag)	Pyroxene (Cpx+Kfs+Plag)	Garnet (Grt + Pyx)	Marble (Cal+Wo+Grt)		Metamorphic	4 cm	Jamtveit *et al.* (1992)
Pelite (Qtz+Bt+Feld)	Diopside	Garnet	Wollastonite	Marble (Cal)		Metamorphic		Brock (1972)
Granite (Hbl+Bt+Plag+Kfs+Qtz)	Garnet (Grt+Wo)	Pyroxene (Cpx)	Pyx-Mont (Cpx+Mtc)	Ol (Fo+Mtc)	Marble (Cal+Fo)	Magmatic	10 cm	Tilley (1951b)
Quartz	Wollastonite	Rankinite	Spurrite	Tilleyite	Marble (Cal)	Metamorphic (chert nodule)	1-2 cm	Joesten (1974)
	Quartz	Tremolite	Calcite	Marble (Dol)		Vein	1 cm	Walther (1983)
Cal + Di	Cal + Tr + Di	Cal + Tr	Calcite	Marble (Dol)		Vein	< 15 cm	Bucher-Nurminen (1981)
	Cal + Tr	Cal + Fo + Tr	Cal + Fo	Marble (Dol+ Cal + Fo)		Vein	< 15 cm	Bucher-Nurminen (1981)
Cal + T + Di	Cal + Tr + Phl	Cal + Tr	Cal +Fo	Marble (Dol+ Cte + Fo)		Vein	< 15 cm	Bucher-Nurminen (1981)
Cal + Di	Cal + Tr + Di	Cal + Fo	Cal + Atg + Fo	Marble (Dol)		Vein	< 15 cm	Bucher-Nurminen (1981)
	Cal + Tr + Phl	Cal + Tr + Tlc + Phl	Do + Cal + Tr + Phl	Marble (Dol)		Vein	< 15 cm	Bucher-Nurminen (1981)
Q. Diorite	Silica-Enriched	Endoskarn (Czo + Plag)	Pyrox. (Cpx+Tr)	Forsterite (Cal + Fo)	Marble (Dol)	Metamorphic	3.5 cm	Frisch and Helgeson (1984)
(same)		Pyrox. (Cpx+Tr)	Antigorite	Forsterite (Cal + Fo)	Marble (Dol)	Metamorphic	3.5 cm	Frisch and Helgeson (1984)
Czo	Di + Tr + Czo + Chl	Pyrox. (Cpx)	Antigorite	Forsterite (Cal + Fo)	Marble (Dol)	Vein	1.5 cm	Frisch and Helgeson (1984)

FIGURE 30.24 Chert nodule in carbonate with layer sequence: calcite|tilleyite|wollastonite|quartz. Christmas Mountains, Texas. From Joesten and Fisher (1988). Copyright © The Geological Society of America, Inc.

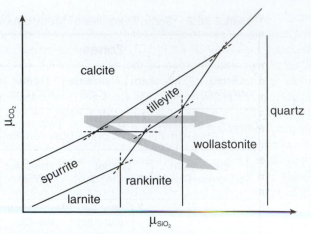

FIGURE 30.25 Schematic isothermal isobaric $\mu_{CO_2}-\mu_{H_2O}$ diagram for fluids in the CaO-SiO$_2$-H$_2$O system at high temperatures. After Joesten (1974). Reprinted by permission of the *American Journal of Science*.

marginal layers that he attributed to diffusion both of CaO into the nodule and SiO$_2$ outward (bimetasomatism). The mineralogy of the layering depends on the temperature (the proximity to the gabbro contact) and the extent of diffusion (the progress of replacement reactions as zones grow). Among the more complex zonal patterns between the marble and the chert nodule core, listed in order of distance from the contact in meters, are:

distance	rim → core zonation
0–5 m:	Cal | spurrite | rankinite | Wo | Qtz
0–16 m:	Cal | tilleyite | spurrite | rankinite | Wo | Qtz
16–31m:	Cal | tilleyite | Wo | Qtz
31–60m:	Cal | Wo | Qtz
> 60 m:	Cal | Qtz

Figure 30.25 is a $\mu_{CO_2}-\mu_{SiO_2}$ diagram Joesten (1974) used to explain the development of the zones. μ-μ diagrams are relatively easy to construct if the reactions are known because the slope of any reaction boundary is simply the ratio of the stoichiometric coefficient of the abscissa component divided by that of the ordinate component as they are involved in the reaction (Korzhinskii, 1959). For example, the spurrite–rankinite reaction can be expressed as:

$$3\,Ca_5Si_2O_8CO_3 + 4\,Si \tag{30.14}$$
$$\text{spurrite}$$
$$= 5\,Ca_3Si_2O_7 + 3\,CO_2$$
$$\text{rankinite}$$

The slope is thus +4/3 (positive because CO$_2$ and SiO$_2$ are on opposite sides of the reaction). The tilleyite to rankinite reaction is:

$$3\,Ca_5Si_2O_7(CO_3)_2 + 4\,SiO_2 \tag{30.15}$$
$$\text{tilleyite}$$
$$= 5\,Ca_3Si_2O_7 + 6\,CO_2$$
$$\text{rankinite}$$

so the slope is only +4/6 (or 2/3). Using these slopes and the Schreinemakers approach (Section 26.10), we can create Figure 30.25 if the stable phases are known. Note that the sequence:

Cal | tilleyite | rankinite | Wo | Qtz

also described by Joesten, can be created by increasing the chemical potential of SiO$_2$ (or decreasing that of CaO in this simple binary system) at constant μ_{CO_2} along the horizontal gray arrow in Figure 30.25. This is analogous to the process of SiO$_2$ diffusing into the marble from the chert, or of CaO diffusing into the nodule from the carbonate. Joesten proposed that spurrite develops by subsequent dehydration of tilleyite at its interior margin to produce the sequence described between 5 and 16 m from the contact above. Alternatively, CO$_2$ may have decreased as SiO$_2$ increased (sloping gray arrow in Figure 30.25) to develop the sequence directly. Joesten (1974) and Joesten and Fisher (1988) proceeded to use quantitative models of diffusion and growth to predict the sequence of zones that formed over time and the migration of zone boundaries as one mineral zone replaces another. Remember from the J. B. Thompson column that a zone, such as the enstatite zone in Figure 30.10, may grow at the expense of forsterite on the left if SiO$_2$ is more mobile and at the expense of quartz on the right if MgO is more mobile. Modeling which boundary moves in a given situation requires quantitative models based on realistic diffusivities (see below).

As discussed above, for systems with multiple components there may be several diffusing constituents. No single unique path is implicated because several components may migrate, and one cannot determine their interactions *a priori* (Brady, 1977). Modeling the zonal sequences in more complex skarns in systems including CaO, SiO$_2$, MgO, Al$_2$O$_3$, and K$_2$O is thus more complicated and variable, typically involving both multimineralic and monomineralic zones.

For example, A. B. Thompson (1975) described the zonal sequence listed in the seventh row in Table 30.2, which developed at the contact between interbedded pelite and carbonate layers in a roof-pendant of metasediments in Vermont. Figure 30.26 is an ACF diagram (projected from K_2O) showing the location of the minerals in the various zones and the trace of whole-rock compositions between adjacent pelite and calcite marble. Remember (in contrast to the application of such diagrams in normal usage), when elements are "mobile," three-phase assemblages tend to be less common than two-phase (tie-line) assemblages or one-phase assemblages. The ideal sequence, if two-phase assemblages predominate, can be determined by the successive tie-lines intersected by a straight line from the pelite (Shale) to the carbonate (Cal) in Figure 30.26:

pelite (Ms, Bt, An) | An + Hbl | Hbl + Czo | Czo +
Cpx | Cpx + Grt | Ves + Cpx | Cal

This sequence, however, is not unique. Although CaO appears to be most mobile, and the traverse from the pelite to the carbonate appears to be dominated by an increase in CaO, the A/F ratio need not be constant and may be controlled by phases stabilized by varying μ_{CaO}. Note that the sequence of analyzed bulk-rock compositions follows a zigzag curve in Figure 30.26 that includes some one-phase (Cpx) and three-phase (Grt + Cpx + Czo) assemblages. It is a common feature of metasomatic zones that the bulk rock composition varies in such an irregular fashion. Remember, smoother chemical gradients in the more mobile components, typically in the *fluid phase*, drive the reactions, causing the rock compositions to conform to the extent that they can, given the limited mobility of other components. Alternative sequences may avoid Grt and Ves between the Cpx

zone and the marble, or they may cross Grt + Ves instead of Ves + Cpx. There is no way to predict which will occur from Figure 30.26 alone.

Figure 30.27 is a diagram developed by Frantz and Mao (1979) for the CaO-MgO-SiO_2-CO_2-H_2O (CMS-CH) system that shows the composition of the solute species in the H_2O-CO_2 fluid that is in equilibrium with several common calc-silicate phases. Because experimental mineral–fluid solution data are not yet available, the diagram is only schematic. This diagram is similar to a ternary μ_{CaO}–μ_{MgO}–μ_{SiO_2} diagram (or log a equivalent) because μ-μ and activity diagrams, such as Figures 30.12, 30.19, and 30.25, refer to the chemical potential of components in the *fluid* phase. Again, because components are mobile, the number of phases is reduced, and because more than two components are involved, no unique path between any two points is predetermined. The fluid is always approaching equilibrium with the solid mineral assemblage, which may also buffer the fluids. If quartz and a dolomitic marble are initially in contact, a number of paths are possible between the quartz and dolomite fields in Figure 30.26, each representing a different sequence of zones.

For example, the direct path (a) in Figure 30.27 would produce a sequence of metasomatic zones:

(a) Qtz | Tr | Fo | Dol

Alternatively, paths (b) and (c) may occur if the solids buffer the fluid and two-phase zones are stabilized, resulting in the following sequences:

(b) Qtz | Tlc + Tr | Tr + Fo | Fo | Fo + Cal | Dol
(c) Qtz | Di + Tr | Tr | Tr + Cal | Cal + Dol | Dol

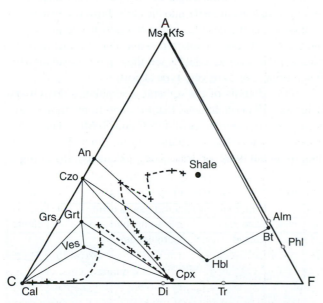

FIGURE 30.26 Al_2O_3-CaO-(FeO + MgO) diagram (projected from K_2O), showing the mineral phases and calculated bulk compositional path for metasomatic zones that develop at the contact between pelitic and carbonate layers near Lake Willoughby, Vermont. Ideal mineral compositions are in gray, real ones in black. After A. B. Thompson (1975). Reprinted by permission of Oxford University Press. Plus signs represent analyzed bulk-rock compositions within zones.

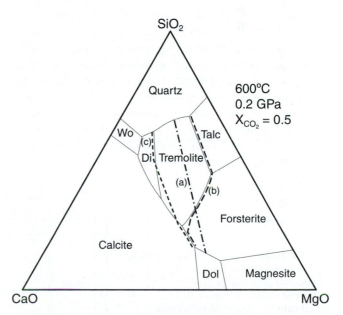

FIGURE 30.27 Schematic CaO-MgO-SiO_2-CO_2-H_2O diagram, showing the composition of the fluid solution in equilibrium with the phases shown at approximately 600°C and 0.2 GPa (projected from H_2O and CO_2 at a constant 1:1 ratio). After Frantz and Mao (1976). Reprinted by permission of the *American Journal of Science*.

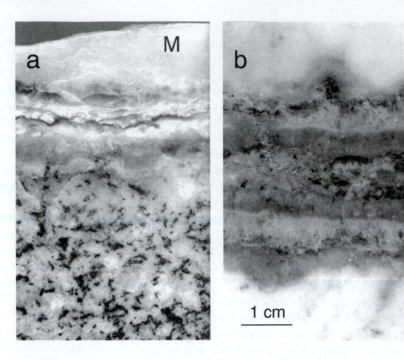

FIGURE 30.28 (a) Metasomatic zones separating quartz diorite (bottom) from marble (top). Zonation corresponds to the third row from the bottom in Table 30.2. (b) Symmetric metasomatic vein in the dolomite. Zonation corresponds to the last row in Table 30.2. Adamello Alps. After Frisch and Helgeson (1984). Reprinted by permission of the *American Journal of Science*. Photos courtesy of Hal Helgeson.

Some of these paths correspond to sequences observed in nature. [Note the similarity to some of the sequences described by Bucher-Nurminen (1981) in Table 30.2.] Numerous other paths involving monomineralic and bimineralic zones are also possible in Figure 30.27, involving wollastonite or talc as well as the phases listed above.

I present one final example because of its careful mineralogical control and theoretical treatment. Frisch and Helgeson (1984) described several zoned skarns, including magmatic and vein types, that develop in dolomitic marbles in the Adamello region of the Italian Alps. Figure 30.28 shows two samples, one at the contact between quartz diorite and marble, and the other a vein within the marble. Both samples illustrate the marked zonation typical of metasomatized rocks. Figure 30.29 shows how the mineralogy of the rocks varies along a traverse across the zones of the magmatic quartz diorite–dolomite contact corresponding to the sample shown in Figure 30.28a. As in the ultramafic examples discussed above, careful petrography reveals that the zones are not monomineralic, but usually mixtures of two or more minerals, reflecting varying degrees of component mobility. Note also that the quartz diorite is also altered to an endoskarn of tremolite, calcite, muscovite and clinozoisite.

Frisch and Helgeson (1984) considered several models for the development of the metasomatic zones in both the magmatic and vein occurrences and attempted material balance calculations for various reference frames (Section 30.2.3) based on each model. Models include (1) reciprocal diffusion (bimetasomatism) between the units, (2) infiltration of fluid from the pluton into the marble, (3) infiltration of both units by a fluid entering along a contact fissure, and (4) infiltration of fluid along the contact fissure accompanied by diffusional transfer into the units. Frisch and Helgeson concluded that model (4) conformed best to the geological and material balance constraints, and that the metasomatic zones at the igneous contact are thus really postmagmatic metamorphic, or even vein-type skarns.

On the basis of the mineral assemblages, Frisch and Helgeson (1984) concluded that the conditions of metamorphism were in the vicinity of 425°C and 50 MPa. The occurrence of clinozoisite + quartz requires that X_{CO_2} in the fluid is about 0.02 or less. They calculated log a-log a

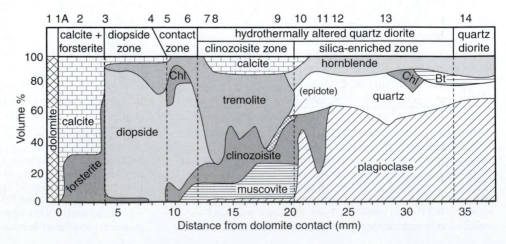

FIGURE 30.29 Mineral zones and modes developed at the contact between quartz diorite and dolomitic marble in Figure 30.28a. Initial contact may be at either side of the contact zone. Index numbers at the top indicate the locations of bulk chemical analyses. After Frisch and Helgeson (1984). Reprinted by permission of the *American Journal of Science*.

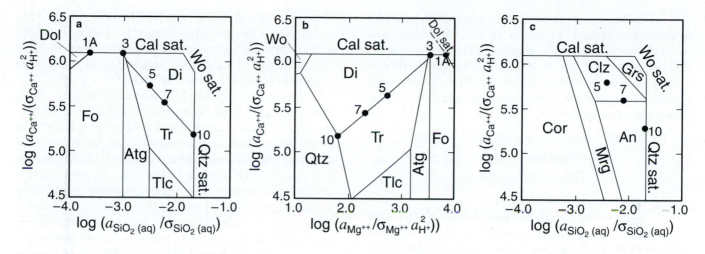

FIGURE 30.30 (a) $\log a_{CaO} - \log a_{SiO_2}$ and (b) $\log a_{CaO} - \log a_{MgO}$ diagrams in the system $CaO-MgO-SiO_2-H_2O-CO_2$. (c) $\log a_{CaO} - \log a_{SiO_2}$ diagram for the same system + Al_2O_3, all at 425°C, 0.05 GPa, and $X_{CO_2} = 0.007$. Numbered points correspond to the index numbers in Figure 30.29. After Frisch and Helgeson (1984). Reprinted by permission of the *American Journal of Science*.

diagrams for fluids in equilibrium with various minerals. Such $\log a$ diagrams are quantitative and can be calculated using the thermodynamic computer program SUPCRT92 (Helgeson et al., 1978; Johnson et al., 1992; see Section 27.5).

Figure 30.30 illustrates Frisch and Helgeson's (1984) $\log a_{Ca}-\log a_{SiO_2}$ and $\log a_{Ca}-\log a_{Mg}$ diagrams. In an attempt to be as rigorous as possible, the expression for the activity of a solute species i with charge z is modified by dividing by σ_i, a factor to account for solvation, and a_H^z to account for hydration (Walther and Helgeson, 1980). Bowers et al. (1984) published a compendium of such activity diagrams for use in analyzing systems involving equilibrium among minerals and aqueous solutions. The traverse of index numbers at the top of Figure 30.29 corresponds to the partial list of numbered dots in Figure 30.30, in which point 1A represents Cal + Dol + Fo equilibrium at the marble boundary, point 3 = Cal + Fo + Di, points 5 and 7 = Di + Tr, and point 10 represents quartz saturation in equilibrium with tremolite within the quartz diorite. In the Al-bearing system of Figure 30.30c, the occurrence of clinozoisite is shown at points 5 and 7, and An + Qtz is shown at point 10.

Note from Figure 30.30 that fairly large changes in the interstitial fluid are responsible for the mineralogy of the various zones and that the gradient of points in Figure 30.30a changes abruptly at the mineral zone boundary, consistent with buffering of activities at boundaries and variation across the boundaries between the buffered points (as in Figure 30.19).

Unlike infiltration of H_2O or CO_2, the nonvolatile migration and zoning seen in ultramafic blackwalls and skarns is rare over large distances and uncommon even over small distances. Conditions favoring mobilization of nonvolatile components include high temperatures, strong μ gradients, the presence of saline (generally chloride) aqueous fluids, and high permeability. For multicomponent systems, simple profiles of bulk rock composition along a traverse are generally irregular and will not directly indicate the metasomatic process, which involves gradients of dissolved species in a fluid phase.

Calculating chemical potential or log activity diagrams for solute species, as constrained by the mineral buffering assemblages (usually at zone boundaries where more phases coexist), provides much better indicators of the processes involved.

30.2.6 Quantitative Models and Experiments of Metasomatism

Several investigators have attempted to develop quantitative models of diffusion, infiltration, or combined processes based on mobile component concentrations and the extent of the mobility of each. The mathematics is complex and I consider these models beyond the scope of this book (and of my poor brain). Modern models expand upon Korzhinskii's (1959, 1970) initial theories and attempt to predict not only the fluxes, but also the widths of the metasomatic zones and the reactions by which they grow as zone boundaries migrate in one direction or another. For an excellent review, see Zaraisky (1993). Numeric solutions based on local equilibrium models include the infiltration models of Hofmann (1972), Fisher (1973), Lichtner (1985), and Lichtner et al. (1986); the diffusion model of Weare et al. (1976); and the combined diffusion–infiltration models of Frantz and Mao (1976, 1979) and Fletcher and Vidale (1975). Joesten and Fisher (1988) applied a diffusion model to the development of the monomineralic zones around the chert nodules discussed above. The growth of reaction rims and coronas in Section 23.6 is an identical diffusion-controlled process, and the theory can be applied to their growth as well. Zaraisky (1993) praised the "macrokinetic" model (Balashov and Lebedeva, 1991) as the current culmination of treating metasomatic zonation in terms of local equilibrium and thermodynamics of reversible reactions, combined with a kinetic description of the approach toward those limits. Given initial and boundary conditions, they developed a set of differential equations that attempts to fully describe the structure of a zoned metasomatic column, with local equilibrium as a limiting case.

The Russians are presently the experts in experimental metasomatic analogs. Plyusnina et al. (1995) performed several experiments with granite–dunite interactions in H_2O-chloride solutions at 400 and 500°C and 0.1 GPa to model ultramafic blackwall zones. At 500°C, a 2-mm zone developed between the contrasting rock types, in which the following subzones formed:

$$(\text{tonalite}) - (\text{Ab} + \text{Na-amphibole} \pm \text{quartz})$$
$$- (\text{oligoclase} + \text{Tlc}) * (\text{Tlc} + \text{Di})$$
$$- (\text{Tlc} + \text{Ol}) - (\text{dunite})$$

The asterisk marks the initial boundary.

At 400°C, the zoning was similar, but a chlorite zone appeared in place of the talc + diopside zone. Si appeared to be most mobile, becoming depleted in the marginal granite. Na was also mobile. Mg migration from the dunite was negligible, creating only some talc with plagioclase on the granitic side of the initial contact. Ca migration into the dunite resulted in the generation of diopside.

Vidale (1969) in the United States performed some experiments on juxtaposed calcite and model pelite, in saline (chloride) solutions, producing zoned skarns at the contact. The bulk of the experimental work on skarns, however, is also Russian (see Zharikov and Zaraisky, 1991; and Zaraisky, 1991, for reviews). Experimental charges are generally set up as an initial column in an open inert platinum capsule with powdered marble at the bottom and a powdered silicate rock at the open top. Capsules are surrounded by an aqueous (commonly alkali halide) solution that can enter the open end of a capsule and permeate through the silicate into the carbonate. Runs generally last two weeks. Results are usually a series of discrete zones through which the composition of the rock adapted consecutively and step-wise to the infiltrating fluids, approaching a state of local equilibrium across the column. Figure 30.31 shows an example of a traverse across a section cut through an experimental charge after two weeks at 600°C and 0.1 GPa (see also the back cover). Notice the similarity between the zoning developed in the charge and some natural examples listed in Table 30.2. Motion of Si, Ca, Fe, Mg, and, to a lesser extent, Al is required to produce the observed zoning.

It is possible to control and vary the rock types, fluids, and conditions in these experiments. Because the initial contact is also known, experimental results can be useful in constraining models of zone growth and development. Of course high temperature and the presence of a fluid phase are both conducive to chemical mobility, but the presence of a chloride brine appears to be equally important in facilitating metasomatism. When a 1.0 molar NaOH solution was used in the experiments, the granodiorite hydrated and albitized due to the Na content, but only a tiny and poorly developed contact zone of wollastonite formed (Zaraisky, 1991). The number of zones that form depends upon both the chemical complexity of the initial contact and the experimental conditions. The accumulated results of these and future experiments will greatly help constrain quantitative models of metasomatism.

The pioneering work of D.S. Korzhinskii has led to a fruitful approach to the study of metasomatic rocks, in the field, in the lab, and on the computer. In the words of Zaraisky (1993):

> Nearly all the principal propositions of Korzhinskii's theory of metasomatic zoning have been proven by experiment and computation. Most important of all is the strong numerical proof of the previously intuitively accepted assumption that the direction of metasomatic processes and the fundamental features of metasomatic zoning are controlled by thermodynamic relationships, and that the structure of any column, including those initially in a state far from equilibrium, in due course asymptotically approaches local equilibrium.

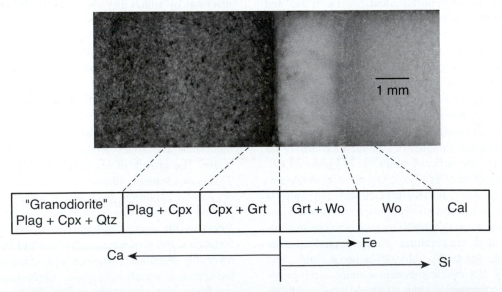

FIGURE 30.31   Zonation in an experimental skarn formed at the contact between granodiorite and limestone at 600°C, $P_{fluid} = 0.1$ GPa ($X_{CO_2} = 0.07$). After Zharikov and Zaraisky (1991). Photo courtesy G. Zaraisky.

Summary

Fluids play a critical role in many geologic processes. Fluids can dissolve material, transport heat and solutes, precipitate minerals, exchange components as they react with minerals, enhance partial melting, and catalyze deformation processes by weakening rocks. Deep fluids are typically supercritical mixtures dominated by C-O-H and typically richest in H_2O and CO_2 (and perhaps CH_4 at low temperatures and in more reducing conditions). Dissolved nonvolatile species tend to be dissociated into ions at lower temperature and higher pressure and more associated at higher grades. Most fluids are chloride brines with $NaCl > KCl > CaCl_2 \gg MgCl_2$. Aqueous SiO_2 is also typically present, as are bicarbonate, sulfate, and bromide.

In general usage, metasomatism is considered to involve the migration of appreciable nonvolatile material into or from a rock during metamorphism. Such migration involves mass transfer via either diffusion of matter through another medium (minerals or intergranular fluid) or as dissolved species carried along by infiltrating fluids (or both). Diffusion is *very* slow through minerals, but may approach a few meters per year through (intergranular) fluids. Estimates of time-integrated fluid fluxes suggest that huge quantities of fluid can pass through crystalline rocks ($\sim 10^3$ m^3 of fluid through each m^2 cross section of rock during regional metamorphism and perhaps 100 times that in channelized flow). Such flow is capable of redistributing heat and matter, thereby affecting the style of metamorphism and the rocks produced.

If migrating species are not in equilibrium with the new host rock, they react with the rock, exchanging components with the solid phases until local equilibrium is attained (or at least approached). As fluid moves, it carries solute species and continuously exchanges them with successive rock volumes that it encounters, generally producing different reactions at different zones along the flow.

Diffusion is driven by chemical potential (μ) gradients: any component will diffuse (if possible) from areas of high μ to areas of low μ. Diffusion (or infiltration) thus tends to reduce μ gradients. When effective, component mobility reduces the number of phases and generates discontinuities in X_{bulk}. The mineralogical phase rule may be expressed as $\phi \leq C_i$ where C_i is the number of "inert" (non-mobile) components. In cases of high mobility, C_i reduces to the lowest possible value (one) and monomineralic zones result. P-μ, T-μ, μ-μ, or log a-log a diagrams are useful in addressing systems with mobile components.

Reactions involving net transfer (i.e., migration) can be written in a number of ways, depending on what species are migrating. When matter is in motion it becomes difficult to determine original contacts between lithologic units, making the determination of a reference frame (what actually moved how far and in what quantity) a problem. Gresens-type and isocon diagrams are two attempts to solve the reference-frame and closure (sum to 100%) problems. We can successfully model some sequences of low-ϕ metasomatic zones that develop across initial contacts between rocks of contrasting composition using paths across μ-μ (or log a-log a) diagrams.

Key Terms

Supercritical fluid *655*
Solubility product (K_{SP}) *656*
Dielectric constant *657*
Dissociation constant (K) *657*
Metasomatism *658*
Diffusion *659*
Bimetasomatism *659*
Fick's law *659*
Diffusion coefficient *659*

Infiltration *660*
Microfractures/hydraulic fracturing *660*
Reaction-enhanced permeability *660*
Flow-focusing *660*
Thermal decompaction *660*
Dilatancy pumping *660*
Fluid:rock ratios *661*
Time-integrated fluid flux (q_{TI}) *661*

Local equilibrium *662*
Chemical potential *662*
Mobile/immobile components *664*
Activity *665*
Isocon *668*
Blackwall *669*
Skarn/tactite *674*
Endoskarn/exoskarn *674*

Review Questions and Problems

Review Questions and Problems are located on the author's web page at the following address: **http://www.prenhall.com/winter**

Important "First Principle" Concepts

- Even dense crystalline rocks are capable of transmitting huge quantities of fluid during a period of regional metamorphism. Such flow can redistribute heat and dissolved constituents in both regional and contact metamorphism, profoundly affecting the style of metamorphism, as well as the composition and texture of metamorphic rocks.

- At equilibrium, the chemical potential of any component is the same in all coexisting phases. If gradients in μ exist, diffusion or infiltration will attempt to reduce the gradient (to zero, if migration is effective).

- A reduction in the number of coexisting phases and accompanying discontinuities in bulk composition (even where none existed before) are the hallmarks of metasomatic processes.

Suggested Further Readings

Metamorphic Fluids

Ague, J. J. (2003). Fluid flow in the deep crust. Chapter 3.06. *Treatise on Geochemistry*, **3**, 195–228. Elsevier. Amsterdam.

Bickle M. J., and D. McKenzie (1987). The transport of heat and matter by fluids during metamorphism. *Contrib. Mineral. Petrol.,* **95**, 384–392.

Bredehoft, J. D., and D. L. Norton (eds.). (1990). *The Role of Fluids in Crustal Processes.* National Academy of Sciences. Washington, DC.

Crawford, M. L., and L. S. Hollister. (1986). Metamorphic fluids: The evidence from fluid inclusions. In: *Fluid-Rock Interactions During Metamorphism* (eds. J. V. Walther and B. J. Wood). Springer-Verlag. New York. pp. 1–35.

Eugster, H. P., and L. Baumgartner. (1987). Mineral solubilities and speciation in supercritical metamorphic fluids. In: *Thermodynamic Modeling of Geological Materials: Minerals, Fluids, Melts* (eds. I. S. E. Carmichael and H. P. Eugster). *Reviews in Mineralogy,* **17**. Mineralogical Society of America. Washington, DC. pp. 367–403.

Ferry, J. M. (1994). A historical review of metamorphic fluid flow. *J. Geophys. Res.,* **99**, 15,487–15,498.

Ferry, J. M., and D. M. Burt. (1982). Characterization of metamorphic fluid composition through mineral equilibria. In: *Characterization of Metamorphism Through Mineral Equilibria* (ed. J. M. Ferry). *Reviews in Mineralogy,* **10**, Mineralogical Society of America. Washington, DC. pp. 207–262.

Ferry, J. M., and G. M. Dipple. (1991). Fluid flow, mineral reactions, and metasomatism. *Geology,* **19**, 211–214.

Fyfe, W. S., N. Price, and A. B. Thompson (1978). *Fluids in the Earth's Crust.* Elsevier. Amsterdam.

Labotka, T. C. (1991). Chemical and physical properties of fluids. In: *Contact Metamorphism* (ed. D. M. Kerrick). *Reviews in Mineralogy,* **26,** Mineralogical Society of America. Washington, DC. pp. 43–104.

Manning, C. E., and S. E. Ingebritsen. (1999). Permeability of the continental crust: implications of geothermal data and metamorphic systems. *Rev. Geophys.,* **37**, 127–150.

Schmulovich, K. I., B. W. D. Yardley, and G. C. Gonchar (eds.). (1995). *Fluids in the Crust: Equilibrium and Transport Properties.* Chapman & Hall. London.

Walther, J. V., and B. J. Wood (eds.). (1986). *Fluid–Rock Interactions During Metamorphism. Advances in Physical Geochemistry 5.* Springer-Verlag. New York.

Metasomatism

Greenwood, H. J. (ed.). (1983). *Studies in Metamorphism and Metasomatism. Amer. J. Sci.,* Special Volume **283-A**.

Helgeson, H. C. (ed.). (1987). *Chemical Transport in Metasomatic Processes.* D. Reidel. Dordrecht, The Netherlands.

Hofmann, A. W., B. J. Giletti, H. S. Yoder, and R. A. Yund (eds.). (1974). *Geochemical Transport and Kinetics.* Publication **634**. Carnegie Institution. Washington, DC.

Korzhinskii, D. S. (1959). *Physicochemical Basis of the Analysis of the Paragenesis of Minerals.* Consultants Bureau. New York.

Korzhinskii, D. S. (1970). *Theory of Metasomatic Zoning.* Clarendon Press. Oxford, UK.

APPENDIX A
Estimating the Density and Viscosity of Silicate Melts

DENSITY

The density calculation below follows the method of Bottinga and Weill (1970), in which the density (ρ) at a particular temperature (chosen so that the system must be entirely molten) is given by:

$$\rho = \sum_i X_i M_i / X_i \overline{V}_i \tag{A.1}$$

where X_i, M_i, and V_i are the mole fraction, molecular weight, and partial molar volume of each component i. The calculation begins with a whole-rock chemical analysis (taken from Table 8.1) in weight percent (wt. %) oxides (column 2 in Table A.1), which are converted to mole proportions (column 4) by dividing each by the molecular weight (column 3). The partial molar volumes, thermal expansions, and thermal compressibilities are listed in columns 5, 6, and 7. I have included the more recent volume data of Lange and Carmichael (1987), along with their coefficients of thermal expansion and the compressibility data of Kress and Carmichael (1991), none of which are in the original formulation of Bottinga and Weill (1970). I have also included the molar volume data for H_2O from Lange (1994), but the data are not particularly precise, and analyses of H_2O are not always available. The calculation thus allows both a hydrous and an anhydrous density; the latter simply omits H_2O.

The molar volume of each oxide at the temperature and pressure of interest is given by:

$$V_{i\,(\text{at P and T})} = V_{i\,(\text{at 0.1 MPa and T})} + \frac{dV_i}{dT}(T - 1400°C)$$
$$+ \frac{dV_i}{dP}(P - 0.1\,\text{MPa}) \tag{A.2}$$

Note that the expansion and compressibility values in columns 6 and 7 in Table A.1 are in 10^{-3} cm (for legibility), so they must be multiplied by 10^3 when used in Equation (A.2) to yield the proper values for volume in cm^3. Of course, one must first choose a temperature and pressure for the calculation (the pressure correction may be ignored by selecting an atmospheric pressure of 0.1 MPa). V_i is then calculated for each oxide in column 8. Because no volume data are available for MnO, I have added the mol proportion of MnO to FeO in column 4. Because Mn and Fe have similar valence and ionic radius this is probably a better approximation than simply omitting MnO. Because column 4 has molecular proportions and Equation (A.1) is the mol fraction (which must sum to 1.0), Equation (A.1) may be approximated by:

$$\rho = 100 / \sum_i Y_i V_i \tag{A.3}$$

where Y_i is the unnormalized mol proportions in column 4. This is so because the values in column 4 are based on 100 grams of sample (column 2) and have the molecular weights as denominators, which cancel the M_i values in Equation (A.1). The final column 9 is then calculated by multiplying the values in column 8 by those in column 4, and the density is simply equal to 100 divided by the sum of column 9 ($= 38.17\ cm^3/100$g in the example). The anhydrous density is calculated by omitting the final H_2O value in column 9.

A spreadsheet for calculating the density by this procedure is available from my web page, **http://www.prenhall.com/winter**.

VISCOSITY

The viscosity calculation below follows the method of Shaw (1972), based on the earlier work of Bottinga and Weill (1972). Because the polymerization of Si-O bonds largely controls the viscosities of silicate liquids, Shaw based his empirical approach on the viscosities of binary experimental systems for each individual oxide mixed with SiO_2 (excluding the minor element oxides MnO and P_2O_5). The viscosity of the full multicomponent system is then determined by:

$$\ln \eta = s \cdot (10^4/T_K) - c_T \cdot s + c_\eta \tag{A.4}$$

where η is the viscosity (in poise), T_K is the temperature in degrees kelvin, s is the slope of $\ln \eta$ versus $10^4/T_K$ calculated for a given multicomponent mixture (rock analysis), and c_T and c_η are the coordinates of the point of intersection of the various Arrhenius slopes. Shaw determined weighted mean averages of these coordinates as $c_T = 1.50$ and $c_\eta = -6.40$.

The method for calculating the slope for a multicomponent mixture begins with a whole-rock chemical analysis in weight percent (wt. %) oxides (column 2 in Table A.2), which are converted to mole proportions (column 4) by dividing each by the molecular weight (column 3). Fe_2O_3 (if present) is converted to FeO by the expedient of doubling the value in column 4 (basing the value on single Fe atoms rather than Fe_2). Al is similarly put on a single-Al basis by doubling the value in column 4. This is *not* done, however, for Na_2O, K_2O, or H_2O. Mol proportions are then converted to mol fractions (X_i) by normalizing the values in column 4 so they sum to 1.0: multiplying each value by $1.0/\Sigma$ (column 4).

Column 6 lists s_o^i, the values of Arrhenius slopes for binary oxide-SiO_2 mixtures extrapolated to pure SiO_2. Note that FeO and MgO behave similarly and have the same values, as do Na_2O and K_2O. I have taken the liberty of assuming that MnO behaves like FeO and MgO and have used the same

TABLE A.1   Magma Density Calculation after the method of Bottinga and Weill (1970)

Oxide	Wt. %	Mol. Wt.	Y_i Mol. Prop.	V_i at 1400°C	dv/dT	dv/dP	T°C = 1200 V_i at T&P	P (Mpa) = 0.1 Y_iV_i
SiO_2	49.20	60.09	0.82	26.90	0.00	−1.89	26.90	22.02
TiO_2	2.03	79.88	0.03	23.16	7.24	−2.31	21.71	0.55
Al_2O_3	16.10	101.96	0.16	37.11	2.62	−2.26	36.59	5.78
Fe_2O_3	2.72	159.70	0.02	42.13	9.09	−2.53	40.31	0.69
FeO	7.77	71.85	0.11	13.65	2.92	−0.45	13.07	1.45
MnO	0.18	70.94					0.00	
MgO	6.44	40.31	0.16	11.45	2.62	0.27	10.93	1.75
CaO	10.50	56.08	0.19	16.57	2.92	0.34	15.99	2.99
Na_2O	3.01	61.95	0.05	28.78	7.41	−2.40	27.30	1.33
K_2O	0.14	94.20	0.00	45.84	11.91	−6.75	43.46	0.06
P_2O_5	0.23	141.94	0.00				0.00	0.00
H_2O	1.65	18.02	0.09	17.00			17.00	1.56
Total	99.97		1.62				$V_{liq} = \Sigma Y_iV_i =$	38.17
Units	g/100g	g/mol	mol/100g	cm^3/mol	$\dfrac{10^{-3}cm^3}{mol\ K}$	$\dfrac{10^{-3}cm^3}{mol\ MPa}$	cm^3/mol	cm^3/100g

$$\rho = 100/\sum_i Y_i V_i$$

Density (g/cm^3) =	**2.62**	hydrous
	2.73	anhydrous

TABLE A.2   Magma Viscosity Calculation after the method of Shaw (1972)

Oxide	Wt. %	Mol. Wt.	Y_i Mol. Prop.	X_i Mol. Frac.	s_i^o	T°C = 800 $Xs_{i^o}*X_{sio2}$	$X_i (s_{i^o}*X_{sio_2})$
SiO_2	71.90	60.09	1.197	0.627			
TiO_2	0.09	79.88	0.001	0.001	4.5	2.82	0.00
Al_2O_3	12.10	101.96	0.237	0.124	6.7	4.20	0.52
Fe_2O_3	0.57	159.70	0.007	0.004	3.4	2.13	0.01
FeO	0.52	71.85	0.007	0.004	3.4	2.13	0.01
MnO	0.00	70.94	0.000	0.000	3.4	2.13	0.00
MgO	0.04	40.31	0.001	0.001	3.4	2.13	0.00
CaO	0.27	56.08	0.005	0.003	4.5	2.82	0.01
Na_2O	3.94	61.98	0.064	0.033	2.8	1.76	0.06
K_2O	4.32	94.20	0.046	0.024	2.8	1.76	0.04
P_2O_5	0.00	141.94	0.000	0.000			
H_2O	6.20	18.02	0.344	0.180	2.0	1.25	0.23
Total	99.95		1.91	1.00			0.87
Units	g/100g	g/mol	mol/100g	mol/1 mol			

Mean slope (s) = $\Sigma Xi(s_{i^o}*X_{Sio2})/(1 - X_{SiO2})$ = 2.34

ln η = slope*($10^4/T_K$) − 1.50*slope − 6.40 = 11.94 | η = 1.53E + 04 Pa sec |

value of 3.4 for it as well. Each of these s_o^i values is then multiplied by the mol fraction of SiO_2 to yield the slopes appropriate for each binary system (column 7). Then multiply each value in column 7 by the mol fraction of each oxide (column 5, excluding SiO_2 of course) to yield the values in the final column, 8. Sum the values in column 8 and divide that sum by $(1-X_{SiO_2})$ to give the mean slope (s) of the mixture. Substitute this value of s into Equation (A.4), using the values of c_T and c_η given, and determine the exponent of the log term to yield the viscosity η in poises (g/cm s) or convert to Pa s (Pascal seconds, S.I. units): 1 Pa s = 10 poise.

McBirney and Murase (1984) and Murase et al. (1985) proposed the following empirical equation for calculating the viscosity of partially crystallized magmas:

$$\log \eta_{eff} = \log \eta_o + \frac{0.019 D_m}{(1/\phi)^{-1/3} - 1}$$

where:

η_{eff} = effective viscosity of the melt/crystal mixture

η_o = viscosity of the liquid alone (as calculated above)

D_m = mean diameter of the crystals (in microns)

ϕ = volume fraction of crystals

A spreadsheet for calculating the viscosity by this procedure is available from my web page, **http://www.prenhall.com/winter**. The Bottinga and Weill (1972) calculation uses more types of molar species and constants, each of which are valid over only a limited range of X_{SiO_2}, making the calculation rather difficult for a spreadsheet. The program Magma by Ken Wohletz at Los Alamos calculates several classification parameters and properties, including the viscosity (using base$_{10}$ logs) according to the models of Botinga and Weill (1972) and Shaw (1972). A pointer to the free download can be found on my web page (**http://www.prenhall.com/winter**).

APPENDIX B
The CIPW Norm

As discussed in Section 8.4, the norm is a method of recalculating a chemical analysis of an igneous rock into a set of ideal minerals (Table B.1). The norm was originally proposed as a method of classifying igneous rocks, but the classification has since given way to the direct chemical classifications presented in Chapter 2. The norm is still used by some petrologists, and it is most useful for volcanic or glassy rocks, where the modal mineralogy is unavailable. In the United States, we commonly use the formulation developed by Cross, Iddings, Pirsson, and Washington (the "CIPW" norm), which expresses the "normative" minerals as *weight* percentages. The method described here is a simplification of the CIPW norm first proposed by Cross et al. (1902) and modified by Johanssen (1939) and Kelsey (1965). In Europe, the Molecular (or "Niggli") norm, which expresses the "normative" minerals as *molecular* percentages, is more common (Niggli, 1936). For the Niggli norm calculation, see Barth (1962). The following mid-ocean ridge basalt analysis will be used as an illustration of the technique. As an exercise, you may want to create a norm spreadsheet and follow the sample calculation below (Table B.2). The process shown here assumes that you are calculating the norm on a spreadsheet. Table B.3 is another CIPW norm calculation for a nepheline basalt, which illustrates some of the less common steps.

Tholeiitic Basalt Mid-Atlantic Ridge	
SiO_2	49.2
TiO_2	2.03
Al_2O_3	16.1
Fe_2O_3	2.72
FeO	7.77
MnO	0.18
MgO	6.44
CaO	10.5
Na_2O	3.01
K_2O	0.14
P_2O_5	0.23
H_2O	1.65
S	0.14
Total	100.11

The norm calculation creates ideal minerals by allocating the various chemical constituents in an analysis to a set of prescribed minerals (Table B.1).

Because the volatile constituents are rarely analyzed, the norm is based on *anhydrous* minerals, so that norms cannot be equivalent to corresponding modes for rocks that contain micas or amphiboles. Norms also account for solid solution in minerals, but only in a cursory fashion. As an example of what a norm does, consider the constituent K_2O. In most igneous rocks, potassium is concentrated in orthoclase, $KAlSi_3O_8$, at least if we are forced to ignore hydrous micas and amphiboles. Each mole of K_2O in a sample can be combined with an equivalent mole of Al_2O_3 and 6 moles of SiO_2 to make a single mole of $K_2Al_2Si_6O_{16}$ ($= 2$ moles of $KAlSi_3O_8$). A norm calculation is thus a process by which oxides are apportioned so as to create idealized minerals, attempting to approximate the "preference" of an oxide for certain minerals over others. It begins with the easiest choices, first allocating oxides that tend to occur in only a single mineral, and then turning to more arbitrary combinations based on what remains after the easier ones have been handled. *SiO_2 balancing is a crucial requirement* in the norm calculation, as you shall see in the "cookbook" process and example described below. In some cases of low-SiO_2 alkaline rocks it can be a complex process to accommodate some components as silicate minerals and still balance the silica. The calculation makes a "first-pass" in which oxides are combined to create common minerals in a "provisional" manner (listed with a prime). If the "first-pass" works, and there is sufficient SiO_2 in the analysis to accommodate the oxides in the analysis and still have enough left over for a bit of free SiO_2 (as normative *q*uartz), the process is complete and the calculation was mercifully simple. If there isn't enough silica to satisfy all of the minerals, you must go back and "borrow" some SiO_2 from some of the provisional minerals you have already calculated (*or'*, *ab'*, *hy'*, etc.) and create other, less siliceous minerals in order to free up SiO_2 and balance the calculation. The process may sound complex, so it is best illustrated by example. Refer to Tables B.1 and B.2 as we apply the procedure below to the basalt analysis above.

1. The *molecular* proportion (amount) of each constituent is determined (column 4 in Table B.2) by reference to a table of molecular weights (column 3 in Table B.2). If any result is less than 0.002 it may be neglected (I included some low-concentration oxides anyway to illustrate the calculation). We convert to *moles* because we combine elements in their *molar* proportions to make minerals (pure forsterite requires two moles of MgO plus one mole of SiO_2 to make a mole of Mg_2SiO_4). Norms ignore hydrous phases and water, so we overlook the OH in column 4 and normalize all the other oxides to add to 100%.

2. The *amounts of MnO and NiO* (if any) are *added to that of FeO* in column 4 because they are only minor constituents and they usually substitute for Fe in mafic minerals. *BaO and SrO are similarly added to CaO* because Ba and Sr are also minor constituents and commonly substitute for Ca in feldspars.

TABLE B.1   Normative Minerals

Mineral	Abbr.	Formula	Formula Wt.
Quartz	*q*	SiO_2	60.08
Corundum	*c*	Al_2O_3	101.96
Zircon	*z*	$ZrO_2 \cdot SiO_2$	183.31
Orthoclase	*or*	$K_2O \cdot Al_2O_3 \cdot 6\,SiO_2$	556.67
Albite	*ab*	$Na_2O \cdot Al_2O_3 \cdot 6\,SiO_2$	524.46
Anorthite	*an*	$CaO \cdot Al_2O_3 \cdot 2\,SiO_2$	278.21
Leucite	*lc*	$K_2O \cdot Al_2O_3 \cdot 4\,SiO_2$	436.5
Nepheline	*ne*	$Na_2O \cdot Al_2O_3 \cdot 2\,SiO_2$	284.11
Kaliophilite	*kp*	$K_2O \cdot Al_2O_3 \cdot 2\,SiO_2$	316.33
Halite	*hl*	$Na_2O \cdot 2\,Cl$	132.89
Thenardite	*th*	$Na_2O \cdot SO_3$	142.04
Sodium Carbonate	*nc*	$Na_2O \cdot CO_2$	105.99
Acmite	*ac*	$Na_2O \cdot Fe_2O_3 \cdot 4\,SiO_2$	462.02
Sodium metasilicate	*ns*	$Na_2O \cdot SiO_2$	122.07
Potassium metasilicate	*ks*	$K_2O \cdot SiO_2$	154.28
Diopside	*di*	$CaO \cdot (Mg,Fe)O \cdot 2\,SiO_2$	216.56–248.10*
Wollastonite	*wo*	$CaO \cdot SiO_2$	116.17
Hypersthene	*hy*	$(Mg,Fe)O \cdot SiO_2$	100.39–131.93*
Olivine	*ol*	$2\,(Mg,Fe)O \cdot SiO_2$	140.69–203.78*
Calcium silicate	*cs*	$2\,CaO \cdot SiO_2$	172.24
Magnetite	*mt*	$FeO \cdot Fe_2O_3$	231.54
Chromite	*cm*	$FeO \cdot Cr_2O_3$	223.84
Ilmenite	*il*	$FeO \cdot TiO_2$	151.75
Hematite	*hm*	Fe_2O_3	159.69
Titanite	*tn*	$CaO \cdot TiO_2 \cdot SiO_2$	196.07
Perovskite	*pf*	$CaO \cdot TiO_2$	135.98
Rutile	*ru*	TiO_2	79.9
Apatite	*ap*	$3.3\,CaO \cdot P_2O_5$	328.68
Fluorite	*fl*	CaF_2	78.07
Pyrite	*pr*	FeS_2	119.98
Calcite	*cc*	$CaO \cdot CO_2$	100.09

* Two numbers = formula weights of Mg and Fe end-members, respectively.

3a. *An amount of CaO equal to 3.33 times that of P_2O_5 (or 3.00 P_2O_5 and 0.33 F, if the latter is present) is allotted for apatite (**ap**).* Because phosphate occurs only in the mineral apatite, we begin by putting the 0.0016 moles of P_2O_5 in the ***ap*** column 5. Because apatite has the formula $Ca_5(PO_4)_3(OH)$, each mole is equivalent to 5 moles CaO + 1.5 moles of P_2O_5 + 0.5 moles of H_2O. Thus for every 1.5 P_2O_5 we must add 5 CaO. If we multiply through by 2/3 we get 3.33 CaO for every single P_2O_5 in apatite. Thus we put $3.33 \times 0.0016 = 0.0053$ CaO in the ***ap*** column as well. Because norms ignore hydrous phases and water, we overlook the OH in ***ap***. To summarize: We've used 0.0016 moles of P_2O_5 + 0.0053 moles of CaO to make 0.0016 moles of apatite (assuming a 2/3 apatite formula = $Ca_{3.33}(PO_4)_2$—as listed in Table B.1). *In this and all subsequent steps, be sure to put the number of moles of the mineral created at the bottom of each column (the row labeled "Mol proportion of normative minerals" in Table B.2).* In this case, it is 0.0016 moles of ***ap***. Be sure that you understand these numbers. *You do not add the numbers* to determine how many moles of apatite are created. We combine 4 wheels + 2 axles + 1 handle + 1 body to make up a *single* wagon, not $4 + 2 + 1 + 1 = 8$ wagons. In our normative apatite we have combined 0.0016 moles of CaO with 0.0053 moles of P_2O_5 to create 0.0016 moles of apatite. Use Table B.1 as a guide for determining the molar combinations.

3b. *An amount of Na_2O equal to that of any Cl_2 is allotted for halite (**hl**).* The oxygen in the normative halite leads to only a minor error, largely compensated by the somewhat bizarre halite formula used in Table B.1. We have no Cl_2 in our analysis, so we can ignore this for our basalt. If Cl_2 is in the analysis, create a column for ***hl***, and put all of the Cl_2 in it, plus an equivalent amount of Na_2O. The number of moles of ***hl*** at the bottom is equivalent to the number of moles of Cl_2 (or Na_2O, because Na_2O and Cl_2 are combined in equal proportions).

3c. *An amount of Na_2O equal to that of the SO_3 (if any) is allotted for thenardite (**th**).* The SO_3 stated in most analyses represents the S of pyrite, so that *step 3c is applicable only when the rock contains minerals of the haüyne group. This is rare, and we ignore it for our basalt. If the haüyne group is

TABLE B.2 CIPW Norm Calculation for Tholeiitic Basalt

Step:				3a	3d	3f	4a	4c	4d	5c	7a	7a	7c	9
Oxide	**Wt. %**	**MW**	**Mol. Prop.**	**ap**	**pr**	**il**	**or'**	**ab'**	**an**	**mt**	**Remainder**	**di'**	**hy'**	**q**
SiO_2	49.2	60.08	0.8189				0.0089	0.2914	0.2157				0.1480	0.1519	0.0031
TiO_2	2.03	79.88	0.0254			0.0254								
Al_2O_3	16.1	101.96	0.1579				0.0015	0.0486	0.1079					
Fe_2O_3	2.72	159.69	0.0170							0.0170				
FeO	7.77	71.85	0.1107		0.0022	0.0254				0.0170	0.0660	0.0216	0.0444	
MnO	0.18	70.94												
MgO	6.44	40.30	0.1598								0.1598	0.0523	0.1075	
CaO	10.5	56.08	0.1872	0.0054					0.1079			0.0740		
Na_2O	3.01	61.98	0.0486					0.0486				0.0486		
K_2O	0.14	94.20	0.0015				0.0015			0.0015				
P_2O_5	0.23	141.94	0.0016	0.0016										
H_2O	1.65													
S	0.14	32.06	0.0044		0.0044									
Total	100.11													
Mol proportion of normative mineral				0.0016	0.0022	0.0254	0.0015	0.0486	0.1079	0.0170	Mg/(Mg + Re) = 0.708	0.0740	0.1519	0.0031
Formula weight				328.68	119.98	151.75	556.67	524.46	278.21	231.54		225.78	109.61	60.08
Wt. % of normative minerals				**0.53**	**0.26**	**3.86**	**0.83**	**25.47**	**30.01**	**3.94**		**16.70**	**16.65**	**0.18**

Y = 0.8158   D = −0.0031

TABLE B.3 CIPW Norm Calculation for orendite, Leucite Hills, Wyoming.

Step:				3a	3f	4a	4c	5a	5c	7a	7a	10b	10b	12b	13b	15b	15b	16	16
Oxide	Wt. %	MW	Mol. Prop.	ap	il	or'	ab'	ac	mt	Remainder	di'	hy'	ol'	ne	lc'	cs	ol	lc	kp
SiO_2	35.6	60.08	0.5925			(0.1771)	(0.5821)	0.1508			(0.2018)	(0.0777)	(0.0389)	0.1940	(0.1180)	0.0504	0.0893	0.0978	0.0101
TiO_2	5.98	79.88	0.0749		0.0749														
Al_2O_3	12.9	101.96	0.1265			(0.0295)	(0.0970)							0.0970	(0.0295)			0.0245	0.0051
Fe_2O_3	7.68	159.69	0.0481					0.0377	0.0104										
FeO	9.28	71.85	0.1299		0.0749				0.0104	0.0446	(0.0252)	(0.0194)	(0.0194)				0.0446		
MnO	0.05	70.94																	
MgO	5.40	40.30	0.1340								(0.0757)	(0.0583)	(0.0583)				0.1340		
CaO	8.46	56.08	0.1509	0.0500							(0.1009)					0.1009			
Na_2O	8.35	61.98	0.1347				(0.0970)	0.0377						0.0970					
K_2O	2.78	94.20	0.0295			(0.0295)									(0.0295)			0.0245	0.0051
P_2O_5	2.13	141.94	0.0150	0.0150															
H_2O																			
S		32.06	0.0000																
Total	98.61																		
Mol proportion of normative mineral				0.0150	0.0749	(0.0295)	(0.0970)	0.0377	0.0104		(0.1009)	(0.0777)	(0.0389)	0.0970	(0.0295)	0.0504	0.0893	0.0245	0.0051
Formula weight				328.68	151.75	(556.67)	(524.46)	462.02	231.54		(224.44)	(108.27)	(156.45)	284.11	(436.50)	172.24	156.45	436.50	316.33
Wt. % of normative minerals				4.93	11.36			17.42	2.40					27.56		8.69	13.97	10.67	1.60

$Mg/(Mg + Fe) = 0.750$

$Y = 1.1895$ $D = 0.5969$ $D1 = D2 = 0.5581$ $D3 = 0.1700$ $D4 = D5 = 0.1110$ $D6 = 0.0101$

Cells with parentheses are provisional normative minerals that were reduced in a subsequent step.

present, create a column for *th* and set the number of moles of *th* equal to the number of moles of Na_2O (see Table B.1).

3d. *An amount of FeO equal to half that of any S (or the S of erroneously stated SO_3 as above) is allotted for pyrite (we have 0.0022 moles of **pr** in column 6 of Table B.2).*

3e. *An amount of FeO equal to that of any Cr_2O_3 is allotted for chromite (**cm**). The number of moles of **cm** at the bottom is equal to the number of moles of Cr_2O_3. There is none in our example analysis, so we ignore this step.*

3f. *An amount of FeO equal to that of the TiO_2 is allotted for ilmenite (**il**, column 7). If there is an excess of TiO_2, an equal amount of CaO will be allotted to the excess TiO_2 for provisional titanite (**tn'**), but only after the allotment of CaO and Al_2O_3 for anorthite (step 4d). If there is still an excess of TiO_2 it is calculated as rutile (**ru**). For our example, we have 0.0254 moles of **il** in column 7, and there is no excess TiO_2 after that).*

3g. *An amount of CaO equal to half that of any remaining F is allotted for fluorite (**fl**). There is none in our analysis.*

3h. *If the rock is not decomposed and contains cancrinite, an amount of Na_2O equal to that of the CO_2 is allotted for sodium carbonate (**nc**). This is rare and usually ignored. If the rock contains calcite, an amount of CaO equal to that of the CO_2 is allotted for calcite. If the modal calcite is secondary, the calculated calcite molecule is to be disregarded as not forming part of the norm.*

3i. *Allot an amount of SiO_2 equal to any ZrO_2 in zircon (**Z**). There is none in our analysis.*

4a. *An amount of Al_2O_3 equal to that of the K_2O is allotted for provisional orthoclase (**or'**). In our case, we combine all of our 0.0015 moles of K_2O with 0.0015 moles of Al_2O_3 and then allocate $6 \cdot 0.0015 = 0.0090$ moles of SiO_2 to create 0.0015 moles of **or'** in column 8 (a double Or molecule, as shown in Table B.1).*

4b. *If there is an excess of K_2O over Al_2O_3 (extremely rare), it is calculated as potassium metasilicate (**ks**). To do this, combine an amount of SiO_2 equal to that of the excess K_2O in a **ks** column. This is not required for our example.*

4c. *An excess of Al_2O_3 over the K_2O in **or'** is allotted to an equal amount of remaining Na_2O and six times as much SiO_2 for provisional albite (**ab'**). In column 9 of Table B.2, we combine 0.0486 moles of Na_2O with 0.0486 moles of Al_2O_3 plus $6 \cdot 0.0486$ moles of SiO_2 to make 0.0486 moles of **ab'** (double molecules as in Table B.1). If there is insufficient Al_2O_3, skip the **ab'** and go to step 4g, as is required for the nepheline basalt in Table B.3.*

4d. *If there is an excess of Al_2O_3 over the $(K_2O + Na_2O)$ used in 4a and 4c, it is allotted to an equal amount of remaining CaO for anorthite (**an**). There is 0.1079 moles of Al_2O_3 remaining in our example after allocation to **or'** and **ab'**. Note that this value is equivalent to the Al_2O_3 that exists in excess of that required to make orthoclase and albite, alumina yet available to form other minerals. Note the similarity between this value and that of A in the calculation of the ACF diagram in Section 24.3.1.1. We can now allocate this leftover Al_2O_3 to **an** by combining it with an equivalent amount of CaO and twice as much SiO_2 in column 10. Because there are 2 molecules of Al in both Al_2O_3 and*

$CaAl_2Si_2O_8$, we can deal with a *single* **An** molecule now, and not a *double* one, as in **or** and **ab** (note the formulas and formula weights in Table B.1).

4e. *If there is an excess of Al_2O_3 over this CaO, it is calculated as corundum (**c**). This is not the case for our examples.*

4f. *If there is an excess of CaO over the Al_2O_3 of 4d, it is reserved for diopside (**di**) and wollastonite (**wo**) in steps 7a and 7b.*

4g. *If in step 4c there is an excess of Na_2O over Al_2O_3, remove all **an** from the norm and proceed to steps 5a and 5b. If there is no excess Na_2O proceed to step 5c. In our example all the Na_2O was consumed in step 4c to make **ab'**, so we proceed to step 5c.*

5a. *If there is an excess of Na_2O over Al_2O_3 after step 4c and any remaining Fe_2O_3, allocate an amount of Na_2O equal to the remaining Fe_2O_3, plus four times as much SiO_2, to acmite (**ac**). This does not apply to our Table B.2 example but is true for Table B.3.*

5b. *If there is still an excess of Na_2O over Fe_2O_3 (rare), it is combined with an equivalent amount of SiO_2, as sodium metasilicate (**ns**). This too does not apply to our Table B.2 example, but does apply to Table B.3.*

5c. *If, as usually happens, there is an excess of Fe_2O_3 over remaining Na_2O, it is assigned to magnetite (**mt**), an equal amount of FeO being allotted to it out of what remains from the formation of pyrite, chromite, and ilmenite (see steps 3d, 3e, 3f). See column 11 in Table B.2.*

5d. *If there is still an excess of Fe_2O_3, it is combined with an equivalent amount of FeO as hematite (**hm**).*

6. *All the MgO and the FeO remaining from the previous allotments (see steps 3d, 3e, 3f, and 5c) are added together as the **Remainder**, and their relative proportions are ascertained. In column 12 in Table B.2 we see that there is 0.0660 moles of FeO and 0.1598 moles of MgO remaining at this point. The proportion, $Mg/(Mg + Fe)$, equals 0.708.*

7a. *To the amount of CaO remaining after allotment in step 4d is allotted provisionally an equal amount of $(MgO + FeO)$ to form diopside (**di'**); the relative proportions of FeO and MgO, as they occur in the **Remainder**, are transferred to diopside (and all subsequent mafics). The norm thus makes no attempt to consider the unequal distribution of Fe and Mg between coexisting phases, which would be the case in natural minerals. In column 13 of Table B.2, we have 0.0740 moles of remaining CaO, which we then combine with an equivalent amount of $(FeO + MgO)$ in the proportions 0.292:0.708. This requires $0.292 \cdot 0.0740 = 0.0216$ moles of FeO and $0.708 \cdot 0.0740 = 0.0524$ moles of MgO because these two quantities are both (a) in the proper proportions, and (b) sum to 0.0740. The spreadsheet carries real numbers, and round-off made the value of MgO in Table B.2 = 0.0523. Twice as much SiO_2 (0.1480 moles) are also required to form **di'**.*

7b. *If there is an excess of CaO, it is combined with an equivalent amount of SiO_2 as provisional wollastonite (**wo'**). This is not necessary in the examples.*

7c. *If there is an excess of $MgO + FeO$ over that needed for diopside (7a), allocate it to provisional hypersthene* *(**hy'**) in the ratio that they occur in the **Remainder**. An amount of SiO$_2$ equal to the sum of MgO and FeO is also allotted.* This is done in column 14 of Table B.2, resulting in the remaining 0.1075 moles of MgO, combined with the remaining 0.0444 moles of FeO and 0.1519 moles of SiO$_2$ creating 0.1519 moles of **hy'**. If we allocated MgO and FeO in the proper proportions to **di'** in step *7a*, the proportions will remain as MgO/(MgO + FeO) = 0.1075/(0.1075 + 0.0444) = 0.708, the proportions calculated in the **Remainder** in step *6*.

All the oxides have now been assigned to actual or provisional mineral molecules and we have next to consider the distribution of the silica.

---

8. Thus far we have allocated silica to silicates as follows. You may want to check to see that it is combined with:

- Any ZrO$_2$ in the ratio 1:1 to form zircon (**Z**, step *3i*)
- Any excess CaO in the ratio 1:1 to form titanite (**tn**, step *3f*)
- Any excess Na$_2$O in the ratio 4:1 to form acmite (**ac**, step *5a*)
- Any excess K$_2$O and Na$_2$O in the ratio 1:1 to form potassium and sodium metasilicates (**ks** and **ns**, steps *4b, 5b*)
- K$_2$O in the ratio 6:1 for provisional orthoclase (**or'**, step *4a*, 6: 1)
- Na$_2$O in the ratio 6:1 for provisional albite (**ab'**, step *4c*)
- CaO in the ratio 2:1 for provisional anorthite (**an'**, step *4d*)
- CaO + (Mg,Fe)O in the ratio 1:1 for diopside (**di**, step *7a*)
- any excess CaO in the ratio 1:1 for wollastonite (**wo**, step *7b*)
- (Mg,Fe)O in the ratio 1:1 for provisional hypersthene (**hy'**, step *7c*).

Set Y = to the sum of all the assigned SiO$_2$ by adding all the SiO$_2$ values to the right of column 4 in row 1 of Table B.2. In our Table B.2 example, Y = 0.8158 moles of allocated SiO$_2$. *Next calculate D = Y − the total SiO$_2$ in column 4*. In our example, SiO$_2$ in column 4 = 0.8189 and our allocated SiO$_2$ (Y) = 0.8158. D is thus equal to 0.8158 − 0.8189 = −0.0031. D is negative, meaning that the SiO$_2$ allocated to normative minerals is 0.0031 moles *less* than the total SiO$_2$ available in column 4. We have thus not over-allocated it.

9. *If* D *is negative* (there is excess SiO$_2$: the total amount of SiO$_2$ in column 4 > Y), as is common, *the excess SiO$_2$ is calculated as quartz* (**q**). This is true for our Table B.2 example, so we have 0.0031 moles of normative **q** in column 15. This near-to-over-saturation in SiO$_2$ is a common feature of tholeiitic basalts (our example rock). *For most common silica-oversaturated (quartz-bearing) rocks, this concludes the allocation of components. Go directly to step 17 to conclude the CIPW norm calculation.* I suggest you do this now if you're following the example in Table B.2. Table B.3, a CIPW norm calculation for a nepheline basalt, illustrates some of the following steps, required if SiO$_2$ is over-allocated.

If *D* is positive (*Y* > the total moles of SiO$_2$ in column 4), there is insufficient SiO$_2$ to accommodate the provisional silicates we have created so far, and we have over-allocated it. We shall thus have to go back, cup in hand, and borrow some SiO$_2$ from the provisional minerals that we have already created. For example, you can reduce **hy'** and produce **ol** + SiO$_2$ (Mg$_2$Si$_2$O$_6$ → Mg$_2$SiO$_4$ + SiO$_2$). If only some of the **hy'** is needed to free up enough SiO$_2$ by this process, the remaining **hy** is no longer provisional (first-pass) and has no hyphen. Thus **hy'** disappears, replaced by a lesser amount of **hy** plus some new **ol** in the norm as SiO$_2$ is liberated to make up the deficit. Similar SiO$_2$-producing reactions are **ab'** = **ne** + SiO$_2$, **or'** = **lc** + SiO$_2$, etc. This is performed in the following steps. Turn to Table B.3 as a guide. In it, we see that *Y* = 1.1895 and *D* = 0.5969 in step *8*, so we have a serious shortage of SiO$_2$. If *D* is positive, proceed to step *10*.

10a. *If* D < **hy'**/2 *(from step 7c) set* **ol** = D *and the final* **hy** = **hy'** − 2D. To do this on a spreadsheet, transfer 2D moles of (FeO + MgO) in the defined proportions from **hy'** to a new **ol** column and allocate D moles of SiO$_2$ to **ol** as well. For each mole of **hy'** converted to **ol** a mole of SiO$_2$ is freed up (for a constant amount of (Mg, Fe)O), so subtract D moles of SiO$_2$ from **hy'**. **hy** is the **hy'** that remains. *If the transfer of 2D moles of SiO$_2$ from* **hy'** *to* **ol** *reduces the SiO$_2$ deficit to zero, you may now proceed to step 17.* This step does not apply to our examples.

10b. *If* D > **hy'**/2, as in Table B.3, *convert all of the* **hy'** *to* **ol** *by setting* **hy** = 0 *and* **ol** = **hy'**/2. This frees up as much SiO$_2$ as possible by this method. Be sure to allot FeO and MgO to **ol** in the same proportions that they are in the **Remainder** column (this should be automatic), and only half of the SiO$_2$ in the original **hy'** column is allocated to **ol**. In Table B.3 **hy'** → 0 (I put the original calculation in parentheses) and **ol** is created with half the SiO$_2$ of **hy'** freeing up **hy'**/2 moles of SiO$_2$. We have reduced the SiO$_2$ deficit (D) by **hy'**/2, but still have not generated enough SiO$_2$ to finish the norm. If a SiO$_2$ deficit still remains, *calculate the remaining SiO$_2$ deficit:* D_1 = D − **hy'**/2 (the original provisional **hy'**) and *proceed to step 11*. In Table B.3, D_1 = 0.1534 − (0.5 · 0.0881) = 0.1094.

11a. *If there is no* **tn'** *in your norm, set* D_2 = D_1 *and proceed to step 12. If* D_1 < **tn'** *(step 3f), set* **tn** = **tn'** − D_1, *and reallocate the difference in CaO and TiO$_2$ (equivalent to* D_1*) to perovskite (**pf**). This frees up an amount of SiO$_2$ equivalent to* D_1, *and SiO$_2$ is balanced. Proceed to step 17.* This step does not apply to our examples.

11b. *If* D_1 > **tn'**, *convert all* **tn'** *to* **pf** *by allocating all of the CaO and TiO$_2$ to a* **pf** *column and free up an amount of SiO$_2$ equivalent to* D_1. *Set* D_2 = D_1/4 − **tn'** *(the original value). Proceed to step 12.* Although this step does not apply to our examples, we must still calculate D_2 = D_1 − 0 in Table B.3.

12a. *If* D_2 < 4**ab'**, *we can free up enough SiO$_2$ to complete the norm by transferring some* **ab'** *to* **ne**. *Set* **ne** = D_2/4 *and* **ab** = **ab'** − D_2/4. On a spreadsheet, this involves transferring D_2/4 moles of Na$_2$O and an equivalent amount of Al$_2$O$_3$ from **ab'** to **ne** and only twice as much SiO$_2$. This releases D_2 moles of SiO_2, which is our deficit. *The silica deficiency is now zero; proceed to step 17.*

12b. *If* $D_2 > 4ab'$ (as is the case for the example in Table B.3), *convert all of the ab' to ne, liberating* D_2 *moles of* SiO_2. *Set* $D_3 = D_2 - 4ab'$ *(original ab') and proceed to step 13.* This is done for the example in Table B.3, where $D_3 = D_2 - 4 \cdot 0.097 = 0.170$. I placed the original **ab'** in parentheses.

13a. *If* $D_3 < 2or'$ *we can free up enough SiO_2 to complete the norm by transferring some or' to lc. Set* $lc = D_3/2$ *and* $or = or' - D_3/2$. This involves transferring $D_3/2$ moles of K_2O and an equivalent amount of Al_2O_3 from *or'* to *lc* and only four times as much SiO_2. This releases D_3 moles of SiO_2, which is our deficit. *The silica deficiency is now zero, so proceed to step 17.*

13b. *If* $D_3 > 4or'$ (as is the case for the example in Table B.3), *convert all of the or' to lc, liberating* D_3 *moles of SiO_2* (see Table B.3). *Set* $D_4 = D_3 - 2or'$ (original *or'*) *and proceed to step 14.* I placed the original **or'** in parentheses.

14a. *If there is no wo' in your norm thus far, set* $D_5 = D_4$ *and proceed to step 15.* This is the case for Table B.3.

14b. *If* $D_4 > wo'/2$, we can free up enough SiO_2 to complete the norm by *transferring some wo' to cs. Set* $cs = D_4$ *and* $wo = wo' - D_4$. This involves transferring D_4 moles of CaO from *wo'* to *cs* and only half as much SiO_2. This releases D_4 moles of SiO_2, which is our deficit. *The silica deficiency is now zero, so proceed to step 17.*

14c. *If* $D_4 > wo'/2$, *convert all of the wo' to cs, liberating* D_4 *moles of SiO_2. Set* $D_5 = D_4 - wo'/2$ *and proceed to step 15.*

15a. *If* $D_5 > di'$, *increase the amounts of cs and ol by* $D_5/2$ *and set* $di = di' - D_5$. On a spreadsheet, add an amount equal to $D_5/2$ to the amounts of *cs* and *ol* already in the norm by transferring D_5 moles of CaO from *di'* to *cs* and D_5 moles of (FeO + MgO) from *di'* to *ol* (with D_5 moles of SiO_2 to each). This liberates D_5 moles of SiO_2, which is our deficit. *The silica deficiency is now zero. Proceed to step 17.*

15b. *If* $D_5 > di'$ (as is the case for the example in Table B.3), *set* $di = 0$ *and* $D_6 = D_5 - di'$ (original). Add an amount equal to $di'/2$ to the amounts of *cs* and *ol* already in the norm by transferring di' moles of CaO from *di'* to *cs* and di' moles of (FeO + MgO) from *di'* to *ol* (with D_5 moles of SiO_2 to each). This has been done in Table B.3. I placed the original **di'** in parentheses. Proceed to step 16.

16. *Set* $kp = D_6/2$ *and* $lc = lc' - D_6/2$. This involves transferring $D_6/2$ moles of K_2O and Al_2O_3 from *lc'* to *kp*, and twice as much (D_6) moles SiO_2. This liberates $D_6/2$ moles of SiO_2. For most rocks, including Table B.3, the silica deficiency should now be zero.

17. The row below the initial total in Table B.2, called "Mol proportion of normative mineral," should now contain the number of moles of each mineral that you have created by allocating and combining oxides to form the normative minerals in the steps above. If you have forgotten to do so in any step, use Table B.1 to guide you as you calculate the number of moles of each normative mineral. Remember, if you have 0.2 moles of CaO to create **an**, for example, you combine it with 0.2 moles of Al_2O_3 and 0.4 moles of SiO_2 to create **0.2** moles of $CaAl_2Si_2O_8$.

*Now add the formula weights of each normative mineral (from Table B.1) in a row below each of the mol proportions. The formula weights for **di**, **hy**, and **ol** are variable and must be calculated on the basis of Mg/(Mg + Fe). For example, because MgO/(MgO + FeO) in the **Remainder** column of Table B.2 = 0.708, the formula weight of **di** is:*

$$
\begin{aligned}
\text{f.w of } \textbf{\textit{di}} &= X_{mg} \cdot (\text{f.w of Mg-Di}) + X_{Fe} \\
&\quad \cdot (\text{f.w. of Fe-Di } (= \text{Hd})) \\
&= 0.708 \cdot 216.56 + 0.292 \cdot 248.1 \\
&= 225.77 \text{ g/mol}
\end{aligned}
$$

*Formula weights of **hy** and **ol** are calculated in the same fashion.*

Next multiply the Mol proportions by the molecular weights to produce the weights of the normative minerals in the next row of Table B.2. This step is omitted in molecular or Niggli norms. These weights should total the same as the anhydrous oxide total, within round-off errors, and should approximate the wt. % of the normative minerals. The final CIPW norm is this bottom row of Table B.2.

The oxidation state of Fe in an analysis can profoundly affect the norm, particularly the degree of silica saturation. The higher the Fe^{3+}/Fe^{2+} (all else being equal), the more magnetite in the norm and thus less mafic silicates, consuming less SiO_2. Oxidation can thus shift an Fe-rich rock from SiO_2-undersaturated (**ne**-normative) to SiO_2-saturated (**q** normative). Often an analysis will not determine or report Fe^{3+}/Fe^{2+}. If total Fe is reported as Fe_2O_3, the norm may seriously over-allocate **mt**, underestimate mafic silicates, and overestimate SiO_2 and **q**. If total Fe is reported as FeO, the norm may seriously under-allocate **mt**, overestimate mafic silicates, and underestimate SiO_2 and **q**. Irvine and Baragar (1971) attempted to mitigate this effect by estimating the Fe_2O_3/FeO for analyses in which it is not determined.

INDEX